Rational (Reciprocal) Function

$f(x) = \dfrac{1}{x}$

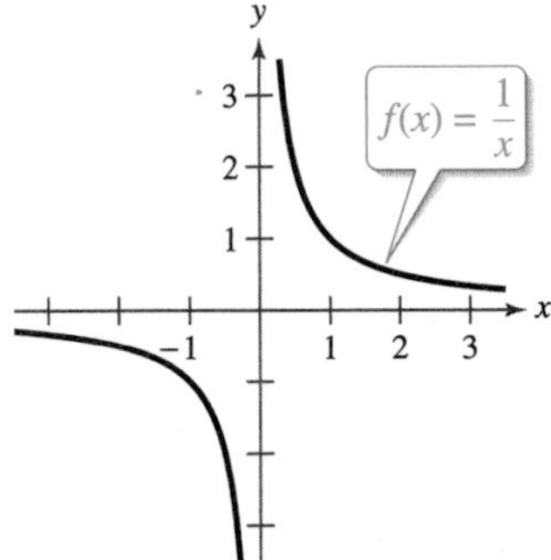

Domain: $(-\infty, 0) \cup (0, \infty)$
Range: $(-\infty, 0) \cup (0, \infty)$
No intercepts
Decreasing on $(-\infty, 0)$ and $(0, \infty)$
Odd function
Origin symmetry
Vertical asymptote: y-axis
Horizontal asymptote: x-axis

Exponential Function

$f(x) = a^x, \ a > 1$

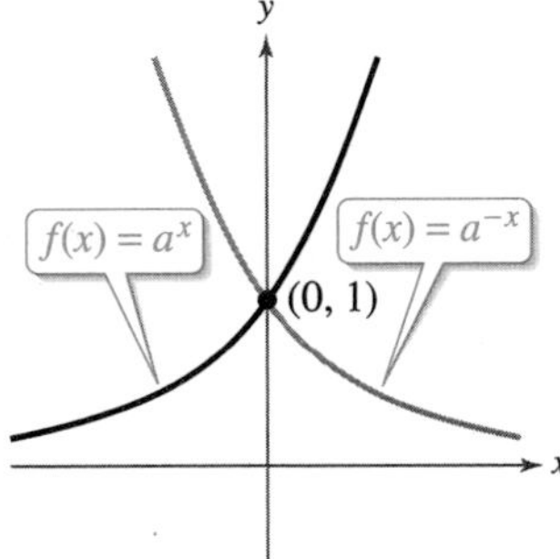

Domain: $(-\infty, \infty)$
Range: $(0, \infty)$
Intercept: $(0, 1)$
Increasing on $(-\infty, \infty)$
for $f(x) = a^x$
Decreasing on $(-\infty, \infty)$
for $f(x) = a^{-x}$
Horizontal asymptote: x-axis
Continuous

Logarithmic Function

$f(x) = \log_a x, \ a > 1$

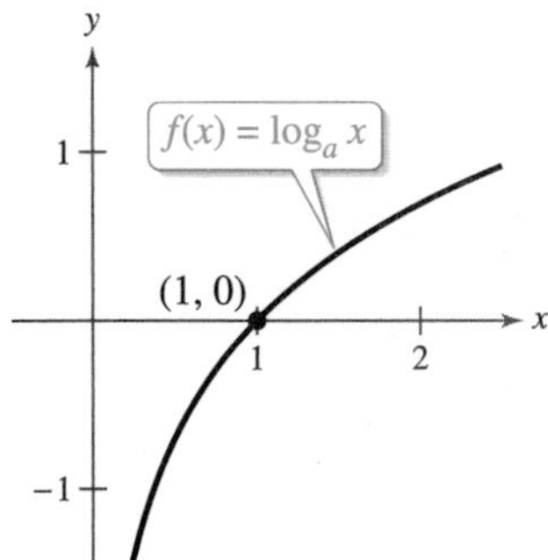

Domain: $(0, \infty)$
Range: $(-\infty, \infty)$
Intercept: $(1, 0)$
Increasing on $(0, \infty)$
Vertical asymptote: y-axis
Continuous
Reflection of graph of $f(x) = a^x$
in the line $y = x$

Sine Function

$f(x) = \sin x$

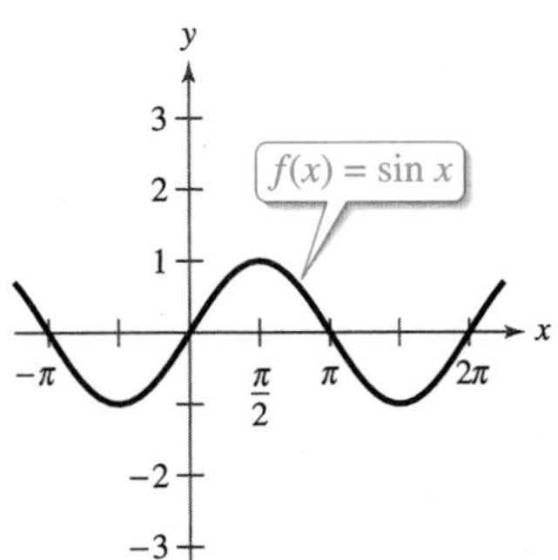

Domain: $(-\infty, \infty)$
Range: $[-1, 1]$
Period: 2π
x-intercepts: $(n\pi, 0)$
y-intercept: $(0, 0)$
Odd function
Origin symmetry

Cosine Function

$f(x) = \cos x$

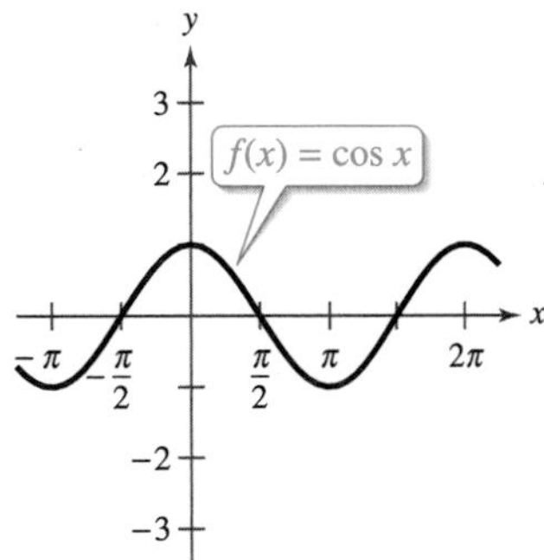

Domain: $(-\infty, \infty)$
Range: $[-1, 1]$
Period: 2π
x-intercepts: $\left(\dfrac{\pi}{2} + n\pi, 0\right)$
y-intercept: $(0, 1)$
Even function
y-axis symmetry

Tangent Function

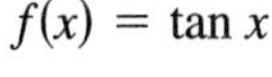

$f(x) = \tan x$

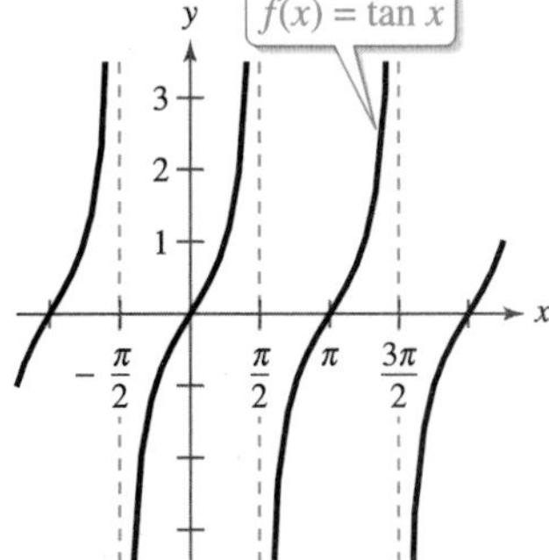

Domain: all $x \neq \dfrac{\pi}{2} + n\pi$
Range: $(-\infty, \infty)$
Period: π
x-intercepts: $(n\pi, 0)$
y-intercept: $(0, 0)$
Vertical asymptotes:
$x = \dfrac{\pi}{2} + n\pi$
Odd function
Origin symmetry

Cosecant Function

$f(x) = \csc x$

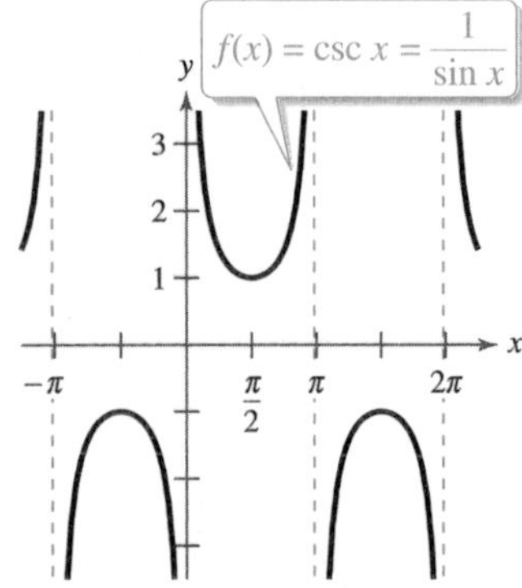

Domain: all $x \neq n\pi$
Range: $(-\infty, -1] \cup [1, \infty)$
Period: 2π
No intercepts
Vertical asymptotes: $x = n\pi$
Odd function
Origin symmetry

Secant Function

$f(x) = \sec x$

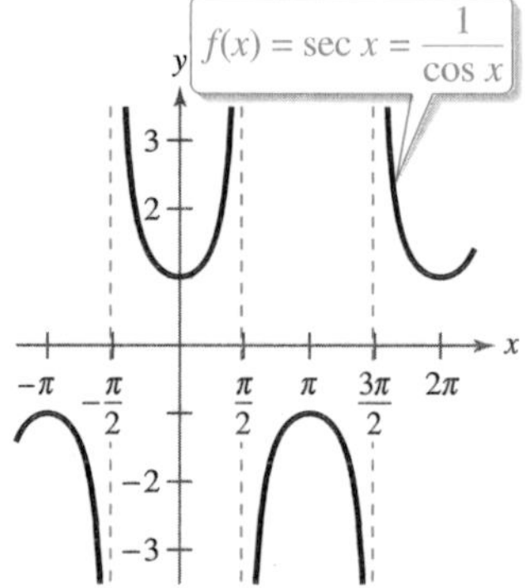

Domain: all $x \neq \frac{\pi}{2} + n\pi$
Range: $(-\infty, -1] \cup [1, \infty)$
Period: 2π
y-intercept: $(0, 1)$
Vertical asymptotes:

$$x = \frac{\pi}{2} + n\pi$$

Even function
y-axis symmetry

Cotangent Function

$f(x) = \cot x$

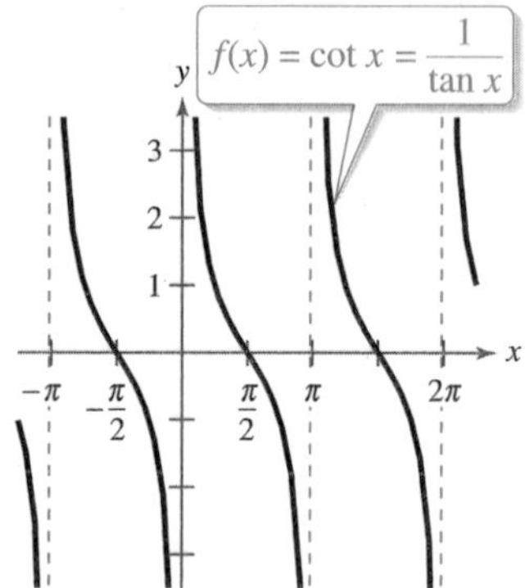

Domain: all $x \neq n\pi$
Range: $(-\infty, \infty)$
Period: π

x-intercepts: $\left(\frac{\pi}{2} + n\pi, 0\right)$

Vertical asymptotes: $x = n\pi$
Odd function
Origin symmetry

Inverse Sine Function

$f(x) = \arcsin x$

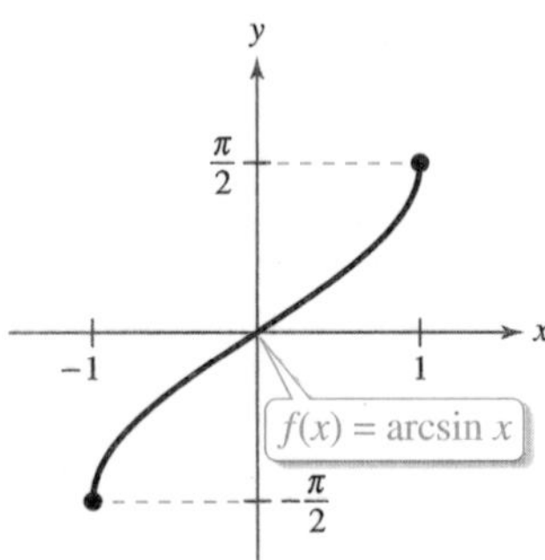

Domain: $[-1, 1]$

Range: $\left[-\frac{\pi}{2}, \frac{\pi}{2}\right]$

Intercept: $(0, 0)$
Odd function
Origin symmetry

Inverse Cosine Function

$f(x) = \arccos x$

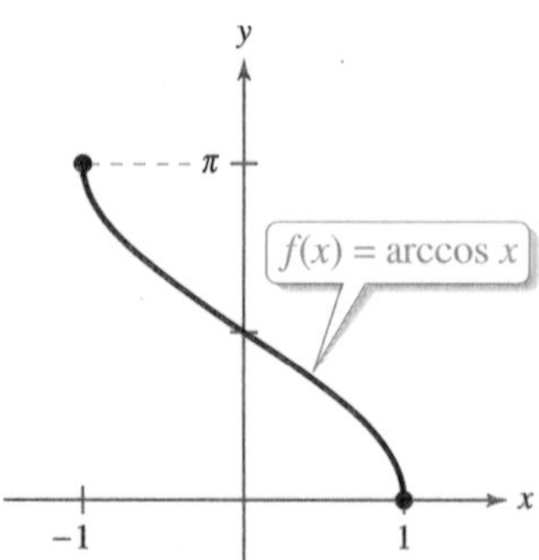

Domain: $[-1, 1]$
Range: $[0, \pi]$

y-intercept: $\left(0, \frac{\pi}{2}\right)$

Inverse Tangent Function

$f(x) = \arctan x$

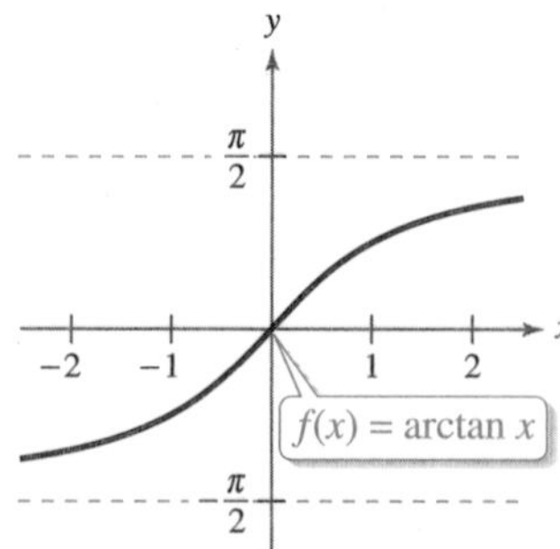

Domain: $(-\infty, \infty)$

Range: $\left(-\frac{\pi}{2}, \frac{\pi}{2}\right)$

Intercept: $(0, 0)$
Horizontal asymptotes:

$$y = \pm\frac{\pi}{2}$$

Odd function
Origin symmetry

TRIGONOMETRY

with

CalcChat® and CalcView®

10E

Ron Larson
The Pennsylvania State University
The Behrend College

With the assistance of David C. Falvo
The Pennsylvania State University
The Behrend College

Australia • Brazil • Mexico • Singapore • United Kingdom • United States

Trigonometry
with CalcChat and CalcView
Tenth Edition

Ron Larson

Product Director: Terry Boyle
Product Manager: Gary Whalen
Senior Content Developer: Stacy Green
Associate Content Developer: Samantha Lugtu
Product Assistant: Katharine Werring
Media Developer: Lynh Pham
Marketing Manager: Ryan Ahern
Content Project Manager: Jennifer Risden
Manufacturing Planner: Doug Bertke
Production Service: Larson Texts, Inc.
Photo Researcher: Lumina Datamatics
Text Researcher: Lumina Datamatics
Illustrator: Larson Texts, Inc.
Text Designer: Larson Texts, Inc.
Cover Designer: Larson Texts, Inc.
Front Cover Image: betibup33/Shutterstock.com
Back Cover Image: Dragonfly22/Shutterstock.com
Compositor: Larson Texts, Inc.

Library of Congress Control Number: 2016944977

Student Edition:
ISBN: 978-1-337-27846-1

Loose-leaf Edition:
ISBN: 978-1-337-29127-9

Cengage Learning
20 Channel Center Street
Boston, MA 02210
USA

Printed in the United States of America
Print Number: 01 Print Year: 2016

Contents

▷ Appendix A

*Available at the text-specific website *www.cengagebrain.com*

Preface

Welcome to *Trigonometry,* Tenth Edition. We are excited to offer you a new edition with even more resources that will help you understand and master trigonometry. This textbook includes features and resources that continue to make *Trigonometry* a valuable learning tool for students and a trustworthy teaching tool for instructors.

Trigonometry provides the clear instruction, precise mathematics, and thorough coverage that you expect for your course. Additionally, this new edition provides you with **free** access to three companion websites:

- **CalcView.com**—video solutions to selected exercises
- **CalcChat.com**—worked-out solutions to odd-numbered exercises and access to online tutors
- **LarsonPrecalculus.com**—companion website with resources to supplement your learning

These websites will help enhance and reinforce your understanding of the material presented in this text and prepare you for future mathematics courses. CalcView® and CalcChat® are also available as free mobile apps.

Features

NEW CalcView®

The website *CalcView.com* contains video solutions of selected exercises. Watch instructors progress step-by-step through solutions, providing guidance to help you solve the exercises. The CalcView mobile app is available for free at the Apple® App Store® or Google Play™ store. The app features an embedded QR Code® reader that can be used to scan the on-page codes and go directly to the videos. You can also access the videos at *CalcView.com.*

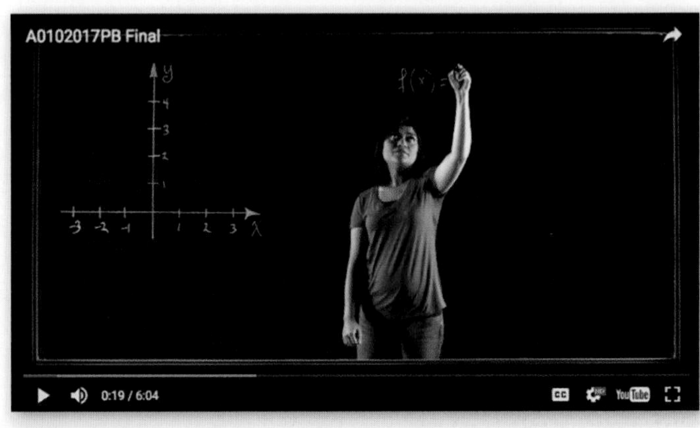

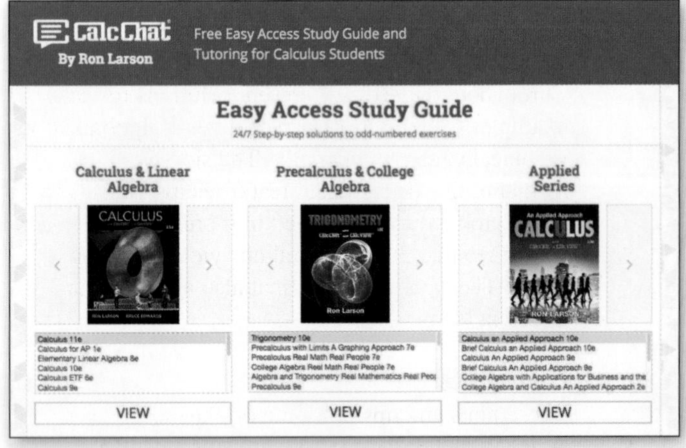

UPDATED CalcChat®

In each exercise set, be sure to notice the reference to *CalcChat.com.* This website provides free step-by-step solutions to all odd-numbered exercises in many of our textbooks. Additionally, you can chat with a tutor, at no charge, during the hours posted at the site. For over 14 years, hundreds of thousands of students have visited this site for help. The CalcChat mobile app is also available as a free download at the Apple® App Store® or Google Play™ store and features an embedded QR Code® reader.

App Store is a service mark of Apple Inc. Google Play is a trademark of Google Inc.
QR Code is a registered trademark of Denso Wave Incorporated.

REVISED LarsonPrecalculus.com

All companion website features have been updated based on this revision, plus we have added a new Collaborative Project feature. Access to these features is free. You can view and listen to worked-out solutions of Checkpoint problems in English or Spanish, explore examples, download data sets, watch lesson videos, and much more.

NEW Collaborative Project

You can find these extended group projects at *LarsonPrecalculus.com.* Check your understanding of the chapter concepts by solving in-depth, real-life problems. These collaborative projects provide an interesting and engaging way for you and other students to work together and investigate ideas.

REVISED Exercise Sets

The exercise sets have been carefully and extensively examined to ensure they are rigorous and relevant, and include topics our users have suggested. The exercises have been reorganized and titled so you can better see the connections between examples and exercises. Multi-step, real-life exercises reinforce problem-solving skills and mastery of concepts by giving you the opportunity to apply the concepts in real-life situations. Error Analysis exercises have been added throughout the text to help you identify common mistakes.

Table of Contents Changes

Based on market research and feedback from users, Section 4.3, The Complex Plane, has been added. In addition, examples on finding the magnitude of a scalar multiple (Section 3.3) and multiplying in the complex plane (Section 4.4) have been added.

Chapter Opener

Each Chapter Opener highlights real-life applications used in the examples and exercises.

Section Objectives

A bulleted list of learning objectives provides you the opportunity to preview what will be presented in the upcoming section.

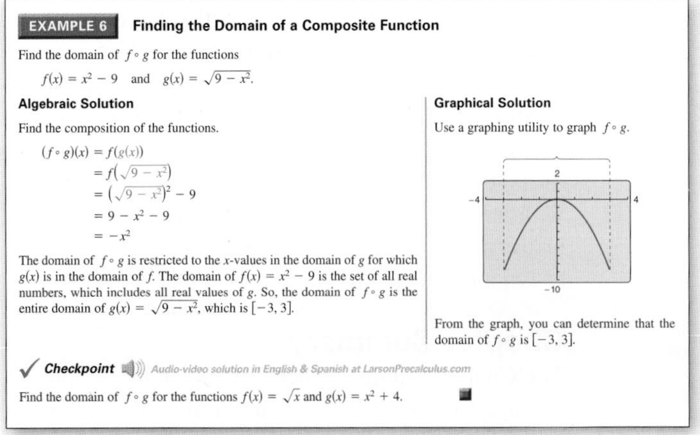

EXAMPLE 6 **Finding the Domain of a Composite Function**

Find the domain of $f \circ g$ for the functions

$f(x) = x^2 - 9$ and $g(x) = \sqrt{9 - x^2}$.

Algebraic Solution

Find the composition of the functions.

$$\begin{aligned}(f \circ g)(x) &= f(g(x)) \\ &= f(\sqrt{9 - x^2}) \\ &= (\sqrt{9 - x^2})^2 - 9 \\ &= 9 - x^2 - 9 \\ &= -x^2\end{aligned}$$

The domain of $f \circ g$ is restricted to the x-values in the domain of g for which $g(x)$ is in the domain of f. The domain of $f(x) = x^2 - 9$ is the set of all real numbers, which includes all real values of g. So, the domain of $f \circ g$ is the entire domain of $g(x) = \sqrt{9 - x^2}$, which is $[-3, 3]$.

Graphical Solution

Use a graphing utility to graph $f \circ g$.

From the graph, you can determine that the domain of $f \circ g$ is $[-3, 3]$.

✓ ***Checkpoint*** *Audio-video solution in English & Spanish at LarsonPrecalculus.com*

Find the domain of $f \circ g$ for the functions $f(x) = \sqrt{x}$ and $g(x) = x^2 + 4$.

Side-By-Side Examples

Throughout the text, we present solutions to many examples from multiple perspectives—algebraically, graphically, and numerically. The side-by-side format of this pedagogical feature helps you to see that a problem can be solved in more than one way and to see that different methods yield the same result. The side-by-side format also addresses many different learning styles.

Remarks

These hints and tips reinforce or expand upon concepts, help you learn how to study mathematics, caution you about common errors, address special cases, or show alternative or additional steps to a solution of an example.

Checkpoints

Accompanying every example, the Checkpoint problems encourage immediate practice and check your understanding of the concepts presented in the example. View and listen to worked-out solutions of the Checkpoint problems in English or Spanish at *LarsonPrecalculus.com.*

Technology

The technology feature gives suggestions for effectively using tools such as calculators, graphing utilities, and spreadsheet programs to help deepen your understanding of concepts, ease lengthy calculations, and provide alternate solution methods for verifying answers obtained by hand.

Historical Notes

These notes provide helpful information regarding famous mathematicians and their work.

Algebra of Calculus

Throughout the text, special emphasis is given to the algebraic techniques used in calculus. Algebra of Calculus examples and exercises are integrated throughout the text and are identified by the symbol ∫.

Summarize

The Summarize feature at the end of each section helps you organize the lesson's key concepts into a concise summary, providing you with a valuable study tool.

Vocabulary Exercises

The vocabulary exercises appear at the beginning of the exercise set for each section. These problems help you review previously learned vocabulary terms that you will use in solving the section exercises.

TECHNOLOGY Use a graphing utility to check the result of Example 2. To do this, enter

$$Y1 = -(\sin(X))^3$$

and

$$Y2 = \sin(X)(\cos(X))^2 - \sin(X).$$

Select the *line* style for Y1 and the *path* style for Y2, then graph both equations in the same viewing window. The two graphs *appear* to coincide, so it is reasonable to assume that their expressions are equivalent. Note that the actual equivalence of the expressions can only be verified algebraically, as in Example 2. This graphical approach is only to check your work.

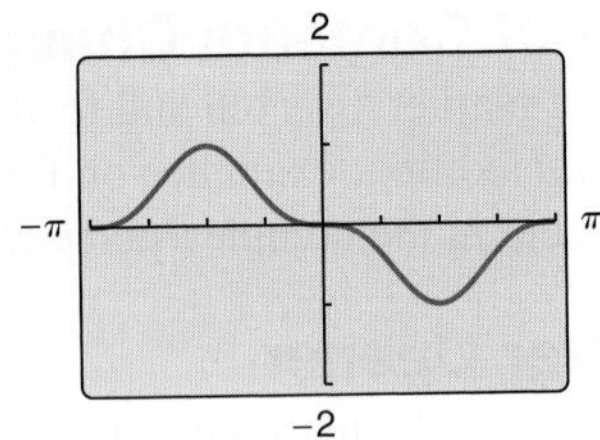

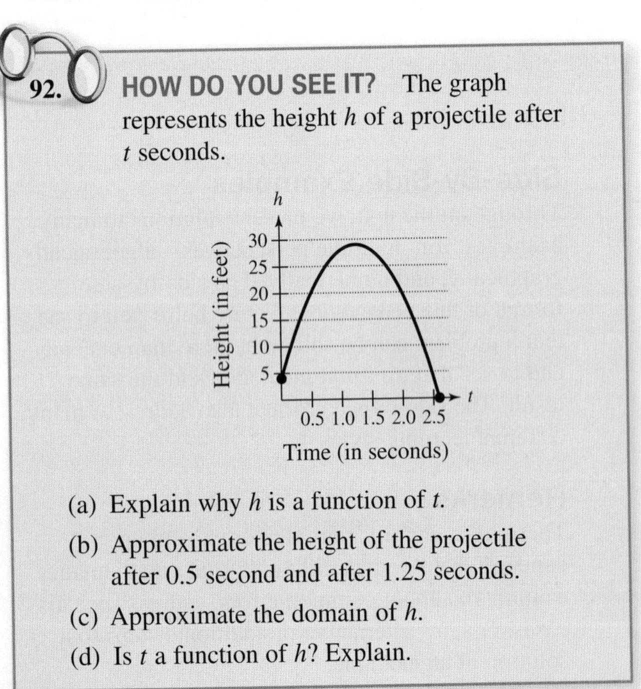

How Do You See It?

The How Do You See It? feature in each section presents a real-life exercise that you will solve by visual inspection using the concepts learned in the lesson. This exercise is excellent for classroom discussion or test preparation.

Project

The projects at the end of selected sections involve in-depth applied exercises in which you will work with large, real-life data sets, often creating or analyzing models. These projects are offered online at *LarsonPrecalculus.com.*

Chapter Summary

The Chapter Summary includes explanations and examples of the objectives taught in each chapter.

Instructor Resources

Annotated Instructor's Edition / ISBN-13: 978-1-337-27847-8
This is the complete student text plus point-of-use annotations for the instructor, including extra projects, classroom activities, teaching strategies, and additional examples. Answers to even-numbered text exercises, Vocabulary Checks, and Explorations are also provided.

Complete Solutions Manual (on instructor companion site)
This manual contains solutions to all exercises from the text, including Chapter Review Exercises and Chapter Tests, and Practice Tests with solutions.

Cengage Learning Testing Powered by Cognero (login.cengage.com)
CLT is a flexible online system that allows you to author, edit, and manage test bank content; create multiple test versions in an instant; and deliver tests from your LMS, your classroom, or wherever you want. This is available online via *www.cengage.com/login.*

Instructor Companion Site
Everything you need for your course in one place! This collection of book-specific lecture and class tools is available online via *www.cengage.com/login.* Access and download PowerPoint® presentations, images, the instructor's manual, and more.

The Test Bank (on instructor companion site)
This contains text-specific multiple-choice and free response test forms.

Lesson Plans (on instructor companion site)
This manual provides suggestions for activities and lessons with notes on time allotment in order to ensure timeliness and efficiency during class.

MindTap for Mathematics
MindTap® is the digital learning solution that helps instructors engage and transform today's students into critical thinkers. Through paths of dynamic assignments and applications that you can personalize, real-time course analytics and an accessible reader, MindTap helps you turn cookie cutter into cutting edge, apathy into engagement, and memorizers into higher-level thinkers.

Enhanced WebAssign® 

Exclusively from Cengage Learning, Enhanced WebAssign combines the exceptional mathematics content that you know and love with the most powerful online homework solution, WebAssign. Enhanced WebAssign engages students with immediate feedback, rich tutorial content, and interactive, fully customizable e-books (YouBook), helping students to develop a deeper conceptual understanding of their subject matter. Quick Prep and Just In Time exercises provide opportunities for students to review prerequisite skills and content, both at the start of the course and at the beginning of each section. Flexible assignment options give instructors the ability to release assignments conditionally on the basis of students' prerequisite assignment scores. Visit us at **www.cengage.com/ewa** to learn more.

Student Resources

Student Study and Solutions Manual / ISBN-13: 978-1-337-27848-5
This guide offers step-by-step solutions for all odd-numbered text exercises, Chapter Tests, and Cumulative Tests. It also contains Practice Tests.

Note-Taking Guide / ISBN-13: 978-1-337-27849-2
This is an innovative study aid, in the form of a notebook organizer, that helps students develop a section-by-section summary of key concepts.

CengageBrain.com
To access additional course materials, please visit *www.cengagebrain.com.* At the *CengageBrain.com* home page, search for the ISBN of your title (from the back cover of your book) using the search box at the top of the page. This will take you to the product page where these resources can be found.

MindTap for Mathematics
MindTap® provides you with the tools you need to better manage your limited time—you can complete assignments whenever and wherever you are ready to learn with course material specially customized for you by your instructor and streamlined in one proven, easy-to-use interface. With an array of tools and apps—from note taking to flashcards—you'll get a true understanding of course concepts, helping you to achieve better grades and setting the groundwork for your future courses. This access code entitles you to one term of usage.

Enhanced WebAssign® ENHANCED WebAssign
Enhanced WebAssign (assigned by the instructor) provides you with instant feedback on homework assignments. This online homework system is easy to use and includes helpful links to textbook sections, video examples, and problem-specific tutorials.

Acknowledgments

I would like to thank the many people who have helped me prepare the text and the supplements package. Their encouragement, criticisms, and suggestions have been invaluable.

Thank you to all of the instructors who took the time to review the changes in this edition and to provide suggestions for improving it. Without your help, this book would not be possible.

Reviewers of the Tenth Edition

Gurdial Arora, *Xavier University of Louisiana*
Russell C. Chappell, *Twinsburg High School, Ohio*
Darlene Martin, *Lawson State Community College*
John Fellers, *North Allegheny School District*
Professor Steven Sikes, *Collin College*
Ann Slate, *Surry Community College*
John Elias, *Glenda Dawson High School*
Kathy Wood, *Lansing Catholic High School*
Darin Bauguess, *Surry Community College*
Brianna Kurtz, *Daytona State College*

Reviewers of the Previous Editions

Timothy Andrew Brown, *South Georgia College;* Blair E. Caboot, *Keystone College;* Shannon Cornell, *Amarillo College;* Gayla Dance, *Millsaps College;* Paul Finster, *El Paso Community College;* Paul A. Flasch, *Pima Community College West Campus;* Vadas Gintautas, *Chatham University;* Lorraine A. Hughes, *Mississippi State University;* Shu-Jen Huang, *University of Florida;* Renyetta Johnson, *East Mississippi Community College;* George Keihany, *Fort Valley State University;* Mulatu Lemma, *Savannah State University;* William Mays Jr., *Salem Community College;* Marcella Melby, *University of Minnesota;* Jonathan Prewett, *University of Wyoming;* Denise Reid, *Valdosta State University;* David L. Sonnier, *Lyon College;* David H. Tseng, *Miami Dade College—Kendall Campus;* Kimberly Walters, *Mississippi State University;* Richard Weil, *Brown College;* Solomon Willis, *Cleveland Community College;* Bradley R. Young, *Darton College*

My thanks to Robert Hostetler, The Behrend College, The Pennsylvania State University, and David Heyd, The Behrend College, The Pennsylvania State University, for their significant contributions to previous editions of this text.

I would also like to thank the staff at Larson Texts, Inc. who assisted with proofreading the manuscript, preparing and proofreading the art package, and checking and typesetting the supplements.

On a personal level, I am grateful to my spouse, Deanna Gilbert Larson, for her love, patience, and support. Also, a special thanks goes to R. Scott O'Neil. If you have suggestions for improving this text, please feel free to write to me. Over the past two decades, I have received many useful comments from both instructors and students, and I value these comments very highly.

Ron Larson, Ph.D.
Professor of Mathematics
Penn State University
www.RonLarson.com

P Prerequisites

Snowstorm *(Exercise 47, page 84)*

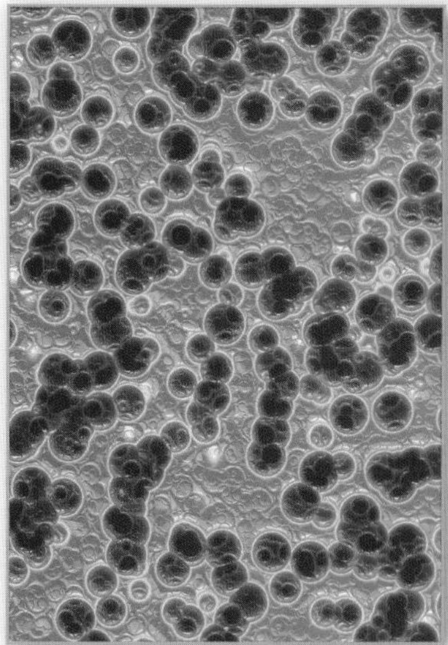

Bacteria *(Example 8, page 98)*

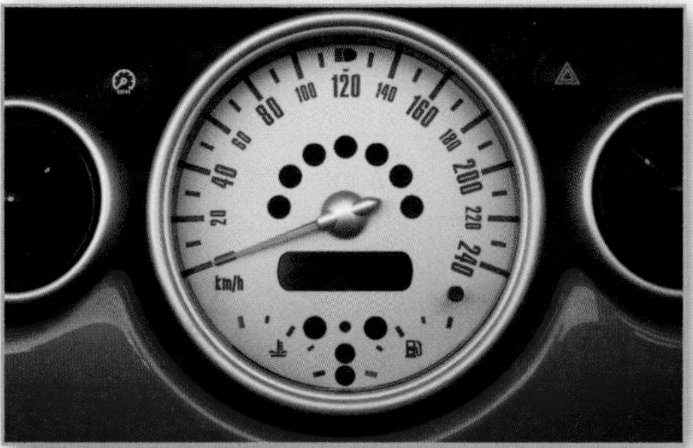

Average Speed *(Example 7, page 72)*

Alternative-Fuel Stations *(Example 10, page 60)*

Americans with Disabilities Act *(page 46)*

P.1 Review of Real Numbers and Their Properties

Real numbers can represent many real-life quantities. For example, in Exercises 49–52 on page 13, you will use real numbers to represent the federal surplus or deficit.

- **Represent and classify real numbers.**
- **Order real numbers and use inequalities.**
- **Find the absolute values of real numbers and find the distance between two real numbers.**
- **Evaluate algebraic expressions.**
- **Use the basic rules and properties of algebra.**

Real Numbers

Real numbers can describe quantities in everyday life such as age, miles per gallon, and population. Symbols such as

$$-5, 9, 0, \tfrac{4}{3}, 0.666\ldots, 28.21, \sqrt{2}, \pi, \text{ and } \sqrt[3]{-32}$$

represent real numbers. Here are some important **subsets** (each member of a subset B is also a member of a set A) of the real numbers. The three dots, or *ellipsis points,* tell you that the pattern continues indefinitely.

$\{1, 2, 3, 4, \ldots\}$ Set of natural numbers

$\{0, 1, 2, 3, 4, \ldots\}$ Set of whole numbers

$\{\ldots, -3, -2, -1, 0, 1, 2, 3, \ldots\}$ Set of integers

A real number is **rational** when it can be written as the ratio p/q of two integers, where $q \neq 0$. For example, the numbers

$$\tfrac{1}{3} = 0.3333\ldots = 0.\overline{3}, \tfrac{1}{8} = 0.125, \text{ and } \tfrac{125}{111} = 1.126126\ldots = 1.\overline{126}$$

are rational. The decimal representation of a rational number either repeats $\left(\text{as in } \tfrac{173}{55} = 3.1\overline{45}\right)$ or terminates $\left(\text{as in } \tfrac{1}{2} = 0.5\right)$. A real number that cannot be written as the ratio of two integers is **irrational.** The decimal representation of an irrational number neither terminates nor repeats. For example, the numbers

$$\sqrt{2} = 1.4142135\ldots \approx 1.41 \quad \text{and} \quad \pi = 3.1415926\ldots \approx 3.14$$

are irrational. (The symbol $\approx$ means "is approximately equal to.") Figure P.1 shows subsets of the real numbers and their relationships to each other.

Real numbers
- Irrational numbers
- Rational numbers
  - Integers
    - Negative integers
    - Whole numbers
      - Natural numbers
      - Zero
  - Noninteger fractions (positive and negative)

Subsets of the real numbers
Figure P.1

EXAMPLE 1 **Classifying Real Numbers**

Determine which numbers in the set $\left\{-13, -\sqrt{5}, -1, -\tfrac{1}{3}, 0, \tfrac{5}{8}, \sqrt{2}, \pi, 7\right\}$ are (a) natural numbers, (b) whole numbers, (c) integers, (d) rational numbers, and (e) irrational numbers.

Solution

a. Natural numbers: $\{7\}$

b. Whole numbers: $\{0, 7\}$

c. Integers: $\{-13, -1, 0, 7\}$

d. Rational numbers: $\left\{-13, -1, -\tfrac{1}{3}, 0, \tfrac{5}{8}, 7\right\}$

e. Irrational numbers: $\left\{-\sqrt{5}, \sqrt{2}, \pi\right\}$

✓ ***Checkpoint*** *Audio-video solution in English & Spanish at LarsonPrecalculus.com*

Repeat Example 1 for the set $\left\{-\pi, -\tfrac{1}{4}, \tfrac{6}{3}, \tfrac{1}{2}\sqrt{2}, -7.5, -1, 8, -22\right\}$.

Michael G Smith/Shutterstock.com

Real numbers are represented graphically on the **real number line.** When you draw a point on the real number line that corresponds to a real number, you are **plotting** the real number. The point representing 0 on the real number line is the **origin.** Numbers to the right of 0 are positive, and numbers to the left of 0 are negative, as shown in the figure below. The term **nonnegative** describes a number that is either positive or zero.

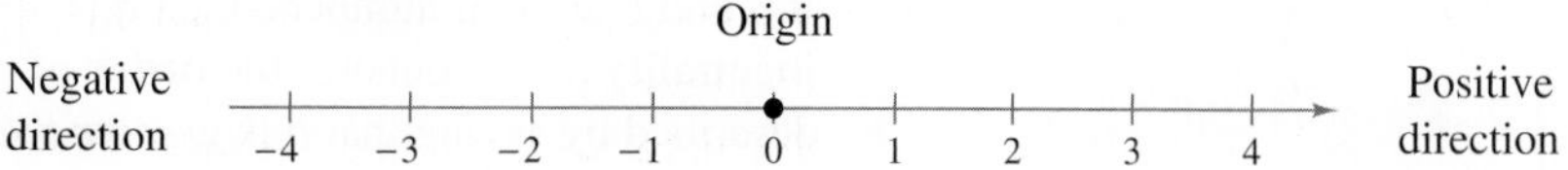

As the next two number lines illustrate, there is a *one-to-one correspondence* between real numbers and points on the real number line.

$-\frac{5}{3}$ 0.75 π

Every real number corresponds to exactly one point on the real number line.

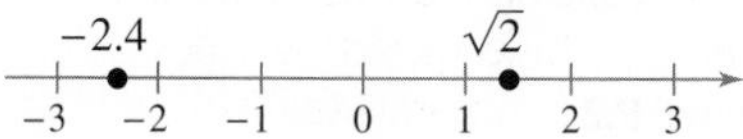

Every point on the real number line corresponds to exactly one real number.

EXAMPLE 2 Plotting Points on the Real Number Line

Plot the real numbers on the real number line.

a. $-\frac{7}{4}$

b. 2.3

c. $\frac{2}{3}$

d. -1.8

Solution The figure below shows all four points.

-1.8 $-\frac{7}{4}$ $\frac{2}{3}$ 2.3

a. The point representing the real number $-\frac{7}{4} = -1.75$ lies between -2 and -1, but closer to -2, on the real number line.

b. The point representing the real number 2.3 lies between 2 and 3, but closer to 2, on the real number line.

c. The point representing the real number $\frac{2}{3} = 0.666\ldots$ lies between 0 and 1, but closer to 1, on the real number line.

d. The point representing the real number -1.8 lies between -2 and -1, but closer to -2, on the real number line. Note that the point representing -1.8 lies slightly to the left of the point representing $-\frac{7}{4}$.

✓ ***Checkpoint*** *Audio-video solution in English & Spanish at LarsonPrecalculus.com*

Plot the real numbers on the real number line.

a. $\frac{5}{2}$ **b.** -1.6

c. $-\frac{3}{4}$ **d.** 0.7

Ordering Real Numbers

One important property of real numbers is that they are *ordered.*

> **Definition of Order on the Real Number Line**
>
> If a and b are real numbers, then a is *less than* b when $b - a$ is positive. The **inequality** $a < b$ denotes the **order** of a and b. This relationship can also be described by saying that b is *greater than* a and writing $b > a$. The inequality $a \leq b$ means that a is *less than or equal to* b, and the inequality $b \geq a$ means that b is *greater than or equal to* a. The symbols $<$, $>$, $\leq$, and $\geq$ are *inequality symbols.*

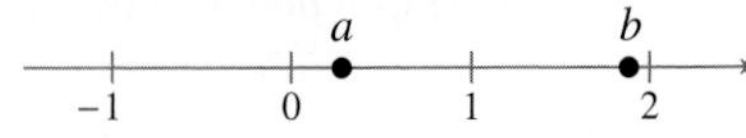

$a < b$ if and only if a lies to the left of b.

Figure P.2

Geometrically, this definition implies that $a < b$ if and only if a lies to the *left* of b on the real number line, as shown in Figure P.2.

EXAMPLE 3 Ordering Real Numbers

Place the appropriate inequality symbol ($<$ or $>$) between the pair of real numbers.

a. $-3, 0$ **b.** $-2, -4$ **c.** $\frac{1}{4}, \frac{1}{3}$

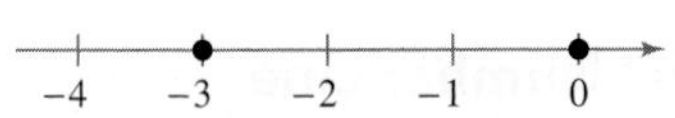

Figure P.3

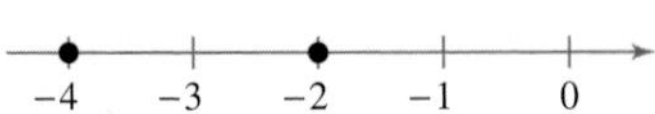

Figure P.4

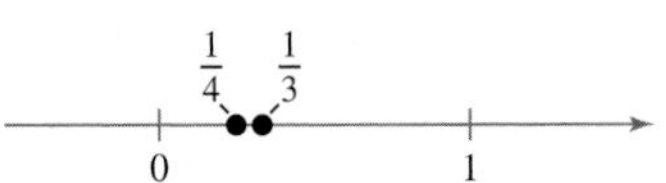

Figure P.5

Solution

a. On the real number line, -3 lies to the left of 0, as shown in Figure P.3. So, you can say that -3 is *less than* 0, and write $-3 < 0$.

b. On the real number line, -2 lies to the right of -4, as shown in Figure P.4. So, you can say that -2 is *greater than* -4, and write $-2 > -4$.

c. On the real number line, $\frac{1}{4}$ lies to the left of $\frac{1}{3}$, as shown in Figure P.5. So, you can say that $\frac{1}{4}$ is *less than* $\frac{1}{3}$, and write $\frac{1}{4} < \frac{1}{3}$.

Place the appropriate inequality symbol ($<$ or $>$) between the pair of real numbers.

a. $1, -5$ **b.** $\frac{3}{2}, 7$ **c.** $-\frac{2}{3}, -\frac{3}{4}$

EXAMPLE 4 Interpreting Inequalities

See LarsonPrecalculus.com for an interactive version of this type of example.

Describe the subset of real numbers that the inequality represents.

a. $x \leq 2$ **b.** $-2 \leq x < 3$

Solution

$x \leq 2$

0 1 2 3 4 x

Figure P.6

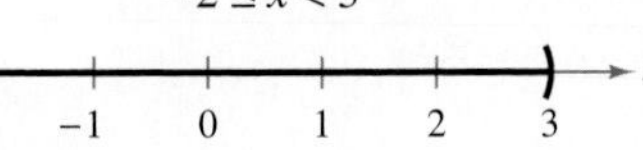

Figure P.7

a. The inequality $x \leq 2$ denotes all real numbers less than or equal to 2, as shown in Figure P.6.

b. The inequality $-2 \leq x < 3$ means that $x \geq -2$ *and* $x < 3$. This "double inequality" denotes all real numbers between -2 and 3, including -2 but not including 3, as shown in Figure P.7.

✓ **Checkpoint** Audio-video solution in English & Spanish at LarsonPrecalculus.com

Describe the subset of real numbers that the inequality represents.

a. $x > -3$ **b.** $0 < x \leq 4$

Inequalities can describe subsets of real numbers called **intervals.** In the bounded intervals below, the real numbers a and b are the **endpoints** of each interval. The endpoints of a closed interval are included in the interval, whereas the endpoints of an open interval are not included in the interval.

REMARK The reason that the four types of intervals at the right are called *bounded* is that each has a finite length. An interval that does not have a finite length is *unbounded* (see below).

Bounded Intervals on the Real Number Line

Notation	Interval Type	Inequality	Graph
$[a, b]$	Closed	$a \le x \le b$	
(a, b)	Open	$a < x < b$	
$[a, b)$		$a \le x < b$	
$(a, b]$		$a < x \le b$	

The symbols ∞, **positive infinity,** and $-\infty$, **negative infinity,** do not represent real numbers. They are convenient symbols used to describe the unboundedness of an interval such as $(1, \infty)$ or $(-\infty, 3]$.

REMARK Whenever you write an interval containing ∞ or $-\infty$, always use a parenthesis and never a bracket next to these symbols. This is because ∞ and $-\infty$ are never included in the interval.

Unbounded Intervals on the Real Number Line

Notation	Interval Type	Inequality	Graph
$[a, \infty)$		$x \ge a$	
(a, ∞)	Open	$x > a$	
$(-\infty, b]$		$x \le b$	
$(-\infty, b)$	Open	$x < b$	
$(-\infty, \infty)$	Entire real line	$-\infty < x < \infty$	

EXAMPLE 5 Interpreting Intervals

a. The interval $(-1, 0)$ consists of all real numbers greater than -1 and less than 0.

b. The interval $[2, \infty)$ consists of all real numbers greater than or equal to 2.

✓ **Checkpoint** Audio-video solution in English & Spanish at LarsonPrecalculus.com

Give a verbal description of the interval $[-2, 5)$.

EXAMPLE 6 Using Inequalities to Represent Intervals

a. The inequality $c \le 2$ can represent the statement "c is at most 2."

b. The inequality $-3 < x \le 5$ can represent "all x in the interval $(-3, 5]$."

✓ **Checkpoint** Audio-video solution in English & Spanish at LarsonPrecalculus.com

Use inequality notation to represent the statement "x is less than 4 and at least -2." ■

Absolute Value and Distance

The **absolute value** of a real number is its *magnitude,* or the distance between the origin and the point representing the real number on the real number line.

Definition of Absolute Value

If a is a real number, then the **absolute value** of a is

$$|a| = \begin{cases} a, & a \geq 0 \\ -a, & a < 0 \end{cases}.$$

Notice in this definition that the absolute value of a real number is never negative. For example, if $a = -5$, then $|-5| = -(-5) = 5$. The absolute value of a real number is either positive or zero. Moreover, 0 is the only real number whose absolute value is 0. So, $|0| = 0$.

Properties of Absolute Values

1. $|a| \geq 0$

2. $|-a| = |a|$

3. $|ab| = |a||b|$

4. $\left|\frac{a}{b}\right| = \frac{|a|}{|b|}, \quad b \neq 0$

EXAMPLE 7 Finding Absolute Values

a. $|-15| = 15$ **b.** $\left|\frac{2}{3}\right| = \frac{2}{3}$

c. $|-4.3| = 4.3$ **d.** $-|-6| = -(6) = -6$

✓ ***Checkpoint***  *Audio-video solution in English & Spanish at LarsonPrecalculus.com*

Evaluate each expression.

a. $|1|$ **b.** $-\left|\frac{3}{4}\right|$ **c.** $\frac{2}{|-3|}$ **d.** $-|0.7|$

EXAMPLE 8 Evaluating an Absolute Value Expression

Evaluate $\frac{|x|}{x}$ for (a) $x > 0$ and (b) $x < 0$.

Solution

a. If $x > 0$, then x is positive and $|x| = x$. So, $\frac{|x|}{x} = \frac{x}{x} = 1$.

b. If $x < 0$, then x is negative and $|x| = -x$. So, $\frac{|x|}{x} = \frac{-x}{x} = -1$.

✓ ***Checkpoint*** *Audio-video solution in English & Spanish at LarsonPrecalculus.com*

Evaluate $\frac{|x+3|}{x+3}$ for (a) $x > -3$ and (b) $x < -3$.

The **Law of Trichotomy** states that for any two real numbers a and b, *precisely* one of three relationships is possible:

$$a = b, \quad a < b, \quad \text{or} \quad a > b. \qquad \text{Law of Trichotomy}$$

EXAMPLE 9 Comparing Real Numbers

Place the appropriate symbol ($<$, $>$, or $=$) between the pair of real numbers.

a. $|-4| \ \square\ |3|$ **b.** $|-10| \ \square\ |10|$ **c.** $-|-7| \ \square\ |-7|$

Solution

a. $|-4| > |3|$ because $|-4| = 4$ and $|3| = 3$, and 4 is greater than 3.

b. $|-10| = |10|$ because $|-10| = 10$ and $|10| = 10$.

c. $-|-7| < |-7|$ because $-|-7| = -7$ and $|-7| = 7$, and -7 is less than 7.

✓ ***Checkpoint*** *Audio-video solution in English & Spanish at LarsonPrecalculus.com*

Place the appropriate symbol ($<$, $>$, or $=$) between the pair of real numbers.

a. $|-3| \ \square\ |4|$

b. $-|-4| \ \square\ -|4|$

c. $|-3| \ \square\ -|-3|$

Absolute value can be used to find the distance between two points on the real number line. For example, the distance between -3 and 4 is

$$|-3 - 4| = |-7|$$
$$= 7$$

as shown in Figure P.8.

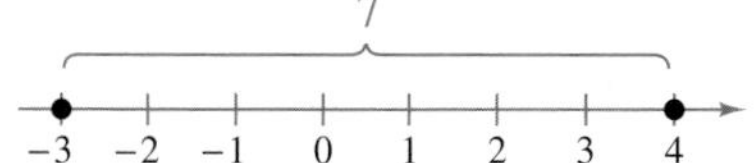

The distance between -3 and 4 is 7.
Figure P.8

Distance Between Two Points on the Real Number Line

Let a and b be real numbers. The **distance between a and b** is

$$d(a, b) = |b - a| = |a - b|.$$

One application of finding the distance between two points on the real number line is finding a change in temperature.

EXAMPLE 10 Finding a Distance

Find the distance between -25 and 13.

Solution

The distance between -25 and 13 is

$$|-25 - 13| = |-38| = 38. \qquad \text{Distance between } -25 \text{ and } 13$$

The distance can also be found as follows.

$$|13 - (-25)| = |38| = 38 \qquad \text{Distance between } -25 \text{ and } 13$$

✓ ***Checkpoint*** *Audio-video solution in English & Spanish at LarsonPrecalculus.com*

a. Find the distance between 35 and -23.

b. Find the distance between -35 and -23.

c. Find the distance between 35 and 23.

Algebraic Expressions

One characteristic of algebra is the use of letters to represent numbers. The letters are **variables,** and combinations of letters and numbers are **algebraic expressions.** Here are a few examples of algebraic expressions.

$$5x, \quad 2x - 3, \quad \frac{4}{x^2 + 2}, \quad 7x + y$$

Definition of an Algebraic Expression

An **algebraic expression** is a collection of letters **(variables)** and real numbers **(constants)** combined using the operations of addition, subtraction, multiplication, division, and exponentiation.

The **terms** of an algebraic expression are those parts that are separated by *addition.* For example, $x^2 - 5x + 8 = x^2 + (-5x) + 8$ has three terms: x^2 and $-5x$ are the **variable terms** and 8 is the **constant term.** For terms such as x^2, $-5x$, and 8, the numerical factor is the **coefficient.** Here, the coefficients are 1, -5, and 8.

EXAMPLE 11 Identifying Terms and Coefficients

Algebraic Expression	Terms	Coefficients
a. $5x - \frac{1}{7}$	$5x, -\frac{1}{7}$	$5, -\frac{1}{7}$
b. $2x^2 - 6x + 9$	$2x^2, -6x, 9$	$2, -6, 9$
c. $\frac{3}{x} + \frac{1}{2}x^4 - y$	$\frac{3}{x}, \frac{1}{2}x^4, -y$	$3, \frac{1}{2}, -1$

Identify the terms and coefficients of $-2x + 4$.

The **Substitution Principle** states, "If $a = b$, then b can replace a in any expression involving a." Use the Substitution Principle to **evaluate** an algebraic expression by substituting numerical values for each of the variables in the expression. The next example illustrates this.

EXAMPLE 12 Evaluating Algebraic Expressions

Expression	Value of Variable	Substitute.	Value of Expression
a. $-3x + 5$	$x = 3$	$-3(3) + 5$	$-9 + 5 = -4$
b. $3x^2 + 2x - 1$	$x = -1$	$3(-1)^2 + 2(-1) - 1$	$3 - 2 - 1 = 0$
c. $\frac{2x}{x + 1}$	$x = -3$	$\frac{2(-3)}{-3 + 1}$	$\frac{-6}{-2} = 3$

Note that you must substitute the value for *each* occurrence of the variable.

Checkpoint Audio-video solution in English & Spanish at LarsonPrecalculus.com

Evaluate $4x - 5$ when $x = 0$.

Basic Rules of Algebra

There are four arithmetic operations with real numbers: *addition, multiplication, subtraction,* and *division,* denoted by the symbols $+$, $\times$ or $\cdot$, $-$, and $\div$ or $/$, respectively. Of these, addition and multiplication are the two primary operations. Subtraction and division are the inverse operations of addition and multiplication, respectively.

Definitions of Subtraction and Division

Subtraction: Add the opposite.

$$a - b = a + (-b)$$

Division: Multiply by the reciprocal.

If $b \neq 0$, then $a/b = a\left(\frac{1}{b}\right) = \frac{a}{b}$.

In these definitions, $-b$ is the **additive inverse** (or opposite) of b, and $1/b$ is the **multiplicative inverse** (or reciprocal) of b. In the fractional form a/b, a is the **numerator** of the fraction and b is the **denominator.**

The properties of real numbers below are true for variables and algebraic expressions as well as for real numbers, so they are often called the **Basic Rules of Algebra.** Formulate a verbal description of each of these properties. For example, the first property states that *the order in which two real numbers are added does not affect their sum.*

Basic Rules of Algebra

Let a, b, and c be real numbers, variables, or algebraic expressions.

Property		Example
Commutative Property of Addition:	$a + b = b + a$	$4x + x^2 = x^2 + 4x$
Commutative Property of Multiplication:	$ab = ba$	$(4 - x)x^2 = x^2(4 - x)$
Associative Property of Addition:	$(a + b) + c = a + (b + c)$	$(x + 5) + x^2 = x + (5 + x^2)$
Associative Property of Multiplication:	$(ab)c = a(bc)$	$(2x \cdot 3y)(8) = (2x)(3y \cdot 8)$
Distributive Properties:	$a(b + c) = ab + ac$	$3x(5 + 2x) = 3x \cdot 5 + 3x \cdot 2x$
	$(a + b)c = ac + bc$	$(y + 8)y = y \cdot y + 8 \cdot y$
Additive Identity Property:	$a + 0 = a$	$5y^2 + 0 = 5y^2$
Multiplicative Identity Property:	$a \cdot 1 = a$	$(4x^2)(1) = 4x^2$
Additive Inverse Property:	$a + (-a) = 0$	$5x^3 + (-5x^3) = 0$
Multiplicative Inverse Property:	$a \cdot \frac{1}{a} = 1, \quad a \neq 0$	$(x^2 + 4)\left(\frac{1}{x^2 + 4}\right) = 1$

Subtraction is defined as "adding the opposite," so the Distributive Properties are also true for subtraction. For example, the "subtraction form" of $a(b + c) = ab + ac$ is $a(b - c) = ab - ac$. Note that the operations of subtraction and division are neither commutative nor associative. The examples

$$7 - 3 \neq 3 - 7 \quad \text{and} \quad 20 \div 4 \neq 4 \div 20$$

show that subtraction and division are not commutative. Similarly

$$5 - (3 - 2) \neq (5 - 3) - 2 \quad \text{and} \quad 16 \div (4 \div 2) \neq (16 \div 4) \div 2$$

demonstrate that subtraction and division are not associative.

EXAMPLE 13 Identifying Rules of Algebra

Identify the rule of algebra illustrated by the statement.

a. $(5x^3)2 = 2(5x^3)$ **b.** $(4x + 3) - (4x + 3) = 0$

c. $7x \cdot \frac{1}{7x} = 1, \quad x \neq 0$ **d.** $(2 + 5x^2) + x^2 = 2 + (5x^2 + x^2)$

Solution

a. This statement illustrates the Commutative Property of Multiplication. In other words, you obtain the same result whether you multiply $5x^3$ by 2, or 2 by $5x^3$.

b. This statement illustrates the Additive Inverse Property. In terms of subtraction, this property states that when any expression is subtracted from itself, the result is 0.

c. This statement illustrates the Multiplicative Inverse Property. Note that x must be a nonzero number. The reciprocal of x is undefined when x is 0.

d. This statement illustrates the Associative Property of Addition. In other words, to form the sum $2 + 5x^2 + x^2$, it does not matter whether 2 and $5x^2$, or $5x^2$ and x^2 are added first.

Identify the rule of algebra illustrated by the statement.

a. $x + 9 = 9 + x$ **b.** $5(x^3 \cdot 2) = (5x^3)2$ **c.** $(2 + 5x^2)y^2 = 2 \cdot y^2 + 5x^2 \cdot y^2$

REMARK Notice the difference between the *opposite of a number* and a *negative number.* If a is already negative, then its opposite, $-a$, is positive. For example, if $a = -5$, then

$$-a = -(-5) = 5.$$

Properties of Negation and Equality

Let a, b, and c be real numbers, variables, or algebraic expressions.

Property	Example
1. $(-1)a = -a$	$(-1)7 = -7$
2. $-(-a) = a$	$-(-6) = 6$
3. $(-a)b = -(ab) = a(-b)$	$(-5)3 = -(5 \cdot 3) = 5(-3)$
4. $(-a)(-b) = ab$	$(-2)(-x) = 2x$
5. $-(a + b) = (-a) + (-b)$	$-(x + 8) = (-x) + (-8)$ $= -x - 8$
6. If $a = b$, then $a \pm c = b \pm c$.	$\frac{1}{2} + 3 = 0.5 + 3$
7. If $a = b$, then $ac = bc$.	$4^2 \cdot 2 = 16 \cdot 2$
8. If $a \pm c = b \pm c$, then $a = b$.	$1.4 - 1 = \frac{7}{5} - 1 \Rightarrow 1.4 = \frac{7}{5}$
9. If $ac = bc$ and $c \neq 0$, then $a = b$.	$3x = 3 \cdot 4 \Rightarrow x = 4$

REMARK The "or" in the Zero-Factor Property includes the possibility that either or both factors may be zero. This is an *inclusive or,* and it is generally the way the word "or" is used in mathematics.

Properties of Zero

Let a and b be real numbers, variables, or algebraic expressions.

1. $a + 0 = a$ and $a - 0 = a$ **2.** $a \cdot 0 = 0$

3. $\frac{0}{a} = 0, \quad a \neq 0$ **4.** $\frac{a}{0}$ is undefined.

5. Zero-Factor Property: If $ab = 0$, then $a = 0$ or $b = 0$.

Properties and Operations of Fractions

Let a, b, c, and d be real numbers, variables, or algebraic expressions such that $b \neq 0$ and $d \neq 0$.

1. **Equivalent Fractions:** $\dfrac{a}{b} = \dfrac{c}{d}$ if and only if $ad = bc$.
2. **Rules of Signs:** $-\dfrac{a}{b} = \dfrac{-a}{b} = \dfrac{a}{-b}$ and $\dfrac{-a}{-b} = \dfrac{a}{b}$
3. **Generate Equivalent Fractions:** $\dfrac{a}{b} = \dfrac{ac}{bc}, \quad c \neq 0$
4. **Add or Subtract with Like Denominators:** $\dfrac{a}{b} \pm \dfrac{c}{b} = \dfrac{a \pm c}{b}$
5. **Add or Subtract with Unlike Denominators:** $\dfrac{a}{b} \pm \dfrac{c}{d} = \dfrac{ad \pm bc}{bd}$
6. **Multiply Fractions:** $\dfrac{a}{b} \cdot \dfrac{c}{d} = \dfrac{ac}{bd}$
7. **Divide Fractions:** $\dfrac{a}{b} \div \dfrac{c}{d} = \dfrac{a}{b} \cdot \dfrac{d}{c} = \dfrac{ad}{bc}, \quad c \neq 0$

REMARK In Property 1, the phrase "if and only if" implies two statements. One statement is: If $a/b = c/d$, then $ad = bc$. The other statement is: If $ad = bc$, where $b \neq 0$ and $d \neq 0$, then $a/b = c/d$.

EXAMPLE 14 Properties and Operations of Fractions

a. $\dfrac{x}{5} = \dfrac{3 \cdot x}{3 \cdot 5} = \dfrac{3x}{15}$ **b.** $\dfrac{7}{x} \div \dfrac{3}{2} = \dfrac{7}{x} \cdot \dfrac{2}{3} = \dfrac{14}{3x}$

✓ ***Checkpoint*** *Audio-video solution in English & Spanish at LarsonPrecalculus.com*

a. Multiply fractions: $\dfrac{3}{5} \cdot \dfrac{x}{6}$ **b.** Add fractions: $\dfrac{x}{10} + \dfrac{2x}{5}$

REMARK The number 1 is neither prime nor composite.

If a, b, and c are integers such that $ab = c$, then a and b are **factors** or **divisors** of c. A **prime number** is an integer that has exactly two positive factors—itself and 1—such as 2, 3, 5, 7, and 11. The numbers 4, 6, 8, 9, and 10 are **composite** because each can be written as the product of two or more prime numbers. The **Fundamental Theorem of Arithmetic** states that every positive integer greater than 1 is a prime number or can be written as the product of prime numbers in precisely one way (disregarding order). For example, the *prime factorization* of 24 is $24 = 2 \cdot 2 \cdot 2 \cdot 3$.

Summarize (Section P.1)

1. Explain how to represent and classify real numbers *(pages 2 and 3)*. For examples of representing and classifying real numbers, see Examples 1 and 2.
2. Explain how to order real numbers and use inequalities *(pages 4 and 5)*. For examples of ordering real numbers and using inequalities, see Examples 3–6.
3. State the definition of the absolute value of a real number *(page 6)*. For examples of using absolute value, see Examples 7–10.
4. Explain how to evaluate an algebraic expression *(page 8)*. For examples involving algebraic expressions, see Examples 11 and 12.
5. State the basic rules and properties of algebra *(pages 9–11)*. For examples involving the basic rules and properties of algebra, see Examples 13 and 14.

P.1 Exercises

See CalcChat.com for tutorial help and worked-out solutions to odd-numbered exercises.

Vocabulary: Fill in the blanks.

1. The decimal representation of an ________ number neither terminates nor repeats.
2. The point representing 0 on the real number line is the ________.
3. The distance between the origin and a point representing a real number on the real number line is the ________ ________ of the real number.
4. A number that can be written as the product of two or more prime numbers is a ________ number.
5. The ________ of an algebraic expression are those parts that are separated by addition.
6. The ________ ________ states that if $ab = 0$, then $a = 0$ or $b = 0$.

Skills and Applications

Classifying Real Numbers In Exercises 7–10, determine which numbers in the set are (a) natural numbers, (b) whole numbers, (c) integers, (d) rational numbers, and (e) irrational numbers.

7. $\left\{-9, -\frac{7}{2}, 5, \frac{2}{3}, \sqrt{2}, 0, 1, -4, 2, -11\right\}$
8. $\left\{\sqrt{5}, -7, -\frac{7}{3}, 0, 3.14, \frac{5}{4}, -3, 12, 5\right\}$
9. $\{2.01, 0.\overline{6}, -13, 0.010110111 \ldots, 1, -6\}$
10. $\left\{25, -17, -\frac{12}{5}, \sqrt{9}, 3.12, \frac{1}{2}\pi, 7, -11.1, 13\right\}$

Plotting Points on the Real Number Line In Exercises 11 and 12, plot the real numbers on the real number line.

11. (a) 3 (b) $\frac{7}{2}$ (c) $-\frac{5}{2}$ (d) -5.2
12. (a) 8.5 (b) $\frac{4}{3}$ (c) -4.75 (d) $-\frac{8}{3}$

Plotting and Ordering Real Numbers In Exercises 13–16, plot the two real numbers on the real number line. Then place the appropriate inequality symbol (< or >) between them.

13. $-4, -8$
14. $1, \frac{16}{3}$
15. $\frac{5}{6}, \frac{2}{3}$
16. $-\frac{8}{7}, -\frac{3}{7}$

Interpreting an Inequality or an Interval In Exercises 17–24, (a) give a verbal description of the subset of real numbers represented by the inequality or the interval, (b) sketch the subset on the real number line, and (c) state whether the subset is bounded or unbounded.

17. $x \leq 5$
18. $x < 0$
19. $-2 < x < 2$
20. $0 < x \leq 6$
21. $[4, \infty)$
22. $(-\infty, 2)$
23. $[-5, 2)$
24. $(-1, 2]$

Using Inequality and Interval Notation In Exercises 25–28, use inequality notation and interval notation to describe the set.

25. y is nonnegative.
26. y is no more than 25.
27. t is at least 10 and at most 22.
28. k is less than 5 but no less than -3.

Evaluating an Absolute Value Expression In Exercises 29–38, evaluate the expression.

29. $|-10|$
30. $|0|$
31. $|3 - 8|$
32. $|6 - 2|$
33. $|-1| - |-2|$
34. $-3 - |-3|$
35. $5|-5|$
36. $-4|-4|$
37. $\dfrac{|x + 2|}{x + 2}, \quad x < -2$
38. $\dfrac{|x - 1|}{x - 1}, \quad x > 1$

Comparing Real Numbers In Exercises 39–42, place the appropriate symbol (<, >, or =) between the pair of real numbers.

39. $|-4| \;\square\; |4|$
40. $-5 \;\square\; -|5|$
41. $-|-6| \;\square\; |-6|$
42. $-|-2| \;\square\; -|2|$

Finding a Distance In Exercises 43–46, find the distance between a and b.

43. $a = 126, b = 75$
44. $a = -20, b = 30$
45. $a = -\frac{5}{2}, b = 0$
46. $a = -\frac{1}{4}, b = -\frac{11}{4}$

Using Absolute Value Notation In Exercises 47 and 48, use absolute value notation to represent the situation.

47. The distance between x and 5 is no more than 3.
48. The distance between x and -10 is at least 6.

The symbol and a red exercise number indicates that a video solution can be seen at *CalcView.com*.

Federal Deficit

In Exercises 49–52, use the bar graph, which shows the receipts of the federal government (in billions of dollars) for selected years from 2008 through 2014. In each exercise, you are given the expenditures of the federal government. Find the magnitude of the surplus or deficit for the year. *(Source: U.S. Office of Management and Budget)*

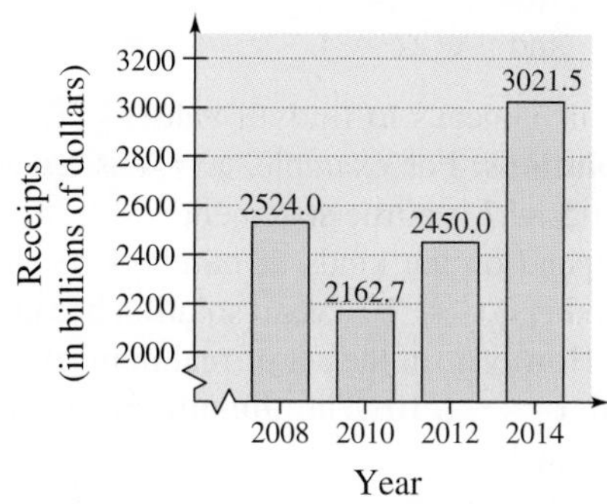

	Year	Receipts, R	Expenditures, E	$\lvert R - E \rvert$
49.	2008		\$2982.5 billion	
50.	2010		\$3457.1 billion	
51.	2012		\$3537.0 billion	
52.	2014		\$3506.1 billion	

Identifying Terms and Coefficients **In Exercises 53–58, identify the terms. Then identify the coefficients of the variable terms of the expression.**

53. $7x + 4$ **54.** $2x - 3$

55. $6x^3 - 5x$ **56.** $4x^3 + 0.5x - 5$

57. $3\sqrt{3}x^2 + 1$ **58.** $2\sqrt{2}x^2 - 3$

Evaluating an Algebraic Expression **In Exercises 59–64, evaluate the expression for each value of x. (If not possible, state the reason.)**

59. $4x - 6$ (a) $x = -1$ (b) $x = 0$

60. $9 - 7x$ (a) $x = -3$ (b) $x = 3$

61. $x^2 - 3x + 2$ (a) $x = 0$ (b) $x = -1$

62. $-x^2 + 5x - 4$ (a) $x = -1$ (b) $x = 1$

63. $\dfrac{x + 1}{x - 1}$ (a) $x = 1$ (b) $x = -1$

64. $\dfrac{x - 2}{x + 2}$ (a) $x = 2$ (b) $x = -2$

Michael G Smith/Shutterstock.com

Identifying Rules of Algebra **In Exercises 65–68, identify the rule(s) of algebra illustrated by the statement.**

65. $\dfrac{1}{h + 6}(h + 6) = 1, \quad h \neq -6$

66. $(x + 3) - (x + 3) = 0$

67. $x(3y) = (x \cdot 3)y = (3x)y$

68. $\frac{1}{7}(7 \cdot 12) = \left(\frac{1}{7} \cdot 7\right)12 = 1 \cdot 12 = 12$

Operations with Fractions **In Exercises 69–72, perform the operation. (Write fractional answers in simplest form.)**

69. $\dfrac{2x}{3} - \dfrac{x}{4}$ **70.** $\dfrac{3x}{4} + \dfrac{x}{5}$

71. $\dfrac{3x}{10} \cdot \dfrac{5}{6}$ **72.** $\dfrac{2x}{3} \div \dfrac{6}{7}$

Exploration

True or False? **In Exercises 73–75, determine whether the statement is true or false. Justify your answer.**

73. Every nonnegative number is positive.

74. If $a > 0$ and $b < 0$, then $ab > 0$.

75. If $a < 0$ and $b < 0$, then $ab > 0$.

76. **HOW DO YOU SEE IT?** Match each description with its graph. Which types of real numbers shown in Figure P.1 on page 2 may be included in a range of prices? a range of lengths? Explain.

(i) [number line with closed interval from 1.87 to 1.93; ticks 1.87 1.88 1.89 1.90 1.91 1.92 1.93]

(ii) [number line with points at 1.87 1.88 1.89 1.90 1.91 1.92 1.93]

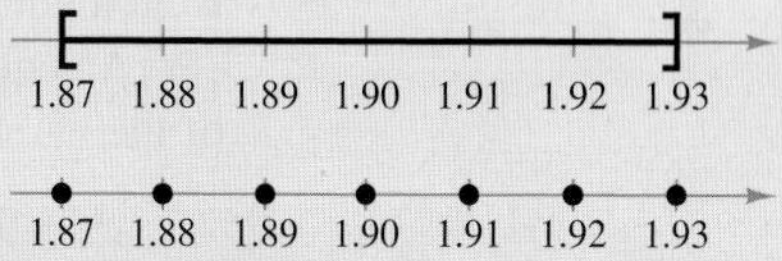

(a) The price of an item is within \$0.03 of \$1.90.

(b) The distance between the prongs of an electric plug may not differ from 1.9 centimeters by more than 0.03 centimeter.

77. **Conjecture**

(a) Use a calculator to complete the table.

n	0.0001	0.01	1	100	10,000
$\dfrac{5}{n}$					

(b) Use the result from part (a) to make a conjecture about the value of $5/n$ as n (i) approaches 0, and (ii) increases without bound.

P.2 Solving Equations

Linear equations have many real-life applications, such as in forensics. For example, in Exercises 107 and 108 on page 25, you will use linear equations to determine height from femur length.

- **Identify different types of equations.**
- **Solve linear equations in one variable and rational equations.**
- **Solve quadratic equations by factoring, extracting square roots, completing the square, and using the Quadratic Formula.**
- **Solve polynomial equations of degree three or greater.**
- **Solve radical equations.**
- **Solve absolute value equations.**

Equations and Solutions of Equations

An **equation** in x is a statement that two algebraic expressions are equal. For example,

$$3x - 5 = 7, \quad x^2 - x - 6 = 0, \quad \text{and} \quad \sqrt{2x} = 4$$

are equations. To **solve** an equation in x means to find all values of x for which the equation is true. Such values are **solutions.** For example, $x = 4$ is a solution of the equation $3x - 5 = 7$ because $3(4) - 5 = 7$ is a true statement.

The solutions of an equation depend on the kinds of numbers being considered. For example, in the set of rational numbers, $x^2 = 10$ has no solution because there is no rational number whose square is 10. However, in the set of real numbers, the equation has the two solutions $x = \sqrt{10}$ and $x = -\sqrt{10}$. The **domain** is the set of all real numbers for which the equation is defined.

An equation that is true for *every* real number in the domain of the variable is an **identity.** For example,

$$x^2 - 9 = (x + 3)(x - 3) \qquad \text{Identity}$$

is an identity because it is a true statement for any real value of x. The equation

$$\frac{x}{3x^2} = \frac{1}{3x} \qquad \text{Identity}$$

is an identity because it is true for any nonzero real value of x.

An equation that is true for just *some* (but not all) of the real numbers in the domain of the variable is a **conditional equation.** For example, the equation

$$x^2 - 9 = 0 \qquad \text{Conditional equation}$$

is conditional because $x = 3$ and $x = -3$ are the only values in the domain that satisfy the equation.

A **contradiction** is an equation that is false for *every* real number in the domain of the variable. For example, the equation

$$2x - 4 = 2x + 1 \qquad \text{Contradiction}$$

is a contradiction because there are no real values of x for which the equation is true.

Linear and Rational Equations

Definition of a Linear Equation in One Variable

A **linear equation in one variable** x is an equation that can be written in the standard form

$$ax + b = 0$$

where a and b are real numbers with $a \neq 0$.

iStockphoto.com/Nikola Nastasic

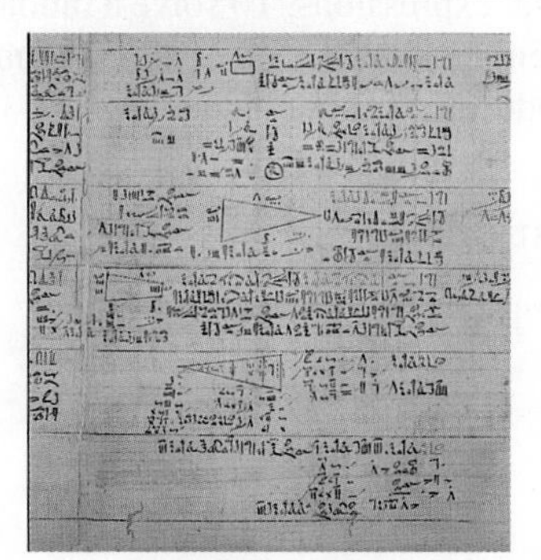

HISTORICAL NOTE

This ancient Egyptian papyrus, discovered in 1858, contains one of the earliest examples of mathematical writing in existence. The papyrus itself dates back to around 1650 B.C., but it is actually a copy of writings from two centuries earlier. The algebraic equations on the papyrus were written in words. Diophantus, a Greek who lived around A.D. 250, is often called the Father of Algebra. He was the first to use abbreviated word forms in equations.

A linear equation in one variable has exactly one solution. To see this, consider the steps below. (Remember that $a \neq 0$.)

$$ax + b = 0 \qquad \text{Write original equation.}$$

$$ax = -b \qquad \text{Subtract } b \text{ from each side.}$$

$$x = -\frac{b}{a} \qquad \text{Divide each side by } a.$$

The above suggests that to solve a conditional equation in x, you isolate x on one side of the equation using a sequence of **equivalent equations,** each having the same solution as the original equation. The operations that yield equivalent equations come from the properties of equality reviewed in Section P.1.

Generating Equivalent Equations

An equation can be transformed into an *equivalent equation* by one or more of the steps listed below.

	Given Equation	Equivalent Equation
1. Remove symbols of grouping, combine like terms, or simplify fractions on one or both sides of the equation.	$2x - x = 4$	$x = 4$
2. Add (or subtract) the same quantity to (from) *each* side of the equation.	$x + 1 = 6$	$x = 5$
3. Multiply (or divide) *each* side of the equation by the same *nonzero* quantity.	$2x = 6$	$x = 3$
4. Interchange the two sides of the equation.	$2 = x$	$x = 2$

In Example 1, you will use these steps to solve linear equations in one variable x.

REMARK After solving an equation, you should check each solution in the original equation. For instance, here is a check of the solution in Example 1(a).

$3x - 6 = 0$ Write original equation.

$3(2) - 6 \stackrel{?}{=} 0$ Substitute 2 for x.

$0 = 0$ Solution checks. ✓

Check the solution in Example 1(b) on your own.

EXAMPLE 1 Solving Linear Equations

a. $3x - 6 = 0$ Original equation

$3x = 6$ Add 6 to each side.

$x = 2$ Divide each side by 3.

b. $5x + 4 = 3x - 8$ Original equation

$2x + 4 = -8$ Subtract $3x$ from each side.

$2x = -12$ Subtract 4 from each side.

$x = -6$ Divide each side by 2.

✓ ***Checkpoint*** *Audio-video solution in English & Spanish at LarsonPrecalculus.com*

Solve each equation.

a. $7 - 2x = 15$

b. $7x - 9 = 5x + 7$

REMARK An equation with a single fraction on each side can be cleared of denominators by *cross multiplying.* To do this, multiply the left numerator by the right denominator and the right numerator by the left denominator.

$\frac{a}{b} = \frac{c}{d}$ Original equation

$ad = cb$ Cross multiply.

A **rational equation** involves one or more rational expressions. To solve a rational equation, multiply every term by the least common denominator (LCD) of all the terms. This clears the original equation of fractions and produces a simpler equation.

EXAMPLE 2 Solving a Rational Equation

Solve $\frac{x}{3} + \frac{3x}{4} = 2.$

Solution

$$\frac{x}{3} + \frac{3x}{4} = 2 \qquad \text{Write original equation.}$$

$$(12)\frac{x}{3} + (12)\frac{3x}{4} = (12)2 \qquad \text{Multiply each term by the LCD.}$$

$$4x + 9x = 24 \qquad \text{Simplify.}$$

$$13x = 24 \qquad \text{Combine like terms.}$$

$$x = \frac{24}{13} \qquad \text{Divide each side by 13.}$$

The solution is $x = \frac{24}{13}$. Check this in the original equation.

Checkpoint Audio-video solution in English & Spanish at LarsonPrecalculus.com

Solve $\frac{4x}{9} - \frac{1}{3} = x + \frac{5}{3}.$

When multiplying or dividing an equation by a *variable expression,* it is possible to introduce an **extraneous solution,** which is a solution that does not satisfy the original equation.

EXAMPLE 3 An Equation with an Extraneous Solution

See LarsonPrecalculus.com for an interactive version of this type of example.

Solve $\frac{1}{x-2} = \frac{3}{x+2} - \frac{6x}{x^2-4}.$

REMARK Recall that the least common denominator of two or more fractions consists of the product of all prime factors in the denominators, with each factor given the highest power of its occurrence in any denominator. For instance, in Example 3, factoring the denominator $x^2 - 4$ shows that the LCD is $(x + 2)(x - 2)$.

Solution The LCD is $x^2 - 4 = (x + 2)(x - 2)$. Multiply each term by the LCD.

$$\frac{1}{x-2}(x+2)(x-2) = \frac{3}{x+2}(x+2)(x-2) - \frac{6x}{x^2-4}(x+2)(x-2)$$

$$x + 2 = 3(x - 2) - 6x, \quad x \neq \pm 2$$

$$x + 2 = 3x - 6 - 6x$$

$$x + 2 = -3x - 6$$

$$4x = -8$$

$$x = -2 \qquad \text{Extraneous solution}$$

In the original equation, $x = -2$ yields a denominator of zero. So, $x = -2$ is an extraneous solution, and the original equation has *no solution.*

Checkpoint Audio-video solution in English & Spanish at LarsonPrecalculus.com

Solve $\frac{3x}{x-4} = 5 + \frac{12}{x-4}.$

Quadratic Equations

A **quadratic equation** in x is an equation that can be written in the general form

$$ax^2 + bx + c = 0$$

where a, b, and c are real numbers with $a \neq 0$. A quadratic equation in x is also called a **second-degree polynomial equation** in x.

You should be familiar with the four methods for solving quadratic equations listed below.

Solving a Quadratic Equation

Factoring

If $ab = 0$, then $a = 0$ or $b = 0$. Zero-Factor Property

Example:
$$x^2 - x - 6 = 0$$
$$(x - 3)(x + 2) = 0$$
$$x - 3 = 0 \Rightarrow x = 3$$
$$x + 2 = 0 \Rightarrow x = -2$$

Extracting Square Roots

If $u^2 = c$, where $c > 0$, then $u = \pm\sqrt{c}$. Square Root Principle

Example:
$$(x + 3)^2 = 16$$
$$x + 3 = \pm 4$$
$$x = -3 \pm 4$$
$$x = 1 \quad \text{or} \quad x = -7$$

Completing the Square

If $x^2 + bx = c$, then

$$x^2 + bx + \left(\frac{b}{2}\right)^2 = c + \left(\frac{b}{2}\right)^2$$ Add $\left(\frac{b}{2}\right)^2$ to each side.

$$\left(x + \frac{b}{2}\right)^2 = c + \frac{b^2}{4}.$$

Example:
$$x^2 + 6x = 5$$
$$x^2 + 6x + 3^2 = 5 + 3^2$$ Add $\left(\frac{6}{2}\right)^2$ to each side.
$$(x + 3)^2 = 14$$
$$x + 3 = \pm\sqrt{14}$$
$$x = -3 \pm \sqrt{14}$$

Quadratic Formula

If $ax^2 + bx + c = 0$, then $x = \dfrac{-b \pm \sqrt{b^2 - 4ac}}{2a}$.

Example: $2x^2 + 3x - 1 = 0$

$$x = \frac{-3 \pm \sqrt{3^2 - 4(2)(-1)}}{2(2)}$$
$$= \frac{-3 \pm \sqrt{17}}{4}$$

REMARK It is possible to solve every quadratic equation by completing the square or using the Quadratic Formula.

EXAMPLE 4 Solving Quadratic Equations by Factoring

a.

$2x^2 + 9x + 7 = 3$	Original equation
$2x^2 + 9x + 4 = 0$	Write in general form.
$(2x + 1)(x + 4) = 0$	Factor.
$2x + 1 = 0 \Rightarrow x = -\frac{1}{2}$	Set 1st factor equal to 0 and solve.
$x + 4 = 0 \Rightarrow x = -4$	Set 2nd factor equal to 0 and solve.

The solutions are $x = -\frac{1}{2}$ and $x = -4$. Check these in the original equation.

b.

$6x^2 - 3x = 0$	Original equation
$3x(2x - 1) = 0$	Factor.
$3x = 0 \Rightarrow x = 0$	Set 1st factor equal to 0 and solve.
$2x - 1 = 0 \Rightarrow x = \frac{1}{2}$	Set 2nd factor equal to 0 and solve.

The solutions are $x = 0$ and $x = \frac{1}{2}$. Check these in the original equation.

Checkpoint *Audio-video solution in English & Spanish at LarsonPrecalculus.com*

Solve $2x^2 - 3x + 1 = 6$ by factoring.

Note that the method of solution in Example 4 is based on the Zero-Factor Property from Section P.1. This property applies *only* to equations written in general form (in which the right side of the equation is zero). So, collect all terms on one side *before* factoring. For example, in the equation $(x - 5)(x + 2) = 8$, it is *incorrect* to set each factor equal to 8. Solve this equation correctly on your own. Then check the solutions in the original equation.

EXAMPLE 5 Extracting Square Roots

Solve each equation by extracting square roots.

a. $4x^2 = 12$

b. $(x - 3)^2 = 7$

Solution

a.

$4x^2 = 12$	Write original equation.
$x^2 = 3$	Divide each side by 4.
$x = \pm\sqrt{3}$	Extract square roots.

The solutions are $x = \sqrt{3}$ and $x = -\sqrt{3}$. Check these in the original equation.

b.

$(x - 3)^2 = 7$	Write original equation.
$x - 3 = \pm\sqrt{7}$	Extract square roots.
$x = 3 \pm \sqrt{7}$	Add 3 to each side.

The solutions are $x = 3 \pm \sqrt{7}$. Check these in the original equation.

Checkpoint *Audio-video solution in English & Spanish at LarsonPrecalculus.com*

Solve each equation by extracting square roots.

a. $3x^2 = 36$

b. $(x - 1)^2 = 10$

When solving quadratic equations by completing the square, you must add $(b/2)^2$ to *each side* in order to maintain equality. When the leading coefficient is *not* 1, divide each side of the equation by the leading coefficient *before* completing the square, as shown in Example 7.

EXAMPLE 6 Completing the Square: Leading Coefficient Is 1

Solve $x^2 + 2x - 6 = 0$ by completing the square.

Solution

$$x^2 + 2x - 6 = 0 \qquad \text{Write original equation.}$$

$$x^2 + 2x = 6 \qquad \text{Add 6 to each side.}$$

$$x^2 + 2x + 1^2 = 6 + 1^2 \qquad \text{Add } 1^2 \text{ to each side.}$$

(Half of 2)^2

$$(x + 1)^2 = 7 \qquad \text{Simplify.}$$

$$x + 1 = \pm\sqrt{7} \qquad \text{Extract square roots.}$$

$$x = -1 \pm \sqrt{7} \qquad \text{Subtract 1 from each side.}$$

The solutions are

$$x = -1 \pm \sqrt{7}.$$

Check these in the original equation.

✓ Checkpoint *Audio-video solution in English & Spanish at LarsonPrecalculus.com*

Solve $x^2 - 4x - 1 = 0$ by completing the square.

EXAMPLE 7 Completing the Square: Leading Coefficient Is Not 1

Solve $3x^2 - 4x - 5 = 0$ by completing the square.

Solution

$$3x^2 - 4x - 5 = 0 \qquad \text{Write original equation.}$$

$$3x^2 - 4x = 5 \qquad \text{Add 5 to each side.}$$

$$x^2 - \frac{4}{3}x = \frac{5}{3} \qquad \text{Divide each side by 3.}$$

$$x^2 - \frac{4}{3}x + \left(-\frac{2}{3}\right)^2 = \frac{5}{3} + \left(-\frac{2}{3}\right)^2 \qquad \text{Add } \left(-\tfrac{2}{3}\right)^2 \text{ to each side.}$$

(Half of $-\frac{4}{3}$)^2

$$\left(x - \frac{2}{3}\right)^2 = \frac{19}{9} \qquad \text{Simplify.}$$

$$x - \frac{2}{3} = \pm\frac{\sqrt{19}}{3} \qquad \text{Extract square roots.}$$

$$x = \frac{2}{3} \pm \frac{\sqrt{19}}{3} \qquad \text{Add } \tfrac{2}{3} \text{ to each side.}$$

✓ Checkpoint *Audio-video solution in English & Spanish at LarsonPrecalculus.com*

Solve $3x^2 - 10x - 2 = 0$ by completing the square.

EXAMPLE 8 The Quadratic Formula: Two Distinct Solutions

Use the Quadratic Formula to solve $x^2 + 3x = 9$.

Solution

$$x^2 + 3x = 9 \qquad \text{Write original equation.}$$

$$x^2 + 3x - 9 = 0 \qquad \text{Write in general form.}$$

$$x = \frac{-b \pm \sqrt{b^2 - 4ac}}{2a} \qquad \text{Quadratic Formula}$$

$$x = \frac{-3 \pm \sqrt{(3)^2 - 4(1)(-9)}}{2(1)} \qquad \text{Substitute } a = 1,\ b = 3, \text{ and } c = -9.$$

$$x = \frac{-3 \pm \sqrt{45}}{2} \qquad \text{Simplify.}$$

$$x = \frac{-3 \pm 3\sqrt{5}}{2} \qquad \text{Simplify.}$$

REMARK When you use the Quadratic Formula, remember that *before* applying the formula, you must first write the quadratic equation in general form.

The two solutions are

$$x = \frac{-3 + 3\sqrt{5}}{2} \quad \text{and} \quad x = \frac{-3 - 3\sqrt{5}}{2}.$$

Check these in the original equation.

✓ ***Checkpoint*** *Audio-video solution in English & Spanish at LarsonPrecalculus.com*

Use the Quadratic Formula to solve $3x^2 + 2x = 10$.

EXAMPLE 9 The Quadratic Formula: One Solution

Use the Quadratic Formula to solve $8x^2 - 24x + 18 = 0$.

Solution

$$8x^2 - 24x + 18 = 0 \qquad \text{Write original equation.}$$

$$4x^2 - 12x + 9 = 0 \qquad \text{Divide out common factor of 2.}$$

$$x = \frac{-b \pm \sqrt{b^2 - 4ac}}{2a} \qquad \text{Quadratic Formula}$$

$$x = \frac{-(-12) \pm \sqrt{(-12)^2 - 4(4)(9)}}{2(4)} \qquad \text{Substitute } a = 4,\ b = -12, \text{ and } c = 9.$$

$$x = \frac{12 \pm \sqrt{0}}{8} \qquad \text{Simplify.}$$

$$x = \frac{3}{2} \qquad \text{Simplify.}$$

This quadratic equation has only one solution: $x = \frac{3}{2}$. Check this in the original equation.

✓ ***Checkpoint*** *Audio-video solution in English & Spanish at LarsonPrecalculus.com*

Use the Quadratic Formula to solve $18x^2 - 48x + 32 = 0$. ■

Note that you could have solved Example 9 without first dividing out a common factor of 2. Substituting $a = 8$, $b = -24$, and $c = 18$ into the Quadratic Formula produces the same result.

Polynomial Equations of Higher Degree

Sometimes, the methods used to solve quadratic equations can be extended to solve polynomial equations of higher degrees.

REMARK A common mistake when solving an equation such as that in Example 10 is to divide each side of the equation by the variable factor x^2. This loses the solution $x = 0$. When solving a polynomial equation, always write the equation in general form, then factor the polynomial and set each factor equal to zero. Do not divide each side of an equation by a variable factor in an attempt to simplify the equation.

EXAMPLE 10 Solving a Polynomial Equation by Factoring

Solve $3x^4 = 48x^2$ and check your solution(s).

Solution First write the polynomial equation in general form. Then factor the polynomial, set each factor equal to zero, and solve.

$3x^4 = 48x^2$ Write original equation.

$3x^4 - 48x^2 = 0$ Write in general form.

$3x^2(x^2 - 16) = 0$ Factor out common factor.

$3x^2(x + 4)(x - 4) = 0$ Factor completely.

$3x^2 = 0 \Rightarrow x = 0$ Set 1st factor equal to 0 and solve.

$x + 4 = 0 \Rightarrow x = -4$ Set 2nd factor equal to 0 and solve.

$x - 4 = 0 \Rightarrow x = 4$ Set 3rd factor equal to 0 and solve.

Check these solutions by substituting in the original equation.

Check

$3(0)^4 \stackrel{?}{=} 48(0)^2 \Rightarrow 0 = 0$ 0 checks. ✓

$3(-4)^4 \stackrel{?}{=} 48(-4)^2 \Rightarrow 768 = 768$ -4 checks. ✓

$3(4)^4 \stackrel{?}{=} 48(4)^2 \Rightarrow 768 = 768$ 4 checks. ✓

So, the solutions are

$x = 0, \quad x = -4, \quad \text{and} \quad x = 4.$

✓ ***Checkpoint*** *Audio-video solution in English & Spanish at LarsonPrecalculus.com*

Solve $9x^4 - 12x^2 = 0$ and check your solution(s).

EXAMPLE 11 Solving a Polynomial Equation by Factoring

Solve $x^3 - 3x^2 - 3x + 9 = 0$.

Solution

$x^3 - 3x^2 - 3x + 9 = 0$ Write original equation.

$x^2(x - 3) - 3(x - 3) = 0$ Group terms and factor.

$(x - 3)(x^2 - 3) = 0$ $(x - 3)$ is a common factor.

$x - 3 = 0 \Rightarrow x = 3$ Set 1st factor equal to 0 and solve.

$x^2 - 3 = 0 \Rightarrow x = \pm\sqrt{3}$ Set 2nd factor equal to 0 and solve.

The solutions are $x = 3$, $x = \sqrt{3}$, and $x = -\sqrt{3}$. Check these in the original equation.

✓ ***Checkpoint*** *Audio-video solution in English & Spanish at LarsonPrecalculus.com*

Solve each equation.

a. $x^3 - 5x^2 - 2x + 10 = 0$

b. $6x^3 - 27x^2 - 54x = 0$

Radical Equations

REMARK When squaring each side of an equation or raising each side of an equation to a rational power, it is possible to introduce extraneous solutions. So when using such operations, checking your solutions is crucial.

A **radical equation** is an equation that involves one or more radical expressions. Examples 12 and 13 demonstrate how to solve radical equations.

EXAMPLE 12 Solving Radical Equations

a. $\sqrt{2x+7} - x = 2$ Original equation

$\sqrt{2x+7} = x + 2$ Isolate radical.

$2x + 7 = x^2 + 4x + 4$ Square each side.

$0 = x^2 + 2x - 3$ Write in general form.

$0 = (x+3)(x-1)$ Factor.

$x + 3 = 0 \implies x = -3$ Set 1st factor equal to 0 and solve.

$x - 1 = 0 \implies x = 1$ Set 2nd factor equal to 0 and solve.

Checking these values shows that the only solution is $x = 1$.

b. $\sqrt{2x-5} - \sqrt{x-3} = 1$ Original equation

$\sqrt{2x-5} = \sqrt{x-3} + 1$ Isolate $\sqrt{2x-5}$.

$2x - 5 = x - 3 + 2\sqrt{x-3} + 1$ Square each side.

$x - 3 = 2\sqrt{x-3}$ Isolate $2\sqrt{x-3}$.

$x^2 - 6x + 9 = 4(x-3)$ Square each side.

$x^2 - 10x + 21 = 0$ Write in general form.

$(x-3)(x-7) = 0$ Factor.

$x - 3 = 0 \implies x = 3$ Set 1st factor equal to 0 and solve.

$x - 7 = 0 \implies x = 7$ Set 2nd factor equal to 0 and solve.

The solutions are $x = 3$ and $x = 7$. Check these in the original equation.

REMARK When an equation contains two radical expressions, it may not be possible to isolate both of them in the first step. In such cases, you may have to isolate radical expressions at *two* different stages in the solution, as shown in Example 12(b).

✓ ***Checkpoint*** *Audio-video solution in English & Spanish at LarsonPrecalculus.com*

Solve $-\sqrt{40 - 9x} + 2 = x$.

EXAMPLE 13 Solving an Equation Involving a Rational Exponent

Solve $(x-4)^{2/3} = 25$.

Solution

$(x-4)^{2/3} = 25$ Write original equation.

$\sqrt[3]{(x-4)^2} = 25$ Rewrite in radical form.

$(x-4)^2 = 15{,}625$ Cube each side.

$x - 4 = \pm 125$ Extract square roots.

$x = 129, \ x = -121$ Add 4 to each side.

The solutions are $x = 129$ and $x = -121$. Check these in the original equation.

✓ ***Checkpoint*** *Audio-video solution in English & Spanish at LarsonPrecalculus.com*

Solve $(x-5)^{2/3} = 16$.

Absolute Value Equations

An **absolute value equation** is an equation that involves one or more absolute value expressions. To solve an absolute value equation, remember that the expression inside the absolute value bars can be positive or negative. This results in *two* separate equations, each of which must be solved. For example, the equation

$$|x - 2| = 3$$

results in the two equations

$$x - 2 = 3 \quad \text{and} \quad -(x - 2) = 3$$

which implies that the original equation has two solutions: $x = 5$ and $x = -1$.

EXAMPLE 14 Solving an Absolute Value Equation

Solve $|x^2 - 3x| = -4x + 6$.

Solution Solve the two equations below.

First Equation

$x^2 - 3x = -4x + 6$	Use positive expression.
$x^2 + x - 6 = 0$	Write in general form.
$(x + 3)(x - 2) = 0$	Factor.
$x + 3 = 0 \implies x = -3$	Set 1st factor equal to 0 and solve.
$x - 2 = 0 \implies x = 2$	Set 2nd factor equal to 0 and solve.

Second Equation

$-(x^2 - 3x) = -4x + 6$	Use negative expression.
$x^2 - 7x + 6 = 0$	Write in general form.
$(x - 1)(x - 6) = 0$	Factor.
$x - 1 = 0 \implies x = 1$	Set 1st factor equal to 0 and solve.
$x - 6 = 0 \implies x = 6$	Set 2nd factor equal to 0 and solve.

Check the values in the original equation to determine that the only solutions are $x = -3$ and $x = 1$.

✓ ***Checkpoint*** *Audio-video solution in English & Spanish at LarsonPrecalculus.com*

Solve $|x^2 + 4x| = 7x + 18$.

Summarize (Section P.2)

1. State the definitions of an identity, a conditional equation, and a contradiction *(page 14)*.
2. State the definition of a linear equation in one variable *(page 14)*. For examples of solving linear equations and rational equations, see Examples 1–3.
3. List the four methods for solving quadratic equations discussed in this section *(page 17)*. For examples of solving quadratic equations, see Examples 4–9.
4. Explain how to solve a polynomial equation of degree three or greater *(page 21)*, a radical equation *(page 22)*, and an absolute value equation *(page 23)*. For examples of solving these types of equations, see Examples 10–14.

P.2 Exercises

See CalcChat.com for tutorial help and worked-out solutions to odd-numbered exercises.

Vocabulary: Fill in the blanks.

1. An ________ is a statement that equates two algebraic expressions.
2. A linear equation in one variable x is an equation that can be written in the standard form ________.
3. An ________ solution is a solution that does not satisfy the original equation.
4. Four methods for solving quadratic equations are ________, extracting ________ ________, ________ the ________, and the ________ ________.

Skills and Applications

Solving a Linear Equation In Exercises 5–12, solve the equation and check your solution. (If not possible, explain why.)

5. $x + 11 = 15$
6. $7 - x = 19$
7. $7 - 2x = 25$
8. $7x + 2 = 23$
9. $3x - 5 = 2x + 7$
10. $4y + 2 - 5y = 7 - 6y$
11. $x - 3(2x + 3) = 8 - 5x$
12. $9x - 10 = 5x + 2(2x - 5)$

Solving a Rational Equation In Exercises 13–24, solve the equation and check your solution. (If not possible, explain why.)

13. $\frac{3x}{8} - \frac{4x}{3} = 4$
14. $\frac{5x}{4} + \frac{1}{2} = x - \frac{1}{2}$
15. $\frac{5x - 4}{5x + 4} = \frac{2}{3}$
16. $\frac{10x + 3}{5x + 6} = \frac{1}{2}$
17. $10 - \frac{13}{x} = 4 + \frac{5}{x}$
18. $\frac{1}{x} + \frac{2}{x - 5} = 0$
19. $\frac{x}{x + 4} + \frac{4}{x + 4} = -2$
20. $\frac{7}{2x + 1} - \frac{8x}{2x - 1} = -4$
21. $\frac{2}{(x - 4)(x - 2)} = \frac{1}{x - 4} + \frac{2}{x - 2}$
22. $\frac{12}{(x - 1)(x + 3)} = \frac{3}{x - 1} + \frac{2}{x + 3}$
23. $\frac{1}{x - 3} + \frac{1}{x + 3} = \frac{10}{x^2 - 9}$
24. $\frac{1}{x - 2} + \frac{3}{x + 3} = \frac{4}{x^2 + x - 6}$

Solving a Quadratic Equation by Factoring In Exercises 25–34, solve the quadratic equation by factoring.

25. $6x^2 + 3x = 0$
26. $8x^2 - 2x = 0$
27. $x^2 + 10x + 25 = 0$
28. $x^2 - 2x - 8 = 0$
29. $3 + 5x - 2x^2 = 0$
30. $4x^2 + 12x + 9 = 0$
31. $16x^2 - 9 = 0$
32. $-x^2 + 8x = 12$
33. $\frac{3}{4}x^2 + 8x + 20 = 0$
34. $\frac{1}{8}x^2 - x - 16 = 0$

Extracting Square Roots In Exercises 35–42, solve the equation by extracting square roots. When a solution is irrational, list both the exact solution *and* its approximation rounded to two decimal places.

35. $x^2 = 49$
36. $x^2 = 43$
37. $3x^2 = 81$
38. $9x^2 = 36$
39. $(x - 4)^2 = 49$
40. $(x + 9)^2 = 24$
41. $(2x - 1)^2 = 18$
42. $(x - 7)^2 = (x + 3)^2$

Completing the Square In Exercises 43–50, solve the quadratic equation by completing the square.

43. $x^2 + 4x - 32 = 0$
44. $x^2 - 2x - 3 = 0$
45. $x^2 + 4x + 2 = 0$
46. $x^2 + 8x + 14 = 0$
47. $6x^2 - 12x = -3$
48. $4x^2 - 4x = 1$
49. $2x^2 + 5x - 8 = 0$
50. $3x^2 - 4x - 7 = 0$

Rewriting an Expression In Exercises 51–54, rewrite the quadratic portion of the algebraic expression as the sum or difference of two squares by completing the square.

51. $\frac{1}{x^2 - 2x + 5}$
52. $\frac{4}{x^2 + 10x + 74}$
53. $\frac{1}{\sqrt{3 + 2x - x^2}}$
54. $\frac{1}{\sqrt{12 + 4x - x^2}}$

Using the Quadratic Formula In Exercises 55–68, use the Quadratic Formula to solve the equation.

55. $2x^2 + x - 1 = 0$
56. $2x^2 - x - 1 = 0$
57. $9x^2 + 30x + 25 = 0$
58. $28x - 49x^2 = 4$
59. $2x^2 - 7x + 1 = 0$
60. $3x + x^2 - 1 = 0$
61. $12x - 9x^2 = -3$
62. $9x^2 - 37 = 6x$
63. $2 + 2x - x^2 = 0$
64. $x^2 + 10 + 8x = 0$
65. $8t = 5 + 2t^2$
66. $25h^2 + 80h = -61$
67. $(y - 5)^2 = 2y$
68. $(z + 6)^2 = -2z$

Using the Quadratic Formula In Exercises 69–72, use the Quadratic Formula to solve the equation. (Round your answer to three decimal places.)

69. $5.1x^2 - 1.7x = 3.2$ **70.** $2x^2 - 2.53x = 0.42$

71. $-0.005x^2 + 0.101x - 0.193 = 0$

72. $-3.22x^2 - 0.08x + 28.651 = 0$

Choosing a Method In Exercises 73–80, solve the equation using any convenient method.

73. $x^2 - 2x - 1 = 0$ **74.** $14x^2 + 42x = 0$

75. $(x + 2)^2 = 64$ **76.** $x^2 - 14x + 49 = 0$

77. $x^2 - x - \frac{11}{4} = 0$ **78.** $x^2 + 3x - \frac{3}{4} = 0$

79. $3x + 4 = 2x^2 - 7$ **80.** $(x + 1)^2 = x^2$

Solving a Rational Equation In Exercises 81–84, solve the equation. Check your solutions.

81. $\frac{1}{x} - \frac{1}{x+1} = 3$ **82.** $\frac{4}{x+1} - \frac{3}{x+2} = 1$

83. $\frac{x}{x^2 - 4} + \frac{1}{x+2} = 3$ **84.** $\frac{x+1}{3} - \frac{x+1}{x+2} = 0$

Solving a Polynomial Equation In Exercises 85–90, solve the equation. Check your solutions.

85. $6x^4 - 54x^2 = 0$ **86.** $5x^3 + 30x^2 + 45x = 0$

87. $x^3 + 2x^2 - 8x = 16$ **88.** $x^3 - 3x^2 - x = -3$

89. $x^4 - 4x^2 + 3 = 0$ **90.** $x^4 - 13x^2 + 36 = 0$

Solving a Radical Equation In Exercises 91–98, solve the equation. Check your solutions.

91. $\sqrt{5x} - 10 = 0$ **92.** $\sqrt{x+8} - 5 = 0$

93. $4 + \sqrt[3]{2x - 9} = 0$ **94.** $\sqrt[3]{12 - x} - 3 = 0$

95. $\sqrt{x+8} = 2 + x$ **96.** $2x = \sqrt{-5x + 24} - 3$

97. $\sqrt{x - 3} + 1 = \sqrt{x}$

98. $2\sqrt{x+1} - \sqrt{2x+3} = 1$

Solving an Equation Involving a Rational Exponent In Exercises 99–102, solve the equation. Check your solutions.

99. $(x - 5)^{3/2} = 8$ **100.** $(x^2 - x - 22)^{3/2} = 27$

101. $3x(x - 1)^{1/2} + 2(x - 1)^{3/2} = 0$

102. $4x^2(x - 1)^{1/3} + 6x(x - 1)^{4/3} = 0$

Solving an Absolute Value Equation In Exercises 103–106, solve the equation. Check your solutions.

103. $|2x - 5| = 11$ **104.** $|3x + 2| = 7$

105. $|x + 1| = x^2 - 5$ **106.** $|x^2 + 6x| = 3x + 18$

Forensics

In Exercises 107 and 108, use the following information. The relationship between the length of an adult's femur (thigh bone) and the height of the adult can be approximated by the linear equations

$y = 0.514x - 14.75$ Female

$y = 0.532x - 17.03$ Male

where y is the length of the femur in inches and x is the height of the adult in inches (see figure).

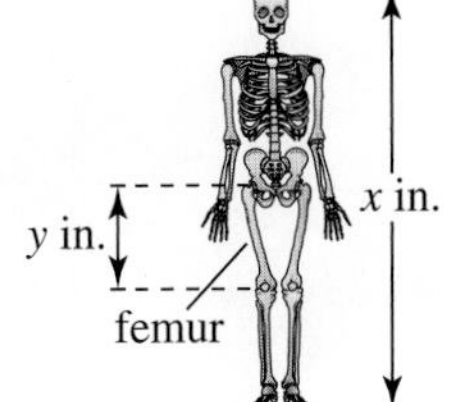

107. A crime scene investigator discovers a femur belonging to an adult human female. The bone is 18 inches long. Estimate the height of the female.

108. Officials search a forest for a missing man who is 6 feet 2 inches tall. They find an adult male femur that is 23 inches long. Is it possible that the femur belongs to the missing man?

Exploration

True or False? In Exercises 109–111, determine whether the statement is true or false. Justify your answer.

109. An equation can never have more than one extraneous solution.

110. The equation $2(x - 3) + 1 = 2x - 5$ has no solution.

111. The equation $\sqrt{x + 10} - \sqrt{x - 10} = 0$ has no solution.

112. **HOW DO YOU SEE IT?** The figure shows a glass cube partially filled with water.

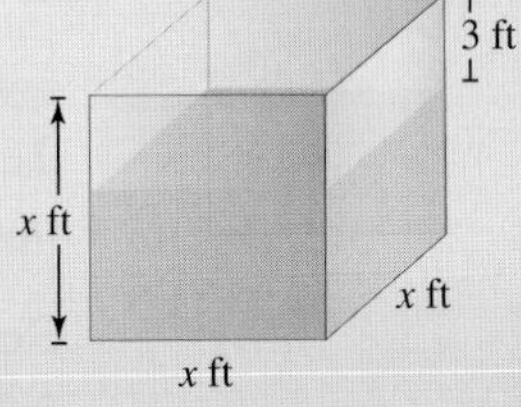

(a) What does the expression $x^2(x - 3)$ represent?

(b) Given $x^2(x - 3) = 320$, explain how to find the capacity of the cube.

113. **Think About It** Are $(3x + 2)/5 = 7$ and $x + 9 = 20$ equivalent equations? Explain.

iStockphoto.com/Nikola Nastasic

P.3 The Cartesian Plane and Graphs of Equations

The Cartesian plane can help you visualize relationships between two variables. For example, in Exercise 37 on page 37, given how far north and west one city is from another, plotting points to represent the cities can help you visualize these distances and determine the flying distance between the cities.

- **Plot points in the Cartesian plane.**
- **Use the Distance Formula to find the distance between two points.**
- **Use the Midpoint Formula to find the midpoint of a line segment.**
- **Use a coordinate plane to model and solve real-life problems.**
- **Sketch graphs of equations.**
- **Find *x*- and *y*-intercepts of graphs of equations.**
- **Use symmetry to sketch graphs of equations.**
- **Write equations of circles.**

The Cartesian Plane

Just as you can represent real numbers by points on a real number line, you can represent ordered pairs of real numbers by points in a plane called the **rectangular coordinate system,** or the **Cartesian plane,** named after the French mathematician René Descartes (1596–1650).

Two real number lines intersecting at right angles form the Cartesian plane, as shown in Figure P.9. The horizontal real number line is usually called the ***x*-axis,** and the vertical real number line is usually called the ***y*-axis.** The point of intersection of these two axes is the **origin,** and the two axes divide the plane into four **quadrants.**

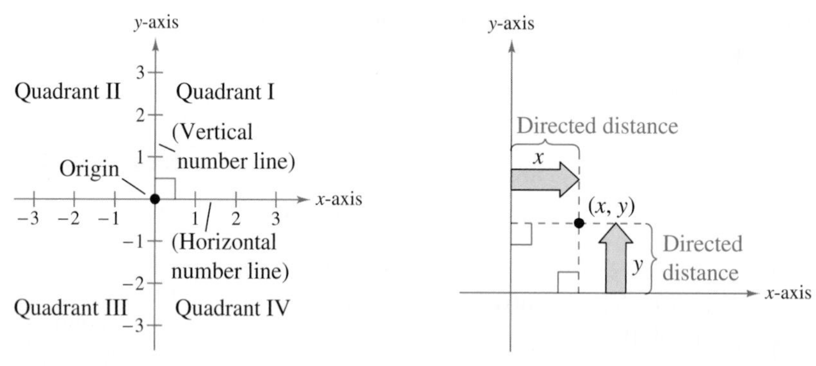

Figure P.9

Figure P.10

Each point in the plane corresponds to an **ordered pair** (x, y) of real numbers x and y, called **coordinates** of the point. The ***x*-coordinate** represents the directed distance from the y-axis to the point, and the ***y*-coordinate** represents the directed distance from the x-axis to the point, as shown in Figure P.10.

The notation (x, y) denotes both a point in the plane and an open interval on the real number line. The context will tell you which meaning is intended.

EXAMPLE 1 Plotting Points in the Cartesian Plane

Plot the points $(-1, 2)$, $(3, 4)$, $(0, 0)$, $(3, 0)$, and $(-2, -3)$.

Solution To plot the point $(-1, 2)$, imagine a vertical line through -1 on the x-axis and a horizontal line through 2 on the y-axis. The intersection of these two lines is the point $(-1, 2)$. Plot the other four points in a similar way, as shown in Figure P.11.

Figure P.11

Checkpoint Audio-video solution in English & Spanish at LarsonPrecalculus.com

Plot the points $(-3, 2)$, $(4, -2)$, $(3, 1)$, $(0, -2)$, and $(-1, -2)$.

The beauty of a rectangular coordinate system is that it allows you to *see* relationships between two variables. It would be difficult to overestimate the importance of Descartes's introduction of coordinates in the plane. Today, his ideas are in common use in virtually every scientific and business-related field.

DATA

Year, t	Subscribers, N
2005	207.9
2006	233.0
2007	255.4
2008	270.3
2009	285.6
2010	296.3
2011	316.0
2012	326.5
2013	335.7
2014	355.4

Spreadsheet at LarsonPrecalculus.com

EXAMPLE 2 Sketching a Scatter Plot

The table shows the numbers N (in millions) of subscribers to a cellular telecommunication service in the United States from 2005 through 2014, where t represents the year. Sketch a scatter plot of the data. *(Source: CTIA-The Wireless Association)*

Solution To sketch a *scatter plot* of the data shown in the table, represent each pair of values by an ordered pair (t, N) and plot the resulting points. For example, let (2005, 207.9) represent the first pair of values. Note that in the scatter plot below, the break in the t-axis indicates omission of the years before 2005, and the break in the N-axis indicates omission of the numbers less than 150 million.

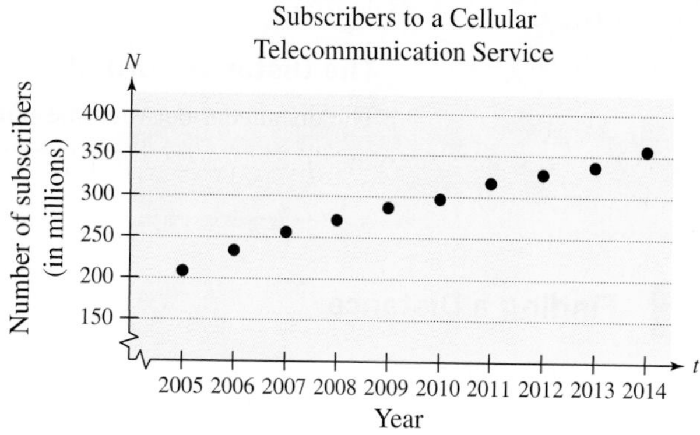

✓ Checkpoint *Audio-video solution in English & Spanish at LarsonPrecalculus.com*

The table shows the numbers N (in thousands) of cellular telecommunication service employees in the United States from 2005 through 2014, where t represents the year. Sketch a scatter plot of the data. *(Source: CTIA-The Wireless Association)*

DATA

t	N
2005	233.1
2006	253.8
2007	266.8
2008	268.5
2009	249.2
2010	250.4
2011	238.1
2012	230.1
2013	230.4
2014	232.2

Spreadsheet at LarsonPrecalculus.com

In Example 2, you could let $t = 1$ represent the year 2005. In that case, there would not be a break in the horizontal axis, and the labels 1 through 10 (instead of 2005 through 2014) would be on the tick marks.

TECHNOLOGY The scatter plot in Example 2 is only one way to represent the data graphically. You could also represent the data using a bar graph or a line graph. Use a graphing utility to represent the data given in Example 2 graphically.

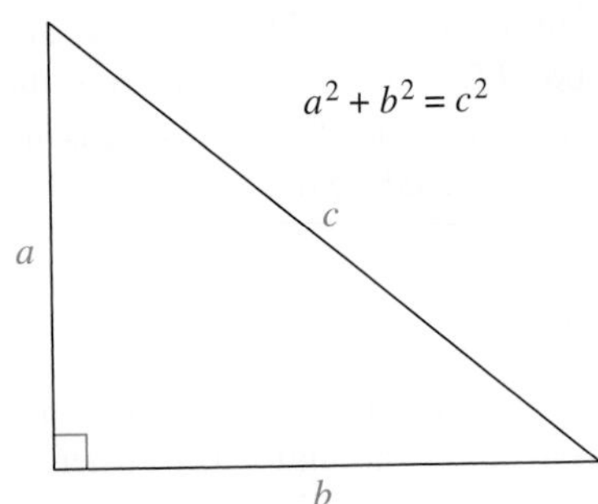

Figure P.12

The Distance Formula

Recall from the Pythagorean Theorem that, for a right triangle with hypotenuse of length c and sides of lengths a and b, you have

$$a^2 + b^2 = c^2 \qquad \text{Pythagorean Theorem}$$

as shown in Figure P.12. (The converse is also true. That is, if $a^2 + b^2 = c^2$, then the triangle is a right triangle.)

Using the points (x_1, y_1) and (x_2, y_2), you can form a right triangle, as shown in Figure P.13. The length of the hypotenuse of the right triangle is the distance d between the two points. The length of the vertical side of the triangle is $|y_2 - y_1|$ and the length of the horizontal side is $|x_2 - x_1|$. By the Pythagorean Theorem,

$$d^2 = |x_2 - x_1|^2 + |y_2 - y_1|^2$$

$$d = \sqrt{|x_2 - x_1|^2 + |y_2 - y_1|^2}$$

$$= \sqrt{(x_2 - x_1)^2 + (y_2 - y_1)^2}.$$

This result is the **Distance Formula.**

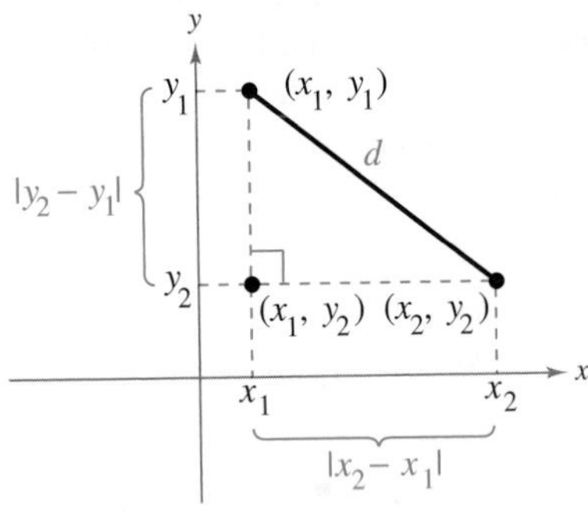

Figure P.13

> **The Distance Formula**
>
> The distance d between the points (x_1, y_1) and (x_2, y_2) in the plane is
>
> $$d = \sqrt{(x_2 - x_1)^2 + (y_2 - y_1)^2}.$$

EXAMPLE 3 Finding a Distance

Find the distance between the points

$(-2, 1)$ and $(3, 4)$.

Algebraic Solution

Let $(x_1, y_1) = (-2, 1)$ and $(x_2, y_2) = (3, 4)$. Then apply the Distance Formula.

$d = \sqrt{(x_2 - x_1)^2 + (y_2 - y_1)^2}$ — Distance Formula

$= \sqrt{[3 - (-2)]^2 + (4 - 1)^2}$ — Substitute for x_1, y_1, x_2, and y_2.

$= \sqrt{(5)^2 + (3)^2}$ — Simplify.

$= \sqrt{34}$ — Simplify.

≈ 5.83 — Use a calculator.

So, the distance between the points is about 5.83 units.

Check

$d^2 \stackrel{?}{=} 5^2 + 3^2$ — Pythagorean Theorem

$(\sqrt{34})^2 \stackrel{?}{=} 5^2 + 3^2$ — Substitute for d.

$34 = 34$ — Distance checks. ✓

Graphical Solution

Use centimeter graph paper to plot the points $A(-2, 1)$ and $B(3, 4)$. Carefully sketch the line segment from A to B. Then use a centimeter ruler to measure the length of the segment.

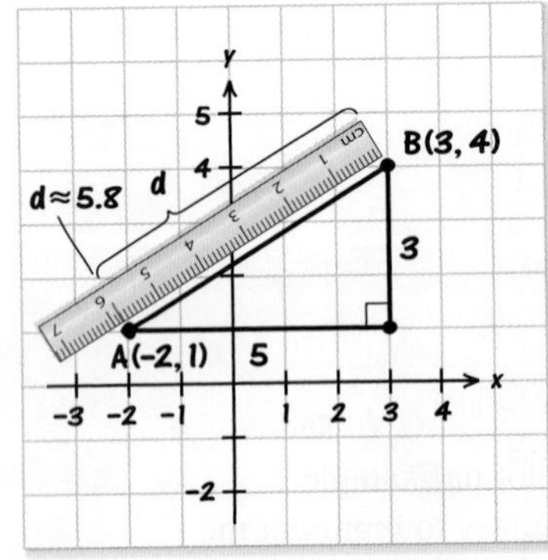

The line segment measures about 5.8 centimeters. So, the distance between the points is about 5.8 units.

✓ **Checkpoint** Audio-video solution in English & Spanish at LarsonPrecalculus.com

Find the distance between the points $(3, 1)$ and $(-3, 0)$.

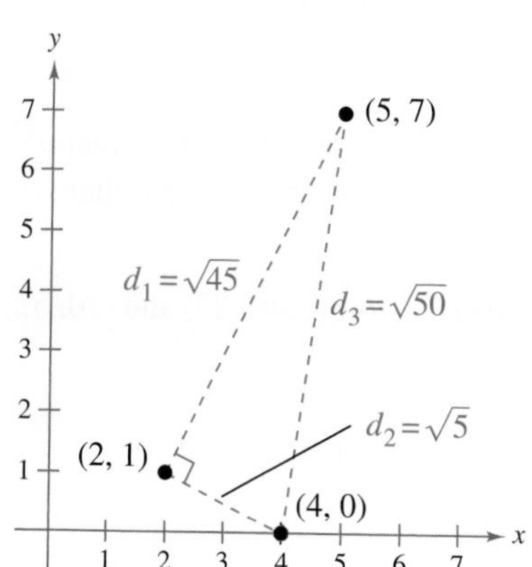

Figure P.14

EXAMPLE 4 **Verifying a Right Triangle**

Show that the points

$$(2, 1), \quad (4, 0), \quad \text{and} \quad (5, 7)$$

are vertices of a right triangle.

Solution The three points are plotted in Figure P.14. Using the Distance Formula, the lengths of the three sides are

$$d_1 = \sqrt{(5-2)^2 + (7-1)^2} = \sqrt{9+36} = \sqrt{45},$$

$$d_2 = \sqrt{(4-2)^2 + (0-1)^2} = \sqrt{4+1} = \sqrt{5}, \text{ and}$$

$$d_3 = \sqrt{(5-4)^2 + (7-0)^2} = \sqrt{1+49} = \sqrt{50}.$$

Because $(d_1)^2 + (d_2)^2 = 45 + 5 = 50 = (d_3)^2$, you can conclude by the converse of the Pythagorean Theorem that the triangle is a right triangle.

✓ ***Checkpoint*** *Audio-video solution in English & Spanish at LarsonPrecalculus.com*

Show that the points $(2, -1)$, $(5, 5)$, and $(6, -3)$ are vertices of a right triangle.

The Midpoint Formula

To find the **midpoint** of the line segment that joins two points in a coordinate plane, find the average values of the respective coordinates of the two endpoints using the **Midpoint Formula.**

> **The Midpoint Formula**
>
> The midpoint of the line segment joining the points (x_1, y_1) and (x_2, y_2) is
>
> $$\text{Midpoint} = \left(\frac{x_1 + x_2}{2}, \frac{y_1 + y_2}{2}\right).$$

For a proof of the Midpoint Formula, see Proofs in Mathematics on page 118.

EXAMPLE 5 **Finding the Midpoint of a Line Segment**

Find the midpoint of the line segment joining the points

$$(-5, -3) \quad \text{and} \quad (9, 3).$$

Solution Let $(x_1, y_1) = (-5, -3)$ and $(x_2, y_2) = (9, 3)$.

$$\text{Midpoint} = \left(\frac{x_1 + x_2}{2}, \frac{y_1 + y_2}{2}\right) \qquad \text{Midpoint Formula}$$

$$= \left(\frac{-5+9}{2}, \frac{-3+3}{2}\right) \qquad \text{Substitute for } x_1, y_1, x_2, \text{ and } y_2.$$

$$= (2, 0) \qquad \text{Simplify.}$$

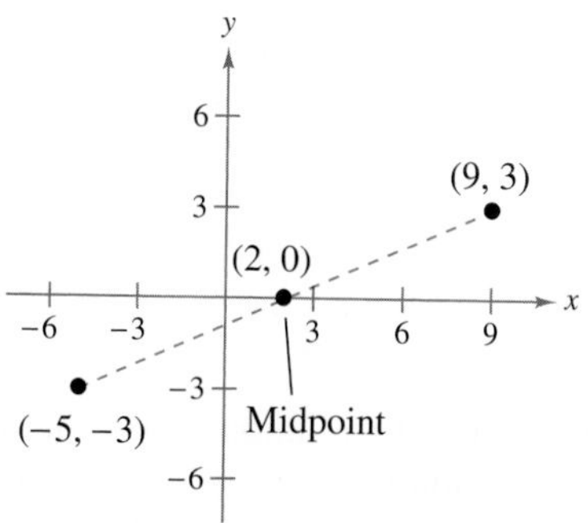

Figure P.15

The midpoint of the line segment is $(2, 0)$, as shown in Figure P.15.

✓ ***Checkpoint*** *Audio-video solution in English & Spanish at LarsonPrecalculus.com*

Find the midpoint of the line segment joining the points $(-2, 8)$ and $(4, -10)$.

Applications

EXAMPLE 6 Finding the Length of a Pass

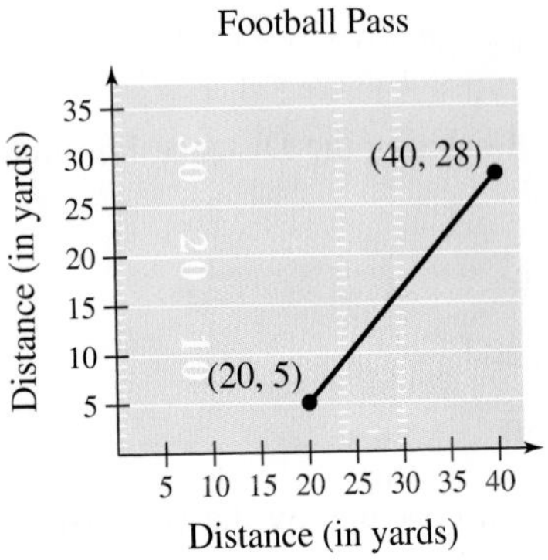

A football quarterback throws a pass from the 28-yard line, 40 yards from the sideline. A wide receiver catches the pass on the 5-yard line, 20 yards from the same sideline, as shown in Figure P.16. How long is the pass?

Football Pass

Figure P.16

Solution The length of the pass is the distance between the points (40, 28) and (20, 5).

$$d = \sqrt{(x_2 - x_1)^2 + (y_2 - y_1)^2}$$ Distance Formula

$$= \sqrt{(40 - 20)^2 + (28 - 5)^2}$$ Substitute for x_1, y_1, x_2, and y_2.

$$= \sqrt{20^2 + 23^2}$$ Simplify.

$$= \sqrt{400 + 529}$$ Simplify.

$$= \sqrt{929}$$ Simplify.

$$\approx 30$$ Use a calculator.

So, the pass is about 30 yards long.

✓ ***Checkpoint*** *Audio-video solution in English & Spanish at LarsonPrecalculus.com*

A football quarterback throws a pass from the 10-yard line, 10 yards from the sideline. A wide receiver catches the pass on the 32-yard line, 25 yards from the same sideline. How long is the pass?

In Example 6, the scale along the goal line does not normally appear on a football field. However, when you use coordinate geometry to solve real-life problems, you are free to place the coordinate system in any way that helps you solve the problem.

EXAMPLE 7 Estimating Annual Sales

Starbucks Corporation had annual sales of approximately \$13.3 billion in 2012 and \$16.4 billion in 2014. Without knowing any additional information, what would you estimate the 2013 sales to have been? *(Source: Starbucks Corporation)*

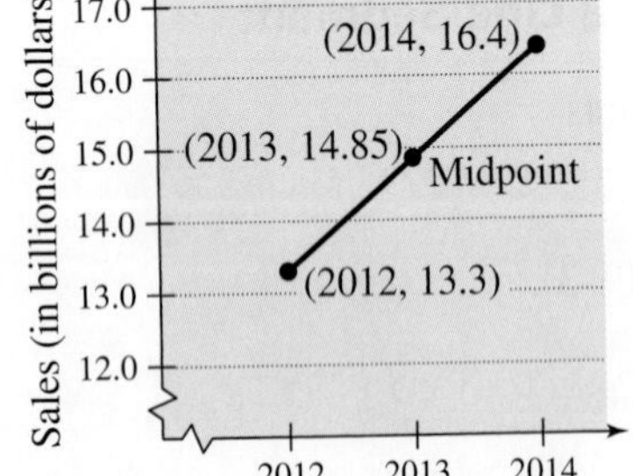

Figure P.17

Solution Assuming that sales followed a linear pattern, you can estimate the 2013 sales by finding the midpoint of the line segment connecting the points (2012, 13.3) and (2014, 16.4).

$$\text{Midpoint} = \left(\frac{x_1 + x_2}{2}, \frac{y_1 + y_2}{2}\right)$$ Midpoint Formula

$$= \left(\frac{2012 + 2014}{2}, \frac{13.3 + 16.4}{2}\right)$$ Substitute for x_1, x_2, y_1, and y_2.

$$= (2013, 14.85)$$ Simplify.

So, you would estimate the 2013 sales to have been about \$14.85 billion, as shown in Figure P.17. (The actual 2013 sales were about \$14.89 billion.)

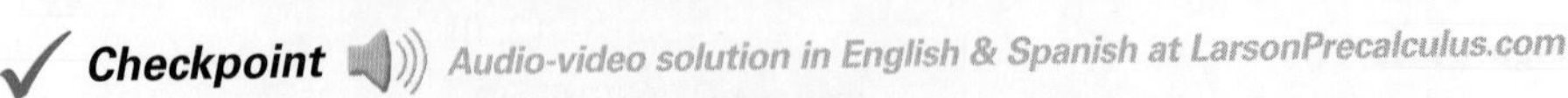

Yahoo! Inc. had annual revenues of approximately \$5.0 billon in 2012 and \$4.6 billion in 2014. Without knowing any additional information, what would you estimate the 2013 revenue to have been? *(Source: Yahoo! Inc.)*

The Graph of an Equation

Earlier in this section, you used a coordinate system to graphically represent the relationship between two quantities as points in a coordinate plane. (See Example 2.)

Frequently, a relationship between two quantities is expressed as an **equation in two variables.** For example, $y = 7 - 3x$ is an equation in x and y. An ordered pair (a, b) is a **solution** or **solution point** of an equation in x and y when the substitutions $x = a$ and $y = b$ result in a true statement. For example, $(1, 4)$ is a solution of $y = 7 - 3x$ because $4 = 7 - 3(1)$ is a true statement.

In the remainder of this section, you will review some basic procedures for sketching the graph of an equation in two variables. The **graph of an equation** is the set of all points that are solutions of the equation. The basic technique used for sketching the graph of an equation is the **point-plotting method.** To sketch a graph using the point-plotting method, first, when possible, isolate one of the variables. Next, construct a table of values showing several solution points. Then, plot the points from your table in a rectangular coordinate system. Finally, connect the points with a smooth curve or line.

EXAMPLE 8 Sketching the Graph of an Equation

See LarsonPrecalculus.com for an interactive version of this type of example.

Sketch the graph of $y = x^2 - 2$.

Solution

The equation is already solved for y, so begin by constructing a table of values.

x	-2	-1	0	1	2	3
$y = x^2 - 2$	2	-1	-2	-1	2	7
(x, y)	$(-2, 2)$	$(-1, -1)$	$(0, -2)$	$(1, -1)$	$(2, 2)$	$(3, 7)$

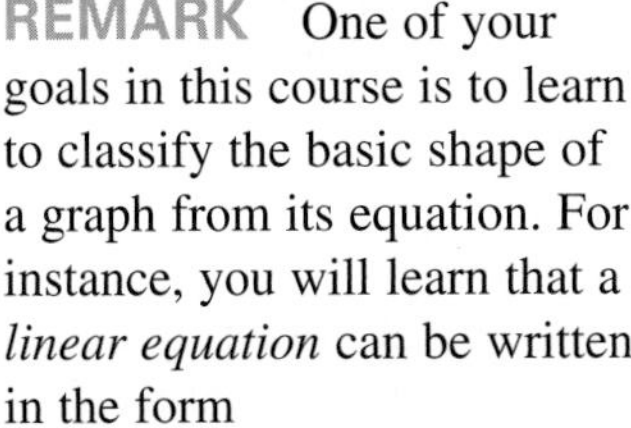

Next, plot the points given in the table, as shown in Figure P.18. Finally, connect the points with a smooth curve, as shown in Figure P.19.

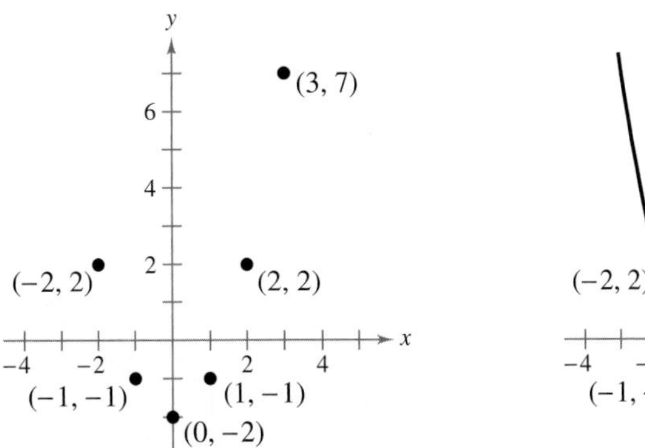

Figure P.18

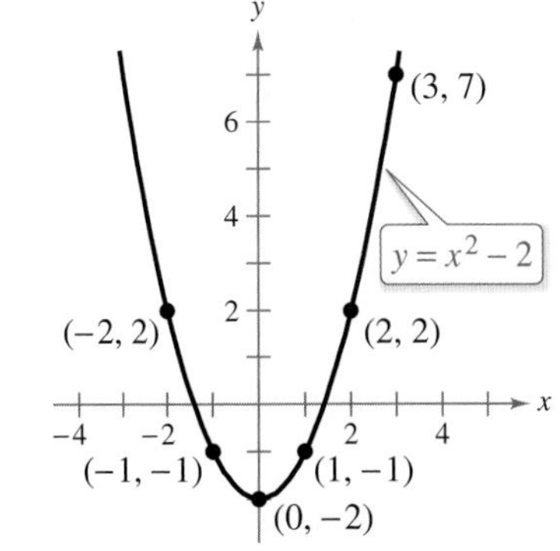

Figure P.19

✓ **Checkpoint** Audio-video solution in English & Spanish at LarsonPrecalculus.com

Sketch the graph of each equation.

a. $y = x^2 + 3$ **b.** $y = 1 - x^2$

TECHNOLOGY To graph an equation involving x and y on a graphing utility, use the procedure below.

1. If necessary, rewrite the equation so that y is isolated on the left side.
2. Enter the equation in the graphing utility.
3. Determine a *viewing window* that shows all important features of the graph.
4. Graph the equation.

Intercepts of a Graph

Solution points of an equation that have zero as either the x-coordinate or the y-coordinate are called **intercepts.** They are the points at which the graph intersects or touches the x- or y-axis. It is possible for a graph to have no intercepts, one intercept, or several intercepts, as shown in the graphs below.

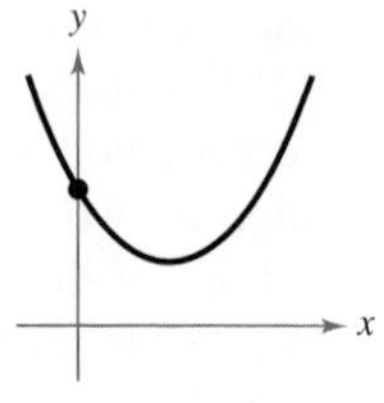

No x-intercepts
One y-intercept

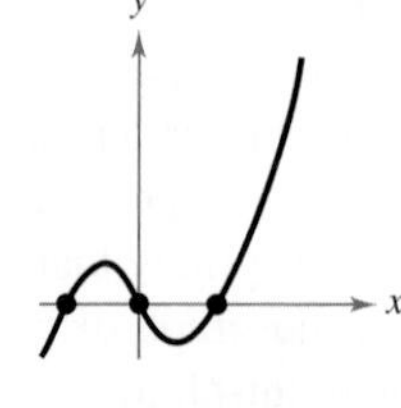

Three x-intercepts
One y-intercept

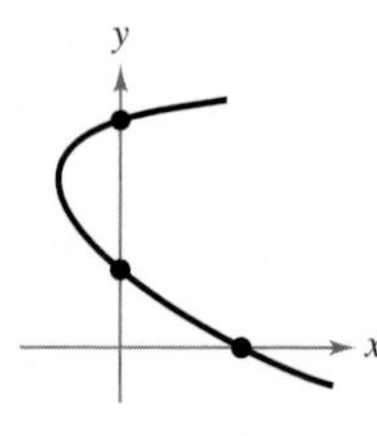

One x-intercept
Two y-intercepts

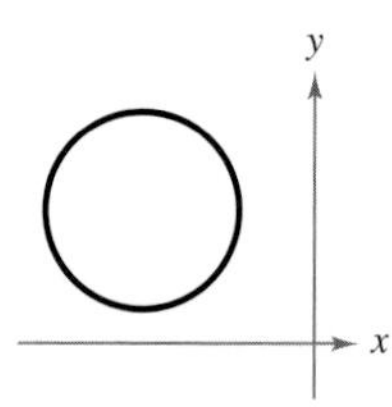

No intercepts

Note that an x-intercept can be written as the ordered pair $(a, 0)$ and a y-intercept can be written as the ordered pair $(0, b)$. Sometimes it is convenient to denote the x-intercept as the x-coordinate a of the point $(a, 0)$ or the y-intercept as the y-coordinate b of the point $(0, b)$. Unless it is necessary to make a distinction, the term *intercept* will refer to either the point or the coordinate.

Finding Intercepts

1. To find x-intercepts, let y be zero and solve the equation for x.
2. To find y-intercepts, let x be zero and solve the equation for y.

EXAMPLE 9 Finding x- and y-Intercepts

Find the x- and y-intercepts of the graph of

$$y = x^3 - 4x.$$

Solution

To find the x-intercepts of the graph of $y = x^3 - 4x$, let $y = 0$. Then

$$0 = x^3 - 4x$$
$$= x(x^2 - 4)$$

has the solutions $x = 0$ and $x = \pm 2$.

x-intercepts: $(0, 0), (2, 0), (-2, 0)$ See figure.

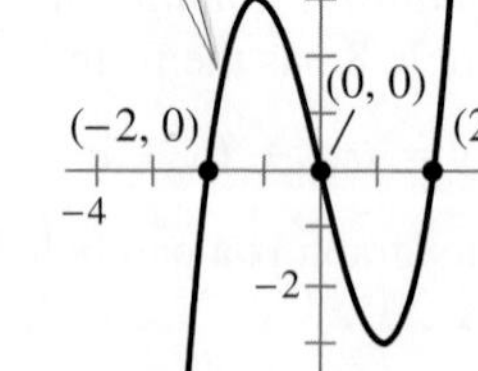

To find the y-intercept of the graph of $y = x^3 - 4x$, let $x = 0$. Then

$$y = (0)^3 - 4(0)$$

has one solution, $y = 0$.

y-intercept: $(0, 0)$ See figure.

Checkpoint Audio-video solution in English & Spanish at LarsonPrecalculus.com

Find the x- and y-intercepts of the graph of

$$y = -x^2 - 5x.$$

Symmetry

Graphs of equations can have **symmetry** with respect to one of the coordinate axes or with respect to the origin. Symmetry with respect to the x-axis means that when you fold the Cartesian plane along the x-axis, the portion of the graph above the x-axis coincides with the portion below the x-axis. Symmetry with respect to the y-axis or the origin can be described in a similar manner. The graphs below show these three types of symmetry.

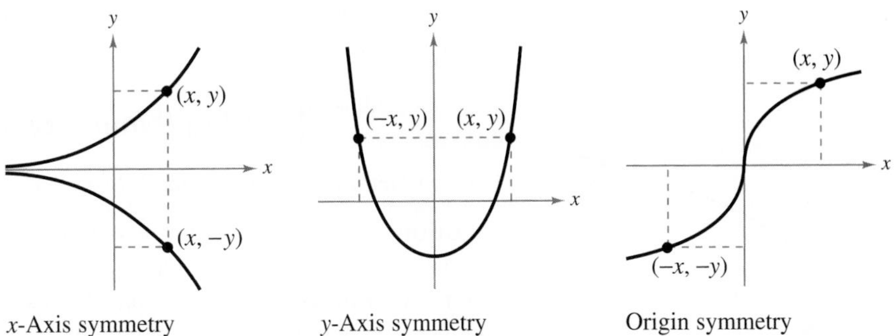

x-Axis symmetry y-Axis symmetry   Origin symmetry

Knowing the symmetry of a graph *before* attempting to sketch it is helpful, because then you need only half as many solution points to sketch the graph.

Graphical Tests for Symmetry

1. A graph is **symmetric with respect to the x-axis** if, whenever (x, y) is on the graph, $(x, -y)$ is also on the graph.
2. A graph is **symmetric with respect to the y-axis** if, whenever (x, y) is on the graph, $(-x, y)$ is also on the graph.
3. A graph is **symmetric with respect to the origin** if, whenever (x, y) is on the graph, $(-x, -y)$ is also on the graph.

EXAMPLE 10 Testing for Symmetry

The graph of $y = x^2 - 2$ is symmetric with respect to the y-axis because (x, y) and $(-x, y)$ are on the graph of $y = x^2 - 2$. (See figure.) The table below illustrates that the graph is symmetric with respect to the y-axis.

x	-3	-2	-1
y	7	2	-1
(x, y)	$(-3, 7)$	$(-2, 2)$	$(-1, -1)$

x	1	2	3
y	-1	2	7
(x, y)	$(1, -1)$	$(2, 2)$	$(3, 7)$

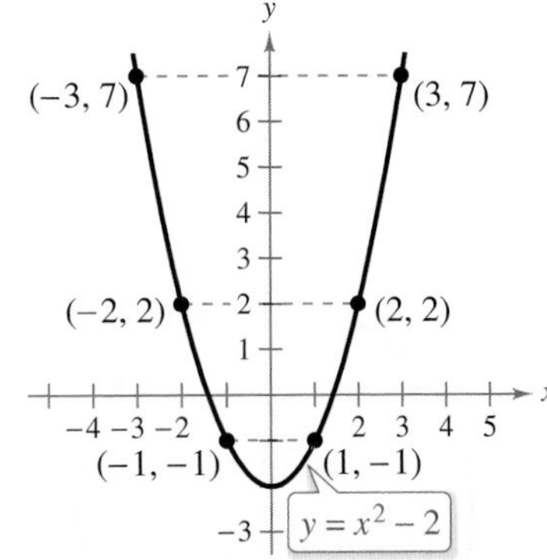

y-Axis symmetry

✓ Checkpoint  *Audio-video solution in English & Spanish at LarsonPrecalculus.com*

Determine the symmetry of the graph of $y^2 = 6 - x$.

Algebraic Tests for Symmetry

1. The graph of an equation is symmetric with respect to the x-axis when replacing y with $-y$ yields an equivalent equation.
2. The graph of an equation is symmetric with respect to the y-axis when replacing x with $-x$ yields an equivalent equation.
3. The graph of an equation is symmetric with respect to the origin when replacing x with $-x$ and y with $-y$ yields an equivalent equation.

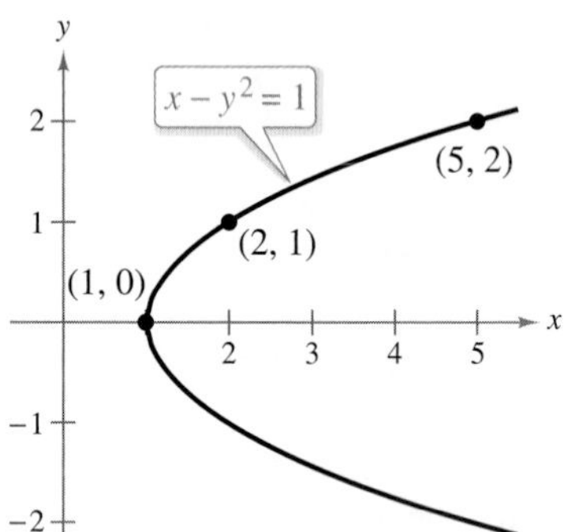

Figure P.20

EXAMPLE 11 Using Symmetry as a Sketching Aid

Use symmetry to sketch the graph of $x - y^2 = 1$.

Solution Of the three tests for symmetry, the test for x-axis symmetry is the only one satisfied, because $x - (-y)^2 = 1$ is equivalent to $x - y^2 = 1$. So, the graph is symmetric with respect to the x-axis. Find solution points above (or below) the x-axis and then use symmetry to obtain the graph, as shown in Figure P.20.

Checkpoint Audio-video solution in English & Spanish at LarsonPrecalculus.com

Use symmetry to sketch the graph of $y = x^2 - 4$.

Figure P.21

EXAMPLE 12 Sketching the Graph of an Equation

Sketch the graph of $y = |x - 1|$.

Solution This equation fails all three tests for symmetry, so its graph is not symmetric with respect to either axis or to the origin. The absolute value bars tell you that y is always nonnegative. Construct a table of values. Then plot and connect the points, as shown in Figure P.21. Notice from the table that $x = 0$ when $y = 1$. So, the y-intercept is $(0, 1)$. Similarly, $y = 0$ when $x = 1$. So, the x-intercept is $(1, 0)$.

x	-2	-1	0	1	2	3	4
$y = \|x - 1\|$	3	2	1	0	1	2	3
(x, y)	$(-2, 3)$	$(-1, 2)$	$(0, 1)$	$(1, 0)$	$(2, 1)$	$(3, 2)$	$(4, 3)$

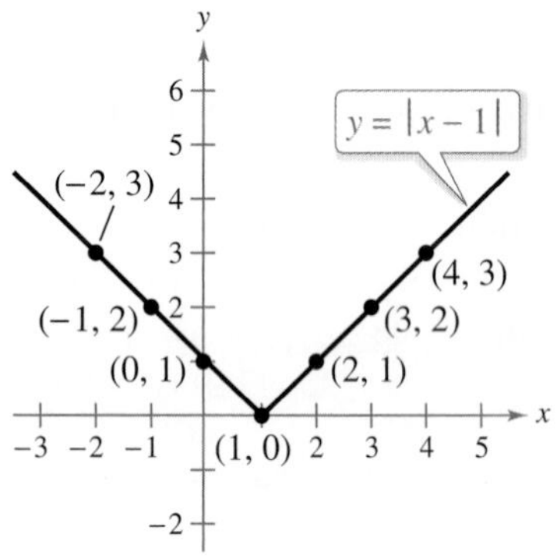

Checkpoint Audio-video solution in English & Spanish at LarsonPrecalculus.com

Sketch the graph of $y = |x - 2|$.

Circles

A **circle** is a set of points (x, y) in a plane that are the same distance r from a point called the center, (h, k), as shown in Figure P.22. By the Distance Formula,

$$\sqrt{(x - h)^2 + (y - k)^2} = r.$$

By squaring each side of this equation, you obtain the **standard form of the equation of a circle.** For example, for a circle with its center at $(h, k) = (1, 3)$ and radius $r = 4$,

$$\sqrt{(x - 1)^2 + (y - 3)^2} = 4 \quad \text{Substitute for } h, k, \text{ and } r.$$

$$(x - 1)^2 + (y - 3)^2 = 16. \quad \text{Square each side.}$$

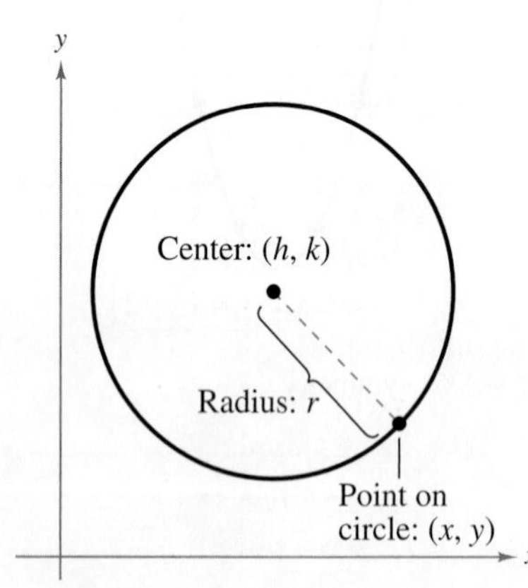

Figure P.22

Standard Form of the Equation of a Circle

A point (x, y) lies on the circle of **radius** r and **center** (h, k) if and only if

$$(x - h)^2 + (y - k)^2 = r^2.$$

REMARK To find h and k from the standard form of the equation of a circle, you may want to rewrite one or both of the quantities in parentheses. For example, $x + 1 = x - (-1)$.

From this result, the standard form of the equation of a circle with radius r *and center at the origin,* $(h, k) = (0, 0)$, is $x^2 + y^2 = r^2$.

EXAMPLE 13 Writing the Equation of a Circle

The point $(3, 4)$ lies on a circle whose center is at $(-1, 2)$, as shown in Figure P.23. Write the standard form of the equation of this circle.

Solution

The radius of the circle is the distance between $(-1, 2)$ and $(3, 4)$.

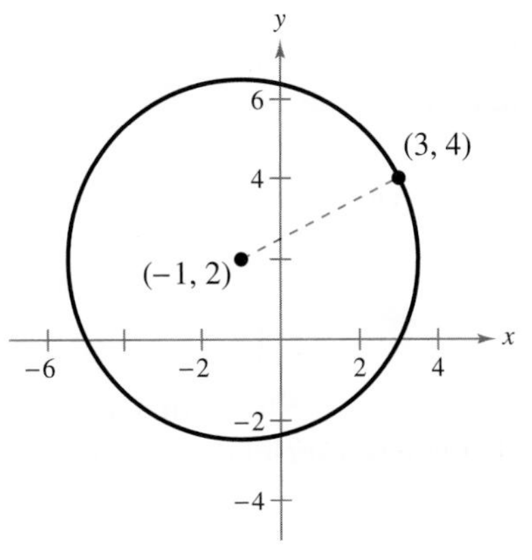

Figure P.23

$r = \sqrt{(x - h)^2 + (y - k)^2}$ — Distance Formula

$= \sqrt{[3 - (-1)]^2 + (4 - 2)^2}$ — Substitute for x, y, h, and k.

$= \sqrt{20}$ — Radius

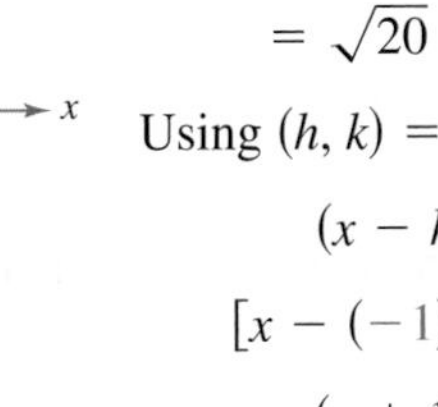

Using $(h, k) = (-1, 2)$ and $r = \sqrt{20}$, the equation of the circle is

$(x - h)^2 + (y - k)^2 = r^2$ — Equation of circle

$[x - (-1)]^2 + (y - 2)^2 = \left(\sqrt{20}\right)^2$ — Substitute for h, k, and r.

$(x + 1)^2 + (y - 2)^2 = 20.$ — Standard form

✓ ***Checkpoint*** *Audio-video solution in English & Spanish at LarsonPrecalculus.com*

The point $(1, -2)$ lies on a circle whose center is at $(-3, -5)$. Write the standard form of the equation of this circle.

Summarize (Section P.3)

1. Describe the Cartesian plane *(page 26)*. For examples of plotting points in the Cartesian plane, see Examples 1 and 2.
2. State the Distance Formula *(page 28)*. For examples of using the Distance Formula to find the distance between two points, see Examples 3 and 4.
3. State the Midpoint Formula *(page 29)*. For an example of using the Midpoint Formula to find the midpoint of a line segment, see Example 5.
4. Describe examples of how to use a coordinate plane to model and solve real-life problems *(page 30, Examples 6 and 7)*.
5. Explain how to sketch the graph of an equation *(page 31)*. For an example of sketching the graph of an equation, see Example 8.
6. Explain how to find the x- and y-intercepts of a graph *(page 32)*. For an example of finding x- and y-intercepts, see Example 9.
7. Explain how to use symmetry to graph an equation *(pages 33 and 34)*. For an example of using symmetry to graph an equation, see Example 11.
8. State the standard form of the equation of a circle *(page 35)*. For an example of writing the standard form of the equation of a circle, see Example 13.

P.3 Exercises

See CalcChat.com for tutorial help and worked-out solutions to odd-numbered exercises.

Vocabulary: Fill in the blanks.

1. An ordered pair of real numbers can be represented in a plane called the rectangular coordinate system or the ________ plane.
2. The ________ ________ is derived from the Pythagorean Theorem.
3. Finding the average values of the respective coordinates of the two endpoints of a line segment in a coordinate plane is also known as using the ________ ________.
4. An ordered pair (a, b) is a ________ of an equation in x and y when the substitutions $x = a$ and $y = b$ result in a true statement.
5. The set of all solutions points of an equation is the ________ of the equation.
6. The points at which a graph intersects or touches an axis are the ________ of the graph.
7. A graph is symmetric with respect to the ________ if, whenever (x, y) is on the graph, $(-x, y)$ is also on the graph.
8. The equation $(x - h)^2 + (y - k)^2 = r^2$ is the standard form of the equation of a ________ with center ________ and radius ________.

Skills and Applications

Approximating Coordinate of Points **In Exercises 9 and 10, approximate the coordinates of the points.**

9. 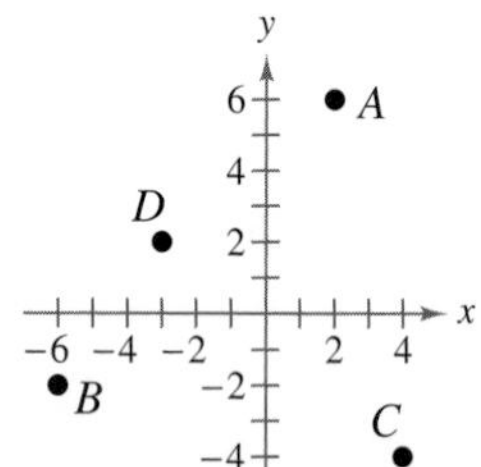

10. 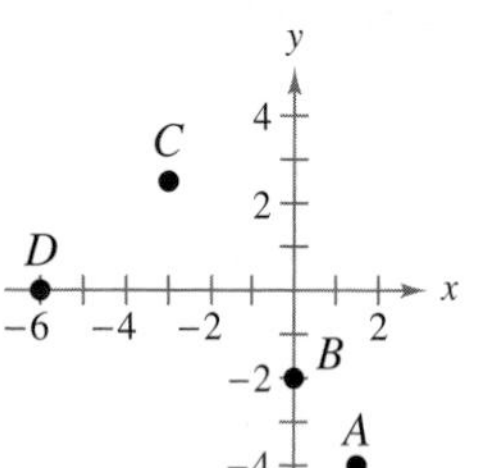

Plotting Points in the Cartesian Plane **In Exercises 11 and 12, plot the points.**

11. $(2, 4), (3, -1), (-6, 2), (-4, 0), (-1, -8), (1.5, -3.5)$
12. $(1, -5), (-2, -7), (3, 3), (-2, 4), (0, 5), \left(\frac{2}{3}, \frac{5}{2}\right)$

Finding the Coordinates of a Point **In Exercises 13 and 14, find the coordinates of the point.**

13. The point is three units to the left of the y-axis and four units above the x-axis.
14. The point is on the x-axis and 12 units to the left of the y-axis.

Determining Quadrant(s) for a Point **In Exercises 15–20, determine the quadrant(s) in which (x, y) could be located.**

15. $x > 0$ and $y < 0$
16. $x < 0$ and $y < 0$
17. $x = -4$ and $y > 0$
18. $x < 0$ and $y = 7$
19. $x + y = 0, x \neq 0, y \neq 0$
20. $xy > 0$

Sketching a Scatter Plot **In Exercises 21 and 22, sketch a scatter plot of the data.**

21. The table shows the number y of Wal-Mart stores for each year x from 2007 through 2014. *(Source: Wal-Mart Stores, Inc.)*

DATA

Spreadsheet at LarsonPrecalculus.com

Year, x	Number of Stores, y
2007	7262
2008	7720
2009	8416
2010	8970
2011	10,130
2012	10,773
2013	10,942
2014	11,453

22. The ordered pairs below give the lowest temperature on record y (in degrees Fahrenheit) in Duluth, Minnesota, for each month x, where $x = 1$ represents January. *(Spreadsheet at LarsonPrecalculus.com)* *(Source: NOAA)*

$(1, -39)$ $(4, -5)$ $(7, 35)$ $(10, 8)$
$(2, -39)$ $(5, 17)$ $(8, 32)$ $(11, -23)$
$(3, -29)$ $(6, 27)$ $(9, 22)$ $(12, -34)$

Finding a Distance **In Exercises 23–26, find the distance between the points.**

23. $(-2, 6), (3, -6)$
24. $(8, 5), (0, 20)$
25. $(1, 4), (-5, -1)$
26. $(9.5, -2.6), (-3.9, 8.2)$

Verifying a Right Triangle In Exercises 27 and 28, (a) find the length of each side of the right triangle, and (b) show that these lengths satisfy the Pythagorean Theorem.

27.

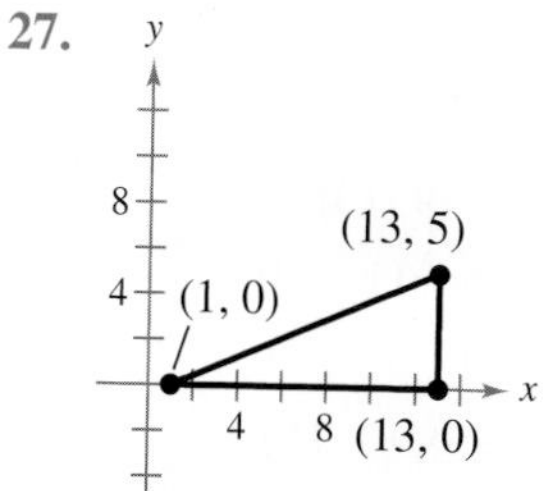

28.

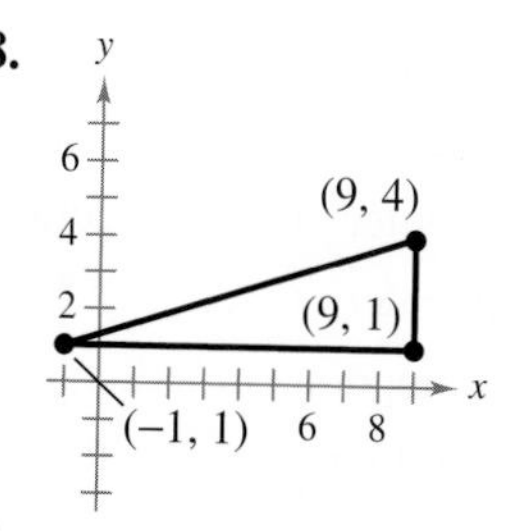

Verifying a Polygon In Exercises 29 and 30, show that the points form the vertices of the indicated polygon.

29. Right triangle: $(4, 0)$, $(2, 1)$, $(-1, -5)$

30. Right triangle: $(-1, 3)$, $(3, 5)$, $(5, 1)$

31. Isosceles triangle: $(1, -3)$, $(3, 2)$, $(-2, 4)$

32. Isosceles triangle: $(2, 3)$, $(4, 9)$, $(-2, 7)$

Plotting, Distance, and Midpoint In Exercises 33–36, (a) plot the points, (b) find the distance between the points, and (c) find the midpoint of the line segment joining the points.

33. $(1, 1)$, $(9, 7)$

34. $(6, -3)$, $(6, 5)$

35. $(-1, 2)$, $(5, 4)$

36. $\left(\frac{1}{2}, 1\right)$, $\left(-\frac{5}{2}, \frac{4}{3}\right)$

37. Flying Distance An airplane flies from Naples, Italy, in a straight line to Rome, Italy, which is 120 kilometers north and 150 kilometers west of Naples. How far does the plane fly?

38. **Sports** A soccer player passes the ball from a point that is 18 yards from the endline and 12 yards from the sideline. A teammate who is 42 yards from the same endline and 50 yards from the same sideline receives the pass. (See figure.) How long is the pass?

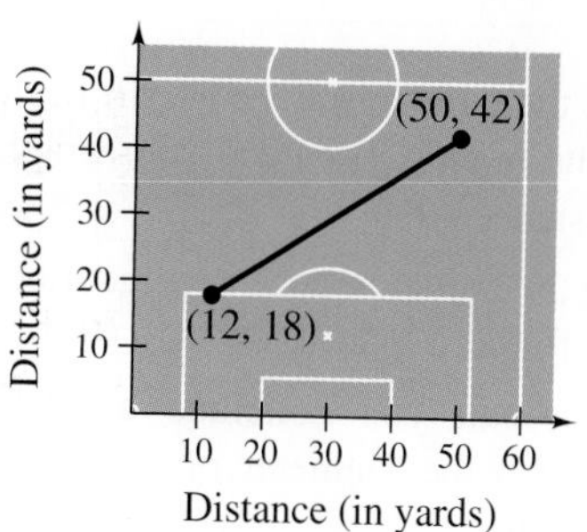

39. **Sales** The Coca-Cola Company had sales of $35,123 million in 2010 and $45,998 million in 2014. Use the Midpoint Formula to estimate the sales in 2012. Assume that the sales followed a linear pattern. *(Source: The Coca-Cola Company)*

40. **Revenue per Share** The revenue per share for Twitter, Inc. was $1.17 in 2013 and $3.25 in 2015. Use the Midpoint Formula to estimate the revenue per share in 2014. Assume that the revenue per share followed a linear pattern. *(Source: Twitter, Inc.)*

Determining Solution Points In Exercises 41–46, determine whether each point lies on the graph of the equation.

	Equation	Points	
41.	$y = \sqrt{x + 4}$	(a) $(0, 2)$	(b) $(5, 3)$
42.	$y = 4 - \lvert x - 2\rvert$	(a) $(1, 5)$	(b) $(6, 0)$
43.	$y = x^2 - 3x + 2$	(a) $(2, 0)$	(b) $(-2, 8)$
44.	$y = 3 - 2x^2$	(a) $(-1, 1)$	(b) $(-2, 11)$
45.	$x^2 + y^2 = 20$	(a) $(3, -2)$	(b) $(-4, 2)$
46.	$2x^2 + 5y^2 = 8$	(a) $(6, 0)$	(b) $(0, 4)$

Sketching the Graph of an Equation In Exercises 47–50, complete the table. Use the resulting solution points to sketch the graph of the equation.

47. $y = -2x + 5$

x	-1	0	1	2	$\frac{5}{2}$
y					
(x, y)					

48. $y + 1 = \frac{3}{4}x$

x	-2	0	1	$\frac{4}{3}$	2
y					
(x, y)					

49. $y + 3x = x^2$

x	-1	0	1	2	3
y					
(x, y)					

50. $y = 5 - x^2$

x	-2	-1	0	1	2
y					
(x, y)					

Finding *x*- and *y*-Intercepts **In Exercises 51–60, find the *x*- and *y*-intercepts of the graph of the equation.**

51. $y = 5x - 6$
52. $y = 8 - 3x$
53. $y = \sqrt{x + 4}$
54. $y = \sqrt{2x - 1}$
55. $y = |3x - 7|$
56. $y = -|x + 10|$
57. $y = 2x^3 - 4x^2$
58. $y = x^4 - 25$
59. $y^2 = 6 - x$
60. $y^2 = x + 1$

Testing for Symmetry **In Exercises 61–68, use the algebraic tests to check for symmetry with respect to both axes and the origin.**

61. $x^2 - y = 0$
62. $x - y^2 = 0$
63. $y = x^3$
64. $y = x^4 - x^2 + 3$
65. $y = \dfrac{x}{x^2 + 1}$
66. $y = \dfrac{1}{x^2 + 1}$
67. $xy^2 + 10 = 0$
68. $xy = 4$

Using Symmetry as a Sketching Aid **In Exercises 69–72, assume that the graph has the given type of symmetry. Complete the graph of the equation. To print an enlarged copy of the graph, go to *MathGraphs.com*.**

69.

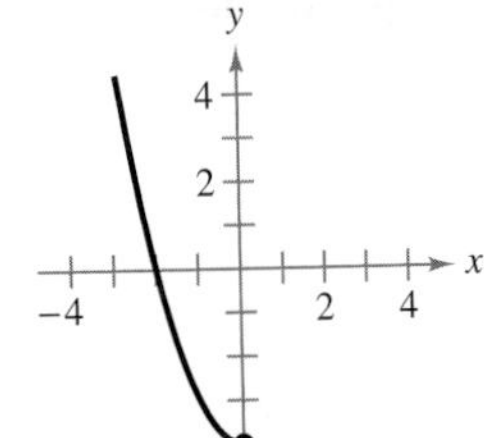

y-Axis symmetry

70.

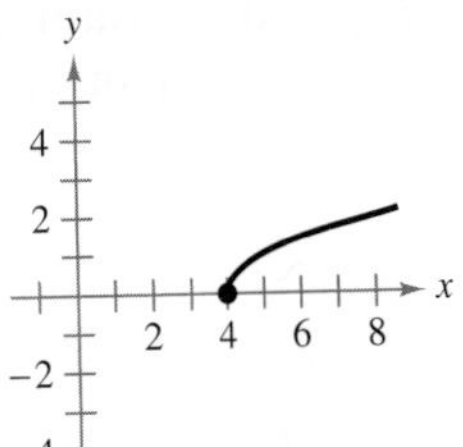

x-Axis symmetry

71.

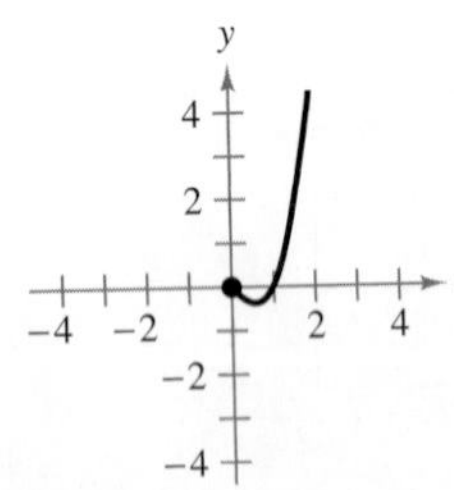

Origin symmetry

72.

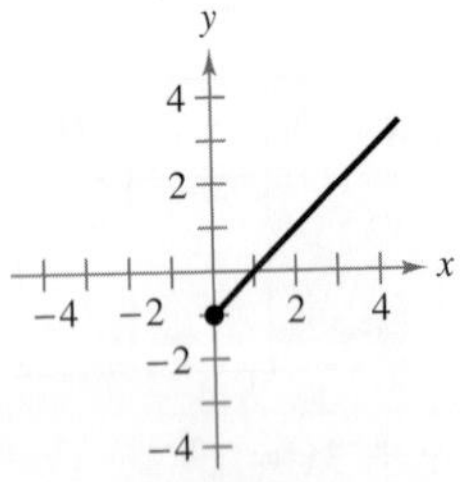

y-Axis symmetry

Sketching the Graph of an Equation **In Exercises 73–82, find any intercepts and test for symmetry. Then sketch the graph of the equation.**

73. $y = -3x + 1$
74. $y = 2x - 3$
75. $y = x^2 - 2x$
76. $x = y^2 - 1$
77. $y = x^3 + 3$
78. $y = x^3 - 1$
79. $y = \sqrt{x - 3}$
80. $y = \sqrt{1 - x}$
81. $y = |x - 6|$
82. $y = 1 - |x|$

Writing the Equation of a Circle **In Exercises 83–88, write the standard form of the equation of the circle with the given characteristics.**

83. Center: $(0, 0)$; Radius: 7
84. Center: $(-4, 5)$; Radius: 2
85. Center: $(3, 8)$; Solution point: $(-9, 13)$
86. Center: $(-2, -6)$; Solution point: $(1, -10)$
87. Endpoints of a diameter: $(3, 2), (-9, -8)$
88. Endpoints of a diameter: $(11, -5), (3, 15)$

Sketching a Circle **In Exercises 89–92, find the center and radius of the circle with the given equation. Then sketch the circle.**

89. $x^2 + y^2 = 25$
90. $x^2 + (y - 1)^2 = 1$
91. $\left(x - \frac{1}{2}\right)^2 + \left(y - \frac{1}{2}\right)^2 = \frac{9}{4}$
92. $(x - 2)^2 + (y + 3)^2 = \frac{16}{9}$

93. **Depreciation** A hospital purchases a new magnetic resonance imaging (MRI) machine for \$1.2 million. The depreciated value y (reduced value) after t years is given by $y = 1{,}200{,}000 - 80{,}000t$, $0 \le t \le 10$. Sketch the graph of the equation.

94. **Depreciation** You purchase an all-terrain vehicle (ATV) for \$9500. The depreciated value y (reduced value) after t years is given by $y = 9500 - 1000t$, $0 \le t \le 6$. Sketch the graph of the equation.

95. **Electronics** The resistance y (in ohms) of 1000 feet of solid copper wire at 68 degrees Fahrenheit is

$$y = \frac{10{,}370}{x^2}$$

where x is the diameter of the wire in mils (0.001 inch).

(a) Complete the table.

x	5	10	20	30	40	50
y						

x	60	70	80	90	100
y					

(b) Use the table of values in part (a) to sketch a graph of the model. Then use your graph to estimate the resistance when $x = 85.5$.

(c) Use the model to confirm algebraically the estimate you found in part (b).

(d) What can you conclude about the relationship between the diameter of the copper wire and the resistance?

96. Population Statistics The table shows the life expectancies of a child (at birth) in the United States for selected years from 1940 through 2010. *(Source: U.S. National Center for Health Statistics)*

DATA — Spreadsheet at LarsonPrecalculus.com

Year	Life Expectancy, y
1940	62.9
1950	68.2
1960	69.7
1970	70.8
1980	73.7
1990	75.4
2000	76.8
2010	78.7

A model for the life expectancy during this period is

$$y = \frac{63.6 + 0.97t}{1 + 0.01t}, \quad 0 \le t \le 70$$

where y represents the life expectancy and t is the time in years, with $t = 0$ corresponding to 1940.

(a) Use a graphing utility to graph the data from the table and the model in the same viewing window. How well does the model fit the data? Explain.

(b) Determine the life expectancy in 1990 both graphically and algebraically.

(c) Use the graph to determine the year when life expectancy was approximately 70.1. Verify your answer algebraically.

(d) Find the y-intercept of the graph of the model. What does it represent in the context of the problem?

(e) Do you think this model can be used to predict the life expectancy of a child 50 years from now? Explain.

Exploration

True or False? **In Exercises 97–100, determine whether the statement is true or false. Justify your answer.**

97. To divide a line segment into 16 equal parts, you have to use the Midpoint Formula 16 times.

98. The points $(-8, 4)$, $(2, 11)$, and $(-5, 1)$ represent the vertices of an isosceles triangle.

99. The graph of a linear equation cannot be symmetric with respect to the origin.

100. A circle can have a total of zero, one, two, three, or four x- and y-intercepts.

101. Think About It When plotting points on the rectangular coordinate system, when should you use different scales for the x- and y-axes? Explain.

102. Think About It What is the y-coordinate of any point on the x-axis? What is the x-coordinate of any point on the y-axis?

103. Proof Prove that the diagonals of the parallelogram in the figure intersect at their midpoints.

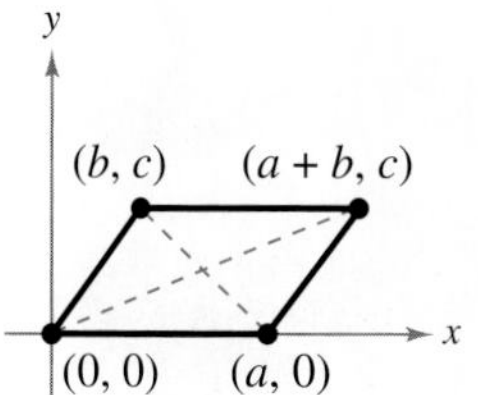

104. HOW DO YOU SEE IT? Use the plot of the point (x_0, y_0) in the figure. Match the transformation of the point with the correct plot. Explain. [The plots are labeled (i), (ii), (iii), and (iv).]

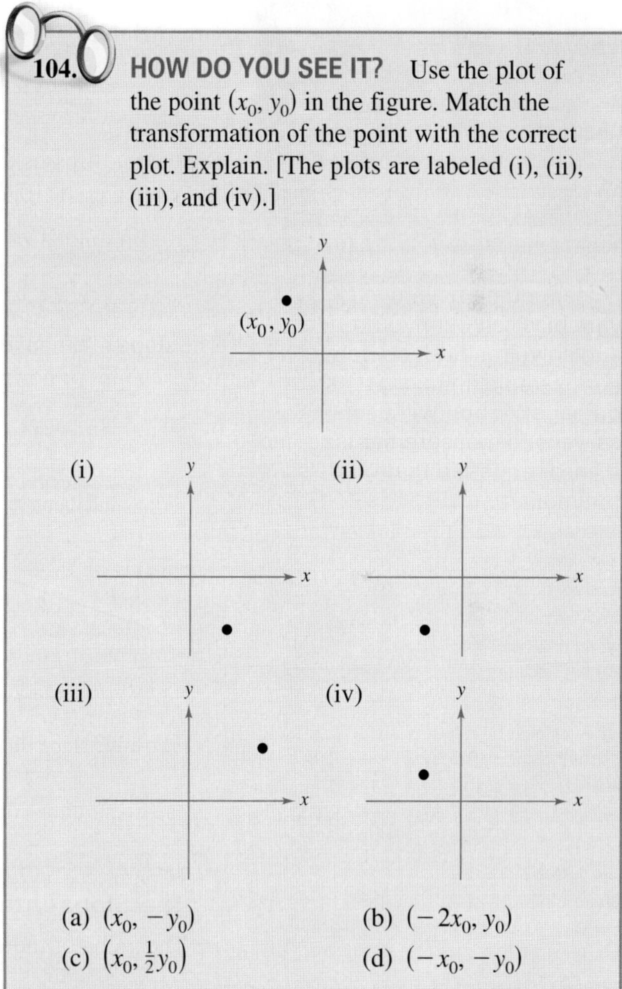

(a) $(x_0, -y_0)$ (b) $(-2x_0, y_0)$

(c) $\left(x_0, \frac{1}{2}y_0\right)$ (d) $(-x_0, -y_0)$

105. Using the Midpoint Formula A line segment has (x_1, y_1) as one endpoint and (x_m, y_m) as its midpoint. Find the other endpoint (x_2, y_2) of the line segment in terms of x_1, y_1, x_m, and y_m. Then use the result to find the missing point.

(a) $(x_1, y_1) = (1, -2)$
$(x_m, y_m) = (4, -1)$
$(x_2, y_2) = \square$

(b) $(x_1, y_1) = (-5, 11)$
$(x_m, y_m) = (2, 4)$
$(x_2, y_2) = \square$

The symbol [graphing utility icon] indicates an exercise or a part of an exercise in which you are instructed to use a graphing utility.

P.4 Linear Equations in Two Variables

Linear equations in two variables can help you model and solve real-life problems. For example, in Exercise 90 on page 51, you will use a surveyor's measurements to find a linear equation that models a mountain road.

- **Use slope to graph linear equations in two variables.**
- **Find the slope of a line given two points on the line.**
- **Write linear equations in two variables.**
- **Use slope to identify parallel and perpendicular lines.**
- **Use slope and linear equations in two variables to model and solve real-life problems.**

Using Slope

The simplest mathematical model for relating two variables is the **linear equation in two variables** $y = mx + b$. The equation is called *linear* because its graph is a line. (In mathematics, the term *line* means *straight line*.) By letting $x = 0$, you obtain

$$y = m(0) + b = b.$$

So, the line crosses the y-axis at $y = b$, as shown in the figures below. In other words, the y-intercept is $(0, b)$. The steepness, or *slope,* of the line is m.

$$y = mx + b$$

Slope ⟶ m; y-Intercept ⟶ b

The **slope** of a nonvertical line is the number of units the line rises (or falls) vertically for each unit of horizontal change from left to right, as shown below.

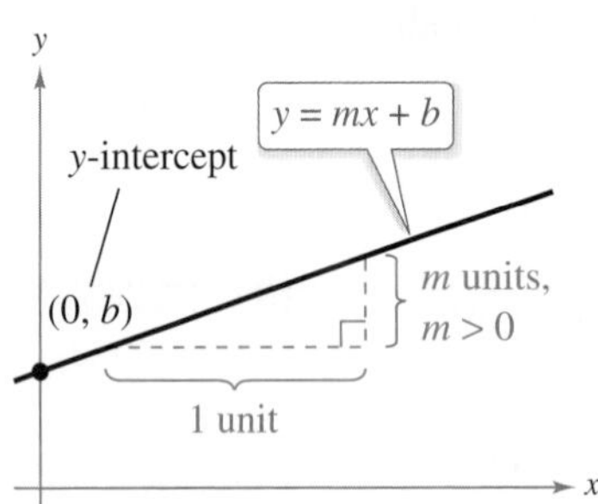

Positive slope, line rises

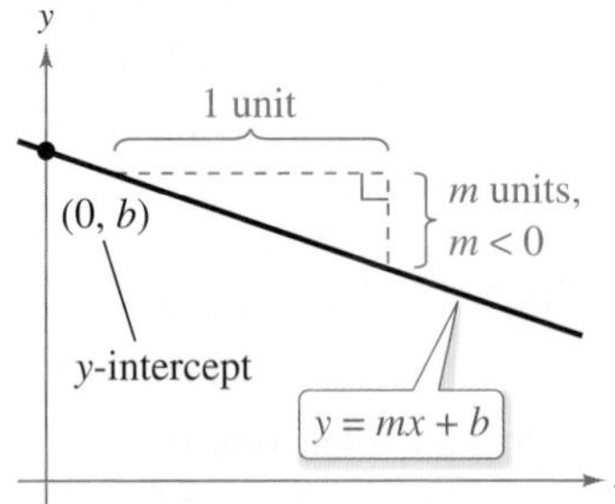

Negative slope, line falls

A linear equation written in **slope-intercept form** has the form $y = mx + b$.

> **The Slope-Intercept Form of the Equation of a Line**
>
> The graph of the equation
>
> $$y = mx + b$$
>
> is a line whose slope is m and whose y-intercept is $(0, b)$.

Once you determine the slope and the y-intercept of a line, it is relatively simple to sketch its graph. In the next example, note that none of the lines is vertical. A vertical line has an equation of the form

$$x = a. \qquad \text{Vertical line}$$

The equation of a vertical line cannot be written in the form $y = mx + b$ because the slope of a vertical line is undefined (see Figure P.24).

Slope is undefined.
Figure P.24

EXAMPLE 1 Graphing Linear Equations

See LarsonPrecalculus.com for an interactive version of this type of example.

Sketch the graph of each linear equation.

a. $y = 2x + 1$

b. $y = 2$

c. $x + y = 2$

Solution

a. Because $b = 1$, the y-intercept is $(0, 1)$. Moreover, the slope is $m = 2$, so the line *rises* two units for each unit the line moves to the right (see figure).

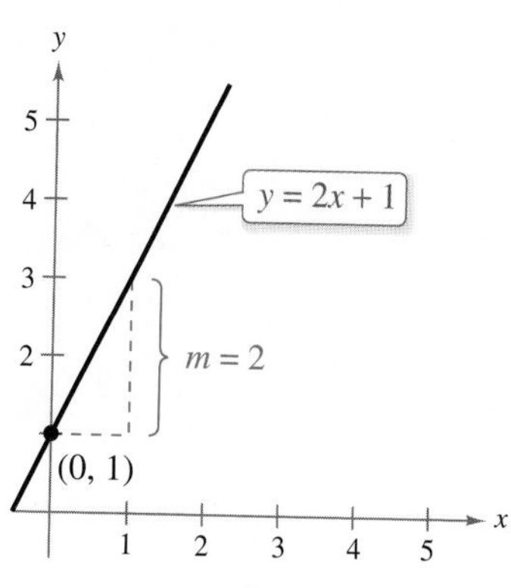

When m is positive, the line rises.

b. By writing this equation in the form $y = (0)x + 2$, you find that the y-intercept is $(0, 2)$ and the slope is $m = 0$. A slope of 0 implies that the line is horizontal—that is, it does not rise *or* fall (see figure).

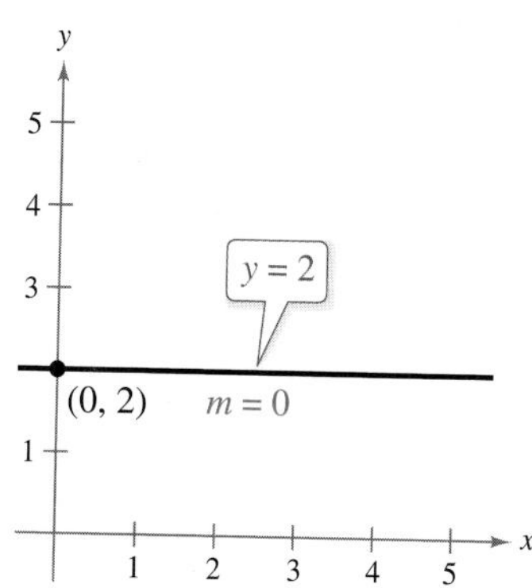

When m is 0, the line is horizontal.

c. By writing this equation in slope-intercept form

$$x + y = 2 \qquad \text{Write original equation.}$$

$$y = -x + 2 \qquad \text{Subtract } x \text{ from each side.}$$

$$y = (-1)x + 2 \qquad \text{Write in slope-intercept form.}$$

you find that the y-intercept is $(0, 2)$. Moreover, the slope is $m = -1$, so the line *falls* one unit for each unit the line moves to the right (see figure).

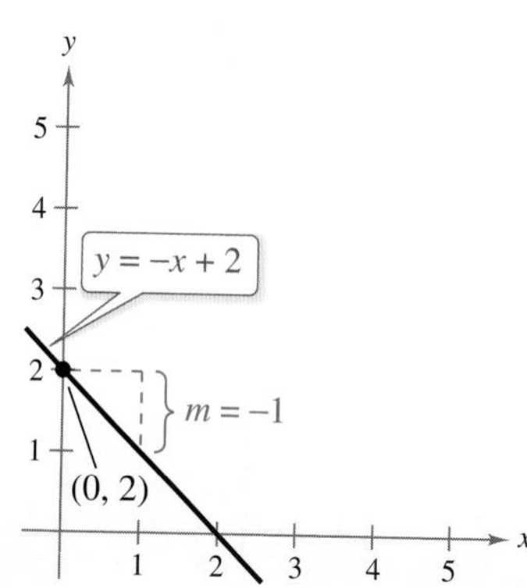

When m is negative, the line falls.

✓ **Checkpoint**  Audio-video solution in English & Spanish at LarsonPrecalculus.com

Sketch the graph of each linear equation.

a. $y = 3x + 2$ **b.** $y = -3$ **c.** $4x + y = 5$

Finding the Slope of a Line

Given an equation of a line, you can find its slope by writing the equation in slope-intercept form. When you are not given an equation, you can still find the slope by using two points on the line. For example, consider the line passing through the points (x_1, y_1) and (x_2, y_2) in the figure below.

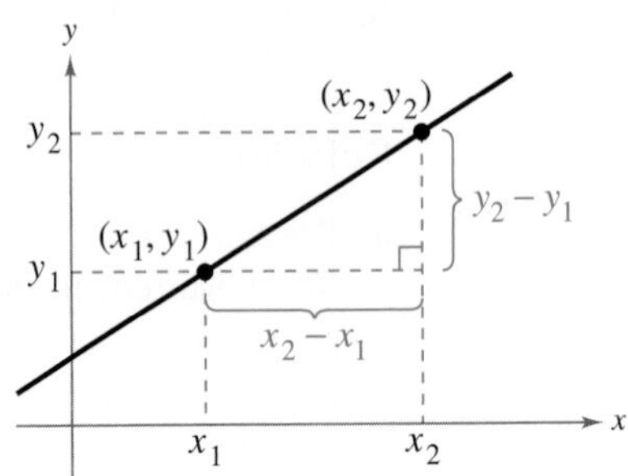

As you move from left to right along this line, a change of $(y_2 - y_1)$ units in the vertical direction corresponds to a change of $(x_2 - x_1)$ units in the horizontal direction.

$$y_2 - y_1 = \text{change in } y = \text{rise}$$

and

$$x_2 - x_1 = \text{change in } x = \text{run}$$

The ratio of $(y_2 - y_1)$ to $(x_2 - x_1)$ represents the slope of the line that passes through the points (x_1, y_1) and (x_2, y_2).

$$\text{Slope} = \frac{\text{change in } y}{\text{change in } x} = \frac{\text{rise}}{\text{run}} = \frac{y_2 - y_1}{x_2 - x_1}$$

The Slope of a Line Passing Through Two Points

The **slope** m of the nonvertical line through (x_1, y_1) and (x_2, y_2) is

$$m = \frac{y_2 - y_1}{x_2 - x_1}$$

where $x_1 \neq x_2$.

When using the formula for slope, the *order of subtraction* is important. Given two points on a line, you are free to label either one of them as (x_1, y_1) and the other as (x_2, y_2). However, once you do this, you must form the numerator and denominator using the same order of subtraction.

$$\underbrace{m = \frac{y_2 - y_1}{x_2 - x_1}}_{\text{Correct}} \qquad \underbrace{m = \frac{y_1 - y_2}{x_1 - x_2}}_{\text{Correct}} \qquad \underbrace{m = \frac{y_2 - y_1}{x_1 - x_2}}_{\text{Incorrect}} \times$$

For example, the slope of the line passing through the points (3, 4) and (5, 7) can be calculated as

$$m = \frac{7 - 4}{5 - 3} = \frac{3}{2}$$

or as

$$m = \frac{4 - 7}{3 - 5} = \frac{-3}{-2} = \frac{3}{2}.$$

EXAMPLE 2 **Finding the Slope of a Line Through Two Points**

Find the slope of the line passing through each pair of points.

a. $(-2, 0)$ and $(3, 1)$ **b.** $(-1, 2)$ and $(2, 2)$

c. $(0, 4)$ and $(1, -1)$ **d.** $(3, 4)$ and $(3, 1)$

Solution

a. Letting $(x_1, y_1) = (-2, 0)$ and $(x_2, y_2) = (3, 1)$, you find that the slope is

$$m = \frac{y_2 - y_1}{x_2 - x_1} = \frac{1 - 0}{3 - (-2)} = \frac{1}{5}.$$ See Figure P.25.

b. The slope of the line passing through $(-1, 2)$ and $(2, 2)$ is

$$m = \frac{2 - 2}{2 - (-1)} = \frac{0}{3} = 0.$$ See Figure P.26.

c. The slope of the line passing through $(0, 4)$ and $(1, -1)$ is

$$m = \frac{-1 - 4}{1 - 0} = \frac{-5}{1} = -5.$$ See Figure P.27.

d. The slope of the line passing through $(3, 4)$ and $(3, 1)$ is

$$m = \frac{1 - 4}{3 - 3} = \frac{-3}{0}.$$ See Figure P.28.

Division by 0 is undefined, so the slope is undefined and the line is vertical.

REMARK In Figures P.25 through P.28, note the relationships between slope and the orientation of the line.

a. Positive slope: line rises from left to right

b. Zero slope: line is horizontal

c. Negative slope: line falls from left to right

d. Undefined slope: line is vertical

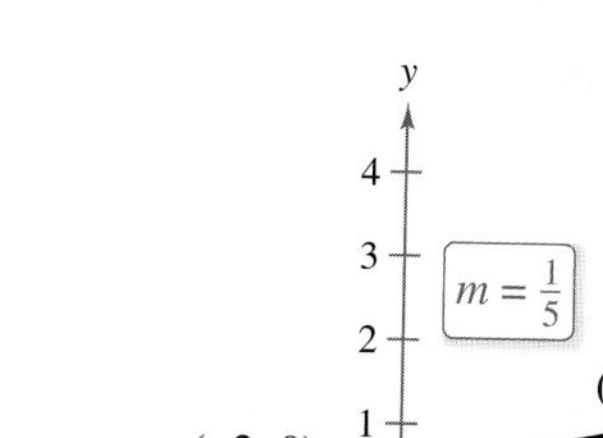

Figure P.25

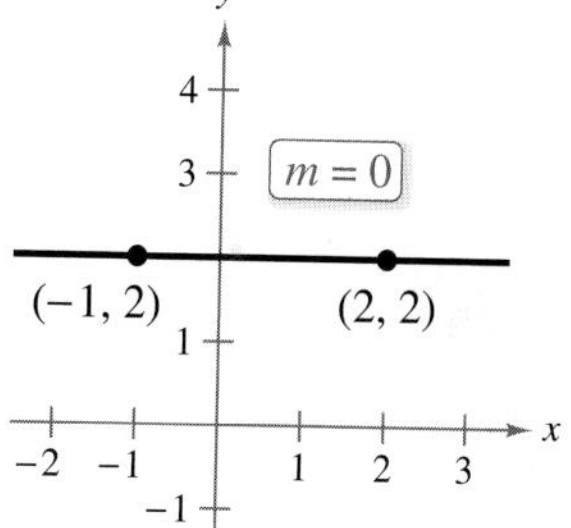

Figure P.26

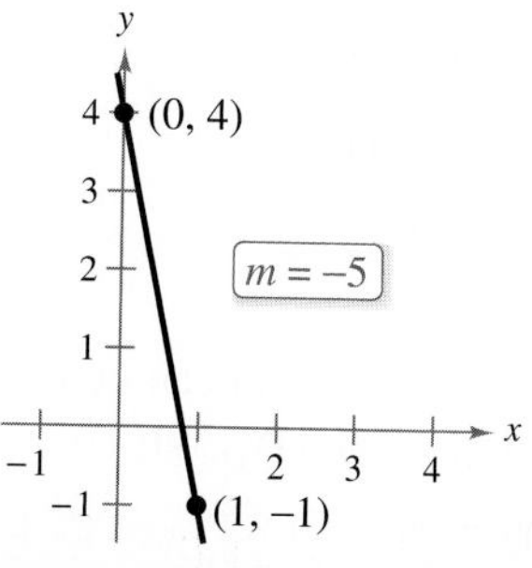

Figure P.27

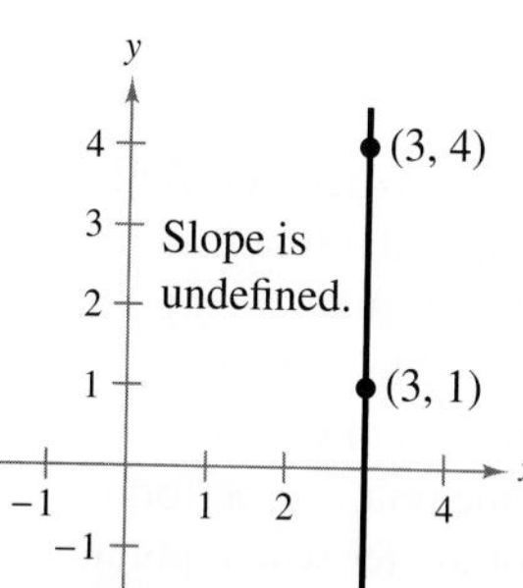

Figure P.28

✓ ***Checkpoint*** *Audio-video solution in English & Spanish at LarsonPrecalculus.com*

Find the slope of the line passing through each pair of points.

a. $(-5, -6)$ and $(2, 8)$ **b.** $(4, 2)$ and $(2, 5)$

c. $(0, 0)$ and $(0, -6)$ **d.** $(0, -1)$ and $(3, -1)$

Writing Linear Equations in Two Variables

If (x_1, y_1) is a point on a line of slope m and (x, y) is *any other* point on the line, then

$$\frac{y - y_1}{x - x_1} = m.$$

This equation in the variables x and y can be rewritten in the **point-slope form** of the equation of a line

$$y - y_1 = m(x - x_1).$$

Point-Slope Form of the Equation of a Line

The equation of the line with slope m passing through the point (x_1, y_1) is

$$y - y_1 = m(x - x_1).$$

The point-slope form is useful for *finding* the equation of a line. You should remember this form.

EXAMPLE 3 Using the Point-Slope Form

Find the slope-intercept form of the equation of the line that has a slope of 3 and passes through the point $(1, -2)$.

Solution Use the point-slope form with $m = 3$ and $(x_1, y_1) = (1, -2)$.

$$y - y_1 = m(x - x_1) \quad \text{Point-slope form}$$
$$y - (-2) = 3(x - 1) \quad \text{Substitute for } m, x_1, \text{ and } y_1.$$
$$y + 2 = 3x - 3 \quad \text{Simplify.}$$
$$y = 3x - 5 \quad \text{Write in slope-intercept form.}$$

The slope-intercept form of the equation of the line is $y = 3x - 5$. Figure P.29 shows the graph of this equation.

Figure P.29

Checkpoint Audio-video solution in English & Spanish at LarsonPrecalculus.com

Find the slope-intercept form of the equation of the line that has the given slope and passes through the given point.

a. $m = 2,\quad (3, -7)$

b. $m = -\frac{2}{3},\quad (1, 1)$

c. $m = 0,\quad (1, 1)$

REMARK When you find an equation of the line that passes through two given points, you only need to substitute the coordinates of one of the points in the point-slope form. It does not matter which point you choose because both points will yield the same result.

The point-slope form can be used to find an equation of the line passing through two points (x_1, y_1) and (x_2, y_2). To do this, first find the slope of the line.

$$m = \frac{y_2 - y_1}{x_2 - x_1}, \quad x_1 \neq x_2$$

Then use the point-slope form to obtain the equation.

$$y - y_1 = \frac{y_2 - y_1}{x_2 - x_1}(x - x_1) \quad \text{Two-point form}$$

This is sometimes called the **two-point form** of the equation of a line.

Parallel and Perpendicular Lines

Slope can tell you whether two nonvertical lines in a plane are parallel, perpendicular, or neither.

> **Parallel and Perpendicular Lines**
>
> 1. Two distinct nonvertical lines are **parallel** if and only if their slopes are equal. That is,
>
> $$m_1 = m_2.$$
>
> 2. Two nonvertical lines are **perpendicular** if and only if their slopes are negative reciprocals of each other. That is,
>
> $$m_1 = \frac{-1}{m_2}.$$

EXAMPLE 4 Finding Parallel and Perpendicular Lines

Find the slope-intercept form of the equations of the lines that pass through the point $(2, -1)$ and are (a) parallel to and (b) perpendicular to the line $2x - 3y = 5$.

Solution Write the equation of the given line in slope-intercept form.

$2x - 3y = 5$	Write original equation.
$-3y = -2x + 5$	Subtract $2x$ from each side.
$y = \frac{2}{3}x - \frac{5}{3}$	Write in slope-intercept form.

Notice that the line has a slope of $m = \frac{2}{3}$.

a. Any line parallel to the given line must also have a slope of $\frac{2}{3}$. Use the point-slope form with $m = \frac{2}{3}$ and $(x_1, y_1) = (2, -1)$.

$y - (-1) = \frac{2}{3}(x - 2)$	Write in point-slope form.
$3(y + 1) = 2(x - 2)$	Multiply each side by 3.
$3y + 3 = 2x - 4$	Distributive Property
$y = \frac{2}{3}x - \frac{7}{3}$	Write in slope-intercept form.

Notice the similarity between the slope-intercept form of this equation and the slope-intercept form of the given equation.

b. Any line perpendicular to the given line must have a slope of $-\frac{3}{2}$ (because $-\frac{3}{2}$ is the negative reciprocal of $\frac{2}{3}$). Use the point-slope form with $m = -\frac{3}{2}$ and $(x_1, y_1) = (2, -1)$.

$y - (-1) = -\frac{3}{2}(x - 2)$	Write in point-slope form.
$2(y + 1) = -3(x - 2)$	Multiply each side by 2.
$2y + 2 = -3x + 6$	Distributive Property
$y = -\frac{3}{2}x + 2$	Write in slope-intercept form.

The graphs of all three equations are shown in Figure P.30.

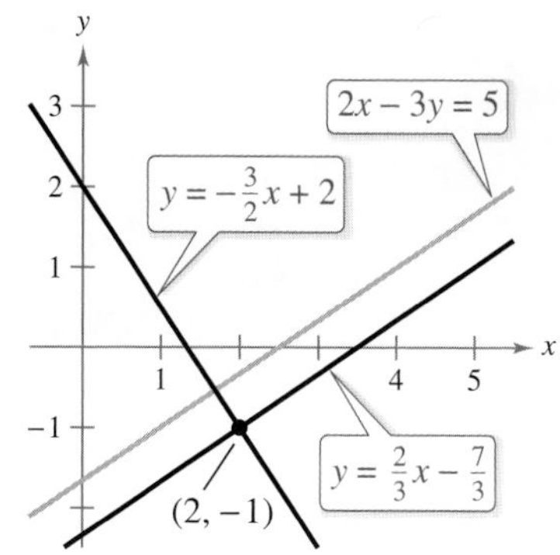

Figure P.30

Find the slope-intercept form of the equations of the lines that pass through the point $(-4, 1)$ and are (a) parallel to and (b) perpendicular to the line $5x - 3y = 8$. ■

TECHNOLOGY On a graphing utility, lines will not appear to have the correct slope unless you use a viewing window that has a square setting. For instance, graph the lines in Example 4 using the standard setting $-10 \le x \le 10$ and $-10 \le y \le 10$. Then reset the viewing window with the square setting $-9 \le x \le 9$ and $-6 \le y \le 6$. On which setting do the lines $y = \frac{2}{3}x - \frac{5}{3}$ and $y = -\frac{3}{2}x + 2$ appear to be perpendicular?

Applications

In real-life problems, the slope of a line can be interpreted as either a *ratio* or a *rate*. When the x-axis and y-axis have the same unit of measure, the slope has no units and is a **ratio.** When the x-axis and y-axis have different units of measure, the slope is a **rate** or **rate of change.**

EXAMPLE 5 Using Slope as a Ratio

The Americans with Disabilities Act (ADA) became law on July 26, 1990. It is the most comprehensive formulation of rights for persons with disabilities in U.S. (and world) history.

The maximum recommended slope of a wheelchair ramp is $\frac{1}{12}$. A business installs a wheelchair ramp that rises 22 inches over a horizontal length of 24 feet. Is the ramp steeper than recommended? *(Source: ADA Standards for Accessible Design)*

Solution The horizontal length of the ramp is 24 feet or $12(24) = 288$ inches (see figure). So, the slope of the ramp is

$$\text{Slope} = \frac{\text{vertical change}}{\text{horizontal change}} = \frac{22 \text{ in.}}{288 \text{ in.}} \approx 0.076.$$

Because $\frac{1}{12} \approx 0.083$, the slope of the ramp is not steeper than recommended.

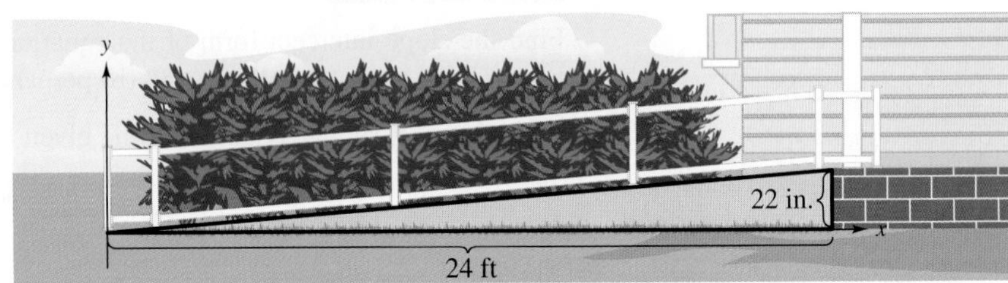

✓ **Checkpoint**  *Audio-video solution in English & Spanish at LarsonPrecalculus.com*

The business in Example 5 installs a second ramp that rises 36 inches over a horizontal length of 32 feet. Is the ramp steeper than recommended?

EXAMPLE 6 Using Slope as a Rate of Change

A kitchen appliance manufacturing company determines that the total cost C (in dollars) of producing x units of a blender is given by

$$C = 25x + 3500. \qquad \text{Cost equation}$$

Interpret the y-intercept and slope of this line.

Solution The y-intercept $(0, 3500)$ tells you that the cost of producing 0 units is \$3500. This is the *fixed cost* of production—it includes costs that must be paid regardless of the number of units produced. The slope of $m = 25$ tells you that the cost of producing each unit is \$25, as shown in Figure P.31. Economists call the cost per unit the *marginal cost*. When the production increases by one unit, the "margin," or extra amount of cost, is \$25. So, the cost increases at a rate of \$25 per unit.

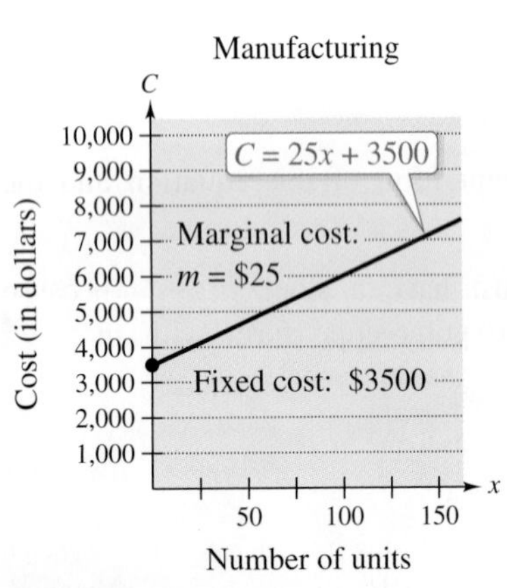

Production cost
Figure P.31

✓ **Checkpoint**  *Audio-video solution in English & Spanish at LarsonPrecalculus.com*

An accounting firm determines that the value V (in dollars) of a copier t years after its purchase is given by

$$V = -300t + 1500.$$

Interpret the y-intercept and slope of this line.

Businesses can deduct most of their expenses in the same year they occur. One exception is the cost of property that has a useful life of more than 1 year. Such costs must be *depreciated* (decreased in value) over the useful life of the property. Depreciating the *same amount* each year is called *linear* or *straight-line depreciation*. The *book value* is the difference between the original value and the total amount of depreciation accumulated to date.

EXAMPLE 7 Straight-Line Depreciation

A college purchased exercise equipment worth \$12,000 for the new campus fitness center. The equipment has a useful life of 8 years. The salvage value at the end of 8 years is \$2000. Write a linear equation that describes the book value of the equipment each year.

Solution Let V represent the value of the equipment at the end of year t. Represent the initial value of the equipment by the data point (0, 12,000) and the salvage value of the equipment by the data point (8, 2000). The slope of the line is

$$m = \frac{2000 - 12{,}000}{8 - 0} = -\$1250$$

which represents the annual depreciation in *dollars per year*. Using the point-slope form, write an equation of the line.

$$V - 12{,}000 = -1250(t - 0) \quad \text{Write in point-slope form.}$$

$$V = -1250t + 12{,}000 \quad \text{Write in slope-intercept form.}$$

The table shows the book value at the end of each year, and Figure P.32 shows the graph of the equation.

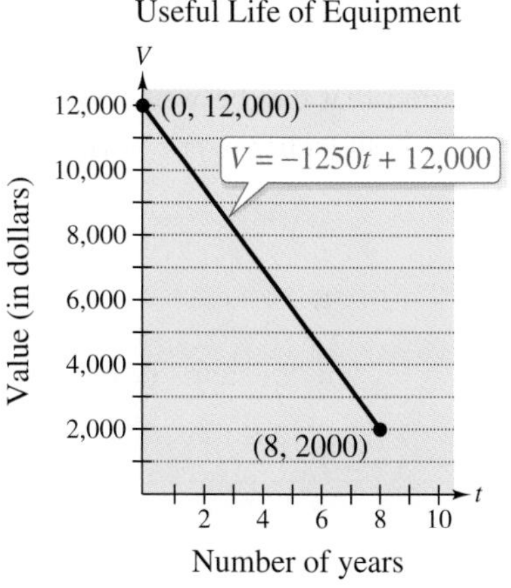

Straight-line depreciation
Figure P.32

Year, t	Value, V
0	12,000
1	10,750
2	9500
3	8250
4	7000
5	5750
6	4500
7	3250
8	2000

✓ **Checkpoint** Audio-video solution in English & Spanish at LarsonPrecalculus.com

A manufacturing firm purchases a machine worth \$24,750. The machine has a useful life of 6 years. After 6 years, the machine will have to be discarded and replaced, because it will have no salvage value. Write a linear equation that describes the book value of the machine each year.

In many real-life applications, the two data points that determine the line are often given in a disguised form. Note how the data points are described in Example 7.

EXAMPLE 8 Predicting Sales

The sales for NIKE were approximately \$25.3 billion in 2013 and \$27.8 billion in 2014. Using only this information, write a linear equation that gives the sales in terms of the year. Then predict the sales in 2017. *(Source: NIKE Inc.)*

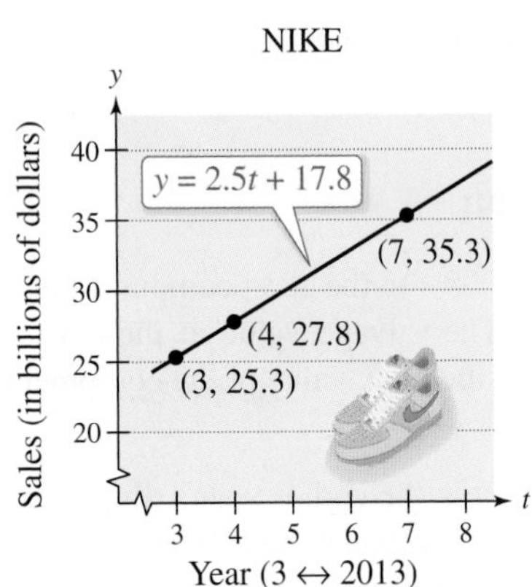

Figure P.33

Solution Let $t = 3$ represent 2013. Then the two given values are represented by the data points (3, 25.3) and (4, 27.8) The slope of the line through these points is

$$m = \frac{27.8 - 25.3}{4 - 3} = 2.5.$$

Use the point-slope form to write an equation that relates the sales y and the year t.

$$y - 25.3 = 2.5(t - 3) \quad \text{Write in point-slope form.}$$

$$y = 2.5t + 17.8 \quad \text{Write in slope-intercept form.}$$

According to this equation, the sales in 2017 will be

$y = 2.5(7) + 17.8 = 17.5 + 17.8 = \$35.3$ billion. (See Figure P.33.)

✓ Checkpoint *Audio-video solution in English & Spanish at LarsonPrecalculus.com*

The sales for Foot Locker were approximately \$6.5 billion in 2013 and \$7.2 billion in 2014. Repeat Example 8 using this information. *(Source: Foot Locker)*

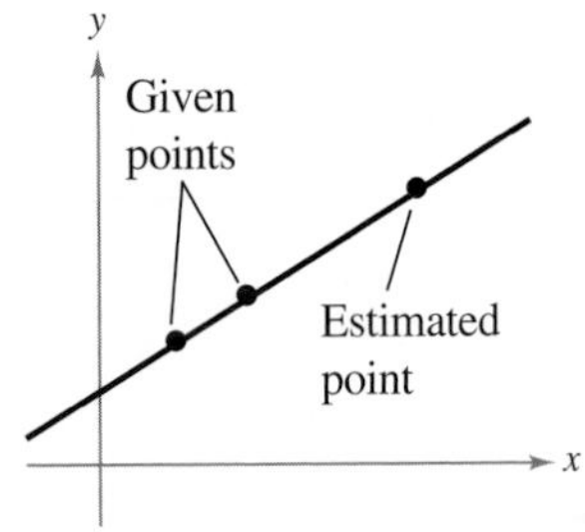

Linear extrapolation
Figure P.34

The prediction method illustrated in Example 8 is called **linear extrapolation.** Note in Figure P.34 that an extrapolated point does not lie between the given points. When the estimated point lies between two given points, as shown in Figure P.35, the procedure is called **linear interpolation.**

The slope of a vertical line is undefined, so its equation cannot be written in slope-intercept form. However, every line has an equation that can be written in the **general form** $Ax + By + C = 0$, where A and B are not both zero.

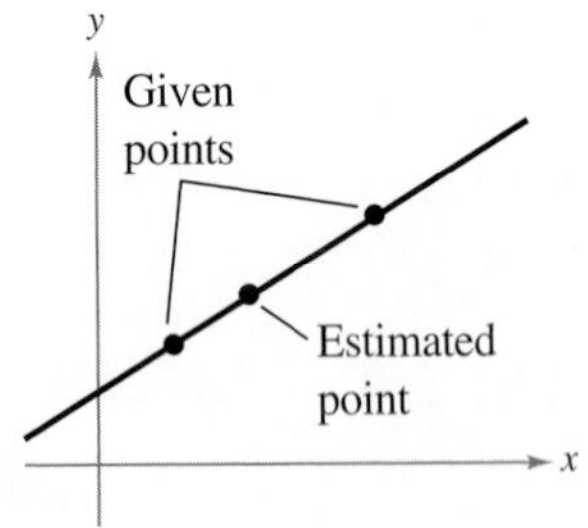

Linear interpolation
Figure P.35

Summary of Equations of Lines

1. General form: $Ax + By + C = 0$
2. Vertical line: $x = a$
3. Horizontal line: $y = b$
4. Slope-intercept form: $y = mx + b$
5. Point-slope form: $y - y_1 = m(x - x_1)$
6. Two-point form: $y - y_1 = \dfrac{y_2 - y_1}{x_2 - x_1}(x - x_1)$

Summarize (Section P.4)

1. Explain how to use slope to graph a linear equation in two variables *(page 40)* and how to find the slope of a line passing through two points *(page 42)*. For examples of using and finding slopes, see Examples 1 and 2.
2. State the point-slope form of the equation of a line *(page 44)*. For an example of using point-slope form, see Example 3.
3. Explain how to use slope to identify parallel and perpendicular lines *(page 45)*. For an example of finding parallel and perpendicular lines, see Example 4.
4. Describe examples of how to use slope and linear equations in two variables to model and solve real-life problems *(pages 46–48, Examples 5–8)*.

P.4 Exercises

See CalcChat.com for tutorial help and worked-out solutions to odd-numbered exercises.

Vocabulary: Fill in the blanks.

1. The simplest mathematical model for relating two variables is the ________ equation in two variables $y = mx + b$.
2. For a line, the ratio of the change in y to the change in x is the ________ of the line.
3. The ________-________ form of the equation of a line with slope m passing through the point (x_1, y_1) is $y - y_1 = m(x - x_1)$.
4. Two distinct nonvertical lines are ________ if and only if their slopes are equal.
5. Two nonvertical lines are ________ if and only if their slopes are negative reciprocals of each other.
6. When the x-axis and y-axis have different units of measure, the slope can be interpreted as a ________.
7. ________ ________ is the prediction method used to estimate a point on a line when the point does not lie between the given points.
8. Every line has an equation that can be written in ________ form.

Skills and Applications

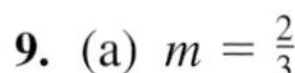

Identifying Lines **In Exercises 9 and 10, identify the line that has each slope.**

9. (a) $m = \frac{2}{3}$
   (b) m is undefined.
   (c) $m = -2$

10. (a) $m = 0$
    (b) $m = -\frac{3}{4}$
    (c) $m = 1$

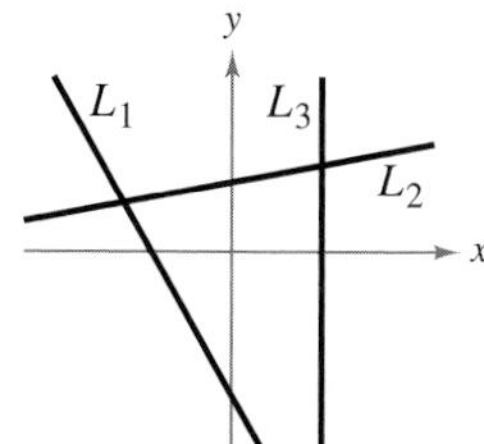

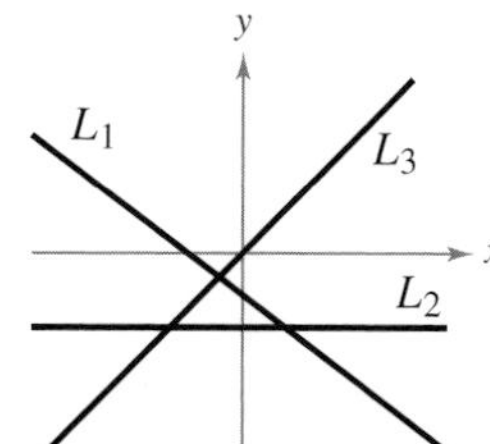

Sketching Lines **In Exercises 11 and 12, sketch the lines through the point with the given slopes on the same set of coordinate axes.**

	Point	Slopes	
11.	(2, 3)	(a) 0	(b) 1
		(c) 2	(d) −3
12.	(−4, 1)	(a) 3	(b) −3
		(c) $\frac{1}{2}$	(d) Undefined

Estimating the Slope of a Line **In Exercises 13 and 14, estimate the slope of the line.**

13.

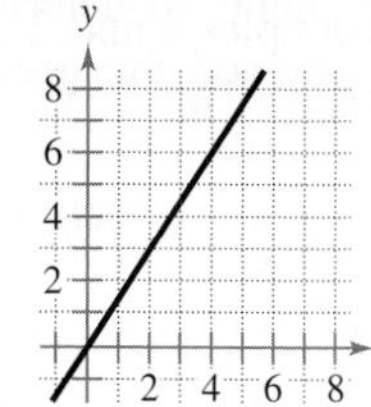

14.

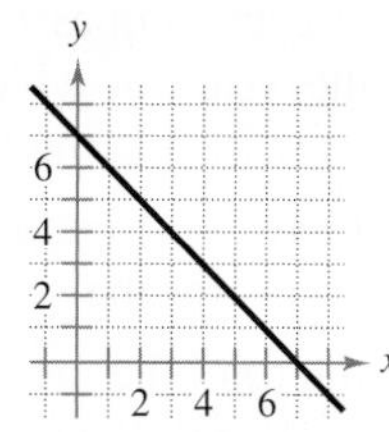

Graphing a Linear Equation **In Exercises 15–24, find the slope and y-intercept (if possible) of the line. Sketch the line.**

15. $y = 5x + 3$
16. $y = -x - 10$
17. $y = -\frac{3}{4}x - 1$
18. $y = \frac{2}{3}x + 2$
19. $y - 5 = 0$
20. $x + 4 = 0$
21. $5x - 2 = 0$
22. $3y + 5 = 0$
23. $7x - 6y = 30$
24. $2x + 3y = 9$

Finding the Slope of a Line Through Two Points **In Exercises 25–34, find the slope of the line passing through the pair of points.**

25. $(0, 9), (6, 0)$
26. $(10, 0), (0, -5)$
27. $(-3, -2), (1, 6)$
28. $(2, -1), (-2, 1)$
29. $(5, -7), (8, -7)$
30. $(-2, 1), (-4, -5)$
31. $(-6, -1), (-6, 4)$
32. $(0, -10), (-4, 0)$
33. $(4.8, 3.1), (-5.2, 1.6)$
34. $\left(\frac{11}{2}, -\frac{4}{3}\right), \left(-\frac{3}{2}, -\frac{1}{3}\right)$

Using the Slope and a Point **In Exercises 35–42, use the slope of the line and the point on the line to find three additional points through which the line passes. (There are many correct answers.)**

35. $m = 0, \quad (5, 7)$
36. $m = 0, \quad (3, -2)$
37. $m = 2, \quad (-5, 4)$
38. $m = -2, \quad (0, -9)$
39. $m = -\frac{1}{3}, \quad (4, 5)$
40. $m = \frac{1}{4}, \quad (3, -4)$
41. m is undefined, $\quad (-4, 3)$
42. m is undefined, $\quad (2, 14)$

Using the Point-Slope Form **In Exercises 43–54, find the slope-intercept form of the equation of the line that has the given slope and passes through the given point. Sketch the line.**

43. $m = 3,\quad (0, -2)$
44. $m = -1,\quad (0, 10)$
45. $m = -2,\quad (-3, 6)$
46. $m = 4,\quad (0, 0)$
47. $m = -\frac{1}{3},\quad (4, 0)$
48. $m = \frac{1}{4},\quad (8, 2)$
49. $m = -\frac{1}{2},\quad (2, -3)$
50. $m = \frac{3}{4},\quad (-2, -5)$
51. $m = 0,\quad \left(4, \frac{5}{2}\right)$
52. $m = 6,\quad \left(2, \frac{3}{2}\right)$
53. $m = 5,\quad (-5.1, 1.8)$
54. $m = 0,\quad (-2.5, 3.25)$

Finding an Equation of a Line **In Exercises 55–64, find an equation of the line passing through the pair of points. Sketch the line.**

55. $(5, -1), (-5, 5)$
56. $(4, 3), (-4, -4)$
57. $(-7, 2), (-7, 5)$
58. $(-6, -3), (2, -3)$
59. $\left(2, \frac{1}{2}\right), \left(\frac{1}{2}, \frac{5}{4}\right)$
60. $(1, 1), \left(6, -\frac{2}{3}\right)$
61. $(1, 0.6), (-2, -0.6)$
62. $(-8, 0.6), (2, -2.4)$
63. $(2, -1), \left(\frac{1}{3}, -1\right)$
64. $\left(\frac{7}{3}, -8\right), \left(\frac{7}{3}, 1\right)$

Parallel and Perpendicular Lines **In Exercises 65–68, determine whether the lines are parallel, perpendicular, or neither.**

65. $L_1\colon y = -\frac{2}{3}x - 3$
$L_2\colon y = -\frac{2}{3}x + 4$

66. $L_1\colon y = \frac{1}{4}x - 1$
$L_2\colon y = 4x + 7$

67. $L_1\colon y = \frac{1}{2}x - 3$
$L_2\colon y = -\frac{1}{2}x + 1$

68. $L_1\colon y = -\frac{4}{5}x - 5$
$L_2\colon y = \frac{5}{4}x + 1$

Parallel and Perpendicular Lines **In Exercises 69–72, determine whether the lines L_1 and L_2 passing through the pairs of points are parallel, perpendicular, or neither.**

69. $L_1\colon (0, -1), (5, 9)$
$L_2\colon (0, 3), (4, 1)$

70. $L_1\colon (-2, -1), (1, 5)$
$L_2\colon (1, 3), (5, -5)$

71. $L_1\colon (-6, -3), (2, -3)$
$L_2\colon \left(3, -\frac{1}{2}\right), \left(6, -\frac{1}{2}\right)$

72. $L_1\colon (4, 8), (-4, 2)$
$L_2\colon (3, -5), \left(-1, \frac{1}{3}\right)$

Finding Parallel and Perpendicular Lines **In Exercises 73–80, find equations of the lines that pass through the given point and are (a) parallel to and (b) perpendicular to the given line.**

73. $4x - 2y = 3,\quad (2, 1)$
74. $x + y = 7,\quad (-3, 2)$
75. $3x + 4y = 7,\quad \left(-\frac{2}{3}, \frac{7}{8}\right)$
76. $5x + 3y = 0,\quad \left(\frac{7}{8}, \frac{3}{4}\right)$
77. $y + 5 = 0,\quad (-2, 4)$
78. $x - 4 = 0,\quad (3, -2)$
79. $x - y = 4,\quad (2.5, 6.8)$
80. $6x + 2y = 9,\quad (-3.9, -1.4)$

Using Intercept Form **In Exercises 81–86, use the *intercept form* to find the general form of the equation of the line with the given intercepts. The intercept form of the equation of a line with intercepts $(a, 0)$ and $(0, b)$ is**

$$\frac{x}{a} + \frac{y}{b} = 1,\quad a \neq 0,\quad b \neq 0.$$

81. x-intercept: $(3, 0)$
y-intercept: $(0, 5)$

82. x-intercept: $(-3, 0)$
y-intercept: $(0, 4)$

83. x-intercept: $\left(-\frac{1}{6}, 0\right)$
y-intercept: $\left(0, -\frac{2}{3}\right)$

84. x-intercept: $\left(\frac{2}{3}, 0\right)$
y-intercept: $(0, -2)$

85. Point on line: $(1, 2)$
x-intercept: $(c, 0),\quad c \neq 0$
y-intercept: $(0, c),\quad c \neq 0$

86. Point on line: $(-3, 4)$
x-intercept: $(d, 0),\quad d \neq 0$
y-intercept: $(0, d),\quad d \neq 0$

87. Sales The slopes of lines representing annual sales y in terms of time x in years are given below. Use the slopes to interpret any change in annual sales for a one-year increase in time.

(a) The line has a slope of $m = 135$.

(b) The line has a slope of $m = 0$.

(c) The line has a slope of $m = -40$.

88. Sales The graph shows the sales (in billions of dollars) for Apple Inc. in the years 2009 through 2015. *(Source: Apple Inc.)*

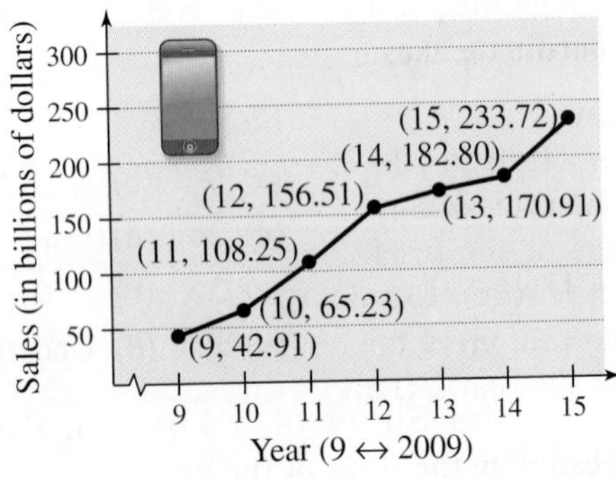

(a) Use the slopes of the line segments to determine the years in which the sales showed the greatest increase and the least increase.

(b) Find the slope of the line segment connecting the points for the years 2009 and 2015.

(c) Interpret the meaning of the slope in part (b) in the context of the problem.

89. Road Grade You are driving on a road that has a 6% uphill grade. This means that the slope of the road is $\frac{6}{100}$. Approximate the amount of vertical change in your position when you drive 200 feet.

90. Road Grade

From the top of a mountain road, a surveyor takes several horizontal measurements x and several vertical measurements y, as shown in the table (x and y are measured in feet).

x	300	600	900	1200
y	−25	−50	−75	−100

x	1500	1800	2100
y	−125	−150	−175

(a) Sketch a scatter plot of the data.

(b) Use a straightedge to sketch the line that you think best fits the data.

(c) Find an equation for the line you sketched in part (b).

(d) Interpret the meaning of the slope of the line in part (c) in the context of the problem.

(e) The surveyor needs to put up a road sign that indicates the steepness of the road. For example, a surveyor would put up a sign that states "8% grade" on a road with a downhill grade that has a slope of $-\frac{8}{100}$. What should the sign state for the road in this problem?

Rate of Change **In Exercises 91 and 92, you are given the dollar value of a product in 2016 and the rate at which the value of the product is expected to change during the next 5 years. Use this information to write a linear equation that gives the dollar value V of the product in terms of the year t. (Let $t = 16$ represent 2016.)**

	2016 Value	Rate
91.	$3000	$150 decrease per year
92.	$200	$6.50 increase per year

93. Cost The cost C of producing n computer laptop bags is given by

$$C = 1.25n + 15{,}750, \quad n > 0.$$

Explain what the C-intercept and the slope represent.

94. Monthly Salary A pharmaceutical salesperson receives a monthly salary of $5000 plus a commission of 7% of sales. Write a linear equation for the salesperson's monthly wage W in terms of monthly sales S.

95. Depreciation A sandwich shop purchases a used pizza oven for $875. After 5 years, the oven will have to be discarded and replaced. Write a linear equation giving the value V of the equipment during the 5 years it will be in use.

96. Depreciation A school district purchases a high-volume printer, copier, and scanner for $24,000. After 10 years, the equipment will have to be replaced. Its value at that time is expected to be $2000. Write a linear equation giving the value V of the equipment during the 10 years it will be in use.

97. Temperature Conversion Write a linear equation that expresses the relationship between the temperature in degrees Celsius C and degrees Fahrenheit F. Use the fact that water freezes at 0°C (32°F) and boils at 100°C (212°F).

98. Neurology The average weight of a male child's brain is 970 grams at age 1 and 1270 grams at age 3. *(Source: American Neurological Association)*

(a) Assuming that the relationship between brain weight y and age t is linear, write a linear model for the data.

(b) What is the slope and what does it tell you about brain weight?

(c) Use your model to estimate the average brain weight at age 2.

(d) Use your school's library, the Internet, or some other reference source to find the actual average brain weight at age 2. How close was your estimate?

(e) Do you think your model could be used to determine the average brain weight of an adult? Explain.

99. Cost, Revenue, and Profit A roofing contractor purchases a shingle delivery truck with a shingle elevator for $42,000. The vehicle requires an average expenditure of $9.50 per hour for fuel and maintenance, and the operator is paid $11.50 per hour.

(a) Write a linear equation giving the total cost C of operating this equipment for t hours. (Include the purchase cost of the equipment.)

(b) Assuming that customers are charged $45 per hour of machine use, write an equation for the revenue R obtained from t hours of use.

(c) Use the formula for profit $P = R - C$ to write an equation for the profit obtained from t hours of use.

(d) Use the result of part (c) to find the break-even point—that is, the number of hours this equipment must be used to yield a profit of 0 dollars.

100. Geometry The length and width of a rectangular garden are 15 meters and 10 meters, respectively. A walkway of width x surrounds the garden.

(a) Draw a diagram that gives a visual representation of the problem.

(b) Write the equation for the perimeter y of the walkway in terms of x.

(c) Use a graphing utility to graph the equation for the perimeter.

(d) Determine the slope of the graph in part (c). For each additional one-meter increase in the width of the walkway, determine the increase in its perimeter.

Exploration

True or False? **In Exercises 101 and 102, determine whether the statement is true or false. Justify your answer.**

101. A line with a slope of $-\frac{5}{7}$ is steeper than a line with a slope of $-\frac{6}{7}$.

102. The line through $(-8, 2)$ and $(-1, 4)$ and the line through $(0, -4)$ and $(-7, 7)$ are parallel.

103. Right Triangle Explain how you can use slope to show that the points $A(-1, 5)$, $B(3, 7)$, and $C(5, 3)$ are the vertices of a right triangle.

104. Vertical Line Explain why the slope of a vertical line is undefined.

105. Error Analysis Describe the error.

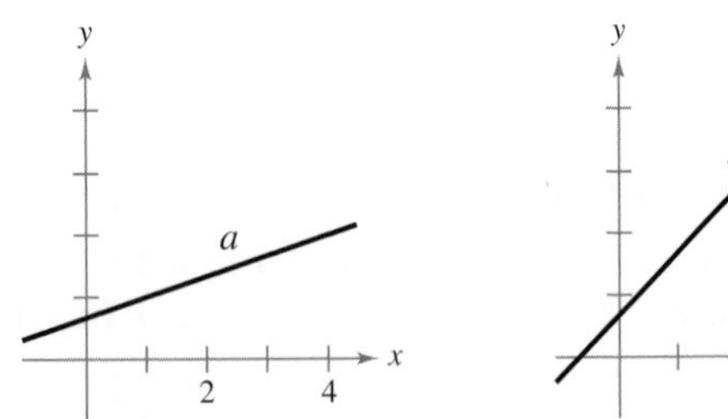

Line b has a greater slope than line a. ✗

106. Perpendicular Segments Find d_1 and d_2 in terms of m_1 and m_2, respectively (see figure). Then use the Pythagorean Theorem to find a relationship between m_1 and m_2.

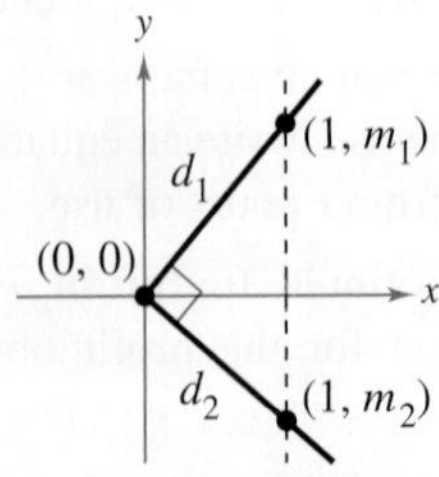

107. Think About It Is it possible for two lines with positive slopes to be perpendicular? Explain.

108. Slope and Steepness The slopes of two lines are -4 and $\frac{5}{2}$. Which is steeper? Explain.

109. Comparing Slopes Use a graphing utility to compare the slopes of the lines $y = mx$, where $m = 0.5$, 1, 2, and 4. Which line rises most quickly? Now, let $m = -0.5$, -1, -2, and -4. Which line falls most quickly? Use a square setting to obtain a true geometric perspective. What can you conclude about the slope and the "rate" at which the line rises or falls?

110. HOW DO YOU SEE IT? Match the description of the situation with its graph. Also determine the slope and y-intercept of each graph and interpret the slope and y-intercept in the context of the situation. [The graphs are labeled (i), (ii), (iii), and (iv).]

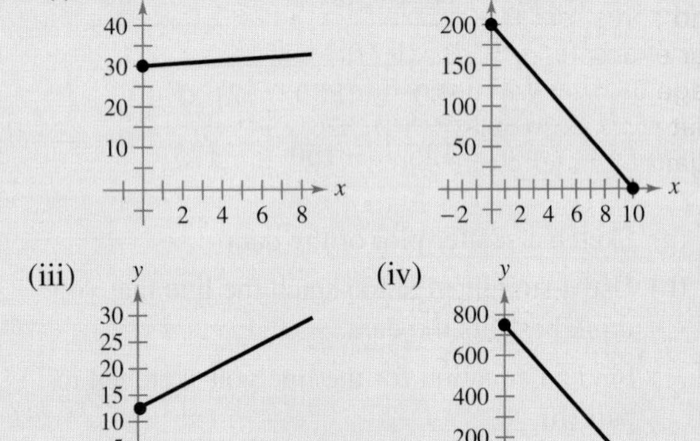

(a) A person is paying \$20 per week to a friend to repay a \$200 loan.

(b) An employee receives \$12.50 per hour plus \$2 for each unit produced per hour.

(c) A sales representative receives \$30 per day for food plus \$0.32 for each mile traveled.

(d) A computer that was purchased for \$750 depreciates \$100 per year.

Finding a Relationship for Equidistance **In Exercises 111–114, find a relationship between x and y such that (x, y) is equidistant (the same distance) from the two points.**

111. $(4, -1), (-2, 3)$

112. $(6, 5), (1, -8)$

113. $\left(3, \frac{5}{2}\right), (-7, 1)$

114. $\left(-\frac{1}{2}, -4\right), \left(\frac{7}{2}, \frac{5}{4}\right)$

Project: Bachelor's Degrees To work an extended application analyzing the numbers of bachelor's degrees earned by women in the United States from 2002 through 2013, visit this text's website at *LarsonPrecalculus.com*. *(Source: National Center for Education Statistics)*

P.5 Functions

Functions are used to model and solve real-life problems. For example, in Exercise 70 on page 65, you will use a function that models the force of water against the face of a dam.

- **Determine whether relations between two variables are functions, and use function notation.**
- **Find the domains of functions.**
- **Use functions to model and solve real-life problems.**
- **Evaluate difference quotients.**

Introduction to Functions and Function Notation

Many everyday phenomena involve two quantities that are related to each other by some rule of correspondence. The mathematical term for such a rule of correspondence is a **relation.** In mathematics, equations and formulas often represent relations. For example, the simple interest I earned on \$1000 for 1 year is related to the annual interest rate r by the formula $I = 1000r$.

The formula $I = 1000r$ represents a special kind of relation that matches each item from one set with *exactly one* item from a different set. Such a relation is a **function.**

> **Definition of Function**
>
> A **function** f from a set A to a set B is a relation that assigns to each element x in the set A exactly one element y in the set B. The set A is the **domain** (or set of inputs) of the function f, and the set B contains the **range** (or set of outputs).

To help understand this definition, look at the function below, which relates the time of day to the temperature.

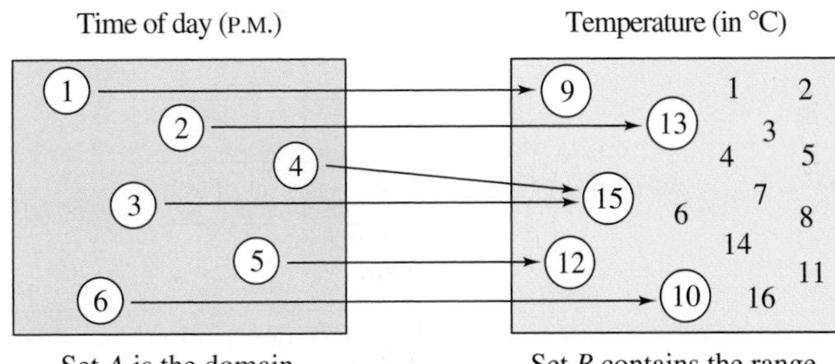

Set A is the domain.
Inputs: 1, 2, 3, 4, 5, 6

Set B contains the range.
Outputs: 9, 10, 12, 13, 15

The ordered pairs below can represent this function. The first coordinate (x-value) is the input and the second coordinate (y-value) is the output.

$$\{(1, 9), (2, 13), (3, 15), (4, 15), (5, 12), (6, 10)\}$$

> **Characteristics of a Function from Set *A* to Set *B***
>
> **1.** Each element in A must be matched with an element in B.
>
> **2.** Some elements in B may not be matched with any element in A.
>
> **3.** Two or more elements in A may be matched with the same element in B.
>
> **4.** An element in A (the domain) cannot be matched with two different elements in B.

iStockphoto.com/Jennifer_Sharp

Here are four common ways to represent functions.

Four Ways to Represent a Function

1. *Verbally* by a sentence that describes how the input variable is related to the output variable
2. *Numerically* by a table or a list of ordered pairs that matches input values with output values
3. *Graphically* by points in a coordinate plane in which the horizontal positions represent the input values and the vertical positions represent the output values
4. *Algebraically* by an equation in two variables

To determine whether a relation is a function, you must decide whether each input value is matched with exactly one output value. When any input value is matched with two or more output values, the relation is not a function.

EXAMPLE 1 Testing for Functions

Determine whether the relation represents y as a function of x.

a. The input value x is the number of representatives from a state, and the output value y is the number of senators.

b.

Input, x	Output, y
2	11
2	10
3	8
4	5
5	1

c.

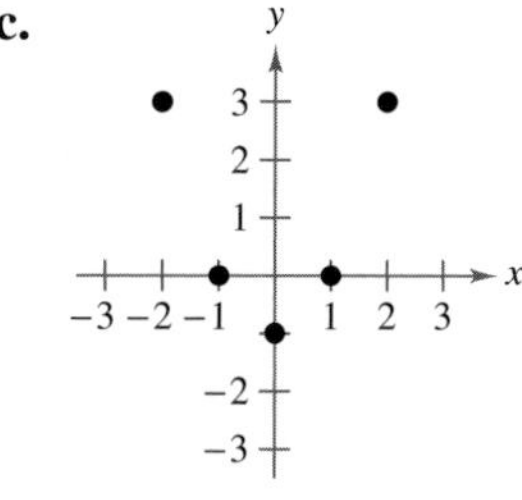

Solution

a. This verbal description *does* describe y as a function of x. Regardless of the value of x, the value of y is always 2. This is an example of a *constant function*.

b. This table *does not* describe y as a function of x. The input value 2 is matched with two different y-values.

c. The graph *does* describe y as a function of x. Each input value is matched with exactly one output value.

✓ Checkpoint *Audio-video solution in English & Spanish at LarsonPrecalculus.com*

Determine whether the relation represents y as a function of x.

a.

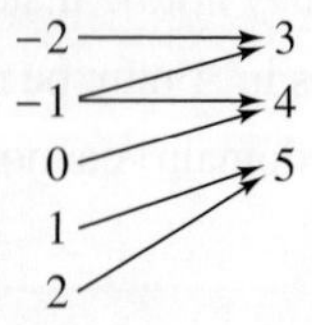

b.

Input, x	0	1	2	3	4
Output, y	-4	-2	0	2	4

HISTORICAL NOTE

Many consider Leonhard Euler (1707–1783), a Swiss mathematician, to be the most prolific and productive mathematician in history. One of his greatest influences on mathematics was his use of symbols, or notation. Euler introduced the function notation $y = f(x)$.

Representing functions by sets of ordered pairs is common in *discrete mathematics.* In algebra, however, it is more common to represent functions by equations or formulas involving two variables. For example, the equation

$y = x^2$ y is a function of x.

represents the variable y as a function of the variable x. In this equation, x is the **independent variable** and y is the **dependent variable.** The domain of the function is the set of all values taken on by the independent variable x, and the range of the function is the set of all values taken on by the dependent variable y.

EXAMPLE 2 Testing for Functions Represented Algebraically

See LarsonPrecalculus.com for an interactive version of this type of example.

Determine whether each equation represents y as a function of x.

a. $x^2 + y = 1$

b. $-x + y^2 = 1$

Solution To determine whether y is a function of x, solve for y in terms of x.

a. Solving for y yields

$x^2 + y = 1$ Write original equation.

$y = 1 - x^2.$ Solve for y.

To each value of x there corresponds exactly one value of y. So, y is a function of x.

b. Solving for y yields

$-x + y^2 = 1$ Write original equation.

$y^2 = 1 + x$ Add x to each side.

$y = \pm\sqrt{1 + x}.$ Solve for y.

The $\pm$ indicates that to a given value of x there correspond two values of y. So, y is not a function of x.

Determine whether each equation represents y as a function of x.

a. $x^2 + y^2 = 8$ **b.** $y - 4x^2 = 36$

When using an equation to represent a function, it is convenient to name the function for easy reference. For example, the equation $y = 1 - x^2$ describes y as a function of x. By renaming this function "f," you can write the input, output, and equation using **function notation.**

Input	Output	Equation
x	$f(x)$	$f(x) = 1 - x^2$

The symbol $f(x)$ is read as *the value of f at x* or simply *f of x*. The symbol $f(x)$ corresponds to the y-value for a given x. So, $y = f(x)$. Keep in mind that f is the *name* of the function, whereas $f(x)$ is the *value* of the function at x. For example, the function $f(x) = 3 - 2x$ has *function values* denoted by $f(-1)$, $f(0)$, $f(2)$, and so on. To find these values, substitute the specified input values into the given equation.

For $x = -1$, $f(-1) = 3 - 2(-1) = 3 + 2 = 5.$

For $x = 0$, $f(0) = 3 - 2(0) = 3 - 0 = 3.$

For $x = 2$, $f(2) = 3 - 2(2) = 3 - 4 = -1.$

AS400 DB/Corbis

Although it is often convenient to use f as a function name and x as the independent variable, other letters may be used as well. For example,

$$f(x) = x^2 - 4x + 7, \quad f(t) = t^2 - 4t + 7, \quad \text{and} \quad g(s) = s^2 - 4s + 7$$

all define the same function. In fact, the role of the independent variable is that of a "placeholder." Consequently, the function can be described by

$$f(\blacksquare) = (\blacksquare)^2 - 4(\blacksquare) + 7.$$

EXAMPLE 3 Evaluating a Function

Let $g(x) = -x^2 + 4x + 1$. Find each function value.

a. $g(2)$ **b.** $g(t)$ **c.** $g(x + 2)$

Solution

a. Replace x with 2 in $g(x) = -x^2 + 4x + 1$.

$$\begin{aligned} g(2) &= -(2)^2 + 4(2) + 1 \\ &= -4 + 8 + 1 \\ &= 5 \end{aligned}$$

b. Replace x with t.

$$\begin{aligned} g(t) &= -(t)^2 + 4(t) + 1 \\ &= -t^2 + 4t + 1 \end{aligned}$$

c. Replace x with $x + 2$.

$$\begin{aligned} g(x + 2) &= -(x + 2)^2 + 4(x + 2) + 1 \\ &= -(x^2 + 4x + 4) + 4x + 8 + 1 \\ &= -x^2 - 4x - 4 + 4x + 8 + 1 \\ &= -x^2 + 5 \end{aligned}$$

REMARK In Example 3(c), note that $g(x + 2)$ is not equal to $g(x) + g(2)$. In general, $g(u + v) \neq g(u) + g(v)$.

✓ **Checkpoint** Audio-video solution in English & Spanish at LarsonPrecalculus.com

Let $f(x) = 10 - 3x^2$. Find each function value.

a. $f(2)$ **b.** $f(-4)$ **c.** $f(x - 1)$

A function defined by two or more equations over a specified domain is called a **piecewise-defined function.**

EXAMPLE 4 A Piecewise-Defined Function

Evaluate the function when $x = -1$, 0, and 1.

$$f(x) = \begin{cases} x^2 + 1, & x < 0 \\ x - 1, & x \geq 0 \end{cases}$$

Solution Because $x = -1$ is less than 0, use $f(x) = x^2 + 1$ to obtain $f(-1) = (-1)^2 + 1 = 2$. For $x = 0$, use $f(x) = x - 1$ to obtain $f(0) = (0) - 1 = -1$. For $x = 1$, use $f(x) = x - 1$ to obtain $f(1) = (1) - 1 = 0$.

✓ **Checkpoint** Audio-video solution in English & Spanish at LarsonPrecalculus.com

Evaluate the function given in Example 4 when $x = -2$, 2, and 3.

EXAMPLE 5 Finding Values for Which $f(x) = 0$

Find all real values of x for which $f(x) = 0$.

a. $f(x) = -2x + 10$ **b.** $f(x) = x^2 - 5x + 6$

Solution For each function, set $f(x) = 0$ and solve for x.

a. $-2x + 10 = 0$ Set $f(x)$ equal to 0.

$-2x = -10$ Subtract 10 from each side.

$x = 5$ Divide each side by -2.

So, $f(x) = 0$ when $x = 5$.

b. $x^2 - 5x + 6 = 0$ Set $f(x)$ equal to 0.

$(x - 2)(x - 3) = 0$ Factor.

$x - 2 = 0 \Longrightarrow x = 2$ Set 1st factor equal to 0 and solve.

$x - 3 = 0 \Longrightarrow x = 3$ Set 2nd factor equal to 0 and solve.

So, $f(x) = 0$ when $x = 2$ or $x = 3$.

Checkpoint *Audio-video solution in English & Spanish at LarsonPrecalculus.com*

Find all real values of x for which $f(x) = 0$, where $f(x) = x^2 - 16$.

EXAMPLE 6 Finding Values for Which $f(x) = g(x)$

Find the values of x for which $f(x) = g(x)$.

a. $f(x) = x^2 + 1$ and $g(x) = 3x - x^2$

b. $f(x) = x^2 - 1$ and $g(x) = -x^2 + x + 2$

Solution

a. $x^2 + 1 = 3x - x^2$ Set $f(x)$ equal to $g(x)$.

$2x^2 - 3x + 1 = 0$ Write in general form.

$(2x - 1)(x - 1) = 0$ Factor.

$2x - 1 = 0 \Longrightarrow x = \frac{1}{2}$ Set 1st factor equal to 0 and solve.

$x - 1 = 0 \Longrightarrow x = 1$ Set 2nd factor equal to 0 and solve.

So, $f(x) = g(x)$ when $x = \dfrac{1}{2}$ or $x = 1$.

b. $x^2 - 1 = -x^2 + x + 2$ Set $f(x)$ equal to $g(x)$.

$2x^2 - x - 3 = 0$ Write in general form.

$(2x - 3)(x + 1) = 0$ Factor.

$2x - 3 = 0 \Longrightarrow x = \frac{3}{2}$ Set 1st factor equal to 0 and solve.

$x + 1 = 0 \Longrightarrow x = -1$ Set 2nd factor equal to 0 and solve.

So, $f(x) = g(x)$ when $x = \dfrac{3}{2}$ or $x = -1$.

Checkpoint *Audio-video solution in English & Spanish at LarsonPrecalculus.com*

Find the values of x for which $f(x) = g(x)$, where $f(x) = x^2 + 6x - 24$ and $g(x) = 4x - x^2$.

The Domain of a Function

> **TECHNOLOGY** Use a graphing utility to graph the functions $y = \sqrt{4 - x^2}$ and $y = \sqrt{x^2 - 4}$. What is the domain of each function? Do the domains of these two functions overlap? If so, for what values do the domains overlap?

The domain of a function can be described explicitly or it can be *implied* by the expression used to define the function. The **implied domain** is the set of all real numbers for which the expression is defined. For example, the function

$$f(x) = \frac{1}{x^2 - 4}$$ Domain excludes x-values that result in division by zero.

has an implied domain consisting of all real x other than $x = \pm 2$. These two values are excluded from the domain because division by zero is undefined. Another common type of implied domain is that used to avoid even roots of negative numbers. For example, the function

$$f(x) = \sqrt{x}$$ Domain excludes x-values that result in even roots of negative numbers.

is defined only for $x \geq 0$. So, its implied domain is the interval $[0, \infty)$. In general, the domain of a function *excludes* values that cause division by zero *or* that result in the even root of a negative number.

EXAMPLE 7 Finding the Domains of Functions

Find the domain of each function.

a. f: $\{(-3, 0), (-1, 4), (0, 2), (2, 2), (4, -1)\}$ **b.** $g(x) = \dfrac{1}{x + 5}$

c. Volume of a sphere: $V = \frac{4}{3}\pi r^3$ **d.** $h(x) = \sqrt{4 - 3x}$

Solution

a. The domain of f consists of all first coordinates in the set of ordered pairs.

$$\text{Domain} = \{-3, -1, 0, 2, 4\}$$

b. Excluding x-values that yield zero in the denominator, the domain of g is the set of all real numbers x except $x = -5$.

c. This function represents the volume of a sphere, so the values of the radius r must be positive. The domain is the set of all real numbers r such that $r > 0$.

d. This function is defined only for x-values for which

$$4 - 3x \geq 0.$$

By solving the inequality, you can conclude that $x \leq \frac{4}{3}$. So, the domain is the interval $\left(-\infty, \frac{4}{3}\right]$.

Find the domain of each function.

a. f: $\{(-2, 2), (-1, 1), (0, 3), (1, 1), (2, 2)\}$ **b.** $g(x) = \dfrac{1}{3 - x}$

c. Circumference of a circle: $C = 2\pi r$ **d.** $h(x) = \sqrt{x - 16}$

In Example 7(c), note that the domain of a function may be implied by the physical context. For example, from the equation

$$V = \frac{4}{3}\pi r^3$$

you have no reason to restrict r to positive values, but the physical context implies that a sphere cannot have a negative or zero radius.

Applications

EXAMPLE 8 The Dimensions of a Container

You work in the marketing department of a soft-drink company and are experimenting with a new can for iced tea that is slightly narrower and taller than a standard can. For your experimental can, the ratio of the height to the radius is 4.

a. Write the volume of the can as a function of the radius r.

b. Write the volume of the can as a function of the height h.

Solution

a. $V(r) = \pi r^2 h = \pi r^2(4r) = 4\pi r^3$ — Write V as a function of r.

b. $V(h) = \pi r^2 h = \pi\left(\frac{h}{4}\right)^2 h = \frac{\pi h^3}{16}$ — Write V as a function of h.

✓ Checkpoint Audio-video solution in English & Spanish at LarsonPrecalculus.com

For the experimental can described in Example 8, write the *surface area* as a function of (a) the radius r and (b) the height h.

EXAMPLE 9 The Path of a Baseball

A batter hits a baseball at a point 3 feet above ground at a velocity of 100 feet per second and an angle of 45°. The path of the baseball is given by the function

$$f(x) = -0.0032x^2 + x + 3$$

where $f(x)$ is the height of the baseball (in feet) and x is the horizontal distance from home plate (in feet). Will the baseball clear a 10-foot fence located 300 feet from home plate?

Algebraic Solution

Find the height of the baseball when $x = 300$.

$f(x) = -0.0032x^2 + x + 3$ — Write original function.

$f(300) = -0.0032(300)^2 + 300 + 3$ — Substitute 300 for x.

$= 15$ — Simplify.

When $x = 300$, the height of the baseball is 15 feet. So, the baseball will clear a 10-foot fence.

Graphical Solution

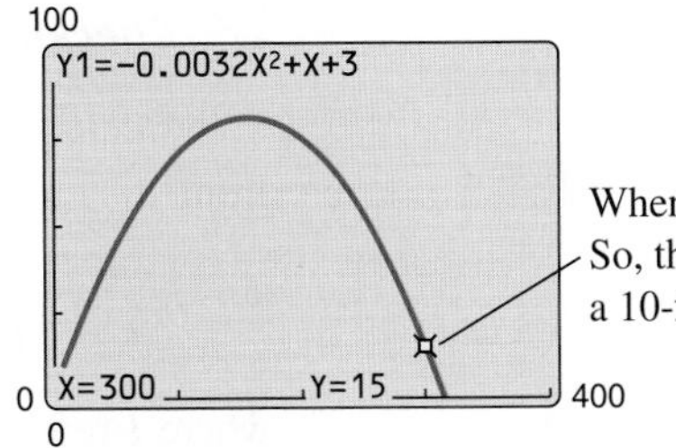

✓ Checkpoint Audio-video solution in English & Spanish at LarsonPrecalculus.com

A second baseman throws a baseball toward the first baseman 60 feet away. The path of the baseball is given by the function

$$f(x) = -0.004x^2 + 0.3x + 6$$

where $f(x)$ is the height of the baseball (in feet) and x is the horizontal distance from the second baseman (in feet). The first baseman can reach 8 feet high. Can the first baseman catch the baseball without jumping?

Flexible-fuel vehicles are designed to operate on gasoline, E85, or a mixture of the two fuels. The concentration of ethanol in E85 fuel ranges from 51% to 83%, depending on where and when the E85 is produced.

EXAMPLE 10 **Alternative-Fuel Stations**

The number S of fuel stations that sold E85 (a gasoline-ethanol blend) in the United States increased in a linear pattern from 2008 through 2011, and then increased in a different linear pattern from 2012 through 2015, as shown in the bar graph. These two patterns can be approximated by the function

$$S(t) = \begin{cases} 260.8t - 439, & 8 \le t \le 11 \\ 151.2t + 714, & 12 \le t \le 15 \end{cases}$$

where t represents the year, with $t = 8$ corresponding to 2008. Use this function to approximate the number of stations that sold E85 each year from 2008 to 2015. *(Source: Alternative Fuels Data Center)*

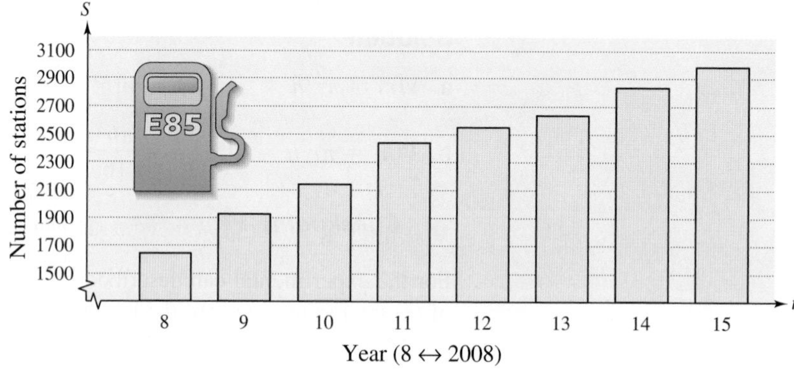

Solution From 2008 through 2011, use $S(t) = 260.8t - 439$.

$$\underbrace{1647}_{2008} \quad \underbrace{1908}_{2009} \quad \underbrace{2169}_{2010} \quad \underbrace{2430}_{2011}$$

From 2012 to 2015, use $S(t) = 151.2t + 714$.

$$\underbrace{2528}_{2012} \quad \underbrace{2680}_{2013} \quad \underbrace{2831}_{2014} \quad \underbrace{2982}_{2015}$$

✓ **Checkpoint** *Audio-video solution in English & Spanish at LarsonPrecalculus.com*

The number S of fuel stations that sold compressed natural gas in the United States from 2009 to 2015 can be approximated by the function

$$S(t) = \begin{cases} 69t + 151, & 9 \le t \le 11 \\ 160t - 803, & 12 \le t \le 15 \end{cases}$$

where t represents the year, with $t = 9$ corresponding to 2009. Use this function to approximate the number of stations that sold compressed natural gas each year from 2009 through 2015. *(Source: Alternative Fuels Data Center)*

Difference Quotients

One of the basic definitions in calculus uses the ratio

$$\frac{f(x + h) - f(x)}{h}, \quad h \neq 0.$$

This ratio is a **difference quotient,** as illustrated in Example 11.

EXAMPLE 11 Evaluating a Difference Quotient

For $f(x) = x^2 - 4x + 7$, find $\dfrac{f(x+h) - f(x)}{h}$.

Solution

$$\frac{f(x+h) - f(x)}{h} = \frac{[(x+h)^2 - 4(x+h) + 7] - (x^2 - 4x + 7)}{h}$$

$$= \frac{x^2 + 2xh + h^2 - 4x - 4h + 7 - x^2 + 4x - 7}{h}$$

$$= \frac{2xh + h^2 - 4h}{h} = \frac{h(2x + h - 4)}{h} = 2x + h - 4, \quad h \neq 0$$

REMARK You may find it easier to calculate the difference quotient in Example 11 by first finding $f(x+h)$, and then substituting the resulting expression into the difference quotient

$$\frac{f(x+h) - f(x)}{h}.$$

Checkpoint Audio-video solution in English & Spanish at LarsonPrecalculus.com

For $f(x) = x^2 + 2x - 3$, find $\dfrac{f(x+h) - f(x)}{h}$.

Summary of Function Terminology

Function: A **function** is a relationship between two variables such that to each value of the independent variable there corresponds exactly one value of the dependent variable.

Function notation: $y = f(x)$

f is the *name* of the function.

y is the **dependent variable.**

x is the **independent variable.**

$f(x)$ is the *value of the function at x.*

Domain: The **domain** of a function is the set of all values (inputs) of the independent variable for which the function is defined. If x is in the domain of f, then f is *defined* at x. If x is not in the domain of f, then f is *undefined* at x.

Range: The **range** of a function is the set of all values (outputs) taken on by the dependent variable (that is, the set of all function values).

Implied domain: If f is defined by an algebraic expression and the domain is not specified, then the **implied domain** consists of all real numbers for which the expression is defined.

Summarize (Section P.5)

1. State the definition of a function and describe function notation *(pages 53–57)*. For examples of determining functions and using function notation, see Examples 1–6.
2. State the definition of the implied domain of a function *(page 58)*. For an example of finding the domains of functions, see Example 7.
3. Describe examples of how functions can model real-life problems *(pages 59 and 60, Examples 8–10)*.
4. State the definition of a difference quotient *(page 60)*. For an example of evaluating a difference quotient, see Example 11.

P.5 Exercises

See CalcChat.com for tutorial help and worked-out solutions to odd-numbered exercises.

Vocabulary: Fill in the blanks.

1. A relation that assigns to each element x from a set of inputs, or ________, exactly one element y in a set of outputs, or ________, is a ________.
2. For an equation that represents y as a function of x, the set of all values taken on by the ________ variable x is the domain, and the set of all values taken on by the ________ variable y is the range.
3. If the domain of the function f is not given, then the set of values of the independent variable for which the expression is defined is the ________ ________.
4. One of the basic definitions in calculus uses the ratio $\frac{f(x+h)-f(x)}{h}$, $h \neq 0$. This ratio is a ________ ________.

Skills and Applications

Testing for Functions In Exercises 5–8, determine whether the relation represents y as a function of x.

5. *Domain, x* *Range, y*

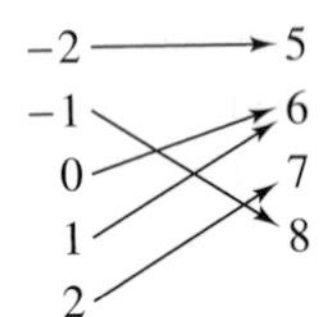

6. *Domain, x* *Range, y*

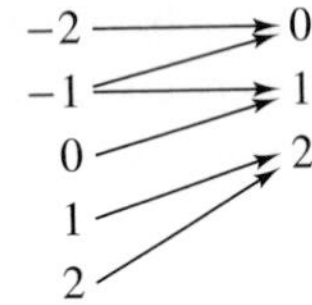

7.

Input, x	10	7	4	7	10
Output, y	3	6	9	12	15

8.

Input, x	−2	0	2	4	6
Output, y	1	1	1	1	1

Testing for Functions In Exercises 9 and 10, which sets of ordered pairs represent functions from A to B? Explain.

9. $A = \{0, 1, 2, 3\}$ and $B = \{-2, -1, 0, 1, 2\}$
   (a) $\{(0, 1), (1, -2), (2, 0), (3, 2)\}$
   (b) $\{(0, -1), (2, 2), (1, -2), (3, 0), (1, 1)\}$
   (c) $\{(0, 0), (1, 0), (2, 0), (3, 0)\}$
   (d) $\{(0, 2), (3, 0), (1, 1)\}$
10. $A = \{a, b, c\}$ and $B = \{0, 1, 2, 3\}$
   (a) $\{(a, 1), (c, 2), (c, 3), (b, 3)\}$
   (b) $\{(a, 1), (b, 2), (c, 3)\}$
   (c) $\{(1, a), (0, a), (2, c), (3, b)\}$
   (d) $\{(c, 0), (b, 0), (a, 3)\}$

Testing for Functions Represented Algebraically In Exercises 11–18, determine whether the equation represents y as a function of x.

11. $x^2 + y^2 = 4$
12. $x^2 - y = 9$
13. $y = \sqrt{16 - x^2}$
14. $y = \sqrt{x + 5}$
15. $y = 4 - |x|$
16. $|y| = 4 - x$
17. $y = -75$
18. $x - 1 = 0$

Evaluating a Function In Exercises 19–30, find each function value, if possible.

19. $f(x) = 3x - 5$
   (a) $f(1)$ (b) $f(-3)$ (c) $f(x + 2)$
20. $V(r) = \frac{4}{3}\pi r^3$
   (a) $V(3)$ (b) $V\left(\frac{3}{2}\right)$ (c) $V(2r)$
21. $g(t) = 4t^2 - 3t + 5$
   (a) $g(2)$ (b) $g(t - 2)$ (c) $g(t) - g(2)$
22. $h(t) = -t^2 + t + 1$
   (a) $h(2)$ (b) $h(-1)$ (c) $h(x + 1)$
23. $f(y) = 3 - \sqrt{y}$
   (a) $f(4)$ (b) $f(0.25)$ (c) $f(4x^2)$
24. $f(x) = \sqrt{x + 8} + 2$
   (a) $f(-8)$ (b) $f(1)$ (c) $f(x - 8)$
25. $q(x) = 1/(x^2 - 9)$
   (a) $q(0)$ (b) $q(3)$ (c) $q(y + 3)$
26. $q(t) = (2t^2 + 3)/t^2$
   (a) $q(2)$ (b) $q(0)$ (c) $q(-x)$
27. $f(x) = |x|/x$
   (a) $f(2)$ (b) $f(-2)$ (c) $f(x - 1)$
28. $f(x) = |x| + 4$
   (a) $f(2)$ (b) $f(-2)$ (c) $f(x^2)$
29. $f(x) = \begin{cases} 2x + 1, & x < 0 \\ 2x + 2, & x \geq 0 \end{cases}$
   (a) $f(-1)$ (b) $f(0)$ (c) $f(2)$
30. $f(x) = \begin{cases} -3x - 3, & x < -1 \\ x^2 + 2x - 1, & x \geq -1 \end{cases}$
   (a) $f(-2)$ (b) $f(-1)$ (c) $f(1)$

Evaluating a Function **In Exercises 31–34, complete the table.**

31. $f(x) = -x^2 + 5$

x	-2	-1	0	1	2
$f(x)$					

32. $h(t) = \frac{1}{2}|t + 3|$

t	-5	-4	-3	-2	-1
$h(t)$					

33. $f(x) = \begin{cases} -\frac{1}{2}x + 4, & x \leq 0 \\ (x - 2)^2, & x > 0 \end{cases}$

x	-2	-1	0	1	2
$f(x)$					

34. $f(x) = \begin{cases} 9 - x^2, & x < 3 \\ x - 3, & x \geq 3 \end{cases}$

x	1	2	3	4	5
$f(x)$					

Finding Values for Which $f(x) = 0$ **In Exercises 35–42, find all real values of x for which $f(x) = 0$.**

35. $f(x) = 15 - 3x$

36. $f(x) = 4x + 6$

37. $f(x) = \dfrac{3x - 4}{5}$

38. $f(x) = \dfrac{12 - x^2}{8}$

39. $f(x) = x^2 - 81$

40. $f(x) = x^2 - 6x - 16$

41. $f(x) = x^3 - x$

42. $f(x) = x^3 - x^2 - 3x + 3$

Finding Values for Which $f(x) = g(x)$ **In Exercises 43–46, find the value(s) of x for which $f(x) = g(x)$.**

43. $f(x) = x^2, \quad g(x) = x + 2$

44. $f(x) = x^2 + 2x + 1, \quad g(x) = 5x + 19$

45. $f(x) = x^4 - 2x^2, \quad g(x) = 2x^2$

46. $f(x) = \sqrt{x} - 4, \quad g(x) = 2 - x$

Finding the Domain of a Function **In Exercises 47–56, find the domain of the function.**

47. $f(x) = 5x^2 + 2x - 1$

48. $g(x) = 1 - 2x^2$

49. $g(y) = \sqrt{y + 6}$

50. $f(t) = \sqrt[3]{t + 4}$

51. $g(x) = \dfrac{1}{x} - \dfrac{3}{x + 2}$

52. $h(x) = \dfrac{6}{x^2 - 4x}$

53. $f(s) = \dfrac{\sqrt{s - 1}}{s - 4}$

54. $f(x) = \dfrac{\sqrt{x + 6}}{6 + x}$

55. $f(x) = \dfrac{x - 4}{\sqrt{x}}$

56. $f(x) = \dfrac{x + 2}{\sqrt{x - 10}}$

57. Maximum Volume An open box of maximum volume is made from a square piece of material 24 centimeters on a side by cutting equal squares from the corners and turning up the sides (see figure).

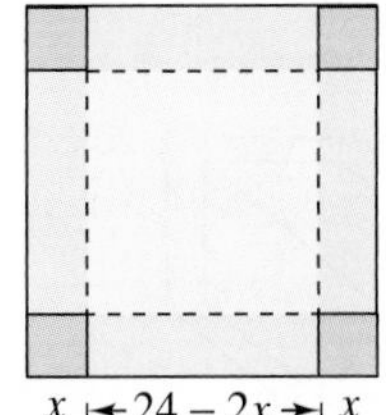

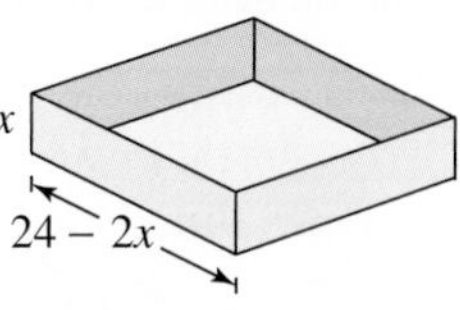

(a) The table shows the volumes V (in cubic centimeters) of the box for various heights x (in centimeters). Use the table to estimate the maximum volume.

Height, x	1	2	3	4	5	6
Volume, V	484	800	972	1024	980	864

(b) Plot the points (x, V) from the table in part (a). Does the relation defined by the ordered pairs represent V as a function of x?

(c) Given that V is a function of x, write the function and determine its domain.

58. Maximum Profit The cost per unit in the production of an MP3 player is \$60. The manufacturer charges \$90 per unit for orders of 100 or less. To encourage large orders, the manufacturer reduces the charge by \$0.15 per MP3 player for each unit ordered in excess of 100 (for example, the charge is reduced to \$87 per MP3 player for an order size of 120).

(a) The table shows the profits P (in dollars) for various numbers of units ordered, x. Use the table to estimate the maximum profit.

Units, x	130	140	150	160	170
Profit, P	3315	3360	3375	3360	3315

(b) Plot the points (x, P) from the table in part (a). Does the relation defined by the ordered pairs represent P as a function of x?

(c) Given that P is a function of x, write the function and determine its domain. (*Note:* $P = R - C$, where R is revenue and C is cost.)

59. Geometry Write the area A of a square as a function of its perimeter P.

60. Geometry Write the area A of a circle as a function of its circumference C.

61. Path of a Ball You throw a baseball to a child 25 feet away. The height y (in feet) of the baseball is given by

$$y = -\frac{1}{10}x^2 + 3x + 6$$

where x is the horizontal distance (in feet) from where you threw the ball. Can the child catch the baseball while holding a baseball glove at a height of 5 feet?

62. Postal Regulations A rectangular package has a combined length and girth (perimeter of a cross section) of 108 inches (see figure).

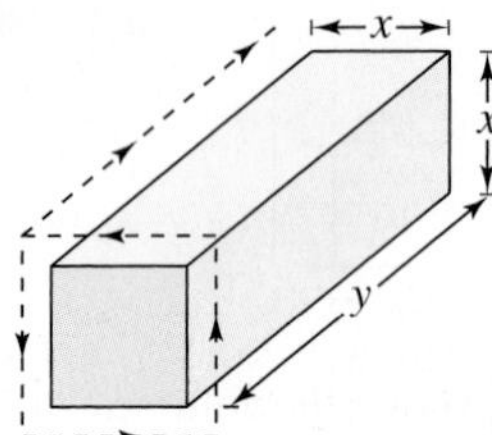

(a) Write the volume V of the package as a function of x. What is the domain of the function?

(b) Use a graphing utility to graph the function. Be sure to use an appropriate window setting.

(c) What dimensions will maximize the volume of the package? Explain.

63. Geometry A right triangle is formed in the first quadrant by the x- and y-axes and a line through the point (2, 1) (see figure). Write the area A of the triangle as a function of x, and determine the domain of the function.

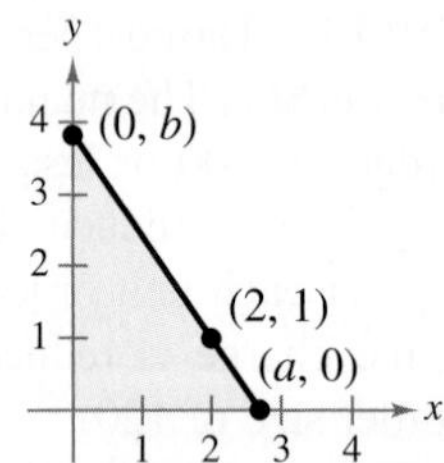

64. Geometry A rectangle is bounded by the x-axis and the semicircle $y = \sqrt{36 - x^2}$ (see figure). Write the area A of the rectangle as a function of x, and graphically determine the domain of the function.

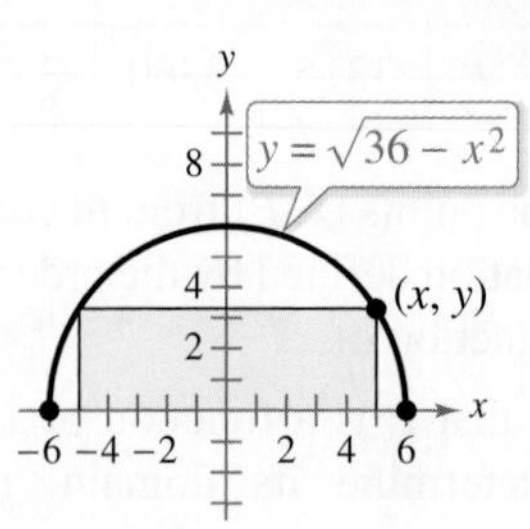

65. Pharmacology The percent p of prescriptions filled with generic drugs at CVS Pharmacies from 2008 through 2014 (see figure) can be approximated by the model

$$p(t) = \begin{cases} 2.77t + 45.2, & 8 \le t \le 11 \\ 1.95t + 55.9, & 12 \le t \le 14 \end{cases}$$

where t represents the year, with $t = 8$ corresponding to 2008. Use this model to find the percent of prescriptions filled with generic drugs in each year from 2008 through 2014. (*Source: CVS Health*)

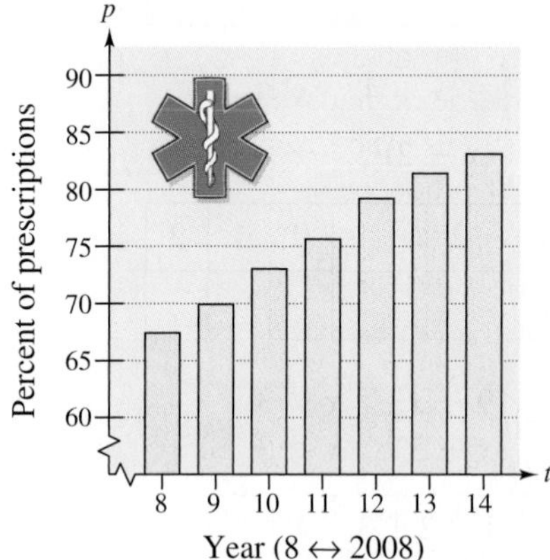

66. Median Sale Price The median sale price p (in thousands of dollars) of an existing one-family home in the United States from 2002 through 2014 (see figure) can be approximated by the model

$$p(t) = \begin{cases} -0.757t^2 + 20.80t + 127.2, & 2 \le t \le 6 \\ 3.879t^2 - 82.50t + 605.8, & 7 \le t \le 11 \\ -4.171t^2 + 124.34t - 714.2, & 12 \le t \le 14 \end{cases}$$

where t represents the year, with $t = 2$ corresponding to 2002. Use this model to find the median sale price of an existing one-family home in each year from 2002 through 2014. (*Source: National Association of Realtors*)

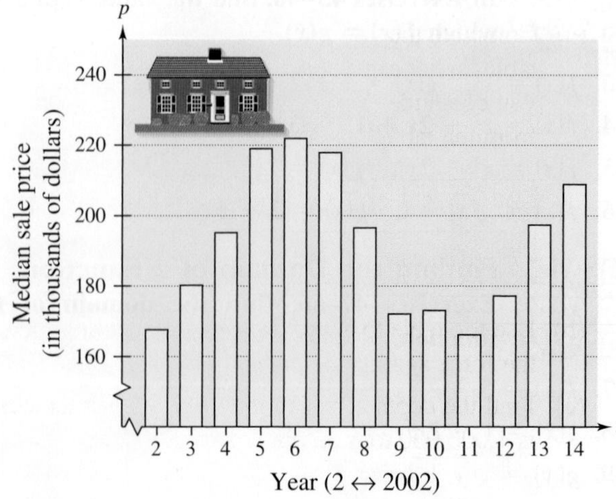

67. Cost, Revenue, and Profit A company produces a product for which the variable cost is \$12.30 per unit and the fixed costs are \$98,000. The product sells for \$17.98. Let x be the number of units produced and sold.

(a) The total cost for a business is the sum of the variable cost and the fixed costs. Write the total cost C as a function of the number of units produced.

(b) Write the revenue R as a function of the number of units sold.

(c) Write the profit P as a function of the number of units sold. (*Note:* $P = R - C$)

68. Average Cost The inventor of a new game believes that the variable cost for producing the game is \$0.95 per unit and the fixed costs are \$6000. The inventor sells each game for \$1.69. Let x be the number of games produced.

(a) The total cost for a business is the sum of the variable cost and the fixed costs. Write the total cost C as a function of the number of games produced.

(b) Write the average cost per unit $\overline{C} = \dfrac{C}{x}$ as a function of x.

69. Height of a Balloon A balloon carrying a transmitter ascends vertically from a point 3000 feet from the receiving station.

(a) Draw a diagram that gives a visual representation of the problem. Let h represent the height of the balloon and let d represent the distance between the balloon and the receiving station.

(b) Write the height of the balloon as a function of d. What is the domain of the function?

70. Physics

The function $F(y) = 149.76\sqrt{10}y^{5/2}$ estimates the force F (in tons) of water against the face of a dam, where y is the depth of the water (in feet).

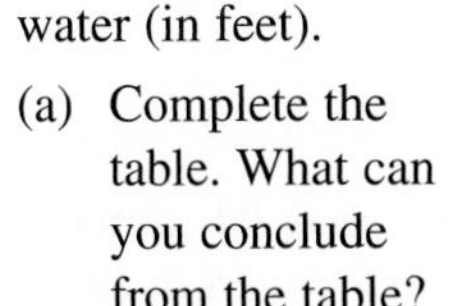

(a) Complete the table. What can you conclude from the table?

y	5	10	20	30	40
$F(y)$					

(b) Use the table to approximate the depth at which the force against the dam is 1,000,000 tons.

(c) Find the depth at which the force against the dam is 1,000,000 tons algebraically.

71. Transportation For groups of 80 or more people, a charter bus company determines the rate per person according to the formula

$$\text{Rate} = 8 - 0.05(n - 80), \quad n \geq 80$$

where the rate is given in dollars and n is the number of people.

(a) Write the revenue R for the bus company as a function of n.

(b) Use the function in part (a) to complete the table. What can you conclude?

n	90	100	110	120	130	140	150
$R(n)$							

72. E-Filing The table shows the numbers of tax returns (in millions) made through e-file from 2007 through 2014. Let $f(t)$ represent the number of tax returns made through e-file in the year t. (*Source: eFile*)

DATA

Year	Number of Tax Returns Made Through E-File
2007	80.0
2008	89.9
2009	95.0
2010	98.7
2011	112.2
2012	112.1
2013	114.4
2014	125.8

Spreadsheet at LarsonPrecalculus.com

(a) Find $\dfrac{f(2014) - f(2007)}{2014 - 2007}$ and interpret the result in the context of the problem.

(b) Make a scatter plot of the data.

(c) Find a linear model for the data algebraically. Let N represent the number of tax returns made through e-file and let $t = 7$ correspond to 2007.

(d) Use the model found in part (c) to complete the table.

t	7	8	9	10	11	12	13	14
N								

(e) Compare your results from part (d) with the actual data.

(f) Use a graphing utility to find a linear model for the data. Let $x = 7$ correspond to 2007. How does the model you found in part (c) compare with the model given by the graphing utility?

Evaluating a Difference Quotient **In Exercises 73–80, find the difference quotient and simplify your answer.**

73. $f(x) = x^2 - 2x + 4, \quad \dfrac{f(2 + h) - f(2)}{h}, \quad h \neq 0$

74. $f(x) = 5x - x^2, \quad \dfrac{f(5 + h) - f(5)}{h}, \quad h \neq 0$

75. $f(x) = x^3 + 3x, \quad \dfrac{f(x + h) - f(x)}{h}, \quad h \neq 0$

76. $f(x) = 4x^3 - 2x, \quad \dfrac{f(x + h) - f(x)}{h}, \quad h \neq 0$

77. $g(x) = \dfrac{1}{x^2}, \quad \dfrac{g(x) - g(3)}{x - 3}, \quad x \neq 3$

78. $f(t) = \dfrac{1}{t - 2}, \quad \dfrac{f(t) - f(1)}{t - 1}, \quad t \neq 1$

79. $f(x) = \sqrt{5x}, \quad \dfrac{f(x) - f(5)}{x - 5}, \quad x \neq 5$

80. $f(x) = x^{2/3} + 1, \quad \dfrac{f(x) - f(8)}{x - 8}, \quad x \neq 8$

Modeling Data **In Exercises 81–84, determine which of the following functions**

$$f(x) = cx,\ g(x) = cx^2,\ h(x) = c\sqrt{|x|},\ \text{and}\ r(x) = \frac{c}{x}$$

can be used to model the data and determine the value of the constant c that will make the function fit the data in the table.

81.

x	-4	-1	0	1	4
y	-32	-2	0	-2	-32

82.

x	-4	-1	0	1	4
y	-1	$-\frac{1}{4}$	0	$\frac{1}{4}$	1

83.

x	-4	-1	0	1	4
y	-8	-32	Undefined	32	8

84.

x	-4	-1	0	1	4
y	6	3	0	3	6

Exploration

True or False? **In Exercises 85–88, determine whether the statement is true or false. Justify your answer.**

85. Every relation is a function.

86. Every function is a relation.

87. For the function

$f(x) = x^4 - 1$

the domain is $(-\infty, \infty)$ and the range is $(0, \infty)$.

88. The set of ordered pairs $\{(-8, -2), (-6, 0), (-4, 0), (-2, 2), (0, 4), (2, -2)\}$ represents a function.

89. **Error Analysis** Describe the error.

The functions

$f(x) = \sqrt{x - 1}$ and $g(x) = \dfrac{1}{\sqrt{x - 1}}$

have the same domain, which is the set of all real numbers x such that $x \geq 1$.

90. **Think About It** Consider

$f(x) = \sqrt{x - 2}$ and $g(x) = \sqrt[3]{x - 2}$.

Why are the domains of f and g different?

91. **Think About It** Given $f(x) = x^2$, is f the independent variable? Why or why not?

92. **HOW DO YOU SEE IT?** The graph represents the height h of a projectile after t seconds.

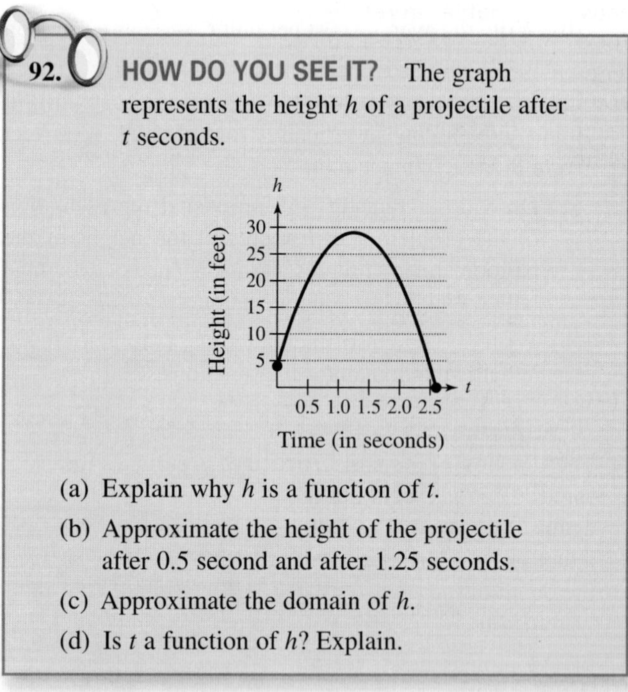

(a) Explain why h is a function of t.

(b) Approximate the height of the projectile after 0.5 second and after 1.25 seconds.

(c) Approximate the domain of h.

(d) Is t a function of h? Explain.

Think About It **In Exercises 93 and 94, determine whether the statements use the word *function* in ways that are mathematically correct. Explain.**

93. (a) The sales tax on a purchased item is a function of the selling price.

(b) Your score on the next algebra exam is a function of the number of hours you study the night before the exam.

94. (a) The amount in your savings account is a function of your salary.

(b) The speed at which a free-falling baseball strikes the ground is a function of the height from which it was dropped.

P.6 Analyzing Graphs of Functions

Graphs of functions can help you visualize relationships between variables in real life. For example, in Exercise 90 on page 77, you will use the graph of a function to visually represent the temperature in a city over a 24-hour period.

- **Use the Vertical Line Test for functions.**
- **Find the zeros of functions.**
- **Determine intervals on which functions are increasing or decreasing.**
- **Determine relative minimum and relative maximum values of functions.**
- **Determine the average rate of change of a function.**
- **Identify even and odd functions.**

The Graph of a Function

In Section P.5, you studied functions from an algebraic point of view. In this section, you will study functions from a graphical perspective.

The **graph of a function** f is the collection of ordered pairs $(x, f(x))$ such that x is in the domain of f. As you study this section, remember that

$x =$ the directed distance from the y-axis

$y = f(x) =$ the directed distance from the x-axis

as shown in the figure at the right.

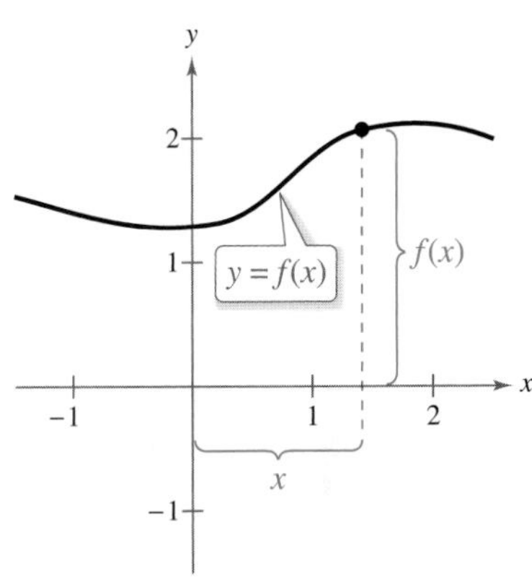

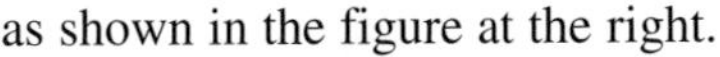

EXAMPLE 1 **Finding the Domain and Range of a Function**

Use the graph of the function f, shown in Figure P.36, to find (a) the domain of f, (b) the function values $f(-1)$ and $f(2)$, and (c) the range of f.

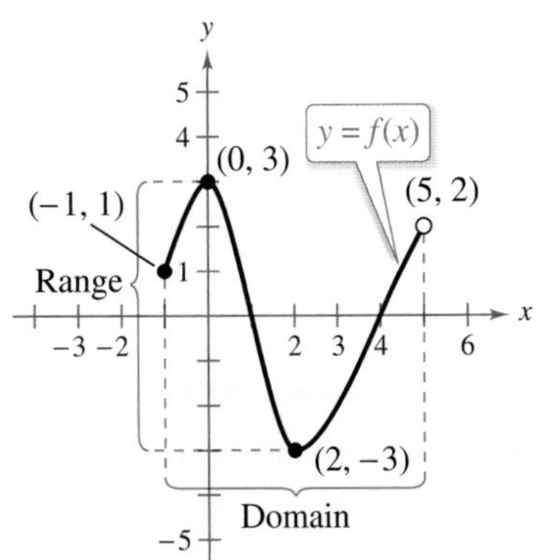

Figure P.36

Solution

a. The closed dot at $(-1, 1)$ indicates that $x = -1$ is in the domain of f, whereas the open dot at $(5, 2)$ indicates that $x = 5$ is not in the domain. So, the domain of f is all x in the interval $[-1, 5)$.

b. One point on the graph of f is $(-1, 1)$, so $f(-1) = 1$. Another point on the graph of f is $(2, -3)$, so $f(2) = -3$.

c. The graph does not extend below $f(2) = -3$ or above $f(0) = 3$, so the range of f is the interval $[-3, 3]$.

REMARK The use of dots (open or closed) at the extreme left and right points of a graph indicates that the graph does not extend beyond these points. If such dots are not on the graph, then assume that the graph extends beyond these points.

✓ ***Checkpoint*** *Audio-video solution in English & Spanish at LarsonPrecalculus.com*

Use the graph of the function f to find (a) the domain of f, (b) the function values $f(0)$ and $f(3)$, and (c) the range of f.

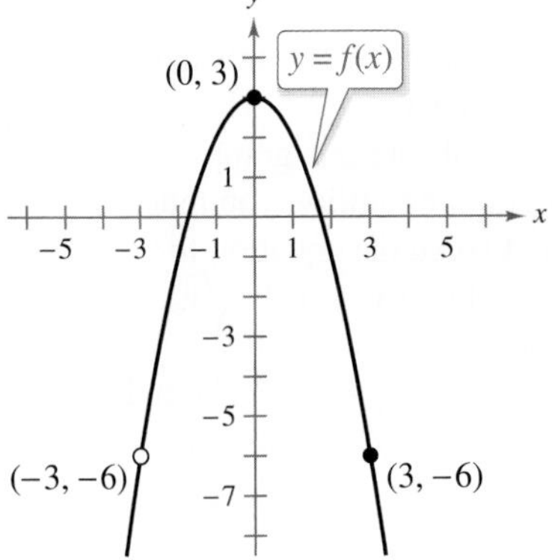

iStockphoto.com/Alija

By the definition of a function, at most one y-value corresponds to a given x-value. So, no two points on the graph of a function have the same x-coordinate, or lie on the same vertical line. It follows, then, that a vertical line can intersect the graph of a function at most once. This observation provides a convenient visual test called the **Vertical Line Test** for functions.

> **Vertical Line Test for Functions**
>
> A set of points in a coordinate plane is the graph of y as a function of x if and only if no *vertical* line intersects the graph at more than one point.

EXAMPLE 2 Vertical Line Test for Functions

Use the Vertical Line Test to determine whether each graph represents y as a function of x.

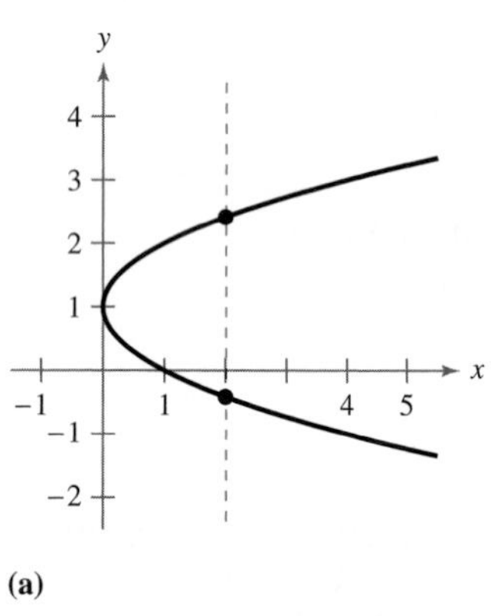
(a)

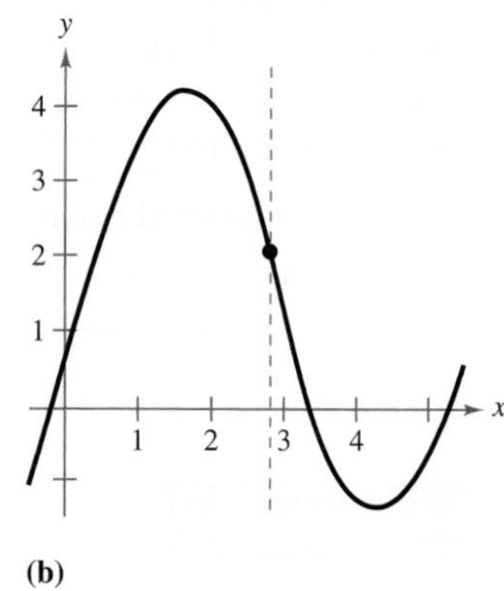
(b)

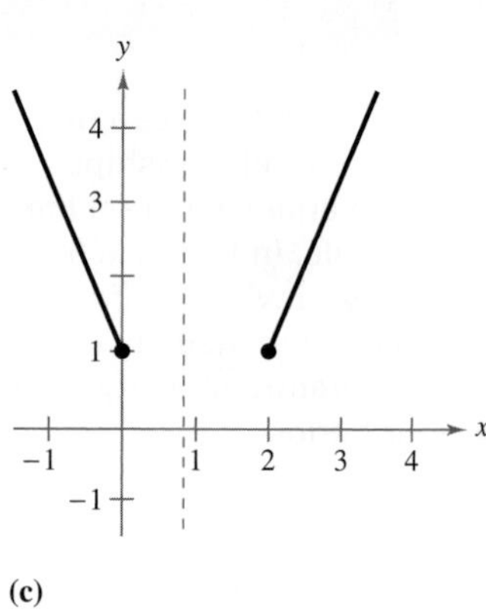
(c)

Solution

a. This *is not* a graph of y as a function of x, because there are vertical lines that intersect the graph twice. That is, for a particular input x, there is more than one output y.

b. This *is* a graph of y as a function of x, because every vertical line intersects the graph at most once. That is, for a particular input x, there is at most one output y.

c. This *is* a graph of y as a function of x, because every vertical line intersects the graph at most once. That is, for a particular input x, there is at most one output y. (Note that when a vertical line does not intersect the graph, it simply means that the function is undefined for that particular value of x.)

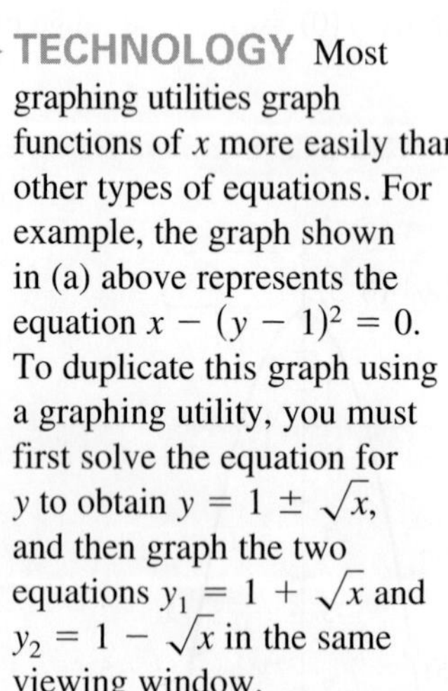
TECHNOLOGY Most graphing utilities graph functions of x more easily than other types of equations. For example, the graph shown in (a) above represents the equation $x - (y - 1)^2 = 0$. To duplicate this graph using a graphing utility, you must first solve the equation for y to obtain $y = 1 \pm \sqrt{x}$, and then graph the two equations $y_1 = 1 + \sqrt{x}$ and $y_2 = 1 - \sqrt{x}$ in the same viewing window.

✓ **Checkpoint** Audio-video solution in English & Spanish at LarsonPrecalculus.com

Use the Vertical Line Test to determine whether the graph represents y as a function of x.

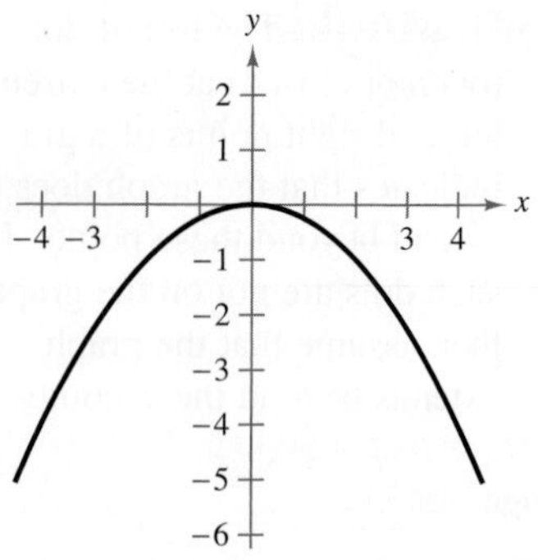

Zeros of a Function

If the graph of a function of x has an x-intercept at $(a, 0)$, then a is a **zero** of the function.

> **Zeros of a Function**
>
> The **zeros of a function** $y = f(x)$ are the x-values for which $f(x) = 0$.

EXAMPLE 3 Finding the Zeros of Functions

Find the zeros of each function algebraically.

a. $f(x) = 3x^2 + x - 10$

b. $g(x) = \sqrt{10 - x^2}$

c. $h(t) = \dfrac{2t - 3}{t + 5}$

Solution To find the zeros of a function, set the function equal to zero and solve for the independent variable.

a.

$3x^2 + x - 10 = 0$	Set $f(x)$ equal to 0.
$(3x - 5)(x + 2) = 0$	Factor.
$3x - 5 = 0 \Rightarrow x = \frac{5}{3}$	Set 1st factor equal to 0 and solve.
$x + 2 = 0 \Rightarrow x = -2$	Set 2nd factor equal to 0 and solve.

The zeros of f are $x = \frac{5}{3}$ and $x = -2$. In Figure P.37, note that the graph of f has $\left(\frac{5}{3}, 0\right)$ and $(-2, 0)$ as its x-intercepts.

Zeros of f: $x = -2$, $x = \frac{5}{3}$
Figure P.37

b.

$\sqrt{10 - x^2} = 0$	Set $g(x)$ equal to 0.
$10 - x^2 = 0$	Square each side.
$10 = x^2$	Add x^2 to each side.
$\pm\sqrt{10} = x$	Extract square roots.

The zeros of g are $x = -\sqrt{10}$ and $x = \sqrt{10}$. In Figure P.38, note that the graph of g has $\left(-\sqrt{10}, 0\right)$ and $\left(\sqrt{10}, 0\right)$ as its x-intercepts.

Zeros of g: $x = \pm\sqrt{10}$
Figure P.38

c.

$\dfrac{2t - 3}{t + 5} = 0$	Set $h(t)$ equal to 0.
$2t - 3 = 0$	Multiply each side by $t + 5$.
$2t = 3$	Add 3 to each side.
$t = \dfrac{3}{2}$	Divide each side by 2.

The zero of h is $t = \frac{3}{2}$. In Figure P.39, note that the graph of h has $\left(\frac{3}{2}, 0\right)$ as its t-intercept.

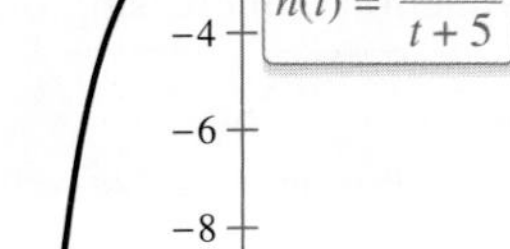

Zero of h: $t = \frac{3}{2}$
Figure P.39

Checkpoint Audio-video solution in English & Spanish at LarsonPrecalculus.com

Find the zeros of each function.

a. $f(x) = 2x^2 + 13x - 24$ **b.** $g(t) = \sqrt{t - 25}$ **c.** $h(x) = \dfrac{x^2 - 2}{x - 1}$

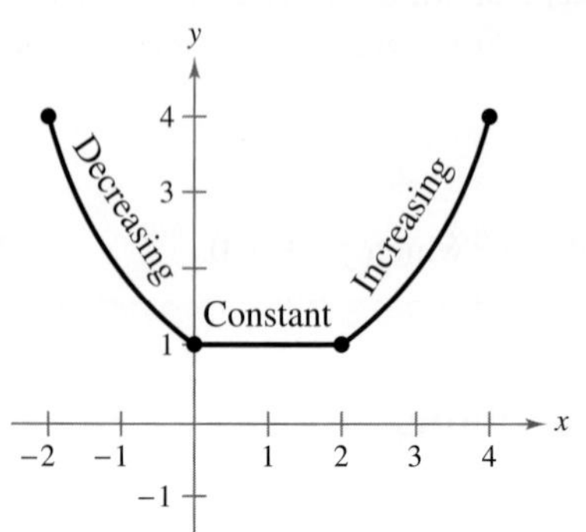

Figure P.40

Increasing and Decreasing Functions

The more you know about the graph of a function, the more you know about the function itself. Consider the graph shown in Figure P.40. As you move from *left to right,* this graph falls from $x = -2$ to $x = 0$, is constant from $x = 0$ to $x = 2$, and rises from $x = 2$ to $x = 4$.

> **Increasing, Decreasing, and Constant Functions**
>
> A function f is **increasing** on an interval when, for any x_1 and x_2 in the interval,
>
> $$x_1 < x_2 \quad \text{implies} \quad f(x_1) < f(x_2).$$
>
> A function f is **decreasing** on an interval when, for any x_1 and x_2 in the interval,
>
> $$x_1 < x_2 \quad \text{implies} \quad f(x_1) > f(x_2).$$
>
> A function f is **constant** on an interval when, for any x_1 and x_2 in the interval,
>
> $$f(x_1) = f(x_2).$$

EXAMPLE 4 Describing Function Behavior

Determine the open intervals on which each function is increasing, decreasing, or constant.

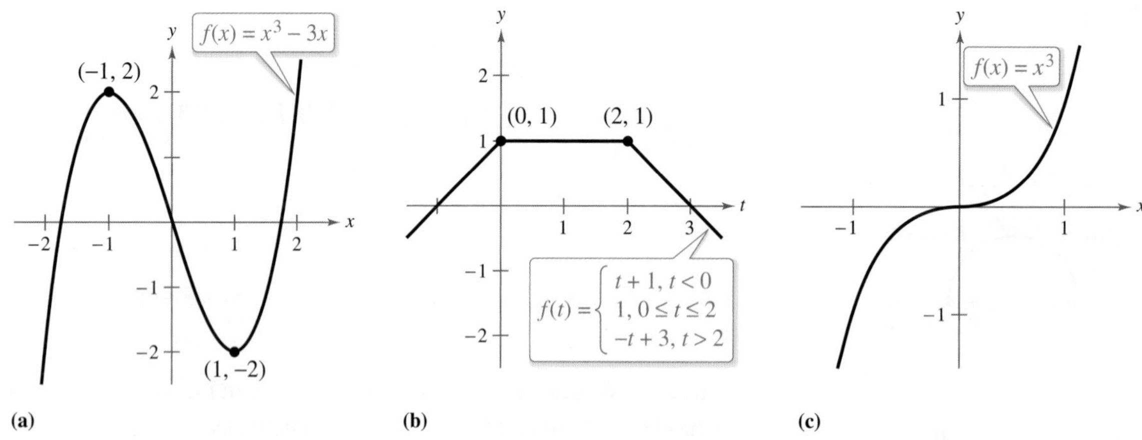

Solution

a. This function is increasing on the interval $(-\infty, -1)$, decreasing on the interval $(-1, 1)$, and increasing on the interval $(1, \infty)$.

b. This function is increasing on the interval $(-\infty, 0)$, constant on the interval $(0, 2)$, and decreasing on the interval $(2, \infty)$.

c. This function may appear to be constant on an interval near $x = 0$, but for all real values of x_1 and x_2, if $x_1 < x_2$, then $(x_1)^3 < (x_2)^3$. So, the function is increasing on the interval $(-\infty, \infty)$.

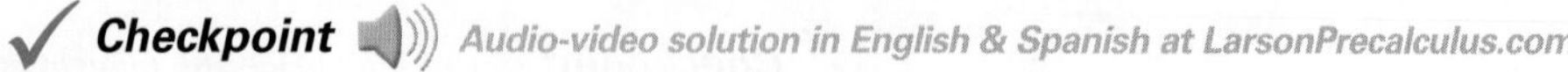

Graph the function

$$f(x) = x^3 + 3x^2 - 1.$$

Then determine the open intervals on which the function is increasing, decreasing, or constant.

Relative Minimum and Relative Maximum Values

The points at which a function changes its increasing, decreasing, or constant behavior are helpful in determining the **relative minimum** or **relative maximum** values of the function.

REMARK A relative minimum or relative maximum is also referred to as a local minimum or local maximum.

> **Definitions of Relative Minimum and Relative Maximum**
>
> A function value $f(a)$ is a **relative minimum** of f when there exists an interval (x_1, x_2) that contains a such that
>
> $$x_1 < x < x_2 \quad \text{implies} \quad f(a) \le f(x).$$
>
> A function value $f(a)$ is a **relative maximum** of f when there exists an interval (x_1, x_2) that contains a such that
>
> $$x_1 < x < x_2 \quad \text{implies} \quad f(a) \ge f(x).$$

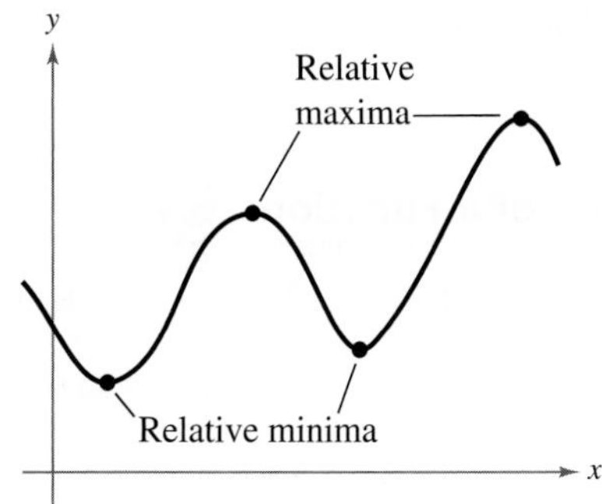

Figure P.41

Figure P.41 shows several different examples of relative minima and relative maxima. By writing a second-degree polynomial function in the form $f(x) = a(x - h)^2 + k$, you can find the *exact point* (h, k) at which it has a relative minimum or relative maximum. For the time being, however, you can use a graphing utility to find reasonable approximations of these points.

EXAMPLE 5 Approximating a Relative Minimum

Use a graphing utility to approximate the relative minimum of the function

$$f(x) = 3x^2 - 4x - 2.$$

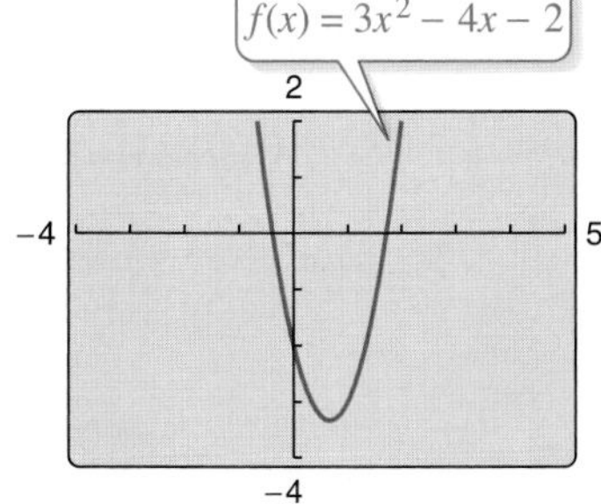

Figure P.42

Solution The graph of f is shown in Figure P.42. By using the *zoom* and *trace* features or the *minimum* feature of a graphing utility, you can approximate that the relative minimum of the function occurs at the point

$$(0.67, -3.33).$$

So, the relative minimum is approximately -3.33. By writing the given function in the form $f(x) = 3\left(x - \frac{2}{3}\right)^2 - \frac{10}{3}$, determine that the exact point at which the relative minimum occurs is $\left(\frac{2}{3}, -\frac{10}{3}\right)$ and the exact relative minimum is $-\frac{10}{3}$.

✓ **Checkpoint** *Audio-video solution in English & Spanish at LarsonPrecalculus.com*

Use a graphing utility to approximate the relative maximum of the function

$$f(x) = -4x^2 - 7x + 3.$$

You can also use the *table* feature of a graphing utility to numerically approximate the relative minimum of the function in Example 5. Using a table that begins at 0.6 and increments the value of x by 0.01, you can approximate that the minimum of

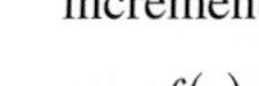

$$f(x) = 3x^2 - 4x - 2$$

occurs at the point $(0.67, -3.33)$.

TECHNOLOGY When you use a graphing utility to approximate the x- and y-values of the point where a relative minimum or relative maximum occurs, the *zoom* feature will often produce graphs that are nearly flat. To overcome this problem, manually change the vertical setting of the viewing window. The graph will stretch vertically when the values of Ymin and Ymax are closer together.

Average Rate of Change

In Section P.4, you learned that the slope of a line can be interpreted as a *rate of change*. For a nonlinear graph, the **average rate of change** between any two points $(x_1, f(x_1))$ and $(x_2, f(x_2))$ is the slope of the line through the two points (see Figure P.43). The line through the two points is called a **secant line,** and the slope of this line is denoted as m_{sec}.

$$\text{Average rate of change of } f \text{ from } x_1 \text{ to } x_2 = \frac{f(x_2) - f(x_1)}{x_2 - x_1} = \frac{\text{change in } y}{\text{change in } x} = m_{sec}$$

Figure P.43

EXAMPLE 6 Average Rate of Change of a Function 

Find the average rates of change of $f(x) = x^3 - 3x$ (a) from $x_1 = -2$ to $x_2 = -1$ and (b) from $x_1 = 0$ to $x_2 = 1$ (see Figure P.44).

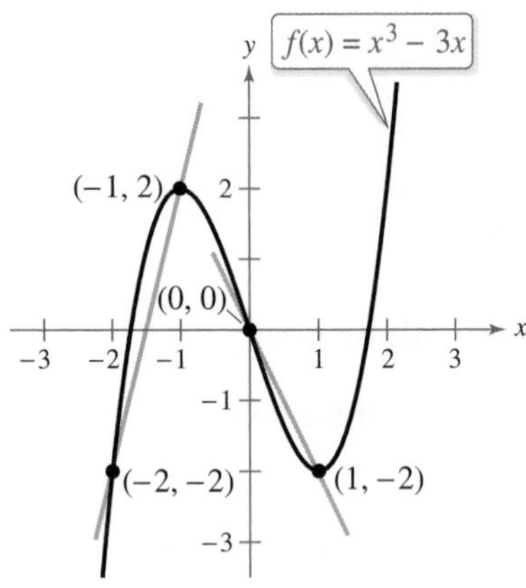

Figure P.44

Solution

a. The average rate of change of f from $x_1 = -2$ to $x_2 = -1$ is

$$\frac{f(x_2) - f(x_1)}{x_2 - x_1} = \frac{f(-1) - f(-2)}{-1 - (-2)} = \frac{2 - (-2)}{1} = 4.$$

Secant line has positive slope.

b. The average rate of change of f from $x_1 = 0$ to $x_2 = 1$ is

$$\frac{f(x_2) - f(x_1)}{x_2 - x_1} = \frac{f(1) - f(0)}{1 - 0} = \frac{-2 - 0}{1} = -2.$$

Secant line has negative slope.

Find the average rates of change of $f(x) = x^2 + 2x$ (a) from $x_1 = -3$ to $x_2 = -2$ and (b) from $x_1 = -2$ to $x_2 = 0$.

EXAMPLE 7 Finding Average Speed 

The distance s (in feet) a moving car is from a stoplight is given by the function

$$s(t) = 20t^{3/2}$$

where t is the time (in seconds). Find the average speed of the car (a) from $t_1 = 0$ to $t_2 = 4$ seconds and (b) from $t_1 = 4$ to $t_2 = 9$ seconds.

Average speed is an average rate of change.

Solution

a. The average speed of the car from $t_1 = 0$ to $t_2 = 4$ seconds is

$$\frac{s(t_2) - s(t_1)}{t_2 - t_1} = \frac{s(4) - s(0)}{4 - 0} = \frac{160 - 0}{4} = 40 \text{ feet per second.}$$

b. The average speed of the car from $t_1 = 4$ to $t_2 = 9$ seconds is

$$\frac{s(t_2) - s(t_1)}{t_2 - t_1} = \frac{s(9) - s(4)}{9 - 4} = \frac{540 - 160}{5} = 76 \text{ feet per second.}$$

✓ **Checkpoint** Audio-video solution in English & Spanish at LarsonPrecalculus.com

In Example 7, find the average speed of the car (a) from $t_1 = 0$ to $t_2 = 1$ second and (b) from $t_1 = 1$ second to $t_2 = 4$ seconds.

Even and Odd Functions

In Section P.3, you studied different types of symmetry of a graph. In the terminology of functions, a function is said to be **even** when its graph is symmetric with respect to the y-axis and **odd** when its graph is symmetric with respect to the origin. The symmetry tests in Section P.3 yield the tests for even and odd functions below.

> **Tests for Even and Odd Functions**
>
> A function $y = f(x)$ is **even** when, for each x in the domain of f, $f(-x) = f(x)$.
>
> A function $y = f(x)$ is **odd** when, for each x in the domain of f, $f(-x) = -f(x)$.

EXAMPLE 8 Even and Odd Functions

See LarsonPrecalculus.com for an interactive version of this type of example.

a. The function $g(x) = x^3 - x$ is odd because $g(-x) = -g(x)$, as follows.

$$g(-x) = (-x)^3 - (-x) \quad \text{Substitute } -x \text{ for } x.$$
$$= -x^3 + x \quad \text{Simplify.}$$
$$= -(x^3 - x) \quad \text{Distributive Property}$$
$$= -g(x) \quad \text{Test for odd function}$$

b. The function $h(x) = x^2 + 1$ is even because $h(-x) = h(x)$, as follows.

$$h(-x) = (-x)^2 + 1 = x^2 + 1 = h(x) \quad \text{Test for even function}$$

Figure P.45 shows the graphs and symmetry of these two functions.

(a) Symmetric to origin: Odd Function

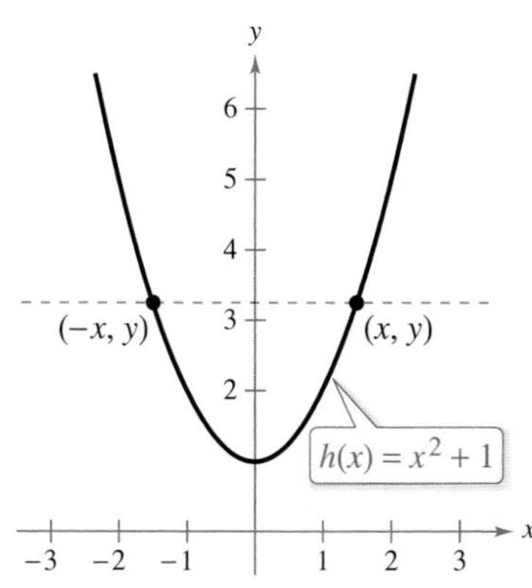

(b) Symmetric to y-axis: Even Function

Figure P.45

✓ Checkpoint Audio-video solution in English & Spanish at LarsonPrecalculus.com

Determine whether each function is even, odd, or neither. Then describe the symmetry.

a. $f(x) = 5 - 3x$ **b.** $g(x) = x^4 - x^2 - 1$ **c.** $h(x) = 2x^3 + 3x$

> ### Summarize (Section P.6)
>
> 1. State the Vertical Line Test for functions *(page 68)*. For an example of using the Vertical Line Test, see Example 2.
> 2. Explain how to find the zeros of a function *(page 69)*. For an example of finding the zeros of functions, see Example 3.
> 3. Explain how to determine intervals on which functions are increasing or decreasing *(page 70)*. For an example of describing function behavior, see Example 4.
> 4. Explain how to determine relative minimum and relative maximum values of functions *(page 71)*. For an example of approximating a relative minimum, see Example 5.
> 5. Explain how to determine the average rate of change of a function *(page 72)*. For examples of determining average rates of change, see Examples 6 and 7.
> 6. State the definitions of an even function and an odd function *(page 73)*. For an example of identifying even and odd functions, see Example 8.

P.6 Exercises

See CalcChat.com for tutorial help and worked-out solutions to odd-numbered exercises.

Vocabulary: Fill in the blanks.

1. The ________ ________ ________ is used to determine whether a graph represents y as a function of x.
2. The ________ of a function $y = f(x)$ are the values of x for which $f(x) = 0$.
3. A function f is ________ on an interval when, for any x_1 and x_2 in the interval, $x_1 < x_2$ implies $f(x_1) > f(x_2)$.
4. A function value $f(a)$ is a relative ________ of f when there exists an interval (x_1, x_2) containing a such that $x_1 < x < x_2$ implies $f(a) \geq f(x)$.
5. The ________ ________ ________ ________ between any two points $(x_1, f(x_1))$ and $(x_2, f(x_2))$ is the slope of the line through the two points, and this line is called the ________ line.
6. A function f is ________ when, for each x in the domain of f, $f(-x) = -f(x)$.

Skills and Applications

Domain, Range, and Values of a Function **In Exercises 7–10, use the graph of the function to find the domain and range of f and each function value.**

7. (a) $f(-1)$ (b) $f(0)$ (c) $f(1)$ (d) $f(2)$

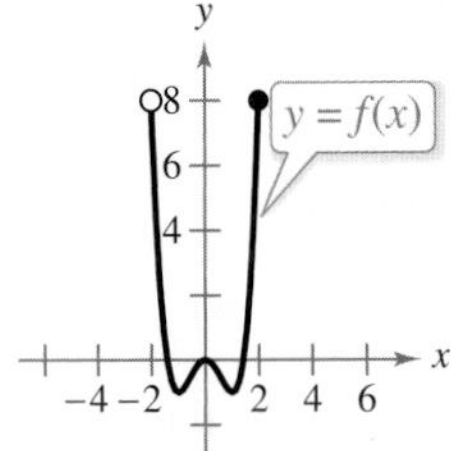

8. (a) $f(-1)$ (b) $f(0)$ (c) $f(1)$ (d) $f(3)$

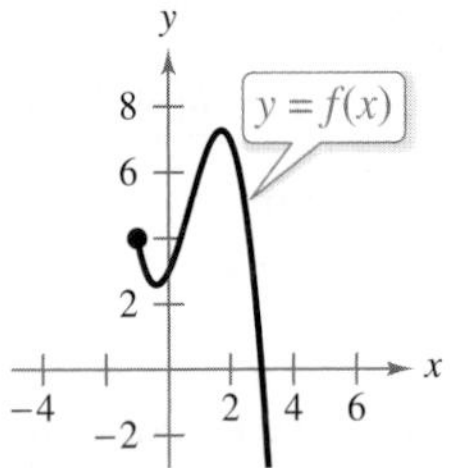

9. (a) $f(2)$ (b) $f(1)$ (c) $f(3)$ (d) $f(-1)$

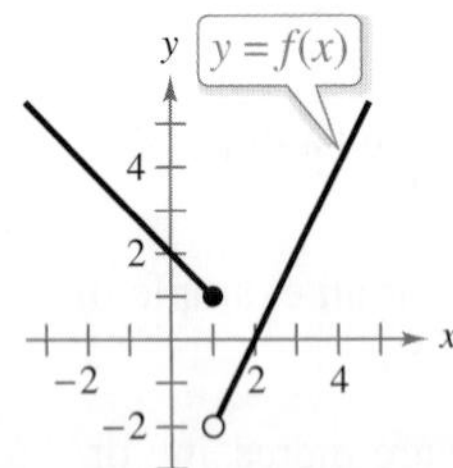

10. (a) $f(-2)$ (b) $f(1)$ (c) $f(0)$ (d) $f(2)$

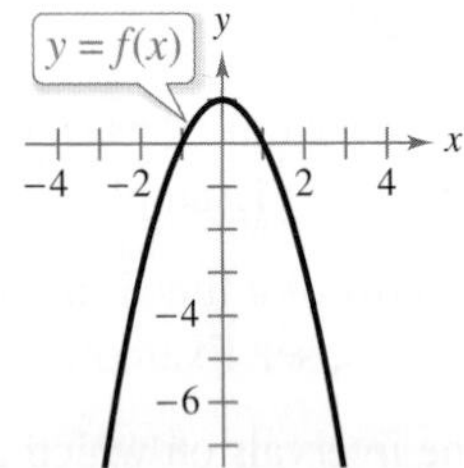

Vertical Line Test for Functions **In Exercises 11–14, use the Vertical Line Test to determine whether the graph represents y as a function of x. To print an enlarged copy of the graph, go to *MathGraphs.com*.**

11.

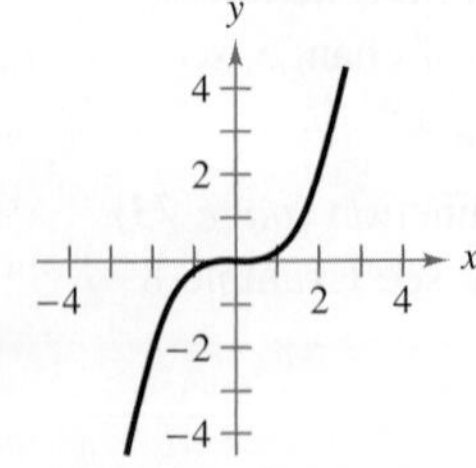

12.

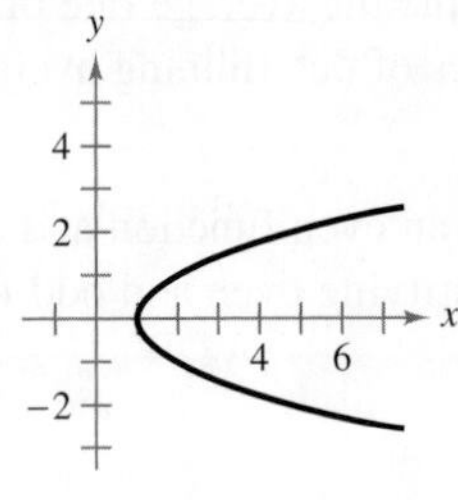

13.

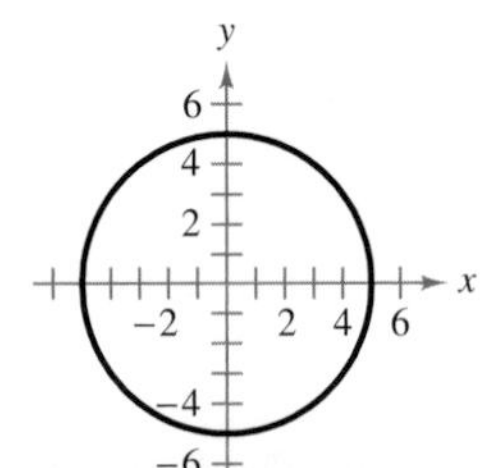

14.

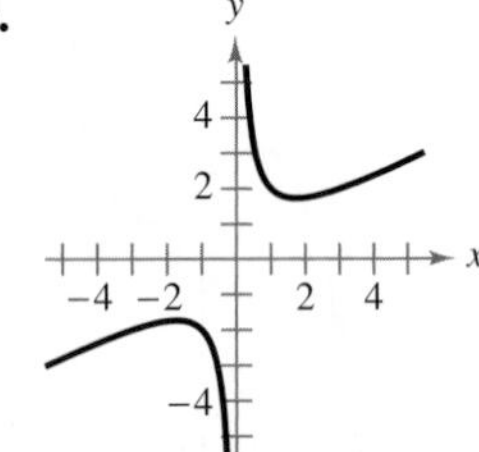

Finding the Zeros of a Function **In Exercises 15–26, find the zeros of the function algebraically.**

15. $f(x) = 3x + 18$
16. $f(x) = 15 - 2x$
17. $f(x) = 2x^2 - 7x - 30$
18. $f(x) = 3x^2 + 22x - 16$
19. $f(x) = \dfrac{x + 3}{2x^2 - 6}$
20. $f(x) = \dfrac{x^2 - 9x + 14}{4x}$
21. $f(x) = \frac{1}{3}x^3 - 2x$
22. $f(x) = -25x^4 + 9x^2$
23. $f(x) = x^3 - 4x^2 - 9x + 36$
24. $f(x) = 4x^3 - 24x^2 - x + 6$
25. $f(x) = \sqrt{2x} - 1$
26. $f(x) = \sqrt{3x + 2}$

Graphing and Finding Zeros **In Exercises 27–32, (a) use a graphing utility to graph the function and find the zeros of the function and (b) verify your results from part (a) algebraically.**

27. $f(x) = x^2 - 6x$
28. $f(x) = 2x^2 - 13x - 7$
29. $f(x) = \sqrt{2x + 11}$
30. $f(x) = \sqrt{3x - 14} - 8$
31. $f(x) = \dfrac{3x - 1}{x - 6}$
32. $f(x) = \dfrac{2x^2 - 9}{3 - x}$

Describing Function Behavior In Exercises 33–40, determine the open intervals on which the function is increasing, decreasing, or constant.

33. $f(x) = -\frac{1}{2}x^3$

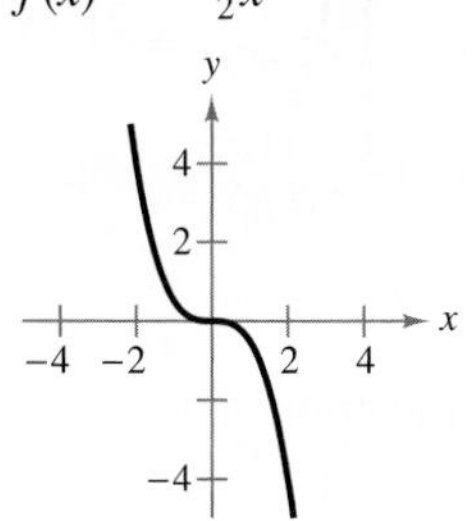

34. $f(x) = x^2 - 4x$

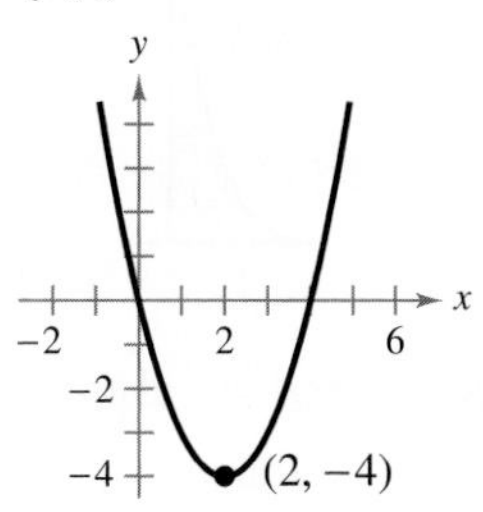

35. $f(x) = \sqrt{x^2 - 1}$

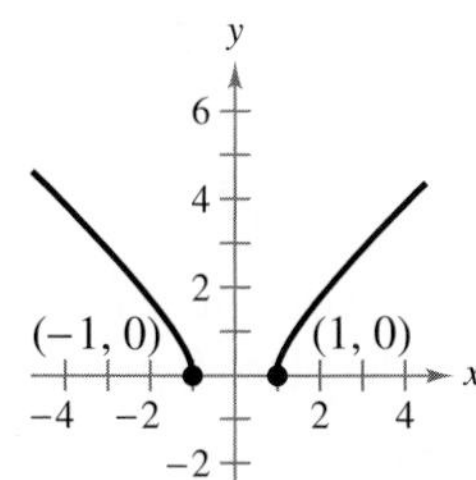

36. $f(x) = x^3 - 3x^2 + 2$

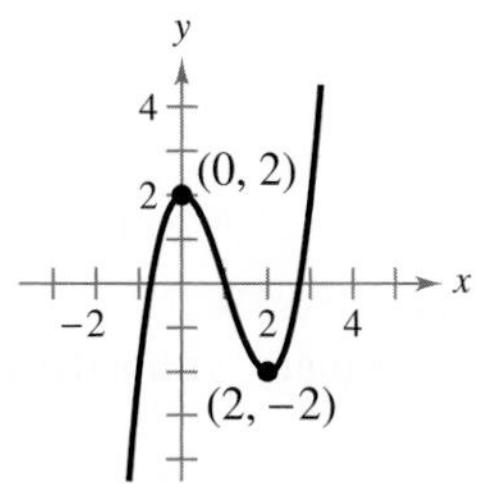

37. $f(x) = |x + 1| + |x - 1|$

(−1, 2) (1, 2)

38. $f(x) = \dfrac{x^2 + x + 1}{x + 1}$

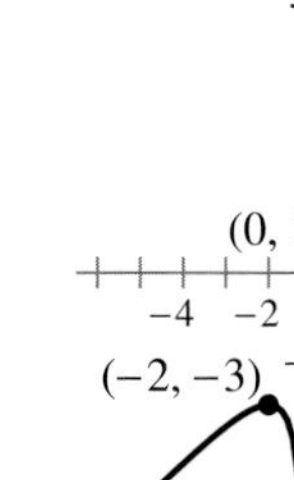

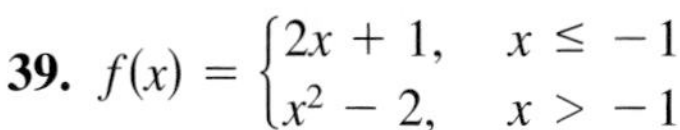

39. $f(x) = \begin{cases} 2x + 1, & x \leq -1 \\ x^2 - 2, & x > -1 \end{cases}$

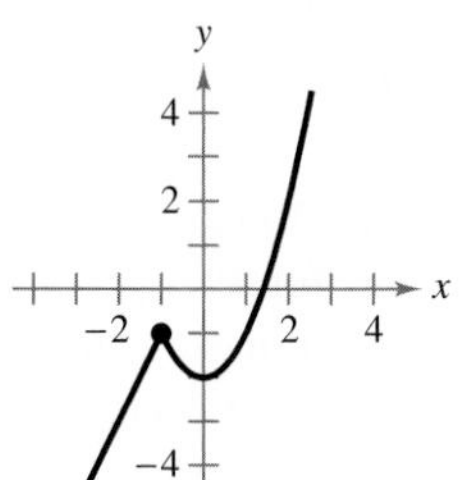

40. $f(x) = \begin{cases} x + 3, & x \leq 0 \\ 3, & 0 < x \leq 2 \\ 2x + 1, & x > 2 \end{cases}$

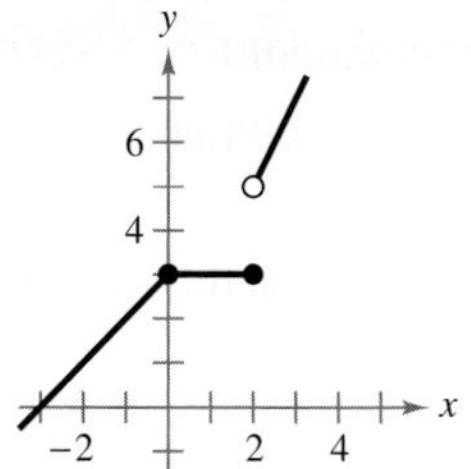

Describing Function Behavior In Exercises 41–48, use a graphing utility to graph the function and visually determine the open intervals on which the function is increasing, decreasing, or constant. Use a table of values to verify your results.

41. $f(x) = 3$
42. $g(x) = x$
43. $g(x) = \frac{1}{2}x^2 - 3$
44. $f(x) = 3x^4 - 6x^2$
45. $f(x) = \sqrt{1 - x}$
46. $f(x) = x\sqrt{x + 3}$
47. $f(x) = x^{3/2}$
48. $f(x) = x^{2/3}$

Approximating Relative Minima or Maxima In Exercises 49–54, use a graphing utility to approximate (to two decimal places) any relative minima or maxima of the function.

49. $f(x) = x(x + 3)$
50. $f(x) = -x^2 + 3x - 2$
51. $h(x) = x^3 - 6x^2 + 15$
52. $f(x) = x^3 - 3x^2 - x + 1$
53. $h(x) = (x - 1)\sqrt{x}$
54. $g(x) = x\sqrt{4 - x}$

Graphical Reasoning In Exercises 55–60, graph the function and determine the interval(s) for which $f(x) \geq 0$.

55. $f(x) = 4 - x$
56. $f(x) = 4x + 2$
57. $f(x) = 9 - x^2$
58. $f(x) = x^2 - 4x$
59. $f(x) = \sqrt{x - 1}$
60. $f(x) = |x + 5|$

Average Rate of Change of a Function In Exercises 61–64, find the average rate of change of the function from x_1 to x_2.

Function	x-Values
61. $f(x) = -2x + 15$	$x_1 = 0, x_2 = 3$
62. $f(x) = x^2 - 2x + 8$	$x_1 = 1, x_2 = 5$
63. $f(x) = x^3 - 3x^2 - x$	$x_1 = -1, x_2 = 2$
64. $f(x) = -x^3 + 6x^2 + x$	$x_1 = 1, x_2 = 6$

65. **Research and Development** The amounts (in billions of dollars) the U.S. federal government spent on research and development for defense from 2010 through 2014 can be approximated by the model

$$y = 0.5079t^2 - 8.168t + 95.08$$

where t represents the year, with $t = 0$ corresponding to 2010. *(Source: American Association for the Advancement of Science)*

(a) Use a graphing utility to graph the model.

(b) Find the average rate of change of the model from 2010 to 2014. Interpret your answer in the context of the problem.

66. Finding Average Speed Use the information in Example 7 to find the average speed of the car from $t_1 = 0$ to $t_2 = 9$ seconds. Explain why the result is less than the value obtained in part (b) of Example 7.

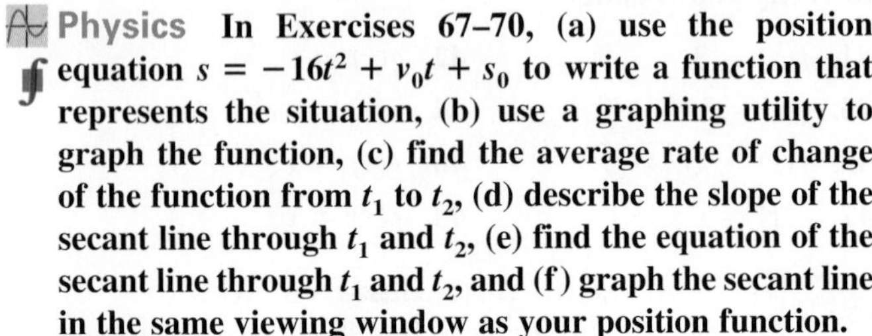

Physics In Exercises 67–70, (a) use the position equation $s = -16t^2 + v_0t + s_0$ to write a function that represents the situation, (b) use a graphing utility to graph the function, (c) find the average rate of change of the function from t_1 to t_2, (d) describe the slope of the secant line through t_1 and t_2, (e) find the equation of the secant line through t_1 and t_2, and (f) graph the secant line in the same viewing window as your position function.

67. An object is thrown upward from a height of 6 feet at a velocity of 64 feet per second.

$t_1 = 0, t_2 = 3$

68. An object is thrown upward from a height of 6.5 feet at a velocity of 72 feet per second.

$t_1 = 0, t_2 = 4$

69. An object is thrown upward from ground level at a velocity of 120 feet per second.

$t_1 = 3, t_2 = 5$

70. An object is dropped from a height of 80 feet.

$t_1 = 1, t_2 = 2$

Even, Odd, or Neither? In Exercises 71–76, determine whether the function is even, odd, or neither. Then describe the symmetry.

71. $f(x) = x^6 - 2x^2 + 3$ **72.** $g(x) = x^3 - 5x$

73. $h(x) = x\sqrt{x+5}$ **74.** $f(x) = x\sqrt{1-x^2}$

75. $f(s) = 4s^{3/2}$ **76.** $g(s) = 4s^{2/3}$

Even, Odd, or Neither? In Exercises 77–82, sketch a graph of the function and determine whether it is even, odd, or neither. Verify your answer algebraically.

77. $f(x) = -9$ **78.** $f(x) = 5 - 3x$

79. $f(x) = -|x - 5|$ **80.** $h(x) = x^2 - 4$

81. $f(x) = \sqrt[3]{4x}$ **82.** $f(x) = \sqrt[3]{x-4}$

Height of a Rectangle In Exercises 83 and 84, write the height h of the rectangle as a function of x.

83.

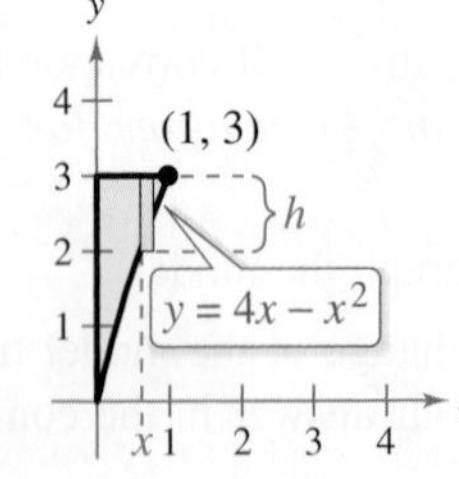

84.

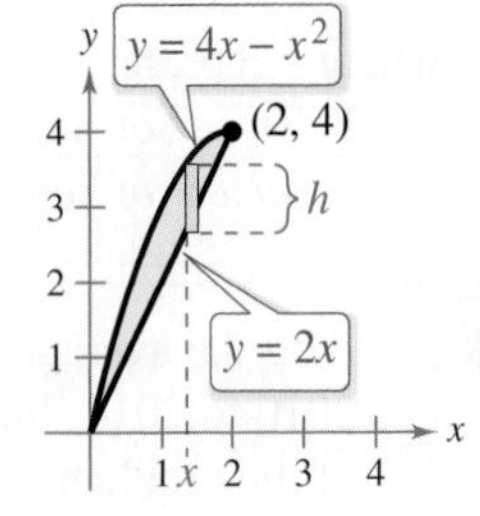

Length of a Rectangle In Exercises 85 and 86, write the length L of the rectangle as a function of y.

85.

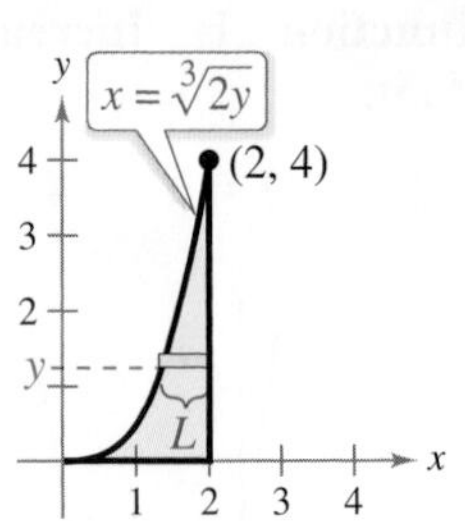

86.

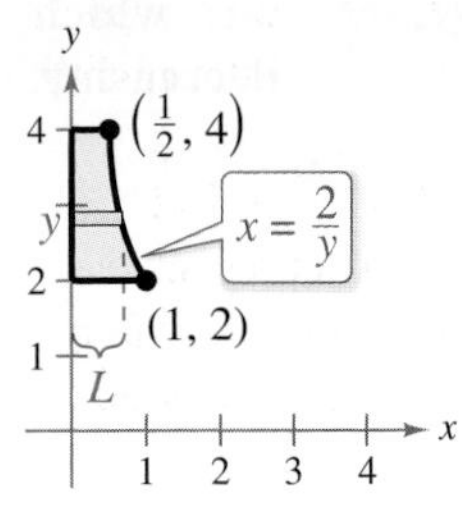

87. Error Analysis Describe the error.

The function $f(x) = 2x^3 - 5$ is odd because $f(-x) = -f(x)$, as follows.

$$f(-x) = 2(-x)^3 - 5$$
$$= -2x^3 - 5$$
$$= -(2x^3 - 5)$$
$$= -f(x)$$

88. Geometry Corners of equal size are cut from a square with sides of length 8 meters (see figure).

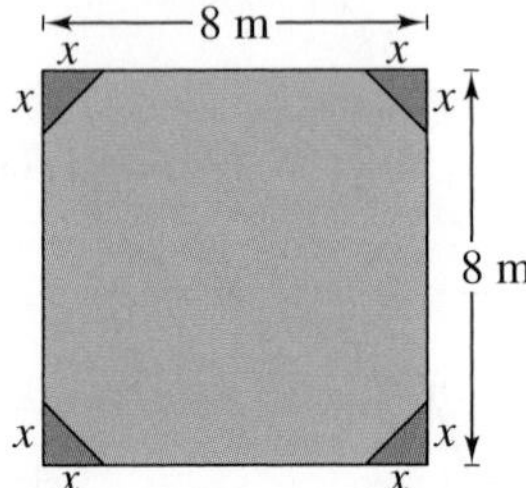

(a) Write the area A of the resulting figure as a function of x. Determine the domain of the function.

(b) Use a graphing utility to graph the area function over its domain. Use the graph to find the range of the function.

(c) Identify the figure that results when x is the maximum value in the domain of the function. What would be the length of each side of the figure?

89. Coordinate Axis Scale Each function described below models the specified data for the years 2006 through 2016, with $t = 6$ corresponding to 2006. Estimate a reasonable scale for the vertical axis (e.g., hundreds, thousands, millions, etc.) of the graph and justify your answer. (There are many correct answers.)

(a) $f(t)$ represents the average salary of college professors.

(b) $f(t)$ represents the U.S. population.

(c) $f(t)$ represents the percent of the civilian workforce that is unemployed.

(d) $f(t)$ represents the number of games a college football team wins.

90. Temperature The table shows the temperatures y (in degrees Fahrenheit) in a city over a 24-hour period. Let x represent the time of day, where $x = 0$ corresponds to 6 A.M.

DATA

Time, x	Temperature, y
0	34
2	50
4	60
6	64
8	63
10	59
12	53
14	46
16	40
18	36
20	34
22	37
24	45

Spreadsheet at LarsonPrecalculus.com

These data can be approximated by the model

$$y = 0.026x^3 - 1.03x^2 + 10.2x + 34, \quad 0 \le x \le 24.$$

(a) Use a graphing utility to create a scatter plot of the data. Then graph the model in the same viewing window.

(b) How well does the model fit the data?

(c) Use the graph to approximate the times when the temperature was increasing and decreasing.

(d) Use the graph to approximate the maximum and minimum temperatures during this 24-hour period.

(e) Could this model predict the temperatures in the city during the next 24-hour period? Why or why not?

Exploration

True or False? **In Exercises 91–93, determine whether the statement is true or false. Justify your answer.**

91. A function with a square root cannot have a domain that is the set of real numbers.

92. It is possible for an odd function to have the interval $[0, \infty)$ as its domain.

93. It is impossible for an even function to be increasing on its entire domain.

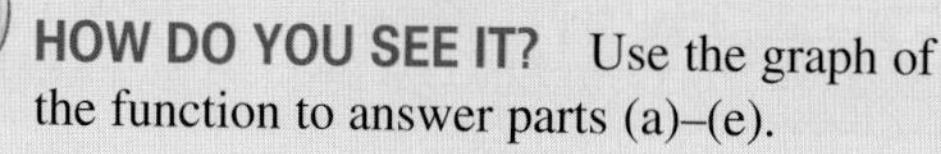

94. HOW DO YOU SEE IT? Use the graph of the function to answer parts (a)–(e).

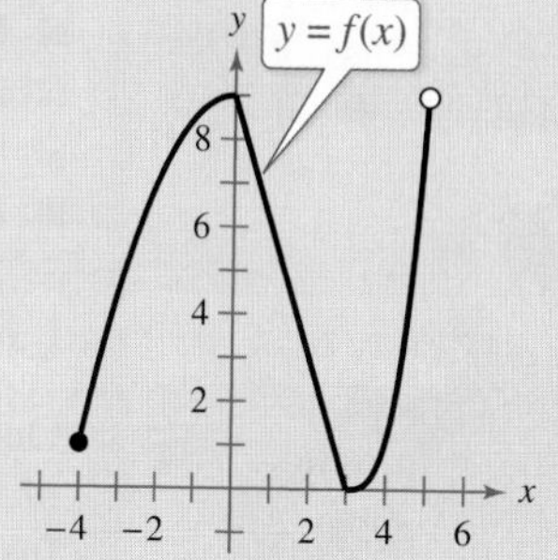

(a) Find the domain and range of f.

(b) Find the zero(s) of f.

(c) Determine the open intervals on which f is increasing, decreasing, or constant.

(d) Approximate any relative minimum or relative maximum values of f.

(e) Is f even, odd, or neither?

Think About It **In Exercises 95 and 96, find the coordinates of a second point on the graph of a function f when the given point is on the graph and the function is (a) even and (b) odd.**

95. $\left(-\frac{5}{3}, -7\right)$ **96.** $(2a, 2c)$

97. Writing Use a graphing utility to graph each function. Write a paragraph describing any similarities and differences you observe among the graphs.

(a) $y = x$ (b) $y = x^2$ (c) $y = x^3$

(d) $y = x^4$ (e) $y = x^5$ (f) $y = x^6$

98. Graphical Reasoning Graph each of the functions with a graphing utility. Determine whether each function is even, odd, or neither.

$f(x) = x^2 - x^4$ $g(x) = 2x^3 + 1$

$h(x) = x^5 - 2x^3 + x$ $j(x) = 2 - x^6 - x^8$

$k(x) = x^5 - 2x^4 + x - 2$ $p(x) = x^9 + 3x^5 - x^3 + x$

What do you notice about the equations of functions that are odd? What do you notice about the equations of functions that are even? Can you describe a way to identify a function as odd or even by inspecting the equation? Can you describe a way to identify a function as neither odd nor even by inspecting the equation?

99. Even, Odd, or Neither? Determine whether g is even, odd, or neither when f is an even function. Explain.

(a) $g(x) = -f(x)$ (b) $g(x) = f(-x)$

(c) $g(x) = f(x) - 2$ (d) $g(x) = f(x - 2)$

iStockphoto.com/Alija

P.7 A Library of Parent Functions

Piecewise-defined functions model many real-life situations. For example, in Exercise 47 on page 84, you will write a piecewise-defined function to model the depth of snow during a snowstorm.

- **Identify and graph linear and squaring functions.**
- **Identify and graph cubic, square root, and reciprocal functions.**
- **Identify and graph step and other piecewise-defined functions.**
- **Recognize graphs of parent functions.**

Linear and Squaring Functions

One of the goals of this text is to enable you to recognize the basic shapes of the graphs of different types of functions. For example, you know that the graph of the **linear function** $f(x) = ax + b$ is a line with slope $m = a$ and y-intercept at $(0, b)$. The graph of a linear function has the characteristics below.

- The domain of the function is the set of all real numbers.
- When $m \neq 0$, the range of the function is the set of all real numbers.
- The graph has an x-intercept at $(-b/m, 0)$ and a y-intercept at $(0, b)$.
- The graph is increasing when $m > 0$, decreasing when $m < 0$, and constant when $m = 0$.

EXAMPLE 1 Writing a Linear Function

Write the linear function f for which $f(1) = 3$ and $f(4) = 0$.

Solution To find the equation of the line that passes through $(x_1, y_1) = (1, 3)$ and $(x_2, y_2) = (4, 0)$, first find the slope of the line.

$$m = \frac{y_2 - y_1}{x_2 - x_1} = \frac{0 - 3}{4 - 1} = \frac{-3}{3} = -1$$

Next, use the point-slope form of the equation of a line.

$$y - y_1 = m(x - x_1) \quad \text{Point-slope form}$$

$$y - 3 = -1(x - 1) \quad \text{Substitute for } x_1, y_1, \text{ and } m.$$

$$y = -x + 4 \quad \text{Simplify.}$$

$$f(x) = -x + 4 \quad \text{Function notation}$$

The figure below shows the graph of this function.

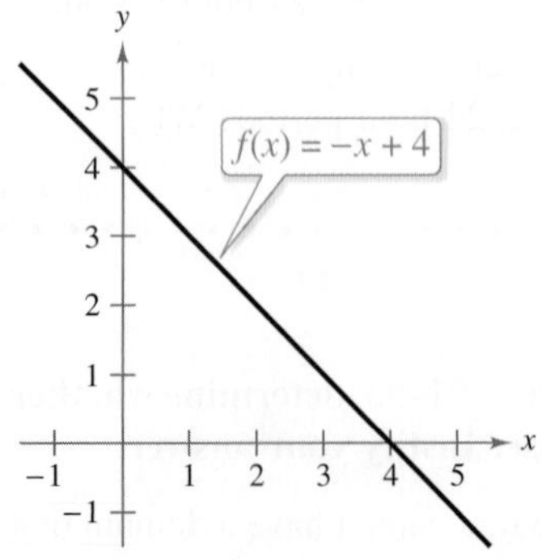

Checkpoint Audio-video solution in English & Spanish at LarsonPrecalculus.com

Write the linear function f for which $f(-2) = 6$ and $f(4) = -9$.

There are two special types of linear functions, the **constant function** and the **identity function.** A constant function has the form

$$f(x) = c$$

and has a domain of all real numbers with a range consisting of a single real number c. The graph of a constant function is a horizontal line, as shown in Figure P.46. The identity function has the form

$$f(x) = x.$$

Its domain and range are the set of all real numbers. The identity function has a slope of $m = 1$ and a y-intercept at $(0, 0)$. The graph of the identity function is a line for which each x-coordinate equals the corresponding y-coordinate. The graph is always increasing, as shown in Figure P.47.

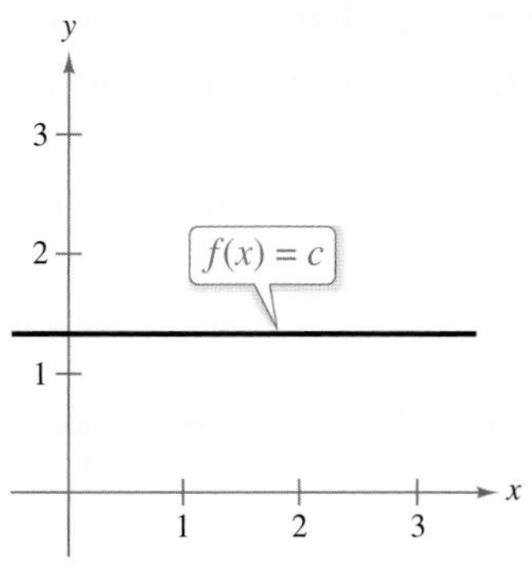

Figure P.46

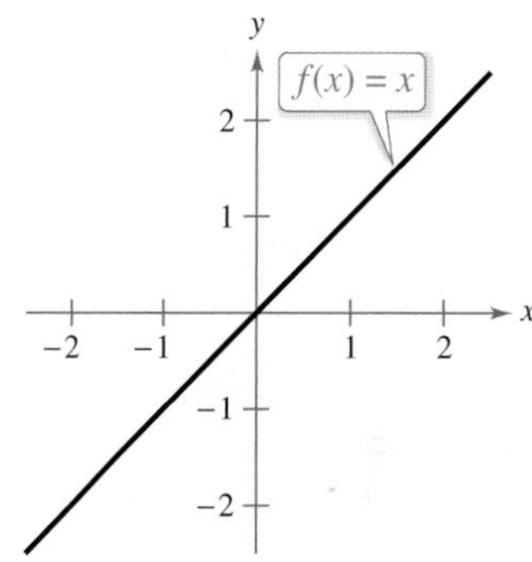

Figure P.47

The graph of the **squaring function**

$$f(x) = x^2$$

is a U-shaped curve with the characteristics below.

- The domain of the function is the set of all real numbers.
- The range of the function is the set of all nonnegative real numbers.
- The function is even.
- The graph has an intercept at $(0, 0)$.
- The graph is decreasing on the interval $(-\infty, 0)$ and increasing on the interval $(0, \infty)$.
- The graph is symmetric with respect to the y-axis.
- The graph has a relative minimum at $(0, 0)$.

The figure below shows the graph of the squaring function.

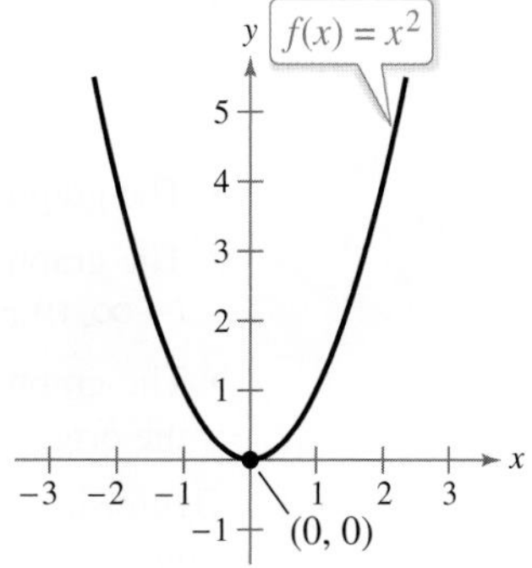

Cubic, Square Root, and Reciprocal Functions

Here are the basic characteristics of the graphs of the **cubic, square root,** and **reciprocal functions.**

1. The graph of the *cubic* function

$$f(x) = x^3$$

has the characteristics below.

- The domain of the function is the set of all real numbers.
- The range of the function is the set of all real numbers.
- The function is odd.
- The graph has an intercept at $(0, 0)$.
- The graph is increasing on the interval $(-\infty, \infty)$.
- The graph is symmetric with respect to the origin.

The figure shows the graph of the cubic function.

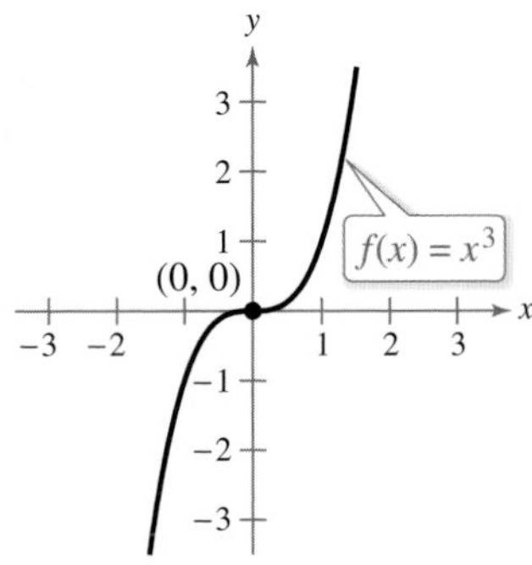

Cubic function

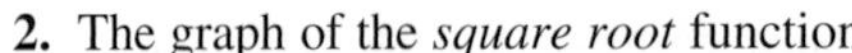

2. The graph of the *square root* function

$$f(x) = \sqrt{x}$$

has the characteristics below.

- The domain of the function is the set of all nonnegative real numbers.
- The range of the function is the set of all nonnegative real numbers.
- The graph has an intercept at $(0, 0)$.
- The graph is increasing on the interval $(0, \infty)$.

The figure shows the graph of the square root function.

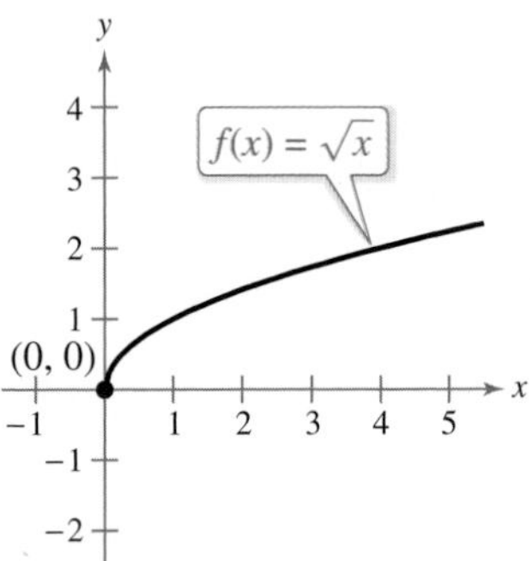

Square root function

3. The graph of the *reciprocal* function

$$f(x) = \frac{1}{x}$$

has the characteristics below.

- The domain of the function is $(-\infty, 0) \cup (0, \infty)$.
- The range of the function is $(-\infty, 0) \cup (0, \infty)$.
- The function is odd.
- The graph does not have any intercepts.
- The graph is decreasing on the intervals $(-\infty, 0)$ and $(0, \infty)$.
- The graph is symmetric with respect to the origin.

The figure shows the graph of the reciprocal function.

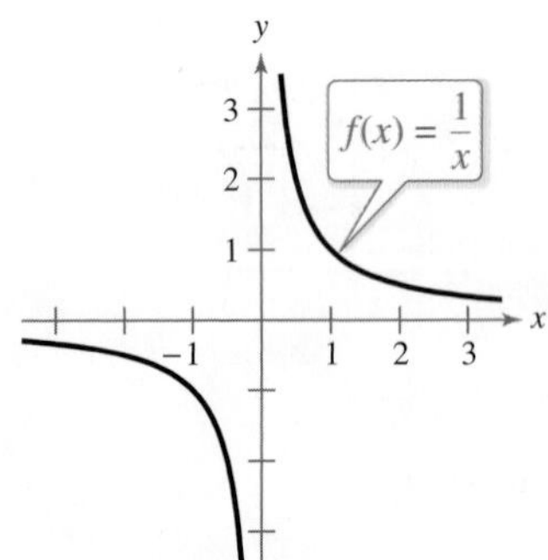

Reciprocal function

Step and Piecewise-Defined Functions

Functions whose graphs resemble sets of stairsteps are known as **step functions.** One common type of step function is the **greatest integer function,** denoted by $[\![x]\!]$ and defined as

$$f(x) = [\![x]\!] = \textit{the greatest integer less than or equal to } x.$$

Here are several examples of evaluating the greatest integer function.

$$[\![-1]\!] = (\text{greatest integer} \leq -1) = -1$$
$$[\![-\tfrac{1}{2}]\!] = \left(\text{greatest integer} \leq -\tfrac{1}{2}\right) = -1$$
$$[\![\tfrac{1}{10}]\!] = \left(\text{greatest integer} \leq \tfrac{1}{10}\right) = 0$$
$$[\![1.5]\!] = (\text{greatest integer} \leq 1.5) = 1$$
$$[\![1.9]\!] = (\text{greatest integer} \leq 1.9) = 1$$

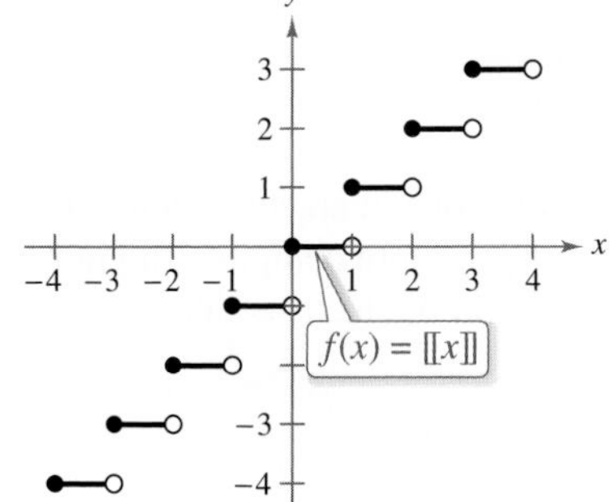

Figure P.48

The graph of the greatest integer function

$$f(x) = [\![x]\!]$$

has the characteristics below, as shown in Figure P.48.

- The domain of the function is the set of all real numbers.
- The range of the function is the set of all integers.
- The graph has a y-intercept at $(0, 0)$ and x-intercepts in the interval $[0, 1)$.
- The graph is constant between each pair of consecutive integer values of x.
- The graph jumps vertically one unit at each integer value of x.

TECHNOLOGY Most graphing utilities display graphs in *connected* mode, which works well for graphs that do not have breaks. For graphs that do have breaks, such as the graph of the greatest integer function, it may be better to use *dot* mode. Graph the greatest integer function [often called Int(x)] in *connected* and *dot* modes, and compare the two results.

EXAMPLE 2 Evaluating a Step Function

Evaluate the function $f(x) = [\![x]\!] + 1$ when $x = -1, 2$, and $\frac{3}{2}$.

Solution For $x = -1$, the greatest integer ≤ -1 is -1, so

$$f(-1) = [\![-1]\!] + 1 = -1 + 1 = 0.$$

For $x = 2$, the greatest integer ≤ 2 is 2, so

$$f(2) = [\![2]\!] + 1 = 2 + 1 = 3.$$

For $x = \frac{3}{2}$, the greatest integer $\leq \frac{3}{2}$ is 1, so

$$f\left(\tfrac{3}{2}\right) = [\![\tfrac{3}{2}]\!] + 1 = 1 + 1 = 2.$$

Verify your answers by examining the graph of $f(x) = [\![x]\!] + 1$ shown in Figure P.49.

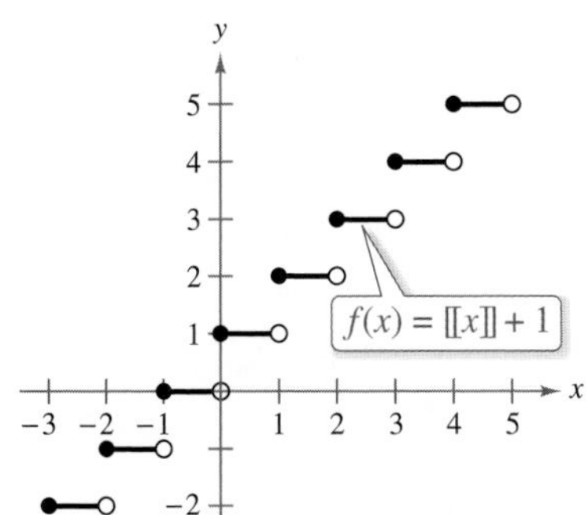

Figure P.49

✓ **Checkpoint** Audio-video solution in English & Spanish at LarsonPrecalculus.com

Evaluate the function $f(x) = [\![x + 2]\!]$ when $x = -\frac{3}{2}, 1$, and $-\frac{5}{2}$.

Recall from Section P.5 that a piecewise-defined function is defined by two or more equations over a specified domain. To graph a piecewise-defined function, graph each equation separately over the specified domain, as shown in Example 3.

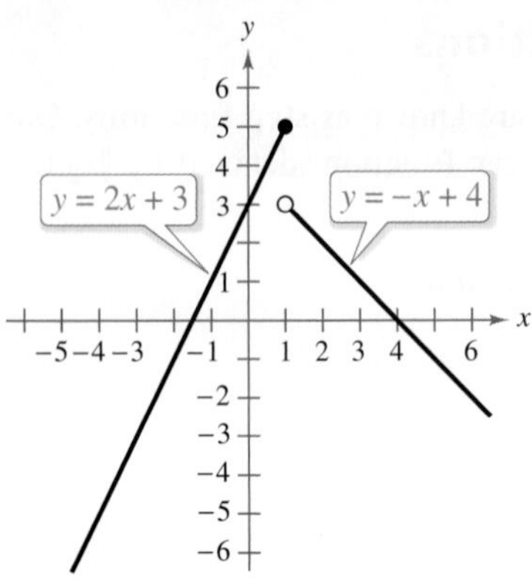

Figure P.50

EXAMPLE 3 Graphing a Piecewise-Defined Function

See LarsonPrecalculus.com for an interactive version of this type of example.

Sketch the graph of $f(x) = \begin{cases} 2x + 3, & x \leq 1 \\ -x + 4, & x > 1 \end{cases}.$

Solution This piecewise-defined function consists of two linear functions. At $x = 1$ and to the left of $x = 1$, the graph is the line $y = 2x + 3$, and to the right of $x = 1$, the graph is the line $y = -x + 4$, as shown in Figure P.50. Notice that the point $(1, 5)$ is a solid dot and the point $(1, 3)$ is an open dot. This is because $f(1) = 2(1) + 3 = 5$.

Checkpoint Audio-video solution in English & Spanish at LarsonPrecalculus.com

Sketch the graph of $f(x) = \begin{cases} -\frac{1}{2}x - 6, & x \leq -4 \\ x + 5, & x > -4 \end{cases}.$

Parent Functions

The graphs below represent the most commonly used functions in algebra. Familiarity with the characteristics of these graphs will help you analyze more complicated graphs obtained from these graphs by the transformations studied in the next section.

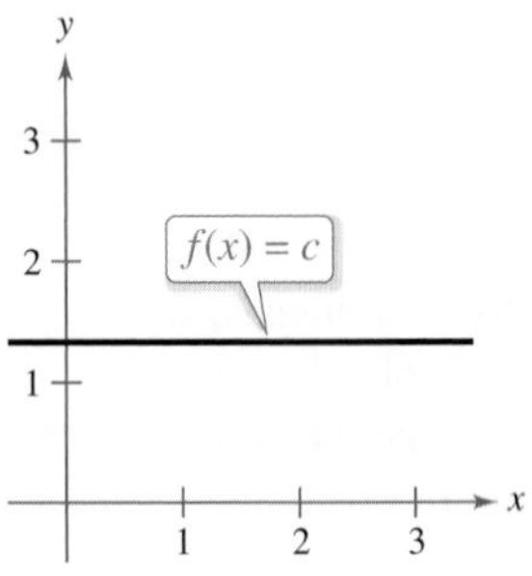

(a) Constant Function

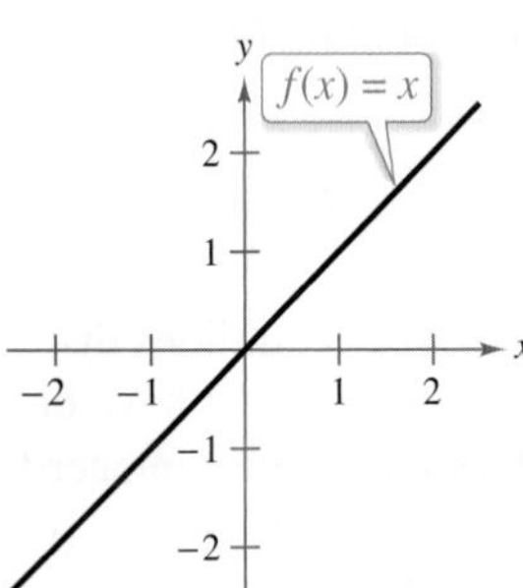

(b) Identity Function

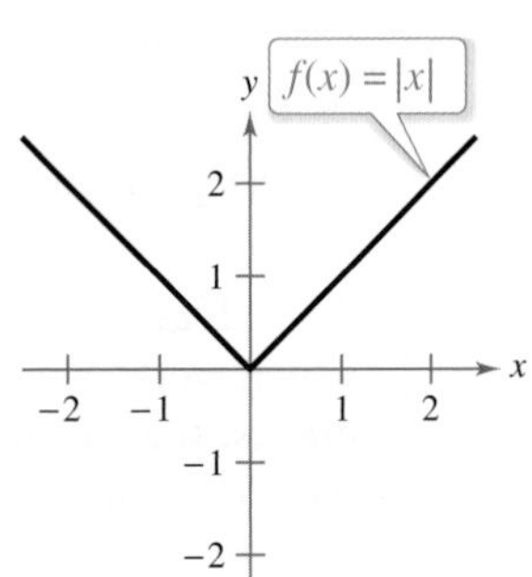

(c) Absolute Value Function

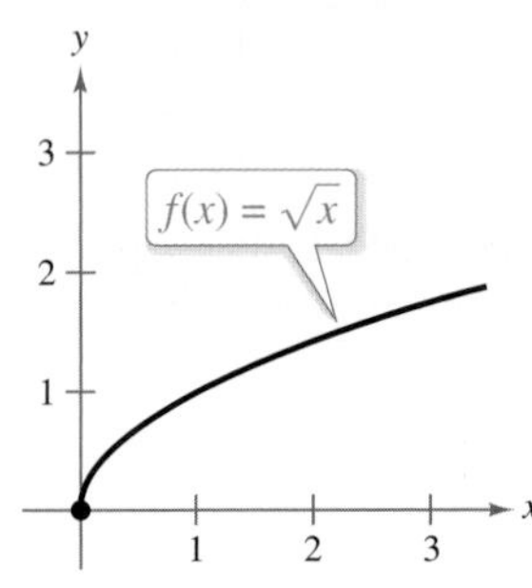

(d) Square Root Function

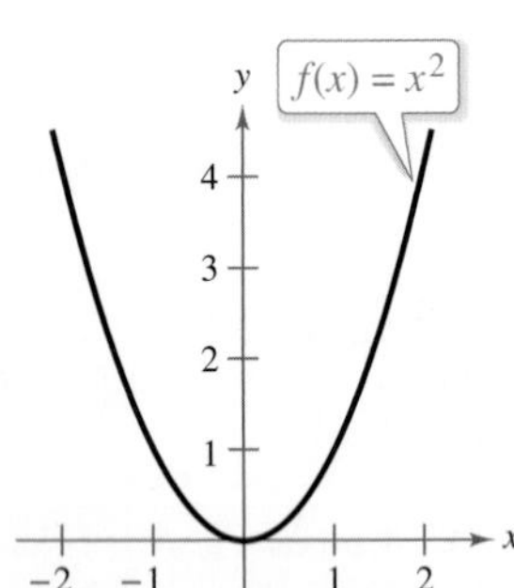

(e) Squaring Function

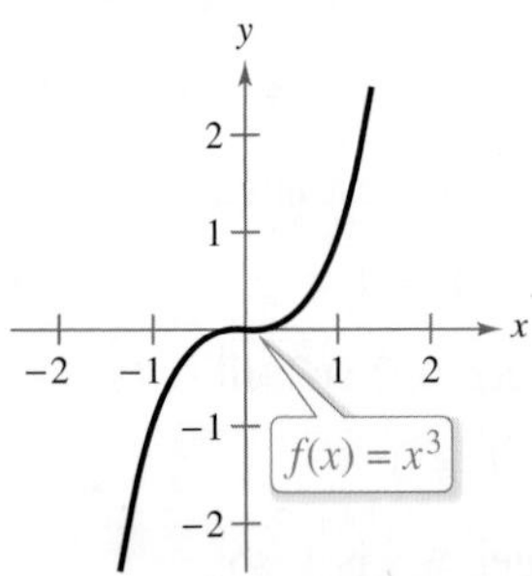

(f) Cubic Function

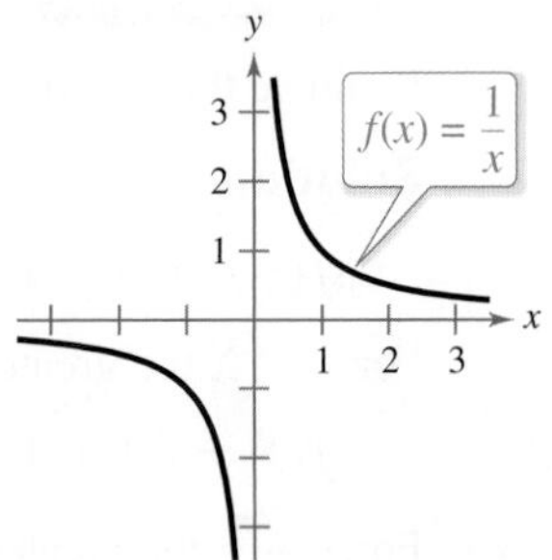

(g) Reciprocal Function

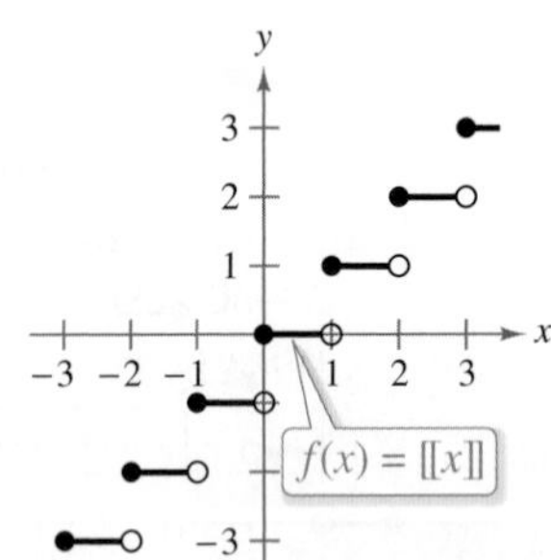

(h) Greatest Integer Function

Summarize (Section P.7)

1. Explain how to identify and graph linear and squaring functions *(pages 78 and 79)*. For an example involving a linear function, see Example 1.
2. Explain how to identify and graph cubic, square root, and reciprocal functions *(page 80)*.
3. Explain how to identify and graph step and other piecewise-defined functions *(page 81)*. For examples involving these functions, see Examples 2 and 3.
4. Identify and sketch the graphs of parent functions *(page 82)*.

P.7 Exercises

See CalcChat.com for tutorial help and worked-out solutions to odd-numbered exercises.

Vocabulary

In Exercises 1–9, write the most specific name of the function.

1. $f(x) = [\![x]\!]$
2. $f(x) = x$
3. $f(x) = 1/x$
4. $f(x) = x^2$
5. $f(x) = \sqrt{x}$
6. $f(x) = c$
7. $f(x) = |x|$
8. $f(x) = x^3$
9. $f(x) = ax + b$

10. Fill in the blank: The constant function and the identity function are two special types of ________ functions.

Skills and Applications

Writing a Linear Function **In Exercises 11–14, (a) write the linear function f that has the given function values and (b) sketch the graph of the function.**

11. $f(1) = 4, \quad f(0) = 6$
12. $f(-3) = -8, \quad f(1) = 2$
13. $f(\frac{1}{2}) = -\frac{5}{3}, \quad f(6) = 2$
14. $f(\frac{3}{5}) = \frac{1}{2}, \quad f(4) = 9$

Graphing a Function **In Exercises 15–26, use a graphing utility to graph the function. Be sure to choose an appropriate viewing window.**

15. $f(x) = 2.5x - 4.25$
16. $f(x) = \frac{5}{6} - \frac{2}{3}x$
17. $g(x) = x^2 + 3$
18. $f(x) = -2x^2 - 1$
19. $f(x) = x^3 - 1$
20. $f(x) = (x - 1)^3 + 2$
21. $f(x) = \sqrt{x} + 4$
22. $h(x) = \sqrt{x + 2} + 3$
23. $f(x) = \dfrac{1}{x - 2}$
24. $k(x) = 3 + \dfrac{1}{x + 3}$
25. $g(x) = |x| - 5$
26. $f(x) = |x - 1|$

Evaluating a Step Function **In Exercises 27–30, evaluate the function for the given values.**

27. $f(x) = [\![x]\!]$
    (a) $f(2.1)$ (b) $f(2.9)$ (c) $f(-3.1)$ (d) $f(\frac{7}{2})$
28. $h(x) = [\![x + 3]\!]$
    (a) $h(-2)$ (b) $h(\frac{1}{2})$ (c) $h(4.2)$ (d) $h(-21.6)$
29. $k(x) = [\![2x + 1]\!]$
    (a) $k(\frac{1}{3})$ (b) $k(-2.1)$ (c) $k(1.1)$ (d) $k(\frac{2}{3})$
30. $g(x) = -7[\![x + 4]\!] + 6$
    (a) $g(\frac{1}{8})$ (b) $g(9)$ (c) $g(-4)$ (d) $g(\frac{3}{2})$

Graphing a Step Function **In Exercises 31–34, sketch the graph of the function.**

31. $g(x) = -[\![x]\!]$
32. $g(x) = 4[\![x]\!]$
33. $g(x) = [\![x]\!] - 1$
34. $g(x) = [\![x - 3]\!]$

Graphing a Piecewise-Defined Function **In Exercises 35–40, sketch the graph of the function.**

35. $g(x) = \begin{cases} x + 6, & x \le -4 \\ \frac{1}{2}x - 4, & x > -4 \end{cases}$

36. $f(x) = \begin{cases} 4 + x, & x \le 2 \\ x^2 + 2, & x > 2 \end{cases}$

37. $f(x) = \begin{cases} 1 - (x - 1)^2, & x \le 2 \\ \sqrt{x - 2}, & x > 2 \end{cases}$

38. $f(x) = \begin{cases} \sqrt{4 + x}, & x < 0 \\ \sqrt{4 - x}, & x \ge 0 \end{cases}$

39. $h(x) = \begin{cases} 4 - x^2, & x < -2 \\ 3 + x, & -2 \le x < 0 \\ x^2 + 1, & x \ge 0 \end{cases}$

40. $k(x) = \begin{cases} 2x + 1, & x \le -1 \\ 2x^2 - 1, & -1 < x \le 1 \\ 1 - x^2, & x > 1 \end{cases}$

Graphing a Function **In Exercises 41 and 42, (a) use a graphing utility to graph the function and (b) state the domain and range of the function.**

41. $s(x) = 2(\frac{1}{4}x - [\![\frac{1}{4}x]\!])$
42. $k(x) = 4(\frac{1}{2}x - [\![\frac{1}{2}x]\!])^2$

43. **Wages** A mechanic's pay is \$14 per hour for regular time and time-and-a-half for overtime. The weekly wage function is

$$W(h) = \begin{cases} 14h, & 0 < h \le 40 \\ 21(h - 40) + 560, & h > 40 \end{cases}$$

where h is the number of hours worked in a week.

(a) Evaluate $W(30)$, $W(40)$, $W(45)$, and $W(50)$.

(b) The company decreases the regular work week to 36 hours. What is the new weekly wage function?

(c) The company increases the mechanic's pay to \$16 per hour. What is the new weekly wage function? Use a regular work week of 40 hours.

44. Revenue The table shows the monthly revenue y (in thousands of dollars) of a landscaping business for each month of the year 2016, with $x = 1$ representing January.

DATA

Month, x	Revenue, y
1	5.2
2	5.6
3	6.6
4	8.3
5	11.5
6	15.8
7	12.8
8	10.1
9	8.6
10	6.9
11	4.5
12	2.7

Spreadsheet at LarsonPrecalculus.com

A mathematical model that represents these data is

$$f(x) = \begin{cases} -1.97x + 26.3 \\ 0.505x^2 - 1.47x + 6.3 \end{cases}.$$

(a) Use a graphing utility to graph the model. What is the domain of each part of the piecewise-defined function? How can you tell?

(b) Find $f(5)$ and $f(11)$ and interpret your results in the context of the problem.

(c) How do the values obtained from the model in part (b) compare with the actual data values?

45. Fluid Flow The intake pipe of a 100-gallon tank has a flow rate of 10 gallons per minute, and two drainpipes have flow rates of 5 gallons per minute each. The figure shows the volume V of fluid in the tank as a function of time t. Determine whether the input pipe and each drainpipe are open or closed in specific subintervals of the 1 hour of time shown in the graph. (There are many correct answers.)

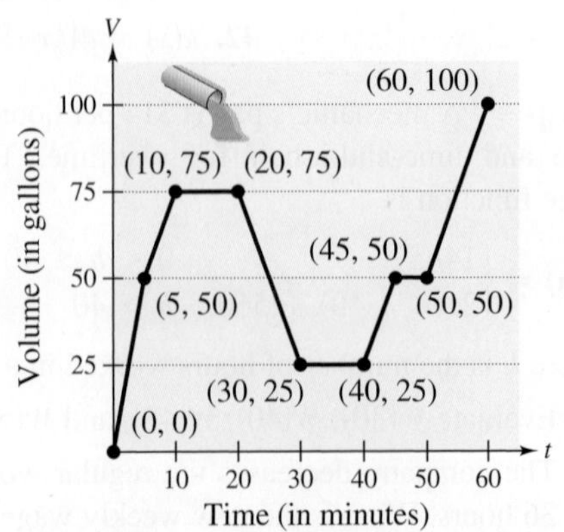

46. Delivery Charges The cost of mailing a package weighing up to, but not including, 1 pound is \$2.72. Each additional pound or portion of a pound costs \$0.50.

(a) Use the greatest integer function to create a model for the cost C of mailing a package weighing x pounds, where $x > 0$.

(b) Sketch the graph of the function.

47. Snowstorm

During a nine-hour snowstorm, it snows at a rate of 1 inch per hour for the first 2 hours, at a rate of 2 inches per hour for the next 6 hours, and at a rate of 0.5 inch per hour for the final hour. Write and graph a piecewise-defined function that gives the depth of the snow during the snowstorm. How many inches of snow accumulated from the storm?

48. HOW DO YOU SEE IT? For each graph of f shown below, answer parts (a)–(d).

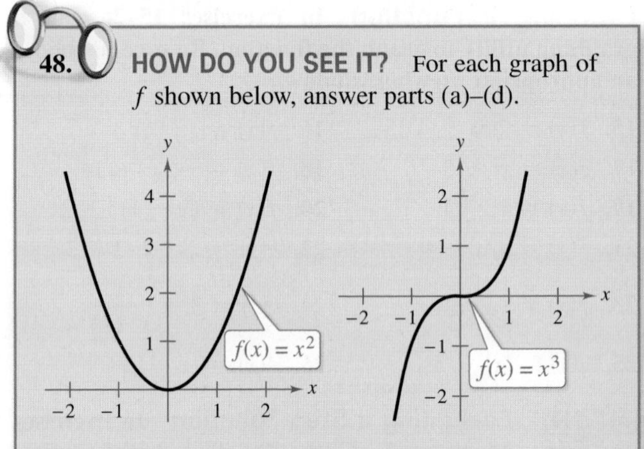

(a) Find the domain and range of f.

(b) Find the x- and y-intercepts of the graph of f.

(c) Determine the open intervals on which f is increasing, decreasing, or constant.

(d) Determine whether f is even, odd, or neither. Then describe the symmetry.

Exploration

True or False? **In Exercises 49 and 50, determine whether the statement is true or false. Justify your answer.**

49. A piecewise-defined function will always have at least one x-intercept or at least one y-intercept.

50. A linear equation will always have an x-intercept and a y-intercept.

Jan_S/Shutterstock.com

P.8 Transformations of Functions

Transformations of functions model many real-life applications. For example, in Exercise 61 on page 92, you will use a transformation of a function to model the number of horsepower required to overcome wind drag on an automobile.

- Use vertical and horizontal shifts to sketch graphs of functions.
- Use reflections to sketch graphs of functions.
- Use nonrigid transformations to sketch graphs of functions.

Shifting Graphs

Many functions have graphs that are transformations of the parent graphs summarized in Section P.7. For example, you obtain the graph of

$$h(x) = x^2 + 2$$

by shifting the graph of $f(x) = x^2$ *up* two units, as shown in Figure P.51. In function notation, h and f are related as follows.

$$h(x) = x^2 + 2 = f(x) + 2$$ Upward shift of two units

Similarly, you obtain the graph of

$$g(x) = (x - 2)^2$$

by shifting the graph of $f(x) = x^2$ to the *right* two units, as shown in Figure P.52. In this case, the functions g and f have the following relationship.

$$g(x) = (x - 2)^2 = f(x - 2)$$ Right shift of two units

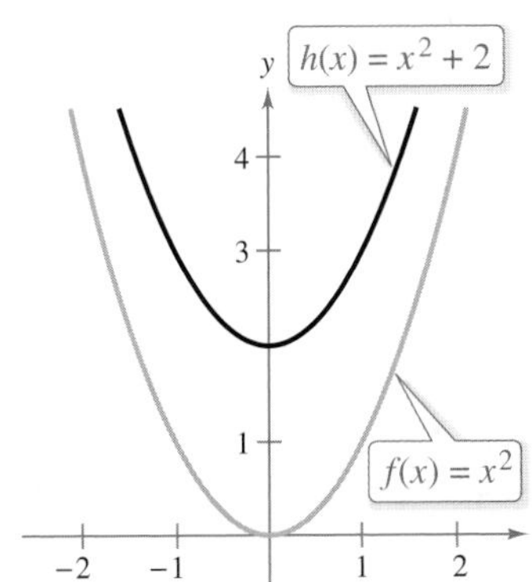

Figure P.51

Figure P.52

The list below summarizes this discussion about horizontal and vertical shifts.

> **REMARK** In items 3 and 4, be sure you see that $h(x) = f(x - c)$ corresponds to a *right* shift and $h(x) = f(x + c)$ corresponds to a *left* shift for $c > 0$.

Vertical and Horizontal Shifts

Let c be a positive real number. **Vertical and horizontal shifts** in the graph of $y = f(x)$ are represented as follows.

1. Vertical shift c units *up*: $h(x) = f(x) + c$
2. Vertical shift c units *down*: $h(x) = f(x) - c$
3. Horizontal shift c units to the *right*: $h(x) = f(x - c)$
4. Horizontal shift c units to the *left*: $h(x) = f(x + c)$

Some graphs are obtained from combinations of vertical and horizontal shifts, as demonstrated in Example 1(b). Vertical and horizontal shifts generate a *family of functions,* each with the same shape but at a different location in the plane.

EXAMPLE 1 Shifting the Graph of a Function

Use the graph of $f(x) = x^3$ to sketch the graph of each function.

a. $g(x) = x^3 - 1$

b. $h(x) = (x + 2)^3 + 1$

Solution

a. Relative to the graph of $f(x) = x^3$, the graph of

$$g(x) = x^3 - 1$$

is a downward shift of one unit, as shown below.

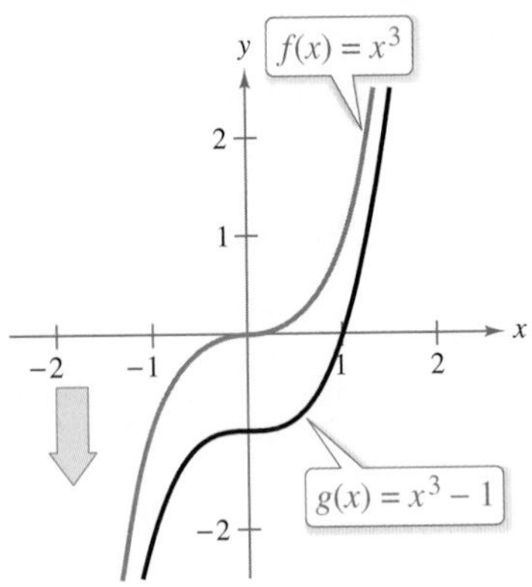

b. Relative to the graph of $f(x) = x^3$, the graph of

$$h(x) = (x + 2)^3 + 1$$

is a left shift of two units and an upward shift of one unit, as shown below.

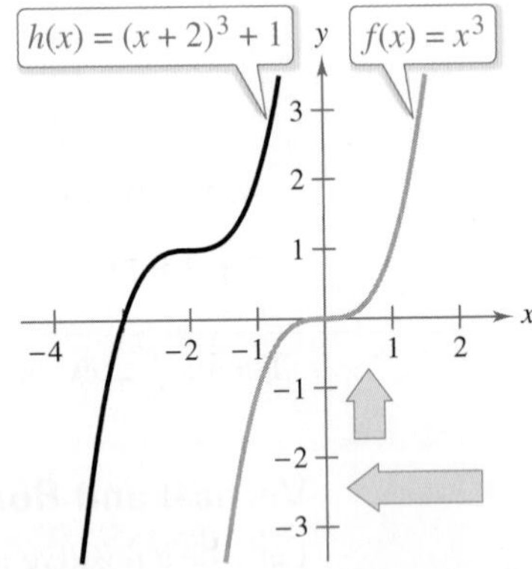

✓ **Checkpoint** Audio-video solution in English & Spanish at LarsonPrecalculus.com

Use the graph of $f(x) = x^3$ to sketch the graph of each function.

a. $h(x) = x^3 + 5$

b. $g(x) = (x - 3)^3 + 2$

In Example 1(a), note that $g(x) = f(x) - 1$ and in Example 1(b), $h(x) = f(x + 2) + 1$. In Example 1(b), you obtain the same result whether the vertical shift precedes the horizontal shift or the horizontal shift precedes the vertical shift.

Reflecting Graphs

Another common type of transformation is a **reflection.** For example, if you consider the x-axis to be a mirror, then the graph of $h(x) = -x^2$ is the mirror image (or reflection) of the graph of $f(x) = x^2$, as shown in Figure P.53.

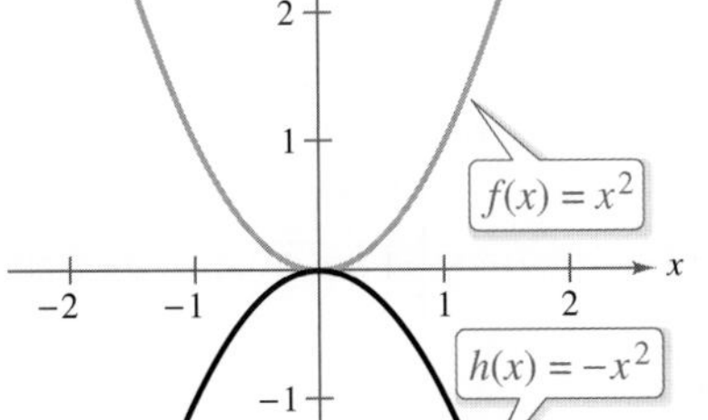

Figure P.53

Reflections in the Coordinate Axes

Reflections in the coordinate axes of the graph of $y = f(x)$ are represented as follows.

1. Reflection in the x-axis: $h(x) = -f(x)$

2. Reflection in the y-axis: $h(x) = f(-x)$

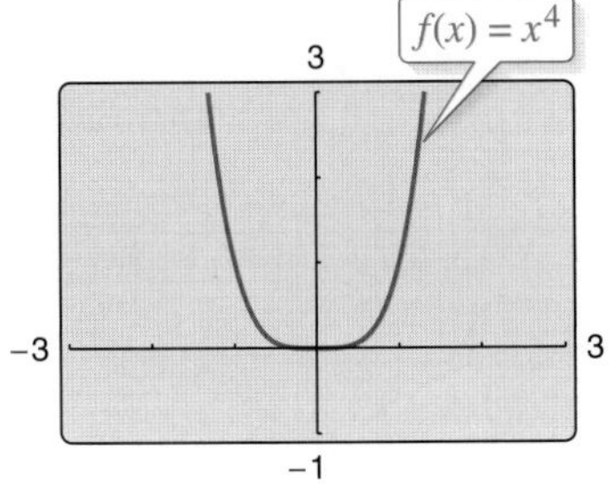

Figure P.54

EXAMPLE 2 Writing Equations from Graphs

The graph of the function

$$f(x) = x^4$$

is shown in Figure P.54. Each graph below is a transformation of the graph of f. Write an equation for the function represented by each graph.

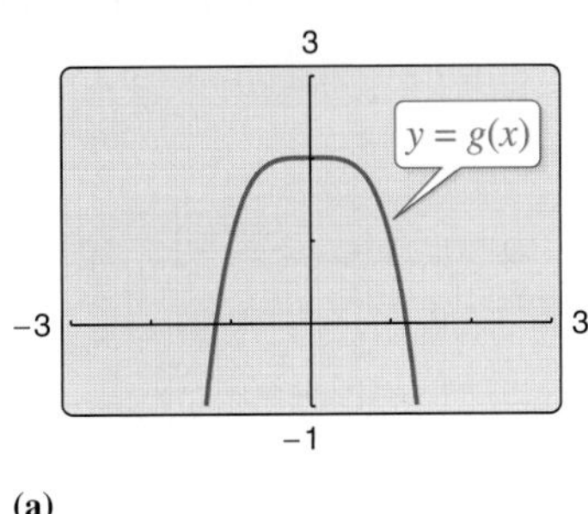

(a)

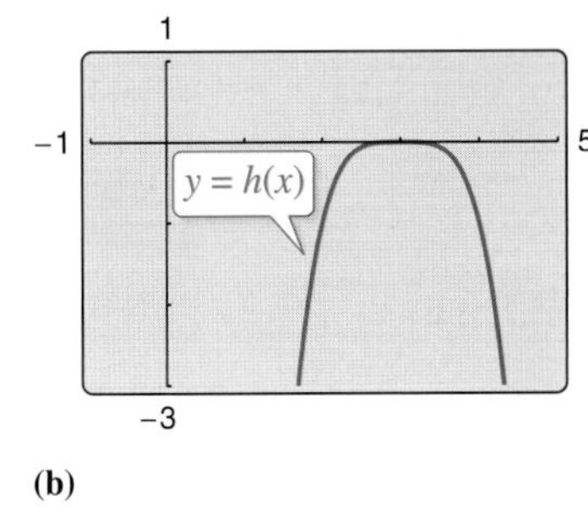

(b)

Solution

a. The graph of g is a reflection in the x-axis *followed by* an upward shift of two units of the graph of $f(x) = x^4$. So, an equation for g is

$$g(x) = -x^4 + 2.$$

b. The graph of h is a right shift of three units *followed by* a reflection in the x-axis of the graph of $f(x) = x^4$. So, an equation for h is

$$h(x) = -(x - 3)^4.$$

✓ ***Checkpoint***  *Audio-video solution in English & Spanish at LarsonPrecalculus.com*

The graph is a transformation of the graph of $f(x) = x^4$. Write an equation for the function represented by the graph.

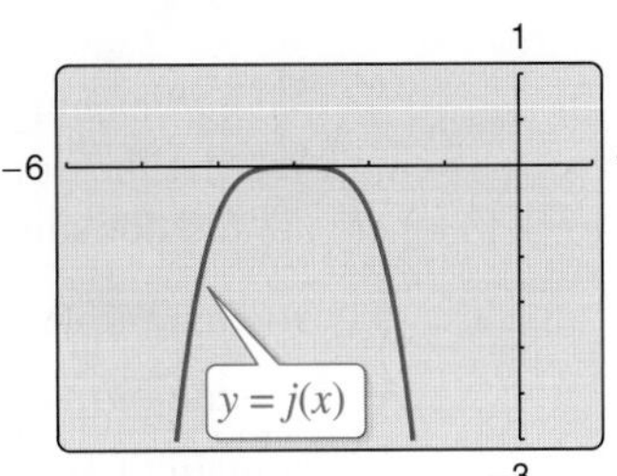

EXAMPLE 3 Reflections and Shifts

Compare the graph of each function with the graph of $f(x) = \sqrt{x}$.

a. $g(x) = -\sqrt{x}$ **b.** $h(x) = \sqrt{-x}$ **c.** $k(x) = -\sqrt{x + 2}$

Algebraic Solution

a. The graph of g is a reflection of the graph of f in the x-axis because

$$\begin{aligned} g(x) &= -\sqrt{x} \\ &= -f(x). \end{aligned}$$

b. The graph of h is a reflection of the graph of f in the y-axis because

$$\begin{aligned} h(x) &= \sqrt{-x} \\ &= f(-x). \end{aligned}$$

c. The graph of k is a left shift of two units followed by a reflection in the x-axis because

$$\begin{aligned} k(x) &= -\sqrt{x + 2} \\ &= -f(x + 2). \end{aligned}$$

Graphical Solution

a. Graph f and g on the same set of coordinate axes. The graph of g is a reflection of the graph of f in the x-axis.

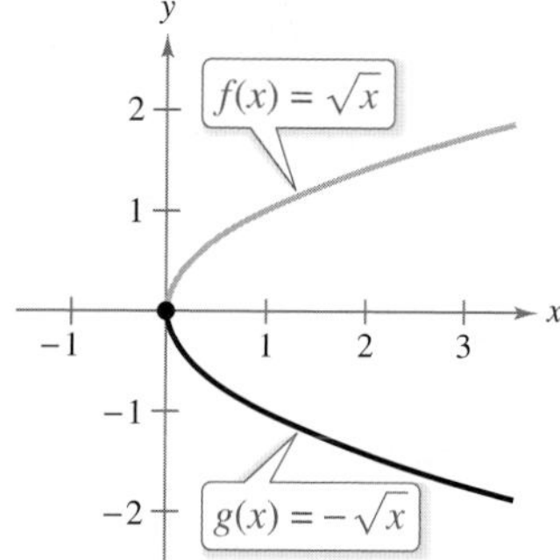

b. Graph f and h on the same set of coordinate axes. The graph of h is a reflection of the graph of f in the y-axis.

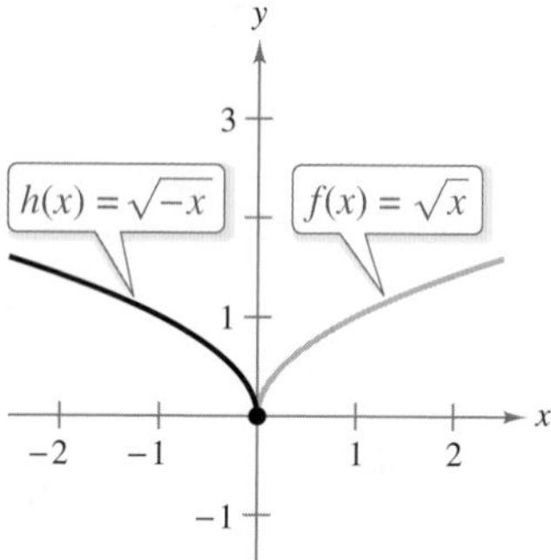

c. Graph f and k on the same set of coordinate axes. The graph of k is a left shift of two units followed by a reflection in the x-axis of the graph of f.

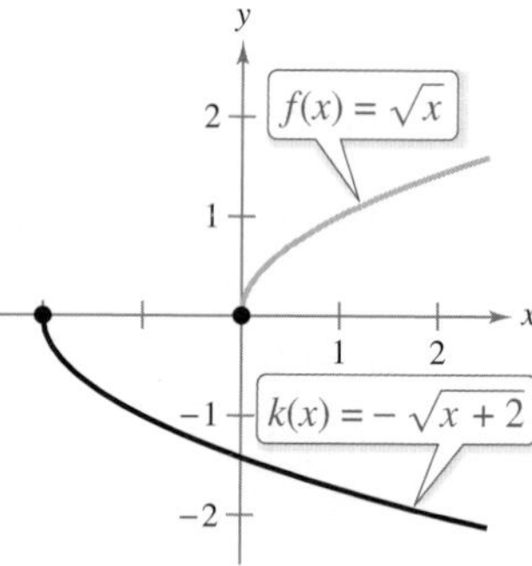

✓ **Checkpoint** Audio-video solution in English & Spanish at LarsonPrecalculus.com

Compare the graph of each function with the graph of

$f(x) = \sqrt{x - 1}$.

a. $g(x) = -\sqrt{x - 1}$ **b.** $h(x) = \sqrt{-x - 1}$

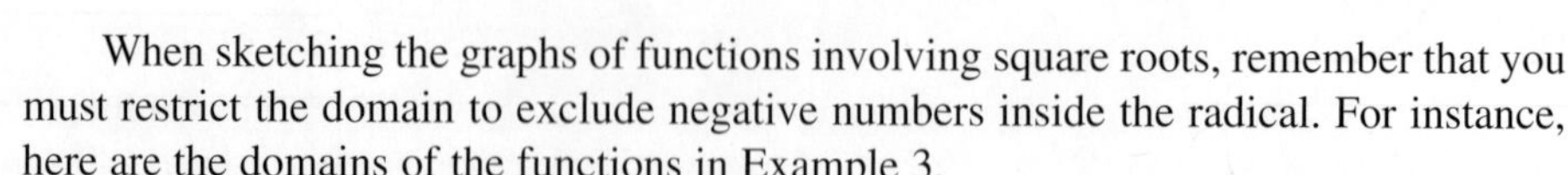

When sketching the graphs of functions involving square roots, remember that you must restrict the domain to exclude negative numbers inside the radical. For instance, here are the domains of the functions in Example 3.

Domain of $g(x) = -\sqrt{x}$: $x \ge 0$

Domain of $h(x) = \sqrt{-x}$: $x \le 0$

Domain of $k(x) = -\sqrt{x + 2}$: $x \ge -2$

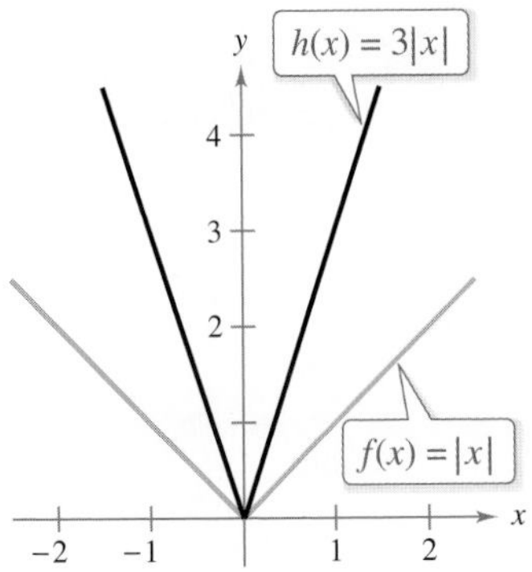

Figure P.55

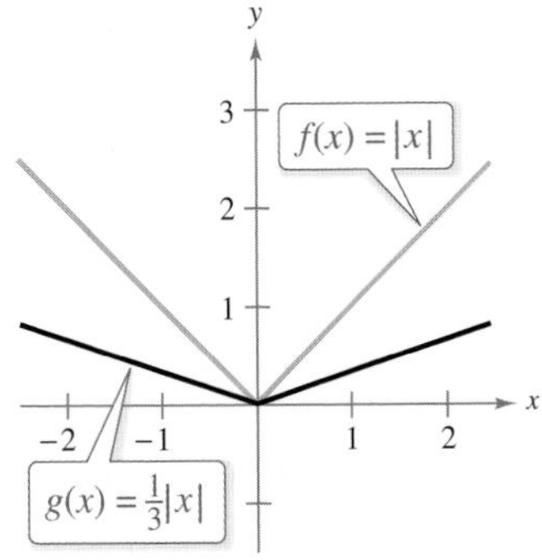

Figure P.56

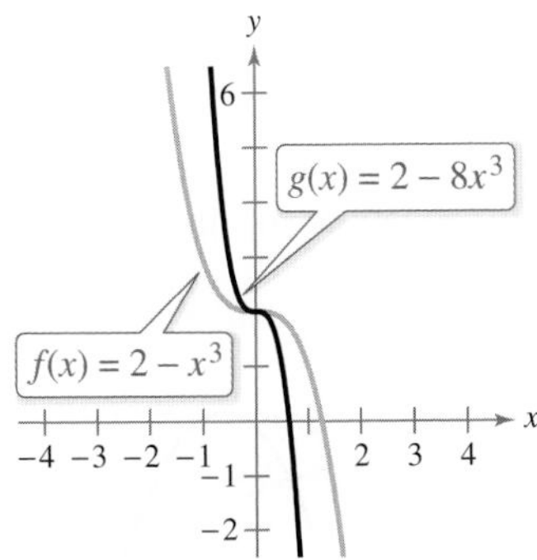

Figure P.57

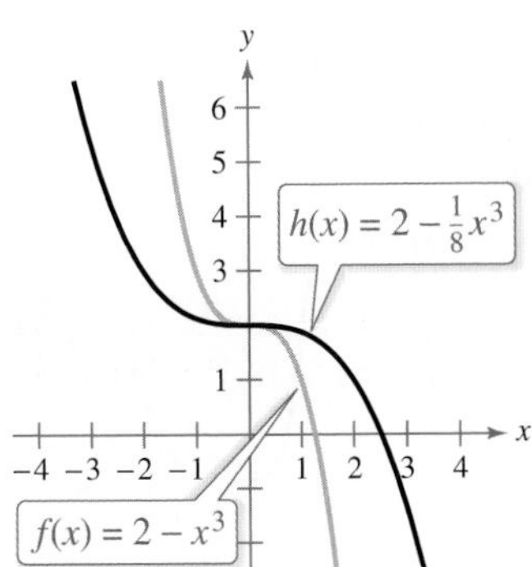

Figure P.58

Nonrigid Transformations

Horizontal shifts, vertical shifts, and reflections are **rigid transformations** because the basic shape of the graph is unchanged. These transformations change only the *position* of the graph in the coordinate plane. **Nonrigid transformations** are those that cause a *distortion*—a change in the shape of the original graph. For example, a nonrigid transformation of the graph of $y = f(x)$ is represented by $g(x) = cf(x)$, where the transformation is a **vertical stretch** when $c > 1$ and a **vertical shrink** when $0 < c < 1$. Another nonrigid transformation of the graph of $y = f(x)$ is represented by $h(x) = f(cx)$, where the transformation is a **horizontal shrink** when $c > 1$ and a **horizontal stretch** when $0 < c < 1$.

EXAMPLE 4 Nonrigid Transformations

Compare the graph of each function with the graph of $f(x) = |x|$.

a. $h(x) = 3|x|$ **b.** $g(x) = \frac{1}{3}|x|$

Solution

a. Relative to the graph of $f(x) = |x|$, the graph of $h(x) = 3|x| = 3f(x)$ is a vertical stretch (each y-value is multiplied by 3). (See Figure P.55.)

b. Similarly, the graph of $g(x) = \frac{1}{3}|x| = \frac{1}{3}f(x)$ is a vertical shrink $\left(\text{each } y\text{-value is multiplied by } \frac{1}{3}\right)$ of the graph of f. (See Figure P.56.)

Compare the graph of each function with the graph of $f(x) = x^2$.

a. $g(x) = 4x^2$ **b.** $h(x) = \frac{1}{4}x^2$

EXAMPLE 5 Nonrigid Transformations

See LarsonPrecalculus.com for an interactive version of this type of example.

Compare the graph of each function with the graph of $f(x) = 2 - x^3$.

a. $g(x) = f(2x)$ **b.** $h(x) = f\left(\frac{1}{2}x\right)$

Solution

a. Relative to the graph of $f(x) = 2 - x^3$, the graph of $g(x) = f(2x) = 2 - (2x)^3 = 2 - 8x^3$ is a horizontal shrink $(c > 1)$. (See Figure P.57.)

b. Similarly, the graph of $h(x) = f\left(\frac{1}{2}x\right) = 2 - \left(\frac{1}{2}x\right)^3 = 2 - \frac{1}{8}x^3$ is a horizontal stretch $(0 < c < 1)$ of the graph of f. (See Figure P.58.)

Checkpoint Audio-video solution in English & Spanish at LarsonPrecalculus.com

Compare the graph of each function with the graph of $f(x) = x^2 + 3$.

a. $g(x) = f(2x)$ **b.** $h(x) = f\left(\frac{1}{2}x\right)$

Summarize (Section P.8)

1. Explain how to shift the graph of a function vertically and horizontally *(page 85)*. For an example of shifting the graph of a function, see Example 1.
2. Explain how to reflect the graph of a function in the x-axis and in the y-axis *(page 87)*. For examples of reflecting graphs of functions, see Examples 2 and 3.
3. Describe nonrigid transformations of the graph of a function *(page 89)*. For examples of nonrigid transformations, see Examples 4 and 5.

P.8 Exercises

See CalcChat.com for tutorial help and worked-out solutions to odd-numbered exercises.

Vocabulary

In Exercises 1–3, fill in the blanks.

1. Horizontal shifts, vertical shifts, and reflections are ________ transformations.
2. A reflection in the x-axis of the graph of $y = f(x)$ is represented by $h(x) =$ ________, while a reflection in the y-axis of the graph of $y = f(x)$ is represented by $h(x) =$ ________.
3. A nonrigid transformation of the graph of $y = f(x)$ represented by $g(x) = cf(x)$ is a ________ ________ when $c > 1$ and a ________ ________ when $0 < c < 1$.

4. Match each function h with the transformation it represents, where $c > 0$.

   (a) $h(x) = f(x) + c$ (i) A horizontal shift of f, c units to the right

   (b) $h(x) = f(x) - c$ (ii) A vertical shift of f, c units down

   (c) $h(x) = f(x + c)$ (iii) A horizontal shift of f, c units to the left

   (d) $h(x) = f(x - c)$ (iv) A vertical shift of f, c units up

Skills and Applications

5. **Shifting the Graph of a Function** For each function, sketch the graphs of the function when $c = -2$, -1, 1, and 2 on the same set of coordinate axes.

   (a) $f(x) = |x| + c$ (b) $f(x) = |x - c|$

6. **Shifting the Graph of a Function** For each function, sketch the graphs of the function when $c = -3$, -2, 2, and 3 on the same set of coordinate axes.

   (a) $f(x) = \sqrt{x} + c$ (b) $f(x) = \sqrt{x - c}$

7. **Shifting the Graph of a Function** For each function, sketch the graphs of the function when $c = -4$, -1, 2, and 5 on the same set of coordinate axes.

   (a) $f(x) = [\![x]\!] + c$ (b) $f(x) = [\![x + c]\!]$

8. **Shifting the Graph of a Function** For each function, sketch the graphs of the function when $c = -3$, -2, 1, and 2 on the same set of coordinate axes.

   (a) $f(x) = \begin{cases} x^2 + c, & x < 0 \\ -x^2 + c, & x \geq 0 \end{cases}$

   (b) $f(x) = \begin{cases} (x + c)^2, & x < 0 \\ -(x + c)^2, & x \geq 0 \end{cases}$

Sketching Transformations **In Exercises 9 and 10, use the graph of f to sketch each graph. To print an enlarged copy of the graph, go to *MathGraphs.com*.**

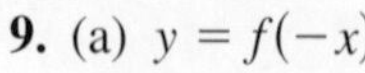

9. (a) $y = f(-x)$

   (b) $y = f(x) + 4$

   (c) $y = 2f(x)$

   (d) $y = -f(x - 4)$

   (e) $y = f(x) - 3$

   (f) $y = -f(x) - 1$

   (g) $y = f(2x)$

10. (a) $y = f(x - 5)$

    (b) $y = -f(x) + 3$

    (c) $y = \frac{1}{3}f(x)$

    (d) $y = -f(x + 1)$

    (e) $y = f(-x)$

    (f) $y = f(x) - 10$

    (g) $y = f\left(\frac{1}{3}x\right)$

11. **Writing Equations from Graphs** Use the graph of $f(x) = x^2$ to write an equation for the function represented by each graph.

    (a)

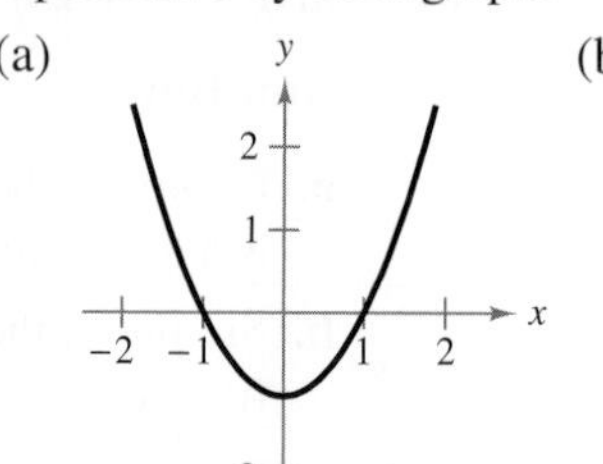

    (b)

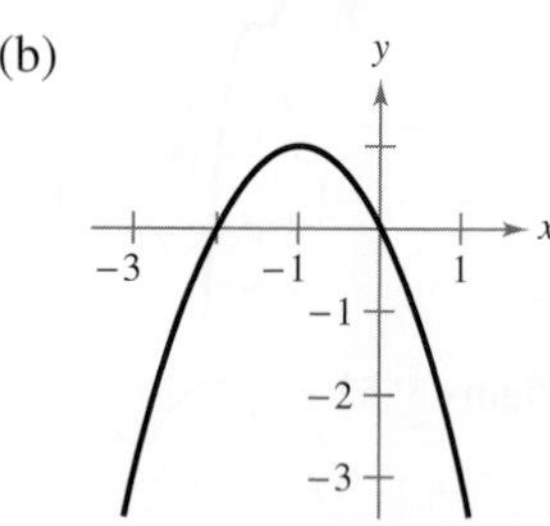

12. **Writing Equations from Graphs** Use the graph of $f(x) = x^3$ to write an equation for the function represented by each graph.

    (a)

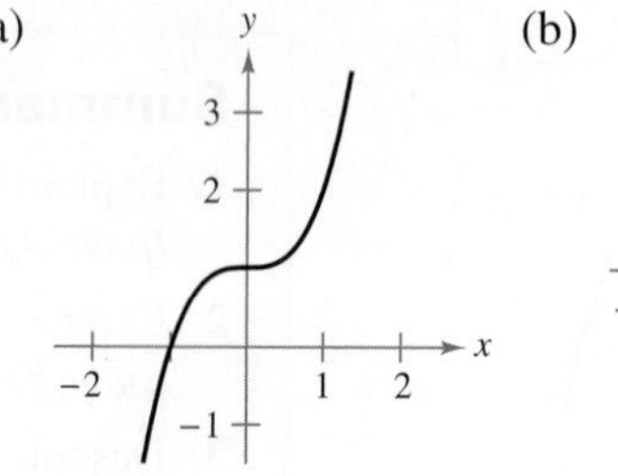

    (b)

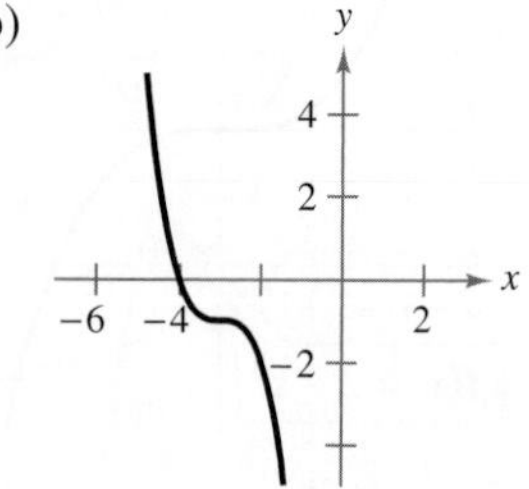

13. Writing Equations from Graphs Use the graph of $f(x) = |x|$ to write an equation for the function represented by each graph.

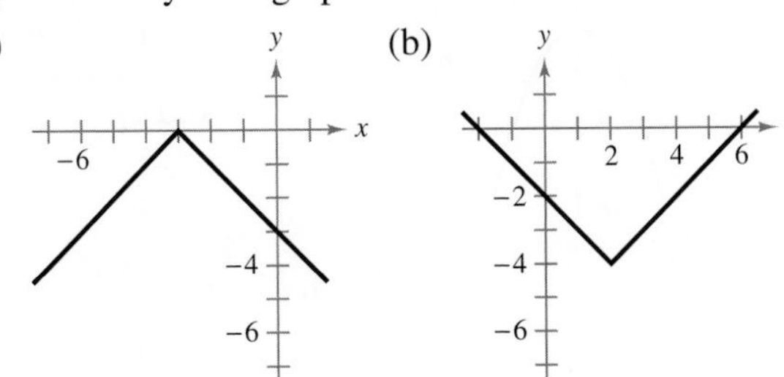

14. Writing Equations from Graphs Use the graph of $f(x) = \sqrt{x}$ to write an equation for the function represented by each graph.

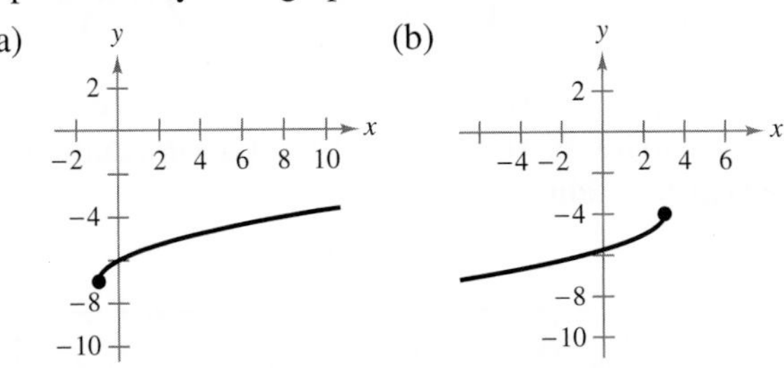

Writing Equations from Graphs In Exercises 15–20, identify the parent function and the transformation represented by the graph. Write an equation for the function represented by the graph.

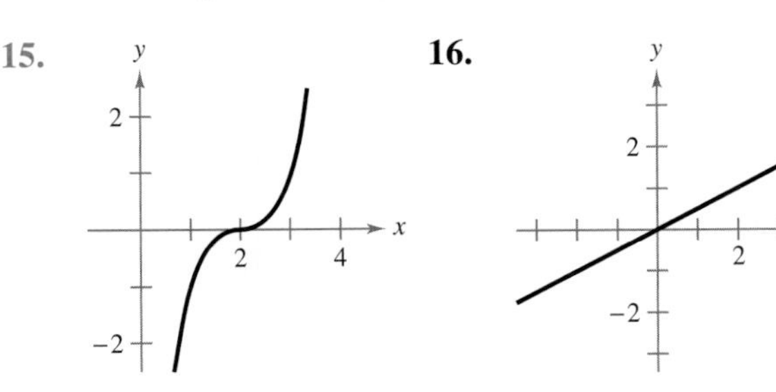

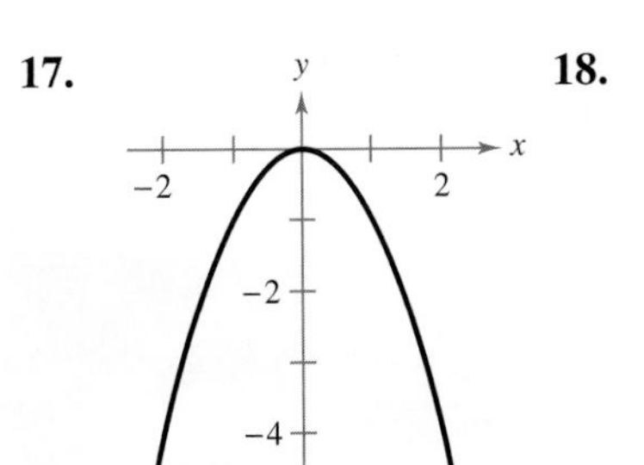

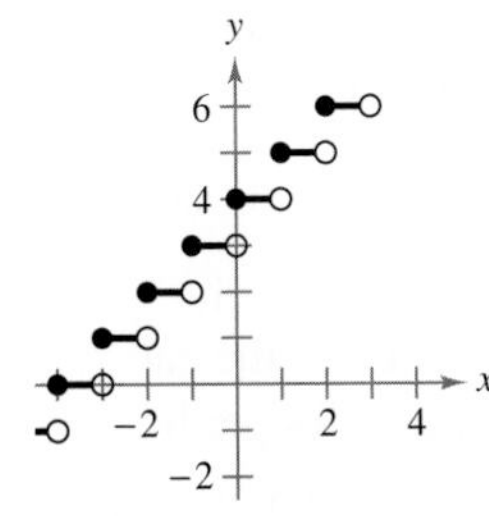

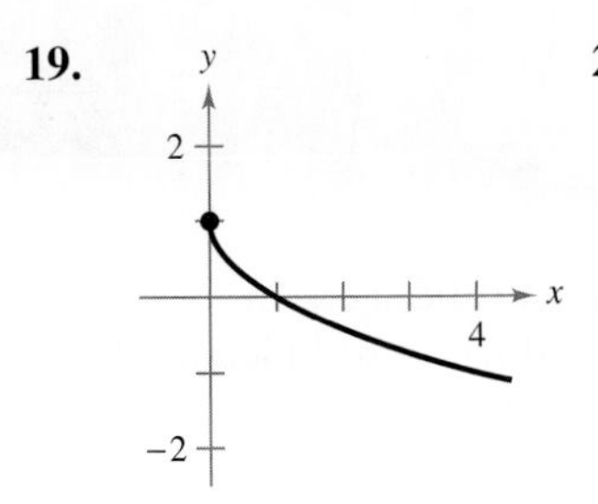

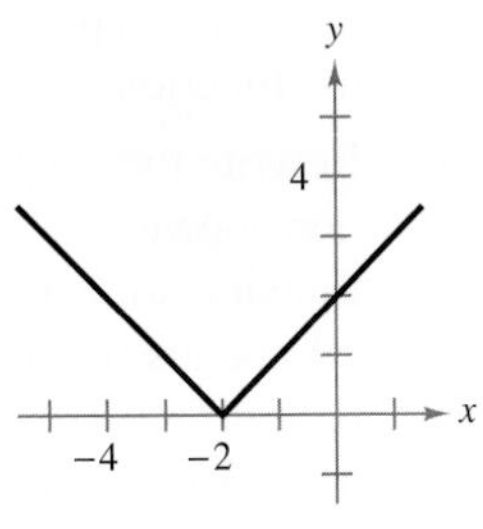

Describing Transformations In Exercises 21–38, g is related to one of the parent functions described in Section P.7. (a) Identify the parent function f. (b) Describe the sequence of transformations from f to g. (c) Sketch the graph of g. (d) Use function notation to write g in terms of f.

21. $g(x) = x^2 + 6$ **22.** $g(x) = x^2 - 2$

23. $g(x) = -(x - 2)^3$ **24.** $g(x) = -(x + 1)^3$

25. $g(x) = -3 - (x + 1)^2$

26. $g(x) = 4 - (x - 2)^2$

27. $g(x) = |x - 1| + 2$ **28.** $g(x) = |x + 3| - 2$

29. $g(x) = 2\sqrt{x}$ **30.** $g(x) = \frac{1}{2}\sqrt{x}$

31. $g(x) = 2[\![x]\!] - 1$ **32.** $g(x) = -[\![x]\!] + 1$

33. $g(x) = |2x|$ **34.** $g(x) = \left|\frac{1}{2}x\right|$

35. $g(x) = -2x^2 + 1$ **36.** $g(x) = \frac{1}{2}x^2 - 2$

37. $g(x) = 3|x - 1| + 2$

38. $g(x) = -2|x + 1| - 3$

Writing an Equation from a Description In Exercises 39–46, write an equation for the function whose graph is described.

39. The shape of $f(x) = x^2$, but shifted three units to the right and seven units down

40. The shape of $f(x) = x^2$, but shifted two units to the left, nine units up, and then reflected in the x-axis

41. The shape of $f(x) = x^3$, but shifted 13 units to the right

42. The shape of $f(x) = x^3$, but shifted six units to the left, six units down, and then reflected in the y-axis

43. The shape of $f(x) = |x|$, but shifted 12 units up and then reflected in the x-axis

44. The shape of $f(x) = |x|$, but shifted four units to the left and eight units down

45. The shape of $f(x) = \sqrt{x}$, but shifted six units to the left and then reflected in both the x-axis and the y-axis

46. The shape of $f(x) = \sqrt{x}$, but shifted nine units down and then reflected in both the x-axis and the y-axis

47. Writing Equations from Graphs Use the graph of $f(x) = x^2$ to write an equation for the function represented by each graph.

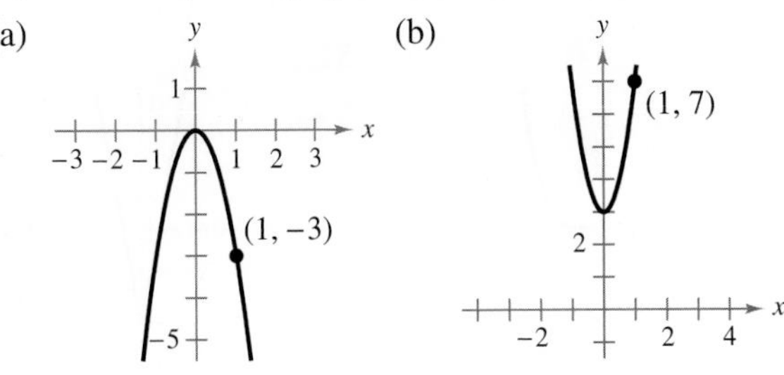

48. Writing Equations from Graphs Use the graph of

$f(x) = x^3$

to write an equation for the function represented by each graph.

(a)

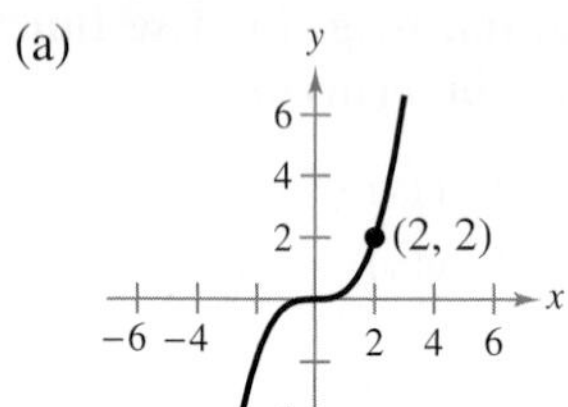

(b)

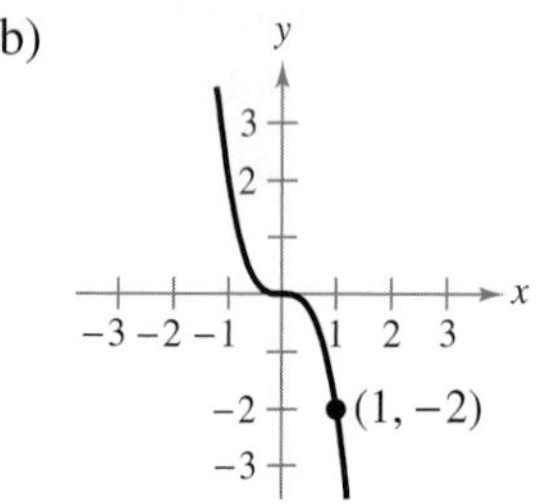

49. Writing Equations from Graphs Use the graph of

$f(x) = |x|$

to write an equation for the function represented by each graph.

(a)

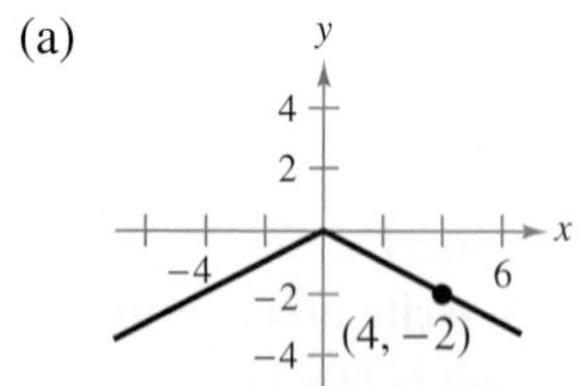

(b)

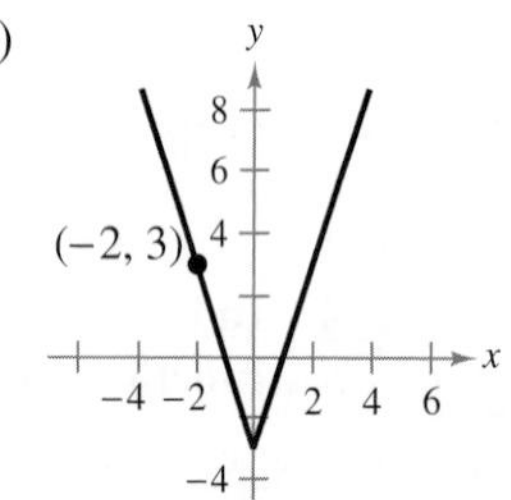

50. Writing Equations from Graphs Use the graph of

$f(x) = \sqrt{x}$

to write an equation for the function represented by each graph.

(a)

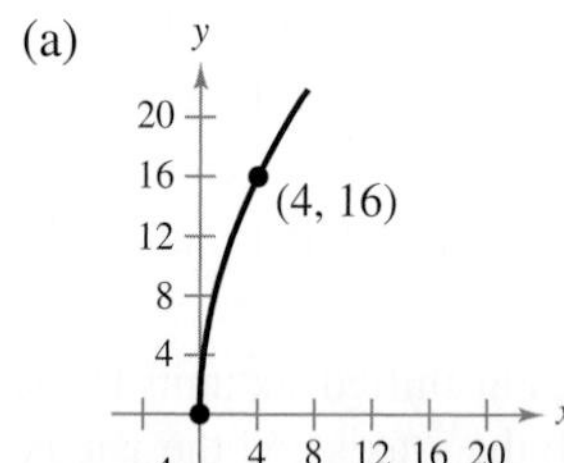

(b)

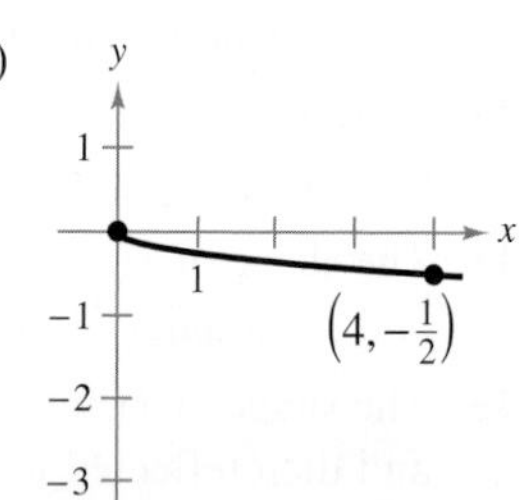

Writing Equations from Graphs **In Exercises 51–56, identify the parent function and the transformation represented by the graph. Write an equation for the function represented by the graph. Then use a graphing utility to verify your answer.**

51.

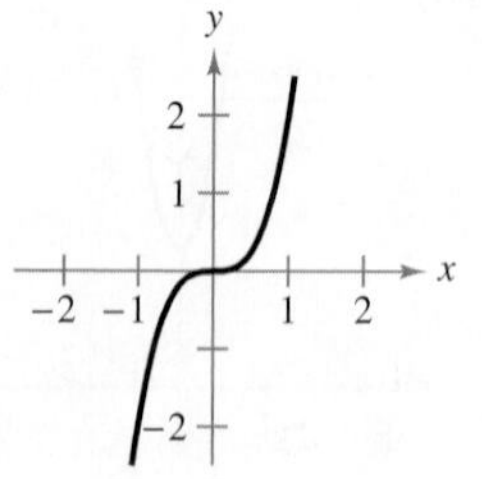

52.

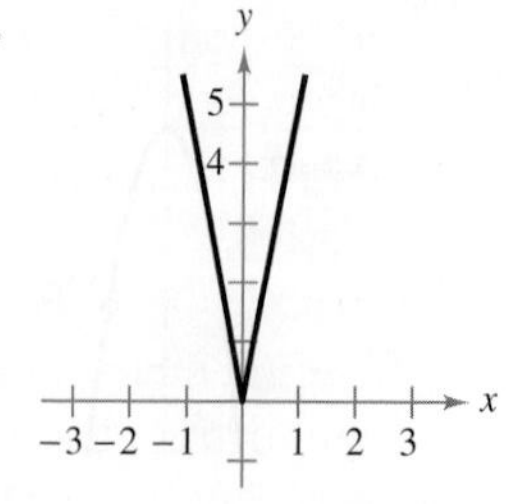

53.

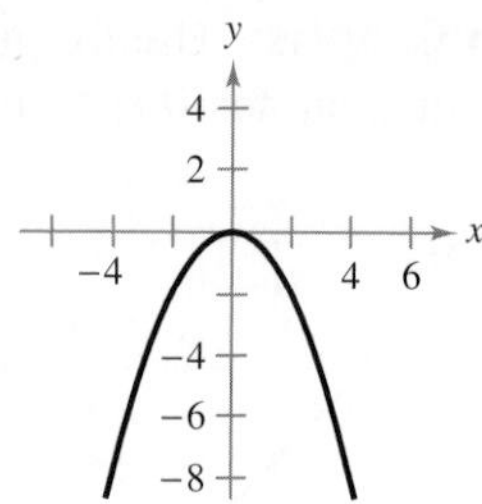

54.

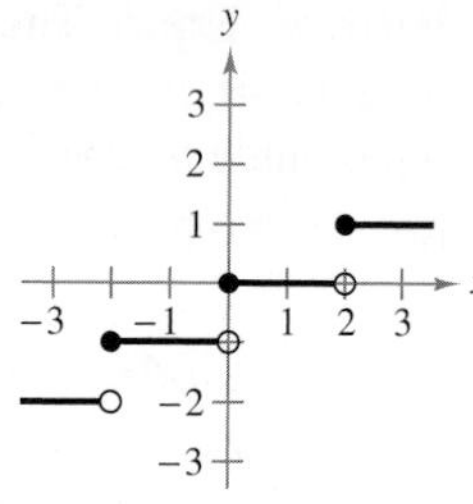

55.

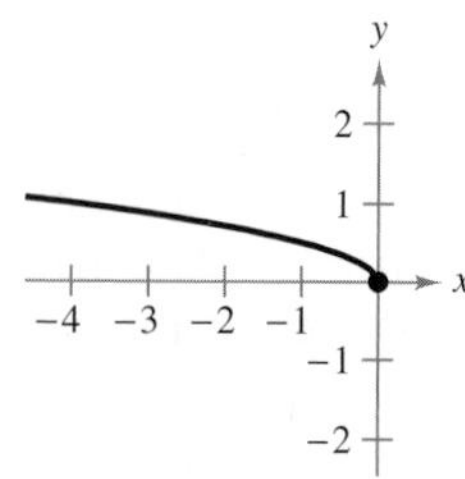

56.

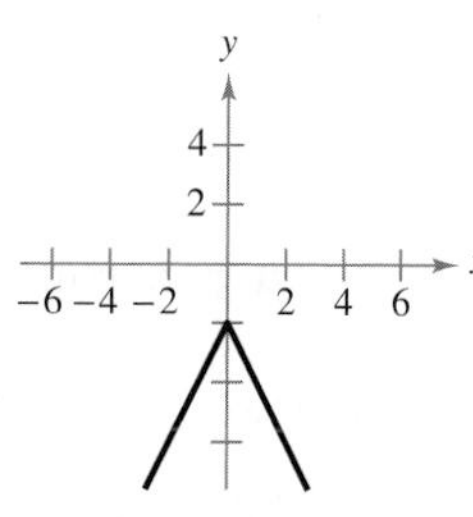

Writing Equations from Graphs **In Exercises 57–60, write an equation for the transformation of the parent function.**

57.

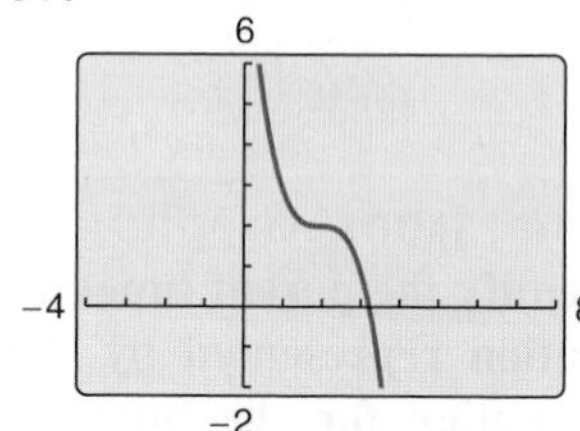

58.

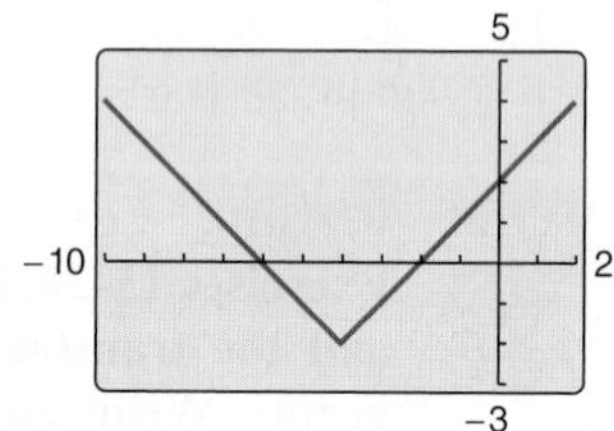

59.

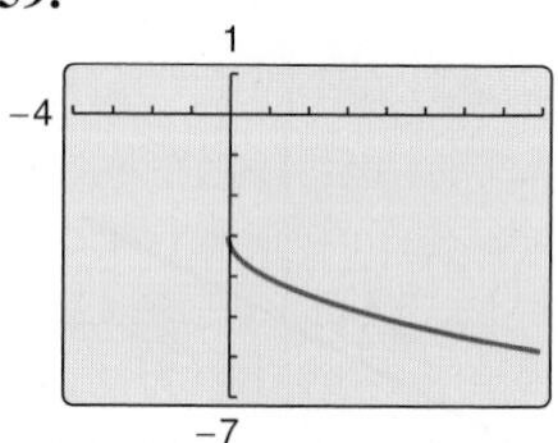

60.

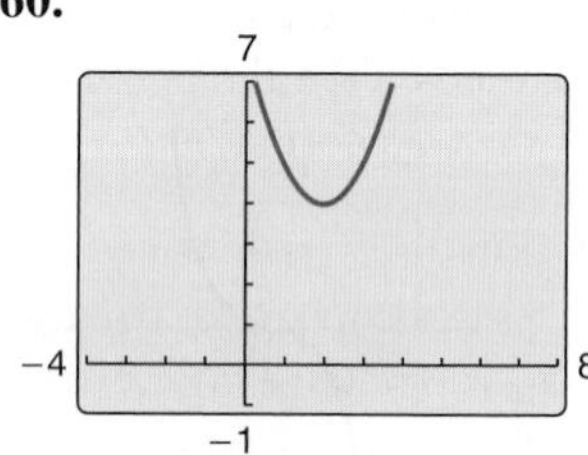

61. Automobile Aerodynamics The horsepower H required to overcome wind drag on a particular automobile is given by

$H(x) = 0.00004636x^3$

where x is the speed of the car (in miles per hour).

(a) Use a graphing utility to graph the function.

(b) Rewrite the horsepower function so that x represents the speed in kilometers per hour. [Find $H(x/1.6)$.] Identify the type of transformation applied to the graph of the horsepower function.

Robert Young/Shutterstock.com

62. Households The number N (in millions) of households in the United States from 2000 through 2014 can be approximated by

$N(x) = -0.023(x - 33.12)^2 + 131, \quad 0 \le t \le 14$

where t represents the year, with $t = 0$ corresponding to 2000. *(Source: U.S. Census Bureau)*

(a) Describe the transformation of the parent function $f(x) = x^2$. Then use a graphing utility to graph the function over the specified domain.

(b) Find the average rate of change of the function from 2000 to 2014. Interpret your answer in the context of the problem.

(c) Use the model to predict the number of households in the United States in 2022. Does your answer seem reasonable? Explain.

Exploration

True or False? **In Exercises 63–66, determine whether the statement is true or false. Justify your answer.**

63. The graph of $y = f(-x)$ is a reflection of the graph of $y = f(x)$ in the x-axis.

64. The graph of $y = -f(x)$ is a reflection of the graph of $y = f(x)$ in the y-axis.

65. The graphs of $f(x) = |x| + 6$ and $f(x) = |-x| + 6$ are identical.

66. If the graph of the parent function $f(x) = x^2$ is shifted six units to the right, three units up, and reflected in the x-axis, then the point $(-2, 19)$ will lie on the graph of the transformation.

67. Finding Points on a Graph The graph of $y = f(x)$ passes through the points $(0, 1)$, $(1, 2)$, and $(2, 3)$. Find the corresponding points on the graph of $y = f(x + 2) - 1$.

68. Think About It Two methods of graphing a function are plotting points and translating a parent function as shown in this section. Which method of graphing do you prefer to use for each function? Explain.

(a) $f(x) = 3x^2 - 4x + 1$ (b) $f(x) = 2(x - 1)^2 - 6$

69. Error Analysis Describe the error.

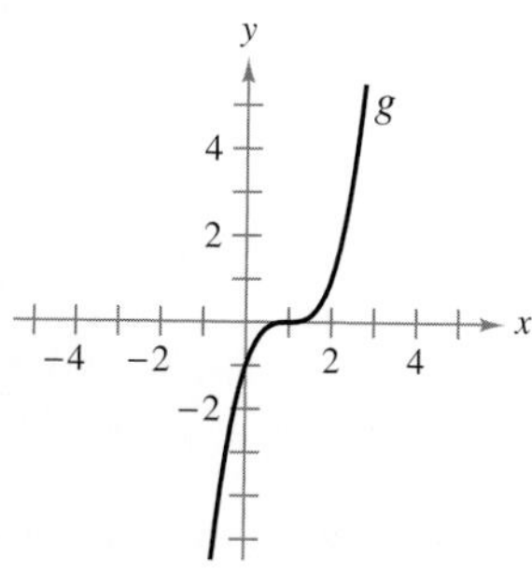

The graph of g is a right shift of one unit of the graph of $f(x) = x^3$. So, an equation for g is $g(x) = (x + 1)^3$.

70. HOW DO YOU SEE IT? Use the graph of $y = f(x)$ to find the open intervals on which the graph of each transformation is increasing and decreasing. If not possible, state the reason.

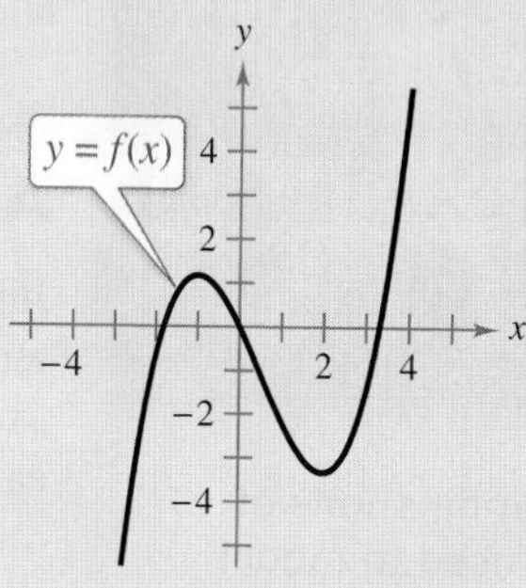

(a) $y = f(-x)$ (b) $y = -f(x)$ (c) $y = \frac{1}{2}f(x)$
(d) $y = -f(x - 1)$ (e) $y = f(x - 2) + 1$

71. Describing Profits Management originally predicted that the profits from the sales of a new product could be approximated by the graph of the function f shown. The actual profits are represented by the graph of the function g along with a verbal description. Use the concepts of transformations of graphs to write g in terms of f.

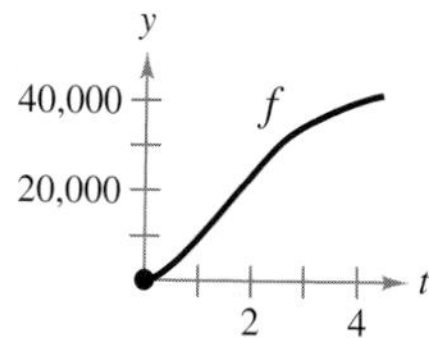

(a) The profits were only three-fourths as large as expected.

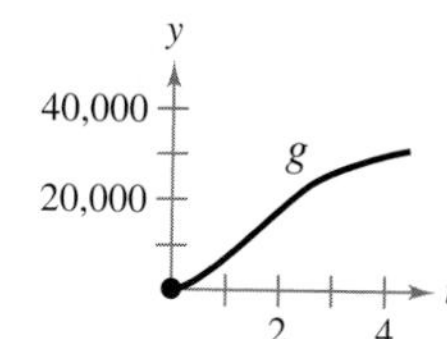

(b) The profits were consistently \$10,000 greater than predicted.

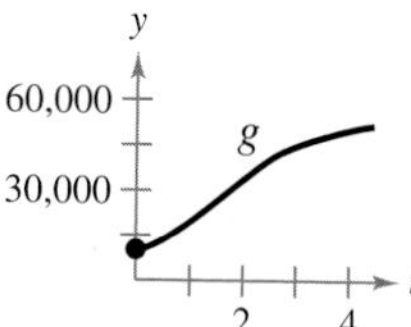

(c) There was a two-year delay in the introduction of the product. After sales began, profits grew as expected.

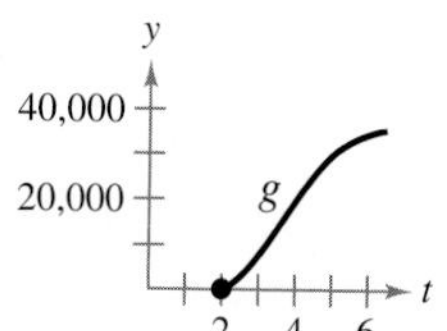

72. Reversing the Order of Transformations Reverse the order of transformations in Example 2(a). Do you obtain the same graph? Do the same for Example 2(b). Do you obtain the same graph? Explain.

P.9 Combinations of Functions: Composite Functions

Arithmetic combinations of functions are used to model and solve real-life problems. For example, in Exercise 60 on page 100, you will use arithmetic combinations of functions to analyze numbers of pets in the United States.

- **Add, subtract, multiply, and divide functions.**
- **Find the composition of one function with another function.**
- **Use combinations and compositions of functions to model and solve real-life problems.**

Arithmetic Combinations of Functions

Just as two real numbers can be combined by the operations of addition, subtraction, multiplication, and division to form other real numbers, two *functions* can be combined to create new functions. For example, the functions $f(x) = 2x - 3$ and $g(x) = x^2 - 1$ can be combined to form the sum, difference, product, and quotient of f and g.

$$f(x) + g(x) = (2x - 3) + (x^2 - 1) = x^2 + 2x - 4 \qquad \text{Sum}$$

$$f(x) - g(x) = (2x - 3) - (x^2 - 1) = -x^2 + 2x - 2 \qquad \text{Difference}$$

$$f(x)g(x) = (2x - 3)(x^2 - 1) = 2x^3 - 3x^2 - 2x + 3 \qquad \text{Product}$$

$$\frac{f(x)}{g(x)} = \frac{2x - 3}{x^2 - 1}, \quad x \neq \pm 1 \qquad \text{Quotient}$$

The domain of an **arithmetic combination** of functions f and g consists of all real numbers that are common to the domains of f and g. In the case of the quotient $f(x)/g(x)$, there is the further restriction that $g(x) \neq 0$.

Sum, Difference, Product, and Quotient of Functions

Let f and g be two functions with overlapping domains. Then, for all x common to both domains, the *sum, difference, product,* and *quotient* of f and g are defined as follows.

1. Sum: $(f + g)(x) = f(x) + g(x)$
2. Difference: $(f - g)(x) = f(x) - g(x)$
3. Product: $(fg)(x) = f(x) \cdot g(x)$
4. Quotient: $\left(\frac{f}{g}\right)(x) = \frac{f(x)}{g(x)}, \quad g(x) \neq 0$

EXAMPLE 1 Finding the Sum of Two Functions

Given $f(x) = 2x + 1$ and $g(x) = x^2 + 2x - 1$, find $(f + g)(x)$. Then evaluate the sum when $x = 3$.

Solution The sum of f and g is

$$(f + g)(x) = f(x) + g(x) = (2x + 1) + (x^2 + 2x - 1) = x^2 + 4x.$$

When $x = 3$, the value of this sum is

$$(f + g)(3) = 3^2 + 4(3) = 21.$$

✓ ***Checkpoint*** *Audio-video solution in English & Spanish at LarsonPrecalculus.com*

Given $f(x) = x^2$ and $g(x) = 1 - x$, find $(f + g)(x)$. Then evaluate the sum when $x = 2$.

Rita Kochmarjova/Shutterstock.com

EXAMPLE 2 Finding the Difference of Two Functions

Given $f(x) = 2x + 1$ and $g(x) = x^2 + 2x - 1$, find $(f - g)(x)$. Then evaluate the difference when $x = 2$.

Solution The difference of f and g is

$$(f - g)(x) = f(x) - g(x) = (2x + 1) - (x^2 + 2x - 1) = -x^2 + 2.$$

When $x = 2$, the value of this difference is

$$(f - g)(2) = -(2)^2 + 2 = -2.$$

✓ ***Checkpoint*** Audio-video solution in English & Spanish at LarsonPrecalculus.com

Given $f(x) = x^2$ and $g(x) = 1 - x$, find $(f - g)(x)$. Then evaluate the difference when $x = 3$.

EXAMPLE 3 Finding the Product of Two Functions

Given $f(x) = x^2$ and $g(x) = x - 3$, find $(fg)(x)$. Then evaluate the product when $x = 4$.

Solution The product of f and g is

$$(fg)(x) = f(x)g(x) = (x^2)(x - 3) = x^3 - 3x^2.$$

When $x = 4$, the value of this product is

$$(fg)(4) = 4^3 - 3(4)^2 = 16.$$

✓ ***Checkpoint*** Audio-video solution in English & Spanish at LarsonPrecalculus.com

Given $f(x) = x^2$ and $g(x) = 1 - x$, find $(fg)(x)$. Then evaluate the product when $x = 3$.

In Examples 1–3, both f and g have domains that consist of all real numbers. So, the domains of $f + g$, $f - g$, and fg are also the set of all real numbers. Remember to consider any restrictions on the domains of f and g when forming the sum, difference, product, or quotient of f and g.

EXAMPLE 4 Finding the Quotients of Two Functions

Find $(f/g)(x)$ and $(g/f)(x)$ for the functions $f(x) = \sqrt{x}$ and $g(x) = \sqrt{4 - x^2}$. Then find the domains of f/g and g/f.

Solution The quotient of f and g is

$$\left(\frac{f}{g}\right)(x) = \frac{f(x)}{g(x)} = \frac{\sqrt{x}}{\sqrt{4 - x^2}}$$

and the quotient of g and f is

$$\left(\frac{g}{f}\right)(x) = \frac{g(x)}{f(x)} = \frac{\sqrt{4 - x^2}}{\sqrt{x}}.$$

The domain of f is $[0, \infty)$ and the domain of g is $[-2, 2]$. The intersection of these domains is $[0, 2]$. So, the domains of f/g and g/f are as follows.

Domain of f/g: $[0, 2)$ Domain of g/f: $(0, 2]$

REMARK Note that the domain of f/g includes $x = 0$, but not $x = 2$, because $x = 2$ yields a zero in the denominator, whereas the domain of g/f includes $x = 2$, but not $x = 0$, because $x = 0$ yields a zero in the denominator.

✓ ***Checkpoint*** Audio-video solution in English & Spanish at LarsonPrecalculus.com

Find $(f/g)(x)$ and $(g/f)(x)$ for the functions $f(x) = \sqrt{x - 3}$ and $g(x) = \sqrt{16 - x^2}$. Then find the domains of f/g and g/f.

Composition of Functions

Another way of combining two functions is to form the **composition** of one with the other. For example, if $f(x) = x^2$ and $g(x) = x + 1$, then the composition of f with g is

$$f(g(x)) = f(x + 1) = (x + 1)^2.$$

This composition is denoted as $f \circ g$ and reads as "f composed with g."

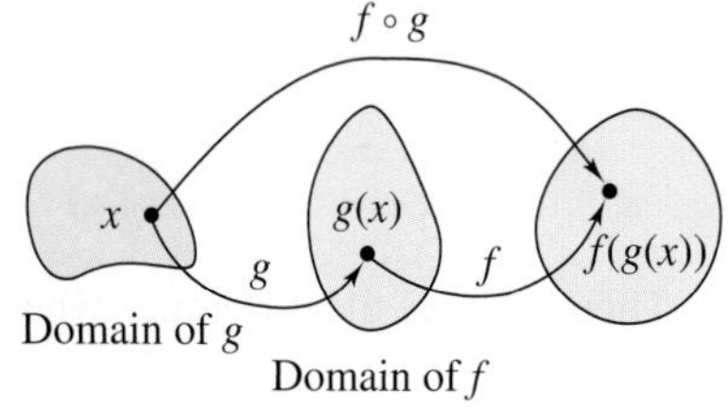

Figure P.59

> **Definition of Composition of Two Functions**
>
> The **composition** of the function f with the function g is
>
> $$(f \circ g)(x) = f(g(x)).$$
>
> The domain of $f \circ g$ is the set of all x in the domain of g such that $g(x)$ is in the domain of f. (See Figure P.59.)

EXAMPLE 5 Compositions of Functions

See LarsonPrecalculus.com for an interactive version of this type of example.

Given $f(x) = x + 2$ and $g(x) = 4 - x^2$, find the following.

a. $(f \circ g)(x)$ **b.** $(g \circ f)(x)$ **c.** $(g \circ f)(-2)$

Solution

a. The composition of f with g is as shown.

$(f \circ g)(x) = f(g(x))$ Definition of $f \circ g$

$= f(4 - x^2)$ Definition of $g(x)$

$= (4 - x^2) + 2$ Definition of $f(x)$

$= -x^2 + 6$ Simplify.

b. The composition of g with f is as shown.

$(g \circ f)(x) = g(f(x))$ Definition of $g \circ f$

$= g(x + 2)$ Definition of $f(x)$

$= 4 - (x + 2)^2$ Definition of $g(x)$

$= 4 - (x^2 + 4x + 4)$ Expand.

$= -x^2 - 4x$ Simplify.

Note that, in this case, $(f \circ g)(x) \neq (g \circ f)(x)$.

c. Evaluate the result of part (b) when $x = -2$.

$(g \circ f)(-2) = -(-2)^2 - 4(-2)$ Substitute.

$= -4 + 8$ Simplify.

$= 4$ Simplify.

REMARK The tables of values below help illustrate the composition $(f \circ g)(x)$ in Example 5(a).

x	0	1	2	3
$g(x)$	4	3	0	−5

$g(x)$	4	3	0	−5
$f(g(x))$	6	5	2	−3

x	0	1	2	3
$f(g(x))$	6	5	2	−3

Note that the first two tables are combined (or "composed") to produce the values in the third table.

Checkpoint *Audio-video solution in English & Spanish at LarsonPrecalculus.com*

Given $f(x) = 2x + 5$ and $g(x) = 4x^2 + 1$, find the following.

a. $(f \circ g)(x)$ **b.** $(g \circ f)(x)$ **c.** $(f \circ g)\left(-\frac{1}{2}\right)$

EXAMPLE 6 Finding the Domain of a Composite Function

Find the domain of $f \circ g$ for the functions

$$f(x) = x^2 - 9 \quad \text{and} \quad g(x) = \sqrt{9 - x^2}.$$

Algebraic Solution

Find the composition of the functions.

$$\begin{aligned}(f \circ g)(x) &= f(g(x)) \\ &= f\left(\sqrt{9 - x^2}\right) \\ &= \left(\sqrt{9 - x^2}\right)^2 - 9 \\ &= 9 - x^2 - 9 \\ &= -x^2\end{aligned}$$

The domain of $f \circ g$ is restricted to the x-values in the domain of g for which $g(x)$ is in the domain of f. The domain of $f(x) = x^2 - 9$ is the set of all real numbers, which includes all real values of g. So, the domain of $f \circ g$ is the entire domain of $g(x) = \sqrt{9 - x^2}$, which is $[-3, 3]$.

Graphical Solution

Use a graphing utility to graph $f \circ g$.

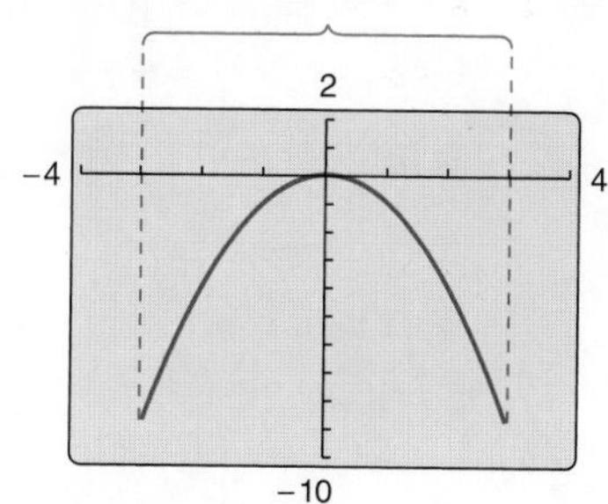

From the graph, you can determine that the domain of $f \circ g$ is $[-3, 3]$.

✓ Checkpoint Audio-video solution in English & Spanish at LarsonPrecalculus.com

Find the domain of $f \circ g$ for the functions $f(x) = \sqrt{x}$ and $g(x) = x^2 + 4$. ■

In Examples 5 and 6, you formed the composition of two given functions. In calculus, it is also important to be able to identify two functions that make up a given composite function. For example, the function $h(x) = (3x - 5)^3$ is the composition of $f(x) = x^3$ and $g(x) = 3x - 5$. That is,

$$h(x) = (3x - 5)^3 = [g(x)]^3 = f(g(x)).$$

Basically, to "decompose" a composite function, look for an "inner" function and an "outer" function. In the function h above, $g(x) = 3x - 5$ is the inner function and $f(x) = x^3$ is the outer function.

EXAMPLE 7 Decomposing a Composite Function 

Write the function $h(x) = \dfrac{1}{(x - 2)^2}$ as a composition of two functions.

Solution Consider $g(x) = x - 2$ as the inner function and $f(x) = \dfrac{1}{x^2} = x^{-2}$ as the outer function. Then write

$$\begin{aligned}h(x) &= \frac{1}{(x - 2)^2} \\ &= (x - 2)^{-2} \\ &= f(x - 2) \\ &= f(g(x)).\end{aligned}$$

✓ Checkpoint Audio-video solution in English & Spanish at LarsonPrecalculus.com

Write the function $h(x) = \dfrac{\sqrt[3]{8 - x}}{5}$ as a composition of two functions. ■

Application

EXAMPLE 8 Bacteria Count

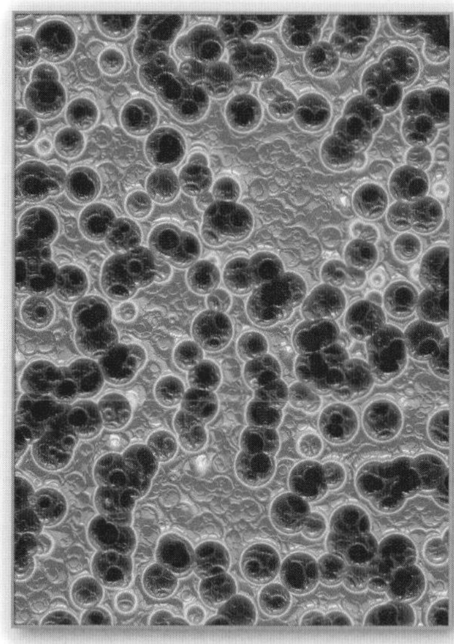

Refrigerated foods can have two types of bacteria: pathogenic bacteria, which can cause foodborne illness, and spoilage bacteria, which give foods an unpleasant look, smell, taste, or texture.

The number N of bacteria in a refrigerated food is given by

$$N(T) = 20T^2 - 80T + 500, \quad 2 \le T \le 14$$

where T is the temperature of the food in degrees Celsius. When the food is removed from refrigeration, the temperature of the food is given by

$$T(t) = 4t + 2, \quad 0 \le t \le 3$$

where t is the time in hours.

a. Find and interpret $(N \circ T)(t)$.

b. Find the time when the bacteria count reaches 2000.

Solution

a.
$$\begin{aligned}(N \circ T)(t) &= N(T(t)) \\ &= 20(4t + 2)^2 - 80(4t + 2) + 500 \\ &= 20(16t^2 + 16t + 4) - 320t - 160 + 500 \\ &= 320t^2 + 320t + 80 - 320t - 160 + 500 \\ &= 320t^2 + 420\end{aligned}$$

The composite function $N \circ T$ represents the number of bacteria in the food as a function of the amount of time the food has been out of refrigeration.

b. The bacteria count reaches 2000 when $320t^2 + 420 = 2000$. By solving this equation algebraically, you find that the count reaches 2000 when $t \approx 2.2$ hours. Note that the negative solution $t \approx -2.2$ hours is rejected because it is not in the domain of the composite function.

✓ Checkpoint  *Audio-video solution in English & Spanish at LarsonPrecalculus.com*

The number N of bacteria in a refrigerated food is given by

$$N(T) = 8T^2 - 14T + 200, \quad 2 \le T \le 12$$

where T is the temperature of the food in degrees Celsius. When the food is removed from refrigeration, the temperature of the food is given by

$$T(t) = 2t + 2, \quad 0 \le t \le 5$$

where t is the time in hours.

a. Find $(N \circ T)(t)$.

b. Find the time when the bacteria count reaches 1000.

Summarize (Section P.9)

1. Explain how to add, subtract, multiply, and divide functions *(page 94)*. For examples of finding arithmetic combinations of functions, see Examples 1–4.
2. Explain how to find the composition of one function with another function *(page 96)*. For examples that use compositions of functions, see Examples 5–7.
3. Describe a real-life example that uses a composition of functions *(page 98, Example 8)*.

iStockphoto.com/Satori13

P.9 Exercises

See **CalcChat.com** for tutorial help and worked-out solutions to odd-numbered exercises.

Vocabulary: Fill in the blanks.

1. Two functions f and g can be combined by the arithmetic operations of ________, ________, ________, and ________ to create new functions.
2. The ________ of the function f with the function g is $(f \circ g)(x) = f(g(x))$.

Skills and Applications

Graphing the Sum of Two Functions In Exercises 3 and 4, use the graphs of f and g to graph $h(x) = (f + g)(x)$. To print an enlarged copy of the graph, go to *MathGraphs.com*.

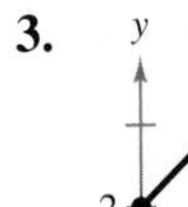

3. 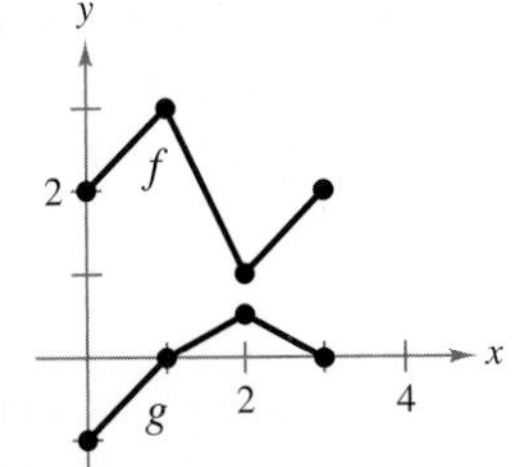

4. 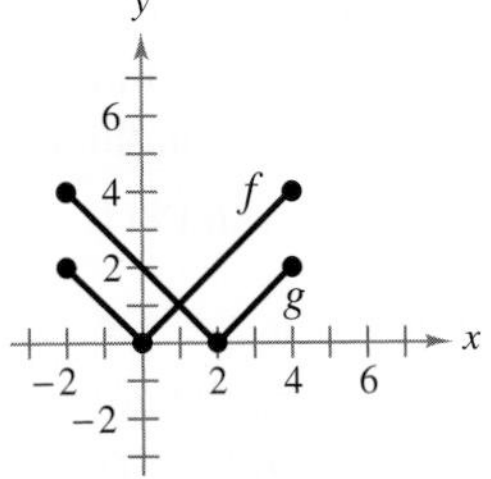

Finding Arithmetic Combinations of Functions In Exercises 5–12, find (a) $(f + g)(x)$, (b) $(f - g)(x)$, (c) $(fg)(x)$, and (d) $(f/g)(x)$. What is the domain of f/g?

5. $f(x) = x + 2, \quad g(x) = x - 2$
6. $f(x) = 2x - 5, \quad g(x) = 2 - x$
7. $f(x) = x^2, \quad g(x) = 4x - 5$
8. $f(x) = 3x + 1, \quad g(x) = x^2 - 16$
9. $f(x) = x^2 + 6, \quad g(x) = \sqrt{1 - x}$
10. $f(x) = \sqrt{x^2 - 4}, \quad g(x) = \dfrac{x^2}{x^2 + 1}$
11. $f(x) = \dfrac{x}{x + 1}, \quad g(x) = x^3$
12. $f(x) = \dfrac{2}{x}, \quad g(x) = \dfrac{1}{x^2 - 1}$

Evaluating an Arithmetic Combination of Functions In Exercises 13–24, evaluate the function for $f(x) = x + 3$ and $g(x) = x^2 - 2$.

13. $(f + g)(2)$
14. $(f + g)(-1)$
15. $(f - g)(0)$
16. $(f - g)(1)$
17. $(f - g)(3t)$
18. $(f + g)(t - 2)$
19. $(fg)(6)$
20. $(fg)(-6)$
21. $(f/g)(5)$
22. $(f/g)(0)$
23. $(f/g)(-1) - g(3)$
24. $(fg)(5) + f(4)$

Graphical Reasoning In Exercises 25–28, use a graphing utility to graph f, g, and $f + g$ in the same viewing window. Which function contributes most to the magnitude of the sum when $0 \le x \le 2$? Which function contributes most to the magnitude of the sum when $x > 6$?

25. $f(x) = 3x, \quad g(x) = -\dfrac{x^3}{10}$
26. $f(x) = \dfrac{x}{2}, \quad g(x) = \sqrt{x}$
27. $f(x) = 3x + 2, \quad g(x) = -\sqrt{x + 5}$
28. $f(x) = x^2 - \frac{1}{2}, \quad g(x) = -3x^2 - 1$

Finding Compositions of Functions In Exercises 29–34, find (a) $f \circ g$, (b) $g \circ f$, and (c) $g \circ g$.

29. $f(x) = x + 8, \quad g(x) = x - 3$
30. $f(x) = -4x, \quad g(x) = x + 7$
31. $f(x) = x^2, \quad g(x) = x - 1$
32. $f(x) = 3x, \quad g(x) = x^4$
33. $f(x) = \sqrt[3]{x - 1}, \quad g(x) = x^3 + 1$
34. $f(x) = x^3, \quad g(x) = \dfrac{1}{x}$

Finding Domains of Functions and Composite Functions In Exercises 35–42, find (a) $f \circ g$ and (b) $g \circ f$. Find the domain of each function and of each composite function.

35. $f(x) = \sqrt{x + 4}, \quad g(x) = x^2$
36. $f(x) = \sqrt[3]{x - 5}, \quad g(x) = x^3 + 1$
37. $f(x) = x^3, \quad g(x) = x^{2/3}$
38. $f(x) = x^5, \quad g(x) = \sqrt[4]{x}$
39. $f(x) = |x|, \quad g(x) = x + 6$
40. $f(x) = |x - 4|, \quad g(x) = 3 - x$
41. $f(x) = \dfrac{1}{x}, \quad g(x) = x + 3$
42. $f(x) = \dfrac{3}{x^2 - 1}, \quad g(x) = x + 1$

Graphing Combinations of Functions In Exercises 43 and 44, on the same set of coordinate axes, (a) graph the functions f, g, and $f + g$ and (b) graph the functions f, g, and $f \circ g$.

43. $f(x) = \frac{1}{2}x, \quad g(x) = x - 4$

44. $f(x) = x + 3, \quad g(x) = x^2$

Evaluating Combinations of Functions In Exercises 45–48, use the graphs of f and g to evaluate the functions.

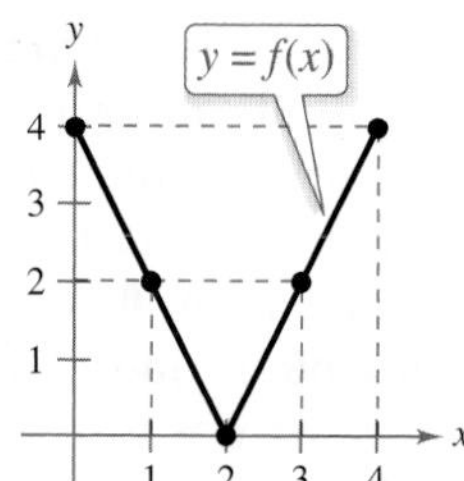

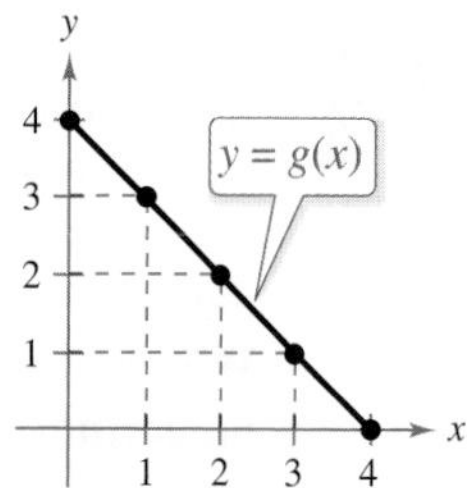

45. (a) $(f + g)(3)$ (b) $(f/g)(2)$

46. (a) $(f - g)(1)$ (b) $(fg)(4)$

47. (a) $(f \circ g)(2)$ (b) $(g \circ f)(2)$

48. (a) $(f \circ g)(1)$ (b) $(g \circ f)(3)$

Decomposing a Composite Function In Exercises 49–56, find two functions f and g such that $(f \circ g)(x) = h(x)$. (There are many correct answers.)

49. $h(x) = (2x + 1)^2$

50. $h(x) = (1 - x)^3$

51. $h(x) = \sqrt[3]{x^2 - 4}$

52. $h(x) = \sqrt{9 - x}$

53. $h(x) = \dfrac{1}{x + 2}$

54. $h(x) = \dfrac{4}{(5x + 2)^2}$

55. $h(x) = \dfrac{-x^2 + 3}{4 - x^2}$

56. $h(x) = \dfrac{27x^3 + 6x}{10 - 27x^3}$

57. Stopping Distance The research and development department of an automobile manufacturer determines that when a driver is required to stop quickly to avoid an accident, the distance (in feet) the car travels during the driver's reaction time is given by $R(x) = \frac{3}{4}x$, where x is the speed of the car in miles per hour. The distance (in feet) the car travels while the driver is braking is given by $B(x) = \frac{1}{15}x^2$.

(a) Find the function that represents the total stopping distance T.

(b) Graph the functions R, B, and T on the same set of coordinate axes for $0 \le x \le 60$.

(c) Which function contributes most to the magnitude of the sum at higher speeds? Explain.

58. Business The annual cost C (in thousands of dollars) and revenue R (in thousands of dollars) for a company each year from 2010 through 2016 can be approximated by the models

$$C = 254 - 9t + 1.1t^2 \quad \text{and} \quad R = 341 + 3.2t$$

where t is the year, with $t = 10$ corresponding to 2010.

(a) Write a function P that represents the annual profit of the company.

(b) Use a graphing utility to graph C, R, and P in the same viewing window.

59. Vital Statistics Let $b(t)$ be the number of births in the United States in year t, and let $d(t)$ represent the number of deaths in the United States in year t, where $t = 10$ corresponds to 2010.

(a) If $p(t)$ is the population of the United States in year t, find the function $c(t)$ that represents the percent change in the population of the United States.

(b) Interpret $c(16)$.

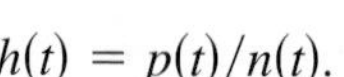

60. Pets

Let $d(t)$ be the number of dogs in the United States in year t, and let $c(t)$ be the number of cats in the United States in year t, where $t = 10$ corresponds to 2010.

(a) Find the function $p(t)$ that represents the total number of dogs and cats in the United States.

(b) Interpret $p(16)$.

(c) Let $n(t)$ represent the population of the United States in year t, where $t = 10$ corresponds to 2010. Find and interpret

$h(t) = p(t)/n(t)$.

61. Geometry A square concrete foundation is a base for a cylindrical tank (see figure).

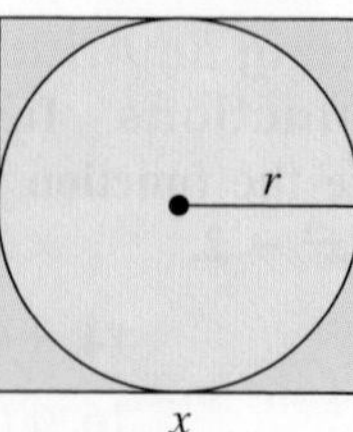

(a) Write the radius r of the tank as a function of the length x of the sides of the square.

(b) Write the area A of the circular base of the tank as a function of the radius r.

(c) Find and interpret $(A \circ r)(x)$.

62. **Biology** The number N of bacteria in a refrigerated food is given by

$$N(T) = 10T^2 - 20T + 600, \quad 2 \le T \le 20$$

where T is the temperature of the food in degrees Celsius. When the food is removed from refrigeration, the temperature of the food is given by

$$T(t) = 3t + 2, \quad 0 \le t \le 6$$

where t is the time in hours.

(a) Find and interpret $(N \circ T)(t)$.

(b) Find the bacteria count after 0.5 hour.

(c) Find the time when the bacteria count reaches 1500.

63. **Salary** You are a sales representative for a clothing manufacturer. You are paid an annual salary, plus a bonus of 3% of your sales over \$500,000. Consider the two functions $f(x) = x - 500{,}000$ and $g(x) = 0.03x$. When x is greater than \$500,000, which of the following represents your bonus? Explain.

(a) $f(g(x)$

(b) $g(f(x))$

64. **Consumer Awareness** The suggested retail price of a new hybrid car is p dollars. The dealership advertises a factory rebate of \$2000 and a 10% discount.

(a) Write a function R in terms of p giving the cost of the hybrid car after receiving the rebate from the factory.

(b) Write a function S in terms of p giving the cost of the hybrid car after receiving the dealership discount.

(c) Find and interpret $(R \circ S)(p)$ and $(S \circ R)(p)$.

(d) Find $(R \circ S)(25{,}795)$ and $(S \circ R)(25{,}795)$. Which yields the lower cost for the hybrid car? Explain.

Exploration

True or False? **In Exercises 65 and 66, determine whether the statement is true or false. Justify your answer.**

65. If $f(x) = x + 1$ and $g(x) = 6x$, then

$(f \circ g)(x) = (g \circ f)(x)$.

66. When you are given two functions f and g and a constant c, you can find $(f \circ g)(c)$ if and only if $g(c)$ is in the domain of f.

Siblings **In Exercises 67 and 68, three siblings are three different ages. The oldest is twice the age of the middle sibling, and the middle sibling is six years older than one-half the age of the youngest.**

67. (a) Write a composite function that gives the oldest sibling's age in terms of the youngest. Explain how you arrived at your answer.

(b) If the oldest sibling is 16 years old, find the ages of the other two siblings.

68. (a) Write a composite function that gives the youngest sibling's age in terms of the oldest. Explain how you arrived at your answer.

(b) If the youngest sibling is 2 years old, find the ages of the other two siblings.

69. **Proof** Prove that the product of two odd functions is an even function, and that the product of two even functions is an even function.

70. **Conjecture** Use examples to hypothesize whether the product of an odd function and an even function is even or odd. Then prove your hypothesis.

71. **Writing Functions** Write two unique functions f and g such that $(f \circ g)(x) = (g \circ f)(x)$ and f and g are (a) linear functions and (b) polynomial functions with degrees greater than one.

72. **HOW DO YOU SEE IT?** The graphs labeled L_1, L_2, L_3, and L_4 represent four different pricing discounts, where p is the original price (in dollars) and S is the sale price (in dollars). Match each function with its graph. Describe the situations in parts (c) and (d).

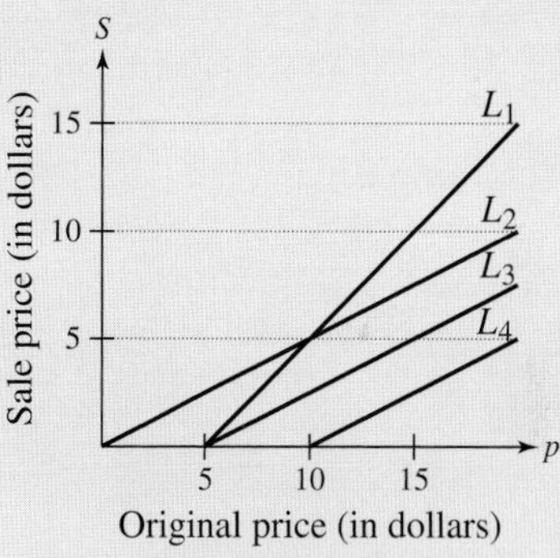

(a) $f(p)$: A 50% discount is applied.

(b) $g(p)$: A \$5 discount is applied.

(c) $(g \circ f)(p)$

(d) $(f \circ g)(p)$

73. **Proof**

(a) Given a function f, prove that g is even and h is odd, where $g(x) = \frac{1}{2}[f(x) + f(-x)]$ and

$h(x) = \frac{1}{2}[f(x) - f(-x)]$.

(b) Use the result of part (a) to prove that any function can be written as a sum of even and odd functions. [*Hint:* Add the two equations in part (a).]

(c) Use the result of part (b) to write each function as a sum of even and odd functions.

$$f(x) = x^2 - 2x + 1, \quad k(x) = \frac{1}{x + 1}$$

P.10 Inverse Functions

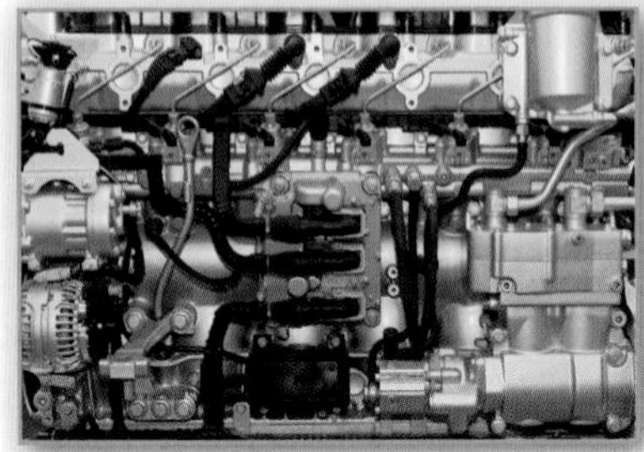

Inverse functions can help you model and solve real-life problems. For example, in Exercise 90 on page 110, you will write an inverse function and use it to determine the percent load interval for a diesel engine.

- **Find inverse functions informally and verify that two functions are inverse functions of each other.**
- **Use graphs to verify that two functions are inverse functions of each other.**
- **Use the Horizontal Line Test to determine whether functions are one-to-one.**
- **Find inverse functions algebraically.**

Inverse Functions

Recall from Section P.5 that a set of ordered pairs can represent a function. For example, the function $f(x) = x + 4$ from the set $A = \{1, 2, 3, 4\}$ to the set $B = \{5, 6, 7, 8\}$ can be written as

$$f(x) = x + 4: \{(1, 5), (2, 6), (3, 7), (4, 8)\}.$$

In this case, by interchanging the first and second coordinates of each of the ordered pairs, you form the **inverse function** of f, which is denoted by f^{-1}. It is a function from the set B to the set A, and can be written as

$$f^{-1}(x) = x - 4: \{(5, 1), (6, 2), (7, 3), (8, 4)\}.$$

Note that the domain of f is equal to the range of f^{-1}, and vice versa, as shown in the figure below. Also note that the functions f and f^{-1} have the effect of "undoing" each other. In other words, when you form the composition of f with f^{-1} or the composition of f^{-1} with f, you obtain the identity function.

$$f(f^{-1}(x)) = f(x - 4) = (x - 4) + 4 = x$$
$$f^{-1}(f(x)) = f^{-1}(x + 4) = (x + 4) - 4 = x$$

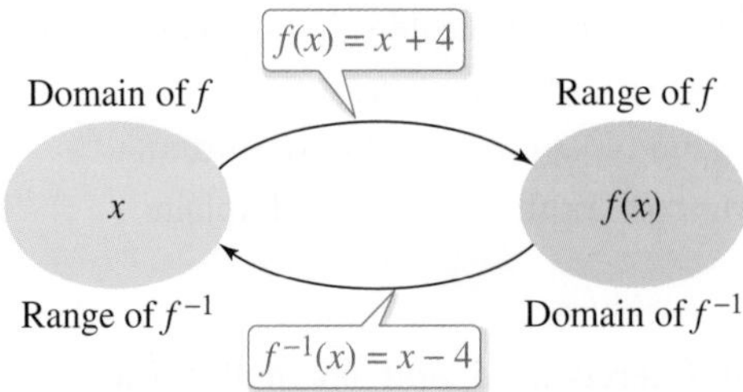

EXAMPLE 1 Finding an Inverse Function Informally

Find the inverse function of $f(x) = 4x$. Then verify that both $f(f^{-1}(x))$ and $f^{-1}(f(x))$ are equal to the identity function.

Solution The function f *multiplies* each input by 4. To "undo" this function, you need to *divide* each input by 4. So, the inverse function of $f(x) = 4x$ is

$$f^{-1}(x) = \frac{x}{4}.$$

Verify that $f(f^{-1}(x)) = x$ and $f^{-1}(f(x)) = x$.

$$f(f^{-1}(x)) = f\left(\frac{x}{4}\right) = 4\left(\frac{x}{4}\right) = x \qquad f^{-1}(f(x)) = f^{-1}(4x) = \frac{4x}{4} = x$$

✓ **Checkpoint** *Audio-video solution in English & Spanish at LarsonPrecalculus.com*

Find the inverse function of $f(x) = \frac{1}{5}x$. Then verify that both $f(f^{-1}(x))$ and $f^{-1}(f(x))$ are equal to the identity function.

Definition of Inverse Function

Let f and g be two functions such that

$$f(g(x)) = x \qquad \text{for every } x \text{ in the domain of } g$$

and

$$g(f(x)) = x \qquad \text{for every } x \text{ in the domain of } f.$$

Under these conditions, the function g is the **inverse function** of the function f. The function g is denoted by f^{-1} (read "f-inverse"). So,

$$f(f^{-1}(x)) = x \quad \text{and} \quad f^{-1}(f(x)) = x.$$

The domain of f must be equal to the range of f^{-1}, and the range of f must be equal to the domain of f^{-1}.

Do not be confused by the use of -1 to denote the inverse function f^{-1}. In this text, whenever f^{-1} is written, it *always* refers to the inverse function of the function f and *not* to the reciprocal of $f(x)$.

If the function g is the inverse function of the function f, then it must also be true that the function f is the inverse function of the function g. So, it is correct to say that the functions f and g are *inverse functions of each other.*

EXAMPLE 2 Verifying Inverse Functions

Which of the functions is the inverse function of $f(x) = \dfrac{5}{x-2}$?

$$g(x) = \frac{x-2}{5} \qquad h(x) = \frac{5}{x} + 2$$

Solution By forming the composition of f with g, you have

$$f(g(x)) = f\left(\frac{x-2}{5}\right) = \frac{5}{\left(\frac{x-2}{5}\right) - 2} = \frac{25}{x-12} \neq x.$$

This composition is not equal to the identity function x, so g *is not* the inverse function of f. By forming the composition of f with h, you have

$$f(h(x)) = f\left(\frac{5}{x} + 2\right) = \frac{5}{\left(\frac{5}{x} + 2\right) - 2} = \frac{5}{\left(\frac{5}{x}\right)} = x.$$

So, it appears that h *is* the inverse function of f. Confirm this by showing that the composition of h with f is also equal to the identity function.

$$h(f(x)) = h\left(\frac{5}{x-2}\right) = \frac{5}{\left(\frac{5}{x-2}\right)} + 2 = x - 2 + 2 = x$$

Check to see that the domain of f is the same as the range of h and vice versa.

✓ **Checkpoint** Audio-video solution in English & Spanish at LarsonPrecalculus.com

Which of the functions is the inverse function of $f(x) = \dfrac{x-4}{7}$?

$$g(x) = 7x + 4 \qquad h(x) = \frac{7}{x-4}$$

The Graph of an Inverse Function

The graphs of a function f and its inverse function f^{-1} are related to each other in this way: If the point (a, b) lies on the graph of f, then the point (b, a) must lie on the graph of f^{-1}, and vice versa. This means that the graph of f^{-1} is a *reflection* of the graph of f in the line $y = x$, as shown in Figure P.60.

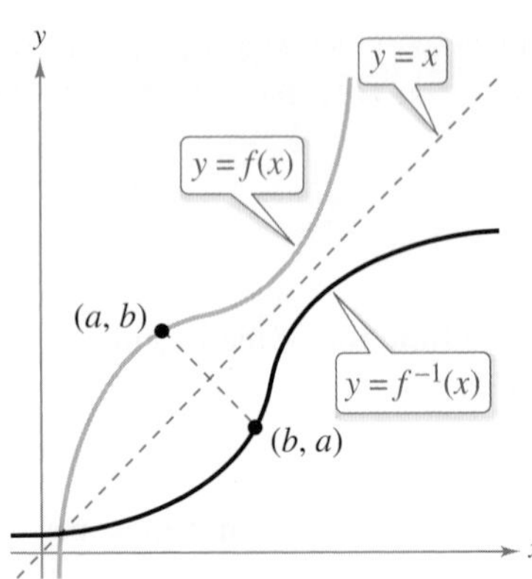

Figure P.60

EXAMPLE 3 Verifying Inverse Functions Graphically

Verify graphically that the functions $f(x) = 2x - 3$ and $g(x) = \frac{1}{2}(x + 3)$ are inverse functions of each other.

Solution Sketch the graphs of f and g on the same rectangular coordinate system, as shown in Figure P.61. It appears that the graphs are reflections of each other in the line $y = x$. Further verify this reflective property by testing a few points on each graph. Note that for each point (a, b) on the graph of f, the point (b, a) is on the graph of g.

Graph of $f(x) = 2x - 3$	Graph of $g(x) = \frac{1}{2}(x + 3)$
$(-1, -5)$	$(-5, -1)$
$(0, -3)$	$(-3, 0)$
$(1, -1)$	$(-1, 1)$
$(2, 1)$	$(1, 2)$
$(3, 3)$	$(3, 3)$

The graphs of f and g are reflections of each other in the line $y = x$. So, f and g are inverse functions of each other.

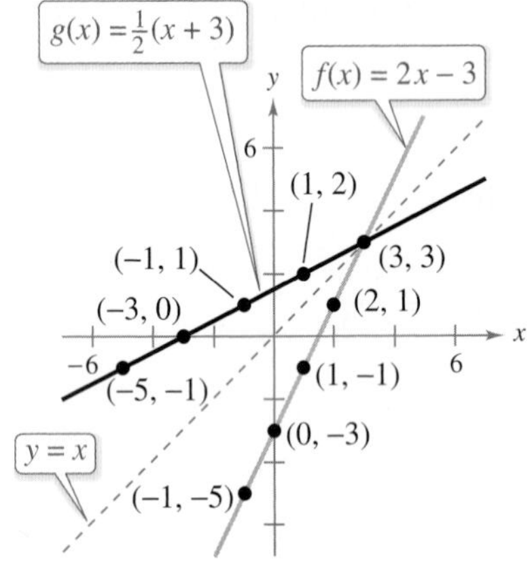

Figure P.61

Checkpoint Audio-video solution in English & Spanish at LarsonPrecalculus.com

Verify graphically that the functions $f(x) = 4x - 1$ and $g(x) = \frac{1}{4}(x + 1)$ are inverse functions of each other.

EXAMPLE 4 Verifying Inverse Functions Graphically

Verify graphically that the functions $f(x) = x^2$ $(x \geq 0)$ and $g(x) = \sqrt{x}$ are inverse functions of each other.

Solution Sketch the graphs of f and g on the same rectangular coordinate system, as shown in Figure P.62. It appears that the graphs are reflections of each other in the line $y = x$. Test a few points on each graph.

Graph of $f(x) = x^2,\ x \geq 0$	Graph of $g(x) = \sqrt{x}$
$(0, 0)$	$(0, 0)$
$(1, 1)$	$(1, 1)$
$(2, 4)$	$(4, 2)$
$(3, 9)$	$(9, 3)$

The graphs of f and g are reflections of each other in the line $y = x$. So, f and g are inverse functions of each other.

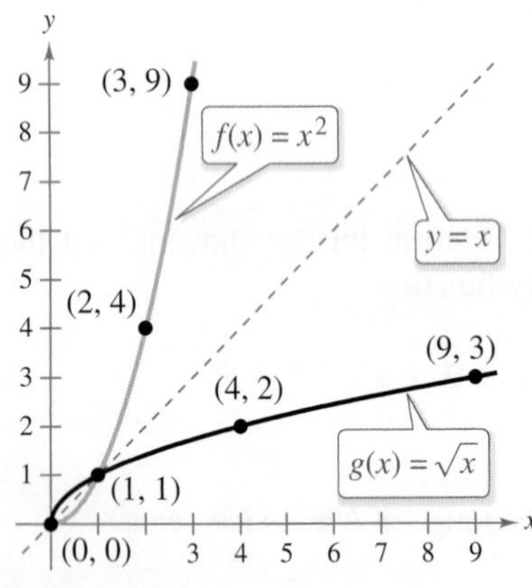

Figure P.62

Checkpoint Audio-video solution in English & Spanish at LarsonPrecalculus.com

Verify graphically that the functions $f(x) = x^2 + 1$ $(x \geq 0)$ and $g(x) = \sqrt{x - 1}$ are inverse functions of each other.

One-to-One Functions

The reflective property of the graphs of inverse functions gives you a graphical test for determining whether a function has an inverse function. This test is the **Horizontal Line Test** for inverse functions.

> **Horizontal Line Test for Inverse Functions**
>
> A function f has an inverse function if and only if no *horizontal* line intersects the graph of f at more than one point.

If no horizontal line intersects the graph of f at more than one point, then no y-value corresponds to more than one x-value. This is the essential characteristic of **one-to-one functions.**

> **One-to-One Functions**
>
> A function f is **one-to-one** when each value of the dependent variable corresponds to exactly one value of the independent variable. A function f has an inverse function if and only if f is one-to-one.

Consider the table of values for the function $f(x) = x^2$ on the left. The output $f(x) = 4$ corresponds to two inputs, $x = -2$ and $x = 2$, so f is not one-to-one. In the table on the right, x and y are interchanged. Here $x = 4$ corresponds to both $y = -2$ and $y = 2$, so this table does not represent a function. So, $f(x) = x^2$ is not one-to-one and does not have an inverse function.

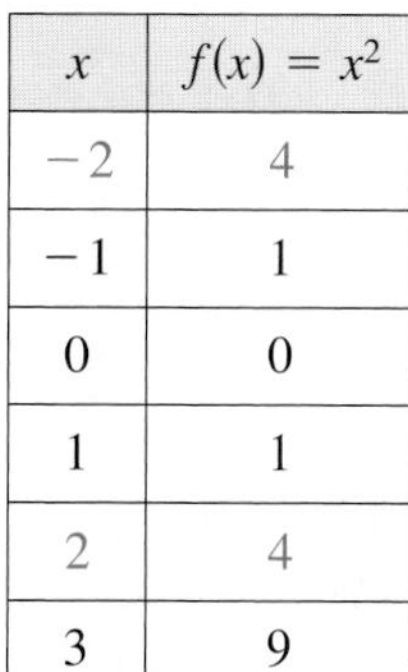

x	$f(x) = x^2$
-2	4
-1	1
0	0
1	1
2	4
3	9

x	y
4	-2
1	-1
0	0
1	1
4	2
9	3

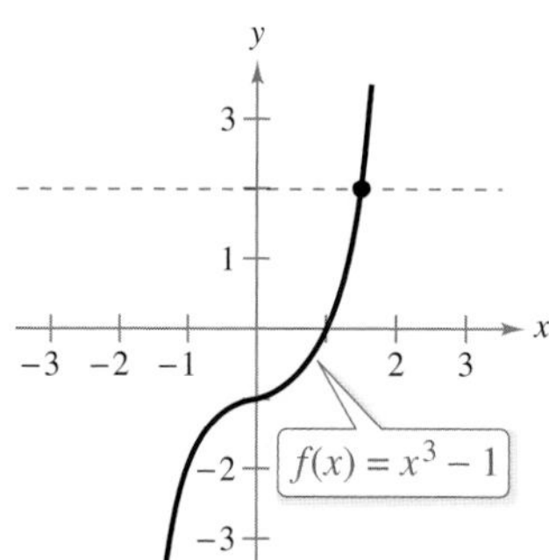

Figure P.63

EXAMPLE 5 Applying the Horizontal Line Test

See LarsonPrecalculus.com for an interactive version of this type of example.

a. The graph of the function $f(x) = x^3 - 1$ is shown in Figure P.63. No horizontal line intersects the graph of f at more than one point, so f *is* a one-to-one function and *does* have an inverse function.

b. The graph of the function $f(x) = x^2 - 1$ is shown in Figure P.64. It is possible to find a horizontal line that intersects the graph of f at more than one point, so f *is not* a one-to-one function and *does not* have an inverse function.

Figure P.64

✓ Checkpoint Audio-video solution in English & Spanish at LarsonPrecalculus.com

Use the graph of f to determine whether the function has an inverse function.

a. $f(x) = \frac{1}{2}(3 - x)$ **b.** $f(x) = |x|$

Finding Inverse Functions Algebraically

For relatively simple functions (such as the one in Example 1), you can find inverse functions by inspection. For more complicated functions, however, it is best to use the guidelines below. The key step in these guidelines is Step 3—interchanging the roles of x and y. This step corresponds to the fact that inverse functions have ordered pairs with the coordinates reversed.

REMARK Note what happens when you try to find the inverse function of a function that is not one-to-one.

$f(x) = x^2 + 1$	Original function
$y = x^2 + 1$	Replace $f(x)$ with y.
$x = y^2 + 1$	Interchange x and y.
$x - 1 = y^2$	Isolate y-term.
$y = \pm\sqrt{x - 1}$	Solve for y.

You obtain two y-values for each x.

Finding an Inverse Function

1. Use the Horizontal Line Test to decide whether f has an inverse function.
2. In the equation for $f(x)$, replace $f(x)$ with y.
3. Interchange the roles of x and y, and solve for y.
4. Replace y with $f^{-1}(x)$ in the new equation.
5. Verify that f and f^{-1} are inverse functions of each other by showing that the domain of f is equal to the range of f^{-1}, the range of f is equal to the domain of f^{-1}, and $f(f^{-1}(x)) = x$ and $f^{-1}(f(x)) = x$.

EXAMPLE 6 **Finding an Inverse Function Algebraically**

Find the inverse function of

$$f(x) = \frac{5 - x}{3x + 2}.$$

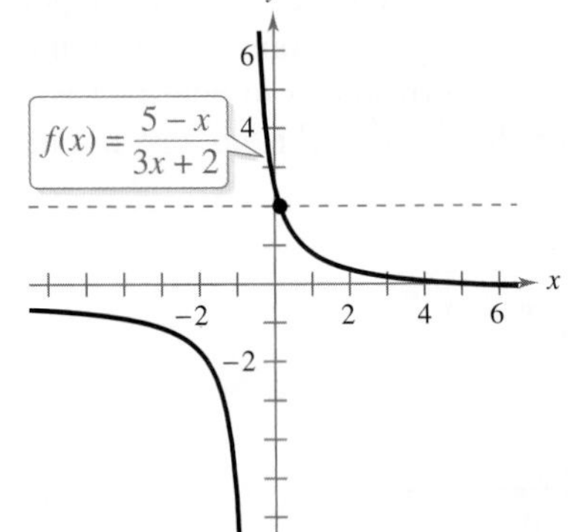

Figure P.65

Solution The graph of f is shown in Figure P.65. This graph passes the Horizontal Line Test. So, you know that f is one-to-one and has an inverse function.

$f(x) = \dfrac{5 - x}{3x + 2}$	Write original function.
$y = \dfrac{5 - x}{3x + 2}$	Replace $f(x)$ with y.
$x = \dfrac{5 - y}{3y + 2}$	Interchange x and y.
$x(3y + 2) = 5 - y$	Multiply each side by $3y + 2$.
$3xy + 2x = 5 - y$	Distributive Property
$3xy + y = 5 - 2x$	Collect terms with y.
$y(3x + 1) = 5 - 2x$	Factor.
$y = \dfrac{5 - 2x}{3x + 1}$	Solve for y.
$f^{-1}(x) = \dfrac{5 - 2x}{3x + 1}$	Replace y with $f^{-1}(x)$.

Check that $f(f^{-1}(x)) = x$ and $f^{-1}(f(x)) = x$.

✓ **Checkpoint** Audio-video solution in English & Spanish at LarsonPrecalculus.com

Find the inverse function of

$$f(x) = \frac{5 - 3x}{x + 2}.$$

EXAMPLE 7 **Finding an Inverse Function Algebraically**

Find the inverse function of

$$f(x) = \sqrt{2x - 3}.$$

Solution The graph of f is shown in the figure below. This graph passes the Horizontal Line Test. So, you know that f is one-to-one and has an inverse function.

$f(x) = \sqrt{2x - 3}$ Write original function.

$y = \sqrt{2x - 3}$ Replace $f(x)$ with y.

$x = \sqrt{2y - 3}$ Interchange x and y.

$x^2 = 2y - 3$ Square each side.

$2y = x^2 + 3$ Isolate y-term.

$y = \dfrac{x^2 + 3}{2}$ Solve for y.

$f^{-1}(x) = \dfrac{x^2 + 3}{2}, \; x \geq 0$ Replace y with $f^{-1}(x)$.

The graph of f^{-1} in the figure is the reflection of the graph of f in the line $y = x$. Note that the range of f is the interval $[0, \infty)$, which implies that the domain of f^{-1} is the interval $[0, \infty)$. Moreover, the domain of f is the interval $\left[\frac{3}{2}, \infty\right)$, which implies that the range of f^{-1} is the interval $\left[\frac{3}{2}, \infty\right)$. Verify that $f(f^{-1}(x)) = x$ and $f^{-1}(f(x)) = x$.

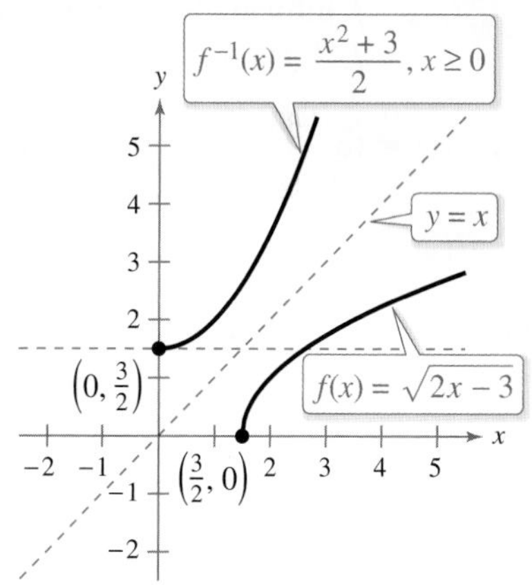

Checkpoint Audio-video solution in English & Spanish at LarsonPrecalculus.com

Find the inverse function of

$$f(x) = \sqrt[3]{10 + x}.$$

Summarize (Section P.10)

1. State the definition of an inverse function *(page 103)*. For examples of finding inverse functions informally and verifying inverse functions, see Examples 1 and 2.
2. Explain how to use graphs to verify that two functions are inverse functions of each other *(page 104)*. For examples of verifying inverse functions graphically, see Examples 3 and 4.
3. Explain how to use the Horizontal Line Test to determine whether a function is one-to-one *(page 105)*. For an example of applying the Horizontal Line Test, see Example 5.
4. Explain how to find an inverse function algebraically *(page 106)*. For examples of finding inverse functions algebraically, see Examples 6 and 7.

P.10 Exercises

See CalcChat.com for tutorial help and worked-out solutions to odd-numbered exercises.

Vocabulary: Fill in the blanks.

1. If $f(g(x))$ and $g(f(x))$ both equal x, then the function g is the ________ function of the function f.
2. The inverse function of f is denoted by ________.
3. The domain of f is the ________ of f^{-1}, and the ________ of f^{-1} is the range of f.
4. The graphs of f and f^{-1} are reflections of each other in the line ________.
5. A function f is ________ when each value of the dependent variable corresponds to exactly one value of the independent variable.
6. A graphical test for the existence of an inverse function of f is the _______ Line Test.

Skills and Applications

Finding an Inverse Function Informally **In Exercises 7–14, find the inverse function of f informally. Verify that $f(f^{-1}(x)) = x$ and $f^{-1}(f(x)) = x$.**

7. $f(x) = 6x$
8. $f(x) = \frac{1}{3}x$
9. $f(x) = 3x + 1$
10. $f(x) = \frac{x - 3}{2}$
11. $f(x) = x^2 - 4,\ x \geq 0$
12. $f(x) = x^2 + 2,\ x \geq 0$
13. $f(x) = x^3 + 1$
14. $f(x) = \frac{x^5}{4}$

Verifying Inverse Functions **In Exercises 15–18, verify that f and g are inverse functions algebraically.**

15. $f(x) = \frac{x - 9}{4}, \quad g(x) = 4x + 9$
16. $f(x) = -\frac{3}{2}x - 4, \quad g(x) = -\frac{2x + 8}{3}$
17. $f(x) = \frac{x^3}{4}, \quad g(x) = \sqrt[3]{4x}$
18. $f(x) = x^3 + 5, \quad g(x) = \sqrt[3]{x - 5}$

Sketching the Graph of an Inverse Function **In Exercises 19 and 20, use the graph of the function to sketch the graph of its inverse function $y = f^{-1}(x)$.**

19.

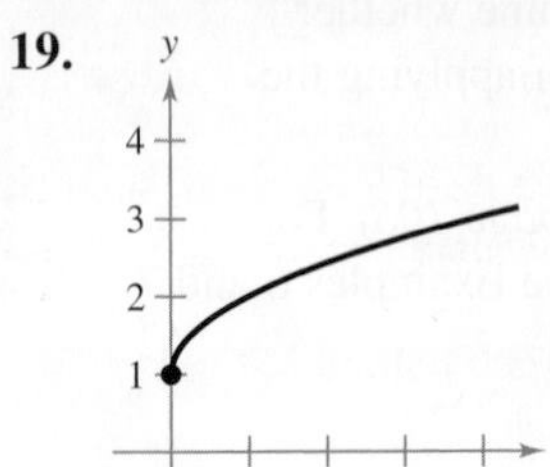

20.

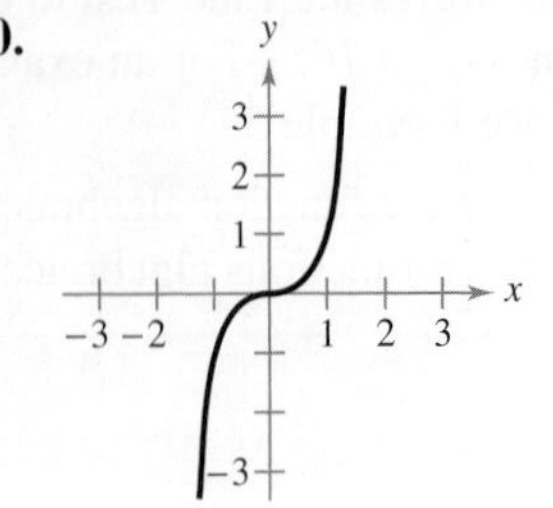

Verifying Inverse Functions **In Exercises 21–32, verify that f and g are inverse functions (a) algebraically and (b) graphically.**

21. $f(x) = x - 5, \quad g(x) = x + 5$
22. $f(x) = 2x, \quad g(x) = \frac{x}{2}$
23. $f(x) = 7x + 1, \quad g(x) = \frac{x - 1}{7}$
24. $f(x) = 3 - 4x, \quad g(x) = \frac{3 - x}{4}$
25. $f(x) = x^3, \quad g(x) = \sqrt[3]{x}$
26. $f(x) = \frac{x^3}{3}, \quad g(x) = \sqrt[3]{3x}$
27. $f(x) = \sqrt{x + 5}, \quad g(x) = x^2 - 5, \quad x \geq 0$
28. $f(x) = 1 - x^3, \quad g(x) = \sqrt[3]{1 - x}$
29. $f(x) = \frac{1}{x}, \quad g(x) = \frac{1}{x}$
30. $f(x) = \frac{1}{1 + x}, \quad x \geq 0, \quad g(x) = \frac{1 - x}{x}, \quad 0 < x \leq 1$
31. $f(x) = \frac{x - 1}{x + 5}, \quad g(x) = -\frac{5x + 1}{x - 1}$
32. $f(x) = \frac{x + 3}{x - 2}, \quad g(x) = \frac{2x + 3}{x - 1}$

Using a Table to Determine an Inverse Function **In Exercises 33 and 34, does the function have an inverse function?**

33.

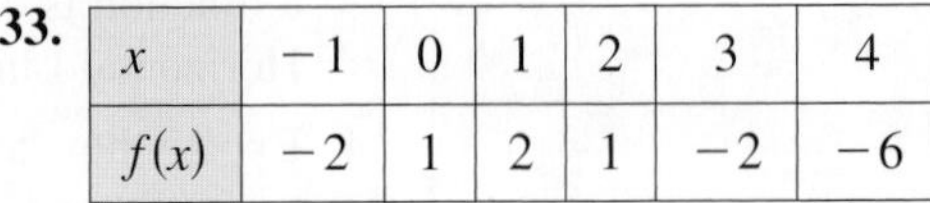

x	-1	0	1	2	3	4
$f(x)$	-2	1	2	1	-2	-6

34.

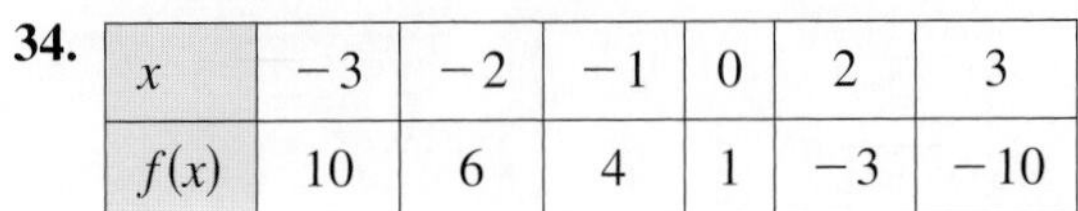

x	-3	-2	-1	0	2	3
$f(x)$	10	6	4	1	-3	-10

Using a Table to Find an Inverse Function In Exercises 35 and 36, use the table of values for $y = f(x)$ to complete a table for $y = f^{-1}(x)$.

35.

x	-1	0	1	2	3	4
$f(x)$	3	5	7	9	11	13

36.

x	-3	-2	-1	0	1	2
$f(x)$	10	5	0	-5	-10	-15

Applying the Horizontal Line Test In Exercises 37–40, does the function have an inverse function?

37.
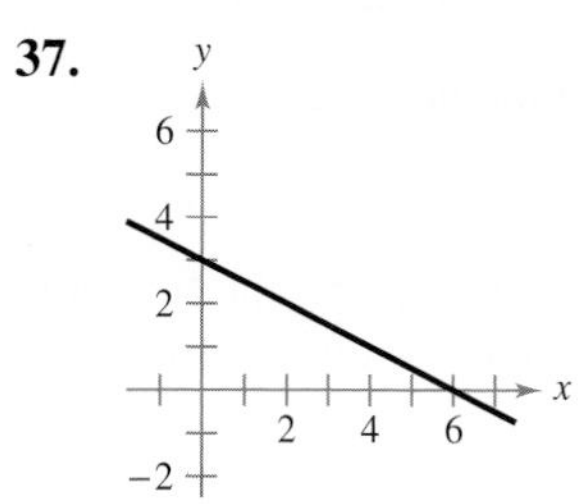

38.
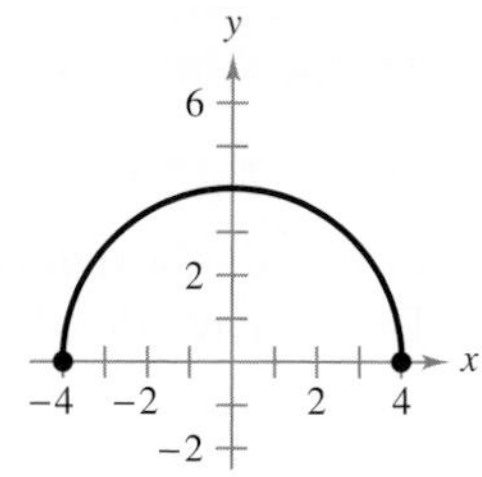

39.
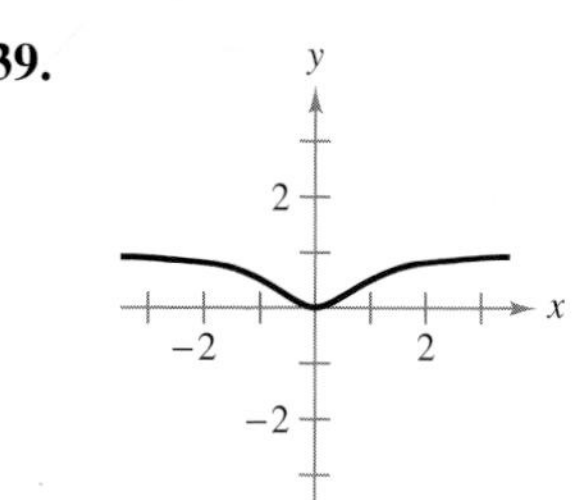

40.
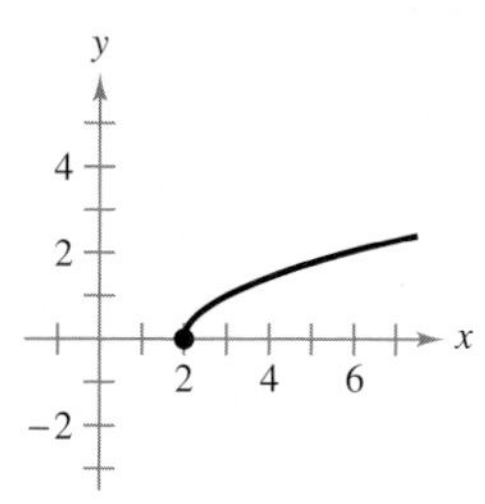

Applying the Horizontal Line Test In Exercises 41–44, use a graphing utility to graph the function, and use the Horizontal Line Test to determine whether the function has an inverse function.

41. $g(x) = (x + 3)^2 + 2$

42. $f(x) = \frac{1}{5}(x + 2)^3$

43. $f(x) = x\sqrt{9 - x^2}$

44. $h(x) = |x| - |x - 4|$

Finding and Analyzing Inverse Functions In Exercises 45–54, (a) find the inverse function of f, (b) graph both f and f^{-1} on the same set of coordinate axes, (c) describe the relationship between the graphs of f and f^{-1}, and (d) state the domains and ranges of f and f^{-1}.

45. $f(x) = x^5 - 2$

46. $f(x) = x^3 + 8$

47. $f(x) = \sqrt{4 - x^2}, \quad 0 \le x \le 2$

48. $f(x) = x^2 - 2, \quad x \le 0$

49. $f(x) = \dfrac{4}{x}$

50. $f(x) = -\dfrac{2}{x}$

51. $f(x) = \dfrac{x + 1}{x - 2}$

52. $f(x) = \dfrac{x - 2}{3x + 5}$

53. $f(x) = \sqrt[3]{x - 1}$

54. $f(x) = x^{3/5}$

Finding an Inverse Function In Exercises 55–70, determine whether the function has an inverse function. If it does, find the inverse function.

55. $f(x) = x^4$

56. $f(x) = \dfrac{1}{x^2}$

57. $g(x) = \dfrac{x + 1}{6}$

58. $f(x) = 3x + 5$

59. $p(x) = -4$

60. $f(x) = 0$

61. $f(x) = (x + 3)^2, \quad x \ge -3$

62. $q(x) = (x - 5)^2$

63. $f(x) = \begin{cases} x + 3, & x < 0 \\ 6 - x, & x \ge 0 \end{cases}$

64. $f(x) = \begin{cases} -x, & x \le 0 \\ x^2 - 3x, & x > 0 \end{cases}$

65. $h(x) = |x + 1| - 1$

66. $f(x) = |x - 2|, \quad x \le 2$

67. $f(x) = \sqrt{2x + 3}$

68. $f(x) = \sqrt{x - 2}$

69. $f(x) = \dfrac{6x + 4}{4x + 5}$

70. $f(x) = \dfrac{5x - 3}{2x + 5}$

Restricting the Domain In Exercises 71–78, restrict the domain of the function f so that the function is one-to-one and has an inverse function. Then find the inverse function f^{-1}. State the domains and ranges of f and f^{-1}. Explain your results. (There are many correct answers.)

71. $f(x) = |x + 2|$

72. $f(x) = |x - 5|$

73. $f(x) = (x + 6)^2$

74. $f(x) = (x - 4)^2$

75. $f(x) = -2x^2 + 5$

76. $f(x) = \frac{1}{2}x^2 - 1$

77. $f(x) = |x - 4| + 1$

78. $f(x) = -|x - 1| - 2$

Composition with Inverses In Exercises 79–84, use the functions $f(x) = \frac{1}{8}x - 3$ and $g(x) = x^3$ to find the value or function.

79. $(f^{-1} \circ g^{-1})(1)$

80. $(g^{-1} \circ f^{-1})(-3)$

81. $(f^{-1} \circ f^{-1})(4)$

82. $(g^{-1} \circ g^{-1})(-1)$

83. $(f \circ g)^{-1}$

84. $g^{-1} \circ f^{-1}$

Composition with Inverses In Exercises 85–88, use the functions $f(x) = x + 4$ and $g(x) = 2x - 5$ to find the function.

85. $g^{-1} \circ f^{-1}$

86. $f^{-1} \circ g^{-1}$

87. $(f \circ g)^{-1}$

88. $(g \circ f)^{-1}$

89. Hourly Wage Your wage is \$10.00 per hour plus \$0.75 for each unit produced per hour. So, your hourly wage y in terms of the number of units produced x is $y = 10 + 0.75x$.

(a) Find the inverse function. What does each variable represent in the inverse function?

(b) Determine the number of units produced when your hourly wage is \$24.25.

90. Diesel Mechanics

The function

$$y = 0.03x^2 + 245.50, \quad 0 < x < 100$$

approximates the exhaust temperature y in degrees Fahrenheit, where x is the percent load for a diesel engine.

(a) Find the inverse function. What does each variable represent in the inverse function?

(b) Use a graphing utility to graph the inverse function.

(c) The exhaust temperature of the engine must not exceed 500 degrees Fahrenheit. What is the percent load interval?

Exploration

True or False? **In Exercises 91 and 92, determine whether the statement is true or false. Justify your answer.**

91. If f is an even function, then f^{-1} exists.

92. If the inverse function of f exists and the graph of f has a y-intercept, then the y-intercept of f is an x-intercept of f^{-1}.

Creating a Table **In Exercises 93 and 94, use the graph of the function f to create a table of values for the given points. Then create a second table that can be used to find f^{-1}, and sketch the graph of f^{-1}, if possible.**

93.

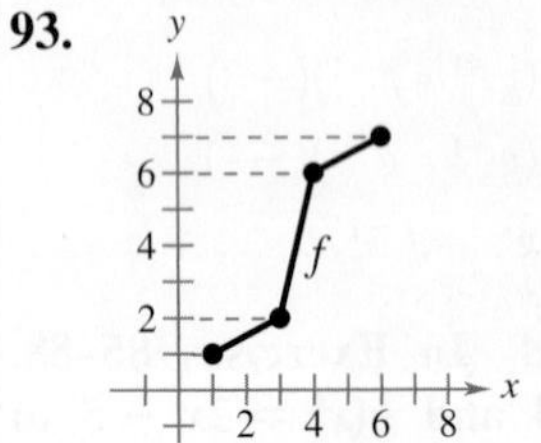

94.

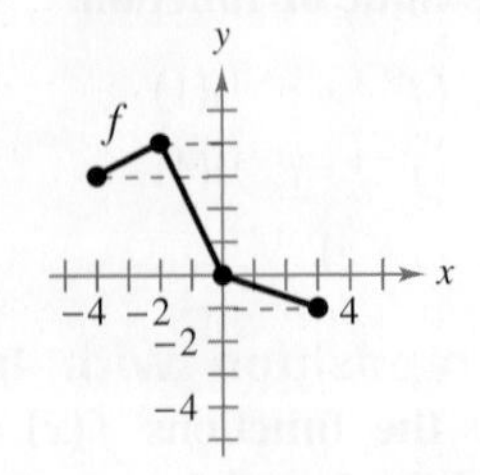

95. Proof Prove that if f and g are one-to-one functions, then $(f \circ g)^{-1}(x) = (g^{-1} \circ f^{-1})(x)$.

96. Proof Prove that if f is a one-to-one odd function, then f^{-1} is an odd function.

97. Think About It The function $f(x) = k(2 - x - x^3)$ has an inverse function, and $f^{-1}(3) = -2$. Find k.

98. Think About It Consider the functions $f(x) = x + 2$ and $f^{-1}(x) = x - 2$. Evaluate $f(f^{-1}(x))$ and $f^{-1}(f(x))$ for the given values of x. What can you conclude about the functions?

x	-10	0	7	45
$f(f^{-1}(x))$				
$f^{-1}(f(x))$				

99. Think About It Restrict the domain of

$$f(x) = x^2 + 1$$

to $x \geq 0$. Use a graphing utility to graph the function. Does the restricted function have an inverse function? Explain.

100. HOW DO YOU SEE IT? The cost C for a business to make personalized T-shirts is given by

$$C(x) = 7.50x + 1500$$

where x represents the number of T-shirts.

(a) The graphs of C and C^{-1} are shown below. Match each function with its graph.

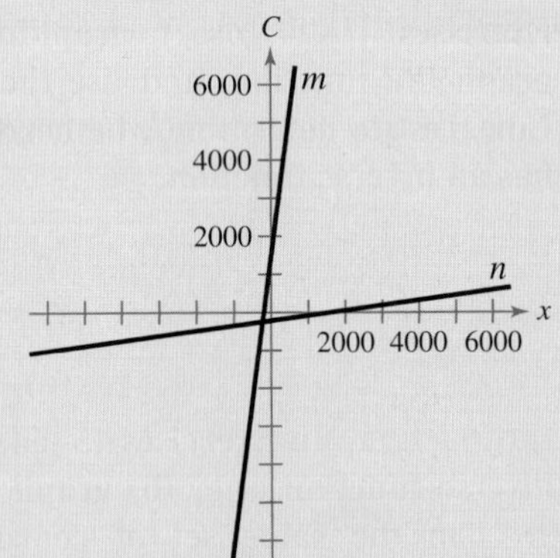

(b) Explain what $C(x)$ and $C^{-1}(x)$ represent in the context of the problem.

One-to-One Function Representation **In Exercises 101 and 102, determine whether the situation can be represented by a one-to-one function. If so, write a statement that best describes the inverse function.**

101. The number of miles n a marathon runner has completed in terms of the time t in hours

102. The depth of the tide d at a beach in terms of the time t over a 24-hour period

Baloncici/Shutterstock.com

Chapter Summary

	What Did You Learn?	Explanation/Examples	Review Exercises
Section P.1	Represent and classify real numbers *(p. 2)*.	Real numbers include both rational and irrational numbers. Real numbers can be represented graphically on the real number line.	1, 2
	Order real numbers and use inequalities *(p. 4)*.	$a < b$: a is less than b. $a > b$: a is greater than b. $a \le b$: a is less than or equal to b. $a \ge b$: a is greater than or equal to b.	3, 4
	Find the absolute values of real numbers and find the distance between two real numbers *(p. 6)*.	**Absolute value of a:** $\lvert a\rvert = \begin{cases} a, & a \ge 0 \\ -a, & a < 0 \end{cases}$ **Distance between a and b:** $d(a, b) = \lvert b - a\rvert = \lvert a - b\rvert$	5–8
	Evaluate algebraic expressions *(p. 8)*.	Evaluate an algebraic expression by substituting numerical values for each of the variables in the expression.	9, 10
	Use the basic rules and properties of algebra *(p. 9)*.	The basic rules of algebra, the properties of negation and equality, the properties of zero, and the properties and operations of fractions can be used to perform operations.	11–22
Section P.2	Identify different types of equations *(p. 14)*, and solve linear equations in one variable and rational equations *(p. 15)*.	**Identity:** true for *every* real number in the domain **Conditional equation:** true for just *some* (but not all) of the real numbers in the domain **Contradiction:** false for *every* real number in the domain	23–26
	Solve quadratic equations *(p. 17)*, polynomial equations of degree three or greater *(p. 21)*, radical equations *(p. 22)*, and absolute value equations *(p. 23)*.	Four methods for solving quadratic equations are factoring, extracting square roots, completing the square, and the Quadratic Formula. Sometimes, these methods can be extended to solve polynomial equations of higher degree. When solving equations involving radicals or absolute values, be sure to check for extraneous solutions.	27–38
Section P.3	Plot points in the Cartesian plane *(p. 26)*, and use the Distance Formula *(p. 28)* and the Midpoint Formula *(p. 29)*.	For an ordered pair (x, y), the x-coordinate represents the directed distance from the y-axis to the point, and the y-coordinate represents the directed distance from the x-axis to the point.	39, 40
	Use a coordinate plane to model and solve real-life problems *(p. 30)*.	The coordinate plane can be used to find the length of a football pass. (See Example 6.)	41, 42
	Sketch graphs of equations *(p. 31)*, and find x- and y-intercepts of graphs of equations *(p. 32)*.	To graph an equation, construct a table of values, plot the points, and connect the points with a smooth curve or line. To find x-intercepts, let y be zero and solve for x. To find y-intercepts, let x be zero and solve for y.	43–48
	Use symmetry to sketch graphs of equations *(p. 33)*.	Graphs can have symmetry with respect to one of the coordinate axes or with respect to the origin. You can test for symmetry graphically and algebraically.	49–52
	Write equations of circles *(p. 34)*.	A point (x, y) lies on the circle of radius r and center (h, k) if and only if $(x - h)^2 + (y - k)^2 = r^2$.	53–56

	What Did You Learn?	Explanation/Examples	Review Exercises
Section P.4	Use slope to graph linear equations in two variables *(p. 40)*.	**The Slope-Intercept Form of the Equation of a Line** The graph of the equation $y = mx + b$ is a line whose slope is m and whose y-intercept is $(0, b)$.	57–60
	Find the slope of a line given two points on the line *(p. 42)*.	The slope m of the nonvertical line through (x_1, y_1) and (x_2, y_2) is $m = \frac{y_2 - y_1}{x_2 - x_1}$ where $x_1 \neq x_2$.	61, 62
	Write linear equations in two variables *(p. 44)*.	**Point-Slope Form of the Equation of a Line** The equation of the line with slope m passing through the point (x_1, y_1) is $y - y_1 = m(x - x_1)$.	63–66
	Use slope to identify parallel and perpendicular lines *(p. 45)*.	**Parallel lines:** Slopes are equal. **Perpendicular lines:** Slopes are negative reciprocals of each other.	67, 68
	Use slope and linear equations in two variables to model and solve real-life problems *(p. 46)*.	A linear equation in two variables can be used to describe the book value of exercise equipment each year. (See Example 7.)	69, 70
Section P.5	Determine whether relations between two variables are functions, and use function notation *(p. 53)*.	A function f from a set A (domain) to a set B (range) is a relation that assigns to each element x in the set A exactly one element y in the set B. **Equation:** $f(x) = 5 - x^2$ **$f(2)$:** $f(2) = 5 - 2^2 = 1$	71–76
	Find the domains of functions *(p. 58)*.	**Domain of $f(x) = 5 - x^2$:** All real numbers	77, 78
	Use functions to model and solve real-life problems *(p. 59)*.	A function can be used to model the path of a baseball. (See Example 9.)	79
	Evaluate difference quotients *(p. 60)*.	**Difference quotient:** $\frac{f(x + h) - f(x)}{h}$, $h \neq 0$	80
Section P.6	Use the Vertical Line Test for functions *(p. 68)*.	A set of points in a coordinate plane is the graph of y as a function of x if and only if no *vertical* line intersects the graph at more than one point.	81, 82
	Find the zeros of functions *(p. 69)*.	**Zeros of $f(x)$:** x-values for which $f(x) = 0$	83, 84
	Determine intervals on which functions are increasing or decreasing *(p. 70)*, determine relative minimum and relative maximum values of functions *(p. 71)*, and determine the average rate of change of a function *(p. 72)*.	To determine whether a function is increasing, decreasing, or constant on an interval, determine whether the graph of the function rises, falls, or is constant from left to right. The points at which the behavior of a function changes can help determine relative minimum or relative maximum values. The average rate of change between any two points is the slope of the line (secant line) through the two points.	85–90
	Identify even and odd functions *(p. 73)*.	**Even:** For each x in the domain of f, $f(-x) = f(x)$. **Odd:** For each x in the domain of f, $f(-x) = -f(x)$.	91–94

	What Did You Learn?	Explanation/Examples	Review Exercises
Section P.7	Identify and graph linear functions *(p. 78)*, squaring functions *(p. 79)*, cubic, square root, and reciprocal functions *(p. 80)*, and step and other piecewise-defined functions *(p. 81)*. Recognize graphs of parent functions *(p. 82)*.	**Linear:** $f(x) = ax + b$ 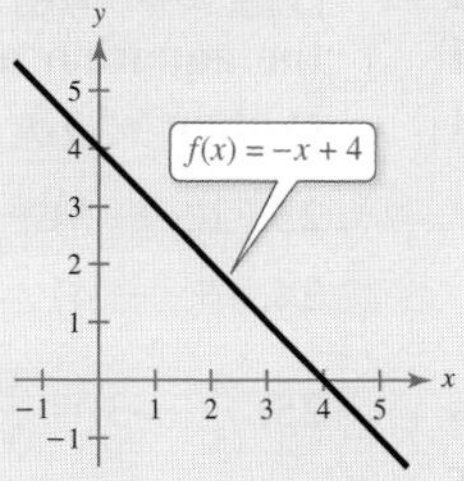 **Squaring:** $f(x) = x^2$ 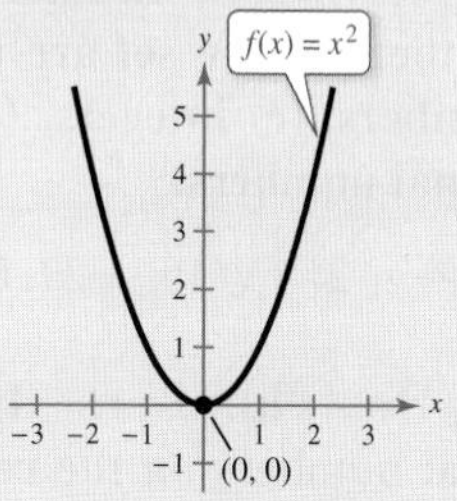 **Square Root:** $f(x) = \sqrt{x}$ 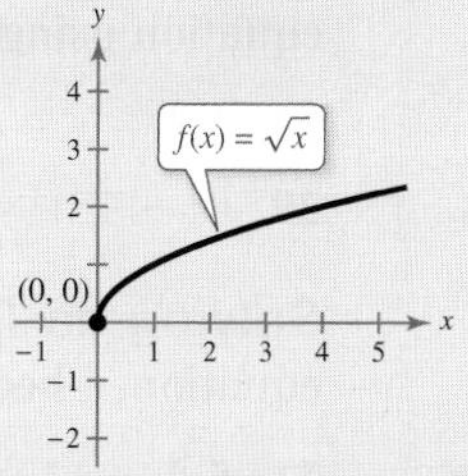 **Step:** $f(x) = [\![x]\!]$ 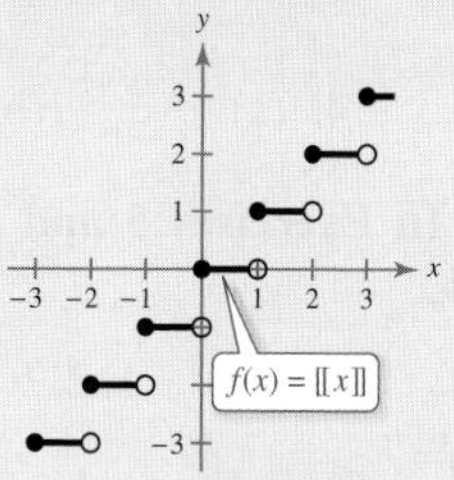 The graphs of eight of the most commonly used functions in algebra are shown on page 82.	95–100
Section P.8	Use vertical and horizontal shifts *(p. 85)*, reflections *(p. 87)*, and nonrigid transformations *(p. 89)* to sketch graphs of functions.	**Vertical shifts:** $h(x) = f(x) + c$ or $h(x) = f(x) - c$ **Horizontal shifts:** $h(x) = f(x - c)$ or $h(x) = f(x + c)$ **Reflection in *x*-axis:** $h(x) = -f(x)$ **Reflection in *y*-axis:** $h(x) = f(-x)$ **Nonrigid transformations:** $h(x) = cf(x)$ or $h(x) = f(cx)$	101–110
Section P.9	Add, subtract, multiply, and divide functions *(p. 94)*.	$(f + g)(x) = f(x) + g(x)$ $(f - g)(x) = f(x) - g(x)$ $(fg)(x) = f(x) \cdot g(x)$ $(f/g)(x) = f(x)/g(x),\ g(x) \neq 0$	111, 112
	Find the composition of one function with another function *(p. 96)*.	The composition of the function f with the function g is $(f \circ g)(x) = f(g(x))$.	113, 114
	Use combinations and compositions of functions to model and solve real-life problems *(p. 98)*.	A composite function can be used to represent the number of bacteria in food as a function of the amount of time the food has been out of refrigeration. (See Example 8.)	115, 116
Section P.10	Find inverse functions informally and verify that two functions are inverse functions of each other *(p. 102)*.	Let f and g be two functions such that $f(g(x)) = x$ for every x in the domain of g and $g(f(x)) = x$ for every x in the domain of f. Under these conditions, the function g is the inverse function of the function f.	117, 118
	Use graphs to verify that two functions are inverses of each other *(p. 104)*, use the Horizontal Line Test to determine whether functions are one-to-one *(p. 105)*, and find inverse functions algebraically *(p. 106)*.	If the point (a, b) lies on the graph of f, then the point (b, a) must lie on the graph of f^{-1}, and vice versa. In short, the graph of f^{-1} is a reflection of the graph of f in the line $y = x$. **Horizontal Line Test for Inverse Functions** A function f has an inverse function if and only if no *horizontal* line intersects the graph of f at more than one point. To find an inverse function, replace $f(x)$ with y, interchange the roles of x and y, solve for y, and then replace y with $f^{-1}(x)$.	119–124

Review Exercises

See CalcChat.com for tutorial help and worked-out solutions to odd-numbered exercises.

P.1 **Classifying Real Numbers** **In Exercises 1 and 2, determine which numbers in the set are (a) natural numbers, (b) whole numbers, (c) integers, (d) rational numbers, and (e) irrational numbers.**

1. $\left\{11, -\frac{8}{9}, \frac{5}{2}, \sqrt{6}, 0.4\right\}$ **2.** $\left\{\sqrt{15}, -22, 0, 5.2, \frac{3}{7}\right\}$

Plotting and Ordering Real Numbers **In Exercises 3 and 4, plot the two real numbers on the real number line. Then place the appropriate inequality symbol (< or >) between them.**

3. (a) $\frac{5}{4}$ (b) $\frac{7}{8}$ **4.** (a) $-\frac{9}{25}$ (b) $-\frac{5}{7}$

Finding a Distance **In Exercises 5 and 6, find the distance between *a* and *b*.**

5. $a = -74, b = 48$ **6.** $a = -112, b = -6$

Using Absolute Value Notation **In Exercises 7 and 8, use absolute value notation to describe the situation.**

7. The distance between x and 7 is at least 4.

8. The distance between x and 25 is no more than 10.

Evaluating an Algebraic Expression **In Exercises 9 and 10, evaluate the expression for each value of *x*. (If not possible, state the reason.)**

	Expression	Values	
9.	$-x^2 + x - 1$	(a) $x = 1$	(b) $x = -1$
10.	$\dfrac{x}{x-3}$	(a) $x = -3$	(b) $x = 3$

Identifying Rules of Algebra **In Exercises 11–16, identify the rule(s) of algebra illustrated by the statement.**

11. $0 + (a - 5) = a - 5$ **12.** $1 \cdot (3x + 4) = 3x + 4$

13. $2x + (3x - 10) = (2x + 3x) - 10$

14. $4(t + 2) = 4 \cdot t + 4 \cdot 2$

15. $(t^2 + 1) + 3 = 3 + (t^2 + 1)$

16. $\dfrac{2}{y+4} \cdot \dfrac{y+4}{2} = 1, \quad y \neq -4$

Performing Operations **In Exercises 17–22, perform the operation(s). (Write fractional answers in simplest form.)**

17. $-6 + 6$ **18.** $2 - (-3)$

19. $(-8)(-4)$ **20.** $5(20 + 7)$

21. $\dfrac{x}{2} - \dfrac{2x}{5}$ **22.** $\dfrac{9}{x} \div \dfrac{1}{6}$

P.2 **Solving an Equation** **In Exercises 23–36, solve the equation and check your solution. (If not possible, explain why.)**

23. $2(x + 5) - 7 = x + 9$

24. $7(x - 4) = 1 - (x + 9)$

25. $\dfrac{x}{5} - 3 = \dfrac{x}{3} + 1$ **26.** $3 + \dfrac{2}{x-5} = \dfrac{2x}{x-5}$

Choosing a Method **In Exercises 27–30, solve the equation using any convenient method.**

27. $2x^2 - x - 28 = 0$ **28.** $6 = 3x^2$

29. $(x + 13)^2 = 25$ **30.** $9x^2 - 12x = 14$

Solving an Equation **In Exercises 31–38, solve the equation. Check your solutions.**

31. $5x^4 - 12x^3 = 0$ **32.** $x^3 + 8x^2 - 2x = 16$

33. $\sqrt{x + 4} = 3$ **34.** $5\sqrt{x} - \sqrt{x - 1} = 6$

35. $(x - 1)^{2/3} - 25 = 0$ **36.** $(x + 2)^{3/4} = 27$

37. $|x - 5| = 10$ **38.** $|x^2 - 3| = 2x$

P.3 **Plotting, Distance, and Midpoint** **In Exercises 39 and 40, (a) plot the points, (b) find the distance between the points, and (c) find the midpoint of the line segment joining the points.**

39. $(-3, 8), (1, 5)$ **40.** $(-2, 6), (4, -3)$

Meteorology **In Exercises 41 and 42, use the following information. The apparent temperature is a measure of relative discomfort to a person from heat and high humidity. The table shows the actual temperatures *x* (in degrees Fahrenheit) versus the apparent temperatures *y* (in degrees Fahrenheit) for a relative humidity of 75%.**

x	70	75	80	85	90	95	100
y	70	77	85	95	109	130	150

41. Sketch a scatter plot of the data shown in the table.

42. Find the change in the apparent temperature when the actual temperature changes from 70°F to 100°F.

Sketching the Graph of an Equation **In Exercises 43–46, construct a table of values that consists of several solution points of the equation. Use the resulting solution points to sketch the graph of the equation.**

43. $y = 2x - 6$ **44.** $y = -\frac{1}{2}x + 2$

45. $y = x^2 + 2x$ **46.** $y = 2x^2 - x - 9$

Finding *x*- and *y*-Intercepts In Exercises 47 and 48, find the *x*- and *y*-intercepts of the graph of the equation.

47. $y = (x - 3)^2 - 4$ 48. $y = |x + 1| - 3$

Testing for Symmetry In Exercises 49–52, use the algebraic tests to check for symmetry with respect to both axes and the origin. Then sketch the graph of the equation.

49. $y = -3x + 7$ 50. $y = 3x^3$

51. $y = -x^4 + 6x^2$ 52. $y = |x| - 4$

Writing the Equation of a Circle In Exercises 53 and 54, write the standard form of the equation of the circle for which the endpoints of a diameter are given.

53. $(0, 0), (4, -6)$ 54. $(-2, -3), (4, -10)$

Sketching the Graph of a Circle In Exercises 55 and 56, find the center and radius of the circle. Then sketch the graph of the circle.

55. $x^2 + y^2 = 9$

56. $(x + 4)^2 + \left(y - \frac{3}{2}\right)^2 = 100$

P.4 Graphing a Linear Equation In Exercises 57–60, find the slope and *y*-intercept (if possible) of the line. Sketch the line.

57. $y = -\frac{1}{2}x + 1$ 58. $2x - 3y = 6$

59. $y = 1$ 60. $x = -6$

Finding the Slope of a Line Through Two Points In Exercises 61 and 62, find the slope of the line passing through the pair of points.

61. $(5, -2), (-1, 4)$ 62. $(-1, 6), (3, -2)$

Using the Point-Slope Form In Exercises 63 and 64, find the slope-intercept form of the equation of the line that has the given slope and passes through the given point. Sketch the line.

63. $m = \frac{1}{3}, \quad (6, -5)$ 64. $m = -\frac{3}{4}, \quad (-4, -2)$

Finding an Equation of a Line In Exercises 65 and 66, find an equation of the line passing through the pair of points. Sketch the line.

65. $(-6, 4), (4, 9)$ 66. $(-9, -3), (-3, -5)$

Finding Parallel and Perpendicular Lines In Exercises 67 and 68, find equations of the lines that pass through the given point and are (a) parallel to and (b) perpendicular to the given line.

67. $5x - 4y = 8, \quad (3, -2)$ 68. $2x + 3y = 5, \quad (-8, 3)$

69. **Sales** A discount outlet offers a 20% discount on all items. Write a linear equation giving the sale price S for an item with a list price L.

70. **Hourly Wage** A manuscript translator charges a starting fee of \$50 plus \$2.50 per page translated. Write a linear equation for the amount A earned for translating p pages.

P.5 Testing for Functions Represented Algebraically In Exercises 71–74, determine whether the equation represents *y* as a function of *x*.

71. $16x - y^4 = 0$ 72. $2x - y - 3 = 0$

73. $y = \sqrt{1 - x}$ 74. $|y| = x + 2$

Evaluating a Function In Exercises 75 and 76, find each function value.

75. $g(x) = x^{4/3}$

(a) $g(8)$ (b) $g(t + 1)$ (c) $g(-27)$ (d) $g(-x)$

76. $h(x) = |x - 2|$

(a) $h(-4)$ (b) $h(-2)$ (c) $h(0)$ (d) $h(-x + 2)$

Finding the Domain of a Function In Exercises 77 and 78, find the domain of the function.

77. $f(x) = \sqrt{25 - x^2}$ 78. $h(x) = \dfrac{x}{x^2 - x - 6}$

79. **Physics** The velocity of a ball projected upward from ground level is given by $v(t) = -32t + 48$, where t is the time in seconds and v is the velocity in feet per second.

(a) Find the velocity when $t = 1$.

(b) Find the time when the ball reaches its maximum height. [*Hint:* Find the time when $v(t) = 0$.]

80. **Evaluating a Difference Quotient** Find the difference quotient and simplify your answer.

$$f(x) = 2x^2 + 3x - 1, \quad \frac{f(x + h) - f(x)}{h}, \quad h \neq 0$$

P.6 Vertical Line Test for Functions In Exercises 81 and 82, sketch the graph of the equation. Then use the Vertical Line Test to determine whether the graph represents *y* as a function of *x*.

81. $y = (x - 3)^2$ 82. $x = -|4 - y|$

Finding the Zeros of a Function In Exercises 83 and 84, find the zeros of the function algebraically.

83. $f(x) = \sqrt{2x + 1}$ 84. $f(x) = \dfrac{x^3 - x^2}{2x + 1}$

Describing Function Behavior In Exercises 85 and 86, use a graphing utility to graph the function and visually determine the open intervals on which the function is increasing, decreasing, or constant.

85. $f(x) = |x| + |x + 1|$ 86. $f(x) = (x^2 - 4)^2$

Approximating Relative Minima or Maxima In Exercises 87 and 88, use a graphing utility to approximate (to two decimal places) any relative minima or maxima of the function.

87. $f(x) = -x^2 + 2x + 1$
88. $f(x) = x^3 - 4x^2 - 1$

Average Rate of Change of a Function In Exercises 89 and 90, find the average rate of change of the function from x_1 to x_2.

89. $f(x) = -x^2 + 8x - 4, \quad x_1 = 0, x_2 = 4$
90. $f(x) = x^3 + 2x + 1, \quad x_1 = 1, x_2 = 3$

Even, Odd, or Neither? In Exercises 91–94, determine whether the function is even, odd, or neither. Then describe the symmetry.

91. $f(x) = x^5 + 4x - 7$
92. $f(x) = x^4 - 20x^2$
93. $f(x) = 2x\sqrt{x^2 + 3}$
94. $f(x) = \sqrt[5]{6x^2}$

P.7 Writing a Linear Function In Exercises 95 and 96, (a) write the linear function f that has the given function values and (b) sketch the graph of the function.

95. $f(2) = -6, \quad f(-1) = 3$
96. $f(0) = -5, \quad f(4) = -8$

Graphing a Function In Exercises 97–100, sketch the graph of the function.

97. $g(x) = [\![x]\!] - 2$
98. $g(x) = [\![x + 4]\!]$
99. $f(x) = \begin{cases} 5x - 3, & x \geq -1 \\ -4x + 5, & x < -1 \end{cases}$
100. $f(x) = \begin{cases} 2x + 1, & x \leq 2 \\ x^2 + 1, & x > 2 \end{cases}$

P.8 Describing Transformations In Exercises 101–110, h is related to one of the parent functions described in this chapter. (a) Identify the parent function f. (b) Describe the sequence of transformations from f to h. (c) Sketch the graph of h. (d) Use function notation to write h in terms of f.

101. $h(x) = x^2 - 9$
102. $h(x) = (x - 2)^3 + 2$
103. $h(x) = -\sqrt{x} + 4$
104. $h(x) = |x + 3| - 5$
105. $h(x) = -(x + 2)^2 + 3$
106. $h(x) = \frac{1}{2}(x - 1)^2 - 2$
107. $h(x) = -[\![x]\!] + 6$
108. $h(x) = -\sqrt{x + 1} + 9$
109. $h(x) = 5[\![x - 9]\!]$
110. $h(x) = -\frac{1}{3}x^3$

P.9 Finding Arithmetic Combinations of Functions In Exercises 111 and 112, find (a) $(f + g)(x)$, (b) $(f - g)(x)$, (c) $(fg)(x)$, and (d) $(f/g)(x)$. What is the domain of f/g?

111. $f(x) = x^2 + 3, \quad g(x) = 2x - 1$
112. $f(x) = x^2 - 4, \quad g(x) = \sqrt{3 - x}$

Finding Domains of Functions and Composite Functions In Exercises 113 and 114, find (a) $f \circ g$ and (b) $g \circ f$. Find the domain of each function and of each composite function.

113. $f(x) = \frac{1}{3}x - 3, \quad g(x) = 3x + 1$
114. $f(x) = \sqrt{x + 1}, \quad g(x) = x^2$

Retail In Exercises 115 and 116, the price of a washing machine is x dollars. The function $f(x) = x - 100$ gives the price of the washing machine after a \$100 rebate. The function $g(x) = 0.95x$ gives the price of the washing machine after a 5% discount.

115. Find and interpret $(f \circ g)(x)$.
116. Find and interpret $(g \circ f)(x)$.

P.10 Finding an Inverse Function Informally In Exercises 117 and 118, find the inverse function of f informally. Verify that $f(f^{-1}(x)) = x$ and $f^{-1}(f(x)) = x$.

117. $f(x) = \dfrac{x - 4}{5}$
118. $f(x) = x^3 - 1$

Applying the Horizontal Line Test In Exercises 119 and 120, use a graphing utility to graph the function, and use the Horizontal Line Test to determine whether the function has an inverse function.

119. $f(x) = (x - 1)^2$
120. $h(t) = \dfrac{2}{t - 3}$

Finding and Analyzing Inverse Functions In Exercises 121 and 122, (a) find the inverse function of f, (b) graph both f and f^{-1} on the same set of coordinate axes, (c) describe the relationship between the graphs of f and f^{-1}, and (d) state the domains and ranges of f and f^{-1}.

121. $f(x) = \frac{1}{2}x - 3$
122. $f(x) = \sqrt{x + 1}$

Restricting the Domain In Exercises 123 and 124, restrict the domain of the function f to an interval on which the function is increasing, and find f^{-1} on that interval.

123. $f(x) = 2(x - 4)^2$
124. $f(x) = |x - 2|$

Exploration

True or False? In Exercises 125 and 126, determine whether the statement is true or false. Justify your answer.

125. Relative to the graph of $f(x) = \sqrt{x}$, the graph of the function $h(x) = -\sqrt{x + 9} - 13$ is shifted 9 units to the left and 13 units down, then reflected in the x-axis.
126. If f and g are two inverse functions, then the domain of g is equal to the range of f.

Chapter Test

See CalcChat.com for tutorial help and worked-out solutions to odd-numbered exercises.

Take this test as you would take a test in class. When you are finished, check your work against the answers given in the back of the book.

1. Place the appropriate inequality symbol ($<$ or $>$) between the real numbers $-\frac{10}{3}$ and $-\frac{5}{3}$.
2. Find the distance between the real numbers $-\frac{7}{4}$ and $\frac{5}{4}$.
3. Identify the rule of algebra illustrated by $(5 - x) + 0 = 5 - x$.

In Exercises 4–7, solve the equation and check your solution. (If not possible, explain why.)

4. $\frac{2}{3}(x - 1) + \frac{1}{4}x = 10$
5. $(x - 4)(x + 2) = 7$
6. $\dfrac{x - 2}{x + 2} + \dfrac{4}{x + 2} + 4 = 0$
7. $|3x - 1| = 7$

8. Plot the points $(-2, 5)$ and $(6, 0)$. Then find the distance between the points and the midpoint of the line segment joining the points.

In Exercises 9–11, find any intercepts and test for symmetry. The sketch the graph of the equation.

9. $y = 4 - \frac{3}{4}x$
10. $y = 4 - \frac{3}{4}|x|$
11. $y = x - x^3$

12. Find the center and radius of the circle given by $(x - 3)^2 + y^2 = 9$. Then sketch the circle.

In Exercises 13 and 14, find an equation of the line passing through the pair of points. Sketch the line.

13. $(-2, 5), (1, -7)$
14. $(-4, -7), \left(1, \frac{4}{3}\right)$

15. Find equations of the lines that pass through the point $(0, 4)$ and are (a) parallel to and (b) perpendicular to the line $5x + 2y = 3$.
16. Let $f(x) = |x + 2| - 15$. Find each function value.
    (a) $f(-8)$ (b) $f(14)$ (c) $f(x - 6)$

In Exercises 17–19, (a) use a graphing utility to graph the function, (b) find the domain of the function, (c) approximate the open intervals on which the function is increasing, decreasing, or constant, and (d) determine whether the function is even, odd, or neither.

17. $f(x) = |x + 5|$
18. $f(x) = 4x\sqrt{3 - x}$
19. $f(x) = 2x^6 + 5x^4 - x^2$

In Exercises 20–22, (a) identify the parent function f in the transformation, (b) describe the sequence of transformations from f to h, and (c) sketch the graph of h.

20. $h(x) = 4[\![x]\!]$
21. $h(x) = -\sqrt{x + 5} + 8$
22. $h(x) = -2(x - 5)^3 + 3$

In Exercises 23 and 24, find (a) $(f + g)(x)$, (b) $(f - g)(x)$, (c) $(fg)(x)$, (d) $(f/g)(x)$, (e) $(f \circ g)(x)$, and (f) $(g \circ f)(x)$.

23. $f(x) = 3x^2 - 7,\quad g(x) = -x^2 - 4x + 5$
24. $f(x) = 1/x,\quad g(x) = 2\sqrt{x}$

In Exercises 25–27, determine whether the function has an inverse function. If it does, find the inverse function.

25. $f(x) = x^3 + 8$
26. $f(x) = |x^2 - 3| + 6$
27. $f(x) = 3x\sqrt{x}$

Proofs in Mathematics

What does the word *proof* mean to you? In mathematics, the word *proof* means a valid argument. When you prove a statement or theorem, you must use facts, definitions, and accepted properties in a logical order. You can also use previously proved theorems in your proof. For example, the proof of the Midpoint Formula below uses the Distance Formula. There are several different proof methods, which you will see in later chapters.

The Midpoint Formula *(p. 29)*

The midpoint of the line segment joining the points (x_1, y_1) and (x_2, y_2) is

$$\text{Midpoint} = \left(\frac{x_1 + x_2}{2}, \frac{y_1 + y_2}{2}\right).$$

THE CARTESIAN PLANE

The Cartesian plane is named after French mathematician René Descartes (1596–1650). According to some accounts, while Descartes was lying in bed, he noticed a fly buzzing around on the ceiling. He realized that he could describe the fly's position by its distance from the bedroon walls. This led to the development of the Cartesian plane. Descartes felt that using a coordinate plane could facilitate descriptions of the positions of objects.

Proof

Using the figure, you must show that $d_1 = d_2$ and $d_1 + d_2 = d_3$.

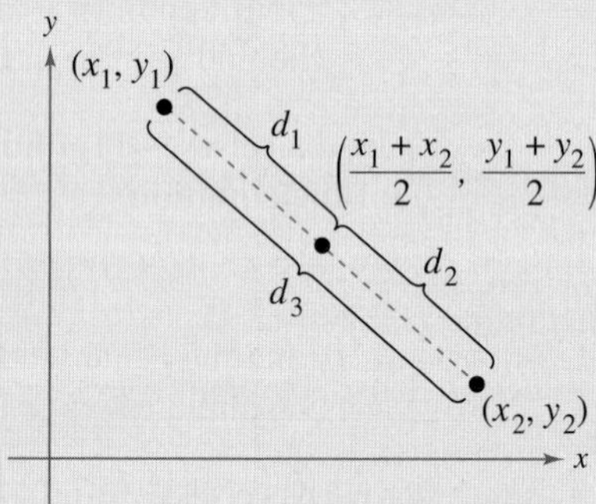

By the Distance Formula, you obtain

$$d_1 = \sqrt{\left(\frac{x_1 + x_2}{2} - x_1\right)^2 + \left(\frac{y_1 + y_2}{2} - y_1\right)^2}$$

$$= \sqrt{\left(\frac{x_2 - x_1}{2}\right)^2 + \left(\frac{y_2 - y_1}{2}\right)^2}$$

$$= \frac{1}{2}\sqrt{(x_2 - x_1)^2 + (y_2 - y_1)^2},$$

$$d_2 = \sqrt{\left(x_2 - \frac{x_1 + x_2}{2}\right)^2 + \left(y_2 - \frac{y_1 + y_2}{2}\right)^2}$$

$$= \sqrt{\left(\frac{x_2 - x_1}{2}\right)^2 + \left(\frac{y_2 - y_1}{2}\right)^2}$$

$$= \frac{1}{2}\sqrt{(x_2 - x_1)^2 + (y_2 - y_1)^2},$$

and

$$d_3 = \sqrt{(x_2 - x_1)^2 + (y_2 - y_1)^2}.$$

So, it follows that $d_1 = d_2$ and $d_1 + d_2 = d_3$. ■

P.S. Problem Solving

1. **Monthly Wages** As a salesperson, you receive a monthly salary of \$2000, plus a commission of 7% of sales. You receive an offer for a new job at \$2300 per month, plus a commission of 5% of sales.
   (a) Write a linear equation for your current monthly wage W_1 in terms of your monthly sales S.
   (b) Write a linear equation for the monthly wage W_2 of your new job offer in terms of the monthly sales S.
   (c) Use a graphing utility to graph both equations in the same viewing window. Find the point of intersection. What does the point of intersection represent?
   (d) You expect sales of \$20,000 per month. Should you change jobs? Explain.

2. **Cellphone Keypad** For the numbers 2 through 9 on a cellphone keypad (see figure), consider two relations: one mapping numbers onto letters, and the other mapping letters onto numbers. Are both relations functions? Explain.

3. **Sums and Differences of Functions** What can be said about the sum and difference of each pair of functions?
   (a) Two even functions
   (b) Two odd functions
   (c) An odd function and an even function

4. **Inverse Functions** The functions

   $f(x) = x$ and $g(x) = -x$

   are their own inverse functions. Graph each function and explain why this is true. Graph other linear functions that are their own inverse functions. Find a formula for a family of linear functions that are their own inverse functions.

5. **Proof** Prove that a function of the form

   $y = a_{2n}x^{2n} + a_{2n-2}x^{2n-2} + \cdots + a_2x^2 + a_0$

   is an even function.

6. **Miniature Golf** A golfer is trying to make a hole-in-one on the miniature golf green shown. The golf ball is at the point (2.5, 2) and the hole is at the point (9.5, 2). The golfer wants to bank the ball off the side wall of the green at the point (x, y). Find the coordinates of the point (x, y). Then write an equation for the path of the ball.

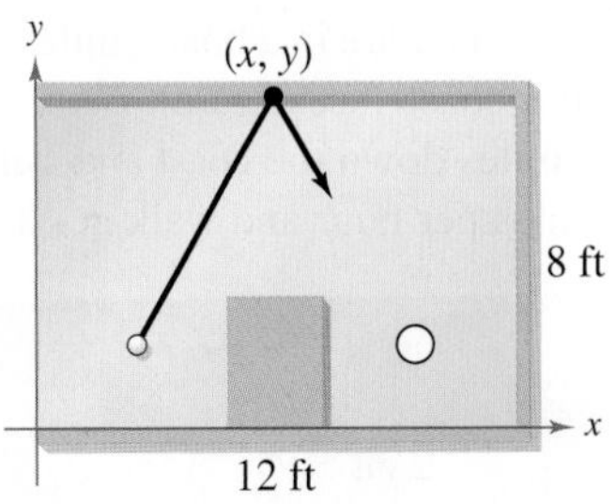

Figure for 6

7. **Titanic** At 2:00 P.M. on April 11, 1912, the *Titanic* left Cobh, Ireland, on her voyage to New York City. At 11:40 P.M. on April 14, the *Titanic* struck an iceberg and sank, having covered only about 2100 miles of the approximately 3400-mile trip.
   (a) What was the total duration of the voyage in hours?
   (b) What was the average speed in miles per hour?
   (c) Write a function relating the distance of the *Titanic* from New York City and the number of hours traveled. Find the domain and range of the function.
   (d) Graph the function in part (c).

8. **Average Rate of Change** Consider the function $f(x) = -x^2 + 4x - 3$. Find the average rate of change of the function from x_1 to x_2.
   (a) $x_1 = 1, x_2 = 2$
   (b) $x_1 = 1, x_2 = 1.5$
   (c) $x_1 = 1, x_2 = 1.25$
   (d) $x_1 = 1, x_2 = 1.125$
   (e) $x_1 = 1, x_2 = 1.0625$
   (f) Does the average rate of change seem to be approaching one value? If so, state the value.
   (g) Find the equations of the secant lines through the points $(x_1, f(x_1))$ and $(x_2, f(x_2))$ for parts (a)–(e).
   (h) Find the equation of the line through the point $(1, f(1))$ using your answer from part (f) as the slope of the line.

9. **Inverse of a Composition** Consider the functions $f(x) = 4x$ and $g(x) = x + 6$.
   (a) Find $(f \circ g)(x)$.
   (b) Find $(f \circ g)^{-1}(x)$.
   (c) Find $f^{-1}(x)$ and $g^{-1}(x)$.
   (d) Find $(g^{-1} \circ f^{-1})(x)$ and compare the result with that of part (b).
   (e) Repeat parts (a) through (d) for $f(x) = x^3 + 1$ and $g(x) = 2x$.
   (f) Write two one-to-one functions f and g, and repeat parts (a) through (d) for these functions.
   (g) Make a conjecture about $(f \circ g)^{-1}(x)$ and $(g^{-1} \circ f^{-1})(x)$.

10. Trip Time You are in a boat 2 miles from the nearest point on the coast (see figure). You plan to travel to point Q, 3 miles down the coast and 1 mile inland. You row at 2 miles per hour and walk at 4 miles per hour.

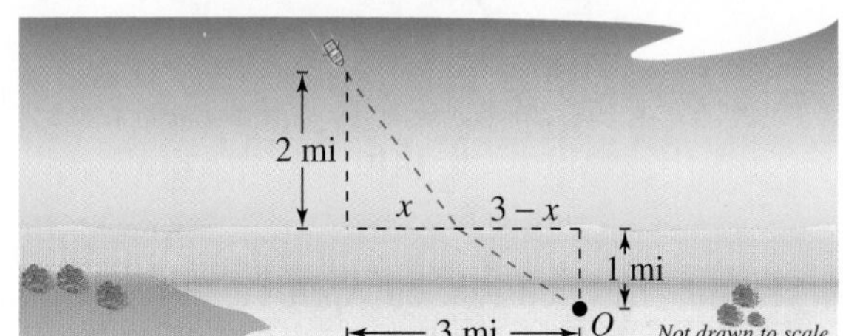

(a) Write the total time T (in hours) of the trip as a function of the distance x (in miles).

(b) Determine the domain of the function.

(c) Use a graphing utility to graph the function. Be sure to choose an appropriate viewing window.

(d) Find the value of x that minimizes T.

(e) Write a brief paragraph interpreting these values.

11. Heaviside Function The **Heaviside function**

$$H(x) = \begin{cases} 1, & x \geq 0 \\ 0, & x < 0 \end{cases}$$

is widely used in engineering applications. (See figure.) To print an enlarged copy of the graph, go to *MathGraphs.com*.

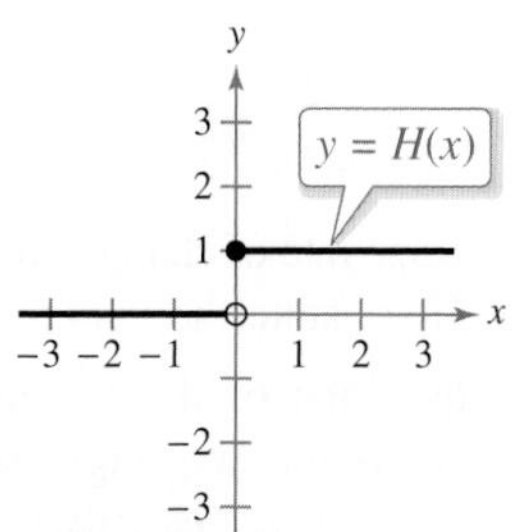

Sketch the graph of each function by hand.

(a) $H(x) - 2$

(b) $H(x - 2)$

(c) $-H(x)$

(d) $H(-x)$

(e) $\frac{1}{2}H(x)$

(f) $-H(x - 2) + 2$

12. Repeated Composition Let $f(x) = \dfrac{1}{1 - x}$.

(a) Find the domain and range of f.

(b) Find $f(f(x))$. What is the domain of this function?

(c) Find $f(f(f(x)))$. Is the graph a line? Why or why not?

13. Associative Property with Compositions Show that the Associative Property holds for compositions of functions—that is,

$$(f \circ (g \circ h))(x) = ((f \circ g) \circ h)(x).$$

14. Graphical Reasoning Use the graph of the function f to sketch the graph of each function. To print an enlarged copy of the graph, go to *MathGraphs.com*.

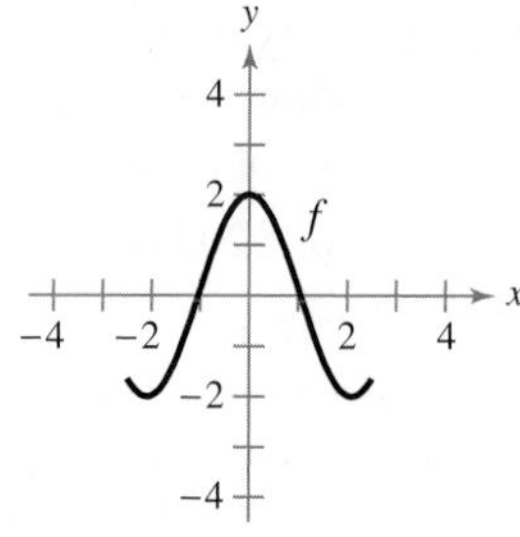

(a) $f(x + 1)$

(b) $f(x) + 1$

(c) $2f(x)$

(d) $f(-x)$

(e) $-f(x)$

(f) $|f(x)|$

(g) $f(|x|)$

15. Graphical Reasoning Use the graphs of f and f^{-1} to complete each table of function values.

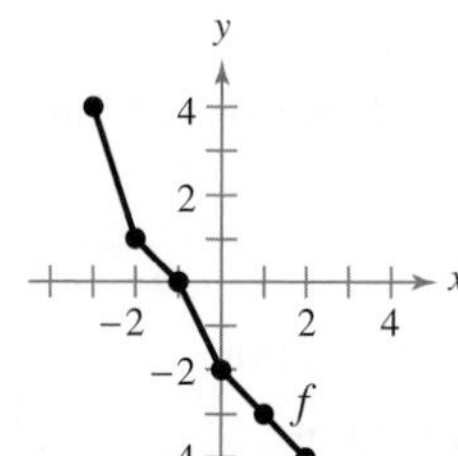

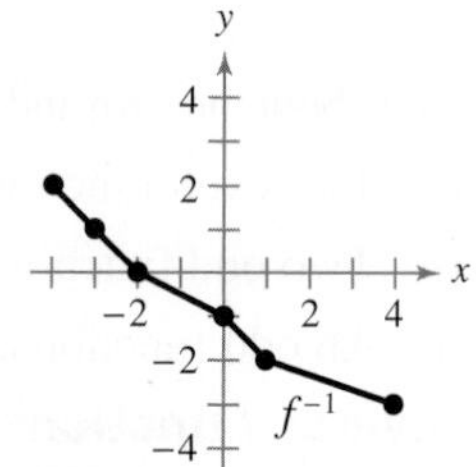

(a)

x	-4	-2	0	4
$(f(f^{-1}(x)))$				

(b)

x	-3	-2	0	1
$(f + f^{-1})(x)$				

(c)

x	-3	-2	0	1
$(f \cdot f^{-1})(x)$				

(d)

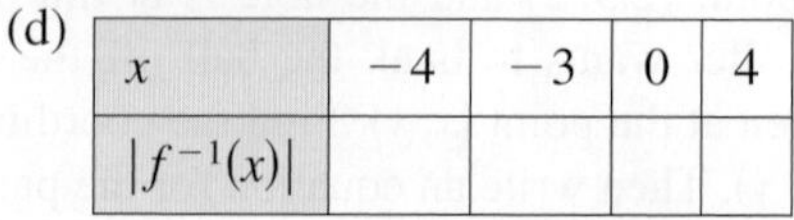

x	-4	-3	0	4
$\lvert f^{-1}(x)\rvert$				

1 Trigonometry

Television Coverage (Exercise 85, page 179)

Waterslide Design (Exercise 30, page 197)

Respiratory Cycle (Exercise 80, page 168)

Temperature of a City (Exercise 99, page 158)

Skateboard Ramp (Example 10, page 145)

1.1 Radian and Degree Measure

Angles and their measure have a wide variety of real-life applications. For example, in Exercise 68 on page 131, you will use angles and their measure to model the distance a cyclist travels.

- **Describe angles.**
- **Use radian measure.**
- **Use degree measure.**
- **Use angles and their measure to model and solve real-life problems.**

Angles

As derived from the Greek language, the word **trigonometry** means "measurement of triangles." Originally, trigonometry dealt with relationships among the sides and angles of triangles and was instrumental in the development of astronomy, navigation, and surveying. With the development of calculus and the physical sciences in the 17th century, a different perspective arose—one that viewed the classic trigonometric relationships as *functions* with the set of real numbers as their domains. Consequently, the applications of trigonometry expanded to include a vast number of physical phenomena, such as sound waves, planetary orbits, vibrating strings, pendulums, and orbits of atomic particles. This text incorporates *both* perspectives, starting with angles and their measure.

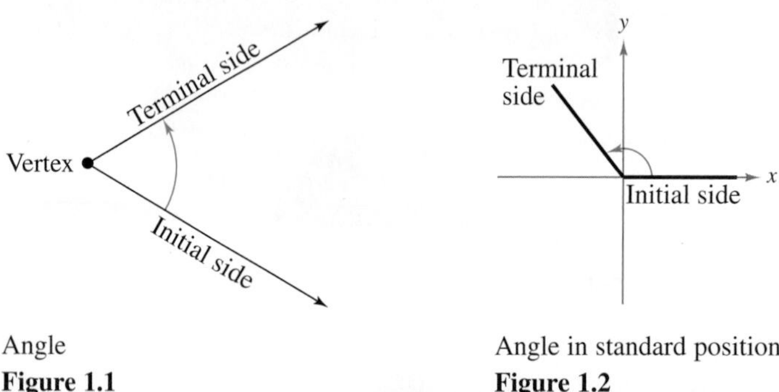

Angle
Figure 1.1

Angle in standard position
Figure 1.2

Rotating a ray (half-line) about its endpoint determines an **angle.** The starting position of the ray is the **initial side** of the angle, and the position after rotation is the **terminal side,** as shown in Figure 1.1. The endpoint of the ray is the **vertex** of the angle. This perception of an angle fits a coordinate system in which the origin is the vertex and the initial side coincides with the positive x-axis. Such an angle is in **standard position,** as shown in Figure 1.2. Counterclockwise rotation generates **positive angles** and clockwise rotation generates **negative angles,** as shown in Figure 1.3. Labels for angles can be Greek letters such as α (alpha), β (beta), and θ (theta) or uppercase letters such as A, B, and C. In Figure 1.4, note that angles α and β have the same initial and terminal sides. Such angles are **coterminal.**

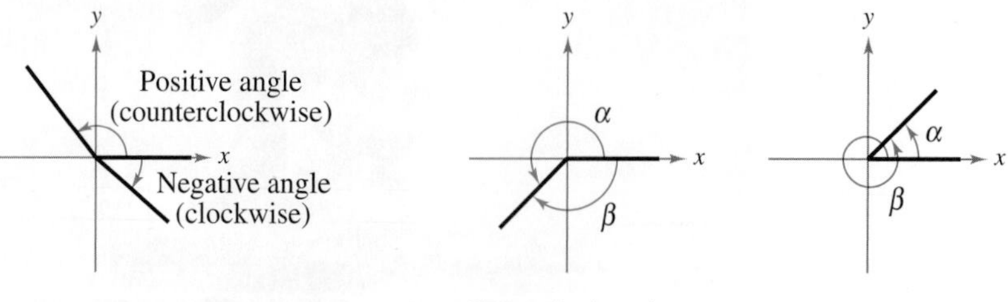

Figure 1.3

Coterminal angles
Figure 1.4

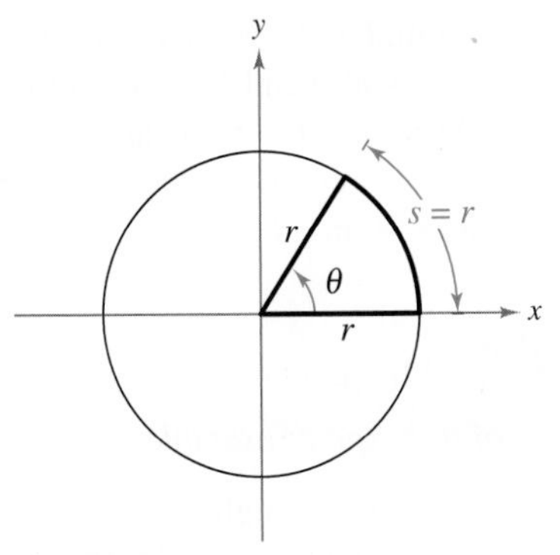

Arc length = radius when $\theta = 1$ radian.
Figure 1.5

Radian Measure

The amount of rotation from the initial side to the terminal side determines the **measure of an angle.** One way to measure angles is in *radians.* This type of measure is especially useful in calculus. To define a radian, use a **central angle** of a circle, which is an angle whose vertex is the center of the circle, as shown in Figure 1.5.

Definition of a Radian

One **radian** (rad) is the measure of a central angle θ that intercepts an arc s equal in length to the radius r of the circle. (See Figure 1.5.) Algebraically, this means that

$$\theta = \frac{s}{r}$$

where θ is measured in radians. (Note that $\theta = 1$ when $s = r$.)

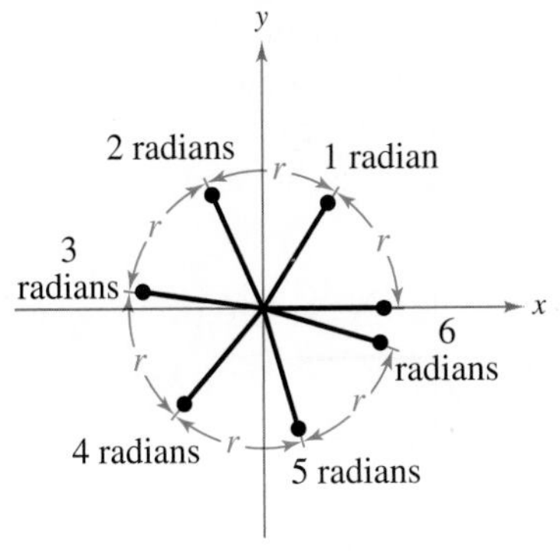

Figure 1.6

The circumference of a circle is $2\pi r$ units, so it follows that a central angle of one full revolution (counterclockwise) corresponds to an arc length of $s = 2\pi r$. Moreover, $2\pi \approx 6.28$, so there are just over six radius lengths in a full circle, as shown in Figure 1.6. The units of measure for s and r are the same, so the ratio s/r has no units—it is a real number.

The measure of an angle of one full revolution is $s/r = 2\pi r/r = 2\pi$ radians, so you can obtain the following.

$$\frac{1}{2} \text{ revolution} = \frac{2\pi}{2} = \pi \text{ radians} \qquad \frac{1}{4} \text{ revolution} = \frac{2\pi}{4} = \frac{\pi}{2} \text{ radians}$$

$$\frac{1}{6} \text{ revolution} = \frac{2\pi}{6} = \frac{\pi}{3} \text{ radians}$$

These and other common angles are shown below.

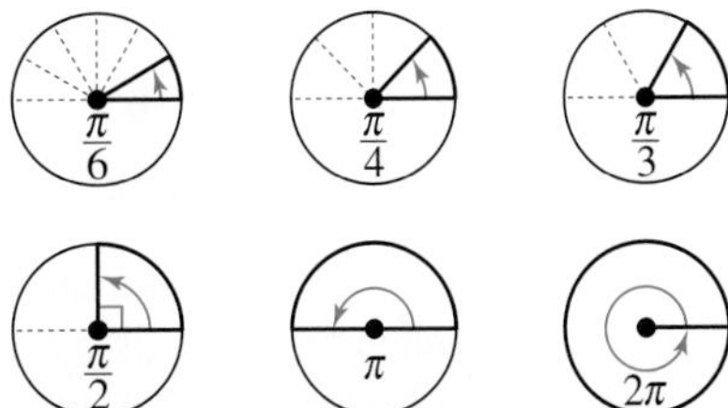

REMARK The phrase "θ lies in a quadrant" is an abbreviation for the phrase "the terminal side of θ lies in a quadrant." The terminal sides of the "quadrantal angles" 0, $\pi/2$, π, and $3\pi/2$ do not lie within quadrants.

Recall that the four quadrants in a coordinate system are numbered I, II, III, and IV. The figure below shows which angles between 0 and 2π lie in each of the four quadrants. Note that angles between 0 and $\pi/2$ are **acute** angles and angles between $\pi/2$ and π are **obtuse** angles.

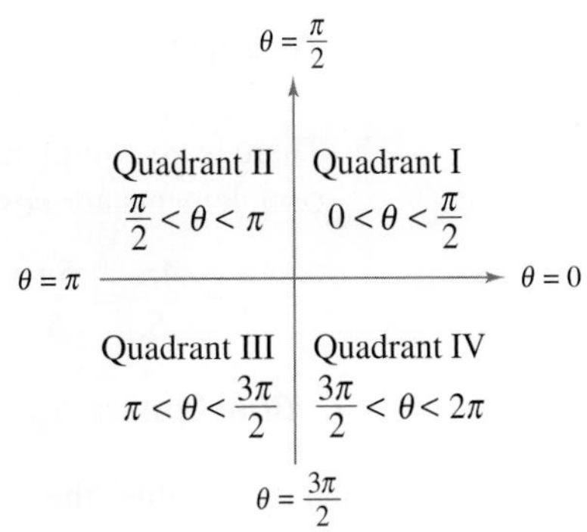

Two angles are coterminal when they have the same initial and terminal sides. For example, the angles 0 and 2π are coterminal, as are the angles $\pi/6$ and $13\pi/6$. To find an angle that is coterminal to a given angle θ, add or subtract 2π (one revolution), as demonstrated in Example 1. A given angle θ has infinitely many coterminal angles. For example, $\theta = \pi/6$ is coterminal with $(\pi/6) + 2n\pi$, where n is an integer.

EXAMPLE 1 Finding Coterminal Angles

See LarsonPrecalculus.com for an interactive version of this type of example.

▷ **ALGEBRA HELP** To review operations involving fractions, see Section P.1.

a. For the positive angle $13\pi/6$, subtract 2π to obtain a coterminal angle.

$$\frac{13\pi}{6} - 2\pi = \frac{\pi}{6}$$ See Figure 1.7.

b. For the negative angle $-2\pi/3$, add 2π to obtain a coterminal angle.

$$-\frac{2\pi}{3} + 2\pi = \frac{4\pi}{3}$$ See Figure 1.8.

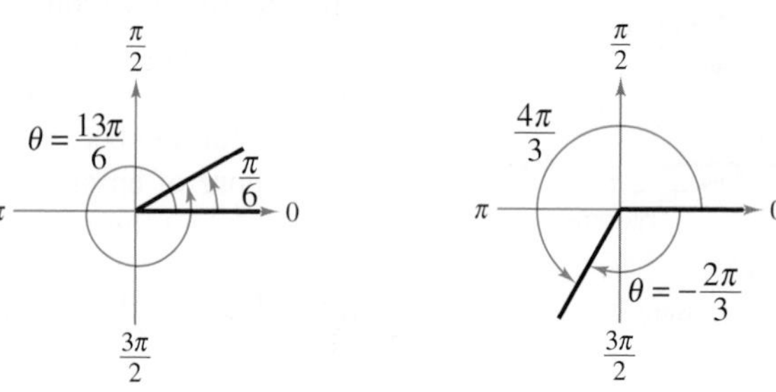

Figure 1.7 **Figure 1.8**

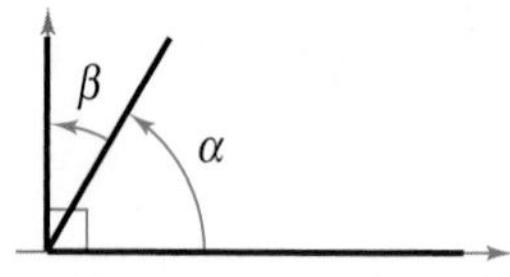

Complementary angles

✓ **Checkpoint** *Audio-video solution in English & Spanish at LarsonPrecalculus.com*

Determine two coterminal angles (one positive and one negative) for each angle.

a. $\theta = \dfrac{9\pi}{4}$ **b.** $\theta = -\dfrac{\pi}{3}$

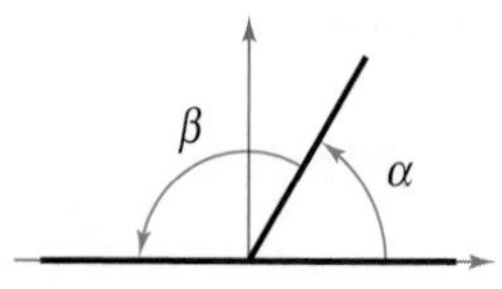

Supplementary angles
Figure 1.9

Two positive angles α and β are **complementary** (complements of each other) when their sum is $\pi/2$. Two positive angles are **supplementary** (supplements of each other) when their sum is π. (See Figure 1.9.)

EXAMPLE 2 Complementary and Supplementary Angles

a. The complement of $\dfrac{2\pi}{5}$ is $\dfrac{\pi}{2} - \dfrac{2\pi}{5} = \dfrac{5\pi}{10} - \dfrac{4\pi}{10} = \dfrac{\pi}{10}$.

The supplement of $\dfrac{2\pi}{5}$ is $\pi - \dfrac{2\pi}{5} = \dfrac{5\pi}{5} - \dfrac{2\pi}{5} = \dfrac{3\pi}{5}$.

b. There is no complement of $4\pi/5$ because $4\pi/5$ is greater than $\pi/2$. (Remember that complements are *positive* angles.) The supplement of $4\pi/5$ is

$$\pi - \frac{4\pi}{5} = \frac{5\pi}{5} - \frac{4\pi}{5} = \frac{\pi}{5}.$$

Find (if possible) the complement and supplement of (a) $\pi/6$ and (b) $5\pi/6$.

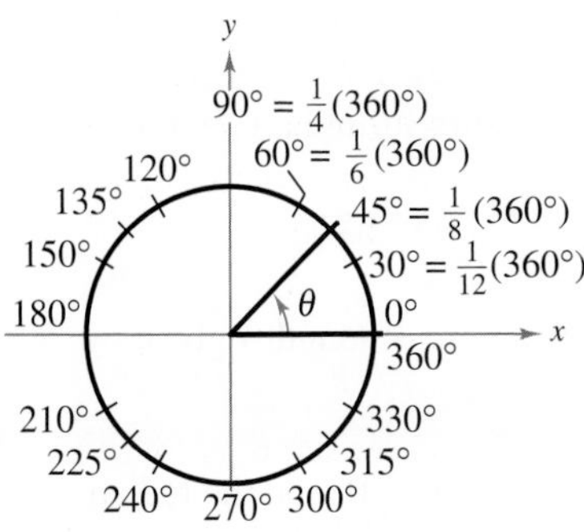

Figure 1.10

Degree Measure

Another way to measure angles is in **degrees,** denoted by the symbol °. A measure of one degree (1°) is equivalent to a rotation of $\frac{1}{360}$ of a complete revolution about the vertex. To measure angles, it is convenient to mark degrees on the circumference of a circle, as shown in Figure 1.10. So, a full revolution (counterclockwise) corresponds to 360°, a half revolution corresponds to 180°, a quarter revolution corresponds to 90°, and so on.

One complete revolution corresponds to 2π radians, so degrees and radians are related by the equations

$$360° = 2\pi \text{ rad} \quad \text{and} \quad 180° = \pi \text{ rad}.$$

From these equations, you obtain

$$1° = \frac{\pi}{180} \text{ rad} \quad \text{and} \quad 1 \text{ rad} = \left(\frac{180}{\pi}\right)^{\circ}$$

which lead to the conversion rules below.

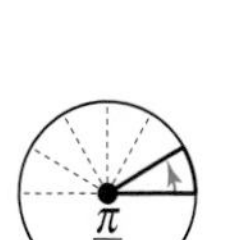

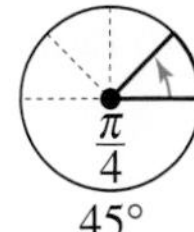

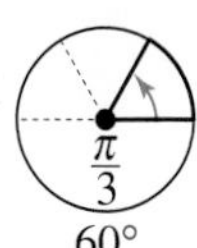

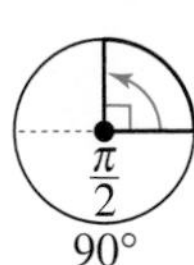

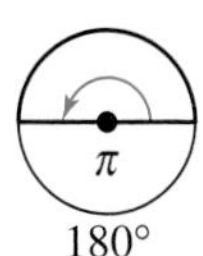

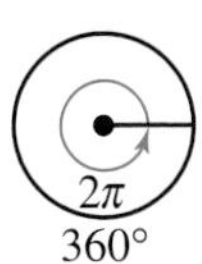

Figure 1.11

Conversions Between Degrees and Radians

1. To convert degrees to radians, multiply degrees by $\dfrac{\pi \text{ rad}}{180°}$.
2. To convert radians to degrees, multiply radians by $\dfrac{180°}{\pi \text{ rad}}$.

To apply these two conversion rules, use the basic relationship π rad = 180°. (See Figure 1.11.)

When no units of angle measure are specified, *radian measure is implied.* For example, $\theta = 2$ implies that $\theta = 2$ radians.

TECHNOLOGY With calculators, it is convenient to use *decimal* degrees to denote fractional parts of degrees. Historically, however, fractional parts of degrees were expressed in *minutes* and *seconds*, using the prime (′) and double prime (″) notations, respectively. That is,

$$1' = \text{one minute} = \tfrac{1}{60}(1°)$$
$$1'' = \text{one second} = \tfrac{1}{3600}(1°).$$

For example, you would write an angle θ of 64 degrees, 32 minutes, and 47 seconds as $\theta = 64° \, 32' \, 47''$.

Many calculators have special keys for converting an angle in degrees, minutes, and seconds (D° M′ S″) to decimal degree form and vice versa.

EXAMPLE 3 **Converting from Degrees to Radians**

a. $135° = (135 \text{ deg})\left(\dfrac{\pi \text{ rad}}{180 \text{ deg}}\right) = \dfrac{3\pi}{4}$ radians Multiply by $\dfrac{\pi \text{ rad}}{180°}$.

b. $540° = (540 \text{ deg})\left(\dfrac{\pi \text{ rad}}{180 \text{ deg}}\right) = 3\pi$ radians Multiply by $\dfrac{\pi \text{ rad}}{180°}$.

Checkpoint Audio-video solution in English & Spanish at LarsonPrecalculus.com

Convert each degree measure to radian measure as a multiple of π. Do not use a calculator.

a. 60° **b.** 320°

EXAMPLE 4 **Converting from Radians to Degrees**

a. $-\dfrac{\pi}{2} \text{ rad} = \left(-\dfrac{\pi}{2} \text{ rad}\right)\left(\dfrac{180 \text{ deg}}{\pi \text{ rad}}\right) = -90°$ Multiply by $\dfrac{180°}{\pi \text{ rad}}$.

b. $2 \text{ rad} = (2 \text{ rad})\left(\dfrac{180 \text{ deg}}{\pi \text{ rad}}\right) = \dfrac{360°}{\pi} \approx 114.59°$ Multiply by $\dfrac{180°}{\pi \text{ rad}}$.

Checkpoint Audio-video solution in English & Spanish at LarsonPrecalculus.com

Convert each radian measure to degree measure. Do not use a calculator.

a. $\pi/6$ **b.** $5\pi/3$

Applications

To measure arc length along a circle, use the radian measure formula, $\theta = s/r$.

Arc Length

For a circle of radius r, a central angle θ intercepts an arc of length s given by

$$s = r\theta \qquad \text{Length of circular arc}$$

where θ is measured in radians. Note that if $r = 1$, then $s = \theta$, and the radian measure of θ equals the arc length.

EXAMPLE 5 Finding Arc Length

A circle has a radius of 4 inches. Find the length of the arc intercepted by a central angle of 240°, as shown in Figure 1.12.

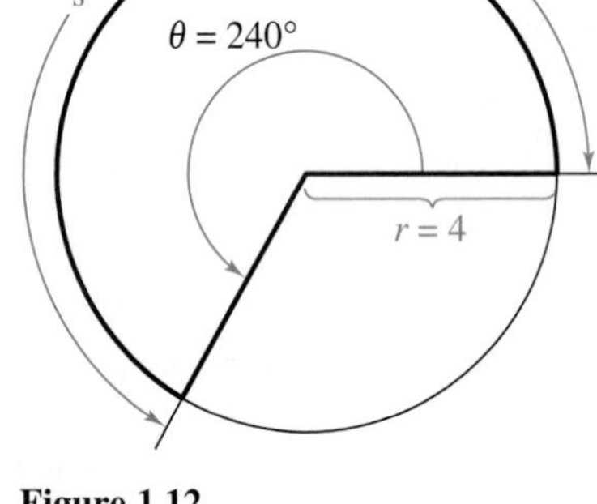

Figure 1.12

Solution To use the formula $s = r\theta$, first convert 240° to radian measure.

$$240° = (240 \text{ deg})\left(\frac{\pi \text{ rad}}{180 \text{ deg}}\right)$$

$$= \frac{4\pi}{3} \text{ radians}$$

Then, using a radius of $r = 4$ inches, find the arc length.

$$s = r\theta \qquad \text{Length of circular arc}$$

$$= 4\left(\frac{4\pi}{3}\right) \qquad \text{Substitute for } r \text{ and } \theta.$$

$$\approx 16.76 \text{ inches} \qquad \text{Use a calculator.}$$

Note that the units for r determine the units for $r\theta$ because θ is in radian measure, which has no units.

✓ ***Checkpoint*** *Audio-video solution in English & Spanish at LarsonPrecalculus.com*

A circle has a radius of 27 inches. Find the length of the arc intercepted by a central angle of 160°.

REMARK Linear speed measures how fast the particle moves, and angular speed measures how fast the angle changes. To establish a relationship between linear speed v and angular speed ω, divide each side of the formula for arc length by t, as shown.

$$s = r\theta$$

$$\frac{s}{t} = \frac{r\theta}{t}$$

$$v = r\omega$$

The formula for the length of a circular arc can be used to analyze the motion of a particle moving at a *constant speed* along a circular path.

Linear and Angular Speeds

Consider a particle moving at a constant speed along a circular arc of radius r. If s is the length of the arc traveled in time t, then the **linear speed** v of the particle is

$$\text{Linear speed } v = \frac{\text{arc length}}{\text{time}} = \frac{s}{t}.$$

Moreover, if θ is the angle (in radian measure) corresponding to the arc length s, then the **angular speed** ω (the lowercase Greek letter omega) of the particle is

$$\text{Angular speed } \omega = \frac{\text{central angle}}{\text{time}} = \frac{\theta}{t}.$$

EXAMPLE 6 Finding Linear Speed

The second hand of a clock is 10.2 centimeters long, as shown at the right. Find the linear speed of the tip of the second hand as it passes around the clock face.

Solution In one revolution, the arc length traveled is

$$s = 2\pi r$$

$$= 2\pi(10.2) \qquad \text{Substitute for } r.$$

$$= 20.4\pi \text{ centimeters.}$$

The time required for the second hand to travel this distance is

$$t = 1 \text{ minute} = 60 \text{ seconds.}$$

So, the linear speed of the tip of the second hand is

$$v = \frac{s}{t}$$

$$= \frac{20.4\pi \text{ centimeters}}{60 \text{ seconds}}$$

$$\approx 1.07 \text{ centimeters per second.}$$

 **Checkpoint** 

The second hand of a clock is 8 centimeters long. Find the linear speed of the tip of the second hand as it passes around the clock face.

EXAMPLE 7 Finding Angular and Linear Speeds

Figure 1.13

The blades of a wind turbine are 116 feet long (see Figure 1.13). The propeller rotates at 15 revolutions per minute.

a. Find the angular speed of the propeller in radians per minute.

b. Find the linear speed of the tips of the blades.

Solution

a. Each revolution corresponds to 2π radians, so the propeller turns $15(2\pi) = 30\pi$ radians per minute. In other words, the angular speed is

$$\omega = \frac{\theta}{t} = \frac{30\pi \text{ radians}}{1 \text{ minute}} = 30\pi \text{ radians per minute.}$$

b. The linear speed is

$$v = \frac{s}{t} = \frac{r\theta}{t} = \frac{116(30\pi) \text{ feet}}{1 \text{ minute}} \approx 10{,}933 \text{ feet per minute.}$$

 **Checkpoint**  Audio-video solution in English & Spanish at LarsonPrecalculus.com

The circular blade on a saw has a radius of 4 inches and it rotates at 2400 revolutions per minute.

a. Find the angular speed of the blade in radians per minute.

b. Find the linear speed of the edge of the blade.

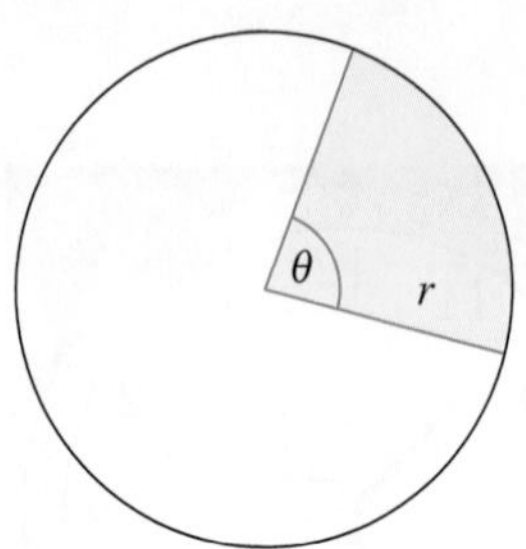

Figure 1.14

A **sector** of a circle is the region bounded by two radii of the circle and their intercepted arc (see Figure 1.14).

Area of a Sector of a Circle

For a circle of radius r, the area A of a sector of the circle with central angle θ is

$$A = \frac{1}{2}r^2\theta$$

where θ is measured in radians.

EXAMPLE 8 Area of a Sector of a Circle

A sprinkler on a golf course fairway sprays water over a distance of 70 feet and rotates through an angle of 120° (see Figure 1.15). Find the area of the fairway watered by the sprinkler.

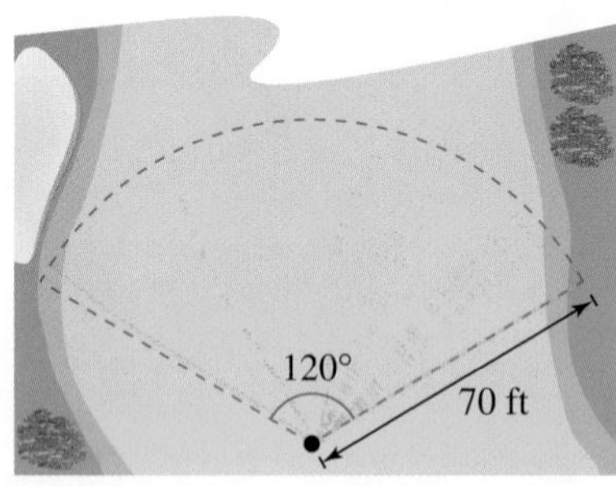

Figure 1.15

Solution

First convert 120° to radian measure.

$$\theta = 120°$$

$$= (120 \text{ deg})\left(\frac{\pi \text{ rad}}{180 \text{ deg}}\right) \qquad \text{Multiply by } \frac{\pi \text{ rad}}{180°}.$$

$$= \frac{2\pi}{3} \text{ radians}$$

Then, using $\theta = 2\pi/3$ and $r = 70$, the area is

$$A = \frac{1}{2}r^2\theta \qquad \text{Formula for the area of a sector of a circle}$$

$$= \frac{1}{2}(70)^2\left(\frac{2\pi}{3}\right) \qquad \text{Substitute for } r \text{ and } \theta.$$

$$= \frac{4900\pi}{3} \qquad \text{Multiply.}$$

$$\approx 5131 \text{ square feet.} \qquad \text{Use a calculator.}$$

A sprinkler sprays water over a distance of 40 feet and rotates through an angle of 80°. Find the area watered by the sprinkler.

Summarize (Section 1.1)

1. Describe an angle *(page 122)*.
2. Explain how to use radian measure *(page 123)*. For examples involving radian measure, see Examples 1 and 2.
3. Explain how to use degree measure *(page 125)*. For examples involving degree measure, see Examples 3 and 4.
4. Describe real-life applications involving angles and their measure *(pages 126–128, Examples 5–8)*.

1.1 Exercises

See CalcChat.com for tutorial help and worked-out solutions to odd-numbered exercises.

Vocabulary: Fill in the blanks.

1. Two angles that have the same initial and terminal sides are ________.
2. One ________ is the measure of a central angle that intercepts an arc equal in length to the radius of the circle.
3. Two positive angles that have a sum of $\pi/2$ are ________ angles, and two positive angles that have a sum of π are ________ angles.
4. The angle measure that is equivalent to a rotation of $\frac{1}{360}$ of a complete revolution about an angle's vertex is one ________.
5. The ________ speed of a particle is the ratio of the change in the central angle to the elapsed time.
6. The area A of a sector of a circle with radius r and central angle θ, where θ is measured in radians, is given by the formula ________.

Skills and Applications

Estimating an Angle **In Exercises 7–10, estimate the angle to the nearest one-half radian.**

7. 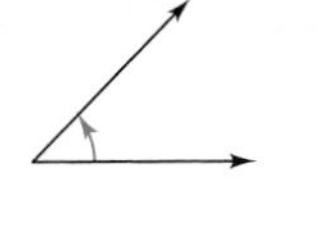

8. 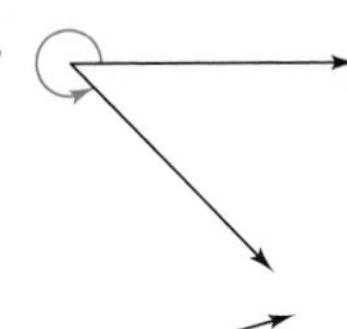

9.

10.

Determining Quadrants **In Exercises 11 and 12, determine the quadrant in which each angle lies.**

11. (a) $\frac{\pi}{4}$ (b) $-\frac{5\pi}{4}$

12. (a) $-\frac{\pi}{6}$ (b) $\frac{11\pi}{9}$

Sketching Angles **In Exercises 13 and 14, sketch each angle in standard position.**

13. (a) $\frac{\pi}{3}$ (b) $-\frac{2\pi}{3}$

14. (a) $\frac{5\pi}{2}$ (b) 4

Finding Coterminal Angles **In Exercises 15 and 16, determine two coterminal angles (one positive and one negative) for each angle. Give your answers in radians.**

15. (a) $\frac{\pi}{6}$ (b) $-\frac{5\pi}{6}$

16. (a) $\frac{2\pi}{3}$ (b) $-\frac{9\pi}{4}$

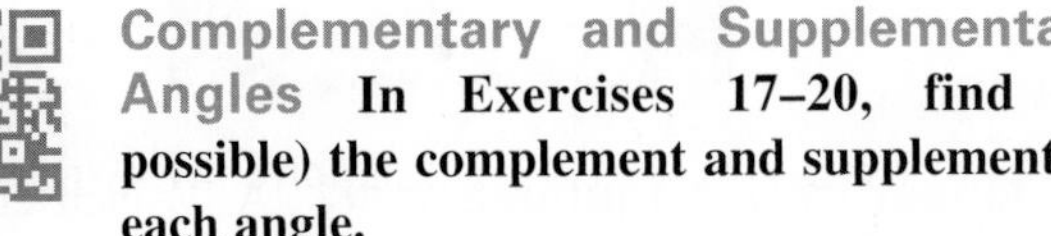

Complementary and Supplementary Angles **In Exercises 17–20, find (if possible) the complement and supplement of each angle.**

17. (a) $\frac{\pi}{12}$ (b) $\frac{11\pi}{12}$

18. (a) $\frac{\pi}{3}$ (b) $\frac{\pi}{4}$

19. (a) 1 (b) 2

20. (a) 3 (b) 1.5

Estimating an Angle **In Exercises 21–24, estimate the number of degrees in the angle.**

21. 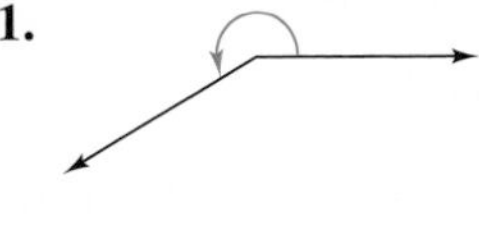

22. 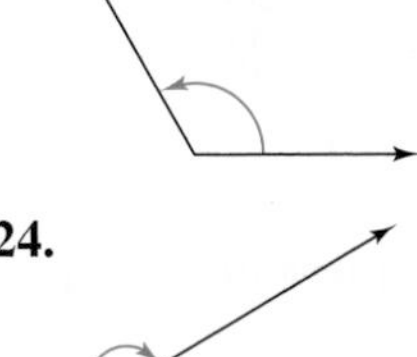

23.

24.

Determining Quadrants **In Exercises 25 and 26, determine the quadrant in which each angle lies.**

25. (a) 130° (b) −8.3°

26. (a) −132° 50′ (b) 3.4°

Sketching Angles **In Exercises 27 and 28, sketch each angle in standard position.**

27. (a) 270° (b) −120°

28. (a) 135° (b) −750°

Finding Coterminal Angles **In Exercises 29 and 30, determine two coterminal angles (one positive and one negative) for each angle. Give your answers in degrees.**

29. (a) 120° (b) −210°

30. (a) 45° (b) −420°

Complementary and Supplementary Angles **In Exercises 31–34, find (if possible) the complement and supplement of each angle.**

31. (a) 18° (b) 85°

32. (a) 46° (b) 93°

33. (a) 24° (b) 126°

34. (a) 130° (b) 170°

The symbol and a red exercise number indicates that a video solution can be seen at *CalcView.com*.

Converting from Degrees to Radians **In Exercises 35 and 36, convert each degree measure to radian measure as a multiple of π. Do not use a calculator.**

35. (a) $120°$ (b) $-20°$
36. (a) $-60°$ (b) $144°$

Converting from Radians to Degrees **In Exercises 37 and 38, convert each radian measure to degree measure. Do not use a calculator.**

37. (a) $\frac{3\pi}{2}$ (b) $-\frac{7\pi}{6}$
38. (a) $-\frac{7\pi}{12}$ (b) $\frac{5\pi}{4}$

Converting from Degrees to Radians **In Exercises 39–42, convert the degree measure to radian measure. Round to three decimal places.**

39. $45°$ 40. $-48.27°$
41. $-0.54°$ 42. $345°$

Converting from Radians to Degrees **In Exercises 43–46, convert the radian measure to degree measure. Round to three decimal places, if necessary.**

43. $\frac{5\pi}{11}$ 44. $\frac{15\pi}{8}$
45. -4.2π 46. -0.57

Converting to Decimal Degree Form **In Exercises 47 and 48, convert each angle measure to decimal degree form.**

47. (a) $54° 45'$ (b) $-128° 30'$
48. (a) $135° 10' 36''$ (b) $-408° 16' 20''$

Converting to D° M′ S″ Form **In Exercises 49 and 50, convert each angle measure to D° M′ S″ form.**

49. (a) $240.6°$ (b) $-145.8°$
50. (a) $345.12°$ (b) $-3.58°$

Finding Arc Length **In Exercises 51 and 52, find the length of the arc on a circle of radius r intercepted by a central angle θ.**

51. $r = 15$ inches, $\theta = 120°$
52. $r = 3$ meters, $\theta = 150°$

Finding the Central Angle **In Exercises 53 and 54, find the radian measure of the central angle of a circle of radius r that intercepts an arc of length s.**

53. $r = 80$ kilometers, $s = 150$ kilometers
54. $r = 14$ feet, $s = 8$ feet

Finding the Central Angle **In Exercises 55 and 56, find the radian measure of the central angle.**

55.
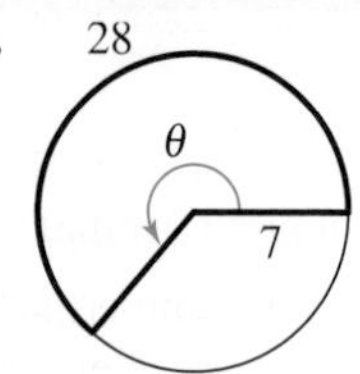

56.
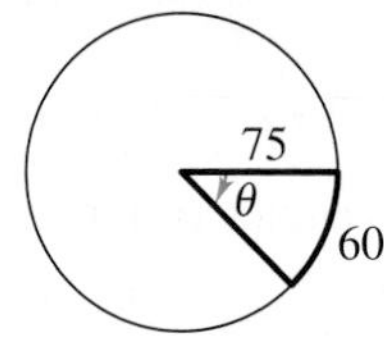

Area of a Sector of a Circle **In Exercises 57 and 58, find the area of the sector of a circle of radius r and central angle θ.**

57. $r = 6$ inches, $\theta = \frac{\pi}{3}$ 58. $r = 2.5$ feet, $\theta = 225°$

Error Analysis **In Exercises 59 and 60, describe the error.**

59. $20° = (20 \text{ deg})\left(\frac{180 \text{ rad}}{\pi \text{ deg}}\right) = \frac{3600}{\pi} \text{ rad}$ ✗

60. A circle has a radius of 6 millimeters. The length of the arc intercepted by a central angle of $72°$ is

$$s = r\theta$$
$$= 6(72)$$
$$= 432 \text{ millimeters.}$$ ✗

Earth-Space Science **In Exercises 61 and 62, find the distance between the cities. Assume that Earth is a sphere of radius 4000 miles and that the cities are on the same longitude (one city is due north of the other).**

	City	Latitude
61.	Dallas, Texas	$32° 47' 9''$ N
	Omaha, Nebraska	$41° 15' 50''$ N
62.	San Francisco, California	$37° 47' 36''$ N
	Seattle, Washington	$47° 37' 18''$ N

63. **Instrumentation** The pointer on a voltmeter is 6 centimeters in length (see figure). Find the number of degrees through which the pointer rotates when it moves 2.5 centimeters on the scale.

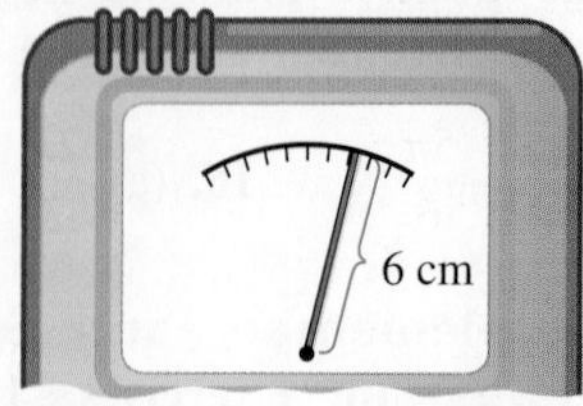

64. **Linear and Angular Speed** A $7\frac{1}{4}$-inch circular power saw blade rotates at 5200 revolutions per minute.
    (a) Find the angular speed of the saw blade in radians per minute.
    (b) Find the linear speed (in feet per minute) of the saw teeth as they contact the wood being cut.

65. **Linear and Angular Speed** A carousel with a 50-foot diameter makes 4 revolutions per minute.
(a) Find the angular speed of the carousel in radians per minute.
(b) Find the linear speed (in feet per minute) of the platform rim of the carousel.

66. **Linear and Angular Speed** A Blu-ray disc is approximately 12 centimeters in diameter. The drive motor of a Blu-ray player is able to rotate up to 10,000 revolutions per minute.
(a) Find the maximum angular speed (in radians per second) of a Blu-ray disc as it rotates.
(b) Find the maximum linear speed (in meters per second) of a point on the outermost track as the disc rotates.

67. **Linear and Angular Speed** A computerized spin balance machine rotates a 25-inch-diameter tire at 480 revolutions per minute.
(a) Find the road speed (in miles per hour) at which the tire is being balanced.
(b) At what rate should the spin balance machine be set so that the tire is being tested for 55 miles per hour?

68. **Speed of a Bicycle**
The radii of the pedal sprocket, the wheel sprocket, and the wheel of the bicycle in the figure are 4 inches, 2 inches, and 14 inches, respectively. A cyclist pedals at a rate of 1 revolution per second.

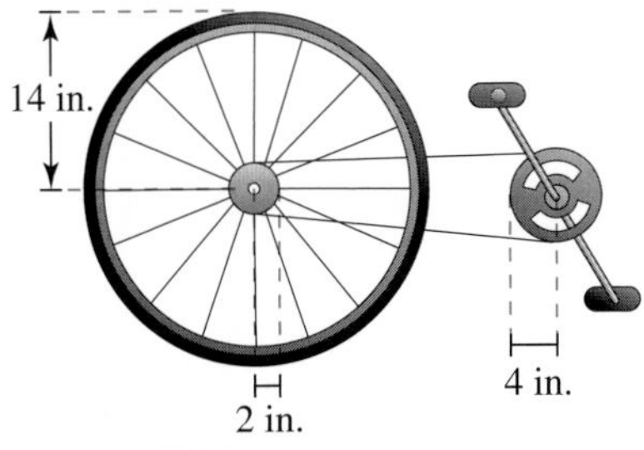

(a) Find the speed of the bicycle in feet per second and miles per hour.
(b) Use your result from part (a) to write a function for the distance d (in miles) a cyclist travels in terms of the number n of revolutions of the pedal sprocket.

(c) Write a function for the distance d (in miles) a cyclist travels in terms of the time t (in seconds). Compare this function with the function from part (b).

69. **Area** A sprinkler on a golf green is set to spray water over a distance of 15 meters and to rotate through an angle of 150°. Draw a diagram that shows the region that can be irrigated with the sprinkler. Find the area of the region.

70. **Area** A car's rear windshield wiper rotates 125°. The total length of the wiper mechanism is 25 inches and the length of the wiper blade is 14 inches. Find the area wiped by the wiper blade.

Exploration

True or False? **In Exercises 71–74, determine whether the statement is true or false. Justify your answer.**

71. An angle measure containing π must be in radian measure.

72. A measurement of 4 radians corresponds to two complete revolutions from the initial side to the terminal side of an angle.

73. The difference between the measures of two coterminal angles is always a multiple of 360° when expressed in degrees and is always a multiple of 2π radians when expressed in radians.

74. An angle that measures $-1260°$ lies in Quadrant III.

75. **Writing** When the radius of a circle increases and the magnitude of a central angle is held constant, how does the length of the intercepted arc change? Explain.

76. **HOW DO YOU SEE IT?** Determine which angles in the figure are coterminal angles with angle A. Explain.

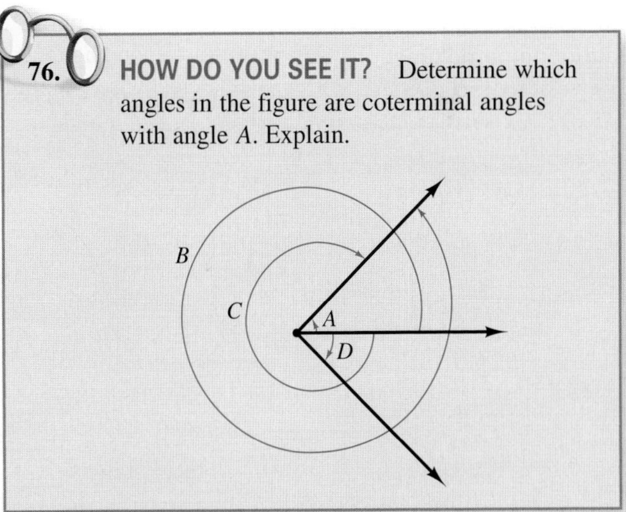

77. **Think About It** A fan motor turns at a given angular speed. How does the speed of the tips of the blades change when a fan of greater diameter is installed on the motor? Explain.

78. **Think About It** Is a degree or a radian the larger unit of measure? Explain.

79. **Proof** Prove that the area of a circular sector of radius r with central angle θ is $A = \frac{1}{2}\theta r^2$, where θ is measured in radians.

Connel/Shutterstock.com

1.2 Trigonometric Functions: The Unit Circle

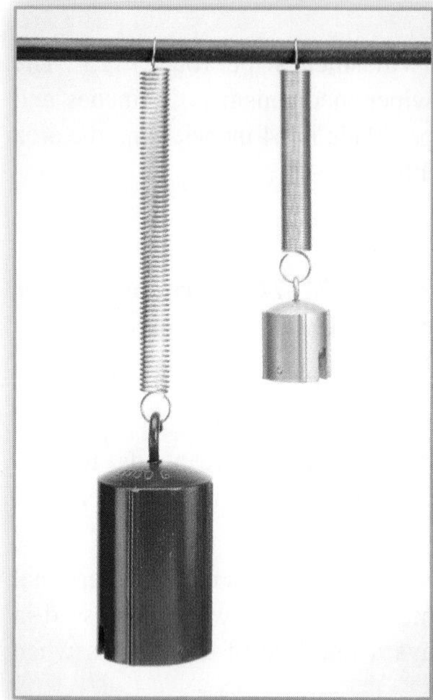

Trigonometric functions can help you analyze the movement of an oscillating weight. For example, in Exercise 50 on page 138, you will analyze the displacement of an oscillating weight suspended by a spring using a model that is a trigonometric function.

- **Identify a unit circle and describe its relationship to real numbers.**
- **Evaluate trigonometric functions using the unit circle.**
- **Use domain and period to evaluate sine and cosine functions, and use a calculator to evaluate trigonometric functions.**

The Unit Circle

The two historical perspectives of trigonometry incorporate different methods for introducing the trigonometric functions. One such perspective is based on the unit circle.

Consider the **unit circle** given by

$$x^2 + y^2 = 1 \qquad \text{Unit circle}$$

as shown in the figure below.

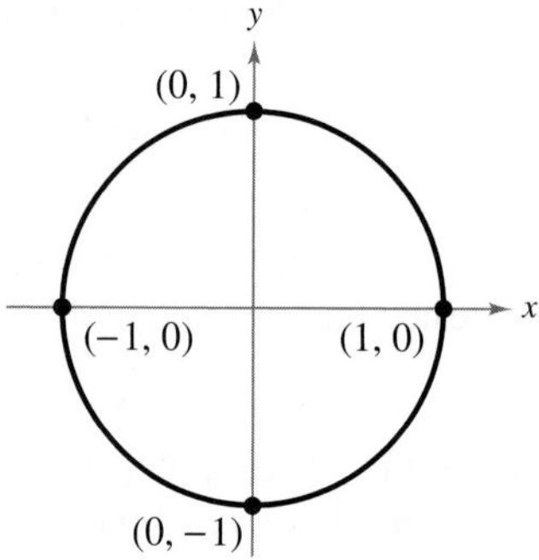

Imagine wrapping the real number line around this circle, with positive numbers corresponding to a counterclockwise wrapping and negative numbers corresponding to a clockwise wrapping, as shown in the figures below.

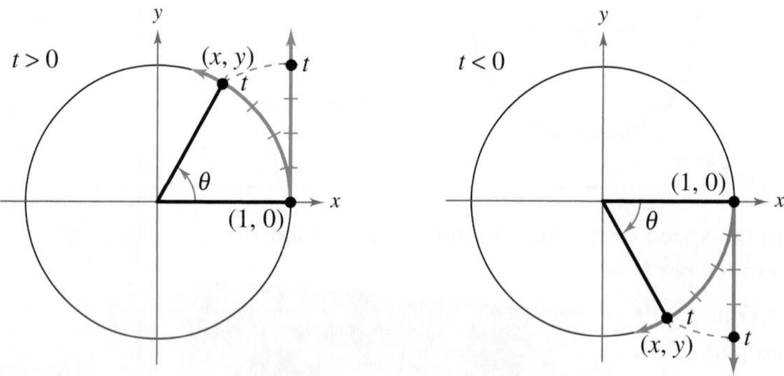

As the real number line wraps around the unit circle, each real number t corresponds to a point (x, y) on the circle. For example, the real number 0 corresponds to the point $(1, 0)$. Moreover, the unit circle has a circumference of 2π, so the real number 2π also corresponds to the point $(1, 0)$.

Each real number t also corresponds to a central angle θ (in standard position) whose radian measure is t. With this interpretation of t, the arc length formula

$$s = r\theta \quad (\text{with } r = 1)$$

indicates that the real number t is the (directional) length of the arc intercepted by the angle θ, given in radians.

The Trigonometric Functions

From the preceding discussion, the coordinates x and y are two functions of the real variable t. These coordinates are used to define the six trigonometric functions of a real number t.

sine **cosecant** **cosine** **secant** **tangent** **cotangent**

Abbreviations for these six functions are sin, csc, cos, sec, tan, and cot, respectively.

REMARK Note that the functions in the second row are the *reciprocals* of the corresponding functions in the first row.

Definitions of Trigonometric Functions

Let t be a real number and let (x, y) be the point on the unit circle corresponding to t.

$$\sin t = y \qquad \cos t = x \qquad \tan t = \frac{y}{x}, \quad x \neq 0$$

$$\csc t = \frac{1}{y}, \quad y \neq 0 \qquad \sec t = \frac{1}{x}, \quad x \neq 0 \qquad \cot t = \frac{x}{y}, \quad y \neq 0$$

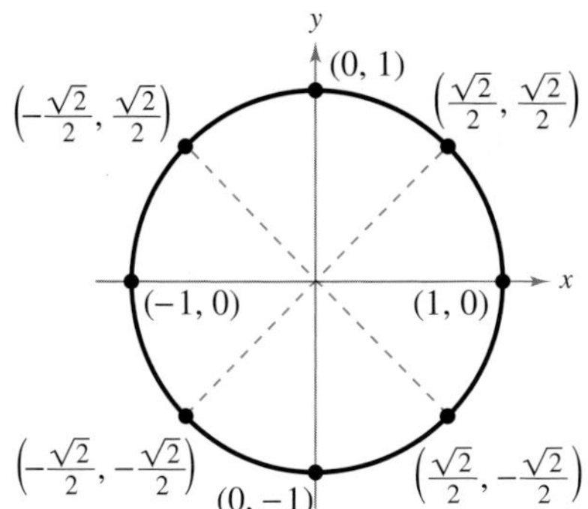

Figure 1.16

In the definitions of the trigonometric functions, note that the tangent and secant are not defined when $x = 0$. For example, $t = \pi/2$ corresponds to $(x, y) = (0, 1)$, so $\tan(\pi/2)$ and $\sec(\pi/2)$ are *undefined.* Similarly, the cotangent and cosecant are not defined when $y = 0$. For example, $t = 0$ corresponds to $(x, y) = (1, 0)$, so cot 0 and csc 0 are *undefined.*

In Figure 1.16, the unit circle is divided into eight equal arcs, corresponding to t-values of

$$0, \frac{\pi}{4}, \frac{\pi}{2}, \frac{3\pi}{4}, \pi, \frac{5\pi}{4}, \frac{3\pi}{2}, \frac{7\pi}{4}, \text{ and } 2\pi.$$

Similarly, in Figure 1.17, the unit circle is divided into 12 equal arcs, corresponding to t-values of

$$0, \frac{\pi}{6}, \frac{\pi}{3}, \frac{\pi}{2}, \frac{2\pi}{3}, \frac{5\pi}{6}, \pi, \frac{7\pi}{6}, \frac{4\pi}{3}, \frac{3\pi}{2}, \frac{5\pi}{3}, \frac{11\pi}{6}, \text{ and } 2\pi.$$

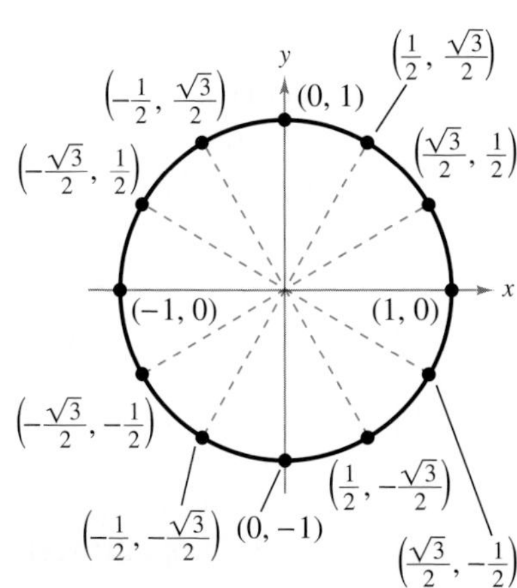

Figure 1.17

To verify the points on the unit circle in Figure 1.16, note that

$$\left(\frac{\sqrt{2}}{2}, \frac{\sqrt{2}}{2}\right)$$

lies on the line $y = x$. So, substituting x for y in the equation of the unit circle produces the following

$$x^2 + x^2 = 1 \implies 2x^2 = 1 \implies x^2 = \frac{1}{2} \implies x = \pm\frac{\sqrt{2}}{2}$$

Because the point is in the first quadrant and $y = x$, you have

$$x = \frac{\sqrt{2}}{2} \quad \text{and} \quad y = \frac{\sqrt{2}}{2}.$$

Similar reasoning can be used to verify the rest of the points in Figure 1.16 and the points in Figure 1.17.

Using the (x, y) coordinates in Figures 1.16 and 1.17, you can evaluate the trigonometric functions for these common t-values. Examples 1 and 2 demonstrate this procedure. You should study and learn these exact function values for common t-values because they will help you perform calculations in later sections.

> **ALGEBRA HELP** To review dividing fractions, see Section P.1.

EXAMPLE 1 Evaluating Trigonometric Functions

See LarsonPrecalculus.com for an interactive version of this type of example.

Evaluate the six trigonometric functions at each real number.

a. $t = \dfrac{\pi}{6}$ **b.** $t = \dfrac{5\pi}{4}$ **c.** $t = \pi$ **d.** $t = -\dfrac{\pi}{3}$

Solution For each t-value, begin by finding the corresponding point (x, y) on the unit circle. Then use the definitions of trigonometric functions listed on page 133.

a. $t = \pi/6$ corresponds to the point $(x, y) = \left(\sqrt{3}/2, 1/2\right)$.

$$\sin\frac{\pi}{6} = y = \frac{1}{2} \qquad \csc\frac{\pi}{6} = \frac{1}{y} = \frac{1}{1/2} = 2$$

$$\cos\frac{\pi}{6} = x = \frac{\sqrt{3}}{2} \qquad \sec\frac{\pi}{6} = \frac{1}{x} = \frac{2}{\sqrt{3}} = \frac{2\sqrt{3}}{3}$$

$$\tan\frac{\pi}{6} = \frac{y}{x} = \frac{1/2}{\sqrt{3}/2} = \frac{1}{\sqrt{3}} = \frac{\sqrt{3}}{3} \qquad \cot\frac{\pi}{6} = \frac{x}{y} = \frac{\sqrt{3}/2}{1/2} = \sqrt{3}$$

b. $t = 5\pi/4$ corresponds to the point $(x, y) = \left(-\sqrt{2}/2, -\sqrt{2}/2\right)$.

$$\sin\frac{5\pi}{4} = y = -\frac{\sqrt{2}}{2} \qquad \csc\frac{5\pi}{4} = \frac{1}{y} = -\frac{2}{\sqrt{2}} = -\sqrt{2}$$

$$\cos\frac{5\pi}{4} = x = -\frac{\sqrt{2}}{2} \qquad \sec\frac{5\pi}{4} = \frac{1}{x} = -\frac{2}{\sqrt{2}} = -\sqrt{2}$$

$$\tan\frac{5\pi}{4} = \frac{y}{x} = \frac{-\sqrt{2}/2}{-\sqrt{2}/2} = 1 \qquad \cot\frac{5\pi}{4} = \frac{x}{y} = \frac{-\sqrt{2}/2}{-\sqrt{2}/2} = 1$$

c. $t = \pi$ corresponds to the point $(x, y) = (-1, 0)$.

$$\sin\pi = y = 0 \qquad \csc\pi = \frac{1}{y} \text{ is undefined.}$$

$$\cos\pi = x = -1 \qquad \sec\pi = \frac{1}{x} = \frac{1}{-1} = -1$$

$$\tan\pi = \frac{y}{x} = \frac{0}{-1} = 0 \qquad \cot\pi = \frac{x}{y} \text{ is undefined.}$$

d. Moving *clockwise* around the unit circle, $t = -\pi/3$ corresponds to the point $(x, y) = \left(1/2, -\sqrt{3}/2\right)$.

$$\sin\left(-\frac{\pi}{3}\right) = y = -\frac{\sqrt{3}}{2} \qquad \csc\left(-\frac{\pi}{3}\right) = \frac{1}{y} = -\frac{2}{\sqrt{3}} = -\frac{2\sqrt{3}}{3}$$

$$\cos\left(-\frac{\pi}{3}\right) = x = \frac{1}{2} \qquad \sec\left(-\frac{\pi}{3}\right) = \frac{1}{x} = \frac{1}{1/2} = 2$$

$$\tan\left(-\frac{\pi}{3}\right) = \frac{y}{x} = \frac{-\sqrt{3}/2}{1/2} = -\sqrt{3}$$

$$\cot\left(-\frac{\pi}{3}\right) = \frac{x}{y} = \frac{1/2}{-\sqrt{3}/2} = -\frac{1}{\sqrt{3}} = -\frac{\sqrt{3}}{3}$$

✓ **Checkpoint** *Audio-video solution in English & Spanish at LarsonPrecalculus.com*

Evaluate the six trigonometric functions at each real number.

a. $t = \pi/2$ **b.** $t = 0$ **c.** $t = -5\pi/6$ **d.** $t = -3\pi/4$

Domain and Period of Sine and Cosine

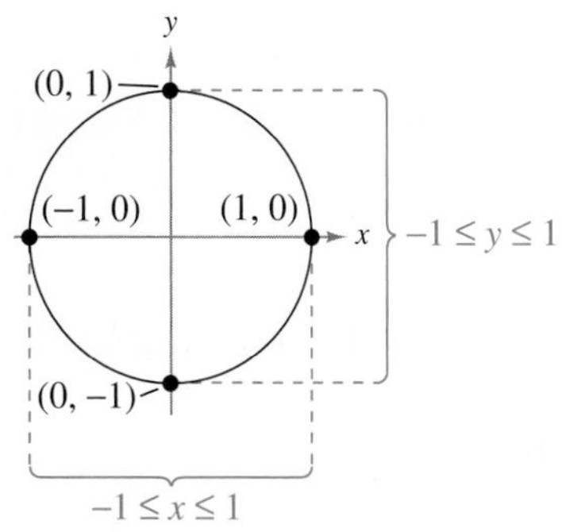

Figure 1.18

The *domain* of the sine and cosine functions is the set of all real numbers. To determine the *range* of these two functions, consider the unit circle shown in Figure 1.18. You know that $\sin t = y$ and $\cos t = x$. Moreover, (x, y) is on the unit circle, so you also know that $-1 \le y \le 1$ and $-1 \le x \le 1$. This means that the values of sine and cosine also range between -1 and 1.

$$-1 \le y \le 1 \qquad \text{and} \qquad -1 \le x \le 1$$
$$-1 \le \sin t \le 1 \qquad\qquad -1 \le \cos t \le 1$$

Adding 2π to each value of t in the interval $[0, 2\pi]$ results in a revolution around the unit circle, as shown in the figure below.

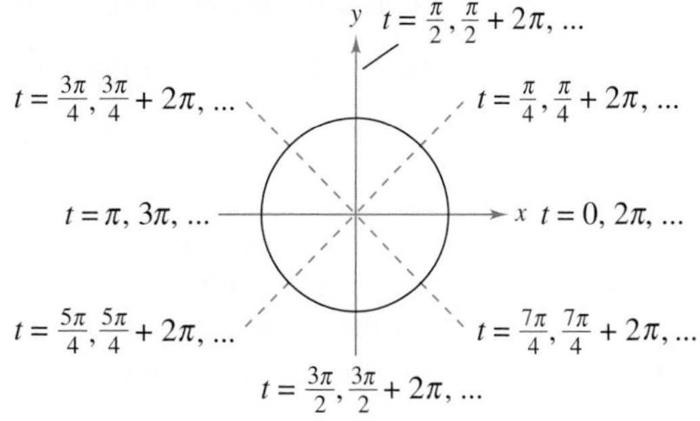

The values of $\sin(t + 2\pi)$ and $\cos(t + 2\pi)$ correspond to those of $\sin t$ and $\cos t$. Repeated revolutions (positive or negative) on the unit circle yield similar results. This leads to the general result

$$\sin(t + 2\pi n) = \sin t \quad \text{and} \quad \cos(t + 2\pi n) = \cos t$$

for any integer n and real number t. Functions that behave in such a repetitive (or cyclic) manner are **periodic.**

REMARK From this definition, it follows that the sine and cosine functions are periodic and have a period of 2π. The other four trigonometric functions are also periodic and will be discussed further in Section 1.6.

Definition of Periodic Function

A function f is **periodic** when there exists a positive real number c such that

$$f(t + c) = f(t)$$

for all t in the domain of f. The smallest number c for which f is periodic is the **period** of f.

Recall from Section P.6 that a function f is *even* when $f(-t) = f(t)$ and is *odd* when $f(-t) = -f(t)$.

Even and Odd Trigonometric Functions

The cosine and secant functions are *even.*

$$\cos(-t) = \cos t \qquad \sec(-t) = \sec t$$

The sine, cosecant, tangent, and cotangent functions are *odd.*

$$\sin(-t) = -\sin t \qquad \csc(-t) = -\csc t$$

$$\tan(-t) = -\tan t \qquad \cot(-t) = -\cot t$$

EXAMPLE 2 Evaluating Sine and Cosine

a. Because $\frac{13\pi}{6} = 2\pi + \frac{\pi}{6}$, you have $\sin \frac{13\pi}{6} = \sin\left(2\pi + \frac{\pi}{6}\right) = \sin \frac{\pi}{6} = \frac{1}{2}$.

b. Because $-\frac{7\pi}{2} = -4\pi + \frac{\pi}{2}$, you have

$$\cos\left(-\frac{7\pi}{2}\right) = \cos\left(-4\pi + \frac{\pi}{2}\right) = \cos \frac{\pi}{2} = 0.$$

c. For $\sin t = \frac{4}{5}$, $\sin(-t) = -\frac{4}{5}$ because the sine function is odd.

✓ ***Checkpoint*** Audio-video solution in English & Spanish at LarsonPrecalculus.com

a. Use the period of the cosine function to evaluate $\cos(9\pi/2)$.

b. Use the period of the sine function to evaluate $\sin(-7\pi/3)$.

c. Evaluate $\cos t$ given that $\cos(-t) = 0.3$.

When evaluating a trigonometric function with a calculator, set the calculator to the desired *mode* of measurement (*degree* or *radian*). Most calculators do not have keys for the cosecant, secant, and cotangent functions. To evaluate these functions, you can use the x^{-1} key with their respective reciprocal functions: sine, cosine, and tangent. For example, to evaluate $\csc(\pi/8)$, use the fact that

$$\csc \frac{\pi}{8} = \frac{1}{\sin(\pi/8)}$$

and enter the keystroke sequence below in *radian* mode.

(SIN (π ÷ 8)) x^{-1} ENTER   Display 2.6131259

▷ TECHNOLOGY When evaluating trigonometric functions with a calculator, remember to enclose all fractional angle measures in parentheses. For example, to evaluate $\sin t$ for $t = \pi/6$, enter

These keystrokes yield the correct value of 0.5. Note that some calculators automatically place a left parenthesis after trigonometric functions.

EXAMPLE 3 Using a Calculator

	Function	Mode	Calculator Keystrokes	Display
a.	$\sin \frac{2\pi}{3}$	Radian	SIN (2 π ÷ 3) ENTER	0.8660254
b.	cot 1.5	Radian	(TAN (1.5)) x^{-1} ENTER	0.0709148

✓ ***Checkpoint*** Audio-video solution in English & Spanish at LarsonPrecalculus.com

Use a calculator to evaluate (a) $\sin(5\pi/7)$ and (b) csc 2.0.

Summarize (Section 1.2)

1. Explain how to identify a unit circle and describe its relationship to real numbers *(page 132)*.
2. State the unit circle definitions of trigonometric functions *(page 133)*. For an example of evaluating trigonometric functions using the unit circle, see Example 1.
3. Explain how to use domain and period to evaluate sine and cosine functions *(page 135)*, and describe how to use a calculator to evaluate trigonometric functions *(page 136)*. For an example of using domain and period to evaluate sine and cosine functions, see Example 2. For an example of using a calculator to evaluate trigonometric functions, see Example 3.

1.2 Exercises

See CalcChat.com for tutorial help and worked-out solutions to odd-numbered exercises.

Vocabulary: Fill in the blanks.

1. Each real number t corresponds to a point (x, y) on the ________ ________.
2. A function f is ________ when there exists a positive real number c such that $f(t + c) = f(t)$ for all t in the domain of f.
3. The smallest number c for which a function f is periodic is the ________ of f.
4. A function f is ________ when $f(-t) = -f(t)$ and ________ when $f(-t) = f(t)$.

Skills and Applications

Evaluating Trigonometric Functions In Exercises 5–8, find the exact values of the six trigonometric functions of the real number t.

5. 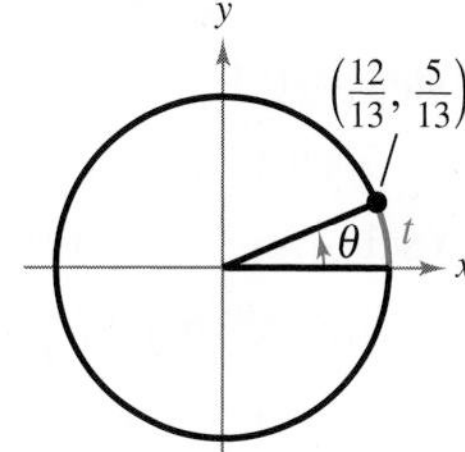

6. 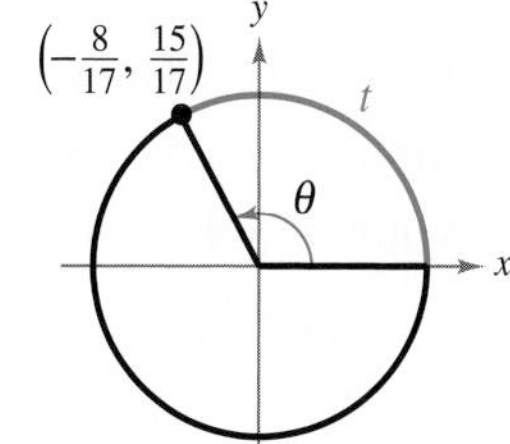

7. 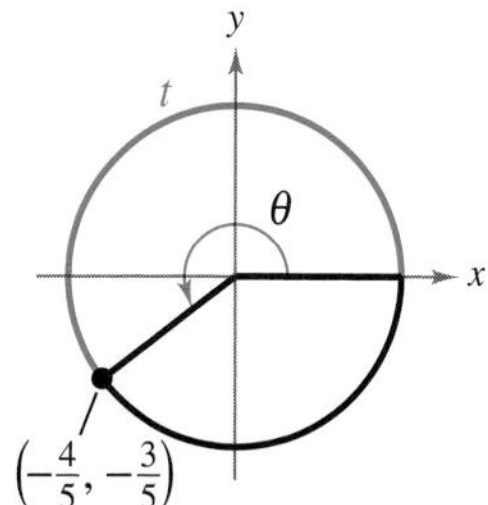

8. 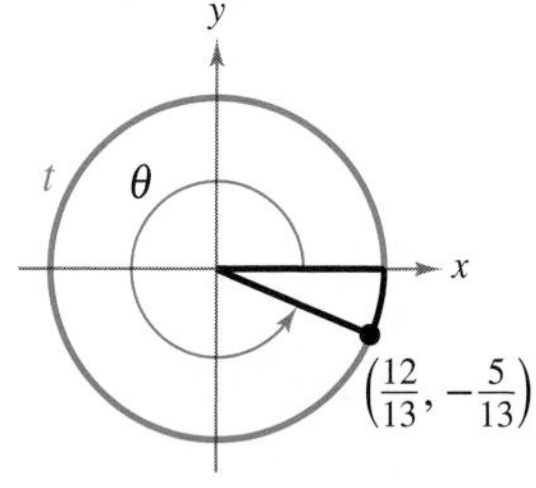

Finding a Point on the Unit Circle In Exercises 9–12, find the point (x, y) on the unit circle that corresponds to the real number t.

9. $t = \pi/2$
10. $t = \pi/4$
11. $t = 5\pi/6$
12. $t = 4\pi/3$

Evaluating Sine, Cosine, and Tangent In Exercises 13–22, evaluate (if possible) the sine, cosine, and tangent at the real number.

13. $t = \dfrac{\pi}{4}$
14. $t = \dfrac{\pi}{3}$
15. $t = -\dfrac{\pi}{6}$
16. $t = -\dfrac{\pi}{4}$
17. $t = -\dfrac{7\pi}{4}$
18. $t = -\dfrac{4\pi}{3}$
19. $t = \dfrac{11\pi}{6}$
20. $t = \dfrac{5\pi}{3}$
21. $t = -\dfrac{3\pi}{2}$
22. $t = -2\pi$

Evaluating Trigonometric Functions In Exercises 23–30, evaluate (if possible) the six trigonometric functions at the real number.

23. $t = 2\pi/3$
24. $t = 5\pi/6$
25. $t = 4\pi/3$
26. $t = 7\pi/4$
27. $t = -5\pi/3$
28. $t = -3\pi/2$
29. $t = -\pi/2$
30. $t = -\pi$

Using Period to Evaluate Sine and Cosine In Exercises 31–36, evaluate the trigonometric function using its period as an aid.

31. $\sin 4\pi$
32. $\cos 3\pi$
33. $\cos(7\pi/3)$
34. $\sin(9\pi/4)$
35. $\sin(19\pi/6)$
36. $\sin(-8\pi/3)$

Using the Value of a Function In Exercises 37–42, use the given value to evaluate each function.

37. $\sin t = \frac{1}{2}$
    (a) $\sin(-t)$
    (b) $\csc(-t)$
38. $\sin(-t) = \frac{3}{8}$
    (a) $\sin t$
    (b) $\csc t$
39. $\cos(-t) = -\frac{1}{5}$
    (a) $\cos t$
    (b) $\sec(-t)$
40. $\cos t = -\frac{3}{4}$
    (a) $\cos(-t)$
    (b) $\sec(-t)$
41. $\sin t = \frac{4}{5}$
    (a) $\sin(\pi - t)$
    (b) $\sin(t + \pi)$
42. $\cos t = \frac{4}{5}$
    (a) $\cos(\pi - t)$
    (b) $\cos(t + \pi)$

Using a Calculator In Exercises 43–48, use a calculator to evaluate the trigonometric function. Round your answer to four decimal places. (Be sure the calculator is in the correct mode.)

43. $\sin 0.6$
44. $\cos(-2.8)$
45. $\tan(\pi/8)$
46. $\tan(5\pi/7)$
47. $\sec 3.1$
48. $\cot(-1.1)$

49. **Harmonic Motion** The displacement from equilibrium of an oscillating weight suspended by a spring is given by

$$y(t) = \frac{1}{2}\cos 6t$$

where y is the displacement in feet and t is the time in seconds. Find the displacement when (a) $t = 0$, (b) $t = \frac{1}{4}$, and (c) $t = \frac{1}{2}$.

50. **Harmonic Motion**

The displacement from equilibrium of an oscillating weight suspended by a spring is given by $y(t) = 3\sin(\pi t/4)$, where y is the displacement in feet and t is the time in seconds.

(a) Complete the table

t	0	$\frac{1}{2}$	1	$\frac{3}{2}$	2
y					

(b) Use the *table* feature of a graphing utility to determine when the displacement is maximum.

(c) Use the *table* feature of the graphing utility to approximate the time t $(0 < t < 8)$ when the weight reaches equilibrium.

Exploration

True of False? **In Exercises 51–54, determine whether the statement is true or false. Justify your answer.**

51. Because $\sin(-t) = -\sin t$, the sine of a negative angle is a negative number.

52. The real number 0 corresponds to the point $(0, 1)$ on the unit circle.

53. $\tan a = \tan(a - 6\pi)$

54. $\cos\left(-\frac{7\pi}{2}\right) = \cos\left(\pi + \frac{\pi}{2}\right)$

55. **Conjecture** Let (x_1, y_1) and (x_2, y_2) be points on the unit circle corresponding to $t = t_1$ and $t = \pi - t_1$, respectively.

(a) Identify the symmetry of the points (x_1, y_1) and (x_2, y_2).

(b) Make a conjecture about any relationship between $\sin t_1$ and $\sin(\pi - t_1)$.

(c) Make a conjecture about any relationship between $\cos t_1$ and $\cos(\pi - t_1)$.

56. **Using the Unit Circle** Use the unit circle to verify that the cosine and secant functions are even and that the sine, cosecant, tangent, and cotangent functions are odd.

57. **Error Analysis** Describe the error.

Your classmate uses a calculator to evaluate $\tan(\pi/2)$ and gets a result of 0.0274224385. ✗

58. **Verifying Expressions Are Not Equal** Verify that

$$\sin(t_1 + t_2) \neq \sin t_1 + \sin t_2$$

by approximating sin 0.25, sin 0.75, and sin 1.

59. **Using Technology** With a graphing utility in *radian* and *parametric* modes, enter the equations

X1T = cos T and Y1T = sin T

and use the settings below.

Tmin = 0, Tmax = 6.3, Tstep = 0.1

Xmin = −1.5, Xmax = 1.5, Xscl = 1

Ymin = −1, Ymax = 1, Yscl = 1

(a) Graph the entered equations and describe the graph.

(b) Use the *trace* feature to move the cursor around the graph. What do the t-values represent? What do the x- and y-values represent?

(c) What are the least and greatest values of x and y?

60. **HOW DO YOU SEE IT?** Use the figure below.

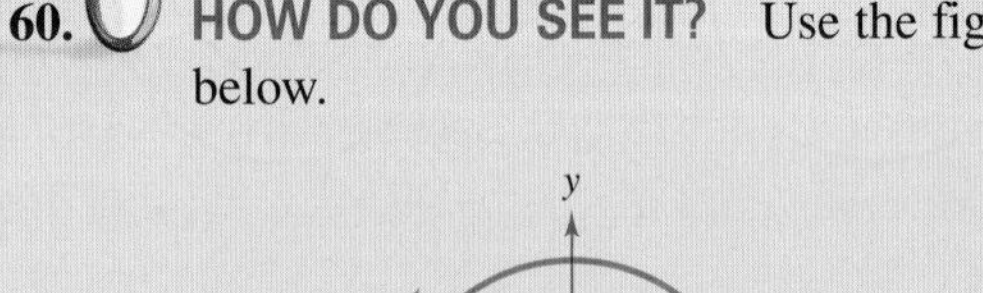

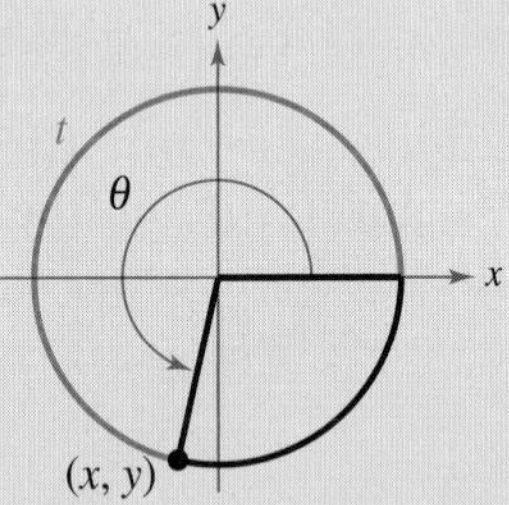

(a) Are all of the trigonometric functions of t defined? Explain.

(b) For those trigonometric functions that are defined, determine whether the sign of the trigonometric function is positive or negative. Explain.

61. **Think About It** Because $f(t) = \sin t$ is an odd function and $g(t) = \cos t$ is an even function, what can be said about the function $h(t) = f(t)g(t)$?

62. **Think About It** Because $f(t) = \sin t$ and $g(t) = \tan t$ are odd functions, what can be said about the function $h(t) = f(t)g(t)$?

The symbol indicates an exercise or a part of an exercise in which you are instructed to use a graphing utility.

1.3 Right Triangle Trigonometry

Right triangle trigonometry has many real-life applications. For example, in Exercise 72 on page 149, you will use right triangle trigonometry to analyze the height of a helium-filled balloon.

- **Evaluate trigonometric functions of acute angles.**
- **Use fundamental trigonometric identities.**
- **Use trigonometric functions to model and solve real-life problems.**

The Six Trigonometric Functions

This section introduces the trigonometric functions from a *right triangle* perspective. Consider the right triangle shown below, in which one acute angle is labeled θ. Relative to the angle θ, the three sides of the triangle are the **hypotenuse,** the **opposite side** (the side opposite the angle θ), and the **adjacent side** (the side adjacent to the angle θ).

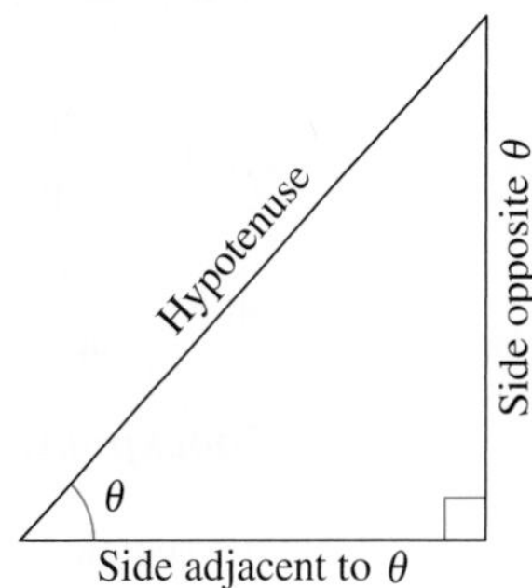

Using the lengths of these three sides, you can form six ratios that define the six trigonometric functions of the acute angle θ.

sine **cosecant** **cosine** **secant** **tangent** **cotangent**

In the definitions below,

$$0^\circ < \theta < 90^\circ$$

(θ lies in the first quadrant). For such angles, the value of each trigonometric function is *positive*.

Right Triangle Definitions of Trigonometric Functions

Let θ be an *acute* angle of a right triangle. The six trigonometric functions of the angle θ are defined below. (Note that the functions in the second row are the *reciprocals* of the corresponding functions in the first row.)

$$\sin\theta = \frac{\text{opp}}{\text{hyp}} \qquad \cos\theta = \frac{\text{adj}}{\text{hyp}} \qquad \tan\theta = \frac{\text{opp}}{\text{adj}}$$

$$\csc\theta = \frac{\text{hyp}}{\text{opp}} \qquad \sec\theta = \frac{\text{hyp}}{\text{adj}} \qquad \cot\theta = \frac{\text{adj}}{\text{opp}}$$

The abbreviations

opp, *adj,* and *hyp*

represent the lengths of the three sides of a right triangle.

opp = the length of the side *opposite* θ

adj = the length of the side *adjacent to* θ

hyp = the length of the *hypotenuse*

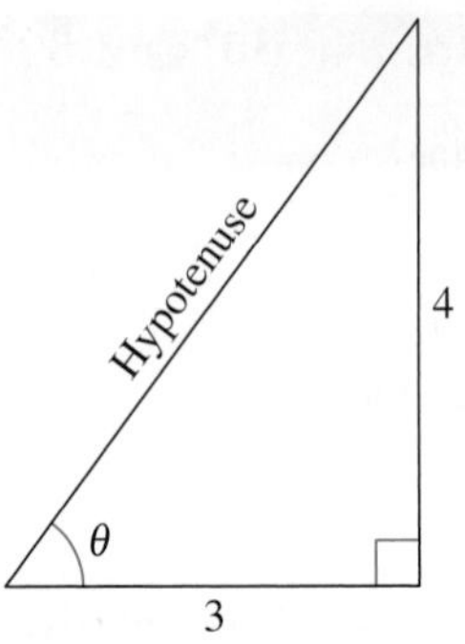

Figure 1.19

EXAMPLE 1 Evaluating Trigonometric Functions

See LarsonPrecalculus.com for an interactive version of this type of example.

Use the triangle in Figure 1.19 to find the values of the six trigonometric functions of θ.

Solution By the Pythagorean Theorem, $(\text{hyp})^2 = (\text{opp})^2 + (\text{adj})^2$, it follows that

$$\begin{aligned}\text{hyp} &= \sqrt{4^2 + 3^2}\\ &= \sqrt{25}\\ &= 5.\end{aligned}$$

So, the six trigonometric functions of θ are

$$\sin\theta = \frac{\text{opp}}{\text{hyp}} = \frac{4}{5} \qquad \csc\theta = \frac{\text{hyp}}{\text{opp}} = \frac{5}{4}$$

$$\cos\theta = \frac{\text{adj}}{\text{hyp}} = \frac{3}{5} \qquad \sec\theta = \frac{\text{hyp}}{\text{adj}} = \frac{5}{3}$$

$$\tan\theta = \frac{\text{opp}}{\text{adj}} = \frac{4}{3} \qquad \cot\theta = \frac{\text{adj}}{\text{opp}} = \frac{3}{4}.$$

Use the triangle below to find the values of the six trigonometric functions of θ.

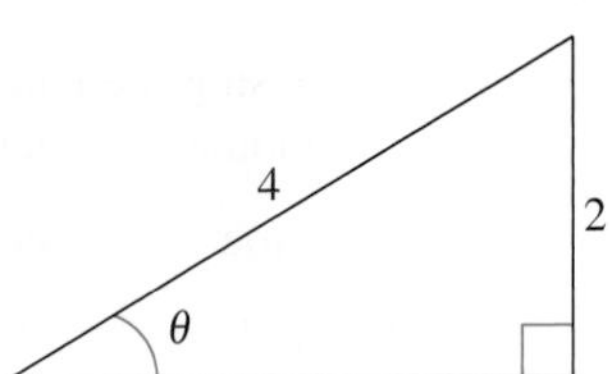

HISTORICAL NOTE

Georg Joachim Rheticus (1514–1576) was the leading Teutonic mathematical astronomer of the 16th century. He was the first to define the trigonometric functions as ratios of the sides of a right triangle.

In Example 1, you were given the lengths of two sides of the right triangle, but not the angle θ. Often, you will be asked to find the trigonometric functions of a *given* acute angle θ. To do this, construct a right triangle having θ as one of its angles.

EXAMPLE 2 Evaluating Trigonometric Functions of 45°

Find the values of sin 45°, cos 45°, and tan 45°.

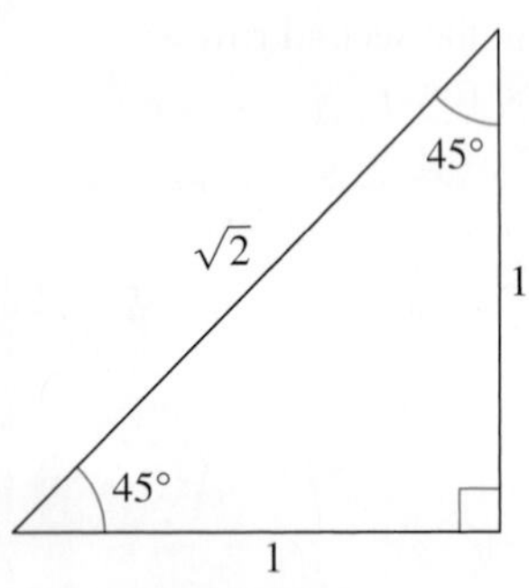

Figure 1.20

Solution Construct a right triangle having 45° as one of its acute angles, as shown in Figure 1.20. Choose 1 as the length of the adjacent side. From geometry, you know that the other acute angle is also 45°. So, the triangle is isosceles and the length of the opposite side is also 1. By the Pythagorean Theorem, the length of the hypotenuse is $\sqrt{2}$.

$$\sin 45^\circ = \frac{\text{opp}}{\text{hyp}} = \frac{1}{\sqrt{2}} = \frac{\sqrt{2}}{2}$$

$$\cos 45^\circ = \frac{\text{adj}}{\text{hyp}} = \frac{1}{\sqrt{2}} = \frac{\sqrt{2}}{2}$$

$$\tan 45^\circ = \frac{\text{opp}}{\text{adj}} = \frac{1}{1} = 1$$

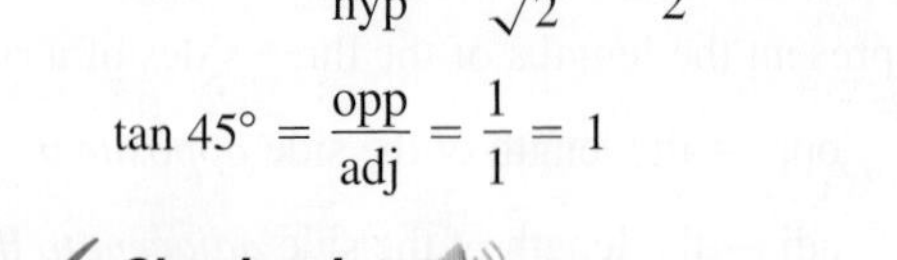

Find the values of cot 45°, sec 45°, and csc 45°.

EXAMPLE 3 Evaluating Trigonometric Functions of 30° and 60°

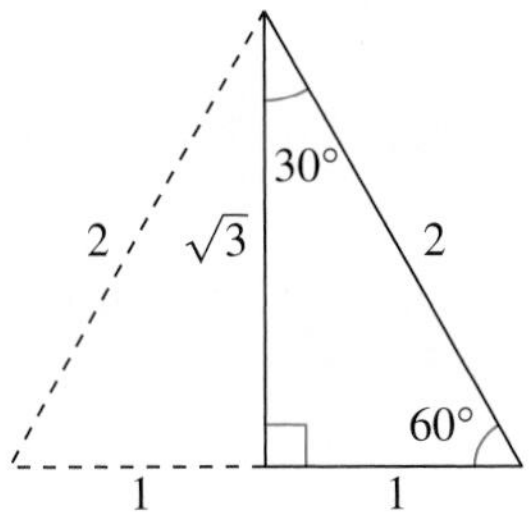

Use the equilateral triangle shown in Figure 1.21 to find the values of sin 60°, cos 60°, sin 30°, and cos 30°.

Solution For $\theta = 60°$, you have adj $= 1$, opp $= \sqrt{3}$, and hyp $= 2$. So,

$$\sin 60° = \frac{\text{opp}}{\text{hyp}} = \frac{\sqrt{3}}{2} \quad \text{and} \quad \cos 60° = \frac{\text{adj}}{\text{hyp}} = \frac{1}{2}.$$

For $\theta = 30°$, adj $= \sqrt{3}$, opp $= 1$, and hyp $= 2$. So,

$$\sin 30° = \frac{\text{opp}}{\text{hyp}} = \frac{1}{2} \quad \text{and} \quad \cos 30° = \frac{\text{adj}}{\text{hyp}} = \frac{\sqrt{3}}{2}.$$

Figure 1.21

Checkpoint Audio-video solution in English & Spanish at LarsonPrecalculus.com

Use the equilateral triangle shown in Figure 1.21 to find the values of tan 60° and tan 30°.

REMARK The angles 30°, 45°, and 60° ($\pi/6$, $\pi/4$, and $\pi/3$ radians, respectively) occur frequently in trigonometry, so you should learn to construct the triangles shown in Figures 1.20 and 1.21.

Sines, Cosines, and Tangents of Special Angles

$$\sin 30° = \sin \frac{\pi}{6} = \frac{1}{2} \qquad \cos 30° = \cos \frac{\pi}{6} = \frac{\sqrt{3}}{2} \qquad \tan 30° = \tan \frac{\pi}{6} = \frac{\sqrt{3}}{3}$$

$$\sin 45° = \sin \frac{\pi}{4} = \frac{\sqrt{2}}{2} \qquad \cos 45° = \cos \frac{\pi}{4} = \frac{\sqrt{2}}{2} \qquad \tan 45° = \tan \frac{\pi}{4} = 1$$

$$\sin 60° = \sin \frac{\pi}{3} = \frac{\sqrt{3}}{2} \qquad \cos 60° = \cos \frac{\pi}{3} = \frac{1}{2} \qquad \tan 60° = \tan \frac{\pi}{3} = \sqrt{3}$$

Note that $\sin 30° = \frac{1}{2} = \cos 60°$. This occurs because 30° and 60° are complementary angles. In general, it can be shown from the right triangle definitions that *cofunctions of complementary angles are equal*. That is, if θ is an acute angle, then the relationships below are true.

$$\sin(90° - \theta) = \cos\theta \qquad \cos(90° - \theta) = \sin\theta \qquad \tan(90° - \theta) = \cot\theta$$

$$\cot(90° - \theta) = \tan\theta \qquad \sec(90° - \theta) = \csc\theta \qquad \csc(90° - \theta) = \sec\theta$$

To use a calculator to evaluate trigonometric functions of angles measured in degrees, remember to set the calculator to *degree* mode.

EXAMPLE 4 Using a Calculator

Use a calculator to evaluate sec 5° 40′ 12″.

Solution Begin by converting to decimal degree form. $\left[\text{Recall that } 1' = \frac{1}{60}(1°) \text{ and } 1'' = \frac{1}{3600}(1°).\right]$

$$5° \, 40' \, 12'' = 5° + \left(\frac{40}{60}\right)^{\circ} + \left(\frac{12}{3600}\right)^{\circ} = 5.67°$$

Then, use a calculator to evaluate sec 5.67°.

Function	Calculator Keystrokes	Display
sec 5° 40′ 12″ = sec 5.67°	(COS (5.67)) x⁻¹ ENTER	1.0049166

Checkpoint Audio-video solution in English & Spanish at LarsonPrecalculus.com

Use a calculator to evaluate csc 34° 30′ 36″.

Trigonometric Identities

Trigonometric identities are relationships between trigonometric functions.

Fundamental Trigonometric Identities

Reciprocal Identities

$$\sin\theta = \frac{1}{\csc\theta} \qquad \cos\theta = \frac{1}{\sec\theta} \qquad \tan\theta = \frac{1}{\cot\theta}$$

$$\csc\theta = \frac{1}{\sin\theta} \qquad \sec\theta = \frac{1}{\cos\theta} \qquad \cot\theta = \frac{1}{\tan\theta}$$

Quotient Identities

$$\tan\theta = \frac{\sin\theta}{\cos\theta} \qquad \cot\theta = \frac{\cos\theta}{\sin\theta}$$

Pythagorean Identities

$$\sin^2\theta + \cos^2\theta = 1$$

$$1 + \tan^2\theta = \sec^2\theta$$

$$1 + \cot^2\theta = \csc^2\theta$$

Note that $\sin^2\theta$ represents $(\sin\theta)^2$, $\cos^2\theta$ represents $(\cos\theta)^2$, and so on.

REMARK Do not confuse, for example, $\sin^2\theta$ with $\sin\theta^2$. With $\sin^2\theta$, you are squaring $\sin\theta$. With $\sin\theta^2$, you are squaring θ and then finding the sine.

EXAMPLE 5 Applying Trigonometric Identities

Let θ be an acute angle such that $\sin\theta = 0.6$. Find the value of (a) $\cos\theta$ and (b) $\tan\theta$ using trigonometric identities.

Solution

a. To find the value of $\cos\theta$, use the Pythagorean identity

$$\sin^2\theta + \cos^2\theta = 1.$$

So, you have

$$(0.6)^2 + \cos^2\theta = 1 \qquad \text{Substitute 0.6 for } \sin\theta.$$

$$\cos^2\theta = 1 - (0.6)^2 \qquad \text{Subtract } (0.6)^2 \text{ from each side.}$$

$$\cos^2\theta = 0.64 \qquad \text{Simplify.}$$

$$\cos\theta = \sqrt{0.64} \qquad \text{Extract positive square root.}$$

$$\cos\theta = 0.8. \qquad \text{Simplify.}$$

b. Now, knowing the sine and cosine of θ, you can find the tangent of θ.

$$\tan\theta = \frac{\sin\theta}{\cos\theta} = \frac{0.6}{0.8} = 0.75$$

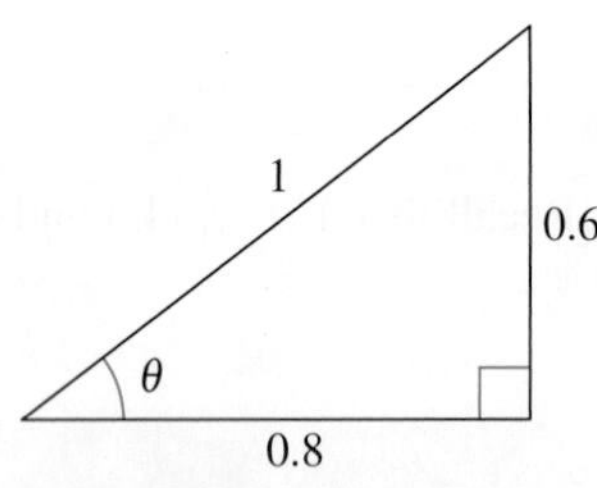

Figure 1.22

Use the definitions of $\cos\theta$ and $\tan\theta$ and the triangle shown in Figure 1.22 to check these results.

Checkpoint Audio-video solution in English & Spanish at LarsonPrecalculus.com

Let θ be an acute angle such that $\cos\theta = 0.96$. Find the value of (a) $\sin\theta$ and (b) $\tan\theta$ using trigonometric identities.

EXAMPLE 6 Applying Trigonometric Identities

Let θ be an acute angle such that $\tan \theta = \frac{1}{3}$. Find the value of (a) $\cot \theta$ and (b) $\sec \theta$ using trigonometric identities.

Solution

a. $\cot \theta = \dfrac{1}{\tan \theta}$ — Reciprocal identity

$= \dfrac{1}{1/3}$ — Substitute $\frac{1}{3}$ for $\tan \theta$.

$= 3$ — Simplify.

b. $\sec^2 \theta = 1 + \tan^2 \theta$ — Pythagorean identity

$\sec^2 \theta = 1 + \left(\dfrac{1}{3}\right)^2$ — Substitute $\frac{1}{3}$ for $\tan \theta$.

$\sec^2 \theta = \dfrac{10}{9}$ — Simplify.

$\sec \theta = \dfrac{\sqrt{10}}{3}$ — Extract positive square root and simplify.

Use the definitions of $\cot \theta$ and $\sec \theta$ and the triangle below to check these results.

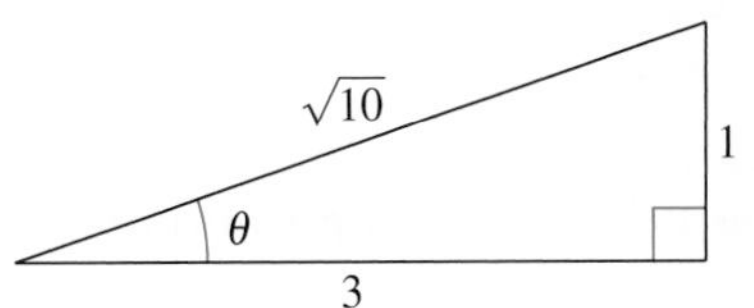

✓ Checkpoint Audio-video solution in English & Spanish at LarsonPrecalculus.com

Let θ be an acute angle such that $\tan \theta = 2$. Find the value of (a) $\cot \theta$ and (b) $\sec \theta$ using trigonometric identities.

EXAMPLE 7 Using Trigonometric Identities

Use trigonometric identities to transform the left side of the equation into the right side $(0 < \theta < \pi/2)$.

a. $\sin \theta \csc \theta = 1$ **b.** $(\csc \theta + \cot \theta)(\csc \theta - \cot \theta) = 1$

Solution

a. $\sin \theta \csc \theta = \left(\dfrac{1}{\cancel{\csc \theta}}\right)\cancel{\csc \theta} = 1$ — Use a reciprocal identity and simplify.

b. $(\csc \theta + \cot \theta)(\csc \theta - \cot \theta)$

$= \csc^2 \theta - \csc \theta \cot \theta + \csc \theta \cot \theta - \cot^2 \theta$ — FOIL Method

$= \csc^2 \theta - \cot^2 \theta$ — Simplify.

$= 1$ — Pythagorean identity

✓ Checkpoint Audio-video solution in English & Spanish at LarsonPrecalculus.com

Use trigonometric identities to transform the left side of the equation into the right side $(0 < \theta < \pi/2)$.

a. $\tan \theta \csc \theta = \sec \theta$ **b.** $(\csc \theta + 1)(\csc \theta - 1) = \cot^2 \theta$ ■

Applications Involving Right Triangles

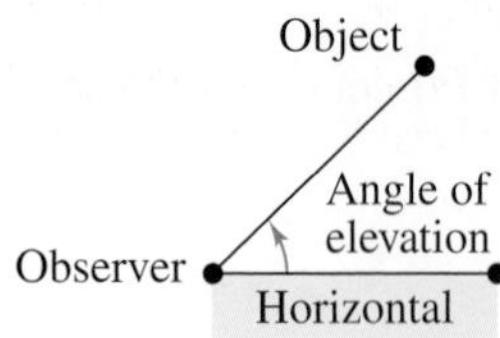

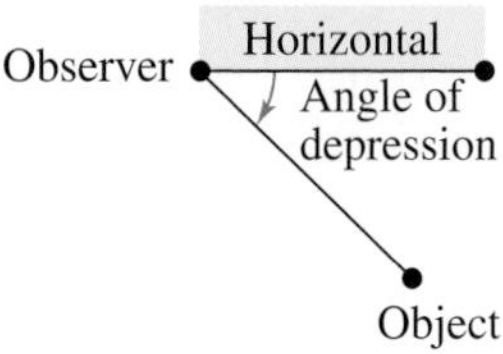

Figure 1.23

Many applications of trigonometry involve **solving right triangles.** In this type of application, you are usually given one side of a right triangle and one of the acute angles and are asked to find one of the other sides, *or* you are given two sides and are asked to find one of the acute angles.

In Example 8, you are given the **angle of elevation,** which represents the angle from the horizontal upward to an object. In other applications you may be given the **angle of depression,** which represents the angle from the horizontal downward to an object. (See Figure 1.23.)

EXAMPLE 8 Solving a Right Triangle

A surveyor stands 115 feet from the base of the Washington Monument, as shown in Figure 1.24. The surveyor measures the angle of elevation to the top of the monument to be 78.3°. How tall is the Washington Monument?

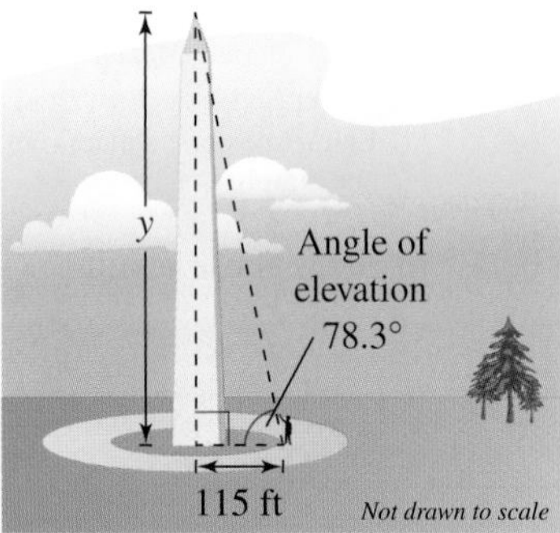

Figure 1.24

Solution From Figure 1.24,

$$\tan 78.3° = \frac{\text{opp}}{\text{adj}} = \frac{y}{115}$$

where y is the height of the monument. So, the height of the Washington Monument is

$$\begin{aligned} y &= 115 \tan 78.3° \\ &\approx 115(4.8288) \\ &\approx 555 \text{ feet.} \end{aligned}$$

✓ **Checkpoint** Audio-video solution in English & Spanish at LarsonPrecalculus.com

The angle of elevation to the top of a flagpole at a distance of 19 feet from its base is 64.6°. How tall is the flagpole?

EXAMPLE 9 Solving a Right Triangle

A lighthouse is 200 yards from a bike path along the edge of a lake. A walkway to the lighthouse is 400 yards long. (See Figure 1.25.) Find the acute angle θ between the bike path and the walkway.

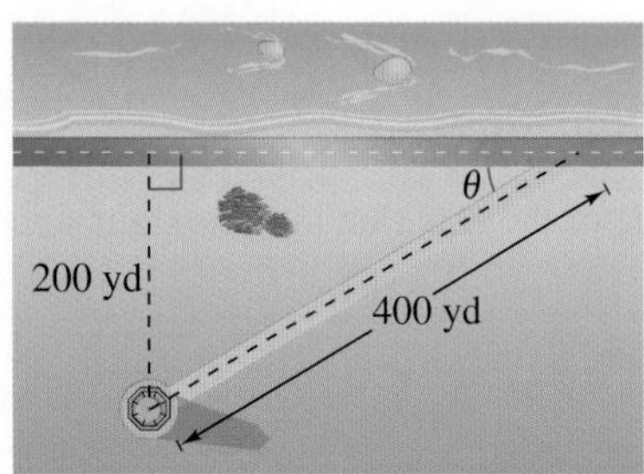

Figure 1.25

Solution From Figure 1.25, the sine of the angle θ is

$$\sin \theta = \frac{\text{opp}}{\text{hyp}} = \frac{200}{400} = \frac{1}{2}.$$

You should recognize that $\theta = 30°$.

✓ **Checkpoint** Audio-video solution in English & Spanish at LarsonPrecalculus.com

Find the acute angle θ between the two paths shown below.

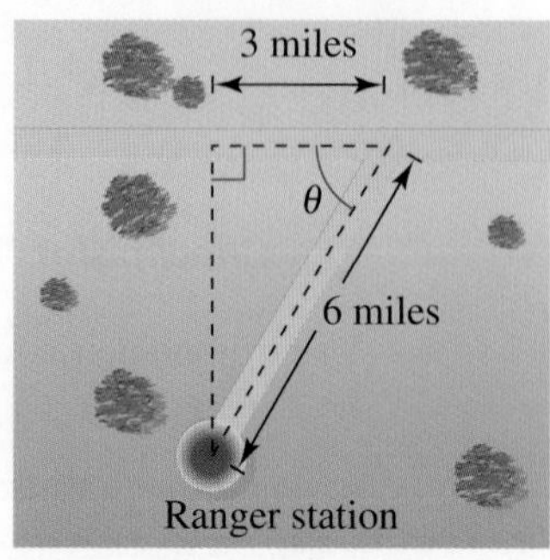

■

In Example 9, you were able to recognize that the special angle $\theta = 30°$ satisfies the equation $\sin \theta = \frac{1}{2}$. However, when θ is not a special angle, you can *estimate* its value. For example, to estimate the acute angle θ in the equation $\sin \theta = 0.6$, you could reason that $\sin 30° = \frac{1}{2} = 0.5000$ and $\sin 45° = 1/\sqrt{2} \approx 0.7071$, so θ lies somewhere between 30° and 45°. In a later section, you will study a method of determining a more precise value of θ.

EXAMPLE 10 Solving a Right Triangle

Skateboarders can go to a skatepark, which is a recreational environment built with many different types of ramps and rails.

Find the length c and the height b of the skateboard ramp below.

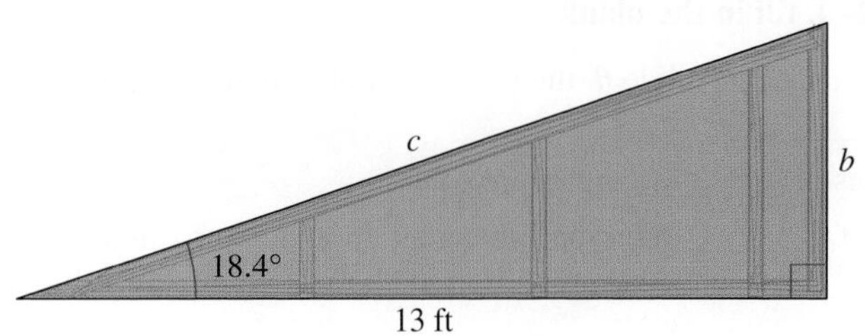

Solution From the figure,

$$\cos 18.4° = \frac{\text{adj}}{\text{hyp}} = \frac{13}{c}.$$

So, the length of the skateboard ramp is

$$c = \frac{13}{\cos 18.4°} \approx \frac{13}{0.9489} \approx 13.7 \text{ feet.}$$

Also from the figure,

$$\tan 18.4° = \frac{\text{opp}}{\text{adj}} = \frac{b}{13}.$$

So, the height is

$$b = 13 \tan 18.4° \approx 13(0.3327) \approx 4.3 \text{ feet.}$$

✓ Checkpoint Audio-video solution in English & Spanish at LarsonPrecalculus.com

Find the length c and the horizontal length a of the loading ramp below.

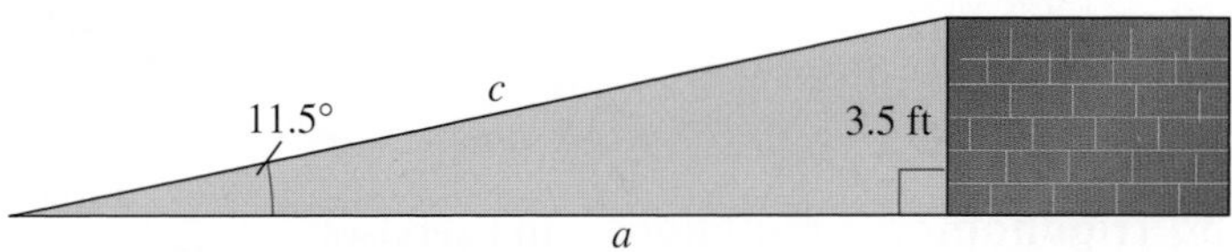

Summarize (Section 1.3)

1. State the right triangle definitions of the six trigonometric functions *(page 139)*. For examples of evaluating trigonometric functions of acute angles, see Examples 1–4.
2. List the reciprocal, quotient, and Pythagorean identities *(page 142)*. For examples of using these identities, see Examples 5–7.
3. Describe real-life applications of trigonometric functions *(pages 144 and 145, Examples 8–10)*.

1.3 Exercises

See CalcChat.com for tutorial help and worked-out solutions to odd-numbered exercises.

Vocabulary

1. Match each trigonometric function with its right triangle definition.

(a) sine (b) cosine (c) tangent (d) cosecant (e) secant (f) cotangent

(i) $\dfrac{\text{hypotenuse}}{\text{adjacent}}$ (ii) $\dfrac{\text{adjacent}}{\text{opposite}}$ (iii) $\dfrac{\text{hypotenuse}}{\text{opposite}}$ (iv) $\dfrac{\text{adjacent}}{\text{hypotenuse}}$ (v) $\dfrac{\text{opposite}}{\text{hypotenuse}}$ (vi) $\dfrac{\text{opposite}}{\text{adjacent}}$

In Exercises 2–4, fill in the blanks.

2. Relative to the acute angle θ, the three sides of a right triangle are the ________ side, the ________ side, and the ________.

3. Cofunctions of ________ angles are equal.

4. An angle of ________ represents the angle from the horizontal upward to an object, whereas an angle of ________ represents the angle from the horizontal downward to an object.

Skills and Applications

Evaluating Trigonometric Functions

In Exercises 5–10, find the exact values of the six trigonometric functions of the angle θ.

5.

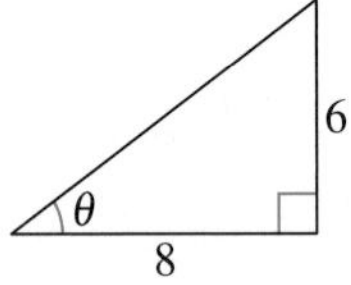

6.

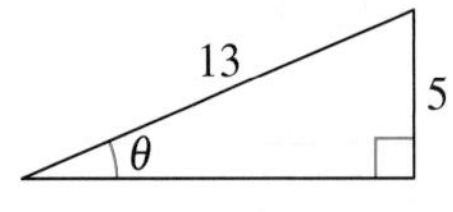

7.

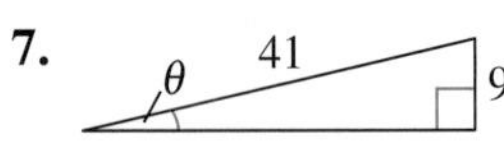

8.

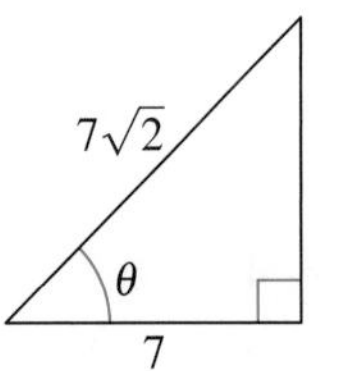

9.

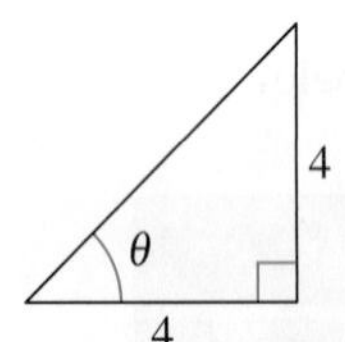

10.

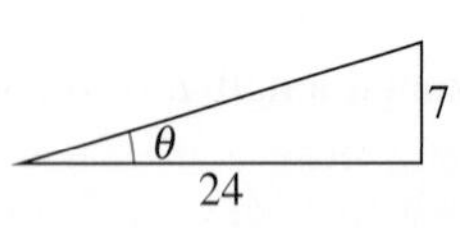

Evaluating Trigonometric Functions

In Exercises 11–14, find the exact values of the six trigonometric functions of the angle θ for each of the two triangles. Explain why the function values are the same.

11.

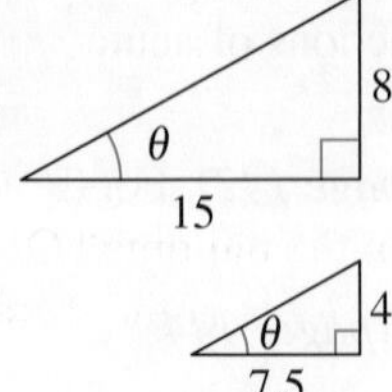

12.

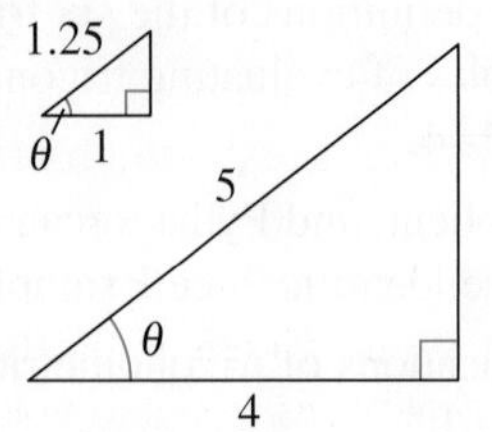

13.

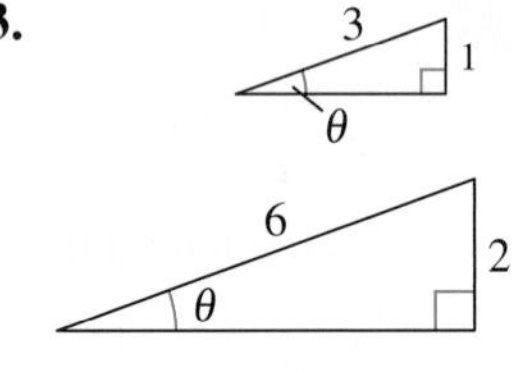

14.

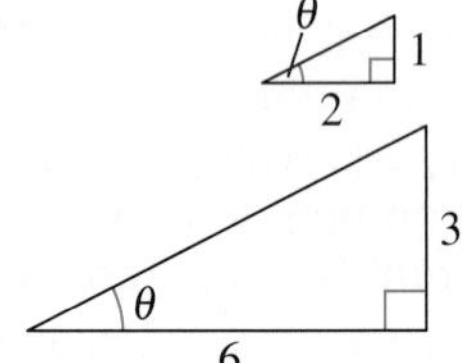

Evaluating Trigonometric Functions

In Exercises 15–22, sketch a right triangle corresponding to the trigonometric function of the acute angle θ. Then find the exact values of the other five trigonometric functions of θ.

15. $\cos\theta = \frac{15}{17}$ **16.** $\sin\theta = \frac{3}{5}$

17. $\sec\theta = \frac{6}{5}$ **18.** $\tan\theta = \frac{4}{5}$

19. $\sin\theta = \frac{1}{5}$ **20.** $\sec\theta = \frac{17}{7}$

21. $\cot\theta = 3$ **22.** $\csc\theta = 9$

Evaluating Trigonometric Functions of 30°, 45°, and 60°

In Exercises 23–28, construct an appropriate triangle to find the missing values. ($0° \le \theta \le 90°$, $0 \le \theta \le \pi/2$)

	Function	θ (deg)	θ (rad)	Function Value
23.	tan	30°		
24.	cos	45°		
25.	sin		$\dfrac{\pi}{4}$	
26.	tan		$\dfrac{\pi}{3}$	
27.	sec		$\dfrac{\pi}{4}$	
28.	csc		$\dfrac{\pi}{6}$	

Using a Calculator **In Exercises 29–36, use a calculator to evaluate each function. Round your answers to four decimal places. (Be sure the calculator is in the correct mode.)**

29. (a) $\sin 20^\circ$ (b) $\cos 70^\circ$

30. (a) $\tan 23.5^\circ$ (b) $\cot 66.5^\circ$

31. (a) $\sin 14.21^\circ$ (b) $\csc 14.21^\circ$

32. (a) $\cot 79.56^\circ$ (b) $\sec 79.56^\circ$

33. (a) $\cos 4^\circ 50' 15''$ (b) $\sec 4^\circ 50' 15''$

34. (a) $\sec 42^\circ 12'$ (b) $\csc 48^\circ 7'$

35. (a) $\cot 17^\circ 15'$ (b) $\tan 17^\circ 15'$

36. (a) $\sec 56^\circ 8' 10''$ (b) $\cos 56^\circ 8' 10''$

Applying Trigonometric Identities **In Exercises 37–42, use the given function value(s) and the trigonometric identities to find the exact value of each indicated trigonometric function.**

37. $\sin 60^\circ = \dfrac{\sqrt{3}}{2}$, $\cos 60^\circ = \dfrac{1}{2}$

(a) $\sin 30^\circ$ (b) $\cos 30^\circ$

(c) $\tan 60^\circ$ (d) $\cot 60^\circ$

38. $\sin 30^\circ = \dfrac{1}{2}$, $\tan 30^\circ = \dfrac{\sqrt{3}}{3}$

(a) $\csc 30^\circ$ (b) $\cot 60^\circ$

(c) $\cos 30^\circ$ (d) $\cot 30^\circ$

39. $\cos \theta = \frac{1}{3}$

(a) $\sin \theta$ (b) $\tan \theta$

(c) $\sec \theta$ (d) $\csc(90^\circ - \theta)$

40. $\sec \theta = 5$

(a) $\cos \theta$ (b) $\cot \theta$

(c) $\cot(90^\circ - \theta)$ (d) $\sin \theta$

41. $\cot \alpha = 3$

(a) $\tan \alpha$ (b) $\csc \alpha$

(c) $\cot(90^\circ - \alpha)$ (d) $\sin \alpha$

42. $\cos \beta = \dfrac{\sqrt{7}}{4}$

(a) $\sec \beta$ (b) $\sin \beta$

(c) $\cot \beta$ (d) $\sin(90^\circ - \beta)$

Using Trigonometric Identities **In Exercises 43–52, use trigonometric identities to transform the left side of the equation into the right side ($0 < \theta < \pi/2$).**

43. $\tan \theta \cot \theta = 1$

44. $\cos \theta \sec \theta = 1$

45. $\tan \alpha \cos \alpha = \sin \alpha$

46. $\cot \alpha \sin \alpha = \cos \alpha$

47. $(1 + \sin \theta)(1 - \sin \theta) = \cos^2 \theta$

48. $(1 + \cos \theta)(1 - \cos \theta) = \sin^2 \theta$

49. $(\sec \theta + \tan \theta)(\sec \theta - \tan \theta) = 1$

50. $\sin^2 \theta - \cos^2 \theta = 2 \sin^2 \theta - 1$

51. $\dfrac{\sin \theta}{\cos \theta} + \dfrac{\cos \theta}{\sin \theta} = \csc \theta \sec \theta$

52. $\dfrac{\tan \beta + \cot \beta}{\tan \beta} = \csc^2 \beta$

Finding Special Angles of a Triangle **In Exercises 53–58, find each value of θ in degrees ($0^\circ < \theta < 90^\circ$) and radians ($0 < \theta < \pi/2$) without using a calculator.**

53. (a) $\sin \theta = \frac{1}{2}$ (b) $\csc \theta = 2$

54. (a) $\cos \theta = \dfrac{\sqrt{2}}{2}$ (b) $\tan \theta = 1$

55. (a) $\sec \theta = 2$ (b) $\cot \theta = 1$

56. (a) $\tan \theta = \sqrt{3}$ (b) $\csc \theta = \sqrt{2}$

57. (a) $\csc \theta = \dfrac{2\sqrt{3}}{3}$ (b) $\sin \theta = \dfrac{\sqrt{2}}{2}$

58. (a) $\cot \theta = \dfrac{\sqrt{3}}{3}$ (b) $\sec \theta = \sqrt{2}$

Finding Side Lengths of a Triangle **In Exercises 59–62, find the exact values of the indicated variables.**

59. Find x and y.

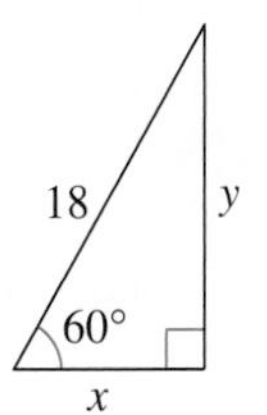

60. Find x and r.

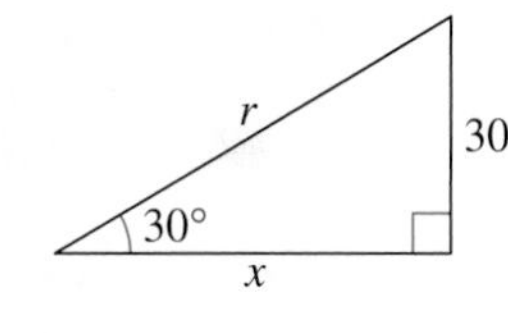

61. Find x and r.

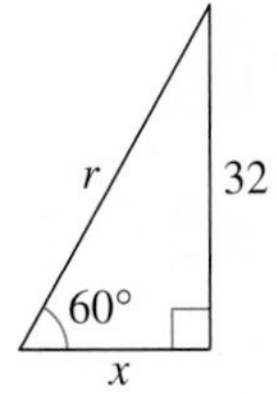

62. Find x and r.

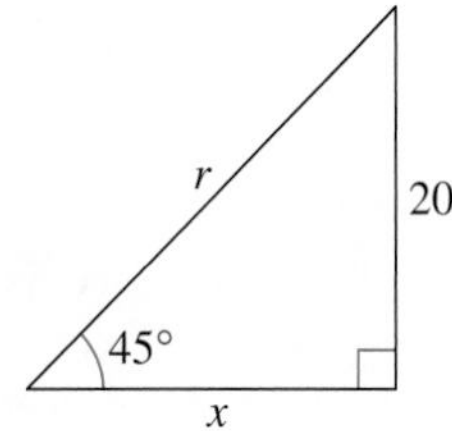

63. Empire State Building You are standing 45 meters from the base of the Empire State Building. You estimate that the angle of elevation to the top of the 86th floor (the observatory) is 82°. The total height of the building is another 123 meters above the 86th floor. What is the approximate height of the building? One of your friends is on the 86th floor. What is the distance between you and your friend?

64. Height of a Tower A six-foot person walks from the base of a broadcasting tower directly toward the tip of the shadow cast by the tower. When the person is 132 feet from the tower and 3 feet from the tip of the shadow, the person's shadow starts to appear beyond the tower's shadow.

(a) Draw a right triangle that gives a visual representation of the problem. Label the known quantities of the triangle and use a variable to represent the height of the tower.

(b) Use a trigonometric function to write an equation involving the unknown quantity.

(c) What is the height of the tower?

65. Angle of Elevation You are skiing down a mountain with a vertical height of 1250 feet. The distance from the top of the mountain to the base is 2500 feet. What is the angle of elevation from the base to the top of the mountain?

66. Biology A biologist wants to know the width w of a river to properly set instruments for an experiment. From point A, the biologist walks downstream 100 feet and sights to point C (see figure). From this sighting, it is determined that $\theta = 54°$. How wide is the river?

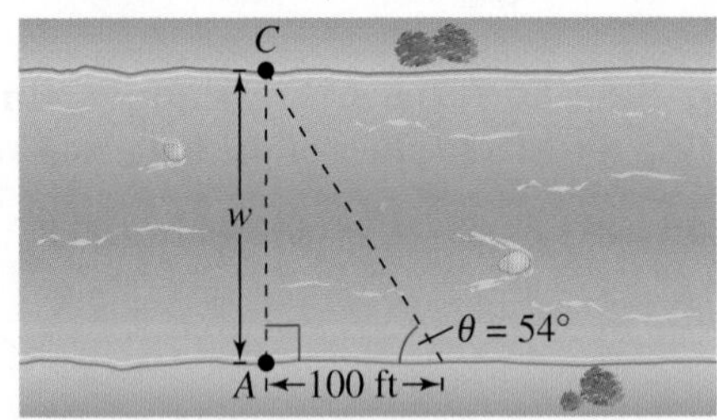

67. Guy Wire A guy wire runs from the ground to a cell tower. The wire is attached to the cell tower 150 feet above the ground. The angle formed between the wire and the ground is 43° (see figure).

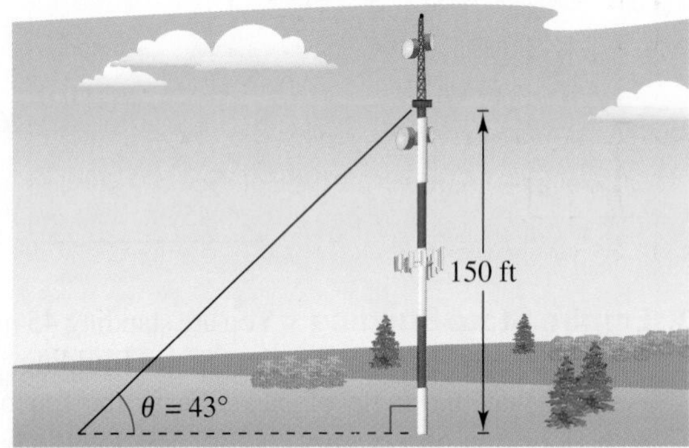

(a) How long is the guy wire?

(b) How far from the base of the tower is the guy wire anchored to the ground?

68. Height of a Mountain In traveling across flat land, you see a mountain directly in front of you. Its angle of elevation (to the peak) is 3.5°. After you drive 13 miles closer to the mountain, the angle of elevation is 9° (see figure). Approximate the height of the mountain.

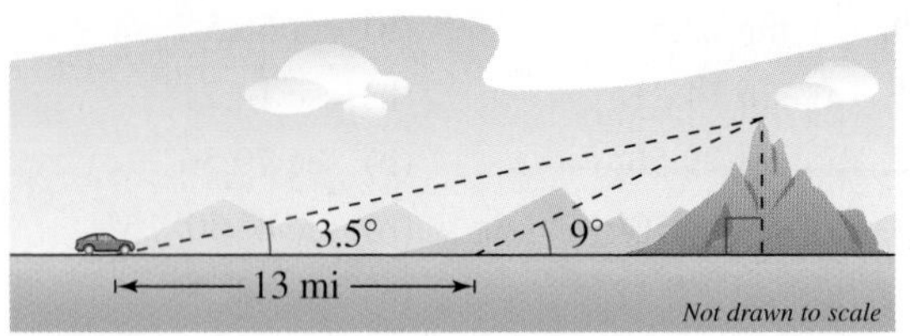

69. Machine Shop Calculations A steel plate has the form of one-fourth of a circle with a radius of 60 centimeters. Two two-centimeter holes are drilled in the plate, positioned as shown in the figure. Find the coordinates of the center of each hole.

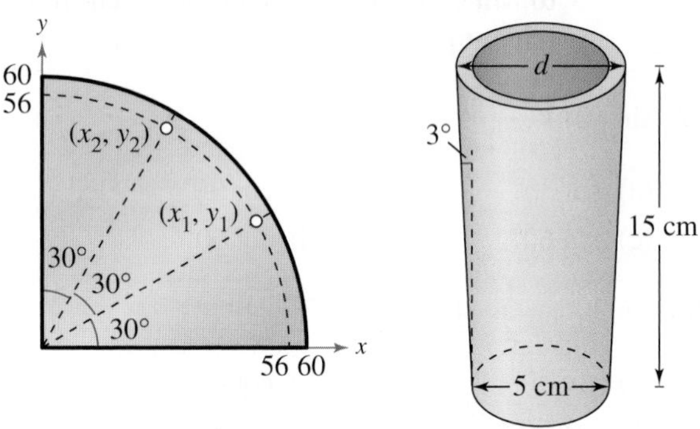

Figure for 69 Figure for 70

70. Machine Shop Calculations A tapered shaft has a diameter of 5 centimeters at the small end and is 15 centimeters long (see figure). The taper is 3°. Find the diameter d of the large end of the shaft.

71. Geometry Use a compass to sketch a quarter of a circle of radius 10 centimeters. Using a protractor, construct an angle of 20° in standard position (see figure). Drop a perpendicular line from the point of intersection of the terminal side of the angle and the arc of the circle. By actual measurement, calculate the coordinates (x, y) of the point of intersection and use these measurements to approximate the six trigonometric functions of a 20° angle.

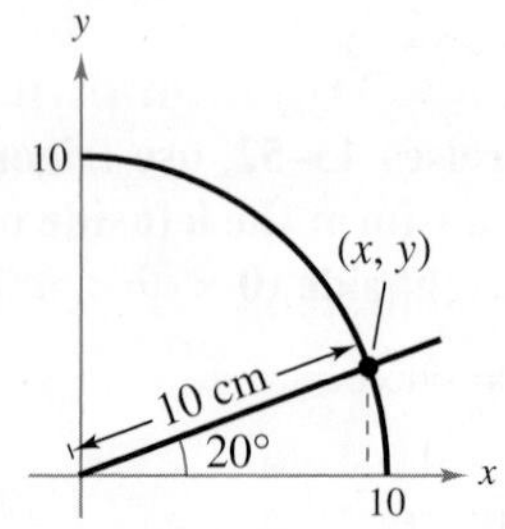

72. Helium-Filled Balloon A 20-meter line is used to tether a helium-filled balloon. The line makes an angle of approximately 85° with the ground because of a breeze.

(a) Draw a right triangle that gives a visual representation of the problem. Label the known quantities of the triangle and use a variable to represent the height of the balloon.

(b) Use a trigonometric function to write and solve an equation for the height of the balloon.

(c) The breeze becomes stronger and the angle the line makes with the ground decreases. How does this affect the triangle you drew in part (a)?

(d) Complete the table, which shows the heights (in meters) of the balloon for decreasing angle measures θ.

Angle, θ	80°	70°	60°	50°
Height				

Angle, θ	40°	30°	20°	10°
Height				

(e) As θ approaches 0°, how does this affect the height of the balloon? Draw a right triangle to explain your reasoning.

73. Johnstown Inclined Plane The Johnstown Inclined Plane in Pennsylvania is one of the longest and steepest hoists in the world. The railway cars travel a distance of 896.5 feet at an angle of approximately 35.4°, rising to a height of 1693.5 feet above sea level.

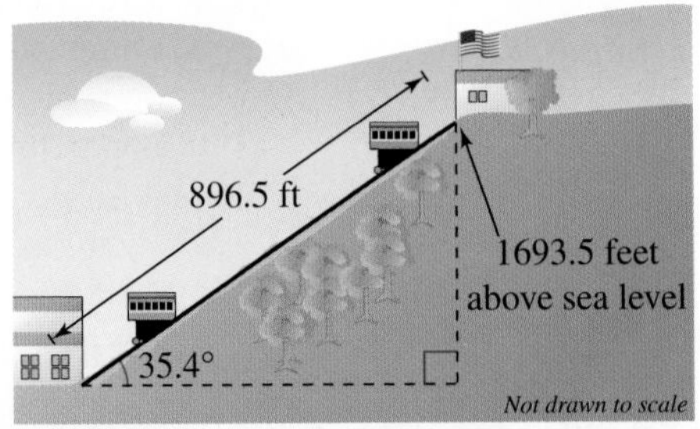

(a) Find the vertical rise of the inclined plane.

(b) Find the elevation of the lower end of the inclined plane.

(c) The cars move up the mountain at a rate of 300 feet per minute. Find the rate at which they rise vertically.

74. Error Analysis Describe the error.

$$\cos 60° = \frac{\text{opp}}{\text{hyp}} = \frac{1}{2} \quad ✗$$

Exploration

True or False? **In Exercises 75–80, determine whether the statement is true or false. Justify your answer.**

75. $\sin 60° \csc 60° = 1$

76. $\sec 30° = \csc 30°$

77. $\sin 45° + \cos 45° = 1$

78. $\cos 60° - \sin 30° = 0$

79. $\dfrac{\sin 60°}{\sin 30°} = \sin 2°$

80. $\tan[(5°)^2] = \tan^2 5°$

81. Think About It You are given the value of $\tan \theta$. Is it possible to find the value of $\sec \theta$ without finding the measure of θ? Explain.

82. HOW DO YOU SEE IT? Use the figure below.

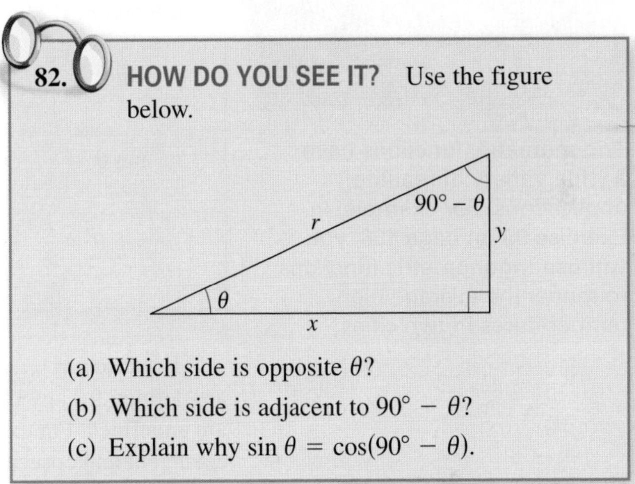

(a) Which side is opposite θ?

(b) Which side is adjacent to $90° - \theta$?

(c) Explain why $\sin \theta = \cos(90° - \theta)$.

83. Think About It Complete the table.

θ	0.1	0.2	0.3	0.4	0.5
$\sin \theta$					

(a) Is θ or $\sin \theta$ greater for θ in the interval (0, 0.5]?

(b) As θ approaches 0, how do θ and $\sin \theta$ compare? Explain.

84. Think About It Complete the table.

θ	0°	18°	36°	54°	72°	90°
$\sin \theta$						
$\cos \theta$						

(a) Discuss the behavior of the sine function for $0° \le \theta \le 90°$.

(b) Discuss the behavior of the cosine function for $0° \le \theta \le 90°$.

(c) Use the definitions of the sine and cosine functions to explain the results of parts (a) and (b).

1.4 Trigonometric Functions of Any Angle

Trigonometric functions have a wide variety of real-life applications. For example, in Exercise 99 on page 158, you will use trigonometric functions to model the average high temperatures in two cities.

- **Evaluate trigonometric functions of any angle.**
- **Find reference angles.**
- **Evaluate trigonometric functions of real numbers.**

Introduction

In Section 1.3, the definitions of trigonometric functions were restricted to acute angles. In this section, the definitions are extended to cover *any* angle. When θ is an *acute* angle, the definitions here coincide with those in the preceding section.

Definitions of Trigonometric Functions of Any Angle

Let θ be an angle in standard position with (x, y) a point on the terminal side of θ and $r = \sqrt{x^2 + y^2} \neq 0$.

$$\sin \theta = \frac{y}{r} \qquad \cos \theta = \frac{x}{r}$$

$$\tan \theta = \frac{y}{x}, \quad x \neq 0 \qquad \cot \theta = \frac{x}{y}, \quad y \neq 0$$

$$\sec \theta = \frac{r}{x}, \quad x \neq 0 \qquad \csc \theta = \frac{r}{y}, \quad y \neq 0$$

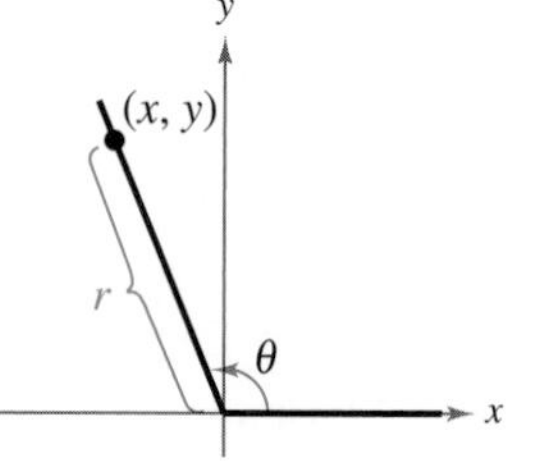

Because $r = \sqrt{x^2 + y^2}$ *cannot* be zero, it follows that the sine and cosine functions are defined for any real value of θ. However, when $x = 0$, the tangent and secant of θ are undefined. For example, the tangent of 90° is undefined. Similarly, when $y = 0$, the cotangent and cosecant of θ are undefined.

EXAMPLE 1 Evaluating Trigonometric Functions

Let $(-3, 4)$ be a point on the terminal side of θ. Find the sine, cosine, and tangent of θ.

Solution Referring to Figure 1.26, $x = -3$, $y = 4$, and

$$r = \sqrt{x^2 + y^2} = \sqrt{(-3)^2 + 4^2} = 5.$$

So, you have

$$\sin \theta = \frac{y}{r} = \frac{4}{5}$$

$$\cos \theta = \frac{x}{r} = -\frac{3}{5}$$

and

$$\tan \theta = \frac{y}{x} = -\frac{4}{3}.$$

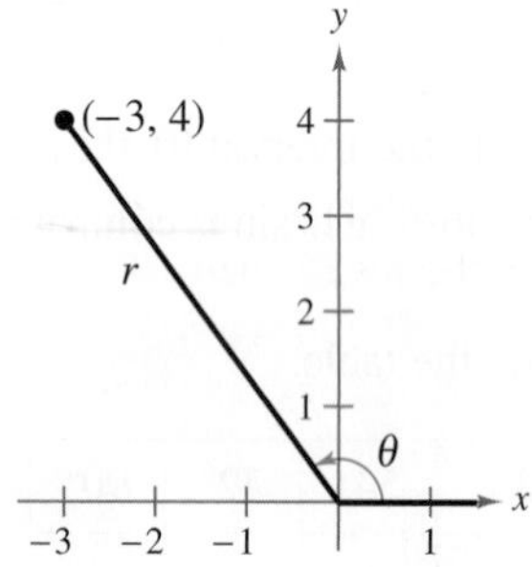

Figure 1.26

ALGEBRA HELP The formula $r = \sqrt{x^2 + y^2}$ is an application of the Distance Formula. To review the Distance Formula, see Section P.3.

Let $(-2, 3)$ be a point on the terminal side of θ. Find the sine, cosine, and tangent of θ.

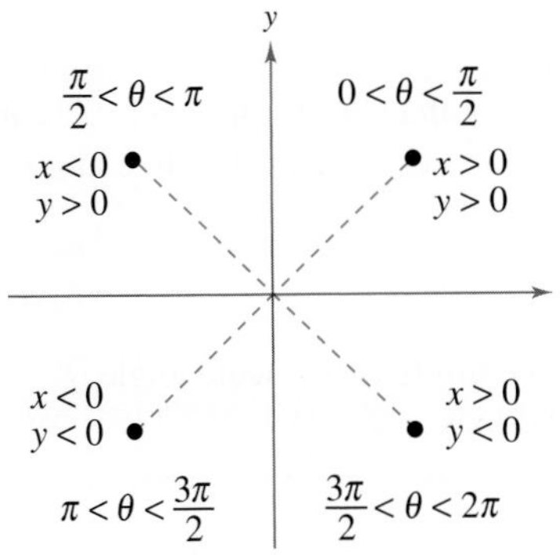

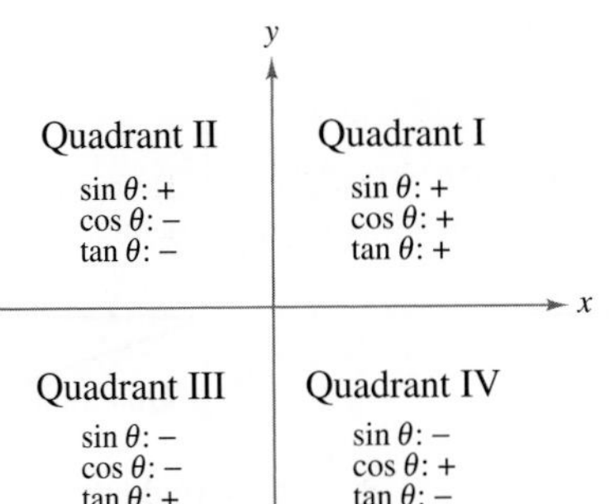

Figure 1.27

The *signs* of the trigonometric functions in the four quadrants can be determined from the definitions of the functions. For example, $\cos\theta = x/r$, so $\cos\theta$ is positive wherever $x > 0$, which is in Quadrants I and IV. (Remember, r is always positive.) Figure 1.27 shows this and other results. Use similar reasoning to verify the other results.

EXAMPLE 2 Evaluating Trigonometric Functions

Given $\tan\theta = -\frac{5}{4}$ and $\cos\theta > 0$, find $\sin\theta$ and $\sec\theta$.

Solution Note that θ lies in Quadrant IV because that is the only quadrant in which the tangent is negative and the cosine is positive. Moreover, using

$$\tan\theta = \frac{y}{x} = -\frac{5}{4}$$

and the fact that y is negative in Quadrant IV, let $y = -5$ and $x = 4$. So, $r = \sqrt{16 + 25} = \sqrt{41}$ and you have the results below.

$$\sin\theta = \frac{y}{r}$$

$$= \frac{-5}{\sqrt{41}} \qquad \text{Exact value}$$

$$\approx -0.7809 \qquad \text{Approximate value}$$

$$\sec\theta = \frac{r}{x}$$

$$= \frac{\sqrt{41}}{4} \qquad \text{Exact value}$$

$$\approx 1.6008 \qquad \text{Approximate value}$$

Checkpoint *Audio-video solution in English & Spanish at LarsonPrecalculus.com*

Given $\sin\theta = \frac{4}{5}$ and $\tan\theta < 0$, find $\cos\theta$ and $\tan\theta$.

EXAMPLE 3 Trigonometric Functions of Quadrantal Angles

Evaluate the cosine and tangent functions at the quadrantal angles $0, \frac{\pi}{2}, \pi$, and $\frac{3\pi}{2}$.

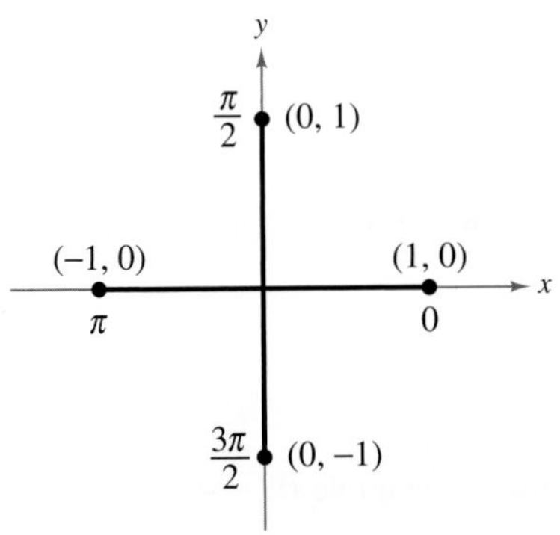

Figure 1.28

Solution To begin, choose a point on the terminal side of each angle, as shown in Figure 1.28. For each of the four points, $r = 1$ and you have the results below.

$$\cos 0 = \frac{x}{r} = \frac{1}{1} = 1 \qquad \tan 0 = \frac{y}{x} = \frac{0}{1} = 0 \qquad (x, y) = (1, 0)$$

$$\cos\frac{\pi}{2} = \frac{x}{r} = \frac{0}{1} = 0 \qquad \tan\frac{\pi}{2} = \frac{y}{x} = \frac{1}{0} \Rightarrow \text{undefined} \qquad (x, y) = (0, 1)$$

$$\cos\pi = \frac{x}{r} = \frac{-1}{1} = -1 \qquad \tan\pi = \frac{y}{x} = \frac{0}{-1} = 0 \qquad (x, y) = (-1, 0)$$

$$\cos\frac{3\pi}{2} = \frac{x}{r} = \frac{0}{1} = 0 \qquad \tan\frac{3\pi}{2} = \frac{y}{x} = \frac{-1}{0} \Rightarrow \text{undefined} \qquad (x, y) = (0, -1)$$

Evaluate the sine and cotangent functions at the quadrantal angle $\frac{3\pi}{2}$.

Reference Angles

The values of the trigonometric functions of angles greater than 90° (or less than 0°) can be determined from their values at corresponding acute angles called **reference angles.**

> **Definition of a Reference Angle**
>
> Let θ be an angle in standard position. Its **reference angle** is the acute angle θ' formed by the terminal side of θ and the horizontal axis.

The three figures below show the reference angles for θ in Quadrants II, III, and IV.

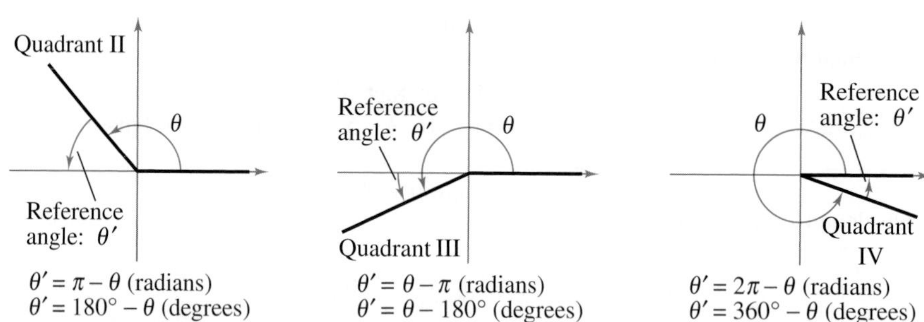

EXAMPLE 4 Finding Reference Angles

Find the reference angle θ'.

a. $\theta = 300°$ **b.** $\theta = 2.3$ **c.** $\theta = -135°$

Solution

a. Because 300° lies in Quadrant IV, the angle it makes with the x-axis is

$$\theta' = 360° - 300°$$
$$= 60°. \quad \text{Degrees}$$

Figure 1.29 shows the angle $\theta = 300°$ and its reference angle $\theta' = 60°$.

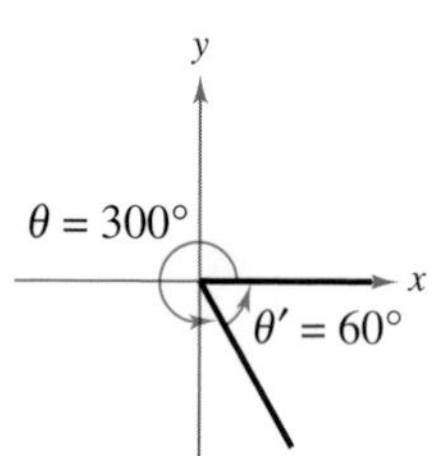

Figure 1.29

b. Because 2.3 lies between $\pi/2 \approx 1.5708$ and $\pi \approx 3.1416$, it follows that it is in Quadrant II and its reference angle is

$$\theta' = \pi - 2.3$$
$$\approx 0.8416. \quad \text{Radians}$$

Figure 1.30 shows the angle $\theta = 2.3$ and its reference angle $\theta' = \pi - 2.3$.

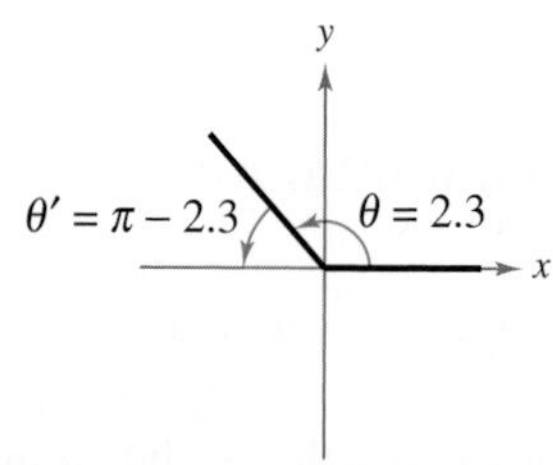

Figure 1.30

c. First, determine that $-135°$ is coterminal with 225°, which lies in Quadrant III. So, the reference angle is

$$\theta' = 225° - 180°$$
$$= 45°. \quad \text{Degrees}$$

Figure 1.31 shows the angle $\theta = -135°$ and its reference angle $\theta' = 45°$.

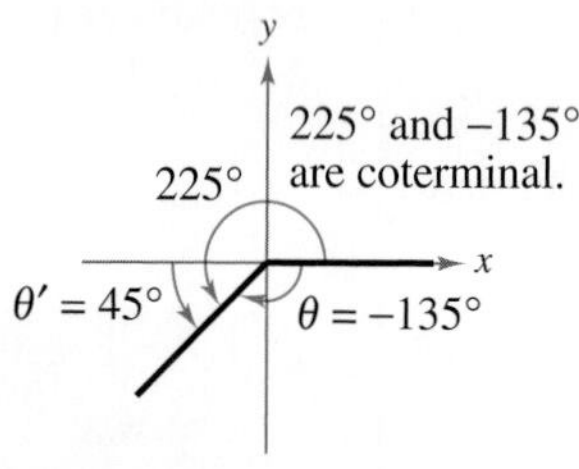

Figure 1.31

✓ **Checkpoint** Audio-video solution in English & Spanish at LarsonPrecalculus.com

Find the reference angle θ'.

a. $\theta = 213°$ **b.** $\theta = \dfrac{14\pi}{9}$ **c.** $\theta = \dfrac{4\pi}{5}$

Trigonometric Functions of Real Numbers

To see how to use a reference angle to evaluate a trigonometric function, consider the point (x, y) on the terminal side of the angle θ, as shown at the right. You know that

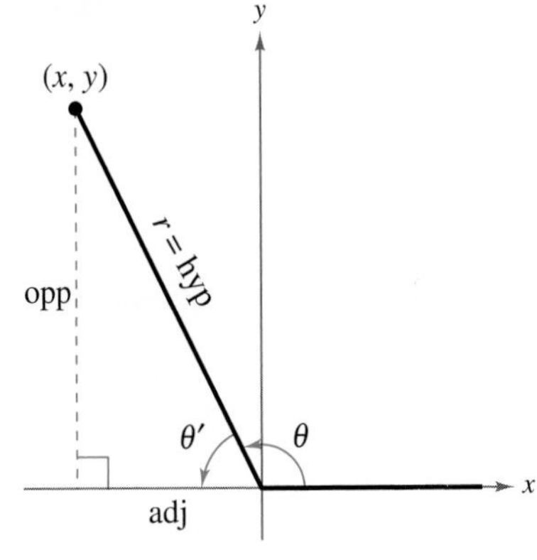

$opp = |y|,\ adj = |x|$

$$\sin\theta = \frac{y}{r}$$

and

$$\tan\theta = \frac{y}{x}.$$

For the right triangle with acute angle θ' and sides of lengths $|x|$ and $|y|$, you have

$$\sin\theta' = \frac{\text{opp}}{\text{hyp}} = \frac{|y|}{r}$$

and

$$\tan\theta' = \frac{\text{opp}}{\text{adj}} = \frac{|y|}{|x|}.$$

So, it follows that $\sin\theta$ and $\sin\theta'$ are equal, *except possibly in sign*. The same is true for $\tan\theta$ and $\tan\theta'$ and for the other four trigonometric functions. In all cases, the quadrant in which θ lies determines the sign of the function value.

Evaluating Trigonometric Functions of Any Angle

To find the value of a trigonometric function of any angle θ:

1. Determine the function value of the associated reference angle θ'.
2. Depending on the quadrant in which θ lies, affix the appropriate sign to the function value.

REMARK Learning the table of values at the right is worth the effort because doing so will increase both your efficiency and your confidence when working in trigonometry. Below is a pattern for the sine function that may help you remember the values.

θ	0°	30°	45°	60°	90°
$\sin\theta$	$\frac{\sqrt{0}}{2}$	$\frac{\sqrt{1}}{2}$	$\frac{\sqrt{2}}{2}$	$\frac{\sqrt{3}}{2}$	$\frac{\sqrt{4}}{2}$

Reverse the order to get cosine values of the same angles.

Using reference angles and the special angles discussed in the preceding section enables you to greatly extend the scope of *exact* trigonometric function values. For example, knowing the function values of 30° means that you know the function values of all angles for which 30° is a reference angle. For convenience, the table below shows the exact values of the sine, cosine, and tangent functions of special angles and quadrantal angles.

Trigonometric Values of Common Angles

θ (degrees)	0°	30°	45°	60°	90°	180°	270°
θ (radians)	0	$\frac{\pi}{6}$	$\frac{\pi}{4}$	$\frac{\pi}{3}$	$\frac{\pi}{2}$	π	$\frac{3\pi}{2}$
$\sin\theta$	0	$\frac{1}{2}$	$\frac{\sqrt{2}}{2}$	$\frac{\sqrt{3}}{2}$	1	0	-1
$\cos\theta$	1	$\frac{\sqrt{3}}{2}$	$\frac{\sqrt{2}}{2}$	$\frac{1}{2}$	0	-1	0
$\tan\theta$	0	$\frac{\sqrt{3}}{3}$	1	$\sqrt{3}$	Undef.	0	Undef.

EXAMPLE 5 Using Reference Angles

See LarsonPrecalculus.com for an interactive version of this type of example.

Evaluate each trigonometric function.

a. $\cos \dfrac{4\pi}{3}$ **b.** $\tan(-210°)$ **c.** $\csc \dfrac{11\pi}{4}$

Solution

a. Because $\theta = 4\pi/3$ lies in Quadrant III, the reference angle is

$$\theta' = \frac{4\pi}{3} - \pi = \frac{\pi}{3}$$

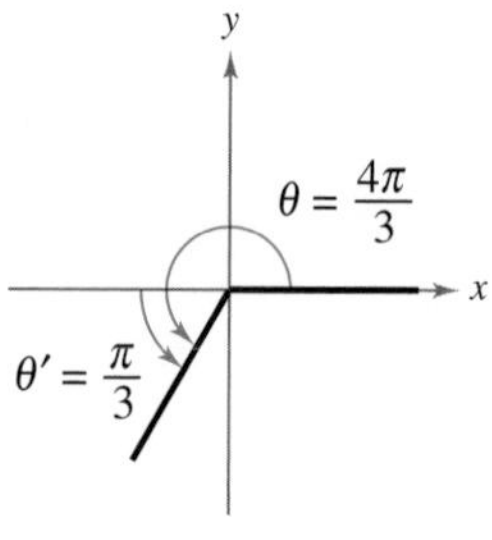

as shown at the right. The cosine is negative in Quadrant III, so

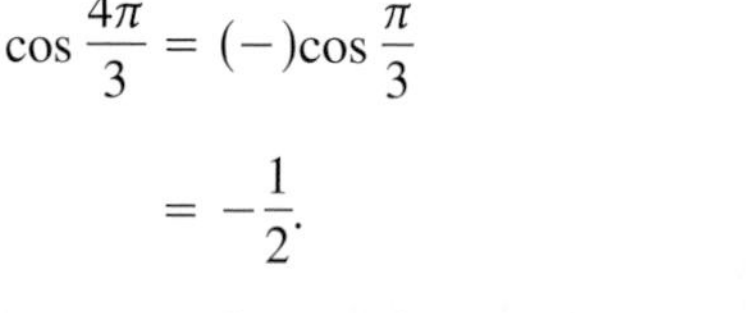

$$\cos \frac{4\pi}{3} = (-)\cos \frac{\pi}{3}$$
$$= -\frac{1}{2}.$$

b. Because $-210° + 360° = 150°$, it follows that $-210°$ is coterminal with the second-quadrant angle $150°$. So, the reference angle is

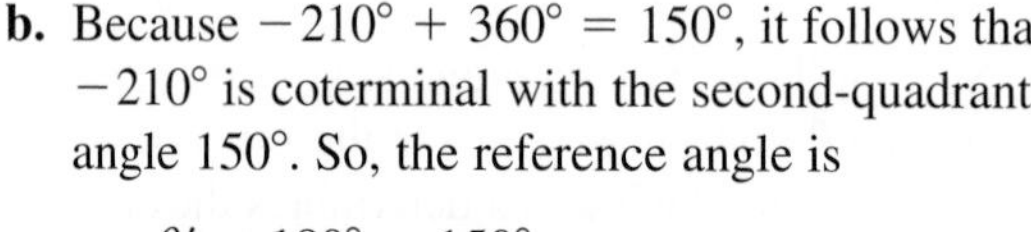

$$\theta' = 180° - 150°$$
$$= 30°$$

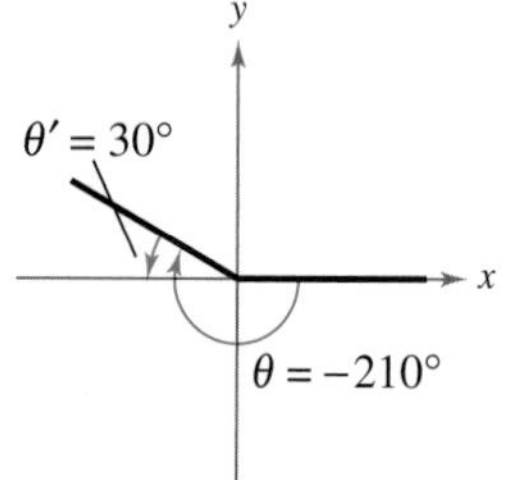

as shown at the right. The tangent is negative in Quadrant II, so

$$\tan(-210°) = (-)\tan 30°$$
$$= -\frac{\sqrt{3}}{3}.$$

c. Because $(11\pi/4) - 2\pi = 3\pi/4$, it follows that $11\pi/4$ is coterminal with the second-quadrant angle $3\pi/4$. So, the reference angle is

$$\theta' = \pi - \frac{3\pi}{4} = \frac{\pi}{4}$$

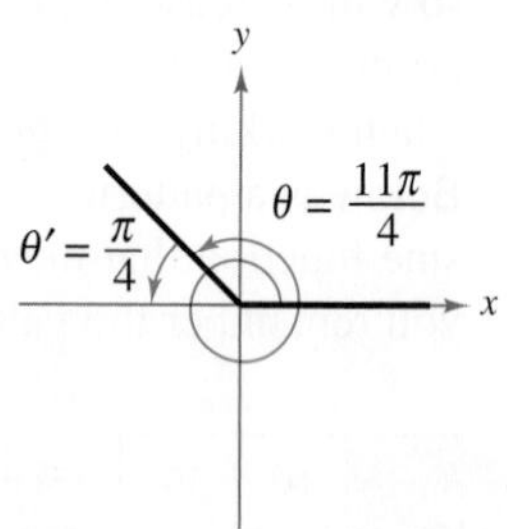

as shown at the right. The cosecant is positive in Quadrant II, so

$$\csc \frac{11\pi}{4} = (+)\csc \frac{\pi}{4}$$
$$= \frac{1}{\sin(\pi/4)}$$
$$= \sqrt{2}.$$

✓ **Checkpoint**  *Audio-video solution in English & Spanish at LarsonPrecalculus.com*

Evaluate each trigonometric function.

a. $\sin \dfrac{7\pi}{4}$ **b.** $\cos(-120°)$ **c.** $\tan \dfrac{11\pi}{6}$

EXAMPLE 6 Using Trigonometric Identities

Let θ be an angle in Quadrant II such that $\sin\theta = \frac{1}{3}$. Find (a) $\cos\theta$ and (b) $\tan\theta$ by using trigonometric identities.

Solution

REMARK The fundamental trigonometric identities listed in the preceding section (for an acute angle θ) are also valid when θ is any angle in the domain of the function.

a. Using the Pythagorean identity $\sin^2\theta + \cos^2\theta = 1$, you obtain

$$\left(\frac{1}{3}\right)^2 + \cos^2\theta = 1 \quad\Longrightarrow\quad \cos^2\theta = 1 - \frac{1}{9} = \frac{8}{9}.$$

You know that $\cos\theta < 0$ in Quadrant II, so use the negative root to obtain

$$\cos\theta = -\frac{\sqrt{8}}{\sqrt{9}} = -\frac{2\sqrt{2}}{3}.$$

b. Using the trigonometric identity $\tan\theta = \dfrac{\sin\theta}{\cos\theta}$, you obtain

$$\tan\theta = \frac{1/3}{-2\sqrt{2}/3} = -\frac{1}{2\sqrt{2}} = -\frac{\sqrt{2}}{4}.$$

Checkpoint Audio-video solution in English & Spanish at LarsonPrecalculus.com

Let θ be an angle in Quadrant III such that $\sin\theta = -\frac{4}{5}$. Find (a) $\cos\theta$ and (b) $\tan\theta$ by using trigonometric identities.

EXAMPLE 7 Using a Calculator

Use a calculator to evaluate each trigonometric function.

a. $\cot 410°$ **b.** $\sin(-7)$ **c.** $\sec\dfrac{\pi}{9}$

Solution

	Function	Mode	Calculator Keystrokes	Display
a.	$\cot 410°$	Degree	(TAN (410)) x⁻¹ ENTER	0.8390996
b.	$\sin(-7)$	Radian	SIN ((−) 7) ENTER	−0.6569866
c.	$\sec(\pi/9)$	Radian	(COS (π ÷ 9)) x⁻¹ ENTER	1.0641778

Checkpoint Audio-video solution in English & Spanish at LarsonPrecalculus.com

Use a calculator to evaluate each trigonometric function.

a. $\tan 119°$ **b.** $\csc 5$ **c.** $\cos\dfrac{\pi}{5}$

Summarize (Section 1.4)

1. State the definitions of the trigonometric functions of any angle *(page 150)*. For examples of evaluating trigonometric functions, see Examples 1–3.
2. Explain how to use a reference angle *(page 152)*. For an example of finding reference angles, see Example 4.
3. Explain how to evaluate a trigonometric function of a real number *(page 153)*. For examples of evaluating trigonometric functions of real numbers, see Examples 5–7.

1.4 Exercises

See CalcChat.com for tutorial help and worked-out solutions to odd-numbered exercises.

Vocabulary: Fill in the blanks.

In Exercises 1–6, let θ be an angle in standard position with (x, y) a point on the terminal side of θ and $r = \sqrt{x^2 + y^2} \neq 0$.

1. $\sin \theta =$ ________
2. $\frac{r}{y} =$ ________
3. $\tan \theta =$ ________
4. $\sec \theta =$ ________
5. $\frac{x}{r} =$ ________
6. $\frac{x}{y} =$ ________
7. Because $r = \sqrt{x^2 + y^2}$ cannot be ________, the sine and cosine functions are ________ for any real value of θ.
8. The acute angle formed by the terminal side of an angle θ in standard position and the horizontal axis is the ________ angle of θ and is denoted by θ'.

Skills and Applications

Evaluating Trigonometric Functions **In Exercises 9–12, find the exact values of the six trigonometric functions of each angle θ.**

9. (a)

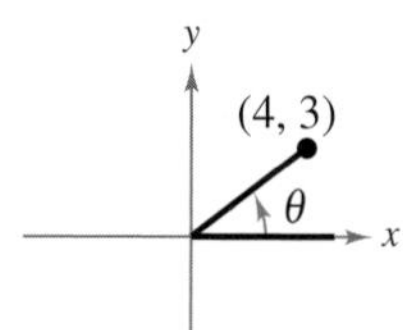

(b)

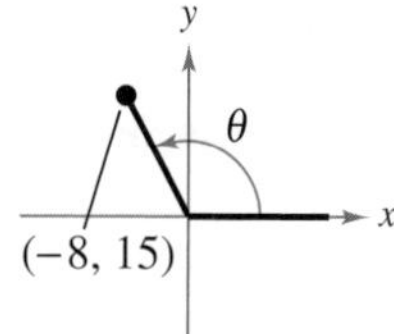

10. (a)

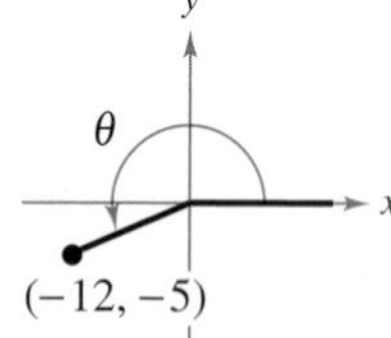

(b)

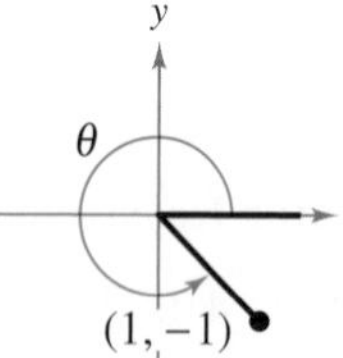

11. (a)

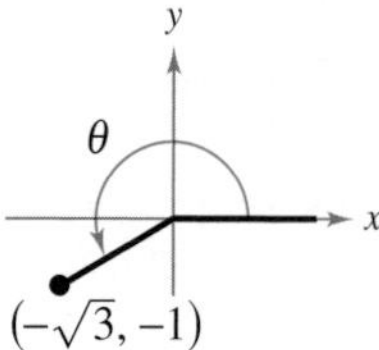

(b)

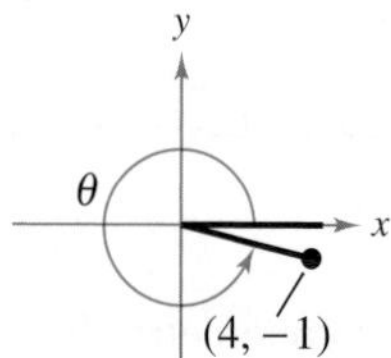

12. (a)

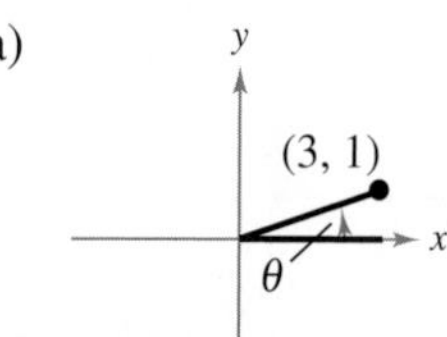

(b)

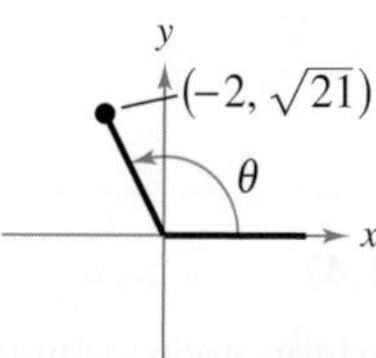

Evaluating Trigonometric Functions **In Exercises 13–18, the point is on the terminal side of an angle in standard position. Find the exact values of the six trigonometric functions of the angle.**

13. $(5, 12)$
14. $(8, 15)$
15. $(-5, -2)$
16. $(-4, 10)$
17. $(-5.4, 7.2)$
18. $\left(3\frac{1}{2}, -2\sqrt{15}\right)$

Determining a Quadrant **In Exercises 19–22, determine the quadrant in which θ lies.**

19. $\sin \theta > 0, \quad \cos \theta > 0$
20. $\sin \theta < 0, \quad \cos \theta < 0$
21. $\csc \theta > 0, \quad \tan \theta < 0$
22. $\sec \theta > 0, \quad \cot \theta < 0$

Evaluating Trigonometric Functions **In Exercises 23–32, find the exact values of the remaining trigonometric functions of θ satisfying the given conditions.**

23. $\tan \theta = \frac{15}{8}, \quad \sin \theta > 0$
24. $\cos \theta = \frac{8}{17}, \quad \tan \theta < 0$
25. $\sin \theta = 0.6, \quad \theta$ lies in Quadrant II.
26. $\cos \theta = -0.8, \quad \theta$ lies in Quadrant III.
27. $\cot \theta = -3, \quad \cos \theta > 0$
28. $\csc \theta = 4, \quad \cot \theta < 0$
29. $\cos \theta = 0, \quad \csc \theta = 1$
30. $\sin \theta = 0, \quad \sec \theta = -1$
31. $\cot \theta$ is undefined, $\frac{\pi}{2} \leq \theta \leq \frac{3\pi}{2}$
32. $\tan \theta$ is undefined, $\pi \leq \theta \leq 2\pi$

An Angle Formed by a Line Through the Origin **In Exercises 33–36, the terminal side of θ lies on the given line in the specified quadrant. Find the exact values of the six trigonometric functions of θ by finding a point on the line.**

	Line	Quadrant
33.	$y = -x$	II
34.	$y = \frac{1}{3}x$	III
35.	$2x - y = 0$	I
36.	$4x + 3y = 0$	IV

Trigonometric Function of a Quadrantal Angle **In Exercises 37–46, evaluate the trigonometric function of the quadrantal angle, if possible.**

37. $\sin 0$ **38.** $\csc \frac{3\pi}{2}$

39. $\sec \frac{3\pi}{2}$ **40.** $\sec \pi$

41. $\sin \frac{\pi}{2}$ **42.** $\cot 0$

43. $\csc \pi$ **44.** $\cot \frac{\pi}{2}$

45. $\cos \frac{9\pi}{2}$ **46.** $\tan\left(-\frac{\pi}{2}\right)$

Finding a Reference Angle **In Exercises 47–54, find the reference angle θ'. Sketch θ in standard position and label θ'.**

47. $\theta = 160°$ **48.** $\theta = 309°$

49. $\theta = -125°$ **50.** $\theta = -215°$

51. $\theta = \frac{2\pi}{3}$ **52.** $\theta = \frac{7\pi}{6}$

53. $\theta = 4.8$ **54.** $\theta = 12.9$

Using a Reference Angle **In Exercises 55–68, evaluate the sine, cosine, and tangent of the angle without using a calculator.**

55. $225°$ **56.** $300°$

57. $750°$ **58.** $675°$

59. $-120°$ **60.** $-570°$

61. $\frac{2\pi}{3}$ **62.** $\frac{3\pi}{4}$

63. $-\frac{\pi}{6}$ **64.** $-\frac{2\pi}{3}$

65. $\frac{11\pi}{4}$ **66.** $\frac{13\pi}{6}$

67. $-\frac{17\pi}{6}$ **68.** $-\frac{23\pi}{4}$

Using a Trigonometric Identity **In Exercises 69–74, use the function value to find the indicated trigonometric value in the specified quadrant.**

	Function Value	Quadrant	Trigonometric Value
69.	$\sin \theta = -\frac{3}{5}$	IV	$\cos \theta$
70.	$\cot \theta = -3$	II	$\csc \theta$
71.	$\tan \theta = \frac{3}{2}$	III	$\sec \theta$
72.	$\csc \theta = -2$	IV	$\cot \theta$
73.	$\cos \theta = \frac{5}{8}$	I	$\csc \theta$
74.	$\sec \theta = -\frac{9}{4}$	III	$\cot \theta$

Using a Calculator **In Exercises 75–90, use a calculator to evaluate the trigonometric function. Round your answer to four decimal places. (Be sure the calculator is in the correct mode.)**

75. $\sin 10°$ **76.** $\tan 304°$

77. $\cos(-110°)$ **78.** $\sin(-330°)$

79. $\cot 178°$ **80.** $\sec 72°$

81. $\csc 405°$ **82.** $\cot(-560°)$

83. $\tan \frac{\pi}{9}$ **84.** $\cos \frac{2\pi}{7}$

85. $\sec \frac{11\pi}{8}$ **86.** $\csc \frac{15\pi}{4}$

87. $\sin(-0.65)$ **88.** $\cos 1.35$

89. $\csc(-10)$ **90.** $\sec(-4.6)$

Solving for θ **In Exercises 91–96, find two solutions of each equation. Give your answers in degrees ($0° \le \theta < 360°$) and in radians ($0 \le \theta < 2\pi$). Do not use a calculator.**

91. (a) $\sin \theta = \frac{1}{2}$ **92.** (a) $\cos \theta = \frac{\sqrt{2}}{2}$

(b) $\sin \theta = -\frac{1}{2}$ (b) $\cos \theta = -\frac{\sqrt{2}}{2}$

93. (a) $\cos \theta = \frac{1}{2}$ **94.** (a) $\sin \theta = \frac{\sqrt{3}}{2}$

(b) $\sec \theta = 2$ (b) $\csc \theta = \frac{2\sqrt{3}}{3}$

95. (a) $\tan \theta = 1$ **96.** (a) $\cot \theta = 0$

(b) $\cot \theta = -\sqrt{3}$ (b) $\sec \theta = -\sqrt{2}$

97. Distance An airplane, flying at an altitude of 6 miles, is on a flight path that passes directly over an observer (see figure). Let θ be the angle of elevation from the observer to the plane. Find the distance d from the observer to the plane when (a) $\theta = 30°$, (b) $\theta = 90°$, and (c) $\theta = 120°$.

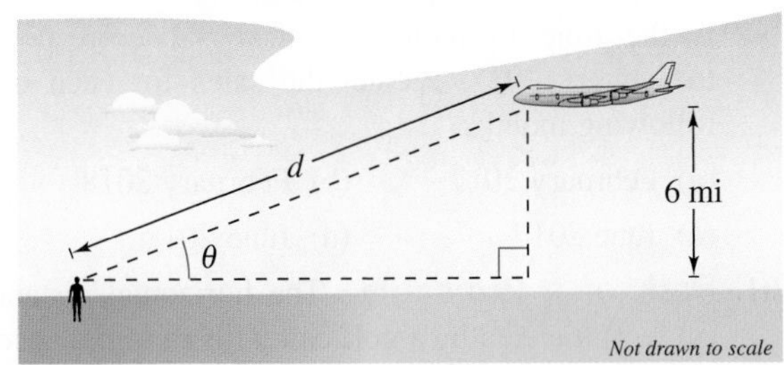

98. Harmonic Motion The displacement from equilibrium of an oscillating weight suspended by a spring is given by $y(t) = 2 \cos 6t$, where y is the displacement in centimeters and t is the time in seconds. Find the displacement when (a) $t = 0$, (b) $t = \frac{1}{4}$, and (c) $t = \frac{1}{2}$.

99. Temperature The table shows the average high temperatures (in degrees Fahrenheit) in Boston, Massachusetts (B), and Fairbanks, Alaska (F), for selected months in 2015. *(Source: U.S. Climate Data)*

DATA

Month	Boston, B	Fairbanks, F
January	33	1
March	41	31
June	72	71
August	83	62
November	56	17

Spreadsheet at LarsonPrecalculus.com

(a) Use the *regression* feature of a graphing utility to find a model of the form

$$y = a\sin(bt + c) + d$$

for each city. Let t represent the month, with $t = 1$ corresponding to January.

(b) Use the models from part (a) to estimate the monthly average high temperatures for the two cities in February, April, May, July, September, October, and December.

(c) Use a graphing utility to graph both models in the same viewing window. Compare the temperatures for the two cities.

100. Sales A company that produces snowboards forecasts monthly sales over the next 2 years to be

$$S = 23.1 + 0.442t + 4.3\cos\frac{\pi t}{6}$$

where S is measured in thousands of units and t is the time in months, with $t = 1$ corresponding to January 2017. Predict the sales for each of the following months.

(a) February 2017 (b) February 2018

(c) June 2017 (d) June 2018

101. Path of a Projectile The horizontal distance d (in feet) traveled by a golf ball with an initial speed of 100 feet per second is modeled by

$$d = 312.5\sin 2\theta$$

where θ is the angle at which the golf ball is hit. Find the horizontal distance traveled by the golf ball when (a) $\theta = 30°$, (b) $\theta = 50°$, and (c) $\theta = 60°$.

102. HOW DO YOU SEE IT? Consider an angle in standard position with $r = 12$ centimeters, as shown in the figure. Describe the changes in the values of x, y, $\sin\theta$, $\cos\theta$, and $\tan\theta$ as θ increases continuously from 0° to 90°.

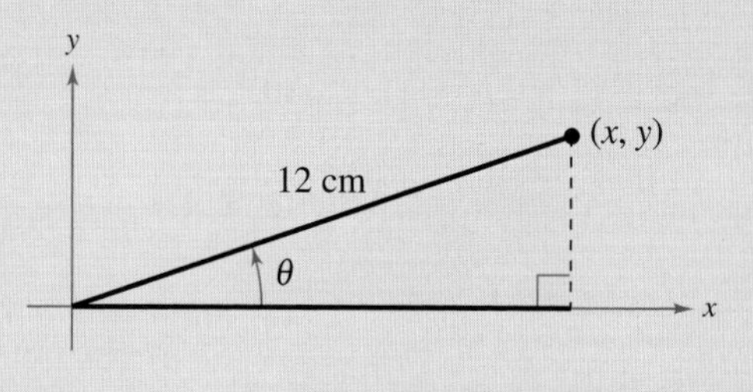

Exploration

True or False? **In Exercises 103 and 104, determine whether the statement is true or false. Justify your answer.**

103. In each of the four quadrants, the signs of the secant function and the sine function are the same.

104. The reference angle for an angle θ (in degrees) is the angle $\theta' = 360°n - \theta$, where n is an integer and $0° \le \theta' \le 360°$.

105. Writing Write a short essay explaining to a classmate how to evaluate the six trigonometric functions of any angle θ in standard position. Include an explanation of reference angles and how to use them, the signs of the functions in each of the four quadrants, and the trigonometric values of common angles. Include figures or diagrams in your essay.

106. Think About It The figure shows point $P(x, y)$ on a unit circle and right triangle OAP.

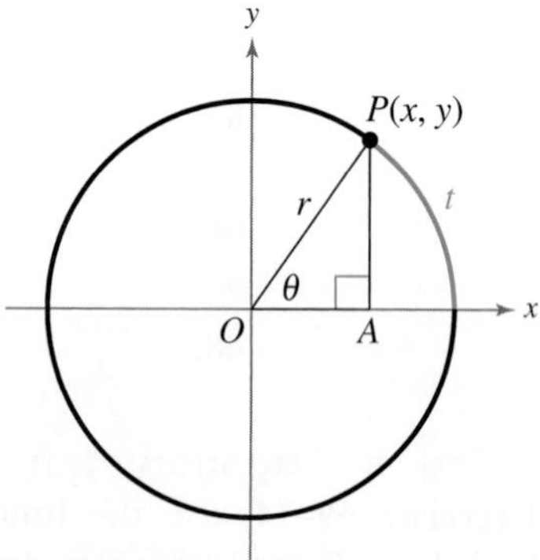

(a) Find $\sin t$ and $\cos t$ using the unit circle definitions of sine and cosine (from Section 1.2).

(b) What is the value of r? Explain.

(c) Use the definitions of sine and cosine given in this section to find $\sin\theta$ and $\cos\theta$. Write your answers in terms of x and y.

(d) Based on your answers to parts (a) and (c), what can you conclude?

1.5 Graphs of Sine and Cosine Functions

Graphs of sine and cosine functions have many scientific applications. For example, in Exercise 80 on page 168, you will use the graph of a sine function to analyze airflow during a respiratory cycle.

- **Sketch the graphs of basic sine and cosine functions.**
- **Use amplitude and period to help sketch the graphs of sine and cosine functions.**
- **Sketch translations of the graphs of sine and cosine functions.**
- **Use sine and cosine functions to model real-life data.**

Basic Sine and Cosine Curves

In this section, you will study techniques for sketching the graphs of the sine and cosine functions. The graph of the sine function, shown in Figure 1.32, is a **sine curve.** In the figure, the black portion of the graph represents one period of the function and is **one cycle** of the sine curve. The gray portion of the graph indicates that the basic sine curve repeats indefinitely to the left and right. Figure 1.33 shows the graph of the cosine function.

Recall from Section 1.2 that the domain of the sine and cosine functions is the set of all real numbers. Moreover, the range of each function is the interval $[-1, 1]$, and each function has a period of 2π. This information is consistent with the basic graphs shown in Figures 1.32 and 1.33.

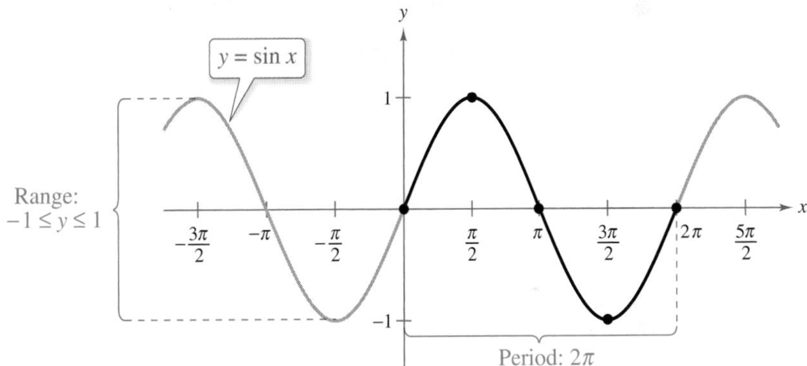

Figure 1.32

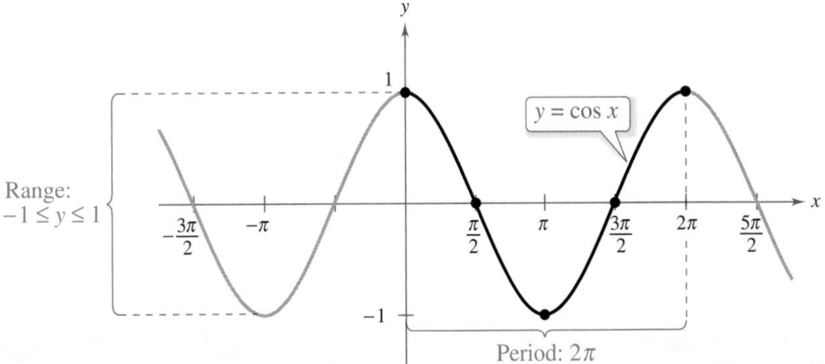

Figure 1.33

Note in Figures 1.32 and 1.33 that the sine curve is symmetric with respect to the *origin*, whereas the cosine curve is symmetric with respect to the *y-axis*. These properties of symmetry follow from the fact that the sine function is odd and the cosine function is even.

Neo2620/Shutterstock.com

To sketch the graphs of the basic sine and cosine functions, it helps to note five **key points** in one period of each graph: the *intercepts, maximum points,* and *minimum points* (see graphs below).

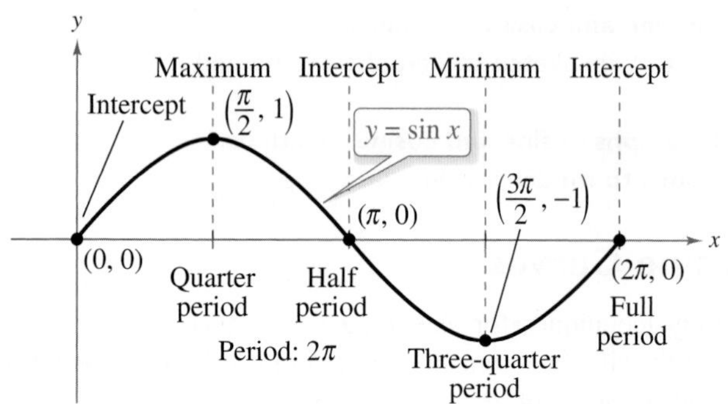

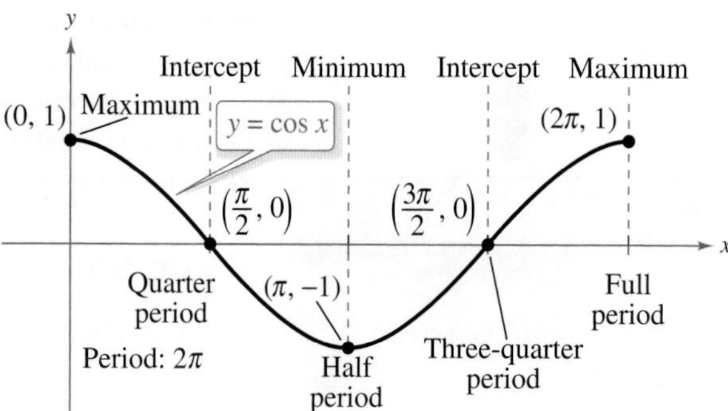

EXAMPLE 1 Using Key Points to Sketch a Sine Curve

See LarsonPrecalculus.com for an interactive version of this type of example.

Sketch the graph of

$$y = 2 \sin x$$

on the interval $[-\pi, 4\pi]$.

Solution Note that

$$y = 2 \sin x$$
$$= 2(\sin x).$$

So, the y-values for the key points have twice the magnitude of those on the graph of $y = \sin x$. Divide the period 2π into four equal parts to obtain the key points

Intercept	Maximum	Intercept	Minimum		Intercept
$(0, 0)$,	$\left(\frac{\pi}{2}, 2\right)$,	$(\pi, 0)$,	$\left(\frac{3\pi}{2}, -2\right)$,	and	$(2\pi, 0)$.

By connecting these key points with a smooth curve and extending the curve in both directions over the interval $[-\pi, 4\pi]$, you obtain the graph below.

TECHNOLOGY When using a graphing utility to graph trigonometric functions, pay special attention to the viewing window you use. For example, graph

$$y = \frac{\sin 10x}{10}$$

in the standard viewing window in *radian* mode. What do you observe? Use the *zoom* feature to find a viewing window that displays a good view of the graph.

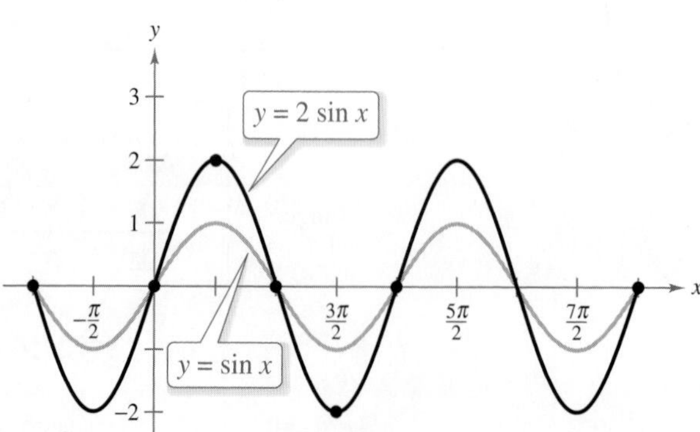

Checkpoint Audio-video solution in English & Spanish at LarsonPrecalculus.com

Sketch the graph of

$$y = 2 \cos x$$

on the interval $\left[-\frac{\pi}{2}, \frac{9\pi}{2}\right]$.

Amplitude and Period

In the rest of this section, you will study the effect of each of the constants a, b, c, and d on the graphs of equations of the forms

$$y = d + a\sin(bx - c)$$

and

$$y = d + a\cos(bx - c).$$

A quick review of the transformations you studied in Section P.8 will help in this investigation.

The constant factor a in $y = a\sin x$ and $y = a\cos x$ acts as a *scaling factor*—a *vertical stretch* or *vertical shrink* of the basic curve. When $|a| > 1$, the basic curve is stretched, and when $0 < |a| < 1$, the basic curve is shrunk. The result is that the graphs of $y = a\sin x$ and $y = a\cos x$ range between $-a$ and a instead of between -1 and 1. The absolute value of a is the **amplitude** of the function. The range of the function for $a > 0$ is $-a \le y \le a$.

Definition of the Amplitude of Sine and Cosine Curves

The **amplitude** of $y = a\sin x$ and $y = a\cos x$ represents half the distance between the maximum and minimum values of the function and is given by

$$\text{Amplitude} = |a|.$$

EXAMPLE 2 Scaling: Vertical Shrinking and Stretching

In the same coordinate plane, sketch the graph of each function.

a. $y = \dfrac{1}{2}\cos x$

b. $y = 3\cos x$

Solution

a. The amplitude of $y = \frac{1}{2}\cos x$ is $\frac{1}{2}$, so the maximum value is $\frac{1}{2}$ and the minimum value is $-\frac{1}{2}$. Divide one cycle, $0 \le x \le 2\pi$, into four equal parts to obtain the key points

Maximum	Intercept	Minimum	Intercept		Maximum
$\left(0, \frac{1}{2}\right)$,	$\left(\frac{\pi}{2}, 0\right)$,	$\left(\pi, -\frac{1}{2}\right)$,	$\left(\frac{3\pi}{2}, 0\right)$,	and	$\left(2\pi, \frac{1}{2}\right)$.

b. A similar analysis shows that the amplitude of $y = 3\cos x$ is 3, and the key points are

Maximum	Intercept	Minimum	Intercept		Maximum
$(0, 3)$,	$\left(\frac{\pi}{2}, 0\right)$,	$(\pi, -3)$,	$\left(\frac{3\pi}{2}, 0\right)$,	and	$(2\pi, 3)$.

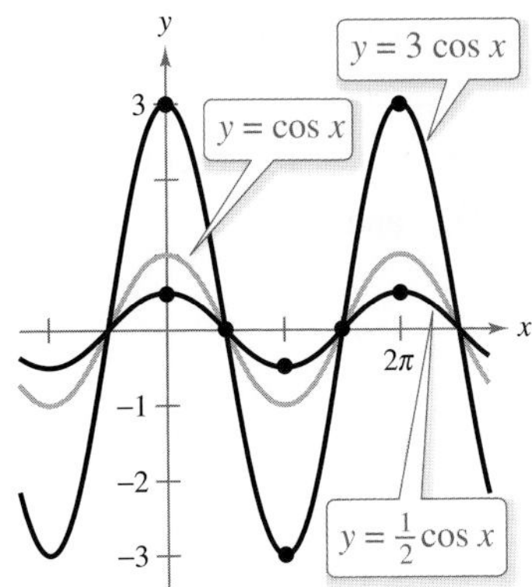

Figure 1.34

Figure 1.34 shows the graphs of these two functions. Notice that the graph of $y = \frac{1}{2}\cos x$ is a vertical *shrink* of the graph of $y = \cos x$ and the graph of $y = 3\cos x$ is a vertical *stretch* of the graph of $y = \cos x$.

In the same coordinate plane, sketch the graph of each function.

a. $y = \dfrac{1}{3}\sin x$

b. $y = 3\sin x$

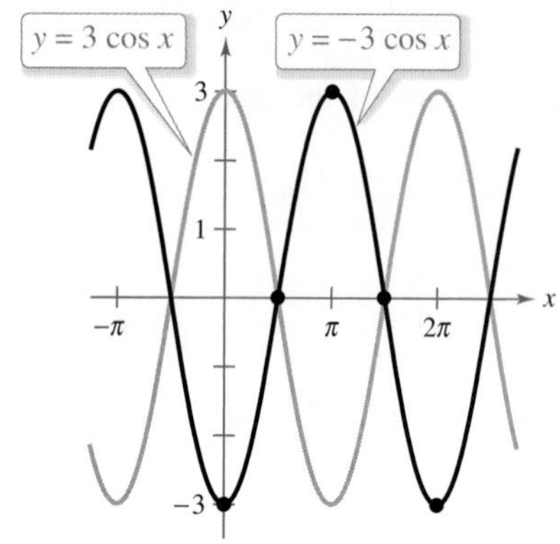

Figure 1.35

You know from Section P.8 that the graph of $y = -f(x)$ is a **reflection** in the x-axis of the graph of $y = f(x)$. For example, the graph of $y = -3 \cos x$ is a reflection of the graph of $y = 3 \cos x$, as shown in Figure 1.35.

Next, consider the effect of the positive real number b on the graphs of $y = a \sin bx$ and $y = a \cos bx$. For example, compare the graphs of $y = a \sin x$ and $y = a \sin bx$. The graph of $y = a \sin x$ completes one cycle from $x = 0$ to $x = 2\pi$, so it follows that the graph of $y = a \sin bx$ completes one cycle from $x = 0$ to $x = 2\pi/b$.

Period of Sine and Cosine Functions

Let b be a positive real number. The **period** of $y = a \sin bx$ and $y = a \cos bx$ is given by

$$\text{Period} = \frac{2\pi}{b}.$$

Note that when $0 < b < 1$, the period of $y = a \sin bx$ is greater than 2π and represents a *horizontal stretch* of the basic curve. Similarly, when $b > 1$, the period of $y = a \sin bx$ is less than 2π and represents a *horizontal shrink* of the basic curve. These two statements are also true for $y = a \cos bx$. When b is negative, rewrite the function using the identity $\sin(-x) = -\sin x$ or $\cos(-x) = \cos x$.

EXAMPLE 3 Scaling: Horizontal Stretching

Sketch the graph of

$$y = \sin \frac{x}{2}.$$

Solution The amplitude is 1. Moreover, $b = \frac{1}{2}$, so the period is

$$\frac{2\pi}{b} = \frac{2\pi}{\frac{1}{2}} = 4\pi. \qquad \text{Substitute for } b.$$

Now, divide the period-interval $[0, 4\pi]$ into four equal parts using the values π, 2π, and 3π to obtain the key points

Intercept	Maximum	Intercept	Minimum		Intercept
$(0, 0)$,	$(\pi, 1)$,	$(2\pi, 0)$,	$(3\pi, -1)$,	and	$(4\pi, 0)$.

The graph is shown below.

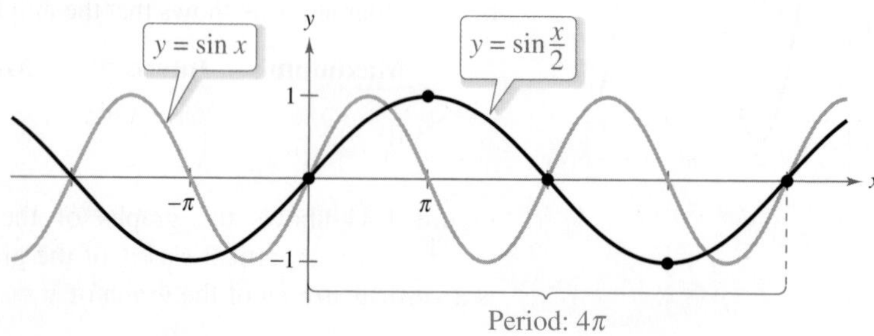

REMARK In general, to divide a period-interval into four equal parts, successively add "period/4," starting with the left endpoint of the interval. For example, for the period-interval $[-\pi/6, \pi/2]$ of length $2\pi/3$, you would successively add

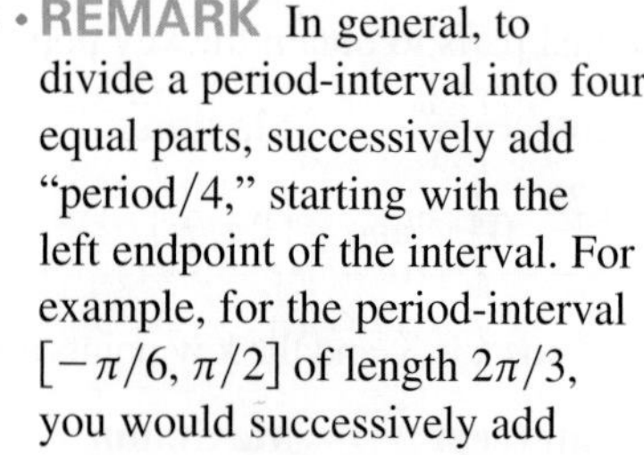

$$\frac{2\pi/3}{4} = \frac{\pi}{6}$$

to obtain $-\pi/6$, 0, $\pi/6$, $\pi/3$, and $\pi/2$ as the x-values for the key points on the graph.

Checkpoint Audio-video solution in English & Spanish at LarsonPrecalculus.com

Sketch the graph of

$$y = \cos \frac{x}{3}.$$

Translations of Sine and Cosine Curves

The constant c in the equations

$$y = a\sin(bx - c) \quad \text{and} \quad y = a\cos(bx - c)$$

results in *horizontal translations* (shifts) of the basic curves. For example, compare the graphs of $y = a\sin bx$ and $y = a\sin(bx - c)$. The graph of $y = a\sin(bx - c)$ completes one cycle from $bx - c = 0$ to $bx - c = 2\pi$. Solve for x to find that the interval for one cycle is

$$\underbrace{\frac{c}{b}}_{\text{Left endpoint}} \le x \le \underbrace{\frac{c}{b} + \overbrace{\frac{2\pi}{b}}^{\text{Period}}}_{\text{Right endpoint}}.$$

This implies that the period of $y = a\sin(bx - c)$ is $2\pi/b$, and the graph of $y = a\sin bx$ is shifted by an amount c/b. The number c/b is the **phase shift.**

Graphs of Sine and Cosine Functions

The graphs of $y = a\sin(bx - c)$ and $y = a\cos(bx - c)$ have the characteristics below. (Assume $b > 0$.)

Amplitude $= |a|$ Period $= \dfrac{2\pi}{b}$

The left and right endpoints of a one-cycle interval can be determined by solving the equations $bx - c = 0$ and $bx - c = 2\pi$.

EXAMPLE 4 Horizontal Translation

Analyze the graph of $y = \dfrac{1}{2}\sin\left(x - \dfrac{\pi}{3}\right)$.

Algebraic Solution

The amplitude is $\frac{1}{2}$ and the period is $2\pi/1 = 2\pi$. Solving the equations

$$x - \frac{\pi}{3} = 0 \quad \Rightarrow \quad x = \frac{\pi}{3}$$

and

$$x - \frac{\pi}{3} = 2\pi \quad \Rightarrow \quad x = \frac{7\pi}{3}$$

shows that the interval $[\pi/3, 7\pi/3]$ corresponds to one cycle of the graph. Dividing this interval into four equal parts produces the key points

Intercept	Maximum	Intercept	Minimum		Intercept
$\left(\frac{\pi}{3}, 0\right)$,	$\left(\frac{5\pi}{6}, \frac{1}{2}\right)$,	$\left(\frac{4\pi}{3}, 0\right)$,	$\left(\frac{11\pi}{6}, -\frac{1}{2}\right)$,	and	$\left(\frac{7\pi}{3}, 0\right)$.

Graphical Solution

Use a graphing utility set in *radian* mode to graph $y = (1/2)\sin[x - (\pi/3)]$, as shown in the figure below. Use the *minimum, maximum,* and *zero* or *root* features of the graphing utility to approximate the key points $(1.05, 0)$, $(2.62, 0.5)$, $(4.19, 0)$, $(5.76, -0.5)$, and $(7.33, 0)$.

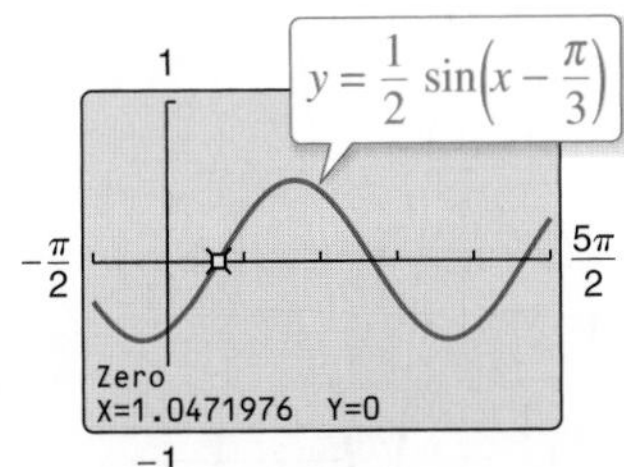

Checkpoint Audio-video solution in English & Spanish at LarsonPrecalculus.com

Analyze the graph of $y = 2\cos\left(x - \dfrac{\pi}{2}\right)$.

EXAMPLE 5 Horizontal Translation

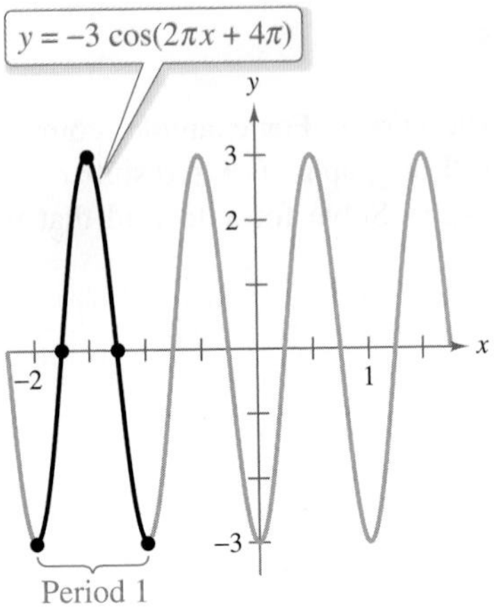

Sketch the graph of

$$y = -3\cos(2\pi x + 4\pi).$$

Solution The amplitude is 3 and the period is $2\pi/2\pi = 1$. Solving the equations

$$2\pi x + 4\pi = 0$$
$$2\pi x = -4\pi$$
$$x = -2$$

and

$$2\pi x + 4\pi = 2\pi$$
$$2\pi x = -2\pi$$
$$x = -1$$

shows that the interval $[-2, -1]$ corresponds to one cycle of the graph. Dividing this interval into four equal parts produces the key points

Minimum	Intercept	Maximum	Intercept		Minimum
$(-2, -3)$,	$\left(-\frac{7}{4}, 0\right)$,	$\left(-\frac{3}{2}, 3\right)$,	$\left(-\frac{5}{4}, 0\right)$,	and	$(-1, -3)$.

Figure 1.36 shows the graph.

Figure 1.36

Sketch the graph of

$$y = -\frac{1}{2}\sin(\pi x + \pi).$$

The constant d in the equations

$$y = d + a\sin(bx - c) \quad \text{and} \quad y = d + a\cos(bx - c)$$

results in *vertical translations* of the basic curves. The shift is d units up for $d > 0$ and d units down for $d < 0$. In other words, the graph oscillates about the horizontal line $y = d$ instead of about the x-axis.

EXAMPLE 6 Vertical Translation

Sketch the graph of

$$y = 2 + 3\cos 2x.$$

Solution The amplitude is 3 and the period is $2\pi/2 = \pi$. The key points over the interval $[0, \pi]$ are

$$(0, 5), \quad \left(\frac{\pi}{4}, 2\right), \quad \left(\frac{\pi}{2}, -1\right), \quad \left(\frac{3\pi}{2}, 2\right), \quad \text{and} \quad (\pi, 5).$$

Figure 1.37 shows the graph. Compared with the graph of $f(x) = 3\cos 2x$, the graph of $y = 2 + 3\cos 2x$ is shifted up two units.

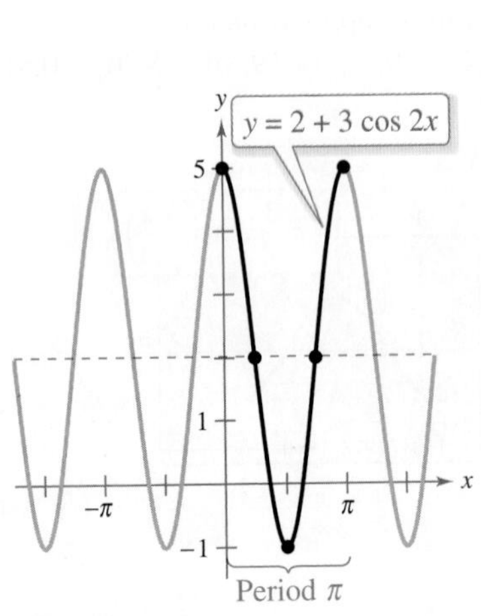

Figure 1.37

Sketch the graph of

$$y = 2\cos x - 5.$$

Mathematical Modeling

EXAMPLE 7 Finding a Trigonometric Model

DATA

Time, t	Depth, y
0	3.4
2	8.7
4	11.3
6	9.1
8	3.8
10	0.1
12	1.2

Spreadsheet at LarsonPrecalculus.com

The table shows the depths (in feet) of the water at the end of a dock every two hours from midnight to noon, where $t = 0$ corresponds to midnight. (a) Use a trigonometric function to model the data. (b) Find the depths at 9 A.M. and 3 P.M. (c) A boat needs at least 10 feet of water to moor at the dock. During what times in the afternoon can it safely dock?

Solution

a. Begin by graphing the data, as shown in Figure 1.38. Use either a sine or cosine model. For example, a cosine model has the form $y = a\cos(bt - c) + d$. The difference between the maximum value and the minimum value is twice the amplitude of the function. So, the amplitude is

$$a = \tfrac{1}{2}[(\text{maximum depth}) - (\text{minimum depth})] = \tfrac{1}{2}(11.3 - 0.1) = 5.6.$$

The cosine function completes one half of a cycle between the times at which the maximum and minimum depths occur. So, the period p is

$$p = 2[(\text{time of min. depth}) - (\text{time of max. depth})] = 2(10 - 4) = 12$$

which implies that $b = 2\pi/p \approx 0.524$. The maximum depth occurs 4 hours after midnight, so consider the left endpoint to be $c/b = 4$, which means that $c \approx 4(0.524) = 2.096$. Moreover, the average depth is $\frac{1}{2}(11.3 + 0.1) = 5.7$, so it follows that $d = 5.7$. Substituting the values of a, b, c, and d into the cosine model yields $y = 5.6\cos(0.524t - 2.096) + 5.7$.

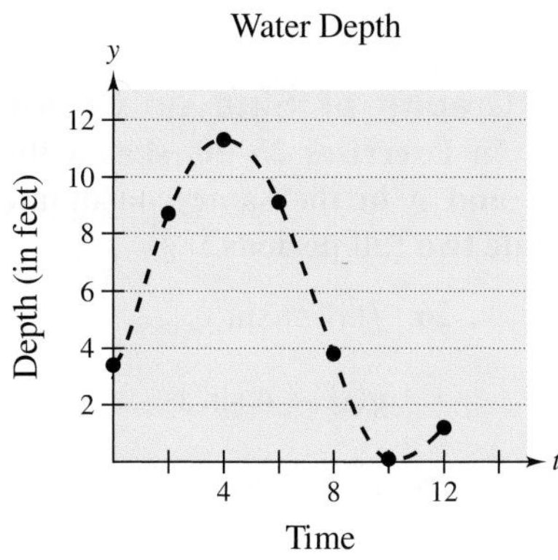

Figure 1.38

b. The depths at 9 A.M. and 3 P.M. are

$$y = 5.6\cos(0.524 \cdot 9 - 2.096) + 5.7 \approx 0.84 \text{ foot} \qquad \text{9 A.M.}$$

and

$$y = 5.6\cos(0.524 \cdot 15 - 2.096) + 5.7 \approx 10.56 \text{ feet.} \qquad \text{3 P.M.}$$

c. Using a graphing utility, graph the model with the line $y = 10$. Using the *intersect* feature, determine that the depth is at least 10 feet between 2:42 P.M. ($t \approx 14.7$) and 5:18 P.M. ($t \approx 17.3$), as shown in Figure 1.39.

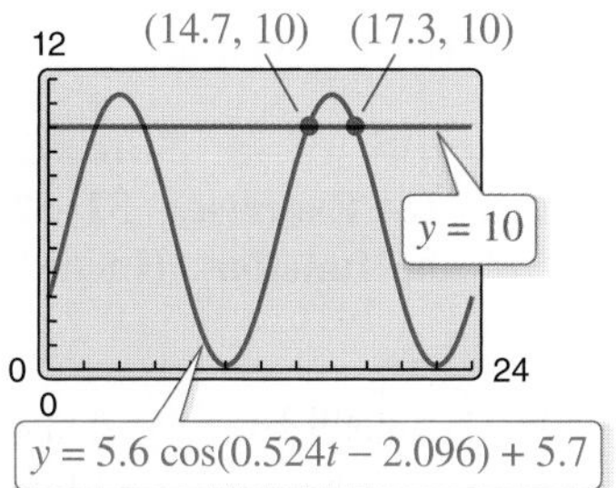

Figure 1.39

Find a sine model for the data in Example 7.

Summarize (Section 1.5)

1. Explain how to sketch the graphs of basic sine and cosine functions *(page 159)*. For an example of sketching the graph of a sine function, see Example 1.
2. Explain how to use amplitude and period to help sketch the graphs of sine and cosine functions *(pages 161 and 162)*. For examples of using amplitude and period to sketch graphs of sine and cosine functions, see Examples 2 and 3.
3. Explain how to sketch translations of the graphs of sine and cosine functions *(page 163)*. For examples of translating the graphs of sine and cosine functions, see Examples 4–6.
4. Give an example of using a sine or cosine function to model real-life data *(page 165, Example 7)*.

1.5 Exercises

See CalcChat.com for tutorial help and worked-out solutions to odd-numbered exercises.

Vocabulary: Fill in the blanks.

1. One period of a sine or cosine function is one ________ of the sine or cosine curve.
2. The ________ of a sine or cosine curve represents half the distance between the maximum and minimum values of the function.
3. For the function $y = a \sin(bx - c)$, $\frac{c}{b}$ represents the ________ ________ of one cycle of the graph of the function.
4. For the function $y = d + a \cos(bx - c)$, d represents a ________ ________ of the basic curve.

Skills and Applications

Finding the Period and Amplitude In Exercises 5–12, find the period and amplitude.

5. $y = 2 \sin 5x$
6. $y = 3 \cos 2x$
7. $y = \frac{3}{4} \cos \frac{\pi x}{2}$
8. $y = -5 \sin \frac{\pi x}{3}$
9. $y = -\frac{1}{2} \sin \frac{5x}{4}$
10. $y = \frac{1}{4} \sin \frac{x}{6}$
11. $y = -\frac{5}{3} \cos \frac{\pi x}{12}$
12. $y = -\frac{2}{5} \cos 10\pi x$

Describing the Relationship Between Graphs In Exercises 13–24, describe the relationship between the graphs of f and g. Consider amplitude, period, and shifts.

13. $f(x) = \cos x$
 $g(x) = \cos 5x$
14. $f(x) = \sin x$
 $g(x) = 2 \sin x$
15. $f(x) = \cos 2x$
 $g(x) = -\cos 2x$
16. $f(x) = \sin 3x$
 $g(x) = \sin(-3x)$
17. $f(x) = \sin x$
 $g(x) = \sin(x - \pi)$
18. $f(x) = \cos x$
 $g(x) = \cos(x + \pi)$
19. $f(x) = \sin 2x$
 $g(x) = 3 + \sin 2x$
20. $f(x) = \cos 4x$
 $g(x) = -2 + \cos 4x$

21.
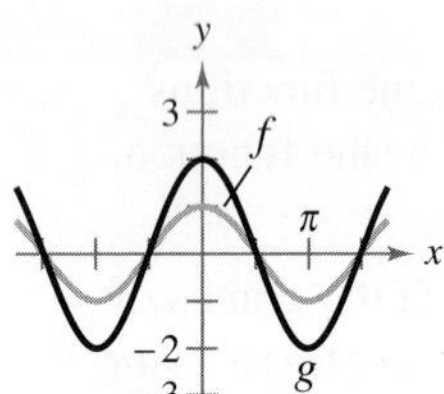

22.
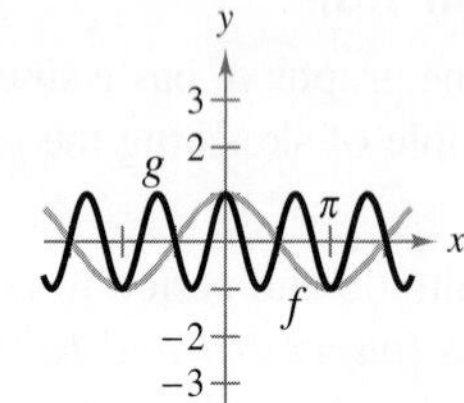

23.
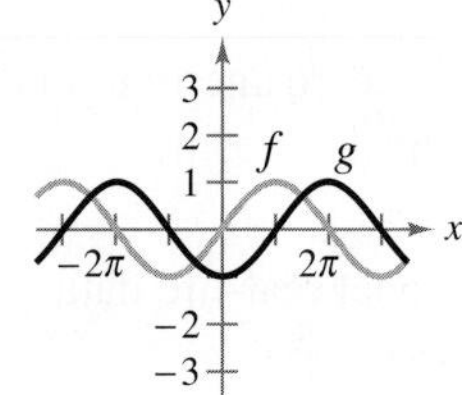

24.
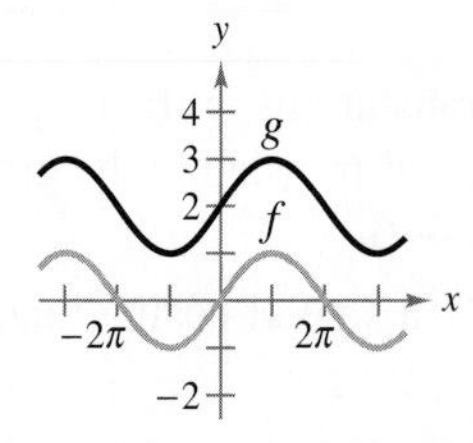

Sketching Graphs of Sine or Cosine Functions In Exercises 25–30, sketch the graphs of f and g in the same coordinate plane. (Include two full periods.)

25. $f(x) = \sin x$
 $g(x) = \sin \frac{x}{3}$
26. $f(x) = \sin x$
 $g(x) = 4 \sin x$
27. $f(x) = \cos x$
 $g(x) = 2 + \cos x$
28. $f(x) = \cos x$
 $g(x) = \cos\left(x + \frac{\pi}{2}\right)$
29. $f(x) = -\cos x$
 $g(x) = -\cos(x - \pi)$
30. $f(x) = -\sin x$
 $g(x) = -3 \sin x$

Sketching the Graph of a Sine or Cosine Function In Exercises 31–52, sketch the graph of the function. (Include two full periods.)

31. $y = 5 \sin x$
32. $y = \frac{1}{4} \sin x$
33. $y = \frac{1}{3} \cos x$
34. $y = 4 \cos x$
35. $y = \cos \frac{x}{2}$
36. $y = \sin 4x$
37. $y = \cos 2\pi x$
38. $y = \sin \frac{\pi x}{4}$
39. $y = -\sin \frac{2\pi x}{3}$
40. $y = 10 \cos \frac{\pi x}{6}$
41. $y = \cos\left(x - \frac{\pi}{2}\right)$
42. $y = \sin(x - 2\pi)$
43. $y = 3 \sin(x + \pi)$
44. $y = -4 \cos\left(x + \frac{\pi}{4}\right)$
45. $y = 2 - \sin \frac{2\pi x}{3}$
46. $y = -3 + 5 \cos \frac{\pi t}{12}$
47. $y = 2 + 5 \cos 6\pi x$
48. $y = 2 \sin 3x + 5$
49. $y = 3 \sin(x + \pi) - 3$
50. $y = -3 \sin(6x + \pi)$
51. $y = \frac{2}{3} \cos\left(\frac{x}{2} - \frac{\pi}{4}\right)$
52. $y = 4 \cos\left(\pi x + \frac{\pi}{2}\right) - 1$

Describing a Transformation **In Exercises 53–58, g is related to a parent function $f(x) = \sin(x)$ or $f(x) = \cos(x)$. (a) Describe the sequence of transformations from f to g. (b) Sketch the graph of g. (c) Use function notation to write g in terms of f.**

53. $g(x) = \sin(4x - \pi)$

54. $g(x) = \sin(2x + \pi)$

55. $g(x) = \cos\left(x - \dfrac{\pi}{2}\right) + 2$

56. $g(x) = 1 + \cos(x + \pi)$

57. $g(x) = 2\sin(4x - \pi) - 3$

58. $g(x) = 4 - \sin\left(2x + \dfrac{\pi}{2}\right)$

Graphing a Sine or Cosine Function **In Exercises 59–64, use a graphing utility to graph the function. (Include two full periods.) Be sure to choose an appropriate viewing window.**

59. $y = -2\sin(4x + \pi)$

60. $y = -4\sin\left(\dfrac{2}{3}x - \dfrac{\pi}{3}\right)$

61. $y = \cos\left(2\pi x - \dfrac{\pi}{2}\right) + 1$

62. $y = 3\cos\left(\dfrac{\pi x}{2} + \dfrac{\pi}{2}\right) - 2$

63. $y = -0.1\sin\left(\dfrac{\pi x}{10} + \pi\right)$

64. $y = \dfrac{1}{100}\cos 120\pi t$

Graphical Reasoning **In Exercises 65–68, find a and d for the function $f(x) = a\cos x + d$ such that the graph of f matches the figure.**

65.

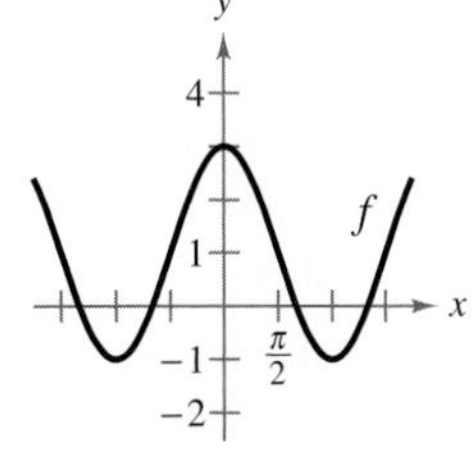

66.

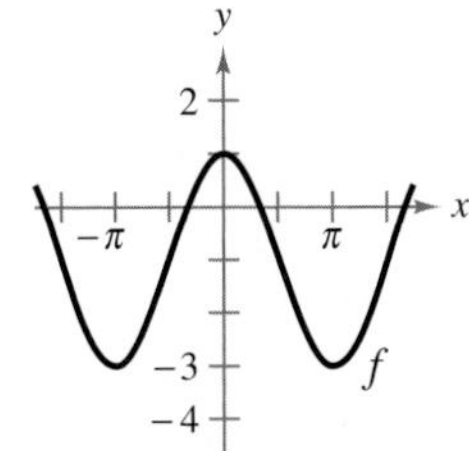

67.

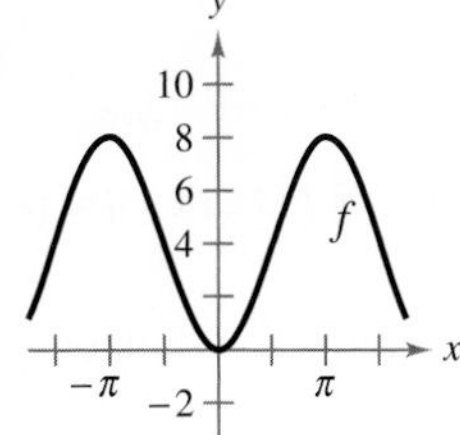

68.

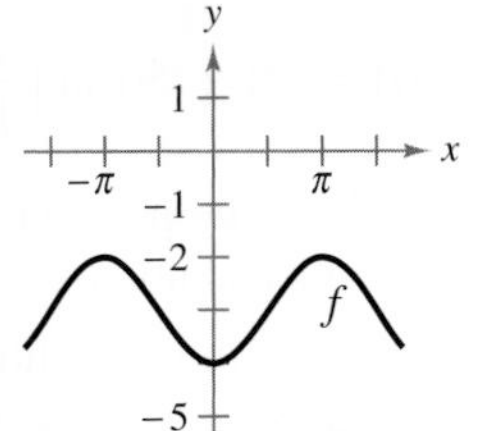

Graphical Reasoning **In Exercises 69–72, find a, b, and c for the function $f(x) = a\sin(bx - c)$ such that the graph of f matches the figure.**

69.

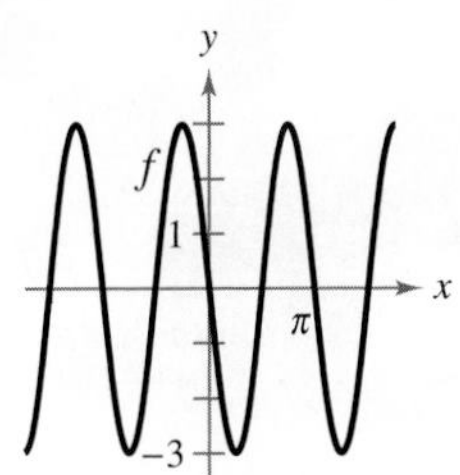

70.

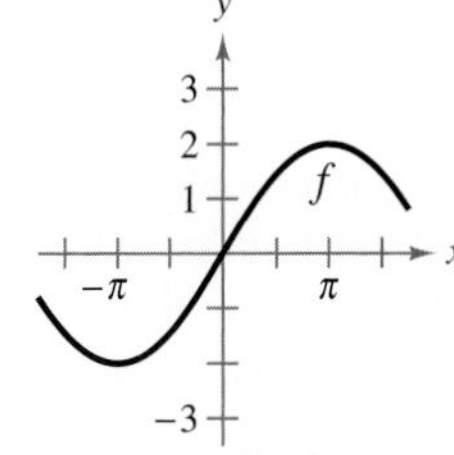

71.

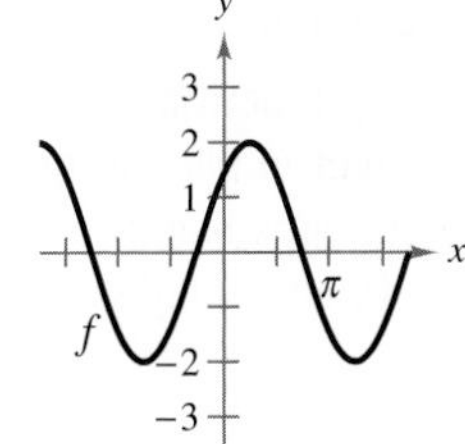

72.

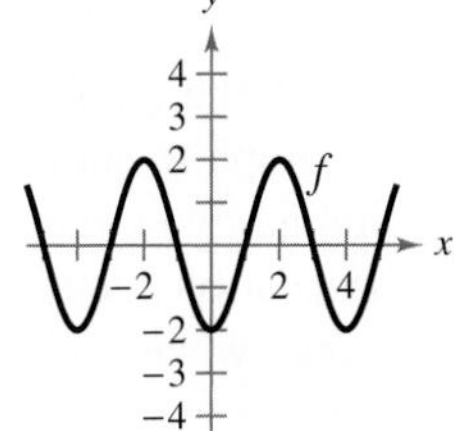

Using Technology **In Exercises 73 and 74, use a graphing utility to graph y_1 and y_2 in the interval $[-2\pi, 2\pi]$. Use the graphs to find real numbers x such that $y_1 = y_2$.**

73. $y_1 = \sin x$, $y_2 = -\frac{1}{2}$

74. $y_1 = \cos x$, $y_2 = -1$

Writing an Equation **In Exercises 75–78, write an equation for a function with the given characteristics.**

75. A sine curve with a period of π, an amplitude of 2, a right phase shift of $\pi/2$, and a vertical translation up 1 unit

76. A sine curve with a period of 4π, an amplitude of 3, a left phase shift of $\pi/4$, and a vertical translation down 1 unit

77. A cosine curve with a period of π, an amplitude of 1, a left phase shift of π, and a vertical translation down $\frac{3}{2}$ units

78. A cosine curve with a period of 4π, an amplitude of 3, a right phase shift of $\pi/2$, and a vertical translation up 2 units

79. **Respiratory Cycle** For a person exercising, the velocity v (in liters per second) of airflow during a respiratory cycle (the time from the beginning of one breath to the beginning of the next) is modeled by

$$v = 1.75\sin(\pi t/2)$$

where t is the time (in seconds). (Inhalation occurs when $v > 0$, and exhalation occurs when $v < 0$.)

(a) Find the time for one full respiratory cycle.

(b) Find the number of cycles per minute.

(c) Sketch the graph of the velocity function.

80. Respiratory Cycle For a person at rest, the velocity v (in liters per second) of airflow during a respiratory cycle (the time from the beginning of one breath to the beginning of the next) is modeled by $v = 0.85 \sin(\pi t/3)$, where t is the time (in seconds).

(a) Find the time for one full respiratory cycle.

(b) Find the number of cycles per minute.

(c) Sketch the graph of the velocity function. Use the graph to confirm your answer in part (a) by finding two times when new breaths begin. (Inhalation occurs when $v > 0$, and exhalation occurs when $v < 0$.)

81. Biology The function $P = 100 - 20 \cos(5\pi t/3)$ approximates the blood pressure P (in millimeters of mercury) at time t (in seconds) for a person at rest.

(a) Find the period of the function.

(b) Find the number of heartbeats per minute.

82. Piano Tuning When tuning a piano, a technician strikes a tuning fork for the A above middle C and sets up a wave motion that can be approximated by $y = 0.001 \sin 880\pi t$, where t is the time (in seconds).

(a) What is the period of the function?

(b) The frequency f is given by $f = 1/p$. What is the frequency of the note?

83. Astronomy The table shows the percent y (in decimal form) of the moon's face illuminated on day x in the year 2018, where $x = 1$ corresponds to January 1. *(Source: U.S. Naval Observatory)*

DATA

x	y
1	1.0
8	0.5
16	0.0
24	0.5
31	1.0
38	0.5

Spreadsheet at LarsonPrecalculus.com

(a) Create a scatter plot of the data.

(b) Find a trigonometric model for the data.

(c) Add the graph of your model in part (b) to the scatter plot. How well does the model fit the data?

(d) What is the period of the model?

(e) Estimate the percent of the moon's face illuminated on March 12, 2018.

84. Meteorology The table shows the maximum daily high temperatures (in degrees Fahrenheit) in Las Vegas L and International Falls I for month t, where $t = 1$ corresponds to January. *(Source: National Climatic Data Center)*

DATA

Month, t	Las Vegas, L	International Falls, I
1	57.1	13.8
2	63.0	22.4
3	69.5	34.9
4	78.1	51.5
5	87.8	66.6
6	98.9	74.2
7	104.1	78.6
8	101.8	76.3
9	93.8	64.7
10	80.8	51.7
11	66.0	32.5
12	57.3	18.1

Spreadsheet at LarsonPrecalculus.com

(a) A model for the temperatures in Las Vegas is

$$L(t) = 80.60 + 23.50 \cos\left(\frac{\pi t}{6} - 3.67\right).$$

Find a trigonometric model for the temperatures in International Falls.

(b) Use a graphing utility to graph the data points and the model for the temperatures in Las Vegas. How well does the model fit the data?

(c) Use the graphing utility to graph the data points and the model for the temperatures in International Falls. How well does the model fit the data?

(d) Use the models to estimate the average maximum temperature in each city. Which value in each model did you use? Explain.

(e) What is the period of each model? Are the periods what you expected? Explain.

(f) Which city has the greater variability in temperature throughout the year? Which value in each model determines this variability? Explain.

85. Ferris Wheel The height h (in feet) above ground of a seat on a Ferris wheel at time t (in seconds) is modeled by

$$h(t) = 53 + 50 \sin\left(\frac{\pi}{10}t - \frac{\pi}{2}\right).$$

(a) Find the period of the model. What does the period tell you about the ride?

(b) Find the amplitude of the model. What does the amplitude tell you about the ride?

(c) Use a graphing utility to graph one cycle of the model.

Neo2620/Shutterstock.com

86. Fuel Consumption The daily consumption C (in gallons) of diesel fuel on a farm is modeled by

$$C = 30.3 + 21.6 \sin\left(\frac{2\pi t}{365} + 10.9\right)$$

where t is the time (in days), with $t = 1$ corresponding to January 1.

(a) What is the period of the model? Is it what you expected? Explain.

(b) What is the average daily fuel consumption? Which value in the model did you use? Explain.

(c) Use a graphing utility to graph the model. Use the graph to approximate the time of the year when consumption exceeds 40 gallons per day.

Exploration

True or False? **In Exercises 87–89, determine whether the statement is true or false. Justify your answer.**

87. The graph of $g(x) = \sin(x + 2\pi)$ is a translation of the graph of $f(x) = \sin x$ exactly one period to the right, and the two graphs look identical.

88. The function $y = \frac{1}{2} \cos 2x$ has an amplitude that is twice that of the function $y = \cos x$.

89. The graph of $y = -\cos x$ is a reflection of the graph of $y = \sin[x + (\pi/2)]$ in the x-axis.

90. HOW DO YOU SEE IT? The figure below shows the graph of $y = \sin(x - c)$ for

$$c = -\frac{\pi}{4},\quad 0,\quad \text{and}\quad \frac{\pi}{4}.$$

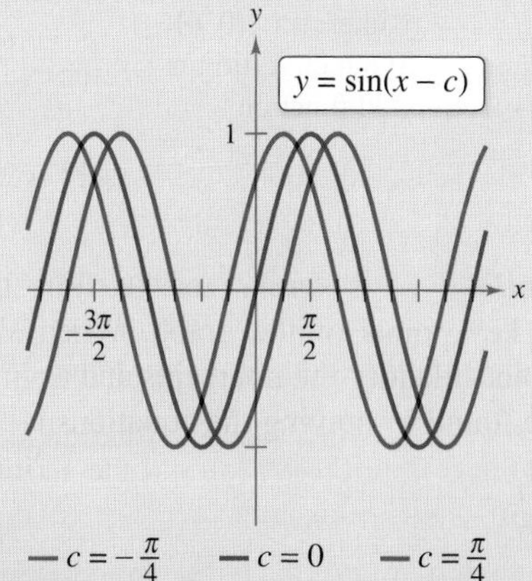

(a) How does the value of c affect the graph?

(b) Which graph is equivalent to that of

$$y = -\cos\left(x + \frac{\pi}{4}\right)?$$

Conjecture **In Exercises 91 and 92, graph f and g in the same coordinate plane. (Include two full periods.) Make a conjecture about the functions.**

91. $f(x) = \sin x, \quad g(x) = \cos\left(x - \frac{\pi}{2}\right)$

92. $f(x) = \sin x, \quad g(x) = -\cos\left(x + \frac{\pi}{2}\right)$

93. Writing Sketch the graph of $y = \cos bx$ for $b = \frac{1}{2}, 2,$ and 3. How does the value of b affect the graph? How many complete cycles of the graph occur between 0 and 2π for each value of b?

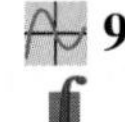
94. Polynomial Approximations Using calculus, it can be shown that the sine and cosine functions can be approximated by the polynomials

$$\sin x \approx x - \frac{x^3}{3!} + \frac{x^5}{5!}$$

and

$$\cos x \approx 1 - \frac{x^2}{2!} + \frac{x^4}{4!}$$

where x is in radians.

(a) Use a graphing utility to graph the sine function and its polynomial approximation in the same viewing window. How do the graphs compare?

(b) Use the graphing utility to graph the cosine function and its polynomial approximation in the same viewing window. How do the graphs compare?

(c) Study the patterns in the polynomial approximations of the sine and cosine functions and predict the next term in each. Then repeat parts (a) and (b). How does the accuracy of the approximations change when an additional term is added?

95. Polynomial Approximations Use the polynomial approximations of the sine and cosine functions in Exercise 94 to approximate each function value. Compare the results with those given by a calculator. Is the error in the approximation the same in each case? Explain.

(a) $\sin \frac{1}{2}$ (b) $\sin 1$

(c) $\sin \frac{\pi}{6}$ (d) $\cos(-0.5)$

(e) $\cos 1$ (f) $\cos \frac{\pi}{4}$

Project: Meteorology To work an extended application analyzing the mean monthly temperature and mean monthly precipitation for Honolulu, Hawaii, visit this text's website at *LarsonPrecalculus.com*. *(Source: National Climatic Data Center)*

1.6 Graphs of Other Trigonometric Functions

Graphs of trigonometric functions have many real-life applications, such as in modeling the distance from a television camera to a unit in a parade, as in Exercise 85 on page 179.

- **Sketch the graphs of tangent functions.**
- **Sketch the graphs of cotangent functions.**
- **Sketch the graphs of secant and cosecant functions.**
- **Sketch the graphs of damped trigonometric functions.**

Graph of the Tangent Function

Recall that the tangent function is odd. That is, $\tan(-x) = -\tan x$. Consequently, the graph of $y = \tan x$ is symmetric with respect to the origin. You also know from the identity $\tan x = (\sin x)/(\cos x)$ that the tangent function is undefined for values at which $\cos x = 0$. Two such values are $x = \pm\pi/2 \approx \pm1.5708$. As shown in the table below, $\tan x$ increases without bound as x approaches $\pi/2$ from the left and decreases without bound as x approaches $-\pi/2$ from the right.

x	$-\frac{\pi}{2}$	-1.57	-1.5	$-\frac{\pi}{4}$	0	$\frac{\pi}{4}$	1.5	1.57	$\frac{\pi}{2}$
$\tan x$	Undef.	-1255.8	-14.1	-1	0	1	14.1	1255.8	Undef.

So, the graph of $y = \tan x$ (shown below) has *vertical asymptotes* at $x = \pi/2$ and $x = -\pi/2$. Moreover, the period of the tangent function is π, so vertical asymptotes also occur at $x = (\pi/2) + n\pi$, where n is an integer. The domain of the tangent function is the set of all real numbers other than $x = (\pi/2) + n\pi$, and the range is the set of all real numbers.

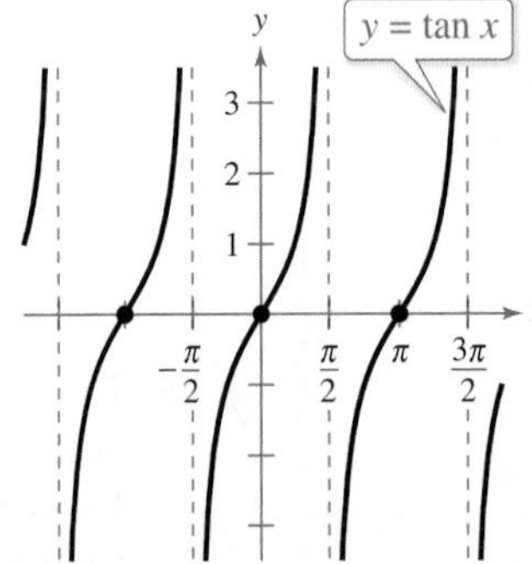

Period: π

Domain: all $x \neq \frac{\pi}{2} + n\pi$

Range: $(-\infty, \infty)$

Vertical asymptotes: $x = \frac{\pi}{2} + n\pi$

x-intercepts: $(n\pi, 0)$

y-intercept: $(0, 0)$

Symmetry: origin

Odd function

ALGEBRA HELP

- To review odd and even functions, see Section P.6.
- To review symmetry of a graph, see Section P.3.
- To review fundamental trigonometric identities, see Section 1.3.
- To review domain and range of a function, see Section P.5.
- To review intercepts of a graph, see Section P.3.

Sketching the graph of $y = a\tan(bx - c)$ is similar to sketching the graph of $y = a\sin(bx - c)$ in that you locate key points of the graph. When sketching the graph of $y = a\tan(bx - c)$, the key points identify the intercepts and asymptotes. Two consecutive vertical asymptotes can be found by solving the equations

$$bx - c = -\frac{\pi}{2} \quad \text{and} \quad bx - c = \frac{\pi}{2}.$$

On the x-axis, the point halfway between two consecutive vertical asymptotes is an x-intercept of the graph. The period of the function $y = a\tan(bx - c)$ is the distance between two consecutive vertical asymptotes. The amplitude of a tangent function is not defined. After plotting two consecutive asymptotes and the x-intercept between them, plot additional points between the asymptotes and sketch one cycle. Finally, sketch one or two additional cycles to the left and right.

EXAMPLE 1 Sketching the Graph of a Tangent Function

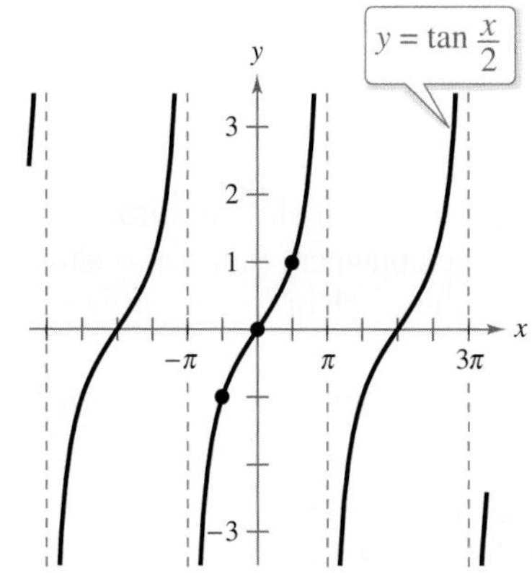

Sketch the graph of $y = \tan \frac{x}{2}$.

Solution

Solving the equations

$$\frac{x}{2} = -\frac{\pi}{2} \quad \text{and} \quad \frac{x}{2} = \frac{\pi}{2}$$

shows that two consecutive vertical asymptotes occur at $x = -\pi$ and $x = \pi$. Between these two asymptotes, find a few points, including the x-intercept, as shown in the table. Figure 1.40 shows three cycles of the graph.

x	$-\pi$	$-\frac{\pi}{2}$	0	$\frac{\pi}{2}$	π
$\tan \frac{x}{2}$	Undef.	-1	0	1	Undef.

Figure 1.40

Checkpoint  Audio-video solution in English & Spanish at LarsonPrecalculus.com

Sketch the graph of $y = \tan \frac{x}{4}$.

EXAMPLE 2 Sketching the Graph of a Tangent Function

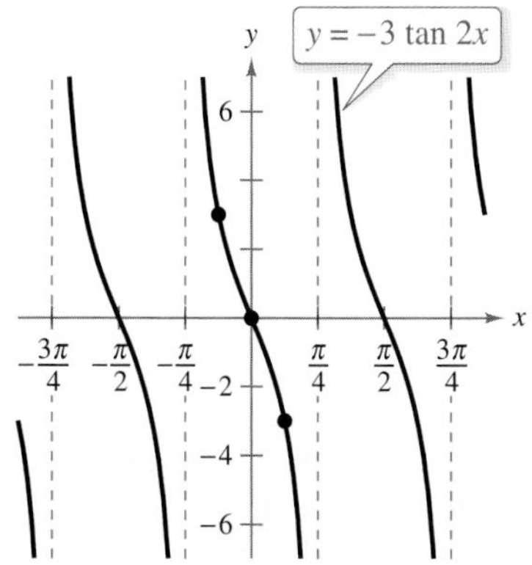

Sketch the graph of $y = -3 \tan 2x$.

Solution

Solving the equations

$$2x = -\frac{\pi}{2} \quad \text{and} \quad 2x = \frac{\pi}{2}$$

shows that two consecutive vertical asymptotes occur at $x = -\pi/4$ and $x = \pi/4$. Between these two asymptotes, find a few points, including the x-intercept, as shown in the table. Figure 1.41 shows three cycles of the graph.

x	$-\frac{\pi}{4}$	$-\frac{\pi}{8}$	0	$\frac{\pi}{8}$	$\frac{\pi}{4}$
$-3 \tan 2x$	Undef.	3	0	-3	Undef.

Figure 1.41

Checkpoint  Audio-video solution in English & Spanish at LarsonPrecalculus.com

Sketch the graph of $y = \tan 2x$.

Compare the graphs in Examples 1 and 2. The graph of $y = a \tan(bx - c)$ increases between consecutive vertical asymptotes when $a > 0$ and decreases between consecutive vertical asymptotes when $a < 0$. In other words, the graph for $a < 0$ is a reflection in the x-axis of the graph for $a > 0$. Also, the period is greater when $0 < b < 1$ than when $b > 1$. In other words, compared with the case where $b = 1$, the period represents a horizontal stretch when $0 < b < 1$ and a horizontal shrink when $b > 1$.

Graph of the Cotangent Function

The graph of the cotangent function is similar to the graph of the tangent function. It also has a period of π. However, the identity

$$y = \cot x = \frac{\cos x}{\sin x}$$

shows that the cotangent function has vertical asymptotes when $\sin x$ is zero, which occurs at $x = n\pi$, where n is an integer. The graph of the cotangent function is shown below. Note that two consecutive vertical asymptotes of the graph of $y = a\cot(bx - c)$ can be found by solving the equations

$$bx - c = 0 \quad \text{and} \quad bx - c = \pi.$$

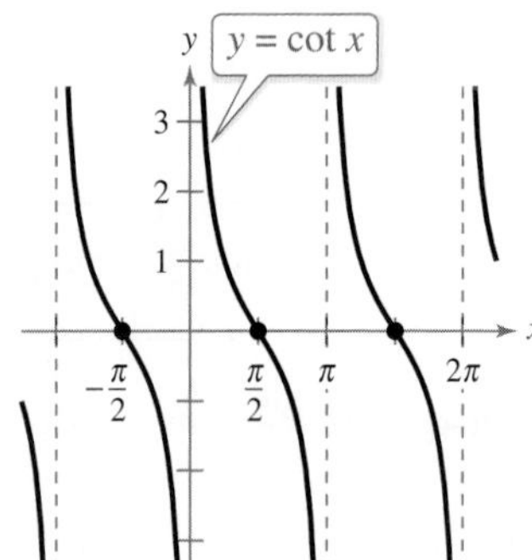

Period: π
Domain: all $x \neq n\pi$
Range: $(-\infty, \infty)$
Vertical asymptotes: $x = n\pi$
x-intercepts: $\left(\frac{\pi}{2} + n\pi, 0\right)$
Symmetry: origin
Odd function

EXAMPLE 3 Sketching the Graph of a Cotangent Function

Sketch the graph of

$$y = 2\cot\frac{x}{3}.$$

Solution

Solving the equations

$$\frac{x}{3} = 0 \quad \text{and} \quad \frac{x}{3} = \pi$$

shows that two consecutive vertical asymptotes occur at $x = 0$ and $x = 3\pi$. Between these two asymptotes, find a few points, including the x-intercept, as shown in the table. Figure 1.42 shows three cycles of the graph. Note that the period is 3π, the distance between consecutive asymptotes.

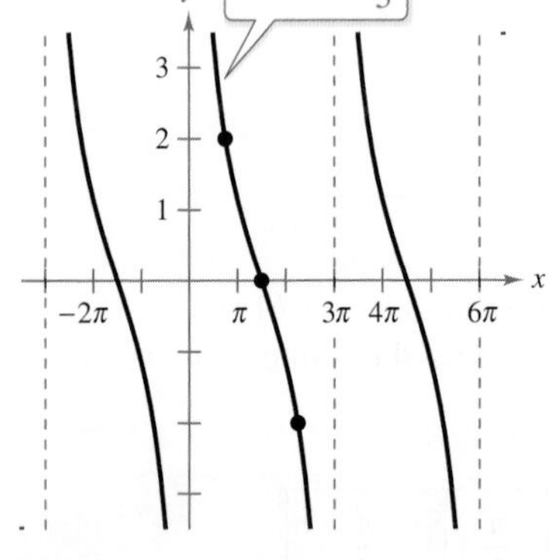

Figure 1.42

x	0	$\frac{3\pi}{4}$	$\frac{3\pi}{2}$	$\frac{9\pi}{4}$	3π
$2\cot\frac{x}{3}$	Undef.	2	0	-2	Undef.

Checkpoint Audio-video solution in English & Spanish at LarsonPrecalculus.com

Sketch the graph of

$$y = \cot\frac{x}{4}.$$

Graphs of the Reciprocal Functions

You can obtain the graphs of the cosecant and secant functions from the graphs of the sine and cosine functions, respectively, using the reciprocal identities

$$\csc x = \frac{1}{\sin x} \quad \text{and} \quad \sec x = \frac{1}{\cos x}.$$

For example, at a given value of x, the y-coordinate of $\sec x$ is the reciprocal of the y-coordinate of $\cos x$. Of course, when $\cos x = 0$, the reciprocal does not exist. Near such values of x, the behavior of the secant function is similar to that of the tangent function. In other words, the graphs of

$$\tan x = \frac{\sin x}{\cos x} \quad \text{and} \quad \sec x = \frac{1}{\cos x}$$

have vertical asymptotes where $\cos x = 0$, that is, at $x = (\pi/2) + n\pi$, where n is an integer. Similarly,

$$\cot x = \frac{\cos x}{\sin x} \quad \text{and} \quad \csc x = \frac{1}{\sin x}$$

have vertical asymptotes where $\sin x = 0$, that is, at $x = n\pi$, where n is an integer.

To sketch the graph of a secant or cosecant function, first make a sketch of its reciprocal function. For example, to sketch the graph of $y = \csc x$, first sketch the graph of $y = \sin x$. Then find reciprocals of the y-coordinates to obtain points on the graph of $y = \csc x$. You can use this procedure to obtain the graphs below.

TECHNOLOGY Some graphing utilities have difficulty graphing trigonometric functions that have vertical asymptotes. In *connected* mode, your graphing utility may connect parts of the graphs of tangent, cotangent, secant, and cosecant functions that are not supposed to be connected. In *dot* mode, the graphs are represented as collections of dots, so the graphs do not resemble solid curves.

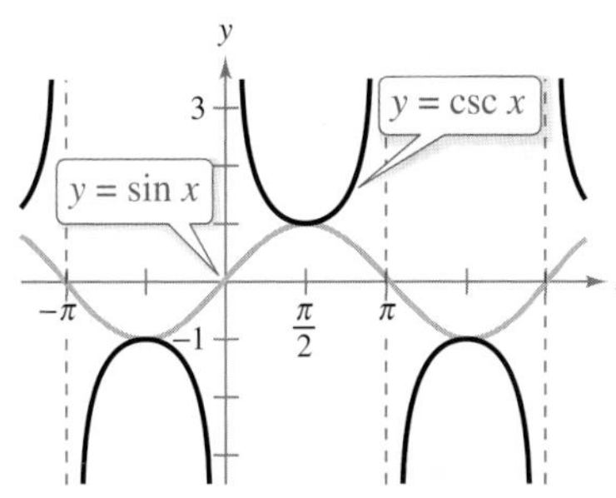

Period: 2π
Domain: all $x \neq n\pi$
Range: $(-\infty, -1] \cup [1, \infty)$
Vertical asymptotes: $x = n\pi$
No intercepts
Symmetry: origin
Odd function

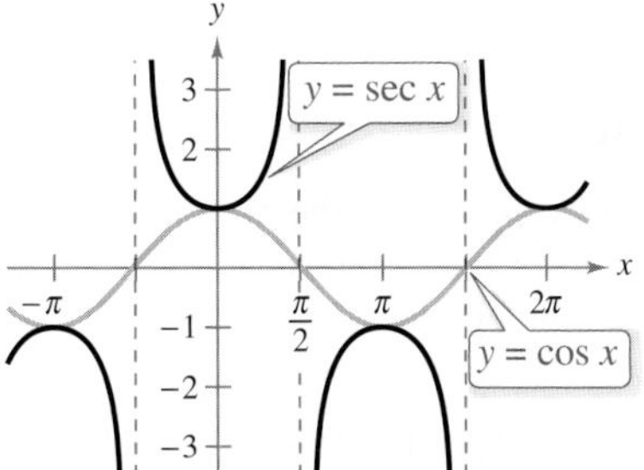

Period: 2π
Domain: all $x \neq \frac{\pi}{2} + n\pi$
Range: $(-\infty, -1] \cup [1, \infty)$
Vertical asymptotes: $x = \frac{\pi}{2} + n\pi$
y-intercept: $(0, 1)$
Symmetry: y-axis
Even function

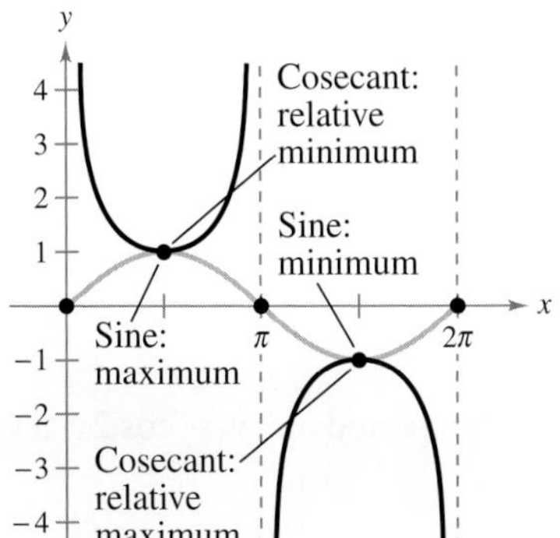

Figure 1.43

In comparing the graphs of the cosecant and secant functions with those of the sine and cosine functions, respectively, note that the "hills" and "valleys" are interchanged. For example, a hill (or maximum point) on the sine curve corresponds to a valley (a relative minimum) on the cosecant curve, and a valley (or minimum point) on the sine curve corresponds to a hill (a relative maximum) on the cosecant curve, as shown in Figure 1.43. Additionally, x-intercepts of the sine and cosine functions become vertical asymptotes of the cosecant and secant functions, respectively (see Figure 1.43).

EXAMPLE 4 Sketching the Graph of a Cosecant Function

Sketch the graph of $y = 2 \csc\left(x + \frac{\pi}{4}\right)$.

Solution

Begin by sketching the graph of

$$y = 2 \sin\left(x + \frac{\pi}{4}\right).$$

For this function, the amplitude is 2 and the period is 2π. Solving the equations

$$x + \frac{\pi}{4} = 0 \quad \text{and} \quad x + \frac{\pi}{4} = 2\pi$$

shows that one cycle of the sine function corresponds to the interval from $x = -\pi/4$ to $x = 7\pi/4$. The gray curve in Figure 1.44 represents the graph of the sine function. At the midpoint and endpoints of this interval, the sine function is zero. So, the corresponding cosecant function

$$\begin{aligned} y &= 2 \csc\left(x + \frac{\pi}{4}\right) \\ &= 2\left(\frac{1}{\sin[x + (\pi/4)]}\right) \end{aligned}$$

has vertical asymptotes at $x = -\pi/4$, $x = 3\pi/4$, $x = 7\pi/4$, and so on. The black curve in Figure 1.44 represents the graph of the cosecant function.

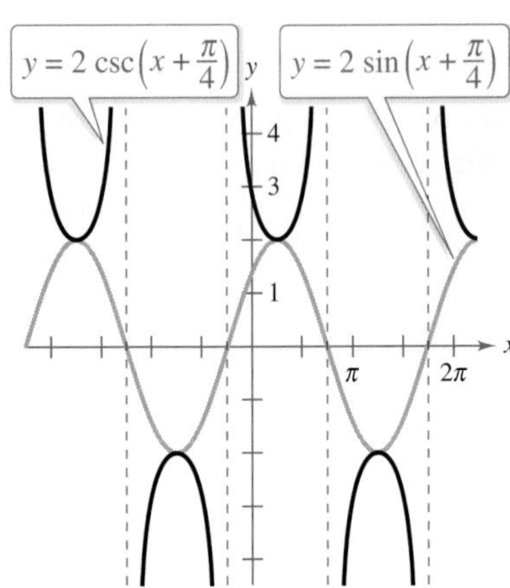

Figure 1.44

Sketch the graph of $y = 2 \csc\left(x + \frac{\pi}{2}\right)$.

EXAMPLE 5 Sketching the Graph of a Secant Function

See LarsonPrecalculus.com for an interactive version of this type of example.

Sketch the graph of $y = \sec 2x$.

Solution

Begin by sketching the graph of $y = \cos 2x$, shown as the gray curve in Figure 1.45. Then, form the graph of $y = \sec 2x$, shown as the black curve in the figure. Note that the x-intercepts of $y = \cos 2x$

$$\left(-\frac{\pi}{4}, 0\right), \left(\frac{\pi}{4}, 0\right), \left(\frac{3\pi}{4}, 0\right), \ldots$$

correspond to the vertical asymptotes

$$x = -\frac{\pi}{4}, \quad x = \frac{\pi}{4}, \quad x = \frac{3\pi}{4}, \ldots$$

of the graph of $y = \sec 2x$. Moreover, notice that the period of $y = \cos 2x$ and $y = \sec 2x$ is π.

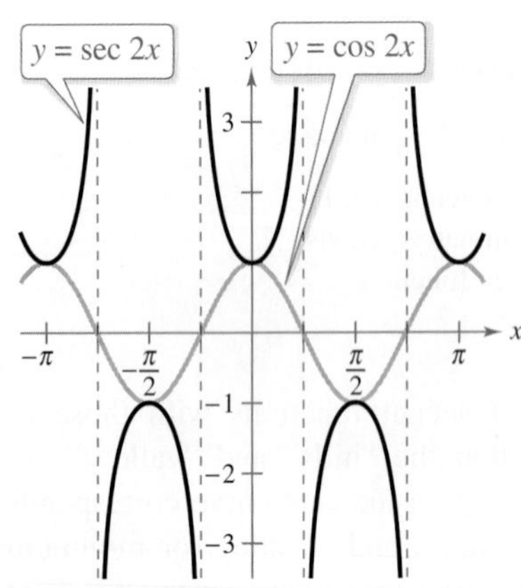

Figure 1.45

Sketch the graph of $y = \sec \frac{x}{2}$.

Damped Trigonometric Graphs

You can graph a *product* of two functions using properties of the individual functions. For example, consider the function

$$f(x) = x \sin x$$

as the product of the functions $y = x$ and $y = \sin x$. Using properties of absolute value and the fact that $|\sin x| \le 1$, you have

$$0 \le |x||\sin x| \le |x|.$$

Consequently,

$$-|x| \le x \sin x \le |x|$$

which means that the graph of $f(x) = x \sin x$ lies between the lines $y = -x$ and $y = x$. Furthermore,

$$f(x) = x \sin x = \pm x \quad \text{at} \quad x = \frac{\pi}{2} + n\pi$$

and

$$f(x) = x \sin x = 0 \quad \text{at} \quad x = n\pi$$

where n is an integer, so the graph of f touches the line $y = x$ or the line $y = -x$ at $x = (\pi/2) + n\pi$ and has x-intercepts at $x = n\pi$. A sketch of f is shown at the right. In the function $f(x) = x \sin x$, the factor x is called the **damping factor.**

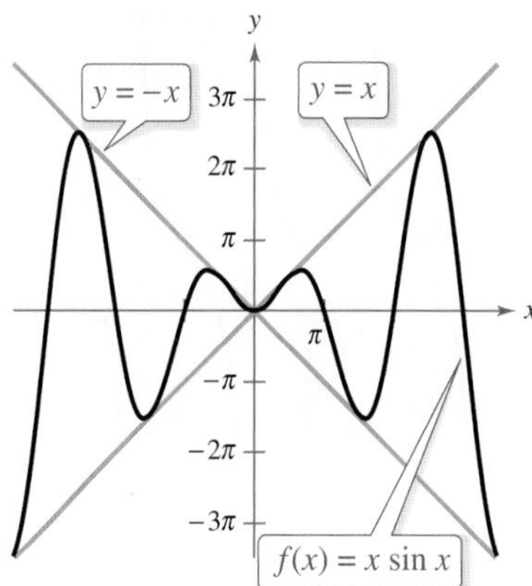

REMARK Do you see why the graph of $f(x) = x \sin x$ touches the lines $y = \pm x$ at $x = (\pi/2) + n\pi$ and why the graph has x-intercepts at $x = n\pi$? Recall that the sine function is equal to ± 1 at odd multiples of $\pi/2$ and is equal to 0 at multiples of π.

EXAMPLE 6 Damped Sine Curve

Sketch the graph of $f(x) = x^2 \sin 3x$.

Solution

Consider f as the product of the two functions $y = x^2$ and $y = \sin 3x$, each of which has the set of real numbers as its domain. For any real number x, you know that $x^2 \ge 0$ and $|\sin 3x| \le 1$. So,

$$x^2|\sin 3x| \le x^2$$

which means that

$$-x^2 \le x^2 \sin 3x \le x^2.$$

Furthermore,

$$f(x) = x^2 \sin 3x = \pm x^2 \quad \text{at} \quad x = \frac{\pi}{6} + \frac{n\pi}{3}$$

and

$$f(x) = x^2 \sin 3x = 0 \quad \text{at} \quad x = \frac{n\pi}{3}$$

so the graph of f touches the curve $y = -x^2$ or the curve $y = x^2$ at $x = (\pi/6) + (n\pi/3)$ and has intercepts at $x = n\pi/3$. Figure 1.46 shows a sketch of f.

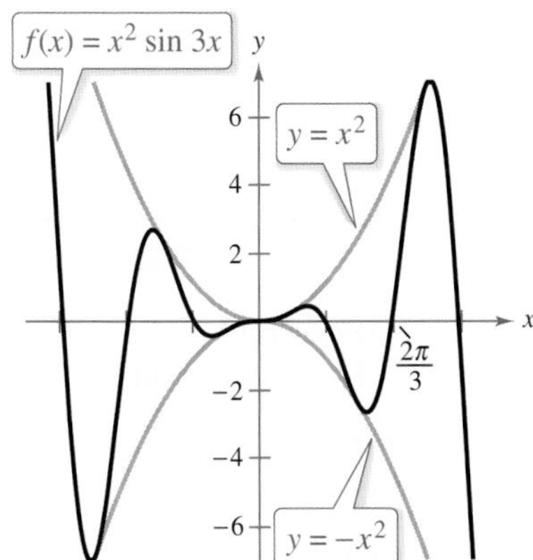

Figure 1.46

Sketch the graph of $f(x) = x^2 \sin 4x$.

Below is a summary of the characteristics of the six basic trigonometric functions.

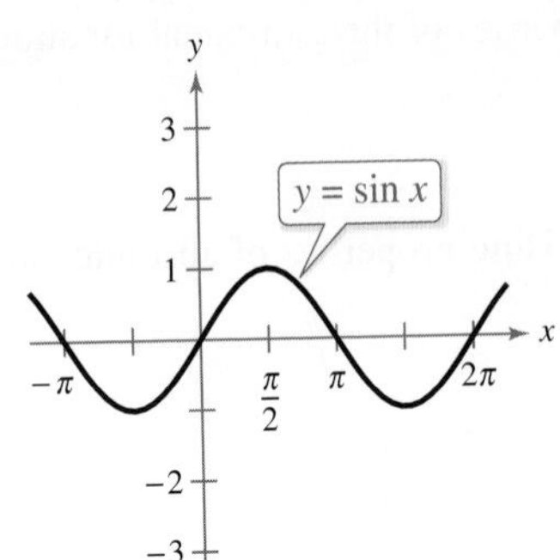

Domain: $(-\infty, \infty)$
Range: $[-1, 1]$
Period: 2π

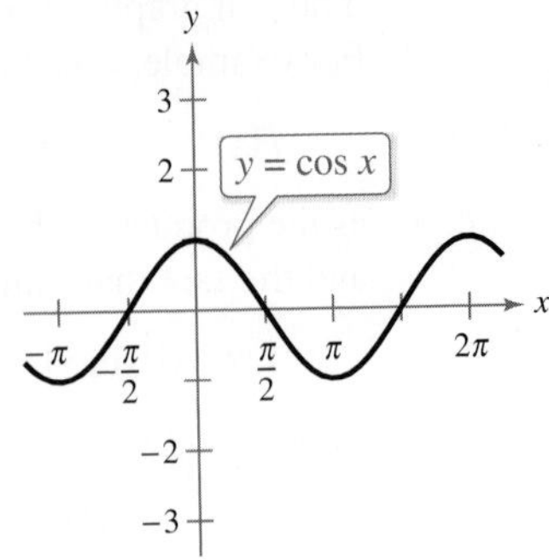

Domain: $(-\infty, \infty)$
Range: $[-1, 1]$
Period: 2π

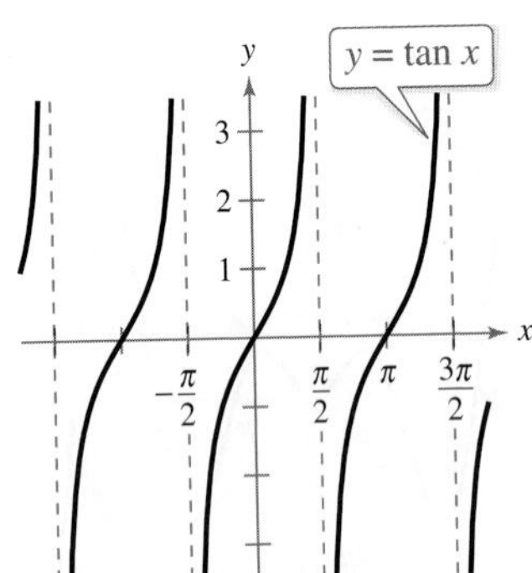

Domain: all $x \neq \frac{\pi}{2} + n\pi$
Range: $(-\infty, \infty)$
Period: π

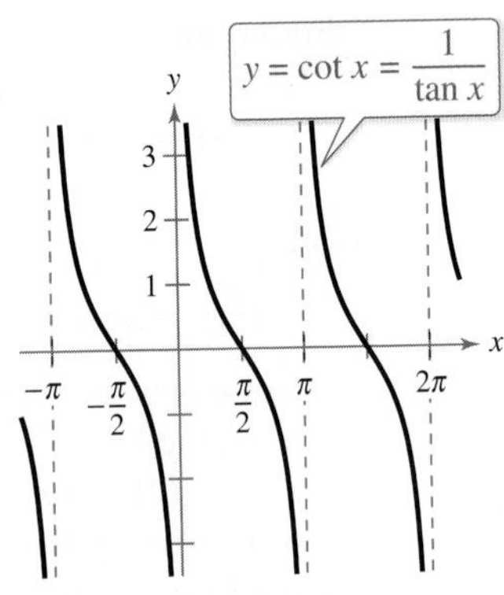

Domain: all $x \neq n\pi$
Range: $(-\infty, \infty)$
Period: π

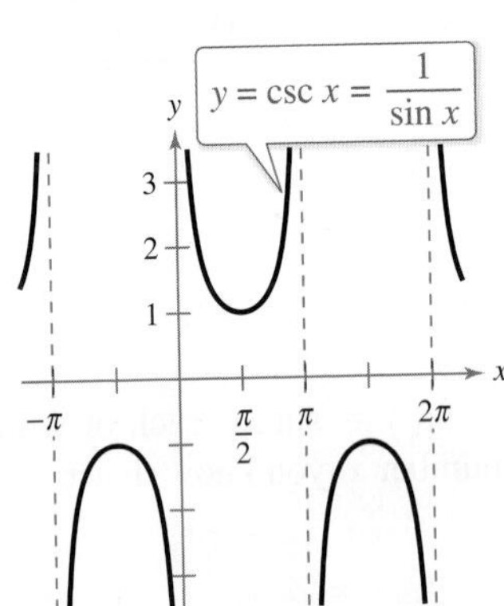

Domain: all $x \neq n\pi$
Range: $(-\infty, -1] \cup [1, \infty)$
Period: 2π

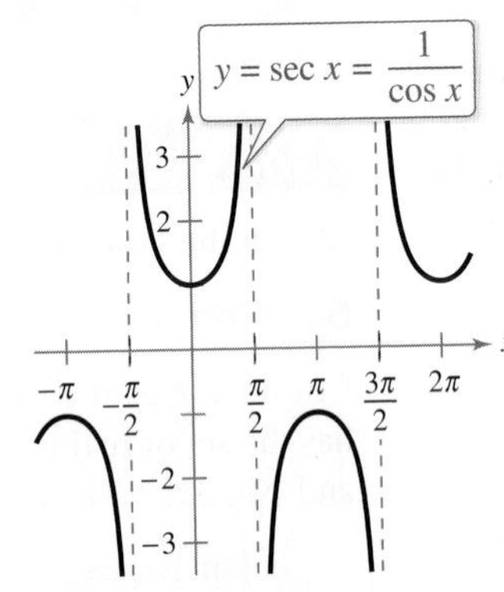

Domain: all $x \neq \frac{\pi}{2} + n\pi$
Range: $(-\infty, -1] \cup [1, \infty)$
Period: 2π

Summarize (Section 1.6)

1. Explain how to sketch the graph of $y = a \tan(bx - c)$ *(page 170)*. For examples of sketching graphs of tangent functions, see Examples 1 and 2.
2. Explain how to sketch the graph of $y = a \cot(bx - c)$ *(page 172)*. For an example of sketching the graph of a cotangent function, see Example 3.
3. Explain how to sketch the graphs of $y = a \csc(bx - c)$ and $y = a \sec(bx - c)$ *(page 173)*. For examples of sketching graphs of cosecant and secant functions, see Examples 4 and 5.
4. Explain how to sketch the graph of a damped trigonometric function *(page 175)*. For an example of sketching the graph of a damped trigonometric function, see Example 6.

1.6 Exercises

See CalcChat.com for tutorial help and worked-out solutions to odd-numbered exercises.

Vocabulary: Fill in the blanks.

1. The tangent, cotangent, and cosecant functions are ________, so the graphs of these functions have symmetry with respect to the ________.
2. The graphs of the tangent, cotangent, secant, and cosecant functions have ________ asymptotes.
3. To sketch the graph of a secant or cosecant function, first make a sketch of its ________ function.
4. For the function $f(x) = g(x) \cdot \sin x$, $g(x)$ is called the ________ factor.
5. The period of $y = \tan x$ is ________.
6. The domain of $y = \cot x$ is all real numbers such that ________.
7. The range of $y = \sec x$ is ________.
8. The period of $y = \csc x$ is ________.

Skills and Applications

Matching In Exercises 9–14, match the function with its graph. State the period of the function. [The graphs are labeled (a), (b), (c), (d), (e), and (f).]

(a)

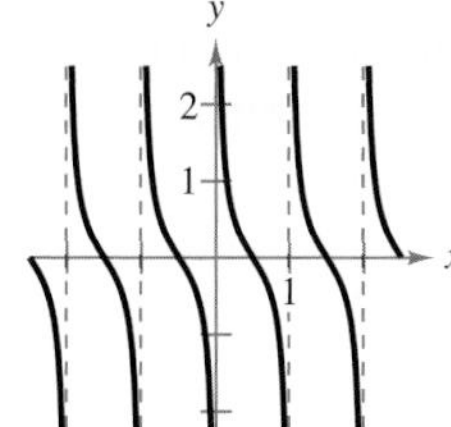

(b)

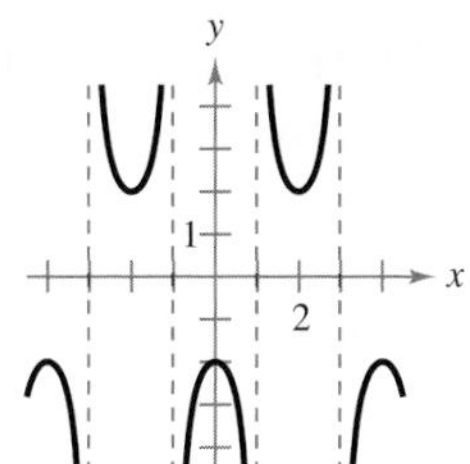

(c)

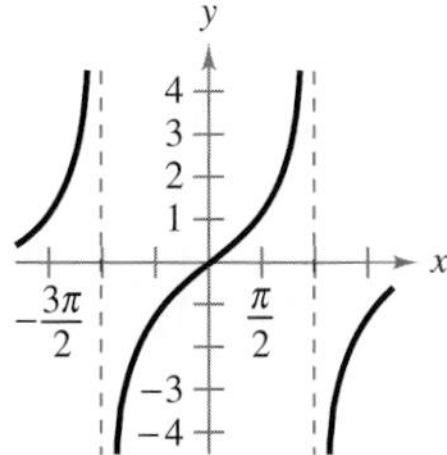

(d)

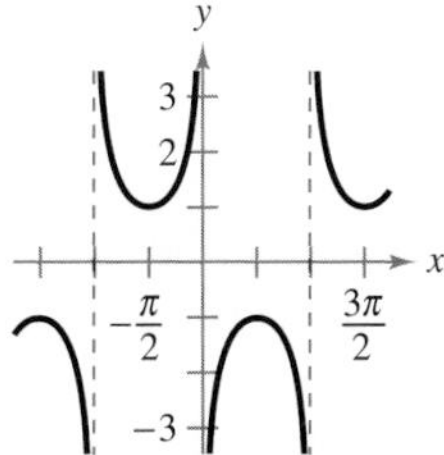

(e)

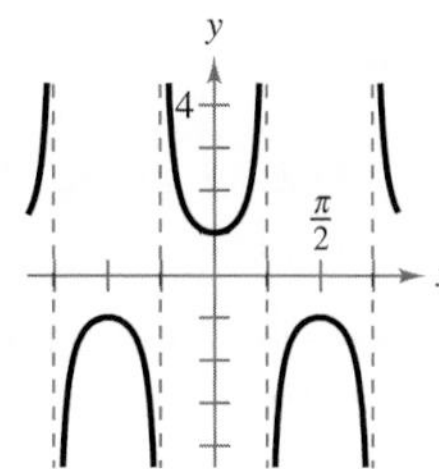

(f)

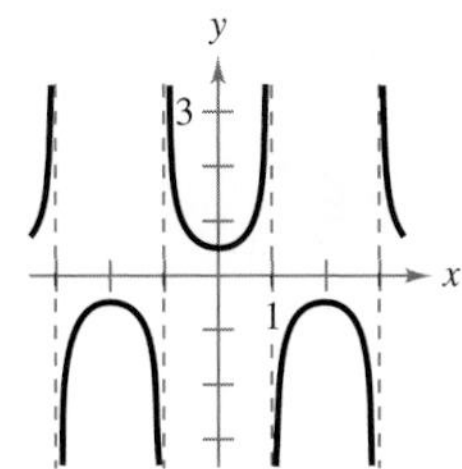

9. $y = \sec 2x$
10. $y = \tan \dfrac{x}{2}$
11. $y = \dfrac{1}{2} \cot \pi x$
12. $y = -\csc x$
13. $y = \dfrac{1}{2} \sec \dfrac{\pi x}{2}$
14. $y = -2 \sec \dfrac{\pi x}{2}$

Sketching the Graph of a Trigonometric Function In Exercises 15–38, sketch the graph of the function. (Include two full periods.)

15. $y = \frac{1}{3} \tan x$
16. $y = -\frac{1}{2} \tan x$
17. $y = -\frac{1}{2} \sec x$
18. $y = \frac{1}{4} \sec x$
19. $y = -2 \tan 3x$
20. $y = -3 \tan \pi x$
21. $y = \csc \pi x$
22. $y = 3 \csc 4x$
23. $y = \frac{1}{2} \sec \pi x$
24. $y = 2 \sec 3x$
25. $y = \csc \dfrac{x}{2}$
26. $y = \csc \dfrac{x}{3}$
27. $y = 3 \cot 2x$
28. $y = 3 \cot \dfrac{\pi x}{2}$
29. $y = \tan \dfrac{\pi x}{4}$
30. $y = \tan 4x$
31. $y = 2 \csc(x - \pi)$
32. $y = \csc(2x - \pi)$
33. $y = 2 \sec(x + \pi)$
34. $y = \tan(x + \pi)$
35. $y = -\sec \pi x + 1$
36. $y = -2 \sec 4x + 2$
37. $y = \dfrac{1}{4} \csc\left(x + \dfrac{\pi}{4}\right)$
38. $y = 2 \cot\left(x + \dfrac{\pi}{2}\right)$

Graphing a Trigonometric Function In Exercises 39–48, use a graphing utility to graph the function. (Include two full periods.)

39. $y = \tan \dfrac{x}{3}$
40. $y = -\tan 2x$
41. $y = -2 \sec 4x$
42. $y = \sec \pi x$
43. $y = \tan\left(x - \dfrac{\pi}{4}\right)$
44. $y = \dfrac{1}{4} \cot\left(x - \dfrac{\pi}{2}\right)$
45. $y = -\csc(4x - \pi)$
46. $y = 2 \sec(2x - \pi)$
47. $y = 0.1 \tan\left(\dfrac{\pi x}{4} + \dfrac{\pi}{4}\right)$
48. $y = \dfrac{1}{3} \sec\left(\dfrac{\pi x}{2} + \dfrac{\pi}{2}\right)$

Solving a Trigonometric Equation **In Exercises 49–56, find the solutions of the equation in the interval $[-2\pi, 2\pi]$. Use a graphing utility to verify your results.**

49. $\tan x = 1$
50. $\tan x = \sqrt{3}$
51. $\cot x = -\sqrt{3}$
52. $\cot x = 1$
53. $\sec x = -2$
54. $\sec x = 2$
55. $\csc x = \sqrt{2}$
56. $\csc x = -2$

Even and Odd Trigonometric Functions **In Exercises 57–64, use the graph of the function to determine whether the function is even, odd, or neither. Verify your answer algebraically.**

57. $f(x) = \sec x$
58. $f(x) = \tan x$
59. $g(x) = \cot x$
60. $g(x) = \csc x$
61. $f(x) = x + \tan x$
62. $f(x) = x^2 - \sec x$
63. $g(x) = x \csc x$
64. $g(x) = x^2 \cot x$

Identifying Damped Trigonometric Functions **In Exercises 65–68, match the function with its graph. Describe the behavior of the function as x approaches zero. [The graphs are labeled (a), (b), (c), and (d).]**

(a)

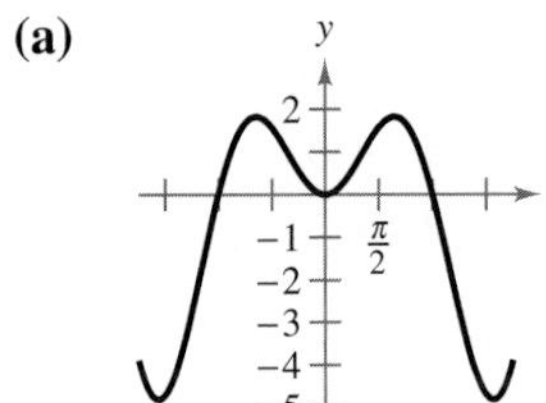

(b)

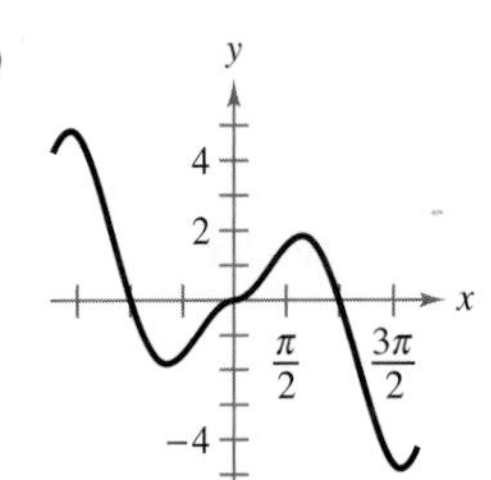

(c)

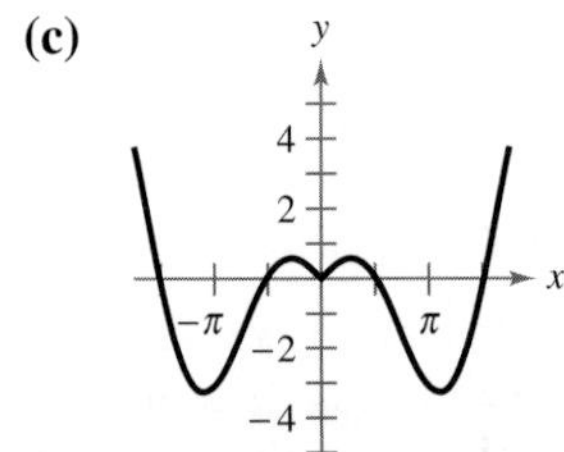

(d)

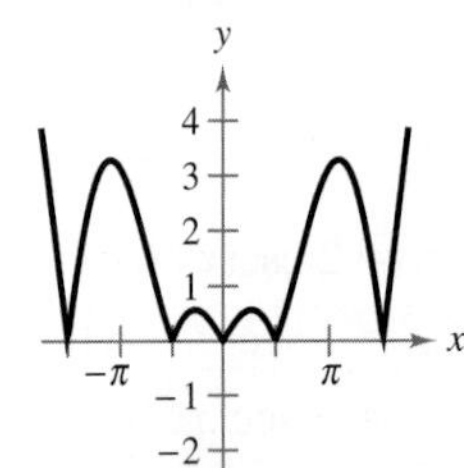

65. $f(x) = |x \cos x|$
66. $f(x) = x \sin x$
67. $g(x) = |x| \sin x$
68. $g(x) = |x| \cos x$

Conjecture **In Exercises 69–72, graph the functions f and g. Use the graphs to make a conjecture about the relationship between the functions.**

69. $f(x) = \sin x + \cos\left(x + \frac{\pi}{2}\right), \quad g(x) = 0$

70. $f(x) = \sin x - \cos\left(x + \frac{\pi}{2}\right), \quad g(x) = 2 \sin x$

71. $f(x) = \sin^2 x, \quad g(x) = \frac{1}{2}(1 - \cos 2x)$

72. $f(x) = \cos^2 \frac{\pi x}{2}, \quad g(x) = \frac{1}{2}(1 + \cos \pi x)$

Analyzing a Damped Trigonometric Graph **In Exercises 73–76, use a graphing utility to graph the function and the damping factor of the function in the same viewing window. Describe the behavior of the function as x increases without bound.**

73. $g(x) = x \cos \pi x$
74. $f(x) = x^2 \cos x$
75. $f(x) = x^3 \sin x$
76. $h(x) = x^3 \cos x$

Analyzing a Trigonometric Graph **In Exercises 77–82, use a graphing utility to graph the function. Describe the behavior of the function as x approaches zero.**

77. $y = \frac{6}{x} + \cos x, \quad x > 0$

78. $y = \frac{4}{x} + \sin 2x, \quad x > 0$

79. $g(x) = \frac{\sin x}{x}$

80. $f(x) = \frac{1 - \cos x}{x}$

81. $f(x) = \sin \frac{1}{x}$

82. $h(x) = x \sin \frac{1}{x}$

83. **Meteorology** The normal monthly high temperatures H (in degrees Fahrenheit) in Erie, Pennsylvania, are approximated by

$$H(t) = 57.54 - 18.53 \cos \frac{\pi t}{6} - 14.03 \sin \frac{\pi t}{6}$$

and the normal monthly low temperatures L are approximated by

$$L(t) = 42.03 - 15.99 \cos \frac{\pi t}{6} - 14.32 \sin \frac{\pi t}{6}$$

where t is the time (in months), with $t = 1$ corresponding to January (see figure). *(Source: NOAA)*

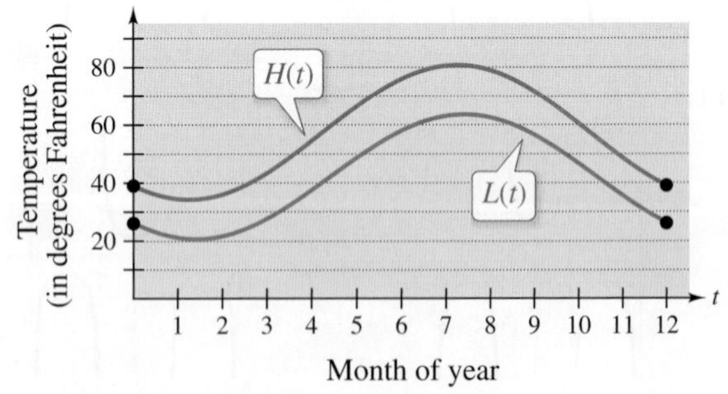

(a) What is the period of each function?

(b) During what part of the year is the difference between the normal high and normal low temperatures greatest? When is it least?

(c) The sun is northernmost in the sky around June 21, but the graph shows the warmest temperatures at a later date. Approximate the lag time of the temperatures relative to the position of the sun.

84. Sales The projected monthly sales S (in thousands of units) of lawn mowers are modeled by

$$S = 74 + 3t - 40 \cos \frac{\pi t}{6}$$

where t is the time (in months), with $t = 1$ corresponding to January.

(a) Graph the sales function over 1 year.

(b) What are the projected sales for June?

85. Television Coverage

A television camera is on a reviewing platform 27 meters from the street on which a parade passes from left to right (see figure). Write the distance d from the camera to a unit in the parade as a function of the angle x, and graph the function over the interval $-\pi/2 < x < \pi/2$. (Consider x as negative when a unit in the parade approaches from the left.)

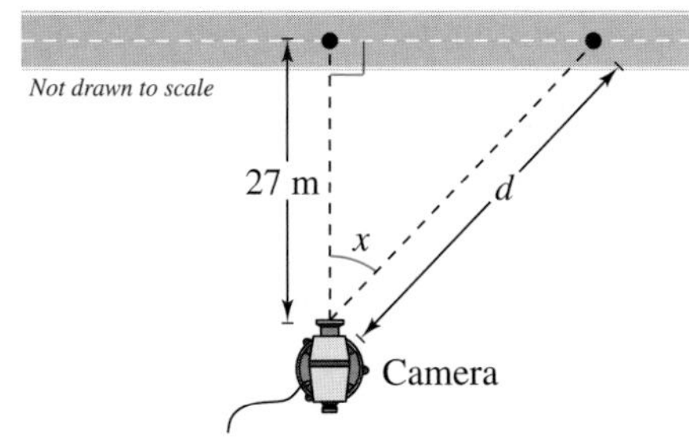

86. Distance A plane flying at an altitude of 7 miles above a radar antenna passes directly over the radar antenna (see figure). Let d be the ground distance from the antenna to the point directly under the plane and let x be the angle of elevation to the plane from the antenna. (d is positive as the plane approaches the antenna.) Write d as a function of x and graph the function over the interval $0 < x < \pi$.

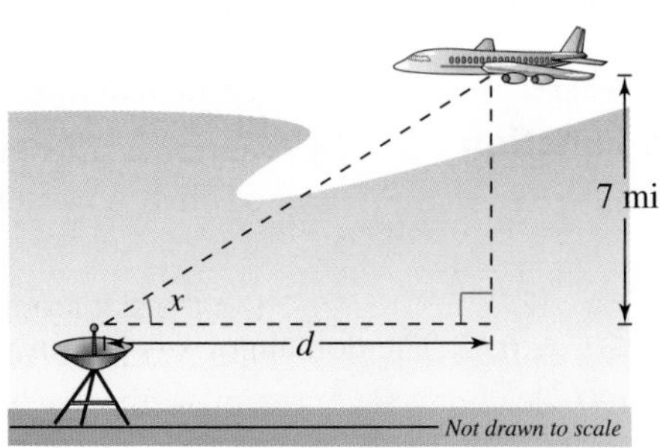

Irin-k/Shutterstock.com

Exploration

True or False? **In Exercises 87 and 88, determine whether the statement is true or false. Justify your answer.**

87. You can obtain the graph of $y = \csc x$ on a calculator by graphing the reciprocal of $y = \sin x$.

88. You can obtain the graph of $y = \sec x$ on a calculator by graphing a translation of the reciprocal of $y = \sin x$.

89. Think About It Consider the function $f(x) = x - \cos x$.

(a) Use a graphing utility to graph the function and verify that there exists a zero between 0 and 1. Use the graph to approximate the zero.

(b) Starting with $x_0 = 1$, generate a sequence x_1, x_2, x_3, . . . , where $x_n = \cos(x_{n-1})$. For example, $x_0 = 1, x_1 = \cos(x_0), x_2 = \cos(x_1), x_3 = \cos(x_2), \ldots$. What value does the sequence approach?

90. HOW DO YOU SEE IT? Determine which function each graph represents. Do not use a calculator. Explain.

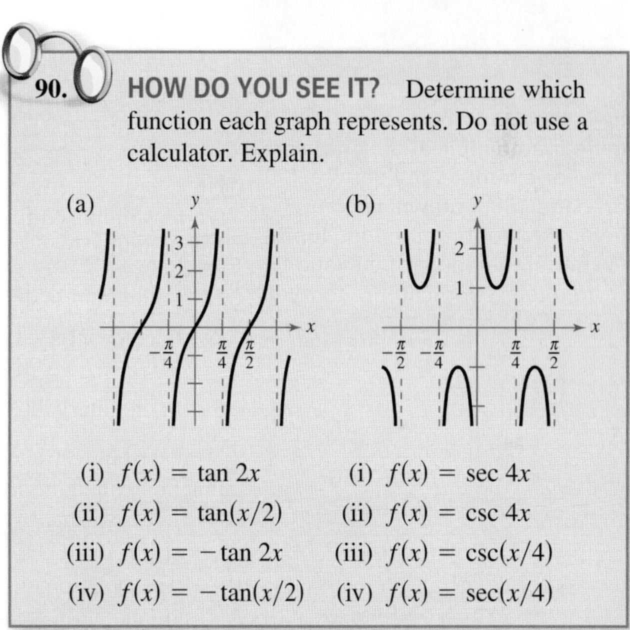

(a)	(b)
(i) $f(x) = \tan 2x$	(i) $f(x) = \sec 4x$
(ii) $f(x) = \tan(x/2)$	(ii) $f(x) = \csc 4x$
(iii) $f(x) = -\tan 2x$	(iii) $f(x) = \csc(x/4)$
(iv) $f(x) = -\tan(x/2)$	(iv) $f(x) = \sec(x/4)$

Graphical Reasoning **In Exercises 91 and 92, use a graphing utility to graph the function. Use the graph to determine the behavior of the function as $x \to c$. (*Note:* The notation $x \to c^+$ indicates that x approaches c from the right and $x \to c^-$ indicates that x approaches c from the left.)**

(a) $x \to 0^+$ **(b)** $x \to 0^-$ **(c)** $x \to \pi^+$ **(d)** $x \to \pi^-$

91. $f(x) = \cot x$

92. $f(x) = \csc x$

Graphical Reasoning **In Exercises 93 and 94, use a graphing utility to graph the function. Use the graph to determine the behavior of the function as $x \to c$.**

(a) $x \to (\pi/2)^+$

(b) $x \to (\pi/2)^-$

(c) $x \to (-\pi/2)^+$

(d) $x \to (-\pi/2)^-$

93. $f(x) = \tan x$

94. $f(x) = \sec x$

1.7 Inverse Trigonometric Functions

Inverse trigonometric functions have many applications in real life. For example, in Exercise 100 on page 188, you will use an inverse trigonometric function to model the angle of elevation from a television camera to a space shuttle.

- Evaluate and graph the inverse sine function.
- Evaluate and graph other inverse trigonometric functions.
- Evaluate compositions with inverse trigonometric functions.

Inverse Sine Function

Recall from Section P.10 that for a function to have an inverse function, it must be one-to-one—that is, it must pass the Horizontal Line Test. Notice in Figure 1.47 that $y = \sin x$ does not pass the test because different values of x yield the same y-value.

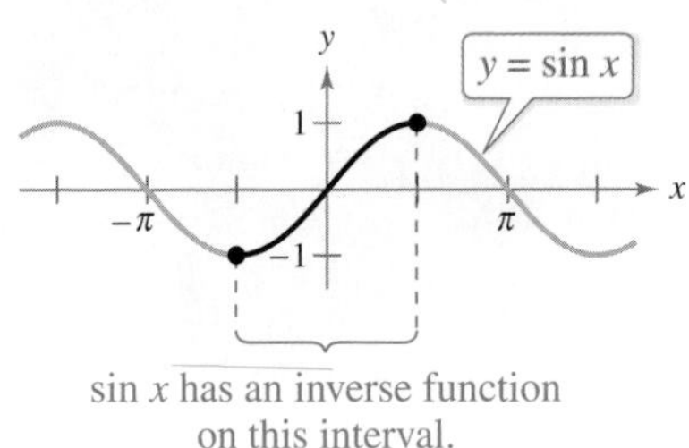

Figure 1.47

However, when you restrict the domain to the interval $-\pi/2 \le x \le \pi/2$ (corresponding to the black portion of the graph in Figure 1.47), the properties listed below hold.

1. On the interval $[-\pi/2, \pi/2]$, the function $y = \sin x$ is increasing.
2. On the interval $[-\pi/2, \pi/2]$, $y = \sin x$ takes on its full range of values, $-1 \le \sin x \le 1$.
3. On the interval $[-\pi/2, \pi/2]$, $y = \sin x$ is one-to-one.

So, on the restricted domain $-\pi/2 \le x \le \pi/2$, $y = \sin x$ has a unique inverse function called the **inverse sine function.** It is denoted by

$$y = \arcsin x \quad \text{or} \quad y = \sin^{-1} x.$$

The notation $\sin^{-1} x$ is consistent with the inverse function notation $f^{-1}(x)$. The $\arcsin x$ notation (read as "the arcsine of x") comes from the association of a central angle with its intercepted *arc length* on a unit circle. So, $\arcsin x$ means the angle (or arc) whose sine is x. Both notations, $\arcsin x$ and $\sin^{-1} x$ are commonly used in mathematics. You must remember that $\sin^{-1} x$ denotes the *inverse* sine function, *not* $1/\sin x$. The values of $\arcsin x$ lie in the interval

$$-\frac{\pi}{2} \le \arcsin x \le \frac{\pi}{2}.$$

Figure 1.48 on the next page shows the graph of $y = \arcsin x$.

REMARK When evaluating the inverse sine function, it helps to remember the phrase "the arcsine of x is the angle (or number) whose sine is x."

Definition of Inverse Sine Function

The **inverse sine function** is defined by

$$y = \arcsin x \quad \text{if and only if} \quad \sin y = x$$

where $-1 \le x \le 1$ and $-\pi/2 \le y \le \pi/2$. The domain of $y = \arcsin x$ is $[-1, 1]$, and the range is $[-\pi/2, \pi/2]$.

REMARK As with trigonometric functions, some of the work with inverse trigonometric functions can be done by *exact* calculations rather than by calculator approximations. Exact calculations help to increase your understanding of inverse functions by relating them to the right triangle definitions of trigonometric functions.

EXAMPLE 1 Evaluating the Inverse Sine Function

If possible, find the exact value of each expression.

a. $\arcsin\left(-\frac{1}{2}\right)$ **b.** $\sin^{-1}\frac{\sqrt{3}}{2}$ **c.** $\sin^{-1} 2$

Solution

a. You know that $\sin\left(-\frac{\pi}{6}\right) = -\frac{1}{2}$ and $-\frac{\pi}{6}$ lies in $\left[-\frac{\pi}{2}, \frac{\pi}{2}\right]$, so

$$\arcsin\left(-\frac{1}{2}\right) = -\frac{\pi}{6}. \qquad \text{Angle whose sine is } -\tfrac{1}{2}$$

b. You know that $\sin\frac{\pi}{3} = \frac{\sqrt{3}}{2}$ and $\frac{\pi}{3}$ lies in $\left[-\frac{\pi}{2}, \frac{\pi}{2}\right]$, so

$$\sin^{-1}\frac{\sqrt{3}}{2} = \frac{\pi}{3}. \qquad \text{Angle whose sine is } \sqrt{3}/2$$

c. It is not possible to evaluate $y = \sin^{-1} x$ when $x = 2$ because there is no angle whose sine is 2. Remember that the domain of the inverse sine function is $[-1, 1]$.

Checkpoint Audio-video solution in English & Spanish at LarsonPrecalculus.com

If possible, find the exact value of each expression.

a. $\arcsin 1$ **b.** $\sin^{-1}(-2)$

EXAMPLE 2 Graphing the Arcsine Function

See LarsonPrecalculus.com for an interactive version of this type of example.

Sketch the graph of $y = \arcsin x$.

Solution

By definition, the equations $y = \arcsin x$ and $\sin y = x$ are equivalent for $-\pi/2 \le y \le \pi/2$. So, their graphs are the same. From the interval $[-\pi/2, \pi/2]$, assign values to y in the equation $\sin y = x$ to make a table of values.

y	$-\frac{\pi}{2}$	$-\frac{\pi}{4}$	$-\frac{\pi}{6}$	0	$\frac{\pi}{6}$	$\frac{\pi}{4}$	$\frac{\pi}{2}$
$x = \sin y$	-1	$-\frac{\sqrt{2}}{2}$	$-\frac{1}{2}$	0	$\frac{1}{2}$	$\frac{\sqrt{2}}{2}$	1

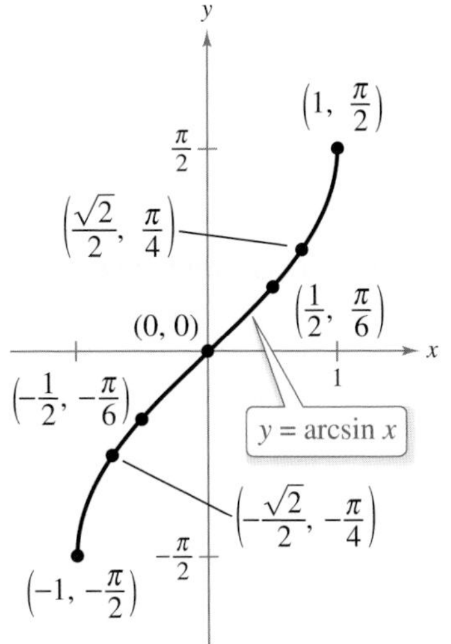

Figure 1.48

Then plot the points and connect them with a smooth curve. Figure 1.48 shows the graph of $y = \arcsin x$. Note that it is the reflection (in the line $y = x$) of the black portion of the graph in Figure 1.47. Be sure you see that Figure 1.48 shows the *entire* graph of the inverse sine function. Remember that the domain of $y = \arcsin x$ is the closed interval $[-1, 1]$ and the range is the closed interval $[-\pi/2, \pi/2]$.

Checkpoint Audio-video solution in English & Spanish at LarsonPrecalculus.com

Use a graphing utility to graph $f(x) = \sin x$, $g(x) = \arcsin x$, and $y = x$ in the same viewing window to verify geometrically that g is the inverse function of f. (Be sure to restrict the domain of f properly.)

Other Inverse Trigonometric Functions

The cosine function is decreasing and one-to-one on the interval $0 \le x \le \pi$, as shown in the graph below.

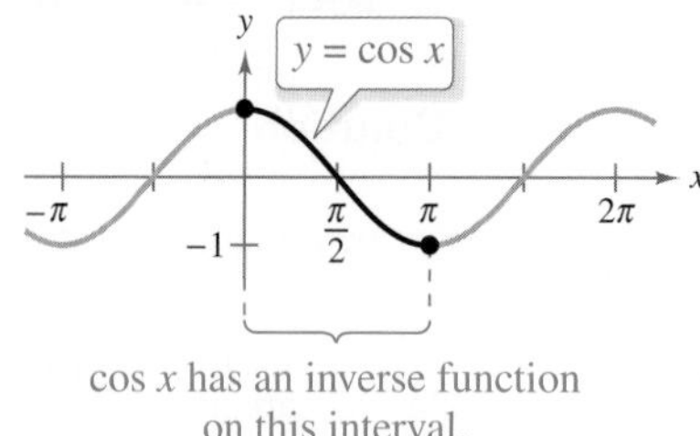

Consequently, on this interval the cosine function has an inverse function—the **inverse cosine function**—denoted by

$$y = \arccos x \quad \text{or} \quad y = \cos^{-1} x.$$

Similarly, to define an **inverse tangent function,** restrict the domain of $y = \tan x$ to the interval $(-\pi/2, \pi/2)$. The inverse tangent function is denoted by

$$y = \arctan x \quad \text{or} \quad y = \tan^{-1} x.$$

The list below summarizes the definitions of the three most common inverse trigonometric functions. Definitions of the remaining three are explored in Exercises 111–113.

Definitions of the Inverse Trigonometric Functions

Function	**Domain**	**Range**
$y = \arcsin x$ if and only if $\sin y = x$	$-1 \le x \le 1$	$-\frac{\pi}{2} \le y \le \frac{\pi}{2}$
$y = \arccos x$ if and only if $\cos y = x$	$-1 \le x \le 1$	$0 \le y \le \pi$
$y = \arctan x$ if and only if $\tan y = x$	$-\infty < x < \infty$	$-\frac{\pi}{2} < y < \frac{\pi}{2}$

The graphs of these three inverse trigonometric functions are shown below.

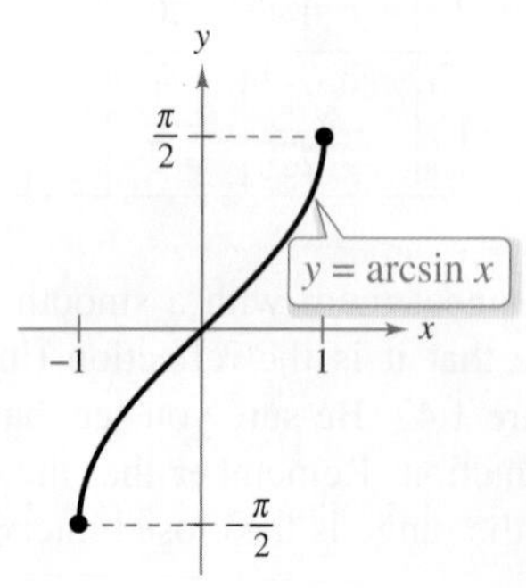

Domain: $[-1, 1]$
Range: $\left[-\frac{\pi}{2}, \frac{\pi}{2}\right]$
Intercept: $(0, 0)$
Symmetry: origin
Odd function

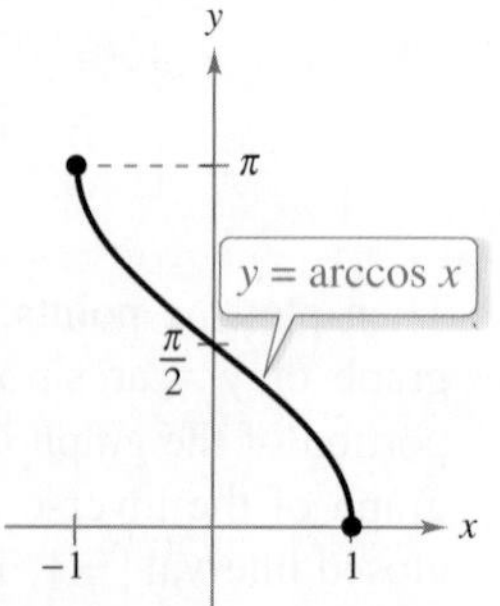

Domain: $[-1, 1]$
Range: $[0, \pi]$
y-intercept: $\left(0, \frac{\pi}{2}\right)$

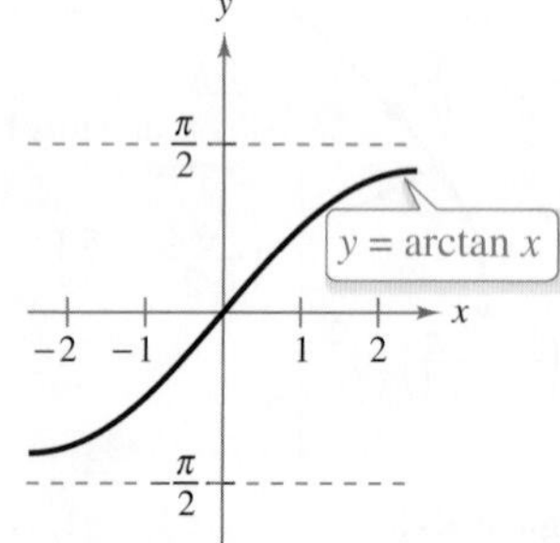

Domain: $(-\infty, \infty)$
Range: $\left(-\frac{\pi}{2}, \frac{\pi}{2}\right)$
Horizontal asymptotes: $y = \pm\frac{\pi}{2}$
Intercept: $(0, 0)$
Symmetry: origin
Odd function

EXAMPLE 3 Evaluating Inverse Trigonometric Functions

Find the exact value of each expression.

a. $\arccos \frac{\sqrt{2}}{2}$

b. $\arctan 0$

c. $\tan^{-1}(-1)$

Solution

a. You know that $\cos(\pi/4) = \sqrt{2}/2$ and $\pi/4$ lies in $[0, \pi]$, so

$$\arccos \frac{\sqrt{2}}{2} = \frac{\pi}{4}.$$ Angle whose cosine is $\sqrt{2}/2$

b. You know that $\tan 0 = 0$ and 0 lies in $(-\pi/2, \pi/2)$, so

$$\arctan 0 = 0.$$ Angle whose tangent is 0

c. You know that $\tan(-\pi/4) = -1$ and $-\pi/4$ lies in $(-\pi/2, \pi/2)$, so

$$\tan^{-1}(-1) = -\frac{\pi}{4}.$$ Angle whose tangent is -1

✓ **Checkpoint** Audio-video solution in English & Spanish at LarsonPrecalculus.com

Find the exact value of $\cos^{-1}(-1)$.

EXAMPLE 4 Calculators and Inverse Trigonometric Functions

Use a calculator to approximate the value of each expression, if possible.

a. $\arctan(-8.45)$

b. $\sin^{-1} 0.2447$

c. $\arccos 2$

Solution

	Function	Mode	Calculator Keystrokes
a.	$\arctan(-8.45)$	Radian	TAN⁻¹ ((−) 8.45) ENTER

From the display, it follows that $\arctan(-8.45) \approx -1.4530010$.

b.	$\sin^{-1} 0.2447$	Radian	SIN⁻¹ (0.2447) ENTER

From the display, it follows that $\sin^{-1} 0.2447 \approx 0.2472103$.

c.	$\arccos 2$	Radian	COS⁻¹ (2) ENTER

The calculator should display an *error message* because the domain of the inverse cosine function is $[-1, 1]$.

✓ **Checkpoint** Audio-video solution in English & Spanish at LarsonPrecalculus.com

Use a calculator to approximate the value of each expression, if possible.

a. $\arctan 4.84$

b. $\arcsin(-1.1)$

c. $\arccos(-0.349)$

In Example 4, had you set the calculator to *degree* mode, the displays would have been in degrees rather than in radians. This convention is peculiar to calculators. By definition, the values of inverse trigonometric functions are *always in radians.*

Compositions with Inverse Trigonometric Functions

▷ **ALGEBRA HELP** To review compositions of functions, see Section P.9.

Recall from Section P.10 that for all x in the domains of f and f^{-1}, inverse functions have the properties

$$f(f^{-1}(x)) = x \quad \text{and} \quad f^{-1}(f(x)) = x.$$

Inverse Properties of Trigonometric Functions

If $-1 \le x \le 1$ and $-\pi/2 \le y \le \pi/2$, then

$$\sin(\arcsin x) = x \quad \text{and} \quad \arcsin(\sin y) = y.$$

If $-1 \le x \le 1$ and $0 \le y \le \pi$, then

$$\cos(\arccos x) = x \quad \text{and} \quad \arccos(\cos y) = y.$$

If x is a real number and $-\pi/2 < y < \pi/2$, then

$$\tan(\arctan x) = x \quad \text{and} \quad \arctan(\tan y) = y.$$

Keep in mind that these inverse properties do not apply for arbitrary values of x and y. For example,

$$\arcsin\left(\sin \frac{3\pi}{2}\right) = \arcsin(-1) = -\frac{\pi}{2} \neq \frac{3\pi}{2}.$$

In other words, the property $\arcsin(\sin y) = y$ is not valid for values of y outside the interval $[-\pi/2, \pi/2]$.

EXAMPLE 5 Using Inverse Properties

If possible, find the exact value of each expression.

a. $\tan[\arctan(-5)]$ **b.** $\arcsin\left(\sin \frac{5\pi}{3}\right)$ **c.** $\cos(\cos^{-1} \pi)$

Solution

a. You know that -5 lies in the domain of the arctangent function, so the inverse property applies, and you have

$$\tan[\arctan(-5)] = -5.$$

b. In this case, $5\pi/3$ does not lie in the range of the arcsine function, $-\pi/2 \le y \le \pi/2$. However, $5\pi/3$ is coterminal with

$$\frac{5\pi}{3} - 2\pi = -\frac{\pi}{3}$$

which does lie in the range of the arcsine function, and you have

$$\arcsin\left(\sin \frac{5\pi}{3}\right) = \arcsin\left[\sin\left(-\frac{\pi}{3}\right)\right] = -\frac{\pi}{3}.$$

c. The expression $\cos(\cos^{-1} \pi)$ is not defined because $\cos^{-1} \pi$ is not defined. Remember that the domain of the inverse cosine function is $[-1, 1]$.

✓ ***Checkpoint*** Audio-video solution in English & Spanish at LarsonPrecalculus.com

If possible, find the exact value of each expression.

a. $\tan[\tan^{-1}(-14)]$ **b.** $\sin^{-1}\left(\sin \frac{7\pi}{4}\right)$ **c.** $\cos(\arccos 0.54)$

EXAMPLE 6 Evaluating Compositions of Functions

Find the exact value of each expression.

a. $\tan\left(\arccos \frac{2}{3}\right)$ **b.** $\cos\left[\arcsin\left(-\frac{3}{5}\right)\right]$

Solution

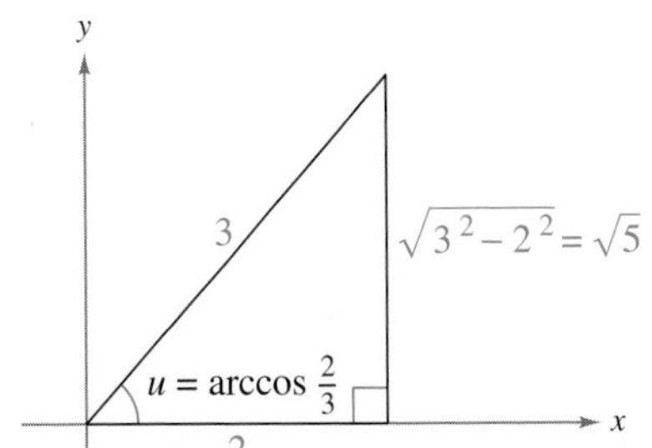

Angle whose cosine is $\frac{2}{3}$
Figure 1.49

a. If you let $u = \arccos \frac{2}{3}$, then $\cos u = \frac{2}{3}$. The range of the inverse cosine function is $[0, \pi]$ and $\cos u$ is positive, so u is a *first*-quadrant angle. Sketch and label a right triangle with acute angle u, as shown in Figure 1.49. Consequently,

$$\tan\left(\arccos \frac{2}{3}\right) = \tan u = \frac{\text{opp}}{\text{adj}} = \frac{\sqrt{5}}{2}.$$

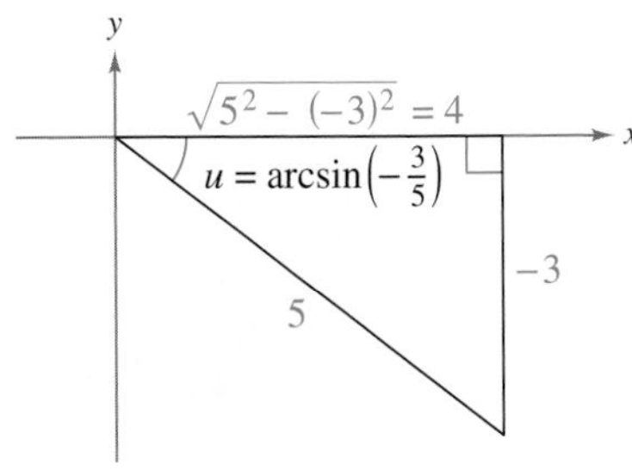

Angle whose sine is $-\frac{3}{5}$
Figure 1.50

b. If you let $u = \arcsin\left(-\frac{3}{5}\right)$, then $\sin u = -\frac{3}{5}$. The range of the inverse sine function is $[-\pi/2, \pi/2]$ and $\sin u$ is negative, so u is a *fourth*-quadrant angle. Sketch and label a right triangle with acute angle u, as shown in Figure 1.50. Consequently,

$$\cos\left[\arcsin\left(-\frac{3}{5}\right)\right] = \cos u = \frac{\text{adj}}{\text{hyp}} = \frac{4}{5}.$$

Checkpoint Audio-video solution in English & Spanish at LarsonPrecalculus.com

Find the exact value of $\cos\left[\arctan\left(-\frac{3}{4}\right)\right]$.

EXAMPLE 7 Some Problems from Calculus

Write an algebraic expression that is equivalent to each expression.

a. $\sin(\arccos 3x)$, $0 \le x \le \frac{1}{3}$ **b.** $\cot(\arccos 3x)$, $0 \le x < \frac{1}{3}$

Solution

If you let $u = \arccos 3x$, then $\cos u = 3x$, where $-1 \le 3x \le 1$. Write

$$\cos u = \frac{\text{adj}}{\text{hyp}} = \frac{3x}{1}$$

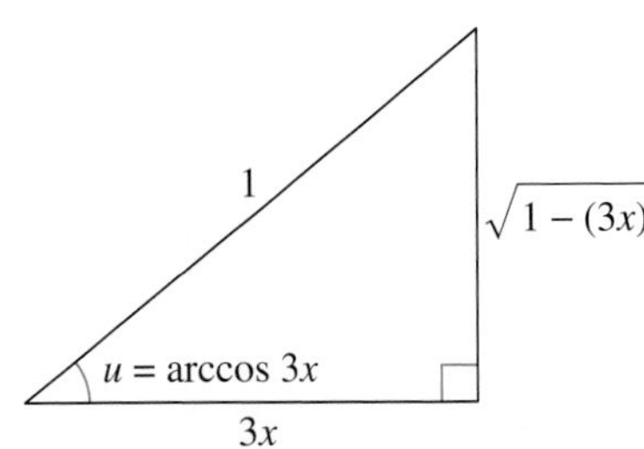

Angle whose cosine is $3x$
Figure 1.51

and sketch a right triangle with acute angle u, as shown in Figure 1.51. From this triangle, convert each expression to algebraic form.

a. $\sin(\arccos 3x) = \sin u = \dfrac{\text{opp}}{\text{hyp}} = \sqrt{1 - 9x^2}, \quad 0 \le x \le \dfrac{1}{3}$

b. $\cot(\arccos 3x) = \cot u = \dfrac{\text{adj}}{\text{opp}} = \dfrac{3x}{\sqrt{1 - 9x^2}}, \quad 0 \le x < \dfrac{1}{3}$

Checkpoint Audio-video solution in English & Spanish at LarsonPrecalculus.com

Write an algebraic expression that is equivalent to $\sec(\arctan x)$.

Summarize (Section 1.7)

1. State the definition of the inverse sine function *(page 180)*. For examples of evaluating and graphing the inverse sine function, see Examples 1 and 2.
2. State the definitions of the inverse cosine and inverse tangent functions *(page 182)*. For examples of evaluating inverse trigonometric functions, see Examples 3 and 4.
3. State the inverse properties of trigonometric functions *(page 184)*. For examples of finding compositions with inverse trigonometric functions, see Examples 5–7.

1.7 Exercises

See CalcChat.com for tutorial help and worked-out solutions to odd-numbered exercises.

Vocabulary: Fill in the blanks.

	Function	Alternative Notation	Domain	Range
1.	$y = \arcsin x$	________	________	$-\frac{\pi}{2} \le y \le \frac{\pi}{2}$
2.	________	$y = \cos^{-1} x$	$-1 \le x \le 1$	________
3.	$y = \arctan x$	________	________	________

4. A trigonometric function has an ________ function only when its domain is restricted.

Skills and Applications

Evaluating an Inverse Trigonometric Function In Exercises 5–18, find the exact value of the expression, if possible.

5. $\arcsin \frac{1}{2}$
6. $\arcsin 0$
7. $\arccos \frac{1}{2}$
8. $\arccos 0$
9. $\arctan \frac{\sqrt{3}}{3}$
10. $\arctan 1$
11. $\arcsin 3$
12. $\arctan \sqrt{3}$
13. $\tan^{-1}(-\sqrt{3})$
14. $\cos^{-1}(-2)$
15. $\arccos\left(-\frac{1}{2}\right)$
16. $\arcsin \frac{\sqrt{2}}{2}$
17. $\sin^{-1}\left(-\frac{\sqrt{3}}{2}\right)$
18. $\tan^{-1}\left(-\frac{\sqrt{3}}{3}\right)$

Graphing an Inverse Trigonometric Function In Exercises 19 and 20, use a graphing utility to graph f, g, and $y = x$ in the same viewing window to verify geometrically that g is the inverse function of f. (Be sure to restrict the domain of f properly.)

19. $f(x) = \cos x, \quad g(x) = \arccos x$
20. $f(x) = \tan x, \quad g(x) = \arctan x$

Calculators and Inverse Trigonometric Functions In Exercises 21–36, use a calculator to approximate the value of the expression, if possible. Round your result to two decimal places.

21. $\arccos 0.37$
22. $\arcsin 0.65$
23. $\arcsin(-0.75)$
24. $\arccos(-0.7)$
25. $\arctan(-3)$
26. $\arctan 25$
27. $\sin^{-1} 1.36$
28. $\cos^{-1} 0.26$
29. $\arccos(-0.41)$
30. $\arcsin(-0.125)$
31. $\arctan 0.92$
32. $\arctan 2.8$
33. $\arcsin \frac{7}{8}$
34. $\arccos\left(-\frac{4}{3}\right)$
35. $\tan^{-1}\left(-\frac{95}{7}\right)$
36. $\tan^{-1}(-\sqrt{372})$

Finding Missing Coordinates In Exercises 37 and 38, determine the missing coordinates of the points on the graph of the function.

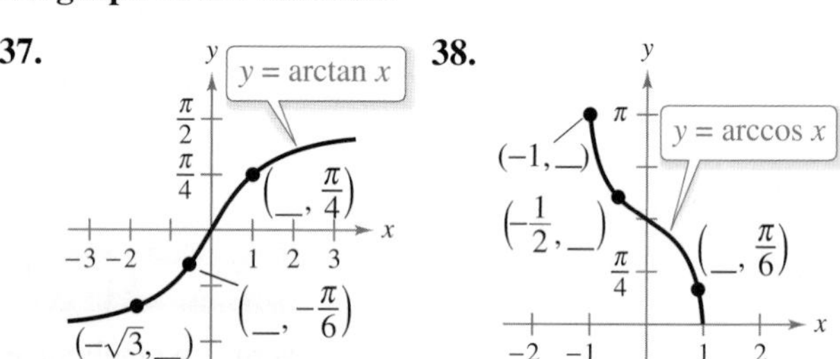

Using an Inverse Trigonometric Function In Exercises 39–44, use an inverse trigonometric function to write θ as a function of x.

39.
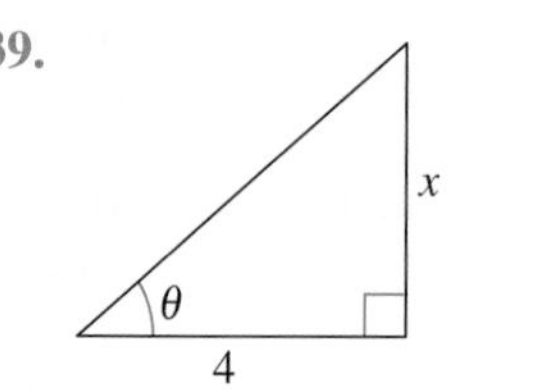

40.
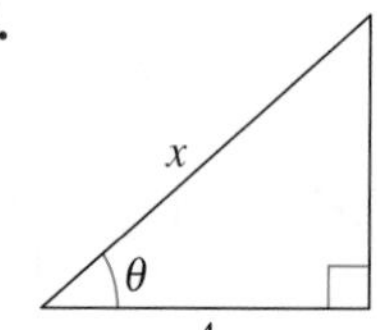

41.
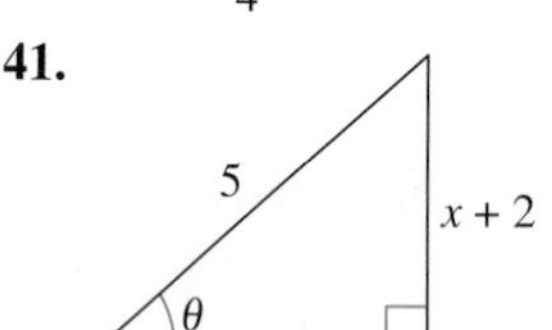

42.
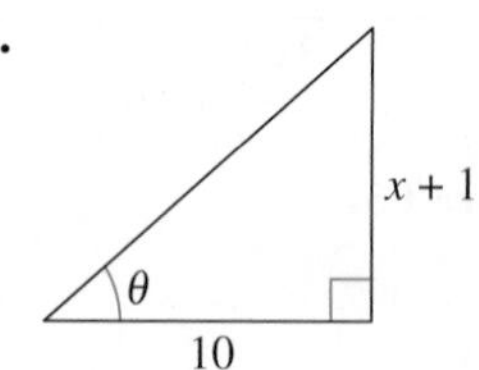

43.
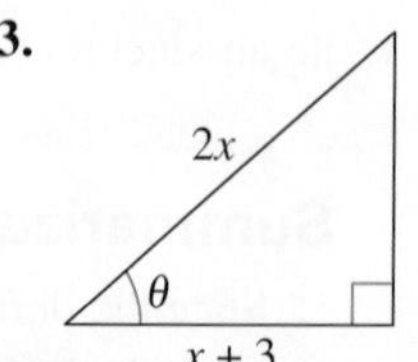

44.
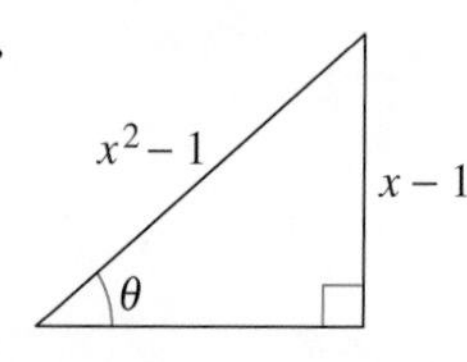

Using Inverse Properties In Exercises 45–50, find the exact value of the expression, if possible.

45. $\sin(\arcsin 0.3)$
46. $\tan(\arctan 45)$
47. $\cos[\arccos(-\sqrt{3})]$
48. $\sin[\arcsin(-0.2)]$
49. $\arcsin[\sin(9\pi/4)]$
50. $\arccos[\cos(-3\pi/2)]$

Evaluating a Composition of Functions **In Exercises 51–62, find the exact value of the expression, if possible.**

51. $\sin\left(\arctan \frac{3}{4}\right)$
52. $\cos\left(\arcsin \frac{4}{5}\right)$
53. $\cos(\tan^{-1} 2)$
54. $\sin\left(\cos^{-1} \sqrt{5}\right)$
55. $\sec\left(\arcsin \frac{5}{13}\right)$
56. $\csc\left[\arctan\left(-\frac{5}{12}\right)\right]$
57. $\cot\left[\arctan\left(-\frac{3}{5}\right)\right]$
58. $\sec\left[\arccos\left(-\frac{3}{4}\right)\right]$
59. $\tan\left[\arccos\left(-\frac{2}{3}\right)\right]$
60. $\cot\left(\arctan \frac{5}{8}\right)$
61. $\csc\left(\cos^{-1} \frac{\sqrt{3}}{2}\right)$
62. $\tan\left[\sin^{-1}\left(-\frac{\sqrt{2}}{2}\right)\right]$

Writing an Expression **In Exercises 63–72, write an algebraic expression that is equivalent to the given expression.**

63. $\cos(\arcsin 2x)$
64. $\sin(\arctan x)$
65. $\cot(\arctan x)$
66. $\sec(\arctan 3x)$
67. $\sin(\arccos x)$
68. $\csc[\arccos(x - 1)]$
69. $\tan\left(\arccos \frac{x}{3}\right)$
70. $\cot\left(\arctan \frac{1}{x}\right)$
71. $\csc\left(\arctan \frac{x}{a}\right)$
72. $\cos\left(\arcsin \frac{x - h}{r}\right)$

Using Technology **In Exercises 73 and 74, use a graphing utility to graph f and g in the same viewing window to verify that the two functions are equal. Explain why they are equal. Identify any asymptotes of the graphs.**

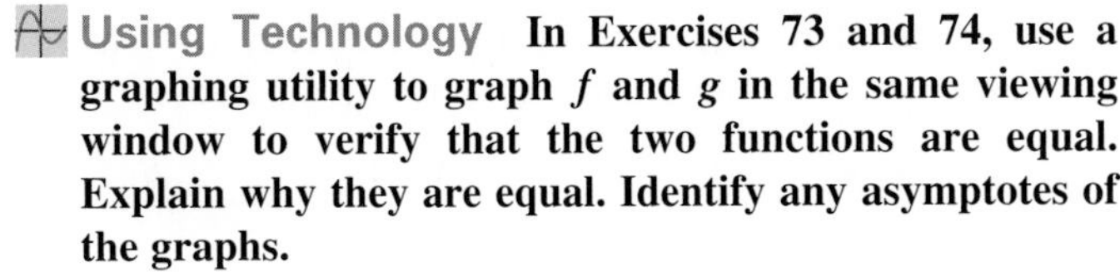

73. $f(x) = \sin(\arctan 2x), \quad g(x) = \dfrac{2x}{\sqrt{1 + 4x^2}}$

74. $f(x) = \tan\left(\arccos \dfrac{x}{2}\right), \quad g(x) = \dfrac{\sqrt{4 - x^2}}{x}$

Completing an Equation **In Exercises 75–78, complete the equation.**

75. $\arctan \dfrac{9}{x} = \arcsin(\quad), \quad x > 0$

76. $\arcsin \dfrac{\sqrt{36 - x^2}}{6} = \arccos(\quad), \quad 0 \le x \le 6$

77. $\arccos \dfrac{3}{\sqrt{x^2 - 2x + 10}} = \arcsin(\quad)$

78. $\arccos \dfrac{x - 2}{2} = \arctan(\quad), \quad 2 < x < 4$

Sketching the Graph of a Function **In Exercises 79–84, sketch the graph of the function and compare the graph to the graph of the parent inverse trigonometric function.**

79. $y = 2 \arcsin x$
80. $f(x) = \arctan 2x$
81. $f(x) = \dfrac{\pi}{2} + \arctan x$
82. $g(t) = \arccos(t + 2)$
83. $h(v) = \arccos \dfrac{v}{2}$
84. $f(x) = \arcsin \dfrac{x}{4}$

Graphing an Inverse Trigonometric Function **In Exercises 85–90, use a graphing utility to graph the function.**

85. $f(x) = 2 \arccos 2x$
86. $f(x) = \pi \arcsin 4x$
87. $f(x) = \arctan(2x - 3)$
88. $f(x) = -3 + \arctan \pi x$
89. $f(x) = \pi - \sin^{-1} \dfrac{2}{3}$
90. $f(x) = \dfrac{\pi}{2} + \cos^{-1} \dfrac{1}{\pi}$

Using a Trigonometric Identity **In Exercises 91 and 92, write the function in terms of the sine function by using the identity**

$$A \cos \omega t + B \sin \omega t = \sqrt{A^2 + B^2} \sin\left(\omega t + \arctan \frac{A}{B}\right).$$

Use a graphing utility to graph both forms of the function. What does the graph imply?

91. $f(t) = 3 \cos 2t + 3 \sin 2t$
92. $f(t) = 4 \cos \pi t + 3 \sin \pi t$

Behavior of an Inverse Trigonometric Function **In Exercises 93–98, fill in the blank. If not possible, state the reason. (*Note:* The notation $x \to c^+$ indicates that x approaches c from the right and $x \to c^-$ indicates that x approaches c from the left.)**

93. As $x \to 1^-$, the value of $\arcsin x \to$ ___.
94. As $x \to 1^-$, the value of $\arccos x \to$ ___.
95. As $x \to \infty$, the value of $\arctan x \to$ ___.
96. As $x \to -1^+$, the value of $\arcsin x \to$ ___.
97. As $x \to -1^+$, the value of $\arccos x \to$ ___.
98. As $x \to -\infty$, the value of $\arctan x \to$ ___.

99. **Docking a Boat** A boat is pulled in by means of a winch located on a dock 5 feet above the deck of the boat (see figure). Let θ be the angle of elevation from the boat to the winch and let s be the length of the rope from the winch to the boat.

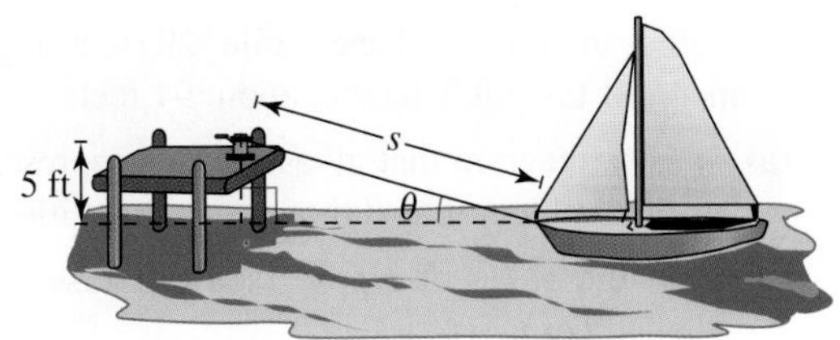

(a) Write θ as a function of s.
(b) Find θ when $s = 40$ feet and $s = 20$ feet.

100. Videography A television camera at ground level films the lift-off of a space shuttle at a point 750 meters from the launch pad (see figure). Let θ be the angle of elevation to the shuttle and let s be the height of the shuttle.

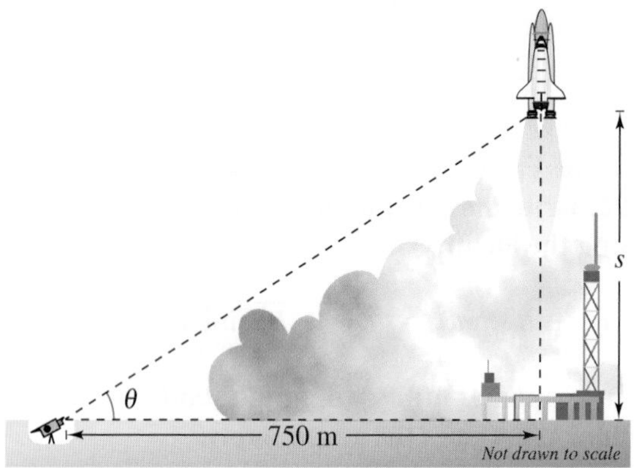

(a) Write θ as a function of s.

(b) Find θ when $s = 300$ meters and $s = 1200$ meters.

101. Granular Angle of Repose Different types of granular substances naturally settle at different angles when stored in cone-shaped piles. This angle θ is called the *angle of repose* (see figure). When rock salt is stored in a cone-shaped pile 5.5 meters high, the diameter of the pile's base is about 17 meters.

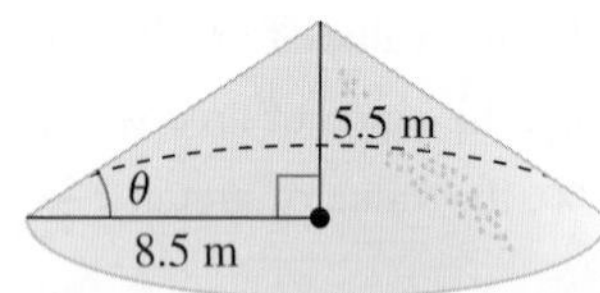

(a) Find the angle of repose for rock salt.

(b) How tall is a pile of rock salt that has a base diameter of 20 meters?

102. Granular Angle of Repose When shelled corn is stored in a cone-shaped pile 20 feet high, the diameter of the pile's base is about 94 feet.

(a) Draw a diagram that gives a visual representation of the problem. Label the known quantities.

(b) Find the angle of repose (see Exercise 101) for shelled corn.

(c) How tall is a pile of shelled corn that has a base diameter of 60 feet?

103. Photography A photographer takes a picture of a three-foot-tall painting hanging in an art gallery. The camera lens is 1 foot below the lower edge of the painting (see figure). The angle β subtended by the camera lens x feet from the painting is given by

$$\beta = \arctan \frac{3x}{x^2 + 4}, \quad x > 0.$$

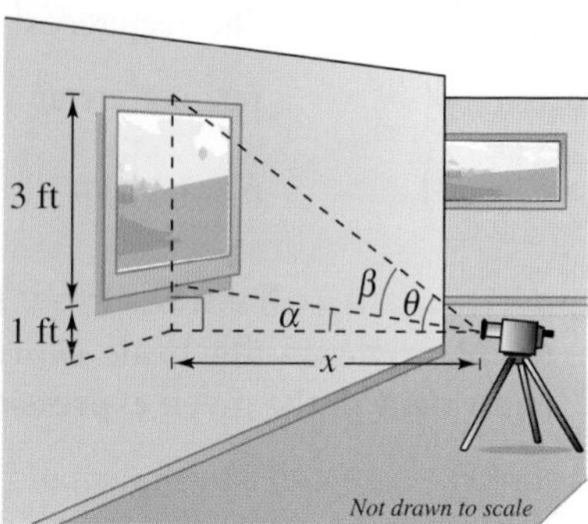

(a) Use a graphing utility to graph β as a function of x.

(b) Use the graph to approximate the distance from the picture when β is maximum.

(c) Identify the asymptote of the graph and interpret its meaning in the context of the problem.

104. Angle of Elevation An airplane flies at an altitude of 6 miles toward a point directly over an observer. Consider θ and x as shown in the figure.

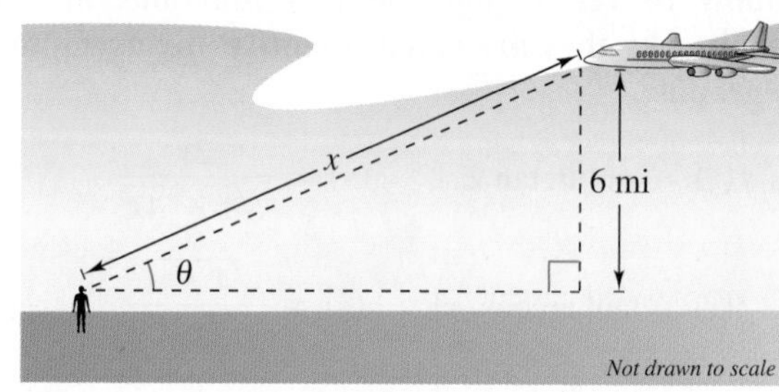

(a) Write θ as a function of x.

(b) Find θ when $x = 12$ miles and $x = 7$ miles.

105. Police Patrol A police car with its spotlight on is parked 20 meters from a warehouse. Consider θ and x as shown in the figure.

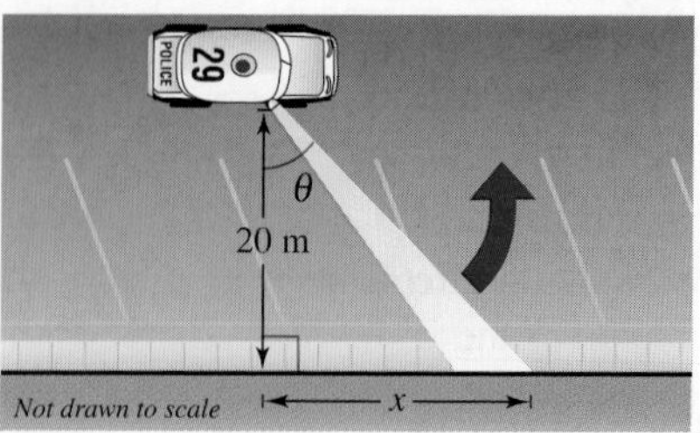

(a) Write θ as a function of x.

(b) Find θ when $x = 5$ meters and $x = 12$ meters.

Exploration

True or False? **In Exercises 106–109, determine whether the statement is true or false. Justify your answer.**

106. $\sin \frac{5\pi}{6} = \frac{1}{2} \Rightarrow \arcsin \frac{1}{2} = \frac{5\pi}{6}$

107. $\tan\left(-\frac{\pi}{4}\right) = -1 \Rightarrow \arctan(-1) = -\frac{\pi}{4}$

108. $\arctan x = \frac{\arcsin x}{\arccos x}$

109. $\sin^{-1} x = \frac{1}{\sin x}$

110. HOW DO YOU SEE IT? Use the figure below to determine the value(s) of x for which each statement is true.

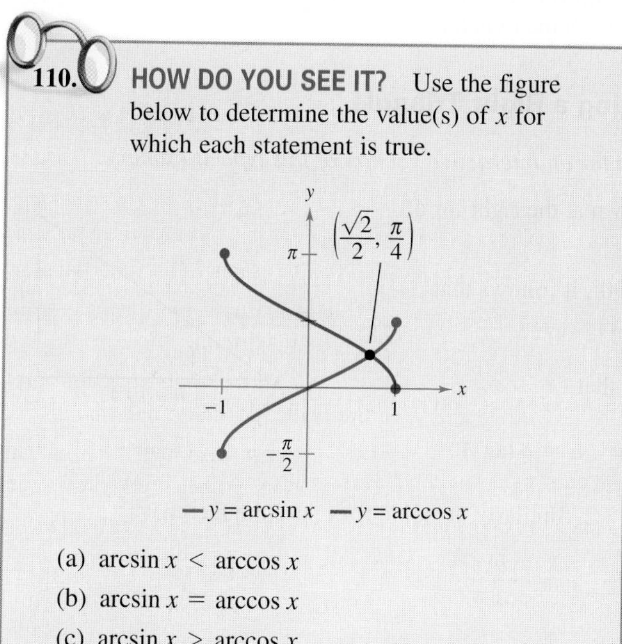

(a) $\arcsin x < \arccos x$

(b) $\arcsin x = \arccos x$

(c) $\arcsin x > \arccos x$

111. Inverse Cotangent Function Define the inverse cotangent function by restricting the domain of the cotangent function to the interval $(0, \pi)$, and sketch the graph of the inverse trigonometric function.

112. Inverse Secant Function Define the inverse secant function by restricting the domain of the secant function to the intervals $[0, \pi/2)$ and $(\pi/2, \pi]$, and sketch the graph of the inverse trigonometric function.

113. Inverse Cosecant Function Define the inverse cosecant function by restricting the domain of the cosecant function to the intervals $[-\pi/2, 0)$ and $(0, \pi/2]$, and sketch the graph of the inverse trigonometric function.

114. Writing Use the results of Exercises 111–113 to explain how to graph (a) the inverse cotangent function, (b) the inverse secant function, and (c) the inverse cosecant function on a graphing utility.

Evaluating an Inverse Trigonometric Function **In Exercises 115–120, use the results of Exercises 111–113 to find the exact value of the expression.**

115. $\text{arcsec}\,\sqrt{2}$

116. $\text{arcsec}\,1$

117. $\text{arccot}(-1)$

118. $\text{arccot}(-\sqrt{3})$

119. $\text{arccsc}(-1)$

120. $\text{arccsc}\,\frac{2\sqrt{3}}{3}$

Calculators and Inverse Trigonometric Functions **In Exercises 121–126, use the results of Exercises 111–113 and a calculator to approximate the value of the expression. Round your result to two decimal places.**

121. $\text{arcsec}\,2.54$

122. $\text{arcsec}(-1.52)$

123. $\text{arccsc}\left(-\frac{25}{3}\right)$

124. $\text{arccsc}(-12)$

125. $\text{arccot}\,5.25$

126. $\text{arccot}\left(-\frac{16}{7}\right)$

127. Area In calculus, it is shown that the area of the region bounded by the graphs of $y = 0$, $y = 1/(x^2 + 1)$, $x = a$, and $x = b$ (see figure) is given by

Area $= \arctan b - \arctan a.$

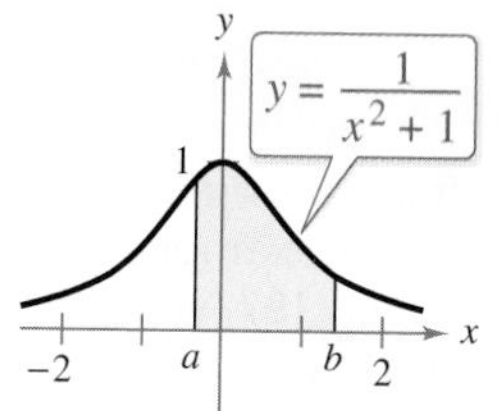

Find the area for each value of a and b.

(a) $a = 0, b = 1$

(b) $a = -1, b = 1$

(c) $a = 0, b = 3$

(d) $a = -1, b = 3$

128. Think About It Use a graphing utility to graph the functions $f(x) = \sqrt{x}$ and $g(x) = 6 \arctan x$. For $x > 0$, it appears that $g > f$. Explain how you know that there exists a positive real number a such that $g < f$ for $x > a$. Approximate the number a.

129. Think About It Consider the functions

$f(x) = \sin x$ and $f^{-1}(x) = \arcsin x.$

(a) Use a graphing utility to graph the composite functions $f \circ f^{-1}$ and $f^{-1} \circ f$.

(b) Explain why the graphs in part (a) are not the graph of the line $y = x$. Why do the graphs of $f \circ f^{-1}$ and $f^{-1} \circ f$ differ?

130. Proof Prove each identity.

(a) $\arcsin(-x) = -\arcsin x$

(b) $\arctan(-x) = -\arctan x$

(c) $\arctan x + \arctan \frac{1}{x} = \frac{\pi}{2}, \quad x > 0$

(d) $\arcsin x + \arccos x = \frac{\pi}{2}$

(e) $\arcsin x = \arctan \frac{x}{\sqrt{1 - x^2}}$

1.8 Applications and Models

Right triangles often occur in real-life situations. For example, in Exercise 30 on page 197, you will use right triangles to analyze the design of a new slide at a water park.

- **Solve real-life problems involving right triangles.**
- **Solve real-life problems involving directional bearings.**
- **Solve real-life problems involving harmonic motion.**

Applications Involving Right Triangles

In this section, the three angles of a right triangle are denoted by A, B, and C (where C is the right angle), and the lengths of the sides opposite these angles are denoted by a, b, and c, respectively (where c is the hypotenuse).

EXAMPLE 1 Solving a Right Triangle

See LarsonPrecalculus.com for an interactive version of this type of example.

Solve the right triangle shown at the right for all unknown sides and angles.

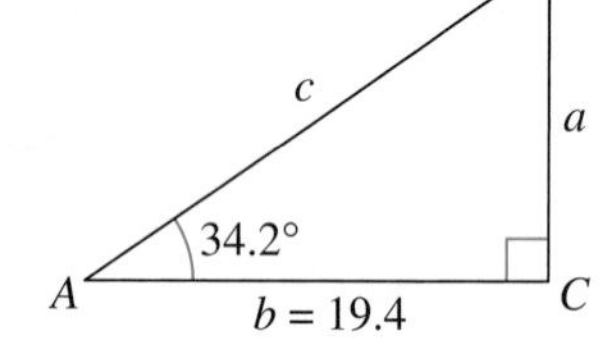

Solution Because $C = 90°$, it follows that

$$A + B = 90° \quad \text{and} \quad B = 90° - 34.2° = 55.8°.$$

To solve for a, use the fact that

$$\tan A = \frac{\text{opp}}{\text{adj}} = \frac{a}{b} \quad \Longrightarrow \quad a = b \tan A.$$

So, $a = 19.4 \tan 34.2° \approx 13.2$. Similarly, to solve for c, use the fact that

$$\cos A = \frac{\text{adj}}{\text{hyp}} = \frac{b}{c} \quad \Longrightarrow \quad c = \frac{b}{\cos A}.$$

So, $c = \dfrac{19.4}{\cos 34.2°} \approx 23.5$.

✓ ***Checkpoint*** *Audio-video solution in English & Spanish at LarsonPrecalculus.com*

Solve the right triangle shown at the right for all unknown sides and angles.

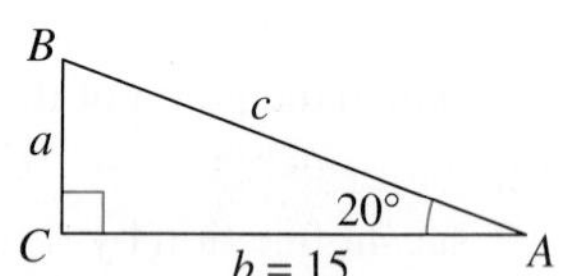

EXAMPLE 2 Finding a Side of a Right Triangle

The height of a mountain is 5000 feet. The distance between its peak and that of an adjacent mountain is 25,000 feet. The angle of elevation between the two peaks is 27°. (See Figure 1.52.) What is the height of the adjacent mountain?

c = 25,000 ft; B; a; A; 27°; b; C; 5,000 ft; *Not drawn to scale*

Figure 1.52

Solution From the figure, $\sin A = a/c$, so

$$a = c \sin A = 25{,}000 \sin 27° \approx 11{,}350.$$

The height of the adjacent mountain is about $11{,}350 + 5000 = 16{,}350$ feet.

✓ ***Checkpoint*** *Audio-video solution in English & Spanish at LarsonPrecalculus.com*

A ladder that is 16 feet long leans against the side of a house. The angle of elevation of the ladder is 80°. Find the height from the top of the ladder to the ground.

István Csak | Dreamstime

EXAMPLE 3 Finding a Side of a Right Triangle

At a point 200 feet from the base of a building, the angle of elevation to the *bottom* of a smokestack is 35°, whereas the angle of elevation to the *top* is 53°, as shown in Figure 1.53. Find the height s of the smokestack alone.

Figure 1.53

Solution

This problem involves two right triangles. For the smaller right triangle, use the fact that

$$\tan 35° = \frac{a}{200}$$

to find that the height of the building is

$$a = 200 \tan 35°.$$

For the larger right triangle, use the equation

$$\tan 53° = \frac{a + s}{200}$$

to find that

$$a + s = 200 \tan 53°.$$

So, the height of the smokestack is

$$\begin{aligned} s &= 200 \tan 53° - a \\ &= 200 \tan 53° - 200 \tan 35° \\ &\approx 125.4 \text{ feet.} \end{aligned}$$

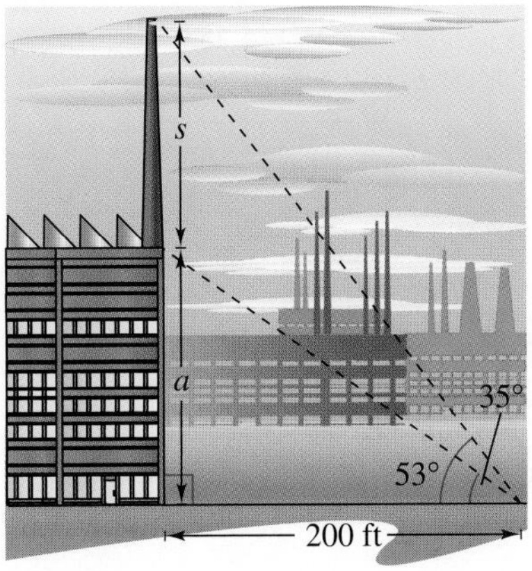

Checkpoint Audio-video solution in English & Spanish at LarsonPrecalculus.com

At a point 65 feet from the base of a church, the angles of elevation to the bottom of the steeple and the top of the steeple are 35° and 43°, respectively. Find the height of the steeple.

EXAMPLE 4 Finding an Angle of Depression

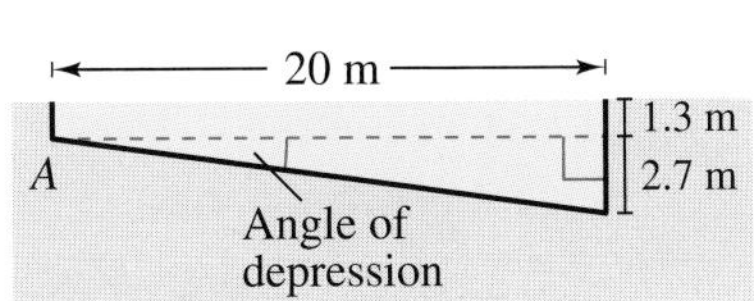

Figure 1.54

A swimming pool is 20 meters long and 12 meters wide. The bottom of the pool is slanted so that the water depth is 1.3 meters at the shallow end and 4 meters at the deep end, as shown in Figure 1.54. Find the angle of depression (in degrees) of the bottom of the pool.

Solution Using the tangent function,

$$\begin{aligned} \tan A &= \frac{\text{opp}}{\text{adj}} \\ &= \frac{2.7}{20} \\ &= 0.135. \end{aligned}$$

So, the angle of depression is

$$A = \arctan 0.135 \approx 0.13419 \text{ radian} \approx 7.69°.$$

Checkpoint Audio-video solution in English & Spanish at LarsonPrecalculus.com

From the time a small airplane is 100 feet high and 1600 ground feet from its landing runway, the plane descends in a straight line to the runway. Determine the angle of descent (in degrees) of the plane.

Trigonometry and Bearings

REMARK In *air navigation*, bearings are measured in degrees *clockwise* from north. The figures below illustrate examples of air navigation bearings

In surveying and navigation, directions can be given in terms of **bearings.** A bearing measures the acute angle that a path or line of sight makes with a fixed north-south line. For example, in the figures below, the bearing S 35° E means 35 degrees east of south, N 80° W means 80 degrees west of north, and N 45° E means 45 degrees east of north.

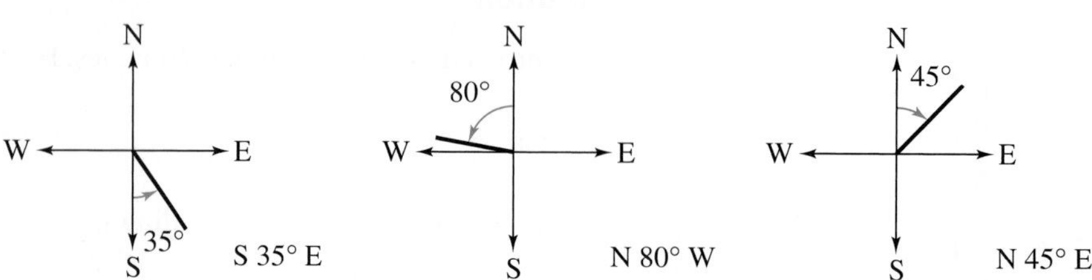

EXAMPLE 5 Finding Directions in Terms of Bearings

A ship leaves port at noon and heads due west at 20 knots, or 20 nautical miles (nmi) per hour. At 2 P.M. the ship changes course to N 54° W, as shown in the figure below. Find the ship's bearing and distance from port at 3 P.M.

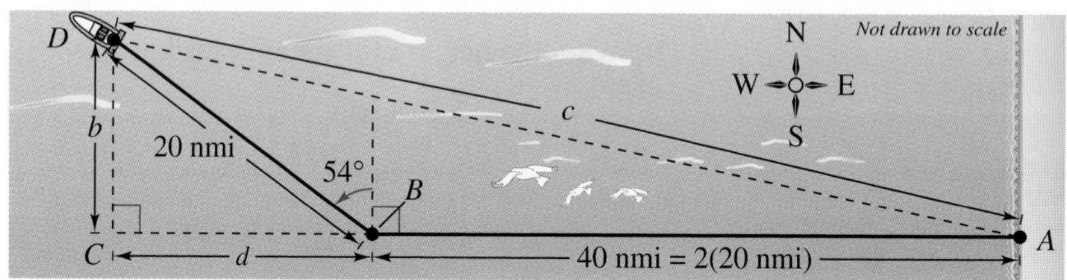

Solution

For triangle BCD, you have

$$B = 90° - 54° = 36°.$$

The two sides of this triangle are

$$b = 20 \sin 36° \quad \text{and} \quad d = 20 \cos 36°.$$

For triangle ACD, find angle A.

$$\tan A = \frac{b}{d + 40} = \frac{20 \sin 36°}{20 \cos 36° + 40} \approx 0.209$$

$$A \approx \arctan 0.209 \approx 0.20603 \text{ radian} \approx 11.80°$$

The angle with the north-south line is $90° - 11.80° = 78.20°$. So, the bearing of the ship is N 78.20° W. Finally, from triangle ACD, you have

$$\sin A = \frac{b}{c}$$

which yields

$$c = \frac{b}{\sin A} = \frac{20 \sin 36°}{\sin 11.80°} \approx 57.5 \text{ nautical miles.}$$ Distance from port

Checkpoint Audio-video solution in English & Spanish at LarsonPrecalculus.com

A sailboat leaves a pier heading due west at 8 knots. After 15 minutes, the sailboat changes course to N 16° W at 10 knots. Find the sailboat's bearing and distance from the pier after 12 minutes on this course.

Harmonic Motion

The periodic nature of the trigonometric functions is useful for describing the motion of a point on an object that vibrates, oscillates, rotates, or is moved by wave motion.

For example, consider a ball that is bobbing up and down on the end of a spring. Assume that the maximum distance the ball moves vertically upward or downward from its equilibrium (at rest) position is 10 centimeters (see figure). Assume further that the time it takes for the ball to move from its maximum displacement above zero to its maximum displacement below zero and back again is $t = 4$ seconds. With the ideal conditions of perfect elasticity and no friction or air resistance, the ball would continue to move up and down in a uniform and regular manner.

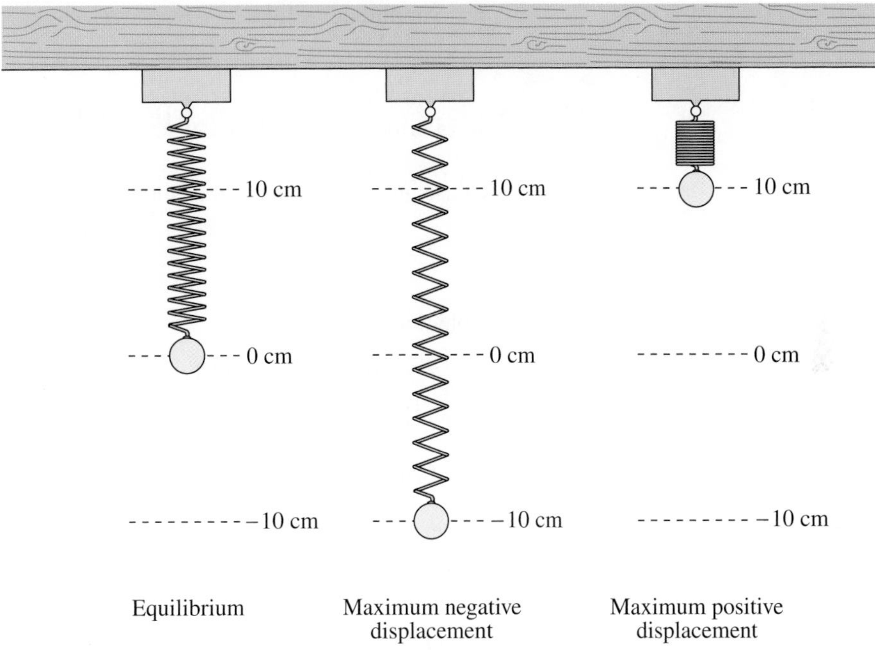

The period (time for one complete cycle) of the motion is

$$\text{Period} = 4 \text{ seconds}$$

the amplitude (maximum displacement from equilibrium) is

$$\text{Amplitude} = 10 \text{ centimeters}$$

and the **frequency** (number of cycles per second) is

$$\text{Frequency} = \frac{1}{4} \text{ cycle per second.}$$

Motion of this nature can be described by a sine or cosine function and is called **simple harmonic motion.**

Definition of Simple Harmonic Motion

A point that moves on a coordinate line is in **simple harmonic motion** when its distance d from the origin at time t is given by either

$$d = a \sin \omega t \quad \text{or} \quad d = a \cos \omega t$$

where a and ω are real numbers such that $\omega > 0$. The motion has amplitude $|a|$, period $\frac{2\pi}{\omega}$, and frequency $\frac{\omega}{2\pi}$.

EXAMPLE 6 Simple Harmonic Motion

Write an equation for the simple harmonic motion of the ball described on the preceding page.

Solution

The spring is at equilibrium ($d = 0$) when $t = 0$, so use the equation

$$d = a \sin \omega t.$$

Moreover, the maximum displacement from zero is 10 and the period is 4. Using this information, you have

$$\text{Amplitude} = |a|$$
$$= 10$$

$$\text{Period} = \frac{2\pi}{\omega} = 4 \quad \Rightarrow \quad \omega = \frac{\pi}{2}.$$

Consequently, an equation of motion is

$$d = 10 \sin \frac{\pi}{2} t.$$

Note that the choice of

$$a = 10 \quad \text{or} \quad a = -10$$

depends on whether the ball initially moves up or down.

 ***Checkpoint***  *Audio-video solution in English & Spanish at LarsonPrecalculus.com*

Write an equation for simple harmonic motion for which $d = 0$ when $t = 0$, the amplitude is 6 centimeters, and the period is 3 seconds.

One illustration of the relationship between sine waves and harmonic motion is in the wave motion that results when you drop a stone into a calm pool of water. The waves move outward in roughly the shape of sine (or cosine) waves, as shown at the right. Now suppose you are fishing in the same pool of water and your fishing bobber does not move horizontally. As the waves move outward from the dropped stone, the fishing bobber moves up and down in simple harmonic motion, as shown below.

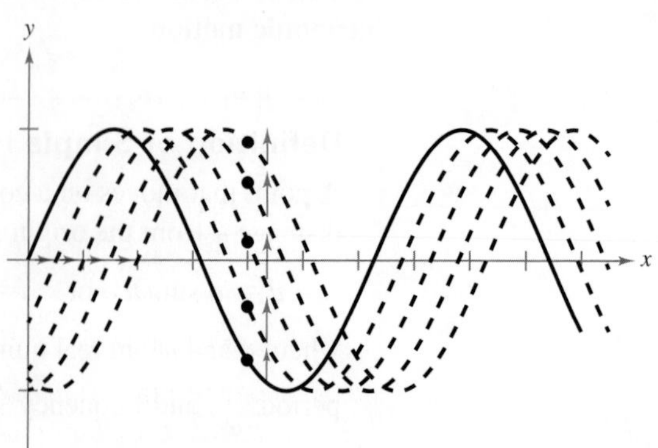

EXAMPLE 7 Simple Harmonic Motion

Consider the equation for simple harmonic motion $d = 6\cos\frac{3\pi}{4}t$. Find (a) the maximum displacement, (b) the frequency, (c) the value of d when $t = 4$, and (d) the least positive value of t for which $d = 0$.

Algebraic Solution

The equation has the form $d = a\cos\omega t$, with $a = 6$ and $\omega = 3\pi/4$.

a. The maximum displacement (from the point of equilibrium) is the amplitude. So, the maximum displacement is 6.

b.
$$\text{Frequency} = \frac{\omega}{2\pi} = \frac{3\pi/4}{2\pi} = \frac{3}{8} \text{ cycle per unit of time}$$

c.
$$d = 6\cos\left[\frac{3\pi}{4}(4)\right] = 6\cos 3\pi = 6(-1) = -6$$

d. To find the least positive value of t for which $d = 0$, solve

$$6\cos\frac{3\pi}{4}t = 0.$$

First divide each side by 6 to obtain

$$\cos\frac{3\pi}{4}t = 0.$$

This equation is satisfied when

$$\frac{3\pi}{4}t = \frac{\pi}{2}, \frac{3\pi}{2}, \frac{5\pi}{2}, \ldots .$$

Multiply these values by $4/(3\pi)$ to obtain

$$t = \frac{2}{3}, 2, \frac{10}{3}, \ldots .$$

So, the least positive value of t is $t = \frac{2}{3}$.

Graphical Solution

Use a graphing utility set in *radian* mode.

a.

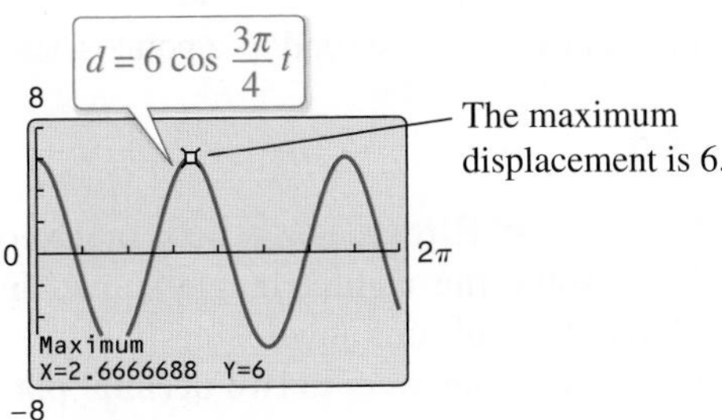

b.

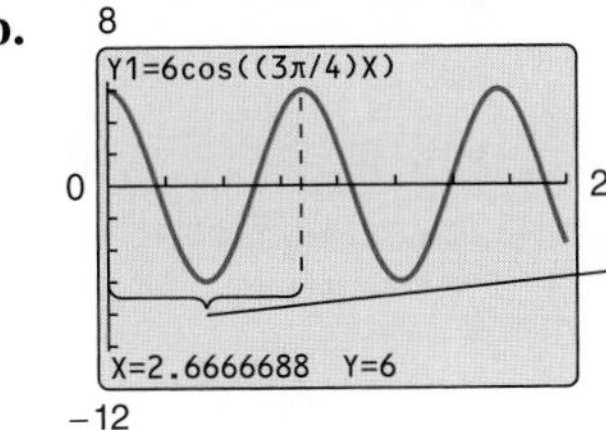

The period is about 2.67. So, the frequency is about $1/2.67 \approx 0.37$ cycle per unit of time.

c.

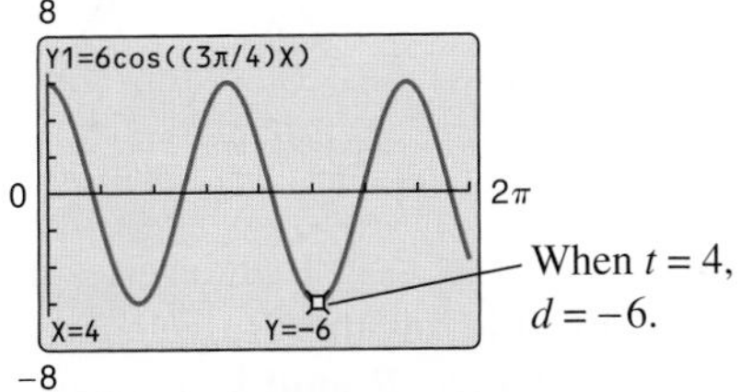

d.

The least positive value of t for which $d = 0$ is $t \approx 0.67$.

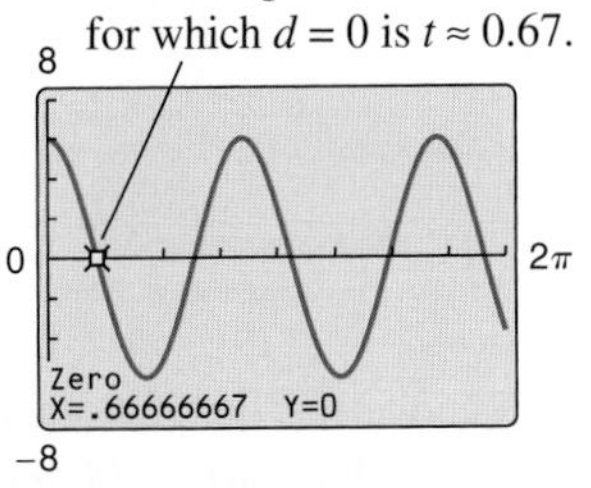

✓ **Checkpoint**  *Audio-video solution in English & Spanish at LarsonPrecalculus.com*

Rework Example 7 for the equation $d = 4\cos 6\pi t$. 

Summarize (Section 1.8)

1. Describe real-life applications of right triangles *(pages 190 and 191, Examples 1–4)*.
2. Describe a real-life application of a directional bearing *(page 192, Example 5)*.
3. Describe real-life applications of simple harmonic motion *(pages 194 and 195, Examples 6 and 7)*.

1.8 Exercises

See CalcChat.com for tutorial help and worked-out solutions to odd-numbered exercises.

Vocabulary: Fill in the blanks.

1. A ________ measures the acute angle that a path or line of sight makes with a fixed north-south line.
2. A point that moves on a coordinate line is in simple ________ ________ when its distance d from the origin at time t is given by either $d = a \sin \omega t$ or $d = a \cos \omega t$.
3. The time for one complete cycle of a point in simple harmonic motion is its ________.
4. The number of cycles per second of a point in simple harmonic motion is its ________.

Skills and Applications

Solving a Right Triangle **In Exercises 5–12, solve the right triangle shown in the figure for all unknown sides and angles. Round your answers to two decimal places.**

5. $A = 60°, \quad c = 12$
6. $B = 25°, \quad b = 4$
7. $B = 72.8°, \quad a = 4.4$
8. $A = 8.4°, \quad a = 40.5$
9. $a = 3, \quad b = 4$
10. $a = 25, \quad c = 35$
11. $b = 15.70, \quad c = 55.16$
12. $b = 1.32, \quad c = 9.45$

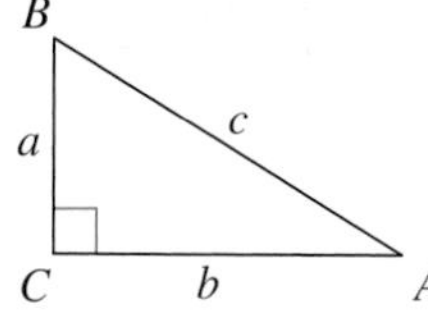

Figure for 5–12

Figure for 13–16

Finding an Altitude **In Exercises 13–16, find the altitude of the isosceles triangle shown in the figure. Round your answers to two decimal places.**

13. $\theta = 45°, \quad b = 6$
14. $\theta = 22°, \quad b = 14$
15. $\theta = 32°, \quad b = 8$
16. $\theta = 27°, \quad b = 11$

17. **Length** The sun is 25° above the horizon. Find the length of a shadow cast by a building that is 100 feet tall (see figure).

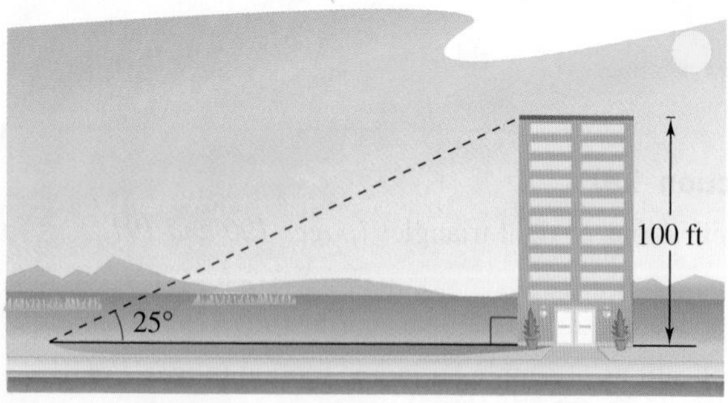

18. **Length** The sun is 20° above the horizon. Find the length of a shadow cast by a park statue that is 12 feet tall.
19. **Height** A ladder that is 20 feet long leans against the side of a house. The angle of elevation of the ladder is 80°. Find the height from the top of the ladder to the ground.
20. **Height** The length of a shadow of a tree is 125 feet when the angle of elevation of the sun is 33°. Approximate the height of the tree.
21. **Height** At a point 50 feet from the base of a church, the angles of elevation to the bottom of the steeple and the top of the steeple are 35° and 48°, respectively. Find the height of the steeple.
22. **Distance** An observer in a lighthouse 350 feet above sea level observes two ships directly offshore. The angles of depression to the ships are 4° and 6.5° (see figure). How far apart are the ships?

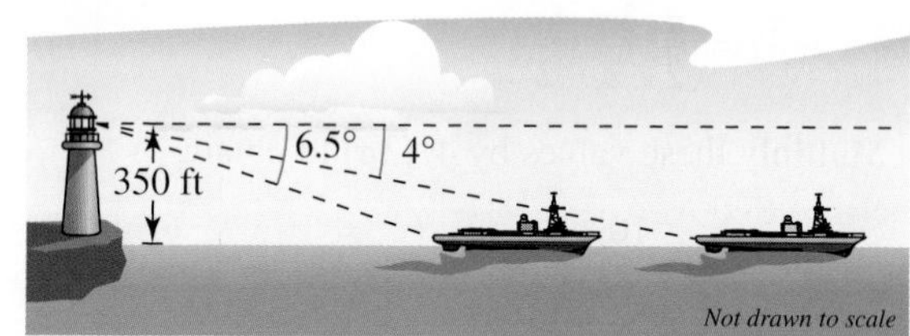

23. **Distance** A passenger in an airplane at an altitude of 10 kilometers sees two towns directly to the east of the plane. The angles of depression to the towns are 28° and 55° (see figure). How far apart are the towns?

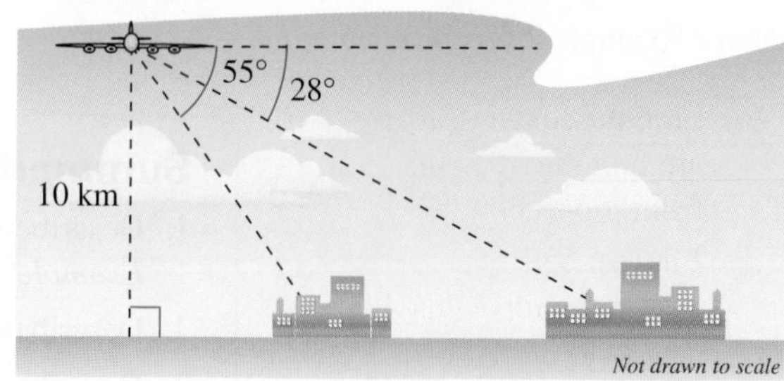

24. Angle of Elevation The height of an outdoor basketball backboard is $12\frac{1}{2}$ feet, and the backboard casts a shadow 17 feet long.

(a) Draw a right triangle that gives a visual representation of the problem. Label the known and unknown quantities.

(b) Use a trigonometric function to write an equation involving the unknown angle of elevation.

(c) Find the angle of elevation.

25. Angle of Elevation An engineer designs a 75-foot cellular telephone tower. Find the angle of elevation to the top of the tower at a point on level ground 50 feet from its base.

26. Angle of Depression A cellular telephone tower that is 120 feet tall is placed on top of a mountain that is 1200 feet above sea level. What is the angle of depression from the top of the tower to a cell phone user who is 5 horizontal miles away and 400 feet above sea level?

27. Angle of Depression A Global Positioning System satellite orbits 12,500 miles above Earth's surface (see figure). Find the angle of depression from the satellite to the horizon. Assume the radius of Earth is 4000 miles.

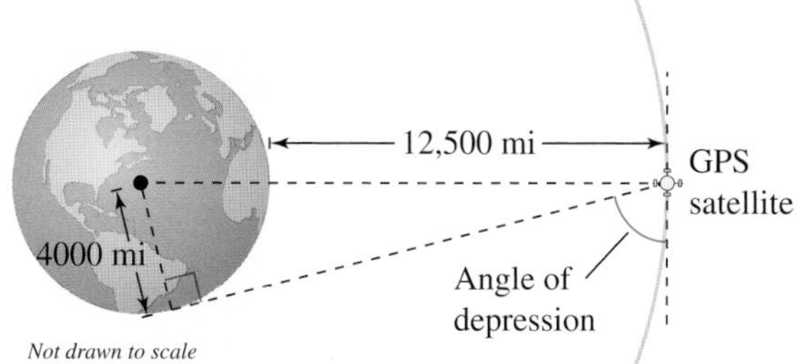

28. Height You are holding one of the tethers attached to the top of a giant character balloon that is floating approximately 20 feet above ground level. You are standing approximately 100 feet ahead of the balloon (see figure).

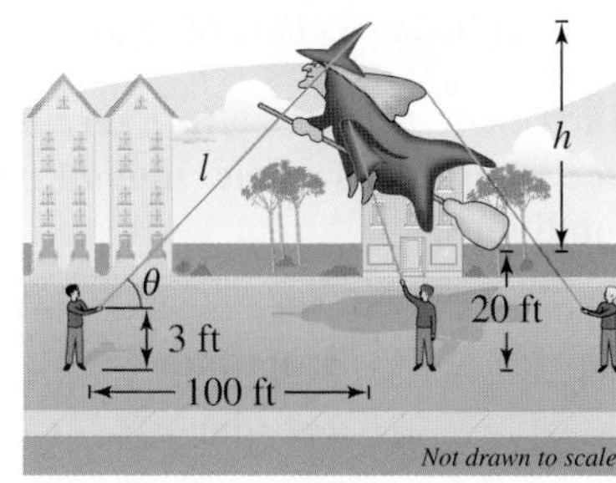

(a) Find an equation for the length l of the tether you are holding in terms of h, the height of the balloon from top to bottom.

(b) Find an equation for the angle of elevation θ from you to the top of the balloon.

(c) The angle of elevation to the top of the balloon is 35°. Find the height h of the balloon.

29. Altitude You observe a plane approaching overhead and assume that its speed is 550 miles per hour. The angle of elevation of the plane is 16° at one time and 57° one minute later. Approximate the altitude of the plane.

30. Waterslide Design

The designers of a water park have sketched a preliminary drawing of a new slide (see figure).

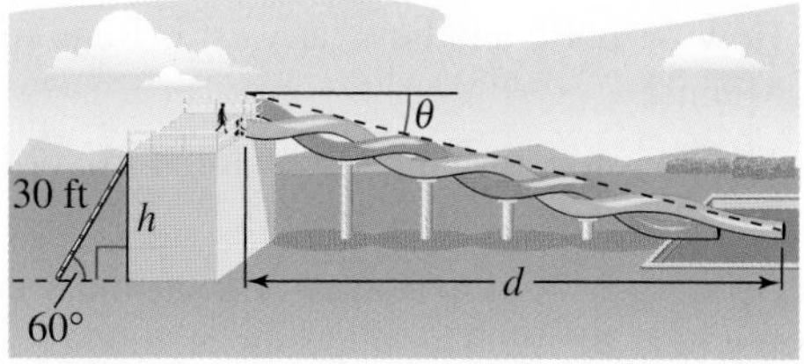

(a) Find the height h of the slide.

(b) Find the angle of depression θ from the top of the slide to the end of the slide at the ground in terms of the horizontal distance d a rider travels.

(c) Safety restrictions require the angle of depression to be no less than 25° and no more than 30°. Find an interval for how far a rider travels horizontally.

31. Speed Enforcement A police department has set up a speed enforcement zone on a straight length of highway. A patrol car is parked parallel to the zone, 200 feet from one end and 150 feet from the other end (see figure).

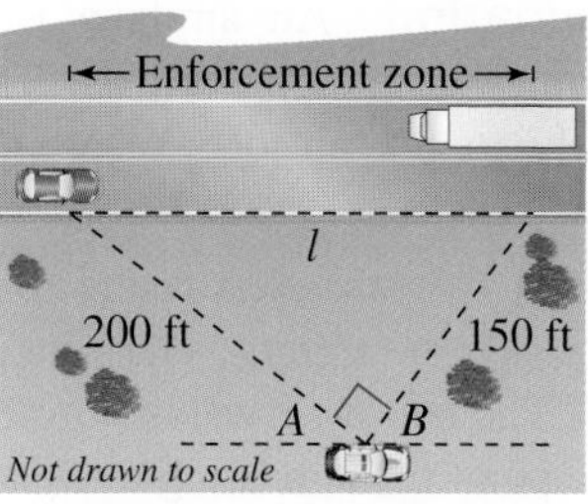

(a) Find the length l of the zone and the measures of angles A and B (in degrees).

(b) Find the minimum amount of time (in seconds) it takes for a vehicle to pass through the zone without exceeding the posted speed limit of 35 miles per hour.

32. Airplane Ascent During takeoff, an airplane's angle of ascent is 18° and its speed is 260 feet per second.

(a) Find the plane's altitude after 1 minute.

(b) How long will it take for the plane to climb to an altitude of 10,000 feet?

33. Air Navigation An airplane flying at 550 miles per hour has a bearing of 52°. After flying for 1.5 hours, how far north and how far east will the plane have traveled from its point of departure?

34. Air Navigation A jet leaves Reno, Nevada, and heads toward Miami, Florida, at a bearing of 100°. The distance between the two cities is approximately 2472 miles.

(a) How far north and how far west is Reno relative to Miami?

(b) The jet is to return directly to Reno from Miami. At what bearing should it travel?

35. Navigation A ship leaves port at noon and has a bearing of S 29° W. The ship sails at 20 knots.

(a) How many nautical miles south and how many nautical miles west will the ship have traveled by 6:00 P.M.?

(b) At 6:00 P.M., the ship changes course to due west. Find the ship's bearing and distance from port at 7:00 P.M.

36. Navigation A privately owned yacht leaves a dock in Myrtle Beach, South Carolina, and heads toward Freeport in the Bahamas at a bearing of S 1.4° E. The yacht averages a speed of 20 knots over the 428-nautical-mile trip.

(a) How long will it take the yacht to make the trip?

(b) How far east and south is the yacht after 12 hours?

(c) A plane leaves Myrtle Beach to fly to Freeport. At what bearing should it travel?

37. Navigation A ship is 45 miles east and 30 miles south of port. The captain wants to sail directly to port. What bearing should the captain take?

38. Air Navigation An airplane is 160 miles north and 85 miles east of an airport. The pilot wants to fly directly to the airport. What bearing should the pilot take?

39. Surveying A surveyor wants to find the distance across a pond (see figure). The bearing from A to B is N 32° W. The surveyor walks 50 meters from A to C, and at the point C the bearing to B is N 68° W.

(a) Find the bearing from A to C.

(b) Find the distance from A to B.

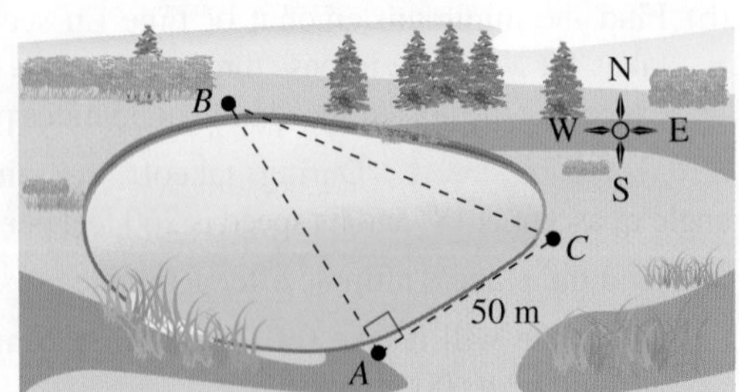

40. Location of a Fire Fire tower A is 30 kilometers due west of fire tower B. A fire is spotted from the towers, and the bearings from A and B are N 76° E and N 56° W, respectively (see figure). Find the distance d of the fire from the line segment AB.

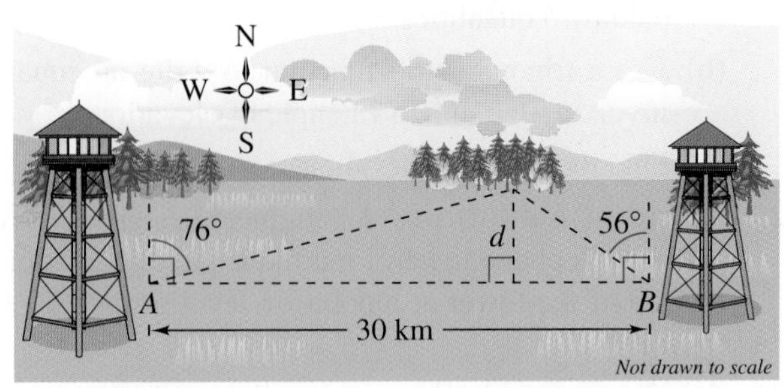

41. Geometry Determine the angle between the diagonal of a cube and the diagonal of its base, as shown in the figure.

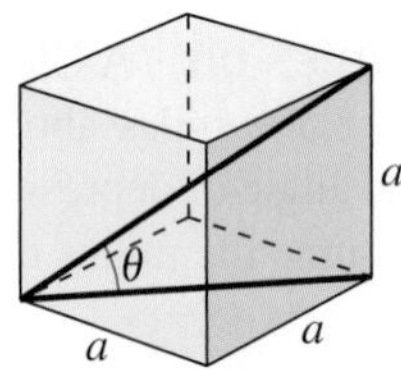

42. Geometry Determine the angle between the diagonal of a cube and its edge, as shown in the figure.

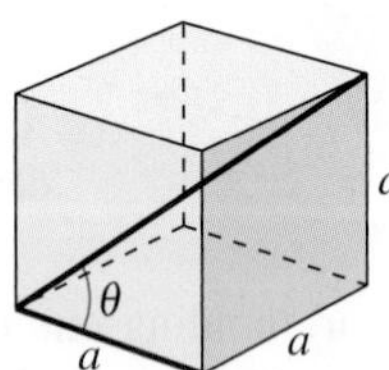

43. Geometry Find the length of the sides of a regular pentagon inscribed in a circle of radius 25 inches.

44. Geometry Find the length of the sides of a regular hexagon inscribed in a circle of radius 25 inches.

Simple Harmonic Motion In Exercises 45–48, find a model for simple harmonic motion satisfying the specified conditions.

	Displacement ($t = 0$)	Amplitude	Period
45.	0	4 centimeters	2 seconds
46.	0	3 meters	6 seconds
47.	3 inches	3 inches	1.5 seconds
48.	2 feet	2 feet	10 seconds

49. Tuning Fork A point on the end of a tuning fork moves in simple harmonic motion described by $d = a \sin \omega t$. Find ω given that the tuning fork for middle C has a frequency of 262 vibrations per second.

50. Wave Motion A buoy oscillates in simple harmonic motion as waves go past. The buoy moves a total of 3.5 feet from its low point to its high point (see figure), and it returns to its high point every 10 seconds. Write an equation that describes the motion of the buoy where the high point corresponds to the time $t = 0$.

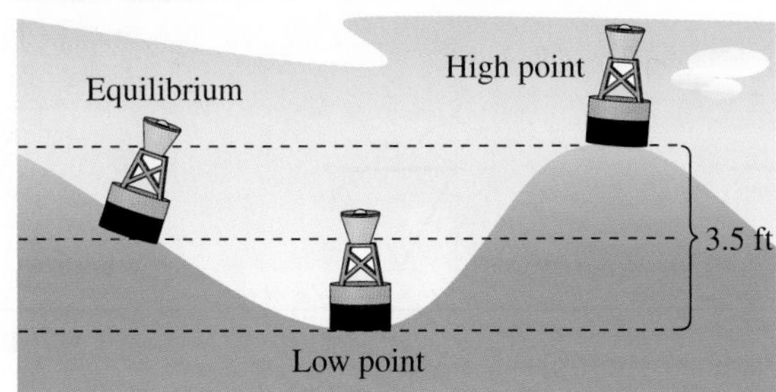

Simple Harmonic Motion In Exercises 51–54, for the simple harmonic motion described by the trigonometric function, find (a) the maximum displacement, (b) the frequency, (c) the value of d when $t = 5$, and (d) the least positive value of t for which $d = 0$. Use a graphing utility to verify your results.

51. $d = 9\cos\frac{6\pi}{5}t$ **52.** $d = \frac{1}{2}\cos 20\pi t$

53. $d = \frac{1}{4}\sin 6\pi t$ **54.** $d = \frac{1}{64}\sin 792\pi t$

55. Oscillation of a Spring A ball that is bobbing up and down on the end of a spring has a maximum displacement of 3 inches. Its motion (in ideal conditions) is modeled by $y = \frac{1}{4}\cos 16t$, $t > 0$, where y is measured in feet and t is the time in seconds.

(a) Graph the function.

(b) What is the period of the oscillations?

(c) Determine the first time the weight passes the point of equilibrium $(y = 0)$.

56. Hours of Daylight The numbers of hours H of daylight in Denver, Colorado, on the 15th of each month starting with January are: 9.68, 10.72, 11.92, 13.25, 14.35, 14.97, 14.72, 13.73, 12.47, 11.18, 10.00, and 9.37. A model for the data is

$$H(t) = 12.13 + 2.77\sin\left(\frac{\pi t}{6} - 1.60\right)$$

where t represents the month, with $t = 1$ corresponding to January. *(Source: United States Navy)*

(a) Use a graphing utility to graph the data and the model in the same viewing window.

(b) What is the period of the model? Is it what you expected? Explain.

(c) What is the amplitude of the model? What does it represent in the context of the problem?

57. Sales The table shows the average sales S (in millions of dollars) of an outerwear manufacturer for each month t, where $t = 1$ corresponds to January.

Time, t	1	2	3	4
Sales, S	13.46	11.15	8.00	4.85

Time, t	5	6	7	8
Sales, S	2.54	1.70	2.54	4.85

Time, t	9	10	11	12
Sales, S	8.00	11.15	13.46	14.30

(a) Create a scatter plot of the data.

(b) Find a trigonometric model that fits the data. Graph the model with your scatter plot. How well does the model fit the data?

(c) What is the period of the model? Do you think it is reasonable given the context? Explain.

(d) Interpret the meaning of the model's amplitude in the context of the problem.

Exploration

58. HOW DO YOU SEE IT? The graph below shows the displacement of an object in simple harmonic motion.

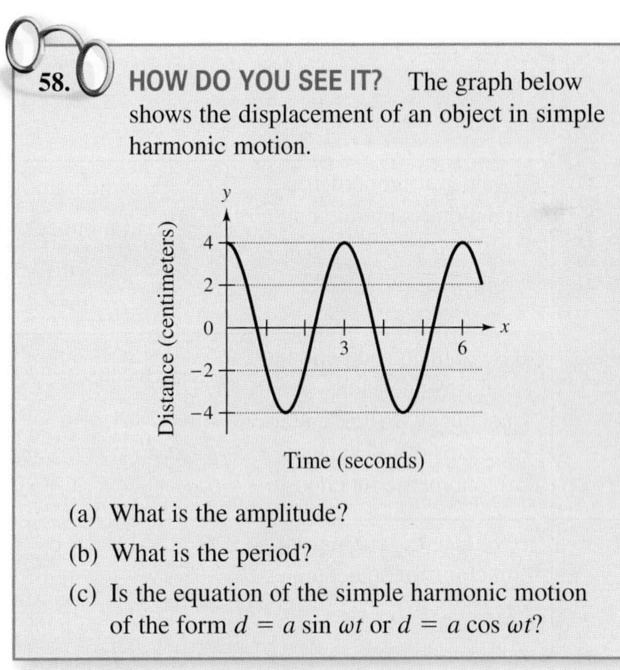

(a) What is the amplitude?

(b) What is the period?

(c) Is the equation of the simple harmonic motion of the form $d = a\sin\omega t$ or $d = a\cos\omega t$?

True or False? In Exercises 59 and 60, determine whether the statement is true or false. Justify your answer.

59. The Leaning Tower of Pisa is not vertical, but when you know the angle of elevation θ to the top of the tower as you stand d feet away from it, its height h can be found using the formula $h = d\tan\theta$.

60. The bearing N 24° E means 24 degrees north of east.

Chapter Summary

	What Did You Learn?	Explanation/Examples	Review Exercises
Section 1.1	Describe angles *(p. 122)*.	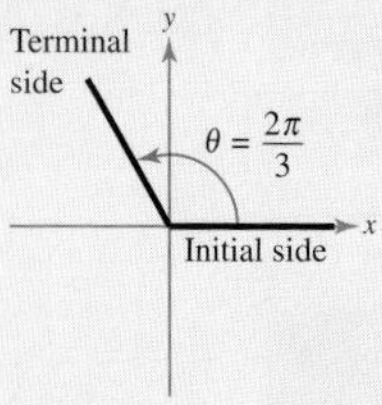 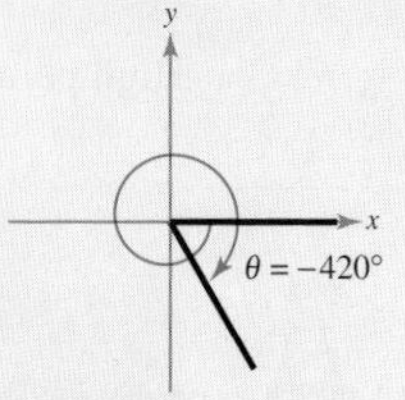	1–4
	Use radian measure *(p. 123)* and degree measure *(p. 125)*.	To convert degrees to radians, multiply degrees by $\frac{\pi \text{ rad}}{180°}$. To convert radians to degrees, multiply radians by $\frac{180°}{\pi \text{ rad}}$.	5–14
	Use angles and their measure to model and solve real-life problems *(p. 126)*.	Angles and their measure can be used to find arc length and the area of a sector of a circle. (See Examples 5 and 8.)	15–18
Section 1.2	Identify a unit circle and describe its relationship to real numbers *(p. 132)*.	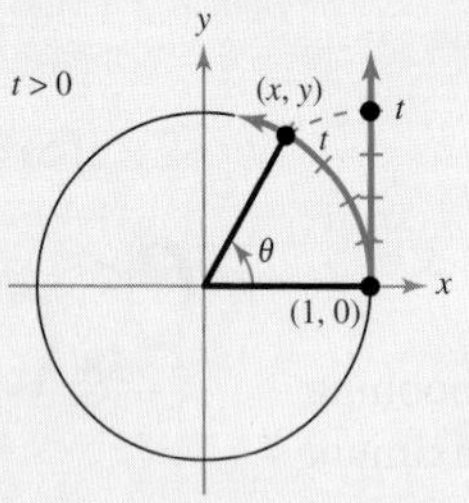 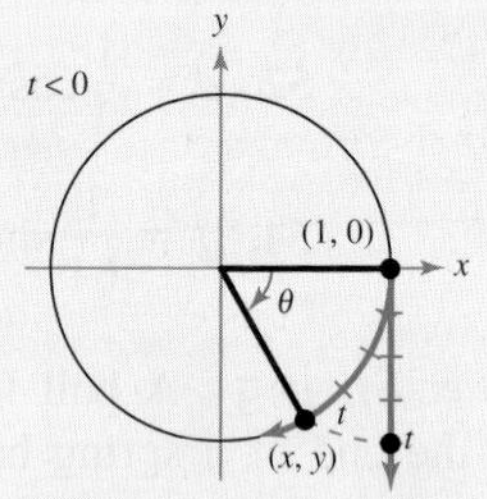	19–22
	Evaluate trigonometric functions using the unit circle *(p. 133)*.	For the point (x, y) on the unit circle corresponding to a real number t: $\sin t = y$; $\cos t = x$; $\tan t = \frac{y}{x}, x \neq 0$; $\csc t = \frac{1}{y}, y \neq 0$; $\sec t = \frac{1}{x}, x \neq 0$; and $\cot t = \frac{x}{y}, y \neq 0$.	23, 24
	Use domain and period to evaluate sine and cosine functions *(p. 135)*, and use a calculator to evaluate trigonometric functions *(p. 136)*.	Because $\frac{13\pi}{6} = 2\pi + \frac{\pi}{6}$, $\sin \frac{13\pi}{6} = \sin \frac{\pi}{6} = \frac{1}{2}$. $\sin \frac{3\pi}{8} \approx 0.9239$, $\cot(-1.2) \approx -0.3888$	25–32
Section 1.3	Evaluate trigonometric functions of acute angles *(p. 139)*.	$\sin \theta = \frac{\text{opp}}{\text{hyp}}$, $\cos \theta = \frac{\text{adj}}{\text{hyp}}$, $\tan \theta = \frac{\text{opp}}{\text{adj}}$ $\csc \theta = \frac{\text{hyp}}{\text{opp}}$, $\sec \theta = \frac{\text{hyp}}{\text{adj}}$, $\cot \theta = \frac{\text{adj}}{\text{opp}}$ $\csc 29° 15' = 1/\sin 29.25° \approx 2.0466$	33–38
	Use fundamental trigonometric identities *(p. 142)*.	$\sin \theta = \frac{1}{\csc \theta}$, $\tan \theta = \frac{\sin \theta}{\cos \theta}$, $\sin^2 \theta + \cos^2 \theta = 1$	39, 40
	Use trigonometric functions to model and solve real-life problems *(p. 144)*.	Trigonometric functions can be used to find the height of a monument, the angle between two paths, and the length and height of a ramp. (See Examples 8–10.)	41, 42

	What Did You Learn?	Explanation/Examples	Review Exercises
Section 1.4	Evaluate trigonometric functions of any angle *(p. 150)*.	Let (3, 4) be a point on the terminal side of θ. Then $\sin\theta = \frac{4}{5}$, $\cos\theta = \frac{3}{5}$, and $\tan\theta = \frac{4}{3}$.	43–50
	Find reference angles *(p. 152)*.	Let θ be an angle in standard position. Its reference angle is the acute angle θ' formed by the terminal side of θ and the horizontal axis.	51–54
	Evaluate trigonometric functions of real numbers *(p. 153)*.	$\cos\frac{7\pi}{3} = \frac{1}{2}$ because $\theta' = \frac{7\pi}{3} - 2\pi = \frac{\pi}{3}$ and $\cos\frac{\pi}{3} = \frac{1}{2}$.	55–62
Section 1.5	Sketch the graphs of sine and cosine functions using amplitude and period *(p. 161)*.	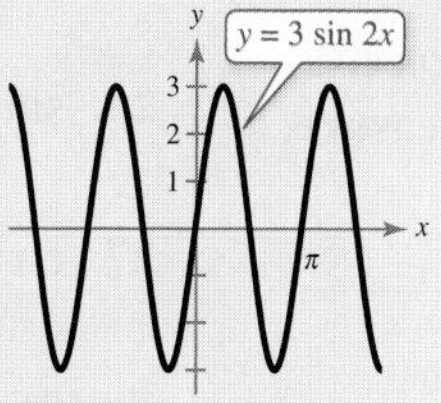 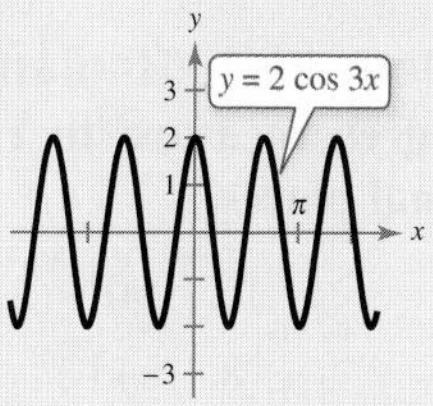 $y = 3\sin 2x$; $y = 2\cos 3x$	63, 64
	Sketch translations of the graphs of sine and cosine functions *(p. 163)*.	For $y = d + a\sin(bx - c)$ and $y = d + a\cos(bx - c)$, the constant c results in horizontal translations and the constant d results in vertical translations. (See Examples 4–6.)	65–68
	Use sine and cosine functions to model real-life data *(p. 165)*.	A cosine function can be used to model the depth of the water at the end of a dock. (See Example 7.)	69, 70
Section 1.6	Sketch the graphs of tangent *(p. 170)*, cotangent *(p. 172)*, secant *(p. 173)*, and cosecant functions *(p. 173)*.	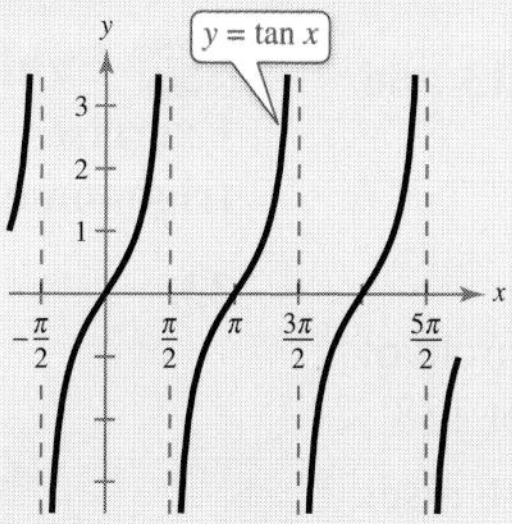 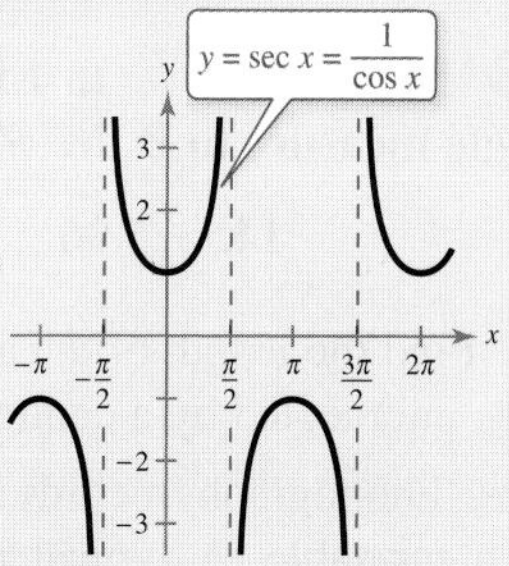 $y = \tan x$; $y = \sec x = \frac{1}{\cos x}$	71–74
	Sketch the graphs of damped trigonometric functions *(p. 175)*.	In $f(x) = x\cos 2x$, the factor x is called the damping factor.	75, 76
Section 1.7	Evaluate and graph inverse trigonometric functions *(p. 180)*.	$\arcsin\left(\frac{1}{2}\right) = \frac{\pi}{6}$, $\cos^{-1}\left(-\frac{\sqrt{2}}{2}\right) = \frac{3\pi}{4}$, $\tan^{-1}\sqrt{3} = \frac{\pi}{3}$	77–86
	Evaluate compositions with inverse trigonometric functions *(p. 184)*.	$\sin(\sin^{-1} 0.4) = 0.4$, $\cos\left(\arctan\frac{5}{12}\right) = \frac{12}{13}$	87–92
Section 1.8	Solve real-life problems involving right triangles *(p. 190)*.	A trigonometric function can be used to find the height of a smokestack on top of a building. (See Example 3.)	93, 94
	Solve real-life problems involving directional bearings *(p. 192)*.	Trigonometric functions can be used to find a ship's bearing and distance from a port at a given time. (See Example 5.)	95
	Solve real-life problems involving harmonic motion *(p. 193)*.	Trigonometric functions can be used to describe the motion of a point on an object that vibrates, oscillates, rotates, or is moved by wave motion. (See Examples 6 and 7.)	96

Review Exercises

See CalcChat.com for tutorial help and worked-out solutions to odd-numbered exercises.

1.1 Using Radian or Degree Measure **In Exercises 1–4, (a) sketch the angle in standard position, (b) determine the quadrant in which the angle lies, and (c) determine two coterminal angles (one positive and one negative).**

1. $\frac{15\pi}{4}$
2. $-\frac{4\pi}{3}$
3. $-110°$
4. $280°$

Converting from Degrees to Radians **In Exercises 5–8, convert the degree measure to radian measure. Round to three decimal places.**

5. $450°$
6. $190°$
7. $-16°$
8. $-112°$

Converting from Radians to Degrees **In Exercises 9–12, convert the radian measure to degree measure. Round to three decimal places, if necessary.**

9. $\frac{3\pi}{10}$
10. $-\frac{11\pi}{6}$
11. -3.5
12. 5.7

Converting to D° M′ S″ Form **In Exercises 13 and 14, convert the angle measure to D° M′ S″ form.**

13. $198.4°$
14. $-5.96°$

15. **Arc Length** Find the length of the arc on a circle of radius 20 inches intercepted by a central angle of 138°.
16. **Phonograph** Phonograph records are vinyl discs that rotate on a turntable. A typical record album is 12 inches in diameter and plays at $33\frac{1}{3}$ revolutions per minute.
    (a) Find the angular speed of a record album.
    (b) Find the linear speed (in inches per minute) of the outer edge of a record album.

Area of a Sector of a Circle **In Exercises 17 and 18, find the area of the sector of a circle of radius *r* and central angle *θ*.**

	Radius r	Central Angle θ
17.	20 inches	150°
18.	7.5 millimeters	$2\pi/3$ radians

1.2 Finding a Point on the Unit Circle **In Exercises 19–22, find the point (x, y) on the unit circle that corresponds to the real number *t*.**

19. $t = 2\pi/3$
20. $t = 7\pi/4$
21. $t = 7\pi/6$
22. $t = -4\pi/3$

Evaluating Trigonometric Functions **In Exercises 23 and 24, evaluate (if possible) the six trigonometric functions at the real number.**

23. $t = \frac{3\pi}{4}$
24. $t = -\frac{2\pi}{3}$

Using Period to Evaluate Sine and Cosine **In Exercises 25–28, evaluate the trigonometric function using its period as an aid.**

25. $\sin \frac{11\pi}{4}$
26. $\cos 4\pi$
27. $\cos\left(-\frac{17\pi}{6}\right)$
28. $\sin\left(-\frac{13\pi}{3}\right)$

Using a Calculator **In Exercises 29–32, use a calculator to evaluate the trigonometric function. Round your answer to four decimal places. (Be sure the calculator is in the correct mode.)**

29. $\sec \frac{12\pi}{5}$
30. $\sin\left(-\frac{\pi}{9}\right)$
31. $\tan 33$
32. $\csc 10.5$

1.3 Evaluating Trigonometric Functions **In Exercises 33 and 34, find the exact values of the six trigonometric functions of the angle *θ*.**

33.

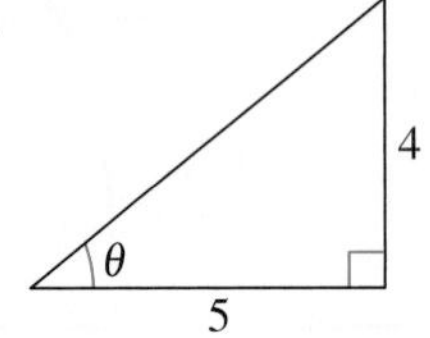

34.

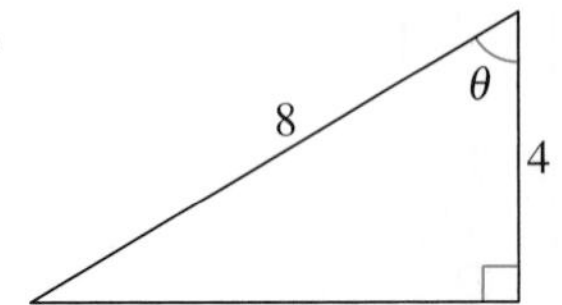

Using a Calculator **In Exercises 35–38, use a calculator to evaluate the trigonometric function. Round your answer to four decimal places. (Be sure the calculator is in the correct mode.)**

35. $\tan 33°$
36. $\sec 79.3°$
37. $\cot 15° 14'$
38. $\cos 78° 11' 58''$

Applying Trigonometric Identities **In Exercises 39 and 40, use the given function value and the trigonometric identities to find the exact value of each indicated trigonometric function.**

39. $\sin \theta = \frac{1}{3}$
    (a) $\csc \theta$ (b) $\cos \theta$
    (c) $\sec \theta$ (d) $\tan \theta$
40. $\csc \theta = 5$
    (a) $\sin \theta$ (b) $\cot \theta$
    (c) $\tan \theta$ (d) $\sec(90° - \theta)$

41. **Railroad Grade** A train travels 3.5 kilometers on a straight track with a grade of 1.2° (see figure). What is the vertical rise of the train in that distance?

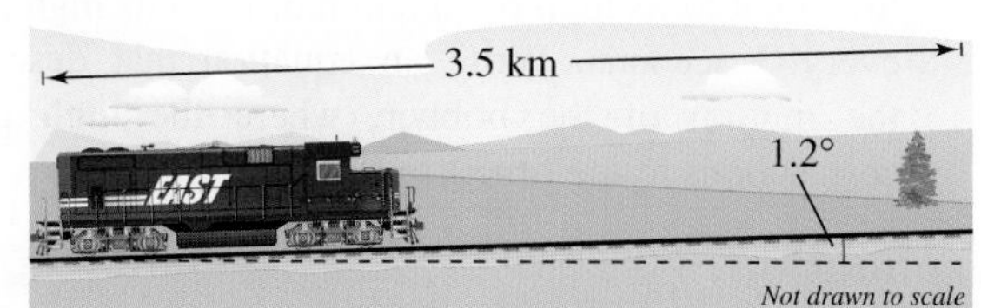

42. **Guy Wire** A guy wire runs from the ground to the top of a 25-foot telephone pole. The angle formed between the wire and the ground is 52°. How far from the base of the pole is the guy wire anchored to the ground? Assume the pole is perpendicular to the ground.

1.4 Evaluating Trigonometric Functions **In Exercises 43–46, the point is on the terminal side of an angle in standard position. Find the exact values of the six trigonometric functions of the angle.**

43. $(12, 16)$ **44.** $(3, -4)$

45. $(0.3, 0.4)$ **46.** $\left(-\frac{10}{3}, -\frac{2}{3}\right)$

Evaluating Trigonometric Functions **In Exercises 47–50, find the exact values of the remaining five trigonometric functions of θ satisfying the given conditions.**

47. $\sec\theta = \frac{6}{5}$, $\tan\theta < 0$

48. $\csc\theta = \frac{3}{2}$, $\cos\theta < 0$

49. $\cos\theta = -\frac{2}{5}$, $\sin\theta > 0$

50. $\sin\theta = -\frac{1}{2}$, $\cos\theta > 0$

Finding a Reference Angle **In Exercises 51–54, find the reference angle θ'. Sketch θ in standard position and label θ'.**

51. $\theta = 264°$ **52.** $\theta = 635°$

53. $\theta = -6\pi/5$ **54.** $\theta = 17\pi/3$

Using a Reference Angle **In Exercises 55–58, evaluate the sine, cosine, and tangent of the angle without using a calculator.**

55. $-150°$ **56.** $495°$

57. $\pi/3$ **58.** $-5\pi/4$

Using a Calculator **In Exercises 59–62, use a calculator to evaluate the trigonometric function. Round your answer to four decimal places. (Be sure the calculator is in the correct mode.)**

59. $\sin 106°$

60. $\tan 37°$

61. $\tan(-17\pi/15)$

62. $\cos(-25\pi/7)$

1.5 Sketching the Graph of a Sine or Cosine Function **In Exercises 63–68, sketch the graph of the function. (Include two full periods.)**

63. $y = \sin 6x$

64. $f(x) = -\cos 3x$

65. $y = 5 + \sin \pi x$

66. $y = -4 - \cos \pi x$

67. $g(t) = \frac{5}{2}\sin(t - \pi)$

68. $g(t) = 3\cos(t + \pi)$

69. **Sound Waves** Sound waves can be modeled using sine functions of the form $y = a \sin bx$, where x is measured in seconds.

(a) Write an equation of a sound wave whose amplitude is 2 and whose period is $\frac{1}{264}$ second.

(b) What is the frequency of the sound wave described in part (a)?

70. **Meteorology** The times S of sunset (Greenwich Mean Time) at 40° north latitude on the 15th of each month starting with January are: 16:59, 17:35, 18:06, 18:38, 19:08, 19:30, 19:28, 18:57, 18:10, 17:21, 16:44, and 16:36. A model (in which minutes have been converted to the decimal parts of an hour) for the data is

$$S(t) = 18.10 - 1.41\sin\left(\frac{\pi t}{6} + 1.55\right)$$

where t represents the month, with $t = 1$ corresponding to January. *(Source: NOAA)*

(a) Use a graphing utility to graph the data and the model in the same viewing window.

(b) What is the period of the model? Is it what you expected? Explain.

(c) What is the amplitude of the model? What does it represent in the context of the problem?

1.6 Sketching the Graph of a Trigonometric Function **In Exercises 71–74, sketch the graph of the function. (Include two full periods.)**

71. $f(t) = \tan\left(t + \frac{\pi}{2}\right)$ **72.** $f(x) = \frac{1}{2}\cot x$

73. $f(x) = \frac{1}{2}\csc\frac{x}{2}$ **74.** $h(t) = \sec\left(t - \frac{\pi}{4}\right)$

Analyzing a Damped Trigonometric Graph **In Exercises 75 and 76, use a graphing utility to graph the function and the damping factor of the function in the same viewing window. Describe the behavior of the function as x increases without bound.**

75. $f(x) = x\cos x$

76. $g(x) = x^4\cos x$

1.7 Evaluating an Inverse Trigonometric Function In Exercises 77–80, find the exact value of the expression.

77. $\arcsin(-1)$

78. $\cos^{-1} 1$

79. $\text{arccot}\ \sqrt{3}$

80. $\text{arcsec}(-\sqrt{2})$

Calculators and Inverse Trigonometric Functions In Exercises 81–84, use a calculator to approximate the value of the expression, if possible. Round your result to two decimal places.

81. $\tan^{-1}(-1.3)$

82. $\arccos 0.372$

83. $\text{arccot}\ 15.5$

84. $\text{arccsc}(-4.03)$

Graphing an Inverse Trigonometric Function In Exercises 85 and 86, use a graphing utility to graph the function.

85. $f(x) = \arctan(x/2)$

86. $f(x) = -\arcsin 2x$

Evaluating a Composition of Functions In Exercises 87–90, find the exact value of the expression.

87. $\cos\left(\arctan \frac{3}{4}\right)$

88. $\tan\left(\arccos \frac{3}{5}\right)$

89. $\sec\left(\arctan \frac{12}{5}\right)$

90. $\cot\left[\arcsin\left(-\frac{12}{13}\right)\right]$

Writing an Expression In Exercises 91 and 92, write an algebraic expression that is equivalent to the given expression.

91. $\tan[\arccos(x/2)]$

92. $\sec[\arcsin(x - 1)]$

1.8

93. **Angle of Elevation** The height of a radio transmission tower is 70 meters, and it casts a shadow of length 30 meters. Draw a right triangle that gives a visual representation of the problem. Label the known and unknown quantities. Then find the angle of elevation.

94. **Height** A football lands at the edge of the roof of your school building. When you are 25 feet from the base of the building, the angle of elevation to the football is 21°. How high off the ground is the football?

95. **Air Navigation** From city A to city B, a plane flies 650 miles at a bearing of 48°. From city B to city C, the plane flies 810 miles at a bearing of 115°. Find the distance from city A to city C and the bearing from city A to city C.

96. **Wave Motion** A fishing bobber oscillates in simple harmonic motion because of the waves in a lake. The bobber moves a total of 1.5 inches from its low point to its high point and returns to its high point every 3 seconds. Write an equation that describes the motion of the bobber, where the high point corresponds to the time $t = 0$.

Exploration

True or False? In Exercises 97 and 98, determine whether the statement is true or false. Justify your answer.

97. The equation $y = \sin\theta$ does not represent y as a function of θ because $\sin 30° = \sin 150°$.

98. Because $\tan(3\pi/4) = -1$, $\arctan(-1) = 3\pi/4$.

99. **Writing** Describe the behavior of $f(\theta) = \sec\theta$ at the zeros of $g(\theta) = \cos\theta$. Explain.

100. **Conjecture**

(a) Use a graphing utility to complete the table.

θ	0.1	0.4	0.7	1.0	1.3
$\tan\left(\theta - \frac{\pi}{2}\right)$					
$-\cot\theta$					

(b) Make a conjecture about the relationship between $\tan[\theta - (\pi/2)]$ and $-\cot\theta$.

101. **Writing** When graphing the sine and cosine functions, determining the amplitude is part of the analysis. Explain why this is not true for the other four trigonometric functions.

102. **Graphical Reasoning** Use the formulas for the area of a circular sector and arc length given in Section 1.1.

(a) For $\theta = 0.8$, write the area and arc length as functions of r. What is the domain of each function? Use a graphing utility to graph the functions. Use the graphs to determine which function changes more rapidly as r increases. Explain.

(b) For $r = 10$ centimeters, write the area and arc length as functions of θ. What is the domain of each function? Use the graphing utility to graph the functions.

103. **Writing** Describe a real-life application that can be represented by a simple harmonic motion model and is different from any that you have seen in this chapter. Explain which equation for simple harmonic motion you would use to model your application and why. Explain how you would determine the amplitude, period, and frequency of the model for your application.

Chapter Test

See CalcChat.com for tutorial help and worked-out solutions to odd-numbered exercises.

Take this test as you would take a test in class. When you are finished, check your work against the answers given in the back of the book.

1. Consider an angle that measures $\frac{5\pi}{4}$ radians.
   (a) Sketch the angle in standard position.
   (b) Determine two coterminal angles (one positive and one negative).
   (c) Convert the radian measure to degree measure.

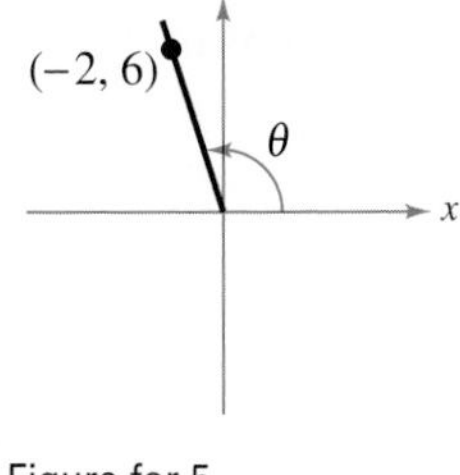

Figure for 5

2. A truck is moving at a rate of 105 kilometers per hour, and the diameter of each of its wheels is 1 meter. Find the angular speed of the wheels in radians per minute.
3. A water sprinkler sprays water on a lawn over a distance of 25 feet and rotates through an angle of 130°. Find the area of the lawn watered by the sprinkler.
4. Given that θ is an acute angle and $\tan\theta = \frac{3}{2}$, find the exact values of the other five trigonometric functions of θ.
5. Find the exact values of the six trigonometric functions of the angle θ shown in the figure.
6. Find the reference angle θ' of the angle $\theta = 205°$. Sketch θ in standard position and label θ'.
7. Determine the quadrant in which θ lies when $\sec\theta < 0$ and $\tan\theta > 0$.
8. Find two exact values of θ in degrees $(0° \le \theta < 360°)$ for which $\cos\theta = -\sqrt{3}/2$. Do not use a calculator.

In Exercises 9 and 10, find the exact values of the remaining five trigonometric functions of θ satisfying the given conditions.

9. $\cos\theta = \frac{3}{5}, \quad \tan\theta < 0$
10. $\sec\theta = -\frac{29}{20}, \quad \sin\theta > 0$

In Exercises 11–13, sketch the graph of the function. (Include two full periods.)

11. $g(x) = -2\sin\left(x - \frac{\pi}{4}\right)$
12. $f(t) = \cos\left(t + \frac{\pi}{2}\right) - 1$
13. $f(x) = \frac{1}{2}\tan 2x$

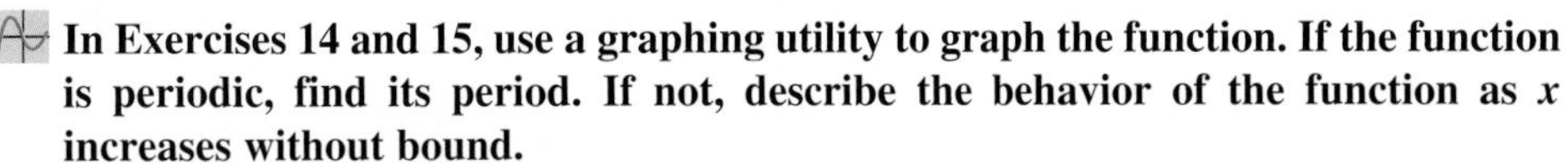

In Exercises 14 and 15, use a graphing utility to graph the function. If the function is periodic, find its period. If not, describe the behavior of the function as x increases without bound.

14. $y = \sin 2\pi x + 2\cos\pi x$
15. $y = 6x\cos(0.25x)$

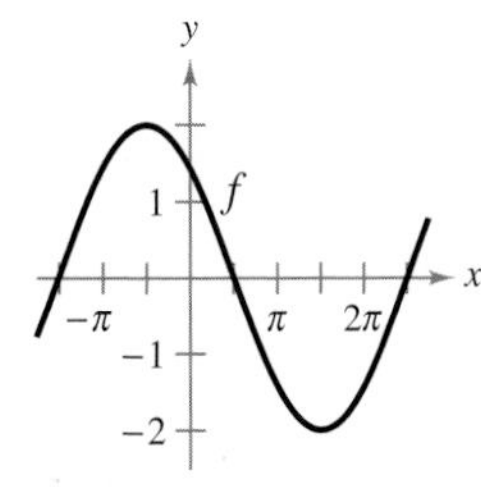

Figure for 16

16. Find a, b, and c for the function $f(x) = a\sin(bx + c)$ such that the graph of f matches the figure.
17. Find the exact value of $\cot\left(\arcsin\frac{3}{8}\right)$.
18. Sketch the graph of the function $f(x) = 2\arcsin\left(\frac{1}{2}x\right)$.
19. An airplane is 90 miles south and 110 miles east of an airport. What bearing should the pilot take to fly directly to the airport?
20. A ball on a spring starts at its lowest point of 6 inches below equilibrium, bounces to its maximum height of 6 inches above equilibrium, and returns to its lowest point in a total of 2 seconds. Write an equation for the simple harmonic motion of the ball.

Proofs in Mathematics

The Pythagorean Theorem

The Pythagorean Theorem is one of the most famous theorems in mathematics. More than 350 different proofs now exist. James A. Garfield, the twentieth president of the United States, developed a proof of the Pythagorean Theorem in 1876. His proof, shown below, involves the fact that two congruent right triangles and an isosceles right triangle can form a trapezoid.

The Pythagorean Theorem

In a right triangle, the sum of the squares of the lengths of the legs is equal to the square of the length of the hypotenuse, where a and b are the lengths of the legs and c is the length of the hypotenuse.

$$a^2 + b^2 = c^2$$

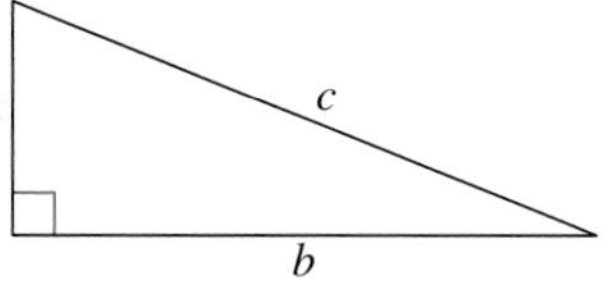

Proof

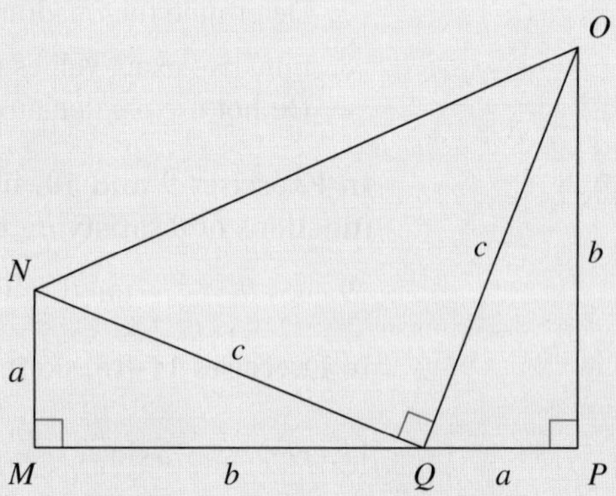

$$\begin{array}{c}\text{Area of} \\ \text{trapezoid } MNOP\end{array} = \begin{array}{c}\text{Area of} \\ \triangle MNQ\end{array} + \begin{array}{c}\text{Area of} \\ \triangle PQO\end{array} + \begin{array}{c}\text{Area of} \\ \triangle NOQ\end{array}$$

$$\frac{1}{2}(a + b)(a + b) = \frac{1}{2}ab + \frac{1}{2}ab + \frac{1}{2}c^2$$

$$\frac{1}{2}(a + b)(a + b) = ab + \frac{1}{2}c^2$$

$$(a + b)(a + b) = 2ab + c^2$$

$$a^2 + 2ab + b^2 = 2ab + c^2$$

$$a^2 + b^2 = c^2$$

P.S. Problem Solving

1. **Angle of Rotation** The restaurant at the top of the Space Needle in Seattle, Washington, is circular and has a radius of 47.25 feet. The dining part of the restaurant revolves, making about one complete revolution every 48 minutes. A dinner party, seated at the edge of the revolving restaurant at 6:45 P.M., finishes at 8:57 P.M.

   (a) Find the angle through which the dinner party rotated.

   (b) Find the distance the party traveled during dinner.

2. **Bicycle Gears** A bicycle's gear ratio is the number of times the freewheel turns for every one turn of the chainwheel (see figure). The table shows the numbers of teeth in the freewheel and chainwheel for the first five gears of an 18-speed touring bicycle. The chainwheel completes one rotation for each gear. Find the angle through which the freewheel turns for each gear. Give your answers in both degrees and radians.

DATA

Spreadsheet at LarsonPrecalculus.com

Gear Number	Number of Teeth in Freewheel	Number of Teeth in Chainwheel
1	32	24
2	26	24
3	22	24
4	32	40
5	19	24

3. **Height of a Ferris Wheel Car** A model for the height h (in feet) of a Ferris wheel car is

   $$h = 50 + 50 \sin 8\pi t$$

   where t is the time (in minutes). (The Ferris wheel has a radius of 50 feet.) This model yields a height of 50 feet when $t = 0$. Alter the model so that the height of the car is 1 foot when $t = 0$.

4. **Periodic Function** The function f is periodic, with period c. So, $f(t + c) = f(t)$. Determine whether each statement is true or false. Explain.

   (a) $f(t - 2c) = f(t)$ (b) $f\left(t + \frac{1}{2}c\right) = f\left(\frac{1}{2}t\right)$

   (c) $f\left(\frac{1}{2}[t + c]\right) = f\left(\frac{1}{2}t\right)$ (d) $f\left(\frac{1}{2}[t + 4c]\right) = f\left(\frac{1}{2}t\right)$

5. **Surveying** A surveyor in a helicopter is determining the width of an island, as shown in the figure.

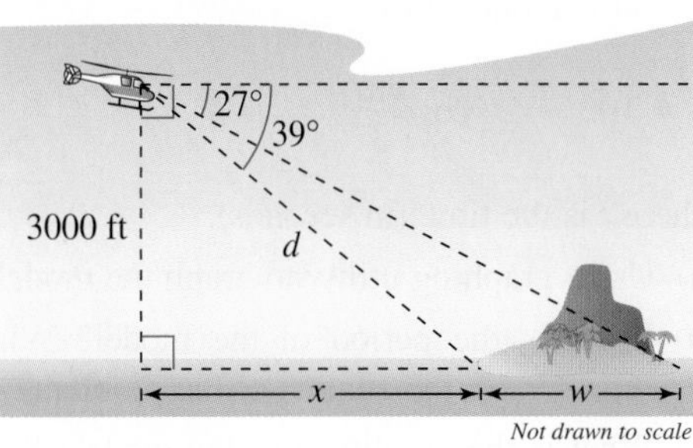

Not drawn to scale

   (a) What is the shortest distance d the helicopter must travel to land on the island?

   (b) What is the horizontal distance x the helicopter must travel before it is directly over the nearer end of the island?

   (c) Find the width w of the island. Explain how you found your answer.

6. **Similar Triangles and Trigonometric Functions** Use the figure below.

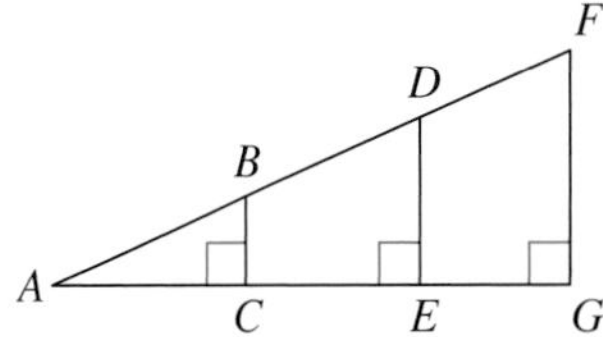

   (a) Explain why $\triangle ABC$, $\triangle ADE$, and $\triangle AFG$ are similar triangles.

   (b) What does similarity imply about the ratios

   $$\frac{BC}{AB}, \frac{DE}{AD}, \text{ and } \frac{FG}{AF}?$$

   (c) Does the value of $\sin A$ depend on which triangle from part (a) is used to calculate it? Does the value of $\sin A$ change when you use a different right triangle similar to the three given triangles?

   (d) Do your conclusions from part (c) apply to the other five trigonometric functions? Explain.

7. **Using Technology** Use a graphing utility to graph h, and use the graph to determine whether h is even, odd, or neither.

   (a) $h(x) = \cos^2 x$ (b) $h(x) = \sin^2 x$

8. **Squares of Even and Odd Functions** Given that f is an even function and g is an odd function, use the results of Exercise 7 to make a conjecture about each function h.

   (a) $h(x) = [f(x)]^2$ (b) $h(x) = [g(x)]^2$

9. Blood Pressure The pressure P (in millimeters of mercury) against the walls of the blood vessels of a patient is modeled by

$$P = 100 - 20 \cos \frac{8\pi t}{3}$$

where t is the time (in seconds).

(a) Use a graphing utility to graph the model.

(b) What is the period of the model? What does it represent in the context of the problem?

(c) What is the amplitude of the model? What does it represent in the context of the problem?

(d) If one cycle of this model is equivalent to one heartbeat, what is the pulse of the patient?

(e) A physician wants the patient's pulse rate to be 64 beats per minute or less. What should the period be? What should the coefficient of t be?

10. Biorhythms A popular theory that attempts to explain the ups and downs of everyday life states that each person has three cycles, called biorhythms, which begin at birth. These three cycles can be modeled by the sine functions below, where t is the number of days since birth.

Physical (23 days): $P = \sin \frac{2\pi t}{23}, \quad t \geq 0$

Emotional (28 days): $E = \sin \frac{2\pi t}{28}, \quad t \geq 0$

Intellectual (33 days): $I = \sin \frac{2\pi t}{33}, \quad t \geq 0$

Consider a person who was born on July 20, 1995.

(a) Use a graphing utility to graph the three models in the same viewing window for $7300 \leq t \leq 7380$.

(b) Describe the person's biorhythms during the month of September 2015.

(c) Calculate the person's three energy levels on September 22, 2015.

11. Graphical Reasoning

(a) Use a graphing utility to graph the functions

$f(x) = 2 \cos 2x + 3 \sin 3x$

and

$g(x) = 2 \cos 2x + 3 \sin 4x.$

(b) Use the graphs from part (a) to find the period of each function.

(c) Is the function $h(x) = A \cos \alpha x + B \sin \beta x$, where α and β are positive integers, periodic? Explain.

12. Analyzing Trigonometric Functions Two trigonometric functions f and g have periods of 2, and their graphs intersect at $x = 5.35$.

(a) Give one positive value of x less than 5.35 and one value of x greater than 5.35 at which the functions have the same value.

(b) Determine one negative value of x at which the graphs intersect.

(c) Is it true that $f(13.35) = g(-4.65)$? Explain.

13. Refraction When you stand in shallow water and look at an object below the surface of the water, the object will look farther away from you than it really is. This is because when light rays pass between air and water, the water refracts, or bends, the light rays. The index of refraction for water is 1.333. This is the ratio of the sine of θ_1 and the sine of θ_2 (see figure).

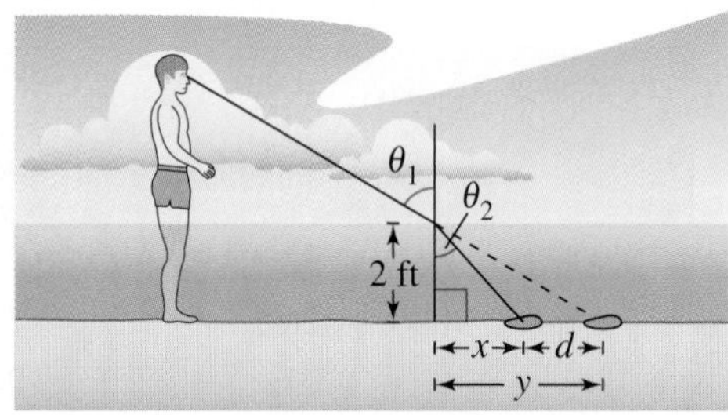

(a) While standing in water that is 2 feet deep, you look at a rock at angle $\theta_1 = 60°$ (measured from a line perpendicular to the surface of the water). Find θ_2.

(b) Find the distances x and y.

(c) Find the distance d between where the rock is and where it appears to be.

(d) What happens to d as you move closer to the rock? Explain.

14. Polynomial Approximation Using calculus, it can be shown that the arctangent function can be approximated by the polynomial

$$\arctan x \approx x - \frac{x^3}{3} + \frac{x^5}{5} - \frac{x^7}{7}$$

where x is in radians.

(a) Use a graphing utility to graph the arctangent function and its polynomial approximation in the same viewing window. How do the graphs compare?

(b) Study the pattern in the polynomial approximation of the arctangent function and predict the next term. Then repeat part (a). How does the accuracy of the approximation change when an additional term is added?

2 Analytic Trigonometry

Projectile Motion (Example 10, page 249)

Standing Waves (Exercise 80, page 241)

Ferris Wheel (Exercise 94, page 235)

Shadow Length (Exercise 62, page 223)

Friction (Exercise 67, page 216)

2.1 Using Fundamental Identities

Fundamental trigonometric identities are useful in simplifying trigonometric expressions. For example, in Exercise 67 on page 216, you will use trigonometric identities to simplify an expression for the coefficient of friction.

- **Recognize and write the fundamental trigonometric identities.**
- **Use the fundamental trigonometric identities to evaluate trigonometric functions, simplify trigonometric expressions, and rewrite trigonometric expressions.**

Introduction

In Chapter 1, you studied the basic definitions, properties, graphs, and applications of the individual trigonometric functions. In this chapter, you will learn how to use the fundamental identities to perform the four tasks listed below.

1. Evaluate trigonometric functions.
2. Simplify trigonometric expressions.
3. Develop additional trigonometric identities.
4. Solve trigonometric equations.

REMARK You should learn the fundamental trigonometric identities well, because you will use them frequently in trigonometry and they will also appear in calculus. Note that u can be an angle, a real number, or a variable.

Fundamental Trigonometric Identities

Reciprocal Identities

$$\sin u = \frac{1}{\csc u} \qquad \cos u = \frac{1}{\sec u} \qquad \tan u = \frac{1}{\cot u}$$

$$\csc u = \frac{1}{\sin u} \qquad \sec u = \frac{1}{\cos u} \qquad \cot u = \frac{1}{\tan u}$$

Quotient Identities

$$\tan u = \frac{\sin u}{\cos u} \qquad \cot u = \frac{\cos u}{\sin u}$$

Pythagorean Identities

$$\sin^2 u + \cos^2 u = 1 \qquad 1 + \tan^2 u = \sec^2 u \qquad 1 + \cot^2 u = \csc^2 u$$

Cofunction Identities

$$\sin\left(\frac{\pi}{2} - u\right) = \cos u \qquad \cos\left(\frac{\pi}{2} - u\right) = \sin u$$

$$\tan\left(\frac{\pi}{2} - u\right) = \cot u \qquad \cot\left(\frac{\pi}{2} - u\right) = \tan u$$

$$\sec\left(\frac{\pi}{2} - u\right) = \csc u \qquad \csc\left(\frac{\pi}{2} - u\right) = \sec u$$

Even/Odd Identities

$$\sin(-u) = -\sin u \qquad \cos(-u) = \cos u \qquad \tan(-u) = -\tan u$$

$$\csc(-u) = -\csc u \qquad \sec(-u) = \sec u \qquad \cot(-u) = -\cot u$$

Pythagorean identities are sometimes used in radical form such as

$$\sin u = \pm\sqrt{1 - \cos^2 u}$$

or

$$\tan u = \pm\sqrt{\sec^2 u - 1}$$

where the sign depends on the choice of u.

chaoss/Shutterstock.com

Using the Fundamental Identities

One common application of trigonometric identities is to use given information about trigonometric functions to evaluate other trigonometric functions.

EXAMPLE 1 Using Identities to Evaluate a Function

Use the conditions $\sec u = -\frac{3}{2}$ and $\tan u > 0$ to find the values of all six trigonometric functions.

Solution Using a reciprocal identity, you have

$$\cos u = \frac{1}{\sec u} = \frac{1}{-3/2} = -\frac{2}{3}.$$

Using a Pythagorean identity, you have

$$\sin^2 u = 1 - \cos^2 u \qquad \text{Pythagorean identity}$$
$$= 1 - \left(-\tfrac{2}{3}\right)^2 \qquad \text{Substitute } -\tfrac{2}{3} \text{ for } \cos u.$$
$$= \tfrac{5}{9}. \qquad \text{Simplify.}$$

Because $\sec u < 0$ and $\tan u > 0$, it follows that u lies in Quadrant III. Moreover, $\sin u$ is negative when u is in Quadrant III, so choose the negative root and obtain $\sin u = -\sqrt{5}/3$. Knowing the values of the sine and cosine enables you to find the values of the remaining trigonometric functions.

$$\sin u = -\frac{\sqrt{5}}{3} \qquad \csc u = \frac{1}{\sin u} = -\frac{3}{\sqrt{5}} = -\frac{3\sqrt{5}}{5}$$

$$\cos u = -\frac{2}{3} \qquad \sec u = -\frac{3}{2}$$

$$\tan u = \frac{\sin u}{\cos u} = \frac{-\sqrt{5}/3}{-2/3} = \frac{\sqrt{5}}{2} \qquad \cot u = \frac{1}{\tan u} = \frac{2}{\sqrt{5}} = \frac{2\sqrt{5}}{5}$$

Use the conditions $\tan x = \frac{1}{3}$ and $\cos x < 0$ to find the values of all six trigonometric functions.

EXAMPLE 2 Simplifying a Trigonometric Expression

Simplify the expression.

$$\sin x \cos^2 x - \sin x$$

Solution First factor out the common monomial factor $\sin x$ and then use a Pythagorean identity.

$$\sin x \cos^2 x - \sin x = \sin x(\cos^2 x - 1) \qquad \text{Factor out common monomial factor.}$$
$$= -\sin x(1 - \cos^2 x) \qquad \text{Factor out } -1.$$
$$= -\sin x(\sin^2 x) \qquad \text{Pythagorean identity}$$
$$= -\sin^3 x \qquad \text{Multiply.}$$

Checkpoint Audio-video solution in English & Spanish at LarsonPrecalculus.com

Simplify the expression.

$$\cos^2 x \csc x - \csc x$$

TECHNOLOGY Use a graphing utility to check the result of Example 2. To do this, enter

$$Y1 = -(\sin(X))^3$$

and

$$Y2 = \sin(X)(\cos(X))^2 - \sin(X).$$

Select the *line* style for Y1 and the *path* style for Y2, then graph both equations in the same viewing window. The two graphs *appear* to coincide, so it is reasonable to assume that their expressions are equivalent. Note that the actual equivalence of the expressions can only be verified algebraically, as in Example 2. This graphical approach is only to check your work.

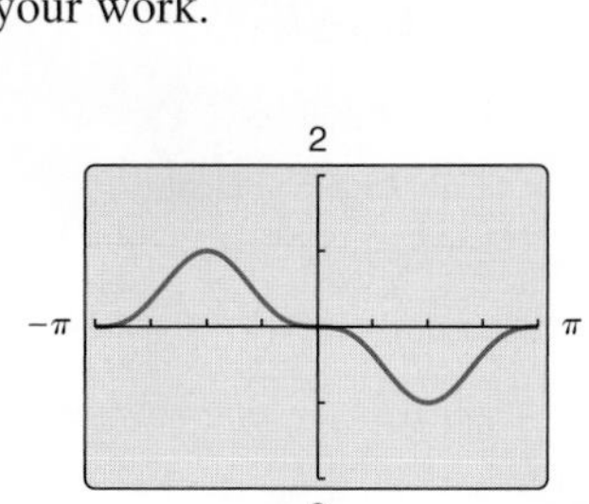

When factoring trigonometric expressions, it is helpful to find a polynomial form that fits the expression, as shown in Example 3.

EXAMPLE 3 Factoring Trigonometric Expressions

Factor each expression.

a. $\sec^2 \theta - 1$ **b.** $4 \tan^2 \theta + \tan \theta - 3$

Solution

a. This expression has the polynomial form $u^2 - v^2$, which is the difference of two squares. It factors as

$$\sec^2 \theta - 1 = (\sec \theta + 1)(\sec \theta - 1).$$

b. This expression has the polynomial form $ax^2 + bx + c$, and it factors as

$$4 \tan^2 \theta + \tan \theta - 3 = (4 \tan \theta - 3)(\tan \theta + 1).$$

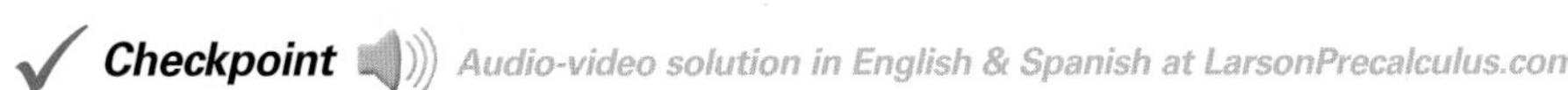

Factor each expression.

a. $1 - \cos^2 \theta$ **b.** $2 \csc^2 \theta - 7 \csc \theta + 6$

In some cases, when factoring or simplifying a trigonometric expression, it is helpful to first rewrite the expression in terms of just *one* trigonometric function or in terms of *sine and cosine only*. These strategies are demonstrated in Examples 4 and 5.

EXAMPLE 4 Factoring a Trigonometric Expression

Factor $\csc^2 x - \cot x - 3$.

Solution Use the identity $\csc^2 x = 1 + \cot^2 x$ to rewrite the expression.

$$\csc^2 x - \cot x - 3 = (1 + \cot^2 x) - \cot x - 3 \qquad \text{Pythagorean identity}$$

$$= \cot^2 x - \cot x - 2 \qquad \text{Combine like terms.}$$

$$= (\cot x - 2)(\cot x + 1) \qquad \text{Factor.}$$

Checkpoint Audio-video solution in English & Spanish at LarsonPrecalculus.com

Factor $\sec^2 x + 3 \tan x + 1$.

REMARK Remember that when adding rational expressions, you must first find the least common denominator (LCD). In Example 5, the LCD is $\sin t$.

EXAMPLE 5 Simplifying a Trigonometric Expression

See LarsonPrecalculus.com for an interactive version of this type of example.

$$\sin t + \cot t \cos t = \sin t + \left(\frac{\cos t}{\sin t}\right) \cos t \qquad \text{Quotient identity}$$

$$= \frac{\sin^2 t + \cos^2 t}{\sin t} \qquad \text{Add fractions.}$$

$$= \frac{1}{\sin t} \qquad \text{Pythagorean identity}$$

$$= \csc t \qquad \text{Reciprocal identity}$$

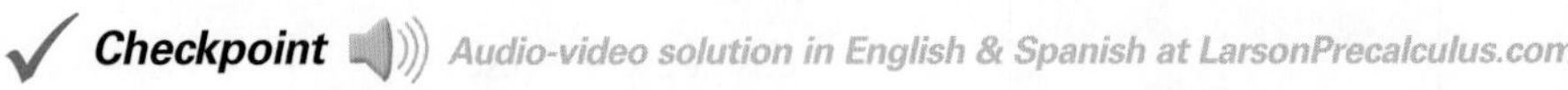

Simplify $\csc x - \cos x \cot x$.

EXAMPLE 6 Adding Trigonometric Expressions

Perform the addition and simplify: $\dfrac{\sin\theta}{1+\cos\theta} + \dfrac{\cos\theta}{\sin\theta}$.

Solution

$$\frac{\sin\theta}{1+\cos\theta} + \frac{\cos\theta}{\sin\theta} = \frac{(\sin\theta)(\sin\theta) + (\cos\theta)(1+\cos\theta)}{(1+\cos\theta)(\sin\theta)}$$

$$= \frac{\sin^2\theta + \cos^2\theta + \cos\theta}{(1+\cos\theta)(\sin\theta)} \qquad \text{Multiply.}$$

$$= \frac{\cancel{1+\cos\theta}}{(\cancel{1+\cos\theta})(\sin\theta)} \qquad \text{Pythagorean identity}$$

$$= \frac{1}{\sin\theta} \qquad \text{Divide out common factor.}$$

$$= \csc\theta \qquad \text{Reciprocal identity}$$

✓ **Checkpoint**  *Audio-video solution in English & Spanish at LarsonPrecalculus.com*

Perform the addition and simplify: $\dfrac{1}{1+\sin\theta} + \dfrac{1}{1-\sin\theta}$.

The next two examples involve techniques for rewriting expressions in forms that are used in calculus.

EXAMPLE 7 Rewriting a Trigonometric Expression 

Rewrite $\dfrac{1}{1+\sin x}$ so that it is *not* in fractional form.

Solution From the Pythagorean identity

$$\cos^2 x = 1 - \sin^2 x = (1-\sin x)(1+\sin x)$$

multiplying both the numerator and the denominator by $(1-\sin x)$ will produce a monomial denominator.

$$\frac{1}{1+\sin x} = \frac{1}{1+\sin x}\cdot\frac{1-\sin x}{1-\sin x} \qquad \text{Multiply numerator and denominator by } (1-\sin x).$$

$$= \frac{1-\sin x}{1-\sin^2 x} \qquad \text{Multiply.}$$

$$= \frac{1-\sin x}{\cos^2 x} \qquad \text{Pythagorean identity}$$

$$= \frac{1}{\cos^2 x} - \frac{\sin x}{\cos^2 x} \qquad \text{Write as separate fractions.}$$

$$= \frac{1}{\cos^2 x} - \frac{\sin x}{\cos x}\cdot\frac{1}{\cos x} \qquad \text{Product of fractions}$$

$$= \sec^2 x - \tan x \sec x \qquad \text{Reciprocal and quotient identities}$$

✓ **Checkpoint**  *Audio-video solution in English & Spanish at LarsonPrecalculus.com*

Rewrite $\dfrac{\cos^2\theta}{1-\sin\theta}$ so that it is *not* in fractional form.

EXAMPLE 8 Trigonometric Substitution 

Use the substitution $x = 2 \tan \theta$, $0 < \theta < \pi/2$, to write

$$\sqrt{4 + x^2}$$

as a trigonometric function of θ.

Solution Begin by letting $x = 2 \tan \theta$. Then, you obtain

$$\begin{aligned}
\sqrt{4 + x^2} &= \sqrt{4 + (2 \tan \theta)^2} && \text{Substitute } 2 \tan \theta \text{ for } x. \\
&= \sqrt{4 + 4 \tan^2 \theta} && \text{Property of exponents} \\
&= \sqrt{4(1 + \tan^2 \theta)} && \text{Factor.} \\
&= \sqrt{4 \sec^2 \theta} && \text{Pythagorean identity} \\
&= 2 \sec \theta. && \sec \theta > 0 \text{ for } 0 < \theta < \frac{\pi}{2}
\end{aligned}$$

✓ ***Checkpoint*** *Audio-video solution in English & Spanish at LarsonPrecalculus.com*

Use the substitution $x = 3 \sin \theta$, $0 < \theta < \pi/2$, to write

$$\sqrt{9 - x^2}$$

as a trigonometric function of θ.

The figure below shows the right triangle illustration of the trigonometric substitution $x = 2 \tan \theta$ in Example 8.

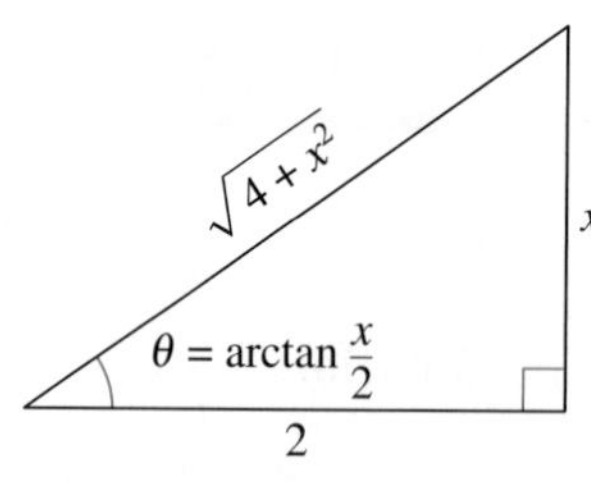

$$2 \tan \theta = x \quad \Longrightarrow \quad \tan \theta = \frac{x}{2}$$

You can use this triangle to check the solution to Example 8. For $0 < \theta < \pi/2$, you have

$$\text{opp} = x, \quad \text{adj} = 2, \quad \text{and} \quad \text{hyp} = \sqrt{4 + x^2}.$$

Using these expressions,

$$\sec \theta = \frac{\text{hyp}}{\text{adj}} = \frac{\sqrt{4 + x^2}}{2}.$$

So, $2 \sec \theta = \sqrt{4 + x^2}$, and the solution checks.

Summarize (Section 2.1)

1. State the fundamental trigonometric identities *(page 210)*.
2. Explain how to use the fundamental trigonometric identities to evaluate trigonometric functions, simplify trigonometric expressions, and rewrite trigonometric expressions *(pages 211–214)*. For examples of these concepts, see Examples 1–8.

2.1 Exercises

See CalcChat.com for tutorial help and worked-out solutions to odd-numbered exercises.

Vocabulary: Fill in the blank to complete the trigonometric identity.

1. $\dfrac{\sin u}{\cos u} = $ ________
2. $\dfrac{1}{\sin u} = $ ________
3. $\dfrac{1}{\tan u} = $ ________
4. $\sec\left(\dfrac{\pi}{2} - u\right) = $ ________
5. $\sin^2 u + \cos^2 u = $ ________
6. $\sin(-u) = $ ________

Skills and Applications

Using Identities to Evaluate a Function In Exercises 7–12, use the given conditions to find the values of all six trigonometric functions.

7. $\sec x = -\frac{5}{2}$, $\tan x < 0$
8. $\csc x = -\frac{7}{6}$, $\tan x > 0$
9. $\sin \theta = -\frac{3}{4}$, $\cos \theta > 0$
10. $\cos \theta = \frac{2}{3}$, $\sin \theta < 0$
11. $\tan x = \frac{2}{3}$, $\cos x > 0$
12. $\cot x = \frac{7}{4}$, $\sin x < 0$

Matching Trigonometric Expressions In Exercises 13–18, match the trigonometric expression with its simplified form.

(a) $\csc x$ (b) -1 (c) 1

(d) $\sin x \tan x$ (e) $\sec^2 x$ (f) $\sec x$

13. $\sec x \cos x$
14. $\cot^2 x - \csc^2 x$
15. $\cos x(1 + \tan^2 x)$
16. $\cot x \sec x$
17. $\dfrac{\sec^2 x - 1}{\sin^2 x}$
18. $\dfrac{\cos^2[(\pi/2) - x]}{\cos x}$

Simplifying a Trigonometric Expression In Exercises 19–22, use the fundamental identities to simplify the expression. (There is more than one correct form of each answer).

19. $\dfrac{\tan \theta \cot \theta}{\sec \theta}$
20. $\cos\left(\dfrac{\pi}{2} - x\right) \sec x$
21. $\tan^2 x - \tan^2 x \sin^2 x$
22. $\sin^2 x \sec^2 x - \sin^2 x$

Factoring a Trigonometric Expression In Exercises 23–32, factor the expression. Use the fundamental identities to simplify, if necessary. (There is more than one correct form of each answer.)

23. $\dfrac{\sec^2 x - 1}{\sec x - 1}$
24. $\dfrac{\cos x - 2}{\cos^2 x - 4}$
25. $1 - 2\cos^2 x + \cos^4 x$
26. $\sec^4 x - \tan^4 x$
27. $\cot^3 x + \cot^2 x + \cot x + 1$
28. $\sec^3 x - \sec^2 x - \sec x + 1$
29. $3\sin^2 x - 5\sin x - 2$
30. $6\cos^2 x + 5\cos x - 6$
31. $\cot^2 x + \csc x - 1$
32. $\sin^2 x + 3\cos x + 3$

Simplifying a Trigonometric Expression In Exercises 33–40, use the fundamental identities to simplify the expression. (There is more than one correct form of each answer.)

33. $\tan \theta \csc \theta$
34. $\tan(-x) \cos x$
35. $\sin \phi(\csc \phi - \sin \phi)$
36. $\cos x(\sec x - \cos x)$
37. $\sin \beta \tan \beta + \cos \beta$
38. $\cot u \sin u + \tan u \cos u$
39. $\dfrac{1 - \sin^2 x}{\csc^2 x - 1}$
40. $\dfrac{\cos^2 y}{1 - \sin y}$

Multiplying Trigonometric Expressions In Exercises 41 and 42, perform the multiplication and use the fundamental identities to simplify. (There is more than one correct form of each answer.)

41. $(\sin x + \cos x)^2$
42. $(2\csc x + 2)(2\csc x - 2)$

Adding or Subtracting Trigonometric Expressions In Exercises 43–48, perform the addition or subtraction and use the fundamental identities to simplify. (There is more than one correct form of each answer.)

43. $\dfrac{1}{1 + \cos x} + \dfrac{1}{1 - \cos x}$
44. $\dfrac{1}{\sec x + 1} - \dfrac{1}{\sec x - 1}$
45. $\dfrac{\cos x}{1 + \sin x} - \dfrac{\cos x}{1 - \sin x}$
46. $\dfrac{\sin x}{1 + \cos x} + \dfrac{\sin x}{1 - \cos x}$
47. $\tan x - \dfrac{\sec^2 x}{\tan x}$
48. $\dfrac{\cos x}{1 + \sin x} + \dfrac{1 + \sin x}{\cos x}$

Rewriting a Trigonometric Expression In Exercises 49 and 50, rewrite the expression so that it is *not* in fractional form. (There is more than one correct form of each answer.)

49. $\dfrac{\sin^2 y}{1 - \cos y}$
50. $\dfrac{5}{\tan x + \sec x}$

Trigonometric Functions and Expressions In Exercises 51–54, use a graphing utility to determine which of the six trigonometric functions is equal to the expression. Verify your answer algebraically.

51. $\frac{1}{2}(\sin x \cot x + \cos x)$
52. $\sec x \csc x - \tan x$
53. $\dfrac{\tan x + 1}{\sec x + \csc x}$
54. $\dfrac{1}{\sin x}\left(\dfrac{1}{\cos x} - \cos x\right)$

Trigonometric Substitution In Exercises 55–58, use the trigonometric substitution to write the algebraic expression as a trigonometric function of θ, where $0 < \theta < \pi/2$.

55. $\sqrt{9 - x^2}$, $x = 3 \cos \theta$
56. $\sqrt{49 - x^2}$, $x = 7 \sin \theta$
57. $\sqrt{x^2 - 4}$, $x = 2 \sec \theta$
58. $\sqrt{9x^2 + 25}$, $3x = 5 \tan \theta$

Trigonometric Substitution In Exercises 59–62, use the trigonometric substitution to write the algebraic equation as a trigonometric equation of θ, where $-\pi/2 < \theta < \pi/2$. Then find $\sin \theta$ and $\cos \theta$.

59. $\sqrt{2} = \sqrt{4 - x^2}$, $x = 2 \sin \theta$
60. $2\sqrt{2} = \sqrt{16 - 4x^2}$, $x = 2 \cos \theta$
61. $3 = \sqrt{36 - x^2}$, $x = 6 \sin \theta$
62. $5\sqrt{3} = \sqrt{100 - x^2}$, $x = 10 \cos \theta$

Solving a Trigonometric Equation In Exercises 63–66, use a graphing utility to solve the equation for θ, where $0 \le \theta < 2\pi$.

63. $\sin \theta = \sqrt{1 - \cos^2 \theta}$
64. $\cos \theta = -\sqrt{1 - \sin^2 \theta}$
65. $\sec \theta = \sqrt{1 + \tan^2 \theta}$
66. $\csc \theta = \sqrt{1 + \cot^2 \theta}$

67. **Friction** The forces acting on an object weighing W units on an inclined plane positioned at an angle of θ with the horizontal (see figure) are modeled by $\mu W \cos \theta = W \sin \theta$, where μ is the coefficient of friction. Solve the equation for μ and simplify the result.

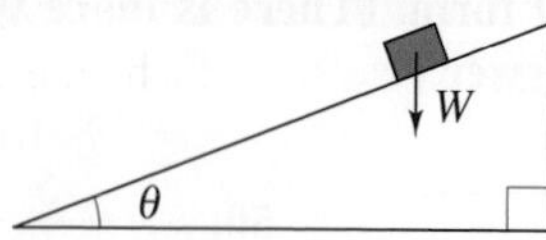

68. **Rate of Change** The rate of change of the function $f(x) = \sec x + \cos x$ is given by the expression $\sec x \tan x - \sin x$. Show that this expression can also be written as $\sin x \tan^2 x$.

Exploration

True or False? In Exercises 69 and 70, determine whether the statement is true or false. Justify your answer.

69. The quotient identities and reciprocal identities can be used to write any trigonometric function in terms of sine and cosine.

70. A cofunction identity can transform a tangent function into a cosecant function.

Analyzing Trigonometric Functions In Exercises 71 and 72, fill in the blanks. (*Note:* The notation $x \to c^+$ indicates that x approaches c from the right and $x \to c^-$ indicates that x approaches c from the left.)

71. As $x \to \left(\frac{\pi}{2}\right)^-$, $\tan x \to$ ____ and $\cot x \to$ ____.

72. As $x \to \pi^+$, $\sin x \to$ ____ and $\csc x \to$ ____.

73. **Error Analysis** Describe the error.

$$\frac{\sin \theta}{\cos(-\theta)} = \frac{\sin \theta}{-\cos \theta} = -\tan \theta$$

74. **Trigonometric Substitution** Use the trigonometric substitution $u = a \tan \theta$, where $-\pi/2 < \theta < \pi/2$ and $a > 0$, to simplify the expression $\sqrt{a^2 + u^2}$.

75. **Writing Trigonometric Functions in Terms of Sine** Write each of the other trigonometric functions of θ in terms of $\sin \theta$.

76. **HOW DO YOU SEE IT?** Explain how to use the figure to derive the Pythagorean identities

$\sin^2 \theta + \cos^2 \theta = 1$,

$1 + \tan^2 \theta = \sec^2 \theta$,

and $1 + \cot^2 \theta = \csc^2 \theta$.

Discuss how to remember these identities and other fundamental trigonometric identities.

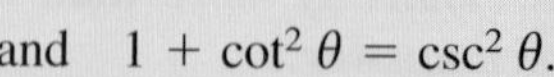

77. **Rewriting a Trigonometric Expression** Rewrite the expression below in terms of $\sin \theta$ and $\cos \theta$.

$$\frac{\sec \theta(1 + \tan \theta)}{\sec \theta + \csc \theta}$$

2.2 Verifying Trigonometric Identities

Trigonometric identities enable you to rewrite trigonometric equations that model real-life situations. For example, in Exercise 62 on page 223, trigonometric identities can help you simplify an equation that models the length of a shadow cast by a gnomon (a device used to tell time).

- **Verify trigonometric identities.**

Verifying Trigonometric Identities

In this section, you will study techniques for verifying trigonometric identities. In the next section, you will study techniques for solving trigonometric equations. The key to both verifying identities *and* solving equations is your ability to use the fundamental identities and the rules of algebra to rewrite trigonometric expressions.

Remember that a *conditional equation* is an equation that is true for only some of the values in the domain of the variable. For example, the conditional equation

$$\sin x = 0 \qquad \text{Conditional equation}$$

is true only for

$$x = n\pi$$

where n is an integer. When you are finding the values of the variable for which the equation is true, you are *solving* the equation.

On the other hand, an equation that is true for all real values in the domain of the variable is an *identity*. For example, the familiar equation

$$\sin^2 x = 1 - \cos^2 x \qquad \text{Identity}$$

is true for all real numbers x. So, it is an identity.

Although there are similarities, verifying that a trigonometric equation is an identity is quite different from solving an equation. There is no well-defined set of rules to follow in verifying trigonometric identities, the process is best learned through practice.

Guidelines for Verifying Trigonometric Identities

1. Work with one side of the equation at a time. It is often better to work with the more complicated side first.
2. Look for opportunities to factor an expression, add fractions, square a binomial, or create a monomial denominator.
3. Look for opportunities to use the fundamental identities. Note which functions are in the final expression you want. Sines and cosines pair up well, as do secants and tangents, and cosecants and cotangents.
4. When the preceding guidelines do not help, try converting all terms to sines and cosines.
5. Always try *something*. Even making an attempt that leads to a dead end can provide insight.

Verifying trigonometric identities is a useful process when you need to convert a trigonometric expression into a form that is more useful algebraically. When you verify an identity, you cannot *assume* that the two sides of the equation are equal because you are trying to verify that they *are* equal. As a result, when verifying identities, you cannot use operations such as adding the same quantity to each side of the equation or cross multiplication.

Leonard Zhukovsky/Shutterstock.com

EXAMPLE 1 Verifying a Trigonometric Identity

Verify the identity $\dfrac{\sec^2 \theta - 1}{\sec^2 \theta} = \sin^2 \theta$.

REMARK Remember that an identity is only true for all real values in the domain of the variable. For instance, in Example 1 the identity is not true when $\theta = \pi/2$ because $\sec^2 \theta$ is undefined when $\theta = \pi/2$.

Solution Start with the left side because it is more complicated.

$$\frac{\sec^2 \theta - 1}{\sec^2 \theta} = \frac{\tan^2 \theta}{\sec^2 \theta} \qquad \text{Pythagorean identity}$$

$$= \tan^2 \theta(\cos^2 \theta) \qquad \text{Reciprocal identity}$$

$$= \frac{\sin^2 \theta}{(\cancel{\cos^2 \theta})}(\cancel{\cos^2 \theta}) \qquad \text{Quotient identity}$$

$$= \sin^2 \theta \qquad \text{Simplify.}$$

Notice that you verify the identity by starting with the left side of the equation (the more complicated side) and using the fundamental trigonometric identities to simplify it until you obtain the right side.

Checkpoint Audio-video solution in English & Spanish at LarsonPrecalculus.com

Verify the identity $\dfrac{\sin^2 \theta + \cos^2 \theta}{\cos^2 \theta \sec^2 \theta} = 1$. ■

There can be more than one way to verify an identity. Here is another way to verify the identity in Example 1.

$$\frac{\sec^2 \theta - 1}{\sec^2 \theta} = \frac{\sec^2 \theta}{\sec^2 \theta} - \frac{1}{\sec^2 \theta} \qquad \text{Write as separate fractions.}$$

$$= 1 - \cos^2 \theta \qquad \text{Reciprocal identity}$$

$$= \sin^2 \theta \qquad \text{Pythagorean identity}$$

EXAMPLE 2 Verifying a Trigonometric Identity

Verify the identity $2 \sec^2 \alpha = \dfrac{1}{1 - \sin \alpha} + \dfrac{1}{1 + \sin \alpha}$.

Algebraic Solution

Start with the right side because it is more complicated.

$$\frac{1}{1 - \sin \alpha} + \frac{1}{1 + \sin \alpha} = \frac{1 + \sin \alpha + 1 - \sin \alpha}{(1 - \sin \alpha)(1 + \sin \alpha)} \qquad \text{Add fractions.}$$

$$= \frac{2}{1 - \sin^2 \alpha} \qquad \text{Simplify.}$$

$$= \frac{2}{\cos^2 \alpha} \qquad \text{Pythagorean identity}$$

$$= 2 \sec^2 \alpha \qquad \text{Reciprocal identity}$$

Numerical Solution

Use a graphing utility to create a table that shows the values of $y_1 = 2/\cos^2 x$ and $y_2 = [1/(1 - \sin x)] + [1/(1 + \sin x)]$ for different values of x.

X	Y1	Y2
-.5	2.5969	2.5969
-.25	2.1304	2.1304
0	2	2
.25	2.1304	2.1304
.5	2.5969	2.5969
.75	3.7357	3.7357
1	6.851	6.851

X=-.5

The values in the table for y_1 and y_2 appear to be identical, so the equation appears to be an identity.

Checkpoint Audio-video solution in English & Spanish at LarsonPrecalculus.com

Verify the identity $2 \csc^2 \beta = \dfrac{1}{1 - \cos \beta} + \dfrac{1}{1 + \cos \beta}$. ■

EXAMPLE 3 Verifying a Trigonometric Identity

Verify the identity

$$(\tan^2 x + 1)(\cos^2 x - 1) = -\tan^2 x.$$

Algebraic Solution

Apply Pythagorean identities before multiplying.

$$\begin{aligned}(\tan^2 x + 1)(\cos^2 x - 1) &= (\sec^2 x)(-\sin^2 x) && \text{Pythagorean identities}\\ &= -\frac{\sin^2 x}{\cos^2 x} && \text{Reciprocal identity}\\ &= -\left(\frac{\sin x}{\cos x}\right)^2 && \text{Property of exponents}\\ &= -\tan^2 x && \text{Quotient identity}\end{aligned}$$

Graphical Solution

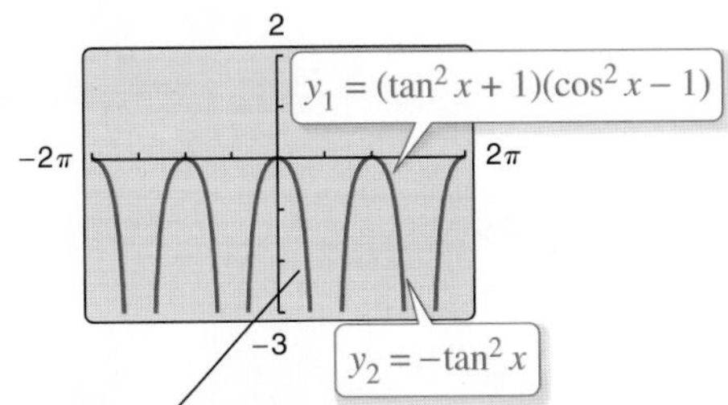

The graphs appear to coincide, so the given equation appears to be an identity.

✓ **Checkpoint** *Audio-video solution in English & Spanish at LarsonPrecalculus.com*

Verify the identity $(\sec^2 x - 1)(\sin^2 x - 1) = -\sin^2 x$.

EXAMPLE 4 Converting to Sines and Cosines

Verify each identity.

a. $\tan x \csc x = \sec x$

b. $\tan x + \cot x = \sec x \csc x$

Solution

a. Convert the left side into sines and cosines.

$$\begin{aligned}\tan x \csc x &= \frac{\sin x}{\cos x} \cdot \frac{1}{\sin x} && \text{Quotient and reciprocal identities}\\ &= \frac{1}{\cos x} && \text{Simplify.}\\ &= \sec x && \text{Reciprocal identity}\end{aligned}$$

b. Convert the left side into sines and cosines.

$$\begin{aligned}\tan x + \cot x &= \frac{\sin x}{\cos x} + \frac{\cos x}{\sin x} && \text{Quotient identities}\\ &= \frac{\sin^2 x + \cos^2 x}{\cos x \sin x} && \text{Add fractions.}\\ &= \frac{1}{\cos x \sin x} && \text{Pythagorean identity}\\ &= \frac{1}{\cos x} \cdot \frac{1}{\sin x} && \text{Product of fractions}\\ &= \sec x \csc x && \text{Reciprocal identities}\end{aligned}$$

Verify each identity.

a. $\cot x \sec x = \csc x$

b. $\csc x - \sin x = \cos x \cot x$

Recall from algebra that *rationalizing the denominator* using conjugates is, on occasion, a powerful simplification technique. A related form of this technique works for simplifying trigonometric expressions as well. For example, to simplify

$$\frac{1}{1 - \cos x}$$

multiply the numerator and the denominator by $1 + \cos x$.

$$\begin{aligned}\frac{1}{1 - \cos x} &= \frac{1}{1 - \cos x}\left(\frac{1 + \cos x}{1 + \cos x}\right)\\ &= \frac{1 + \cos x}{1 - \cos^2 x}\\ &= \frac{1 + \cos x}{\sin^2 x}\\ &= \csc^2 x(1 + \cos x)\end{aligned}$$

The expression $\csc^2 x(1 + \cos x)$ is considered a simplified form of

$$\frac{1}{1 - \cos x}$$

because $\csc^2 x(1 + \cos x)$ does not contain fractions.

EXAMPLE 5 Verifying a Trigonometric Identity

See LarsonPrecalculus.com for an interactive version of this type of example.

Verify the identity $\sec x + \tan x = \dfrac{\cos x}{1 - \sin x}$.

Algebraic Solution

Begin with the *right* side and create a monomial denominator by multiplying the numerator and the denominator by $1 + \sin x$.

$$\begin{aligned}\frac{\cos x}{1 - \sin x} &= \frac{\cos x}{1 - \sin x}\left(\frac{1 + \sin x}{1 + \sin x}\right) && \text{Multiply numerator and denominator by } 1 + \sin x.\\ &= \frac{\cos x + \cos x \sin x}{1 - \sin^2 x} && \text{Multiply.}\\ &= \frac{\cos x + \cos x \sin x}{\cos^2 x} && \text{Pythagorean identity}\\ &= \frac{\cos x}{\cos^2 x} + \frac{\cos x \sin x}{\cos^2 x} && \text{Write as separate fractions.}\\ &= \frac{1}{\cos x} + \frac{\sin x}{\cos x} && \text{Simplify.}\\ &= \sec x + \tan x && \text{Identities}\end{aligned}$$

Graphical Solution

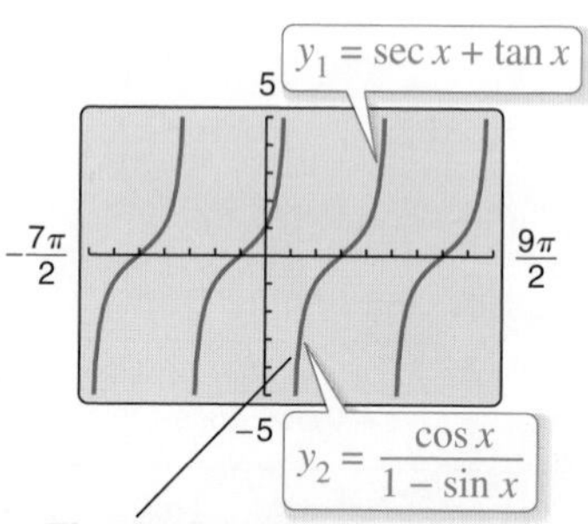

The graphs appear to coincide, so the given equation appears to be an identity.

Checkpoint *Audio-video solution in English & Spanish at LarsonPrecalculus.com*

Verify the identity $\csc x + \cot x = \dfrac{\sin x}{1 - \cos x}$.

In Examples 1 through 5, you have been verifying trigonometric identities by working with one side of the equation and converting to the form given on the other side. On occasion, it is practical to work with each side *separately*, to obtain one common form that is equivalent to both sides. This is illustrated in Example 6.

EXAMPLE 6 Working with Each Side Separately

Verify the identity $\dfrac{\cot^2 \theta}{1 + \csc \theta} = \dfrac{1 - \sin \theta}{\sin \theta}$.

Algebraic Solution

Working with the left side, you have

$$\frac{\cot^2 \theta}{1 + \csc \theta} = \frac{\csc^2 \theta - 1}{1 + \csc \theta} \qquad \text{Pythagorean identity}$$

$$= \frac{(\csc \theta - 1)\cancel{(\csc \theta + 1)}}{\cancel{1 + \csc \theta}} \qquad \text{Factor.}$$

$$= \csc \theta - 1. \qquad \text{Simplify.}$$

Now, simplifying the right side, you have

$$\frac{1 - \sin \theta}{\sin \theta} = \frac{1}{\sin \theta} - \frac{\sin \theta}{\sin \theta} = \csc \theta - 1.$$

This verifies the identity because both sides are equal to $\csc \theta - 1$.

Numerical Solution

Use a graphing utility to create a table that shows the values of

$$y_1 = \frac{\cot^2 x}{1 + \csc x} \quad \text{and} \quad y_2 = \frac{1 - \sin x}{\sin x}$$

for different values of x.

X	Y1	Y2
-.5	-3.086	-3.086
-.25	-5.042	-5.042
0	ERROR	ERROR
.25	3.042	3.042
.5	1.0858	1.0858
.75	.46705	.46705
1	.1884	.1884
X=1		

The values for y_1 and y_2 appear to be identical, so the equation appears to be an identity.

✓ **Checkpoint** *Audio-video solution in English & Spanish at LarsonPrecalculus.com*

Verify the identity $\dfrac{\tan^2 \theta}{1 + \sec \theta} = \dfrac{1 - \cos \theta}{\cos \theta}$.

Example 7 shows powers of trigonometric functions rewritten as more complicated sums of products of trigonometric functions. This is a common procedure used in calculus.

EXAMPLE 7 Two Examples from Calculus

Verify each identity.

a. $\tan^4 x = \tan^2 x \sec^2 x - \tan^2 x$ **b.** $\csc^4 x \cot x = \csc^2 x(\cot x + \cot^3 x)$

Solution

a. $\tan^4 x = (\tan^2 x)(\tan^2 x)$ Write as separate factors.

$= \tan^2 x(\sec^2 x - 1)$ Pythagorean identity

$= \tan^2 x \sec^2 x - \tan^2 x$ Multiply.

b. $\csc^4 x \cot x = \csc^2 x \csc^2 x \cot x$ Write as separate factors.

$= \csc^2 x(1 + \cot^2 x) \cot x$ Pythagorean identity

$= \csc^2 x(\cot x + \cot^3 x)$ Multiply.

✓ **Checkpoint** *Audio-video solution in English & Spanish at LarsonPrecalculus.com*

Verify each identity.

a. $\tan^3 x = \tan x \sec^2 x - \tan x$ **b.** $\sin^3 x \cos^4 x = (\cos^4 x - \cos^6 x) \sin x$

Summarize (Section 2.2)

1. State the guidelines for verifying trigonometric identities *(page 217)*. For examples of verifying trigonometric identities, see Examples 1–7.

2.2 Exercises

See **CalcChat.com** for tutorial help and worked-out solutions to odd-numbered exercises.

Vocabulary

In Exercises 1 and 2, fill in the blanks.

1. An equation that is true for all real values in the domain of the variable is an ________.

2. An equation that is true for only some values in the domain of the variable is a ________ ________.

In Exercises 3–8, fill in the blank to complete the fundamental trigonometric identity.

3. $\dfrac{1}{\cot u} =$ ______ **4.** $\dfrac{\cos u}{\sin u} =$ ______ **5.** $\cos\left(\dfrac{\pi}{2} - u\right) =$ ______

6. $1 +$ ______ $= \csc^2 u$ **7.** $\csc(-u) =$ ______ **8.** $\sec(-u) =$ ______

Skills and Applications

Verifying a Trigonometric Identity

In Exercises 9–18, verify the identity.

9. $\tan t \cot t = 1$

10. $\dfrac{\tan x \cot x}{\cos x} = \sec x$

11. $(1 + \sin \alpha)(1 - \sin \alpha) = \cos^2 \alpha$

12. $\cos^2 \beta - \sin^2 \beta = 2\cos^2 \beta - 1$

13. $\cos^2 \beta - \sin^2 \beta = 1 - 2\sin^2 \beta$

14. $\sin^2 \alpha - \sin^4 \alpha = \cos^2 \alpha - \cos^4 \alpha$

15. $\tan\left(\dfrac{\pi}{2} - \theta\right)\tan \theta = 1$

16. $\dfrac{\cos[(\pi/2) - x]}{\sin[(\pi/2) - x]} = \tan x$

17. $\sin t \csc\left(\dfrac{\pi}{2} - t\right) = \tan t$

18. $\sec^2 y - \cot^2\left(\dfrac{\pi}{2} - y\right) = 1$

Verifying a Trigonometric Identity

In Exercises 19–24, verify the identity algebraically. Use the *table* feature of a graphing utility to check your result numerically.

19. $\dfrac{1}{\tan x} + \dfrac{1}{\cot x} = \tan x + \cot x$

20. $\dfrac{1}{\sin x} - \dfrac{1}{\csc x} = \csc x - \sin x$

21. $\dfrac{1 + \sin \theta}{\cos \theta} + \dfrac{\cos \theta}{1 + \sin \theta} = 2\sec \theta$

22. $\dfrac{\cos \theta \cot \theta}{1 - \sin \theta} - 1 = \csc \theta$

23. $\dfrac{1}{\cos x + 1} + \dfrac{1}{\cos x - 1} = -2\csc x \cot x$

24. $\cos x - \dfrac{\cos x}{1 - \tan x} = \dfrac{\sin x \cos x}{\sin x - \cos x}$

Verifying a Trigonometric Identity

In Exercises 25–30, verify the identity algebraically. Use a graphing utility to check your result graphically.

25. $\sec y \cos y = 1$

26. $\cot^2 y(\sec^2 y - 1) = 1$

27. $\dfrac{\tan^2 \theta}{\sec \theta} = \sin \theta \tan \theta$

28. $\dfrac{\cot^3 t}{\csc t} = \cos t(\csc^2 t - 1)$

29. $\dfrac{1}{\tan \beta} + \tan \beta = \dfrac{\sec^2 \beta}{\tan \beta}$

30. $\dfrac{\sec \theta - 1}{1 - \cos \theta} = \sec \theta$

Converting to Sines and Cosines

In Exercises 31–36, verify the identity by converting the left side into sines and cosines.

31. $\dfrac{\cot^2 t}{\csc t} = \dfrac{1 - \sin^2 t}{\sin t}$

32. $\cos x + \sin x \tan x = \sec x$

33. $\sec x - \cos x = \sin x \tan x$

34. $\cot x - \tan x = \sec x(\csc x - 2\sin x)$

35. $\dfrac{\cot x}{\sec x} = \csc x - \sin x$

36. $\dfrac{\csc(-x)}{\sec(-x)} = -\cot x$

Verifying a Trigonometric Identity

In Exercises 37–42, verify the identity.

37. $\sin^{1/2} x \cos x - \sin^{5/2} x \cos x = \cos^3 x\sqrt{\sin x}$

38. $\sec^6 x(\sec x \tan x) - \sec^4 x(\sec x \tan x) = \sec^5 x \tan^3 x$

39. $(1 + \sin y)[1 + \sin(-y)] = \cos^2 y$

40. $\dfrac{\tan x + \tan y}{1 - \tan x \tan y} = \dfrac{\cot x + \cot y}{\cot x \cot y - 1}$

41. $\sqrt{\dfrac{1 + \sin \theta}{1 - \sin \theta}} = \dfrac{1 + \sin \theta}{|\cos \theta|}$

42. $\dfrac{\cos x - \cos y}{\sin x + \sin y} + \dfrac{\sin x - \sin y}{\cos x + \cos y} = 0$

Error Analysis **In Exercises 43 and 44, describe the error(s).**

43. $\frac{1}{\tan x} + \cot(-x) = \cot x + \cot x = 2 \cot x$

44. $\frac{1 + \sec(-\theta)}{\sin(-\theta) + \tan(-\theta)} = \frac{1 - \sec\theta}{\sin\theta - \tan\theta}$
$= \frac{1 - \sec\theta}{(\sin\theta)[1 - (1/\cos\theta)]}$
$= \frac{1 - \sec\theta}{\sin\theta(1 - \sec\theta)}$
$= \frac{1}{\sin\theta}$
$= \csc\theta$

Determining Trigonometric Identities **In Exercises 45–50, (a) use a graphing utility to graph each side of the equation to determine whether the equation is an identity, (b) use the *table* feature of the graphing utility to determine whether the equation is an identity, and (c) confirm the results of parts (a) and (b) algebraically.**

45. $(1 + \cot^2 x)(\cos^2 x) = \cot^2 x$

46. $\csc x(\csc x - \sin x) + \frac{\sin x - \cos x}{\sin x} + \cot x = \csc^2 x$

47. $2 + \cos^2 x - 3\cos^4 x = \sin^2 x(3 + 2\cos^2 x)$

48. $\tan^4 x + \tan^2 x - 3 = \sec^2 x(4\tan^2 x - 3)$

49. $\frac{1 + \cos x}{\sin x} = \frac{\sin x}{1 - \cos x}$

50. $\frac{\cot\alpha}{\csc\alpha + 1} = \frac{\csc\alpha + 1}{\cot\alpha}$

Verifying a Trigonometric Identity **In Exercises 51–54, verify the identity.**

51. $\tan^5 x = \tan^3 x \sec^2 x - \tan^3 x$

52. $\sec^4 x \tan^2 x = (\tan^2 x + \tan^4 x)\sec^2 x$

53. $\cos^3 x \sin^2 x = (\sin^2 x - \sin^4 x)\cos x$

54. $\sin^4 x + \cos^4 x = 1 - 2\cos^2 x + 2\cos^4 x$

Using Cofunction Identities **In Exercises 55 and 56, use the cofunction identities to evaluate the expression without using a calculator.**

55. $\sin^2 25° + \sin^2 65°$

56. $\tan^2 63° + \cot^2 16° - \sec^2 74° - \csc^2 27°$

Verifying a Trigonometric Identity **In Exercises 57–60, verify the identity.**

57. $\tan(\sin^{-1} x) = \frac{x}{\sqrt{1 - x^2}}$

58. $\cos(\sin^{-1} x) = \sqrt{1 - x^2}$

59. $\tan\left(\sin^{-1}\frac{x - 1}{4}\right) = \frac{x - 1}{\sqrt{16 - (x - 1)^2}}$

60. $\tan\left(\cos^{-1}\frac{x + 1}{2}\right) = \frac{\sqrt{4 - (x + 1)^2}}{x + 1}$

61. **Rate of Change** The rate of change of the function $f(x) = \sin x + \csc x$ is given by the expression $\cos x - \csc x \cot x$. Show that the expression for the rate of change can also be written as $-\cos x \cot^2 x$.

62. **Shadow Length** The length s of a shadow cast by a vertical gnomon (a device used to tell time) of height h when the angle of the sun above the horizon is θ can be modeled by the equation

$$s = \frac{h \sin(90° - \theta)}{\sin\theta},$$

$0° < \theta \le 90°$.

(a) Verify that the expression for s is equal to $h \cot\theta$.

(b) Use a graphing utility to create a table of the lengths s for different values of θ. Let $h = 5$ feet.

(c) Use your table from part (b) to determine the angle of the sun that results in the minimum length of the shadow.

(d) Based on your results from part (c), what time of day do you think it is when the angle of the sun above the horizon is 90°?

Exploration

True or False? **In Exercises 63–65, determine whether the statement is true or false. Justify your answer.**

63. $\tan x^2 = \tan^2 x$

64. $\cos\left(\theta - \frac{\pi}{2}\right) = \sin\theta$

65. The equation $\sin^2\theta + \cos^2\theta = 1 + \tan^2\theta$ is an identity because $\sin^2(0) + \cos^2(0) = 1$ and $1 + \tan^2(0) = 1$.

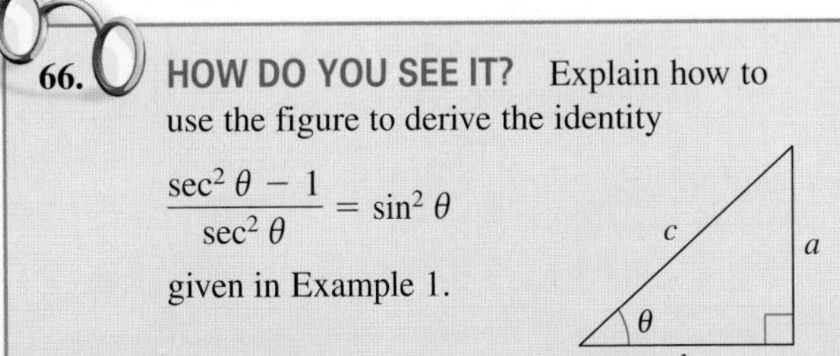

66. **HOW DO YOU SEE IT?** Explain how to use the figure to derive the identity

$$\frac{\sec^2\theta - 1}{\sec^2\theta} = \sin^2\theta$$

given in Example 1.

Think About It **In Exercises 67–70, explain why the equation is not an identity and find one value of the variable for which the equation is not true.**

67. $\sin\theta = \sqrt{1 - \cos^2\theta}$

68. $\tan\theta = \sqrt{\sec^2\theta - 1}$

69. $1 - \cos\theta = \sin\theta$

70. $1 + \tan\theta = \sec\theta$

2.3 Solving Trigonometric Equations

Trigonometric equations have many applications in circular motion. For example, in Exercise 94 on page 235, you will solve a trigonometric equation to determine when a person riding a Ferris wheel will be at certain heights above the ground.

- **Use standard algebraic techniques to solve trigonometric equations.**
- **Solve trigonometric equations of quadratic type.**
- **Solve trigonometric equations involving multiple angles.**
- **Use inverse trigonometric functions to solve trigonometric equations.**

Introduction

To solve a trigonometric equation, use standard algebraic techniques (when possible) such as collecting like terms, extracting square roots, and factoring. Your preliminary goal in solving a trigonometric equation is to *isolate* the trigonometric function on one side of the equation. For example, to solve the equation $2 \sin x = 1$, divide each side by 2 to obtain

$$\sin x = \frac{1}{2}.$$

To solve for x, note in the graph of $y = \sin x$ below that the equation $\sin x = \frac{1}{2}$ has solutions $x = \pi/6$ and $x = 5\pi/6$ in the interval $[0, 2\pi)$. Moreover, because $\sin x$ has a period of 2π, there are infinitely many other solutions, which can be written as

$$x = \frac{\pi}{6} + 2n\pi \quad \text{and} \quad x = \frac{5\pi}{6} + 2n\pi \qquad \text{General solution}$$

where n is an integer. Notice the solutions for $n = \pm 1$ in the graph of $y = \sin x$.

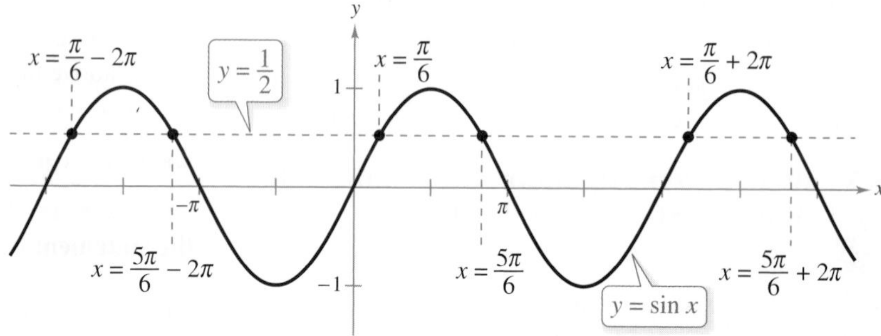

The figure below illustrates another way to show that the equation $\sin x = \frac{1}{2}$ has infinitely many solutions. Any angles that are coterminal with $\pi/6$ or $5\pi/6$ are also solutions of the equation.

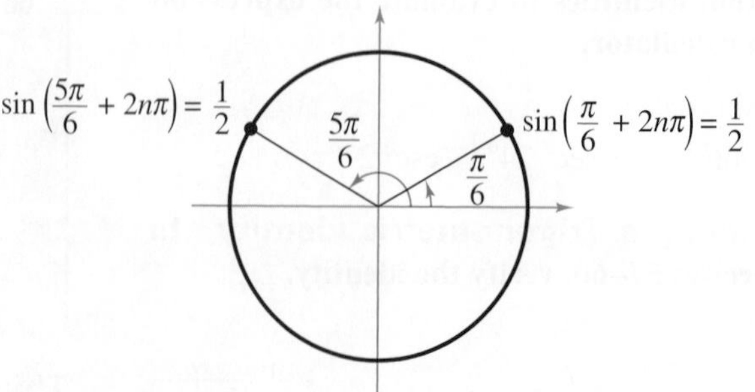

When solving trigonometric equations, write your answer(s) using exact values (when possible) rather than decimal approximations.

EXAMPLE 1 Collecting Like Terms

Solve

$$\sin x + \sqrt{2} = -\sin x.$$

Solution Begin by isolating $\sin x$ on one side of the equation.

$$\sin x + \sqrt{2} = -\sin x \qquad \text{Write original equation.}$$

$$\sin x + \sin x + \sqrt{2} = 0 \qquad \text{Add } \sin x \text{ to each side.}$$

$$\sin x + \sin x = -\sqrt{2} \qquad \text{Subtract } \sqrt{2} \text{ from each side.}$$

$$2 \sin x = -\sqrt{2} \qquad \text{Combine like terms.}$$

$$\sin x = -\frac{\sqrt{2}}{2} \qquad \text{Divide each side by 2.}$$

The period of $\sin x$ is 2π, so first find all solutions in the interval $[0, 2\pi)$. These solutions are $x = 5\pi/4$ and $x = 7\pi/4$. Finally, add multiples of 2π to each of these solutions to obtain the general form

$$x = \frac{5\pi}{4} + 2n\pi \quad \text{and} \quad x = \frac{7\pi}{4} + 2n\pi \qquad \text{General solution}$$

where n is an integer.

✓ ***Checkpoint*** *Audio-video solution in English & Spanish at LarsonPrecalculus.com*

Solve $\sin x - \sqrt{2} = -\sin x$.

EXAMPLE 2 Extracting Square Roots

Solve

$$3 \tan^2 x - 1 = 0.$$

Solution Begin by isolating $\tan x$ on one side of the equation.

$$3 \tan^2 x - 1 = 0 \qquad \text{Write original equation.}$$

$$3 \tan^2 x = 1 \qquad \text{Add 1 to each side.}$$

$$\tan^2 x = \frac{1}{3} \qquad \text{Divide each side by 3.}$$

$$\tan x = \pm\frac{1}{\sqrt{3}} \qquad \text{Extract square roots.}$$

$$\tan x = \pm\frac{\sqrt{3}}{3} \qquad \text{Rationalize the denominator.}$$

REMARK When you extract square roots, make sure you account for both the positive and negative solutions.

The period of $\tan x$ is π, so first find all solutions in the interval $[0, \pi)$. These solutions are $x = \pi/6$ and $x = 5\pi/6$. Finally, add multiples of π to each of these solutions to obtain the general form

$$x = \frac{\pi}{6} + n\pi \quad \text{and} \quad x = \frac{5\pi}{6} + n\pi \qquad \text{General solution}$$

where n is an integer.

Solve $4 \sin^2 x - 3 = 0$.

The equations in Examples 1 and 2 involved only one trigonometric function. When two or more functions occur in the same equation, collect all terms on one side and try to separate the functions by factoring or by using appropriate identities. This may produce factors that yield no solutions, as illustrated in Example 3.

EXAMPLE 3 Factoring

Solve $\cot x \cos^2 x = 2 \cot x$.

Solution Begin by collecting all terms on one side of the equation and factoring.

$$\cot x \cos^2 x = 2 \cot x \qquad \text{Write original equation.}$$

$$\cot x \cos^2 x - 2 \cot x = 0 \qquad \text{Subtract } 2 \cot x \text{ from each side.}$$

$$\cot x(\cos^2 x - 2) = 0 \qquad \text{Factor.}$$

Set each factor equal to zero and isolate the trigonometric function, if necessary.

$$\cot x = 0 \quad \text{or} \quad \cos^2 x - 2 = 0$$

$$\cos^2 x = 2$$

$$\cos x = \pm\sqrt{2}$$

In the interval $(0, \pi)$, the equation $\cot x = 0$ has the solution

$$x = \frac{\pi}{2}.$$

No solution exists for $\cos x = \pm\sqrt{2}$ because $\pm\sqrt{2}$ are outside the range of the cosine function. The period of $\cot x$ is π, so add multiples of π to $x = \pi/2$ to get the general form

$$x = \frac{\pi}{2} + n\pi \qquad \text{General solution}$$

where n is an integer. Confirm this graphically by sketching the graph of $y = \cot x \cos^2 x - 2 \cot x$.

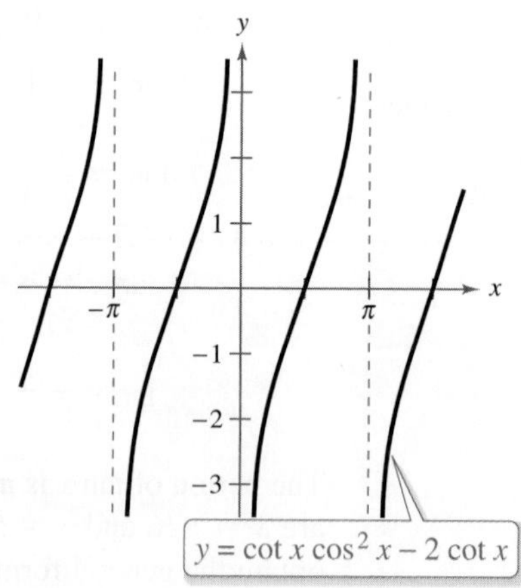

Notice that the x-intercepts occur at

$$-\frac{3\pi}{2}, \quad -\frac{\pi}{2}, \quad \frac{\pi}{2}, \quad \frac{3\pi}{2}$$

and so on. These x-intercepts correspond to the solutions of $\cot x \cos^2 x = 2 \cot x$.

✓ Checkpoint *Audio-video solution in English & Spanish at LarsonPrecalculus.com*

Solve $\sin^2 x = 2 \sin x$.

▷ **ALGEBRA HELP** To review the techniques for solving quadratic equations, see Section P.2.

Equations of Quadratic Type

Below are two examples of trigonometric equations of quadratic type

$ax^2 + bx + c = 0.$

To solve equations of this type, use factoring (when possible) or use the Quadratic Formula.

Quadratic in sin *x*	**Quadratic in sec *x***
$2\sin^2 x - \sin x - 1 = 0$	$\sec^2 x - 3\sec x - 2 = 0$
$2(\sin x)^2 - (\sin x) - 1 = 0$	$(\sec x)^2 - 3(\sec x) - 2 = 0$

EXAMPLE 4 Solving an Equation of Quadratic Type

Find all solutions of $2\sin^2 x - \sin x - 1 = 0$ in the interval $[0, 2\pi)$.

Algebraic Solution

Treat the equation as quadratic in $\sin x$ and factor.

$$2\sin^2 x - \sin x - 1 = 0 \quad \text{Write original equation.}$$

$$(2\sin x + 1)(\sin x - 1) = 0 \quad \text{Factor.}$$

Setting each factor equal to zero, you obtain the following solutions in the interval $[0, 2\pi)$.

$$2\sin x + 1 = 0 \quad \text{or} \quad \sin x - 1 = 0$$

$$\sin x = -\frac{1}{2} \qquad \sin x = 1$$

$$x = \frac{7\pi}{6}, \frac{11\pi}{6} \qquad x = \frac{\pi}{2}$$

Graphical Solution

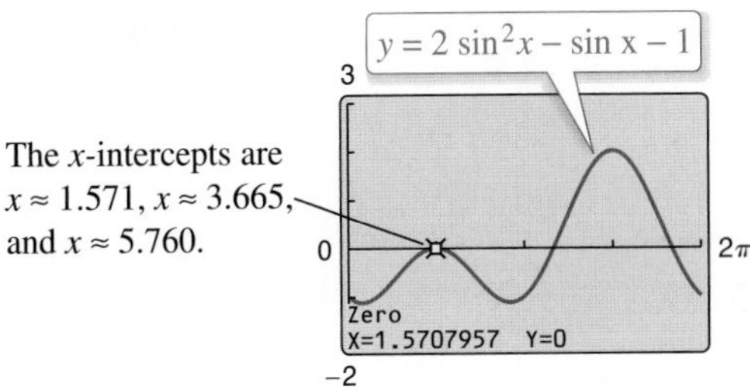

Use the x-intercepts to conclude that the approximate solutions of $2\sin^2 x - \sin x - 1 = 0$ in the interval $[0, 2\pi)$ are

$$x \approx 1.571 \approx \frac{\pi}{2}, \; x \approx 3.665 \approx \frac{7\pi}{6}, \text{ and } x \approx 5.760 \approx \frac{11\pi}{6}.$$

✓ ***Checkpoint*** *Audio-video solution in English & Spanish at LarsonPrecalculus.com*

Find all solutions of $2\sin^2 x - 3\sin x + 1 = 0$ in the interval $[0, 2\pi)$.

EXAMPLE 5 Rewriting with a Single Trigonometric Function

Solve $2\sin^2 x + 3\cos x - 3 = 0$.

Solution This equation contains both sine and cosine functions. Rewrite the equation so that it has only cosine functions by using the identity $\sin^2 x = 1 - \cos^2 x$.

$$2\sin^2 x + 3\cos x - 3 = 0 \quad \text{Write original equation.}$$

$$2(1 - \cos^2 x) + 3\cos x - 3 = 0 \quad \text{Pythagorean identity}$$

$$2\cos^2 x - 3\cos x + 1 = 0 \quad \text{Multiply each side by } -1.$$

$$(2\cos x - 1)(\cos x - 1) = 0 \quad \text{Factor.}$$

Setting each factor equal to zero, you obtain the solutions $x = 0$, $x = \pi/3$, and $x = 5\pi/3$ in the interval $[0, 2\pi)$. Because $\cos x$ has a period of 2π, the general solution is

$$x = 2n\pi, \quad x = \frac{\pi}{3} + 2n\pi, \quad \text{and} \quad x = \frac{5\pi}{3} + 2n\pi \quad \text{General solution}$$

where n is an integer.

✓ ***Checkpoint*** *Audio-video solution in English & Spanish at LarsonPrecalculus.com*

Solve $3\sec^2 x - 2\tan^2 x - 4 = 0$. ■

Sometimes you square each side of an equation to obtain an equation of quadratic type, as demonstrated in the next example. This procedure can introduce extraneous solutions, so check any solutions in the original equation to determine whether they are valid or extraneous.

REMARK You square each side of the equation in Example 6 because the squares of the sine and cosine functions are related by a Pythagorean identity. The same is true for the squares of the secant and tangent functions and for the squares of the cosecant and cotangent functions.

EXAMPLE 6 Squaring and Converting to Quadratic Type

See LarsonPrecalculus.com for an interactive version of this type of example.

Find all solutions of $\cos x + 1 = \sin x$ in the interval $[0, 2\pi)$.

Solution It is not clear how to rewrite this equation in terms of a single trigonometric function. Notice what happens when you square each side of the equation.

$$\cos x + 1 = \sin x \qquad \text{Write original equation.}$$

$$\cos^2 x + 2\cos x + 1 = \sin^2 x \qquad \text{Square each side.}$$

$$\cos^2 x + 2\cos x + 1 = 1 - \cos^2 x \qquad \text{Pythagorean identity}$$

$$\cos^2 x + \cos^2 x + 2\cos x + 1 - 1 = 0 \qquad \text{Rewrite equation.}$$

$$2\cos^2 x + 2\cos x = 0 \qquad \text{Combine like terms.}$$

$$2\cos x(\cos x + 1) = 0 \qquad \text{Factor.}$$

Set each factor equal to zero and solve for x.

$$2\cos x = 0 \qquad \text{or} \qquad \cos x + 1 = 0$$

$$\cos x = 0 \qquad\qquad \cos x = -1$$

$$x = \frac{\pi}{2}, \frac{3\pi}{2} \qquad\qquad x = \pi$$

Because you squared the original equation, check for extraneous solutions.

Check $x = \frac{\pi}{2}$

$$\cos\frac{\pi}{2} + 1 \stackrel{?}{=} \sin\frac{\pi}{2} \qquad \text{Substitute } \frac{\pi}{2} \text{ for } x.$$

$$0 + 1 = 1 \qquad \text{Solution checks. ✓}$$

Check $x = \frac{3\pi}{2}$

$$\cos\frac{3\pi}{2} + 1 \stackrel{?}{=} \sin\frac{3\pi}{2} \qquad \text{Substitute } \frac{3\pi}{2} \text{ for } x.$$

$$0 + 1 \neq -1 \qquad \text{Solution does not check.}$$

Check $x = \pi$

$$\cos\pi + 1 \stackrel{?}{=} \sin\pi \qquad \text{Substitute } \pi \text{ for } x.$$

$$-1 + 1 = 0 \qquad \text{Solution checks. ✓}$$

Of the three possible solutions, $x = 3\pi/2$ is extraneous. So, in the interval $[0, 2\pi)$, the only two solutions are

$$x = \frac{\pi}{2} \quad \text{and} \quad x = \pi.$$

✓ ***Checkpoint*** *Audio-video solution in English & Spanish at LarsonPrecalculus.com*

Find all solutions of $\sin x + 1 = \cos x$ in the interval $[0, 2\pi)$.

Functions Involving Multiple Angles

The next two examples involve trigonometric functions of multiple angles of the forms cos ku and tan ku. To solve equations involving these forms, first solve the equation for ku, and then divide your result by k.

EXAMPLE 7 Solving a Multiple-Angle Equation

Solve $2 \cos 3t - 1 = 0$.

Solution

$$2 \cos 3t - 1 = 0 \quad \text{Write original equation.}$$

$$2 \cos 3t = 1 \quad \text{Add 1 to each side.}$$

$$\cos 3t = \frac{1}{2} \quad \text{Divide each side by 2.}$$

In the interval $[0, 2\pi)$, you know that $3t = \pi/3$ and $3t = 5\pi/3$ are the only solutions, so, in general, you have

$$3t = \frac{\pi}{3} + 2n\pi \quad \text{and} \quad 3t = \frac{5\pi}{3} + 2n\pi.$$

Dividing these results by 3, you obtain the general solution

$$t = \frac{\pi}{9} + \frac{2n\pi}{3} \quad \text{and} \quad t = \frac{5\pi}{9} + \frac{2n\pi}{3} \quad \text{General solution}$$

where n is an integer.

✓ **Checkpoint** Audio-video solution in English & Spanish at LarsonPrecalculus.com

Solve $2 \sin 2t - \sqrt{3} = 0$.

EXAMPLE 8 Solving a Multiple-Angle Equation

$$3 \tan \frac{x}{2} + 3 = 0 \quad \text{Original equation}$$

$$3 \tan \frac{x}{2} = -3 \quad \text{Subtract 3 from each side.}$$

$$\tan \frac{x}{2} = -1 \quad \text{Divide each side by 3.}$$

In the interval $[0, \pi)$, you know that $x/2 = 3\pi/4$ is the only solution, so, in general, you have

$$\frac{x}{2} = \frac{3\pi}{4} + n\pi.$$

Multiplying this result by 2, you obtain the general solution

$$x = \frac{3\pi}{2} + 2n\pi \quad \text{General solution}$$

where n is an integer.

✓ **Checkpoint** Audio-video solution in English & Spanish at LarsonPrecalculus.com

Solve $2 \tan \frac{x}{2} - 2 = 0$.

Using Inverse Functions

EXAMPLE 9 Using Inverse Functions

$$\sec^2 x - 2\tan x = 4 \quad \text{Original equation}$$

$$1 + \tan^2 x - 2\tan x - 4 = 0 \quad \text{Pythagorean identity}$$

$$\tan^2 x - 2\tan x - 3 = 0 \quad \text{Combine like terms.}$$

$$(\tan x - 3)(\tan x + 1) = 0 \quad \text{Factor.}$$

Setting each factor equal to zero, you obtain two solutions in the interval $(-\pi/2, \pi/2)$. [Recall that the range of the inverse tangent function is $(-\pi/2, \pi/2)$.]

$$x = \arctan 3 \quad \text{and} \quad x = \arctan(-1) = -\pi/4$$

Finally, $\tan x$ has a period of π, so add multiples of π to obtain

$$x = \arctan 3 + n\pi \quad \text{and} \quad x = (-\pi/4) + n\pi \quad \text{General solution}$$

where n is an integer. You can use a calculator to approximate the value of arctan 3.

Checkpoint *Audio-video solution in English & Spanish at LarsonPrecalculus.com*

Solve $4\tan^2 x + 5\tan x - 6 = 0$.

EXAMPLE 10 Using the Quadratic Formula

Find all solutions of $\sin^2 x - 3\sin x - 2 = 0$ in the interval $[0, 2\pi)$.

Solution

The expression $\sin^2 x - 3\sin x - 2$ cannot be factored, so use the Quadratic Formula.

$$\sin^2 x - 3\sin x - 2 = 0 \quad \text{Write original equation.}$$

$$\sin x = \frac{-(-3) \pm \sqrt{(-3)^2 - 4(1)(-2)}}{2(1)} \quad \text{Quadratic Formula}$$

$$\sin x = \frac{3 \pm \sqrt{17}}{2} \quad \text{Simplify.}$$

So, $\sin x = \frac{3 + \sqrt{17}}{2} \approx 3.5616$ or $\sin x = \frac{3 - \sqrt{17}}{2} \approx -0.5616$. The range of the sine function is $[-1, 1]$, so $\sin x = \frac{3 + \sqrt{17}}{2}$ has no solution for x. Use a calculator to approximate a solution of $\sin x = \frac{3 - \sqrt{17}}{2}$.

$$x = \arcsin\left(\frac{3 - \sqrt{17}}{2}\right) \approx -0.5963$$

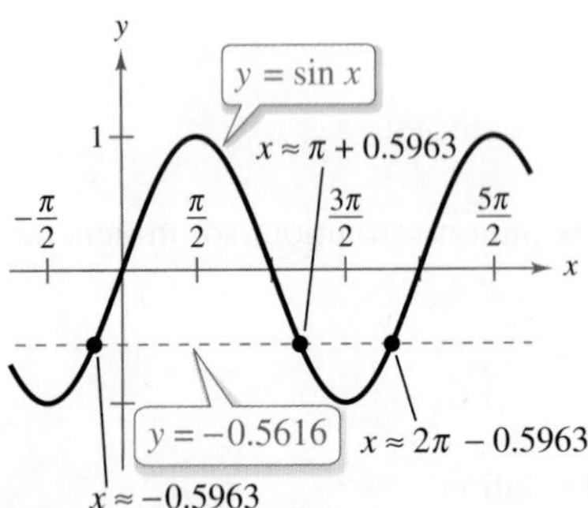

Figure 2.1

Note that this solution is not in the interval $[0, 2\pi)$. To find the solutions in $[0, 2\pi)$, sketch the graphs of $y = \sin x$ and $y = -0.5616$, as shown in Figure 2.1. From the graph, it appears that $\sin x \approx -0.5616$ on the interval $[0, 2\pi)$ when

$$x \approx \pi + 0.5963 \approx 3.7379 \quad \text{and} \quad x \approx 2\pi - 0.5963 \approx 5.6869.$$

So, the solutions of $\sin^2 x - 3\sin x - 2 = 0$ in $[0, 2\pi)$ are $x \approx 3.7379$ and $x \approx 5.6869$.

Checkpoint *Audio-video solution in English & Spanish at LarsonPrecalculus.com*

Find all solutions of $\sin^2 x + 2\sin x - 1 = 0$ in the interval $[0, 2\pi)$.

EXAMPLE 11 Surface Area of a Honeycomb Cell

The surface area S (in square inches) of a honeycomb cell is given by

$$S = 6hs + 1.5s^2\left(\frac{\sqrt{3} - \cos\theta}{\sin\theta}\right), \quad 0° < \theta \leq 90°$$

where $h = 2.4$ inches, $s = 0.75$ inch, and θ is the angle shown in the figure at the right. What value of θ gives the minimum surface area?

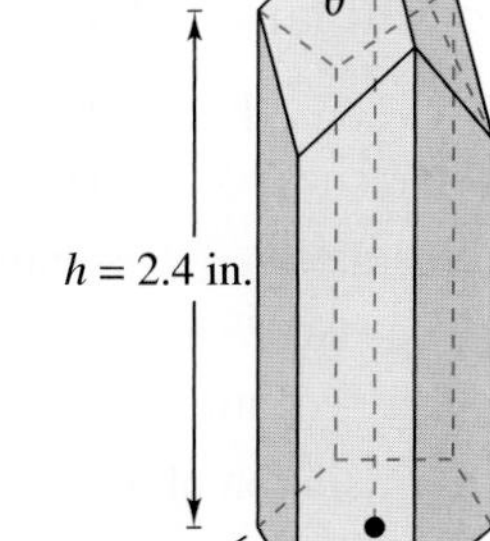

Solution

Letting $h = 2.4$ and $s = 0.75$, you obtain

$$S = 10.8 + 0.84375\left(\frac{\sqrt{3} - \cos\theta}{\sin\theta}\right).$$

Graph this function using a graphing utility set in *degree* mode. Use the *minimum* feature to approximate the minimum point on the graph, as shown in the figure below.

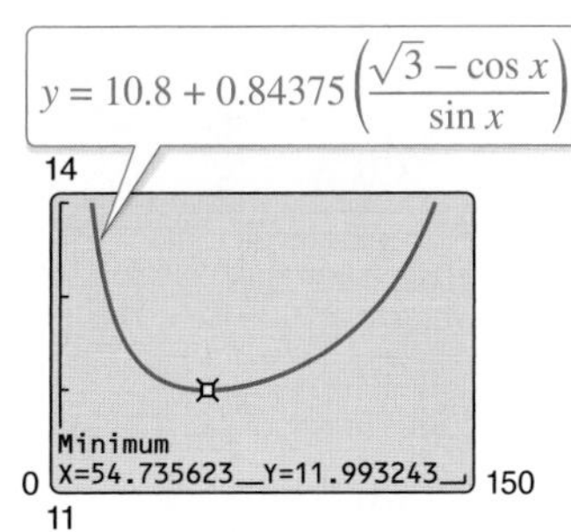

So, the minimum surface area occurs when

$$\theta \approx 54.7356°.$$

REMARK By using calculus, it can be shown that the *exact* minimum surface area occurs when

$$\theta = \arccos\left(\frac{1}{\sqrt{3}}\right).$$

Checkpoint *Audio-video solution in English & Spanish at LarsonPrecalculus.com*

Use the equation for the surface area of a honeycomb cell given in Example 11 with $h = 3.2$ inches and $s = 0.75$ inch. What value of θ gives the minimum surface area?

Summarize (Section 2.3)

1. Explain how to use standard algebraic techniques to solve trigonometric equations *(page 224)*. For examples of using standard algebraic techniques to solve trigonometric equations, see Examples 1–3.
2. Explain how to solve a trigonometric equation of quadratic type *(page 227)*. For examples of solving trigonometric equations of quadratic type, see Examples 4–6.
3. Explain how to solve a trigonometric equation involving multiple angles *(page 229)*. For examples of solving trigonometric equations involving multiple angles, see Examples 7 and 8.
4. Explain how to use inverse trigonometric functions to solve trigonometric equations *(page 230)*. For examples of using inverse trigonometric functions to solve trigonometric equations, see Examples 9–11.

2.3 Exercises

See CalcChat.com for tutorial help and worked-out solutions to odd-numbered exercises.

Vocabulary: Fill in the blanks.

1. When solving a trigonometric equation, the preliminary goal is to ________ the trigonometric function on one side of the equation.
2. The ________ solution of the equation $2 \sin \theta + 1 = 0$ is $\theta = \frac{7\pi}{6} + 2n\pi$ and $\theta = \frac{11\pi}{6} + 2n\pi$, where n is an integer.
3. The equation $2 \tan^2 x - 3 \tan x + 1 = 0$ is a trigonometric equation of ________ type.
4. A solution of an equation that does not satisfy the original equation is an ________ solution.

Skills and Applications

Verifying Solutions In Exercises 5–10, verify that each x-value is a solution of the equation.

5. $\tan x - \sqrt{3} = 0$
   (a) $x = \frac{\pi}{3}$
   (b) $x = \frac{4\pi}{3}$
6. $\sec x - 2 = 0$
   (a) $x = \frac{\pi}{3}$
   (b) $x = \frac{5\pi}{3}$
7. $3 \tan^2 2x - 1 = 0$
   (a) $x = \frac{\pi}{12}$
   (b) $x = \frac{5\pi}{12}$
8. $2 \cos^2 4x - 1 = 0$
   (a) $x = \frac{\pi}{16}$
   (b) $x = \frac{3\pi}{16}$
9. $2 \sin^2 x - \sin x - 1 = 0$
   (a) $x = \frac{\pi}{2}$
   (b) $x = \frac{7\pi}{6}$
10. $\csc^4 x - 4 \csc^2 x = 0$
   (a) $x = \frac{\pi}{6}$
   (b) $x = \frac{5\pi}{6}$

Solving a Trigonometric Equation In Exercises 11–28, solve the equation.

11. $\sqrt{3} \csc x - 2 = 0$
12. $\tan x + \sqrt{3} = 0$
13. $\cos x + 1 = -\cos x$
14. $3 \sin x + 1 = \sin x$
15. $3 \sec^2 x - 4 = 0$
16. $3 \cot^2 x - 1 = 0$
17. $4 \cos^2 x - 1 = 0$
18. $2 - 4 \sin^2 x = 0$
19. $\sin x(\sin x + 1) = 0$
20. $(2 \sin^2 x - 1)(\tan^2 x - 3) = 0$
21. $\cos^3 x - \cos x = 0$
22. $\sec^2 x - 1 = 0$
23. $3 \tan^3 x = \tan x$
24. $\sec x \csc x = 2 \csc x$
25. $2 \cos^2 x + \cos x - 1 = 0$
26. $2 \sin^2 x + 3 \sin x + 1 = 0$
27. $\sec^2 x - \sec x = 2$
28. $\csc^2 x + \csc x = 2$

Solving a Trigonometric Equation In Exercises 29–38, find all solutions of the equation in the interval $[0, 2\pi)$.

29. $\sin x - 2 = \cos x - 2$
30. $\cos x + \sin x \tan x = 2$
31. $2 \sin^2 x = 2 + \cos x$
32. $\tan^2 x = \sec x - 1$
33. $\sin^2 x = 3 \cos^2 x$
34. $2 \sec^2 x + \tan^2 x - 3 = 0$
35. $2 \sin x + \csc x = 0$
36. $3 \sec x - 4 \cos x = 0$
37. $\csc x + \cot x = 1$
38. $\sec x + \tan x = 1$

Solving a Multiple-Angle Equation In Exercises 39–46, solve the multiple-angle equation.

39. $2 \cos 2x - 1 = 0$
40. $2 \sin 2x + \sqrt{3} = 0$
41. $\tan 3x - 1 = 0$
42. $\sec 4x - 2 = 0$
43. $2 \cos \frac{x}{2} - \sqrt{2} = 0$
44. $2 \sin \frac{x}{2} + \sqrt{3} = 0$
45. $3 \tan \frac{x}{2} - \sqrt{3} = 0$
46. $\tan \frac{x}{2} + \sqrt{3} = 0$

Finding x-Intercepts In Exercises 47 and 48, find the x-intercepts of the graph.

47. $y = \sin \frac{\pi x}{2} + 1$

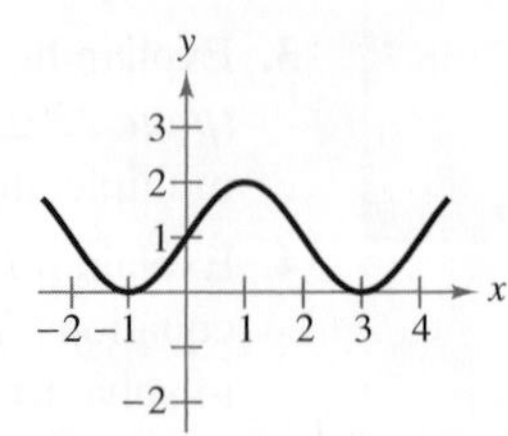

48. $y = \sin \pi x + \cos \pi x$

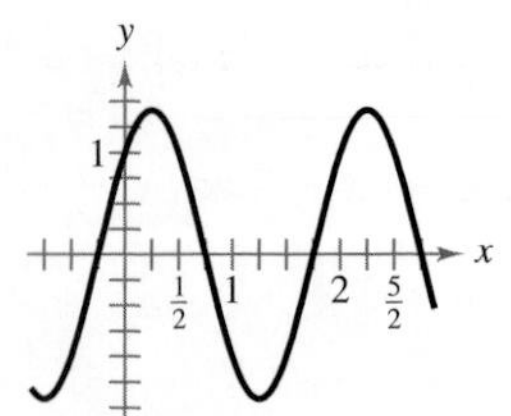

Approximating Solutions **In Exercises 49–58, use a graphing utility to approximate (to three decimal places) the solutions of the equation in the interval $[0, 2\pi)$.**

49. $5 \sin x + 2 = 0$

50. $2 \tan x + 7 = 0$

51. $\sin x - 3 \cos x = 0$

52. $\sin x + 4 \cos x = 0$

53. $\cos x = x$

54. $\tan x = \csc x$

55. $\sec^2 x - 3 = 0$

56. $\csc^2 x - 5 = 0$

57. $2 \tan^2 x = 15$

58. $6 \sin^2 x = 5$

Using Inverse Functions **In Exercises 59–70, solve the equation.**

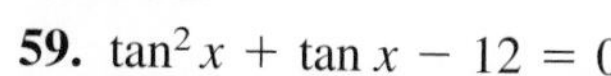

59. $\tan^2 x + \tan x - 12 = 0$

60. $\tan^2 x - \tan x - 2 = 0$

61. $\sec^2 x - 6 \tan x = -4$

62. $\sec^2 x + \tan x = 3$

63. $2 \sin^2 x + 5 \cos x = 4$

64. $2 \cos^2 x + 7 \sin x = 5$

65. $\cot^2 x - 9 = 0$

66. $\cot^2 x - 6 \cot x + 5 = 0$

67. $\sec^2 x - 4 \sec x = 0$

68. $\sec^2 x + 2 \sec x - 8 = 0$

69. $\csc^2 x + 3 \csc x - 4 = 0$

70. $\csc^2 x - 5 \csc x = 0$

Using the Quadratic Formula **In Exercises 71–74, use the Quadratic Formula to find all solutions of the equation in the interval $[0, 2\pi)$. Round your result to four decimal places.**

71. $12 \sin^2 x - 13 \sin x + 3 = 0$

72. $3 \tan^2 x + 4 \tan x - 4 = 0$

73. $\tan^2 x + 3 \tan x + 1 = 0$

74. $4 \cos^2 x - 4 \cos x - 1 = 0$

Approximating Solutions **In Exercises 75–78, use a graphing utility to approximate (to three decimal places) the solutions of the equation in the given interval.**

75. $3 \tan^2 x + 5 \tan x - 4 = 0, \quad \left[-\frac{\pi}{2}, \frac{\pi}{2}\right]$

76. $\cos^2 x - 2 \cos x - 1 = 0, \quad [0, \pi]$

77. $4 \cos^2 x - 2 \sin x + 1 = 0, \quad \left[-\frac{\pi}{2}, \frac{\pi}{2}\right]$

78. $2 \sec^2 x + \tan x - 6 = 0, \quad \left[-\frac{\pi}{2}, \frac{\pi}{2}\right]$

Approximating Maximum and Minimum Points **In Exercises 79–84, (a) use a graphing utility to graph the function and approximate the maximum and minimum points on the graph in the interval $[0, 2\pi)$, and (b) solve the trigonometric equation and verify that its solutions are the x-coordinates of the maximum and minimum points of f. (Calculus is required to find the trigonometric equation.)**

	Function	Trigonometric Equation
79.	$f(x) = \sin^2 x + \cos x$	$2 \sin x \cos x - \sin x = 0$
80.	$f(x) = \cos^2 x - \sin x$	$-2 \sin x \cos x - \cos x = 0$
81.	$f(x) = \sin x + \cos x$	$\cos x - \sin x = 0$
82.	$f(x) = 2 \sin x + \cos 2x$	$2 \cos x - 4 \sin x \cos x = 0$
83.	$f(x) = \sin x \cos x$	$-\sin^2 x + \cos^2 x = 0$
84.	$f(x) = \sec x + \tan x - x$	$\sec x \tan x + \sec^2 x = 1$

Number of Points of Intersection **In Exercises 85 and 86, use the graph to approximate the number of points of intersection of the graphs of y_1 and y_2.**

85. $y_1 = 2 \sin x$
$y_2 = 3x + 1$

86. $y_1 = 2 \sin x$

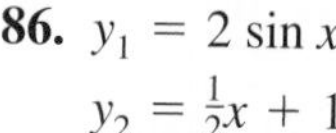

$y_2 = \frac{1}{2}x + 1$

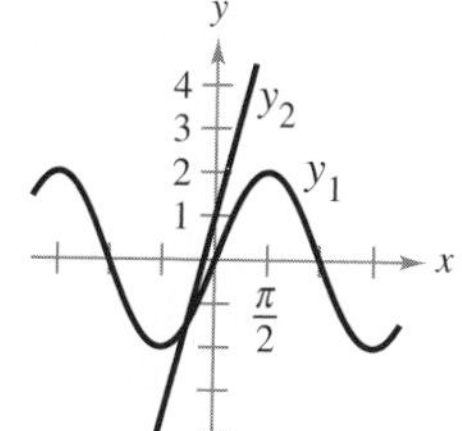

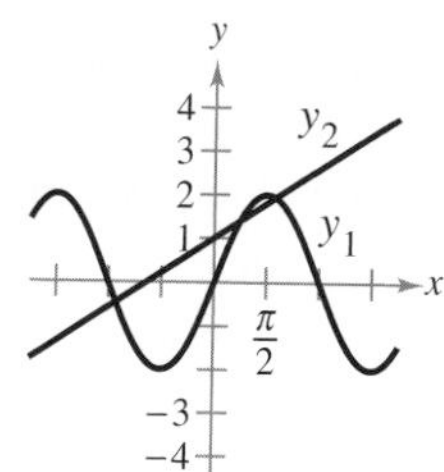

87. **Graphical Reasoning** Consider the function

$$f(x) = \frac{\sin x}{x}$$

and its graph, shown in the figure below.

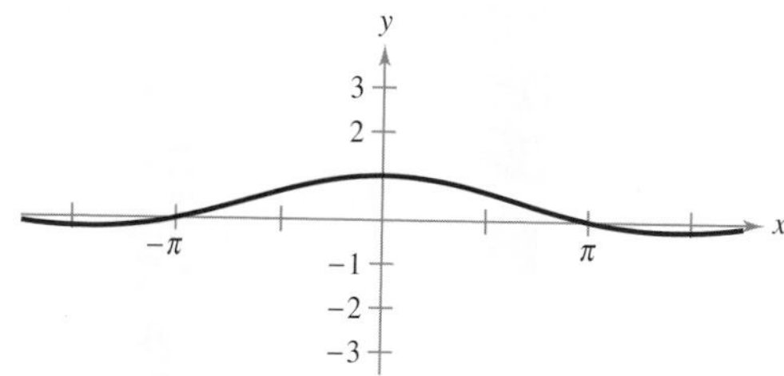

(a) What is the domain of the function?

(b) Identify any symmetry and any asymptotes of the graph.

(c) Describe the behavior of the function as $x \to 0$.

(d) How many solutions does the equation

$$\frac{\sin x}{x} = 0$$

have in the interval $[-8, 8]$? Find the solutions.

88. Graphical Reasoning Consider the function

$$f(x) = \cos \frac{1}{x}$$

and its graph, shown in the figure below.

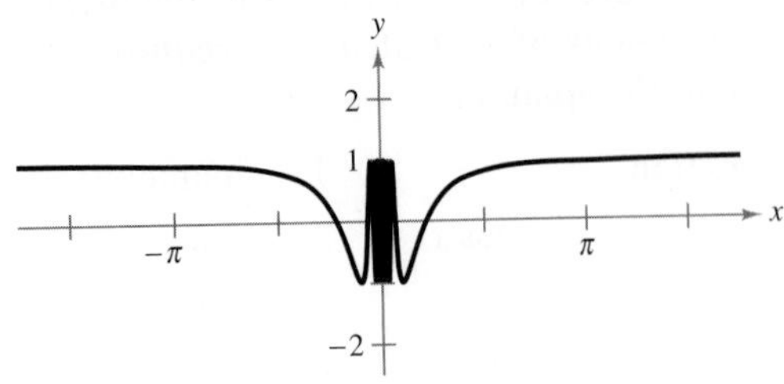

(a) What is the domain of the function?

(b) Identify any symmetry and any asymptotes of the graph.

(c) Describe the behavior of the function as $x \to 0$.

(d) How many solutions does the equation

$$\cos \frac{1}{x} = 0$$

have in the interval $[-1, 1]$? Find the solutions.

(e) Does the equation $\cos(1/x) = 0$ have a greatest solution? If so, then approximate the solution. If not, then explain why.

89. Harmonic Motion A weight is oscillating on the end of a spring (see figure). The displacement from equilibrium of the weight relative to the point of equilibrium is given by

$$y = \tfrac{1}{12}(\cos 8t - 3 \sin 8t)$$

where y is the displacement (in meters) and t is the time (in seconds). Find the times when the weight is at the point of equilibrium ($y = 0$) for $0 \le t \le 1$.

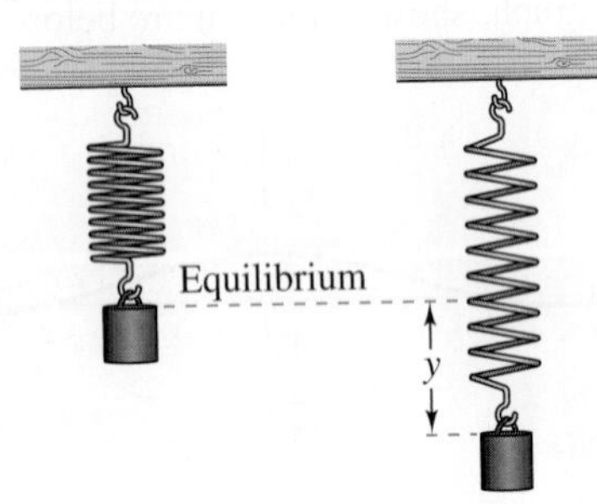

90. Damped Harmonic Motion The displacement from equilibrium of a weight oscillating on the end of a spring is given by

$$y = 1.56t^{-1/2} \cos 1.9t$$

where y is the displacement (in feet) and t is the time (in seconds). Use a graphing utility to graph the displacement function for $0 < t \le 10$. Find the time beyond which the distance between the weight and equilibrium does not exceed 1 foot.

91. Equipment Sales The monthly sales S (in hundreds of units) of skiing equipment at a sports store are approximated by

$$S = 58.3 + 32.5 \cos \frac{\pi t}{6}$$

where t is the time (in months), with $t = 1$ corresponding to January. Determine the months in which sales exceed 7500 units.

92. Projectile Motion A baseball is hit at an angle of θ with the horizontal and with an initial velocity of $v_0 = 100$ feet per second. An outfielder catches the ball 300 feet from home plate (see figure). Find θ when the range r of a projectile is given by

$$r = \frac{1}{32} v_0^2 \sin 2\theta.$$

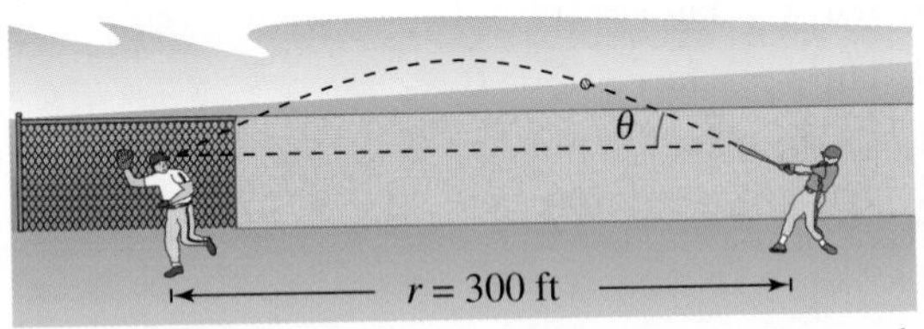

Not drawn to scale

93. Meteorology The table shows the normal daily high temperatures C in Chicago (in degrees Fahrenheit) for month t, with $t = 1$ corresponding to January. *(Source: NOAA)*

DATA

Month, t	Chicago, C
1	31.0
2	35.3
3	46.6
4	59.0
5	70.0
6	79.7
7	84.1
8	81.9
9	74.8
10	62.3
11	48.2
12	34.8

Spreadsheet at LarsonPrecalculus.com

(a) Use a graphing utility to create a scatter plot of the data.

(b) Find a cosine model for the temperatures.

(c) Graph the model and the scatter plot in the same viewing window. How well does the model fit the data?

(d) What is the overall normal daily high temperature?

(e) Use the graphing utility to determine the months during which the normal daily high temperature is above 72°F and below 72°F.

94. Ferris Wheel The height h (in feet) above ground of a seat on a Ferris wheel at time t (in minutes) can be modeled by

$$h(t) = 53 + 50 \sin\left(\frac{\pi}{16}t - \frac{\pi}{2}\right).$$

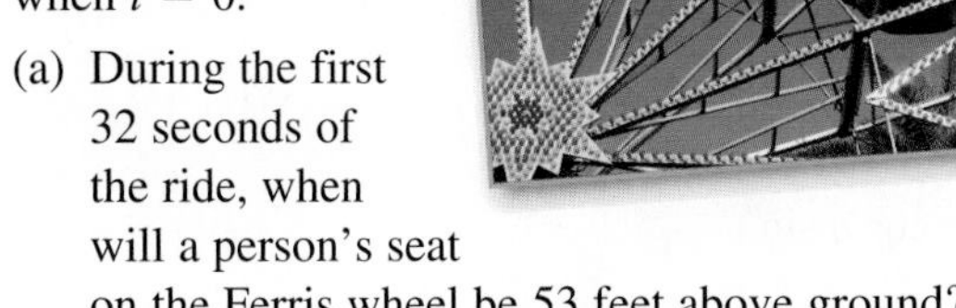

The wheel makes one revolution every 32 seconds. The ride begins when $t = 0$.

(a) During the first 32 seconds of the ride, when will a person's seat on the Ferris wheel be 53 feet above ground?

(b) When will a person's seat be at the top of the Ferris wheel for the first time during the ride? For a ride that lasts 160 seconds, how many times will a person's seat be at the top of the ride, and at what times?

95. Geometry The area of a rectangle inscribed in one arc of the graph of $y = \cos x$ (see figure) is given by

$A = 2x \cos x,\ 0 < x < \pi/2.$

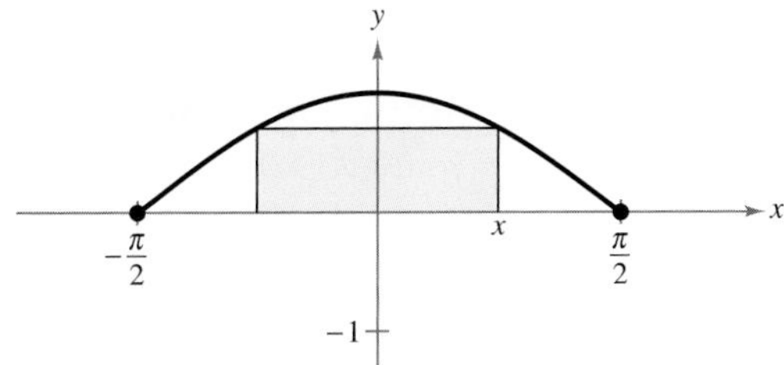

(a) Use a graphing utility to graph the area function, and approximate the area of the largest inscribed rectangle.

(b) Determine the values of x for which $A \geq 1$.

96. Quadratic Approximation Consider the function

$f(x) = 3 \sin(0.6x - 2).$

(a) Approximate the zero of the function in the interval $[0, 6]$.

(b) A quadratic approximation agreeing with f at $x = 5$ is

$g(x) = -0.45x^2 + 5.52x - 13.70.$

Use a graphing utility to graph f and g in the same viewing window. Describe the result.

(c) Use the Quadratic Formula to find the zeros of g. Compare the zero of g in the interval $[0, 6]$ with the result of part (a).

Fixed Point In Exercises 97 and 98, find the least positive fixed point of the function f. [A *fixed point* of a function f is a real number c such that $f(c) = c$.]

97. $f(x) = \tan(\pi x/4)$

98. $f(x) = \cos x$

Exploration

True or False? In Exercises 99 and 100, determine whether the statement is true or false. Justify your answer.

99. The equation $2 \sin 4t - 1 = 0$ has four times the number of solutions in the interval $[0, 2\pi)$ as the equation $2 \sin t - 1 = 0$.

100. The trigonometric equation $\sin x = 3.4$ can be solved using an inverse trigonometric function.

101. Think About It Explain what happens when you divide each side of the equation $\cot x \cos^2 x = 2 \cot x$ by $\cot x$. Is this a correct method to use when solving equations?

102. HOW DO YOU SEE IT? Explain how to use the figure to solve the equation $2 \cos x - 1 = 0$.

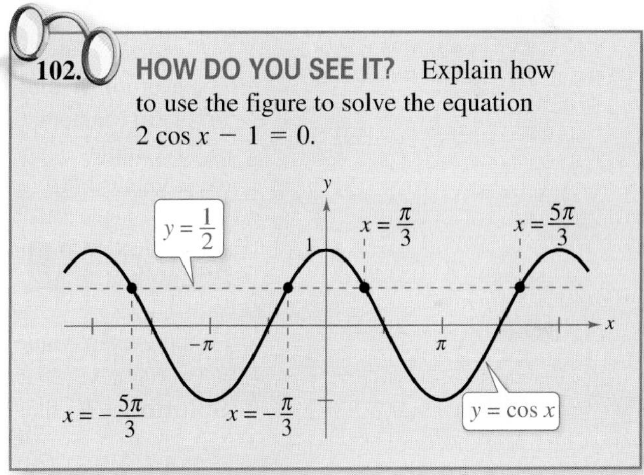

103. Graphical Reasoning Use a graphing utility to confirm the solutions found in Example 6 in two different ways.

(a) Graph both sides of the equation and find the x-coordinates of the points at which the graphs intersect.

Left side: $y = \cos x + 1$

Right side: $y = \sin x$

(b) Graph the equation $y = \cos x + 1 - \sin x$ and find the x-intercepts of the graph.

(c) Do both methods produce the same x-values? Which method do you prefer? Explain.

Project: Meteorology To work an extended application analyzing the normal daily high temperatures in Phoenix, Arizona, and in Seattle, Washington, visit this text's website at *LarsonPrecalculus.com.* *(Source: NOAA)*

2.4 Sum and Difference Formulas

Sum and difference formulas are used to model standing waves, such as those produced in a guitar string. For example, in Exercise 80 on page 241, you will use a sum formula to write the equation of a standing wave.

- **Use sum and difference formulas to evaluate trigonometric functions, verify identities, and solve trigonometric equations.**

Using Sum and Difference Formulas

In this section and the next, you will study the uses of several trigonometric identities and formulas.

Sum and Difference Formulas

$$\sin(u + v) = \sin u \cos v + \cos u \sin v$$

$$\sin(u - v) = \sin u \cos v - \cos u \sin v$$

$$\cos(u + v) = \cos u \cos v - \sin u \sin v$$

$$\cos(u - v) = \cos u \cos v + \sin u \sin v$$

$$\tan(u + v) = \frac{\tan u + \tan v}{1 - \tan u \tan v} \qquad \tan(u - v) = \frac{\tan u - \tan v}{1 + \tan u \tan v}$$

For a proof of the sum and difference formulas for $\cos(u \pm v)$ and $\tan(u \pm v)$, see Proofs in Mathematics on page 257.

Examples 1 and 2 show how **sum and difference formulas** enable you to find exact values of trigonometric functions involving sums or differences of special angles.

EXAMPLE 1 Evaluating a Trigonometric Function

Find the exact value of $\sin \frac{\pi}{12}$.

Solution To find the *exact* value of $\sin(\pi/12)$, use the fact that

$$\frac{\pi}{12} = \frac{\pi}{3} - \frac{\pi}{4}$$

with the formula for $\sin(u - v)$.

$$\begin{aligned}\sin \frac{\pi}{12} &= \sin\left(\frac{\pi}{3} - \frac{\pi}{4}\right) \\ &= \sin \frac{\pi}{3} \cos \frac{\pi}{4} - \cos \frac{\pi}{3} \sin \frac{\pi}{4} \\ &= \frac{\sqrt{3}}{2}\left(\frac{\sqrt{2}}{2}\right) - \frac{1}{2}\left(\frac{\sqrt{2}}{2}\right) \\ &= \frac{\sqrt{6} - \sqrt{2}}{4}\end{aligned}$$

Check this result on a calculator by comparing its value to $\sin(\pi/12) \approx 0.2588$.

✓ **Checkpoint** Audio-video solution in English & Spanish at LarsonPrecalculus.com

Find the exact value of $\cos \frac{\pi}{12}$.

Brian A Jackson/Shutterstock.com

REMARK Another way to solve Example 2 is to use the fact that $75^\circ = 120^\circ - 45^\circ$ with the formula for $\cos(u - v)$.

EXAMPLE 2 Evaluating a Trigonometric Function

Find the exact value of $\cos 75^\circ$.

Solution Use the fact that $75^\circ = 30^\circ + 45^\circ$ with the formula for $\cos(u + v)$.

$$\begin{aligned}\cos 75^\circ &= \cos(30^\circ + 45^\circ)\\ &= \cos 30^\circ \cos 45^\circ - \sin 30^\circ \sin 45^\circ\\ &= \frac{\sqrt{3}}{2}\left(\frac{\sqrt{2}}{2}\right) - \frac{1}{2}\left(\frac{\sqrt{2}}{2}\right)\\ &= \frac{\sqrt{6} - \sqrt{2}}{4}\end{aligned}$$

✓ **Checkpoint** Audio-video solution in English & Spanish at LarsonPrecalculus.com

Find the exact value of $\sin 75^\circ$.

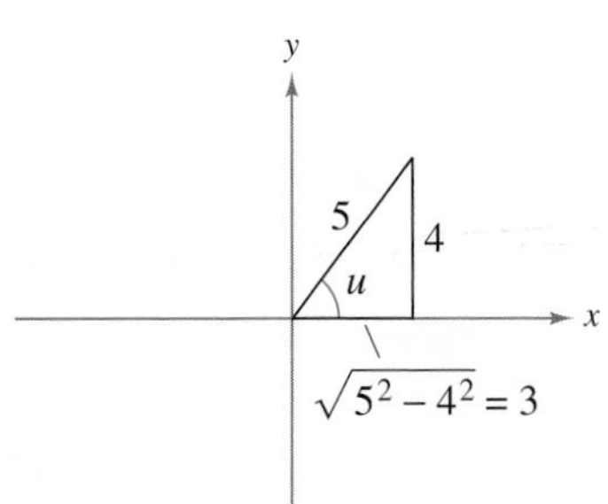

Figure 2.2

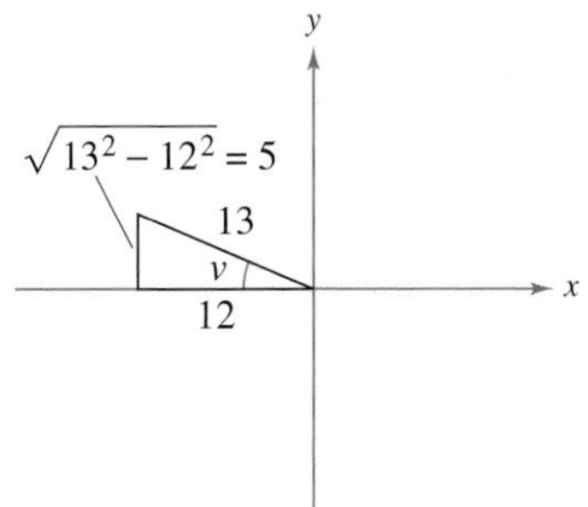

Figure 2.3

EXAMPLE 3 Evaluating a Trigonometric Expression

Find the exact value of $\sin(u + v)$ given $\sin u = 4/5$, where $0 < u < \pi/2$, and $\cos v = -12/13$, where $\pi/2 < v < \pi$.

Solution Because $\sin u = 4/5$ and u is in Quadrant I, $\cos u = 3/5$, as shown in Figure 2.2. Because $\cos v = -12/13$ and v is in Quadrant II, $\sin v = 5/13$, as shown in Figure 2.3. Use these values in the formula for $\sin(u + v)$.

$$\begin{aligned}\sin(u + v) &= \sin u \cos v + \cos u \sin v\\ &= \frac{4}{5}\left(-\frac{12}{13}\right) + \frac{3}{5}\left(\frac{5}{13}\right)\\ &= -\frac{33}{65}\end{aligned}$$

✓ **Checkpoint** Audio-video solution in English & Spanish at LarsonPrecalculus.com

Find the exact value of $\cos(u + v)$ given $\sin u = 12/13$, where $0 < u < \pi/2$, and $\cos v = -3/5$, where $\pi/2 < v < \pi$.

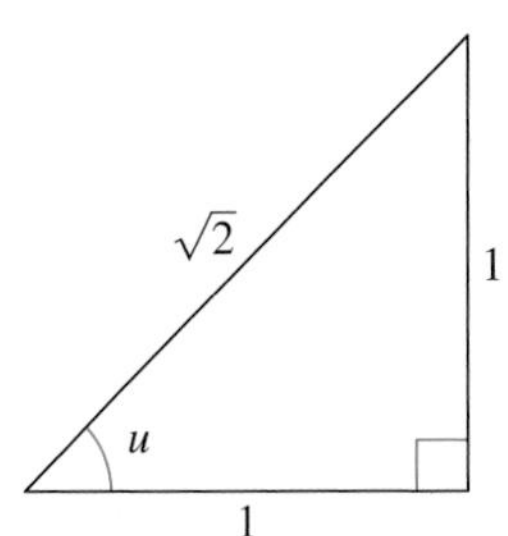

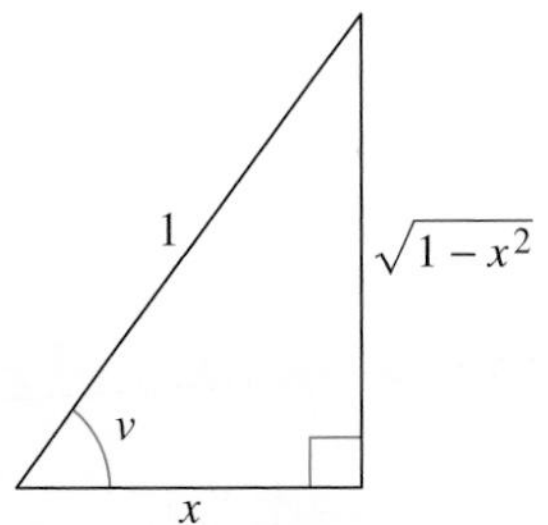

Figure 2.4

EXAMPLE 4 An Application of a Sum Formula

Write $\cos(\arctan 1 + \arccos x)$ as an algebraic expression.

Solution This expression fits the formula for $\cos(u + v)$. Figure 2.4 shows angles $u = \arctan 1$ and $v = \arccos x$.

$$\begin{aligned}\cos(u + v) &= \cos(\arctan 1)\cos(\arccos x) - \sin(\arctan 1)\sin(\arccos x)\\ &= \frac{1}{\sqrt{2}} \cdot x - \frac{1}{\sqrt{2}} \cdot \sqrt{1 - x^2}\\ &= \frac{x - \sqrt{1 - x^2}}{\sqrt{2}}\\ &= \frac{\sqrt{2}x - \sqrt{2 - 2x^2}}{2}\end{aligned}$$

✓ **Checkpoint** Audio-video solution in English & Spanish at LarsonPrecalculus.com

Write $\sin(\arctan 1 + \arccos x)$ as an algebraic expression.

Hipparchus, considered the most important of the Greek astronomers, was born about 190 B.C. in Nicaea. He is credited with the invention of trigonometry, and his work contributed to the derivation of the sum and difference formulas for $\sin(A \pm B)$ and $\cos(A \pm B)$.

EXAMPLE 5 Verifying a Cofunction Identity

See LarsonPrecalculus.com for an interactive version of this type of example.

Verify the cofunction identity $\cos\left(\frac{\pi}{2} - x\right) = \sin x$.

Solution Use the formula for $\cos(u - v)$.

$$\cos\left(\frac{\pi}{2} - x\right) = \cos\frac{\pi}{2}\cos x + \sin\frac{\pi}{2}\sin x$$
$$= (0)(\cos x) + (1)(\sin x)$$
$$= \sin x$$

✓ ***Checkpoint*** *Audio-video solution in English & Spanish at LarsonPrecalculus.com*

Verify the cofunction identity $\sin\left(x - \frac{\pi}{2}\right) = -\cos x$.

Sum and difference formulas can be used to derive **reduction formulas** for rewriting expressions such as

$$\sin\left(\theta + \frac{n\pi}{2}\right) \quad \text{and} \quad \cos\left(\theta + \frac{n\pi}{2}\right), \quad \text{where } n \text{ is an integer}$$

as trigonometric functions of only θ.

EXAMPLE 6 Deriving Reduction Formulas

Write each expression as a trigonometric function of only θ.

a. $\cos\left(\theta - \frac{3\pi}{2}\right)$

b. $\tan(\theta + 3\pi)$

Solution

a. Use the formula for $\cos(u - v)$.

$$\cos\left(\theta - \frac{3\pi}{2}\right) = \cos\theta\cos\frac{3\pi}{2} + \sin\theta\sin\frac{3\pi}{2}$$
$$= (\cos\theta)(0) + (\sin\theta)(-1)$$
$$= -\sin\theta$$

b. Use the formula for $\tan(u + v)$.

$$\tan(\theta + 3\pi) = \frac{\tan\theta + \tan 3\pi}{1 - \tan\theta\tan 3\pi}$$
$$= \frac{\tan\theta + 0}{1 - (\tan\theta)(0)}$$
$$= \tan\theta$$

✓ ***Checkpoint*** *Audio-video solution in English & Spanish at LarsonPrecalculus.com*

Write each expression as a trigonometric function of only θ.

a. $\sin\left(\frac{3\pi}{2} - \theta\right)$ **b.** $\tan\left(\theta - \frac{\pi}{4}\right)$

EXAMPLE 7 Solving a Trigonometric Equation

Find all solutions of $\sin[x + (\pi/4)] + \sin[x - (\pi/4)] = -1$ in the interval $[0, 2\pi)$.

Algebraic Solution

Use sum and difference formulas to rewrite the equation.

$$\sin x \cos \frac{\pi}{4} + \cos x \sin \frac{\pi}{4} + \sin x \cos \frac{\pi}{4} - \cos x \sin \frac{\pi}{4} = -1$$

$$2 \sin x \cos \frac{\pi}{4} = -1$$

$$2(\sin x)\left(\frac{\sqrt{2}}{2}\right) = -1$$

$$\sin x = -\frac{1}{\sqrt{2}}$$

$$\sin x = -\frac{\sqrt{2}}{2}$$

So, the solutions in the interval $[0, 2\pi)$ are $x = \dfrac{5\pi}{4}$ and $x = \dfrac{7\pi}{4}$.

Graphical Solution

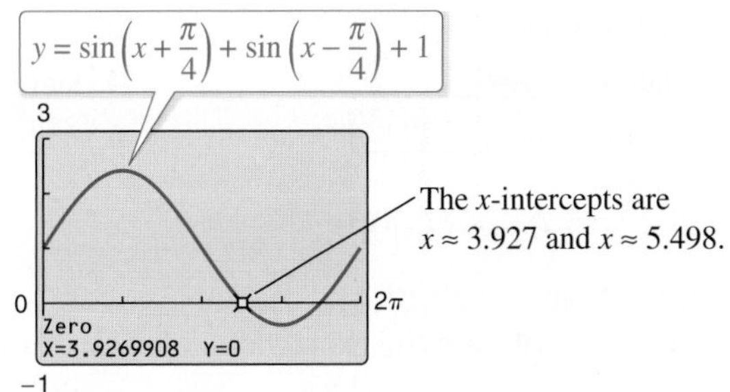

Use the x-intercepts of

$$y = \sin[x + (\pi/4)] + \sin[x - (\pi/4)] + 1$$

to conclude that the approximate solutions in the interval $[0, 2\pi)$ are

$$x \approx 3.927 \approx \frac{5\pi}{4} \quad \text{and} \quad x \approx 5.498 \approx \frac{7\pi}{4}.$$

✓ *Checkpoint* *Audio-video solution in English & Spanish at LarsonPrecalculus.com*

Find all solutions of $\sin[x + (\pi/2)] + \sin[x - (3\pi/2)] = 1$ in the interval $[0, 2\pi)$. ■

The next example is an application from calculus.

EXAMPLE 8 An Application from Calculus

Verify that $\dfrac{\sin(x + h) - \sin x}{h} = (\cos x)\left(\dfrac{\sin h}{h}\right) - (\sin x)\left(\dfrac{1 - \cos h}{h}\right)$, where $h \neq 0$.

Solution Use the formula for $\sin(u + v)$.

$$\frac{\sin(x + h) - \sin x}{h} = \frac{\sin x \cos h + \cos x \sin h - \sin x}{h}$$

$$= \frac{\cos x \sin h - \sin x(1 - \cos h)}{h}$$

$$= (\cos x)\left(\frac{\sin h}{h}\right) - (\sin x)\left(\frac{1 - \cos h}{h}\right)$$

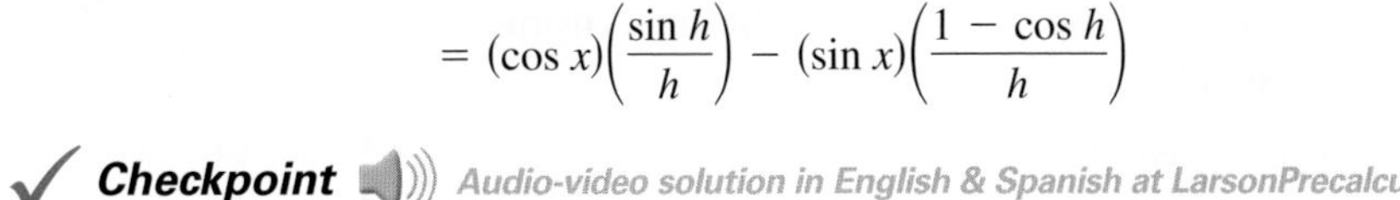

✓ *Checkpoint* *Audio-video solution in English & Spanish at LarsonPrecalculus.com*

Verify that $\dfrac{\cos(x + h) - \cos x}{h} = (\cos x)\left(\dfrac{\cos h - 1}{h}\right) - (\sin x)\left(\dfrac{\sin h}{h}\right)$, where $h \neq 0$.

■

Summarize (Section 2.4)

1. State the sum and difference formulas for sine, cosine, and tangent *(page 236)*. For examples of using the sum and difference formulas to evaluate trigonometric functions, verify identities, and solve trigonometric equations, see Examples 1–8.

2.4 Exercises

See **CalcChat.com** for tutorial help and worked-out solutions to odd-numbered exercises.

Vocabulary: Fill in the blank.

1. $\sin(u - v) =$ ______
2. $\cos(u + v) =$ ______
3. $\tan(u + v) =$ ______
4. $\sin(u + v) =$ ______
5. $\cos(u - v) =$ ______
6. $\tan(u - v) =$ ______

Skills and Applications

Evaluating Trigonometric Expressions In Exercises 7–10, find the exact value of each expression.

7. (a) $\cos\left(\frac{\pi}{4} + \frac{\pi}{3}\right)$ (b) $\cos\frac{\pi}{4} + \cos\frac{\pi}{3}$
8. (a) $\sin\left(\frac{7\pi}{6} - \frac{\pi}{3}\right)$ (b) $\sin\frac{7\pi}{6} - \sin\frac{\pi}{3}$
9. (a) $\sin(135° - 30°)$ (b) $\sin 135° - \cos 30°$
10. (a) $\cos(120° + 45°)$ (b) $\cos 120° + \cos 45°$

Evaluating Trigonometric Functions In Exercises 11–26, find the exact values of the sine, cosine, and tangent of the angle.

11. $\frac{11\pi}{12} = \frac{3\pi}{4} + \frac{\pi}{6}$
12. $\frac{7\pi}{12} = \frac{\pi}{3} + \frac{\pi}{4}$
13. $\frac{17\pi}{12} = \frac{9\pi}{4} - \frac{5\pi}{6}$
14. $-\frac{\pi}{12} = \frac{\pi}{6} - \frac{\pi}{4}$
15. $105° = 60° + 45°$
16. $165° = 135° + 30°$
17. $-195° = 30° - 225°$
18. $255° = 300° - 45°$
19. $\frac{13\pi}{12}$
20. $\frac{19\pi}{12}$
21. $-\frac{5\pi}{12}$
22. $-\frac{7\pi}{12}$
23. $285°$
24. $15°$
25. $-165°$
26. $-105°$

Rewriting a Trigonometric Expression In Exercises 27–34, write the expression as the sine, cosine, or tangent of an angle.

27. $\sin 3 \cos 1.2 - \cos 3 \sin 1.2$
28. $\cos\frac{\pi}{7}\cos\frac{\pi}{5} - \sin\frac{\pi}{7}\sin\frac{\pi}{5}$
29. $\sin 60° \cos 15° + \cos 60° \sin 15°$
30. $\cos 130° \cos 40° - \sin 130° \sin 40°$
31. $\dfrac{\tan(\pi/15) + \tan(2\pi/5)}{1 - \tan(\pi/15)\tan(2\pi/5)}$
32. $\dfrac{\tan 1.1 - \tan 4.6}{1 + \tan 1.1 \tan 4.6}$
33. $\cos 3x \cos 2y + \sin 3x \sin 2y$
34. $\sin x \cos 2x + \cos x \sin 2x$

Evaluating a Trigonometric Expression In Exercises 35–40, find the exact value of the expression.

35. $\sin\frac{\pi}{12}\cos\frac{\pi}{4} + \cos\frac{\pi}{12}\sin\frac{\pi}{4}$
36. $\cos\frac{\pi}{16}\cos\frac{3\pi}{16} - \sin\frac{\pi}{16}\sin\frac{3\pi}{16}$
37. $\cos 130° \cos 10° + \sin 130° \sin 10°$
38. $\sin 100° \cos 40° - \cos 100° \sin 40°$
39. $\dfrac{\tan(9\pi/8) - \tan(\pi/8)}{1 + \tan(9\pi/8)\tan(\pi/8)}$
40. $\dfrac{\tan 25° + \tan 110°}{1 - \tan 25° \tan 110°}$

Evaluating a Trigonometric Expression In Exercises 41–46, find the exact value of the trigonometric expression given that $\sin u = -\frac{3}{5}$, where $3\pi/2 < u < 2\pi$, and $\cos v = \frac{15}{17}$, where $0 < v < \pi/2$.

41. $\sin(u + v)$
42. $\cos(u - v)$
43. $\tan(u + v)$
44. $\csc(u - v)$
45. $\sec(v - u)$
46. $\cot(u + v)$

Evaluating a Trigonometric Expression In Exercises 47–52, find the exact value of the trigonometric expression given that $\sin u = -\frac{7}{25}$ and $\cos v = -\frac{4}{5}$. (Both u and v are in Quadrant III.)

47. $\cos(u + v)$
48. $\sin(u + v)$
49. $\tan(u - v)$
50. $\cot(v - u)$
51. $\csc(u - v)$
52. $\sec(v - u)$

An Application of a Sum or Difference Formula In Exercises 53–56, write the trigonometric expression as an algebraic expression.

53. $\sin(\arcsin x + \arccos x)$
54. $\sin(\arctan 2x - \arccos x)$
55. $\cos(\arccos x + \arcsin x)$
56. $\cos(\arccos x - \arctan x)$

Verifying a Trigonometric Identity In Exercises 57–64, verify the identity.

57. $\sin\left(\frac{\pi}{2} - x\right) = \cos x$

58. $\sin\left(\frac{\pi}{2} + x\right) = \cos x$

59. $\sin\left(\frac{\pi}{6} + x\right) = \frac{1}{2}(\cos x + \sqrt{3}\sin x)$

60. $\cos\left(\frac{5\pi}{4} - x\right) = -\frac{\sqrt{2}}{2}(\cos x + \sin x)$

61. $\tan(\theta + \pi) = \tan\theta$

62. $\tan\left(\frac{\pi}{4} - \theta\right) = \frac{1 - \tan\theta}{1 + \tan\theta}$

63. $\cos(\pi - \theta) + \sin\left(\frac{\pi}{2} + \theta\right) = 0$

64. $\cos(x + y)\cos(x - y) = \cos^2 x - \sin^2 y$

Deriving a Reduction Formula In Exercises 65–68, write the expression as a trigonometric function of only θ, and use a graphing utility to confirm your answer graphically.

65. $\cos\left(\frac{3\pi}{2} - \theta\right)$

66. $\sin(\pi + \theta)$

67. $\csc\left(\frac{3\pi}{2} + \theta\right)$

68. $\cot(\theta - \pi)$

Solving a Trigonometric Equation In Exercises 69–74, find all solutions of the equation in the interval $[0, 2\pi)$.

69. $\sin(x + \pi) - \sin x + 1 = 0$

70. $\cos(x + \pi) - \cos x - 1 = 0$

71. $\cos\left(x + \frac{\pi}{4}\right) - \cos\left(x - \frac{\pi}{4}\right) = 1$

72. $\sin\left(x + \frac{\pi}{6}\right) - \sin\left(x - \frac{7\pi}{6}\right) = \frac{\sqrt{3}}{2}$

73. $\tan(x + \pi) + 2\sin(x + \pi) = 0$

74. $\sin\left(x + \frac{\pi}{2}\right) - \cos^2 x = 0$

Approximating Solutions In Exercises 75–78, use a graphing utility to approximate the solutions of the equation in the interval $[0, 2\pi)$.

75. $\cos\left(x + \frac{\pi}{4}\right) + \cos\left(x - \frac{\pi}{4}\right) = 1$

76. $\tan(x + \pi) - \cos\left(x + \frac{\pi}{2}\right) = 0$

77. $\sin\left(x + \frac{\pi}{2}\right) + \cos^2 x = 0$

78. $\cos\left(x - \frac{\pi}{2}\right) - \sin^2 x = 0$

79. **Harmonic Motion** A weight is attached to a spring suspended vertically from a ceiling. When a driving force is applied to the system, the weight moves vertically from its equilibrium position, and this motion is modeled by

$$y = \frac{1}{3}\sin 2t + \frac{1}{4}\cos 2t$$

where y is the displacement (in feet) from equilibrium of the weight and t is the time (in seconds).

(a) Use the identity

$$a\sin B\theta + b\cos B\theta = \sqrt{a^2 + b^2}\sin(B\theta + C)$$

where $C = \arctan(b/a)$, $a > 0$, to write the model in the form

$$y = \sqrt{a^2 + b^2}\sin(Bt + C).$$

(b) Find the amplitude of the oscillations of the weight.

(c) Find the frequency of the oscillations of the weight.

80. **Standing Waves** The equation of a standing wave is obtained by adding the displacements of two waves traveling in opposite directions (see figure). Assume that each of the waves has amplitude A, period T, and wavelength λ. The models for two such waves are

$$y_1 = A\cos 2\pi\left(\frac{t}{T} - \frac{x}{\lambda}\right) \text{ and } y_2 = A\cos 2\pi\left(\frac{t}{T} + \frac{x}{\lambda}\right).$$

Show that

$$y_1 + y_2 = 2A\cos\frac{2\pi t}{T}\cos\frac{2\pi x}{\lambda}.$$

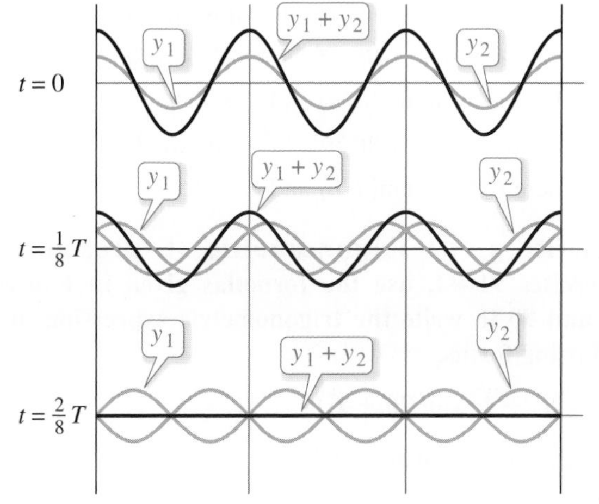

Exploration

True or False? **In Exercises 81–84, determine whether the statement is true or false. Justify your answer.**

81. $\sin(u \pm v) = \sin u \cos v \pm \cos u \sin v$

82. $\cos(u \pm v) = \cos u \cos v \pm \sin u \sin v$

83. When α and β are supplementary,

$\sin \alpha \cos \beta = \cos \alpha \sin \beta$.

84. When A, B, and C form $\triangle ABC$, $\cos(A + B) = -\cos C$.

85. **Error Analysis** Describe the error.

$$\tan\left(x - \frac{\pi}{4}\right) = \frac{\tan x - \tan(\pi/4)}{1 - \tan x \tan(\pi/4)}$$
$$= \frac{\tan x - 1}{1 - \tan x}$$
$$= -1$$

86. **HOW DO YOU SEE IT?** Explain how to use the figure to justify each statement.

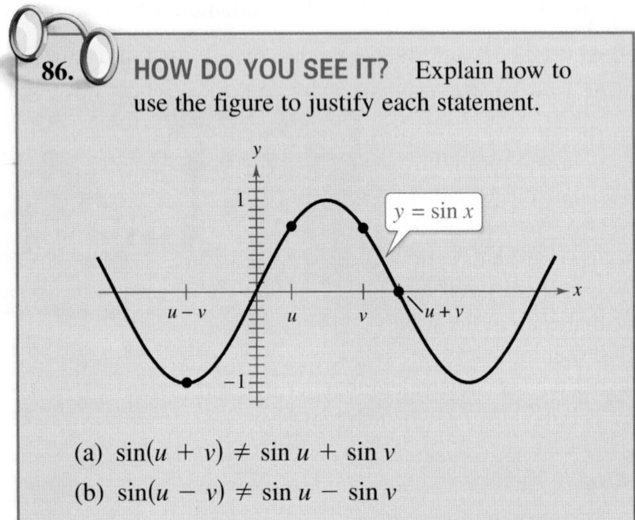

(a) $\sin(u + v) \neq \sin u + \sin v$

(b) $\sin(u - v) \neq \sin u - \sin v$

Verifying an Identity **In Exercises 87–90, verify the identity.**

87. $\cos(n\pi + \theta) = (-1)^n \cos \theta$, n is an integer

88. $\sin(n\pi + \theta) = (-1)^n \sin \theta$, n is an integer

89. $a \sin B\theta + b \cos B\theta = \sqrt{a^2 + b^2} \sin(B\theta + C)$,
where $C = \arctan(b/a)$ and $a > 0$

90. $a \sin B\theta + b \cos B\theta = \sqrt{a^2 + b^2} \cos(B\theta - C)$,
where $C = \arctan(a/b)$ and $b > 0$

Rewriting a Trigonometric Expression **In Exercises 91–94, use the formulas given in Exercises 89 and 90 to write the trigonometric expression in the following forms.**

(a) $\sqrt{a^2 + b^2} \sin(B\theta + C)$

(b) $\sqrt{a^2 + b^2} \cos(B\theta - C)$

91. $\sin \theta + \cos \theta$

92. $3 \sin 2\theta + 4 \cos 2\theta$

93. $12 \sin 3\theta + 5 \cos 3\theta$

94. $\sin 2\theta + \cos 2\theta$

Rewriting a Trigonometric Expression **In Exercises 95 and 96, use the formulas given in Exercises 89 and 90 to write the trigonometric expression in the form $a \sin B\theta + b \cos B\theta$.**

95. $2 \sin[\theta + (\pi/4)]$

96. $5 \cos[\theta - (\pi/4)]$

Angle Between Two Lines **In Exercises 97 and 98, use the figure, which shows two lines whose equations are $y_1 = m_1x + b_1$ and $y_2 = m_2x + b_2$. Assume that both lines have positive slopes. Derive a formula for the angle between the two lines. Then use your formula to find the angle between the given pair of lines.**

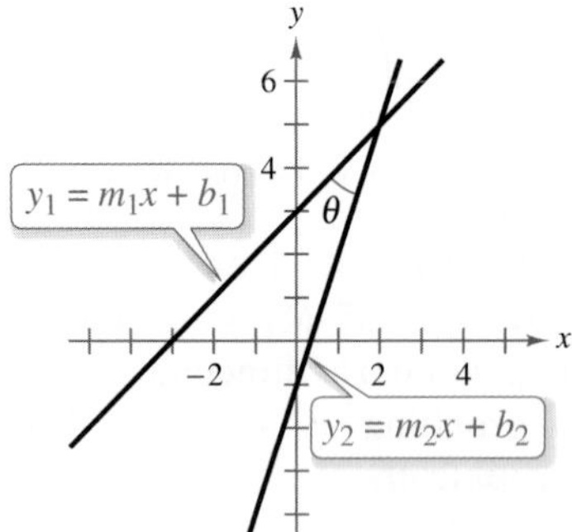

97. $y = x$ and $y = \sqrt{3}x$

98. $y = x$ and $y = x/\sqrt{3}$

Graphical Reasoning **In Exercises 99 and 100, use a graphing utility to graph y_1 and y_2 in the same viewing window. Use the graphs to determine whether $y_1 = y_2$. Explain your reasoning.**

99. $y_1 = \cos(x + 2)$, $y_2 = \cos x + \cos 2$

100. $y_1 = \sin(x + 4)$, $y_2 = \sin x + \sin 4$

101. **Proof** Write a proof of the formula for $\sin(u + v)$. Write a proof of the formula for $\sin(u - v)$.

102. **An Application from Calculus** Let $x = \pi/3$ in the identity in Example 8 and define the functions f and g as follows.

$$f(h) = \frac{\sin[(\pi/3) + h] - \sin(\pi/3)}{h}$$

$$g(h) = \cos\frac{\pi}{3}\left(\frac{\sin h}{h}\right) - \sin\frac{\pi}{3}\left(\frac{1 - \cos h}{h}\right)$$

(a) What are the domains of the functions f and g?

(b) Use a graphing utility to complete the table.

h	0.5	0.2	0.1	0.05	0.02	0.01
$f(h)$						
$g(h)$						

(c) Use the graphing utility to graph the functions f and g.

(d) Use the table and the graphs to make a conjecture about the values of the functions f and g as $h \to 0^+$.

2.5 Multiple-Angle and Product-to-Sum Formulas

A variety of trigonometric formulas enable you to rewrite trigonometric equations in more convenient forms. For example, in Exercise 71 on page 251, you will use a half-angle formula to rewrite an equation relating the Mach number of a supersonic airplane to the apex angle of the cone formed by the sound waves behind the airplane.

- **Use multiple-angle formulas to rewrite and evaluate trigonometric functions.**
- **Use power-reducing formulas to rewrite trigonometric expressions.**
- **Use half-angle formulas to rewrite and evaluate trigonometric functions.**
- **Use product-to-sum and sum-to-product formulas to rewrite and evaluate trigonometric expressions.**
- **Use trigonometric formulas to rewrite real-life models.**

Multiple-Angle Formulas

In this section, you will study four other categories of trigonometric identities.

1. The first category involves *functions of multiple angles* such as $\sin ku$ and $\cos ku$.
2. The second category involves *squares of trigonometric functions* such as $\sin^2 u$.
3. The third category involves *functions of half-angles* such as $\sin(u/2)$.
4. The fourth category involves *products of trigonometric functions* such as $\sin u \cos v$.

You should learn the **double-angle formulas** because they are used often in trigonometry and calculus. For proofs of these formulas, see Proofs in Mathematics on page 257.

Double-Angle Formulas

$$\sin 2u = 2 \sin u \cos u$$

$$\tan 2u = \frac{2 \tan u}{1 - \tan^2 u}$$

$$\begin{aligned} \cos 2u &= \cos^2 u - \sin^2 u \\ &= 2\cos^2 u - 1 \\ &= 1 - 2\sin^2 u \end{aligned}$$

EXAMPLE 1 Solving a Multiple-Angle Equation

Solve $2 \cos x + \sin 2x = 0$.

Solution Begin by rewriting the equation so that it involves trigonometric functions of only x. Then factor and solve.

$$2 \cos x + \sin 2x = 0 \quad \text{Write original equation.}$$

$$2 \cos x + 2 \sin x \cos x = 0 \quad \text{Double-angle formula}$$

$$2 \cos x(1 + \sin x) = 0 \quad \text{Factor.}$$

$$2 \cos x = 0 \quad \text{and} \quad 1 + \sin x = 0 \quad \text{Set factors equal to zero.}$$

$$x = \frac{\pi}{2}, \frac{3\pi}{2} \qquad x = \frac{3\pi}{2} \quad \text{Solutions in } [0, 2\pi)$$

So, the general solution is

$$x = \frac{\pi}{2} + 2n\pi \quad \text{and} \quad x = \frac{3\pi}{2} + 2n\pi$$

where n is an integer. Verify these solutions graphically.

✓ ***Checkpoint*** *Audio-video solution in English & Spanish at LarsonPrecalculus.com*

Solve $\cos 2x + \cos x = 0$.

EXAMPLE 2 Evaluating Functions Involving Double Angles

Use the conditions below to find $\sin 2\theta$, $\cos 2\theta$, and $\tan 2\theta$.

$$\cos \theta = \frac{5}{13}, \quad \frac{3\pi}{2} < \theta < 2\pi$$

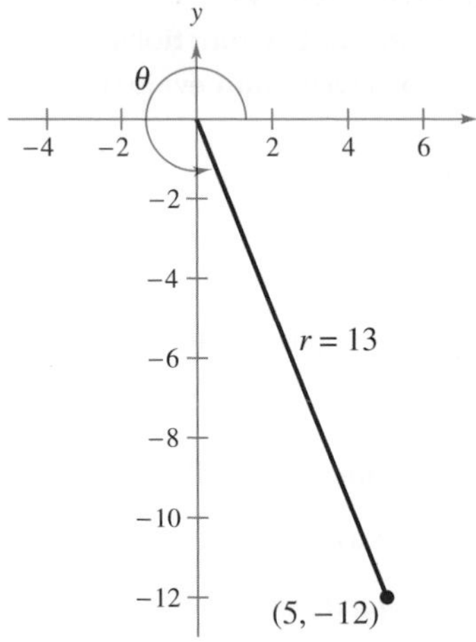

Figure 2.5

Solution From Figure 2.5,

$$\sin \theta = \frac{y}{r} = -\frac{12}{13} \quad \text{and} \quad \tan \theta = \frac{y}{x} = -\frac{12}{5}.$$

Use these values with each of the double-angle formulas.

$$\sin 2\theta = 2 \sin \theta \cos \theta = 2\left(-\frac{12}{13}\right)\left(\frac{15}{13}\right) = -\frac{120}{169}$$

$$\cos 2\theta = 2 \cos^2 \theta - 1 = 2\left(\frac{25}{169}\right) - 1 = -\frac{119}{169}$$

$$\tan 2\theta = \frac{2 \tan \theta}{1 - \tan^2 \theta} = \frac{2\left(-\frac{12}{5}\right)}{1 - \left(-\frac{12}{5}\right)^2} = \frac{120}{119}$$

✓ **Checkpoint** *Audio-video solution in English & Spanish at LarsonPrecalculus.com*

Use the conditions below to find $\sin 2\theta$, $\cos 2\theta$, and $\tan 2\theta$.

$$\sin \theta = \frac{3}{5}, \quad 0 < \theta < \frac{\pi}{2}$$

The double-angle formulas are not restricted to the angles 2θ and θ. Other *double* combinations, such as 4θ and 2θ or 6θ and 3θ, are also valid. Here are two examples.

$$\sin 4\theta = 2 \sin 2\theta \cos 2\theta \quad \text{and} \quad \cos 6\theta = \cos^2 3\theta - \sin^2 3\theta$$

By using double-angle formulas together with the sum formulas given in the preceding section, you can derive other multiple-angle formulas.

EXAMPLE 3 Deriving a Triple-Angle Formula

Rewrite $\sin 3x$ in terms of $\sin x$.

Solution

$$\begin{aligned}
\sin 3x &= \sin(2x + x) && \text{Rewrite the angle as a sum.}\\
&= \sin 2x \cos x + \cos 2x \sin x && \text{Sum formula}\\
&= 2 \sin x \cos x \cos x + (1 - 2 \sin^2 x) \sin x && \text{Double-angle formulas}\\
&= 2 \sin x \cos^2 x + \sin x - 2 \sin^3 x && \text{Distributive Property}\\
&= 2 \sin x(1 - \sin^2 x) + \sin x - 2 \sin^3 x && \text{Pythagorean identity}\\
&= 2 \sin x - 2 \sin^3 x + \sin x - 2 \sin^3 x && \text{Distributive Property}\\
&= 3 \sin x - 4 \sin^3 x && \text{Simplify.}
\end{aligned}$$

✓ **Checkpoint** *Audio-video solution in English & Spanish at LarsonPrecalculus.com*

Rewrite $\cos 3x$ in terms of $\cos x$.

Power-Reducing Formulas

The double-angle formulas can be used to obtain the **power-reducing formulas.**

Power-Reducing Formulas

$$\sin^2 u = \frac{1 - \cos 2u}{2}$$

$$\cos^2 u = \frac{1 + \cos 2u}{2}$$

$$\tan^2 u = \frac{1 - \cos 2u}{1 + \cos 2u}$$

For a proof of the power-reducing formulas, see Proofs in Mathematics on page 258.

Example 4 shows a typical power reduction used in calculus.

EXAMPLE 4 Reducing a Power 

Rewrite $\sin^4 x$ in terms of first powers of the cosines of multiple angles.

Solution Note the repeated use of power-reducing formulas.

$$\sin^4 x = (\sin^2 x)^2 \qquad \text{Property of exponents}$$

$$= \left(\frac{1 - \cos 2x}{2}\right)^2 \qquad \text{Power-reducing formula}$$

$$= \frac{1}{4}(1 - 2\cos 2x + \cos^2 2x) \qquad \text{Expand.}$$

$$= \frac{1}{4}\left(1 - 2\cos 2x + \frac{1 + \cos 4x}{2}\right) \qquad \text{Power-reducing formula}$$

$$= \frac{1}{4} - \frac{1}{2}\cos 2x + \frac{1}{8} + \frac{1}{8}\cos 4x \qquad \text{Distributive Property}$$

$$= \frac{3}{8} - \frac{1}{2}\cos 2x + \frac{1}{8}\cos 4x \qquad \text{Simplify.}$$

$$= \frac{1}{8}(3 - 4\cos 2x + \cos 4x) \qquad \text{Factor out common factor.}$$

Use a graphing utility to check this result, as shown below. Notice that the graphs coincide.

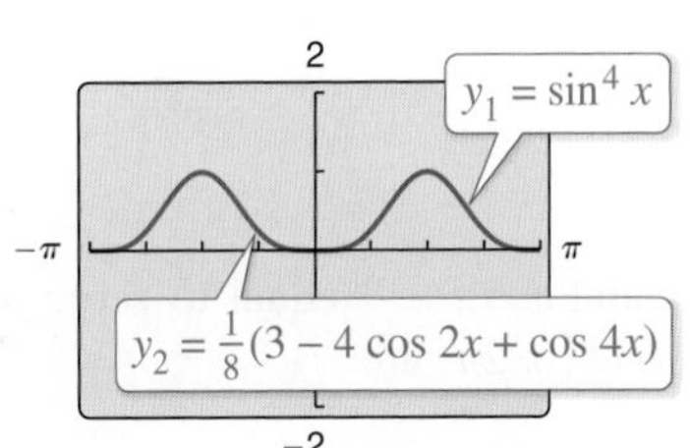

✓ **Checkpoint**  Audio-video solution in English & Spanish at LarsonPrecalculus.com

Rewrite $\tan^4 x$ in terms of first powers of the cosines of multiple angles.

Half-Angle Formulas

You can derive some useful alternative forms of the power-reducing formulas by replacing u with $u/2$. The results are called **half-angle formulas.**

REMARK To find the exact value of a trigonometric function with an angle measure in D° M′ S″ form using a half-angle formula, first convert the angle measure to decimal degree form. Then multiply the resulting angle measure by 2.

Half-Angle Formulas

$$\sin\frac{u}{2} = \pm\sqrt{\frac{1-\cos u}{2}} \qquad \cos\frac{u}{2} = \pm\sqrt{\frac{1+\cos u}{2}}$$

$$\tan\frac{u}{2} = \frac{1-\cos u}{\sin u} = \frac{\sin u}{1+\cos u}$$

The signs of $\sin\frac{u}{2}$ and $\cos\frac{u}{2}$ depend on the quadrant in which $\frac{u}{2}$ lies.

REMARK Use your calculator to verify the result obtained in Example 5. That is, evaluate $\sin 105°$ and $\left(\sqrt{2+\sqrt{3}}\right)/2$. Note that both values are approximately 0.9659258.

EXAMPLE 5 Using a Half-Angle Formula

Find the exact value of $\sin 105°$.

Solution Begin by noting that 105° is half of 210°. Then, use the half-angle formula for $\sin(u/2)$ and the fact that 105° lies in Quadrant II.

$$\sin 105° = \sqrt{\frac{1-\cos 210°}{2}} = \sqrt{\frac{1+\left(\sqrt{3}/2\right)}{2}} = \frac{\sqrt{2+\sqrt{3}}}{2}$$

The positive square root is chosen because $\sin\theta$ is positive in Quadrant II.

✓ **Checkpoint** Audio-video solution in English & Spanish at LarsonPrecalculus.com

Find the exact value of $\cos 105°$.

EXAMPLE 6 Solving a Trigonometric Equation

Find all solutions of $1+\cos^2 x = 2\cos^2\frac{x}{2}$ in the interval $[0, 2\pi)$.

Algebraic Solution

$$1+\cos^2 x = 2\cos^2\frac{x}{2} \qquad \text{Write original equation.}$$

$$1+\cos^2 x = 2\left(\pm\sqrt{\frac{1+\cos x}{2}}\right)^2 \qquad \text{Half-angle formula}$$

$$1+\cos^2 x = 1+\cos x \qquad \text{Simplify.}$$

$$\cos^2 x - \cos x = 0 \qquad \text{Simplify.}$$

$$\cos x(\cos x - 1) = 0 \qquad \text{Factor.}$$

By setting the factors $\cos x$ and $\cos x - 1$ equal to zero, you find that the solutions in the interval $[0, 2\pi)$ are

$$x = \frac{\pi}{2},\quad x = \frac{3\pi}{2},\quad \text{and} \quad x = 0.$$

Graphical Solution

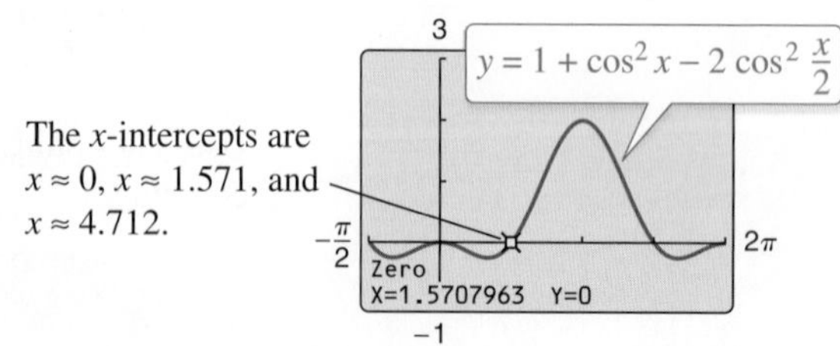

Use the x-intercepts of $y = 1+\cos^2 x - 2\cos^2(x/2)$ to conclude that the approximate solutions of $1+\cos^2 x = 2\cos^2(x/2)$ in the interval $[0, 2\pi)$ are

$$x = 0,\quad x \approx 1.571 \approx \frac{\pi}{2},\quad \text{and} \quad x \approx 4.712 \approx \frac{3\pi}{2}.$$

✓ **Checkpoint** Audio-video solution in English & Spanish at LarsonPrecalculus.com

Find all solutions of $\cos^2 x = \sin^2(x/2)$ in the interval $[0, 2\pi)$. ■

Product-to-Sum and Sum-to-Product Formulas

Each of the **product-to-sum formulas** can be proved using the sum and difference formulas discussed in the preceding section.

Product-to-Sum Formulas

$$\sin u \sin v = \frac{1}{2}[\cos(u - v) - \cos(u + v)]$$

$$\cos u \cos v = \frac{1}{2}[\cos(u - v) + \cos(u + v)]$$

$$\sin u \cos v = \frac{1}{2}[\sin(u + v) + \sin(u - v)]$$

$$\cos u \sin v = \frac{1}{2}[\sin(u + v) - \sin(u - v)]$$

Product-to-sum formulas are used in calculus to solve problems involving the products of sines and cosines of two different angles.

EXAMPLE 7 Writing Products as Sums

Rewrite the product $\cos 5x \sin 4x$ as a sum or difference.

Solution Using the appropriate product-to-sum formula, you obtain

$$\cos 5x \sin 4x = \frac{1}{2}[\sin(5x + 4x) - \sin(5x - 4x)]$$

$$= \frac{1}{2}\sin 9x - \frac{1}{2}\sin x.$$

✓ **Checkpoint** *Audio-video solution in English & Spanish at LarsonPrecalculus.com*

Rewrite the product $\sin 5x \cos 3x$ as a sum or difference.

Occasionally, it is useful to reverse the procedure and write a sum of trigonometric functions as a product. This can be accomplished with the **sum-to-product formulas.**

Sum-to-Product Formulas

$$\sin u + \sin v = 2 \sin\left(\frac{u + v}{2}\right) \cos\left(\frac{u - v}{2}\right)$$

$$\sin u - \sin v = 2 \cos\left(\frac{u + v}{2}\right) \sin\left(\frac{u - v}{2}\right)$$

$$\cos u + \cos v = 2 \cos\left(\frac{u + v}{2}\right) \cos\left(\frac{u - v}{2}\right)$$

$$\cos u - \cos v = -2 \sin\left(\frac{u + v}{2}\right) \sin\left(\frac{u - v}{2}\right)$$

For a proof of the sum-to-product formulas, see Proofs in Mathematics on page 258.

EXAMPLE 8 Using a Sum-to-Product Formula

Find the exact value of $\cos 195° + \cos 105°$.

Solution Use the appropriate sum-to-product formula.

$$\begin{aligned}\cos 195° + \cos 105° &= 2\cos\left(\frac{195° + 105°}{2}\right)\cos\left(\frac{195° - 105°}{2}\right)\\ &= 2\cos 150° \cos 45°\\ &= 2\left(-\frac{\sqrt{3}}{2}\right)\left(\frac{\sqrt{2}}{2}\right)\\ &= -\frac{\sqrt{6}}{2}\end{aligned}$$

✓ ***Checkpoint*** *Audio-video solution in English & Spanish at LarsonPrecalculus.com*

Find the exact value of $\sin 195° + \sin 105°$.

EXAMPLE 9 Solving a Trigonometric Equation

See LarsonPrecalculus.com for an interactive version of this type of example.

Solve $\sin 5x + \sin 3x = 0$.

Solution

$$\sin 5x + \sin 3x = 0 \qquad \text{Write original equation.}$$

$$2\sin\left(\frac{5x + 3x}{2}\right)\cos\left(\frac{5x - 3x}{2}\right) = 0 \qquad \text{Sum-to-product formula}$$

$$2\sin 4x \cos x = 0 \qquad \text{Simplify.}$$

Set the factor $2\sin 4x$ equal to zero. The solutions in the interval $[0, 2\pi)$ are

$$x = 0, \frac{\pi}{4}, \frac{\pi}{2}, \frac{3\pi}{4}, \pi, \frac{5\pi}{4}, \frac{3\pi}{2}, \frac{7\pi}{4}.$$

The equation $\cos x = 0$ yields no additional solutions, so the solutions are of the form $x = n\pi/4$, where n is an integer. To confirm this graphically, sketch the graph of $y = \sin 5x + \sin 3x$, as shown below.

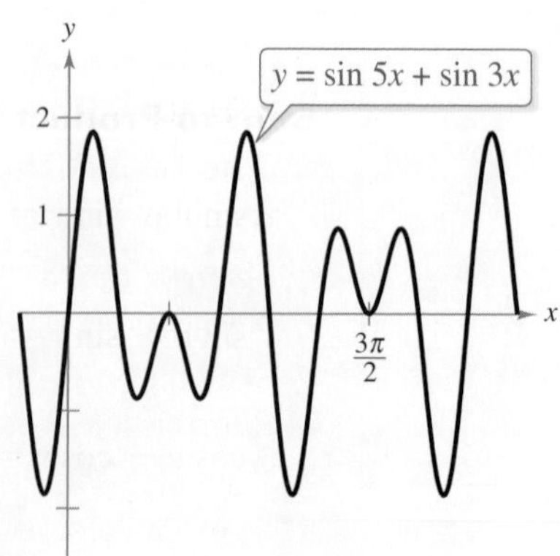

Notice from the graph that the x-intercepts occur at multiples of $\pi/4$.

✓ ***Checkpoint*** *Audio-video solution in English & Spanish at LarsonPrecalculus.com*

Solve $\sin 4x - \sin 2x = 0$.

Application

EXAMPLE 10 Projectile Motion

Kicking a football with an initial velocity of 80 feet per second at an angle of 45° with the horizontal results in a distance traveled of 200 feet.

Ignoring air resistance, the range of a projectile fired at an angle θ with the horizontal and with an initial velocity of v_0 feet per second is given by

$$r = \frac{1}{16} v_0^2 \sin \theta \cos \theta$$

where r is the horizontal distance (in feet) that the projectile travels. A football player can kick a football from ground level with an initial velocity of 80 feet per second.

a. Rewrite the projectile motion model in terms of the first power of the sine of a multiple angle.

b. At what angle must the player kick the football so that the football travels 200 feet?

Solution

a. Use a double-angle formula to rewrite the projectile motion model as

$$r = \frac{1}{32} v_0{}^2 (2 \sin \theta \cos \theta) \qquad \text{Write original model.}$$

$$= \frac{1}{32} v_0^2 \sin 2\theta. \qquad \text{Double-angle formula}$$

b.

$$r = \frac{1}{32} v_0^2 \sin 2\theta \qquad \text{Write projectile motion model.}$$

$$200 = \frac{1}{32}(80)^2 \sin 2\theta \qquad \text{Substitute 200 for } r \text{ and 80 for } v_0.$$

$$200 = 200 \sin 2\theta \qquad \text{Simplify.}$$

$$1 = \sin 2\theta \qquad \text{Divide each side by 200.}$$

You know that $2\theta = \pi/2$. Dividing this result by 2 produces $\theta = \pi/4$, or 45°. So, the player must kick the football at an angle of 45° so that the football travels 200 feet.

✓ ***Checkpoint*** *Audio-video solution in English & Spanish at LarsonPrecalculus.com*

In Example 10, for what angle is the horizontal distance the football travels a maximum?

Summarize (Section 2.5)

1. State the double-angle formulas *(page 243)*. For examples of using multiple-angle formulas to rewrite and evaluate trigonometric functions, see Examples 1–3.
2. State the power-reducing formulas *(page 245)*. For an example of using power-reducing formulas to rewrite a trigonometric expression, see Example 4.
3. State the half-angle formulas *(page 246)*. For examples of using half-angle formulas to rewrite and evaluate trigonometric functions, see Examples 5 and 6.
4. State the product-to-sum and sum-to-product formulas *(page 247)*. For an example of using a product-to-sum formula to rewrite a trigonometric expression, see Example 7. For examples of using sum-to-product formulas to rewrite and evaluate trigonometric functions, see Examples 8 and 9.
5. Describe an example of how to use a trigonometric formula to rewrite a real-life model *(page 249, Example 10)*.

2.5 Exercises

See CalcChat.com for tutorial help and worked-out solutions to odd-numbered exercises.

Vocabulary: Fill in the blank to complete the trigonometric formula.

1. $\sin 2u =$ ______
2. $\cos 2u =$ ______
3. $\sin u \cos v =$ ______
4. $\dfrac{1 - \cos 2u}{1 + \cos 2u} =$ ______
5. $\sin \dfrac{u}{2} =$ ______
6. $\cos u - \cos v =$ ______

Skills and Applications

Solving a Multiple-Angle Equation In Exercises 7–14, solve the equation.

7. $\sin 2x - \sin x = 0$
8. $\sin 2x \sin x = \cos x$
9. $\cos 2x - \cos x = 0$
10. $\cos 2x + \sin x = 0$
11. $\sin 4x = -2 \sin 2x$
12. $(\sin 2x + \cos 2x)^2 = 1$
13. $\tan 2x - \cot x = 0$
14. $\tan 2x - 2 \cos x = 0$

Using a Double-Angle Formula In Exercises 15–20, use a double-angle formula to rewrite the expression.

15. $6 \sin x \cos x$
16. $\sin x \cos x$
17. $6 \cos^2 x - 3$
18. $\cos^2 x - \frac{1}{2}$
19. $4 - 8 \sin^2 x$
20. $10 \sin^2 x - 5$

Evaluating Functions Involving Double Angles In Exercises 21–24, use the given conditions to find the exact values of $\sin 2u$, $\cos 2u$, and $\tan 2u$ using the double-angle formulas.

21. $\sin u = -3/5, \quad 3\pi/2 < u < 2\pi$
22. $\cos u = -4/5, \quad \pi/2 < u < \pi$
23. $\tan u = 3/5, \quad 0 < u < \pi/2$
24. $\sec u = -2, \quad \pi < u < 3\pi/2$

25. **Deriving a Multiple-Angle Formula** Rewrite $\cos 4x$ in terms of $\cos x$.

26. **Deriving a Multiple-Angle Formula** Rewrite $\tan 3x$ in terms of $\tan x$.

Reducing Powers In Exercises 27–34, use the power-reducing formulas to rewrite the expression in terms of first powers of the cosines of multiple angles.

27. $\cos^4 x$
28. $\sin^8 x$
29. $\sin^4 2x$
30. $\cos^4 2x$
31. $\tan^4 2x$
32. $\tan^2 2x \cos^4 2x$
33. $\sin^2 2x \cos^2 2x$
34. $\sin^4 x \cos^2 x$

Using Half-Angle Formulas In Exercises 35–40, use the half-angle formulas to determine the exact values of the sine, cosine, and tangent of the angle.

35. $75°$
36. $165°$
37. $112° 30'$
38. $67° 30'$
39. $\pi/8$
40. $7\pi/12$

Using Half-Angle Formulas In Exercises 41–44, use the given conditions to (a) determine the quadrant in which $u/2$ lies, and (b) find the exact values of $\sin(u/2)$, $\cos(u/2)$, and $\tan(u/2)$ using the half-angle formulas.

41. $\cos u = 7/25, \quad 0 < u < \pi/2$
42. $\sin u = 5/13, \quad \pi/2 < u < \pi$
43. $\tan u = -5/12, \quad 3\pi/2 < u < 2\pi$
44. $\cot u = 3, \quad \pi < u < 3\pi/2$

Solving a Trigonometric Equation In Exercises 45–48, find all solutions of the equation in the interval $[0, 2\pi)$. Use a graphing utility to graph the equation and verify the solutions.

45. $\sin \dfrac{x}{2} + \cos x = 0$
46. $\sin \dfrac{x}{2} + \cos x - 1 = 0$
47. $\cos \dfrac{x}{2} - \sin x = 0$
48. $\tan \dfrac{x}{2} - \sin x = 0$

Using Product-to-Sum Formulas In Exercises 49–52, use the product-to-sum formulas to rewrite the product as a sum or difference.

49. $\sin 5\theta \sin 3\theta$
50. $7 \cos(-5\beta) \sin 3\beta$
51. $\cos 2\theta \cos 4\theta$
52. $\sin(x + y) \cos(x - y)$

Using Sum-to-Product Formulas In Exercises 53–56, use the sum-to-product formulas to rewrite the sum or difference as a product.

53. $\sin 5\theta - \sin 3\theta$
54. $\sin 3\theta + \sin \theta$
55. $\cos 6x + \cos 2x$
56. $\cos x + \cos 4x$

Using Sum-to-Product Formulas **In Exercises 57–60, use the sum-to-product formulas to find the exact value of the expression.**

57. $\sin 75° + \sin 15°$

58. $\cos 120° + \cos 60°$

59. $\cos \frac{3\pi}{4} - \cos \frac{\pi}{4}$

60. $\sin \frac{5\pi}{4} - \sin \frac{3\pi}{4}$

Solving a Trigonometric Equation **In Exercises 61–64, find all solutions of the equation in the interval $[0, 2\pi)$. Use a graphing utility to graph the equation and verify the solutions.**

61. $\sin 6x + \sin 2x = 0$

62. $\cos 2x - \cos 6x = 0$

63. $\frac{\cos 2x}{\sin 3x - \sin x} - 1 = 0$

64. $\sin^2 3x - \sin^2 x = 0$

Verifying a Trigonometric Identity **In Exercises 65–70, verify the identity.**

65. $\csc 2\theta = \frac{\csc \theta}{2 \cos \theta}$

66. $\cos^4 x - \sin^4 x = \cos 2x$

67. $(\sin x + \cos x)^2 = 1 + \sin 2x$

68. $\tan \frac{u}{2} = \csc u - \cot u$

69. $\frac{\sin x \pm \sin y}{\cos x + \cos y} = \tan \frac{x \pm y}{2}$

70. $\cos\left(\frac{\pi}{3} + x\right) + \cos\left(\frac{\pi}{3} - x\right) = \cos x$

71. Mach Number

The Mach number M of a supersonic airplane is the ratio of its speed to the speed of sound. When an airplane travels faster than the speed of sound, the sound waves form a cone behind the airplane. The Mach number is related to the apex angle θ of the cone by $\sin(\theta/2) = 1/M$.

(a) Use a half-angle formula to rewrite the equation in terms of $\cos \theta$.

(b) Find the angle θ that corresponds to a Mach number of 2.

(c) Find the angle θ that corresponds to a Mach number of 4.5.

(d) The speed of sound is about 760 miles per hour. Determine the speed of an object with the Mach numbers from parts (b) and (c).

Chris Parypa Photography/Shutterstock.com

72. Projectile Motion The range of a projectile fired at an angle θ with the horizontal and with an initial velocity of v_0 feet per second is

$$r = \frac{1}{32} v_0^2 \sin 2\theta$$

where r is the horizontal distance (in feet) the projectile travels. An athlete throws a javelin at 75 feet per second. At what angle must the athlete throw the javelin so that the javelin travels 130 feet?

73. Railroad Track When two railroad tracks merge, the overlapping portions of the tracks are in the shapes of circular arcs (see figure). The radius r (in feet) of each arc and the angle θ are related by

$$\frac{x}{2} = 2r \sin^2 \frac{\theta}{2}.$$

Write a formula for x in terms of $\cos \theta$.

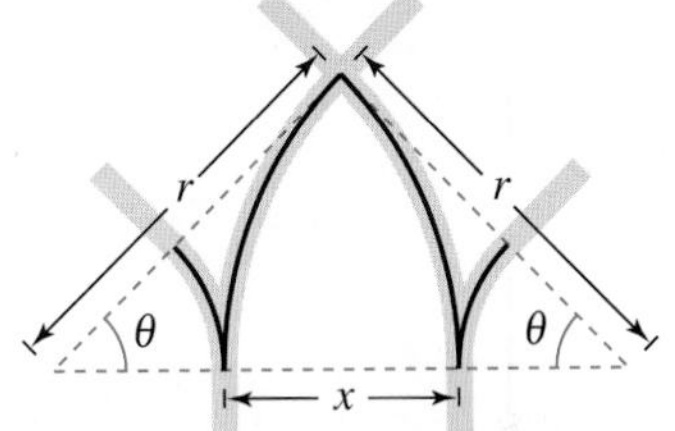

74. **HOW DO YOU SEE IT?** Explain how to use the figure to verify the double-angle formulas (a) $\sin 2u = 2 \sin u \cos u$ and (b) $\cos 2u = \cos^2 u - \sin^2 u$.

Exploration

True or False? **In Exercises 75 and 76, determine whether the statement is true or false. Justify your answer.**

75. The sine function is an odd function, so

$$\sin(-2x) = -2 \sin x \cos x.$$

76. $\sin \frac{u}{2} = -\sqrt{\frac{1 - \cos u}{2}}$ when u is in the second quadrant.

77. Complementary Angles Verify each identity for complementary angles ϕ and θ.

(a) $\sin(\phi - \theta) = \cos 2\theta$

(b) $\cos(\phi - \theta) = \sin 2\theta$

Chapter Summary

	What Did You Learn?	Explanation/Examples	Review Exercises
Section 2.1	Recognize and write the fundamental trigonometric identities *(p. 210).*	**Reciprocal Identities** $\sin u = 1/\csc u$ $\quad \cos u = 1/\sec u$ $\quad \tan u = 1/\cot u$ $\csc u = 1/\sin u$ $\quad \sec u = 1/\cos u$ $\quad \cot u = 1/\tan u$ **Quotient Identities:** $\tan u = \dfrac{\sin u}{\cos u}$, $\cot u = \dfrac{\cos u}{\sin u}$ **Pythagorean Identities:** $\sin^2 u + \cos^2 u = 1$, $1 + \tan^2 u = \sec^2 u$, $\quad 1 + \cot^2 u = \csc^2 u$ **Cofunction Identities** $\sin[(\pi/2) - u] = \cos u$ $\quad \cos[(\pi/2) - u] = \sin u$ $\tan[(\pi/2) - u] = \cot u$ $\quad \cot[(\pi/2) - u] = \tan u$ $\sec[(\pi/2) - u] = \csc u$ $\quad \csc[(\pi/2) - u] = \sec u$ **Even/Odd Identities** $\sin(-u) = -\sin u$ $\quad \cos(-u) = \cos u$ $\quad \tan(-u) = -\tan u$ $\csc(-u) = -\csc u$ $\quad \sec(-u) = \sec u$ $\quad \cot(-u) = -\cot u$	1–4
	Use the fundamental trigonometric identities to evaluate trigonometric functions, simplify trigonometric expressions, and rewrite trigonometric expressions *(p. 211).*	In some cases, when factoring or simplifying a trigonometric expression, it is helpful to rewrite the expression in terms of just *one* trigonometric function or in terms of *sine and cosine only.*	5–18
Section 2.2	Verify trigonometric identities *(p. 217).*	**Guidelines for Verifying Trigonometric Identities** **1.** Work with one side of the equation at a time. **2.** Look to factor an expression, add fractions, square a binomial, or create a monomial denominator. **3.** Look to use the fundamental identities. Note which functions are in the final expression you want. Sines and cosines pair up well, as do secants and tangents, and cosecants and cotangents. **4.** When the preceding guidelines do not help, try converting all terms to sines and cosines. **5.** Always try *something.*	19–26
Section 2.3	Use standard algebraic techniques to solve trigonometric equations *(p. 224).*	Use standard algebraic techniques (when possible) such as collecting like terms, extracting square roots, and factoring to solve trigonometric equations.	27–32
	Solve trigonometric equations of quadratic type *(p. 227).*	To solve trigonometric equations of quadratic type $ax^2 + bx + c = 0$, use factoring (when possible) or use the Quadratic Formula.	33–36
	Solve trigonometric equations involving multiple angles *(p. 229).*	To solve equations that contain forms such as $\sin ku$ or $\cos ku$, first solve the equation for ku, and then divide your result by k.	37–42
	Use inverse trigonometric functions to solve trigonometric equations *(p. 230).*	After factoring an equation, you may get an equation such as $(\tan x - 3)(\tan x + 1) = 0$. In such cases, use inverse trigonometric functions to solve. (See Example 9.)	43–46

	What Did You Learn?	Explanation/Examples	Review Exercises
Section 2.4	Use sum and difference formulas to evaluate trigonometric functions, verify identities, and solve trigonometric equations *(p. 236)*.	**Sum and Difference Formulas** $\sin(u + v) = \sin u \cos v + \cos u \sin v$ $\sin(u - v) = \sin u \cos v - \cos u \sin v$ $\cos(u + v) = \cos u \cos v - \sin u \sin v$ $\cos(u - v) = \cos u \cos v + \sin u \sin v$ $\tan(u + v) = \dfrac{\tan u + \tan v}{1 - \tan u \tan v}$ $\tan(u - v) = \dfrac{\tan u - \tan v}{1 + \tan u \tan v}$	47–62
Section 2.5	Use multiple-angle formulas to rewrite and evaluate trigonometric functions *(p. 243)*.	**Double-Angle Formulas** $\sin 2u = 2 \sin u \cos u$ $\cos 2u = \cos^2 u - \sin^2 u$ $\tan 2u = \dfrac{2 \tan u}{1 - \tan^2 u}$ $= 2\cos^2 u - 1$ $= 1 - 2\sin^2 u$	63–66
	Use power-reducing formulas to rewrite trigonometric expressions *(p. 245)*.	**Power-Reducing Formulas** $\sin^2 u = \dfrac{1 - \cos 2u}{2}$, $\cos^2 u = \dfrac{1 + \cos 2u}{2}$ $\tan^2 u = \dfrac{1 - \cos 2u}{1 + \cos 2u}$	67, 68
	Use half-angle formulas to rewrite and evaluate trigonometric functions *(p. 246)*.	**Half-Angle Formulas** $\sin \dfrac{u}{2} = \pm\sqrt{\dfrac{1 - \cos u}{2}}$, $\cos \dfrac{u}{2} = \pm\sqrt{\dfrac{1 + \cos u}{2}}$ $\tan \dfrac{u}{2} = \dfrac{1 - \cos u}{\sin u} = \dfrac{\sin u}{1 + \cos u}$ The signs of $\sin \dfrac{u}{2}$ and $\cos \dfrac{u}{2}$ depend on the quadrant in which $u/2$ lies.	69–74
	Use product-to-sum and sum-to-product formulas to rewrite and evaluate trigonometric expressions *(p. 247)*.	**Product-to-Sum Formulas** $\sin u \sin v = (1/2)[\cos(u - v) - \cos(u + v)]$ $\cos u \cos v = (1/2)[\cos(u - v) + \cos(u + v)]$ $\sin u \cos v = (1/2)[\sin(u + v) + \sin(u - v)]$ $\cos u \sin v = (1/2)[\sin(u + v) - \sin(u - v)]$ **Sum-to-Product Formulas** $\sin u + \sin v = 2 \sin\left(\dfrac{u + v}{2}\right) \cos\left(\dfrac{u - v}{2}\right)$ $\sin u - \sin v = 2 \cos\left(\dfrac{u + v}{2}\right) \sin\left(\dfrac{u - v}{2}\right)$ $\cos u + \cos v = 2 \cos\left(\dfrac{u + v}{2}\right) \cos\left(\dfrac{u - v}{2}\right)$ $\cos u - \cos v = -2 \sin\left(\dfrac{u + v}{2}\right) \sin\left(\dfrac{u - v}{2}\right)$	75–78
	Use trigonometric formulas to rewrite real-life models *(p. 249)*.	A trigonometric formula can be used to rewrite the projectile motion model $r = (1/16)\, v_0^2 \sin \theta \cos \theta$. (See Example 10.)	79, 80

Review Exercises

See CalcChat.com for tutorial help and worked-out solutions to odd-numbered exercises.

2.1 Recognizing a Fundamental Identity **In Exercises 1–4, name the trigonometric function that is equivalent to the expression.**

1. $\dfrac{\cos x}{\sin x}$
2. $\dfrac{1}{\cos x}$
3. $\sin\left(\dfrac{\pi}{2} - x\right)$
4. $\sqrt{\cot^2 x + 1}$

Using Identities to Evaluate a Function **In Exercises 5 and 6, use the given conditions and fundamental trigonometric identities to find the values of all six trigonometric functions.**

5. $\cos\theta = -\frac{2}{5}$, $\tan\theta > 0$
6. $\cot x = -\frac{2}{3}$, $\cos x < 0$

Simplifying a Trigonometric Expression **In Exercises 7–16, use the fundamental trigonometric identities to simplify the expression. (There is more than one correct form of each answer.)**

7. $\dfrac{1}{\cot^2 x + 1}$
8. $\dfrac{\tan\theta}{1 - \cos^2\theta}$
9. $\tan^2 x(\csc^2 x - 1)$
10. $\cot^2 x(\sin^2 x)$
11. $\dfrac{\cot\left(\dfrac{\pi}{2} - u\right)}{\cos u}$
12. $\dfrac{\sec^2(-\theta)}{\csc^2\theta}$
13. $\cos^2 x + \cos^2 x \cot^2 x$
14. $(\tan x + 1)^2 \cos x$
15. $\dfrac{1}{\csc\theta + 1} - \dfrac{1}{\csc\theta - 1}$
16. $\dfrac{\tan^2 x}{1 + \sec x}$

Trigonometric Substitution **In Exercises 17 and 18, use the trigonometric substitution to write the algebraic expression as a trigonometric function of θ, where $0 < \theta < \pi/2$.**

17. $\sqrt{25 - x^2}$, $x = 5\sin\theta$
18. $\sqrt{x^2 - 16}$, $x = 4\sec\theta$

2.2 Verifying a Trigonometric Identity **In Exercises 19–26, verify the identity.**

19. $\cos x(\tan^2 x + 1) = \sec x$
20. $\sec^2 x \cot x - \cot x = \tan x$
21. $\sin\left(\dfrac{\pi}{2} - \theta\right)\tan\theta = \sin\theta$
22. $\cot\left(\dfrac{\pi}{2} - x\right)\csc x = \sec x$
23. $\dfrac{1}{\tan\theta\csc\theta} = \cos\theta$
24. $\dfrac{1}{\tan x \csc x \sin x} = \cot x$
25. $\sin^5 x \cos^2 x = (\cos^2 x - 2\cos^4 x + \cos^6 x)\sin x$
26. $\cos^3 x \sin^2 x = (\sin^2 x - \sin^4 x)\cos x$

2.3 Solving a Trigonometric Equation **In Exercises 27–32, solve the equation.**

27. $\sin x = \sqrt{3} - \sin x$
28. $4\cos\theta = 1 + 2\cos\theta$
29. $3\sqrt{3}\tan u = 3$
30. $\frac{1}{2}\sec x - 1 = 0$
31. $3\csc^2 x = 4$
32. $4\tan^2 u - 1 = \tan^2 u$

Solving a Trigonometric Equation **In Exercises 33–42, find all solutions of the equation in the interval $[0, 2\pi)$.**

33. $\sin^3 x = \sin x$
34. $2\cos^2 x + 3\cos x = 0$
35. $\cos^2 x + \sin x = 1$
36. $\sin^2 x + 2\cos x = 2$
37. $2\sin 2x - \sqrt{2} = 0$
38. $2\cos\dfrac{x}{2} + 1 = 0$
39. $3\tan^2\left(\dfrac{x}{3}\right) - 1 = 0$
40. $\sqrt{3}\tan 3x = 0$
41. $\cos 4x(\cos x - 1) = 0$
42. $3\csc^2 5x = -4$

Using Inverse Functions **In Exercises 43–46, solve the equation.**

43. $\tan^2 x - 2\tan x = 0$
44. $2\tan^2 x - 3\tan x = -1$
45. $\tan^2\theta + \tan\theta - 6 = 0$
46. $\sec^2 x + 6\tan x + 4 = 0$

2.4 Evaluating Trigonometric Functions **In Exercises 47–50, find the exact values of the sine, cosine, and tangent of the angle.**

47. $75° = 120° - 45°$
48. $375° = 135° + 240°$
49. $\dfrac{25\pi}{12} = \dfrac{11\pi}{6} + \dfrac{\pi}{4}$
50. $\dfrac{19\pi}{12} = \dfrac{11\pi}{6} - \dfrac{\pi}{4}$

Rewriting a Trigonometric Expression **In Exercises 51 and 52, write the expression as the sine, cosine, or tangent of an angle.**

51. $\sin 60° \cos 45° - \cos 60° \sin 45°$
52. $\dfrac{\tan 68° - \tan 115°}{1 + \tan 68° \tan 115°}$

Evaluating a Trigonometric Expression **In Exercises 53–56, find the exact value of the trigonometric expression given that $\tan u = \frac{3}{4}$ and $\cos v = -\frac{4}{5}$. (u is in Quadrant I and v is in Quadrant III.)**

53. $\sin(u + v)$
54. $\tan(u + v)$
55. $\cos(u - v)$
56. $\sin(u - v)$

Verifying a Trigonometric Identity **In Exercises 57–60, verify the identity.**

57. $\cos\left(x + \frac{\pi}{2}\right) = -\sin x$

58. $\tan\left(x - \frac{\pi}{2}\right) = -\cot x$

59. $\tan(\pi - x) = -\tan x$

60. $\sin(x - \pi) = -\sin x$

Solving a Trigonometric Equation **In Exercises 61 and 62, find all solutions of the equation in the interval $[0, 2\pi)$.**

61. $\sin\left(x + \frac{\pi}{4}\right) - \sin\left(x - \frac{\pi}{4}\right) = 1$

62. $\cos\left(x + \frac{\pi}{6}\right) - \cos\left(x - \frac{\pi}{6}\right) = 1$

2.5 Evaluating Functions Involving Double Angles **In Exercises 63 and 64, use the given conditions to find the exact values of $\sin 2u$, $\cos 2u$, and $\tan 2u$ using the double-angle formulas.**

63. $\sin u = \frac{4}{5}, \quad 0 < u < \pi/2$

64. $\cos u = -2/\sqrt{5}, \quad \pi/2 < u < \pi$

Verifying a Trigonometric Identity **In Exercises 65 and 66, use the double-angle formulas to verify the identity algebraically and use a graphing utility to confirm your result graphically.**

65. $\sin 4x = 8\cos^3 x \sin x - 4\cos x \sin x$

66. $\tan^2 x = \dfrac{1 - \cos 2x}{1 + \cos 2x}$

Reducing Powers **In Exercises 67 and 68, use the power-reducing formulas to rewrite the expression in terms of first powers of the cosines of multiple angles.**

67. $\tan^2 3x$

68. $\sin^2 x \cos^2 x$

Using Half-Angle Formulas **In Exercises 69 and 70, use the half-angle formulas to determine the exact values of the sine, cosine, and tangent of the angle.**

69. $-75°$

70. $5\pi/12$

Using Half-Angle Formulas **In Exercises 71–74, use the given conditions to (a) determine the quadrant in which $u/2$ lies, and (b) find the exact values of $\sin(u/2)$, $\cos(u/2)$, and $\tan(u/2)$ using the half-angle formulas.**

71. $\tan u = \frac{4}{3}, \quad \pi < u < \frac{3\pi}{2}$

72. $\sin u = \frac{3}{5}, \quad 0 < u < \frac{\pi}{2}$

73. $\cos u = -\frac{2}{7}, \quad \frac{\pi}{2} < u < \pi$

74. $\tan u = -\frac{\sqrt{21}}{2}, \quad \frac{3\pi}{2} < u < 2\pi$

Using Product-to-Sum Formulas **In Exercises 75 and 76, use the product-to-sum formulas to rewrite the product as a sum or difference.**

75. $\cos 4\theta \sin 6\theta$

76. $2\sin 7\theta \cos 3\theta$

Using Sum-to-Product Formulas **In Exercises 77 and 78, use the sum-to-product formulas to rewrite the sum or difference as a product.**

77. $\cos 6\theta + \cos 5\theta$

78. $\sin 3x - \sin x$

79. **Projectile Motion** A baseball leaves the hand of a player at first base at an angle of θ with the horizontal and at an initial velocity of $v_0 = 80$ feet per second. A player at second base 100 feet away catches the ball. Find θ when the range r of a projectile is

$$r = \frac{1}{32}v_0^2 \sin 2\theta.$$

80. **Geometry** A trough for feeding cattle is 4 meters long and its cross sections are isosceles triangles with the two equal sides being $\frac{1}{2}$ meter (see figure). The angle between the two sides is θ.

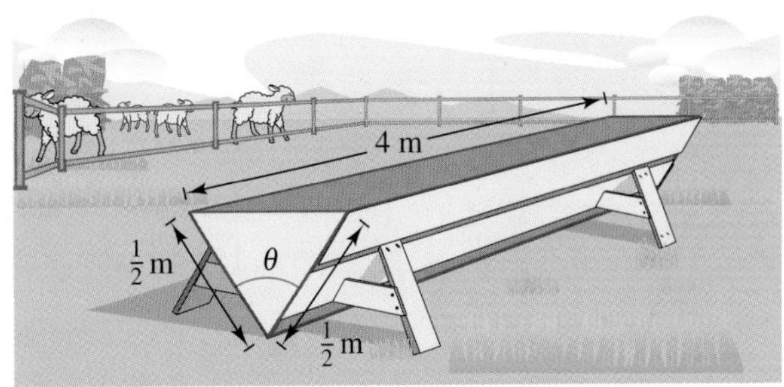

(a) Write the volume of the trough as a function of $\theta/2$.

(b) Write the volume of the trough as a function of θ and determine the value of θ such that the volume is maximized.

Exploration

True or False? **In Exercises 81–84, determine whether the statement is true or false. Justify your answer.**

81. If $\frac{\pi}{2} < \theta < \pi$, then $\cos\frac{\theta}{2} < 0$.

82. $\cot x \sin^2 x = \cos x \sin x$

83. $4\sin(-x)\cos(-x) = -2\sin 2x$

84. $4\sin 45° \cos 15° = 1 + \sqrt{3}$

85. **Think About It** Is it possible for a trigonometric equation that is not an identity to have an infinite number of solutions? Explain.

Chapter Test

See CalcChat.com for tutorial help and worked-out solutions to odd-numbered exercises.

Take this test as you would take a test in class. When you are finished, check your work against the answers given in the back of the book.

1. Use the conditions $\csc\theta = \frac{5}{2}$ and $\tan\theta < 0$ to find the values of all six trigonometric functions.

2. Use the fundamental identities to simplify $\csc^2\beta(1 - \cos^2\beta)$.

3. Factor and simplify $\dfrac{\sec^4 x - \tan^4 x}{\sec^2 x + \tan^2 x}$.

4. Add and simplify $\dfrac{\cos\theta}{\sin\theta} + \dfrac{\sin\theta}{\cos\theta}$.

In Exercises 5–10, verify the identity.

5. $\sin\theta\sec\theta = \tan\theta$

6. $\sec^2 x\tan^2 x + \sec^2 x = \sec^4 x$

7. $\dfrac{\csc\alpha + \sec\alpha}{\sin\alpha + \cos\alpha} = \cot\alpha + \tan\alpha$

8. $\tan\left(x + \dfrac{\pi}{2}\right) = -\cot x$

9. $1 + \cos 10y = 2\cos^2 5y$

10. $\sin\dfrac{\alpha}{3}\cos\dfrac{\alpha}{3} = \dfrac{1}{2}\sin\dfrac{2\alpha}{3}$

11. Rewrite $4\sin 3\theta\cos 2\theta$ as a sum or difference.

12. Rewrite $\cos\left(\theta + \dfrac{\pi}{2}\right) - \cos\left(\theta - \dfrac{\pi}{2}\right)$ as a product.

In Exercises 13–16, find all solutions of the equation in the interval $[0, 2\pi)$.

13. $\tan^2 x + \tan x = 0$

14. $\sin 2\alpha - \cos\alpha = 0$

15. $4\cos^2 x - 3 = 0$

16. $\csc^2 x - \csc x - 2 = 0$

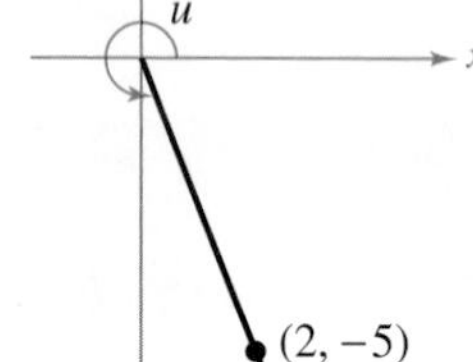

Figure for 19

17. Use a graphing utility to approximate (to three decimal places) the solutions of $5\sin x - x = 0$ in the interval $[0, 2\pi)$.

18. Find the exact value of $\cos 105°$ using the fact that $105° = 135° - 30°$.

19. Use the figure to find the exact values of $\sin 2u$, $\cos 2u$, and $\tan 2u$.

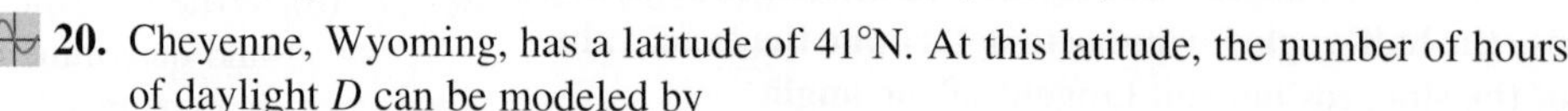

20. Cheyenne, Wyoming, has a latitude of 41°N. At this latitude, the number of hours of daylight D can be modeled by

$$D = 2.914\sin(0.017t - 1.321) + 12.134$$

where t represents the day, with $t = 1$ corresponding to January 1. Use a graphing utility to determine the days on which there are more than 10 hours of daylight. *(Source: U.S. Naval Observatory)*

21. The heights h_1 and h_2 (in feet) above ground of two people in different seats on a Ferris wheel can be modeled by

$$h_1 = 28\cos 10t + 38$$

and

$$h_2 = 28\cos\left[10\left(t - \frac{\pi}{6}\right)\right] + 38, \quad 0 \le t \le 2$$

where t represents the time (in minutes). When are the two people at the same height?

Proofs in Mathematics

Sum and Difference Formulas *(p. 236)*

$$\sin(u + v) = \sin u \cos v + \cos u \sin v$$

$$\sin(u - v) = \sin u \cos v - \cos u \sin v$$

$$\cos(u + v) = \cos u \cos v - \sin u \sin v$$

$$\cos(u - v) = \cos u \cos v + \sin u \sin v$$

$$\tan(u + v) = \frac{\tan u + \tan v}{1 - \tan u \tan v}$$

$$\tan(u - v) = \frac{\tan u - \tan v}{1 + \tan u \tan v}$$

Proof

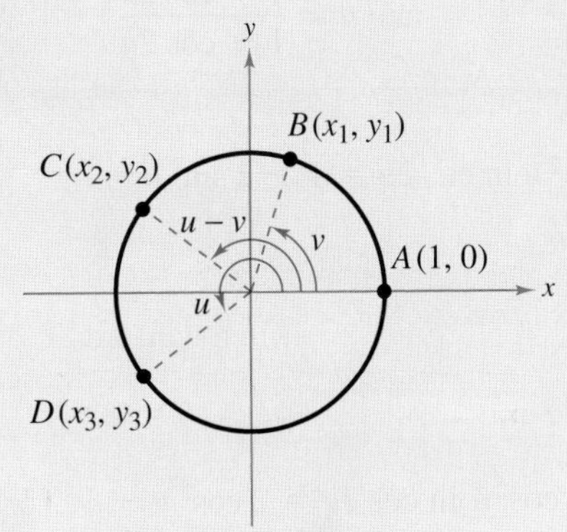

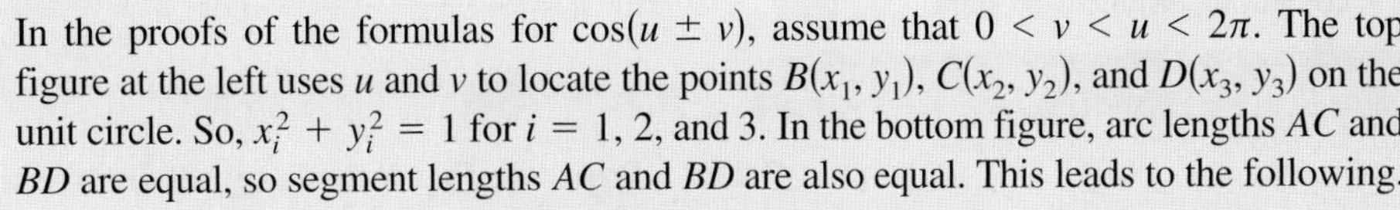

In the proofs of the formulas for $\cos(u \pm v)$, assume that $0 < v < u < 2\pi$. The top figure at the left uses u and v to locate the points $B(x_1, y_1)$, $C(x_2, y_2)$, and $D(x_3, y_3)$ on the unit circle. So, $x_i^2 + y_i^2 = 1$ for $i = 1$, 2, and 3. In the bottom figure, arc lengths AC and BD are equal, so segment lengths AC and BD are also equal. This leads to the following.

$$\sqrt{(x_2 - 1)^2 + (y_2 - 0)^2} = \sqrt{(x_3 - x_1)^2 + (y_3 - y_1)^2}$$

$$x_2^2 - 2x_2 + 1 + y_2^2 = x_3^2 - 2x_1x_3 + x_1^2 + y_3^2 - 2y_1y_3 + y_1^2$$

$$(x_2^2 + y_2^2) + 1 - 2x_2 = (x_3^2 + y_3^2) + (x_1^2 + y_1^2) - 2x_1x_3 - 2y_1y_3$$

$$1 + 1 - 2x_2 = 1 + 1 - 2x_1x_3 - 2y_1y_3$$

$$x_2 = x_3x_1 + y_3y_1$$

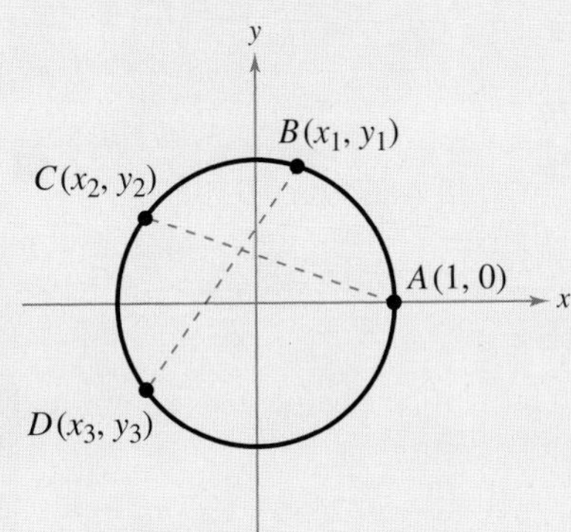

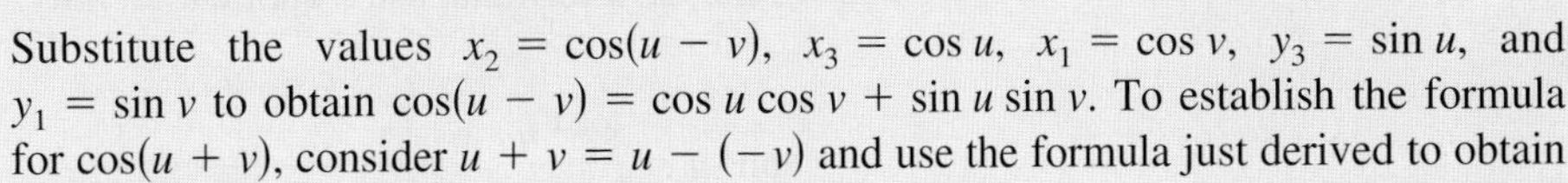

Substitute the values $x_2 = \cos(u - v)$, $x_3 = \cos u$, $x_1 = \cos v$, $y_3 = \sin u$, and $y_1 = \sin v$ to obtain $\cos(u - v) = \cos u \cos v + \sin u \sin v$. To establish the formula for $\cos(u + v)$, consider $u + v = u - (-v)$ and use the formula just derived to obtain

$$\begin{aligned}\cos(u + v) &= \cos[u - (-v)] \\ &= \cos u \cos(-v) + \sin u \sin(-v) \\ &= \cos u \cos v - \sin u \sin v.\end{aligned}$$

You can use the sum and difference formulas for sine and cosine to prove the formulas for $\tan(u \pm v)$.

$$\tan(u \pm v) = \frac{\sin(u \pm v)}{\cos(u \pm v)} = \frac{\sin u \cos v \pm \cos u \sin v}{\cos u \cos v \mp \sin u \sin v}$$

$$= \frac{\dfrac{\sin u \cos v \pm \cos u \sin v}{\cos u \cos v}}{\dfrac{\cos u \cos v \mp \sin u \sin v}{\cos u \cos v}} = \frac{\dfrac{\sin u \cos v}{\cos u \cos v} \pm \dfrac{\cos u \sin v}{\cos u \cos v}}{\dfrac{\cos u \cos v}{\cos u \cos v} \mp \dfrac{\sin u \sin v}{\cos u \cos v}}$$

$$= \frac{\dfrac{\sin u}{\cos u} \pm \dfrac{\sin v}{\cos v}}{1 \mp \dfrac{\sin u}{\cos u} \cdot \dfrac{\sin v}{\cos v}} = \frac{\tan u \pm \tan v}{1 \mp \tan u \tan v}$$

Double-Angle Formulas *(p. 243)*

$$\sin 2u = 2 \sin u \cos u$$

$$\tan 2u = \frac{2 \tan u}{1 - \tan^2 u}$$

$$\begin{aligned}\cos 2u &= \cos^2 u - \sin^2 u \\ &= 2\cos^2 u - 1 \\ &= 1 - 2\sin^2 u\end{aligned}$$

> **TRIGONOMETRY AND ASTRONOMY**
>
> Early astronomers used trigonometry to calculate measurements in the universe. For instance, they used trigonometry to calculate the circumference of Earth and the distance from Earth to the moon. Another major accomplishment in astronomy using trigonometry was computing distances to stars.

Proof Prove each Double-Angle Formula by letting $v = u$ in the corresponding sum formula.

$$\sin 2u = \sin(u + u) = \sin u \cos u + \cos u \sin u = 2 \sin u \cos u$$

$$\cos 2u = \cos(u + u) = \cos u \cos u - \sin u \sin u = \cos^2 u - \sin^2 u$$

$$\tan 2u = \tan(u + u) = \frac{\tan u + \tan u}{1 - \tan u \tan u} = \frac{2 \tan u}{1 - \tan^2 u}$$ ■

Power-Reducing Formulas *(p. 245)*

$$\sin^2 u = \frac{1 - \cos 2u}{2} \qquad \cos^2 u = \frac{1 + \cos 2u}{2} \qquad \tan^2 u = \frac{1 - \cos 2u}{1 + \cos 2u}$$

Proof Prove the first formula by solving for $\sin^2 u$ in $\cos 2u = 1 - 2 \sin^2 u$.

$\cos 2u = 1 - 2 \sin^2 u$ Write double-angle formula.

$2 \sin^2 u = 1 - \cos 2u$ Subtract $\cos 2u$ from, and add $2 \sin^2 u$ to, each side.

$\sin^2 u = \dfrac{1 - \cos 2u}{2}$ Divide each side by 2.

Similarly, to prove the second formula, solve for $\cos^2 u$ in $\cos 2u = 2 \cos^2 u - 1$. To prove the third formula, use a quotient identity.

$$\tan^2 u = \frac{\sin^2 u}{\cos^2 u} = \frac{\dfrac{1 - \cos 2u}{2}}{\dfrac{1 + \cos 2u}{2}} = \frac{1 - \cos 2u}{1 + \cos 2u}$$ ■

Sum-to-Product Formulas *(p. 247)*

$$\sin u + \sin v = 2 \sin\left(\frac{u + v}{2}\right) \cos\left(\frac{u - v}{2}\right)$$

$$\sin u - \sin v = 2 \cos\left(\frac{u + v}{2}\right) \sin\left(\frac{u - v}{2}\right)$$

$$\cos u + \cos v = 2 \cos\left(\frac{u + v}{2}\right) \cos\left(\frac{u - v}{2}\right)$$

$$\cos u - \cos v = -2 \sin\left(\frac{u + v}{2}\right) \sin\left(\frac{u - v}{2}\right)$$

Proof To prove the first formula, let $x = u + v$ and $y = u - v$. Then substitute $u = (x + y)/2$ and $v = (x - y)/2$ in the product-to-sum formula.

$$\sin u \cos v = \frac{1}{2}[\sin(u + v) + \sin(u - v)]$$

$$\sin\left(\frac{x + y}{2}\right) \cos\left(\frac{x - y}{2}\right) = \frac{1}{2}(\sin x + \sin y)$$

$$2 \sin\left(\frac{x + y}{2}\right) \cos\left(\frac{x - y}{2}\right) = \sin x + \sin y$$

The other sum-to-product formulas can be proved in a similar manner. ■

P.S. Problem Solving

1. **Writing Trigonometric Functions in Terms of Cosine** Write each of the other trigonometric functions of θ in terms of $\cos \theta$.

2. **Verifying a Trigonometric Identity** Verify that for all integers n,

$$\cos\left[\frac{(2n + 1)\pi}{2}\right] = 0.$$

3. **Verifying a Trigonometric Identity** Verify that for all integers n,

$$\sin\left[\frac{(12n + 1)\pi}{6}\right] = \frac{1}{2}.$$

4. **Sound Wave** A sound wave is modeled by

$$p(t) = \frac{1}{4\pi}[p_1(t) + 30p_2(t) + p_3(t) + p_5(t) + 30p_6(t)]$$

where $p_n(t) = \frac{1}{n}\sin(524n\pi t)$, and t represents the time (in seconds).

(a) Find the sine components $p_n(t)$ and use a graphing utility to graph the components. Then verify the graph of p shown below.

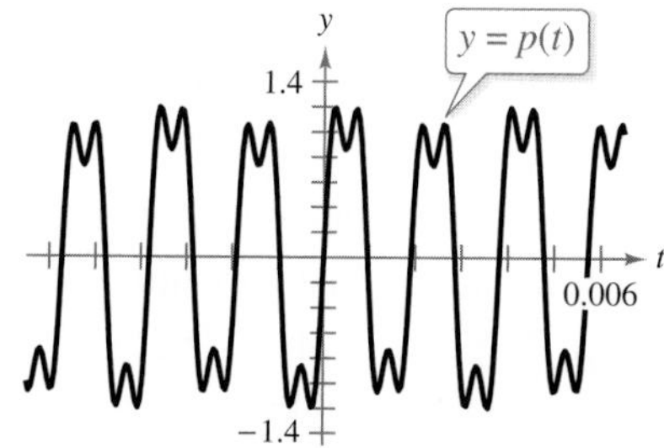

(b) Find the period of each sine component of p. Is p periodic? If so, then what is its period?

(c) Use the graphing utility to find the t-intercepts of the graph of p over one cycle.

(d) Use the graphing utility to approximate the absolute maximum and absolute minimum values of p over one cycle.

5. **Geometry** Three squares of side length s are placed side by side (see figure). Make a conjecture about the relationship between the sum $u + v$ and w. Prove your conjecture by using the identity for the tangent of the sum of two angles.

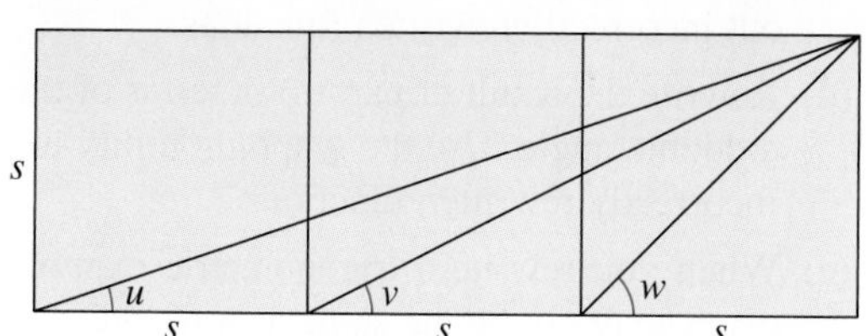

6. **Projectile Motion** The path traveled by an object (neglecting air resistance) that is projected at an initial height of h_0 feet, an initial velocity of v_0 feet per second, and an initial angle θ is given by

$$y = -\frac{16}{v_0^2 \cos^2 \theta}x^2 + (\tan \theta)x + h_0$$

where the horizontal distance x and the vertical distance y are measured in feet. Find a formula for the maximum height of an object projected from ground level at velocity v_0 and angle θ. To do this, find half of the horizontal distance

$$\frac{1}{32}v_0^2 \sin 2\theta$$

and then substitute it for x in the model for the path of a projectile (where $h_0 = 0$).

7. **Geometry** The length of each of the two equal sides of an isosceles triangle is 10 meters (see figure). The angle between the two sides is θ.

(a) Write the area of the triangle as a function of $\theta/2$.

(b) Write the area of the triangle as a function of θ. Determine the value of θ such that the area is a maximum.

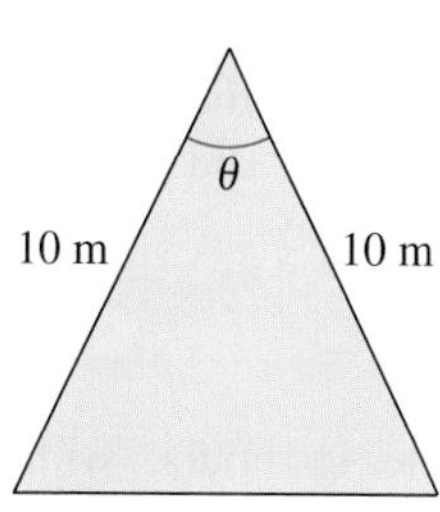

Figure for 7

Figure for 8

8. **Geometry** Use the figure to derive the formulas for

$$\sin\frac{\theta}{2},\ \cos\frac{\theta}{2},\ \text{and}\ \tan\frac{\theta}{2}$$

where θ is an acute angle.

9. **Force** The force F (in pounds) on a person's back when he or she bends over at an angle θ from an upright position is modeled by

$$F = \frac{0.6W\sin(\theta + 90^\circ)}{\sin 12^\circ}$$

where W represents the person's weight (in pounds).

(a) Simplify the model.

(b) Use a graphing utility to graph the model, where $W = 185$ and $0^\circ < \theta < 90^\circ$.

(c) At what angle is the force maximized? At what angle is the force minimized?

10. **Hours of Daylight** The number of hours of daylight that occur at any location on Earth depends on the time of year and the latitude of the location. The equations below model the numbers of hours of daylight in Seward, Alaska (60° latitude), and New Orleans, Louisiana (30° latitude).

$$D = 12.2 - 6.4\cos\left[\frac{\pi(t + 0.2)}{182.6}\right] \quad \text{Seward}$$

$$D = 12.2 - 1.9\cos\left[\frac{\pi(t + 0.2)}{182.6}\right] \quad \text{New Orleans}$$

In these models, D represents the number of hours of daylight and t represents the day, with $t = 0$ corresponding to January 1.

(a) Use a graphing utility to graph both models in the same viewing window. Use a viewing window of $0 \le t \le 365$.

(b) Find the days of the year on which both cities receive the same amount of daylight.

(c) Which city has the greater variation in the number of hours of daylight? Which constant in each model would you use to determine the difference between the greatest and least numbers of hours of daylight?

(d) Determine the period of each model.

11. **Ocean Tide** The tide, or depth of the ocean near the shore, changes throughout the day. The water depth d (in feet) of a bay can be modeled by

$$d = 35 - 28\cos\frac{\pi}{6.2}t$$

where t represents the time in hours, with $t = 0$ corresponding to 12:00 A.M.

(a) Algebraically find the times at which the high and low tides occur.

(b) If possible, algebraically find the time(s) at which the water depth is 3.5 feet.

(c) Use a graphing utility to verify your results from parts (a) and (b).

12. **Piston Heights** The heights h (in inches) of pistons 1 and 2 in an automobile engine can be modeled by

$$h_1 = 3.75\sin 733t + 7.5$$

and

$$h_2 = 3.75\sin 733\left(t + \frac{4\pi}{3}\right) + 7.5$$

respectively, where t is measured in seconds.

(a) Use a graphing utility to graph the heights of these pistons in the same viewing window for $0 \le t \le 1$.

(b) How often are the pistons at the same height?

13. **Index of Refraction** The index of refraction n of a transparent material is the ratio of the speed of light in a vacuum to the speed of light in the material. Some common materials and their indices of refraction are air (1.00), water (1.33), and glass (1.50). Triangular prisms are often used to measure the index of refraction based on the formula

$$n = \frac{\sin\left(\frac{\theta}{2} + \frac{\alpha}{2}\right)}{\sin\frac{\theta}{2}}.$$

For the prism shown in the figure, $\alpha = 60°$.

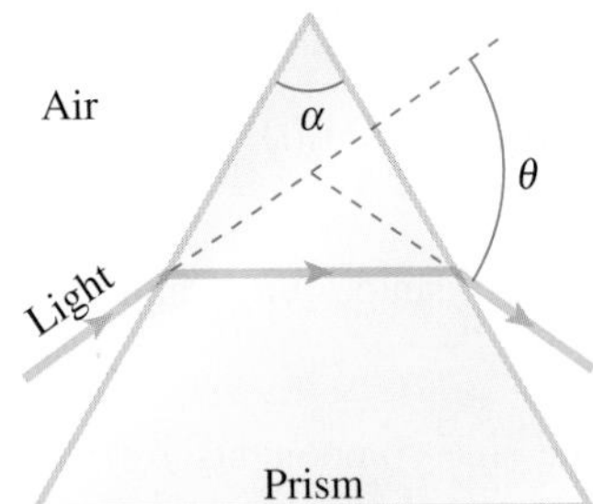

(a) Write the index of refraction as a function of $\cot(\theta/2)$.

(b) Find θ for a prism made of glass.

14. **Sum Formulas**

(a) Write a sum formula for $\sin(u + v + w)$.

(b) Write a sum formula for $\tan(u + v + w)$.

15. **Solving Trigonometric Inequalities** Find the solution of each inequality in the interval $[0, 2\pi)$.

(a) $\sin x \ge 0.5$ (b) $\cos x \le -0.5$

(c) $\tan x < \sin x$ (d) $\cos x \ge \sin x$

16. **Sum of Fourth Powers** Consider the function $f(x) = \sin^4 x + \cos^4 x$.

(a) Use the power-reducing formulas to write the function in terms of cosine to the first power.

(b) Determine another way of rewriting the original function. Use a graphing utility to rule out incorrectly rewritten functions.

(c) Add a trigonometric term to the original function so that it becomes a perfect square trinomial. Rewrite the function as a perfect square trinomial minus the term that you added. Use the graphing utility to rule out incorrectly rewritten functions.

(d) Rewrite the result of part (c) in terms of the sine of a double angle. Use the graphing utility to rule out incorrectly rewritten functions.

(e) When you rewrite a trigonometric expression, the result may not be the same as a friend's. Does this mean that one of you is wrong? Explain.

3 Additional Topics in Trigonometry

Work *(page 296)*

Braking Load
(Exercise 74, page 298)

Air Navigation *(Example 12, page 286)*

Mechanical Engineering
(Exercise 56, page 277)

Surveying *(page 263)*

Clockwise from top left, Vince Clements/Shutterstock.com; Anthony Berenyi/Shutterstock.com; Smart-foto/Shutterstock.com; iStockphoto.com/Andrew Ilyasov/isoft; iStockphoto.com/ggyykk

3.1 Law of Sines

The Law of Sines is a useful tool for solving real-life problems involving oblique triangles. For example, in Exercise 46 on page 269, you will use the Law of Sines to determine the distance from a boat to a shoreline.

- **Use the Law of Sines to solve oblique triangles (AAS or ASA).**
- **Use the Law of Sines to solve oblique triangles (SSA).**
- **Find the areas of oblique triangles.**
- **Use the Law of Sines to model and solve real-life problems.**

Introduction

In Chapter 1, you studied techniques for solving right triangles. In this section and the next, you will solve **oblique triangles**—triangles that have no right angles. As standard notation, the angles of a triangle are labeled A, B, and C, and their opposite sides are labeled a, b, and c, as shown in the figure.

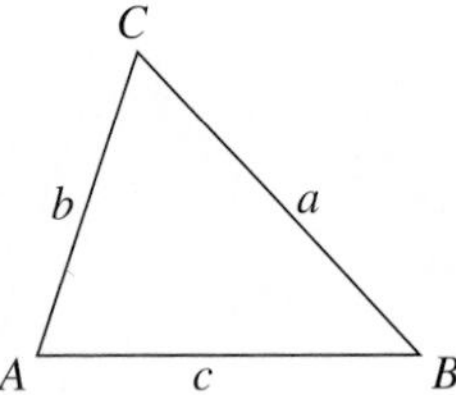

To solve an oblique triangle, you need to know the measure of at least one side and any two other measures of the triangle—the other two sides, two angles, or one angle and one other side. So, there are four cases.

1. Two angles and any side (AAS or ASA)
2. Two sides and an angle opposite one of them (SSA)
3. Three sides (SSS)
4. Two sides and their included angle (SAS)

The first two cases can be solved using the **Law of Sines,** whereas the last two cases require the Law of Cosines (see Section 3.2).

Law of Sines

If ABC is a triangle with sides a, b, and c, then

$$\frac{a}{\sin A} = \frac{b}{\sin B} = \frac{c}{\sin C}.$$

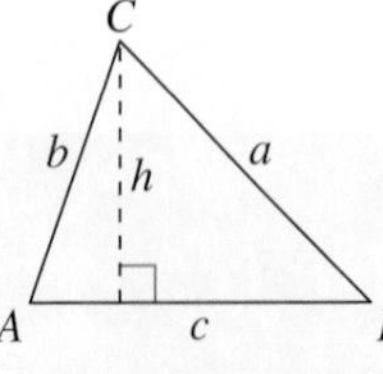

A is acute.

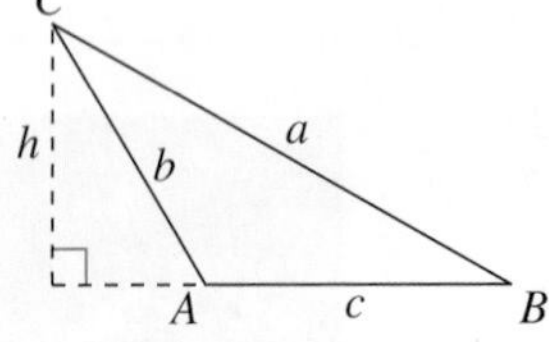

A is obtuse.

The Law of Sines can also be written in the reciprocal form

$$\frac{\sin A}{a} = \frac{\sin B}{b} = \frac{\sin C}{c}.$$

For a proof of the Law of Sines, see Proofs in Mathematics on page 308.

karamysh/Shutterstock.com

EXAMPLE 1 Given Two Angles and One Side—AAS

For the triangle in Figure 3.1, $C = 102°$, $B = 29°$, and $b = 28$ feet. Find the remaining angle and sides.

Solution The third angle of the triangle is

$$A = 180° - B - C = 180° - 29° - 102° = 49°.$$

By the Law of Sines, you have

$$\frac{a}{\sin A} = \frac{b}{\sin B} = \frac{c}{\sin C}.$$

Using $b = 28$ produces

$$a = \frac{b}{\sin B}(\sin A) = \frac{28}{\sin 29°}(\sin 49°) \approx 43.59 \text{ feet}$$

and

$$c = \frac{b}{\sin B}(\sin C) = \frac{28}{\sin 29°}(\sin 102°) \approx 56.49 \text{ feet.}$$

Figure 3.1

In the 1850s, surveyors used the Law of Sines to calculate the height of Mount Everest. Their calculation was within 30 feet of the currently accepted value.

Checkpoint Audio-video solution in English & Spanish at LarsonPrecalculus.com

For the triangle shown, $A = 30°$, $B = 45°$, and $a = 32$ centimeters. Find the remaining angle and sides.

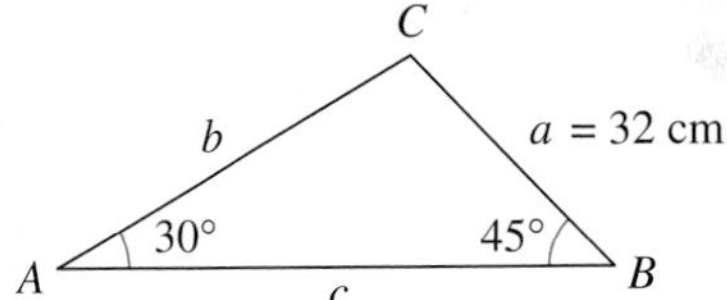

EXAMPLE 2 Given Two Angles and One Side—ASA

A pole tilts toward the sun at an 8° angle from the vertical, and it casts a 22-foot shadow. (See Figure 3.2.) The angle of elevation from the tip of the shadow to the top of the pole is 43°. How tall is the pole?

Solution In Figure 3.2, $A = 43°$ and

$$B = 90° + 8° = 98°.$$

So, the third angle is

$$C = 180° - A - B = 180° - 43° - 98° = 39°.$$

By the Law of Sines, you have

$$\frac{a}{\sin A} = \frac{c}{\sin C}.$$

The shadow length c is $c = 22$ feet, so the height of the pole is

$$a = \frac{c}{\sin C}(\sin A) = \frac{22}{\sin 39°}(\sin 43°) \approx 23.84 \text{ feet.}$$

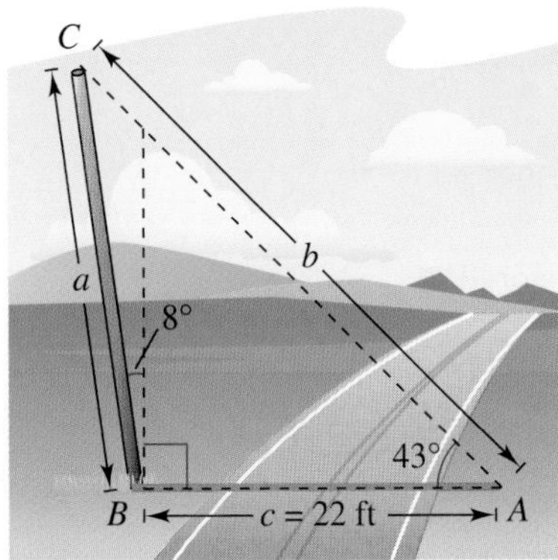

Figure 3.2

Checkpoint Audio-video solution in English & Spanish at LarsonPrecalculus.com

Find the height of the tree shown in the figure.

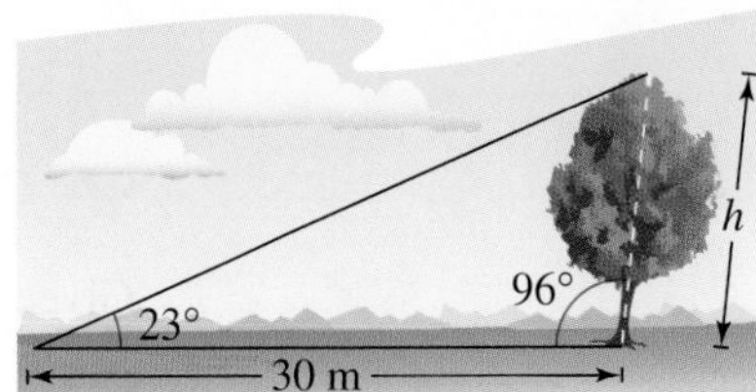

The Ambiguous Case (SSA)

In Examples 1 and 2, you saw that two angles and one side determine a unique triangle. However, if two sides and one opposite angle are given, then three possible situations can occur: (1) no such triangle exists, (2) one such triangle exists, or (3) two distinct triangles exist that satisfy the conditions.

The Ambiguous Case (SSA)

Consider a triangle in which a, b, and A are given. $(h = b \sin A)$

	A is acute.	A is acute.	A is acute.	A is acute.	A is obtuse.	A is obtuse.
Sketch	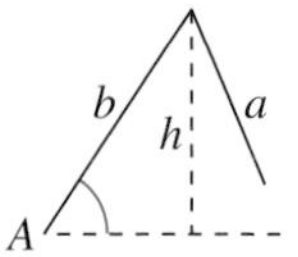		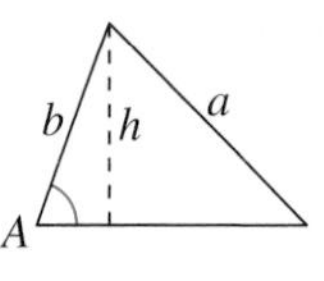	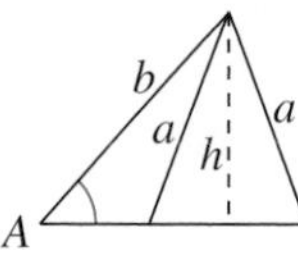	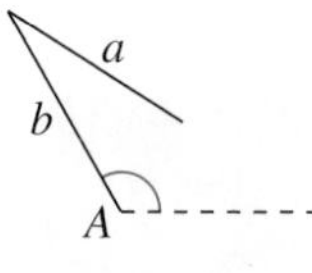	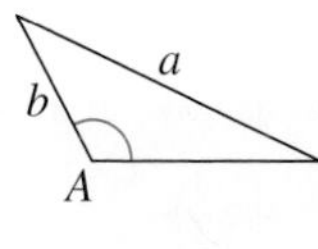
Necessary condition	$a < h$	$a = h$	$a \geq b$	$h < a < b$	$a \leq b$	$a > b$
Triangles possible	None	One	One	Two	None	One

EXAMPLE 3 Single-Solution Case—SSA

See LarsonPrecalculus.com for an interactive version of this type of example.

For the triangle in Figure 3.3, $a = 22$ inches, $b = 12$ inches, and $A = 42°$. Find the remaining side and angles.

One solution: $a \geq b$
Figure 3.3

Solution By the Law of Sines, you have

$$\frac{\sin B}{b} = \frac{\sin A}{a} \qquad \text{Reciprocal form}$$

$$\sin B = b\left(\frac{\sin A}{a}\right) \qquad \text{Multiply each side by } b.$$

$$\sin B = 12\left(\frac{\sin 42°}{22}\right) \qquad \text{Substitute for } A, a, \text{ and } b.$$

$$B \approx 21.41°. \qquad \text{Solve for acute angle } B.$$

Next, subtract to determine that $C \approx 180° - 42° - 21.41° = 116.59°$. Then find the remaining side.

$$\frac{c}{\sin C} = \frac{a}{\sin A} \qquad \text{Law of Sines}$$

$$c = \frac{a}{\sin A}(\sin C) \qquad \text{Multiply each side by } \sin C.$$

$$c \approx \frac{22}{\sin 42°}(\sin 116.59°) \qquad \text{Substitute for } a, A, \text{ and } C.$$

$$c \approx 29.40 \text{ inches} \qquad \text{Simplify.}$$

✓ ***Checkpoint*** *Audio-video solution in English & Spanish at LarsonPrecalculus.com*

Given $A = 31°$, $a = 12$ inches, and $b = 5$ inches, find the remaining side and angles of the triangle.

EXAMPLE 4 No-Solution Case—SSA

Show that there is no triangle for which $a = 15$ feet, $b = 25$ feet, and $A = 85°$.

No solution: $a < h$

Figure 3.4

Solution Begin by making the sketch shown in Figure 3.4. From this figure, it appears that no triangle is possible. Verify this using the Law of Sines.

$$\frac{\sin B}{b} = \frac{\sin A}{a} \quad \text{Reciprocal form}$$

$$\sin B = b\left(\frac{\sin A}{a}\right) \quad \text{Multiply each side by } b.$$

$$\sin B = 25\left(\frac{\sin 85°}{15}\right) \approx 1.6603 > 1$$

This contradicts the fact that $|\sin B| \le 1$. So, no triangle can be formed with sides $a = 15$ feet and $b = 25$ feet and angle $A = 85°$.

 **Checkpoint** Audio-video solution in English & Spanish at LarsonPrecalculus.com

Show that there is no triangle for which $a = 4$ feet, $b = 14$ feet, and $A = 60°$.

EXAMPLE 5 Two-Solution Case—SSA

Find two triangles for which $a = 12$ meters, $b = 31$ meters, and $A = 20.50°$.

Solution Because $h = b \sin A = 31(\sin 20.50°) \approx 10.86$ meters and $h < a < b$, there are two possible triangles. By the Law of Sines, you have

$$\frac{\sin B}{b} = \frac{\sin A}{a} \quad \text{Reciprocal form}$$

$$\sin B = b\left(\frac{\sin A}{a}\right) = 31\left(\frac{\sin 20.50°}{12}\right) \approx 0.9047.$$

There are two angles between 0° and 180° whose sine is approximately 0.9047, $B_1 \approx 64.78°$ and $B_2 \approx 180° - 64.78° = 115.22°$. For $B_1 \approx 64.78°$, you obtain

$$C \approx 180° - 20.50° - 64.78° = 94.72°$$

$$c = \frac{a}{\sin A}(\sin C) \approx \frac{12}{\sin 20.50°}(\sin 94.72°) \approx 34.15 \text{ meters.}$$

For $B_2 \approx 115.22°$, you obtain

$$C \approx 180° - 20.50° - 115.22° = 44.28°$$

$$c = \frac{a}{\sin A}(\sin C) \approx \frac{12}{\sin 20.50°}(\sin 44.28°) \approx 23.92 \text{ meters.}$$

The resulting triangles are shown below.

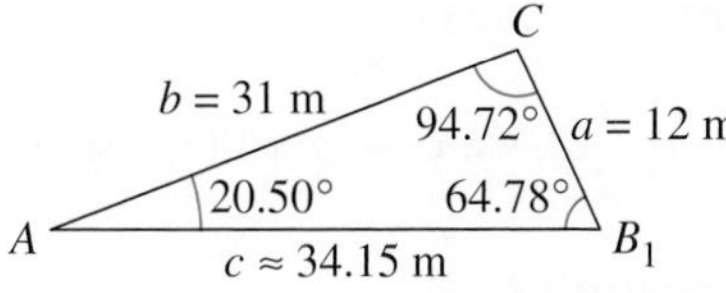

Two solutions: $h < a < b$

Checkpoint Audio-video solution in English & Spanish at LarsonPrecalculus.com

Find two triangles for which $a = 4.5$ feet, $b = 5$ feet, and $A = 58°$.

Area of an Oblique Triangle

The procedure used to prove the Law of Sines leads to a formula for the area of an oblique triangle. Consider the two triangles below.

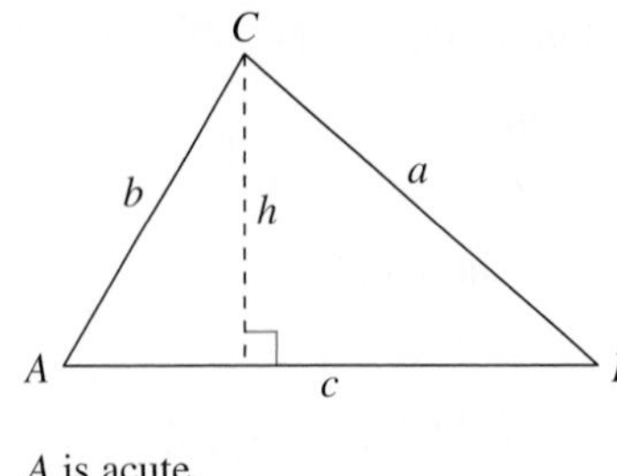

A is acute.

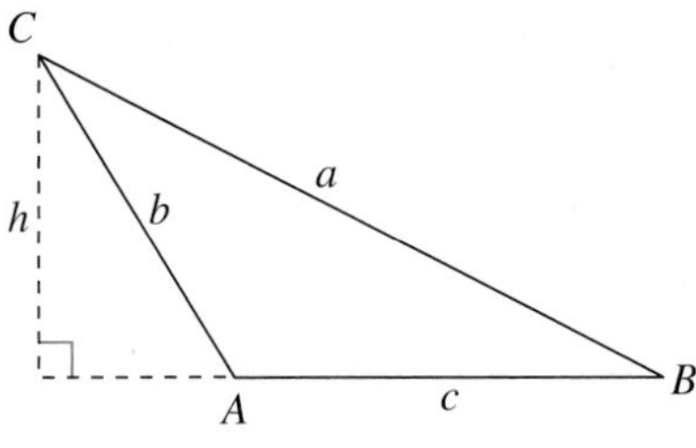

A is obtuse.

REMARK To obtain the height of the obtuse triangle, use the reference angle $180° - A$ and the difference formula for sine:

$$\begin{aligned} h &= b\sin(180° - A) \\ &= b(\sin 180° \cos A \\ &\qquad - \cos 180° \sin A) \\ &= b[0 \cdot \cos A - (-1) \cdot \sin A] \\ &= b \sin A. \end{aligned}$$

Note that each triangle has a height of $h = b \sin A$. Consequently, the area of each triangle is

$$\begin{aligned} \text{Area} &= \frac{1}{2}(\text{base})(\text{height}) \\ &= \frac{1}{2}(c)(b \sin A) \\ &= \frac{1}{2}bc \sin A. \end{aligned}$$

By similar arguments, you can develop the other two forms shown below.

> **Area of an Oblique Triangle**
>
> The area of any triangle is one-half the product of the lengths of two sides times the sine of their included angle. That is,
>
> $$\text{Area} = \frac{1}{2}bc \sin A = \frac{1}{2}ab \sin C = \frac{1}{2}ac \sin B.$$

Note that when angle *A* is 90°, the formula gives the area of a right triangle:

$$\text{Area} = \frac{1}{2}bc(\sin 90°) = \frac{1}{2}bc = \frac{1}{2}(\text{base})(\text{height}). \qquad \sin 90° = 1$$

You obtain similar results for angles *C* and *B* equal to 90°.

EXAMPLE 6 Finding the Area of a Triangular Lot

Find the area of a triangular lot with two sides of lengths 90 meters and 52 meters and an included angle of 102°, as shown in Figure 3.5.

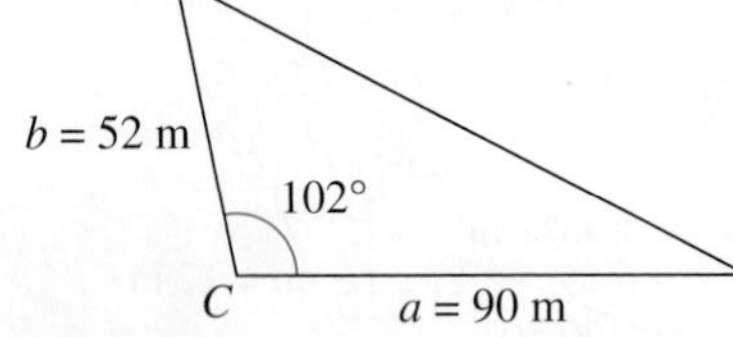

Figure 3.5

Solution The area is

$$\text{Area} = \frac{1}{2}ab \sin C = \frac{1}{2}(90)(52)(\sin 102°) \approx 2289 \text{ square meters.}$$

✓ ***Checkpoint*** *Audio-video solution in English & Spanish at LarsonPrecalculus.com*

Find the area of a triangular lot with two sides of lengths 24 yards and 18 yards and an included angle of 80°.

Application

EXAMPLE 7 An Application of the Law of Sines

The course for a boat race starts at point A and proceeds in the direction S 52° W to point B, then in the direction S 40° E to point C, and finally back to point A, as shown in Figure 3.6. Point C lies 8 kilometers directly south of point A. Approximate the total distance of the race course.

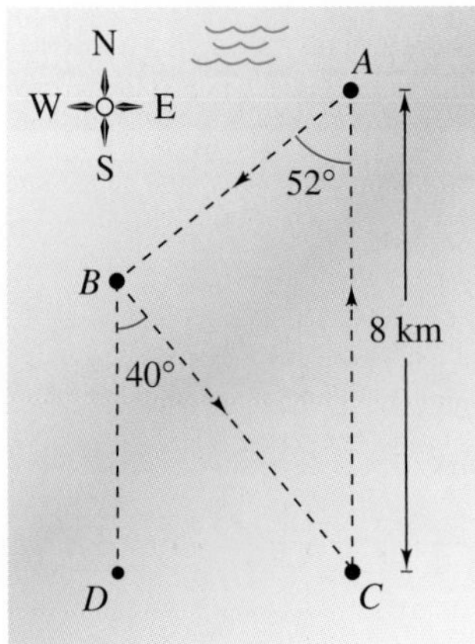

Figure 3.6

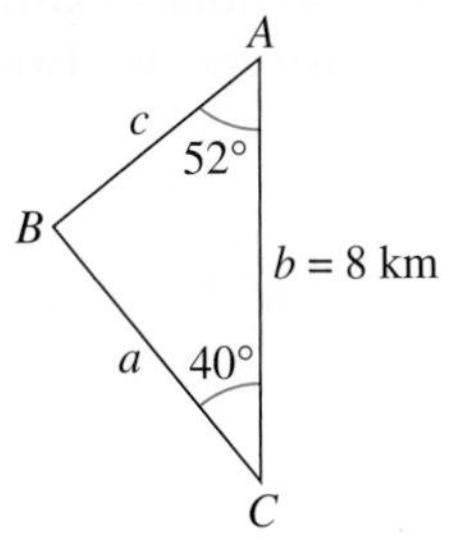

Figure 3.7

Solution The lines BD and AC are parallel, so $\angle BCA \cong \angle CBD$. Consequently, triangle ABC has the measures shown in Figure 3.7. The measure of angle B is $180° - 52° - 40° = 88°$. Using the Law of Sines,

$$\frac{a}{\sin 52°} = \frac{8}{\sin 88°} = \frac{c}{\sin 40°}.$$

Solving for a and c, you have

$$a = \frac{8}{\sin 88°}(\sin 52°) \approx 6.31 \quad \text{and} \quad c = \frac{8}{\sin 88°}(\sin 40°) \approx 5.15.$$

So, the total distance of the course is approximately

$$8 + 6.31 + 5.15 = 19.46 \text{ kilometers}.$$

Checkpoint Audio-video solution in English & Spanish at LarsonPrecalculus.com

On a small lake, you swim from point A to point B at a bearing of N 28° E, then to point C at a bearing of N 58° W, and finally back to point A, as shown in the figure below. Point C lies 800 meters directly north of point A. Approximate the total distance that you swim.

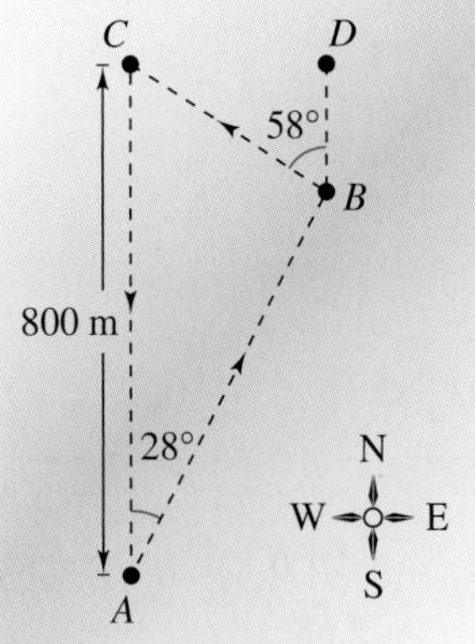

Summarize (Section 3.1)

1. State the Law of Sines *(page 262)*. For examples of using the Law of Sines to solve oblique triangles (AAS or ASA), see Examples 1 and 2.
2. List the necessary conditions and the corresponding numbers of possible triangles for the ambiguous case (SSA) *(page 264)*. For examples of using the Law of Sines to solve oblique triangles (SSA), see Examples 3–5.
3. State the formulas for the area of an oblique triangle *(page 266)*. For an example of finding the area of an oblique triangle, see Example 6.
4. Describe a real-life application of the Law of Sines *(page 267, Example 7)*.

3.1 Exercises

See **CalcChat.com** for tutorial help and worked-out solutions to odd-numbered exercises.

Vocabulary: Fill in the blanks.

1. An ________ triangle is a triangle that has no right angle.
2. For triangle ABC, the Law of Sines is $\dfrac{a}{\sin A} =$ ________ $= \dfrac{c}{\sin C}$.
3. Two ________ and one ________ determine a unique triangle.
4. The area of an oblique triangle ABC is $\frac{1}{2}bc \sin A = \frac{1}{2}ab \sin C =$ ________.

Skills and Applications

Using the Law of Sines **In Exercises 5–22, use the Law of Sines to solve the triangle. Round your answers to two decimal places.**

5.

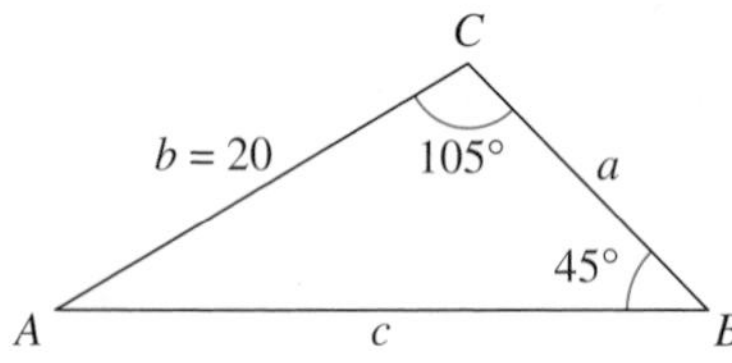

6.

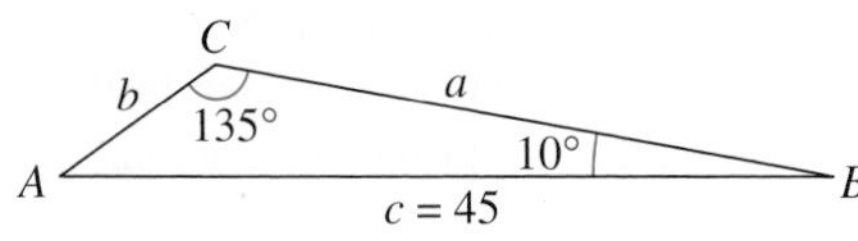

7.

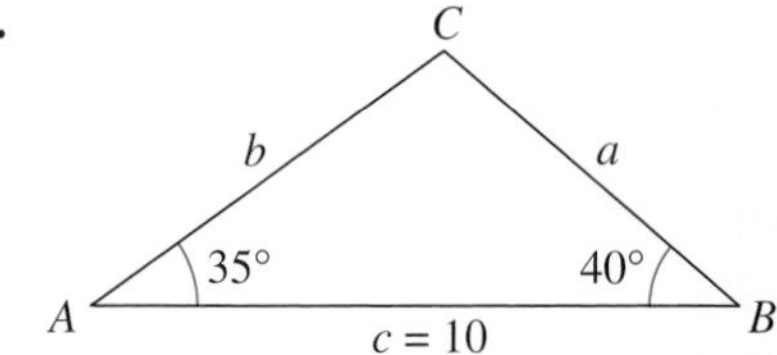

8.

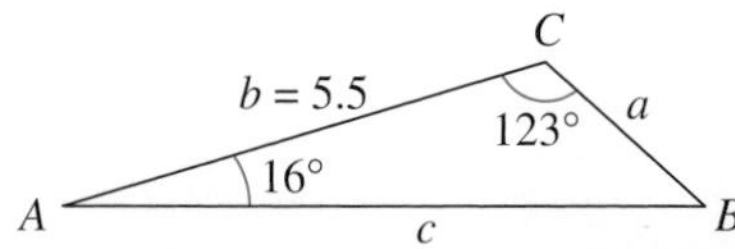

9. $A = 102.4°, \quad C = 16.7°, \quad a = 21.6$

10. $A = 24.3°, \quad C = 54.6°, \quad c = 2.68$

11. $A = 83° 20', \quad C = 54.6°, \quad c = 18.1$

12. $A = 5° 40', \quad B = 8° 15', \quad b = 4.8$

13. $A = 35°, \quad B = 65°, \quad c = 10$

14. $A = 120°, \quad B = 45°, \quad c = 16$

15. $A = 55°, \quad B = 42°, \quad c = \frac{3}{4}$

16. $B = 28°, \quad C = 104°, \quad a = 3\frac{5}{8}$

17. $A = 36°, \quad a = 8, \quad b = 5$

18. $A = 60°, \quad a = 9, \quad c = 7$

19. $A = 145°, \quad a = 14, \quad b = 4$

20. $A = 100°, \quad a = 125, \quad c = 10$

21. $B = 15° 30', \quad a = 4.5, \quad b = 6.8$

22. $B = 2° 45', \quad b = 6.2, \quad c = 5.8$

Using the Law of Sines **In Exercises 23–32, use the Law of Sines to solve (if possible) the triangle. If two solutions exist, find both. Round your answers to two decimal places.**

23. $A = 110°, \quad a = 125, \quad b = 100$

24. $A = 110°, \quad a = 125, \quad b = 200$

25. $A = 76°, \quad a = 18, \quad b = 20$

26. $A = 76°, \quad a = 34, \quad b = 21$

27. $A = 58°, \quad a = 11.4, \quad b = 12.8$

28. $A = 58°, \quad a = 4.5, \quad b = 12.8$

29. $A = 120°, \quad a = b = 25$

30. $A = 120°, \quad a = 25, \quad b = 24$

31. $A = 45°, \quad a = b = 1$

32. $A = 25° 4', \quad a = 9.5, \quad b = 22$

Using the Law of Sines **In Exercises 33–36, find values for b such that the triangle has (a) one solution, (b) two solutions (if possible), and (c) no solution.**

33. $A = 36°, \quad a = 5$

34. $A = 60°, \quad a = 10$

35. $A = 105°, \quad a = 80$

36. $A = 132°, \quad a = 215$

Finding the Area of a Triangle **In Exercises 37–44, find the area of the triangle. Round your answers to one decimal place.**

37. $A = 125°, \quad b = 9, \quad c = 6$

38. $C = 150°, \quad a = 17, \quad b = 10$

39. $B = 39°, \quad a = 25, \quad c = 12$

40. $A = 72°, \quad b = 31, \quad c = 44$

41. $C = 103° 15', \quad a = 16, \quad b = 28$

42. $B = 54° 30', \quad a = 62, \quad c = 35$

43. $A = 67°, \quad B = 43°, \quad a = 8$

44. $B = 118°, \quad C = 29°, \quad a = 52$

45. Height A tree grows at an angle of 4° from the vertical due to prevailing winds. At a point 40 meters from the base of the tree, the angle of elevation to the top of the tree is 30° (see figure).

(a) Write an equation that you can use to find the height h of the tree.

(b) Find the height of the tree.

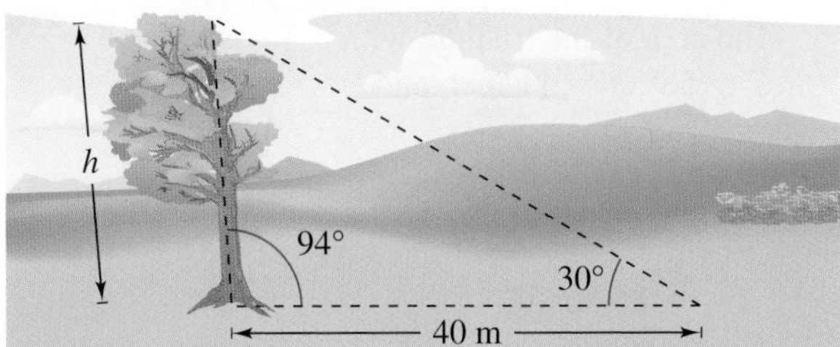

46. Distance A boat is traveling due east parallel to the shoreline at a speed of 10 miles per hour. At a given time, the bearing to a lighthouse is S 70° E, and 15 minutes later the bearing is S 63° E (see figure). The lighthouse is located at the shoreline. What is the distance from the boat to the shoreline?

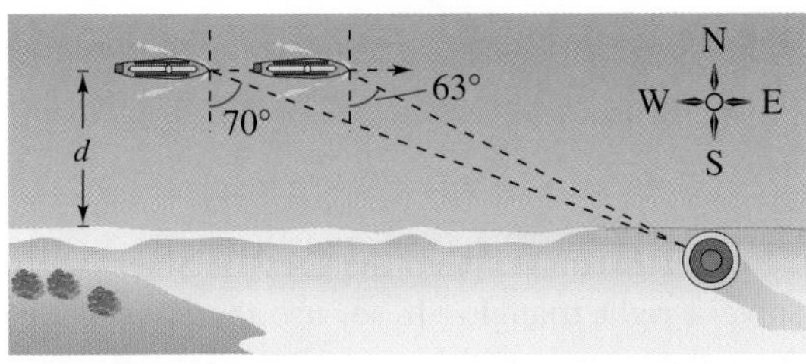

47. Environmental Science The bearing from the Pine Knob fire tower to the Colt Station fire tower is N 65° E, and the two towers are 30 kilometers apart. A fire spotted by rangers in each tower has a bearing of N 80° E from Pine Knob and S 70° E from Colt Station (see figure). Find the distance of the fire from each tower.

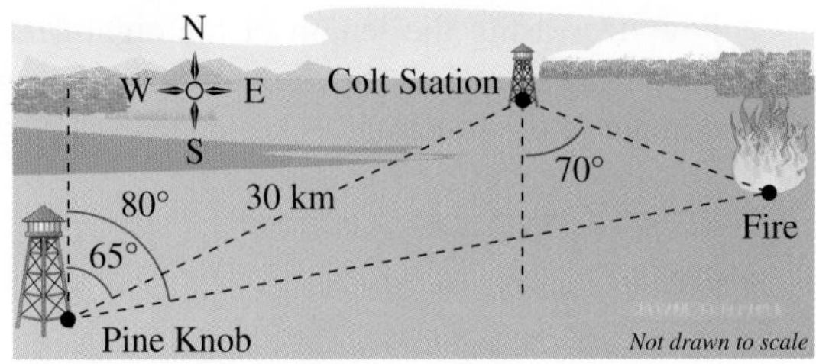

48. Bridge Design A bridge is built across a small lake from a gazebo to a dock (see figure). The bearing from the gazebo to the dock is S 41° W. From a tree 100 meters from the gazebo, the bearings to the gazebo and the dock are S 74° E and S 28° E, respectively. Find the distance from the gazebo to the dock.

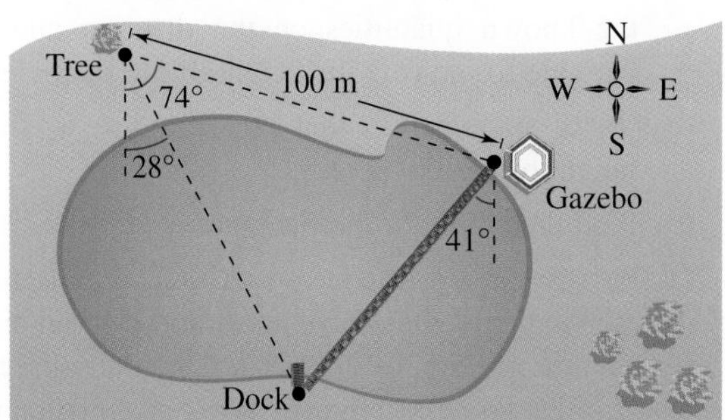

49. Angle of Elevation A 10-meter utility pole casts a 17-meter shadow directly down a slope when the angle of elevation of the sun is 42° (see figure). Find θ, the angle of elevation of the ground.

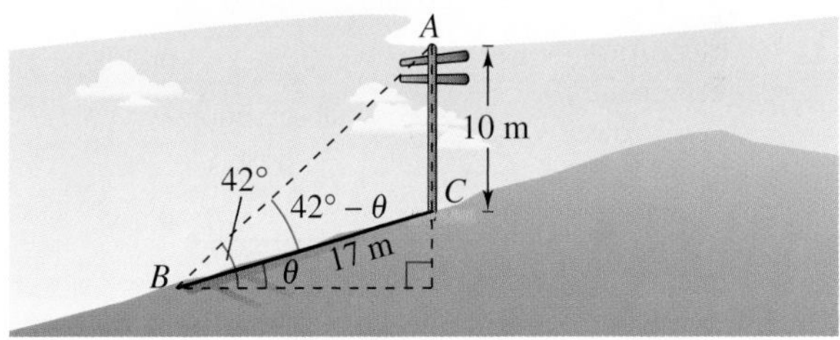

50. Flight Path A plane flies 500 kilometers with a bearing of 316° from Naples to Elgin (see figure). The plane then flies 720 kilometers from Elgin to Canton (Canton is due west of Naples). Find the bearing of the flight from Elgin to Canton.

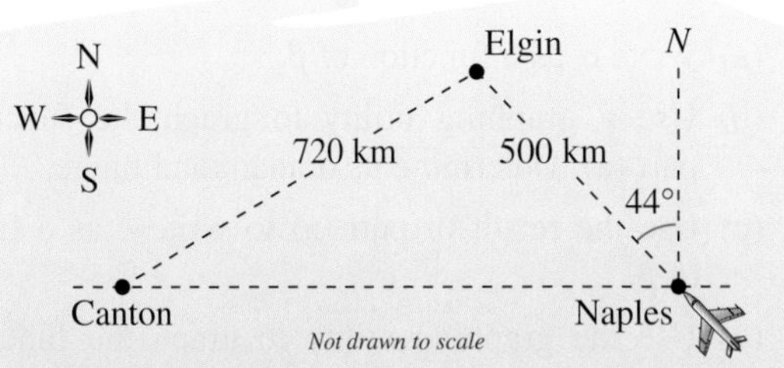

51. Altitude The angles of elevation to an airplane from two points A and B on level ground are 55° and 72°, respectively. The points A and B are 2.2 miles apart, and the airplane is east of both points in the same vertical plane.

(a) Draw a diagram that represents the problem. Show the known quantities on the diagram.

(b) Find the distance between the plane and point B.

(c) Find the altitude of the plane.

(d) Find the distance the plane must travel before it is directly above point A.

52. Height A flagpole at a right angle to the horizontal is located on a slope that makes an angle of 12° with the horizontal. The flagpole's shadow is 16 meters long and points directly up the slope. The angle of elevation from the tip of the shadow to the sun is 20°.

(a) Draw a diagram that represents the problem. Show the known quantities on the diagram and use a variable to indicate the height of the flagpole.

(b) Write an equation that you can use to find the height of the flagpole.

(c) Find the height of the flagpole.

53. Distance Air traffic controllers continuously monitor the angles of elevation θ and ϕ to an airplane from an airport control tower and from an observation post 2 miles away (see figure). Write an equation giving the distance d between the plane and the observation post in terms of θ and ϕ.

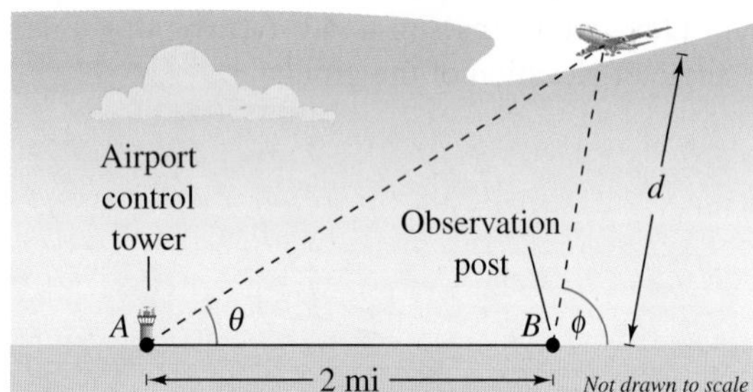

54. Numerical Analysis In the figure, α and β are positive angles.

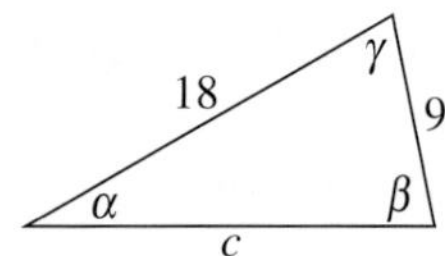

(a) Write α as a function of β.

(b) Use a graphing utility to graph the function in part (a). Determine its domain and range.

(c) Use the result of part (a) to write c as a function of β.

(d) Use the graphing utility to graph the function in part (c). Determine its domain and range.

(e) Complete the table. What can you infer?

β	0.4	0.8	1.2	1.6	2.0	2.4	2.8
α							
c							

Exploration

True or False? **In Exercises 55–58, determine whether the statement is true or false. Justify your answer.**

55. If a triangle contains an obtuse angle, then it must be oblique.

56. Two angles and one side of a triangle do not necessarily determine a unique triangle.

57. When you know the three angles of an oblique triangle, you can solve the triangle.

58. The ratio of any two sides of a triangle is equal to the ratio of the sines of the opposite angles of the two sides.

59. Error Analysis Describe the error.

The area of the triangle with $C = 58°$, $b = 11$ feet, and $c = 16$ feet is

$$\text{Area} = \frac{1}{2}(11)(16)(\sin 58°)$$
$$= 88(\sin 58°)$$
$$\approx 74.63 \text{ square feet.}$$

60. HOW DO YOU SEE IT? In the figure, a triangle is to be formed by drawing a line segment of length a from (4, 3) to the positive x-axis. For what value(s) of a can you form (a) one triangle, (b) two triangles, and (c) no triangles? Explain.

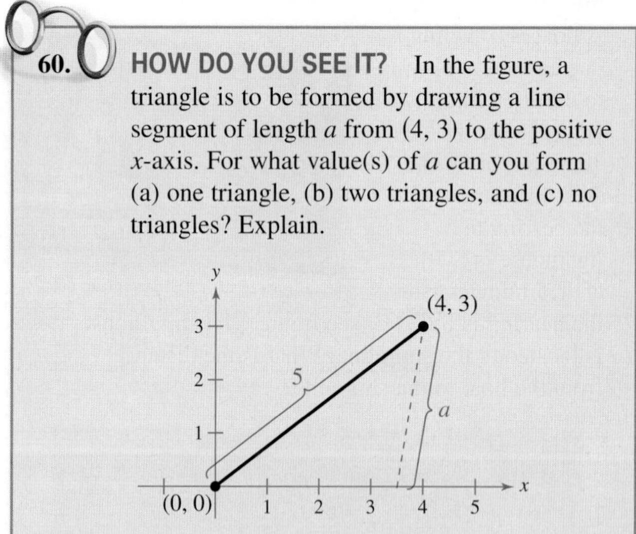

61. Think About It Can the Law of Sines be used to solve a right triangle? If so, use the Law of Sines to solve the triangle with

$$B = 50°,\quad C = 90°,\quad \text{and}\quad a = 10.$$

Is there another way to solve the triangle? Explain.

62. Using Technology

(a) Write the area A of the shaded region in the figure as a function of θ.

(b) Use a graphing utility to graph the function.

(c) Determine the domain of the function. Explain how decreasing the length of the eight-centimeter line segment affects the area of the region and the domain of the function.

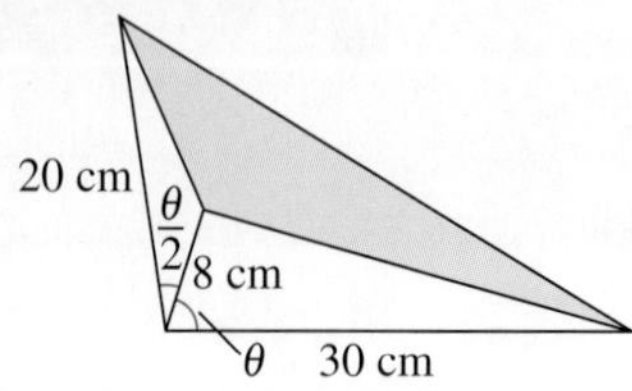

3.2 Law of Cosines

The Law of Cosines is a useful tool for solving real-life problems involving oblique triangles. For example, in Exercise 56 on page 277, you will use the Law of Cosines to determine the total distance a piston moves in an engine.

- **Use the Law of Cosines to solve oblique triangles (SSS or SAS).**
- **Use the Law of Cosines to model and solve real-life problems.**
- **Use Heron's Area Formula to find areas of triangles.**

Introduction

Two cases remain in the list of conditions needed to solve an oblique triangle—SSS and SAS. When you are given three sides (SSS), or two sides and their included angle (SAS), you cannot solve the triangle using the Law of Sines alone. In such cases, use the **Law of Cosines.**

> **Law of Cosines**
>
> | **Standard Form** | **Alternative Form** |
> |---|---|
> | $a^2 = b^2 + c^2 - 2bc\cos A$ | $\cos A = \dfrac{b^2 + c^2 - a^2}{2bc}$ |
> | $b^2 = a^2 + c^2 - 2ac\cos B$ | $\cos B = \dfrac{a^2 + c^2 - b^2}{2ac}$ |
> | $c^2 = a^2 + b^2 - 2ab\cos C$ | $\cos C = \dfrac{a^2 + b^2 - c^2}{2ab}$ |

For a proof of the Law of Cosines, see Proofs in Mathematics on page 308.

EXAMPLE 1 Given Three Sides—SSS

Find the three angles of the triangle shown below.

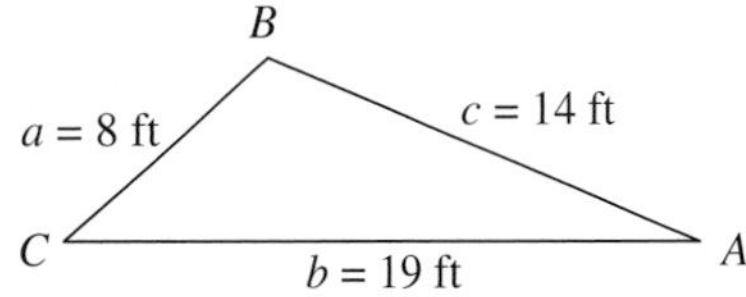

Solution It is a good idea to find the angle opposite the longest side first—side b in this case. Using the alternative form of the Law of Cosines,

$$\cos B = \frac{a^2 + c^2 - b^2}{2ac} = \frac{8^2 + 14^2 - 19^2}{2(8)(14)} \approx -0.4509.$$

Because $\cos B$ is negative, B is an *obtuse* angle given by $B \approx 116.80°$. At this point, use the Law of Sines to determine A.

$$\sin A = a\left(\frac{\sin B}{b}\right) \approx 8\left(\frac{\sin 116.80°}{19}\right) \approx 0.3758$$

The angle B is obtuse and a triangle can have at most one obtuse angle, so you know that A must be acute. So, $A \approx 22.07°$ and $C \approx 180° - 22.07° - 116.80° = 41.13°$.

✓ ***Checkpoint*** Audio-video solution in English & Spanish at LarsonPrecalculus.com

Find the three angles of the triangle whose sides have lengths $a = 6$ centimeters, $b = 8$ centimeters, and $c = 12$ centimeters.

Smart-foto/Shutterstock.com

Do you see why it was wise to find the largest angle *first* in Example 1? Knowing the cosine of an angle, you can determine whether the angle is acute or obtuse. That is,

$$\cos \theta > 0 \quad \text{for} \quad 0° < \theta < 90° \qquad \text{Acute}$$

$$\cos \theta < 0 \quad \text{for} \quad 90° < \theta < 180°. \qquad \text{Obtuse}$$

So, in Example 1, after you find that angle B is obtuse, you know that angles A and C must both be acute. Furthermore, if the largest angle is acute, then the remaining two angles must also be acute.

REMARK When solving an oblique triangle given three sides, use the alternative form of the Law of Cosines to solve for an angle. When solving an oblique triangle given two sides and their included angle, use the standard form of the Law of Cosines to solve for the remaining side.

EXAMPLE 2 Given Two Sides and Their Included Angle—SAS

See LarsonPrecalculus.com for an interactive version of this type of example.

Find the remaining angles and side of the triangle shown below.

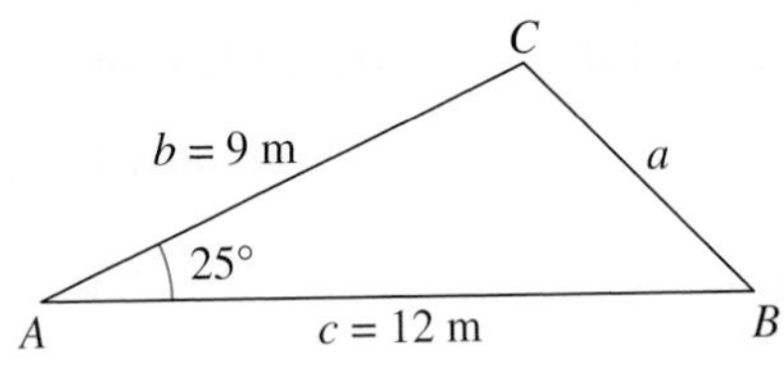

Solution Use the standard form of the Law of Cosines to find side a.

$$a^2 = b^2 + c^2 - 2bc \cos A$$

$$a^2 = 9^2 + 12^2 - 2(9)(12) \cos 25°$$

$$a^2 \approx 29.2375$$

$$a \approx 5.4072 \text{ meters}$$

Next, use the ratio $(\sin A)/a$, the given value of b, and the reciprocal form of the Law of Sines to find B.

$$\frac{\sin B}{b} = \frac{\sin A}{a} \qquad \text{Reciprocal form}$$

$$\sin B = b\left(\frac{\sin A}{a}\right) \qquad \text{Multiply each side by } b.$$

$$\sin B \approx 9\left(\frac{\sin 25°}{5.4072}\right) \qquad \text{Substitute for } A, a, \text{ and } b.$$

$$\sin B \approx 0.7034 \qquad \text{Use a calculator.}$$

There are two angles between 0° and 180° whose sine is approximately 0.7034, $B_1 \approx 44.70°$ and $B_2 \approx 180° - 44.70° = 135.30°$.

For $B_1 \approx 44.70°$,

$$C_1 \approx 180° - 25° - 44.70° = 110.30°.$$

For $B_2 \approx 135.30°$,

$$C_2 \approx 180° - 25° - 135.30° = 19.70°.$$

Side c is the longest side of the triangle, which means that angle C is the largest angle of the triangle. So, $C \approx 110.30°$ and $B \approx 44.70°$.

✓ **Checkpoint** Audio-video solution in English & Spanish at LarsonPrecalculus.com

Given $A = 80°$, $b = 16$ meters, and $c = 12$ meters, find the remaining angles and side of the triangle.

Applications

EXAMPLE 3 An Application of the Law of Cosines

The pitcher's mound on a women's softball field is 43 feet from home plate and the distance between the bases is 60 feet, as shown in Figure 3.8. (The pitcher's mound is *not* halfway between home plate and second base.) How far is the pitcher's mound from first base?

Figure 3.8

Solution In triangle HPF, $H = 45°$ (line segment HP bisects the right angle at H), $f = 43$, and $p = 60$. Using the standard form of the Law of Cosines for this SAS case,

$$h^2 = f^2 + p^2 - 2fp \cos H$$

$$= 43^2 + 60^2 - 2(43)(60) \cos 45°$$

$$\approx 1800.3290.$$

So, the approximate distance from the pitcher's mound to first base is

$$h \approx \sqrt{1800.3290} \approx 42.43 \text{ feet.}$$

Checkpoint Audio-video solution in English & Spanish at LarsonPrecalculus.com

In a softball game, a batter hits a ball to dead center field, a distance of 240 feet from home plate. The center fielder then throws the ball to third base and gets a runner out. The distance between the bases is 60 feet. How far is the center fielder from third base?

EXAMPLE 4 An Application of the Law of Cosines

A ship travels 60 miles due north and then adjusts its course, as shown in Figure 3.9. After traveling 80 miles in this new direction, the ship is 139 miles from its point of departure. Describe the bearing from point B to point C.

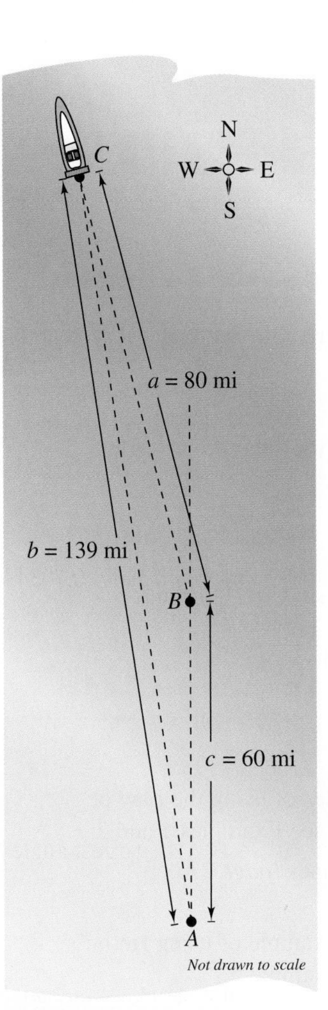

Figure 3.9

Solution You have $a = 80$, $b = 139$, and $c = 60$. So, using the alternative form of the Law of Cosines,

$$\cos B = \frac{a^2 + c^2 - b^2}{2ac}$$

$$= \frac{80^2 + 60^2 - 139^2}{2(80)(60)}$$

$$\approx -0.9709.$$

So, $B \approx 166.14°$, and the bearing measured from due north from point B to point C is approximately $180° - 166.14° = 13.86°$, or N 13.86° W.

Checkpoint Audio-video solution in English & Spanish at LarsonPrecalculus.com

A ship travels 40 miles due east and then changes direction, as shown at the right. After traveling 30 miles in this new direction, the ship is 56 miles from its point of departure. Describe the bearing from point B to point C.

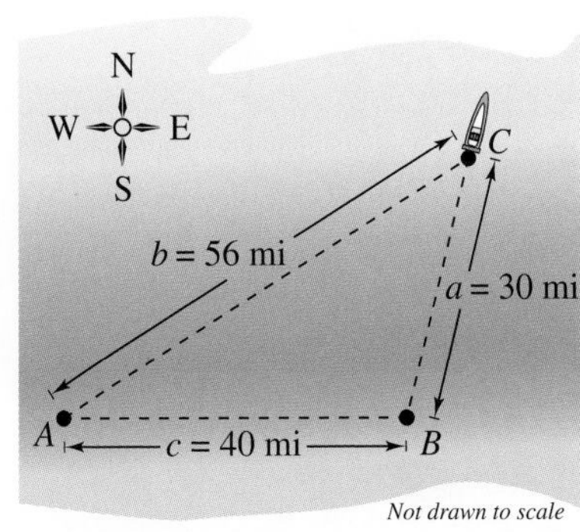

> **HISTORICAL NOTE**
>
> Heron of Alexandria (10–75 A.D.) was a Greek geometer and inventor. His works describe how to find areas of triangles, quadrilaterals, regular polygons with 3 to 12 sides, and circles, as well as surface areas and volumes of three-dimensional objects.

Heron's Area Formula

The Law of Cosines can be used to establish a formula for the area of a triangle. This formula is called **Heron's Area Formula** after the Greek mathematician Heron (ca. 10–75 A.D.).

> **Heron's Area Formula**
>
> Given any triangle with sides of lengths a, b, and c, the area of the triangle is
>
> $$\text{Area} = \sqrt{s(s-a)(s-b)(s-c)}$$
>
> where
>
> $$s = \frac{a+b+c}{2}.$$

For a proof of Heron's Area Formula, see Proofs in Mathematics on page 309.

EXAMPLE 5 Using Heron's Area Formula

Use Heron's Area Formula to find the area of a triangle with sides of lengths $a = 43$ meters, $b = 53$ meters, and $c = 72$ meters.

Solution First, determine that $s = (a + b + c)/2 = 168/2 = 84$. Then Heron's Area Formula yields

$$\begin{aligned}\text{Area} &= \sqrt{s(s-a)(s-b)(s-c)}\\ &= \sqrt{84(84-43)(84-53)(84-72)}\\ &= \sqrt{84(41)(31)(12)}\\ &\approx 1131.89 \text{ square meters.}\end{aligned}$$

Use Heron's Area Formula to find the area of a triangle with sides of lengths $a = 5$ inches, $b = 9$ inches, and $c = 8$ inches.

You have now studied three different formulas for the area of a triangle.

Standard Formula: $\text{Area} = \frac{1}{2}bh$

Oblique Triangle: $\text{Area} = \frac{1}{2}bc \sin A = \frac{1}{2}ab \sin C = \frac{1}{2}ac \sin B$

Heron's Area Formula: $\text{Area} = \sqrt{s(s-a)(s-b)(s-c)}$

> **Summarize** (Section 3.2)
>
> 1. State the Law of Cosines *(page 271)*. For examples of using the Law of Cosines to solve oblique triangles (SSS or SAS), see Examples 1 and 2.
> 2. Describe real-life applications of the Law of Cosines *(page 273, Examples 3 and 4)*.
> 3. State Heron's Area Formula *(page 274)*. For an example of using Heron's Area Formula to find the area of a triangle, see Example 5.

3.2 Exercises

See CalcChat.com for tutorial help and worked-out solutions to odd-numbered exercises.

Vocabulary: Fill in the blanks.

1. The standard form of the Law of Cosines for $\cos B = \dfrac{a^2 + c^2 - b^2}{2ac}$ is ________.
2. When solving an oblique triangle given three sides, use the ________ form of the Law of Cosines to solve for an angle.
3. When solving an oblique triangle given two sides and their included angle, use the ________ form of the Law of Cosines to solve for the remaining side.
4. The Law of Cosines can be used to establish a formula for the area of a triangle called ________ ________ Formula.

Skills and Applications

Using the Law of Cosines **In Exercises 5–24, use the Law of Cosines to solve the triangle. Round your answers to two decimal places.**

5. 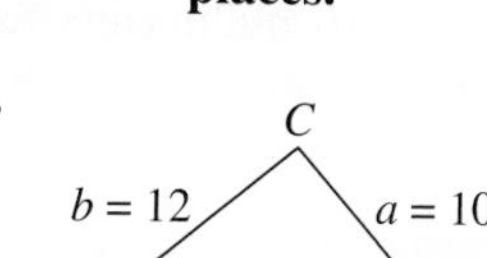

6. 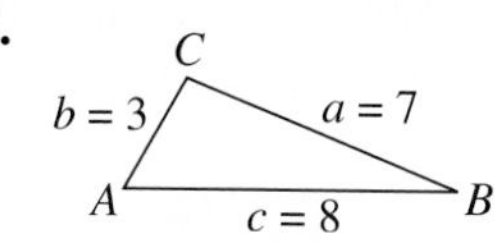

7. 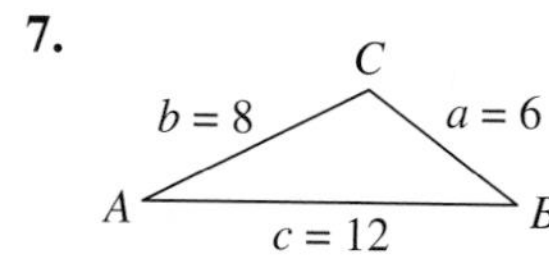

8. 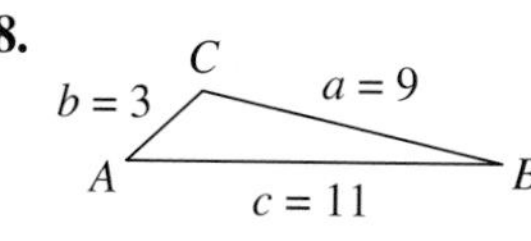

9. 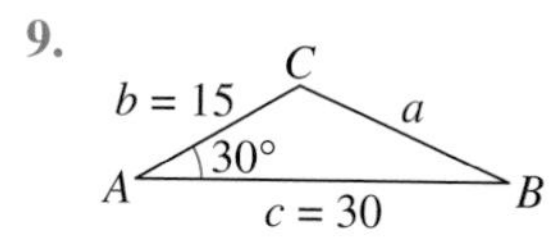

10. 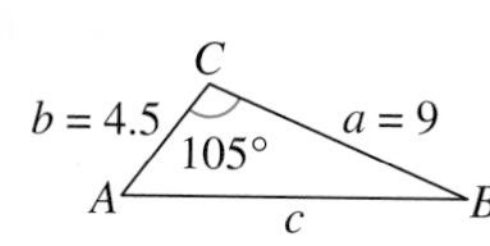

11. 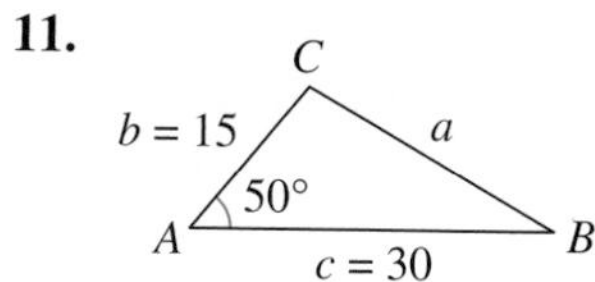

12. 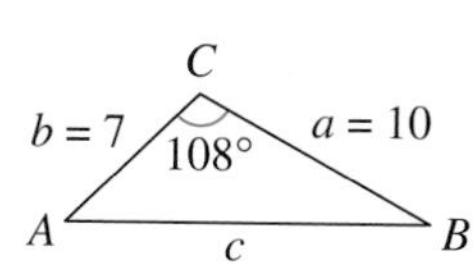

13. $a = 11, \quad b = 15, \quad c = 21$
14. $a = 55, \quad b = 25, \quad c = 72$
15. $a = 2.5, \quad b = 1.8, \quad c = 0.9$
16. $a = 75.4, \quad b = 52.5, \quad c = 52.5$
17. $A = 120°, \quad b = 6, \quad c = 7$
18. $A = 48°, \quad b = 3, \quad c = 14$
19. $B = 10° \, 35', \quad a = 40, \quad c = 30$
20. $B = 75° \, 20', \quad a = 9, \quad c = 6$
21. $B = 125° \, 40', \quad a = 37, \quad c = 37$
22. $C = 15° \, 15', \quad a = 7.45, \quad b = 2.15$
23. $C = 43°, \quad a = \frac{4}{9}, \quad b = \frac{7}{9}$
24. $C = 101°, \quad a = \frac{3}{8}, \quad b = \frac{3}{4}$

Finding Measures in a Parallelogram **In Exercises 25–30, find the missing values by solving the parallelogram shown in the figure. (The lengths of the diagonals are given by c and d.)**

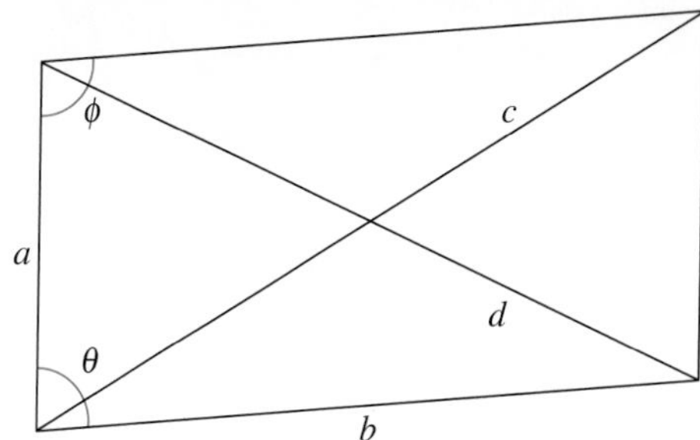

	a	b	c	d	θ	ϕ
25.	5	8			45°	
26.	25	35				120°
27.	10	14	20			
28.	40	60		80		
29.	15		25	20		
30.		25	50	35		

Solving a Triangle **In Exercises 31–36, determine whether the Law of Cosines is needed to solve the triangle. Then solve (if possible) the triangle. If two solutions exist, find both. Round your answers to two decimal places.**

31. $a = 8, \quad c = 5, \quad B = 40°$
32. $a = 10, \quad b = 12, \quad C = 70°$
33. $A = 24°, \quad a = 4, \quad b = 18$
34. $a = 11, \quad b = 13, \quad c = 7$
35. $A = 42°, \quad B = 35°, \quad c = 1.2$
36. $B = 12°, \quad a = 160, \quad b = 63$

Using Heron's Area Formula In Exercises 37–44, use Heron's Area Formula to find the area of the triangle.

37. $a = 6, \quad b = 12, \quad c = 17$

38. $a = 33, \quad b = 36, \quad c = 21$

39. $a = 2.5, \quad b = 10.2, \quad c = 8$

40. $a = 12.32, \quad b = 8.46, \quad c = 15.9$

41. $a = 1, \quad b = \frac{1}{2}, \quad c = \frac{5}{4}$

42. $a = \frac{3}{5}, \quad b = \frac{4}{3}, \quad c = \frac{7}{8}$

43. $A = 80°, \quad b = 75, \quad c = 41$

44. $C = 109°, \quad a = 16, \quad b = 3.5$

45. **Surveying** To approximate the length of a marsh, a surveyor walks 250 meters from point A to point B, then turns 75° and walks 220 meters to point C (see figure). Approximate the length AC of the marsh.

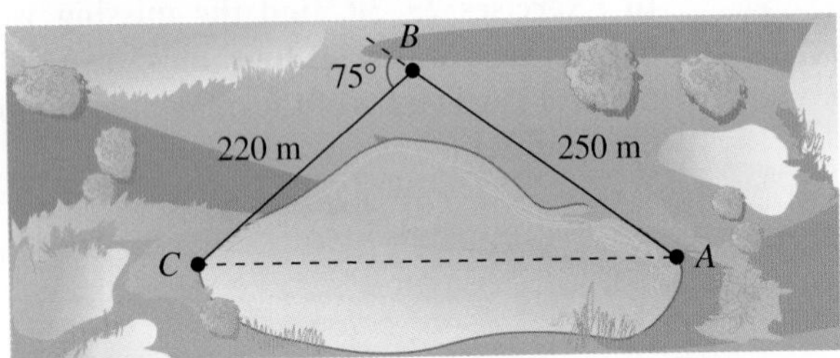

46. **Streetlight Design** Determine the angle θ in the design of the streetlight shown in the figure.

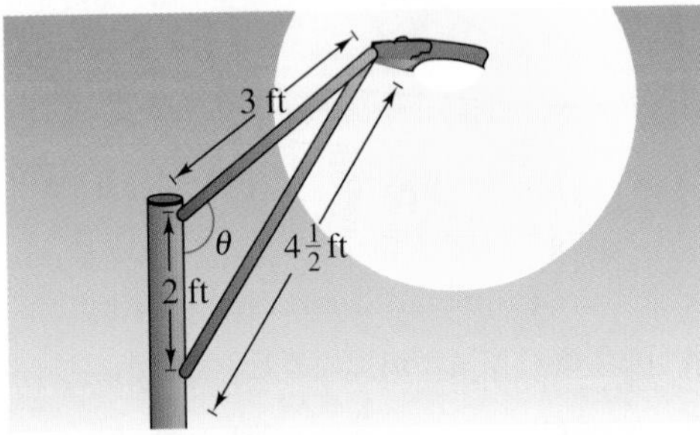

47. **Baseball** A baseball player in center field is approximately 330 feet from a television camera that is behind home plate. A batter hits a fly ball that goes to the wall 420 feet from the camera (see figure). The camera turns 8° to follow the play. Approximately how far does the center fielder have to run to make the catch?

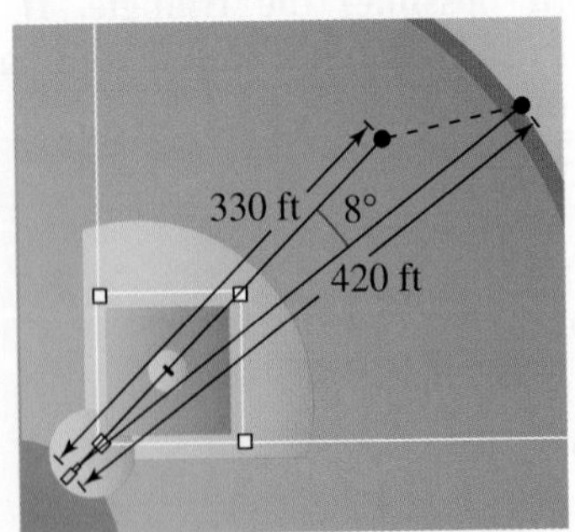

48. **Baseball** On a baseball diamond with 90-foot sides, the pitcher's mound is 60.5 feet from home plate. How far is the pitcher's mound from third base?

49. **Length** A 100-foot vertical tower is built on the side of a hill that makes a 6° angle with the horizontal (see figure). Find the length of each of the two guy wires that are anchored 75 feet uphill and downhill from the base of the tower.

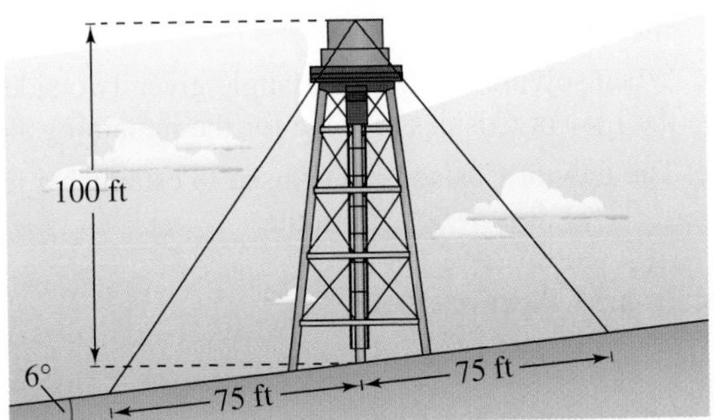

50. **Navigation** On a map, Minneapolis is 165 millimeters due west of Albany, Phoenix is 216 millimeters from Minneapolis, and Phoenix is 368 millimeters from Albany (see figure).

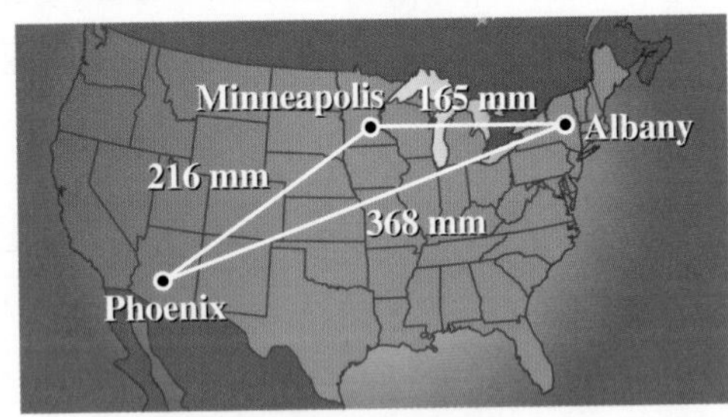

(a) Find the bearing of Minneapolis from Phoenix.

(b) Find the bearing of Albany from Phoenix.

51. **Navigation** A boat race runs along a triangular course marked by buoys A, B, and C. The race starts with the boats headed west for 3700 meters. The other two sides of the course lie to the north of the first side, and their lengths are 1700 meters and 3000 meters. Draw a diagram that gives a visual representation of the problem. Then find the bearings for the last two legs of the race.

52. **Air Navigation** A plane flies 810 miles from Franklin to Centerville with a bearing of 75°. Then it flies 648 miles from Centerville to Rosemount with a bearing of 32°. Draw a diagram that gives a visual representation of the problem. Then find the straight-line distance and bearing from Franklin to Rosemount.

53. **Surveying** A triangular parcel of land has 115 meters of frontage, and the other boundaries have lengths of 76 meters and 92 meters. What angles does the frontage make with the two other boundaries?

54. **Surveying** A triangular parcel of ground has sides of lengths 725 feet, 650 feet, and 575 feet. Find the measure of the largest angle.

55. **Distance** Two ships leave a port at 9 A.M. One travels at a bearing of N 53° W at 12 miles per hour, and the other travels at a bearing of S 67° W at s miles per hour.
   (a) Use the Law of Cosines to write an equation that relates s and the distance d between the two ships at noon.
   (b) Find the speed s that the second ship must travel so that the ships are 43 miles apart at noon.

56. **Mechanical Engineering**
An engine has a seven-inch connecting rod fastened to a crank (see figure).

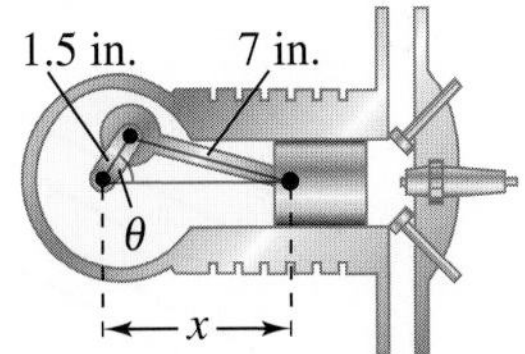

   (a) Use the Law of Cosines to write an equation giving the relationship between x and θ.
   (b) Write x as a function of θ. (Select the sign that yields positive values of x.)
   (c) Use a graphing utility to graph the function in part (b).
   (d) Use the graph in part (c) to determine the total distance the piston moves in one cycle.

57. **Geometry** A triangular parcel of land has sides of lengths 200 feet, 500 feet, and 600 feet. Find the area of the parcel.

58. **Geometry** A parking lot has the shape of a parallelogram (see figure). The lengths of two adjacent sides are 70 meters and 100 meters. The angle between the two sides is 70°. What is the area of the parking lot?

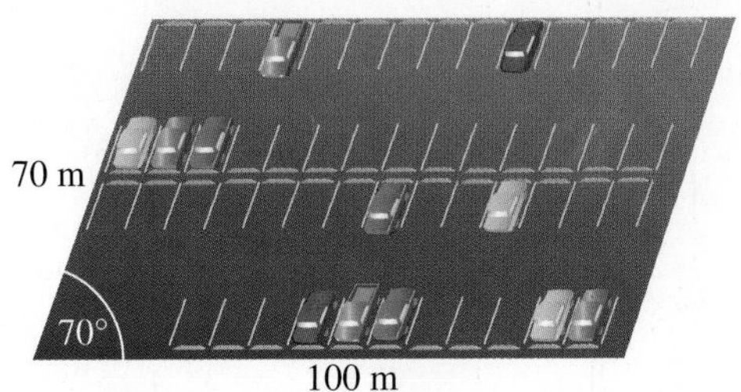

Smart-foto/Shutterstock.com

59. **Geometry** You want to buy a triangular lot measuring 510 yards by 840 yards by 1120 yards. The price of the land is \$2000 per acre. How much does the land cost? (*Hint:* 1 acre = 4840 square yards)

60. **Geometry** You want to buy a triangular lot measuring 1350 feet by 1860 feet by 2490 feet. The price of the land is \$2200 per acre. How much does the land cost? (*Hint:* 1 acre = 43,560 square feet)

Exploration

True or False? **In Exercises 61 and 62, determine whether the statement is true or false. Justify your answer.**

61. In Heron's Area Formula, s is the average of the lengths of the three sides of the triangle.

62. In addition to SSS and SAS, the Law of Cosines can be used to solve triangles with AAS conditions.

63. **Think About It** What familiar formula do you obtain when you use the standard form of the Law of Cosines, $c^2 = a^2 + b^2 - 2ab \cos C$, and you let $C = 90°$? What is the relationship between the Law of Cosines and this formula?

64. **Writing** Describe how the Law of Cosines can be used to solve the ambiguous case of the oblique triangle ABC, where $a = 12$ feet, $b = 30$ feet, and $A = 20°$. Is the result the same as when the Law of Sines is used to solve the triangle? Describe the advantages and the disadvantages of each method.

65. **Writing** In Exercise 64, the Law of Cosines was used to solve a triangle in the two-solution case of SSA. Can the Law of Cosines be used to solve the no-solution and single-solution cases of SSA? Explain.

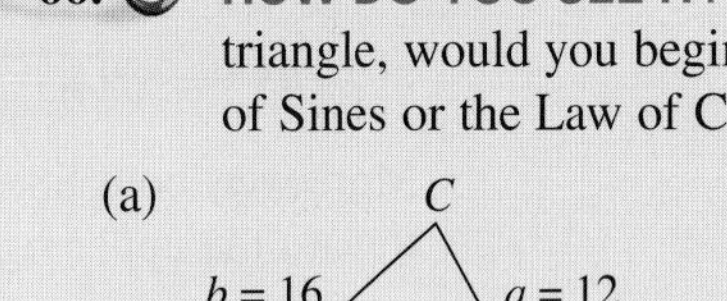

66. **HOW DO YOU SEE IT?** To solve the triangle, would you begin by using the Law of Sines or the Law of Cosines? Explain.

(a)

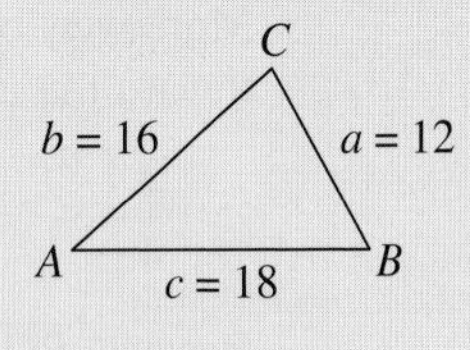

(b)

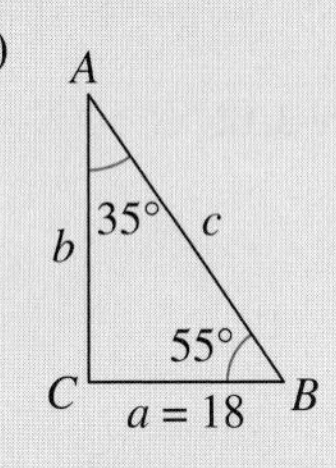

67. **Proof** Use the Law of Cosines to prove each identity.
   (a) $\frac{1}{2}bc(1 + \cos A) = \frac{a + b + c}{2} \cdot \frac{-a + b + c}{2}$
   (b) $\frac{1}{2}bc(1 - \cos A) = \frac{a - b + c}{2} \cdot \frac{a + b - c}{2}$

3.3 Vectors in the Plane

Vectors are useful tools for modeling and solving real-life problems involving magnitude and direction. For instance, in Exercise 94 on page 290, you will use vectors to determine the speed and true direction of a commercial jet.

- Represent vectors as directed line segments.
- Write component forms of vectors.
- Perform basic vector operations and represent vector operations graphically.
- Write vectors as linear combinations of unit vectors.
- Find direction angles of vectors.
- Use vectors to model and solve real-life problems.

Introduction

Quantities such as force and velocity involve both *magnitude* and *direction* and cannot be completely characterized by a single real number. To represent such a quantity, you can use a **directed line segment,** as shown in Figure 3.10. The directed line segment $\overrightarrow{PQ}$ has **initial point** P and **terminal point** Q. Its **magnitude** (or **length**) is denoted by $\|\overrightarrow{PQ}\|$ and can be found using the Distance Formula.

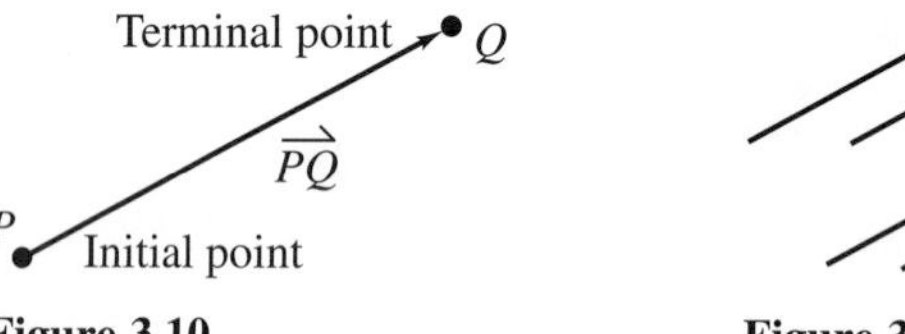

Figure 3.10 Figure 3.11

Two directed line segments that have the same magnitude and direction are *equivalent*. For example, the directed line segments in Figure 3.11 are all equivalent. The set of all directed line segments that are equivalent to the directed line segment $\overrightarrow{PQ}$ is a **vector v in the plane,** written $\mathbf{v} = \overrightarrow{PQ}$. Vectors are denoted by lowercase, boldface letters such as **u**, **v**, and **w**.

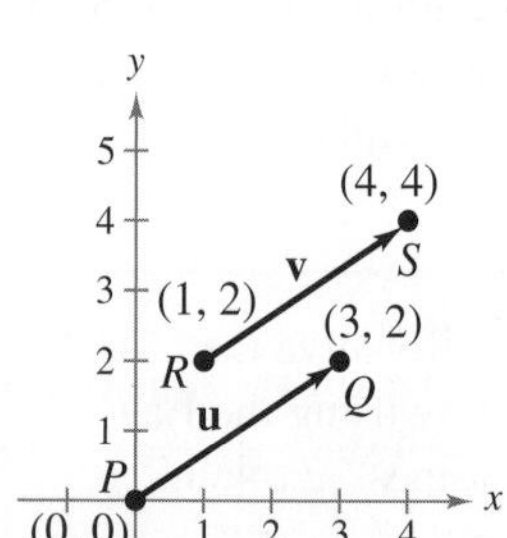

Figure 3.12

EXAMPLE 1 Showing That Two Vectors Are Equivalent

Show that **u** and **v** in Figure 3.12 are equivalent.

Solution From the Distance Formula, $\overrightarrow{PQ}$ and $\overrightarrow{RS}$ have the *same magnitude.*

$$\|\overrightarrow{PQ}\| = \sqrt{(3-0)^2 + (2-0)^2} = \sqrt{13}$$

$$\|\overrightarrow{RS}\| = \sqrt{(4-1)^2 + (4-2)^2} = \sqrt{13}$$

Moreover, both line segments have the *same direction* because they are both directed toward the upper right on lines with a slope of

$$\frac{4-2}{4-1} = \frac{2-0}{3-0} = \frac{2}{3}.$$

Because $\overrightarrow{PQ}$ and $\overrightarrow{RS}$ have the same magnitude and direction, **u** and **v** are equivalent.

✓ Checkpoint Audio-video solution in English & Spanish at LarsonPrecalculus.com

Show that **u** and **v** in the figure at the right are equivalent.

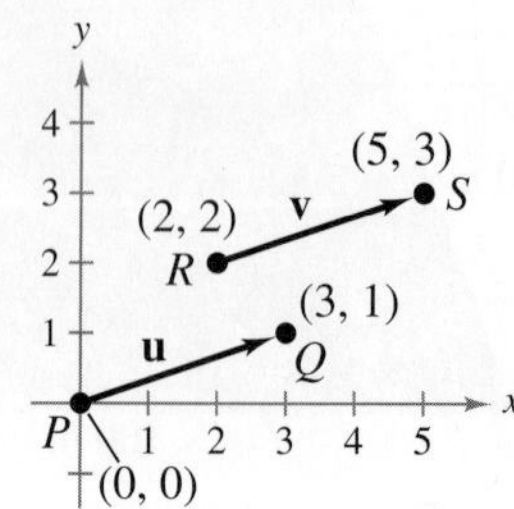

Component Form of a Vector

The directed line segment whose initial point is the origin is often the most convenient representative of a set of equivalent directed line segments. This representative of the vector **v** is in **standard position.**

A vector whose initial point is the origin (0, 0) can be uniquely represented by the coordinates of its terminal point (v_1, v_2). This is the **component form of a vector v,** written as $\mathbf{v} = \langle v_1, v_2 \rangle$. The coordinates v_1 and v_2 are the *components* of **v**. If both the initial point and the terminal point lie at the origin, then **v** is the **zero vector** and is denoted by $\mathbf{0} = \langle 0, 0 \rangle$.

▷ TECHNOLOGY Consult the user's guide for your graphing utility for specific instructions on how to use your graphing utility to graph vectors.

Component Form of a Vector

The component form of the vector with initial point $P(p_1, p_2)$ and terminal point $Q(q_1, q_2)$ is given by

$$\overrightarrow{PQ} = \langle q_1 - p_1, q_2 - p_2 \rangle = \langle v_1, v_2 \rangle = \mathbf{v}.$$

The **magnitude** (or **length**) of **v** is given by

$$\|\mathbf{v}\| = \sqrt{(q_1 - p_1)^2 + (q_2 - p_2)^2} = \sqrt{v_1^2 + v_2^2}.$$

If $\|\mathbf{v}\| = 1$, then **v** is a **unit vector.** Moreover, $\|\mathbf{v}\| = 0$ if and only if **v** is the zero vector **0**.

Two vectors $\mathbf{u} = \langle u_1, u_2 \rangle$ and $\mathbf{v} = \langle v_1, v_2 \rangle$ are *equal* if and only if $u_1 = v_1$ and $u_2 = v_2$. For instance, in Example 1, the vector **u** from $P(0, 0)$ to $Q(3, 2)$ is $\mathbf{u} = \overrightarrow{PQ} = \langle 3 - 0, 2 - 0 \rangle = \langle 3, 2 \rangle$, and the vector **v** from $R(1, 2)$ to $S(4, 4)$ is $\mathbf{v} = \overrightarrow{RS} = \langle 4 - 1, 4 - 2 \rangle = \langle 3, 2 \rangle$. So, the vectors **u** and **v** in Example 1 are equal.

EXAMPLE 2 Finding the Component Form of a Vector

Find the component form and magnitude of the vector **v** that has initial point $(4, -7)$ and terminal point $(-1, 5)$.

Algebraic Solution

Let

$$P(4, -7) = (p_1, p_2)$$

and

$$Q(-1, 5) = (q_1, q_2).$$

Then, the components of $\mathbf{v} = \langle v_1, v_2 \rangle$ are

$$v_1 = q_1 - p_1 = -1 - 4 = -5$$

$$v_2 = q_2 - p_2 = 5 - (-7) = 12.$$

So, $\mathbf{v} = \langle -5, 12 \rangle$ and the magnitude of **v** is

$$\|\mathbf{v}\| = \sqrt{(-5)^2 + 12^2} = \sqrt{169} = 13.$$

Graphical Solution

Use centimeter graph paper to plot the points $P(4, -7)$ and $Q(-1, 5)$. Carefully sketch the vector **v**. Use the sketch to find the components of $\mathbf{v} = \langle v_1, v_2 \rangle$. Then use a centimeter ruler to find the magnitude of **v**. The figure at the right shows that the components of **v** are $v_1 = -5$ and $v_2 = 12$, so $\mathbf{v} = \langle -5, 12 \rangle$. The figure also shows that the magnitude of **v** is $\|\mathbf{v}\| = 13$.

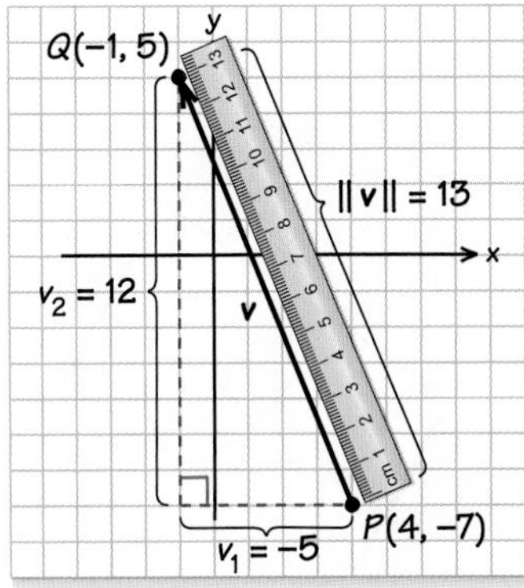

✓ **Checkpoint** Audio-video solution in English & Spanish at LarsonPrecalculus.com

Find the component form and magnitude of the vector **v** that has initial point $(-2, 3)$ and terminal point $(-7, 9)$.

Vector Operations

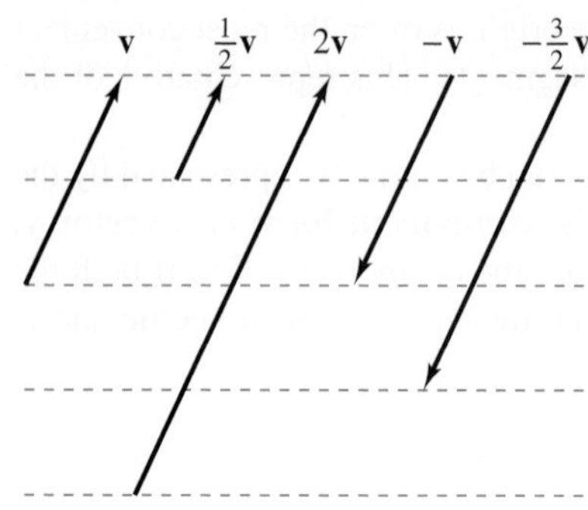

Figure 3.13

The two basic vector operations are **scalar multiplication** and **vector addition.** In operations with vectors, numbers are usually referred to as **scalars.** In this text, scalars will always be real numbers. Geometrically, the product of a vector $\mathbf{v}$ and a scalar k is the vector that is $|k|$ times as long as $\mathbf{v}$. When k is positive, $k\mathbf{v}$ has the same direction as $\mathbf{v}$, and when k is negative, $k\mathbf{v}$ has the direction opposite that of $\mathbf{v}$, as shown in Figure 3.13.

To add two vectors $\mathbf{u}$ and $\mathbf{v}$ geometrically, first position them (without changing their lengths or directions) so that the initial point of the second vector $\mathbf{v}$ coincides with the terminal point of the first vector $\mathbf{u}$. The sum $\mathbf{u} + \mathbf{v}$ is the vector formed by joining the initial point of the first vector $\mathbf{u}$ with the terminal point of the second vector $\mathbf{v}$, as shown in the next two figures. This technique is called the **parallelogram law** for vector addition because the vector $\mathbf{u} + \mathbf{v}$, often called the **resultant** of vector addition, is the diagonal of a parallelogram with adjacent sides $\mathbf{u}$ and $\mathbf{v}$.

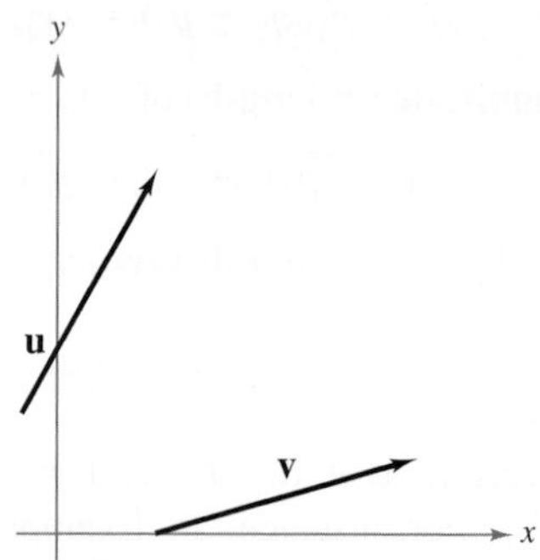

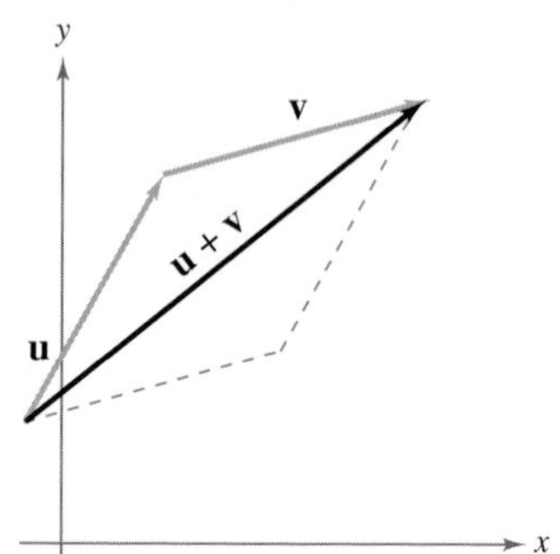

> **Definitions of Vector Addition and Scalar Multiplication**
>
> Let $\mathbf{u} = \langle u_1, u_2 \rangle$ and $\mathbf{v} = \langle v_1, v_2 \rangle$ be vectors and let k be a scalar (a real number). Then the **sum** of $\mathbf{u}$ and $\mathbf{v}$ is the vector
>
> $\mathbf{u} + \mathbf{v} = \langle u_1 + v_1, u_2 + v_2 \rangle$ Sum
>
> and the **scalar multiple** of k times $\mathbf{u}$ is the vector
>
> $k\mathbf{u} = k\langle u_1, u_2 \rangle = \langle ku_1, ku_2 \rangle.$ Scalar multiple

The **negative** of $\mathbf{v} = \langle v_1, v_2 \rangle$ is

$-\mathbf{v} = (-1)\mathbf{v}$

$= \langle -v_1, -v_2 \rangle$ Negative

and the **difference** of $\mathbf{u}$ and $\mathbf{v}$ is

$\mathbf{u} - \mathbf{v} = \mathbf{u} + (-\mathbf{v})$ Add ($-\mathbf{v}$). See Figure 3.14.

$= \langle u_1 - v_1, u_2 - v_2 \rangle.$ Difference

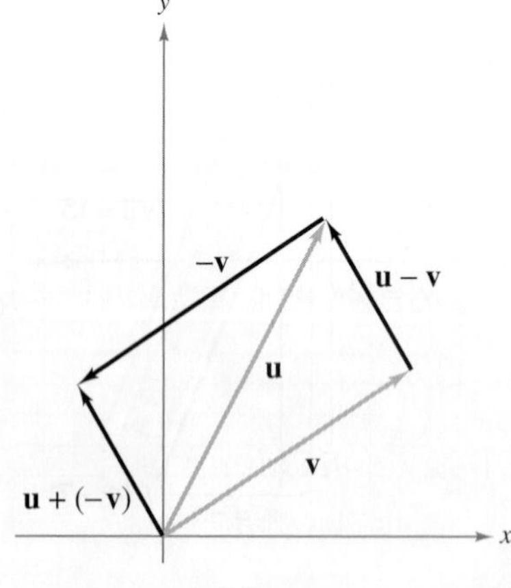

$\mathbf{u} - \mathbf{v} = \mathbf{u} + (-\mathbf{v})$
Figure 3.14

To represent $\mathbf{u} - \mathbf{v}$ geometrically, use directed line segments with the *same* initial point. The difference $\mathbf{u} - \mathbf{v}$ is the vector from the terminal point of $\mathbf{v}$ to the terminal point of $\mathbf{u}$, which is equal to

$\mathbf{u} + (-\mathbf{v})$

as shown in Figure 3.14.

Example 3 illustrates the component definitions of vector addition and scalar multiplication. In this example, note the geometrical interpretations of each of the vector operations.

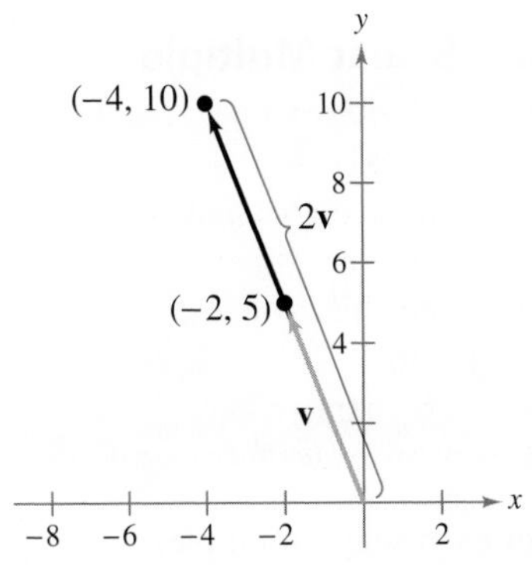

Figure 3.15

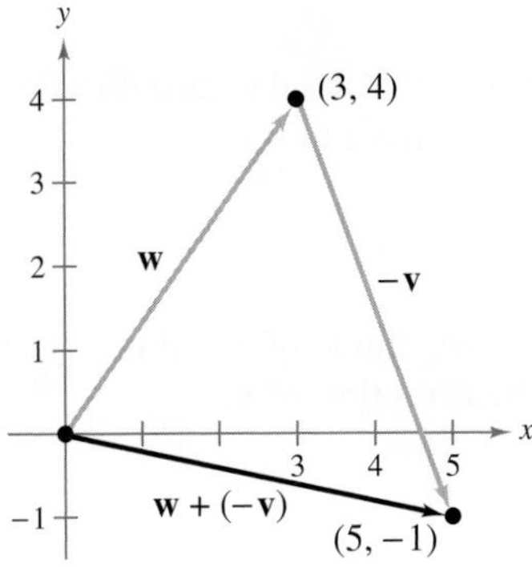

Figure 3.16

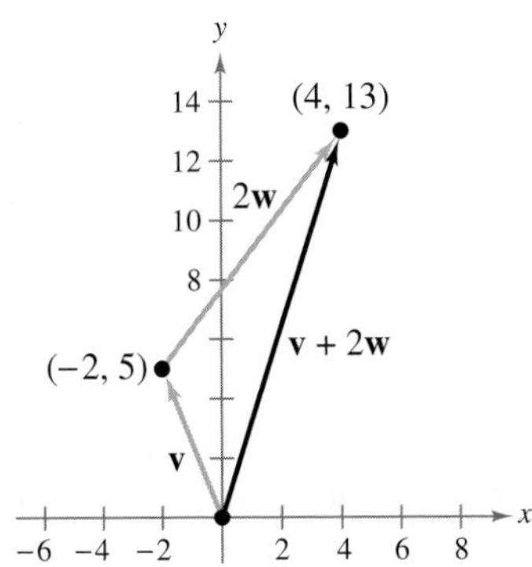

Figure 3.17

EXAMPLE 3 Vector Operations

See LarsonPrecalculus.com for an interactive version of this type of example.

Let $\mathbf{v} = \langle -2, 5\rangle$ and $\mathbf{w} = \langle 3, 4\rangle$. Find each vector.

a. $2\mathbf{v}$ **b.** $\mathbf{w} - \mathbf{v}$ **c.** $\mathbf{v} + 2\mathbf{w}$

Solution

a. Multiplying $\mathbf{v} = \langle -2, 5\rangle$ by the scalar 2, you have

$$\begin{aligned} 2\mathbf{v} &= 2\langle -2, 5\rangle \\ &= \langle 2(-2), 2(5)\rangle \\ &= \langle -4, 10\rangle. \end{aligned}$$

Figure 3.15 shows a sketch of $2\mathbf{v}$.

b. The difference of $\mathbf{w}$ and $\mathbf{v}$ is

$$\begin{aligned} \mathbf{w} - \mathbf{v} &= \langle 3, 4\rangle - \langle -2, 5\rangle \\ &= \langle 3 - (-2), 4 - 5\rangle \\ &= \langle 5, -1\rangle. \end{aligned}$$

Figure 3.16 shows a sketch of $\mathbf{w} - \mathbf{v}$. Note that the figure shows the vector difference $\mathbf{w} - \mathbf{v}$ as the sum $\mathbf{w} + (-\mathbf{v})$.

c. The sum of $\mathbf{v}$ and $2\mathbf{w}$ is

$$\begin{aligned} \mathbf{v} + 2\mathbf{w} &= \langle -2, 5\rangle + 2\langle 3, 4\rangle \\ &= \langle -2, 5\rangle + \langle 2(3), 2(4)\rangle \\ &= \langle -2, 5\rangle + \langle 6, 8\rangle \\ &= \langle -2 + 6, 5 + 8\rangle \\ &= \langle 4, 13\rangle. \end{aligned}$$

Figure 3.17 shows a sketch of $\mathbf{v} + 2\mathbf{w}$.

✓ ***Checkpoint*** *Audio-video solution in English & Spanish at LarsonPrecalculus.com*

Let $\mathbf{u} = \langle 1, 4\rangle$ and $\mathbf{v} = \langle 3, 2\rangle$. Find each vector.

a. $\mathbf{u} + \mathbf{v}$ **b.** $\mathbf{u} - \mathbf{v}$ **c.** $2\mathbf{u} - 3\mathbf{v}$

Vector addition and scalar multiplication share many of the properties of ordinary arithmetic.

Properties of Vector Addition and Scalar Multiplication

Let $\mathbf{u}$, $\mathbf{v}$, and $\mathbf{w}$ be vectors and let c and d be scalars. Then the properties listed below are true.

1. $\mathbf{u} + \mathbf{v} = \mathbf{v} + \mathbf{u}$

2. $(\mathbf{u} + \mathbf{v}) + \mathbf{w} = \mathbf{u} + (\mathbf{v} + \mathbf{w})$

3. $\mathbf{u} + \mathbf{0} = \mathbf{u}$

4. $\mathbf{u} + (-\mathbf{u}) = \mathbf{0}$

5. $c(d\mathbf{u}) = (cd)\mathbf{u}$

6. $(c + d)\mathbf{u} = c\mathbf{u} + d\mathbf{u}$

7. $c(\mathbf{u} + \mathbf{v}) = c\mathbf{u} + c\mathbf{v}$

8. $1(\mathbf{u}) = \mathbf{u}, \quad 0(\mathbf{u}) = \mathbf{0}$

9. $\|c\mathbf{v}\| = |c|\,\|\mathbf{v}\|$

REMARK Property 9 can be stated as: The magnitude of a scalar multiple $c\mathbf{v}$ is the absolute value of c times the magnitude of $\mathbf{v}$.

William Rowan Hamilton (1805–1865), an Irish mathematician, did some of the earliest work with vectors. Hamilton spent many years developing a system of vector-like quantities called quaternions. Although Hamilton was convinced of the benefits of quaternions, the operations he defined did not produce good models for physical phenomena. It was not until the latter half of the nineteenth century that the Scottish physicist James Maxwell (1831–1879) restructured Hamilton's quaternions in a form that is useful for representing physical quantities such as force, velocity, and acceleration.

EXAMPLE 4 Finding the Magnitude of a Scalar Multiple

Let $\mathbf{u} = \langle 1, 3\rangle$ and $\mathbf{v} = \langle -2, 5\rangle$.

a. $\|2\mathbf{u}\| = |2|\|\mathbf{u}\| = |2|\|\langle 1, 3\rangle\| = |2|\sqrt{1^2 + 3^2} = 2\sqrt{10}$

b. $\|-5\mathbf{u}\| = |-5|\|\mathbf{u}\| = |-5|\|\langle 1, 3\rangle\| = |-5|\sqrt{1^2 + 3^2} = 5\sqrt{10}$

c. $\|3\mathbf{v}\| = |3|\|\mathbf{v}\| = |3|\|\langle -2, 5\rangle\| = |3|\sqrt{(-2)^2 + 5^2} = 3\sqrt{29}$

✓ ***Checkpoint*** Audio-video solution in English & Spanish at LarsonPrecalculus.com

Let $\mathbf{u} = \langle 4, -1\rangle$ and $\mathbf{v} = \langle 3, 2\rangle$. Find the magnitude of each scalar multiple.

a. $\|3\mathbf{u}\|$ **b.** $\|-2\mathbf{v}\|$ **c.** $\|5\mathbf{v}\|$

Unit Vectors

In many applications of vectors, it is useful to find a unit vector that has the same direction as a given nonzero vector **v**. To do this, divide **v** by its magnitude to obtain

$$\mathbf{u} = \text{unit vector} = \frac{\mathbf{v}}{\|\mathbf{v}\|} = \left(\frac{1}{\|\mathbf{v}\|}\right)\mathbf{v}.$$ Unit vector in direction of **v**

Note that **u** is a scalar multiple of **v**. The vector **u** has a magnitude of 1 and the same direction as **v**. The vector **u** is called a **unit vector in the direction of v.**

EXAMPLE 5 Finding a Unit Vector

Find a unit vector **u** in the direction of $\mathbf{v} = \langle -2, 5\rangle$. Verify that $\|\mathbf{u}\| = 1$.

Solution The unit vector **u** in the direction of **v** is

$$\frac{\mathbf{v}}{\|\mathbf{v}\|} = \frac{\langle -2, 5\rangle}{\sqrt{(-2)^2 + 5^2}} = \frac{1}{\sqrt{29}}\langle -2, 5\rangle = \left\langle \frac{-2}{\sqrt{29}}, \frac{5}{\sqrt{29}}\right\rangle.$$

This vector has a magnitude of 1 because

$$\sqrt{\left(\frac{-2}{\sqrt{29}}\right)^2 + \left(\frac{5}{\sqrt{29}}\right)^2} = \sqrt{\frac{4}{29} + \frac{25}{29}} = \sqrt{\frac{29}{29}} = 1.$$

✓ ***Checkpoint*** Audio-video solution in English & Spanish at LarsonPrecalculus.com

Find a unit vector **u** in the direction of $\mathbf{v} = \langle 6, -1\rangle$. Verify that $\|\mathbf{u}\| = 1$.

To find a vector **w** with magnitude $\|\mathbf{w}\| = c$ and the same direction as a nonzero vector **v**, multiply the unit vector **u** in the direction of **v** by the scalar c to obtain $\mathbf{w} = c\mathbf{u}$.

EXAMPLE 6 Finding a Vector

Find the vector **w** with magnitude $\|\mathbf{w}\| = 5$ and the same direction as $\mathbf{v} = \langle -2, 3\rangle$.

Solution

$$\mathbf{w} = 5\left(\frac{1}{\|\mathbf{v}\|}\mathbf{v}\right) = 5\left(\frac{1}{\sqrt{(-2)^2 + 3^2}}\langle -2, 3\rangle\right) = \frac{5}{\sqrt{13}}\langle -2, 3\rangle = \left\langle \frac{-10}{\sqrt{13}}, \frac{15}{\sqrt{13}}\right\rangle$$

✓ ***Checkpoint*** Audio-video solution in English & Spanish at LarsonPrecalculus.com

Find the vector **w** with magnitude $\|\mathbf{w}\| = 6$ and the same direction as $\mathbf{v} = \langle 2, -4\rangle$.

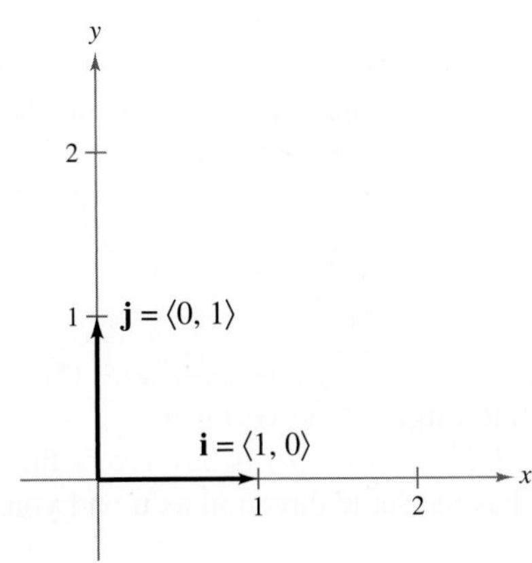

Figure 3.18

The unit vectors $\langle 1, 0\rangle$ and $\langle 0, 1\rangle$ are the **standard unit vectors** and are denoted by

$$\mathbf{i} = \langle 1, 0\rangle \quad \text{and} \quad \mathbf{j} = \langle 0, 1\rangle$$

as shown in Figure 3.18. (Note that the lowercase letter **i** is in boldface and not italicized to distinguish it from the imaginary unit $i = \sqrt{-1}$.) These vectors can be used to represent any vector $\mathbf{v} = \langle v_1, v_2\rangle$, because

$$\mathbf{v} = \langle v_1, v_2\rangle = v_1\langle 1, 0\rangle + v_2\langle 0, 1\rangle = v_1\mathbf{i} + v_2\mathbf{j}.$$

The scalars v_1 and v_2 are the **horizontal** and **vertical components of v,** respectively. The vector sum

$$v_1\mathbf{i} + v_2\mathbf{j}$$

is a **linear combination** of the vectors **i** and **j**. Any vector in the plane can be written as a linear combination of the standard unit vectors **i** and **j**.

EXAMPLE 7 Writing a Linear Combination of Unit Vectors

Let **u** be the vector with initial point $(2, -5)$ and terminal point $(-1, 3)$. Write **u** as a linear combination of the standard unit vectors **i** and **j**.

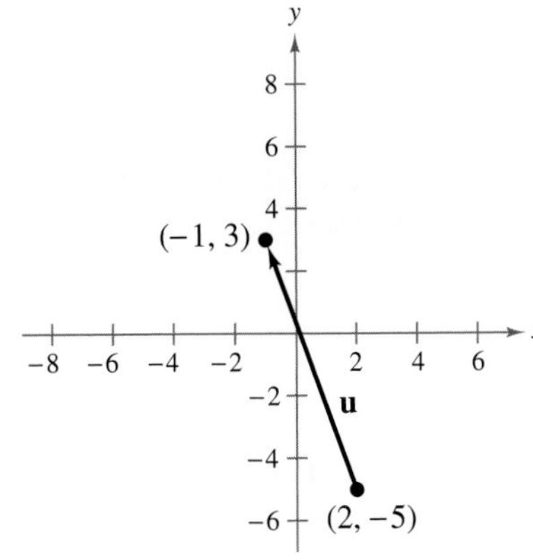

Figure 3.19

Solution Begin by writing the component form of the vector **u**. Then write the component form in terms of **i** and **j**.

$$\mathbf{u} = \langle -1 - 2, 3 - (-5)\rangle = \langle -3, 8\rangle = -3\mathbf{i} + 8\mathbf{j}$$

Figure 3.19 shows the vector **u**.

Checkpoint Audio-video solution in English & Spanish at LarsonPrecalculus.com

Let **u** be the vector with initial point $(-2, 6)$ and terminal point $(-8, 3)$. Write **u** as a linear combination of the standard unit vectors **i** and **j**.

EXAMPLE 8 Vector Operations

Let $\mathbf{u} = -3\mathbf{i} + 8\mathbf{j}$ and $\mathbf{v} = 2\mathbf{i} - \mathbf{j}$. Find $2\mathbf{u} - 3\mathbf{v}$.

Solution It is not necessary to convert **u** and **v** to component form to solve this problem. Just perform the operations with the vectors in unit vector form.

$$\begin{aligned} 2\mathbf{u} - 3\mathbf{v} &= 2(-3\mathbf{i} + 8\mathbf{j}) - 3(2\mathbf{i} - \mathbf{j}) \\ &= -6\mathbf{i} + 16\mathbf{j} - 6\mathbf{i} + 3\mathbf{j} \\ &= -12\mathbf{i} + 19\mathbf{j} \end{aligned}$$

Checkpoint Audio-video solution in English & Spanish at LarsonPrecalculus.com

Let $\mathbf{u} = \mathbf{i} - 2\mathbf{j}$ and $\mathbf{v} = -3\mathbf{i} + 2\mathbf{j}$. Find $5\mathbf{u} - 2\mathbf{v}$.

In Example 8, you could perform the operations in component form by writing

$$\mathbf{u} = -3\mathbf{i} + 8\mathbf{j} = \langle -3, 8\rangle \quad \text{and} \quad \mathbf{v} = 2\mathbf{i} - \mathbf{j} = \langle 2, -1\rangle.$$

The difference of $2\mathbf{u}$ and $3\mathbf{v}$ is

$$\begin{aligned} 2\mathbf{u} - 3\mathbf{v} &= 2\langle -3, 8\rangle - 3\langle 2, -1\rangle \\ &= \langle -6, 16\rangle - \langle 6, -3\rangle \\ &= \langle -6 - 6, 16 - (-3)\rangle \\ &= \langle -12, 19\rangle. \end{aligned}$$

Compare this result with the solution to Example 8.

Direction Angles

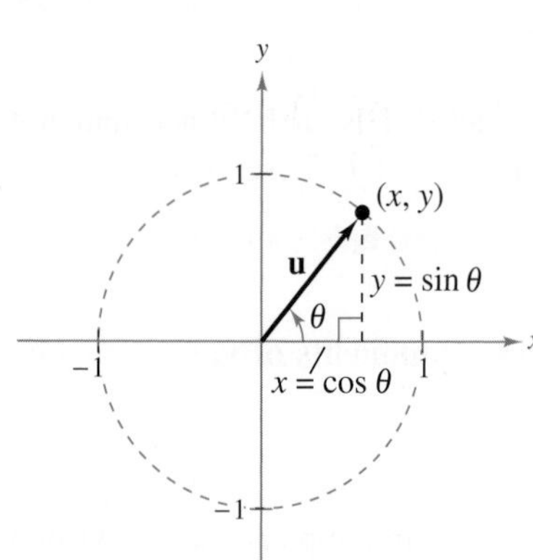

$\|\mathbf{u}\| = 1$
Figure 3.20

If **u** is a unit vector such that θ is the angle (measured counterclockwise) from the positive x-axis to **u**, then the terminal point of **u** lies on the unit circle and you have

$$\begin{aligned}\mathbf{u} &= \langle x, y \rangle \\ &= \langle \cos\theta, \sin\theta \rangle \\ &= (\cos\theta)\mathbf{i} + (\sin\theta)\mathbf{j}\end{aligned}$$

as shown in Figure 3.20. The angle θ is the **direction angle** of the vector **u**.

Consider a unit vector **u** with direction angle θ. If $\mathbf{v} = a\mathbf{i} + b\mathbf{j}$ is any vector that makes an angle θ with the positive x-axis, then it has the same direction as **u** and you can write

$$\begin{aligned}\mathbf{v} &= \|\mathbf{v}\|\langle \cos\theta, \sin\theta \rangle \\ &= \|\mathbf{v}\|(\cos\theta)\mathbf{i} + \|\mathbf{v}\|(\sin\theta)\mathbf{j}.\end{aligned}$$

Because $\mathbf{v} = a\mathbf{i} + b\mathbf{j} = \|\mathbf{v}\|(\cos\theta)\mathbf{i} + \|\mathbf{v}\|(\sin\theta)\mathbf{j}$, it follows that the direction angle θ for **v** is determined from

$$\tan\theta = \frac{\sin\theta}{\cos\theta} \qquad \text{Quotient identity}$$

$$= \frac{\|\mathbf{v}\|\sin\theta}{\|\mathbf{v}\|\cos\theta} \qquad \text{Multiply numerator and denominator by } \|\mathbf{v}\|.$$

$$= \frac{b}{a}. \qquad \text{Simplify.}$$

EXAMPLE 9 Finding Direction Angles of Vectors

Find the direction angle of each vector.

a. $\mathbf{u} = 3\mathbf{i} + 3\mathbf{j}$ **b.** $\mathbf{v} = 3\mathbf{i} - 4\mathbf{j}$

Solution

a. The direction angle is determined from

$$\tan\theta = \frac{b}{a} = \frac{3}{3} = 1.$$

So, $\theta = 45°$, as shown in Figure 3.21.

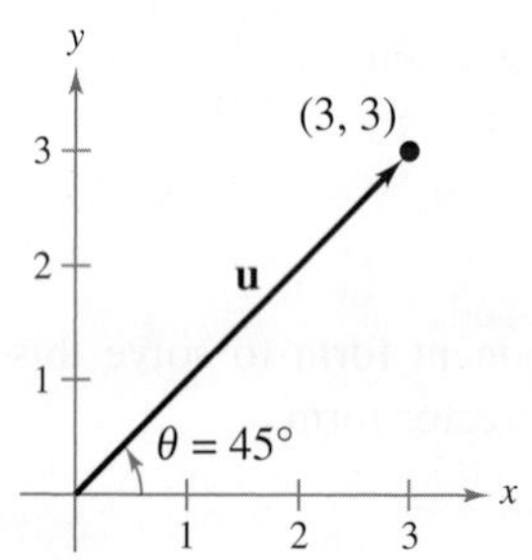

Figure 3.21

b. The direction angle is determined from

$$\tan\theta = \frac{b}{a} = \frac{-4}{3}.$$

Moreover, $\mathbf{v} = 3\mathbf{i} - 4\mathbf{j}$ lies in Quadrant IV, so θ lies in Quadrant IV, and its reference angle is

$$\theta' = \left|\arctan\left(-\frac{4}{3}\right)\right| \approx |-0.9273 \text{ radian}| \approx |-53.13°| = 53.13°.$$

It follows that $\theta \approx 360° - 53.13° = 306.87°$, as shown in Figure 3.22.

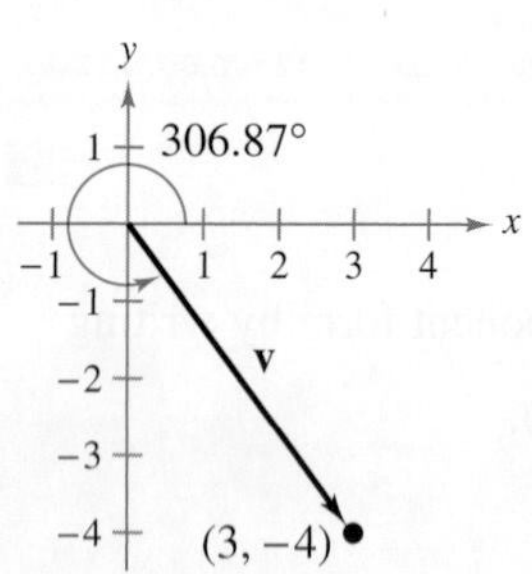

Figure 3.22

✓ **Checkpoint** Audio-video solution in English & Spanish at LarsonPrecalculus.com

Find the direction angle of each vector.

a. $\mathbf{v} = -6\mathbf{i} + 6\mathbf{j}$ **b.** $\mathbf{v} = -7\mathbf{i} - 4\mathbf{j}$

Applications

EXAMPLE 10 Finding the Component Form of a Vector

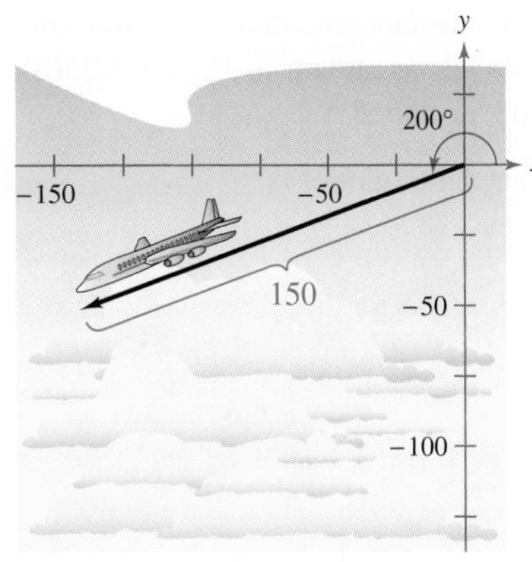

Figure 3.23

Find the component form of the vector that represents the velocity of an airplane descending at a speed of 150 miles per hour at an angle 20° below the horizontal, as shown in Figure 3.23.

Solution The velocity vector **v** has a magnitude of 150 and a direction angle of $\theta = 200°$.

$$\begin{aligned}\mathbf{v} &= \|\mathbf{v}\|(\cos\theta)\mathbf{i} + \|\mathbf{v}\|(\sin\theta)\mathbf{j}\\ &= 150(\cos 200°)\mathbf{i} + 150(\sin 200°)\mathbf{j}\\ &\approx 150(-0.9397)\mathbf{i} + 150(-0.3420)\mathbf{j}\\ &\approx -140.96\mathbf{i} - 52.30\mathbf{j}\\ &= \langle -140.96, -52.30\rangle\end{aligned}$$

Check that **v** has a magnitude of 150.

$$\|\mathbf{v}\| \approx \sqrt{(-140.96)^2 + (-51.30)^2} \approx \sqrt{22{,}501.41} \approx 150$$ Solution checks.

✓ **Checkpoint** Audio-video solution in English & Spanish at LarsonPrecalculus.com

Find the component form of the vector that represents the velocity of an airplane descending at a speed of 100 miles per hour at an angle 15° below the horizontal ($\theta = 195°$).

EXAMPLE 11 Using Vectors to Determine Weight

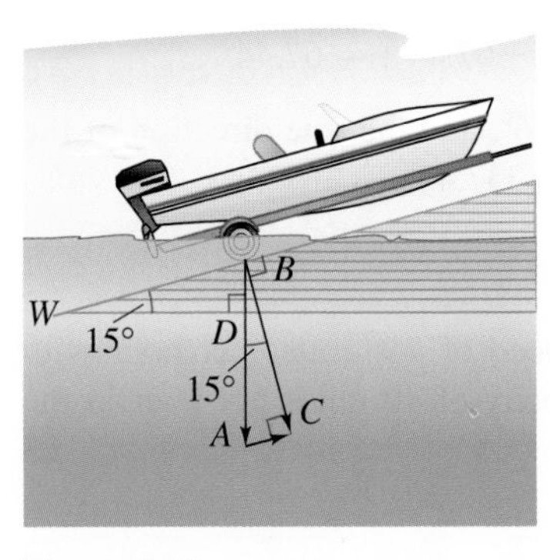

Figure 3.24

A force of 600 pounds is required to pull a boat and trailer up a ramp inclined at 15° from the horizontal. Find the combined weight of the boat and trailer.

Solution Use Figure 3.24 to make the observations below.

$\|\overrightarrow{BA}\|$ = force of gravity = combined weight of boat and trailer

$\|\overrightarrow{BC}\|$ = force against ramp

$\|\overrightarrow{AC}\|$ = force required to move boat up ramp = 600 pounds

Note that $\overline{AC}$ is parallel to the ramp. So, by construction, triangles *BWD* and *ABC* are similar and angle *ABC* is 15°. In triangle *ABC*, you have

$$\begin{aligned}\sin 15° &= \frac{\|\overrightarrow{AC}\|}{\|\overrightarrow{BA}\|}\\ \sin 15° &= \frac{600}{\|\overrightarrow{BA}\|}\\ \|\overrightarrow{BA}\| &= \frac{600}{\sin 15°}\\ \|\overrightarrow{BA}\| &\approx 2318.\end{aligned}$$

So, the combined weight is approximately 2318 pounds.

✓ **Checkpoint** Audio-video solution in English & Spanish at LarsonPrecalculus.com

A force of 500 pounds is required to pull a boat and trailer up a ramp inclined at 12° from the horizontal. Find the combined weight of the boat and trailer.

REMARK Recall from Section 1.8 that in air navigation, bearings are measured in degrees clockwise from north.

EXAMPLE 12 Using Vectors to Find Speed and Direction

An airplane travels at a speed of 500 miles per hour with a bearing of 330° at a fixed altitude with a negligible wind velocity, as shown in Figure 3.25(a). (Note that a bearing of 330° corresponds to a direction angle of 120°.) The airplane encounters a wind with a velocity of 70 miles per hour in the direction N 45° E, as shown in Figure 3.25(b). What are the resultant speed and true direction of the airplane?

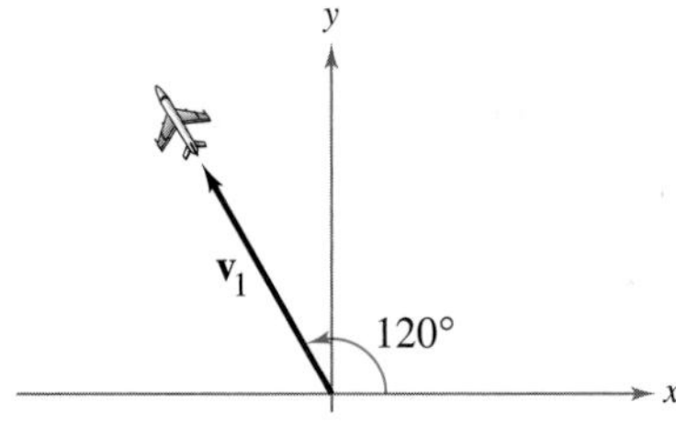

(a)

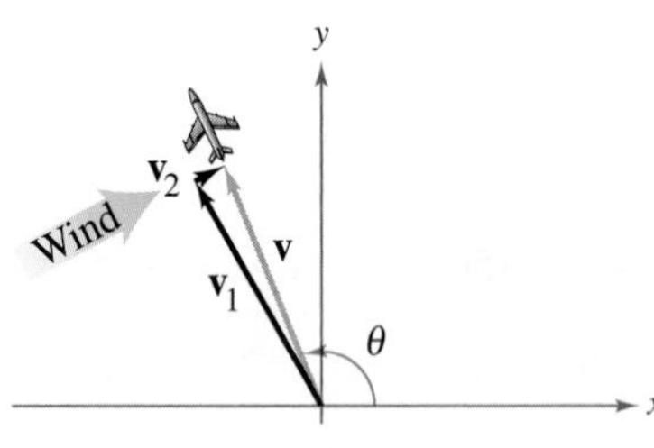

(b)
Figure 3.25

Solution Using Figure 3.25, the velocity of the airplane (alone) is

$$\mathbf{v}_1 = 500\langle \cos 120°, \sin 120° \rangle = \langle -250, 250\sqrt{3} \rangle$$

and the velocity of the wind is

$$\mathbf{v}_2 = 70\langle \cos 45°, \sin 45° \rangle = \langle 35\sqrt{2}, 35\sqrt{2} \rangle.$$

So, the velocity of the airplane (in the wind) is

$$\begin{aligned} \mathbf{v} &= \mathbf{v}_1 + \mathbf{v}_2 \\ &= \langle -250 + 35\sqrt{2}, 250\sqrt{3} + 35\sqrt{2} \rangle \\ &\approx \langle -200.5, 482.5 \rangle \end{aligned}$$

and the resultant speed of the airplane is

$$\|\mathbf{v}\| \approx \sqrt{(-200.5)^2 + (482.5)^2} \approx 522.5 \text{ miles per hour.}$$

To find the direction angle θ of the flight path, you have

$$\tan\theta \approx \frac{482.5}{-200.5} \approx -2.4065.$$

The flight path lies in Quadrant II, so θ lies in Quadrant II, and its reference angle is

$$\theta' \approx |\arctan(-2.4065)| \approx |-1.1770 \text{ radians}| \approx |-67.44°| = 67.44°.$$

So, the direction angle is $\theta \approx 180° - 67.44° = 112.56°$, and the true direction of the airplane is approximately $270° + (180° - 112.56°) = 337.44°$.

Pilots can take advantage of fast-moving air currents called jet streams to decrease travel time.

Checkpoint Audio-video solution in English & Spanish at LarsonPrecalculus.com

Repeat Example 11 for an airplane traveling at a speed of 450 miles per hour with a bearing of 300° that encounters a wind with a velocity of 40 miles per hour in the direction N 30° E.

Summarize (Section 3.3)

1. Explain how to represent a vector as a directed line segment *(page 278)*. For an example involving vectors represented as directed line segments, see Example 1.
2. Explain how to find the component form of a vector *(page 279)*. For an example of finding the component form of a vector, see Example 2.
3. Explain how to perform basic vector operations *(page 280)*. For an example of performing basic vector operations, see Example 3.
4. Explain how to write a vector as a linear combination of unit vectors *(page 282)*. For examples involving unit vectors, see Examples 5–8.
5. Explain how to find the direction angle of a vector *(page 284)*. For an example of finding direction angles of vectors, see Example 9.
6. Describe real-life applications of vectors *(pages 285 and 286, Examples 10–12)*.

3.3 Exercises

See **CalcChat.com** for tutorial help and worked-out solutions to odd-numbered exercises.

Vocabulary: Fill in the blanks.

1. You can use a ________ ________ ________ to represent a quantity that involves both magnitude and direction.
2. The directed line segment $\overrightarrow{PQ}$ has ________ point P and ________ point Q.
3. The set of all directed line segments that are equivalent to a given directed line segment $\overrightarrow{PQ}$ is a ________ **v** in the plane.
4. Two vectors are equivalent when they have the same ________ and the same ________.
5. The directed line segment whose initial point is the origin is in ________ ________.
6. A vector that has a magnitude of 1 is a ________ ________.
7. The two basic vector operations are scalar ________ and vector ________.
8. The vector sum $v_1\mathbf{i} + v_2\mathbf{j}$ is a ________ ________ of the vectors **i** and **j**, and the scalars v_1 and v_2 are the ________ and ________ components of **v**, respectively.

Skills and Applications

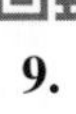

Determining Whether Two Vectors Are Equivalent In Exercises 9–14, determine whether u and v are equivalent. Explain.

9.

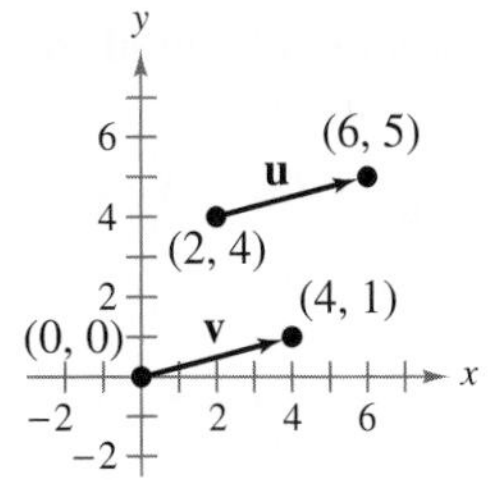

10.

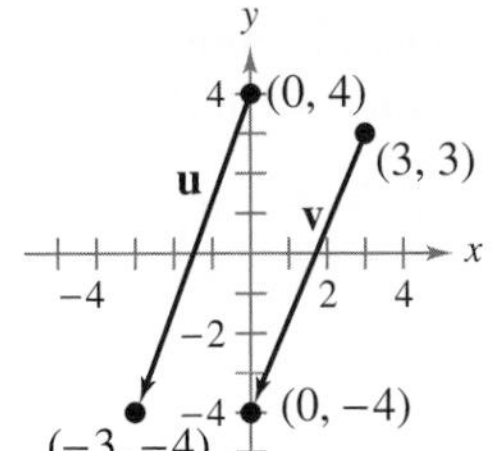

	Vector	Initial Point	Terminal Point
11.	**u**	$(2, 2)$	$(-1, 4)$
	v	$(-3, -1)$	$(-5, 2)$
12.	**u**	$(2, 0)$	$(7, 4)$
	v	$(-8, 1)$	$(2, 9)$
13.	**u**	$(2, -1)$	$(5, -10)$
	v	$(6, 1)$	$(9, -8)$
14.	**u**	$(8, 1)$	$(13, -1)$
	v	$(-2, 4)$	$(-7, 6)$

Finding the Component Form of a Vector In Exercises 15–24, find the component form and magnitude of the vector v.

15.

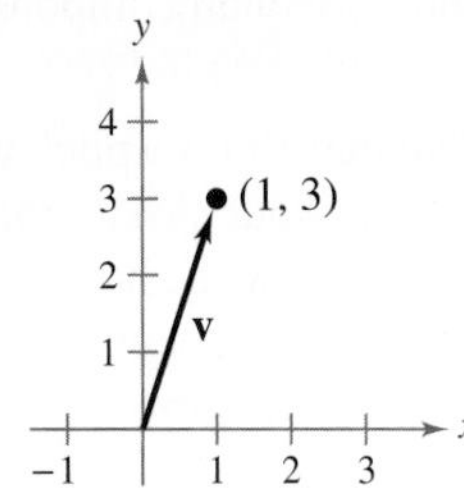

16.

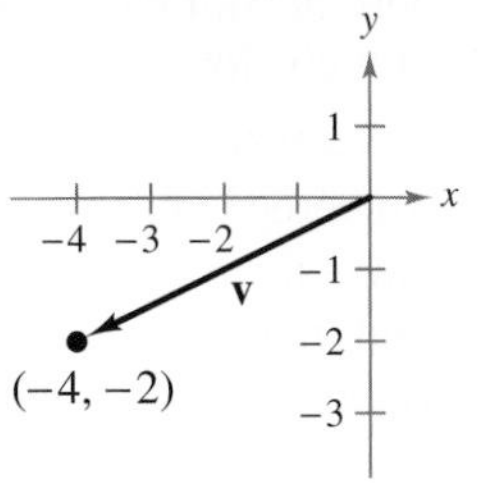

17.

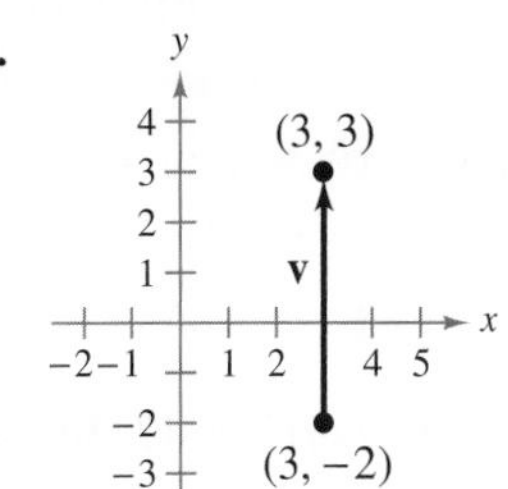

18.

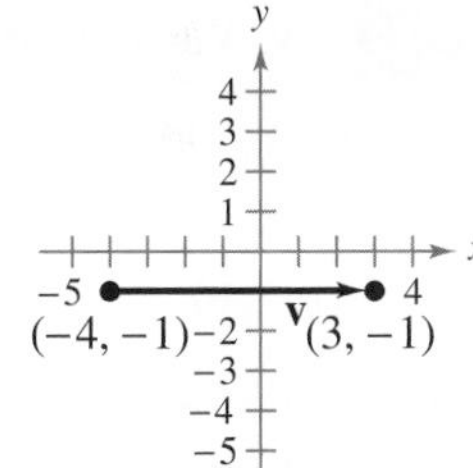

	Initial Point	Terminal Point
19.	$(-3, -5)$	$(-11, 1)$
20.	$(-2, 7)$	$(5, -17)$
21.	$(1, 3)$	$(-8, -9)$
22.	$(17, -5)$	$(9, 3)$
23.	$(-1, 5)$	$(15, -21)$
24.	$(-3, 11)$	$(9, 40)$

Sketching the Graph of a Vector In Exercises 25–30, use the figure to sketch a graph of the specified vector. To print an enlarged copy of the graph, go to *MathGraphs.com*.

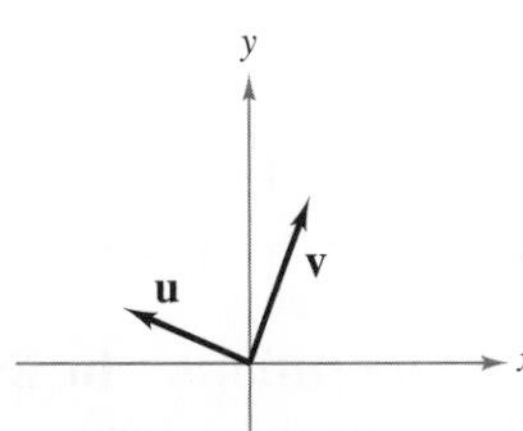

25. $-\mathbf{v}$

26. $5\mathbf{v}$

27. $\mathbf{u} + \mathbf{v}$

28. $\mathbf{u} + 2\mathbf{v}$

29. $\mathbf{u} - \mathbf{v}$

30. $\mathbf{v} - \frac{1}{2}\mathbf{u}$

Vector Operations In Exercises 31–36, find (a) $\mathbf{u} + \mathbf{v}$, (b) $\mathbf{u} - \mathbf{v}$, and (c) $2\mathbf{u} - 3\mathbf{v}$. Then sketch each resultant vector.

31. $\mathbf{u} = \langle 2, 1 \rangle$, $\mathbf{v} = \langle 1, 3 \rangle$
32. $\mathbf{u} = \langle 2, 3 \rangle$, $\mathbf{v} = \langle 4, 0 \rangle$
33. $\mathbf{u} = \langle -5, 3 \rangle$, $\mathbf{v} = \langle 0, 0 \rangle$
34. $\mathbf{u} = \langle 0, 0 \rangle$, $\mathbf{v} = \langle 2, 1 \rangle$
35. $\mathbf{u} = \langle 0, -7 \rangle$, $\mathbf{v} = \langle 1, -2 \rangle$
36. $\mathbf{u} = \langle -3, 1 \rangle$, $\mathbf{v} = \langle 2, -5 \rangle$

Finding the Magnitude of a Scalar Multiple In Exercises 37–40, find the magnitude of the scalar multiple, where $\mathbf{u} = \langle 2, 0 \rangle$ and $\mathbf{v} = \langle -3, 6 \rangle$.

37. $\|5\mathbf{u}\|$
38. $\|4\mathbf{v}\|$
39. $\|-3\mathbf{v}\|$
40. $\left\|-\frac{3}{4}\mathbf{u}\right\|$

Finding a Unit Vector In Exercises 41–46, find a unit vector u in the direction of v. Verify that $\|\mathbf{u}\| = 1$.

41. $\mathbf{v} = \langle 3, 0 \rangle$
42. $\mathbf{v} = \langle 0, -2 \rangle$
43. $\mathbf{v} = \langle -2, 2 \rangle$
44. $\mathbf{v} = \langle -5, 12 \rangle$
45. $\mathbf{v} = \langle 1, -6 \rangle$
46. $\mathbf{v} = \langle -8, -4 \rangle$

Finding a Vector In Exercises 47–50, find the vector w with the given magnitude and the same direction as v.

47. $\|\mathbf{w}\| = 10$, $\mathbf{v} = \langle -3, 4 \rangle$
48. $\|\mathbf{w}\| = 3$, $\mathbf{v} = \langle -12, -5 \rangle$
49. $\|\mathbf{w}\| = 9$, $\mathbf{v} = \langle 2, 5 \rangle$
50. $\|\mathbf{w}\| = 8$, $\mathbf{v} = \langle 3, 3 \rangle$

Writing a Linear Combination of Unit Vectors In Exercises 51–54, the initial and terminal points of a vector are given. Write the vector as a linear combination of the standard unit vectors i and j.

	Initial Point	Terminal Point
51.	$(-2, 1)$	$(3, -2)$
52.	$(0, -2)$	$(3, 6)$
53.	$(0, 1)$	$(-6, 4)$
54.	$(2, 3)$	$(-1, -5)$

Vector Operations In Exercises 55–60, find the component form of v and sketch the specified vector operations geometrically, where $\mathbf{u} = 2\mathbf{i} - \mathbf{j}$ and $\mathbf{w} = \mathbf{i} + 2\mathbf{j}$.

55. $\mathbf{v} = \frac{3}{2}\mathbf{u}$
56. $\mathbf{v} = \frac{3}{4}\mathbf{w}$
57. $\mathbf{v} = \mathbf{u} + 2\mathbf{w}$
58. $\mathbf{v} = -\mathbf{u} + \mathbf{w}$
59. $\mathbf{v} = \mathbf{u} - 2\mathbf{w}$
60. $\mathbf{v} = \frac{1}{2}(3\mathbf{u} + \mathbf{w})$

Finding the Direction Angle of a Vector In Exercises 61–64, find the magnitude and direction angle of the vector v.

61. $\mathbf{v} = 6\mathbf{i} - 6\mathbf{j}$
62. $\mathbf{v} = -5\mathbf{i} + 4\mathbf{j}$
63. $\mathbf{v} = 3(\cos 60°\mathbf{i} + \sin 60°\mathbf{j})$
64. $\mathbf{v} = 8(\cos 135°\mathbf{i} + \sin 135°\mathbf{j})$

Finding the Component Form of a Vector In Exercises 65–70, find the component form of v given its magnitude and the angle it makes with the positive x-axis. Then sketch v.

	Magnitude	Angle
65.	$\|\mathbf{v}\| = 3$	$\theta = 0°$
66.	$\|\mathbf{v}\| = 4\sqrt{3}$	$\theta = 90°$
67.	$\|\mathbf{v}\| = \frac{7}{2}$	$\theta = 150°$
68.	$\|\mathbf{v}\| = 2\sqrt{3}$	$\theta = 45°$
69.	$\|\mathbf{v}\| = 3$	v in the direction $3\mathbf{i} + 4\mathbf{j}$
70.	$\|\mathbf{v}\| = 2$	v in the direction $\mathbf{i} + 3\mathbf{j}$

Finding the Component Form of a Vector In Exercises 71 and 72, find the component form of the sum of u and v with direction angles $\theta_{\mathbf{u}}$ and $\theta_{\mathbf{v}}$.

71. $\|\mathbf{u}\| = 4$, $\theta_{\mathbf{u}} = 60°$
$\|\mathbf{v}\| = 4$, $\theta_{\mathbf{v}} = 90°$

72. $\|\mathbf{u}\| = 20$, $\theta_{\mathbf{u}} = 45°$
$\|\mathbf{v}\| = 50$, $\theta_{\mathbf{v}} = 180°$

Using the Law of Cosines In Exercises 73 and 74, use the Law of Cosines to find the angle α between the vectors. (Assume $0° \le \alpha \le 180°$.)

73. $\mathbf{v} = \mathbf{i} + \mathbf{j}$, $\mathbf{w} = 2\mathbf{i} - 2\mathbf{j}$
74. $\mathbf{v} = \mathbf{i} + 2\mathbf{j}$, $\mathbf{w} = 2\mathbf{i} - \mathbf{j}$

Resultant Force In Exercises 75 and 76, find the angle between the forces given the magnitude of their resultant. (*Hint:* Write force 1 as a vector in the direction of the positive x-axis and force 2 as a vector at an angle θ with the positive x-axis.)

	Force 1	Force 2	Resultant Force
75.	45 pounds	60 pounds	90 pounds
76.	3000 pounds	1000 pounds	3750 pounds

77. **Velocity** A gun with a muzzle velocity of 1200 feet per second is fired at an angle of 6° above the horizontal. Find the vertical and horizontal components of the velocity.

78. **Velocity** Pitcher Aroldis Chapman threw a pitch with a recorded velocity of 105 miles per hour. Assuming he threw the pitch at an angle of 3.5° below the horizontal, find the vertical and horizontal components of the velocity. *(Source: Guinness World Records)*

79. Resultant Force Forces with magnitudes of 125 newtons and 300 newtons act on a hook (see figure). The angle between the two forces is 45°. Find the direction and magnitude of the resultant of these forces. (*Hint:* Write the vector representing each force in component form, then add the vectors.)

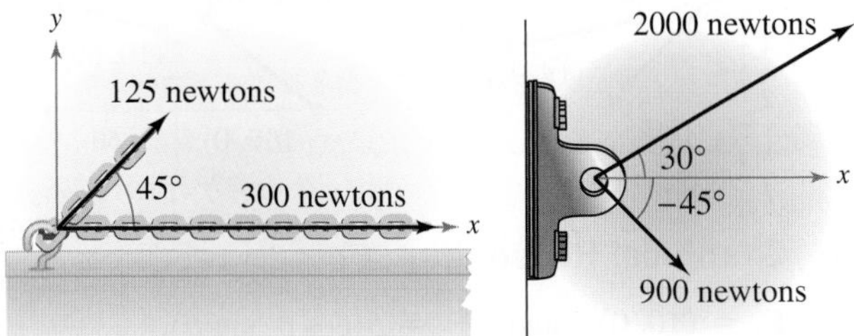

Figure for 79

Figure for 80

80. Resultant Force Forces with magnitudes of 2000 newtons and 900 newtons act on a machine part at angles of 30° and −45°, respectively, with the positive x-axis (see figure). Find the direction and magnitude of the resultant of these forces.

81. Resultant Force Three forces with magnitudes of 75 pounds, 100 pounds, and 125 pounds act on an object at angles of 30°, 45°, and 120°, respectively, with the positive x-axis. Find the direction and magnitude of the resultant of these forces.

82. Resultant Force Three forces with magnitudes of 70 pounds, 40 pounds, and 60 pounds act on an object at angles of −30°, 45°, and 135°, respectively, with the positive x-axis. Find the direction and magnitude of the resultant of these forces.

83. Cable Tension The cranes shown in the figure are lifting an object that weighs 20,240 pounds. Find the tension (in pounds) in the cable of each crane.

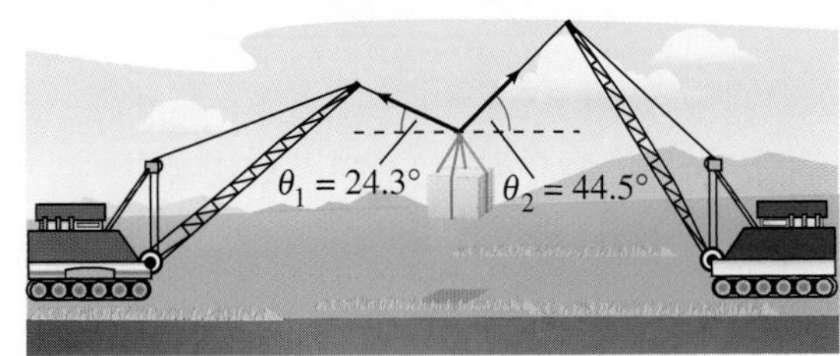

84. Cable Tension Repeat Exercise 83 for $\theta_1 = 35.6°$ and $\theta_2 = 40.4°$.

85. Rope Tension A tetherball weighing 1 pound is pulled outward from the pole by a horizontal force **u** until the rope makes a 45° angle with the pole (see figure). Determine the resulting tension (in pounds) in the rope and the magnitude of **u**.

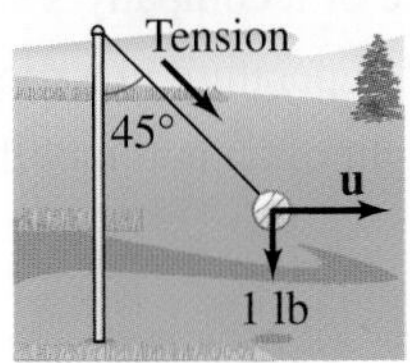

86. Physics Use the figure to determine the tension (in pounds) in each cable supporting the load.

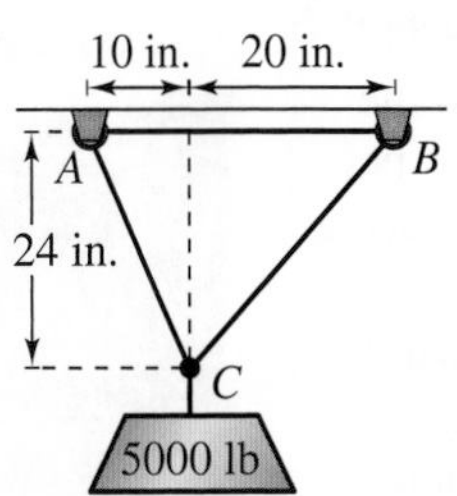

87. Tow Line Tension Two tugboats are towing a loaded barge and the magnitude of the resultant is 6000 pounds directed along the axis of the barge (see figure). Find the tension (in pounds) in the tow lines when they each make an 18° angle with the axis of the barge.

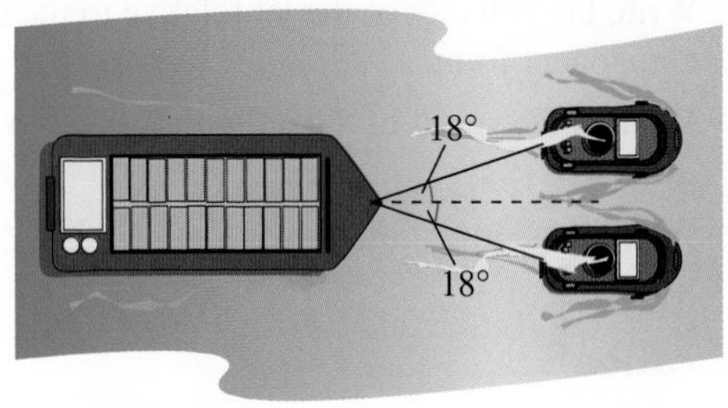

88. Rope Tension To carry a 100-pound cylindrical weight, two people lift on the ends of short ropes that are tied to an eyelet on the top center of the cylinder. Each rope makes a 20° angle with the vertical. Draw a diagram that gives a visual representation of the problem. Then find the tension (in pounds) in the ropes.

Inclined Ramp **In Exercises 89–92, a force of F pounds is required to pull an object weighing W pounds up a ramp inclined at θ degrees from the horizontal.**

89. Find F when $W = 100$ pounds and $\theta = 12°$.

90. Find W when $F = 600$ pounds and $\theta = 14°$.

91. Find θ when $F = 5000$ pounds and $W = 15{,}000$ pounds.

92. Find F when $W = 5000$ pounds and $\theta = 26°$.

93. Air Navigation An airplane travels in the direction of 148° with an airspeed of 875 kilometers per hour. Due to the wind, its groundspeed and direction are 800 kilometers per hour and 140°, respectively (see figure). Find the direction and speed of the wind.

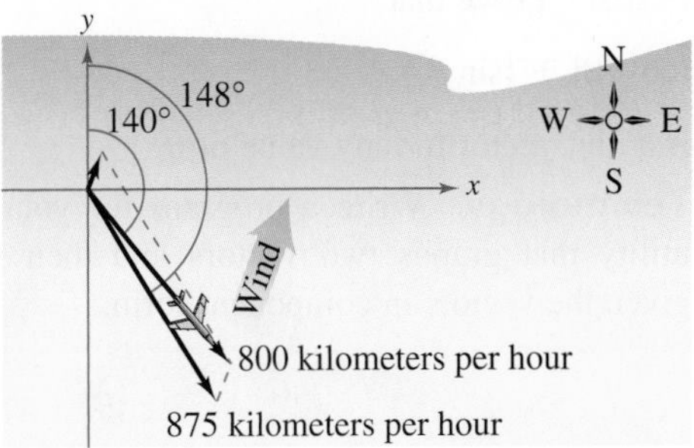

94. Air Navigation A commercial jet travels from Miami to Seattle. The jet's velocity with respect to the air is 580 miles per hour, and its bearing is 332°. The jet encounters a wind with a velocity of 60 miles per hour from the southwest.

(a) Draw a diagram that gives a visual representation of the problem.

(b) Write the velocity of the wind as a vector in component form.

(c) Write the velocity of the jet relative to the air in component form.

(d) What is the speed of the jet with respect to the ground?

(e) What is the true direction of the jet?

Exploration

True or False? **In Exercises 95–98, determine whether the statement is true or false. Justify your answer.**

95. If **u** and **v** have the same magnitude and direction, then **u** and **v** are equivalent.

96. If **u** is a unit vector in the direction of **v**, then $\mathbf{v} = \|\mathbf{v}\|\mathbf{u}$.

97. If $\mathbf{v} = a\mathbf{i} + b\mathbf{j} = \mathbf{0}$, then $a = -b$.

98. If $\mathbf{u} = a\mathbf{i} + b\mathbf{j}$ is a unit vector, then $a^2 + b^2 = 1$.

99. Error Analysis Describe the error in finding the component form of the vector **u** that has initial point $(-3, 4)$ and terminal point $(6, -1)$.

The components are $u_1 = -3 - 6 = -9$ and $u_2 = 4 - (-1) = 5$. So, $\mathbf{u} = \langle -9, 5 \rangle$. ✗

100. Error Analysis Describe the error in finding the direction angle θ of the vector $\mathbf{v} = -5\mathbf{i} + 8\mathbf{j}$.

Because $\tan \theta = \dfrac{b}{a} = \dfrac{8}{-5}$, the reference angle is $\theta' = \left|\arctan\left(-\dfrac{8}{5}\right)\right| \approx |-57.99°| = 57.99°$ and $\theta \approx 360° - 57.99° = 302.01°$. ✗

101. Proof Prove that

$$(\cos \theta)\mathbf{i} + (\sin \theta)\mathbf{j}$$

is a unit vector for any value of θ.

102. Technology Write a program for your graphing utility that graphs two vectors and their difference given the vectors in component form.

Finding the Difference of Two Vectors **In Exercises 103 and 104, use the program in Exercise 102 to find the difference of the vectors shown in the figure.**

103. (1, 6), (5, 2), (4, 5), (9, 4)

104.

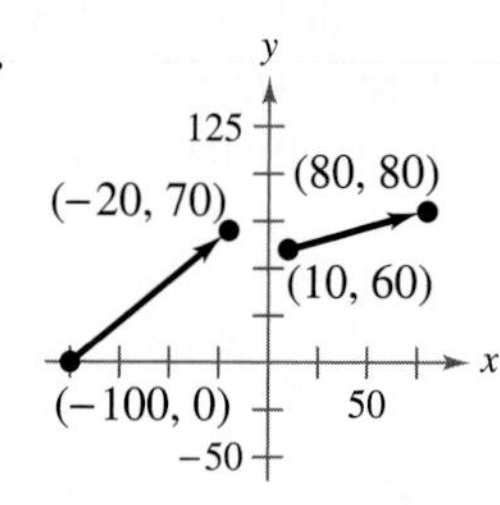

105. Graphical Reasoning Consider two forces

$$\mathbf{F}_1 = \langle 10, 0 \rangle \quad \text{and} \quad \mathbf{F}_2 = 5\langle \cos \theta, \sin \theta \rangle.$$

(a) Find $\|\mathbf{F}_1 + \mathbf{F}_2\|$ as a function of θ.

(b) Use a graphing utility to graph the function in part (a) for $0 \le \theta < 2\pi$.

(c) Use the graph in part (b) to determine the range of the function. What is its maximum, and for what value of θ does it occur? What is its minimum, and for what value of θ does it occur?

(d) Explain why the magnitude of the resultant is never 0.

106. HOW DO YOU SEE IT? Use the figure to determine whether each statement is true or false. Justify your answer.

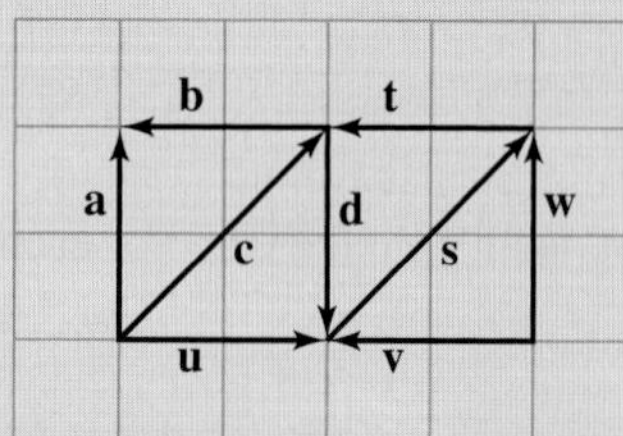

(a) $\mathbf{a} = -\mathbf{d}$

(b) $\mathbf{c} = \mathbf{s}$

(c) $\mathbf{a} + \mathbf{u} = \mathbf{c}$

(d) $\mathbf{v} + \mathbf{w} = -\mathbf{s}$

(e) $\mathbf{a} + \mathbf{w} = -2\mathbf{d}$

(f) $\mathbf{a} + \mathbf{d} = \mathbf{0}$

(g) $\mathbf{u} - \mathbf{v} = -2(\mathbf{b} + \mathbf{t})$

(h) $\mathbf{t} - \mathbf{w} = \mathbf{b} - \mathbf{a}$

107. Writing Give geometric descriptions of (a) vector addition and (b) scalar multiplication.

108. Writing Identify the quantity as a scalar or as a vector. Explain.

(a) The muzzle velocity of a bullet

(b) The price of a company's stock

(c) The air temperature in a room

(d) The weight of an automobile

3.4 Vectors and Dot Products

The dot product of two vectors has many real-life applications. For example, in Exercise 74 on page 298, you will use the dot product to find the force necessary to keep a sport utility vehicle from rolling down a hill.

- **Find the dot product of two vectors and use the properties of the dot product.**
- **Find the angle between two vectors and determine whether two vectors are orthogonal.**
- **Write a vector as the sum of two vector components.**
- **Use vectors to determine the work done by a force.**

The Dot Product of Two Vectors

So far, you have studied two vector operations—vector addition and multiplication by a scalar—each of which yields another vector. In this section, you will study a third vector operation, the **dot product.** This operation yields a scalar, rather than a vector.

Definition of the Dot Product

The **dot product** of $\mathbf{u} = \langle u_1, u_2 \rangle$ and $\mathbf{v} = \langle v_1, v_2 \rangle$ is $\mathbf{u} \cdot \mathbf{v} = u_1v_1 + u_2v_2$.

Properties of the Dot Product

Let **u**, **v**, and **w** be vectors in the plane or in space and let c be a scalar.

1. $\mathbf{u} \cdot \mathbf{v} = \mathbf{v} \cdot \mathbf{u}$
2. $\mathbf{0} \cdot \mathbf{v} = 0$
3. $\mathbf{u} \cdot (\mathbf{v} + \mathbf{w}) = \mathbf{u} \cdot \mathbf{v} + \mathbf{u} \cdot \mathbf{w}$
4. $\mathbf{v} \cdot \mathbf{v} = \|\mathbf{v}\|^2$
5. $c(\mathbf{u} \cdot \mathbf{v}) = c\mathbf{u} \cdot \mathbf{v} = \mathbf{u} \cdot c\mathbf{v}$

For proofs of the properties of the dot product, see Proofs in Mathematics on page 310.

REMARK In Example 1, be sure you see that the dot product of two vectors is a scalar (a real number), not a vector. Moreover, notice that the dot product can be positive, zero, or negative.

EXAMPLE 1 Finding Dot Products

a. $\langle 4, 5 \rangle \cdot \langle 2, 3 \rangle = 4(2) + 5(3)$
$= 8 + 15$
$= 23$

b. $\langle 2, -1 \rangle \cdot \langle 1, 2 \rangle = 2(1) + (-1)(2)$
$= 2 - 2$
$= 0$

c. $\langle 0, 3 \rangle \cdot \langle 4, -2 \rangle = 0(4) + 3(-2)$
$= 0 - 6$
$= -6$

✓ ***Checkpoint*** *Audio-video solution in English & Spanish at LarsonPrecalculus.com*

Find each dot product.

a. $\langle 3, 4 \rangle \cdot \langle 2, -3 \rangle$ **b.** $\langle -3, -5 \rangle \cdot \langle 1, -8 \rangle$ **c.** $\langle -6, 5 \rangle \cdot \langle 5, 6 \rangle$

Anthony Berenyi/Shutterstock.com

EXAMPLE 2 Using Properties of the Dot Product

Let $\mathbf{u} = \langle -1, 3\rangle$, $\mathbf{v} = \langle 2, -4\rangle$, and $\mathbf{w} = \langle 1, -2\rangle$. Find each quantity.

a. $(\mathbf{u} \cdot \mathbf{v})\mathbf{w}$ **b.** $\mathbf{u} \cdot 2\mathbf{v}$ **c.** $\|\mathbf{u}\|$

Solution Begin by finding the dot product of **u** and **v** and the dot product of **u** and **u**.

$$\mathbf{u} \cdot \mathbf{v} = \langle -1, 3\rangle \cdot \langle 2, -4\rangle = -1(2) + 3(-4) = -14$$

$$\mathbf{u} \cdot \mathbf{u} = \langle -1, 3\rangle \cdot \langle -1, 3\rangle = -1(-1) + 3(3) = 10$$

a. $(\mathbf{u} \cdot \mathbf{v})\mathbf{w} = -14\langle 1, -2\rangle = \langle -14, 28\rangle$

b. $\mathbf{u} \cdot 2\mathbf{v} = 2(\mathbf{u} \cdot \mathbf{v}) = 2(-14) = -28$

c. Because $\|\mathbf{u}\|^2 = \mathbf{u} \cdot \mathbf{u} = 10$, it follows that $\|\mathbf{u}\| = \sqrt{\mathbf{u} \cdot \mathbf{u}} = \sqrt{10}$.

Notice that the product in part (a) is a vector, whereas the product in part (b) is a scalar. Can you see why?

✓ ***Checkpoint*** *Audio-video solution in English & Spanish at LarsonPrecalculus.com*

Let $\mathbf{u} = \langle 3, 4\rangle$ and $\mathbf{v} = \langle -2, 6\rangle$. Find each quantity.

a. $(\mathbf{u} \cdot \mathbf{v})\mathbf{v}$ **b.** $\mathbf{u} \cdot (\mathbf{u} + \mathbf{v})$ **c.** $\|\mathbf{v}\|$

The Angle Between Two Vectors

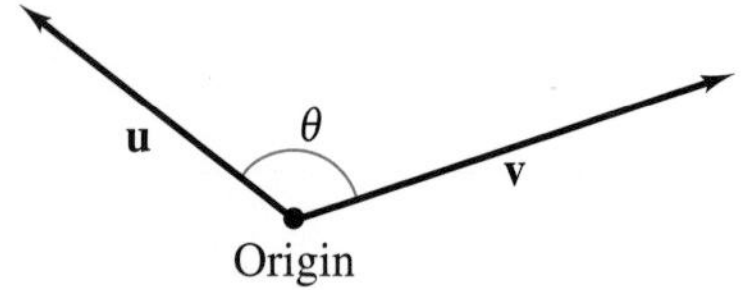

Figure 3.26

The **angle between two nonzero vectors** is the angle θ, $0 \le \theta \le \pi$, between their respective standard position vectors, as shown in Figure 3.26. This angle can be found using the dot product.

> **Angle Between Two Vectors**
>
> If θ is the angle between two nonzero vectors **u** and **v**, then
>
> $$\cos\theta = \frac{\mathbf{u} \cdot \mathbf{v}}{\|\mathbf{u}\|\,\|\mathbf{v}\|}.$$

For a proof of the angle between two vectors, see Proofs in Mathematics on page 310.

EXAMPLE 3 Finding the Angle Between Two Vectors

See LarsonPrecalculus.com for an interactive version of this type of example.

Find the angle θ between $\mathbf{u} = \langle 4, 3\rangle$ and $\mathbf{v} = \langle 3, 5\rangle$ (see Figure 3.27).

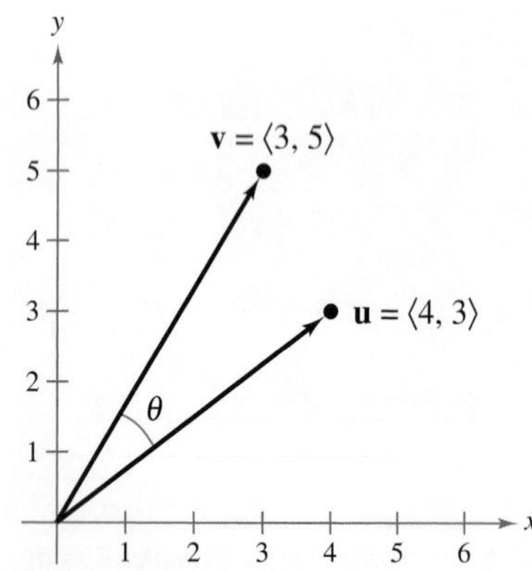

Figure 3.27

Solution

$$\cos\theta = \frac{\mathbf{u} \cdot \mathbf{v}}{\|\mathbf{u}\|\,\|\mathbf{v}\|} = \frac{\langle 4, 3\rangle \cdot \langle 3, 5\rangle}{\|\langle 4, 3\rangle\|\,\|\langle 3, 5\rangle\|} = \frac{4(3) + 3(5)}{\sqrt{4^2 + 3^2}\sqrt{3^2 + 5^2}} = \frac{27}{5\sqrt{34}}$$

This implies that the angle between the two vectors is

$$\theta = \cos^{-1}\frac{27}{5\sqrt{34}} \approx 0.3869 \text{ radian} \approx 22.17°.$$ Use a calculator.

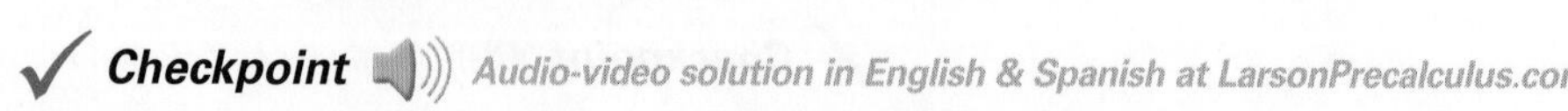

Find the angle θ between $\mathbf{u} = \langle 2, 1\rangle$ and $\mathbf{v} = \langle 1, 3\rangle$.

Rewriting the expression for the angle between two vectors in the form

$$\mathbf{u} \cdot \mathbf{v} = \|\mathbf{u}\| \|\mathbf{v}\| \cos \theta \qquad \text{Alternative form of dot product}$$

produces an alternative way to calculate the dot product. This form shows that $\mathbf{u} \cdot \mathbf{v}$ and $\cos \theta$ always have the same sign, because $\|\mathbf{u}\|$ and $\|\mathbf{v}\|$ are always positive. The figures below show the five possible orientations of two vectors.

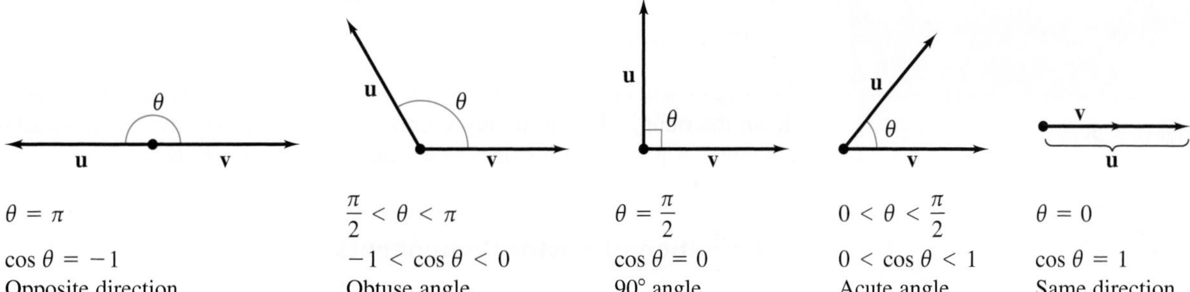

Definition of Orthogonal Vectors

The vectors **u** and **v** are **orthogonal** if and only if $\mathbf{u} \cdot \mathbf{v} = 0$.

The terms *orthogonal* and *perpendicular* have essentially the same meaning—meeting at right angles. Even though the angle between the zero vector and another vector is not defined, it is convenient to extend the definition of orthogonality to include the zero vector. In other words, the zero vector is orthogonal to every vector **u**, because $\mathbf{0} \cdot \mathbf{u} = 0$.

TECHNOLOGY

A graphing utility program that graphs two vectors and finds the angle between them is available at *CengageBrain.com*. Use this program, called "Finding the Angle Between Two Vectors," to verify the solutions to Examples 3 and 4.

EXAMPLE 4 Determining Orthogonal Vectors

Determine whether the vectors $\mathbf{u} = \langle 2, -3 \rangle$ and $\mathbf{v} = \langle 6, 4 \rangle$ are orthogonal.

Solution Find the dot product of the two vectors.

$$\mathbf{u} \cdot \mathbf{v} = \langle 2, -3 \rangle \cdot \langle 6, 4 \rangle = 2(6) + (-3)(4) = 0$$

The dot product is 0, so the two vectors are orthogonal (see figure below).

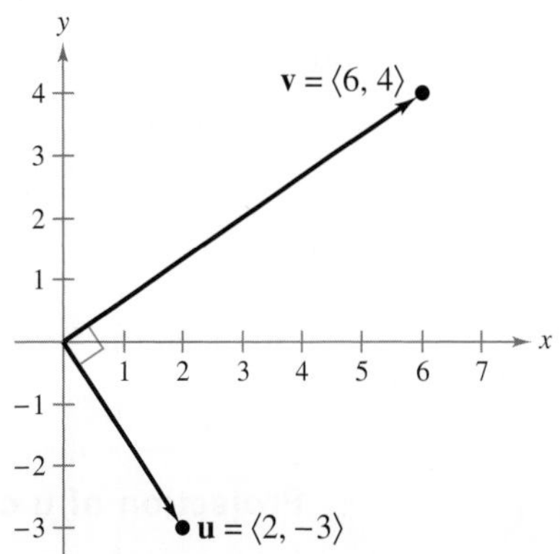

Determine whether the vectors $\mathbf{u} = \langle 6, 10 \rangle$ and $\mathbf{v} = \left\langle -\frac{1}{3}, \frac{1}{5} \right\rangle$ are orthogonal.

Finding Vector Components

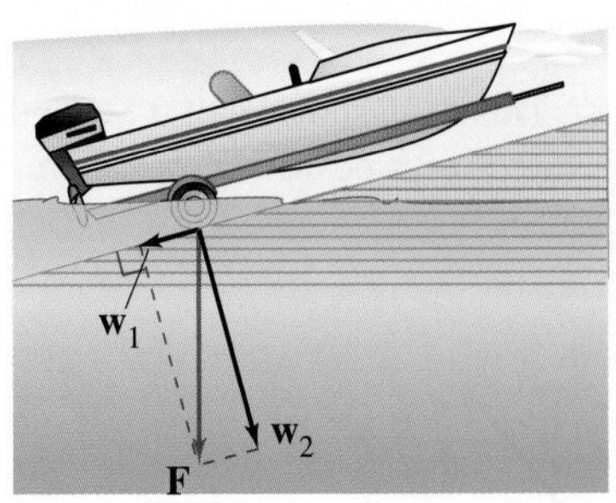

Figure 3.28

You have seen applications in which you add two vectors to produce a resultant vector. Many applications in physics and engineering pose the reverse problem—decomposing a given vector into the sum of two **vector components.**

Consider a boat on an inclined ramp, as shown in Figure 3.28. The force $\mathbf{F}$ due to gravity pulls the boat *down* the ramp and *against* the ramp. These two orthogonal forces $\mathbf{w}_1$ and $\mathbf{w}_2$ are vector components of $\mathbf{F}$. That is,

$$\mathbf{F} = \mathbf{w}_1 + \mathbf{w}_2. \qquad \text{Vector components of } \mathbf{F}$$

The negative of component $\mathbf{w}_1$ represents the force needed to keep the boat from rolling down the ramp, whereas $\mathbf{w}_2$ represents the force that the tires must withstand against the ramp. A procedure for finding $\mathbf{w}_1$ and $\mathbf{w}_2$ is developed below.

Definition of Vector Components

Let $\mathbf{u}$ and $\mathbf{v}$ be nonzero vectors such that

$$\mathbf{u} = \mathbf{w}_1 + \mathbf{w}_2$$

where $\mathbf{w}_1$ and $\mathbf{w}_2$ are orthogonal and $\mathbf{w}_1$ is parallel to (or a scalar multiple of) $\mathbf{v}$, as shown in Figure 3.29. The vectors $\mathbf{w}_1$ and $\mathbf{w}_2$ are **vector components** of $\mathbf{u}$. The vector $\mathbf{w}_1$ is the **projection** of $\mathbf{u}$ onto $\mathbf{v}$ and is denoted by

$$\mathbf{w}_1 = \text{proj}_{\mathbf{v}}\mathbf{u}.$$

The vector $\mathbf{w}_2$ is given by

$$\mathbf{w}_2 = \mathbf{u} - \mathbf{w}_1.$$

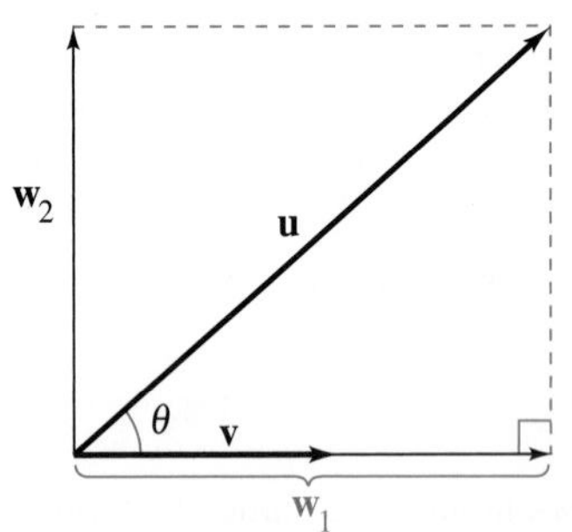

θ is acute.

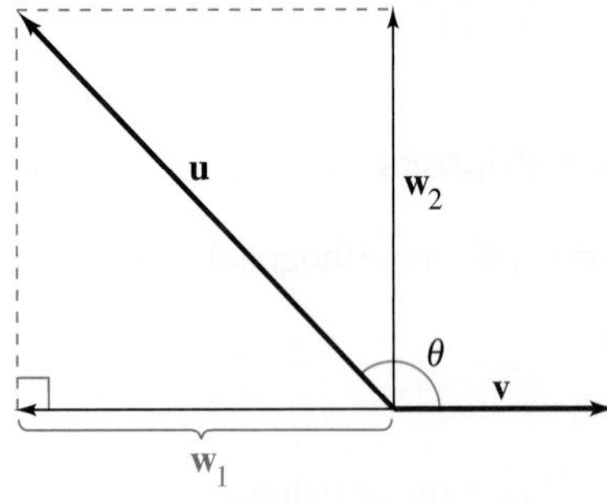

θ is obtuse.
Figure 3.29

To find the component $\mathbf{w}_2$, first find the projection of $\mathbf{u}$ onto $\mathbf{v}$. To find the projection, use the dot product.

$$\mathbf{u} = \mathbf{w}_1 + \mathbf{w}_2$$

$$\mathbf{u} = c\mathbf{v} + \mathbf{w}_2 \qquad \mathbf{w}_1 \text{ is a scalar multiple of } \mathbf{v}.$$

$$\mathbf{u} \cdot \mathbf{v} = (c\mathbf{v} + \mathbf{w}_2) \cdot \mathbf{v} \qquad \text{Dot product of each side with } \mathbf{v}$$

$$\mathbf{u} \cdot \mathbf{v} = c\mathbf{v} \cdot \mathbf{v} + \mathbf{w}_2 \cdot \mathbf{v} \qquad \text{Property 3 of the dot product}$$

$$\mathbf{u} \cdot \mathbf{v} = c\|\mathbf{v}\|^2 + 0 \qquad \mathbf{w}_2 \text{ and } \mathbf{v} \text{ are orthogonal.}$$

So,

$$c = \frac{\mathbf{u} \cdot \mathbf{v}}{\|\mathbf{v}\|^2}$$

and

$$\mathbf{w}_1 = \text{proj}_{\mathbf{v}}\mathbf{u} = c\mathbf{v} = \left(\frac{\mathbf{u} \cdot \mathbf{v}}{\|\mathbf{v}\|^2}\right)\mathbf{v}.$$

Projection of u onto v

Let $\mathbf{u}$ and $\mathbf{v}$ be nonzero vectors. The projection of $\mathbf{u}$ onto $\mathbf{v}$ is given by

$$\text{proj}_{\mathbf{v}}\mathbf{u} = \left(\frac{\mathbf{u} \cdot \mathbf{v}}{\|\mathbf{v}\|^2}\right)\mathbf{v}.$$

EXAMPLE 5 Decomposing a Vector into Components

Find the projection of $\mathbf{u} = \langle 3, -5 \rangle$ onto $\mathbf{v} = \langle 6, 2 \rangle$. Then write $\mathbf{u}$ as the sum of two orthogonal vectors, one of which is $\text{proj}_{\mathbf{v}}\mathbf{u}$.

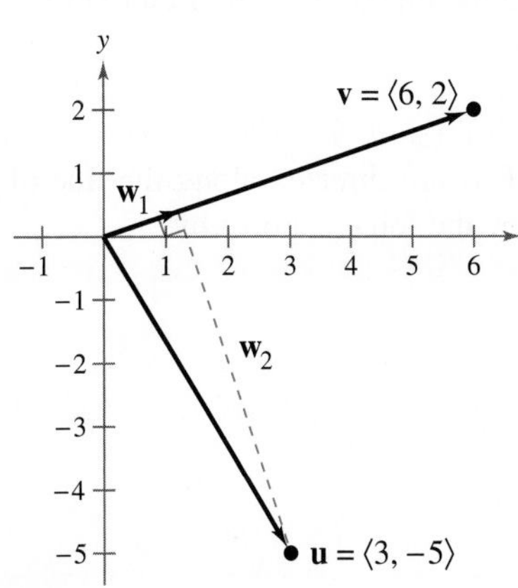

Figure 3.30

Solution The projection of $\mathbf{u}$ onto $\mathbf{v}$ is

$$\mathbf{w}_1 = \text{proj}_{\mathbf{v}}\mathbf{u} = \left(\frac{\mathbf{u} \cdot \mathbf{v}}{\|\mathbf{v}\|^2}\right)\mathbf{v} = \left(\frac{8}{40}\right)\langle 6, 2 \rangle = \left\langle \frac{6}{5}, \frac{2}{5} \right\rangle$$

as shown in Figure 3.30. The component $\mathbf{w}_2$ is

$$\mathbf{w}_2 = \mathbf{u} - \mathbf{w}_1 = \langle 3, -5 \rangle - \left\langle \frac{6}{5}, \frac{2}{5} \right\rangle = \left\langle \frac{9}{5}, -\frac{27}{5} \right\rangle.$$

So,

$$\mathbf{u} = \mathbf{w}_1 + \mathbf{w}_2 = \left\langle \frac{6}{5}, \frac{2}{5} \right\rangle + \left\langle \frac{9}{5}, -\frac{27}{5} \right\rangle = \langle 3, -5 \rangle.$$

Checkpoint Audio-video solution in English & Spanish at LarsonPrecalculus.com

Find the projection of $\mathbf{u} = \langle 3, 4 \rangle$ onto $\mathbf{v} = \langle 8, 2 \rangle$. Then write $\mathbf{u}$ as the sum of two orthogonal vectors, one of which is $\text{proj}_{\mathbf{v}}\mathbf{u}$.

EXAMPLE 6 Finding a Force

A 200-pound cart is on a ramp inclined at 30°, as shown in Figure 3.31. What force is required to keep the cart from rolling down the ramp?

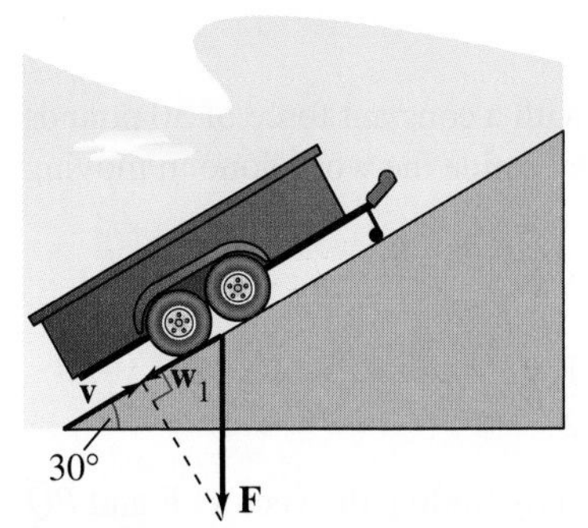

Figure 3.31

Solution The force due to gravity is vertical and downward, so use the vector

$$\mathbf{F} = -200\mathbf{j} \qquad \text{Force due to gravity}$$

to represent the gravitational force. To find the force required to keep the cart from rolling down the ramp, project $\mathbf{F}$ onto a unit vector $\mathbf{v}$ in the direction of the ramp, where

$$\mathbf{v} = (\cos 30°)\mathbf{i} + (\sin 30°)\mathbf{j}$$
$$= \frac{\sqrt{3}}{2}\mathbf{i} + \frac{1}{2}\mathbf{j}. \qquad \text{Unit vector along ramp}$$

So, the projection of $\mathbf{F}$ onto $\mathbf{v}$ is

$$\mathbf{w}_1 = \text{proj}_{\mathbf{v}}\mathbf{F}$$
$$= \left(\frac{\mathbf{F} \cdot \mathbf{v}}{\|\mathbf{v}\|^2}\right)\mathbf{v}$$
$$= (\mathbf{F} \cdot \mathbf{v})\mathbf{v} \qquad \|\mathbf{v}\|^2 = 1$$
$$= (-200)\left(\frac{1}{2}\right)\mathbf{v}$$
$$= -100\left(\frac{\sqrt{3}}{2}\mathbf{i} + \frac{1}{2}\mathbf{j}\right).$$

The magnitude of this force is 100. So, a force of 100 pounds is required to keep the cart from rolling down the ramp.

Checkpoint Audio-video solution in English & Spanish at LarsonPrecalculus.com

Rework Example 6 for a 150-pound cart that is on a ramp inclined at 15°.

Work

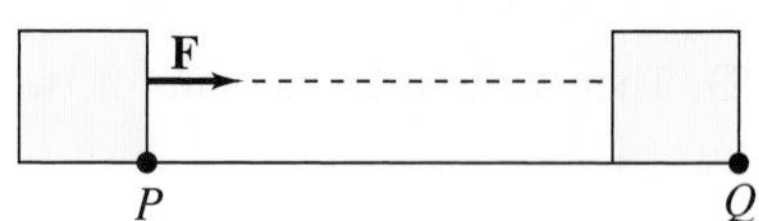

Force acts along the line of motion.
Figure 3.32

The work W done by a constant force $\mathbf{F}$ acting along the line of motion of an object is given by

$$W = (\text{magnitude of force})(\text{distance}) = \|\mathbf{F}\|\|\overrightarrow{PQ}\|$$

as shown in Figure 3.32. When the constant force $\mathbf{F}$ is *not* directed along the line of motion, as shown in Figure 3.33, the work W done by the force is given by

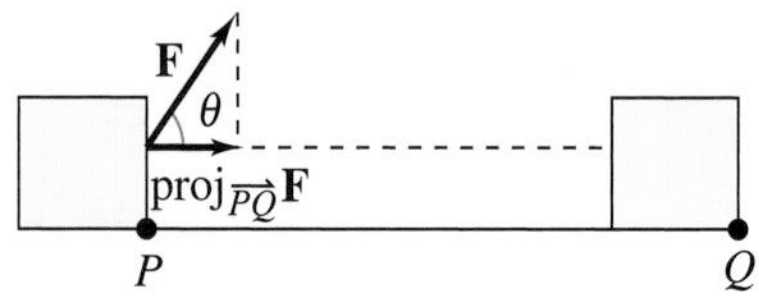

Force acts at angle θ with the line of motion.
Figure 3.33

$$\begin{aligned} W &= \|\text{proj}_{\overrightarrow{PQ}}\mathbf{F}\|\,\|\overrightarrow{PQ}\| && \text{Projection form for work} \\ &= (\cos\theta)\|\mathbf{F}\|\,\|\overrightarrow{PQ}\| && \|\text{proj}_{\overrightarrow{PQ}}\mathbf{F}\| = (\cos\theta)\|\mathbf{F}\| \\ &= \mathbf{F}\cdot\overrightarrow{PQ}. && \text{Alternate form of dot product} \end{aligned}$$

The definition below summarizes the concept of work.

Definition of Work

The **work** W done by a constant force $\mathbf{F}$ as its point of application moves along the vector $\overrightarrow{PQ}$ is given by either formula below.

1. $W = \|\text{proj}_{\overrightarrow{PQ}}\mathbf{F}\|\,\|\overrightarrow{PQ}\|$ — Projection form
2. $W = \mathbf{F}\cdot\overrightarrow{PQ}$ — Dot product form

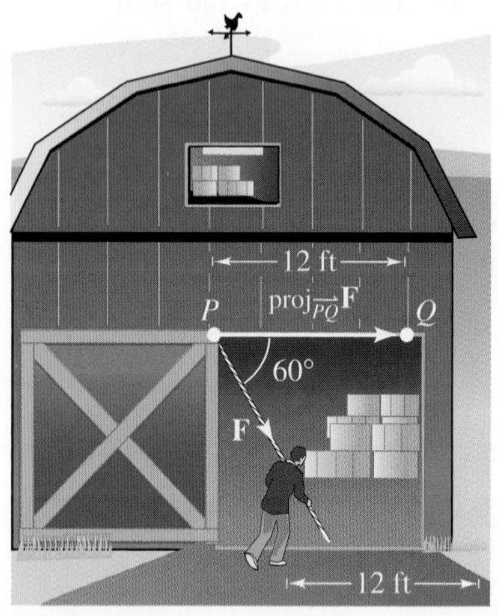

Figure 3.34

EXAMPLE 7 Determining Work

To close a sliding barn door, a person pulls on a rope with a constant force of 50 pounds at a constant angle of 60°, as shown in Figure 3.34. Determine the work done in moving the barn door 12 feet to its closed position.

Solution Use a projection to find the work.

$$W = \|\text{proj}_{\overrightarrow{PQ}}\mathbf{F}\|\,\|\overrightarrow{PQ}\| = (\cos 60°)\|\mathbf{F}\|\,\|\overrightarrow{PQ}\| = \frac{1}{2}(50)(12) = 300 \text{ foot-pounds}$$

So, the work done is 300 foot-pounds. Verify this result by finding the vectors $\mathbf{F}$ and $\overrightarrow{PQ}$ and calculating their dot product.

A person pulls a wagon by exerting a constant force of 35 pounds on a handle that makes a 30° angle with the horizontal. Determine the work done in pulling the wagon 40 feet.

Work is done only when an object is moved. It does not matter how much force is applied—if an object does not move, then no work is done.

Summarize (Section 3.4)

1. State the definition of the dot product and list the properties of the dot product *(page 291)*. For examples of finding dot products and using the properties of the dot product, see Examples 1 and 2.
2. Explain how to find the angle between two vectors and how to determine whether two vectors are orthogonal *(page 292)*. For examples involving the angle between two vectors, see Examples 3 and 4.
3. Explain how to write a vector as the sum of two vector components *(page 294)*. For examples involving vector components, see Examples 5 and 6.
4. State the definition of work *(page 296)*. For an example of determining work, see Example 7.

3.4 Exercises

See CalcChat.com for tutorial help and worked-out solutions to odd-numbered exercises.

Vocabulary: Fill in the blanks.

1. The ________ ________ of two vectors yields a scalar, rather than a vector.
2. The dot product of $\mathbf{u} = \langle u_1, u_2 \rangle$ and $\mathbf{v} = \langle v_1, v_2 \rangle$ is $\mathbf{u} \cdot \mathbf{v} =$ ________.
3. If θ is the angle between two nonzero vectors **u** and **v**, then $\cos \theta =$ ________.
4. The vectors **u** and **v** are ________ if and only if $\mathbf{u} \cdot \mathbf{v} = 0$.
5. The projection of **u** onto **v** is given by $\text{proj}_{\mathbf{v}}\mathbf{u} =$ ________.
6. The work W done by a constant force **F** as its point of application moves along the vector $\overrightarrow{PQ}$ is given by $W =$ ________ or $W =$ ________.

Skills and Applications

Finding a Dot Product In Exercises 7–12, find $\mathbf{u} \cdot \mathbf{v}$.

7. $\mathbf{u} = \langle 7, 1 \rangle$, $\mathbf{v} = \langle -3, 2 \rangle$
8. $\mathbf{u} = \langle 6, 10 \rangle$, $\mathbf{v} = \langle -2, 3 \rangle$
9. $\mathbf{u} = \langle -6, 2 \rangle$, $\mathbf{v} = \langle 1, 3 \rangle$
10. $\mathbf{u} = \langle -2, 5 \rangle$, $\mathbf{v} = \langle -1, -8 \rangle$
11. $\mathbf{u} = 4\mathbf{i} - 2\mathbf{j}$, $\mathbf{v} = \mathbf{i} - \mathbf{j}$
12. $\mathbf{u} = \mathbf{i} - 2\mathbf{j}$, $\mathbf{v} = -2\mathbf{i} - \mathbf{j}$

Using Properties of the Dot Product In Exercises 13–22, use the vectors $\mathbf{u} = \langle 3, 3 \rangle$, $\mathbf{v} = \langle -4, 2 \rangle$, and $\mathbf{w} = \langle 3, -1 \rangle$ to find the quantity. State whether the result is a vector or a scalar.

13. $\mathbf{u} \cdot \mathbf{u}$
14. $3\mathbf{u} \cdot \mathbf{v}$
15. $(\mathbf{u} \cdot \mathbf{v})\mathbf{v}$
16. $(\mathbf{u} \cdot 2\mathbf{v})\mathbf{w}$
17. $(\mathbf{v} \cdot \mathbf{0})\mathbf{w}$
18. $(\mathbf{u} + \mathbf{v}) \cdot \mathbf{0}$
19. $\|\mathbf{w}\| - 1$
20. $2 - \|\mathbf{u}\|$
21. $(\mathbf{u} \cdot \mathbf{v}) - (\mathbf{u} \cdot \mathbf{w})$
22. $(\mathbf{v} \cdot \mathbf{u}) - (\mathbf{w} \cdot \mathbf{v})$

Finding the Magnitude of a Vector In Exercises 23–28, use the dot product to find the magnitude of **u**.

23. $\mathbf{u} = \langle -8, 15 \rangle$
24. $\mathbf{u} = \langle 4, -6 \rangle$
25. $\mathbf{u} = 20\mathbf{i} + 25\mathbf{j}$
26. $\mathbf{u} = 12\mathbf{i} - 16\mathbf{j}$
27. $\mathbf{u} = 6\mathbf{j}$
28. $\mathbf{u} = -21\mathbf{i}$

Finding the Angle Between Two Vectors In Exercises 29–38, find the angle θ (in radians) between the vectors.

29. $\mathbf{u} = \langle 1, 0 \rangle$, $\mathbf{v} = \langle 0, -2 \rangle$
30. $\mathbf{u} = \langle 3, 2 \rangle$, $\mathbf{v} = \langle 4, 0 \rangle$
31. $\mathbf{u} = 3\mathbf{i} + 4\mathbf{j}$, $\mathbf{v} = -2\mathbf{j}$
32. $\mathbf{u} = 2\mathbf{i} - 3\mathbf{j}$, $\mathbf{v} = \mathbf{i} - 2\mathbf{j}$
33. $\mathbf{u} = 2\mathbf{i} - \mathbf{j}$, $\mathbf{v} = 6\mathbf{i} - 3\mathbf{j}$
34. $\mathbf{u} = 5\mathbf{i} + 5\mathbf{j}$, $\mathbf{v} = -6\mathbf{i} + 6\mathbf{j}$
35. $\mathbf{u} = -6\mathbf{i} - 3\mathbf{j}$, $\mathbf{v} = -8\mathbf{i} + 4\mathbf{j}$
36. $\mathbf{u} = 2\mathbf{i} - 3\mathbf{j}$, $\mathbf{v} = 4\mathbf{i} + 3\mathbf{j}$
37. $\mathbf{u} = \cos\left(\frac{\pi}{3}\right)\mathbf{i} + \sin\left(\frac{\pi}{3}\right)\mathbf{j}$, $\mathbf{v} = \cos\left(\frac{3\pi}{4}\right)\mathbf{i} + \sin\left(\frac{3\pi}{4}\right)\mathbf{j}$
38. $\mathbf{u} = \cos\left(\frac{\pi}{4}\right)\mathbf{i} + \sin\left(\frac{\pi}{4}\right)\mathbf{j}$, $\mathbf{v} = \cos\left(\frac{5\pi}{4}\right)\mathbf{i} + \sin\left(\frac{5\pi}{4}\right)\mathbf{j}$

Finding the Angle Between Two Vectors In Exercises 39–42, find the angle θ (in degrees) between the vectors.

39. $\mathbf{u} = 3\mathbf{i} + 4\mathbf{j}$, $\mathbf{v} = -7\mathbf{i} + 5\mathbf{j}$
40. $\mathbf{u} = 6\mathbf{i} - 3\mathbf{j}$, $\mathbf{v} = -4\mathbf{i} - 4\mathbf{j}$
41. $\mathbf{u} = -5\mathbf{i} - 5\mathbf{j}$, $\mathbf{v} = -8\mathbf{i} + 8\mathbf{j}$
42. $\mathbf{u} = 2\mathbf{i} - 3\mathbf{j}$, $\mathbf{v} = 8\mathbf{i} + 3\mathbf{j}$

Finding the Angles in a Triangle In Exercises 43–46, use vectors to find the interior angles of the triangle with the given vertices.

43. $(1, 2), (3, 4), (2, 5)$
44. $(-3, -4), (1, 7), (8, 2)$
45. $(-3, 0), (2, 2), (0, 6)$
46. $(-3, 5), (-1, 9), (7, 9)$

Using the Angle Between Two Vectors In Exercises 47–50, find $\mathbf{u} \cdot \mathbf{v}$, where θ is the angle between **u** and **v**.

47. $\|\mathbf{u}\| = 4$, $\|\mathbf{v}\| = 10$, $\theta = 2\pi/3$
48. $\|\mathbf{u}\| = 4$, $\|\mathbf{v}\| = 12$, $\theta = \pi/3$
49. $\|\mathbf{u}\| = 100$, $\|\mathbf{v}\| = 250$, $\theta = \pi/6$
50. $\|\mathbf{u}\| = 9$, $\|\mathbf{v}\| = 36$, $\theta = 3\pi/4$

Determining Orthogonal Vectors In Exercises 51–56, determine whether u and v are orthogonal.

51. $\mathbf{u} = \langle 3, 15 \rangle$, $\mathbf{v} = \langle -1, 5 \rangle$

52. $\mathbf{u} = \langle 30, 12 \rangle$, $\mathbf{v} = \left\langle \frac{1}{2}, -\frac{5}{4} \right\rangle$

53. $\mathbf{u} = 2\mathbf{i} - 2\mathbf{j}$, $\mathbf{v} = -\mathbf{i} - \mathbf{j}$

54. $\mathbf{u} = \frac{1}{4}(3\mathbf{i} - \mathbf{j})$, $\mathbf{v} = 5\mathbf{i} + 6\mathbf{j}$

55. $\mathbf{u} = \mathbf{i}$, $\mathbf{v} = -2\mathbf{i} + 2\mathbf{j}$

56. $\mathbf{u} = \langle \cos\theta, \sin\theta \rangle$, $\mathbf{v} = \langle \sin\theta, -\cos\theta \rangle$

Decomposing a Vector into Components In Exercises 57–60, find the projection of u onto v. Then write u as the sum of two orthogonal vectors, one of which is $\text{proj}_{\mathbf{v}}\mathbf{u}$.

57. $\mathbf{u} = \langle 2, 2 \rangle$, $\mathbf{v} = \langle 6, 1 \rangle$

58. $\mathbf{u} = \langle 0, 3 \rangle$, $\mathbf{v} = \langle 2, 15 \rangle$

59. $\mathbf{u} = \langle 4, 2 \rangle$, $\mathbf{v} = \langle 1, -2 \rangle$

60. $\mathbf{u} = \langle -3, -2 \rangle$, $\mathbf{v} = \langle -4, -1 \rangle$

Finding the Projection of u onto v In Exercises 61–64, use the graph to find the projection of u onto v. (The terminal points of the vectors in standard position are given.) Use the formula for the projection of u onto v to verify your result.

61.

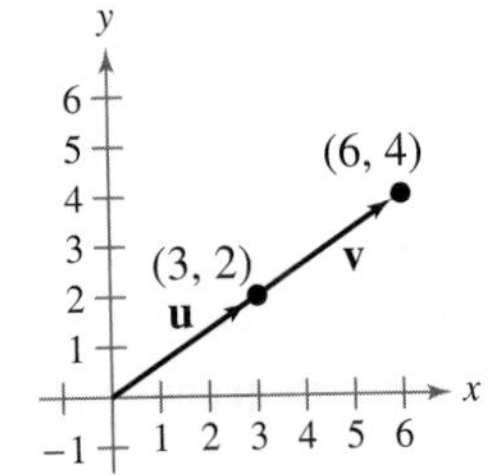

62.

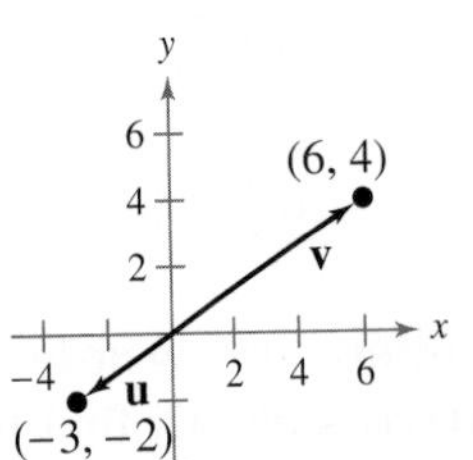

63.

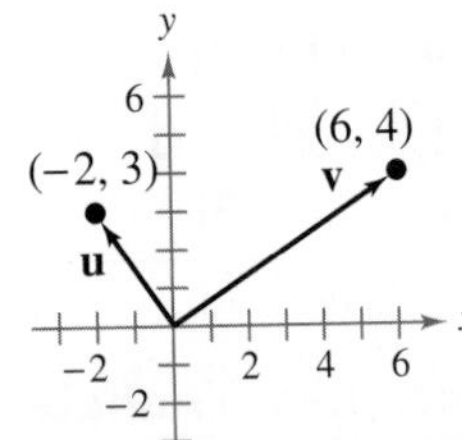

64.

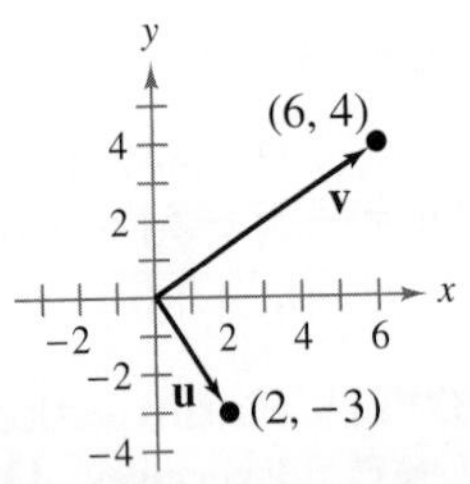

Finding Orthogonal Vectors In Exercises 65–68, find two vectors in opposite directions that are orthogonal to the vector u. (There are many correct answers.)

65. $\mathbf{u} = \langle 3, 5 \rangle$

66. $\mathbf{u} = \langle -8, 3 \rangle$

67. $\mathbf{u} = \frac{1}{2}\mathbf{i} - \frac{2}{3}\mathbf{j}$

68. $\mathbf{u} = -\frac{5}{2}\mathbf{i} - 3\mathbf{j}$

Work In Exercises 69 and 70, determine the work done in moving a particle from P to Q when the magnitude and direction of the force are given by v.

69. $P(0, 0)$, $Q(4, 7)$, $\mathbf{v} = \langle 1, 4 \rangle$

70. $P(1, 3)$, $Q(-3, 5)$, $\mathbf{v} = -2\mathbf{i} + 3\mathbf{j}$

71. **Business** The vector $\mathbf{u} = \langle 1225, 2445 \rangle$ gives the numbers of hours worked by employees of a temporary work agency at two pay levels. The vector $\mathbf{v} = \langle 12.20, 8.50 \rangle$ gives the hourly wage (in dollars) paid at each level, respectively.
    (a) Find the dot product $\mathbf{u} \cdot \mathbf{v}$ and interpret the result in the context of the problem.
    (b) Identify the vector operation used to increase wages by 2%.

72. **Revenue** The vector $\mathbf{u} = \langle 3140, 2750 \rangle$ gives the numbers of hamburgers and hot dogs, respectively, sold at a fast-food stand in one month. The vector $\mathbf{v} = \langle 2.25, 1.75 \rangle$ gives the prices (in dollars) of the food items, respectively.
    (a) Find the dot product $\mathbf{u} \cdot \mathbf{v}$ and interpret the result in the context of the problem.
    (b) Identify the vector operation used to increase the prices by 2.5%.

73. **Physics** A truck with a gross weight of 30,000 pounds is parked on a slope of $d°$ (see figure). Assume that the only force to overcome is the force of gravity.

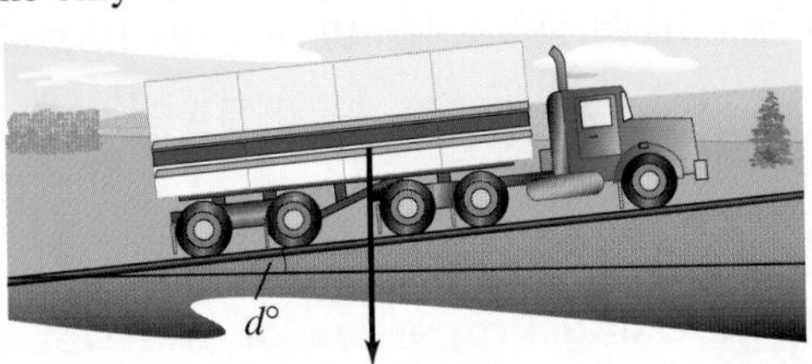

(a) Find the force required to keep the truck from rolling down the hill in terms of d.
(b) Use a graphing utility to complete the table.

d	0°	1°	2°	3°	4°	5°
Force						

d	6°	7°	8°	9°	10°
Force					

(c) Find the force perpendicular to the hill when $d = 5°$.

74. **Braking Load** A sport utility vehicle with a gross weight of 5400 pounds is parked on a slope of 10°. Assume that the only force to overcome is the force of gravity. Find the force required to keep the vehicle from rolling down the hill. Find the force perpendicular to the hill.

Anthony Berenyi/Shutterstock.com

75. **Work** Determine the work done by a person lifting a 245-newton bag of sugar 3 meters.

76. **Work** Determine the work done by a crane lifting a 2400-pound car 5 feet.

77. **Work** A constant force of 45 pounds, exerted at an angle of 30° with the horizontal, is required to slide a table across a floor. Determine the work done in sliding the table 20 feet.

78. **Work** A constant force of 50 pounds, exerted at an angle of 25° with the horizontal, is required to slide a desk across a floor. Determine the work done in sliding the desk 15 feet.

79. **Work** A tractor pulls a log 800 meters, and the tension in the cable connecting the tractor and the log is approximately 15,691 newtons. The direction of the constant force is 35° above the horizontal. Determine the work done in pulling the log.

80. **Work** One of the events in a strength competition is to pull a cement block 100 feet. One competitor pulls the block by exerting a constant force of 250 pounds on a rope attached to the block at an angle of 30° with the horizontal (see figure). Determine the work done in pulling the block.

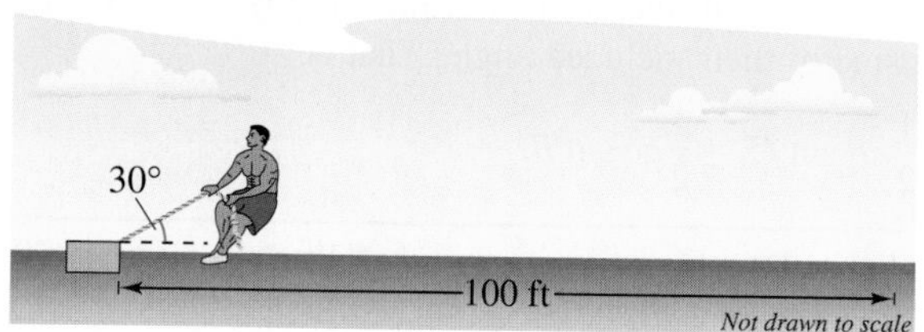

81. **Work** A child pulls a toy wagon by exerting a constant force of 25 pounds on a handle that makes a 20° angle with the horizontal (see figure). Determine the work done in pulling the wagon 50 feet.

82. **Work** A ski patroller pulls a rescue toboggan across a flat snow surface by exerting a constant force of 35 pounds on a handle that makes a 22° angle with the horizontal (see figure). Determine the work done in pulling the toboggan 200 feet.

Exploration

True or False? **In Exercises 83 and 84, determine whether the statement is true or false. Justify your answer.**

83. The work W done by a constant force $\mathbf{F}$ acting along the line of motion of an object is represented by a vector.

84. A sliding door moves along the line of vector $\overrightarrow{PQ}$. If a force is applied to the door along a vector that is orthogonal to $\overrightarrow{PQ}$, then no work is done.

Error Analysis **In Exercises 85 and 86, describe the error in finding the quantity when $\mathbf{u} = \langle 2, -1 \rangle$ and $\mathbf{v} = \langle -3, 5 \rangle$.**

85. $\mathbf{v} \cdot \mathbf{0} = \langle 0, 0 \rangle$ ✗

86. $\mathbf{u} \cdot 2\mathbf{v} = \langle 2, -1 \rangle \cdot \langle -6, 10 \rangle$
$= 2(-6) - (-1)(10)$
$= -12 + 10$
$= -2$ ✗

Finding an Unknown Vector Component **In Exercises 87 and 88, find the value of k such that vectors u and v are orthogonal.**

87. $\mathbf{u} = 8\mathbf{i} + 4\mathbf{j}$
$\mathbf{v} = 2\mathbf{i} - k\mathbf{j}$

88. $\mathbf{u} = -3k\mathbf{i} + 5\mathbf{j}$
$\mathbf{v} = 2\mathbf{i} - 4\mathbf{j}$

89. **Think About It** Let $\mathbf{u}$ be a unit vector. What is the value of $\mathbf{u} \cdot \mathbf{u}$? Explain.

90. **HOW DO YOU SEE IT?** What is known about θ, the angle between two nonzero vectors $\mathbf{u}$ and $\mathbf{v}$, under each condition (see figure)?

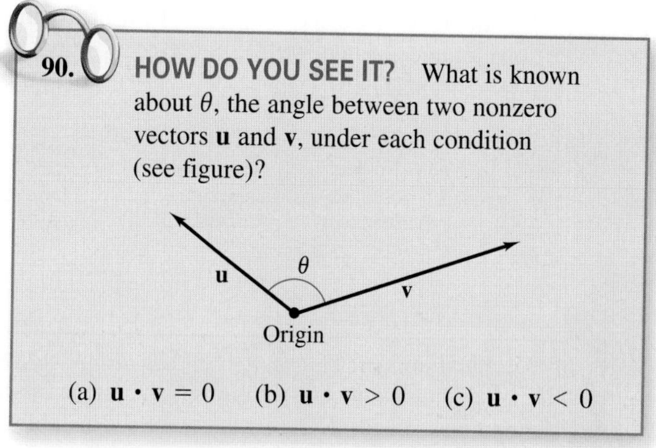

(a) $\mathbf{u} \cdot \mathbf{v} = 0$ (b) $\mathbf{u} \cdot \mathbf{v} > 0$ (c) $\mathbf{u} \cdot \mathbf{v} < 0$

91. **Think About It** What can be said about the vectors $\mathbf{u}$ and $\mathbf{v}$ under each condition?

(a) The projection of $\mathbf{u}$ onto $\mathbf{v}$ equals $\mathbf{u}$.

(b) The projection of $\mathbf{u}$ onto $\mathbf{v}$ equals $\mathbf{0}$.

92. **Proof** Use vectors to prove that the diagonals of a rhombus are perpendicular.

93. **Proof** Prove that

$$\|\mathbf{u} - \mathbf{v}\|^2 = \|\mathbf{u}\|^2 + \|\mathbf{v}\|^2 - 2\mathbf{u} \cdot \mathbf{v}.$$

Chapter Summary

	What Did You Learn?	Explanation/Examples	Review Exercises
Section 3.1	Use the Law of Sines to solve oblique triangles (AAS or ASA) *(p. 262).*	**Law of Sines** If ABC is a triangle with sides a, b, and c, then $\frac{a}{\sin A} = \frac{b}{\sin B} = \frac{c}{\sin C}.$ *A* is acute. *A* is obtuse.	1–12
	Use the Law of Sines to solve oblique triangles (SSA) *(p. 264).*	If two sides and one opposite angle are given, then three possible situations can occur: (1) no such triangle exists (see Example 4), (2) one such triangle exists (see Example 3), or (3) two distinct triangles exist that satisfy the conditions (see Example 5).	1–12
	Find the areas of oblique triangles *(p. 266).*	The area of any triangle is one-half the product of the lengths of two sides times the sine of their included angle. That is, $\text{Area} = \frac{1}{2}bc \sin A = \frac{1}{2}ab \sin C = \frac{1}{2}ac \sin B.$	13–18
	Use the Law of Sines to model and solve real-life problems *(p. 267).*	The Law of Sines can be used to approximate the total distance of a boat race course. (See Example 7.)	19, 20
Section 3.2	Use the Law of Cosines to solve oblique triangles (SSS or SAS) *(p. 271).*	**Law of Cosines** **Standard Form** / **Alternative Form** $a^2 = b^2 + c^2 - 2bc \cos A$ $\cos A = \frac{b^2 + c^2 - a^2}{2bc}$ $b^2 = a^2 + c^2 - 2ac \cos B$ $\cos B = \frac{a^2 + c^2 - b^2}{2ac}$ $c^2 = a^2 + b^2 - 2ab \cos C$ $\cos C = \frac{a^2 + b^2 - c^2}{2ab}$	21–36
	Use the Law of Cosines to model and solve real-life problems *(p. 273).*	The Law of Cosines can be used to find the distance between the pitcher's mound and first base on a women's softball field. (See Example 3.)	37–40
	Use Heron's Area Formula to find areas of triangles *(p. 274).*	**Heron's Area Formula:** Given any triangle with sides of lengths a, b, and c, the area of the triangle is $\text{Area} = \sqrt{s(s-a)(s-b)(s-c)}$ where $s = \frac{a+b+c}{2}.$	41–44

	What Did You Learn?	Explanation/Examples	Review Exercises
Section 3.3	Represent vectors as directed line segments *(p. 278).*	Terminal point Q; $\overrightarrow{PQ}$; P Initial point	45, 46
	Write component forms of vectors *(p. 279).*	The component form of the vector with initial point $P(p_1, p_2)$ and terminal point $Q(q_1, q_2)$ is given by $\overrightarrow{PQ} = \langle q_1 - p_1, q_2 - p_2 \rangle = \langle v_1, v_2 \rangle = \mathbf{v}$.	47–50
	Perform basic vector operations and represent vector operations graphically *(p. 280).*	Let $\mathbf{u} = \langle u_1, u_2 \rangle$ and $\mathbf{v} = \langle v_1, v_2 \rangle$ be vectors and let k be a scalar (a real number). $\mathbf{u} + \mathbf{v} = \langle u_1 + v_1, u_2 + v_2 \rangle$ $k\mathbf{u} = \langle ku_1, ku_2 \rangle$ $-\mathbf{v} = \langle -v_1, -v_2 \rangle$ $\mathbf{u} - \mathbf{v} = \langle u_1 - v_1, u_2 - v_2 \rangle$	51–58, 63–68
	Write vectors as linear combinations of unit vectors *(p. 282).*	The vector sum $\mathbf{v} = \langle v_1, v_2 \rangle = v_1\langle 1, 0 \rangle + v_2\langle 0, 1 \rangle = v_1\mathbf{i} + v_2\mathbf{j}$ is a linear combination of the vectors **i** and **j**.	59–62
	Find direction angles of vectors *(p. 284).*	If $\mathbf{u} = a\mathbf{i} + b\mathbf{j}$, then the direction angle is determined from $\tan \theta = \frac{b}{a}$.	69–74
	Use vectors to model and solve real-life problems *(p. 285).*	Vectors can be used to find the resultant speed and true direction of an airplane. (See Example 12.)	75, 76
Section 3.4	Find the dot product of two vectors and use the properties of the dot product *(p. 291).*	The dot product of $\mathbf{u} = \langle u_1, u_2 \rangle$ and $\mathbf{v} = \langle v_1, v_2 \rangle$ is $\mathbf{u} \cdot \mathbf{v} = u_1v_1 + u_2v_2$.	77–88
	Find the angle between two vectors and determine whether two vectors are orthogonal *(p. 292).*	If θ is the angle between two nonzero vectors **u** and **v**, then $\cos \theta = \frac{\mathbf{u} \cdot \mathbf{v}}{\|\mathbf{u}\|\|\mathbf{v}\|}$. The vectors **u** and **v** are orthogonal if and only if $\mathbf{u} \cdot \mathbf{v} = 0$.	89–96
	Write a vector as the sum of two vector components *(p. 294).*	Many applications in physics and engineering require the decomposition of a given vector into the sum of two vector components. (See Example 6.)	97–100
	Use vectors to determine the work done by a force *(p. 296).*	The work W done by a constant force **F** as its point of application moves along the vector $\overrightarrow{PQ}$ is given by either formula below. **1.** $W = \|\text{proj}_{\overrightarrow{PQ}} \mathbf{F}\|\|\overrightarrow{PQ}\|$ **2.** $W = \mathbf{F} \cdot \overrightarrow{PQ}$	101–104

Review Exercises

See CalcChat.com for tutorial help and worked-out solutions to odd-numbered exercises.

3.1 Using the Law of Sines **In Exercises 1–12, use the Law of Sines to solve (if possible) the triangle. If two solutions exist, find both. Round your answers to two decimal places.**

1. 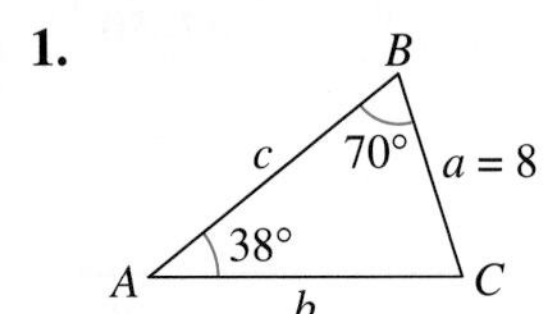

2. 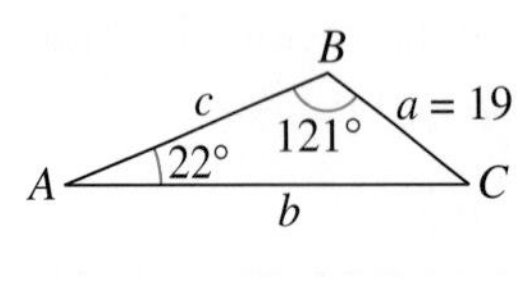

3. $B = 72°, \quad C = 82°, \quad b = 54$
4. $B = 10°, \quad C = 20°, \quad c = 33$
5. $A = 16°, \quad B = 98°, \quad c = 8.4$
6. $A = 95°, \quad B = 45°, \quad c = 104.8$
7. $A = 24°, \quad C = 48°, \quad b = 27.5$
8. $B = 64°, \quad C = 36°, \quad a = 367$
9. $B = 150°, \quad b = 30, \quad c = 10$
10. $B = 150°, \quad a = 10, \quad b = 3$
11. $A = 75°, \quad a = 51.2, \quad b = 33.7$
12. $B = 25°, \quad a = 6.2, \quad b = 4$

Finding the Area of a Triangle **In Exercises 13–18, find the area of the triangle. Round your answers to one decimal place.**

13. $A = 33°, \quad b = 7, \quad c = 10$
14. $B = 80°, \quad a = 4, \quad c = 8$
15. $C = 119°, \quad a = 18, \quad b = 6$
16. $A = 11°, \quad b = 22, \quad c = 21$
17. $B = 72° \, 30', \quad a = 105, \quad c = 64$
18. $C = 84° \, 30', \quad a = 16, \quad b = 20$

19. **Height** A tree stands on a hillside of slope 28° from the horizontal. From a point 75 feet down the hill, the angle of elevation to the top of the tree is 45° (see figure). Find the height of the tree.

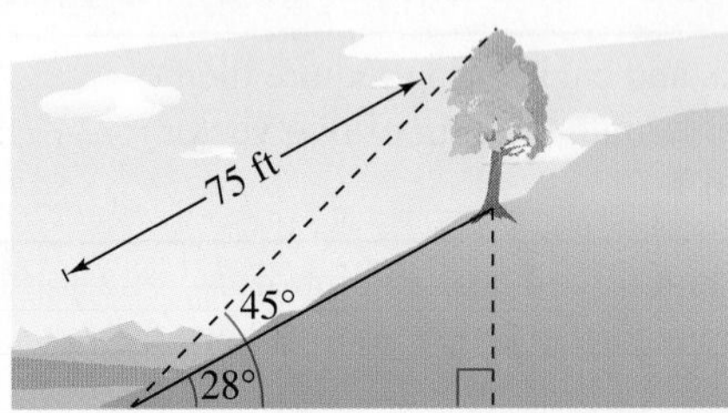

20. **River Width** A surveyor finds that a pier on the opposite bank of a river flowing due east has a bearing of N 22° 30′ E from a certain point and a bearing of N 15° W from a point 400 feet downstream. Find the width of the river.

3.2 Using the Law of Cosines **In Exercises 21–30, use the Law of Cosines to solve the triangle. Round your answers to two decimal places.**

21.

22. 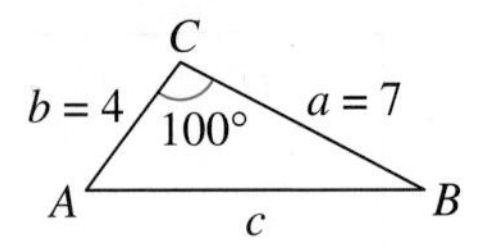

23. $a = 6, \quad b = 9, \quad c = 14$
24. $a = 75, \quad b = 50, \quad c = 110$
25. $a = 2.5, \quad b = 5.0, \quad c = 4.5$
26. $a = 16.4, \quad b = 8.8, \quad c = 12.2$
27. $B = 108°, \quad a = 11, \quad c = 11$
28. $B = 150°, \quad a = 10, \quad c = 20$
29. $C = 43°, \quad a = 22.5, \quad b = 31.4$
30. $A = 62°, \quad b = 11.34, \quad c = 19.52$

Solving a Triangle **In Exercises 31–36, determine whether the Law of Cosines is needed to solve the triangle. Then solve (if possible) the triangle. If two solutions exist, find both. Round your answers to two decimal places.**

31. $C = 64°, \quad b = 9, \quad c = 13$
32. $B = 52°, \quad a = 4, \quad c = 5$
33. $a = 13, \quad b = 15, \quad c = 24$
34. $A = 44°, \quad B = 31°, \quad c = 2.8$
35. $B = 12°, \quad C = 7°, \quad a = 160$
36. $A = 33°, \quad b = 7, \quad c = 10$

37. **Geometry** The lengths of the sides of a parallelogram are 10 feet and 16 feet. Find the lengths of the diagonals of the parallelogram when the sides intersect at an angle of 28°.

38. **Geometry** The lengths of the diagonals of a parallelogram are 30 meters and 40 meters. Find the lengths of the sides of the parallelogram when the diagonals intersect at an angle of 34°.

39. **Surveying** To approximate the length of a marsh, a surveyor walks 425 meters from point A to point B. Then the surveyor turns 65° and walks 300 meters to point C (see figure). Find the length AC of the marsh.

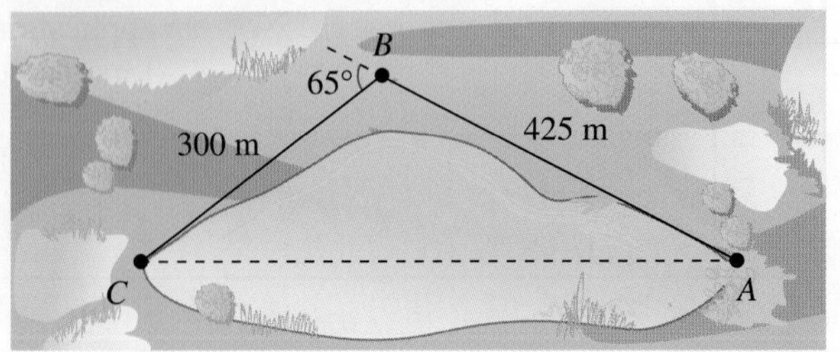

40. Air Navigation Two planes leave an airport at approximately the same time. One plane flies 425 miles per hour at a bearing of 355°, and the other plane flies 530 miles per hour at a bearing of 67° (see figure). Determine the distance between the planes after they fly for 2 hours.

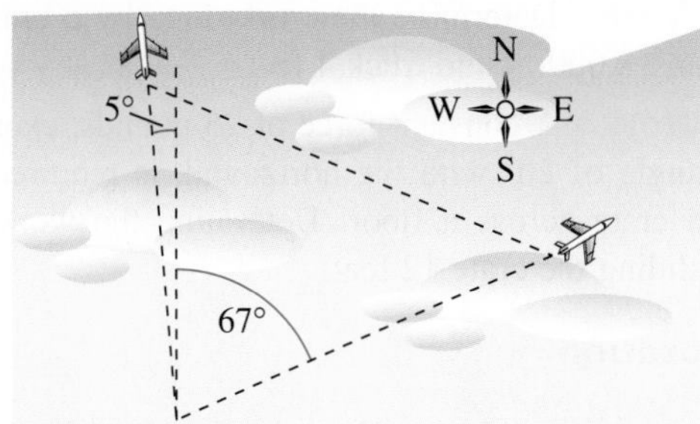

Using Heron's Area Formula **In Exercises 41–44, use Heron's Area Formula to find the area of the triangle.**

41. $a = 3,\quad b = 6,\quad c = 8$

42. $a = 15,\quad b = 8,\quad c = 10$

43. $a = 12.3,\quad b = 15.8,\quad c = 3.7$

44. $a = \frac{4}{5},\quad b = \frac{3}{4},\quad c = \frac{5}{8}$

 **3.3 Determining Whether Two Vectors Are Equivalent** **In Exercises 45 and 46, determine whether u and v are equivalent. Explain.**

45.

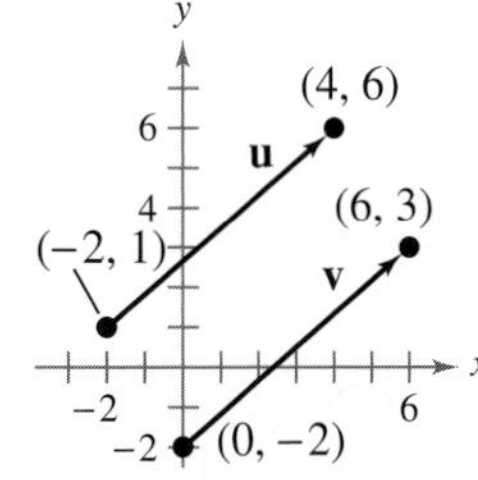

46.

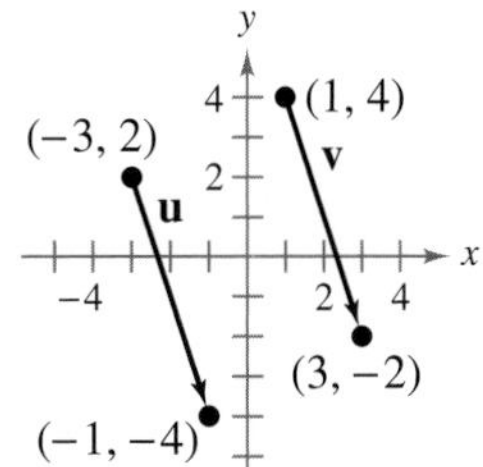

Finding the Component Form of a Vector **In Exercises 47–50, find the component form and magnitude of the vector v.**

47.

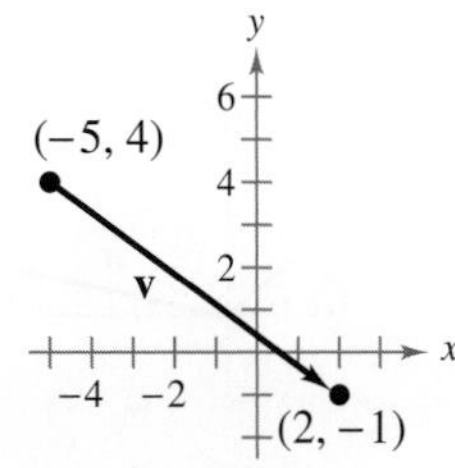

48.

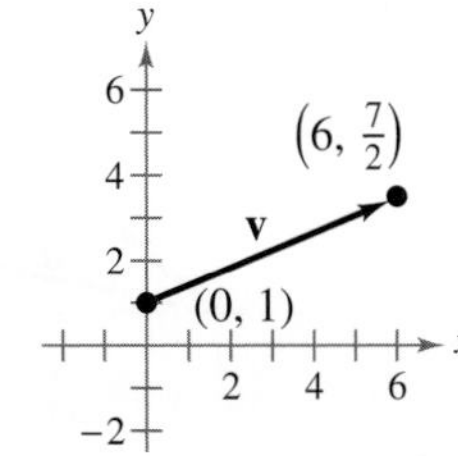

49. Initial point: $(0, 10)$

Terminal point: $(7, 3)$

50. Initial point: $(1, 5)$

Terminal point: $(15, 9)$

Vector Operations **In Exercises 51–58, find (a) u + v, (b) u − v, (c) 4u, and (d) 3v + 5u. Then sketch each resultant vector.**

51. $\mathbf{u} = \langle -1, -3 \rangle,\quad \mathbf{v} = \langle -3, 6 \rangle$

52. $\mathbf{u} = \langle 4, 5 \rangle,\quad \mathbf{v} = \langle 0, -1 \rangle$

53. $\mathbf{u} = \langle -5, 2 \rangle,\quad \mathbf{v} = \langle 4, 4 \rangle$

54. $\mathbf{u} = \langle 1, -8 \rangle,\quad \mathbf{v} = \langle 3, -2 \rangle$

55. $\mathbf{u} = 2\mathbf{i} - \mathbf{j},\quad \mathbf{v} = 5\mathbf{i} + 3\mathbf{j}$

56. $\mathbf{u} = -7\mathbf{i} - 3\mathbf{j},\quad \mathbf{v} = 4\mathbf{i} - \mathbf{j}$

57. $\mathbf{u} = 4\mathbf{i},\quad \mathbf{v} = -\mathbf{i} + 6\mathbf{j}$

58. $\mathbf{u} = -6\mathbf{j},\quad \mathbf{v} = \mathbf{i} + \mathbf{j}$

Writing a Linear Combination of Unit Vectors **In Exercises 59–62, the initial and terminal points of a vector are given. Write the vector as a linear combination of the standard unit vectors i and j.**

	Initial Point	Terminal Point
59.	$(2, 3)$	$(1, 8)$
60.	$(4, -2)$	$(-2, -10)$
61.	$(3, 4)$	$(9, 8)$
62.	$(-2, 7)$	$(5, -9)$

Vector Operations **In Exercises 63–68, find the component form of w and sketch the specified vector operations geometrically, where u = 6i − 5j and v = 10i + 3j.**

63. $\mathbf{w} = 3\mathbf{v}$

64. $\mathbf{w} = \frac{1}{2}\mathbf{v}$

65. $\mathbf{w} = 2\mathbf{u} + \mathbf{v}$

66. $\mathbf{w} = 4\mathbf{u} - 5\mathbf{v}$

67. $\mathbf{w} = 5\mathbf{u} - 4\mathbf{v}$

68. $\mathbf{w} = -3\mathbf{u} + 2\mathbf{v}$

Finding the Direction Angle of a Vector **In Exercises 69–72, find the magnitude and direction angle of the vector v.**

69. $\mathbf{v} = 5\mathbf{i} + 4\mathbf{j}$

70. $\mathbf{v} = -4\mathbf{i} + 7\mathbf{j}$

71. $\mathbf{v} = -3\mathbf{i} - 3\mathbf{j}$

72. $\mathbf{v} = 8\mathbf{i} - \mathbf{j}$

Finding the Component Form of a Vector **In Exercises 73 and 74, find the component form of v given its magnitude ‖v‖ and the angle θ it makes with the positive x-axis. Then sketch v.**

73. $\|\mathbf{v}\| = 8,\quad \theta = 120°$

74. $\|\mathbf{v}\| = \frac{1}{2},\quad \theta = 225°$

75. Rope Tension Two ropes support a 180-pound weight, as shown in the figure. Find the tension in each rope.

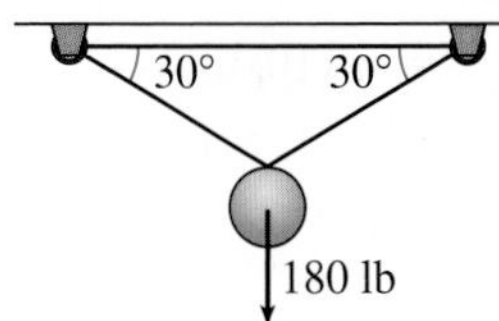

76. **Resultant Force** Forces with magnitudes of 85 pounds and 50 pounds act on a single point at angles of 45° and 60°, respectively, with the positive x-axis. Find the direction and magnitude of the resultant of these forces.

3.4 Finding a Dot Product **In Exercises 77–80, find u · v.**

77. $\mathbf{u} = \langle 6, 7\rangle$, $\mathbf{v} = \langle -3, 9\rangle$

78. $\mathbf{u} = \langle -7, 12\rangle$, $\mathbf{v} = \langle -4, -14\rangle$

79. $\mathbf{u} = 3\mathbf{i} + 7\mathbf{j}$, $\mathbf{v} = 11\mathbf{i} - 5\mathbf{j}$

80. $\mathbf{u} = -7\mathbf{i} + 2\mathbf{j}$, $\mathbf{v} = 16\mathbf{i} - 12\mathbf{j}$

Using Properties of the Dot Product **In Exercises 81–88, use the vectors $\mathbf{u} = \langle -4, 2\rangle$ and $\mathbf{v} = \langle 5, 1\rangle$ to find the quantity. State whether the result is a vector or a scalar.**

81. $2\mathbf{u} \cdot \mathbf{u}$

82. $3\mathbf{u} \cdot \mathbf{v}$

83. $4 - \|\mathbf{u}\|$

84. $\|\mathbf{v}\|^2$

85. $\mathbf{u}(\mathbf{u} \cdot \mathbf{v})$

86. $(\mathbf{u} \cdot \mathbf{v})\mathbf{v}$

87. $(\mathbf{u} \cdot \mathbf{u}) - (\mathbf{u} \cdot \mathbf{v})$

88. $(\mathbf{v} \cdot \mathbf{v}) - (\mathbf{v} \cdot \mathbf{u})$

Finding the Angle Between Two Vectors **In Exercises 89–92, find the angle θ (in degrees) between the vectors.**

89. $\mathbf{u} = \cos\frac{7\pi}{4}\mathbf{i} + \sin\frac{7\pi}{4}\mathbf{j}$
$\mathbf{v} = \cos\frac{5\pi}{6}\mathbf{i} + \sin\frac{5\pi}{6}\mathbf{j}$

90. $\mathbf{u} = \cos 45°\mathbf{i} + \sin 45°\mathbf{j}$
$\mathbf{v} = \cos 300°\mathbf{i} + \sin 300°\mathbf{j}$

91. $\mathbf{u} = \langle 2\sqrt{2}, -4\rangle$, $\mathbf{v} = \langle -\sqrt{2}, 1\rangle$

92. $\mathbf{u} = \langle 3, \sqrt{3}\rangle$, $\mathbf{v} = \langle 4, 3\sqrt{3}\rangle$

Determining Orthogonal Vectors **In Exercises 93–96, determine whether u and v are orthogonal.**

93. $\mathbf{u} = \langle -3, 8\rangle$, $\mathbf{v} = \langle 8, 3\rangle$

94. $\mathbf{u} = \langle \frac{1}{4}, -\frac{1}{2}\rangle$, $\mathbf{v} = \langle -2, 4\rangle$

95. $\mathbf{u} = -\mathbf{i}$, $\mathbf{v} = \mathbf{i} + 2\mathbf{j}$

96. $\mathbf{u} = -2\mathbf{i} + \mathbf{j}$, $\mathbf{v} = 3\mathbf{i} + 6\mathbf{j}$

Decomposing a Vector into Components **In Exercises 97–100, find the projection of u onto v. Then write u as the sum of two orthogonal vectors, one of which is $\text{proj}_{\mathbf{v}}\mathbf{u}$.**

97. $\mathbf{u} = \langle -4, 3\rangle$, $\mathbf{v} = \langle -8, -2\rangle$

98. $\mathbf{u} = \langle 5, 6\rangle$, $\mathbf{v} = \langle 10, 0\rangle$

99. $\mathbf{u} = \langle 2, 7\rangle$, $\mathbf{v} = \langle 1, -1\rangle$

100. $\mathbf{u} = \langle -3, 5\rangle$, $\mathbf{v} = \langle -5, 2\rangle$

Work **In Exercises 101 and 102, determine the work done in moving a particle from P to Q when the magnitude and direction of the force are given by v.**

101. $P(5, 3)$, $Q(8, 9)$, $\mathbf{v} = \langle 2, 7\rangle$

102. $P(-2, -9)$, $Q(-12, 8)$, $\mathbf{v} = 3\mathbf{i} - 6\mathbf{j}$

103. **Work** Determine the work done by a crane lifting an 18,000-pound truck 4 feet.

104. **Work** A constant force of 25 pounds, exerted at an angle of 20° with the horizontal, is required to slide a crate across a floor. Determine the work done in sliding the crate 12 feet.

Exploration

True or False? **In Exercises 105–108, determine whether the statement is true or false. Justify your answer.**

105. Two sides and their included angle determine a unique triangle.

106. If any three sides or angles of an oblique triangle are known, then the triangle can be solved.

107. The Law of Sines is true when one of the angles in the triangle is a right angle.

108. When the Law of Sines is used, the solution is always unique.

109. **Think About It** Which vectors in the figure appear to be equivalent?

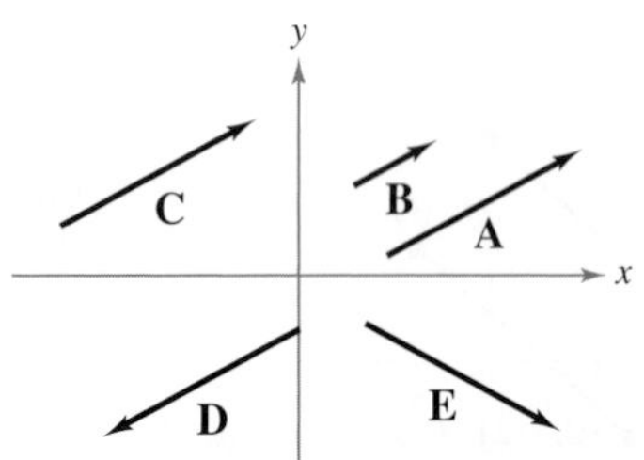

110. **Think About It** The vectors **u** and **v** have the same magnitudes in the two figures. In which figure is the magnitude of the sum greater? Explain.

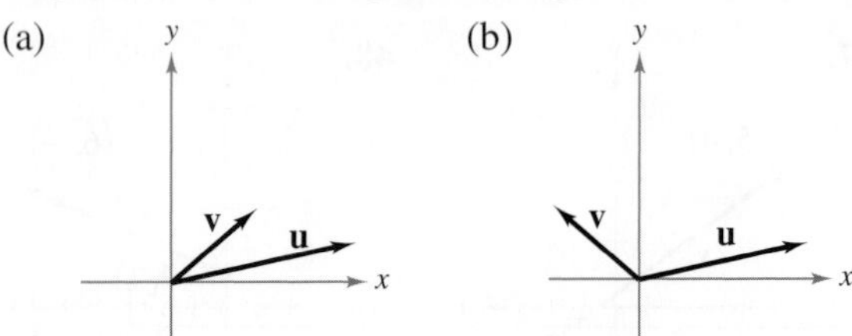

111. **Geometry** Describe geometrically the scalar multiple $k\mathbf{u}$ of the vector $\mathbf{u}$, for $k > 0$ and $k < 0$.

112. **Geometry** Describe geometrically the difference of the vectors **u** and **v**.

113. **Writing** What characterizes a vector in the plane?

Chapter Test

See CalcChat.com for tutorial help and worked-out solutions to odd-numbered exercises.

Take this test as you would take a test in class. When you are finished, check your work against the answers given in the back of the book.

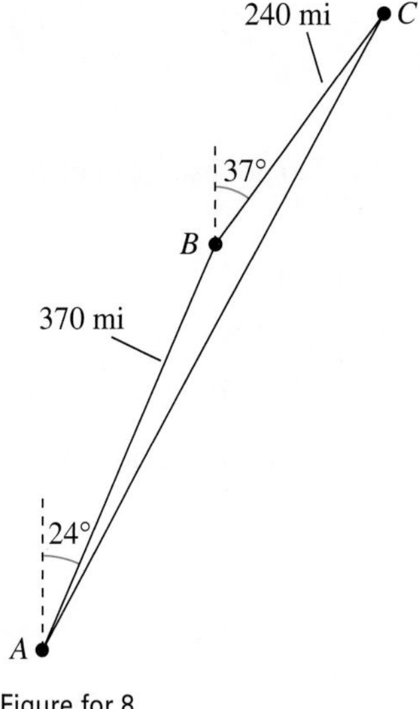

Figure for 8

In Exercises 1–6, determine whether the Law of Cosines is needed to solve the triangle. Then solve (if possible) the triangle. If two solutions exist, find both. Round your answers to two decimal places.

1. $A = 24°, \quad B = 68°, \quad a = 12.2$

2. $B = 110°, \quad C = 28°, \quad a = 15.6$

3. $A = 24°, \quad a = 11.2, \quad b = 13.4$

4. $a = 6.0, \quad b = 7.3, \quad c = 12.4$

5. $B = 100°, \quad a = 23, \quad b = 15$

6. $C = 121°, \quad a = 34, \quad b = 55$

7. A triangular parcel of land has sides of lengths 60 meters, 70 meters, and 82 meters. Find the area of the parcel of land.

8. An airplane flies 370 miles from point A to point B with a bearing of 24°. Then it flies 240 miles from point B to point C with a bearing of 37° (see figure). Find the straight-line distance and bearing from point A to point C.

In Exercises 9 and 10, find the component form of the vector v.

9. Initial point of $\mathbf{v}$: $(-3, 7)$

Terminal point of $\mathbf{v}$: $(11, -16)$

10. Magnitude of $\mathbf{v}$: $\|\mathbf{v}\| = 12$

Direction of $\mathbf{v}$: $\mathbf{u} = \langle 3, -5 \rangle$

In Exercises 11–14, $\mathbf{u} = \langle 2, 7 \rangle$ and $\mathbf{v} = \langle -6, 5 \rangle$. Find the resultant vector and sketch its graph.

11. $\mathbf{u} + \mathbf{v}$

12. $\mathbf{u} - \mathbf{v}$

13. $5\mathbf{u} - 3\mathbf{v}$

14. $4\mathbf{u} + 2\mathbf{v}$

15. Find a unit vector in the direction of $\mathbf{u} = \langle 24, -7 \rangle$.

16. Forces with magnitudes of 250 pounds and 130 pounds act on an object at angles of 45° and −60°, respectively, with the positive x-axis. Find the direction and magnitude of the resultant of these forces.

17. Find the dot product of $\mathbf{u} = \langle -9, 4 \rangle$ and $\mathbf{v} = \langle 1, 2 \rangle$.

18. Find the angle θ (in degrees) between the vectors $\mathbf{u} = \langle -1, 5 \rangle$ and $\mathbf{v} = \langle 3, -2 \rangle$.

19. Determine whether the vectors $\mathbf{u} = \langle 6, -10 \rangle$ and $\mathbf{v} = \langle 5, 3 \rangle$ are orthogonal.

20. Determine whether the vectors $\mathbf{u} = \langle \cos\theta, -\sin\theta \rangle$ and $\mathbf{v} = \langle \sin\theta, \cos\theta \rangle$ are orthogonal, parallel, or neither.

21. Find the projection of $\mathbf{u} = \langle 6, 7 \rangle$ onto $\mathbf{v} = \langle -5, -1 \rangle$. Then write $\mathbf{u}$ as the sum of two orthogonal vectors, one of which is $\text{proj}_{\mathbf{v}}\mathbf{u}$.

22. A 500-pound motorcycle is stopped at a red light on a hill inclined at 12°. Find the force required to keep the motorcycle from rolling down the hill.

Cumulative Test for Chapters 1–3

See CalcChat.com for tutorial help and worked-out solutions to odd-numbered exercises.

Take this test as you would take a test in class. When you are finished, check your work against the answers given in the back of the book.

1. Consider the angle $\theta = -120°$.
   (a) Sketch the angle in standard position.
   (b) Determine a coterminal angle in the interval $[0°, 360°)$.
   (c) Rewrite the angle in radian measure as a multiple of π. Do not use a calculator.
   (d) Find the reference angle θ'.
   (e) Find the exact values of the six trigonometric functions of θ.
2. Convert -1.45 radians to degrees. Round to three decimal places.
3. Find $\cos\theta$ when $\tan\theta = -\frac{21}{20}$ and $\sin\theta < 0$.

In Exercises 4–6, sketch the graph of the function. (Include two full periods.)

4. $f(x) = 3 - 2\sin\pi x$
5. $g(x) = \dfrac{1}{2}\tan\left(x - \dfrac{\pi}{2}\right)$
6. $h(x) = -\sec(x + \pi)$

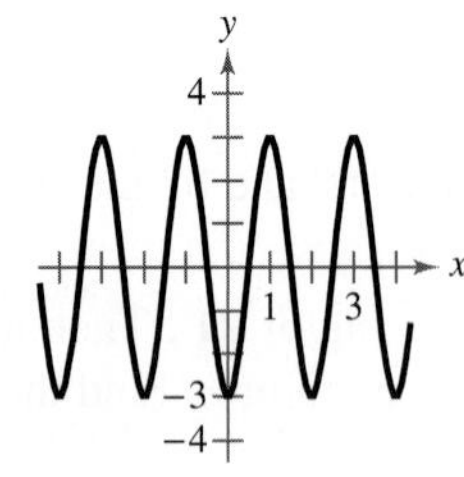

Figure for 7

7. Find a, b, and c for the function $h(x) = a\cos(bx + c)$ such that the graph of h matches the figure.
8. Sketch the graph of the function $f(x) = \frac{1}{2}x\sin x$ on the interval $[-3\pi, 3\pi]$.

In Exercises 9 and 10, find the exact value of the expression.

9. $\tan(\arctan 4.9)$
10. $\tan\left(\arcsin\frac{3}{5}\right)$

11. Write an algebraic expression that is equivalent to $\sin(\arccos 2x)$.
12. Use the fundamental identities to simplify: $\cos\left(\dfrac{\pi}{2} - x\right)\csc x$.
13. Subtract and simplify: $\dfrac{\sin\theta - 1}{\cos\theta} - \dfrac{\cos\theta}{\sin\theta - 1}$.

In Exercises 14–16, verify the identity.

14. $\cot^2\alpha(\sec^2\alpha - 1) = 1$
15. $\sin(x + y)\sin(x - y) = \sin^2 x - \sin^2 y$
16. $\sin^2 x\cos^2 x = \frac{1}{8}(1 - \cos 4x)$

In Exercises 17 and 18, find all solutions of the equation in the interval $[0, 2\pi)$.

17. $2\cos^2\beta - \cos\beta = 0$
18. $3\tan\theta - \cot\theta = 0$

19. Use the Quadratic Formula to find all solutions of the equation in the interval $[0, 2\pi)$: $\sin^2 x + 2\sin x + 1 = 0$.
20. Given that $\sin u = \frac{12}{13}$, $\cos v = \frac{3}{5}$, and angles u and v are both in Quadrant I, find $\tan(u - v)$.

21. Given that $\tan u = \frac{1}{2}$ and $0 < u < \frac{\pi}{2}$, find the exact value of $\tan(2u)$.

22. Given that $\tan u = \frac{4}{3}$ and $0 < u < \frac{\pi}{2}$, find the exact value of $\sin \frac{u}{2}$.

23. Rewrite $5 \sin \frac{3\pi}{4} \cdot \cos \frac{7\pi}{4}$ as a sum or difference.

24. Rewrite $\cos 9x - \cos 7x$ as a product.

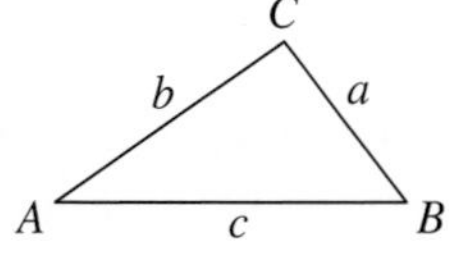

Figure for 25–30

In Exercises 25–30, determine whether the Law of Cosines is needed to solve the triangle at the left, then solve the triangle. Round your answers to two decimal places.

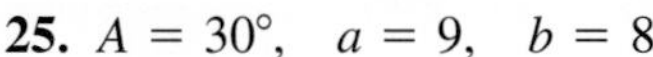

25. $A = 30°, \quad a = 9, \quad b = 8$

26. $A = 30°, \quad b = 8, \quad c = 10$

27. $A = 30°, \quad C = 90°, \quad b = 10$

28. $a = 4.7, \quad b = 8.1, \quad c = 10.3$

29. $A = 45°, \quad B = 26°, \quad c = 20$

30. $C = 80°, \quad a = 1.2, \quad b = 10$

31. Find the area of a triangle with two sides of lengths 7 inches and 12 inches and an included angle of 99°.

32. Use Heron's Area Formula to find the area of a triangle with sides of lengths 30 meters, 41 meters, and 45 meters.

33. Write the vector with initial point $(-1, 2)$ and terminal point $(6, 10)$ as a linear combination of the standard unit vectors **i** and **j**.

34. Find a unit vector **u** in the direction of $\mathbf{v} = \mathbf{i} + \mathbf{j}$.

35. Find $\mathbf{u} \cdot \mathbf{v}$ for $\mathbf{u} = 3\mathbf{i} + 4\mathbf{j}$ and $\mathbf{v} = \mathbf{i} - 2\mathbf{j}$.

36. Find the projection of $\mathbf{u} = \langle 8, -2 \rangle$ onto $\mathbf{v} = \langle 1, 5 \rangle$. Then write **u** as the sum of two orthogonal vectors, one of which is $\text{proj}_{\mathbf{v}}\mathbf{u}$.

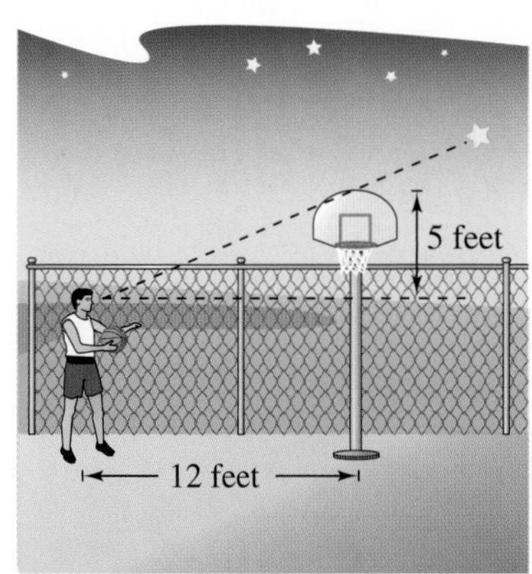

Figure for 40

37. A ceiling fan with 21-inch blades makes 63 revolutions per minute. Find the angular speed of the fan in radians per minute. Find the linear speed (in inches per minute) of the tips of the blades.

38. Find the area of the sector of a circle with a radius of 12 yards and a central angle of 105°.

39. From a point 200 feet from a flagpole, the angles of elevation to the bottom and top of the flag are 16° 45′ and 18°, respectively. Approximate the height of the flag to the nearest foot.

40. To determine the angle of elevation of a star in the sky, you align the star and the top of the backboard of a basketball hoop that is 5 feet higher than your eyes in your line of vision (see figure). Your horizontal distance from the backboard is 12 feet. What is the angle of elevation of the star?

41. Find a model for a particle in simple harmonic motion with a displacement (at $t = 0$) of 4 inches, an amplitude of 4 inches, and a period of 8 seconds.

42. An airplane has a speed of 500 kilometers per hour at a bearing of 30°. The wind velocity is 50 kilometers per hour in the direction N 60° E. Find the resultant speed and true direction of the airplane.

43. A constant force of 85 pounds, exerted at an angle of 60° with the horizontal, is required to slide an object across a floor. Determine the work done in sliding the object 10 feet.

Proofs in Mathematics

LAW OF TANGENTS

Besides the Law of Sines and the Law of Cosines, there is also a Law of Tangents, developed by French mathematician François Viète (1540–1603). The Law of Tangents follows from the Law of Sines and the sum-to-product formulas for sine and is defined as

$$\frac{a+b}{a-b} = \frac{\tan[(A+B)/2]}{\tan[(A-B)/2]}.$$

The Law of Tangents can be used to solve a triangle when two sides and the included angle (SAS) are given. Before the invention of calculators, it was easier to use the Law of Tangents to solve the SAS case instead of the Law of Cosines because the computations by hand were not as tedious.

Law of Sines ***(p. 262)***

If ABC is a triangle with sides a, b, and c, then

$$\frac{a}{\sin A} = \frac{b}{\sin B} = \frac{c}{\sin C}.$$

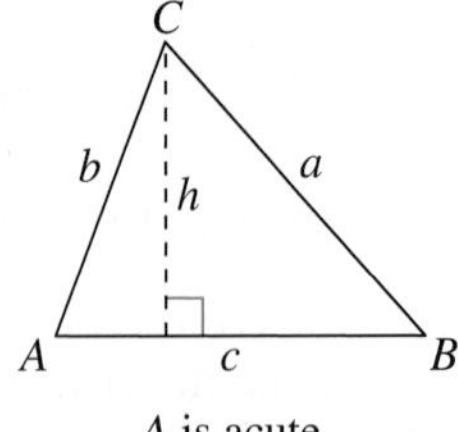

A is acute.

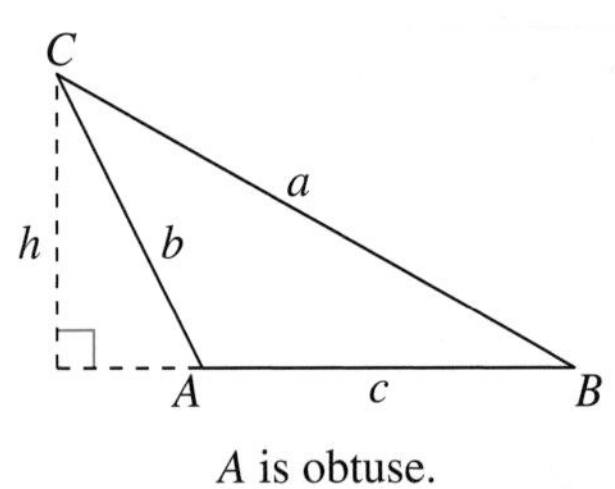

A is obtuse.

Proof

For either triangle shown above, you have

$$\sin A = \frac{h}{b} \implies h = b \sin A \quad \text{and} \quad \sin B = \frac{h}{a} \implies h = a \sin B$$

where h is an altitude. Equating these two values of h, you have

$$a \sin B = b \sin A \quad \text{or} \quad \frac{a}{\sin A} = \frac{b}{\sin B}.$$

Note that $\sin A \neq 0$ and $\sin B \neq 0$ because no angle of a triangle can have a measure of 0° or 180°. In a similar manner, construct an altitude h from vertex B to side AC (extended in the obtuse triangle), as shown at the left. Then you have

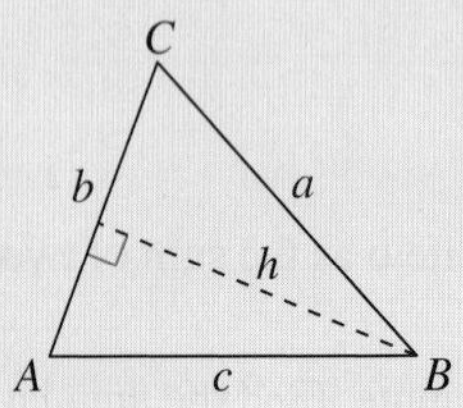

A is acute.

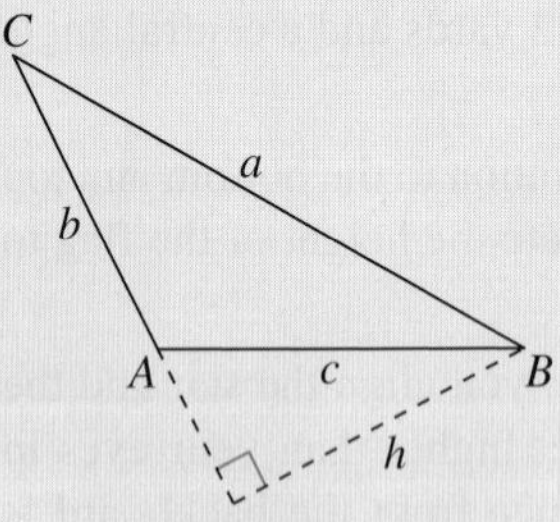

A is obtuse.

$$\sin A = \frac{h}{c} \implies h = c \sin A \quad \text{and} \quad \sin C = \frac{h}{a} \implies h = a \sin C.$$

Equating these two values of h, you have

$$a \sin C = c \sin A \quad \text{or} \quad \frac{a}{\sin A} = \frac{c}{\sin C}.$$

By the Transitive Property of Equality,

$$\frac{a}{\sin A} = \frac{b}{\sin B} = \frac{c}{\sin C}.$$

Law of Cosines ***(p. 271)***

Standard Form	Alternative Form
$a^2 = b^2 + c^2 - 2bc \cos A$	$\cos A = \dfrac{b^2 + c^2 - a^2}{2bc}$
$b^2 = a^2 + c^2 - 2ac \cos B$	$\cos B = \dfrac{a^2 + c^2 - b^2}{2ac}$
$c^2 = a^2 + b^2 - 2ab \cos C$	$\cos C = \dfrac{a^2 + b^2 - c^2}{2ab}$

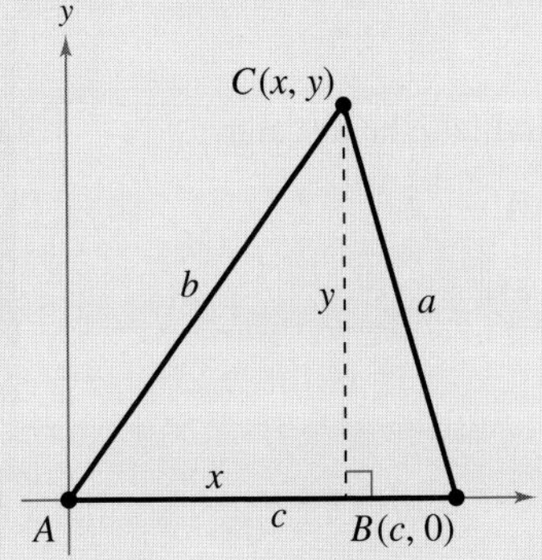

Proof

To prove the first formula, consider the triangle at the left, which has three acute angles. Note that vertex B has coordinates $(c, 0)$. Furthermore, C has coordinates (x, y), where $x = b \cos A$ and $y = b \sin A$. Because a is the distance from C to B, it follows that

$$a = \sqrt{(x - c)^2 + (y - 0)^2} \qquad \text{Distance Formula}$$

$$a^2 = (x - c)^2 + (y - 0)^2 \qquad \text{Square each side.}$$

$$a^2 = (b \cos A - c)^2 + (b \sin A)^2 \qquad \text{Substitute for } x \text{ and } y.$$

$$a^2 = b^2 \cos^2 A - 2bc \cos A + c^2 + b^2 \sin^2 A \qquad \text{Expand.}$$

$$a^2 = b^2(\sin^2 A + \cos^2 A) + c^2 - 2bc \cos A \qquad \text{Factor out } b^2.$$

$$a^2 = b^2 + c^2 - 2bc \cos A. \qquad \sin^2 A + \cos^2 A = 1$$

Similar arguments are used to establish the second and third formulas.

Heron's Area Formula *(p. 274)*

Given any triangle with sides of lengths a, b, and c, the area of the triangle is

$$\text{Area} = \sqrt{s(s - a)(s - b)(s - c)}, \text{ where } s = \frac{a + b + c}{2}.$$

REMARK

$$\frac{1}{2}bc(1 + \cos A)$$
$$= \frac{1}{2}bc\left(1 + \frac{b^2 + c^2 - a^2}{2bc}\right)$$
$$= \frac{1}{2}bc\left(\frac{2bc + b^2 + c^2 - a^2}{2bc}\right)$$
$$= \frac{1}{4}(2bc + b^2 + c^2 - a^2)$$
$$= \frac{1}{4}(b^2 + 2bc + c^2 - a^2)$$
$$= \frac{1}{4}[(b + c)^2 - a^2]$$
$$= \frac{1}{4}(b + c + a)(b + c - a)$$
$$= \frac{a + b + c}{2} \cdot \frac{-a + b + c}{2}$$

Proof

From Section 3.1, you know that

$$\text{Area} = \frac{1}{2}bc \sin A \qquad \text{Formula for the area of an oblique triangle}$$

$$= \sqrt{\frac{1}{4}b^2c^2 \sin^2 A} \qquad \text{Square each side and then take the square root of each side.}$$

$$= \sqrt{\frac{1}{4}b^2c^2(1 - \cos^2 A)} \qquad \text{Pythagorean identity}$$

$$= \sqrt{\left[\frac{1}{2}bc(1 + \cos A)\right]\left[\frac{1}{2}bc(1 - \cos A)\right]}. \qquad \text{Factor.}$$

Using the alternate form of the Law of Cosines,

$$\frac{1}{2}bc(1 + \cos A) = \frac{a + b + c}{2} \cdot \frac{-a + b + c}{2}$$

and

$$\frac{1}{2}bc(1 - \cos A) = \frac{a - b + c}{2} \cdot \frac{a + b - c}{2}.$$

Letting $s = (a + b + c)/2$, rewrite these two equations as

$$\frac{1}{2}bc(1 + \cos A) = s(s - a) \quad \text{and} \quad \frac{1}{2}bc(1 - \cos A) = (s - b)(s - c).$$

Substitute into the last formula for area to conclude that

$$\text{Area} = \sqrt{s(s - a)(s - b)(s - c)}.$$

Properties of the Dot Product *(p. 291)*

Let **u**, **v**, and **w** be vectors in the plane or in space and let c be a scalar.

1. $\mathbf{u} \cdot \mathbf{v} = \mathbf{v} \cdot \mathbf{u}$
2. $\mathbf{0} \cdot \mathbf{v} = 0$
3. $\mathbf{u} \cdot (\mathbf{v} + \mathbf{w}) = \mathbf{u} \cdot \mathbf{v} + \mathbf{u} \cdot \mathbf{w}$
4. $\mathbf{v} \cdot \mathbf{v} = \|\mathbf{v}\|^2$
5. $c(\mathbf{u} \cdot \mathbf{v}) = c\mathbf{u} \cdot \mathbf{v} = \mathbf{u} \cdot c\mathbf{v}$

Proof

Let $\mathbf{u} = \langle u_1, u_2 \rangle$, $\mathbf{v} = \langle v_1, v_2 \rangle$, $\mathbf{w} = \langle w_1, w_2 \rangle$, $\mathbf{0} = \langle 0, 0 \rangle$, and let c be a scalar.

1. $\mathbf{u} \cdot \mathbf{v} = u_1v_1 + u_2v_2 = v_1u_1 + v_2u_2 = \mathbf{v} \cdot \mathbf{u}$
2. $\mathbf{0} \cdot \mathbf{v} = 0 \cdot v_1 + 0 \cdot v_2 = 0$
3. $$\begin{aligned}\mathbf{u} \cdot (\mathbf{v} + \mathbf{w}) &= \mathbf{u} \cdot \langle v_1 + w_1, v_2 + w_2 \rangle \\ &= u_1(v_1 + w_1) + u_2(v_2 + w_2) \\ &= u_1v_1 + u_1w_1 + u_2v_2 + u_2w_2 \\ &= (u_1v_1 + u_2v_2) + (u_1w_1 + u_2w_2) \\ &= \mathbf{u} \cdot \mathbf{v} + \mathbf{u} \cdot \mathbf{w}\end{aligned}$$
4. $\mathbf{v} \cdot \mathbf{v} = v_1^2 + v_2^2 = \left(\sqrt{v_1^2 + v_2^2}\right)^2 = \|\mathbf{v}\|^2$
5. $$\begin{aligned}c(\mathbf{u} \cdot \mathbf{v}) &= c(\langle u_1, u_2 \rangle \cdot \langle v_1, v_2 \rangle) \\ &= c(u_1v_1 + u_2v_2) \\ &= (cu_1)v_1 + (cu_2)v_2 \\ &= \langle cu_1, cu_2 \rangle \cdot \langle v_1, v_2 \rangle \\ &= c\mathbf{u} \cdot \mathbf{v}\end{aligned}$$

Angle Between Two Vectors *(p. 292)*

If θ is the angle between two nonzero vectors **u** and **v**, then $\cos\theta = \dfrac{\mathbf{u} \cdot \mathbf{v}}{\|\mathbf{u}\|\|\mathbf{v}\|}$.

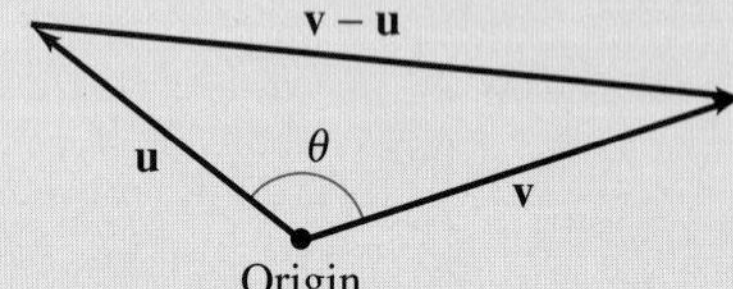

Proof

Consider the triangle determined by vectors **u**, **v**, and **v** − **u**, as shown at the left. By the Law of Cosines,

$$\begin{aligned}\|\mathbf{v} - \mathbf{u}\|^2 &= \|\mathbf{u}\|^2 + \|\mathbf{v}\|^2 - 2\|\mathbf{u}\|\|\mathbf{v}\|\cos\theta \\ (\mathbf{v} - \mathbf{u}) \cdot (\mathbf{v} - \mathbf{u}) &= \|\mathbf{u}\|^2 + \|\mathbf{v}\|^2 - 2\|\mathbf{u}\|\|\mathbf{v}\|\cos\theta \\ (\mathbf{v} - \mathbf{u}) \cdot \mathbf{v} - (\mathbf{v} - \mathbf{u}) \cdot \mathbf{u} &= \|\mathbf{u}\|^2 + \|\mathbf{v}\|^2 - 2\|\mathbf{u}\|\|\mathbf{v}\|\cos\theta \\ \mathbf{v} \cdot \mathbf{v} - \mathbf{u} \cdot \mathbf{v} - \mathbf{v} \cdot \mathbf{u} + \mathbf{u} \cdot \mathbf{u} &= \|\mathbf{u}\|^2 + \|\mathbf{v}\|^2 - 2\|\mathbf{u}\|\|\mathbf{v}\|\cos\theta \\ \|\mathbf{v}\|^2 - 2\mathbf{u} \cdot \mathbf{v} + \|\mathbf{u}\|^2 &= \|\mathbf{u}\|^2 + \|\mathbf{v}\|^2 - 2\|\mathbf{u}\|\|\mathbf{v}\|\cos\theta\end{aligned}$$

$$\cos\theta = \frac{\mathbf{u} \cdot \mathbf{v}}{\|\mathbf{u}\|\|\mathbf{v}\|}.$$

P.S. Problem Solving

1. **Distance** In the figure, a beam of light is directed at the blue mirror, reflected to the red mirror, and then reflected back to the blue mirror. Find PT, the distance that the light travels from the red mirror back to the blue mirror.

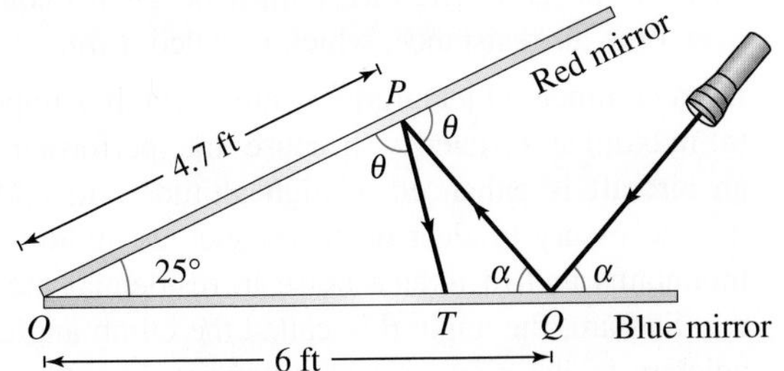

2. **Correcting a Course** A triathlete sets a course to swim S 25° E from a point on shore to a buoy $\frac{3}{4}$ mile away. After swimming 300 yards through a strong current, the triathlete is off course at a bearing of S 35° E. Find the bearing and distance the triathlete needs to swim to correct her course.

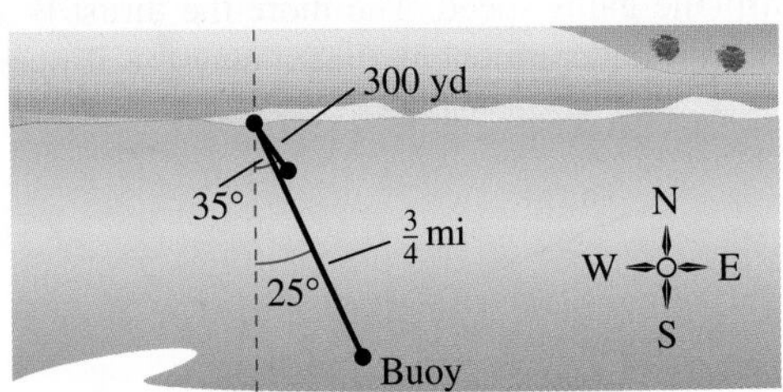

3. **Locating Lost Hikers** A group of hikers is lost in a national park. Two ranger stations receive an emergency SOS signal from the hikers. Station B is 75 miles due east of station A. The bearing from station A to the signal is S 60° E and the bearing from station B to the signal is S 75° W.
   (a) Draw a diagram that gives a visual representation of the problem.
   (b) Find the distance from each station to the SOS signal.
   (c) A rescue party is in the park 20 miles from station A at a bearing of S 80° E. Find the distance and the bearing the rescue party must travel to reach the lost hikers.

4. **Seeding a Courtyard** You are seeding a triangular courtyard. One side of the courtyard is 52 feet long and another side is 46 feet long. The angle opposite the 52-foot side is 65°.
   (a) Draw a diagram that gives a visual representation of the problem.
   (b) How long is the third side of the courtyard?
   (c) One bag of grass seed covers an area of 50 square feet. How many bags of grass seed do you need to cover the courtyard?

5. **Finding Magnitudes** For each pair of vectors, find the value of each expression.

   (i) $\|\mathbf{u}\|$ (ii) $\|\mathbf{v}\|$ (iii) $\|\mathbf{u}+\mathbf{v}\|$

   (iv) $\left\|\frac{\mathbf{u}}{\|\mathbf{u}\|}\right\|$ (v) $\left\|\frac{\mathbf{v}}{\|\mathbf{v}\|}\right\|$ (vi) $\left\|\frac{\mathbf{u}+\mathbf{v}}{\|\mathbf{u}+\mathbf{v}\|}\right\|$

   (a) $\mathbf{u} = \langle 1, -1\rangle$, $\mathbf{v} = \langle -1, 2\rangle$
   (b) $\mathbf{u} = \langle 0, 1\rangle$, $\mathbf{v} = \langle 3, -3\rangle$
   (c) $\mathbf{u} = \left\langle 1, \frac{1}{2}\right\rangle$, $\mathbf{v} = \langle 2, 3\rangle$
   (d) $\mathbf{u} = \langle 2, -4\rangle$, $\mathbf{v} = \langle 5, 5\rangle$

6. **Writing a Vector in Terms of Other Vectors** Write the vector $\mathbf{w}$ in terms of $\mathbf{u}$ and $\mathbf{v}$, given that the terminal point of $\mathbf{w}$ bisects the line segment (see figure).

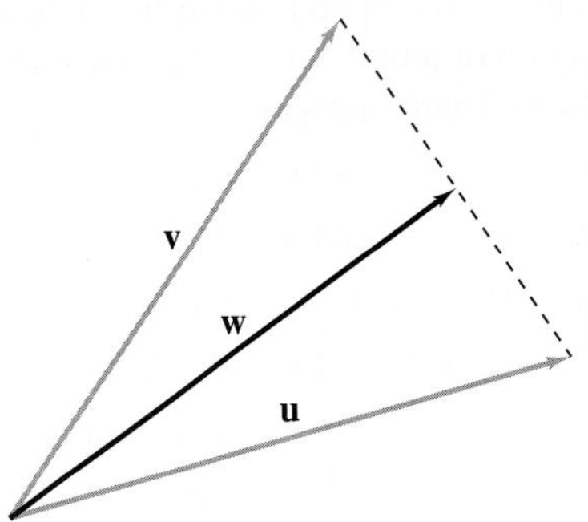

7. **Think About It** Consider two forces of equal magnitude acting on a point.
   (a) If the magnitude of the resultant is the sum of the magnitudes of the two forces, make a conjecture about the angle between the forces.
   (b) If the resultant of the forces is $\mathbf{0}$, make a conjecture about the angle between the forces.
   (c) Can the magnitude of the resultant be greater than the sum of the magnitudes of the two forces? Explain.

8. **Comparing Work** Two forces of the same magnitude $\mathbf{F}_1$ and $\mathbf{F}_2$ act at angles θ_1 and θ_2, respectively. Use a diagram to compare the work done by $\mathbf{F}_1$ with the work done by $\mathbf{F}_2$ in moving along the vector $\overrightarrow{PQ}$ when
   (a) $\theta_1 = -\theta_2$
   (b) $\theta_1 = 60°$ and $\theta_2 = 30°$
   (c) $\theta_1 = 45°$ and $\theta_2 = 60°$.

9. **Proof** Use a half-angle formula and the Law of Cosines to show that, for any triangle,

   (a) $\cos\frac{C}{2} = \sqrt{\frac{s(s-c)}{ab}}$ and

   (b) $\sin\frac{C}{2} = \sqrt{\frac{(s-a)(s-b)}{ab}}$

   where $s = \frac{1}{2}(a+b+c)$.

10. Proof Prove that if **u** is orthogonal to **v** and **w**, then **u** is orthogonal to

$c\mathbf{v} + d\mathbf{w}$

for any scalars c and d.

11. Technology Given vectors **u** and **v** in component form, write a program for your graphing utility that produces each output.

(a) $\|\mathbf{u}\|$

(b) $\|\mathbf{v}\|$

(c) The angle between **u** and **v**

(d) The component form of the projection of **u** onto **v**

12. Technology Use the program you wrote in Exercise 11 to find the angle between **u** and **v** and the projection of **u** onto **v** for the given vectors. Verify your results by hand on paper.

(a) $\mathbf{u} = \langle 8, -4 \rangle$ and $\mathbf{v} = \langle 2, 5 \rangle$

(b) $\mathbf{u} = \langle 2, -6 \rangle$ and $\mathbf{v} = \langle 4, 1 \rangle$

(c) $\mathbf{u} = \langle 5, 6 \rangle$ and $\mathbf{v} = \langle -1, 3 \rangle$

(d) $\mathbf{u} = \langle 3, -2 \rangle$ and $\mathbf{v} = \langle -2, 1 \rangle$

13. Skydiving A skydiver falls at a constant downward velocity of 120 miles per hour. In the figure, vector **u** represents the skydiver's velocity. A steady breeze pushes the skydiver to the east at 40 miles per hour. Vector **v** represents the wind velocity.

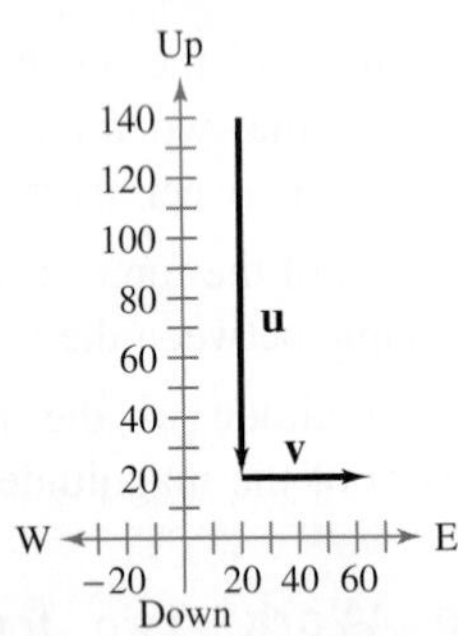

(a) Write the vectors **u** and **v** in component form.

(b) Let $\mathbf{s} = \mathbf{u} + \mathbf{v}$. Use the figure to sketch **s**. To print an enlarged copy of the graph, go to *MathGraphs.com*.

(c) Find the magnitude of **s**. What information does the magnitude give you about the skydiver's fall?

(d) Without wind, the skydiver would fall in a path perpendicular to the ground. At what angle to the ground is the path of the skydiver when affected by the 40-mile-per-hour wind from due west?

(e) The next day, the skydiver falls at a constant downward velocity of 120 miles per hour and a steady breeze pushes the skydiver to the west at 30 miles per hour. Draw a new figure that gives a visual representation of the problem and find the skydiver's new velocity.

14. Speed and Velocity of an Airplane Four basic forces are in action during flight: weight, lift, thrust, and drag. To fly through the air, an object must overcome its own *weight*. To do this, it must create an upward force called *lift*. To generate lift, a forward motion called *thrust* is needed. The thrust must be great enough to overcome air resistance, which is called *drag*.

For a commercial jet aircraft, a quick climb is important to maximize efficiency because the performance of an aircraft is enhanced at high altitudes. In addition, it is necessary to clear obstacles such as buildings and mountains and to reduce noise in residential areas. In the diagram, the angle θ is called the climb angle. The velocity of the plane can be represented by a vector **v** with a vertical component $\|\mathbf{v}\| \sin \theta$ (called climb speed) and a horizontal component $\|\mathbf{v}\| \cos \theta$, where $\|\mathbf{v}\|$ is the speed of the plane.

When taking off, a pilot must decide how much of the thrust to apply to each component. The more the thrust is applied to the horizontal component, the faster the airplane gains speed. The more the thrust is applied to the vertical component, the quicker the airplane climbs.

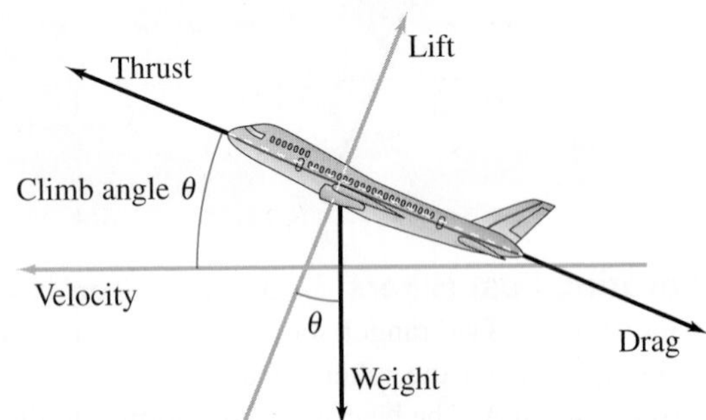

(a) Complete the table for an airplane that has a speed of $\|\mathbf{v}\| = 100$ miles per hour.

θ	0.5°	1.0°	1.5°	2.0°	2.5°	3.0°
$\|\mathbf{v}\| \sin \theta$						
$\|\mathbf{v}\| \cos \theta$						

(b) Does an airplane's speed equal the sum of the vertical and horizontal components of its velocity? If not, how could you find the speed of an airplane whose velocity components were known?

(c) Use the result of part (b) to find the speed of an airplane with the given velocity components.

(i) $\|\mathbf{v}\| \sin \theta = 5.235$ miles per hour
$\|\mathbf{v}\| \cos \theta = 149.909$ miles per hour

(ii) $\|\mathbf{v}\| \sin \theta = 10.463$ miles per hour
$\|\mathbf{v}\| \cos \theta = 149.634$ miles per hour

4 Complex Numbers

Fractals (Exercise 73, page 347)

Ohm's Law (Exercise 69, page 341)

Sailing (Exercise 49, page 335)

Projectile Motion (page 323)

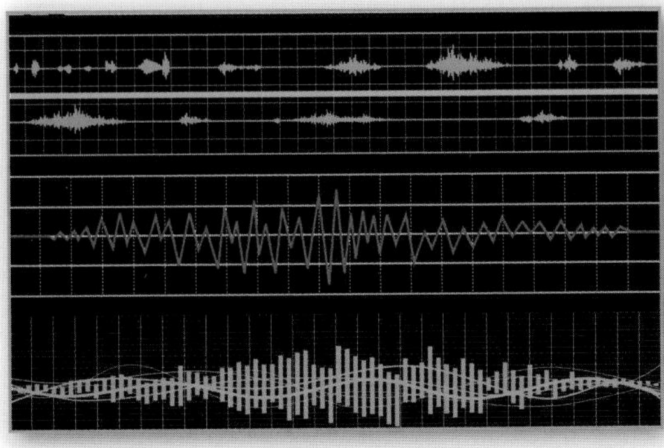

Digital Signal Processing (page 315)

4.1 Complex Numbers

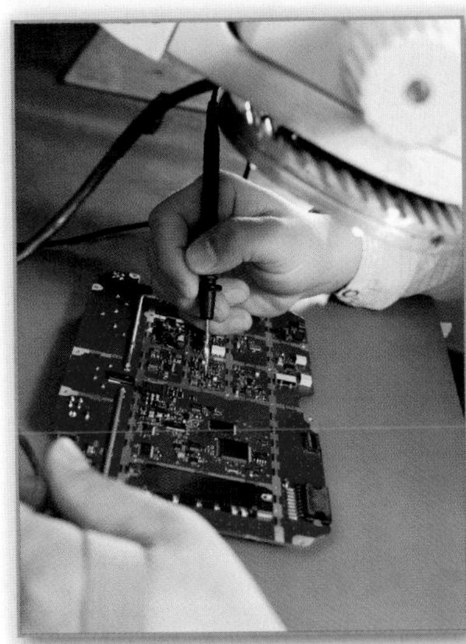
Complex numbers are often used in electrical engineering. For example, in Exercise 87 on page 320, you will use complex numbers to find the impedance of an electrical circuit.

- **Use the imaginary unit *i* to write complex numbers.**
- **Add, subtract, and multiply complex numbers.**
- **Use complex conjugates to write the quotient of two complex numbers in standard form.**
- **Find complex solutions of quadratic equations.**

The Imaginary Unit *i*

Some quadratic equations have no real solutions. For example, the quadratic equation

$$x^2 + 1 = 0$$

has no real solution because there is no real number x that can be squared to produce -1. To overcome this deficiency, mathematicians created an expanded system of numbers using the **imaginary unit *i*,** defined as

$$i = \sqrt{-1} \qquad \text{Imaginary unit}$$

where $i^2 = -1$. By adding real numbers to real multiples of this imaginary unit, you obtain the set of **complex numbers.** Each complex number can be written in the **standard form *a* + *bi*.** For example, the standard form of the complex number $-5 + \sqrt{-9}$ is $-5 + 3i$ because

$$-5 + \sqrt{-9} = -5 + \sqrt{3^2(-1)} = -5 + 3\sqrt{-1} = -5 + 3i.$$

Definition of a Complex Number

Let a and b be real numbers. The number $a + bi$ is a **complex number** written in **standard form.** The real number a is the **real part** and the number bi (where b is a real number) is the **imaginary part** of the complex number.

When $b = 0$, the number $a + bi$ is a real number. When $b \neq 0$, the number $a + bi$ is an **imaginary number.** A number of the form bi, where $b \neq 0$, is a **pure imaginary number.**

Every real number a can be written as a complex number using $b = 0$. That is, for every real number a, $a = a + 0i$. So, the set of real numbers is a subset of the set of complex numbers, as shown in the figure below.

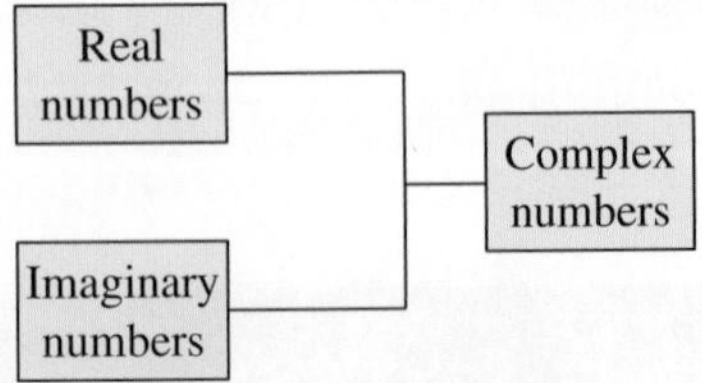

Equality of Complex Numbers

Two complex numbers $a + bi$ and $c + di$, written in standard form, are equal to each other

$$a + bi = c + di \qquad \text{Equality of two complex numbers}$$

if and only if $a = c$ and $b = d$.

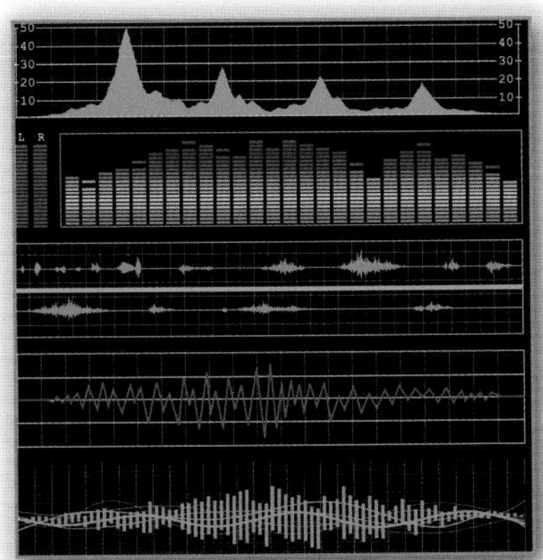

The fast Fourier transform (FFT), which has important applications in digital signal processing, involves operations with complex numbers.

Operations with Complex Numbers

To add (or subtract) two complex numbers, add (or subtract) the real and imaginary parts of the numbers separately.

Addition and Subtraction of Complex Numbers

For two complex numbers $a + bi$ and $c + di$ written in standard form, the sum and difference are

Sum: $(a + bi) + (c + di) = (a + c) + (b + d)i$

Difference: $(a + bi) - (c + di) = (a - c) + (b - d)i.$

The **additive identity** in the complex number system is zero (the same as in the real number system). Furthermore, the **additive inverse** of the complex number $a + bi$ is

$-(a + bi) = -a - bi.$ Additive inverse

So, you have

$(a + bi) + (-a - bi) = 0 + 0i = 0.$

EXAMPLE 1 Adding and Subtracting Complex Numbers

a. $(4 + 7i) + (1 - 6i) = 4 + 7i + 1 - 6i$ Remove parentheses.

$= (4 + 1) + (7 - 6)i$ Group like terms.

$= 5 + i$ Write in standard form.

b. $(1 + 2i) + (3 - 2i) = 1 + 2i + 3 - 2i$ Remove parentheses.

$= (1 + 3) + (2 - 2)i$ Group like terms.

$= 4 + 0i$ Simplify.

$= 4$ Write in standard form.

REMARK Note that the sum of two complex numbers can be a real number.

c. $3i - (-2 + 3i) - (2 + 5i) = 3i + 2 - 3i - 2 - 5i$

$= (2 - 2) + (3 - 3 - 5)i$

$= 0 - 5i$

$= -5i$

d. $(3 + 2i) + (4 - i) - (7 + i) = 3 + 2i + 4 - i - 7 - i$

$= (3 + 4 - 7) + (2 - 1 - 1)i$

$= 0 + 0i$

$= 0$

✓ **Checkpoint** *Audio-video solution in English & Spanish at LarsonPrecalculus.com*

Perform each operation and write the result in standard form.

a. $(7 + 3i) + (5 - 4i)$

b. $(3 + 4i) - (5 - 3i)$

c. $2i + (-3 - 4i) - (-3 - 3i)$

d. $(5 - 3i) + (3 + 5i) - (8 + 2i)$

zphoto/Shutterstock.com

Many of the properties of real numbers are valid for complex numbers as well. Here are some examples.

Associative Properties of Addition and Multiplication

Commutative Properties of Addition and Multiplication

Distributive Property of Multiplication Over Addition

Note the use of these properties when multiplying two complex numbers.

$$(a + bi)(c + di) = a(c + di) + bi(c + di) \quad \text{Distributive Property}$$
$$= ac + (ad)i + (bc)i + (bd)i^2 \quad \text{Distributive Property}$$
$$= ac + (ad)i + (bc)i + (bd)(-1) \quad i^2 = -1$$
$$= ac - bd + (ad)i + (bc)i \quad \text{Commutative Property}$$
$$= (ac - bd) + (ad + bc)i \quad \text{Associative Property}$$

The procedure shown above is similar to multiplying two binomials and combining like terms, as in the FOIL method. So, you do not need to memorize this procedure.

EXAMPLE 2 Multiplying Complex Numbers

See LarsonPrecalculus.com for an interactive version of this type of example.

a. $4(-2 + 3i) = 4(-2) + 4(3i)$ Distributive Property

$= -8 + 12i$ Simplify.

b. $(2 - i)(4 + 3i) = 8 + 6i - 4i - 3i^2$ FOIL Method

$= 8 + 6i - 4i - 3(-1)$ $i^2 = -1$

$= (8 + 3) + (6 - 4)i$ Group like terms.

$= 11 + 2i$ Write in standard form.

c. $(3 + 2i)(3 - 2i) = 9 - 6i + 6i - 4i^2$ FOIL Method

$= 9 - 6i + 6i - 4(-1)$ $i^2 = -1$

$= 9 + 4$ Simplify.

$= 13$ Write in standard form.

d. $(3 + 2i)^2 = (3 + 2i)(3 + 2i)$ Square of a binomial

$= 9 + 6i + 6i + 4i^2$ FOIL Method

$= 9 + 6i + 6i + 4(-1)$ $i^2 = -1$

$= 9 + 12i - 4$ Simplify.

$= 5 + 12i$ Write in standard form.

✓ **Checkpoint** *Audio-video solution in English & Spanish at LarsonPrecalculus.com*

Perform each operation and write the result in standard form.

a. $-5(3 - 2i)$

b. $(2 - 4i)(3 + 3i)$

c. $(4 + 5i)(4 - 5i)$

d. $(4 + 2i)^2$

Complex Conjugates

Notice in Example 2(c) that the product of two complex numbers can be a real number. This occurs with pairs of complex numbers of the form $a + bi$ and $a - bi$, called **complex conjugates.**

$$\begin{aligned}(a + bi)(a - bi) &= a^2 - abi + abi - b^2i^2 \\ &= a^2 - b^2(-1) \\ &= a^2 + b^2\end{aligned}$$

EXAMPLE 3 Multiplying Conjugates

Multiply each complex number by its complex conjugate.

a. $1 + i$ **b.** $4 - 3i$

Solution

a. The complex conjugate of $1 + i$ is $1 - i$.

$$(1 + i)(1 - i) = 1^2 - i^2 = 1 - (-1) = 2$$

b. The complex conjugate of $4 - 3i$ is $4 + 3i$.

$$(4 - 3i)(4 + 3i) = 4^2 - (3i)^2 = 16 - 9i^2 = 16 - 9(-1) = 25$$

✓ **Checkpoint** Audio-video solution in English & Spanish at LarsonPrecalculus.com

Multiply each complex number by its complex conjugate.

a. $3 + 6i$ **b.** $2 - 5i$

To write the quotient of $a + bi$ and $c + di$ in standard form, where c and d are not both zero, multiply the numerator and denominator by the complex conjugate of the *denominator* to obtain

$$\frac{a + bi}{c + di} = \frac{a + bi}{c + di}\left(\frac{c - di}{c - di}\right) = \frac{(ac + bd) + (bc - ad)i}{c^2 + d^2} = \frac{ac + bd}{c^2 + d^2} + \left(\frac{bc - ad}{c^2 + d^2}\right)i.$$

REMARK Note that when you multiply a quotient of complex numbers by

$$\frac{c - di}{c - di}$$

you are multiplying the quotient by a form of 1. So, you are not changing the original expression, you are only writing an equivalent expression.

EXAMPLE 4 A Quotient of Complex Numbers in Standard Form

$$\frac{2 + 3i}{4 - 2i} = \frac{2 + 3i}{4 - 2i}\left(\frac{4 + 2i}{4 + 2i}\right) \quad \text{Multiply numerator and denominator by complex conjugate of denominator.}$$

$$= \frac{8 + 4i + 12i + 6i^2}{16 - 4i^2} \quad \text{Expand.}$$

$$= \frac{8 - 6 + 16i}{16 + 4} \quad i^2 = -1$$

$$= \frac{2 + 16i}{20} \quad \text{Simplify.}$$

$$= \frac{1}{10} + \frac{4}{5}i \quad \text{Write in standard form.}$$

✓ **Checkpoint** Audio-video solution in English & Spanish at LarsonPrecalculus.com

Write $\dfrac{2 + i}{2 - i}$ in standard form.

Complex Solutions of Quadratic Equations

You can write a number such as $\sqrt{-3}$ in standard form by factoring out $i = \sqrt{-1}$.

$$\sqrt{-3} = \sqrt{3(-1)} = \sqrt{3}\sqrt{-1} = \sqrt{3}i$$

The number $\sqrt{3}i$ is the *principal square root* of -3.

> **REMARK** The definition of principal square root uses the rule
>
> $$\sqrt{ab} = \sqrt{a}\sqrt{b}$$
>
> for $a > 0$ and $b < 0$. This rule is not valid when *both* a and b are negative. For example,
>
> $$\begin{aligned}\sqrt{-5}\sqrt{-5} &= \sqrt{5(-1)}\sqrt{5(-1)} \\ &= \sqrt{5}i\sqrt{5}i \\ &= \sqrt{25}i^2 \\ &= 5i^2 \\ &= -5\end{aligned}$$
>
> whereas
>
> $$\sqrt{(-5)(-5)} = \sqrt{25} = 5.$$
>
> Be sure to convert complex numbers to standard form *before* performing any operations.

Principal Square Root of a Negative Number

When a is a positive real number, the **principal square root** of $-a$ is defined as

$$\sqrt{-a} = \sqrt{a}i.$$

EXAMPLE 5 Writing Complex Numbers in Standard Form

a. $\sqrt{-3}\sqrt{-12} = \sqrt{3}i\sqrt{12}i = \sqrt{36}i^2 = 6(-1) = -6$

b. $\sqrt{-48} - \sqrt{-27} = \sqrt{48}i - \sqrt{27}i = 4\sqrt{3}i - 3\sqrt{3}i = \sqrt{3}i$

c. $\left(-1 + \sqrt{-3}\right)^2 = \left(-1 + \sqrt{3}i\right)^2 = 1 - 2\sqrt{3}i + 3(-1)$
$= -2 - 2\sqrt{3}i$

✓ **Checkpoint** *Audio-video solution in English & Spanish at LarsonPrecalculus.com*

Write $\sqrt{-14}\sqrt{-2}$ in standard form.

EXAMPLE 6 Complex Solutions of a Quadratic Equation

Solve $3x^2 - 2x + 5 = 0$.

Solution

$$x = \frac{-(-2) \pm \sqrt{(-2)^2 - 4(3)(5)}}{2(3)} \quad \text{Quadratic Formula}$$

$$= \frac{2 \pm \sqrt{-56}}{6} \quad \text{Simplify.}$$

$$= \frac{2 \pm 2\sqrt{14}i}{6} \quad \text{Write } \sqrt{-56} \text{ in standard form.}$$

$$= \frac{1}{3} \pm \frac{\sqrt{14}}{3}i \quad \text{Write solution in standard form.}$$

✓ **Checkpoint** *Audio-video solution in English & Spanish at LarsonPrecalculus.com*

Solve $8x^2 + 14x + 9 = 0$.

> ▷ **ALGEBRA HELP** To review the Quadratic Formula, see Section P.2.

Summarize (Section 4.1)

1. Explain how to write complex numbers using the imaginary unit i *(page 314)*.
2. Explain how to add, subtract, and multiply complex numbers *(pages 315 and 316, Examples 1 and 2)*.
3. Explain how to use complex conjugates to write the quotient of two complex numbers in standard form *(page 317, Example 4)*.
4. Explain how to find complex solutions of a quadratic equation *(page 318, Example 6)*.

4.1 Exercises

See CalcChat.com for tutorial help and worked-out solutions to odd-numbered exercises.

Vocabulary: Fill in the blanks.

1. A ________ number has the form $a + bi$, where $a \neq 0, b = 0$.
2. An ________ number has the form $a + bi$, where $a \neq 0, b \neq 0$.
3. A ________ ________ number has the form $a + bi$, where $a = 0, b \neq 0$.
4. The imaginary unit i is defined as $i =$ ________, where $i^2 =$ ________.
5. When a is a positive real number, the ________ ________ root of $-a$ is defined as $\sqrt{-a} = \sqrt{a}i$.
6. The numbers $a + bi$ and $a - bi$ are called ________ ________, and their product is a real number $a^2 + b^2$.

Skills and Applications

Equality of Complex Numbers In Exercises 7–10, find real numbers a and b such that the equation is true.

7. $a + bi = 9 + 8i$
8. $a + bi = 10 - 5i$
9. $(a - 2) + (b + 1)i = 6 + 5i$
10. $(a + 2) + (b - 3)i = 4 + 7i$

Writing a Complex Number in Standard Form In Exercises 11–22, write the complex number in standard form.

11. $2 + \sqrt{-25}$
12. $4 + \sqrt{-49}$
13. $1 - \sqrt{-12}$
14. $2 - \sqrt{-18}$
15. $\sqrt{-40}$
16. $\sqrt{-27}$
17. 23
18. 50
19. $-6i + i^2$
20. $-2i^2 + 4i$
21. $\sqrt{-0.04}$
22. $\sqrt{-0.0025}$

Adding or Subtracting Complex Numbers In Exercises 23–30, perform the operation and write the result in standard form.

23. $(5 + i) + (2 + 3i)$
24. $(13 - 2i) + (-5 + 6i)$
25. $(9 - i) - (8 - i)$
26. $(3 + 2i) - (6 + 13i)$
27. $(-2 + \sqrt{-8}) + (5 - \sqrt{-50})$
28. $(8 + \sqrt{-18}) - (4 + 3\sqrt{2}i)$
29. $13i - (14 - 7i)$
30. $25 + (-10 + 11i) + 15i$

Multiplying Complex Numbers In Exercises 31–38, perform the operation and write the result in standard form.

31. $(1 + i)(3 - 2i)$
32. $(7 - 2i)(3 - 5i)$
33. $12i(1 - 9i)$
34. $-8i(9 + 4i)$
35. $(\sqrt{2} + 3i)(\sqrt{2} - 3i)$
36. $(4 + \sqrt{7}i)(4 - \sqrt{7}i)$
37. $(6 + 7i)^2$
38. $(5 - 4i)^2$

Multiplying Conjugates In Exercises 39–46, write the complex conjugate of the complex number. Then multiply the number by its complex conjugate.

39. $9 + 2i$
40. $8 - 10i$
41. $-1 - \sqrt{5}i$
42. $-3 + \sqrt{2}i$
43. $\sqrt{-20}$
44. $\sqrt{-15}$
45. $\sqrt{6}$
46. $1 + \sqrt{8}$

A Quotient of Complex Numbers in Standard Form In Exercises 47–54, write the quotient in standard form.

47. $\dfrac{2}{4 - 5i}$
48. $\dfrac{13}{1 - i}$
49. $\dfrac{5 + i}{5 - i}$
50. $\dfrac{6 - 7i}{1 - 2i}$
51. $\dfrac{9 - 4i}{i}$
52. $\dfrac{8 + 16i}{2i}$
53. $\dfrac{3i}{(4 - 5i)^2}$
54. $\dfrac{5i}{(2 + 3i)^2}$

Performing Operations with Complex Numbers In Exercises 55–58, perform the operation and write the result in standard form.

55. $\dfrac{2}{1 + i} - \dfrac{3}{1 - i}$
56. $\dfrac{2i}{2 + i} + \dfrac{5}{2 - i}$
57. $\dfrac{i}{3 - 2i} + \dfrac{2i}{3 + 8i}$
58. $\dfrac{1 + i}{i} - \dfrac{3}{4 - i}$

Writing a Complex Number in Standard Form In Exercises 59–66, write the complex number in standard form.

59. $\sqrt{-6}\sqrt{-2}$
60. $\sqrt{-5}\sqrt{-10}$
61. $(\sqrt{-15})^2$
62. $(\sqrt{-75})^2$
63. $\sqrt{-8} + \sqrt{-50}$
64. $\sqrt{-45} - \sqrt{-5}$
65. $(3 + \sqrt{-5})(7 - \sqrt{-10})$
66. $(2 - \sqrt{-6})^2$

Complex Solutions of a Quadratic Equation **In Exercises 67–76, use the Quadratic Formula to solve the quadratic equation.**

67. $x^2 - 2x + 2 = 0$
68. $x^2 + 6x + 10 = 0$
69. $4x^2 + 16x + 17 = 0$
70. $9x^2 - 6x + 37 = 0$
71. $4x^2 + 16x + 21 = 0$
72. $16t^2 - 4t + 3 = 0$
73. $\frac{3}{2}x^2 - 6x + 9 = 0$
74. $\frac{7}{8}x^2 - \frac{3}{4}x + \frac{5}{16} = 0$
75. $1.4x^2 - 2x + 10 = 0$
76. $4.5x^2 - 3x + 12 = 0$

Simplifying a Complex Number **In Exercises 77–86, simplify the complex number and write it in standard form.**

77. $-6i^3 + i^2$
78. $4i^2 - 2i^3$
79. $-14i^5$
80. $(-i)^3$
81. $\left(\sqrt{-72}\right)^3$
82. $\left(\sqrt{-2}\right)^6$
83. $\dfrac{1}{i^3}$
84. $\dfrac{1}{(2i)^3}$
85. $(3i)^4$
86. $(-i)^6$

87. Impedance of a Circuit

The opposition to current in an electrical circuit is called its impedance. The impedance z in a parallel circuit with two pathways satisfies the equation

$$\frac{1}{z} = \frac{1}{z_1} + \frac{1}{z_2}$$

where z_1 is the impedance (in ohms) of pathway 1 and z_2 is the impedance (in ohms) of pathway 2.

(a) The impedance of each pathway in a parallel circuit is found by adding the impedances of all components in the pathway. Use the table to find z_1 and z_2.

	Resistor	Inductor	Capacitor
Symbol	$a\ \Omega$	$b\ \Omega$	$c\ \Omega$
Impedance	a	bi	$-ci$

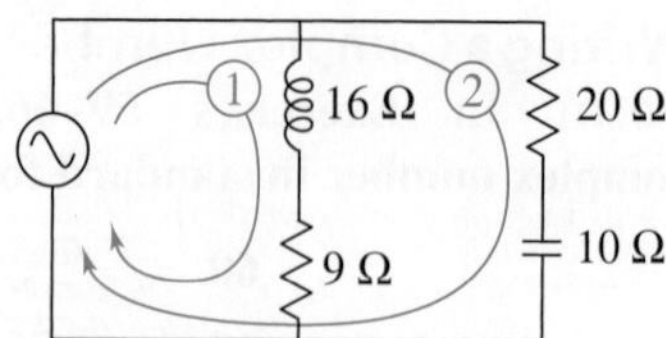

(b) Find the impedance z.

88. **Cube of a Complex Number** Cube each complex number.
(a) $-1 + \sqrt{3}i$ (b) $-1 - \sqrt{3}i$

Exploration

True or False? **In Exercises 89–92, determine whether the statement is true or false. Justify your answer.**

89. The sum of two complex numbers is always a real number.
90. There is no complex number that is equal to its complex conjugate.
91. $-i\sqrt{6}$ is a solution of $x^4 - x^2 + 14 = 56$.
92. $i^{44} + i^{150} - i^{74} - i^{109} + i^{61} = -1$

93. **Pattern Recognition** Find the missing values.

$i^1 = i$ $i^2 = -1$ $i^3 = -i$ $i^4 = 1$
$i^5 = \square$ $i^6 = \square$ $i^7 = \square$ $i^8 = \square$
$i^9 = \square$ $i^{10} = \square$ $i^{11} = \square$ $i^{12} = \square$

What pattern do you see? Write a brief description of how you would find i raised to any positive integer power.

94. **HOW DO YOU SEE IT?** Use the diagram below.

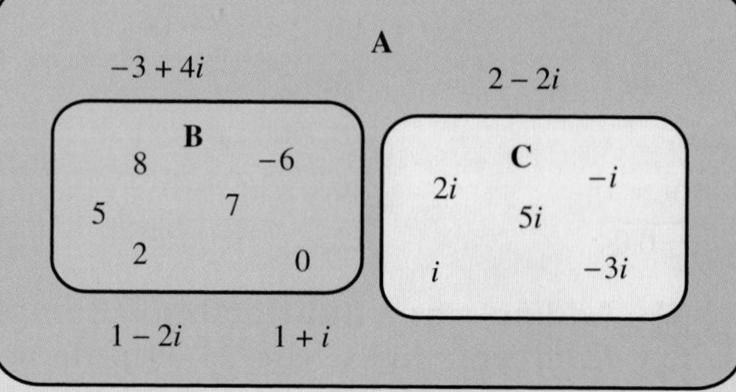

(a) Match each label with its corresponding letter in the diagram.
(i) Pure imaginary numbers
(ii) Real numbers
(iii) Complex numbers

(b) What part of the diagram represents the imaginary numbers? Explain your reasoning.

95. **Error Analysis** Describe the error.

$$\sqrt{-6}\sqrt{-6} = \sqrt{(-6)(-6)} = \sqrt{36} = 6 \quad ✗$$

96. **Proof** Prove that the complex conjugate of the product of two complex numbers $a_1 + b_1 i$ and $a_2 + b_2 i$ is the product of their complex conjugates.

97. **Proof** Prove that the complex conjugate of the sum of two complex numbers $a_1 + b_1 i$ and $a_2 + b_2 i$ is the sum of their complex conjugates.

4.2 Complex Solutions of Equations

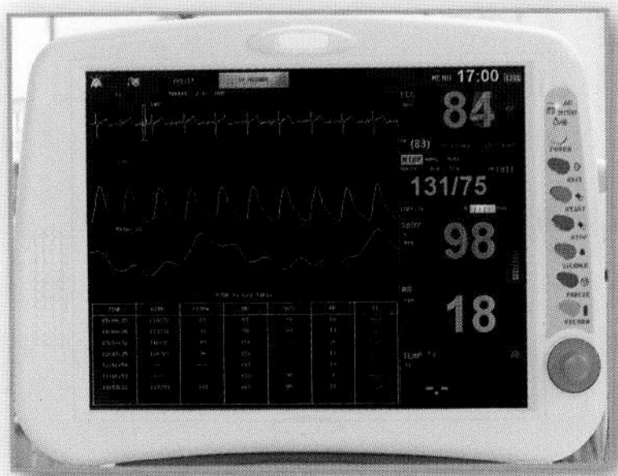

Polynomial equations have many real-life applications. For example, in Exercise 80 on page 328, you will use a quadratic equation to analyze a patient's blood oxygen level.

- **Determine the numbers of solutions of polynomial equations.**
- **Find solutions of polynomial equations.**
- **Find zeros of polynomial functions and find polynomial functions given the zeros of the functions.**

The Number of Solutions of a Polynomial Equation

The **Fundamental Theorem of Algebra** implies that a polynomial equation of degree n has *precisely* n solutions in the complex number system. These solutions can be real or imaginary and may be repeated. The Fundamental Theorem of Algebra and the **Linear Factorization Theorem** are given below for your review. For a proof of the Linear Factorization Theorem, see Proofs in Mathematics on page 354.

The Fundamental Theorem of Algebra

If $f(x)$ is a polynomial of degree n, where $n > 0$, then f has at least one zero in the complex number system.

Note that finding zeros of a polynomial function f is equivalent to finding solutions of the polynomial equation $f(x) = 0$.

Linear Factorization Theorem

If $f(x)$ is a polynomial of degree n, where $n > 0$, then $f(x)$ has precisely n linear factors

$$f(x) = a_n(x - c_1)(x - c_2) \cdot \cdot \cdot (x - c_n)$$

where $c_1, c_2, \ldots, c_n$ are complex numbers.

EXAMPLE 1 Solutions of Polynomial Equations

See LarsonPrecalculus.com for an interactive version of this type of example.

a. The first-degree equation $x - 2 = 0$ has exactly *one* solution: $x = 2$.

b. The second-degree equation

$$x^2 - 6x + 9 = (x - 3)(x - 3) = 0$$

has exactly *two* solutions: $x = 3$ and $x = 3$ (a *repeated solution*).

c. The fourth-degree equation

$$x^4 - 1 = (x - 1)(x + 1)(x - i)(x + i) = 0$$

has exactly *four* solutions: $x = 1$, $x = -1$, $x = i$, and $x = -i$.

Checkpoint Audio-video solution in English & Spanish at LarsonPrecalculus.com

Determine the number of solutions of the equation $x^3 + 9x = 0$.

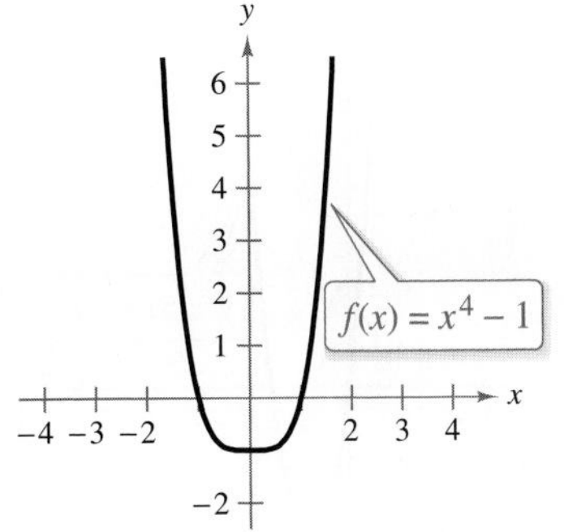

Figure 4.1

You can use a graph to check the number of real solutions of an equation. As shown in Figure 4.1, the graph of $f(x) = x^4 - 1$ has two x-intercepts, which implies that the equation has two real solutions.

Every second-degree equation, $ax^2 + bx + c = 0$, has precisely two solutions given by the Quadratic Formula

$$x = \frac{-b \pm \sqrt{b^2 - 4ac}}{2a}.$$

The quantity under the radical sign, $b^2 - 4ac$, is the **discriminant,** and can be used to determine whether the solutions are real, repeated, or imaginary.

1. If $b^2 - 4ac < 0$, then the equation has two imaginary solutions.
2. If $b^2 - 4ac = 0$, then the equation has one repeated real solution.
3. If $b^2 - 4ac > 0$, then the equation has two distinct real solutions.

EXAMPLE 2 Using the Discriminant

Use the discriminant to determine the numbers of real and imaginary solutions of each equation.

a. $4x^2 - 20x + 25 = 0$ **b.** $13x^2 + 7x + 2 = 0$ **c.** $5x^2 - 8x = 0$

Solution

a. For this equation, $a = 4$, $b = -20$, and $c = 25$.

$$b^2 - 4ac = (-20)^2 - 4(4)(25) = 400 - 400 = 0$$

The discriminant is zero, so there is one repeated real solution.

b. For this equation, $a = 13$, $b = 7$, and $c = 2$.

$$b^2 - 4ac = 7^2 - 4(13)(2) = 49 - 104 = -55$$

The discriminant is negative, so there are two imaginary solutions.

c. For this equation, $a = 5$, $b = -8$, and $c = 0$.

$$b^2 - 4ac = (-8)^2 - 4(5)(0) = 64 - 0 = 64$$

The discriminant is positive, so there are two distinct real solutions.

Use the discriminant to determine the numbers of real and imaginary solutions of each equation.

a. $3x^2 + 2x - 1 = 0$ **b.** $9x^2 + 6x + 1 = 0$ **c.** $9x^2 + 2x + 1 = 0$

The figures below show the graphs of the functions corresponding to the equations in Example 2. Notice that with one repeated solution, the graph *touches* the x-axis at its x-intercept. With two imaginary solutions, the graph has no x-intercepts. With two real solutions, the graph *crosses* the x-axis at its x-intercepts.

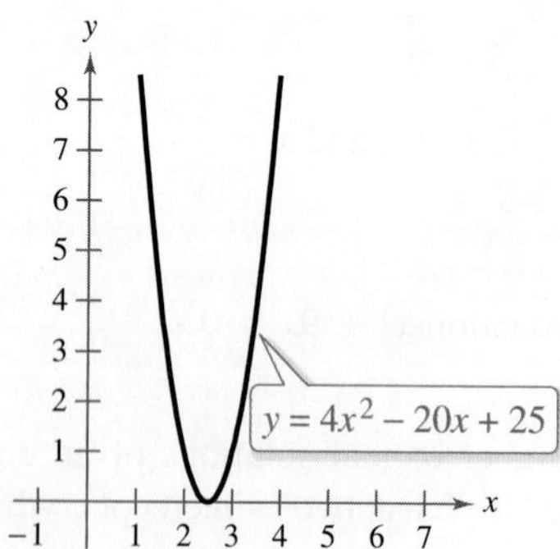

a. Repeated real solution

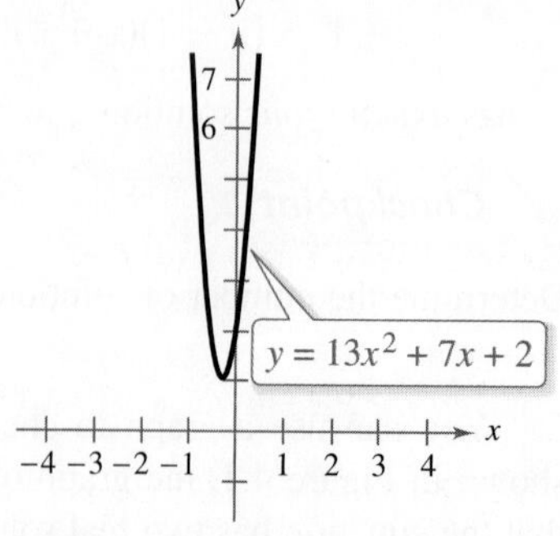

b. No real solution

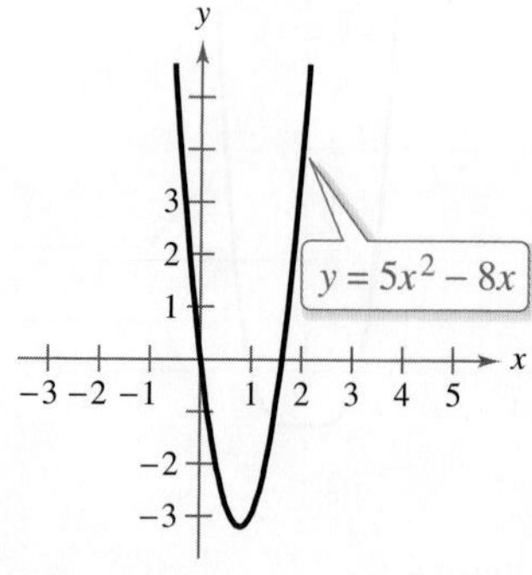

c. Two distinct real solutions

Finding Solutions of Polynomial Equations

To determine whether an object in vertical projectile motion reaches a specific height, solve the quadratic equation that corresponds to the object's position. You will explore this concept further in Exercises 77 and 78 on page 327.

EXAMPLE 3 Solving a Quadratic Equation

Solve $x^2 + 2x + 2 = 0$. Write complex solutions in standard form.

Solution You have

$$a = 1, \quad b = 2, \quad \text{and} \quad c = 2$$

so by the Quadratic Formula,

$$x = \frac{-b \pm \sqrt{b^2 - 4ac}}{2a}$$ Quadratic Formula

$$= \frac{-2 \pm \sqrt{2^2 - 4(1)(2)}}{2(1)}$$ Substitute 1 for a, 2 for b, and 2 for c.

$$= \frac{-2 \pm \sqrt{-4}}{2}$$ Simplify.

$$= \frac{-2 \pm 2i}{2}$$ Write $\sqrt{-4}$ in standard form.

$$= -1 \pm i.$$ Write solution in standard form.

✓ ***Checkpoint*** *Audio-video solution in English & Spanish at LarsonPrecalculus.com*

Solve $x^2 - 4x + 5 = 0$. Write complex solutions in standard form.

In Example 3, the two complex solutions are *complex conjugates.* That is, they are of the form $a \pm bi$. This is not a coincidence, as stated in the theorem below.

> **Complex Solutions Occur in Conjugate Pairs**
>
> If $a + bi$, $b \neq 0$, is a solution of a polynomial equation with *real coefficients,* then the complex conjugate $a - bi$ is also a solution of the equation.

Be sure you see that this result is true only when the polynomial has *real coefficients.* For example, the result applies to the equation $x^2 + 1 = 0$, but not to the equation $x - i = 0$.

EXAMPLE 4 Solving a Polynomial Equation

Solve $x^4 - x^2 - 20 = 0$.

Solution

$$x^4 - x^2 - 20 = 0$$ Write original equation.

$$(x^2 - 5)(x^2 + 4) = 0$$ Factor trinomial.

$$\left(x + \sqrt{5}\right)\left(x - \sqrt{5}\right)(x + 2i)(x - 2i) = 0$$ Factor completely.

Setting each factor equal to zero and solving the resulting equations yields the solutions $x = -\sqrt{5}$, $x = \sqrt{5}$, $x = -2i$, and $x = 2i$.

✓ ***Checkpoint*** *Audio-video solution in English & Spanish at LarsonPrecalculus.com*

Solve $x^4 + 7x^2 - 18 = 0$.

Zeros of Polynomial Functions

Finding the *zeros* of a polynomial function is essentially the same as finding the solutions of the corresponding polynomial equation. For example, the zeros of the polynomial function

$$f(x) = 3x^2 - 4x + 5$$

are the solutions of the polynomial equation

$$3x^2 - 4x + 5 = 0.$$

EXAMPLE 5 Finding the Zeros of a Polynomial Function

Find all the zeros of

$$f(x) = x^4 - 3x^3 + 6x^2 + 2x - 60$$

given that $1 + 3i$ is a zero of f.

Algebraic Solution

Complex zeros occur in conjugate pairs, so you know that $1 - 3i$ is also a zero of f. This means that both

$$[x - (1 + 3i)]$$

and

$$[x - (1 - 3i)]$$

are factors of $f(x)$. Multiplying these two factors produces

$$\begin{aligned}[x - (1 + 3i)][x - (1 - 3i)] &= [(x - 1) - 3i][(x - 1) + 3i] \\ &= (x - 1)^2 - 9i^2 \\ &= x^2 - 2x + 10.\end{aligned}$$

Using long division, divide $x^2 - 2x + 10$ into $f(x)$.

$$\begin{array}{r} x^2 - x - 6 \\ x^2 - 2x + 10 \overline{)\, x^4 - 3x^3 + 6x^2 + 2x - 60} \\ \underline{x^4 - 2x^3 + 10x^2} \\ -x^3 - 4x^2 + 2x \\ \underline{-x^3 + 2x^2 - 10x} \\ -6x^2 + 12x - 60 \\ \underline{-6x^2 + 12x - 60} \\ 0 \end{array}$$

So, you have

$$\begin{aligned} f(x) &= (x^2 - 2x + 10)(x^2 - x - 6) \\ &= (x^2 - 2x + 10)(x - 3)(x + 2) \end{aligned}$$

and can conclude that the zeros of f are

$$x = 1 + 3i,\ x = 1 - 3i,\ x = 3, \text{ and } x = -2.$$

Graphical Solution

Complex zeros occur in conjugate pairs, so you know that $1 - 3i$ is also a zero of f. The polynomial is a fourth-degree polynomial, so you know that there are two other zeros of the function. Use a graphing utility to graph

$$y = x^4 - 3x^3 + 6x^2 + 2x - 60.$$

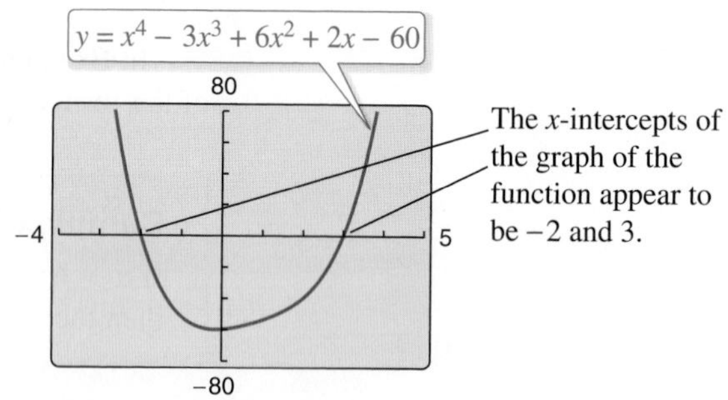

Use the *zero* or *root* feature of the graphing utility to confirm that $x = -2$ and $x = 3$ are x-intercepts of the graph. So, the zeros of f are

$$x = 1 + 3i,\ x = 1 - 3i,\ x = 3, \text{ and } x = -2.$$

✓ **Checkpoint** Audio-video solution in English & Spanish at LarsonPrecalculus.com

Find all the zeros of

$$f(x) = 3x^3 - 2x^2 + 48x - 32$$

given that $4i$ is a zero of f.

EXAMPLE 6 Finding a Polynomial Function with Given Zeros

Find a fourth-degree polynomial function f with real coefficients that has -1, -1, and $3i$ as zeros.

Solution You are given that $3i$ is a zero of f *and* the polynomial has real coefficients, so you know that the complex conjugate $-3i$ must also be a zero. Using the Linear Factorization Theorem, write $f(x)$ as

$$f(x) = a(x + 1)(x + 1)(x - 3i)(x + 3i).$$

For simplicity, let $a = 1$ to obtain

$$f(x) = (x^2 + 2x + 1)(x^2 + 9) = x^4 + 2x^3 + 10x^2 + 18x + 9.$$

✓ **Checkpoint**  Audio-video solution in English & Spanish at LarsonPrecalculus.com

Find a fourth-degree polynomial function f with real coefficients that has 2, -2, and $-7i$ as zeros.

EXAMPLE 7 Finding a Polynomial Function with Given Zeros

Find the cubic polynomial function f with real coefficients that has 2 and $1 - i$ as zeros, and $f(1) = 3$.

Solution You are given that $1 - i$ is a zero of f, so the complex conjugate $1 + i$ is also a zero.

$$\begin{aligned} f(x) &= a(x - 2)[x - (1 - i)][x - (1 + i)] \\ &= a(x - 2)[(x - 1) + i][(x - 1) - i] \\ &= a(x - 2)[(x - 1)^2 + 1] \\ &= a(x - 2)(x^2 - 2x + 2) \\ &= a(x^3 - 4x^2 + 6x - 4) \end{aligned}$$

To find the value of a, use the fact that $f(1) = 3$ to obtain

$$\begin{aligned} f(1) &= a[(1)^3 - 4(1)^2 + 6(1) - 4] \\ 3 &= -a. \end{aligned}$$

So, $a = -3$ and

$$f(x) = -3(x^3 - 4x^2 + 6x - 4) = -3x^3 + 12x^2 - 18x + 12.$$

✓ **Checkpoint**  Audio-video solution in English & Spanish at LarsonPrecalculus.com

Find the cubic polynomial function f with real coefficients that has 1 and $2 + i$ as zeros, and $f(2) = 2$. ■

Summarize (Section 4.2)

1. Explain how to use the Fundamental Theorem of Algebra to determine the number of solutions of a polynomial equation *(page 321, Example 1).*
2. Explain how to use the discriminant to determine the number of real solutions of a quadratic equation *(page 322, Example 2).*
3. Explain how to solve a polynomial equation *(page 323, Examples 3 and 4).*
4. Explain how to find the zeros of a polynomial function and how to find a polynomial function with given zeros *(pages 324 and 325, Examples 5–7).*

4.2 Exercises

See **CalcChat.com** for tutorial help and worked-out solutions to odd-numbered exercises.

Vocabulary: Fill in the blanks.

1. The ________ ________ of ________ states that if $f(x)$ is a polynomial of degree n $(n > 0)$, then f has at least one zero in the complex number system.
2. The ________ ________ ________ states that if $f(x)$ is a polynomial of degree n $(n > 0)$, then $f(x)$ has precisely n linear factors, $f(x) = a_n(x - c_1)(x - c_2) \cdot \cdot \cdot (x - c_n)$, where $c_1, c_2, \ldots, c_n$ are complex numbers.
3. Two complex solutions of the form $a \pm bi$ of a polynomial equation with real coefficients are ________ ________.
4. The quantity under the radical sign of the Quadratic Formula, $b^2 - 4ac$, is the ________.

Skills and Applications

Solutions of a Polynomial Equation In Exercises 5–8, determine the number of solutions of the equation in the complex number system.

5. $2x^3 + 3x + 1 = 0$
6. $x^6 + 4x^2 + 12 = 0$
7. $50 - 2x^4 = 0$
8. $14 - x + 4x^2 - 7x^5 = 0$

Using the Discriminant In Exercises 9–14, use the discriminant to find the number of real and imaginary solutions of the quadratic equation.

9. $2x^2 - 5x + 5 = 0$
10. $\frac{1}{4}x^2 - 5x + 25 = 0$
11. $4x^2 + 12x + 9 = 0$
12. $x^2 - 4x + 53 = 0$
13. $\frac{1}{5}x^2 + \frac{6}{5}x - 8 = 0$
14. $-2x^2 + 11x - 2 = 0$

Solving a Quadratic Equation In Exercises 15–24, solve the quadratic equation. Write complex solutions in standard form.

15. $x^2 - 5 = 0$
16. $3x^2 - 1 = 0$
17. $2 - 2x - x^2 = 0$
18. $x^2 + 10 + 8x = 0$
19. $x^2 - 8x + 16 = 0$
20. $4x^2 + 4x + 1 = 0$
21. $x^2 + 2x + 5 = 0$
22. $x^2 + 16x + 65 = 0$
23. $4x^2 - 4x + 5 = 0$
24. $4x^2 - 4x + 21 = 0$

Solving a Polynomial Equation In Exercises 25–28, solve the polynomial equation. Write complex solutions in standard form.

25. $x^4 - 6x^2 - 7 = 0$
26. $x^4 + 2x^2 - 8 = 0$
27. $x^4 - 5x^2 - 6 = 0$
28. $x^4 + x^2 - 72 = 0$

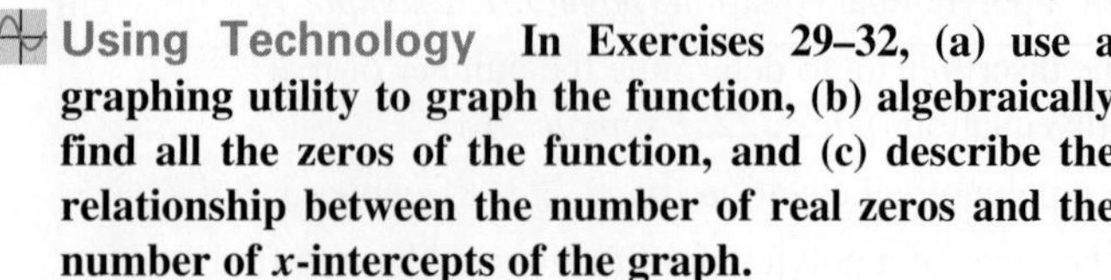

Using Technology In Exercises 29–32, (a) use a graphing utility to graph the function, (b) algebraically find all the zeros of the function, and (c) describe the relationship between the number of real zeros and the number of x-intercepts of the graph.

29. $f(x) = x^3 - 4x^2 + x - 4$
30. $f(x) = x^3 - 4x^2 - 4x + 16$
31. $f(x) = x^4 + 4x^2 + 4$
32. $f(x) = x^4 - 3x^2 - 4$

Finding the Zeros of a Polynomial Function In Exercises 33–48, write the polynomial as a product of linear factors and list all the zeros of the function.

33. $f(x) = x^2 + 36$
34. $f(t) = t^3 + 25t$
35. $f(x) = x^4 - 81$
36. $f(y) = y^4 - 256$
37. $h(x) = x^2 - 2x + 17$
38. $g(x) = x^2 + 10x + 17$
39. $h(x) = x^2 - 6x - 10$
40. $f(z) = z^2 - 2z + 2$
41. $g(x) = x^3 + 3x^2 - 3x - 9$
42. $h(x) = x^3 - 4x^2 + 16x - 64$
43. $f(x) = 2x^3 - x^2 + 36x - 18$
44. $g(x) = 4x^3 + 3x^2 + 96x + 72$
45. $g(x) = x^4 - 6x^3 + 16x^2 - 96x$
46. $h(x) = x^4 + x^3 + 100x^2 + 100x$
47. $f(x) = x^4 + 10x^2 + 9$
48. $f(x) = x^4 + 29x^2 + 100$

Finding the Zeros of a Polynomial Function In Exercises 49–58, use the given zero to find all the zeros of the function.

	Function	Zero
49.	$f(x) = x^3 - x^2 + 4x - 4$	$2i$
50.	$f(x) = x^3 + x^2 + 9x + 9$	$3i$
51.	$f(x) = 2x^4 - x^3 + 7x^2 - 4x - 4$	$2i$
52.	$f(x) = x^4 - 4x^3 + 6x^2 - 4x + 5$	i
53.	$f(x) = x^3 - 2x^2 - 14x + 40$	$3 - i$
54.	$g(x) = 4x^3 + 23x^2 + 34x - 10$	$-3 + i$
55.	$g(x) = x^3 - 8x^2 + 25x - 26$	$3 + 2i$
56.	$f(x) = x^3 + 4x^2 + 14x + 20$	$-1 - 3i$
57.	$h(x) = x^4 - 2x^3 + 8x^2 - 8x + 16$	$1 + \sqrt{3}i$
58.	$h(x) = x^4 - 6x^3 + 14x^2 - 18x + 9$	$1 - \sqrt{2}i$

Finding a Polynomial Function with Given Zeros In Exercises 59–64, find a polynomial function with real coefficients that has the given zeros. (There are many correct answers.)

59. $1, 5i$

60. $4, -3i$

61. $2, 2, 1 + i$

62. $-1, 5, 3 - 2i$

63. $\frac{2}{3}, -1, 3 + \sqrt{2}i$

64. $-\frac{5}{2}, -5, 1 + \sqrt{3}i$

Finding a Polynomial Function with Given Zeros In Exercises 65–70, find the polynomial function f with real coefficients that has the given degree, zeros, and function value.

	Degree	Zeros	Function Value
65.	3	$1, 2i$	$f(-1) = 10$
66.	3	$2, i$	$f(-1) = 6$
67.	4	$-2, 1, i$	$f(0) = -4$
68.	4	$-1, 2, \sqrt{2}i$	$f(1) = 12$
69.	3	$-3, 1 + \sqrt{3}i$	$f(-2) = 12$
70.	3	$-2, 1 - \sqrt{2}i$	$f(-1) = -12$

Finding a Polynomial Function In Exercises 71–74, find a cubic polynomial function f with real coefficients that has the given complex zeros and x-intercept. (There are many correct answers.)

	Complex Zeros	x-Intercept
71.	$x = 4 \pm 2i$	$(-2, 0)$
72.	$x = 3 \pm i$	$(1, 0)$
73.	$x = 2 \pm \sqrt{5}i$	$(2, 0)$
74.	$x = -1 \pm \sqrt{3}i$	$(-4, 0)$

75. **Writing an Equation** The graph of a fourth-degree polynomial function $y = f(x)$ is shown. The equation has $\pm\sqrt{2}i$ as zeros. Write an equation for f.

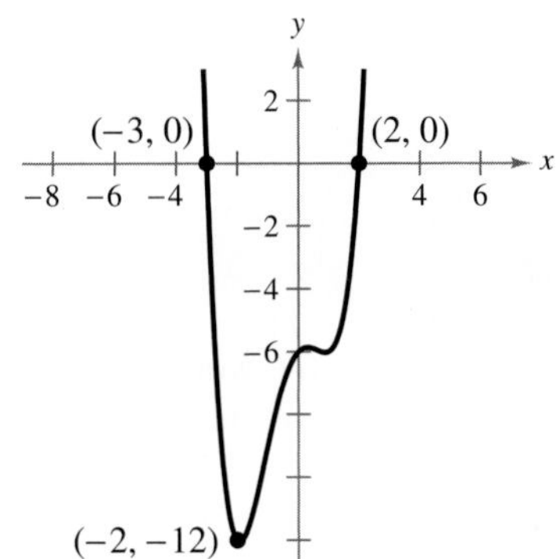

76. **Writing an Equation** The graph of a fourth-degree polynomial function $y = f(x)$ is shown. The equation has $\pm\sqrt{5}i$ as zeros. Write an equation for f.

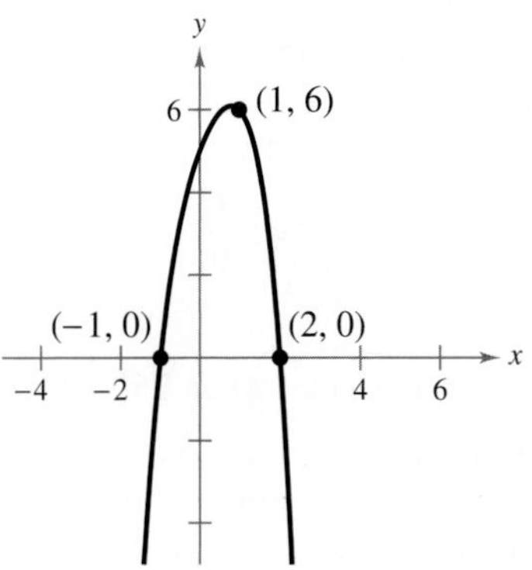

77. **Height of a Ball** A ball is kicked upward from ground level with an initial velocity of 48 feet per second. The height h (in feet) of the ball is modeled by $h(t) = -16t^2 + 48t$ for $0 \le t \le 3$, where t represents the time (in seconds).

(a) Complete the table to find the heights h of the ball for the given times t. Does it appear that the ball reaches a height of 64 feet?

t	0	0.5	1	1.5	2	2.5	3
h							

(b) Algebraically determine whether the ball reaches a height of 64 feet.

(c) Use a graphing utility to graph the function. Graphically determine whether the ball reaches a height of 64 feet.

(d) Compare your results from parts (a), (b), and (c).

78. **Height of a Baseball** A baseball is thrown upward from a height of 5 feet with an initial velocity of 79 feet per second. The height h (in feet) of the baseball is modeled by $h = -16t^2 + 79t + 5$ for $0 \le t \le 5$, where t represents the time (in seconds).

(a) Complete the table to find the heights h of the baseball for the given times t. Does it appear that the baseball reaches a height of 110 feet?

t	0	1	2	3	4	5
h						

(b) Algebraically determine whether the baseball reaches a height of 110 feet.

(c) Use a graphing utility to graph the function. Graphically determine whether the baseball reaches a height of 110 feet.

(d) Compare your results from parts (a), (b), and (c).

79. Profit The demand equation for a microwave oven is given by $p = 140 - 0.0001x$, where p is the unit price (in dollars) of the microwave oven and x is the number of units sold. The cost equation for the microwave oven is $C = 80x + 150{,}000$, where C is the total cost (in dollars) and x is the number of units produced. The total profit P obtained by producing and selling x units is modeled by $P = xp - C$.

(a) Find the profit function P in terms of x.

(b) Find the profit when 250,000 units are sold.

(c) Find the unit price when 250,000 units are sold.

(d) Find (if possible) the unit price that will yield a profit of 10 million dollars. If not possible, explain why.

80. Physiology

Doctors treated a patient at an emergency room from 2:00 P.M. to 7:00 P.M. The patient's blood oxygen level L (in percent form) during this time period can be modeled by

$$L = -0.270t^2 + 3.59t + 83.1, \quad 2 \le t \le 7$$

where t represents the time of day, with $t = 2$ corresponding to 2:00 P.M. Use the model to estimate the time (rounded to the nearest hour) when the patient's blood oxygen level was 93%.

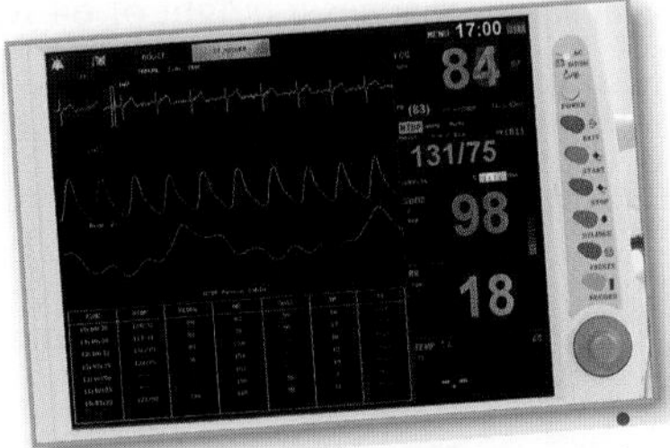

Exploration

True or False **In Exercises 81 and 82, decide whether the statement is true or false. Justify your answer.**

81. It is possible for a third-degree polynomial function with integer coefficients to have no real zeros.

82. If $x = -i$ is a zero of the function

$$f(x) = x^3 + ix^2 + ix - 1$$

then $x = i$ must also be a zero of f.

83. Writing Write a paragraph explaining the relationships among the solutions of a polynomial equation, the zeros of a polynomial function, and the x-intercepts of the graph of a polynomial function. Include examples in your paragraph.

84. Error Analysis Describe the error in finding a polynomial function f with real coefficients that has -2, 3.5, and i as zeros.

A function is $f(x) = (x + 2)(x - 3.5)(x - i)$. ✗

85. Finding a Quadratic Function Find a quadratic function f (with integer coefficients) that has the given zeros. Assume that b is a positive integer and a is an integer not equal to zero.

(a) $\pm\sqrt{b}i$ (b) $a \pm bi$

86. HOW DO YOU SEE IT? In parts (a)–(c) use the graph to determine whether the discriminant of the given equation is *positive, zero,* or *negative.* Explain.

(a) $x^2 - 2x = 0$

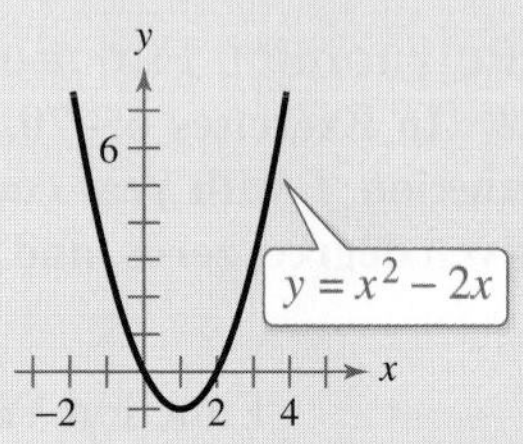

(b) $x^2 - 2x + 1 = 0$

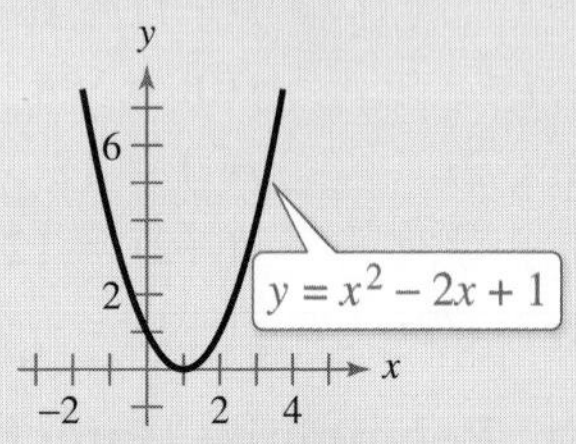

(c) $x^2 - 2x + 2 = 0$

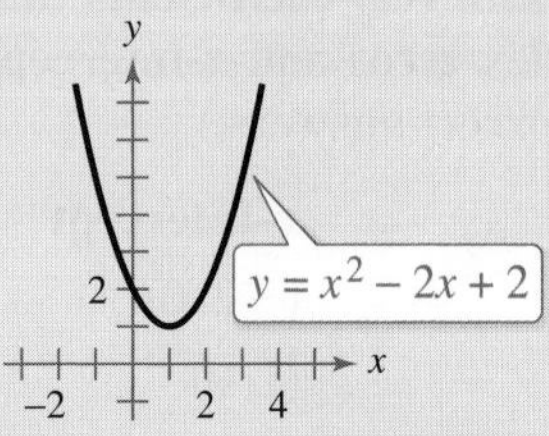

Think About It **In Exercises 87–92, determine (if possible) the zeros of the function g when the function f has zeros at $x = r_1$, $x = r_2$, and $x = r_3$.**

87. $g(x) = -f(x)$

88. $g(x) = 4f(x)$

89. $g(x) = f(x - 5)$

90. $g(x) = f(2x)$

91. $g(x) = 3 + f(x)$

92. $g(x) = f(-x)$

Project: Population To work an extended application analyzing the population of the United States, visit this text's website at *LarsonPrecalculus.com.* *(Source: U.S. Census Bureau)*

4.3 The Complex Plane

The complex plane has many practical applications. For example, in Exercise 49 on page 335, you will use the complex plane to write complex numbers that represent the positions of two ships.

- **Plot complex numbers in the complex plane and find absolute values of complex numbers.**
- **Perform operations with complex numbers in the complex plane.**
- **Use the Distance and Midpoint Formulas in the complex plane.**

The Complex Plane

Just as a real number can be represented by a point on the real number line, a complex number $z = a + bi$ can be represented by the point (a, b) in a coordinate plane (the **complex plane**). In the complex plane, the horizontal axis is the **real axis** and the vertical axis is the **imaginary axis,** as shown in the figure below.

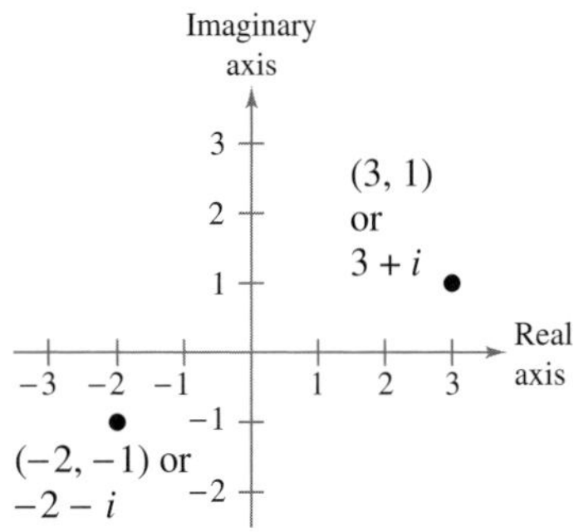

The **absolute value,** or **modulus,** of the complex number $z = a + bi$ is the distance between the origin $(0, 0)$ and the point (a, b). (The plural of modulus is *moduli.*)

Definition of the Absolute Value of a Complex Number

The **absolute value** of the complex number $z = a + bi$ is

$$|a + bi| = \sqrt{a^2 + b^2}.$$

When the complex number $z = a + bi$ is a real number (that is, when $b = 0$), this definition agrees with that given for the absolute value of a real number

$$|a + 0i| = \sqrt{a^2 + 0^2}$$
$$= |a|.$$

EXAMPLE 1 Finding the Absolute Value of a Complex Number

See LarsonPrecalculus.com for an interactive version of this type of example.

Plot $z = -2 + 5i$ in the complex plane and find its absolute value.

Solution The number is plotted in Figure 4.2. It has an absolute value of

$$|z| = \sqrt{(-2)^2 + 5^2}$$
$$= \sqrt{29}.$$

Figure 4.2

Checkpoint Audio-video solution in English & Spanish at LarsonPrecalculus.com

Plot $z = 3 - 4i$ in the complex plane and find its absolute value.

Michael C. Gray/Shutterstock.com

Operations with Complex Numbers in the Complex Plane

In Section 3.3, you learned how to add and subtract vectors geometrically in the coordinate plane. In a similar way, you can add and subtract complex numbers geometrically in the complex plane.

The complex number $z = a + bi$ can be represented by the vector $\mathbf{u} = \langle a, b \rangle$. For example, the complex number $z = 1 + 2i$ can be represented by the vector $\mathbf{u} = \langle 1, 2 \rangle$. To add two complex numbers geometrically, first represent them as vectors $\mathbf{u}$ and $\mathbf{v}$. Then add the vectors, as shown in the next two figures. The sum of the vectors represents the sum of the complex numbers.

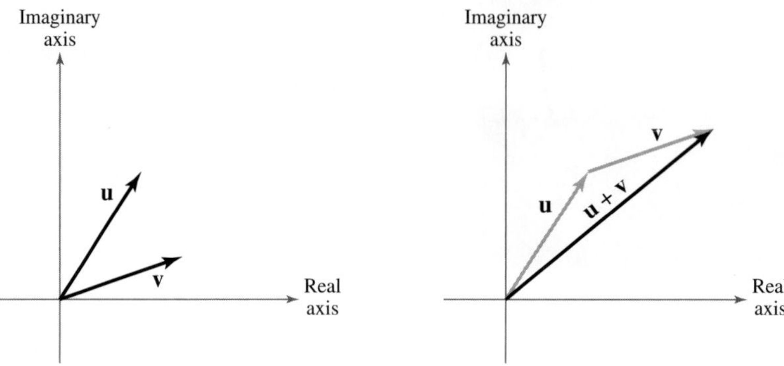

EXAMPLE 2 Adding in the Complex Plane

Find $(1 + 3i) + (2 + i)$ in the complex plane.

Solution

Let the vectors $\mathbf{u} = \langle 1, 3 \rangle$ and $\mathbf{v} = \langle 2, 1 \rangle$ represent the complex numbers $1 + 3i$ and $2 + i$, respectively. Graph the vectors $\mathbf{u}$, $\mathbf{v}$, and $\mathbf{u} + \mathbf{v}$, as shown at the right. From the graph, $\mathbf{u} + \mathbf{v} = \langle 3, 4 \rangle$, which implies that

$$(1 + 3i) + (2 + i) = 3 + 4i.$$

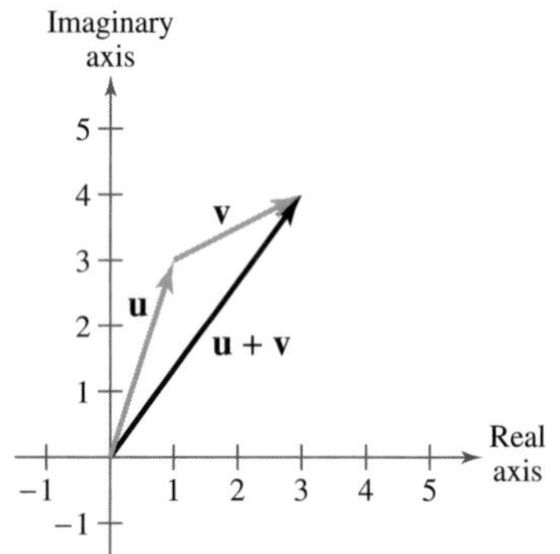

Checkpoint Audio-video solution in English & Spanish at LarsonPrecalculus.com

Find $(3 + i) + (1 + 2i)$ in the complex plane.

To subtract two complex numbers geometrically, first represent them as vectors $\mathbf{u}$ and $\mathbf{v}$. Then subtract the vectors, as shown in the figure below. The difference of the vectors represents the difference of the complex numbers.

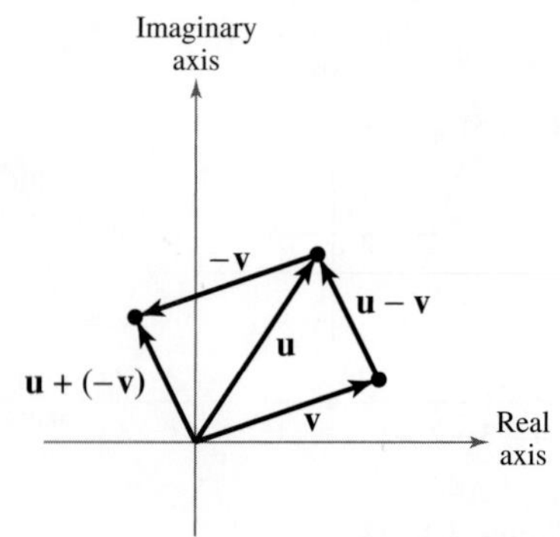

EXAMPLE 3 Subtracting in the Complex Plane

Find $(4 + 2i) - (3 - i)$ in the complex plane.

Solution

Let the vectors $\mathbf{u} = \langle 4, 2 \rangle$ and $\mathbf{v} = \langle 3, -1 \rangle$ represent the complex numbers $4 + 2i$ and $3 - i$, respectively. Graph the vectors $\mathbf{u}$, $-\mathbf{v}$, and $\mathbf{u} + (-\mathbf{v})$, as shown in Figure 4.3. From the graph, $\mathbf{u} - \mathbf{v} = \mathbf{u} + (-\mathbf{v}) = \langle 1, 3 \rangle$, which implies that

$$(4 + 2i) - (3 - i) = 1 + 3i.$$

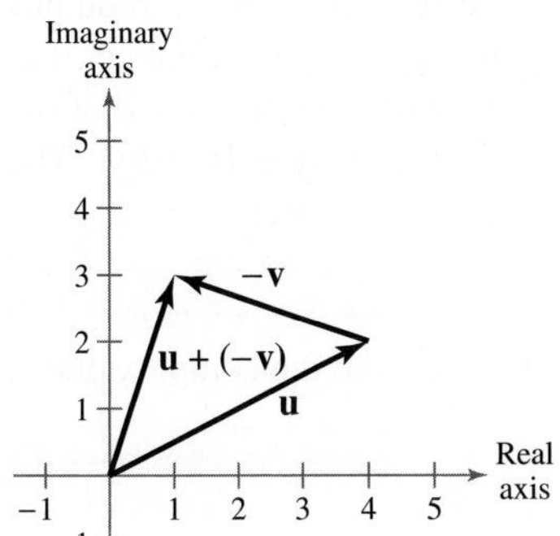

Figure 4.3

Checkpoint Audio-video solution in English & Spanish at LarsonPrecalculus.com

Find $(2 - 4i) - (1 + i)$ in the complex plane.

Recall that the complex numbers $a + bi$ and $a - bi$ are *complex conjugates*. The points (a, b) and $(a, -b)$ are reflections of each other in the real axis, as shown in the figure below. This information enables you to find a complex conjugate geometrically.

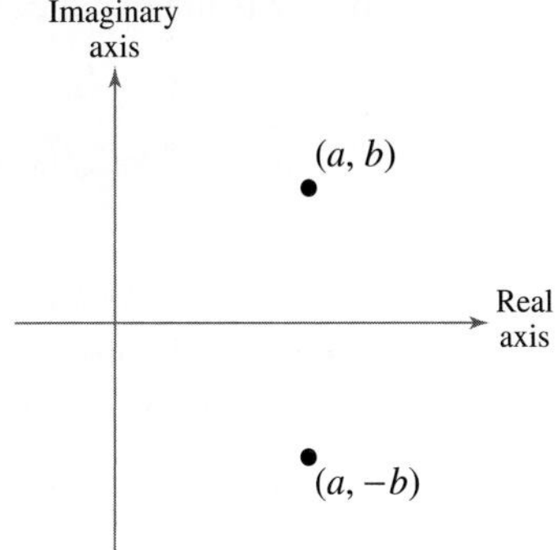

EXAMPLE 4 Complex Conjugates in the Complex Plane

Plot $z = -3 + i$ and its complex conjugate in the complex plane. Write the conjugate as a complex number.

Solution

The figure below shows the point $(-3, 1)$ and its reflection in the real axis, $(-3, -1)$. So, the complex conjugate of $-3 + i$ is $-3 - i$.

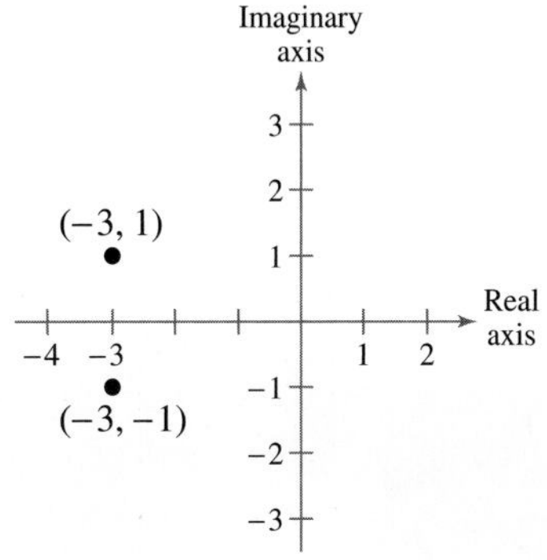

Checkpoint Audio-video solution in English & Spanish at LarsonPrecalculus.com

Plot $z = 2 - 3i$ and its complex conjugate in the complex plane. Write the conjugate as a complex number.

Distance and Midpoint Formulas in the Complex Plane

For two points in the complex plane, the distance between the points is the modulus (or absolute value) of the difference of the two corresponding complex numbers. Let (a, b) and (s, t) be points in the complex plane. One way to write the difference of the corresponding complex numbers is $(s + ti) - (a + bi) = (s - a) + (t - b)i$. The modulus of the difference is

$$|(s - a) + (t - b)i| = \sqrt{(s - a)^2 + (t - b)^2}.$$

So, $d = \sqrt{(s - a)^2 + (t - b)^2}$ is the distance between the points in the complex plane.

Distance Formula in the Complex Plane

The distance d between the points (a, b) and (s, t) in the complex plane is

$$d = \sqrt{(s - a)^2 + (t - b)^2}.$$

Figure 4.4

Figure 4.4 shows the points represented as vectors. The magnitude of the vector $\mathbf{u} - \mathbf{v}$ is the distance between (a, b) and (s, t).

$$\mathbf{u} - \mathbf{v} = \langle s - a, t - b \rangle$$

$$\|\mathbf{u} - \mathbf{v}\| = \sqrt{(s - a)^2 + (t - b)^2}$$

EXAMPLE 5 Finding Distance in the Complex Plane

Find the distance between $2 + 3i$ and $5 - 2i$ in the complex plane.

Solution

Let $a + bi = 2 + 3i$ and $s + ti = 5 - 2i$. The distance is

$$\begin{aligned} d &= \sqrt{(s - a)^2 + (t - b)^2} \\ &= \sqrt{(5 - 2)^2 + (-2 - 3)^2} \\ &= \sqrt{3^2 + (-5)^2} \\ &= \sqrt{34} \\ &\approx 5.83 \text{ units} \end{aligned}$$

as shown in the figure below.

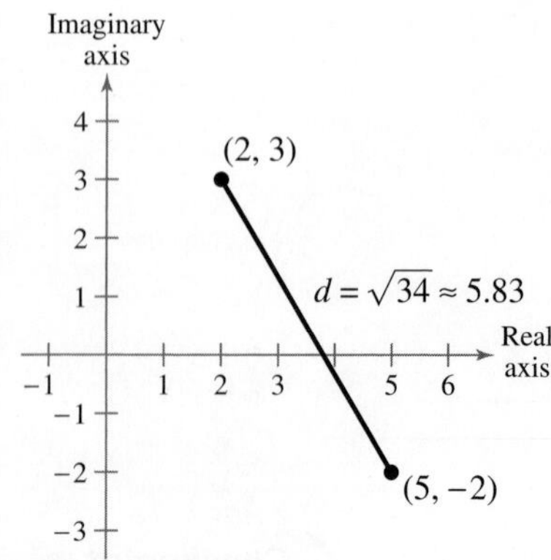

✓ **Checkpoint** Audio-video solution in English & Spanish at LarsonPrecalculus.com

Find the distance between $5 - 4i$ and $6 + 5i$ in the complex plane.

To find the midpoint of the line segment joining two points in the complex plane, find the average values of the respective coordinates of the two endpoints.

Midpoint Formula in the Complex Plane

The midpoint of the line segment joining the points (a, b) and (s, t) in the complex plane is

$$\text{Midpoint} = \left(\frac{a + s}{2}, \frac{b + t}{2}\right).$$

EXAMPLE 6 **Finding a Midpoint in the Complex Plane**

Find the midpoint of the line segment joining the points corresponding to $4 - 3i$ and $2 + 2i$ in the complex plane.

Solution

Let the points $(4, -3)$ and $(2, 2)$ represent the complex numbers $4 - 3i$ and $2 + 2i$, respectively. Apply the Midpoint Formula.

$$\text{Midpoint} = \left(\frac{a + s}{2}, \frac{b + t}{2}\right) = \left(\frac{4 + 2}{2}, \frac{-3 + 2}{2}\right) = \left(3, -\frac{1}{2}\right)$$

The midpoint is $\left(3, -\frac{1}{2}\right)$, as shown in the figure below.

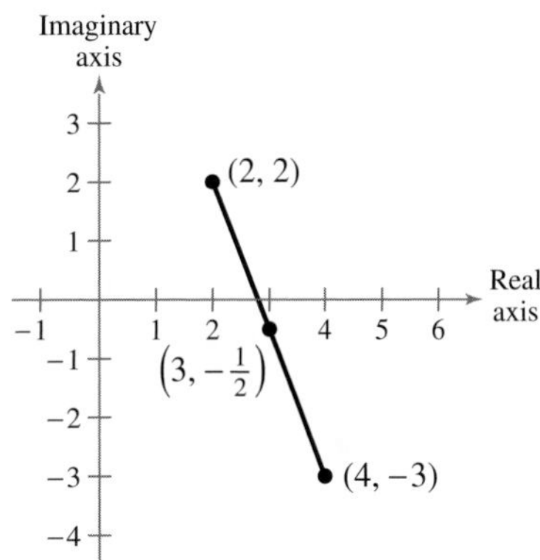

Checkpoint *Audio-video solution in English & Spanish at LarsonPrecalculus.com*

Find the midpoint of the line segment joining the points corresponding to $2 + i$ and $5 - 5i$ in the complex plane.

Summarize (Section 4.3)

1. State the definition of the absolute value, or modulus, of a complex number *(page 329)*. For an example of finding the absolute value of a complex number, see Example 1.
2. Explain how to add, subtract, and find complex conjugates of complex numbers in the complex plane *(page 330)*. For examples of performing operations with complex numbers in the complex plane, see Examples 2–4.
3. Explain how to use the Distance and Midpoint Formulas in the complex plane *(page 332)*. For examples of using the Distance and Midpoint Formulas in the complex plane, see Examples 5 and 6.

4.3 Exercises

See **CalcChat.com** for tutorial help and worked-out solutions to odd-numbered exercises.

Vocabulary: Fill in the blanks.

1. In the complex plane, the horizontal axis is the ________ axis.
2. In the complex plane, the vertical axis is the ________ axis.
3. The ________ ________ of the complex number $a + bi$ is the distance between the origin and (a, b).
4. To subtract two complex numbers geometrically, first represent them as ________.
5. The points that represent a complex number and its complex conjugate are ________ of each other in the real axis.
6. The distance between two points in the complex plane is the ________ of the difference of the two corresponding complex numbers.

Skills and Applications

Matching In Exercises 7–14, match the complex number with its representation in the complex plane. [The representations are labeled (a)–(h).]

(a)

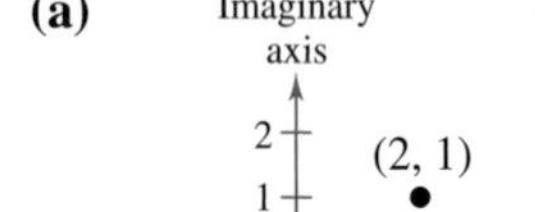

(b)

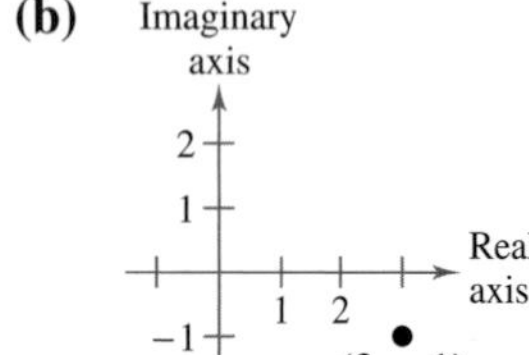

(c)

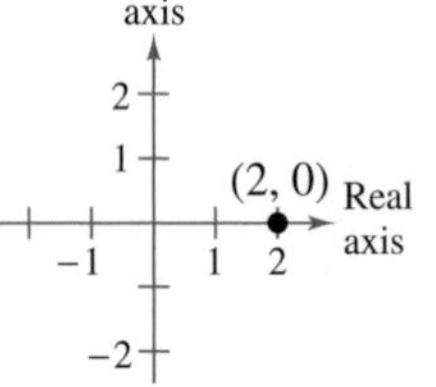

(d)

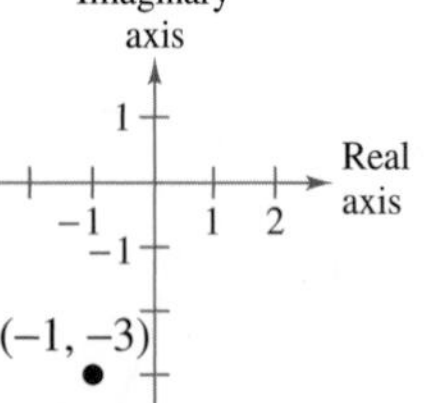

(e)

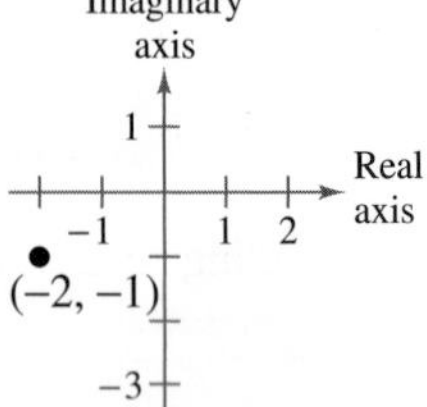

(f)

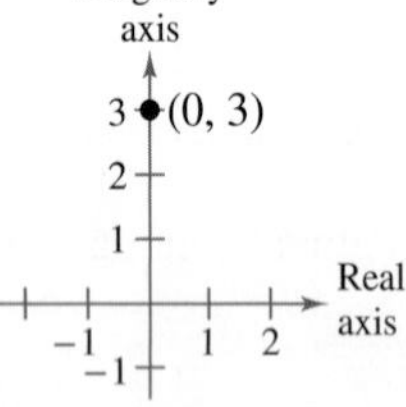

(g)

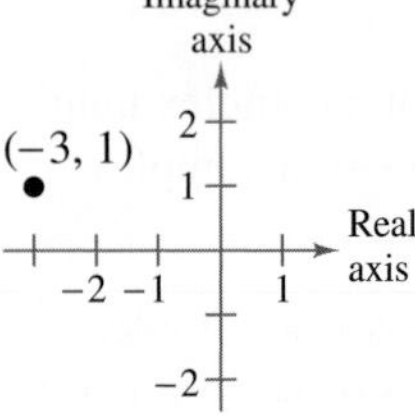

(h)

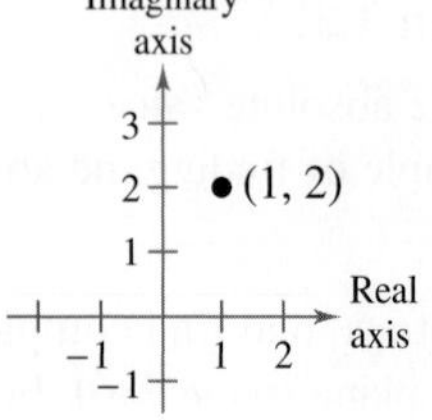

7. 2
8. $3i$
9. $1 + 2i$
10. $2 + i$
11. $3 - i$
12. $-3 + i$
13. $-2 - i$
14. $-1 - 3i$

Finding the Absolute Value of a Complex Number In Exercises 15–20, plot the complex number and find its absolute value.

15. $-7i$
16. -7
17. $-6 + 8i$
18. $5 - 12i$
19. $4 - 6i$
20. $-8 + 3i$

Adding in the Complex Plane In Exercises 21–28, find the sum of the complex numbers in the complex plane.

21. $(3 + i) + (2 + 5i)$
22. $(5 + 2i) + (3 + 4i)$
23. $(8 - 2i) + (2 + 6i)$
24. $(3 - i) + (-1 + 2i)$

25.

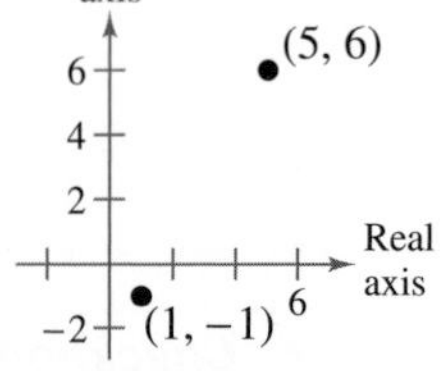

26. Imaginary axis; Real axis; (2, 4); (−1, 3)

27.

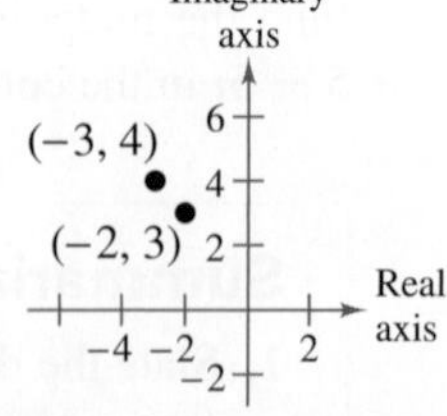

28.

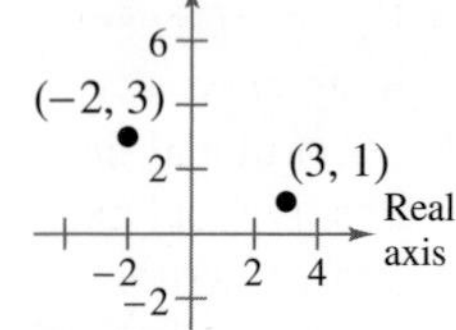

Subtracting in the Complex Plane In Exercises 29–36, find the difference of the complex numbers in the complex plane.

29. $(4 + 2i) - (6 + 4i)$
30. $(-3 + i) - (3 + i)$
31. $(5 - i) - (-5 + 2i)$
32. $(2 - 3i) - (3 + 2i)$
33. $2 - (2 + 6i)$
34. $-3 - (2 + 2i)$
35. $-2i - (3 - 5i)$
36. $3i - (-3 + 7i)$

Complex Conjugates in the Complex Plane **In Exercises 37–40, plot the complex number and its complex conjugate. Write the conjugate as a complex number.**

37. $2 + 3i$
38. $5 - 4i$
39. $-1 - 2i$
40. $-7 + 3i$

Finding Distance in the Complex Plane **In Exercises 41–44, find the distance between the complex numbers in the complex plane.**

41. $1 + 2i, -1 + 4i$
42. $-5 + i, -2 + 5i$
43. $6i, 3 - 4i$
44. $-7 - 3i, 3 + 5i$

Finding a Midpoint in the Complex Plane **In Exercises 45–48, find the midpoint of the line segment joining the points corresponding to the complex numbers in the complex plane.**

45. $2 + i, 6 + 5i$
46. $-3 + 4i, 1 - 2i$
47. $7i, 9 - 10i$
48. $-1 - \frac{3}{4}i, \frac{1}{2} + \frac{1}{4}i$

49. Sailing

Ship A is 3 miles east and 4 miles north of port. Ship B is 5 miles west and 2 miles north of port (see figure).

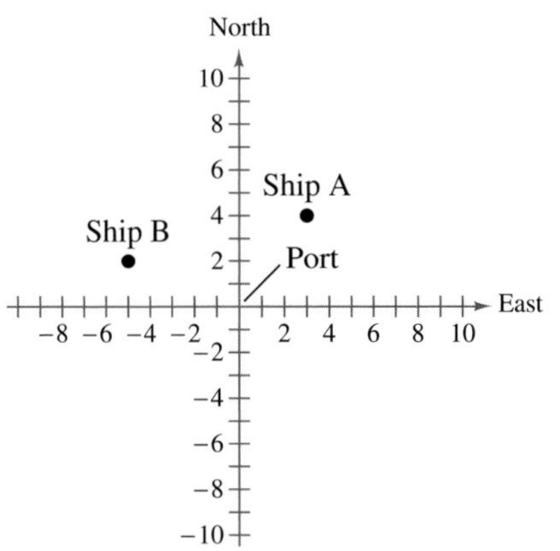

(a) Using the positive imaginary axis as north and the positive real axis as east, write complex numbers that represent the positions of Ship A and Ship B relative to port.

(b) How can you use the complex numbers in part (a) to find the distance between Ship A and Ship B?

50. **Force** Two forces are acting on a point. The first force has a horizontal component of 5 newtons and a vertical component of 3 newtons. The second force has a horizontal component of 4 newtons and a vertical component of 2 newtons.
    (a) Plot the vectors that represent the two forces in the complex plane.
    (b) Find the horizontal and vertical components of the resultant force acting on the point using the complex plane.

Exploration

True or False? **In Exercises 51–54, determine whether the statement is true or false. Justify your answer.**

51. The modulus of a complex number can be real or imaginary.
52. The distance between two points in the complex plane is always real.
53. The modulus of the sum of two complex numbers is equal to the sum of their moduli.
54. The modulus of the difference of two complex numbers is equal to the difference of their moduli.
55. **Think About It** What does the set of all points with the same modulus represent in the complex plane? Explain.

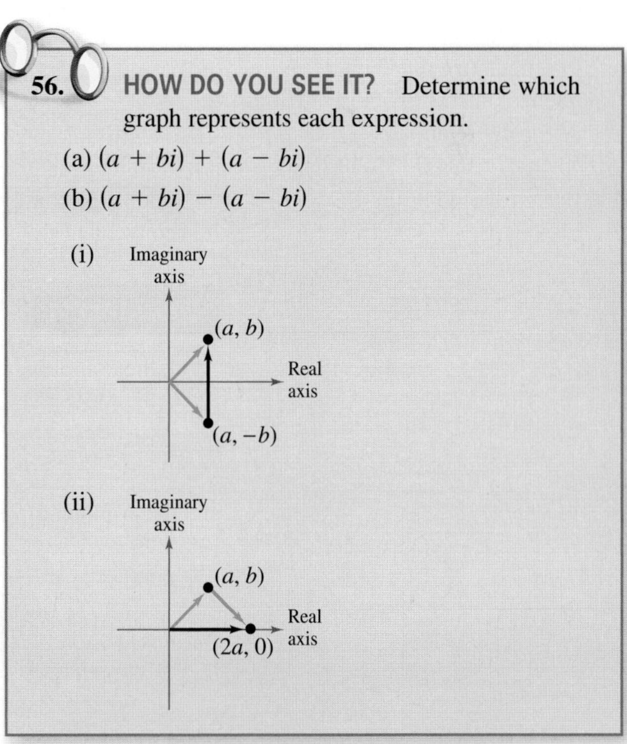

56. **HOW DO YOU SEE IT?** Determine which graph represents each expression.
    (a) $(a + bi) + (a - bi)$
    (b) $(a + bi) - (a - bi)$

57. **Think About It** The points corresponding to a complex number and its complex conjugate are plotted in the complex plane. What type of triangle do these points form with the origin?

4.4 Trigonometric Form of a Complex Number

Trigonometric forms of complex numbers have applications in circuit analysis. For example, in Exercise 69 on page 341, you will use trigonometric forms of complex numbers to find the voltage of an alternating current circuit.

- Write trigonometric forms of complex numbers.
- Multiply and divide complex numbers written in trigonometric form.

Trigonometric Form of a Complex Number

In Section 4.1, you learned how to add, subtract, multiply, and divide complex numbers. To work effectively with *powers* and *roots* of complex numbers, it is helpful to write complex numbers in trigonometric form. Consider the nonzero complex number $a + bi$, plotted at the right. By letting θ be the angle from the positive real axis (measured counterclockwise) to the line segment connecting the origin and the point (a, b), you can write $a = r\cos\theta$ and $b = r\sin\theta$, where $r = \sqrt{a^2 + b^2}$. Consequently, you have $a + bi = (r\cos\theta) + (r\sin\theta)i$, from which you can obtain the **trigonometric form of a complex number.**

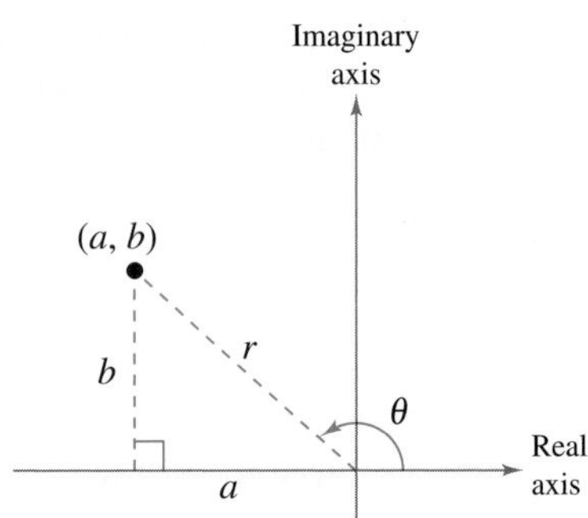

Trigonometric Form of a Complex Number

The **trigonometric form** of the complex number $z = a + bi$ is

$$z = r(\cos\theta + i\sin\theta)$$

where $a = r\cos\theta$, $b = r\sin\theta$, $r = \sqrt{a^2 + b^2}$, and $\tan\theta = b/a$. The number r is the **modulus** of z, and θ is an **argument** of z.

REMARK For $0 \le \theta < 2\pi$, use the guidelines below. When z lies in Quadrant I, $\theta = \arctan(b/a)$. When z lies in Quadrant II or Quadrant III, $\theta = \pi + \arctan(b/a)$. When z lies in Quadrant IV, $\theta = 2\pi + \arctan(b/a)$.

The trigonometric form of a complex number is also called the *polar form.* There are infinitely many choices for θ, so the trigonometric form of a complex number is not unique. Normally, θ is restricted to the interval $0 \le \theta < 2\pi$, although on occasion it is convenient to use $\theta < 0$.

EXAMPLE 1 Trigonometric Form of a Complex Number

Write the complex number $z = -2 - 2\sqrt{3}i$ in trigonometric form.

Solution The modulus of z is

$$r = \left|-2 - 2\sqrt{3}i\right| = \sqrt{(-2)^2 + \left(-2\sqrt{3}\right)^2} = \sqrt{16} = 4$$

and the argument θ is determined from

$$\tan\theta = \frac{b}{a} = \frac{-2\sqrt{3}}{-2} = \sqrt{3}.$$

Because $z = -2 - 2\sqrt{3}i$ lies in Quadrant III, as shown in Figure 4.5, you have $\theta = \pi + \arctan\sqrt{3} = \pi + (\pi/3) = 4\pi/3$. So, the trigonometric form of z is

$$z = r(\cos\theta + i\sin\theta) = 4\left(\cos\frac{4\pi}{3} + i\sin\frac{4\pi}{3}\right).$$

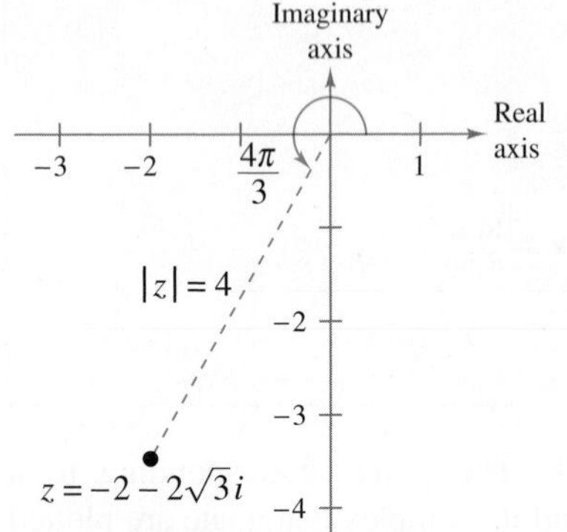

Figure 4.5

Write the complex number $z = 6 - 6i$ in trigonometric form.

EXAMPLE 2 **Trigonometric Form of a Complex Number**

See LarsonPrecalculus.com for an interactive version of this type of example.

Write the complex number $z = 6 + 2i$ in trigonometric form.

Solution The modulus of z is

$$r = |6 + 2i| = \sqrt{6^2 + 2^2} = \sqrt{40} = 2\sqrt{10}$$

and the argument θ is determined from

$$\tan\theta = \frac{b}{a} = \frac{2}{6} = \frac{1}{3}.$$

Because $z = 6 + 2i$ is in Quadrant I, as shown at the right, you have

$$\theta = \arctan\frac{1}{3}$$

$$\approx 0.32175 \text{ radian}$$

$$\approx 18.4°.$$

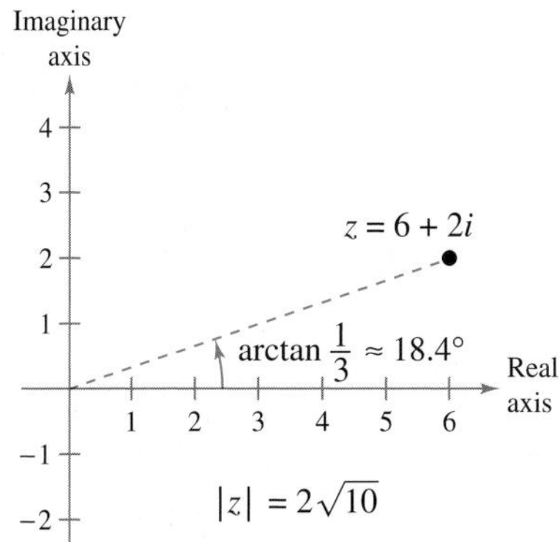

So, the trigonometric form of z is

$$z = r(\cos\theta + i\sin\theta)$$

$$\approx 2\sqrt{10}(\cos 18.4° + i\sin 18.4°).$$

✓ ***Checkpoint*** *Audio-video solution in English & Spanish at LarsonPrecalculus.com*

Write the complex number $z = 3 + 4i$ in trigonometric form.

▷ TECHNOLOGY A graphing utility can be used to convert a complex number in trigonometric form to standard form. For specific keystrokes, see the user's manual for your graphing utility.

EXAMPLE 3 **Writing a Complex Number in Standard Form**

Write $z = 4(\cos 120° + i\sin 120°)$ in standard form $a + bi$.

Solution Because $\cos 120° = -\frac{1}{2}$ and $\sin 120° = \frac{\sqrt{3}}{2}$, you can write

$$z = 4(\cos 120° + i\sin 120°) = 4\left(-\frac{1}{2} + \frac{\sqrt{3}}{2}i\right) = -2 + 2\sqrt{3}i.$$

✓ ***Checkpoint*** *Audio-video solution in English & Spanish at LarsonPrecalculus.com*

Write $z = 2(\cos 150° + i\sin 150°)$ in standard form $a + bi$.

EXAMPLE 4 **Writing a Complex Number in Standard Form**

Write $z = \sqrt{8}\left[\cos\left(-\frac{\pi}{3}\right) + i\sin\left(-\frac{\pi}{3}\right)\right]$ in standard form $a + bi$.

Solution Because $\cos\left(-\frac{\pi}{3}\right) = \frac{1}{2}$ and $\sin\left(-\frac{\pi}{3}\right) = -\frac{\sqrt{3}}{2}$, you can write

$$z = \sqrt{8}\left[\cos\left(-\frac{\pi}{3}\right) + i\sin\left(-\frac{\pi}{3}\right)\right] = 2\sqrt{2}\left(\frac{1}{2} - \frac{\sqrt{3}}{2}i\right) = \sqrt{2} - \sqrt{6}i.$$

✓ ***Checkpoint*** *Audio-video solution in English & Spanish at LarsonPrecalculus.com*

Write $z = 8\left(\cos\frac{2\pi}{3} + i\sin\frac{2\pi}{3}\right)$ in standard form $a + bi$.

Multiplication and Division of Complex Numbers

The trigonometric form adapts nicely to multiplication and division of complex numbers. Consider two complex numbers

$$z_1 = r_1(\cos\theta_1 + i\sin\theta_1) \quad \text{and} \quad z_2 = r_2(\cos\theta_2 + i\sin\theta_2).$$

The product of z_1 and z_2 is

$$\begin{aligned} z_1 z_2 &= r_1 r_2(\cos\theta_1 + i\sin\theta_1)(\cos\theta_2 + i\sin\theta_2) \\ &= r_1 r_2[(\cos\theta_1\cos\theta_2 - \sin\theta_1\sin\theta_2) + i(\sin\theta_1\cos\theta_2 + \cos\theta_1\sin\theta_2)]. \end{aligned}$$

Using the sum and difference formulas for cosine and sine, this equation is equivalent to

$$z_1 z_2 = r_1 r_2[\cos(\theta_1 + \theta_2) + i\sin(\theta_1 + \theta_2)].$$

This establishes the first part of the rule below. The second part is left for you to verify (see Exercise 73).

Product and Quotient of Two Complex Numbers

Let $z_1 = r_1(\cos\theta_1 + i\sin\theta_1)$ and $z_2 = r_2(\cos\theta_2 + i\sin\theta_2)$ be complex numbers.

$$z_1 z_2 = r_1 r_2[\cos(\theta_1 + \theta_2) + i\sin(\theta_1 + \theta_2)] \qquad \text{Product}$$

$$\frac{z_1}{z_2} = \frac{r_1}{r_2}[\cos(\theta_1 - \theta_2) + i\sin(\theta_1 - \theta_2)], \quad z_2 \neq 0 \qquad \text{Quotient}$$

Note that this rule states that to *multiply* two complex numbers you multiply moduli and add arguments, whereas to *divide* two complex numbers you divide moduli and subtract arguments.

EXAMPLE 5 Multiplying Complex Numbers

Find the product $z_1 z_2$ of

$$z_1 = 2(\cos 120° + i\sin 120°) \quad \text{and} \quad z_2 = 8(\cos 330° + i\sin 330°).$$

Solution

$$\begin{aligned} z_1 z_2 &= 2(\cos 120° + i\sin 120°) \cdot 8(\cos 330° + i\sin 330°) \\ &= 16[\cos(120° + 330°) + i\sin(120° + 330°)] && \text{Multiply moduli and add arguments.} \\ &= 16(\cos 450° + i\sin 450°) \\ &= 16(\cos 90° + i\sin 90°) && \text{450° and 90° are coterminal.} \\ &= 16[0 + i(1)] \\ &= 16i \end{aligned}$$

✓ Checkpoint *Audio-video solution in English & Spanish at LarsonPrecalculus.com*

Find the product $z_1 z_2$ of

$$z_1 = 3(\cos 60° + i\sin 60°) \quad \text{and} \quad z_2 = 4(\cos 30° + i\sin 30°).$$

Check the solution to Example 5 by first converting the complex numbers to the standard forms $z_1 = -1 + \sqrt{3}i$ and $z_2 = 4\sqrt{3} - 4i$ and then multiplying algebraically, as in Section 4.1.

$$z_1 z_2 = (-1 + \sqrt{3}i)(4\sqrt{3} - 4i) = -4\sqrt{3} + 4i + 12i + 4\sqrt{3} = 16i$$

TECHNOLOGY Some graphing utilities can multiply and divide complex numbers in trigonometric form. If you have access to such a graphing utility, use it to check the solutions to Examples 5 and 6.

EXAMPLE 6 Dividing Complex Numbers

Find the quotient $\dfrac{z_1}{z_2}$ of $z_1 = 24\left(\cos\dfrac{5\pi}{3} + i\sin\dfrac{5\pi}{3}\right)$ and $z_2 = 8\left(\cos\dfrac{5\pi}{12} + i\sin\dfrac{5\pi}{12}\right)$.

Solution

$$\frac{z_1}{z_2} = \frac{24[\cos(5\pi/3) + i\sin(5\pi/3)]}{8[\cos(5\pi/12) + i\sin(5\pi/12)]}$$

$$= 3\left[\cos\left(\frac{5\pi}{3} - \frac{5\pi}{12}\right) + i\sin\left(\frac{5\pi}{3} - \frac{5\pi}{12}\right)\right] \qquad \text{Divide moduli and subtract arguments.}$$

$$= 3\left(\cos\frac{5\pi}{4} + i\sin\frac{5\pi}{4}\right)$$

$$= -\frac{3\sqrt{2}}{2} - \frac{3\sqrt{2}}{2}i$$

✓ **Checkpoint** *Audio-video solution in English & Spanish at LarsonPrecalculus.com*

Find the quotient $\dfrac{z_1}{z_2}$ of $z_1 = \cos\dfrac{2\pi}{9} + i\sin\dfrac{2\pi}{9}$ and $z_2 = \cos\dfrac{\pi}{18} + i\sin\dfrac{\pi}{18}$.

In Section 4.3, you added, subtracted, and found complex conjugates of complex numbers geometrically in the complex plane. In a similar way, you can multiply complex numbers geometrically in the complex plane.

EXAMPLE 7 Multiplying in the Complex Plane

Find the product z_1z_2 of $z_1 = 2\left(\cos\dfrac{\pi}{6} + i\sin\dfrac{\pi}{6}\right)$ and $z_2 = 2\left(\cos\dfrac{\pi}{3} + i\sin\dfrac{\pi}{3}\right)$ in the complex plane.

Solution

Let $\mathbf{u} = 2\langle\cos(\pi/6), \sin(\pi/6)\rangle = \langle\sqrt{3}, 1\rangle$ and $\mathbf{v} = 2\langle\cos(\pi/3), \sin(\pi/3)\rangle = \langle 1, \sqrt{3}\rangle$. Then $\|\mathbf{u}\| = \sqrt{(\sqrt{3})^2 + 1^2} = \sqrt{4} = 2$ and $\|\mathbf{v}\| = \sqrt{1^2 + (\sqrt{3})^2} = \sqrt{4} = 2$. So, the magnitude of the product vector is $2(2) = 4$. The sum of the direction angles is $(\pi/6) + (\pi/3) = \pi/2$. So, the product vector lies on the imaginary axis and is represented in vector form as $\langle 0, 4\rangle$, as shown in Figure 4.6. This implies that $z_1z_2 = 4i$.

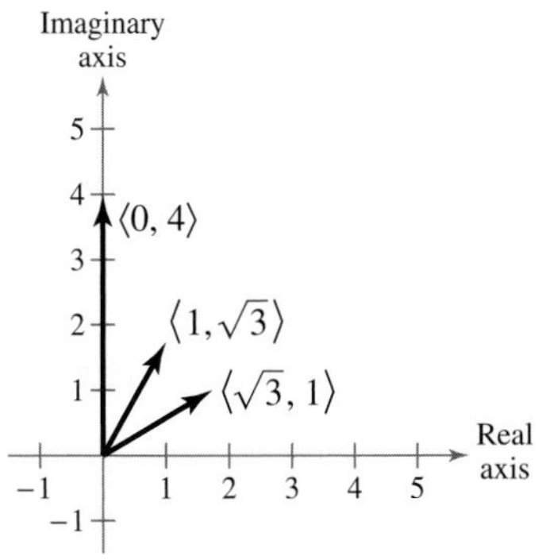

Figure 4.6

✓ **Checkpoint**  *Audio-video solution in English & Spanish at LarsonPrecalculus.com*

Find the product z_1z_2 of $z_1 = 2\left(\cos\dfrac{\pi}{4} + i\sin\dfrac{\pi}{4}\right)$ and $z_2 = 4\left(\cos\dfrac{3\pi}{4} + i\sin\dfrac{3\pi}{4}\right)$ in the complex plane.

Summarize (Section 4.4)

1. State the trigonometric form of a complex number *(page 336)*. For examples of writing complex numbers in trigonometric form and standard form, see Examples 1–4.
2. Explain how to multiply and divide complex numbers written in trigonometric form *(page 338)*. For examples of multiplying and dividing complex numbers written in trigonometric form, see Examples 5–7.

4.4 Exercises

See **CalcChat.com** for tutorial help and worked-out solutions to odd-numbered exercises.

Vocabulary: Fill in the blanks.

1. The ________ ________ of the complex number $z = a + bi$ is $z = r(\cos\theta + i\sin\theta)$, where r is the ________ of z and θ is an ________ of z.
2. Let $z_1 = r_1(\cos\theta_1 + i\sin\theta_1)$ and $z_2 = r_2(\cos\theta_2 + i\sin\theta_2)$ be complex numbers, then the product $z_1z_2 =$ ________ and the quotient $\dfrac{z_1}{z_2} =$ ________ $(z_1 \neq 0)$.

Skills and Applications

Trigonometric Form of a Complex Number

In Exercises 3–6, write the complex number in trigonometric form.

3.

Imaginary axis
4
3
$z = 3i$
2
1
−2 −1
1 2
Real axis

4.

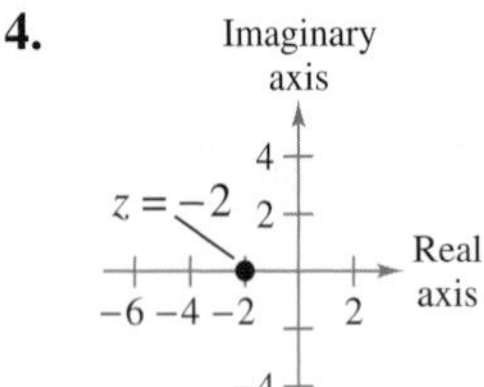

5.

Imaginary axis
Real axis
−3 −2
−2
−3
$z = -3 - 3i$

6.

Imaginary axis
3
$z = -1 + \sqrt{3}i$
Real axis
−3 −2 −1

Trigonometric Form of a Complex Number

In Exercises 7–26, plot the complex number. Then write the trigonometric form of the complex number.

7. $1 + i$
8. $5 - 5i$
9. $1 - \sqrt{3}i$
10. $4 - 4\sqrt{3}i$
11. $-2(1 + \sqrt{3}i)$
12. $\frac{5}{2}(\sqrt{3} - i)$
13. $-5i$
14. $12i$
15. 2
16. 4
17. $-7 + 4i$
18. $3 - i$
19. $3 + \sqrt{3}i$
20. $2\sqrt{2} - i$
21. $-3 - i$
22. $1 + 3i$
23. $5 + 2i$
24. $8 + 3i$
25. $-8 - 5\sqrt{3}i$
26. $-9 - 2\sqrt{10}i$

Writing a Complex Number in Standard Form

In Exercises 27–36, write the standard form of the complex number. Then plot the complex number.

27. $2(\cos 60^\circ + i\sin 60^\circ)$
28. $5(\cos 135^\circ + i\sin 135^\circ)$
29. $\dfrac{9}{4}\left(\cos\dfrac{3\pi}{4} + i\sin\dfrac{3\pi}{4}\right)$
30. $6\left(\cos\dfrac{5\pi}{12} + i\sin\dfrac{5\pi}{12}\right)$
31. $\sqrt{48}[\cos(-30^\circ) + i\sin(-30^\circ)]$
32. $\sqrt{8}(\cos 225^\circ + i\sin 225^\circ)$
33. $7(\cos 0 + i\sin 0)$
34. $8\left(\cos\dfrac{\pi}{2} + i\sin\dfrac{\pi}{2}\right)$
35. $5[\cos(198^\circ 45') + i\sin(198^\circ 45')]$
36. $9.75[\cos(280^\circ 30') + i\sin(280^\circ 30')]$

Writing a Complex Number in Standard Form

In Exercises 37–40, use a graphing utility to write the complex number in standard form.

37. $5\left(\cos\dfrac{\pi}{9} + i\sin\dfrac{\pi}{9}\right)$
38. $10\left(\cos\dfrac{2\pi}{5} + i\sin\dfrac{2\pi}{5}\right)$
39. $2(\cos 155^\circ + i\sin 155^\circ)$
40. $9(\cos 58^\circ + i\sin 58^\circ)$

Multiplying Complex Numbers

In Exercises 41–46, find the product. Leave the result in trigonometric form.

41. $\left[2\left(\cos\dfrac{\pi}{4} + i\sin\dfrac{\pi}{4}\right)\right]\left[6\left(\cos\dfrac{\pi}{12} + i\sin\dfrac{\pi}{12}\right)\right]$
42. $\left[\dfrac{3}{4}\left(\cos\dfrac{\pi}{3} + i\sin\dfrac{\pi}{3}\right)\right]\left[4\left(\cos\dfrac{3\pi}{4} + i\sin\dfrac{3\pi}{4}\right)\right]$
43. $\left[\frac{5}{3}(\cos 120^\circ + i\sin 120^\circ)\right]\left[\frac{2}{3}(\cos 30^\circ + i\sin 30^\circ)\right]$
44. $\left[\frac{1}{2}(\cos 100^\circ + i\sin 100^\circ)\right]\left[\frac{4}{5}(\cos 300^\circ + i\sin 300^\circ)\right]$
45. $(\cos 80^\circ + i\sin 80^\circ)(\cos 330^\circ + i\sin 330^\circ)$
46. $(\cos 5^\circ + i\sin 5^\circ)(\cos 20^\circ + i\sin 20^\circ)$

Dividing Complex Numbers **In Exercises 47–52, find the quotient. Leave the result in trigonometric form.**

47. $\dfrac{3(\cos 50^\circ + i \sin 50^\circ)}{9(\cos 20^\circ + i \sin 20^\circ)}$

48. $\dfrac{\cos 120^\circ + i \sin 120^\circ}{2(\cos 40^\circ + i \sin 40^\circ)}$

49. $\dfrac{\cos \pi + i \sin \pi}{\cos(\pi/3) + i \sin(\pi/3)}$

50. $\dfrac{5(\cos 4.3 + i \sin 4.3)}{4(\cos 2.1 + i \sin 2.1)}$

51. $\dfrac{12(\cos 92^\circ + i \sin 92^\circ)}{2(\cos 122^\circ + i \sin 122^\circ)}$

52. $\dfrac{6(\cos 40^\circ + i \sin 40^\circ)}{7(\cos 100^\circ + i \sin 100^\circ)}$

Multiplying or Dividing Complex Numbers **In Exercises 53–60, (a) write the trigonometric forms of the complex numbers, (b) perform the operation using the trigonometric forms, and (c) perform the indicated operation using the standard forms, and check your result with that of part (b).**

53. $(2 + 2i)(1 - i)$

54. $(\sqrt{3} + i)(1 + i)$

55. $-2i(1 + i)$

56. $3i(1 - \sqrt{2}i)$

57. $\dfrac{3 + 4i}{1 - \sqrt{3}i}$

58. $\dfrac{1 + \sqrt{3}i}{6 - 3i}$

59. $\dfrac{5}{2 + 3i}$

60. $\dfrac{4i}{-4 + 2i}$

Multiplying in the Complex Plane **In Exercises 61–64, find the product in the complex plane.**

61. $\left[2\left(\cos \frac{2\pi}{3} + i \sin \frac{2\pi}{3}\right)\right]\left[\frac{1}{2}\left(\cos \frac{\pi}{3} + i \sin \frac{\pi}{3}\right)\right]$

62. $\left[2\left(\cos \frac{\pi}{4} + i \sin \frac{\pi}{4}\right)\right]\left[3\left(\cos \frac{\pi}{4} + i \sin \frac{\pi}{4}\right)\right]$

63. $\left[4\left(\cos \frac{\pi}{4} + i \sin \frac{\pi}{4}\right)\right]\left[5\left(\cos \frac{\pi}{2} + i \sin \frac{\pi}{2}\right)\right]$

64. $\left[\frac{1}{3}\left(\cos \frac{\pi}{6} + i \sin \frac{\pi}{6}\right)\right]\left[6\left(\cos \frac{\pi}{6} + i \sin \frac{\pi}{6}\right)\right]$

Graphing Complex Numbers **In Exercises 65–68, sketch the graph of all complex numbers z satisfying the given condition.**

65. $|z| = 2$

66. $|z| = 3$

67. $\theta = \dfrac{\pi}{6}$

68. $\theta = \dfrac{5\pi}{4}$

69. Ohm's Law

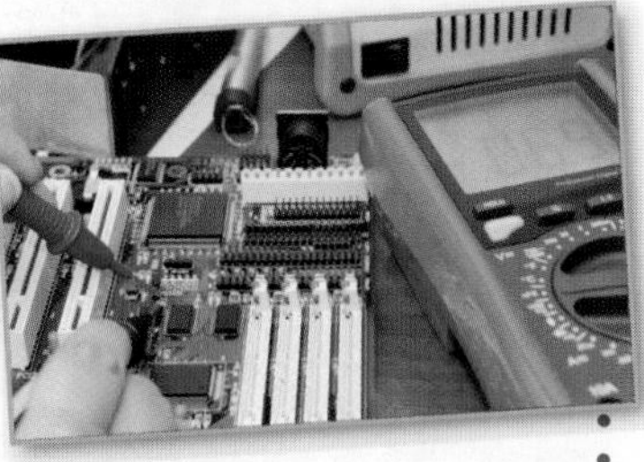

Ohm's law for alternating current circuits is $E = IZ$, where E is the voltage in volts, I is the current in amperes, and Z is the impedance in ohms.

(a) Write E in trigonometric form when $I = 6(\cos 41^\circ + i \sin 41^\circ)$ amperes and $Z = 4[\cos(-11^\circ) + i \sin(-11^\circ)]$ ohms.

(b) Write the voltage from part (a) in standard form.

(c) A voltmeter measures the magnitude of the voltage in a circuit. What would be the reading on a voltmeter for the circuit described in part (a)?

Exploration

True or False? **In Exercises 70 and 71, determine whether the statement is true or false. Justify your answer.**

70. When the argument of a complex number is π, the complex number is a real number.

71. The product of two complex numbers is zero only when the modulus of one (or both) of the numbers is zero.

72. **HOW DO YOU SEE IT?** Match each complex number with its corresponding point.

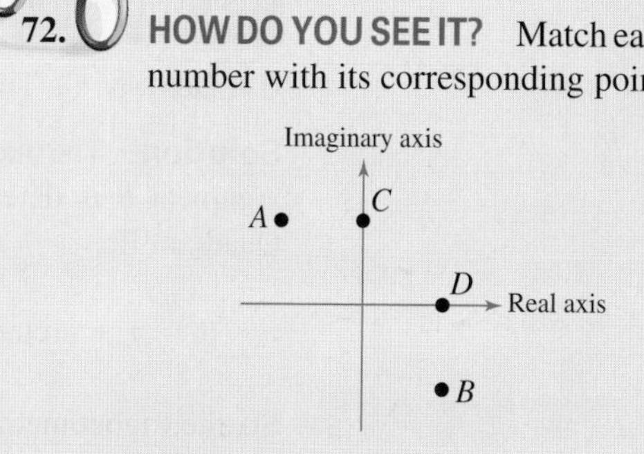

(i) $3(\cos 0^\circ + i \sin 0^\circ)$

(ii) $3(\cos 90^\circ + i \sin 90^\circ)$

(iii) $3\sqrt{2}(\cos 135^\circ + i \sin 135^\circ)$

(iv) $3\sqrt{2}(\cos 315^\circ + i \sin 315^\circ)$

73. **Quotient of Two Complex Numbers** Given two complex numbers $z_1 = r_1(\cos \theta_1 + i \sin \theta_1)$ and $z_2 = r_2(\cos \theta_2 + i \sin \theta_2)$, $z_2 \neq 0$, show that

$z_1/z_2 = (r_1/r_2)[\cos(\theta_1 - \theta_2) + i \sin(\theta_1 - \theta_2)]$.

74. **Complex Conjugates** Show that

$\bar{z} = r[\cos(-\theta) + i \sin(-\theta)]$

is the complex conjugate of $z = r(\cos \theta + i \sin \theta)$. Then find (a) $z\bar{z}$ and (b) $z/\bar{z}$, $\bar{z} \neq 0$.

4.5 DeMoivre's Theorem

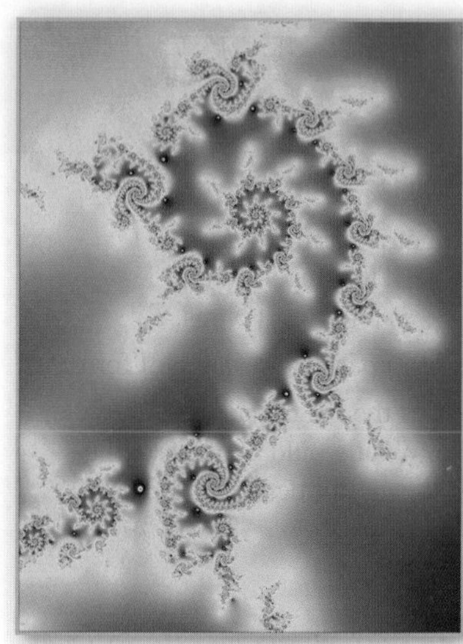

Applications of DeMoivre's Theorem include solving problems that involve powers of complex numbers. For example, in Exercise 73 on page 347, you will use DeMoivre's Theorem in an application related to computer-generated fractals.

- **Use DeMoivre's Theorem to find powers of complex numbers.**
- **Find *n*th roots of complex numbers.**

Powers of Complex Numbers

The trigonometric form of a complex number is used to raise a complex number to a power. To accomplish this, consider repeated use of the multiplication rule.

$$z = r(\cos\theta + i\sin\theta)$$

$$z^2 = r(\cos\theta + i\sin\theta)r(\cos\theta + i\sin\theta) = r^2(\cos 2\theta + i\sin 2\theta)$$

$$z^3 = r^2(\cos 2\theta + i\sin 2\theta)r(\cos\theta + i\sin\theta) = r^3(\cos 3\theta + i\sin 3\theta)$$

$$z^4 = r^4(\cos 4\theta + i\sin 4\theta)$$

$$\vdots$$

This pattern leads to **DeMoivre's Theorem,** which is named after the French mathematician Abraham DeMoivre (1667–1754).

> **DeMoivre's Theorem**
>
> If $z = r(\cos\theta + i\sin\theta)$ is a complex number and n is a positive integer, then
>
> $$z^n = [r(\cos\theta + i\sin\theta)]^n = r^n(\cos n\theta + i\sin n\theta).$$

EXAMPLE 1 Finding a Power of a Complex Number

Use DeMoivre's Theorem to find $(-1 + \sqrt{3}i)^{12}$.

Solution The modulus of $z = -1 + \sqrt{3}i$ is $r = \sqrt{(-1)^2 + (\sqrt{3})^2} = 2$ and the argument θ is determined from $\tan\theta = \sqrt{3}/(-1)$. Because $z = -1 + \sqrt{3}i$ lies in Quadrant II,

$$\theta = \pi + \arctan\frac{\sqrt{3}}{-1} = \pi + \left(-\frac{\pi}{3}\right) = \frac{2\pi}{3}.$$

So, the trigonometric form of z is

$$z = -1 + \sqrt{3}i = 2\left(\cos\frac{2\pi}{3} + i\sin\frac{2\pi}{3}\right).$$

Then, by DeMoivre's Theorem, you have

$$\begin{aligned}(-1 + \sqrt{3}i)^{12} &= \left[2\left(\cos\frac{2\pi}{3} + i\sin\frac{2\pi}{3}\right)\right]^{12}\\ &= 2^{12}\left[\cos\frac{12(2\pi)}{3} + i\sin\frac{12(2\pi)}{3}\right]\\ &= 4096(\cos 8\pi + i\sin 8\pi)\\ &= 4096.\end{aligned}$$

✓ ***Checkpoint*** *Audio-video solution in English & Spanish at LarsonPrecalculus.com*

Use DeMoivre's Theorem to find $(-1 - i)^4$.

Bill Heller/Shutterstock.com

Roots of Complex Numbers

Recall that a consequence of the Fundamental Theorem of Algebra is that a polynomial equation of degree n has n solutions in the complex number system. For example, the equation $x^6 = 1$ has six solutions. To find these solutions, use factoring and the Quadratic Formula.

$$x^6 - 1 = 0$$

$$(x^3 - 1)(x^3 + 1) = 0$$

$$(x - 1)(x^2 + x + 1)(x + 1)(x^2 - x + 1) = 0$$

Consequently, the solutions are

$$x = \pm 1, \quad x = \frac{-1 \pm \sqrt{3}i}{2}, \quad \text{and} \quad x = \frac{1 \pm \sqrt{3}i}{2}.$$

Each of these numbers is a sixth root of 1. In general, an **nth root of a complex number** is defined as follows.

Abraham DeMoivre (1667–1754) is remembered for his work in probability theory and DeMoivre's Theorem. His book *The Doctrine of Chances* (published in 1718) includes the theory of recurring series and the theory of partial fractions.

Definition of an nth Root of a Complex Number

The complex number $u = a + bi$ is an **nth root** of the complex number z when

$$z = u^n$$
$$= (a + bi)^n.$$

To find a formula for an nth root of a complex number, let u be an nth root of z, where

$$u = s(\cos \beta + i \sin \beta)$$

and

$$z = r(\cos \theta + i \sin \theta).$$

By DeMoivre's Theorem and the fact that $u^n = z$, you have

$$s^n(\cos n\beta + i \sin n\beta) = r(\cos \theta + i \sin \theta).$$

Taking the absolute value of each side of this equation, it follows that $s^n = r$. Substituting back into the previous equation and dividing by r gives

$$\cos n\beta + i \sin n\beta = \cos \theta + i \sin \theta.$$

So, it follows that

$$\cos n\beta = \cos \theta$$

and

$$\sin n\beta = \sin \theta.$$

Both sine and cosine have a period of 2π, so these last two equations have solutions if and only if the angles differ by a multiple of 2π. Consequently, there must exist an integer k such that

$$n\beta = \theta + 2\pi k$$

$$\beta = \frac{\theta + 2\pi k}{n}.$$

Substituting this value of β and $s = \sqrt[n]{r}$ into the trigonometric form of u gives the result stated on the next page.

Finding nth Roots of a Complex Number

For a positive integer n, the complex number $z = r(\cos\theta + i\sin\theta)$ has exactly n distinct nth roots given by

$$z_k = \sqrt[n]{r}\left(\cos\frac{\theta + 2\pi k}{n} + i\sin\frac{\theta + 2\pi k}{n}\right)$$

where $k = 0, 1, 2, \ldots, n - 1$.

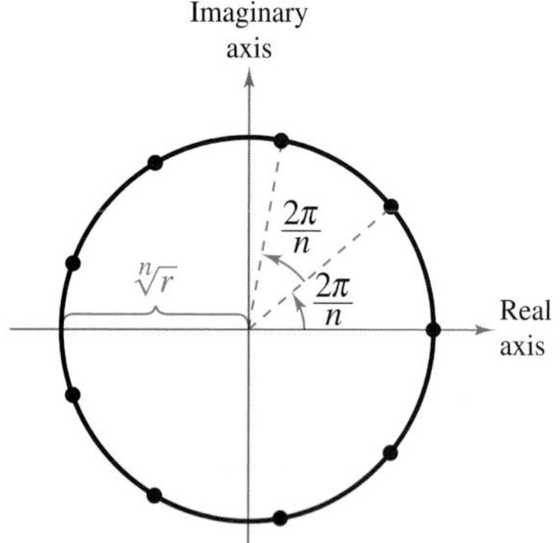

Figure 4.7

When $k > n - 1$, the roots begin to repeat. For example, when $k = n$, the angle

$$\frac{\theta + 2\pi n}{n} = \frac{\theta}{n} + 2\pi$$

is coterminal with θ/n, which is also obtained when $k = 0$.

The formula for the nth roots of a complex number z has a geometrical interpretation, as shown in Figure 4.7. Note that the nth roots of z all have the same magnitude $\sqrt[n]{r}$, so they all lie on a circle of radius $\sqrt[n]{r}$ with center at the origin. Furthermore, successive nth roots have arguments that differ by $2\pi/n$, so the n roots are equally spaced around the circle.

You have already found the sixth roots of 1 by factoring and using the Quadratic Formula. Example 2 shows how to solve the same problem with the formula for nth roots.

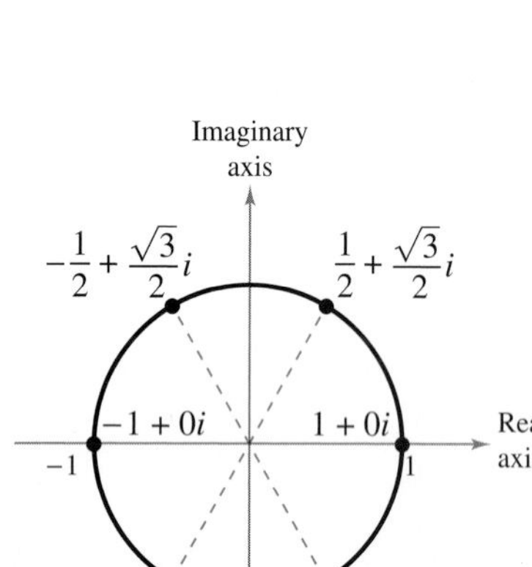

Figure 4.8

EXAMPLE 2 Finding the nth Roots of a Real Number

Find all sixth roots of 1.

Solution First, write 1 in the trigonometric form $z = 1(\cos 0 + i\sin 0)$. Then, by the nth root formula with $n = 6$, $r = 1$, and $\theta = 0$, the roots have the form

$$z_k = \sqrt[6]{1}\left(\cos\frac{0 + 2\pi k}{6} + i\sin\frac{0 + 2\pi k}{6}\right) = \cos\frac{\pi k}{3} + i\sin\frac{\pi k}{3}.$$

So, for $k = 0, 1, 2, 3, 4$, and 5, the roots are as listed below. (See Figure 4.8.)

$$z_0 = \cos 0 + i\sin 0 = 1$$

$$z_1 = \cos\frac{\pi}{3} + i\sin\frac{\pi}{3} = \frac{1}{2} + \frac{\sqrt{3}}{2}i$$

Increment by $\frac{2\pi}{n} = \frac{2\pi}{6} = \frac{\pi}{3}$

$$z_2 = \cos\frac{2\pi}{3} + i\sin\frac{2\pi}{3} = -\frac{1}{2} + \frac{\sqrt{3}}{2}i$$

$$z_3 = \cos\pi + i\sin\pi = -1$$

$$z_4 = \cos\frac{4\pi}{3} + i\sin\frac{4\pi}{3} = -\frac{1}{2} - \frac{\sqrt{3}}{2}i$$

$$z_5 = \cos\frac{5\pi}{3} + i\sin\frac{5\pi}{3} = \frac{1}{2} - \frac{\sqrt{3}}{2}i$$

✓ ***Checkpoint*** Audio-video solution in English & Spanish at LarsonPrecalculus.com

Find all fourth roots of 1. ■

In Figure 4.8, notice that the roots obtained in Example 2 all have a magnitude of 1 and are equally spaced around the unit circle. Also notice that the complex roots occur in conjugate pairs, as discussed in Section 4.2. The n distinct nth roots of 1 are called the ***n*th roots of unity.**

EXAMPLE 3 Finding the *n*th Roots of a Complex Number

See LarsonPrecalculus.com for an interactive version of this type of example.

Find the three cube roots of

$$z = -2 + 2i.$$

Solution The modulus of z is

$$r = \sqrt{(-2)^2 + 2^2}$$
$$= \sqrt{8}$$

and the argument θ is determined from

$$\tan\theta = \frac{b}{a} = \frac{2}{-2} = -1.$$

REMARK In Example 3, $r = \sqrt{8}$, so it follows that

$$\sqrt[n]{r} = \sqrt[3]{\sqrt{8}}$$
$$= \sqrt[3 \cdot 2]{8}$$
$$= \sqrt[6]{8}.$$

Because z lies in Quadrant II, the trigonometric form of z is

$$z = -2 + 2i$$
$$= \sqrt{8}(\cos 135^\circ + i \sin 135^\circ). \qquad \theta = \pi + \arctan(-1) = 3\pi/4 = 135^\circ$$

By the nth root formula, the roots have the form

$$z_k = \sqrt[6]{8}\left(\cos\frac{135^\circ + 360^\circ k}{3} + i \sin\frac{135^\circ + 360^\circ k}{3}\right).$$

So, for $k = 0$, 1, and 2, the roots are as listed below. (See Figure 4.9.)

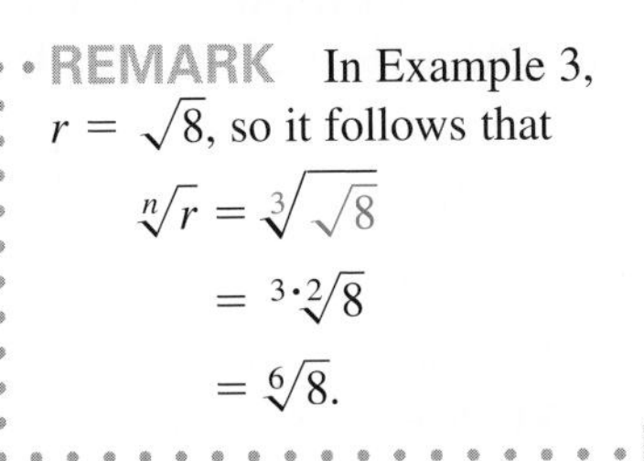

Figure 4.9

$$z_0 = \sqrt[6]{8}\left(\cos\frac{135^\circ + 360^\circ(0)}{3} + i \sin\frac{135^\circ + 360^\circ(0)}{3}\right)$$
$$= \sqrt{2}(\cos 45^\circ + i \sin 45^\circ)$$
$$= 1 + i$$

$$z_1 = \sqrt[6]{8}\left(\cos\frac{135^\circ + 360^\circ(1)}{3} + i \sin\frac{135^\circ + 360^\circ(1)}{3}\right)$$
$$= \sqrt{2}(\cos 165^\circ + i \sin 165^\circ)$$
$$\approx -1.3660 + 0.3660i$$

$$z_2 = \sqrt[6]{8}\left(\cos\frac{135^\circ + 360^\circ(2)}{3} + i \sin\frac{135^\circ + 360^\circ(2)}{3}\right)$$
$$= \sqrt{2}(\cos 285^\circ + i \sin 285^\circ)$$
$$\approx 0.3660 - 1.3660i.$$

✓ ***Checkpoint*** *Audio-video solution in English & Spanish at LarsonPrecalculus.com*

Find the three cube roots of

$$z = -6 + 6i.$$

Summarize (Section 4.5)

1. Explain how to use DeMoivre's Theorem to find a power of a complex number *(page 342)*. For an example of using DeMoivre's Theorem, see Example 1.
2. Explain how to find the nth roots of a complex number *(page 343)*. For examples of finding nth roots of complex numbers, see Examples 2 and 3.

4.5 Exercises

See CalcChat.com for tutorial help and worked-out solutions to odd-numbered exercises.

Vocabulary: Fill in the blanks.

1. ________ Theorem states that if $z = r(\cos\theta + i\sin\theta)$ is a complex number and n is a positive integer, then $z^n = r^n(\cos n\theta + i\sin n\theta)$.
2. The complex number $u = a + bi$ is an ________ ________ of the complex number z when $z = u^n = (a + bi)^n$.
3. Successive nth roots of a complex number have arguments that differ by ________.
4. The n distinct nth roots of 1 are called the nth roots of ________.

Skills and Applications

Finding a Power of a Complex Number **In Exercises 5–28, use DeMoivre's Theorem to find the power of the complex number. Write the result in standard form.**

5. $[5(\cos 20° + i\sin 20°)]^3$
6. $[3(\cos 60° + i\sin 60°)]^4$
7. $\left(\cos\frac{\pi}{4} + i\sin\frac{\pi}{4}\right)^{12}$
8. $\left[2\left(\cos\frac{\pi}{2} + i\sin\frac{\pi}{2}\right)\right]^8$
9. $[5(\cos 3.2 + i\sin 3.2)]^4$
10. $(\cos 0 + i\sin 0)^{20}$
11. $[3(\cos 15° + i\sin 15°)]^4$
12. $[2(\cos 10° + i\sin 10°)]^8$
13. $[5(\cos 95° + i\sin 95°)]^3$
14. $[4(\cos 110° + i\sin 110°)]^4$
15. $\left[2\left(\cos\frac{\pi}{10} + i\sin\frac{\pi}{10}\right)\right]^5$
16. $\left[2\left(\cos\frac{\pi}{8} + i\sin\frac{\pi}{8}\right)\right]^6$
17. $\left[3\left(\cos\frac{2\pi}{3} + i\sin\frac{2\pi}{3}\right)\right]^3$
18. $\left[3\left(\cos\frac{\pi}{12} + i\sin\frac{\pi}{12}\right)\right]^5$
19. $(1 + i)^5$
20. $(2 + 2i)^6$
21. $(-1 + i)^6$
22. $(3 - 2i)^8$
23. $2(\sqrt{3} + i)^{10}$
24. $4(1 - \sqrt{3}i)^3$
25. $(3 - 2i)^5$
26. $(2 + 5i)^6$
27. $(\sqrt{5} - 4i)^3$
28. $(\sqrt{3} + 2i)^4$

Graphing Powers of a Complex Number **In Exercises 29 and 30, represent the powers z, z^2, z^3, and z^4 graphically. Describe the pattern.**

29. $z = \frac{\sqrt{2}}{2}(1 + i)$
30. $z = \frac{1}{2}(1 + \sqrt{3}i)$

Finding the Square Roots of a Complex Number **In Exercises 31–38, find the square roots of the complex number.**

31. $2i$
32. $5i$
33. $-3i$
34. $-6i$
35. $2 - 2i$
36. $2 + 2i$
37. $1 + \sqrt{3}i$
38. $1 - \sqrt{3}i$

Finding the nth Roots of a Complex Number **In Exercises 39–56, (a) use the formula on page 344 to find the roots of the complex number, (b) write each of the roots in standard form, and (c) represent each of the roots graphically.**

39. Square roots of $5(\cos 120° + i\sin 120°)$
40. Square roots of $16(\cos 60° + i\sin 60°)$
41. Cube roots of $8\left(\cos\frac{2\pi}{3} + i\sin\frac{2\pi}{3}\right)$
42. Cube roots of $64\left(\cos\frac{\pi}{3} + i\sin\frac{\pi}{3}\right)$
43. Fifth roots of $243\left(\cos\frac{\pi}{6} + i\sin\frac{\pi}{6}\right)$
44. Fifth roots of $32\left(\cos\frac{5\pi}{6} + i\sin\frac{5\pi}{6}\right)$
45. Fourth roots of $81i$
46. Fourth roots of $625i$
47. Cube roots of $-\frac{125}{2}(1 + \sqrt{3}i)$
48. Cube roots of $-4\sqrt{2}(-1 + i)$
49. Fourth roots of 16
50. Fourth roots of i
51. Fifth roots of 1
52. Cube roots of 1000
53. Cube roots of -125
54. Fourth roots of -4
55. Fifth roots of $4(1 - i)$
56. Sixth roots of $64i$

Solving an Equation **In Exercises 57–72, use the formula on page 344 to find all solutions of the equation and represent the solutions graphically.**

57. $x^4 + i = 0$
58. $x^3 - i = 0$
59. $x^6 + 1 = 0$
60. $x^3 + 1 = 0$
61. $x^5 + 32 = 0$
62. $x^3 + 125 = 0$
63. $x^3 - 27 = 0$
64. $x^5 - 243 = 0$
65. $x^4 + 16i = 0$
66. $x^4 - 256i = 0$
67. $x^4 - 16i = 0$
68. $x^6 + 64i = 0$
69. $x^3 - (1 - i) = 0$
70. $x^5 - (1 - i) = 0$
71. $x^6 + (1 + i) = 0$
72. $x^4 + (1 + i) = 0$

73. **Computer-Generated Fractals** The *prisoner* set and *escape* set of a function play a role in the study of computer-generated *fractals.* A **fractal** is a geometric figure that consists of a pattern that is repeated infinitely on a smaller and smaller scale. To determine whether a complex number z_0 is in the prisoner set or the escape set of a function, consider the following sequence.

$$z_1 = f(z_0),\ z_2 = f(z_1),\ z_3 = f(z_2), . . .$$

If the sequence is bounded (the absolute value of each number in the sequence is less than some fixed number N), then the complex number z_0 is in the prisoner set, and if the sequence is unbounded (the absolute value of the terms of the sequence become infinitely large), then the complex number z_0 is in the escape set. Determine whether each complex number is in the prisoner set or the escape set of the function $f(z) = z^2 - 1$.

(a) $\frac{1}{2}(\cos 0° + i \sin 0°)$

(b) $\sqrt{2}(\cos 30° + i \sin 30°)$

(c) $\sqrt[4]{2}\left(\cos \frac{\pi}{8} + i \sin \frac{\pi}{8}\right)$

(d) $\sqrt{2}(\cos \pi + i \sin \pi)$

74. **HOW DO YOU SEE IT?** The figure shows one of the fourth roots of a complex number z.

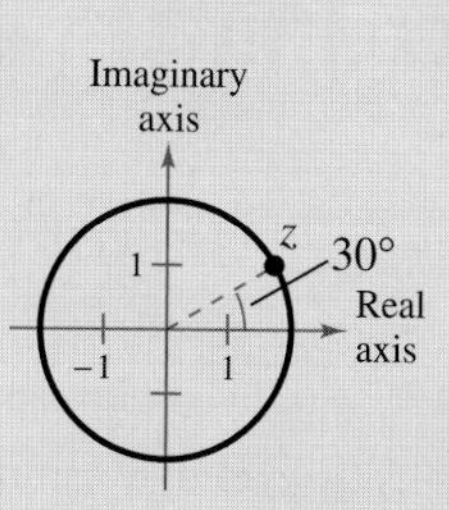

(a) How many roots are not shown?

(b) Describe the other roots.

Exploration

True or False? **In Exercises 75–78, determine whether the statement is true or false. Justify your answer.**

75. Geometrically, the nth roots of any complex number z are all equally spaced around the unit circle.

76. By DeMoivre's Theorem,

$$(4 + \sqrt{6}i)^8 = \cos(32) + i \sin(8\sqrt{6}).$$

77. The complex numbers i and $-i$ are each a cube root of the other.

78. $\sqrt{3} + i$ is a solution of the equation $x^2 - 8i = 0$.

79. **Reasoning** Show that $\frac{1}{2}(1 - \sqrt{3}i)$ is a ninth root of -1.

80. **Reasoning** Show that $2^{-1/4}(1 - i)$ is a fourth root of -2.

Solutions of Quadratic Equations **In Exercises 81 and 82, (a) show that the given value of x is a solution of the quadratic equation, (b) find the other solution and write it in trigonometric form, (c) explain how you obtained your answer to part (b), and (d) show that the solution in part (b) satisfies the quadratic equation.**

81. $x^2 - 4x + 8 = 0;\ x = 2\sqrt{2}(\cos 45° + i \sin 45°)$

82. $x^2 + 2x + 4 = 0;\ x = 2\left(\cos \frac{2\pi}{3} + i \sin \frac{2\pi}{3}\right)$

83. **Solving Quadratic Equations** Use the Quadratic Formula and, if necessary, the theorem on page 344 to solve each equation.

(a) $x^2 + ix + 2 = 0$

(b) $x^2 + 2ix + 1 = 0$

(c) $x^2 + 2ix + \sqrt{3}i = 0$

84. **Reasoning** Show that $2\left(\cos \frac{2\pi}{5} + i \sin \frac{2\pi}{5}\right)$ is a fifth root of 32. Then find the other fifth roots of 32, and verify your results.

85. **Reasoning** Show that $\sqrt{2}(\cos 7.5° + i \sin 7.5°)$ is a fourth root of $2\sqrt{3} + 2i$. Then find the other fourth roots of $2\sqrt{3} + 2i$, and verify your results.

Chapter Summary

	What Did You Learn?	Explanation/Examples	Review Exercises
Section 4.1	Use the imaginary unit i to write complex numbers *(p. 314)*.	Let a and b be real numbers. The number $a + bi$ is a complex number written in standard form. **Equality of Complex Numbers** Two complex numbers $a + bi$ and $c + di$, written in standard form, are equal to each other, $a + bi = c + di$, if and only if $a = c$ and $b = d$.	1–4, 21–24
	Add, subtract, and multiply complex numbers *(p. 315)*.	**Sum:** $(a + bi) + (c + di) = (a + c) + (b + d)i$ **Difference:** $(a + bi) - (c + di) = (a - c) + (b - d)i$ Use the Distributive Property or the FOIL method to multiply two complex numbers.	5–10
	Use complex conjugates to write the quotient of two complex numbers in standard form *(p. 317)*.	Complex numbers of the form $a + bi$ and $a - bi$ are complex conjugates. To write $(a + bi)/(c + di)$ in standard form, where c and d are not both zero, multiply the numerator and denominator by the complex conjugate of the denominator, $c - di$.	11–16
	Find complex solutions of quadratic equations *(p. 318)*.	**Principal Square Root of a Negative Number** When a is a positive number, the principal square root of $-a$ is defined as $\sqrt{-a} = \sqrt{a}i$.	17–20
Section 4.2	Determine the numbers of solutions of polynomial equations *(p. 321)*.	**The Fundamental Theorem of Algebra** If $f(x)$ is a polynomial of degree n, where $n > 0$, then f has at least one zero in the complex number system. **Linear Factorization Theorem** If $f(x)$ is a polynomial of degree n, where $n > 0$, then $f(x)$ has precisely n linear factors $f(x) = a_n(x - c_1)(x - c_2) \cdot \cdot \cdot (x - c_n)$, where $c_1, c_2, \ldots, c_n$ are complex numbers. Every second-degree equation, $ax^2 + bx + c = 0$, has precisely two solutions given by the Quadratic Formula $x = \frac{-b \pm \sqrt{b^2 - 4ac}}{2a}$. The quantity under the radical sign, $b^2 - 4ac$, is the discriminant. **1.** If $b^2 - 4ac < 0$, then the equation has two imaginary solutions. **2.** If $b^2 - 4ac = 0$, then the equation has one repeated real solution. **3.** If $b^2 - 4ac > 0$, then the equation has two distinct real solutions.	25–30
	Find solutions of polynomial equations *(p. 323)*.	If $a + bi$, $b \neq 0$, is a solution of a polynomial equation with real coefficients, then the complex conjugate $a - bi$ is also a solution of the equation.	31–40
	Find zeros of polynomial functions and find polynomial functions given the zeros of the functions *(p 324)*.	Finding the zeros of a polynomial function is essentially the same as finding the solutions of the corresponding polynomial equation.	41–58

	What Did You Learn?	Explanation/Examples	Review Exercises
Section 4.3	Plot complex numbers in the complex plane and find absolute values of complex numbers *(p. 329)*.	A complex number $z = a + bi$ can be represented by the point (a, b) in the complex plane. The horizontal axis is the real axis and the vertical axis is the imaginary axis. The absolute value, or modulus, of $z = a + bi$ is $\lvert a + bi \rvert = \sqrt{a^2 + b^2}$.	59–62
	Perform operations with complex numbers in the complex plane *(p. 330)*.	Complex numbers can be added and subtracted geometrically in the complex plane. The points representing the complex conjugates $a + bi$ and $a - bi$ are reflections of each other in the real axis.	63–68
	Use the Distance and Midpoint Formulas in the complex plane *(p. 332)*.	Let (a, b) and (s, t) be points in the complex plane. **Distance Formula** $d = \sqrt{(s - a)^2 + (t - b)^2}$ **Midpoint Formula** $\text{Midpoint} = \left(\frac{a + s}{2}, \frac{b + t}{2}\right)$	69–72
Section 4.4	Write trigonometric forms of complex numbers *(p. 336)*.	The trigonometric form of the complex number $z = a + bi$ is $z = r(\cos\theta + i\sin\theta)$ where $a = r\cos\theta,\ b = r\sin\theta,\ r = \sqrt{a^2 + b^2}$, and $\tan\theta = b/a$. The number r is the modulus of z, and θ is an argument of z.	73–84
	Multiply and divide complex numbers written in trigonometric form *(p. 338)*.	**Product and Quotient of Two Complex Numbers** Let $z_1 = r_1(\cos\theta_1 + i\sin\theta_1)$ and $z_2 = r_2(\cos\theta_2 + i\sin\theta_2)$ be complex numbers. $z_1z_2 = r_1r_2[\cos(\theta_1 + \theta_2) + i\sin(\theta_1 + \theta_2)]$ $\frac{z_1}{z_2} = \frac{r_1}{r_2}[\cos(\theta_1 - \theta_2) + i\sin(\theta_1 - \theta_2)],\quad z_2 \neq 0$	85–90
Section 4.5	Use DeMoivre's Theorem to find powers of complex numbers *(p. 342)*.	**DeMoivre's Theorem** If $z = r(\cos\theta + i\sin\theta)$ is a complex number and n is a positive integer, then $z^n = [r(\cos\theta + i\sin\theta)]^n = r^n(\cos n\theta + i\sin n\theta)$.	91–96
	Find nth roots of complex numbers *(p. 343)*.	**Definition of an nth Root of a Complex Number** The complex number $u = a + bi$ is an nth root of the complex number z when $z = u^n = (a + bi)^n$. **Finding nth Roots of a Complex Number** For a positive integer n, the complex number $z = r(\cos\theta + i\sin\theta)$ has exactly n distinct nth roots given by $z_k = \sqrt[n]{r}\left(\cos\frac{\theta + 2\pi k}{n} + i\sin\frac{\theta + 2\pi k}{n}\right)$ where $k = 0, 1, 2, \ldots, n - 1$.	97–104

Review Exercises

See CalcChat.com for tutorial help and worked-out solutions to odd-numbered exercises.

4.1 Writing a Complex Number in Standard Form **In Exercises 1–4, write the complex number in standard form.**

1. $6 + \sqrt{-4}$
2. $3 - \sqrt{-25}$
3. $i^2 + 3i$
4. $-5i + i^2$

Performing Operations with Complex Numbers **In Exercises 5–10, perform the operation and write the result in standard form.**

5. $(6 - 4i) + (-9 + i)$
6. $(7 - 2i) - (3 - 8i)$
7. $-3i(-2 + 5i)$
8. $(4 + i)(3 - 10i)$
9. $(1 + 7i)(1 - 7i)$
10. $(5 - 9i)^2$

Quotient of Complex Numbers in Standard Form **In Exercises 11–14, write the quotient in standard form.**

11. $\dfrac{4}{1 - 2i}$
12. $\dfrac{6 - 5i}{i}$
13. $\dfrac{3 + 2i}{5 + i}$
14. $\dfrac{7i}{(3 + 2i)^2}$

Performing Operations with Complex Numbers **In Exercises 15 and 16, perform the operation and write the result in standard form.**

15. $\dfrac{4}{2 - 3i} + \dfrac{2}{1 + i}$
16. $\dfrac{1}{2 + i} - \dfrac{5}{1 + 4i}$

Complex Solutions of a Quadratic Equation **In Exercises 17–20, use the Quadratic Formula to solve the quadratic equation.**

17. $x^2 - 2x + 10 = 0$
18. $x^2 + 6x + 34 = 0$
19. $4x^2 + 4x + 7 = 0$
20. $6x^2 + 3x + 27 = 0$

Simplifying a Complex Number **In Exercises 21–24, simplify the complex number and write the result in standard form.**

21. $10i^2 - i^3$
22. $-8i^6 + i^2$
23. $\dfrac{1}{i^7}$
24. $\dfrac{1}{(4i)^3}$

4.2 Solutions of a Polynomial Equation **In Exercises 25 and 26, determine the number of solutions of the equation in the complex number system.**

25. $-2x^6 + 7x^3 + x^2 + 4x - 19 = 0$
26. $\frac{3}{4}x^3 + \frac{1}{2}x^2 + \frac{3}{2}x + 2 = 0$

Using the Discriminant **In Exercises 27–30, use the discriminant to find the number of real and imaginary solutions of the quadratic equation.**

27. $6x^2 + x - 2 = 0$
28. $9x^2 - 12x + 4 = 0$
29. $0.13x^2 - 0.45x + 0.65 = 0$
30. $4x^2 + \frac{4}{3}x + \frac{1}{9} = 0$

Solving a Polynomial Equation **In Exercises 31–38, solve the polynomial equation. Write complex solutions in standard form.**

31. $x^2 - 2x = 0$
32. $6x - x^2 = 0$
33. $x^2 - 3x + 5 = 0$
34. $x^2 - 4x + 9 = 0$
35. $2x^2 + 3x + 6 = 0$
36. $4x^2 - x + 10 = 0$
37. $x^4 + 8x^2 + 7 = 0$
38. $21 + 4x^2 - x^4 = 0$

39. **Biology** The metabolic rate of an ectothermic organism increases with increasing temperature within a certain range. Experimental data for the oxygen consumption C (in microliters per gram per hour) of a beetle at certain temperatures can be approximated by the model

$$C = 0.45x^2 - 1.65x + 50.75, \quad 10 \le x \le 25$$

where x is the air temperature in degree Celsius. The oxygen consumption is 150 microliters per gram per hour. What is the air temperature?

40. **Profit** The demand equation for a Blu-ray player is $p = 140 - 0.0001x$, where p is the unit price (in dollars) of the Blu-ray player and x is the number of units produced and sold. The cost equation for the Blu-ray player is $C = 75x + 100{,}000$, where C is the total cost (in dollars) and x is the number of units produced. The total profit obtained by producing and selling x units is modeled by

$$P = xp - C.$$

If possible, determine a price p that would yield a profit of 9 million dollars. If not possible, explain.

Finding the Zeros of a Polynomial Function In Exercises 41–46, write the polynomial as a product of linear factors. Then find all the zeros of the function.

41. $r(x) = 2x^2 + 2x + 3$
42. $s(x) = 2x^2 + 5x + 4$
43. $f(x) = 2x^3 - 3x^2 + 50x - 75$
44. $f(x) = 4x^3 - x^2 + 128x - 32$
45. $f(x) = 4x^4 + 3x^2 - 10$
46. $f(x) = 5x^4 + 126x^2 + 25$

Finding the Zeros of a Polynomial Function In Exercises 47–52, use the given zero to find all the zeros of the function.

	Function	Zero
47.	$f(x) = x^3 + 3x^2 - 24x + 28$	2
48.	$f(x) = x^3 + 3x^2 - 5x + 25$	-5
49.	$h(x) = -x^3 + 2x^2 - 16x + 32$	$-4i$
50.	$f(x) = 5x^3 - 4x^2 + 20x - 16$	$2i$
51.	$g(x) = 2x^4 - 3x^3 - 13x^2 + 37x - 15$	$2 + i$
52.	$f(x) = x^4 + 5x^3 + 2x^2 - 50x - 84$	$-3 + \sqrt{5}i$

Finding a Polynomial Function with Given Zeros In Exercises 53–56, find a polynomial function with real coefficients that has the given zeros. (There are many correct answers.)

53. $\frac{2}{3}, 4, \sqrt{3}i$
54. $2, -3, 1 - 2i$
55. $\sqrt{2}i, -5i$
56. $-2i, -4i$

Finding a Polynomial Function with Given Zeros In Exercises 57 and 58, find the polynomial function f with real coefficients that has the given degree, zeros, and function value.

	Degree	Zeros	Function Value
57.	3	$5, 1 - i$	$f(1) = -8$
58.	4	$-3, 0, \sqrt{2}i$	$f(-2) = 24$

4.3 Finding the Absolute Value of a Complex Number In Exercises 59–62, plot the complex number and find its absolute value.

59. $7i$
60. $-6i$
61. $5 + 3i$
62. $-10 - 4i$

Adding in the Complex Plane In Exercises 63 and 64, find the sum of the complex numbers in the complex plane.

63. $(2 + 3i) + (1 - 2i)$
64. $(-4 + 2i) + (2 + i)$

Subtracting in the Complex Plane In Exercises 65 and 66, find the difference of the complex numbers in the complex plane.

65. $(1 + 2i) - (3 + i)$
66. $(-2 + i) - (1 + 4i)$

Complex Conjugates in the Complex Plane In Exercises 67 and 68, plot the complex number and its complex conjugate. Write the conjugate as a complex number.

67. $3 + i$
68. $2 - 5i$

Finding Distance in the Complex Plane In Exercises 69 and 70, find the distance between the complex numbers in the complex plane.

69. $3 + 2i, 2 - i$
70. $1 + 5i, -1 + 3i$

Finding a Midpoint in the Complex Plane In Exercises 71 and 72, find the midpoint of the line segment joining the points corresponding to the complex numbers in the complex plane.

71. $1 + i, 4 + 3i$
72. $2 - i, 1 + 4i$

4.4 Trigonometric Form of a Complex Number In Exercises 73–78, plot the complex number. Then write the trigonometric form of the complex number.

73. $4i$
74. -7
75. $7 - 7i$
76. $5 + 12i$
77. $5 - 12i$
78. $-3\sqrt{3} + 3i$

Writing a Complex Number in Standard Form In Exercises 79–84, write the standard form of the complex number. Then plot the complex number.

79. $2(\cos 30° + i \sin 30°)$
80. $4(\cos 210° + i \sin 210°)$
81. $\sqrt{2}[\cos(-45°) + i \sin(-45°)]$
82. $\sqrt{8}(\cos 315° + i \sin 315°)$
83. $2\left(\cos \frac{5\pi}{6} + i \sin \frac{5\pi}{6}\right)$
84. $4\left(\cos \frac{4\pi}{3} + i \sin \frac{4\pi}{3}\right)$

Multiplying Complex Numbers In Exercises 85 and 86, find the product. Leave the result in trigonometric form.

85. $\left[2\left(\cos \frac{\pi}{4} + i \sin \frac{\pi}{4}\right)\right]\left[2\left(\cos \frac{\pi}{3} + i \sin \frac{\pi}{3}\right)\right]$
86. $\left[4\left(\cos \frac{\pi}{3} + i \sin \frac{\pi}{3}\right)\right]\left[3\left(\cos \frac{5\pi}{6} + i \sin \frac{5\pi}{6}\right)\right]$

Dividing Complex Numbers **In Exercises 87 and 88, find the quotient. Leave the result in trigonometric form.**

87. $\dfrac{2(\cos 60° + i \sin 60°)}{3(\cos 15° + i \sin 15°)}$

88. $\dfrac{\cos 150° + i \sin 150°}{2(\cos 50° + i \sin 50°)}$

Multiplying in the Complex Plane **In Exercises 89 and 90, find the product in the complex plane.**

89. $\left[4\left(\cos \dfrac{2\pi}{3} + i \sin \dfrac{2\pi}{3}\right)\right]\left[2\left(\cos \dfrac{5\pi}{6} + i \sin \dfrac{5\pi}{6}\right)\right]$

90. $[3(\cos \pi + i \sin \pi)]\left[3\left(\cos \dfrac{7\pi}{6} + i \sin \dfrac{7\pi}{6}\right)\right]$

4.5 Finding a Power of a Complex Number **In Exercises 91–96, use DeMoivre's Theorem to find the power of the complex number. Write the result in standard form.**

91. $\left[5\left(\cos \dfrac{\pi}{12} + i \sin \dfrac{\pi}{12}\right)\right]^4$

92. $\left[2\left(\cos \dfrac{4\pi}{15} + i \sin \dfrac{4\pi}{15}\right)\right]^5$

93. $(2 + 3i)^6$

94. $(1 - i)^8$

95. $(-1 + i)^7$

96. $(\sqrt{3} - i)^4$

Finding the *n*th Roots of a Complex Number **In Exercises 97–100, (a) use the formula on page 344 to find the roots of the complex number, (b) write each of the roots in standard form, and (c) represent each of the roots graphically.**

97. Sixth roots of $-729i$

98. Fourth roots of $256i$

99. Cube roots of 8

100. Fifth roots of -1024

Solving an Equation **In Exercises 101–104, use the formula on page 344 to find all solutions of the equation and represent the solutions graphically.**

101. $x^4 + 81 = 0$

102. $x^5 - 32 = 0$

103. $x^3 + 8i = 0$

104. $x^4 - 64i = 0$

Exploration

True or False? **In Exercises 105–107, determine whether the statement is true or false. Justify your answer.**

105. $\sqrt{-18}\sqrt{-2} = \sqrt{(-18)(-2)}$

106. The equation

$$325x^2 - 717x + 398 = 0$$

has no solution.

107. A fourth-degree polynomial with real coefficients can have -5, $-8i$, $4i$, and 5 as its zeros.

108. Writing Quadratic Equations Write quadratic equations that have

(a) two distinct real solutions,

(b) two imaginary solutions, and

(c) no real solution.

Graphical Reasoning **In Exercises 109 and 110, use the graph of the roots of a complex number.**

(a) Write each of the roots in trigonometric form.

(b) Identify the complex number whose roots are given. Use a graphing utility to verify your results.

109.

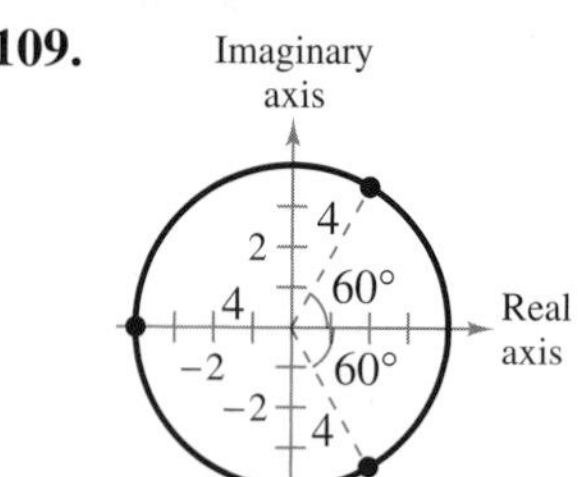

110.

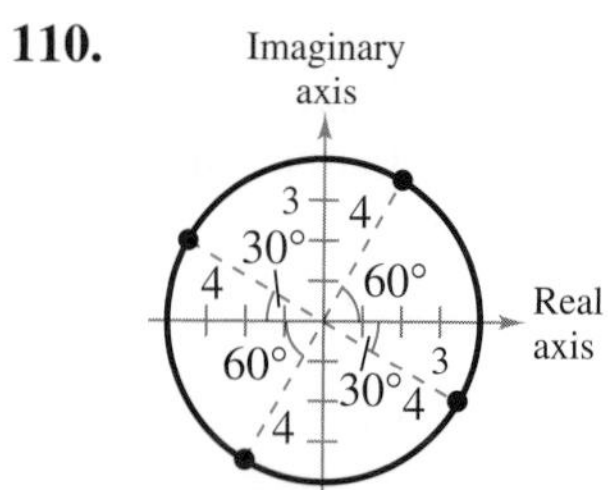

111. Graphical Reasoning The figure shows z_1 and z_2. Describe $z_1 z_2$ and z_1/z_2.

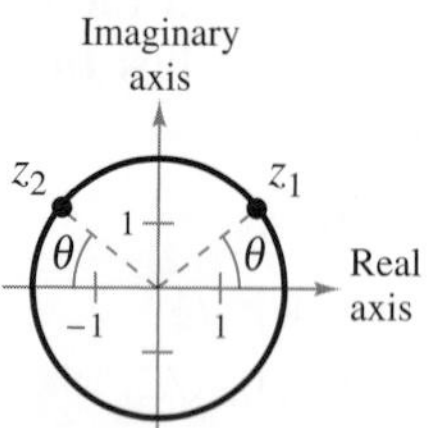

112. Graphical Reasoning The figure shows one of the sixth roots of a complex number z.

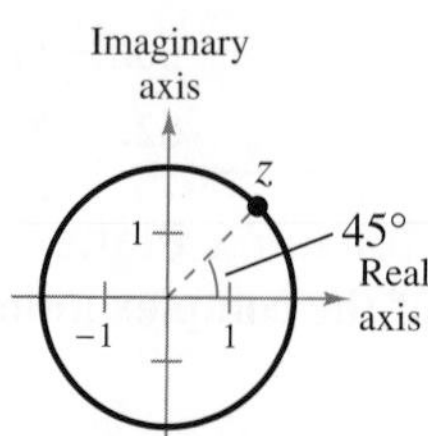

(a) How many roots are not shown?

(b) Describe the other roots.

Chapter Test

See CalcChat.com for tutorial help and worked-out solutions to odd-numbered exercises.

Take this test as you would take a test in class. When you are finished, check your work against the answers given in the back of the book.

1. Write the complex number $-5 + \sqrt{-100}$ in standard form.

In Exercises 2–4, perform the operations and write the result in standard form.

2. $\sqrt{-16} - 2(7 + 2i)$
3. $(4 + 9i)^2$
4. $(6 + \sqrt{7}i)(6 - \sqrt{7}i)$
5. Write the quotient in standard form: $\dfrac{8}{1 + 2i}$.
6. Use the Quadratic Formula to solve the equation $2x^2 - 2x + 3 = 0$.
7. Determine the number of solutions of the equation $x^5 + x^3 - x + 1 = 0$ in the complex number system.

In Exercises 8 and 9, write the polynomial as a product of linear factors. Then find all the zeros of the function.

8. $f(x) = x^3 - 6x^2 + 5x - 30$
9. $f(x) = x^4 - 2x^2 - 24$

In Exercises 10 and 11, use the given zero(s) to find all the zeros of the function.

	Function	**Zero(s)**
10.	$h(x) = x^4 - 2x^2 - 8$	$-2, 2$
11.	$g(v) = 2v^3 - 11v^2 + 22v - 15$	$3/2$

In Exercises 12 and 13, find a polynomial function with real coefficients that has the given zeros. (There are many correct answers.)

12. $0, 2, 3i$
13. $1, 1, 2 + \sqrt{3}i$
14. It is possible for a polynomial function with integer coefficients to have exactly one imaginary zero? Explain.
15. Find the distance between $4 + 3i$ and $1 - i$ in the complex plane.
16. Write the complex number $z = 4 - 4i$ in trigonometric form.
17. Write the complex number $z = 6(\cos 120° + i \sin 120°)$ in standard form.

In Exercises 18 and 19, use DeMoivre's Theorem to find the power of the complex number. Write the results in standard form.

18. $\left[3\left(\cos \dfrac{7\pi}{6} + i \sin \dfrac{7\pi}{6}\right)\right]^8$
19. $(3 - 3i)^6$
20. Find the fourth roots of 256.
21. Find all solutions for the equation $x^3 - 27i = 0$ and represent the solutions graphically.
22. A projectile is fired upward from ground level with an initial velocity of 88 feet per second. The height h (in feet) of the projectile is given by

$$h = -16t^2 + 88t, \quad 0 \le t \le 5.5$$

where t is the time (in seconds). Is it possible for the projectile to reach a height of 125 feet? Explain.

Proofs in Mathematics

The Fundamental Theorem of Algebra, which is closely related to the Linear Factorization Theorem, has a long and interesting history. In the early work with polynomial equations, the Fundamental Theorem of Algebra was thought to have been false, because imaginary solutions were not considered. In fact, in the very early work by mathematicians such as Abu al-Khwarizmi (c. 800 A.D.), negative solutions were also not considered.

Once imaginary numbers were considered, several mathematicians attempted to give a general proof of the Fundamental Theorem of Algebra. These included Jean Le Rond D'Alembert (1746), Leonhard Euler (1749), Joseph-Louis Lagrange (1772), and Pierre Simon Laplace (1795). The mathematician usually credited with the first complete and correct proof of the Fundamental Theorem of Algebra is Carl Friedrich Gauss, who published the proof in 1816.

Linear Factorization Theorem *(p. 321)*

If $f(x)$ is a polynomial of degree n, where $n > 0$, then $f(x)$ has precisely n linear factors

$$f(x) = a_n(x - c_1)(x - c_2) \cdot \cdot \cdot (x - c_n)$$

where $c_1, c_2, \ldots, c_n$ are complex numbers.

Proof

Using the Fundamental Theorem of Algebra, you know that f must have at least one zero, c_1. Consequently, $(x - c_1)$ is a factor of $f(x)$, and you have

$$f(x) = (x - c_1)f_1(x).$$

If the degree of $f_1(x)$ is greater than zero, then you again apply the Fundamental Theorem to conclude that f_1 must have a zero c_2, which implies that

$$f(x) = (x - c_1)(x - c_2)f_2(x).$$

It is clear that the degree of $f_1(x)$ is $n - 1$, that the degree of $f_2(x)$ is $n - 2$, and that you can repeatedly apply the Fundamental Theorem n times until you obtain

$$f(x) = a_n(x - c_1)(x - c_2) \cdot \cdot \cdot (x - c_n)$$

where a_n is the leading coefficient of the polynomial $f(x)$. ■

P.S. Problem Solving

1. **Cube Roots**

(a) The complex numbers

$$z = 2,\ z = \frac{-2 + 2\sqrt{3}i}{2},\ \text{and}\ z = \frac{-2 - 2\sqrt{3}i}{2}$$

are represented graphically (see figure). Evaluate the expression z^3 for each complex number. What do you observe?

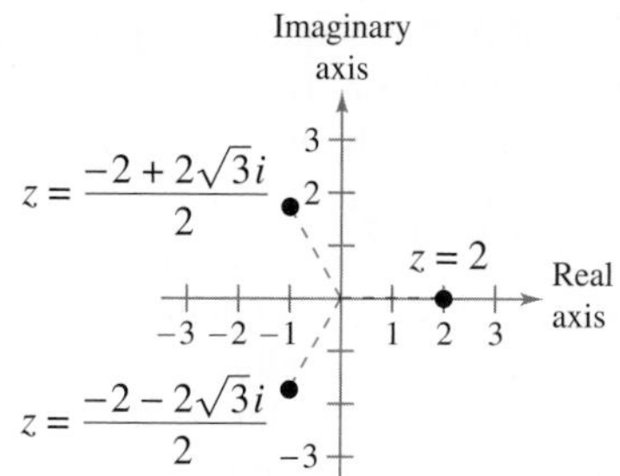

(b) The complex numbers

$$z = 3,\ z = \frac{-3 + 3\sqrt{3}i}{2},\ \text{and}\ z = \frac{-3 - 3\sqrt{3}i}{2}$$

are represented graphically (see figure). Evaluate the expression z^3 for each complex number. What do you observe?

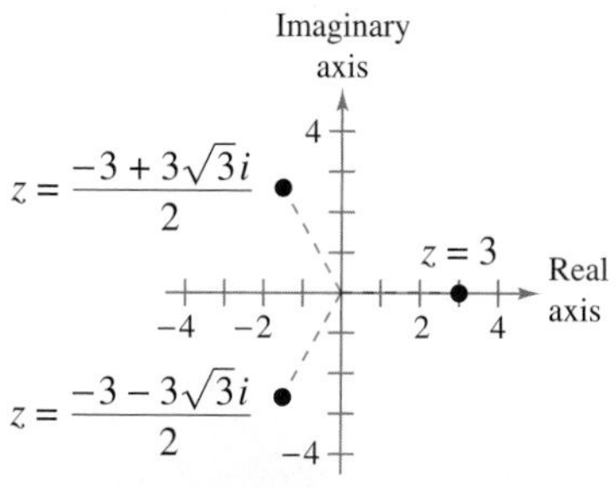

(c) Use your results from parts (a) and (b) to generalize your findings.

2. **Multiplicative Inverse of a Complex Number** The multiplicative inverse of a complex number z is a complex number z_m such that $z \cdot z_m = 1$. Find the multiplicative inverse of each complex number.

(a) $z = 1 + i$ (b) $z = 3 - i$ (c) $z = -2 + 8i$

3. **Writing an Equation** A third-degree polynomial function f has real zeros -2, $\frac{1}{2}$, and 3, and its leading coefficient is negative.

(a) Write an equation for f.

(b) Sketch the graph of the equation from part (a).

(c) How many different polynomial functions are possible for f?

4. **Proof** Prove that the product of a complex number $a + bi$ and its conjugate is a real number.

5. **Proof** Let

$$z = a + bi,\ \bar{z} = a - bi,\ w = c + di,\ \text{and}\ \bar{w} = c - di.$$

Prove each statement.

(a) $\overline{z + w} = \bar{z} + \bar{w}$ (b) $\overline{z - w} = \bar{z} - \bar{w}$

(c) $\overline{zw} = \bar{z} \cdot \bar{w}$ (d) $\overline{z/w} = \bar{z}/\bar{w}$

(e) $(\bar{z})^2 = \overline{z^2}$ (f) $\bar{\bar{z}} = z$

(g) $\bar{z} = z$ when z is real

6. **Finding Values** Find the values of k such that the equation $x^2 - 2kx + k = 0$ has (a) two real solutions and (b) two imaginary solutions.

7. **Finding Values** Use a graphing utility to graph the function $f(x) = x^4 - 4x^2 + k$ for different values of k. Find the values of k such that the zeros of f satisfy the specified characteristics. (Some parts do not have unique answers.)

(a) Four real zeros

(b) Two real zeros and two imaginary zeros

(c) Four imaginary zeros

8. **Finding Values** Will the answers to Exercise 7 change for each function g?

(a) $g(x) = f(x - 2)$ (b) $g(x) = f(2x)$

9. **Reasoning** The graph of one of the following functions is shown below. Identify the function shown in the graph. Explain why each of the others is not the correct function. Use a graphing utility to verify your result.

(a) $f(x) = x^2(x + 2)(x - 3.5)$

(b) $g(x) = (x + 2)(x - 3.5)$

(c) $h(x) = (x + 2)(x - 3.5)(x^2 + 1)$

(d) $k(x) = (x + 1)(x + 2)(x - 3.5)$

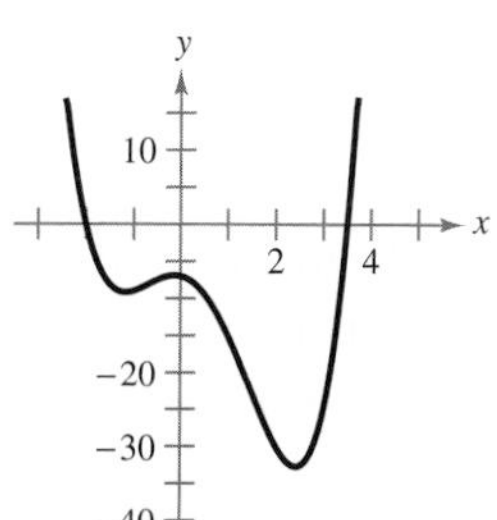

10. **Quadratic Equations with Complex Coefficients** Use the Quadratic Formula and, if necessary, the formula on page 344 to solve each equation with complex coefficients.

(a) $x^2 - (4 + 2i)x + 2 + 4i = 0$

(b) $x^2 - (3 + 2i)x + 5 + i = 0$

(c) $2x^2 + (5 - 8i)x - 13 - i = 0$

(d) $3x^2 - (11 + 14i)x + 1 - 9i = 0$

11. **Reasoning** Use the information in the table to answer each question.

Interval	Value of $f(x)$
$(-\infty, -2)$	Positive
$(-2, 1)$	Negative
$(1, 4)$	Negative
$(4, \infty)$	Positive

(a) What are the three real zeros of the polynomial function f?

(b) What can be said about the behavior of the graph of f at $x = 1$?

(c) What is the least possible degree of f? Explain. Can the degree of f ever be odd? Explain.

(d) Is the leading coefficient of f positive or negative? Explain.

(e) Write an equation for f.

(f) Sketch a graph of the function you wrote in part (e).

12. **Sums and Products of Zeros**

(a) Complete the table.

Function	Zeros	Sum of Zeros	Product of Zeros
$f_1(x) = x^2 - 5x + 6$			
$f_2(x) = x^3 - 7x + 6$			
$f_3(x) = x^4 + 2x^3 + x^2 + 8x - 12$			
$f_4(x) = x^5 - 3x^4 - 9x^3 + 25x^2 - 6x$			

(b) Use the table to make a conjecture relating the sum of the zeros of a polynomial function to the coefficients of the polynomial function.

(c) Use the table to make a conjecture relating the product of the zeros of a polynomial function to the coefficients of the polynomial function.

13. **Reasoning** Let $z = a + bi$ and $\bar{z} = a - bi$, where $a \neq 0$. Show that the equation

$$z^2 - \bar{z}^2 = 0$$

has only real solutions, whereas the equation

$$z^2 + \bar{z}^2 = 0$$

has imaginary solutions.

14. **The Mandelbrot Set** A **fractal** is a geometric figure that consists of a pattern that is repeated infinitely on a smaller and smaller scale. The most famous fractal is the **Mandelbrot Set,** named after the Polish-born mathematician Benoit Mandelbrot. To draw the Mandelbrot Set, consider the sequence of numbers below.

$$c, c^2 + c, (c^2 + c)^2 + c, [(c^2 + c)^2 + c]^2 + c, . . .$$

The behavior of this sequence depends on the value of the complex number c. If the sequence is bounded (the absolute value of each number in the sequence

$$|a + bi| = \sqrt{a^2 + b^2}$$

is less than some fixed number N), then the complex number c is in the Mandelbrot Set, and if the sequence is unbounded (the absolute value of the terms of the sequence become infinitely large), then the complex number c is not in the Mandelbrot Set. Determine whether each complex number c is in the Mandelbrot Set.

(a) $c = i$

(b) $c = 1 + i$

(c) $c = -2$

The figure below shows a graph of the Mandelbrot Set, where the horizontal and vertical axes represent the real and imaginary parts of c, respectively.

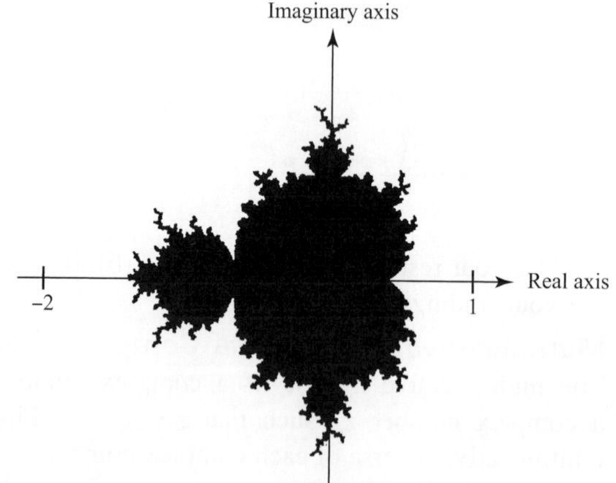

15. **Reasoning** Let z and $\bar{z}$ be complex conjugates. Show that the solutions of

$$|z - 1| \cdot |\bar{z} - 1| = 1$$

are the points (x, y) in the complex plane such that

$$(x - 1)^2 + y^2 = 1.$$

Identify the graph of the solution set. (*Hint:* Let $z = x + yi$.)

sciencephotos/Alamy Stock Photo

5 Exponential and Logarithmic Functions

Earthquakes (Example 6, page 402)

Beaver Population (Exercise 83, page 394)

Sound Intensity (Exercises 79–82, page 384)

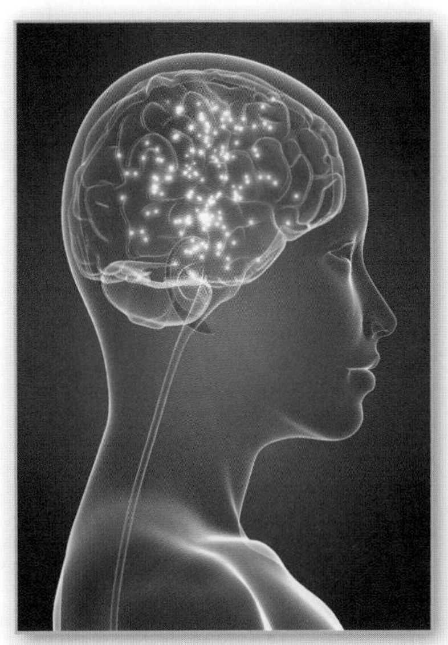

Human Memory Model (Exercise 83, page 378)

Nuclear Reactor Accident (Example 9, page 365)

5.1 Exponential Functions and Their Graphs

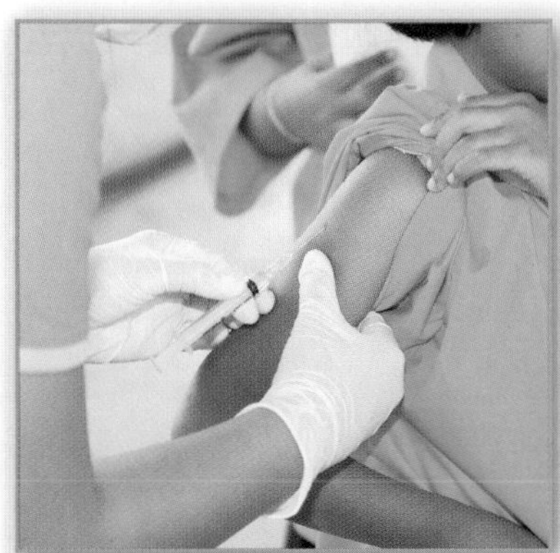

Exponential functions can help you model and solve real-life problems. For example, in Exercise 66 on page 368, you will use an exponential function to model the concentration of a drug in the bloodstream.

- **Recognize and evaluate exponential functions with base a.**
- **Graph exponential functions and use the One-to-One Property.**
- **Recognize, evaluate, and graph exponential functions with base e.**
- **Use exponential functions to model and solve real-life problems.**

Exponential Functions

So far, this text has dealt mainly with **algebraic functions,** which include polynomial functions and rational functions. In this chapter, you will study two types of nonalgebraic functions—*exponential functions* and *logarithmic functions*. These functions are examples of **transcendental functions.** This section will focus on exponential functions.

> **Definition of Exponential Function**
>
> The **exponential function f with base a** is denoted by
>
> $$f(x) = a^x$$
>
> where $a > 0$, $a \neq 1$, and x is any real number.

The base a of an exponential function cannot be 1 because $a = 1$ yields $f(x) = 1^x = 1$. This is a constant function, not an exponential function.

You have evaluated a^x for integer and rational values of x. For example, you know that $4^3 = 64$ and $4^{1/2} = 2$. However, to evaluate 4^x for any real number x, you need to interpret forms with *irrational* exponents. For the purposes of this text, it is sufficient to think of $a^{\sqrt{2}}$ (where $\sqrt{2} \approx 1.41421356$) as the number that has the successively closer approximations

$$a^{1.4}, a^{1.41}, a^{1.414}, a^{1.4142}, a^{1.41421}, \ldots .$$

EXAMPLE 1 Evaluating Exponential Functions

Use a calculator to evaluate each function at the given value of x.

	Function	Value
a.	$f(x) = 2^x$	$x = -3.1$
b.	$f(x) = 2^{-x}$	$x = \pi$
c.	$f(x) = 0.6^x$	$x = \frac{3}{2}$

Solution

	Function Value	Calculator Keystrokes	Display
a.	$f(-3.1) = 2^{-3.1}$	2 [^] [(−)] 3.1 [ENTER]	0.1166291
b.	$f(\pi) = 2^{-\pi}$	2 [^] [(−)] π [ENTER]	0.1133147
c.	$f\left(\frac{3}{2}\right) = (0.6)^{3/2}$	.6 [^] [(] 3 [÷] 2 [)] [ENTER]	0.4647580

✓ Checkpoint Audio-video solution in English & Spanish at LarsonPrecalculus.com

Use a calculator to evaluate $f(x) = 8^{-x}$ at $x = \sqrt{2}$.

When evaluating exponential functions with a calculator, it may be necessary to enclose fractional exponents in parentheses. Some calculators do not correctly interpret an exponent that consists of an expression unless parentheses are used.

Graphs of Exponential Functions

The graphs of all exponential functions have similar characteristics, as shown in Examples 2, 3, and 5.

EXAMPLE 2 Graphs of $y = a^x$

In the same coordinate plane, sketch the graph of each function.

a. $f(x) = 2^x$ **b.** $g(x) = 4^x$

Solution Begin by constructing a table of values.

x	-3	-2	-1	0	1	2
2^x	$\frac{1}{8}$	$\frac{1}{4}$	$\frac{1}{2}$	1	2	4
4^x	$\frac{1}{64}$	$\frac{1}{16}$	$\frac{1}{4}$	1	4	16

To sketch the graph of each function, plot the points from the table and connect them with a smooth curve, as shown in Figure 5.1. Note that both graphs are increasing. Moreover, the graph of $g(x) = 4^x$ is increasing more rapidly than the graph of $f(x) = 2^x$.

Checkpoint Audio-video solution in English & Spanish at LarsonPrecalculus.com

In the same coordinate plane, sketch the graph of each function.

a. $f(x) = 3^x$ **b.** $g(x) = 9^x$

▶ **ALGEBRA HELP** To review the techniques for sketching the graph of an equation, see Section P.3.

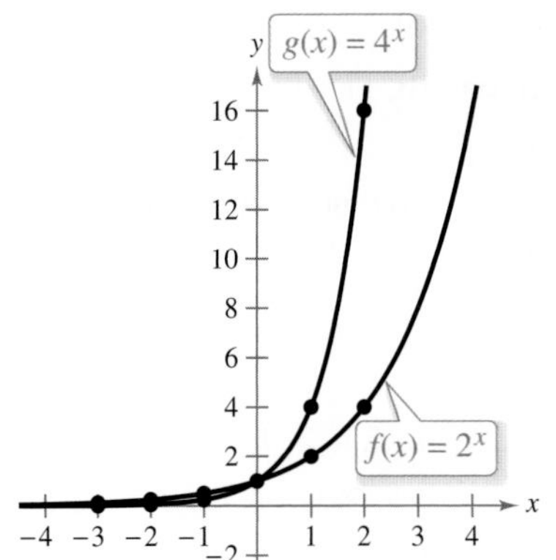

Figure 5.1

The table in Example 2 was evaluated by hand for integer values of x. You can also evaluate $f(x)$ and $g(x)$ for noninteger values of x by using a calculator.

EXAMPLE 3 Graphs of $y = a^{-x}$

In the same coordinate plane, sketch the graph of each function.

a. $F(x) = 2^{-x}$ **b.** $G(x) = 4^{-x}$

Solution Begin by constructing a table of values.

x	-2	-1	0	1	2	3
2^{-x}	4	2	1	$\frac{1}{2}$	$\frac{1}{4}$	$\frac{1}{8}$
4^{-x}	16	4	1	$\frac{1}{4}$	$\frac{1}{16}$	$\frac{1}{64}$

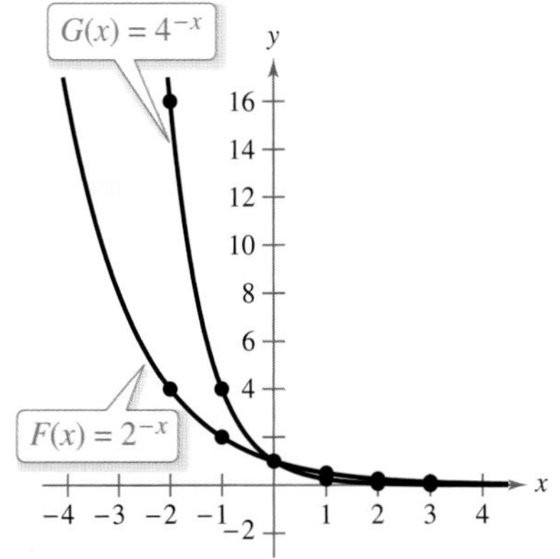

Figure 5.2

To sketch the graph of each function, plot the points from the table and connect them with a smooth curve, as shown in Figure 5.2. Note that both graphs are decreasing. Moreover, the graph of $G(x) = 4^{-x}$ is decreasing more rapidly than the graph of $F(x) = 2^{-x}$.

Checkpoint Audio-video solution in English & Spanish at LarsonPrecalculus.com

In the same coordinate plane, sketch the graph of each function.

a. $f(x) = 3^{-x}$ **b.** $g(x) = 9^{-x}$

Note that it is possible to use one of the properties of exponents to rewrite the functions in Example 3 with positive exponents.

$$F(x) = 2^{-x} = \frac{1}{2^x} = \left(\frac{1}{2}\right)^x \quad \text{and} \quad G(x) = 4^{-x} = \frac{1}{4^x} = \left(\frac{1}{4}\right)^x$$

Comparing the functions in Examples 2 and 3, observe that

$$F(x) = 2^{-x} = f(-x) \quad \text{and} \quad G(x) = 4^{-x} = g(-x).$$

Consequently, the graph of F is a reflection (in the y-axis) of the graph of f. The graphs of G and g have the same relationship. The graphs in Figures 5.1 and 5.2 are typical of the exponential functions $y = a^x$ and $y = a^{-x}$. They have one y-intercept and one horizontal asymptote (the x-axis), and they are continuous. Here is a summary of the basic characteristics of the graphs of these exponential functions.

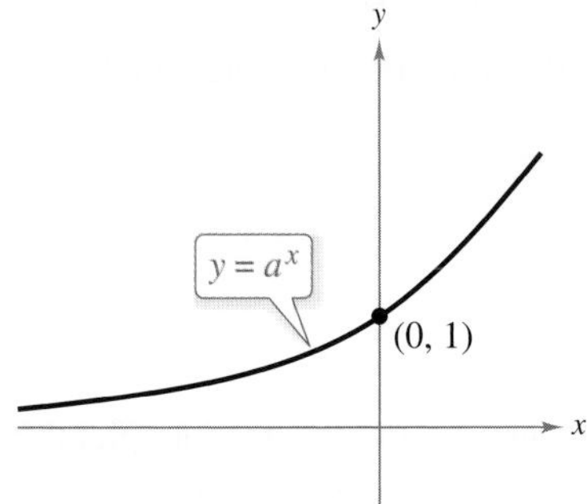

Graph of $y = a^x$, $a > 1$

- Domain: $(-\infty, \infty)$
- Range: $(0, \infty)$
- y-intercept: $(0, 1)$
- Increasing
- x-axis is a horizontal asymptote ($a^x \to 0$ as $x \to -\infty$).
- Continuous

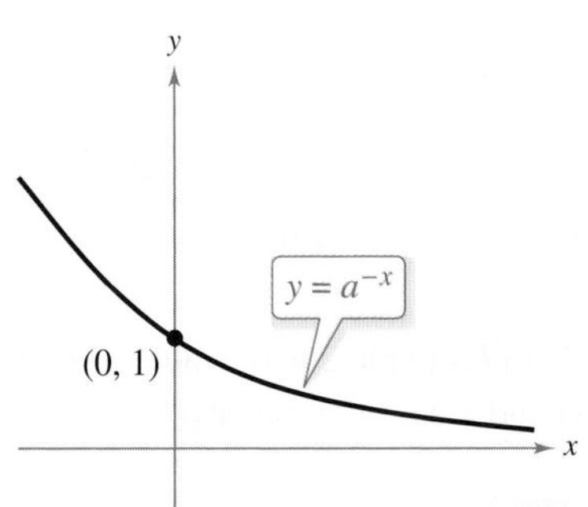

Graph of $y = a^{-x}$, $a > 1$

- Domain: $(-\infty, \infty)$
- Range: $(0, \infty)$
- y-intercept: $(0, 1)$
- Decreasing
- x-axis is a horizontal asymptote ($a^{-x} \to 0$ as $x \to \infty$).
- Continuous

Notice that the graph of an exponential function is always increasing or always decreasing, so the graph passes the Horizontal Line Test. Therefore, an exponential function is a one-to-one function. You can use the following **One-to-One Property** to solve simple exponential equations.

For $a > 0$ and $a \neq 1$, $a^x = a^y$ if and only if $x = y$. One-to-One Property

EXAMPLE 4 Using the One-to-One Property

a. $9 = 3^{x+1}$ Original equation

$3^2 = 3^{x+1}$ $9 = 3^2$

$2 = x + 1$ One-to-One Property

$1 = x$ Solve for x.

b. $\left(\frac{1}{2}\right)^x = 8$ Original equation

$2^{-x} = 2^3$ $\left(\frac{1}{2}\right)^x = 2^{-x}$, $8 = 2^3$

$x = -3$ One-to-One Property

✓ **Checkpoint** *Audio-video solution in English & Spanish at LarsonPrecalculus.com*

Use the One-to-One Property to solve the equation for x.

a. $8 = 2^{2x-1}$ **b.** $\left(\frac{1}{3}\right)^{-x} = 27$

In Example 5, notice how the graph of $y = a^x$ can be used to sketch the graphs of functions of the form $f(x) = b \pm a^{x+c}$.

▷ **ALGEBRA HELP** To review the techniques for transforming the graph of a function, see Section P.8.

EXAMPLE 5 Transformations of Graphs of Exponential Functions

See LarsonPrecalculus.com for an interactive version of this type of example.

Describe the transformation of the graph of $f(x) = 3^x$ that yields each graph.

a.

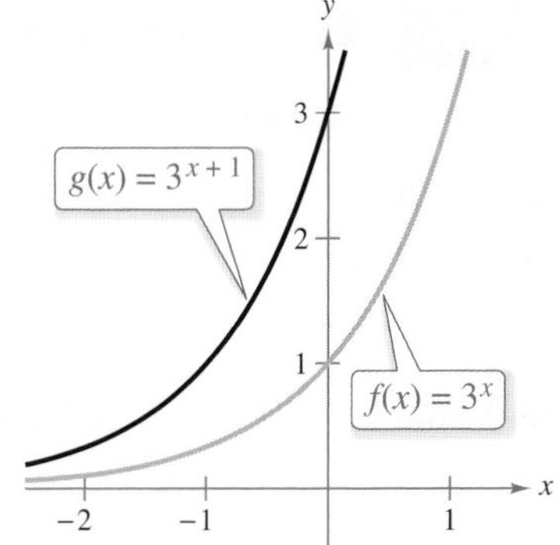

b.

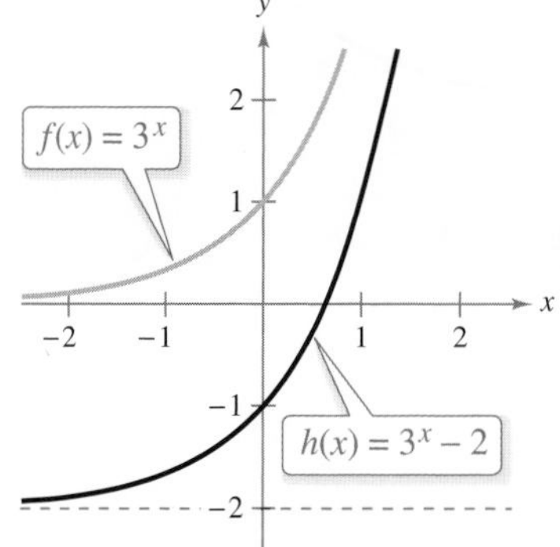

c.

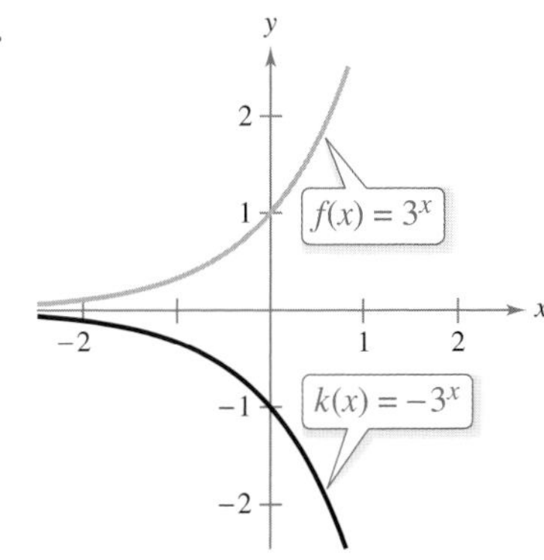

d.

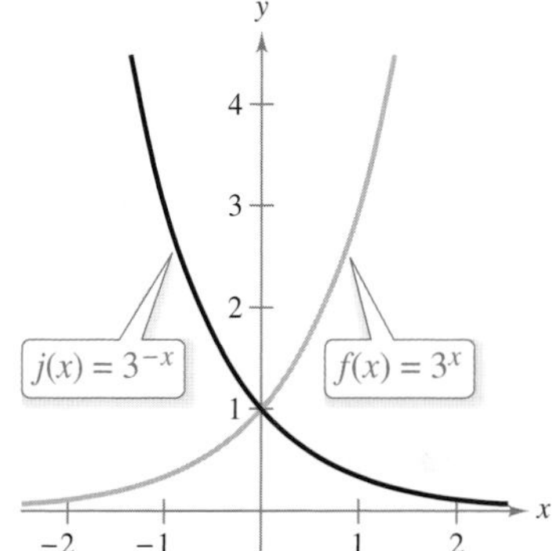

Solution

a. Because $g(x) = 3^{x+1} = f(x + 1)$, the graph of g is obtained by shifting the graph of f one unit to the *left*.

b. Because $h(x) = 3^x - 2 = f(x) - 2$, the graph of h is obtained by shifting the graph of f *down* two units.

c. Because $k(x) = -3^x = -f(x)$, the graph of k is obtained by *reflecting* the graph of f in the x-axis.

d. Because $j(x) = 3^{-x} = f(-x)$, the graph of j is obtained by *reflecting* the graph of f in the y-axis.

Describe the transformation of the graph of $f(x) = 4^x$ that yields the graph of each function.

a. $g(x) = 4^{x-2}$ **b.** $h(x) = 4^x + 3$ **c.** $k(x) = 4^{-x} - 3$

Note how each transformation in Example 5 affects the y-intercept and the horizontal asymptote.

The Natural Base e

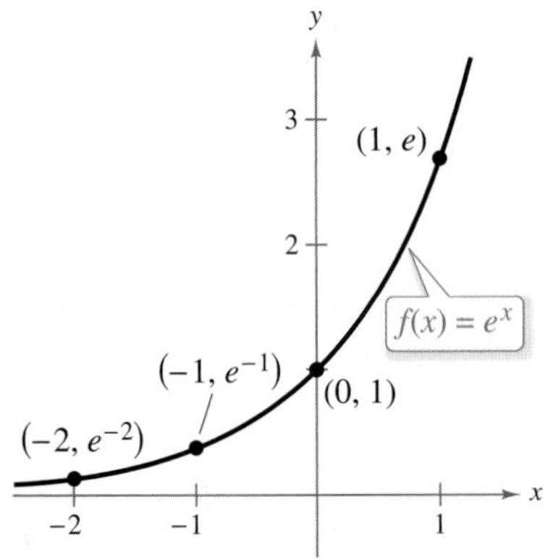

Figure 5.3

In many applications, the most convenient choice for a base is the irrational number

$$e \approx 2.718281828\ldots.$$

This number is called the **natural base.** The function $f(x) = e^x$ is called the **natural exponential function.** Figure 5.3 shows its graph. Be sure you see that for the exponential function $f(x) = e^x$, e is the constant $2.718281828\ldots$, whereas x is the variable.

EXAMPLE 6 Evaluating the Natural Exponential Function

Use a calculator to evaluate the function $f(x) = e^x$ at each value of x.

a. $x = -2$ **b.** $x = -1$

c. $x = 0.25$ **d.** $x = -0.3$

Solution

Function Value	Calculator Keystrokes	Display
a. $f(-2) = e^{-2}$	[eˣ] [(−)] 2 [ENTER]	0.1353353
b. $f(-1) = e^{-1}$	[eˣ] [(−)] 1 [ENTER]	0.3678794
c. $f(0.25) = e^{0.25}$	[eˣ] 0.25 [ENTER]	1.2840254
d. $f(-0.3) = e^{-0.3}$	[eˣ] [(−)] 0.3 [ENTER]	0.7408182

  **Checkpoint** Audio-video solution in English & Spanish at LarsonPrecalculus.com

Use a calculator to evaluate the function $f(x) = e^x$ at each value of x.

a. $x = 0.3$

b. $x = -1.2$

c. $x = 6.2$

EXAMPLE 7 Graphing Natural Exponential Functions

Sketch the graph of each natural exponential function.

a. $f(x) = 2e^{0.24x}$

b. $g(x) = \frac{1}{2}e^{-0.58x}$

Solution Begin by using a graphing utility to construct a table of values.

x	-3	-2	-1	0	1	2	3
$f(x)$	0.974	1.238	1.573	2.000	2.542	3.232	4.109
$g(x)$	2.849	1.595	0.893	0.500	0.280	0.157	0.088

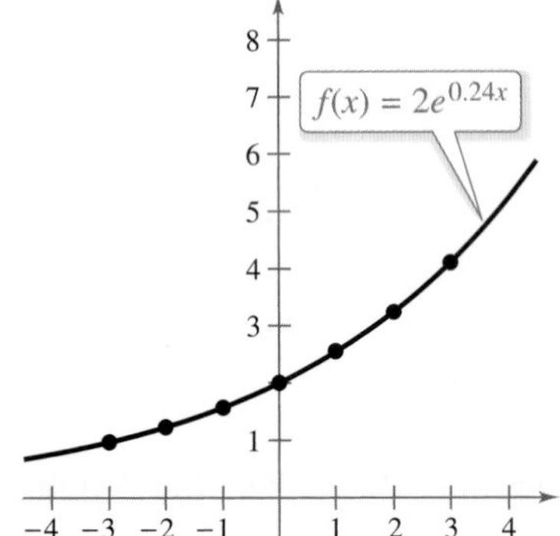

Figure 5.4

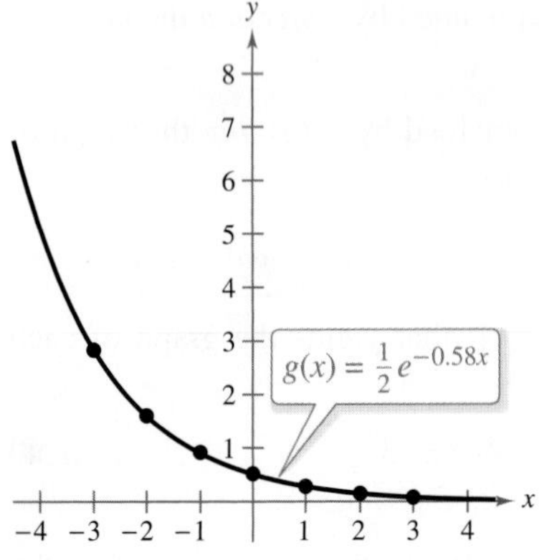

Figure 5.5

To graph each function, plot the points from the table and connect them with a smooth curve, as shown in Figures 5.4 and 5.5. Note that the graph in Figure 5.4 is increasing, whereas the graph in Figure 5.5 is decreasing.

  **Checkpoint** Audio-video solution in English & Spanish at LarsonPrecalculus.com

Sketch the graph of $f(x) = 5e^{0.17x}$.

Applications

One of the most familiar examples of exponential growth is an investment earning *continuously compounded interest*. The formula for *interest compounded n times per year* is

$$A = P\left(1 + \frac{r}{n}\right)^{nt}.$$

In this formula, A is the balance in the account, P is the principal (or original deposit), r is the annual interest rate (in decimal form), n is the number of compoundings per year, and t is the time in years. Exponential functions can be used to *develop* this formula and show how it leads to continuous compounding.

Consider a principal P invested at an annual interest rate r, compounded once per year. When the interest is added to the principal at the end of the first year, the new balance P_1 is

$$\begin{aligned} P_1 &= P + Pr \\ &= P(1 + r). \end{aligned}$$

This pattern of multiplying the balance by $1 + r$ repeats each successive year, as shown here.

Year	**Balance After Each Compounding**
0	$P = P$
1	$P_1 = P(1 + r)$
2	$P_2 = P_1(1 + r) = P(1 + r)(1 + r) = P(1 + r)^2$
3	$P_3 = P_2(1 + r) = P(1 + r)^2(1 + r) = P(1 + r)^3$
⋮	⋮
t	$P_t = P(1 + r)^t$

To accommodate more frequent (quarterly, monthly, or daily) compounding of interest, let n be the number of compoundings per year and let t be the number of years. Then the rate per compounding is r/n, and the account balance after t years is

$$A = P\left(1 + \frac{r}{n}\right)^{nt}. \qquad \text{Amount (balance) with } n \text{ compoundings per year}$$

When the number of compoundings n increases without bound, the process approaches what is called **continuous compounding.** In the formula for n compoundings per year, let $m = n/r$. This yields a new expression.

$$\begin{aligned} A &= P\left(1 + \frac{r}{n}\right)^{nt} && \text{Amount with } n \text{ compoundings per year} \\ &= P\left(1 + \frac{r}{mr}\right)^{mrt} && \text{Substitute } mr \text{ for } n. \\ &= P\left(1 + \frac{1}{m}\right)^{mrt} && \text{Simplify.} \\ &= P\left[\left(1 + \frac{1}{m}\right)^{m}\right]^{rt} && \text{Property of exponents} \end{aligned}$$

m	$\left(1 + \frac{1}{m}\right)^m$
1	2
10	2.59374246
100	2.704813829
1,000	2.716923932
10,000	2.718145927
100,000	2.718268237
1,000,000	2.718280469
10,000,000	2.718281693
↓	↓
∞	e

As m increases without bound (that is, as $m \to \infty$), the table at the left shows that $[1 + (1/m)]^m \to e$. This allows you to conclude that the formula for continuous compounding is

$$A = Pe^{rt}. \qquad \text{Substitute } e \text{ for } [1 + (1/m)]^m.$$

REMARK Be sure you see that, when using the formulas for compound interest, you must write the annual interest rate in decimal form. For example, you must write 6% as 0.06.

Formulas for Compound Interest

After t years, the balance A in an account with principal P and annual interest rate r (in decimal form) is given by one of these two formulas.

1. For n compoundings per year: $A = P\left(1 + \frac{r}{n}\right)^{nt}$

2. For continuous compounding: $A = Pe^{rt}$

EXAMPLE 8 Compound Interest

You invest \$12,000 at an annual rate of 3%. Find the balance after 5 years for each type of compounding.

a. Quarterly

b. Monthly

c. Continuous

Solution

a. For quarterly compounding, use $n = 4$ to find the balance after 5 years.

$$A = P\left(1 + \frac{r}{n}\right)^{nt} \qquad \text{Formula for compound interest}$$

$$= 12{,}000\left(1 + \frac{0.03}{4}\right)^{4(5)} \qquad \text{Substitute for } P, r, n, \text{ and } t.$$

$$\approx 13{,}934.21 \qquad \text{Use a calculator.}$$

b. For monthly compounding, use $n = 12$ to find the balance after 5 years.

$$A = P\left(1 + \frac{r}{n}\right)^{nt} \qquad \text{Formula for compound interest}$$

$$= 12{,}000\left(1 + \frac{0.03}{12}\right)^{12(5)} \qquad \text{Substitute for } P, r, n, \text{ and } t.$$

$$\approx \$13{,}939.40 \qquad \text{Use a calculator.}$$

c. Use the formula for continuous compounding to find the balance after 5 years.

$$A = Pe^{rt} \qquad \text{Formula for continuous compounding}$$

$$= 12{,}000e^{0.03(5)} \qquad \text{Substitute for } P, r, \text{ and } t.$$

$$\approx \$13{,}942.01 \qquad \text{Use a calculator.}$$

✓ ***Checkpoint*** *Audio-video solution in English & Spanish at LarsonPrecalculus.com*

You invest \$6000 at an annual rate of 4%. Find the balance after 7 years for each type of compounding.

a. Quarterly **b.** Monthly **c.** Continuous

In Example 8, note that continuous compounding yields more than quarterly and monthly compounding. This is typical of the two types of compounding. That is, for a given principal, interest rate, and time, continuous compounding will always yield a larger balance than compounding n times per year.

EXAMPLE 9 Radioactive Decay

In 1986, a nuclear reactor accident occurred in Chernobyl in what was then the Soviet Union. The explosion spread highly toxic radioactive chemicals, such as plutonium $\left({}^{239}\text{Pu}\right)$, over hundreds of square miles, and the government evacuated the city and the surrounding area. To see why the city is now uninhabited, consider the model

$$P = 10\left(\frac{1}{2}\right)^{t/24,100}$$

which represents the amount of plutonium P that remains (from an initial amount of 10 pounds) after t years. Sketch the graph of this function over the interval from $t = 0$ to $t = 100,000$, where $t = 0$ represents 1986. How much of the 10 pounds will remain in the year 2020? How much of the 10 pounds will remain after 100,000 years?

The International Atomic Energy Authority ranks nuclear incidents and accidents by severity using a scale from 1 to 7 called the International Nuclear and Radiological Event Scale (INES). A level 7 ranking is the most severe. To date, the Chernobyl accident and an accident at Japan's Fukushima Daiichi power plant in 2011 are the only two disasters in history to be given an INES level 7 ranking.

Solution The graph of this function is shown in the figure at the right. Note from this graph that plutonium has a *half-life* of about 24,100 years. That is, after 24,100 years, *half* of the original amount will remain. After another 24,100 years, one-quarter of the original amount will remain, and so on. In the year 2020 ($t = 34$), there will still be

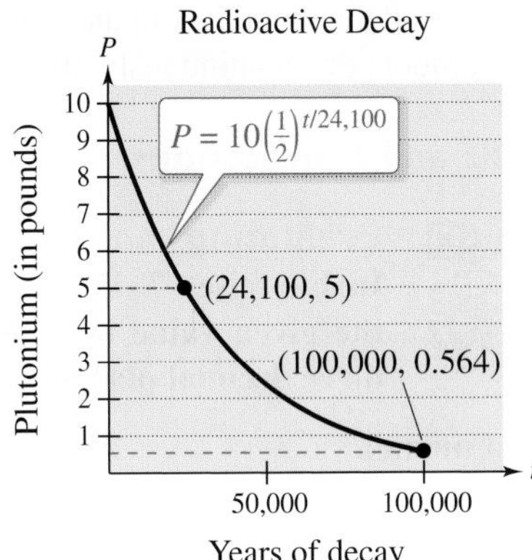

$$\begin{aligned} P &= 10\left(\frac{1}{2}\right)^{34/24,100} \\ &\approx 10\left(\frac{1}{2}\right)^{0.0014108} \\ &\approx 9.990 \text{ pounds} \end{aligned}$$

of plutonium remaining. After 100,000 years, there will still be

$$\begin{aligned} P &= 10\left(\frac{1}{2}\right)^{100,000/24,100} \\ &\approx 0.564 \text{ pound} \end{aligned}$$

of plutonium remaining.

Checkpoint Audio-video solution in English & Spanish at LarsonPrecalculus.com

In Example 9, how much of the 10 pounds will remain in the year 2089? How much of the 10 pounds will remain after 125,000 years?

Summarize (Section 5.1)

1. State the definition of the exponential function f with base a *(page 358)*. For an example of evaluating exponential functions, see Example 1.
2. Describe the basic characteristics of the graphs of the exponential functions $y = a^x$ and $y = a^{-x}$, $a > 1$ *(page 360)*. For examples of graphing exponential functions, see Examples 2, 3, and 5.
3. State the definitions of the natural base and the natural exponential function *(page 362)*. For examples of evaluating and graphing natural exponential functions, see Examples 6 and 7.
4. Describe real-life applications involving exponential functions *(pages 364 and 365, Examples 8 and 9)*.

5.1 Exercises

See CalcChat.com for tutorial help and worked-out solutions to odd-numbered exercises.

Vocabulary: Fill in the blanks.

1. Polynomial and rational functions are examples of ________ functions.
2. Exponential and logarithmic functions are examples of nonalgebraic functions, also called ________ functions.
3. The ________ Property can be used to solve simple exponential equations.
4. The exponential function $f(x) = e^x$ is called the ________ ________ function, and the base e is called the ________ base.
5. To find the amount A in an account after t years with principal P and an annual interest rate r (in decimal form) compounded n times per year, use the formula ________.
6. To find the amount A in an account after t years with principal P and an annual interest rate r (in decimal form) compounded continuously, use the formula ________.

Skills and Applications

Evaluating an Exponential Function In Exercises 7–12, evaluate the function at the given value of x. Round your result to three decimal places.

	Function	Value
7.	$f(x) = 0.9^x$	$x = 1.4$
8.	$f(x) = 4.7^x$	$x = -\pi$
9.	$f(x) = 3^x$	$x = \frac{2}{5}$
10.	$f(x) = \left(\frac{2}{3}\right)^{5x}$	$x = \frac{3}{10}$
11.	$f(x) = 5000(2^x)$	$x = -1.5$
12.	$f(x) = 200(1.2)^{12x}$	$x = 24$

Matching an Exponential Function with Its Graph In Exercises 13–16, match the exponential function with its graph. [The graphs are labeled (a), (b), (c), and (d).]

(a)

(b)

(c)

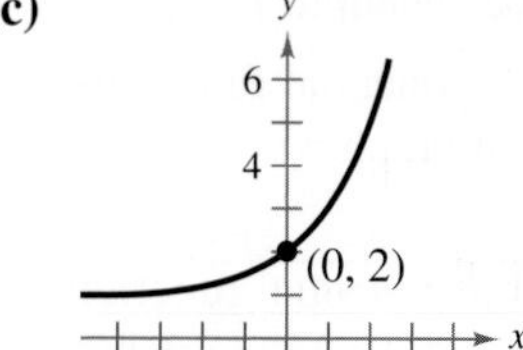

(d)

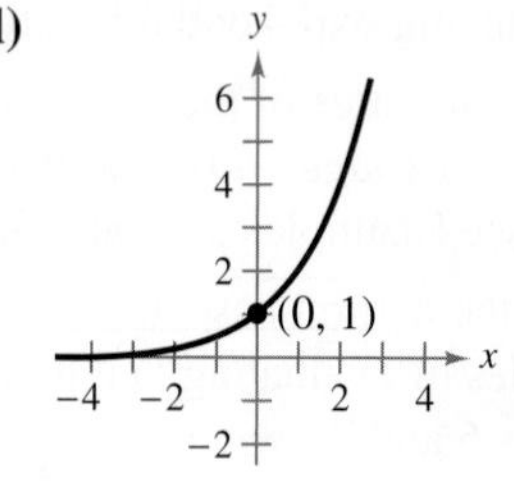

13. $f(x) = 2^x$
14. $f(x) = 2^x + 1$
15. $f(x) = 2^{-x}$
16. $f(x) = 2^{x-2}$

Graphing an Exponential Function In Exercises 17–24, use a graphing utility to construct a table of values for the function. Then sketch the graph of the function.

17. $f(x) = 7^x$
18. $f(x) = 7^{-x}$
19. $f(x) = \left(\frac{1}{4}\right)^{-x}$
20. $f(x) = \left(\frac{1}{4}\right)^x$
21. $f(x) = 4^{x-1}$
22. $f(x) = 4^{x+1}$
23. $f(x) = 2^{x+1} + 3$
24. $f(x) = 3^{x-2} + 1$

Using the One-to-One Property In Exercises 25–28, use the One-to-One Property to solve the equation for x.

25. $3^{x+1} = 27$
26. $2^{x-2} = 64$
27. $\left(\frac{1}{2}\right)^x = 32$
28. $5^{x-2} = \frac{1}{125}$

Transformations of the Graph of an Exponential Function In Exercises 29–32, describe the transformation(s) of the graph of f that yield(s) the graph of g.

29. $f(x) = 3^x, \quad g(x) = 3^x + 1$
30. $f(x) = \left(\frac{7}{2}\right)^x, \quad g(x) = -\left(\frac{7}{2}\right)^{-x}$
31. $f(x) = 10^x, \quad g(x) = 10^{-x+3}$
32. $f(x) = 0.3^x, \quad g(x) = -0.3^x + 5$

Evaluating a Natural Exponential Function In Exercises 33–36, evaluate the function at the given value of x. Round your result to three decimal places.

	Function	Value
33.	$f(x) = e^x$	$x = 1.9$
34.	$f(x) = 1.5e^{x/2}$	$x = 240$
35.	$f(x) = 5000e^{0.06x}$	$x = 6$
36.	$f(x) = 250e^{0.05x}$	$x = 20$

Graphing a Natural Exponential Function In Exercises 37–40, use a graphing utility to construct a table of values for the function. Then sketch the graph of the function.

37. $f(x) = 3e^{x+4}$
38. $f(x) = 2e^{-1.5x}$
39. $f(x) = 2e^{x-2} + 4$
40. $f(x) = 2 + e^{x-5}$

Graphing a Natural Exponential Function In Exercises 41–44, use a graphing utility to graph the exponential function.

41. $s(t) = 2e^{0.5t}$
42. $s(t) = 3e^{-0.2t}$
43. $g(x) = 1 + e^{-x}$
44. $h(x) = e^{x-2}$

Using the One-to-One Property In Exercises 45–48, use the One-to-One Property to solve the equation for x.

45. $e^{3x+2} = e^3$
46. $e^{2x-1} = e^4$
47. $e^{x^2-3} = e^{2x}$
48. $e^{x^2+6} = e^{5x}$

Compound Interest In Exercises 49–52, complete the table by finding the balance A when P dollars is invested at rate r for t years and compounded n times per year.

n	1	2	4	12	365	Continuous
A						

49. $P = \$1500$, $r = 2\%$, $t = 10$ years
50. $P = \$2500$, $r = 3.5\%$, $t = 10$ years
51. $P = \$2500$, $r = 4\%$, $t = 20$ years
52. $P = \$1000$, $r = 6\%$, $t = 40$ years

Compound Interest In Exercises 53–56, complete the table by finding the balance A when \$12,000 is invested at rate r for t years, compounded continuously.

t	10	20	30	40	50
A					

53. $r = 4\%$
54. $r = 6\%$
55. $r = 6.5\%$
56. $r = 3.5\%$

57. **Trust Fund** On the day of a child's birth, a parent deposits \$30,000 in a trust fund that pays 5% interest, compounded continuously. Determine the balance in this account on the child's 25th birthday.

58. **Trust Fund** A philanthropist deposits \$5000 in a trust fund that pays 7.5% interest, compounded continuously. The balance will be given to the college from which the philanthropist graduated after the money has earned interest for 50 years. How much will the college receive?

59. **Inflation** Assuming that the annual rate of inflation averages 4% over the next 10 years, the approximate costs C of goods or services during any year in that decade can be modeled by $C(t) = P(1.04)^t$, where t is the time in years and P is the present cost. The price of an oil change for your car is presently \$29.88. Estimate the price 10 years from now.

60. **Computer Virus** The number V of computers infected by a virus increases according to the model $V(t) = 100e^{4.6052t}$, where t is the time in hours. Find the number of computers infected after (a) 1 hour, (b) 1.5 hours, and (c) 2 hours.

61. **Population Growth** The projected population of the United States for the years 2025 through 2055 can be modeled by $P = 307.58e^{0.0052t}$, where P is the population (in millions) and t is the time (in years), with $t = 25$ corresponding to 2025. *(Source: U.S. Census Bureau)*
    (a) Use a graphing utility to graph the function for the years 2025 through 2055.
    (b) Use the *table* feature of the graphing utility to create a table of values for the same time period as in part (a).
    (c) According to the model, during what year will the population of the United States exceed 430 million?

62. **Population** The population P (in millions) of Italy from 2003 through 2015 can be approximated by the model $P = 57.59e^{0.0051t}$, where t represents the year, with $t = 3$ corresponding to 2003. *(Source: U.S. Census Bureau)*
    (a) According to the model, is the population of Italy increasing or decreasing? Explain.
    (b) Find the populations of Italy in 2003 and 2015.
    (c) Use the model to predict the populations of Italy in 2020 and 2025.

63. **Radioactive Decay** Let Q represent a mass (in grams) of radioactive plutonium (^{239}Pu), whose half-life is 24,100 years. The quantity of plutonium present after t years is $Q = 16\left(\frac{1}{2}\right)^{t/24,100}$.
    (a) Determine the initial quantity (when $t = 0$).
    (b) Determine the quantity present after 75,000 years.
    (c) Use a graphing utility to graph the function over the interval $t = 0$ to $t = 150,000$.

64. **Radioactive Decay** Let Q represent a mass (in grams) of carbon (^{14}C), whose half-life is 5715 years. The quantity of carbon 14 present after t years is $Q = 10\left(\frac{1}{2}\right)^{t/5715}$.
    (a) Determine the initial quantity (when $t = 0$).
    (b) Determine the quantity present after 2000 years.
    (c) Sketch the graph of the function over the interval $t = 0$ to $t = 10,000$.

65. Depreciation The value of a wheelchair conversion van that originally cost \$49,810 depreciates so that each year it is worth $\frac{7}{8}$ of its value for the previous year.

(a) Find a model for $V(t)$, the value of the van after t years.

(b) Determine the value of the van 4 years after it was purchased.

66. Chemistry

Immediately following an injection, the concentration of a drug in the bloodstream is 300 milligrams per milliliter. After t hours, the concentration is 75% of the level of the previous hour.

(a) Find a model for $C(t)$, the concentration of the drug after t hours.

(b) Determine the concentration of the drug after 8 hours.

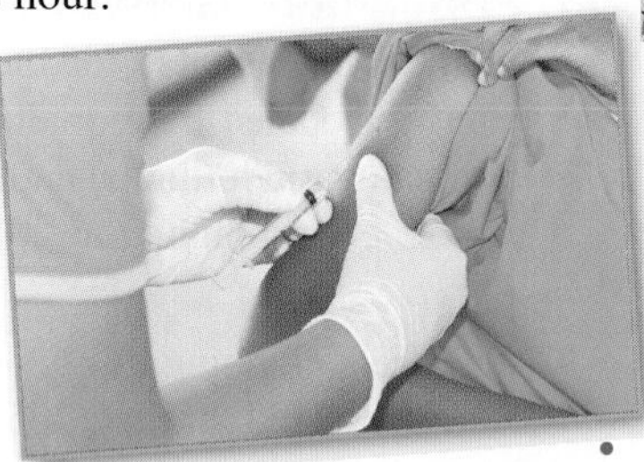

Exploration

True or False? **In Exercises 67 and 68, determine whether the statement is true or false. Justify your answer.**

67. The line $y = -2$ is an asymptote for the graph of $f(x) = 10^x - 2$.

68. $e = \dfrac{271{,}801}{99{,}990}$

Think About It **In Exercises 69–72, use properties of exponents to determine which functions (if any) are the same.**

69. $f(x) = 3^{x-2}$
$g(x) = 3^x - 9$
$h(x) = \frac{1}{9}(3^x)$

70. $f(x) = 4^x + 12$
$g(x) = 2^{2x+6}$
$h(x) = 64(4^x)$

71. $f(x) = 16(4^{-x})$
$g(x) = \left(\frac{1}{4}\right)^{x-2}$
$h(x) = 16(2^{-2x})$

72. $f(x) = e^{-x} + 3$
$g(x) = e^{3-x}$
$h(x) = -e^{x-3}$

73. Solving Inequalities Graph the functions $y = 3^x$ and $y = 4^x$ and use the graphs to solve each inequality.

(a) $4^x < 3^x$ (b) $4^x > 3^x$

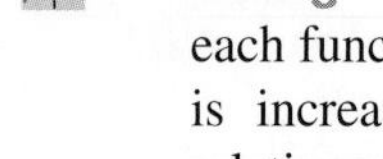

74. Using Technology Use a graphing utility to graph each function. Use the graph to find where the function is increasing and decreasing, and approximate any relative maximum or minimum values.

(a) $f(x) = x^2e^{-x}$

(b) $g(x) = x2^{3-x}$

75. Graphical Reasoning Use a graphing utility to graph $y_1 = [1 + (1/x)]^x$ and $y_2 = e$ in the same viewing window. Using the *trace* feature, explain what happens to the graph of y_1 as x increases.

76. Graphical Reasoning Use a graphing utility to graph

$$f(x) = \left(1 + \frac{0.5}{x}\right)^x \quad \text{and} \quad g(x) = e^{0.5}$$

in the same viewing window. What is the relationship between f and g as x increases and decreases without bound?

77. Comparing Graphs Use a graphing utility to graph each pair of functions in the same viewing window. Describe any similarities and differences in the graphs.

(a) $y_1 = 2^x, y_2 = x^2$

(b) $y_1 = 3^x, y_2 = x^3$

78. HOW DO YOU SEE IT? The figure shows the graphs of $y = 2^x$, $y = e^x$, $y = 10^x$, $y = 2^{-x}$, $y = e^{-x}$, and $y = 10^{-x}$. Match each function with its graph. [The graphs are labeled (a) through (f).] Explain your reasoning.

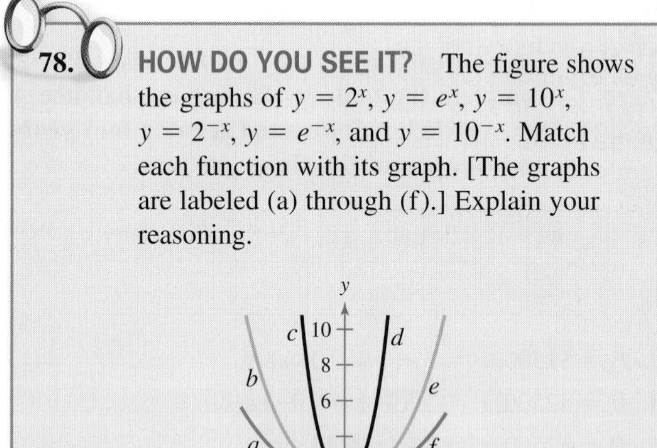

79. Think About It Which functions are exponential?

(a) $f(x) = 3x$ (b) $g(x) = 3x^2$

(c) $h(x) = 3^x$ (d) $k(x) = 2^{-x}$

80. Compound Interest Use the formula

$$A = P\left(1 + \frac{r}{n}\right)^{nt}$$

to calculate the balance A of an investment when $P = \$3000$, $r = 6\%$, and $t = 10$ years, and compounding is done (a) by the day, (b) by the hour, (c) by the minute, and (d) by the second. Does increasing the number of compoundings per year result in unlimited growth of the balance? Explain.

Project: Population per Square Mile To work an extended application analyzing the population per square mile of the United States, visit this text's website at *LarsonPrecalculus.com.* *(Source: U.S. Census Bureau)*

5.2 Logarithmic Functions and Their Graphs

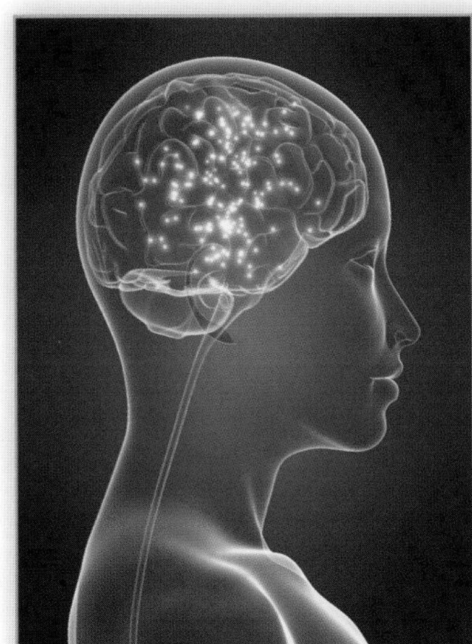

Logarithmic functions can often model scientific observations. For example, in Exercise 83 on page 378, you will use a logarithmic function that models human memory.

- **Recognize and evaluate logarithmic functions with base *a*.**
- **Graph logarithmic functions.**
- **Recognize, evaluate, and graph natural logarithmic functions.**
- **Use logarithmic functions to model and solve real-life problems.**

Logarithmic Functions

In Section 5.1, you learned that the exponential function $f(x) = a^x$ is one-to-one. It follows that $f(x) = a^x$ must have an inverse function. This inverse function is the **logarithmic function with base *a*.**

> **Definition of Logarithmic Function with Base *a***
>
> For $x > 0$, $a > 0$, and $a \neq 1$,
>
> $y = \log_a x$ if and only if $x = a^y$.
>
> The function
>
> $f(x) = \log_a x$ Read as "log base a of x."
>
> is the **logarithmic function with base *a*.**

The equations $y = \log_a x$ and $x = a^y$ are equivalent. For example, $2 = \log_3 9$ is equivalent to $9 = 3^2$, and $5^3 = 125$ is equivalent to $\log_5 125 = 3$.

When evaluating logarithms, remember that *a logarithm is an exponent.* This means that $\log_a x$ is the exponent to which a must be raised to obtain x. For example, $\log_2 8 = 3$ because 2 raised to the third power is 8.

EXAMPLE 1 Evaluating Logarithms

Evaluate each logarithm at the given value of x.

a. $f(x) = \log_2 x,\ x = 32$ **b.** $f(x) = \log_3 x,\ x = 1$

c. $f(x) = \log_4 x,\ x = 2$ **d.** $f(x) = \log_{10} x,\ x = \frac{1}{100}$

Solution

a. $f(32) = \log_2 32 = 5$ because $2^5 = 32.$

b. $f(1) = \log_3 1 = 0$ because $3^0 = 1.$

c. $f(2) = \log_4 2 = \frac{1}{2}$ because $4^{1/2} = \sqrt{4} = 2.$

d. $f\left(\frac{1}{100}\right) = \log_{10} \frac{1}{100} = -2$ because $10^{-2} = \frac{1}{10^2} = \frac{1}{100}.$

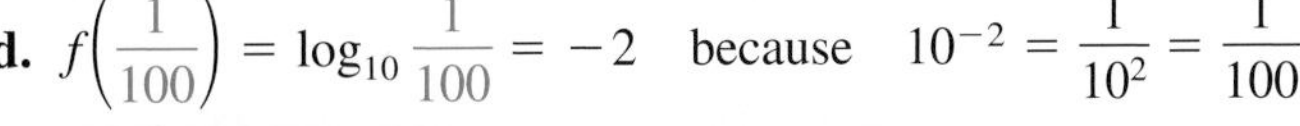

✓ ***Checkpoint*** *Audio-video solution in English & Spanish at LarsonPrecalculus.com*

Evaluate each logarithm at the given value of x.

a. $f(x) = \log_6 x, x = 1$ **b.** $f(x) = \log_5 x, x = \frac{1}{125}$ **c.** $f(x) = \log_7 x, x = 343$

The logarithmic function with base 10 is called the **common logarithmic function.** It is denoted by $\log_{10}$ or simply log. On most calculators, it is denoted by [LOG]. Example 2 shows how to use a calculator to evaluate common logarithmic functions. You will learn how to use a calculator to calculate logarithms with any base in Section 5.3.

EXAMPLE 2 Evaluating Common Logarithms on a Calculator

Use a calculator to evaluate the function $f(x) = \log x$ at each value of x.

a. $x = 10$ **b.** $x = \frac{1}{3}$ **c.** $x = -2$

Solution

	Function Value	Calculator Keystrokes	Display
a.	$f(10) = \log 10$	LOG 10 ENTER	1
b.	$f\left(\frac{1}{3}\right) = \log \frac{1}{3}$	LOG (1 ÷ 3) ENTER	-0.4771213
c.	$f(-2) = \log(-2)$	LOG (−) 2 ENTER	ERROR

Note that the calculator displays an error message (or a complex number) when you try to evaluate $\log(-2)$. This occurs because there is no real number power to which 10 can be raised to obtain -2.

✓ **Checkpoint** Audio-video solution in English & Spanish at LarsonPrecalculus.com

Use a calculator to evaluate the function $f(x) = \log x$ at each value of x.

a. $x = 275$ **b.** $x = -\frac{1}{2}$ **c.** $x = \frac{1}{2}$

The definition of the logarithmic function with base a leads to several properties.

Properties of Logarithms

1. $\log_a 1 = 0$ because $a^0 = 1$.
2. $\log_a a = 1$ because $a^1 = a$.
3. $\log_a a^x = x$ and $a^{\log_a x} = x$ — Inverse Properties
4. If $\log_a x = \log_a y$, then $x = y$. — One-to-One Property

EXAMPLE 3 Using Properties of Logarithms

a. Simplify $\log_4 1$. **b.** Simplify $\log_{\sqrt{7}} \sqrt{7}$. **c.** Simplify $6^{\log_6 20}$.

Solution

a. $\log_4 1 = 0$ — Property 1

b. $\log_{\sqrt{7}} \sqrt{7} = 1$ — Property 2

c. $6^{\log_6 20} = 20$ — Property 3 (Inverse Property)

✓ **Checkpoint** Audio-video solution in English & Spanish at LarsonPrecalculus.com

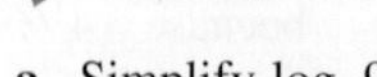

a. Simplify $\log_9 9$. **b.** Simplify $20^{\log_{20} 3}$. **c.** Simplify $\log_{\sqrt{3}} 1$.

EXAMPLE 4 Using the One-to-One Property

a. $\log_3 x = \log_3 12$ — Original equation

$x = 12$ — One-to-One Property

b. $\log(2x + 1) = \log 3x \implies 2x + 1 = 3x \implies 1 = x$

c. $\log_4(x^2 - 6) = \log_4 10 \implies x^2 - 6 = 10 \implies x^2 = 16 \implies x = \pm 4$

✓ **Checkpoint** Audio-video solution in English & Spanish at LarsonPrecalculus.com

Solve $\log_5 (x^2 + 3) = \log_5 12$ for x.

Graphs of Logarithmic Functions

To sketch the graph of $y = \log_a x$, use the fact that the graphs of inverse functions are reflections of each other in the line $y = x$.

EXAMPLE 5 Graphing Exponential and Logarithmic Functions

In the same coordinate plane, sketch the graph of each function.

a. $f(x) = 2^x$ **b.** $g(x) = \log_2 x$

Solution

a. For $f(x) = 2^x$, construct a table of values. By plotting these points and connecting them with a smooth curve, you obtain the graph shown in Figure 5.6.

x	-2	-1	0	1	2	3
$f(x) = 2^x$	$\frac{1}{4}$	$\frac{1}{2}$	1	2	4	8

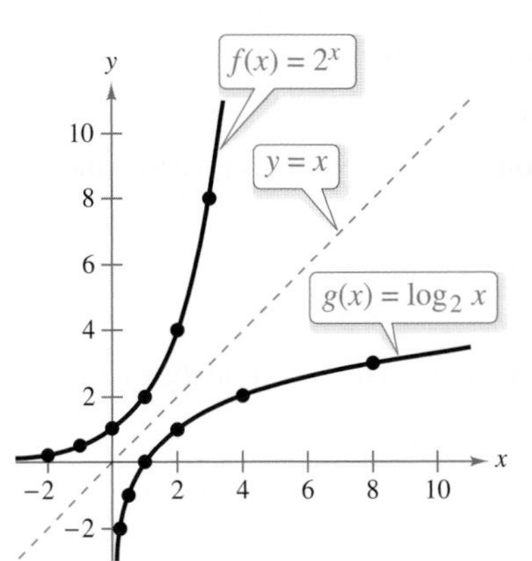

Figure 5.6

b. Because $g(x) = \log_2 x$ is the inverse function of $f(x) = 2^x$, the graph of g is obtained by plotting the points $(f(x), x)$ and connecting them with a smooth curve. The graph of g is a reflection of the graph of f in the line $y = x$, as shown in Figure 5.6.

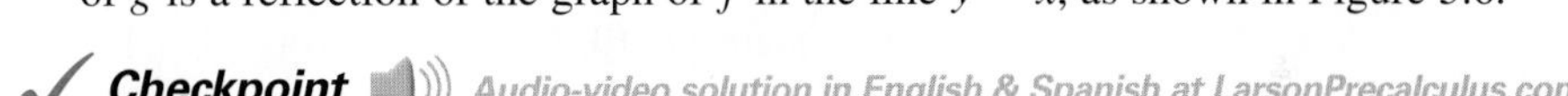

In the same coordinate plane, sketch the graphs of (a) $f(x) = 8^x$ and (b) $g(x) = \log_8 x$.

EXAMPLE 6 Sketching the Graph of a Logarithmic Function

Sketch the graph of $f(x) = \log x$. Identify the vertical asymptote.

Solution Begin by constructing a table of values. Note that some of the values can be obtained without a calculator by using the properties of logarithms. Others require a calculator.

	Without calculator				With calculator		
x	$\frac{1}{100}$	$\frac{1}{10}$	1	10	2	5	8
$f(x) = \log x$	-2	-1	0	1	0.301	0.699	0.903

Next, plot the points and connect them with a smooth curve, as shown in the figure below. The vertical asymptote is $x = 0$ (y-axis).

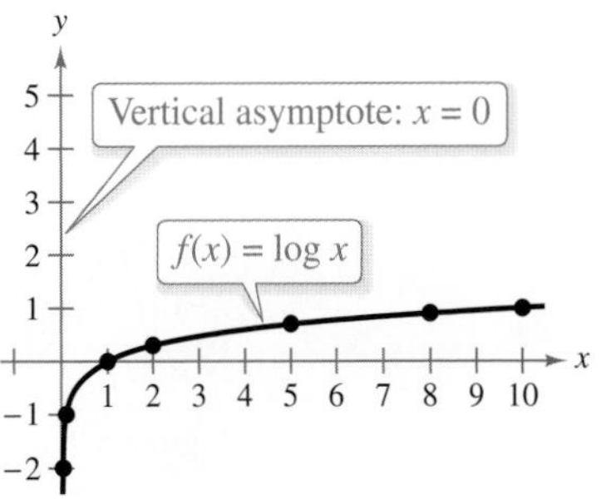

Checkpoint Audio-video solution in English & Spanish at LarsonPrecalculus.com

Sketch the graph of $f(x) = \log_3 x$ by constructing a table of values without using a calculator. Identify the vertical asymptote.

The graph in Example 6 is typical for functions of the form $f(x) = \log_a x$, $a > 1$. They have one x-intercept and one vertical asymptote. Notice how slowly the graph rises for $x > 1$. Here are the basic characteristics of logarithmic graphs.

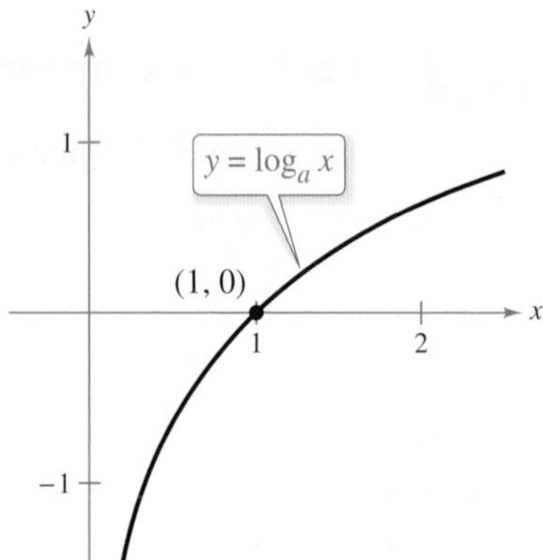

Graph of $y = \log_a x, a > 1$

- Domain: $(0, \infty)$
- Range: $(-\infty, \infty)$
- x-intercept: $(1, 0)$
- Increasing
- One-to-one, therefore has an inverse function
- y-axis is a vertical asymptote ($\log_a x \to -\infty$ as $x \to 0^+$).
- Continuous
- Reflection of graph of $y = a^x$ in the line $y = x$

Some basic characteristics of the graph of $f(x) = a^x$ are listed below to illustrate the inverse relation between $f(x) = a^x$ and $g(x) = \log_a x$.

- Domain: $(-\infty, \infty)$
- Range: $(0, \infty)$
- y-intercept: $(0, 1)$
- x-axis is a horizontal asymptote ($a^x \to 0$ as $x \to -\infty$).

The next example uses the graph of $y = \log_a x$ to sketch the graphs of functions of the form $f(x) = b \pm \log_a(x + c)$.

EXAMPLE 7 Shifting Graphs of Logarithmic Functions

See LarsonPrecalculus.com for an interactive version of this type of example.

Use the graph of $f(x) = \log x$ to sketch the graph of each function.

a. $g(x) = \log(x - 1)$ **b.** $h(x) = 2 + \log x$

Solution

a. Because $g(x) = \log(x - 1) = f(x - 1)$, the graph of g can be obtained by shifting the graph of f one unit to the right, as shown in Figure 5.7.

b. Because $h(x) = 2 + \log x = 2 + f(x)$, the graph of h can be obtained by shifting the graph of f two units up, as shown in Figure 5.8.

REMARK Notice that the vertical transformation in Figure 5.8 keeps the y-axis as the vertical asymptote, but the horizontal transformation in Figure 5.7 yields a new vertical asymptote of $x = 1$.

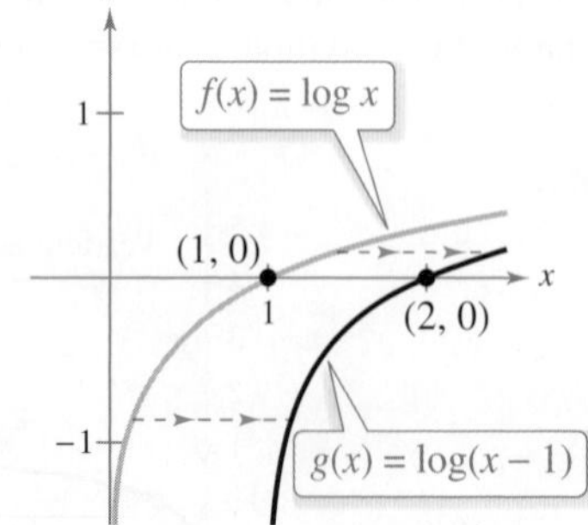

Figure 5.7

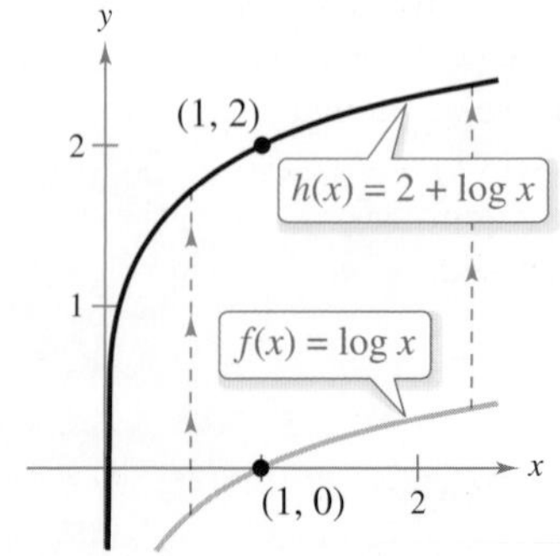

Figure 5.8

ALGEBRA HELP To review the techniques for shifting, reflecting, and stretching graphs, see Section P.8.

✓ ***Checkpoint*** Audio-video solution in English & Spanish at LarsonPrecalculus.com

Use the graph of $f(x) = \log_3 x$ to sketch the graph of each function.

a. $g(x) = -1 + \log_3 x$ **b.** $h(x) = \log_3(x + 3)$

The Natural Logarithmic Function

By looking back at the graph of the natural exponential function introduced on page 362 in Section 5.1, you will see that $f(x) = e^x$ is one-to-one and so has an inverse function. This inverse function is called the **natural logarithmic function** and is denoted by the special symbol ln x, read as "the natural log of x" or "el en of x."

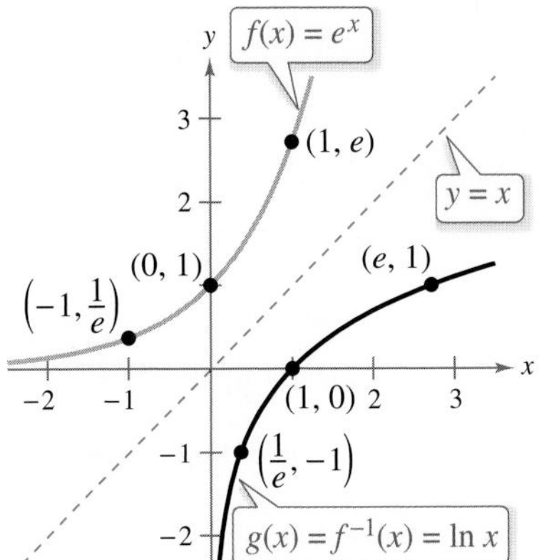

Reflection of graph of $f(x) = e^x$ in the line $y = x$

Figure 5.9

> **TECHNOLOGY** On most calculators, the natural logarithm is denoted by [LN] as illustrated in Example 8.

> **The Natural Logarithmic Function**
>
> The function
>
> $$f(x) = \log_e x = \ln x, \quad x > 0$$
>
> is called the **natural logarithmic function.**

The equations $y = \ln x$ and $x = e^y$ are equivalent. Note that the natural logarithm $\ln x$ is written without a base. The base is understood to be e.

Because the functions $f(x) = e^x$ and $g(x) = \ln x$ are inverse functions of each other, their graphs are reflections of each other in the line $y = x$, as shown in Figure 5.9.

EXAMPLE 8 Evaluating the Natural Logarithmic Function

Use a calculator to evaluate the function $f(x) = \ln x$ at each value of x.

a. $x = 2$

b. $x = 0.3$

c. $x = -1$

d. $x = 1 + \sqrt{2}$

Solution

	Function Value	Calculator Keystrokes	Display
a.	$f(2) = \ln 2$	[LN] 2 [ENTER]	0.6931472
b.	$f(0.3) = \ln 0.3$	[LN] .3 [ENTER]	–1.2039728
c.	$f(-1) = \ln(-1)$	[LN] [(−)] 1 [ENTER]	ERROR
d.	$f(1 + \sqrt{2}) = \ln(1 + \sqrt{2})$	[LN] [(] 1 [+] [√] 2 [)] [ENTER]	0.8813736

> **REMARK** In Example 8(c), be sure you see that $\ln(-1)$ gives an error message on most calculators. This occurs because the domain of ln x is the set of *positive real numbers* (see Figure 5.9). So, $\ln(-1)$ is undefined.

Checkpoint Audio-video solution in English & Spanish at LarsonPrecalculus.com

Use a calculator to evaluate the function $f(x) = \ln x$ at each value of x.

a. $x = 0.01$ **b.** $x = 4$

c. $x = \sqrt{3} + 2$ **d.** $x = \sqrt{3} - 2$

The properties of logarithms on page 370 are also valid for natural logarithms.

> **Properties of Natural Logarithms**
>
> **1.** $\ln 1 = 0$ because $e^0 = 1$.
>
> **2.** $\ln e = 1$ because $e^1 = e$.
>
> **3.** $\ln e^x = x$ and $e^{\ln x} = x$ — Inverse Properties
>
> **4.** If $\ln x = \ln y$, then $x = y$. — One-to-One Property

EXAMPLE 9 Using Properties of Natural Logarithms

Use the properties of natural logarithms to simplify each expression.

a. $\ln \frac{1}{e}$ **b.** $e^{\ln 5}$ **c.** $\frac{\ln 1}{3}$ **d.** $2 \ln e$

Solution

a. $\ln \frac{1}{e} = \ln e^{-1} = -1$ Property 3 (Inverse Property)

b. $e^{\ln 5} = 5$ Property 3 (Inverse Property)

c. $\frac{\ln 1}{3} = \frac{0}{3} = 0$ Property 1

d. $2 \ln e = 2(1) = 2$ Property 2

Checkpoint *Audio-video solution in English & Spanish at LarsonPrecalculus.com*

Use the properties of natural logarithms to simplify each expression.

a. $\ln e^{1/3}$ **b.** $5 \ln 1$ **c.** $\frac{3}{4} \ln e$ **d.** $e^{\ln 7}$

EXAMPLE 10 Finding the Domains of Logarithmic Functions

Find the domain of each function.

a. $f(x) = \ln(x - 2)$ **b.** $g(x) = \ln(2 - x)$ **c.** $h(x) = \ln x^2$

Solution

a. Because $\ln(x - 2)$ is defined only when

$$x - 2 > 0$$

it follows that the domain of f is $(2, \infty)$, as shown in Figure 5.10.

b. Because $\ln(2 - x)$ is defined only when

$$2 - x > 0$$

it follows that the domain of g is $(-\infty, 2)$, as shown in Figure 5.11.

c. Because $\ln x^2$ is defined only when

$$x^2 > 0$$

it follows that the domain of h is all real numbers except $x = 0$, as shown in Figure 5.12.

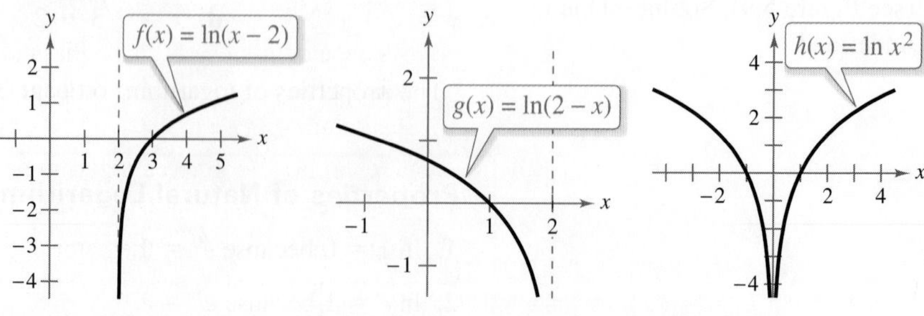

Figure 5.10 Figure 5.11 Figure 5.12

Checkpoint *Audio-video solution in English & Spanish at LarsonPrecalculus.com*

Find the domain of $f(x) = \ln(x + 3)$.

Application

EXAMPLE 11 Human Memory Model

Students participating in a psychology experiment attended several lectures on a subject and took an exam. Every month for a year after the exam, the students took a retest to see how much of the material they remembered. The average scores for the group are given by the *human memory model* $f(t) = 75 - 6\ln(t+1),\ 0 \le t \le 12$, where t is the time in months.

a. What was the average score on the original exam ($t = 0$)?

b. What was the average score at the end of $t = 2$ months?

c. What was the average score at the end of $t = 6$ months?

Algebraic Solution

a. The original average score was

$$\begin{aligned} f(0) &= 75 - 6\ln(0+1) && \text{Substitute 0 for } t. \\ &= 75 - 6\ln 1 && \text{Simplify.} \\ &= 75 - 6(0) && \text{Property of natural logarithms} \\ &= 75. && \text{Solution} \end{aligned}$$

b. After 2 months, the average score was

$$\begin{aligned} f(2) &= 75 - 6\ln(2+1) && \text{Substitute 2 for } t. \\ &= 75 - 6\ln 3 && \text{Simplify.} \\ &\approx 75 - 6(1.0986) && \text{Use a calculator.} \\ &\approx 68.41. && \text{Solution} \end{aligned}$$

c. After 6 months, the average score was

$$\begin{aligned} f(6) &= 75 - 6\ln(6+1) && \text{Substitute 6 for } t. \\ &= 75 - 6\ln 7 && \text{Simplify.} \\ &\approx 75 - 6(1.9459) && \text{Use a calculator.} \\ &\approx 63.32. && \text{Solution} \end{aligned}$$

Graphical Solution

a.

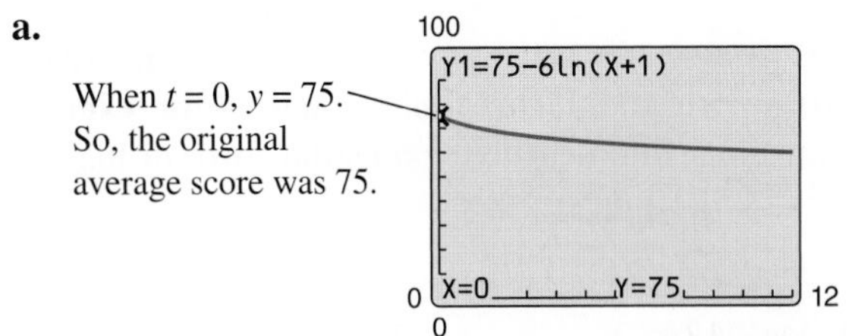

b.

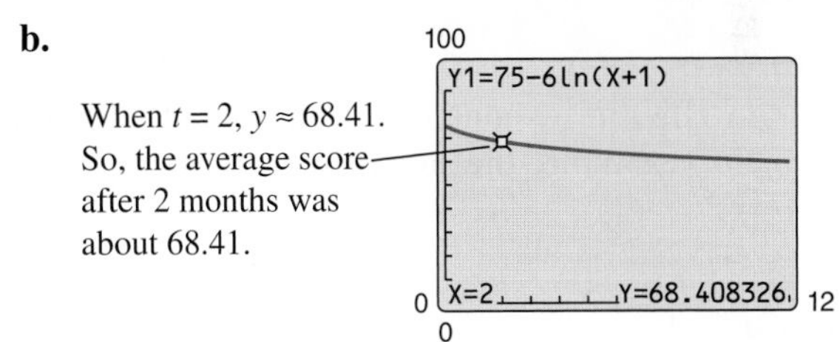

c.

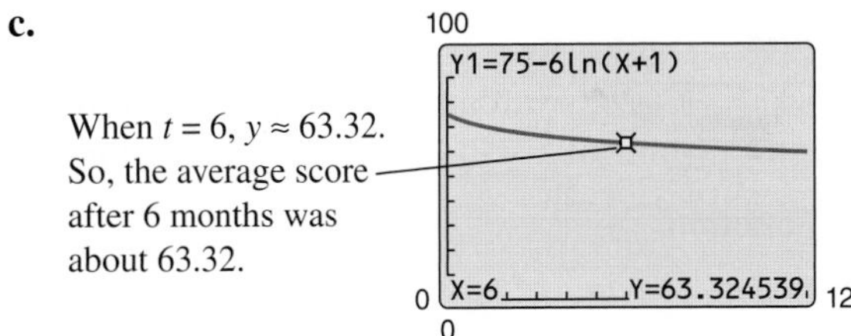

✓ **Checkpoint** *Audio-video solution in English & Spanish at LarsonPrecalculus.com*

In Example 11, find the average score at the end of (a) $t = 1$ month, (b) $t = 9$ months, and (c) $t = 12$ months.

Summarize (Section 5.2)

1. State the definition of the logarithmic function with base *a* *(page 369)* and make a list of the properties of logarithms *(page 370)*. For examples of evaluating logarithmic functions and using the properties of logarithms, see Examples 1–4.
2. Explain how to graph a logarithmic function *(pages 371 and 372)*. For examples of graphing logarithmic functions, see Examples 5–7.
3. State the definition of the natural logarithmic function and make a list of the properties of natural logarithms *(page 373)*. For examples of evaluating natural logarithmic functions and using the properties of natural logarithms, see Examples 8 and 9.
4. Describe a real-life application that uses a logarithmic function to model and solve a problem *(page 375, Example 11)*.

5.2 Exercises

See CalcChat.com for tutorial help and worked-out solutions to odd-numbered exercises.

Vocabulary: Fill in the blanks.

1. The inverse function of the exponential function $f(x) = a^x$ is the ________ function with base a.
2. The common logarithmic function has base ________.
3. The logarithmic function $f(x) = \ln x$ is the ________ logarithmic function and has base ________.
4. The Inverse Properties of logarithms state that $\log_a a^x = x$ and ________.
5. The One-to-One Property of natural logarithms states that if $\ln x = \ln y$, then ________.
6. The domain of the natural logarithmic function is the set of ________ ________ ________.

Skills and Applications

Writing an Exponential Equation **In Exercises 7–10, write the logarithmic equation in exponential form. For example, the exponential form of $\log_5 25 = 2$ is $5^2 = 25$.**

7. $\log_4 16 = 2$
8. $\log_9 \frac{1}{81} = -2$
9. $\log_{12} 12 = 1$
10. $\log_{32} 4 = \frac{2}{5}$

Writing a Logarithmic Equation **In Exercises 11–14, write the exponential equation in logarithmic form. For example, the logarithmic form of $2^3 = 8$ is $\log_2 8 = 3$.**

11. $5^3 = 125$
12. $9^{3/2} = 27$
13. $4^{-3} = \frac{1}{64}$
14. $24^0 = 1$

Evaluating a Logarithm **In Exercises 15–20, evaluate the logarithm at the given value of x without using a calculator.**

Function	Value
15. $f(x) = \log_2 x$	$x = 64$
16. $f(x) = \log_{25} x$	$x = 5$
17. $f(x) = \log_8 x$	$x = 1$
18. $f(x) = \log x$	$x = 10$
19. $g(x) = \log_a x$	$x = a^{-2}$
20. $g(x) = \log_b x$	$x = \sqrt{b}$

Evaluating a Common Logarithm on a Calculator **In Exercises 21–24, use a calculator to evaluate $f(x) = \log x$ at the given value of x. Round your result to three decimal places.**

21. $x = \frac{7}{8}$
22. $x = \frac{1}{500}$
23. $x = 12.5$
24. $x = 96.75$

Using Properties of Logarithms **In Exercises 25–28, use the properties of logarithms to simplify the expression.**

25. $\log_8 8$
26. $\log_\pi \pi^2$
27. $\log_{7.5} 1$
28. $5^{\log_5 3}$

Using the One-to-One Property **In Exercises 29–32, use the One-to-One Property to solve the equation for x.**

29. $\log_5(x + 1) = \log_5 6$
30. $\log_2(x - 3) = \log_2 9$
31. $\log 11 = \log(x^2 + 7)$
32. $\log(x^2 + 6x) = \log 27$

Graphing Exponential and Logarithmic Functions **In Exercises 33–36, sketch the graphs of f and g in the same coordinate plane.**

33. $f(x) = 7^x,\ g(x) = \log_7 x$
34. $f(x) = 5^x,\ g(x) = \log_5 x$
35. $f(x) = 6^x,\ g(x) = \log_6 x$
36. $f(x) = 10^x,\ g(x) = \log x$

Matching a Logarithmic Function with Its Graph **In Exercises 37–40, use the graph of $g(x) = \log_3 x$ to match the given function with its graph. Then describe the relationship between the graphs of f and g. [The graphs are labeled (a), (b), (c), and (d).]**

(a)

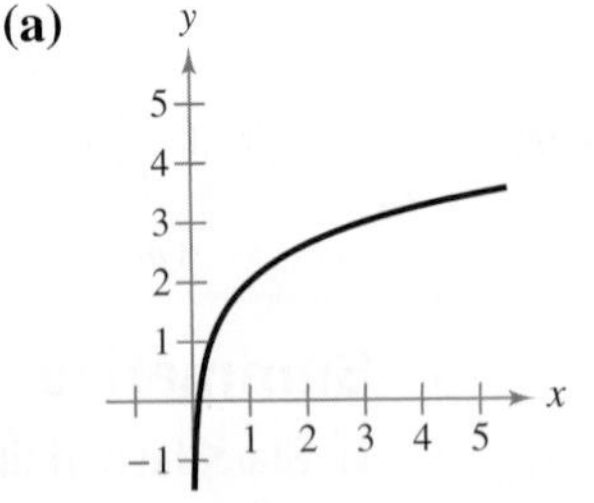

(b)

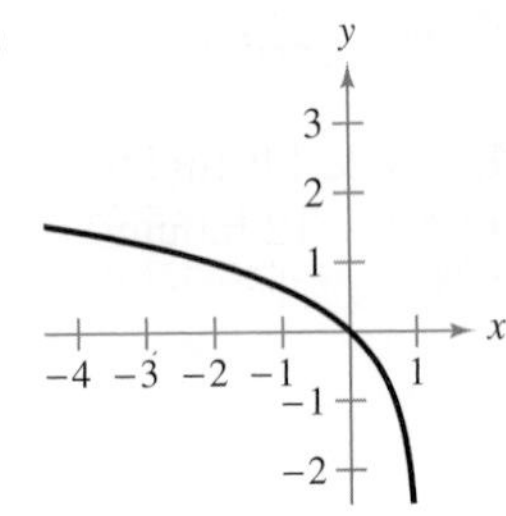

(c)

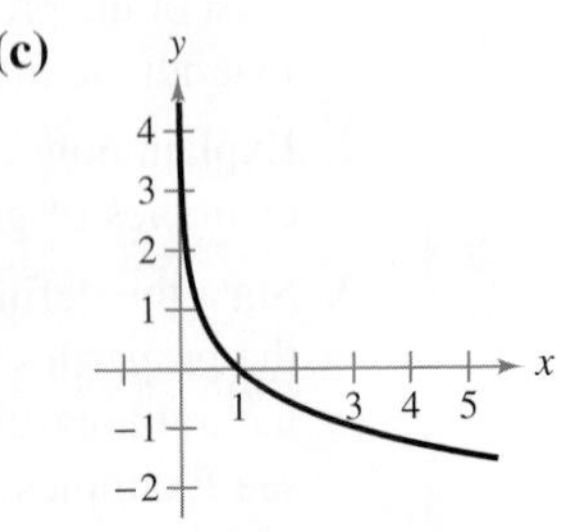

(d)

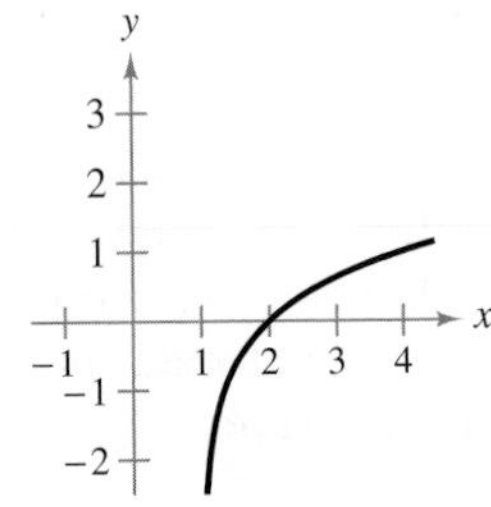

37. $f(x) = \log_3 x + 2$
38. $f(x) = \log_3(x - 1)$

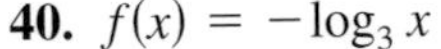

39. $f(x) = \log_3(1 - x)$
40. $f(x) = -\log_3 x$

Sketching the Graph of a Logarithmic Function In Exercises 41–48, find the domain, x-intercept, and vertical asymptote of the logarithmic function and sketch its graph.

41. $f(x) = \log_4 x$
42. $g(x) = \log_6 x$
43. $y = \log_3 x + 1$
44. $h(x) = \log_4(x - 3)$
45. $f(x) = -\log_6(x + 2)$
46. $y = \log_5(x - 1) + 4$
47. $y = \log \frac{x}{7}$
48. $y = \log(-2x)$

Writing a Natural Exponential Equation In Exercises 49–52, write the logarithmic equation in exponential form.

49. $\ln \frac{1}{2} = -0.693 \ldots$
50. $\ln 7 = 1.945 \ldots$
51. $\ln 250 = 5.521 \ldots$
52. $\ln 1 = 0$

Writing a Natural Logarithmic Equation In Exercises 53–56, write the exponential equation in logarithmic form.

53. $e^2 = 7.3890 \ldots$
54. $e^{-3/4} = 0.4723 \ldots$
55. $e^{-4x} = \frac{1}{2}$
56. $e^{2x} = 3$

Evaluating a Logarithmic Function In Exercises 57–60, use a calculator to evaluate the function at the given value of x. Round your result to three decimal places.

	Function	Value
57.	$f(x) = \ln x$	$x = 18.42$
58.	$f(x) = 3 \ln x$	$x = 0.74$
59.	$g(x) = 8 \ln x$	$x = \sqrt{5}$
60.	$g(x) = -\ln x$	$x = \frac{1}{2}$

Using Properties of Natural Logarithms In Exercises 61–66, use the properties of natural logarithms to simplify the expression.

61. $e^{\ln 4}$
62. $\ln \frac{1}{e^2}$
63. $2.5 \ln 1$
64. $\frac{\ln e}{\pi}$
65. $\ln e^{\ln e}$
66. $e^{\ln(1/e)}$

Graphing a Natural Logarithmic Function In Exercises 67–70, find the domain, x-intercept, and vertical asymptote of the logarithmic function and sketch its graph.

67. $f(x) = \ln(x - 4)$
68. $h(x) = \ln(x + 5)$
69. $g(x) = \ln(-x)$
70. $f(x) = \ln(3 - x)$

Graphing a Natural Logarithmic Function In Exercises 71–74, use a graphing utility to graph the function. Be sure to use an appropriate viewing window.

71. $f(x) = \ln(x - 1)$
72. $f(x) = \ln(x + 2)$
73. $f(x) = -\ln x + 8$
74. $f(x) = 3 \ln x - 1$

Using the One-to-One Property In Exercises 75–78, use the One-to-One Property to solve the equation for x.

75. $\ln(x + 4) = \ln 12$
76. $\ln(x - 7) = \ln 7$
77. $\ln(x^2 - x) = \ln 6$
78. $\ln(x^2 - 2) = \ln 23$

79. **Monthly Payment** The model

$$t = 16.625 \ln \frac{x}{x - 750}, \quad x > 750$$

approximates the length of a home mortgage of \$150,000 at 6% in terms of the monthly payment. In the model, t is the length of the mortgage in years and x is the monthly payment in dollars.

(a) Approximate the lengths of a \$150,000 mortgage at 6% when the monthly payment is \$897.72 and when the monthly payment is \$1659.24.

(b) Approximate the total amounts paid over the term of the mortgage with a monthly payment of \$897.72 and with a monthly payment of \$1659.24. What amount of the total is interest costs in each case?

(c) What is the vertical asymptote for the model? Interpret its meaning in the context of the problem.

80. **Telephone Service** The percent P of households in the United States with wireless-only telephone service from 2005 through 2014 can be approximated by the model

$$P = -3.42 + 1.297t \ln t, \quad 5 \le t \le 14$$

where t represents the year, with $t = 5$ corresponding to 2005. *(Source: National Center for Health Statistics)*

(a) Approximate the percents of households with wireless-only telephone service in 2008 and 2012.

(b) Use a graphing utility to graph the function.

(c) Can the model be used to predict the percent of households with wireless-only telephone service in 2020? in 2030? Explain.

81. **Population** The time t (in years) for the world population to double when it is increasing at a continuous rate r (in decimal form) is given by $t = (\ln 2)/r$.

(a) Complete the table and interpret your results.

r	0.005	0.010	0.015	0.020	0.025	0.030
t						

(b) Use a graphing utility to graph the function.

82. Compound Interest A principal P, invested at $5\frac{1}{2}\%$ and compounded continuously, increases to an amount K times the original principal after t years, where $t = (\ln K)/0.055$.

(a) Complete the table and interpret your results.

K	1	2	4	6	8	10	12
t							

(b) Sketch a graph of the function.

83. Human Memory Model

Students in a mathematics class took an exam and then took a retest monthly with an equivalent exam. The average scores for the class are given by the human memory model

$f(t) = 80 - 17\log(t + 1), \quad 0 \le t \le 12$

where t is the time in months.

(a) Use a graphing utility to graph the model over the specified domain.

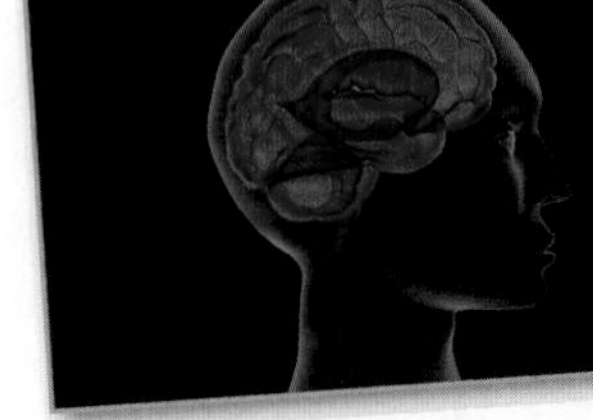

(b) What was the average score on the original exam ($t = 0$)?

(c) What was the average score after 4 months?

(d) What was the average score after 10 months?

84. Sound Intensity The relationship between the number of decibels β and the intensity of a sound I (in watts per square meter) is

$$\beta = 10\log\frac{I}{10^{-12}}.$$

(a) Determine the number of decibels of a sound with an intensity of 1 watt per square meter.

(b) Determine the number of decibels of a sound with an intensity of 10^{-2} watt per square meter.

(c) The intensity of the sound in part (a) is 100 times as great as that in part (b). Is the number of decibels 100 times as great? Explain.

Exploration

True or False? **In Exercises 85 and 86, determine whether the statement is true or false. Justify your answer.**

85. The graph of $f(x) = \log_6 x$ is a reflection of the graph of $g(x) = 6^x$ in the x-axis.

86. The graph of $f(x) = \ln(-x)$ is a reflection of the graph of $h(x) = e^{-x}$ in the line $y = -x$.

87. Graphical Reasoning Use a graphing utility to graph f and g in the same viewing window and determine which is increasing at the greater rate as x approaches $+\infty$. What can you conclude about the rate of growth of the natural logarithmic function?

(a) $f(x) = \ln x, \quad g(x) = \sqrt{x}$

(b) $f(x) = \ln x, \quad g(x) = \sqrt[4]{x}$

88. HOW DO YOU SEE IT? The figure shows the graphs of $f(x) = 3^x$ and $g(x) = \log_3 x$. [The graphs are labeled m and n.]

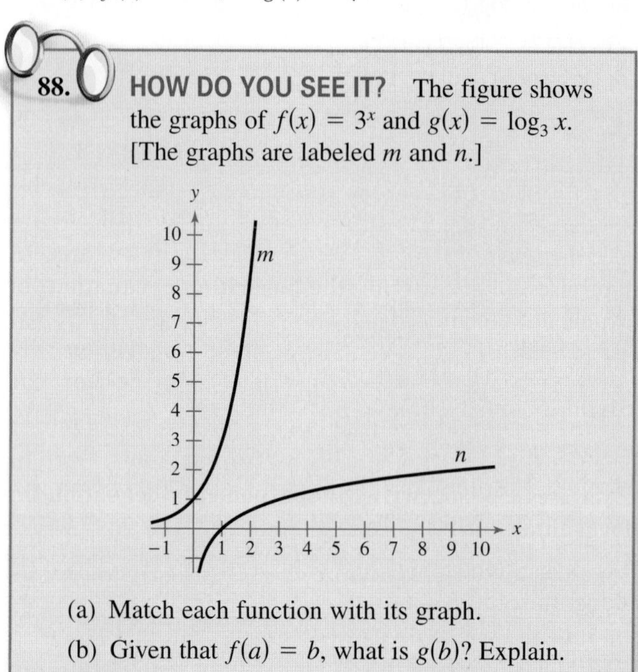

(a) Match each function with its graph.

(b) Given that $f(a) = b$, what is $g(b)$? Explain.

Error Analysis **In Exercises 89 and 90, describe the error.**

89.

x	1	2	8
y	0	1	3

From the table, you can conclude that y is an exponential function of x. ✗

90.

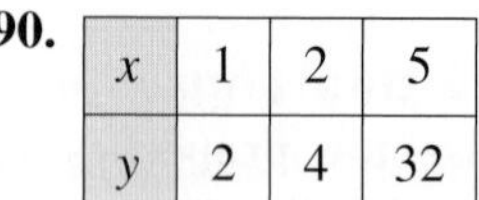

x	1	2	5
y	2	4	32

From the table, you can conclude that y is a logarithmic function of x. 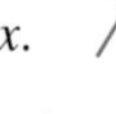

91. Numerical Analysis

(a) Complete the table for the function $f(x) = (\ln x)/x$.

x	1	5	10	10^2	10^4	10^6
$f(x)$						

(b) Use the table in part (a) to determine what value $f(x)$ approaches as x increases without bound.

(c) Use a graphing utility to confirm the result of part (b).

92. Writing Explain why $\log_a x$ is defined only for $0 < a < 1$ and $a > 1$.

5.3 Properties of Logarithms

Logarithmic functions have many real-life applications. For example, in Exercises 79–82 on page 384, you will use a logarithmic function that models the relationship between the number of decibels and the intensity of a sound.

- **Use the change-of-base formula to rewrite and evaluate logarithmic expressions.**
- **Use properties of logarithms to evaluate or rewrite logarithmic expressions.**
- **Use properties of logarithms to expand or condense logarithmic expressions.**
- **Use logarithmic functions to model and solve real-life problems.**

Change of Base

Most calculators have only two types of log keys, LOG for common logarithms (base 10) and LN for natural logarithms (base e). Although common logarithms and natural logarithms are the most frequently used, you may occasionally need to evaluate logarithms with other bases. To do this, use the **change-of-base formula.**

Change-of-Base Formula

Let a, b, and x be positive real numbers such that $a \neq 1$ and $b \neq 1$. Then $\log_a x$ can be converted to a different base as follows.

Base b	**Base 10**	**Base e**
$\log_a x = \dfrac{\log_b x}{\log_b a}$	$\log_a x = \dfrac{\log x}{\log a}$	$\log_a x = \dfrac{\ln x}{\ln a}$

One way to look at the change-of-base formula is that logarithms with base a are *constant multiples* of logarithms with base b. The constant multiplier is

$$\frac{1}{\log_b a}.$$

EXAMPLE 1 Changing Bases Using Common Logarithms

$$\log_4 25 = \frac{\log 25}{\log 4} \qquad \log_a x = \frac{\log x}{\log a}$$

$$\approx \frac{1.39794}{0.60206} \qquad \text{Use a calculator.}$$

$$\approx 2.3219 \qquad \text{Simplify.}$$

✓ ***Checkpoint*** *Audio-video solution in English & Spanish at LarsonPrecalculus.com*

Evaluate $\log_2 12$ using the change-of-base formula and common logarithms.

EXAMPLE 2 Changing Bases Using Natural Logarithms

$$\log_4 25 = \frac{\ln 25}{\ln 4} \qquad \log_a x = \frac{\ln x}{\ln a}$$

$$\approx \frac{3.21888}{1.38629} \qquad \text{Use a calculator.}$$

$$\approx 2.3219 \qquad \text{Simplify.}$$

✓ ***Checkpoint*** *Audio-video solution in English & Spanish at LarsonPrecalculus.com*

Evaluate $\log_2 12$ using the change-of-base formula and natural logarithms. ■

Properties of Logarithms

You know from the preceding section that the logarithmic function with base a is the *inverse function* of the exponential function with base a. So, it makes sense that the properties of exponents have corresponding properties involving logarithms. For example, the exponential property $a^m a^n = a^{m+n}$ has the corresponding logarithmic property $\log_a(uv) = \log_a u + \log_a v$.

REMARK There is no property that can be used to rewrite $\log_a(u \pm v)$. Specifically, $\log_a(u + v)$ is *not* equal to $\log_a u + \log_a v$.

Properties of Logarithms

Let a be a positive number such that $a \neq 1$, let n be a real number, and let u and v be positive real numbers.

	Logarithm with Base a	Natural Logarithm
1. Product Property:	$\log_a(uv) = \log_a u + \log_a v$	$\ln(uv) = \ln u + \ln v$
2. Quotient Property:	$\log_a \frac{u}{v} = \log_a u - \log_a v$	$\ln \frac{u}{v} = \ln u - \ln v$
3. Power Property:	$\log_a u^n = n \log_a u$	$\ln u^n = n \ln u$

For proofs of the properties listed above, see Proofs in Mathematics on page 414.

EXAMPLE 3 Using Properties of Logarithms

Write each logarithm in terms of ln 2 and ln 3.

a. $\ln 6$ **b.** $\ln \frac{2}{27}$

Solution

a. $\ln 6 = \ln(2 \cdot 3)$ Rewrite 6 as $2 \cdot 3$.

$= \ln 2 + \ln 3$ Product Property

b. $\ln \frac{2}{27} = \ln 2 - \ln 27$ Quotient Property

$= \ln 2 - \ln 3^3$ Rewrite 27 as 3^3.

$= \ln 2 - 3 \ln 3$ Power Property

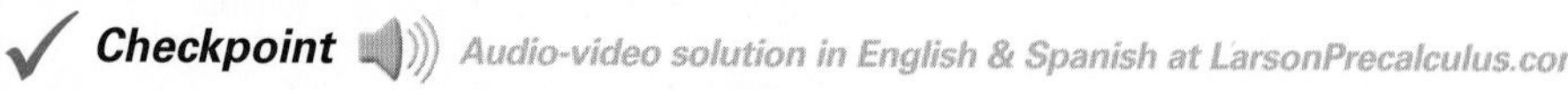

Write each logarithm in terms of log 3 and log 5.

a. $\log 75$ **b.** $\log \frac{9}{125}$

HISTORICAL NOTE

John Napier, a Scottish mathematician, developed logarithms as a way to simplify tedious calculations. Napier worked about 20 years on the development of logarithms before publishing his work is 1614. Napier only partially succeeded in his quest to simplify tedious calculations. Nonetheless, the development of logarithms was a step forward and received immediate recognition.

EXAMPLE 4 Using Properties of Logarithms

Find the exact value of $\log_5 \sqrt[3]{5}$ without using a calculator.

Solution

$$\log_5 \sqrt[3]{5} = \log_5 5^{1/3} = \tfrac{1}{3} \log_5 5 = \tfrac{1}{3}(1) = \tfrac{1}{3}$$

Checkpoint Audio-video solution in English & Spanish at LarsonPrecalculus.com

Find the exact value of $\ln e^6 - \ln e^2$ without using a calculator.

Rewriting Logarithmic Expressions

The properties of logarithms are useful for rewriting logarithmic expressions in forms that simplify the operations of algebra. This is true because these properties convert complicated products, quotients, and exponential forms into simpler sums, differences, and products, respectively.

EXAMPLE 5 Expanding Logarithmic Expressions

Expand each logarithmic expression.

a. $\log_4 5x^3y$ **b.** $\ln \dfrac{\sqrt{3x-5}}{7}$

Solution

a. $\log_4 5x^3y = \log_4 5 + \log_4 x^3 + \log_4 y$ Product Property

$= \log_4 5 + 3\log_4 x + \log_4 y$ Power Property

b. $\ln \dfrac{\sqrt{3x-5}}{7} = \ln \dfrac{(3x-5)^{1/2}}{7}$ Rewrite using rational exponent.

$= \ln(3x-5)^{1/2} - \ln 7$ Quotient Property

$= \dfrac{1}{2}\ln(3x-5) - \ln 7$ Power Property

✓ **Checkpoint** Audio-video solution in English & Spanish at LarsonPrecalculus.com

Expand the expression $\log_3 \dfrac{4x^2}{\sqrt{y}}$.

Example 5 uses the properties of logarithms to *expand* logarithmic expressions. Example 6 reverses this procedure and uses the properties of logarithms to *condense* logarithmic expressions.

EXAMPLE 6 Condensing Logarithmic Expressions

See LarsonPrecalculus.com for an interactive version of this type of example.

Condense each logarithmic expression.

a. $\frac{1}{2}\log x + 3\log(x+1)$ **b.** $2\ln(x+2) - \ln x$ **c.** $\frac{1}{3}[\log_2 x + \log_2(x+1)]$

Solution

a. $\frac{1}{2}\log x + 3\log(x+1) = \log x^{1/2} + \log(x+1)^3$ Power Property

$= \log\left[\sqrt{x}\,(x+1)^3\right]$ Product Property

b. $2\ln(x+2) - \ln x = \ln(x+2)^2 - \ln x$ Power Property

$= \ln \dfrac{(x+2)^2}{x}$ Quotient Property

c. $\frac{1}{3}[\log_2 x + \log_2(x+1)] = \frac{1}{3}\log_2[x(x+1)]$ Product Property

$= \log_2[x(x+1)]^{1/3}$ Power Property

$= \log_2 \sqrt[3]{x(x+1)}$ Rewrite with a radical.

✓ **Checkpoint** Audio-video solution in English & Spanish at LarsonPrecalculus.com

Condense the expression $2[\log(x+3) - 2\log(x-2)]$.

Application

One way to determine a possible relationship between the x- and y-values of a set of nonlinear data is to take the natural logarithm of each x-value and each y-value. If the plotted points $(\ln x, \ln y)$ lie on a line, then x and y are related by the equation $\ln y = m \ln x$, where m is the slope of the line.

EXAMPLE 7 Finding a Mathematical Model

The table shows the mean distance x from the sun and the period y (the time it takes a planet to orbit the sun, in years) for each of the six planets that are closest to the sun. In the table, the mean distance is given in astronomical units (where one astronomical unit is defined as Earth's mean distance from the sun). The points from the table are plotted in Figure 5.13. Find an equation that relates y and x.

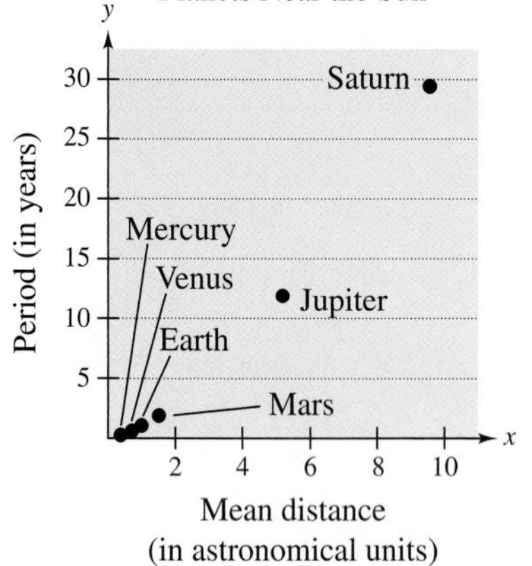

Figure 5.13

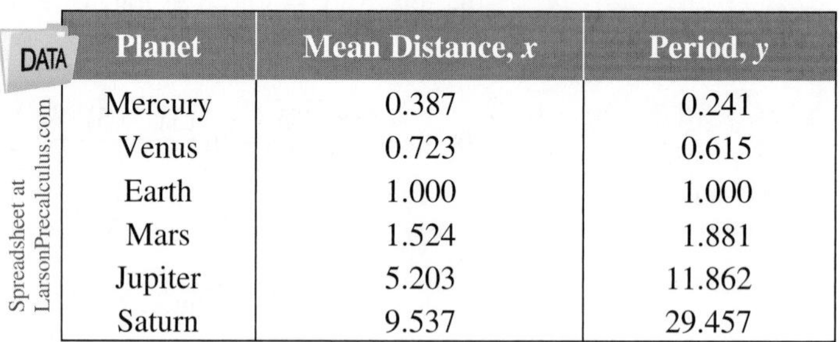

Planet	Mean Distance, x	Period, y
Mercury	0.387	0.241
Venus	0.723	0.615
Earth	1.000	1.000
Mars	1.524	1.881
Jupiter	5.203	11.862
Saturn	9.537	29.457

Solution From Figure 5.13, it is not clear how to find an equation that relates y and x. To solve this problem, make a table of values giving the natural logarithms of all x- and y-values of the data (see the table at the left). Plot each point $(\ln x, \ln y)$. These points appear to lie on a line (see Figure 5.14). Choose two points to determine the slope of the line. Using the points $(0.421, 0.632)$ and $(0, 0)$, the slope of the line is

Planet	$\ln x$	$\ln y$
Mercury	−0.949	−1.423
Venus	−0.324	−0.486
Earth	0.000	0.000
Mars	0.421	0.632
Jupiter	1.649	2.473
Saturn	2.255	3.383

$$m = \frac{0.632 - 0}{0.421 - 0} \approx 1.5 = \frac{3}{2}.$$

By the point-slope form, the equation of the line is $Y = \frac{3}{2}X$, where $Y = \ln y$ and $X = \ln x$. So, an equation that relates y and x is $\ln y = \frac{3}{2} \ln x$.

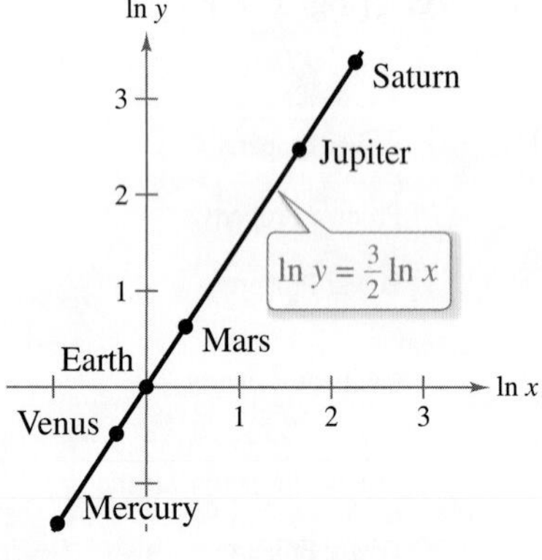

Figure 5.14

✓ ***Checkpoint*** *Audio-video solution in English & Spanish at LarsonPrecalculus.com*

Find a logarithmic equation that relates y and x for the following ordered pairs.

$(0.37, 0.51), (1.00, 1.00), (2.72, 1.95), (7.39, 3.79), (20.09, 7.39)$

Summarize (Section 5.3)

1. State the change-of-base formula *(page 379)*. For examples of using the change-of-base formula to rewrite and evaluate logarithmic expressions, see Examples 1 and 2.
2. Make a list of the properties of logarithms *(page 380)*. For examples of using the properties of logarithms to evaluate or rewrite logarithmic expressions, see Examples 3 and 4.
3. Explain how to use the properties of logarithms to expand or condense logarithmic expressions *(page 381)*. For examples of expanding and condensing logarithmic expressions, see Examples 5 and 6.
4. Describe an example of how to use a logarithmic function to model and solve a real-life problem *(page 382, Example 7)*.

5.3 Exercises

See **CalcChat.com** for tutorial help and worked-out solutions to odd-numbered exercises.

Vocabulary

In Exercises 1–3, fill in the blanks.

1. To evaluate a logarithm to any base, use the ________ formula.
2. The change-of-base formula for base e is $\log_a x =$ ________.
3. When you consider $\log_a x$ to be a constant multiple of $\log_b x$, the constant multiplier is ________.
4. Name the property of logarithms illustrated by each statement.

(a) $\ln(uv) = \ln u + \ln v$ (b) $\log_a u^n = n \log_a u$ (c) $\ln \frac{u}{v} = \ln u - \ln v$

Skills and Applications

Changing Bases **In Exercises 5–8, rewrite the logarithm as a ratio of (a) common logarithms and (b) natural logarithms.**

5. $\log_5 16$
6. $\log_{1/5} 4$
7. $\log_x \frac{3}{10}$
8. $\log_{2.6} x$

Using the Change-of-Base Formula **In Exercises 9–12, evaluate the logarithm using the change-of-base formula. Round your result to three decimal places.**

9. $\log_3 17$
10. $\log_{0.4} 12$
11. $\log_\pi 0.5$
12. $\log_{2/3} 0.125$

Using Properties of Logarithms **In Exercises 13–18, use the properties of logarithms to write the logarithm in terms of $\log_3 5$ and $\log_3 7$.**

13. $\log_3 35$
14. $\log_3 \frac{5}{7}$
15. $\log_3 \frac{7}{25}$
16. $\log_3 175$
17. $\log_3 \frac{21}{5}$
18. $\log_3 \frac{45}{49}$

Using Properties of Logarithms **In Exercises 19–32, find the exact value of the logarithmic expression without using a calculator. (If this is not possible, state the reason.)**

19. $\log_3 9$
20. $\log_5 \frac{1}{125}$
21. $\log_6 \sqrt[3]{\frac{1}{6}}$
22. $\log_2 \sqrt[4]{8}$
23. $\log_2(-2)$
24. $\log_3(-27)$
25. $\ln \sqrt[4]{e^3}$
26. $\ln(1/\sqrt{e})$
27. $\ln e^2 + \ln e^5$
28. $2 \ln e^6 - \ln e^5$
29. $\log_5 75 - \log_5 3$
30. $\log_4 2 + \log_4 32$
31. $\log_4 8$
32. $\log_8 16$

Using Properties of Logarithms **In Exercises 33–40, approximate the logarithm using the properties of logarithms, given $\log_b 2 \approx 0.3562$, $\log_b 3 \approx 0.5646$, and $\log_b 5 \approx 0.8271$.**

33. $\log_b 10$
34. $\log_b \frac{2}{3}$
35. $\log_b 0.04$
36. $\log_b \sqrt{2}$
37. $\log_b 45$
38. $\log_b(3b^2)$
39. $\log_b(2b)^{-2}$
40. $\log_b \sqrt[3]{3b}$

Expanding a Logarithmic Expression **In Exercises 41–60, use the properties of logarithms to expand the expression as a sum, difference, and/or constant multiple of logarithms. (Assume all variables are positive.)**

41. $\ln 7x$
42. $\log_3 13z$
43. $\log_8 x^4$
44. $\ln(xy)^3$
45. $\log_5 \frac{5}{x}$
46. $\log_6 \frac{w^2}{v}$
47. $\ln \sqrt{z}$
48. $\ln \sqrt[3]{t}$
49. $\ln xyz^2$
50. $\log_4 11b^2c$
51. $\ln z(z-1)^2, \quad z > 1$
52. $\ln \frac{x^2 - 1}{x^3}, \quad x > 1$
53. $\log_2 \frac{\sqrt{a^2 - 4}}{7}, \quad a > 2$
54. $\ln \frac{3}{\sqrt{x^2 + 1}}$
55. $\log_5 \frac{x^2}{y^2 z^3}$
56. $\log_{10} \frac{xy^4}{z^5}$
57. $\ln \sqrt[3]{\frac{yz}{x^2}}$
58. $\log_2 x^4 \sqrt{\frac{y}{z^3}}$
59. $\ln \sqrt[4]{x^3(x^2 + 3)}$
60. $\ln \sqrt{x^2(x + 2)}$

Condensing a Logarithmic Expression **In Exercises 61–76, condense the expression to the logarithm of a single quantity.**

61. $\ln 3 + \ln x$
62. $\log_5 8 - \log_5 t$
63. $\frac{2}{3}\log_7(z - 2)$
64. $-4 \ln 3x$
65. $\log_3 5x - 4\log_3 x$
66. $2\log_2 x + 4\log_2 y$
67. $\log x + 2\log(x + 1)$
68. $2\ln 8 - 5\ln(z - 4)$
69. $\log x - 2\log y + 3\log z$
70. $3\log_3 x + \frac{1}{4}\log_3 y - 4\log_3 z$
71. $\ln x - [\ln(x + 1) + \ln(x - 1)]$
72. $4[\ln z + \ln(z + 5)] - 2\ln(z - 5)$
73. $\frac{1}{2}[2\ln(x + 3) + \ln x - \ln(x^2 - 1)]$
74. $2[3\ln x - \ln(x + 1) - \ln(x - 1)]$
75. $\frac{1}{3}[\log_8 y + 2\log_8(y + 4)] - \log_8(y - 1)$
76. $\frac{1}{2}[\log_4(x + 1) + 2\log_4(x - 1)] + 6\log_4 x$

Comparing Logarithmic Quantities **In Exercises 77 and 78, determine which (if any) of the logarithmic expressions are equal. Justify your answer.**

77. $\dfrac{\log_2 32}{\log_2 4}$, $\log_2 \dfrac{32}{4}$, $\log_2 32 - \log_2 4$

78. $\log_7\sqrt{70}$, $\log_7 35$, $\frac{1}{2} + \log_7\sqrt{10}$

Sound Intensity

In Exercises 79–82, use the following information. The relationship between the number of decibels β and the intensity of a sound I (in watts per square meter) is

$$\beta = 10 \log \frac{I}{10^{-12}}.$$

79. Use the properties of logarithms to write the formula in a simpler form. Then determine the number of decibels of a sound with an intensity of 10^{-6} watt per square meter.

80. Find the difference in loudness between an average office with an intensity of 1.26×10^{-7} watt per square meter and a broadcast studio with an intensity of 3.16×10^{-10} watt per square meter.

81. Find the difference in loudness between a vacuum cleaner with an intensity of 10^{-4} watt per square meter and rustling leaves with an intensity of 10^{-11} watt per square meter.

82. You and your roommate are playing your stereos at the same time and at the same intensity. How much louder is the music when both stereos are playing compared with just one stereo playing?

Curve Fitting **In Exercises 83–86, find a logarithmic equation that relates y and x.**

83.

x	1	2	3	4	5	6
y	1	1.189	1.316	1.414	1.495	1.565

84.

x	1	2	3	4	5	6
y	1	0.630	0.481	0.397	0.342	0.303

85.

x	1	2	3	4	5	6
y	2.5	2.102	1.9	1.768	1.672	1.597

86.

x	1	2	3	4	5	6
y	0.5	2.828	7.794	16	27.951	44.091

87. Stride Frequency of Animals Four-legged animals run with two different types of motion: trotting and galloping. An animal that is trotting has at least one foot on the ground at all times, whereas an animal that is galloping has all four feet off the ground at some point in its stride. The number of strides per minute at which an animal breaks from a trot to a gallop depends on the weight of the animal. Use the table to find a logarithmic equation that relates an animal's weight x (in pounds) and its lowest stride frequency while galloping y (in strides per minute).

DATA

Weight, x	Stride Frequency, y
25	191.5
35	182.7
50	173.8
75	164.2
500	125.9
1000	114.2

Spreadsheet at LarsonPrecalculus.com

88. Nail Length The approximate lengths and diameters (in inches) of bright common wire nails are shown in the table. Find a logarithmic equation that relates the diameter y of a bright common wire nail to its length x.

Length, x	Diameter, y
2	0.113
3	0.148
4	0.192
5	0.225
6	0.262

Titima Ongkantong/Shutterstock.com

89. **Comparing Models** A cup of water at an initial temperature of 78°C is placed in a room at a constant temperature of 21°C. The temperature of the water is measured every 5 minutes during a half-hour period. The results are recorded as ordered pairs of the form (t, T), where t is the time (in minutes) and T is the temperature (in degrees Celsius).

$(0, 78.0°)$, $(5, 66.0°)$, $(10, 57.5°)$, $(15, 51.2°)$, $(20, 46.3°)$, $(25, 42.4°)$, $(30, 39.6°)$

(a) Subtract the room temperature from each of the temperatures in the ordered pairs. Use a graphing utility to plot the data points (t, T) and $(t, T - 21)$.

(b) An exponential model for the data $(t, T - 21)$ is $T - 21 = 54.4(0.964)^t$. Solve for T and graph the model. Compare the result with the plot of the original data.

(c) Use the graphing utility to plot the points $(t, \ln(T - 21))$ and observe that the points appear to be linear. Use the *regression* feature of the graphing utility to fit a line to these data. This resulting line has the form $\ln(T - 21) = at + b$, which is equivalent to $e^{\ln(T-21)} = e^{at+b}$. Solve for T, and verify that the result is equivalent to the model in part (b).

(d) Fit a rational model to the data. Take the reciprocals of the y-coordinates of the revised data points to generate the points

$$\left(t, \frac{1}{T - 21}\right).$$

Use the graphing utility to graph these points and observe that they appear to be linear. Use the *regression* feature of the graphing utility to fit a line to these data. The resulting line has the form

$$\frac{1}{T - 21} = at + b.$$

Solve for T, and use the graphing utility to graph the rational function and the original data points.

90. **Writing** Write a short paragraph explaining why the transformations of the data in Exercise 89 were necessary to obtain the models. Why did taking the logarithms of the temperatures lead to a linear scatter plot? Why did taking the reciprocals of the temperatures lead to a linear scatter plot?

Exploration

True or False? **In Exercises 91–96, determine whether the statement is true or false given that $f(x) = \ln x$. Justify your answer.**

91. $f(0) = 0$

92. $f(ax) = f(a) + f(x), \quad a > 0, \quad x > 0$

93. $f(x - 2) = f(x) - f(2), \quad x > 2$

94. $\sqrt{f(x)} = \frac{1}{2}f(x)$

95. If $f(u) = 2f(v)$, then $v = u^2$.

96. If $f(x) < 0$, then $0 < x < 1$.

Using the Change-of-Base Formula **In Exercises 97–100, use the change-of-base formula to rewrite the logarithm as a ratio of logarithms. Then use a graphing utility to graph the ratio.**

97. $f(x) = \log_2 x$

98. $f(x) = \log_{1/2} x$

99. $f(x) = \log_{1/4} x$

100. $f(x) = \log_{11.8} x$

Error Analysis **In Exercises 101 and 102, describe the error.**

101. $(\ln e)^2 = 2(\ln e) = 2(1) = 2$ ✗

102. $\log_2 8 = \log_2(4 + 4)$
$= \log_2 4 + \log_2 4$
$= \log_2 2^2 + \log_2 2^2$
$= 2 + 2$
$= 4$ ✗

103. **Graphical Reasoning** Use a graphing utility to graph the functions $y_1 = \ln x - \ln(x - 3)$ and $y_2 = \ln \dfrac{x}{x - 3}$ in the same viewing window. Does the graphing utility show the functions with the same domain? If not, explain why some numbers are in the domain of one function but not the other.

104. **HOW DO YOU SEE IT?** The figure shows the graphs of $y = \ln x$, $y = \ln x^2$, $y = \ln 2x$, and $y = \ln 2$. Match each function with its graph. (The graphs are labeled A through D.) Explain.

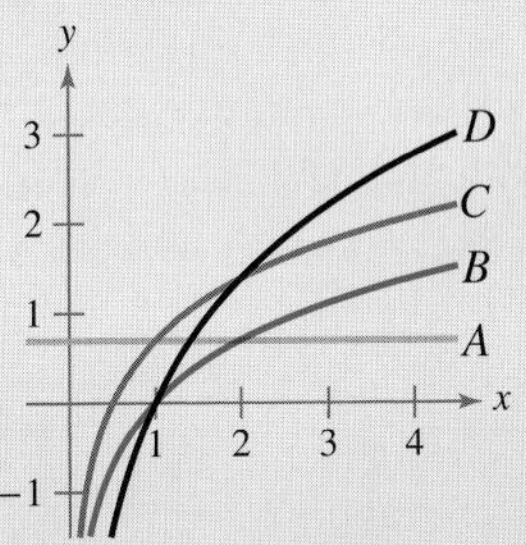

105. **Think About It** For which integers between 1 and 20 can you approximate natural logarithms, given the values $\ln 2 \approx 0.6931$, $\ln 3 \approx 1.0986$, and $\ln 5 \approx 1.6094$? Approximate these logarithms. (Do not use a calculator.)

5.4 Exponential and Logarithmic Equations

Exponential and logarithmic equations have many life science applications. For example, Exercise 83 on page 394 uses an exponential function to model the beaver population in a given area.

- **Solve simple exponential and logarithmic equations.**
- **Solve more complicated exponential equations.**
- **Solve more complicated logarithmic equations.**
- **Use exponential and logarithmic equations to model and solve real-life problems.**

Introduction

So far in this chapter, you have studied the definitions, graphs, and properties of exponential and logarithmic functions. In this section, you will study procedures for *solving equations* involving exponential and logarithmic expressions.

There are two basic strategies for solving exponential or logarithmic equations. The first is based on the One-to-One Properties and was used to solve simple exponential and logarithmic equations in Sections 5.1 and 5.2. The second is based on the Inverse Properties. For $a > 0$ and $a \neq 1$, the properties below are true for all x and y for which $\log_a x$ and $\log_a y$ are defined.

One-to-One Properties	**Inverse Properties**
$a^x = a^y$ if and only if $x = y$.	$a^{\log_a x} = x$
$\log_a x = \log_a y$ if and only if $x = y$.	$\log_a a^x = x$

EXAMPLE 1 **Solving Simple Equations**

	Original Equation	**Rewritten Equation**	**Solution**	**Property**
a.	$2^x = 32$	$2^x = 2^5$	$x = 5$	One-to-One
b.	$\ln x - \ln 3 = 0$	$\ln x = \ln 3$	$x = 3$	One-to-One
c.	$\left(\frac{1}{3}\right)^x = 9$	$3^{-x} = 3^2$	$x = -2$	One-to-One
d.	$e^x = 7$	$\ln e^x = \ln 7$	$x = \ln 7$	Inverse
e.	$\ln x = -3$	$e^{\ln x} = e^{-3}$	$x = e^{-3}$	Inverse
f.	$\log x = -1$	$10^{\log x} = 10^{-1}$	$x = 10^{-1} = \frac{1}{10}$	Inverse
g.	$\log_3 x = 4$	$3^{\log_3 x} = 3^4$	$x = 81$	Inverse

✓ ***Checkpoint***  *Audio-video solution in English & Spanish at LarsonPrecalculus.com*

Solve each equation for x.

a. $2^x = 512$ **b.** $\log_6 x = 3$ **c.** $5 - e^x = 0$ **d.** $9^x = \frac{1}{3}$

Strategies for Solving Exponential and Logarithmic Equations

1. Rewrite the original equation in a form that allows the use of the One-to-One Properties of exponential or logarithmic functions.
2. Rewrite an *exponential* equation in logarithmic form and apply the Inverse Property of logarithmic functions.
3. Rewrite a *logarithmic* equation in exponential form and apply the Inverse Property of exponential functions.

Solving Exponential Equations

EXAMPLE 2 Solving Exponential Equations

Solve each equation and approximate the result to three decimal places, if necessary.

a. $e^{-x^2} = e^{-3x-4}$ **b.** $3(2^x) = 42$

Solution

a.

$e^{-x^2} = e^{-3x-4}$	Write original equation.
$-x^2 = -3x - 4$	One-to-One Property
$x^2 - 3x - 4 = 0$	Write in general form.
$(x + 1)(x - 4) = 0$	Factor.
$x + 1 = 0 \Rightarrow x = -1$	Set 1st factor equal to 0.
$x - 4 = 0 \Rightarrow x = 4$	Set 2nd factor equal to 0.

The solutions are $x = -1$ and $x = 4$. Check these in the original equation.

b.

$3(2^x) = 42$	Write original equation.
$2^x = 14$	Divide each side by 3.
$\log_2 2^x = \log_2 14$	Take log (base 2) of each side.
$x = \log_2 14$	Inverse Property
$x = \dfrac{\ln 14}{\ln 2} \approx 3.807$	Change-of-base formula

The solution is $x = \log_2 14 \approx 3.807$. Check this in the original equation.

REMARK Another way to solve Example 2(b) is by taking the natural log of each side and then applying the Power Property.

$$3(2^x) = 42$$
$$2^x = 14$$
$$\ln 2^x = \ln 14$$
$$x \ln 2 = \ln 14$$
$$x = \frac{\ln 14}{\ln 2} \approx 3.807$$

Notice that you obtain the same result as in Example 2(b).

Checkpoint Audio-video solution in English & Spanish at LarsonPrecalculus.com

Solve each equation and approximate the result to three decimal places, if necessary.

a. $e^{2x} = e^{x^2-8}$ **b.** $2(5^x) = 32$

In Example 2(b), the exact solution is $x = \log_2 14$, and the approximate solution is $x \approx 3.807$. An exact answer is preferred when the solution is an intermediate step in a larger problem. For a final answer, an approximate solution is more practical.

EXAMPLE 3 Solving an Exponential Equation

Solve $e^x + 5 = 60$ and approximate the result to three decimal places.

Solution

$e^x + 5 = 60$	Write original equation.
$e^x = 55$	Subtract 5 from each side.
$\ln e^x = \ln 55$	Take natural log of each side.
$x = \ln 55 \approx 4.007$	Inverse Property

The solution is $x = \ln 55 \approx 4.007$. Check this in the original equation.

REMARK Remember that the natural logarithmic function has a base of e.

Checkpoint Audio-video solution in English & Spanish at LarsonPrecalculus.com

Solve $e^x - 7 = 23$ and approximate the result to three decimal places.

EXAMPLE 4 Solving an Exponential Equation

Solve $2(3^{2t-5}) - 4 = 11$ and approximate the result to three decimal places.

Solution

$2(3^{2t-5}) - 4 = 11$	Write original equation.
$2(3^{2t-5}) = 15$	Add 4 to each side.
$3^{2t-5} = \dfrac{15}{2}$	Divide each side by 2.
$\log_3 3^{2t-5} = \log_3 \dfrac{15}{2}$	Take log (base 3) of each side.
$2t - 5 = \log_3 \dfrac{15}{2}$	Inverse Property
$2t = 5 + \log_3 7.5$	Add 5 to each side.
$t = \dfrac{5}{2} + \dfrac{1}{2}\log_3 7.5$	Divide each side by 2.
$t \approx 3.417$	Use a calculator.

REMARK Remember that to evaluate a logarithm such as $\log_3 7.5$, you need to use the change-of-base formula.

$$\log_3 7.5 = \frac{\ln 7.5}{\ln 3} \approx 1.834$$

The solution is $t = \frac{5}{2} + \frac{1}{2}\log_3 7.5 \approx 3.417$. Check this in the original equation.

Checkpoint Audio-video solution in English & Spanish at LarsonPrecalculus.com

Solve $6(2^{t+5}) + 4 = 11$ and approximate the result to three decimal places.

When an equation involves two or more exponential expressions, you can still use a procedure similar to that demonstrated in Examples 2, 3, and 4. However, it may include additional algebraic techniques.

EXAMPLE 5 Solving an Exponential Equation of Quadratic Type

Solve $e^{2x} - 3e^x + 2 = 0$.

Algebraic Solution

$e^{2x} - 3e^x + 2 = 0$	Write original equation.
$(e^x)^2 - 3e^x + 2 = 0$	Write in quadratic form.
$(e^x - 2)(e^x - 1) = 0$	Factor.
$e^x - 2 = 0$	Set 1st factor equal to 0.
$x = \ln 2$	Solve for x.
$e^x - 1 = 0$	Set 2nd factor equal to 0.
$x = 0$	Solve for x.

The solutions are $x = \ln 2 \approx 0.693$ and $x = 0$. Check these in the original equation.

Graphical Solution

Use a graphing utility to graph $y = e^{2x} - 3e^x + 2$ and then find the zeros.

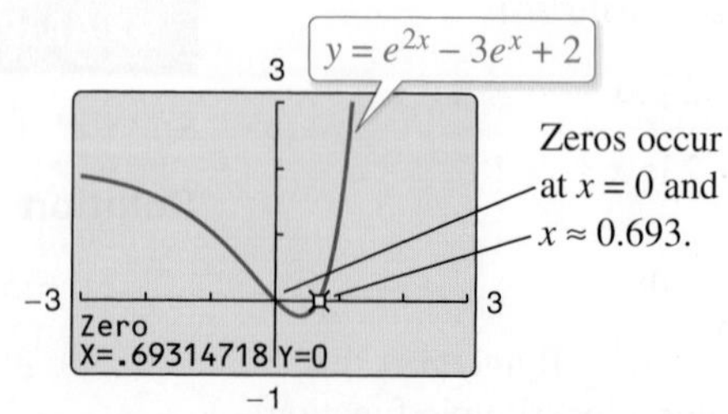

So, the solutions are $x = 0$ and $x \approx 0.693$.

Checkpoint Audio-video solution in English & Spanish at LarsonPrecalculus.com

Solve $e^{2x} - 7e^x + 12 = 0$.

Solving Logarithmic Equations

To solve a logarithmic equation, write it in exponential form. This procedure is called *exponentiating* each side of an equation.

$\ln x = 3$ — Logarithmic form

$e^{\ln x} = e^3$ — Exponentiate each side.

$x = e^3$ — Exponential form

REMARK When solving equations, remember to check your solutions in the original equation to verify that the answer is correct and to make sure that the answer is in the domain of the original equation.

EXAMPLE 6 Solving Logarithmic Equations

a. $\ln x = 2$ — Original equation

$e^{\ln x} = e^2$ — Exponentiate each side.

$x = e^2$ — Inverse Property

b. $\log_3(5x - 1) = \log_3(x + 7)$ — Original equation

$5x - 1 = x + 7$ — One-to-One Property

$x = 2$ — Solve for x.

c. $\log_6(3x + 14) - \log_6 5 = \log_6 2x$ — Original equation

$\log_6\left(\frac{3x + 14}{5}\right) = \log_6 2x$ — Quotient Property of Logarithms

$\frac{3x + 14}{5} = 2x$ — One-to-One Property

$3x + 14 = 10x$ — Multiply each side by 5.

$x = 2$ — Solve for x.

Solve each equation.

a. $\ln x = \frac{2}{3}$ **b.** $\log_2(2x - 3) = \log_2(x + 4)$ **c.** $\log 4x - \log(12 + x) = \log 2$

EXAMPLE 7 Solving a Logarithmic Equation

Solve $5 + 2 \ln x = 4$ and approximate the result to three decimal places.

Algebraic Solution

$5 + 2 \ln x = 4$ — Write original equation.

$2 \ln x = -1$ — Subtract 5 from each side.

$\ln x = -\frac{1}{2}$ — Divide each side by 2.

$e^{\ln x} = e^{-1/2}$ — Exponentiate each side.

$x = e^{-1/2}$ — Inverse Property

$x \approx 0.607$ — Use a calculator.

Graphical Solution

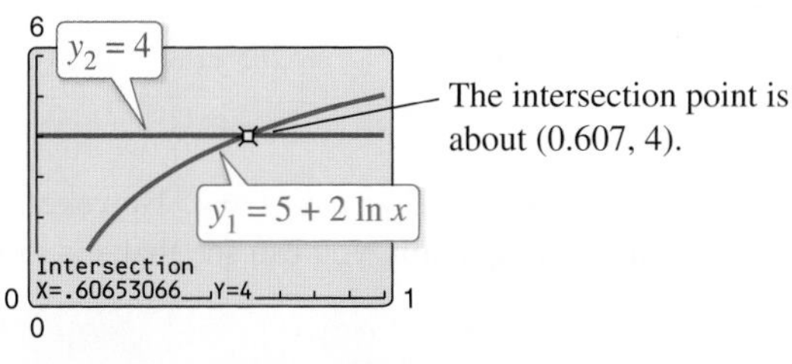

So, the solution is $x \approx 0.607$.

Checkpoint Audio-video solution in English & Spanish at LarsonPrecalculus.com

Solve $7 + 3 \ln x = 5$ and approximate the result to three decimal places.

EXAMPLE 8 Solving a Logarithmic Equation

Solve $2 \log_5 3x = 4$.

Solution

$$
\begin{aligned}
2 \log_5 3x &= 4 && \text{Write original equation.} \\
\log_5 3x &= 2 && \text{Divide each side by 2.} \\
5^{\log_5 3x} &= 5^2 && \text{Exponentiate each side (base 5).} \\
3x &= 25 && \text{Inverse Property} \\
x &= \frac{25}{3} && \text{Divide each side by 3.}
\end{aligned}
$$

The solution is $x = \frac{25}{3}$. Check this in the original equation.

Solve $3 \log_4 6x = 9$.

The domain of a logarithmic function generally does not include all real numbers, so you should be sure to check for extraneous solutions of logarithmic equations.

EXAMPLE 9 Checking for Extraneous Solutions

Solve

$$\log 5x + \log(x - 1) = 2.$$

Algebraic Solution

$$
\begin{aligned}
\log 5x + \log(x - 1) &= 2 && \text{Write original equation.} \\
\log[5x(x - 1)] &= 2 && \text{Product Property of Logarithms} \\
10^{\log(5x^2 - 5x)} &= 10^2 && \text{Exponentiate each side (base 10).} \\
5x^2 - 5x &= 100 && \text{Inverse Property} \\
x^2 - x - 20 &= 0 && \text{Write in general form.} \\
(x - 5)(x + 4) &= 0 && \text{Factor.} \\
x - 5 &= 0 && \text{Set 1st factor equal to 0.} \\
x &= 5 && \text{Solve for } x. \\
x + 4 &= 0 && \text{Set 2nd factor equal to 0.} \\
x &= -4 && \text{Solve for } x.
\end{aligned}
$$

The solutions appear to be $x = 5$ and $x = -4$. However, when you check these in the original equation, you can see that $x = 5$ is the only solution.

Graphical Solution

First, rewrite the original equation as

$$\log 5x + \log(x - 1) - 2 = 0.$$

Then use a graphing utility to graph the equation

$$y = \log 5x + \log(x - 1) - 2$$

and find the zero(s).

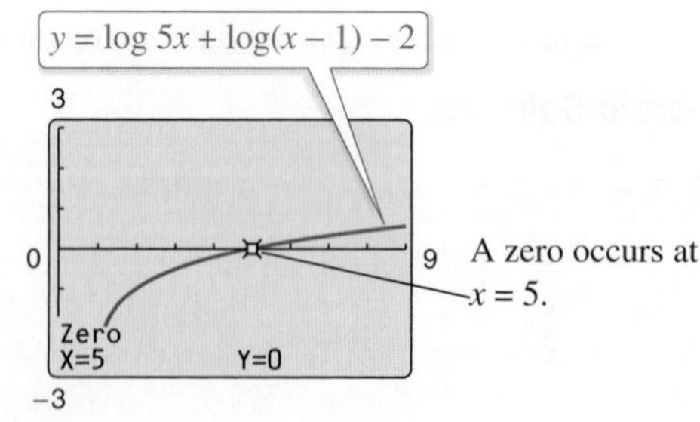

So, the solution is $x = 5$.

✓ **Checkpoint** Audio-video solution in English & Spanish at LarsonPrecalculus.com

Solve $\log x + \log(x - 9) = 1$.

In Example 9, the domain of $\log 5x$ is $x > 0$ and the domain of $\log(x - 1)$ is $x > 1$, so the domain of the original equation is $x > 1$. This means that the solution $x = -4$ is extraneous. The graphical solution verifies this conclusion.

Applications

EXAMPLE 10 Doubling an Investment

See LarsonPrecalculus.com for an interactive version of this type of example.

You invest \$500 at an annual interest rate of 6.75%, compounded continuously. How long will it take your money to double?

Solution Using the formula for continuous compounding, the balance is

$$A = Pe^{rt}$$

$$A = 500e^{0.0675t}.$$

To find the time required for the balance to double, let $A = 1000$ and solve the resulting equation for t.

$500e^{0.0675t} = 1000$	Let $A = 1000$.
$e^{0.0675t} = 2$	Divide each side by 500.
$\ln e^{0.0675t} = \ln 2$	Take natural log of each side.
$0.0675t = \ln 2$	Inverse Property
$t = \dfrac{\ln 2}{0.0675}$	Divide each side by 0.0675.
$t \approx 10.27$	Use a calculator.

The balance in the account will double after approximately 10.27 years. This result is demonstrated graphically below.

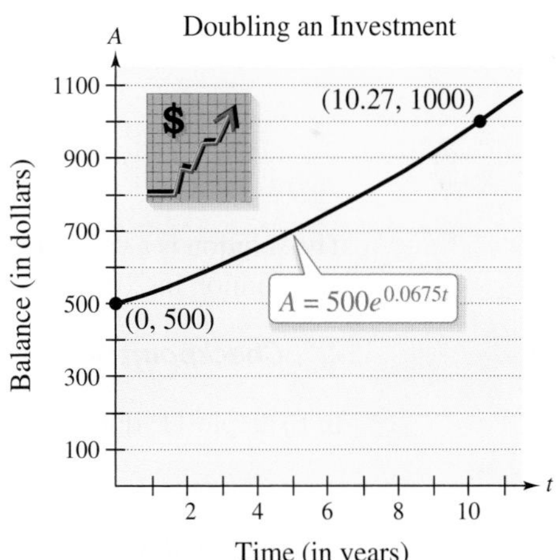

✓ **Checkpoint** Audio-video solution in English & Spanish at LarsonPrecalculus.com

You invest \$500 at an annual interest rate of 5.25%, compounded continuously. How long will it take your money to double? Compare your result with that of Example 10.

In Example 10, an approximate answer of 10.27 years is given. Within the context of the problem, the exact solution

$$t = \frac{\ln 2}{0.0675}$$

does not make sense as an answer.

EXAMPLE 11 Retail Sales

The retail sales y (in billions of dollars) of e-commerce companies in the United States from 2009 through 2014 can be modeled by

$$y = -614 + 342.2 \ln t, \quad 9 \leq t \leq 14$$

where t represents the year, with $t = 9$ corresponding to 2009 (see figure). During which year did the sales reach \$240 billion? *(Source: U.S. Census Bureau)*

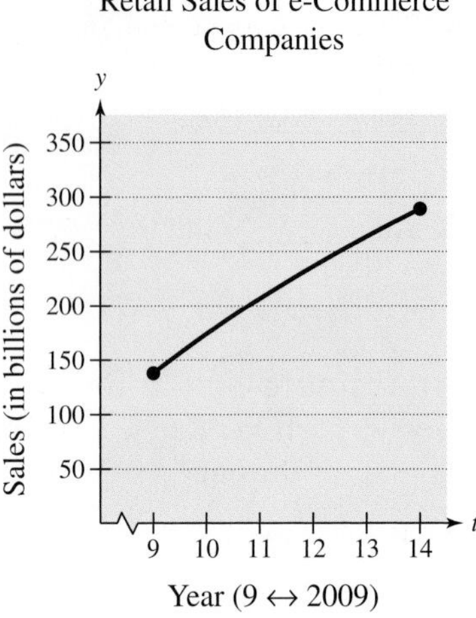

Solution

$-614 + 342.2 \ln t = y$	Write original equation.
$-614 + 342.2 \ln t = 240$	Substitute 240 for y.
$342.2 \ln t = 854$	Add 614 to each side.
$\ln t = \dfrac{854}{342.2}$	Divide each side by 342.2.
$e^{\ln t} = e^{854/342.2}$	Exponentiate each side.
$t = e^{854/342.2}$	Inverse Property
$t \approx 12$	Use a calculator.

The solution is $t \approx 12$. Because $t = 9$ represents 2009, it follows that the sales reached \$240 billion in 2012.

✓ Checkpoint *Audio-video solution in English & Spanish at LarsonPrecalculus.com*

In Example 11, during which year did the sales reach \$180 billion?

Summarize (Section 5.4)

1. State the One-to-One Properties and the Inverse Properties that are used to solve simple exponential and logarithmic equations *(page 386)*. For an example of solving simple exponential and logarithmic equations, see Example 1.
2. Describe strategies for solving exponential equations *(pages 387 and 388)*. For examples of solving exponential equations, see Examples 2–5.
3. Describe strategies for solving logarithmic equations *(pages 389 and 390)*. For examples of solving logarithmic equations, see Examples 6–9.
4. Describe examples of how to use exponential and logarithmic equations to model and solve real-life problems *(pages 391 and 392, Examples 10 and 11)*.

5.4 Exercises

See CalcChat.com for tutorial help and worked-out solutions to odd-numbered exercises.

Vocabulary: Fill in the blanks.

1. To solve exponential and logarithmic equations, you can use the One-to-One and Inverse Properties below.
   (a) $a^x = a^y$ if and only if ________.
   (b) $\log_a x = \log_a y$ if and only if ________.
   (c) $a^{\log_a x} =$ ________
   (d) $\log_a a^x =$ ________
2. An ________ solution does not satisfy the original equation.

Skills and Applications

Determining Solutions In Exercises 3–6, determine whether each x-value is a solution (or an approximate solution) of the equation.

3. $4^{2x-7} = 64$
   (a) $x = 5$
   (b) $x = 2$
   (c) $x = \frac{1}{2}(\log_4 64 + 7)$
4. $4e^{x-1} = 60$
   (a) $x = 1 + \ln 15$
   (b) $x \approx 1.708$
   (c) $x = \ln 16$
5. $\log_2(x + 3) = 10$
   (a) $x = 1021$
   (b) $x = 17$
   (c) $x = 10^2 - 3$
6. $\ln(2x + 3) = 5.8$
   (a) $x = \frac{1}{2}(-3 + \ln 5.8)$
   (b) $x = \frac{1}{2}(-3 + e^{5.8})$
   (c) $x \approx 163.650$

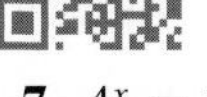

Solving a Simple Equation In Exercises 7–16, solve for x.

7. $4^x = 16$
8. $\left(\frac{1}{2}\right)^x = 32$
9. $\ln x - \ln 2 = 0$
10. $\log x - \log 10 = 0$
11. $e^x = 2$
12. $e^x = \frac{1}{3}$
13. $\ln x = -1$
14. $\log x = -2$
15. $\log_4 x = 3$
16. $\log_5 x = \frac{1}{2}$

Approximating a Point of Intersection In Exercises 17 and 18, approximate the point of intersection of the graphs of f and g. Then solve the equation $f(x) = g(x)$ algebraically to verify your approximation.

17. $f(x) = 2^x,\ g(x) = 8$

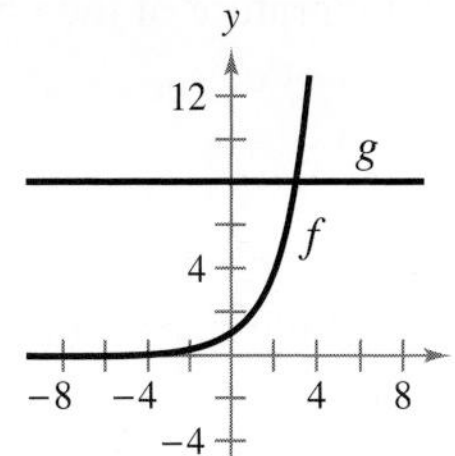

18. $f(x) = \log_3 x,\ g(x) = 2$

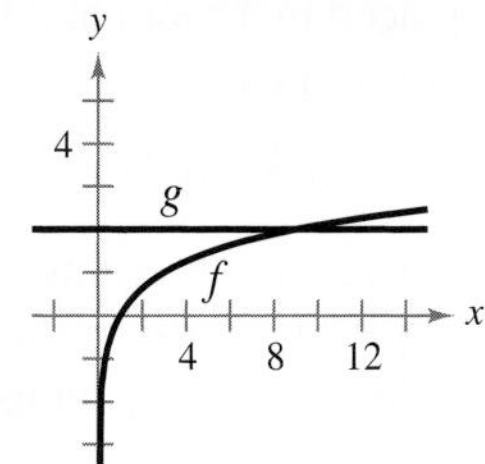

Solving an Exponential Equation In Exercises 19–46, solve the exponential equation algebraically. Approximate the result to three decimal places, if necessary.

19. $e^x = e^{x^2-2}$
20. $e^{x^2-3} = e^{x-2}$
21. $4(3^x) = 20$
22. $4e^x = 91$
23. $e^x - 8 = 31$
24. $5^x + 8 = 26$
25. $3^{2x} = 80$
26. $4^{-3t} = 0.10$
27. $3^{2-x} = 400$
28. $7^{-3-x} = 242$
29. $8(10^{3x}) = 12$
30. $8(3^{6-x}) = 40$
31. $e^{3x} = 12$
32. $500e^{-2x} = 125$
33. $7 - 2e^x = 5$
34. $-14 + 3e^x = 11$
35. $6(2^{3x-1}) - 7 = 9$
36. $8(4^{6-2x}) + 13 = 41$
37. $3^x = 2^{x-1}$
38. $e^{x+1} = 2^{x+2}$
39. $4^x = 5^{x^2}$
40. $3^{x^2} = 7^{6-x}$
41. $e^{2x} - 4e^x - 5 = 0$
42. $e^{2x} - 5e^x + 6 = 0$
43. $\dfrac{1}{1 - e^x} = 5$
44. $\dfrac{100}{1 + e^{2x}} = 1$
45. $\left(1 + \dfrac{0.065}{365}\right)^{365t} = 4$
46. $\left(1 + \dfrac{0.10}{12}\right)^{12t} = 2$

Solving a Logarithmic Equation In Exercises 47–62, solve the logarithmic equation algebraically. Approximate the result to three decimal places, if necessary.

47. $\ln x = -3$
48. $\ln x - 7 = 0$
49. $2.1 = \ln 6x$
50. $\log 3z = 2$
51. $3 - 4\ln x = 11$
52. $3 + 8\ln x = 7$
53. $6\log_3 0.5x = 11$
54. $4\log(x - 6) = 11$
55. $\ln x - \ln(x + 1) = 2$
56. $\ln x + \ln(x + 1) = 1$
57. $\ln(x + 5) = \ln(x - 1) - \ln(x + 1)$
58. $\ln(x + 1) - \ln(x - 2) = \ln x$
59. $\log(3x + 4) = \log(x - 10)$
60. $\log_2 x + \log_2(x + 2) = \log_2(x + 6)$
61. $\log_4 x - \log_4(x - 1) = \frac{1}{2}$
62. $\log 8x - \log(1 + \sqrt{x}) = 2$

Using Technology In Exercises 63–70, use a graphing utility to graphically solve the equation. Approximate the result to three decimal places. Verify your result algebraically.

63. $5^x = 212$
64. $6e^{1-x} = 25$
65. $8e^{-2x/3} = 11$
66. $e^{0.09t} = 3$
67. $3 - \ln x = 0$
68. $10 - 4\ln(x - 2) = 0$
69. $2\ln(x + 3) = 3$
70. $\ln(x + 1) = 2 - \ln x$

Compound Interest In Exercises 71 and 72, you invest \$2500 in an account at interest rate r, compounded continuously. Find the time required for the amount to (a) double and (b) triple.

71. $r = 0.025$
72. $r = 0.0375$

Algebra of Calculus In Exercises 73–80, solve the equation algebraically. Round your result to three decimal places, if necessary. Verify your answer using a graphing utility.

73. $2x^2e^{2x} + 2xe^{2x} = 0$
74. $-x^2e^{-x} + 2xe^{-x} = 0$
75. $-xe^{-x} + e^{-x} = 0$
76. $e^{-2x} - 2xe^{-2x} = 0$
77. $\dfrac{1 + \ln x}{2} = 0$
78. $\dfrac{1 - \ln x}{x^2} = 0$
79. $2x \ln x + x = 0$
80. $2x \ln\left(\dfrac{1}{x}\right) - x = 0$

81. **Average Heights** The percent m of American males between the ages of 20 and 29 who are under x inches tall is modeled by

$$m(x) = \frac{100}{1 + e^{-0.5536(x-69.51)}}, \quad 64 \le x \le 78$$

and the percent f of American females between the ages of 20 and 29 who are under x inches tall is modeled by

$$f(x) = \frac{100}{1 + e^{-0.5834(x-64.49)}}, \quad 60 \le x \le 78.$$

(Source: U.S. National Center for Health Statistics)

(a) Use the graph to determine any horizontal asymptotes of the graphs of the functions. Interpret the meaning in the context of the problem.

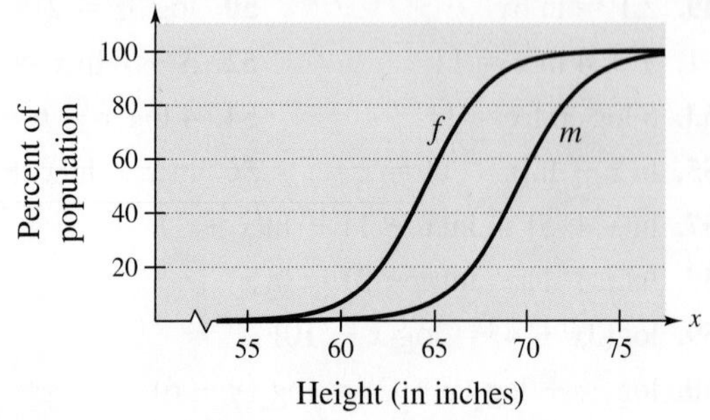

(b) What is the average height of each sex?

82. **Demand** The demand equation for a smartphone is

$$p = 5000\left(1 - \frac{4}{4 + e^{-0.002x}}\right).$$

Find the demand x for each price.

(a) $p = \$169$

(b) $p = \$299$

83. **Ecology** The number N of beavers in a given area after x years can be approximated by

$$N = 5.5 \cdot 10^{0.23x}, \quad 0 \le x \le 10.$$

Use the model to approximate how many years it will take for the beaver population to reach 78.

84. **Ecology** The number N of trees of a given species per acre is approximated by the model

$$N = 3500(10^{-0.12x}), \quad 3 \le x \le 30$$

where x is the average diameter of the trees (in inches) 4.5 feet above the ground. Use the model to approximate the average diameter of the trees in a test plot when $N = 22$.

85. **Population** The population P (in thousands) of Alaska in the years 2005 through 2015 can be modeled by

$$P = 75 \ln t + 540, \quad 5 \le t \le 15$$

where t represents the year, with $t = 5$ corresponding to 2005. During which year did the population of Alaska exceed 720 thousand? *(Source: U.S. Census Bureau)*

86. **Population** The population P (in thousands) of Montana in the years 2005 through 2015 can be modeled by

$$P = 81 \ln t + 807, \quad 5 \le t \le 15$$

where t represents the year, with $t = 5$ corresponding to 2005. During which year did the population of Montana exceed 965 thousand? *(Source: U.S. Census Bureau)*

87. **Temperature** An object at a temperature of 80°C is placed in a room at 20°C. The temperature of the object is given by

$$T = 20 + 60e^{-0.06m}$$

where m represents the number of minutes after the object is placed in the room. How long does it take the object to reach a temperature of 70°C?

88. Temperature An object at a temperature of 160°C was removed from a furnace and placed in a room at 20°C. The temperature T of the object was measured each hour h and recorded in the table. A model for the data is

$T = 20 + 140e^{-0.68h}$.

DATA

Hour, h	Temperature, T
0	160°
1	90°
2	56°
3	38°
4	29°
5	24°

Spreadsheet at LarsonPrecalculus.com

(a) The figure below shows the graph of the model. Use the graph to identify the horizontal asymptote of the model and interpret the asymptote in the context of the problem.

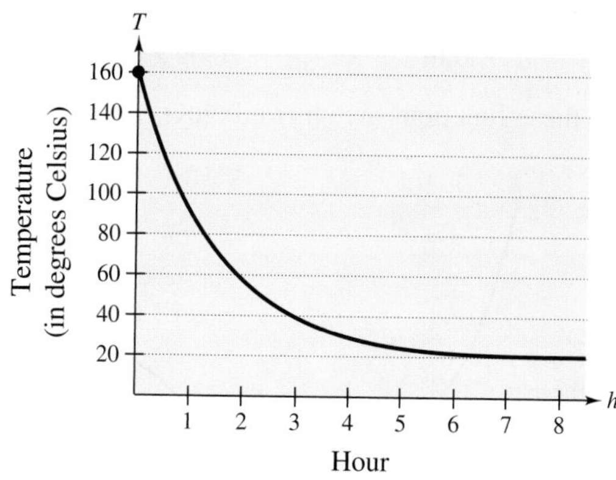

(b) Use the model to approximate the time it took for the object to reach a temperature of 100°C.

Exploration

True or False? **In Exercises 89–92, rewrite each verbal statement as an equation. Then decide whether the statement is true or false. Justify your answer.**

89. The logarithm of the product of two numbers is equal to the sum of the logarithms of the numbers.

90. The logarithm of the sum of two numbers is equal to the product of the logarithms of the numbers.

91. The logarithm of the difference of two numbers is equal to the difference of the logarithms of the numbers.

92. The logarithm of the quotient of two numbers is equal to the difference of the logarithms of the numbers.

93. Think About It Is it possible for a logarithmic equation to have more than one extraneous solution? Explain.

94. HOW DO YOU SEE IT? Solving $\log_3 x + \log_3(x - 8) = 2$ algebraically, the solutions appear to be $x = 9$ and $x = -1$. Use the graph of

$y = \log_3 x + \log_3(x - 8) - 2$

to determine whether each value is an actual solution of the equation. Explain.

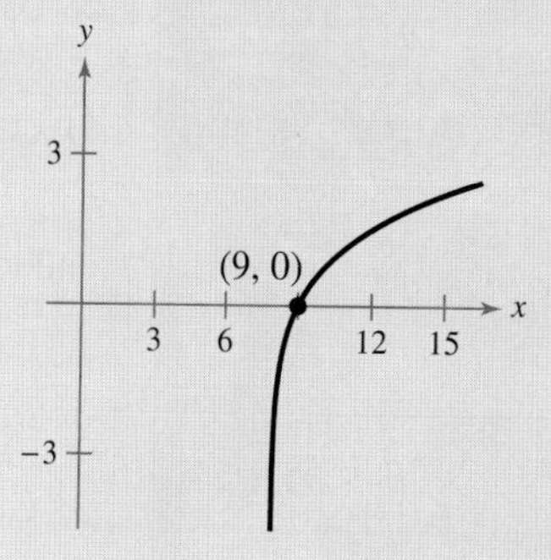

95. Finance You are investing P dollars at an annual interest rate of r, compounded continuously, for t years. Which change below results in the highest value of the investment? Explain.

(a) Double the amount you invest.

(b) Double your interest rate.

(c) Double the number of years.

96. Think About It Are the times required for the investments in Exercises 71 and 72 to quadruple twice as long as the times for them to double? Give a reason for your answer and verify your answer algebraically.

97. Effective Yield The *effective yield* of an investment plan is the percent increase in the balance after 1 year. Find the effective yield for each investment plan. Which investment plan has the greatest effective yield? Which investment plan will have the highest balance after 5 years?

(a) 7% annual interest rate, compounded annually

(b) 7% annual interest rate, compounded continuously

(c) 7% annual interest rate, compounded quarterly

(d) 7.25% annual interest rate, compounded quarterly

98. Graphical Reasoning Let $f(x) = \log_a x$ and $g(x) = a^x$, where $a > 1$.

(a) Let $a = 1.2$ and use a graphing utility to graph the two functions in the same viewing window. What do you observe? Approximate any points of intersection of the two graphs.

(b) Determine the value(s) of a for which the two graphs have one point of intersection.

(c) Determine the value(s) of a for which the two graphs have two points of intersection.

5.5 Exponential and Logarithmic Models

Exponential growth and decay models can often represent populations. For example, in Exercise 30 on page 404, you will use exponential growth and decay models to compare the populations of several countries.

- **Recognize the five most common types of models involving exponential and logarithmic functions.**
- **Use exponential growth and decay functions to model and solve real-life problems.**
- **Use Gaussian functions to model and solve real-life problems.**
- **Use logistic growth functions to model and solve real-life problems.**
- **Use logarithmic functions to model and solve real-life problems.**

Introduction

The five most common types of mathematical models involving exponential functions and logarithmic functions are listed below.

1. **Exponential growth model:** $y = ae^{bx}, \quad b > 0$
2. **Exponential decay model:** $y = ae^{-bx}, \quad b > 0$
3. **Gaussian model:** $y = ae^{-(x-b)^2/c}$
4. **Logistic growth model:** $y = \dfrac{a}{1 + be^{-rx}}$
5. **Logarithmic models:** $y = a + b \ln x, \quad y = a + b \log x$

The basic shapes of the graphs of these functions are shown below.

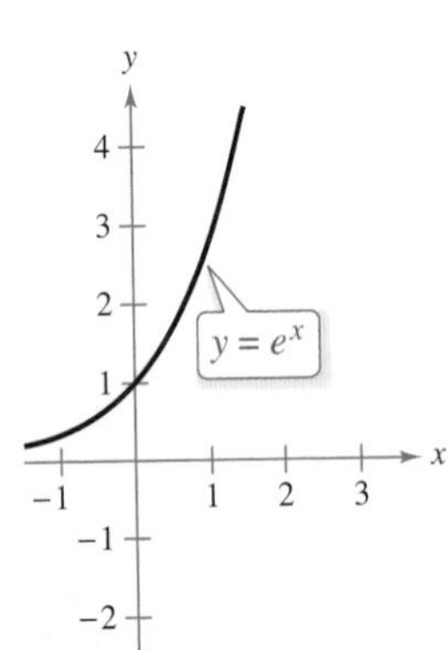

Exponential growth model

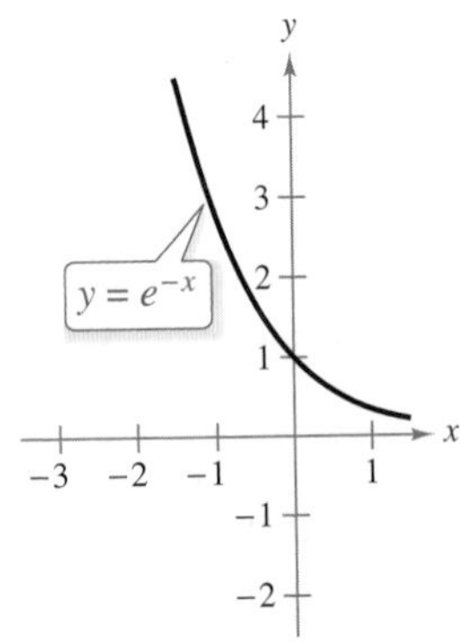

Exponential decay model

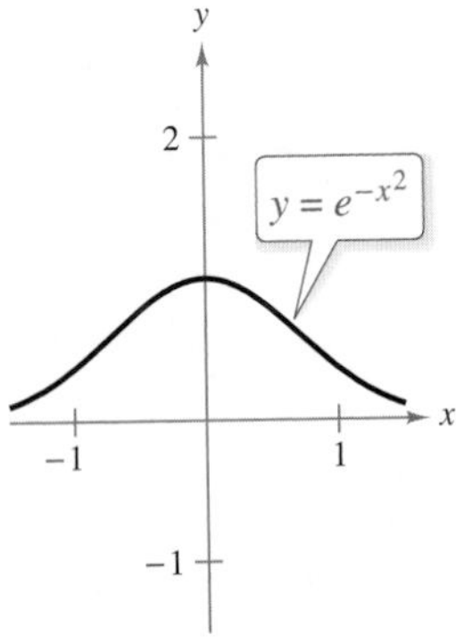

Gaussian model

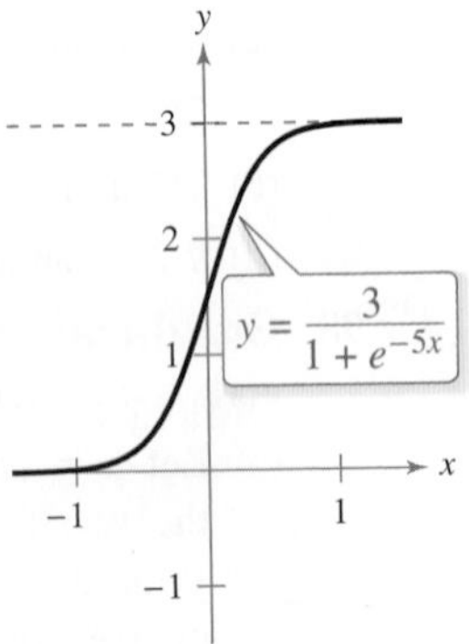

Logistic growth model

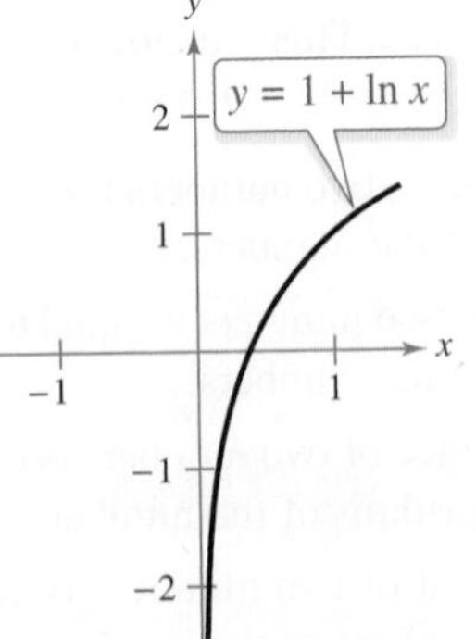

Natural logarithmic model

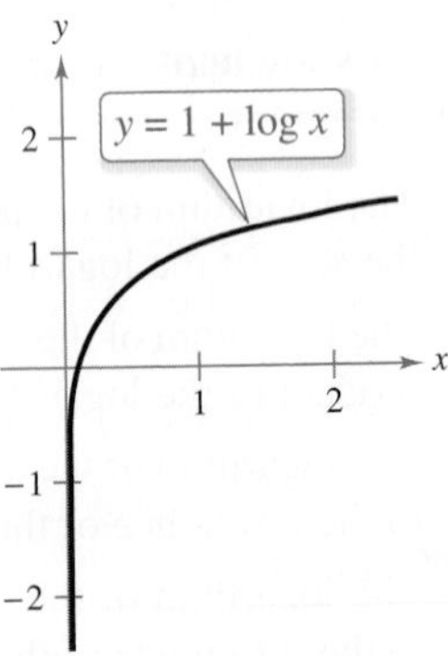

Common logarithmic model

You often gain insight into a situation modeled by an exponential or logarithmic function by identifying and interpreting the asymptotes of the graph of the function. Identify the asymptote(s) of the graph of each function shown above.

Exponential Growth and Decay

EXAMPLE 1 Online Advertising

The amounts S (in billions of dollars) spent in the United States on mobile online advertising in the years 2010 through 2014 are shown in the table. A scatter plot of the data is shown at the right. *(Source: IAB/Price Waterhouse Coopers)*

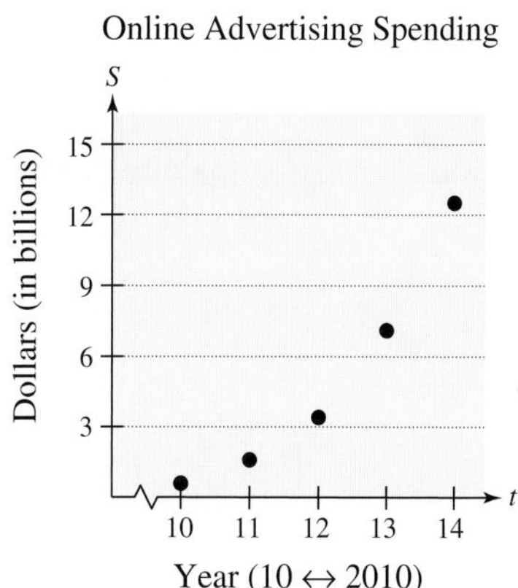

Year	2010	2011	2012	2013	2014
Advertising Spending	0.6	1.6	3.4	7.1	12.5

An exponential growth model that approximates the data is

$$S = 0.00036e^{0.7563t}, \quad 10 \le t \le 14$$

where t represents the year, with $t = 10$ corresponding to 2010. Compare the values found using the model with the amounts shown in the table. According to this model, in what year will the amount spent on mobile online advertising be approximately \$65 billion?

Algebraic Solution

The table compares the actual amounts with the values found using the model.

Year	2010	2011	2012	2013	2014
Advertising Spending	0.6	1.6	3.4	7.1	12.5
Model	0.7	1.5	3.1	6.7	14.3

To find when the amount spent on mobile online advertising is about \$65 billion, let $S = 65$ in the model and solve for t.

$$0.00036e^{0.7563t} = S \quad \text{Write original model.}$$

$$0.00036e^{0.7563t} = 65 \quad \text{Substitute 65 for } S.$$

$$e^{0.7563t} \approx 180{,}556 \quad \text{Divide each side by 0.00036.}$$

$$\ln e^{0.7563t} \approx \ln 180{,}556 \quad \text{Take natural log of each side.}$$

$$0.7563t \approx 12.1038 \quad \text{Inverse Property}$$

$$t \approx 16 \quad \text{Divide each side by 0.7563.}$$

According to the model, the amount spent on mobile online advertising will be about \$65 billion in 2016.

Graphical Solution

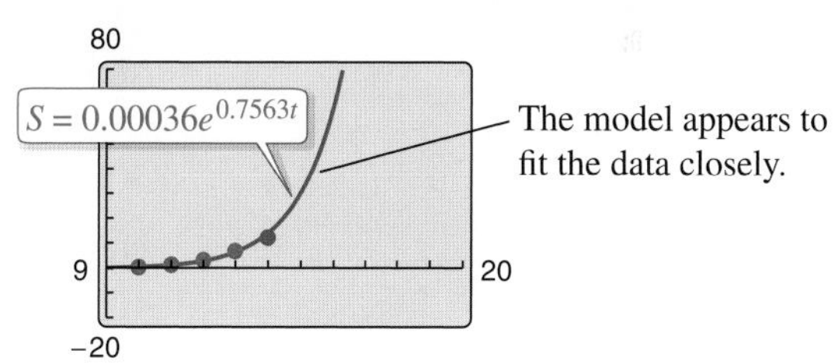

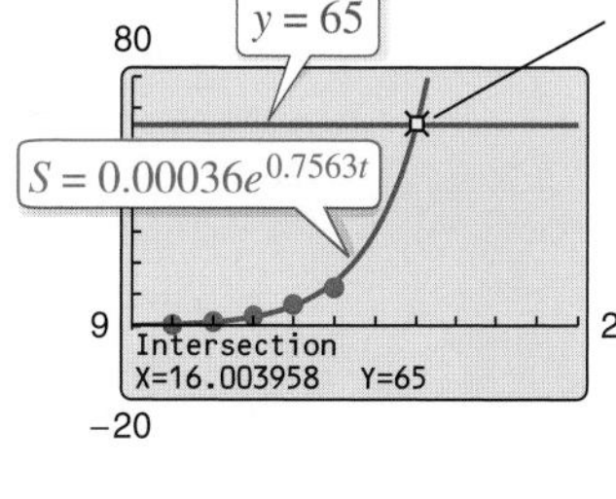

The intersection point of the model and the line $y = 65$ is about (16, 65). So, according to the model, the amount spent on mobile online advertising will be about \$65 billion in 2016.

✓ Checkpoint *Audio-video solution in English & Spanish at LarsonPrecalculus.com*

In Example 1, in what year will the amount spent on mobile online advertising be about \$300 billion?

TECHNOLOGY Some graphing utilities have an *exponential regression* feature that can help you find exponential models to represent data. If you have such a graphing utility, use it to find an exponential model for the data given in Example 1. How does your model compare with the model given in Example 1?

In Example 1, the exponential growth model is given. Sometimes you must find such a model. One technique for doing this is shown in Example 2.

EXAMPLE 2 Modeling Population Growth

In a research experiment, a population of fruit flies is increasing according to the law of exponential growth. After 2 days there are 100 flies, and after 4 days there are 300 flies. How many flies will there be after 5 days?

Solution Let y be the number of flies at time t (in days). From the given information, you know that $y = 100$ when $t = 2$ and $y = 300$ when $t = 4$. Substituting this information into the model $y = ae^{bt}$ produces

$$100 = ae^{2b} \quad \text{and} \quad 300 = ae^{4b}.$$

To solve for b, solve for a in the first equation.

$$100 = ae^{2b} \qquad \text{Write first equation.}$$

$$\frac{100}{e^{2b}} = a \qquad \text{Solve for } a.$$

Then substitute the result into the second equation.

$$300 = ae^{4b} \qquad \text{Write second equation.}$$

$$300 = \left(\frac{100}{e^{2b}}\right)e^{4b} \qquad \text{Substitute } \frac{100}{e^{2b}} \text{ for } a.$$

$$300 = 100e^{2b} \qquad \text{Simplify.}$$

$$\frac{300}{100} = e^{2b} \qquad \text{Divide each side by 100.}$$

$$\ln 3 = 2b \qquad \text{Take natural log of each side.}$$

$$\frac{1}{2}\ln 3 = b \qquad \text{Solve for } b.$$

Now substitute $\frac{1}{2}\ln 3$ for b in the expression you found for a.

$$a = \frac{100}{e^{2[(1/2)\ln 3]}} \qquad \text{Substitute } \tfrac{1}{2}\ln 3 \text{ for } b.$$

$$= \frac{100}{e^{\ln 3}} \qquad \text{Simplify.}$$

$$= \frac{100}{3} \qquad \text{Inverse Property}$$

$$\approx 33.33 \qquad \text{Divide.}$$

So, with $a \approx 33.33$ and $b = \frac{1}{2}\ln 3 \approx 0.5493$, the exponential growth model is

$$y = 33.33e^{0.5493t}$$

as shown in Figure 5.15. After 5 days, the population will be

$$y = 33.33e^{0.5493(5)}$$

$$\approx 520 \text{ flies.}$$

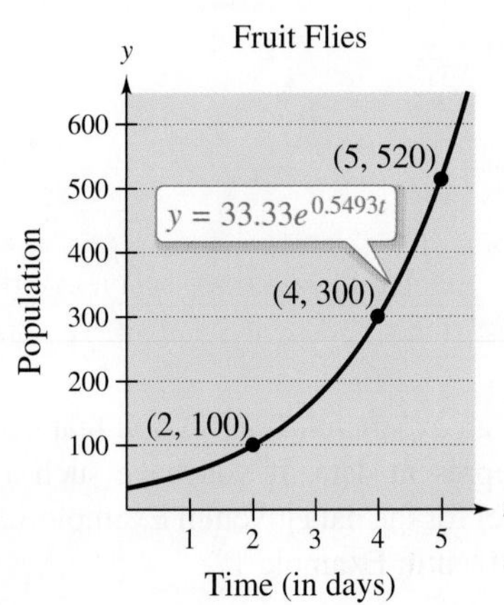

Figure 5.15

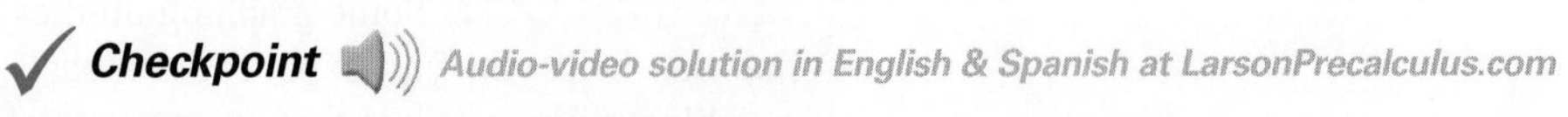

The number of bacteria in a culture is increasing according to the law of exponential growth. After 1 hour there are 100 bacteria, and after 2 hours there are 200 bacteria. How many bacteria will there be after 3 hours?

In living organic material, the ratio of the number of radioactive carbon isotopes (carbon-14) to the number of nonradioactive carbon isotopes (carbon-12) is about 1 to 10^{12}. When organic material dies, its carbon-12 content remains fixed, whereas its radioactive carbon-14 begins to decay with a half-life of about 5700 years. To estimate the age (the number of years since death) of organic material, scientists use the formula

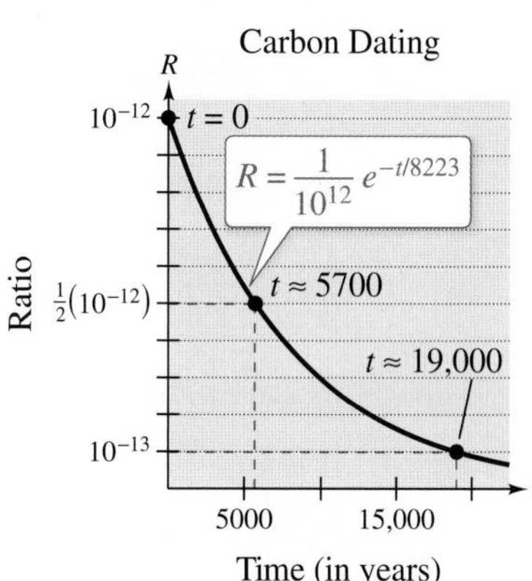

$$R = \frac{1}{10^{12}}e^{-t/8223} \qquad \text{Carbon dating model}$$

where R represents the ratio of carbon-14 to carbon-12 of organic material t years after death. The graph of R is shown at the right. Note that R decreases as t increases.

EXAMPLE 3 Carbon Dating

Estimate the age of a newly discovered fossil for which the ratio of carbon-14 to carbon-12 is $R = \dfrac{1}{10^{13}}$.

Algebraic Solution

In the carbon dating model, substitute the given value of R to obtain the following.

$$\frac{1}{10^{12}}e^{-t/8223} = R \qquad \text{Write original model.}$$

$$\frac{e^{-t/8223}}{10^{12}} = \frac{1}{10^{13}} \qquad \text{Substitute } \frac{1}{10^{13}} \text{ for } R.$$

$$e^{-t/8223} = \frac{1}{10} \qquad \text{Multiply each side by } 10^{12}.$$

$$\ln e^{-t/8223} = \ln\frac{1}{10} \qquad \text{Take natural log of each side.}$$

$$-\frac{t}{8223} \approx -2.3026 \qquad \text{Inverse Property}$$

$$t \approx 18{,}934 \qquad \text{Multiply each side by } -8223.$$

So, to the nearest thousand years, the age of the fossil is about 19,000 years.

Graphical Solution

Use a graphing utility to graph

$$y_1 = \frac{1}{10^{12}}e^{-x/8223} \quad \text{and} \quad y_2 = \frac{1}{10^{13}}$$

in the same viewing window.

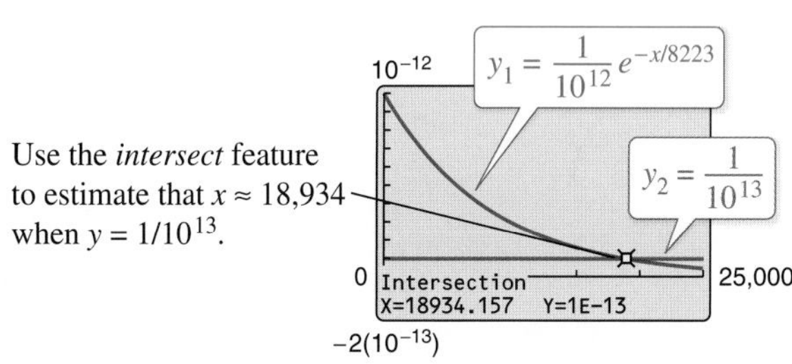

So, to the nearest thousand years, the age of the fossil is about 19,000 years.

✓ Checkpoint *Audio-video solution in English & Spanish at LarsonPrecalculus.com*

Estimate the age of a newly discovered fossil for which the ratio of carbon-14 to carbon-12 is $R = 1/10^{14}$.

The value of b in the exponential decay model $y = ae^{-bt}$ determines the *decay* of radioactive isotopes. For example, to find how much of an initial 10 grams of ^{226}Ra isotope with a half-life of 1599 years is left after 500 years, substitute this information into the model $y = ae^{-bt}$.

$$\frac{1}{2}(10) = 10e^{-b(1599)} \quad \Longrightarrow \quad \ln\frac{1}{2} = -1599b \quad \Longrightarrow \quad b = -\frac{\ln\frac{1}{2}}{1599}$$

Using the value of b found above and $a = 10$, the amount left is

$$y = 10e^{-[-\ln(1/2)/1599](500)} \approx 8.05 \text{ grams.}$$

Gaussian Models

As mentioned at the beginning of this section, Gaussian models are of the form

$$y = ae^{-(x-b)^2/c}.$$

This type of model is commonly used in probability and statistics to represent populations that are **normally distributed.** For *standard* normal distributions, the model takes the form

$$y = \frac{1}{\sqrt{2\pi}}e^{-x^2/2}.$$

The graph of a Gaussian model is called a **bell-shaped curve.** Use a graphing utility to graph the standard normal distribution curve. Can you see why it is called a bell-shaped curve?

The **average value** of a population can be found from the bell-shaped curve by observing where the maximum y-value of the function occurs. The x-value corresponding to the maximum y-value of the function represents the average value of the independent variable—in this case, x.

EXAMPLE 4 SAT Scores

See LarsonPrecalculus.com for an interactive version of this type of example.

In 2015, the SAT mathematics scores for college-bound seniors in the United States roughly followed the normal distribution

$$y = 0.0033e^{-(x-511)^2/28,800}, \quad 200 \le x \le 800$$

where x is the SAT score for mathematics. Use a graphing utility to graph this function and estimate the average SAT mathematics score. *(Source: The College Board)*

Solution The graph of the function is shown below. On this bell-shaped curve, the maximum value of the curve corresponds to the average score. Using the *maximum* feature of the graphing utility, you find that the average mathematics score for college-bound seniors in 2015 was about 511.

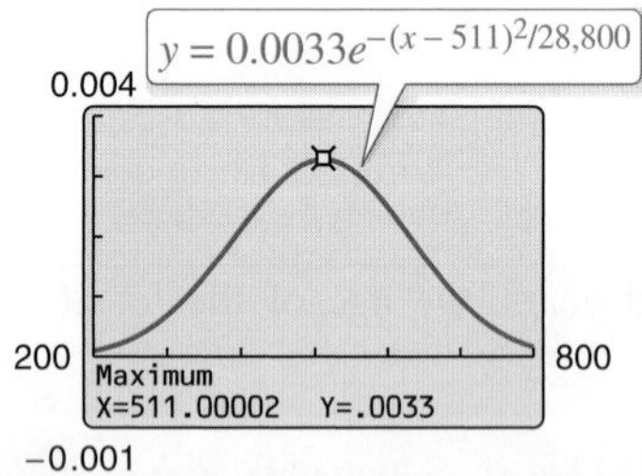

Checkpoint Audio-video solution in English & Spanish at LarsonPrecalculus.com

In 2015, the SAT critical reading scores for college-bound seniors in the United States roughly followed the normal distribution

$$y = 0.0034e^{-(x-495)^2/26,912}, \quad 200 \le x \le 800$$

where x is the SAT score for critical reading. Use a graphing utility to graph this function and estimate the average SAT critical reading score. *(Source: The College Board)*

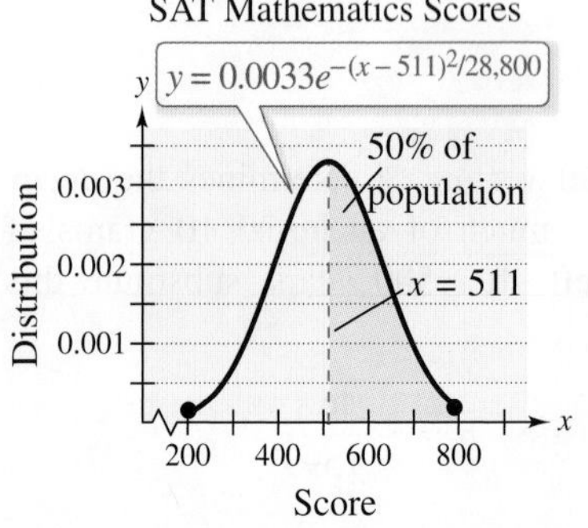

Figure 5.16

In Example 4, note that 50% of the seniors who took the test earned scores greater than 511 (see Figure 5.16).

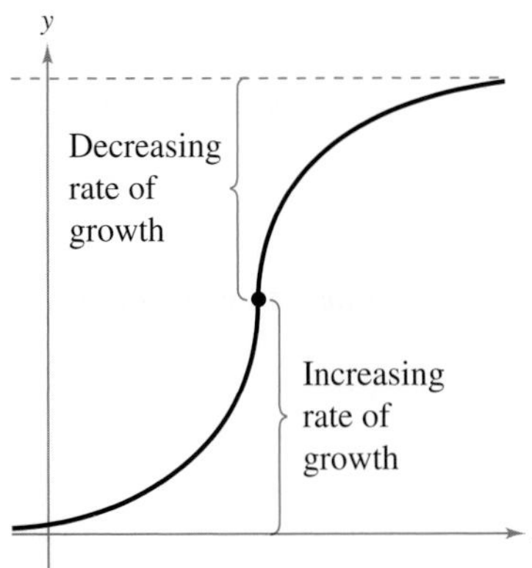

Figure 5.17

Logistic Growth Models

Some populations initially have rapid growth, followed by a declining rate of growth, as illustrated by the graph in Figure 5.17. One model for describing this type of growth pattern is the **logistic curve** given by the function

$$y = \frac{a}{1 + be^{-rx}}$$

where y is the population size and x is the time. An example is a bacteria culture that is initially allowed to grow under ideal conditions and then under less favorable conditions that inhibit growth. A logistic growth curve is also called a **sigmoidal curve.**

EXAMPLE 5 Spread of a Virus

On a college campus of 5000 students, one student returns from vacation with a contagious and long-lasting flu virus. The spread of the virus is modeled by

$$y = \frac{5000}{1 + 4999e^{-0.8t}}, \quad t \geq 0$$

where y is the total number of students infected after t days. The college will cancel classes when 40% or more of the students are infected.

a. How many students are infected after 5 days?

b. After how many days will the college cancel classes?

Algebraic Solution

a. After 5 days, the number of students infected is

$$y = \frac{5000}{1 + 4999e^{-0.8(5)}} = \frac{5000}{1 + 4999e^{-4}} \approx 54.$$

b. The college will cancel classes when the number of infected students is $(0.40)(5000) = 2000$.

$$2000 = \frac{5000}{1 + 4999e^{-0.8t}}$$

$$1 + 4999e^{-0.8t} = 2.5$$

$$e^{-0.8t} = \frac{1.5}{4999}$$

$$-0.8t = \ln \frac{1.5}{4999}$$

$$t = -\frac{1}{0.8} \ln \frac{1.5}{4999}$$

$$t \approx 10.14$$

So, after about 10 days, at least 40% of the students will be infected, and the college will cancel classes.

Graphical Solution

a.

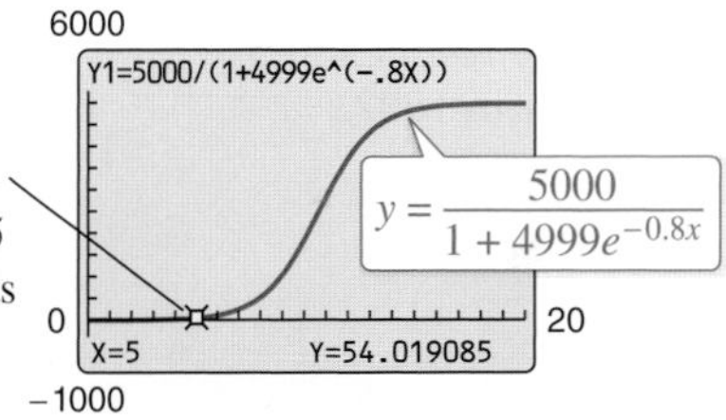

Use the *value* feature to estimate that $y \approx 54$ when $x = 5$. So, after 5 days, about 54 students are infected.

b. The college will cancel classes when the number of infected students is $(0.40)(5000) = 2000$. Use a graphing utility to graph

$$y_1 = \frac{5000}{1 + 4999e^{-0.8x}} \quad \text{and} \quad y_2 = 2000$$

in the same viewing window. Use the *intersect* feature of the graphing utility to find the point of intersection of the graphs.

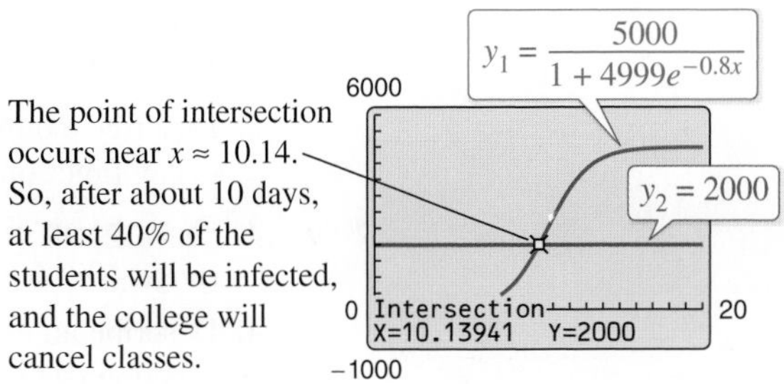

The point of intersection occurs near $x \approx 10.14$. So, after about 10 days, at least 40% of the students will be infected, and the college will cancel classes.

Checkpoint Audio-video solution in English & Spanish at LarsonPrecalculus.com

In Example 5, after how many days are 250 students infected?

Logarithmic Models

On the Richter scale, the magnitude R of an earthquake of intensity I is given by

$$R = \log \frac{I}{I_0}$$

where $I_0 = 1$ is the minimum intensity used for comparison. (Intensity is a measure of the wave energy of an earthquake.)

On April 25, 2015, an earthquake of magnitude 7.8 struck in Nepal. The city of Kathmandu took extensive damage, including the collapse of the 203-foot Dharahara Tower, built by Nepal's first prime minister in 1832.

EXAMPLE 6 Magnitudes of Earthquakes

Find the intensity of each earthquake.

a. Piedmont, California, in 2015: $R = 4.0$ **b.** Nepal in 2015: $R = 7.8$

Solution

a. Because $I_0 = 1$ and $R = 4.0$, you have

$$4.0 = \log \frac{I}{1}$$ Substitute 1 for I_0 and 4.0 for R.

$$10^{4.0} = 10^{\log I}$$ Exponentiate each side.

$$10^{4.0} = I$$ Inverse Property

$$10{,}000 = I.$$ Simplify.

b. For $R = 7.8$, you have

$$7.8 = \log \frac{I}{1}$$ Substitute 1 for I_0 and 7.8 for R.

$$10^{7.8} = 10^{\log I}$$ Exponentiate each side.

$$10^{7.8} = I$$ Inverse Property

$$63{,}000{,}000 \approx I.$$ Use a calculator.

Note that an increase of 3.8 units on the Richter scale (from 4.0 to 7.8) represents an increase in intensity by a factor of $10^{7.8}/10^4 \approx 63{,}000{,}000/10{,}000 = 6300$. In other words, the intensity of the earthquake in Nepal was about 6300 times as great as that of the earthquake in Piedmont, California.

Checkpoint *Audio-video solution in English & Spanish at LarsonPrecalculus.com*

Find the intensities of earthquakes whose magnitudes are (a) $R = 6.0$ and (b) $R = 7.9$.

Summarize (Section 5.5)

1. State the five most common types of models involving exponential and logarithmic functions *(page 396)*.
2. Describe examples of real-life applications that use exponential growth and decay functions *(pages 397–399, Examples 1–3)*.
3. Describe an example of a real-life application that uses a Gaussian function *(page 400, Example 4)*.
4. Describe an example of a real-life application that uses a logistic growth function *(page 401, Example 5)*.
5. Describe an example of a real-life application that uses a logarithmic function *(page 402, Example 6)*.

5.5 Exercises

See CalcChat.com for tutorial help and worked-out solutions to odd-numbered exercises.

Vocabulary: Fill in the blanks.

1. An exponential growth model has the form ________, and an exponential decay model has the form ________.
2. A logarithmic model has the form ________ or ________.
3. In probability and statistics, Gaussian models commonly represent populations that are ________ ________.
4. A logistic growth model has the form ________.

Skills and Applications

Solving for a Variable **In Exercises 5 and 6, (a) solve for *P* and (b) solve for *t*.**

5. $A = Pe^{rt}$ 6. $A = P\left(1 + \frac{r}{n}\right)^{nt}$

Compound Interest **In Exercises 7–12, find the missing values assuming continuously compounded interest.**

	Initial Investment	Annual % Rate	Time to Double	Amount After 10 Years
7.	\$1000	3.5%		
8.	\$750	$10\frac{1}{2}\%$		
9.	\$750		$7\frac{3}{4}$ yr	
10.	\$500			\$1505.00
11.		4.5%		\$10,000.00
12.			12 yr	\$2000.00

Compound Interest **In Exercises 13 and 14, determine the principal *P* that must be invested at rate *r*, compounded monthly, so that \$500,000 will be available for retirement in *t* years.**

13. $r = 5\%, t = 10$ 14. $r = 3\frac{1}{2}\%, t = 15$

Compound Interest **In Exercises 15 and 16, determine the time necessary for *P* dollars to double when it is invested at interest rate *r* compounded (a) annually, (b) monthly, (c) daily, and (d) continuously.**

15. $r = 10\%$ 16. $r = 6.5\%$

17. **Compound Interest** Complete the table for the time *t* (in years) necessary for *P* dollars to triple when it is invested at an interest rate *r* compounded (a) continuously and (b) annually.

r	2%	4%	6%	8%	10%	12%
t						

18. **Modeling Data** Draw scatter plots of the data in Exercise 17. Use the *regression* feature of a graphing utility to find models for the data.

19. **Comparing Models** If \$1 is invested over a 10-year period, then the balance A after t years is given by either $A = 1 + 0.075[\![t]\!]$ or $A = e^{0.07t}$ depending on whether the interest is simple interest at $7\frac{1}{2}\%$ or continuous compound interest at 7%. Graph each function on the same set of axes. Which grows at a greater rate? (Remember that $[\![t]\!]$ is the greatest integer function discussed in Section P.7.)

20. **Comparing Models** If \$1 is invested over a 10-year period, then the balance A after t years is given by either $A = 1 + 0.06[\![t]\!]$ or $A = [1 + (0.055/365)]^{[\![365t]\!]}$ depending on whether the interest is simple interest at 6% or compound interest at $5\frac{1}{2}\%$ compounded daily. Use a graphing utility to graph each function in the same viewing window. Which grows at a greater rate?

Radioactive Decay **In Exercises 21–24, find the missing value for the radioactive isotope.**

	Isotope	Half-life (years)	Initial Quantity	Amount After 1000 Years
21.	^{226}Ra	1599	10 g	
22.	^{14}C	5715	6.5 g	
23.	^{14}C	5715		2 g
24.	^{239}Pu	24,100		0.4 g

Finding an Exponential Model **In Exercises 25–28, find the exponential model that fits the points shown in the graph or table.**

25.

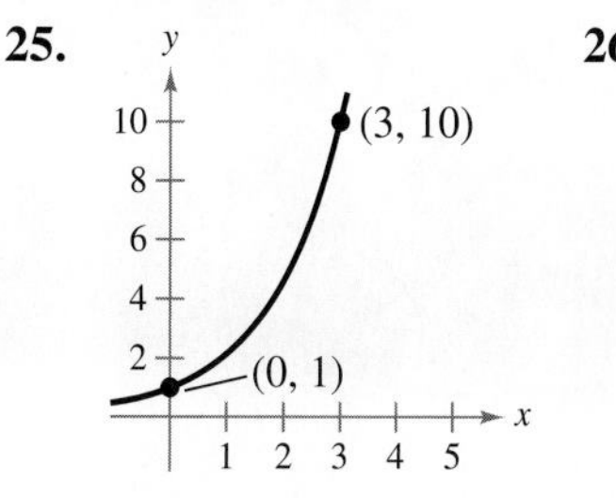

26.

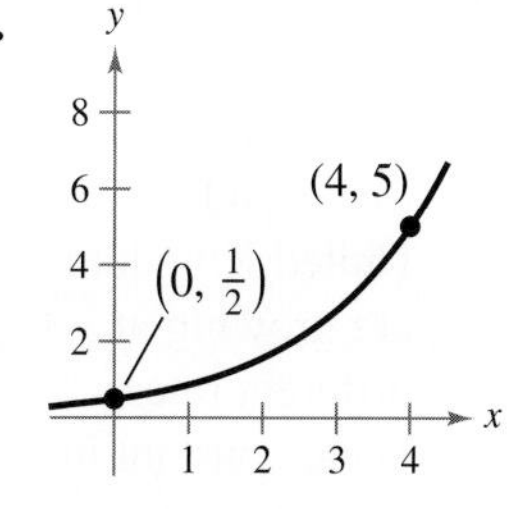

27.

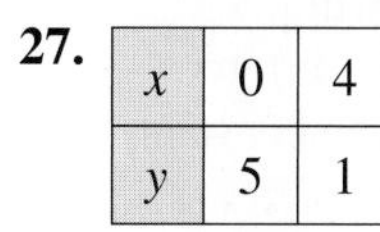

x	0	4
y	5	1

28.

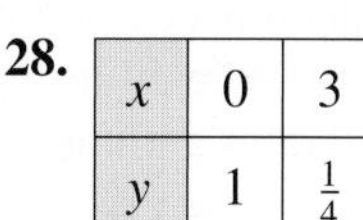

x	0	3
y	1	$\frac{1}{4}$

29. Population The populations P (in thousands) of Horry County, South Carolina, from 1971 through 2014 can be modeled by

$$P = 76.6e^{0.0313t}$$

where t represents the year, with $t = 1$ corresponding to 1971. *(Source: U.S. Census Bureau)*

(a) Use the model to complete the table.

Year	Population
1980	
1990	
2000	
2010	

(b) According to the model, when will the population of Horry County reach 360,000?

(c) Do you think the model is valid for long-term predictions of the population? Explain.

30. Population

The table shows the mid-year populations (in millions) of five countries in 2015 and the projected populations (in millions) for the year 2025. *(Source: U.S. Census Bureau)*

Country	2015	2025
Bulgaria	7.2	6.7
Canada	35.1	37.6
China	1367.5	1407.0
United Kingdom	64.1	67.2
United States	321.4	347.3

(a) Find the exponential growth or decay model $y = ae^{bt}$ or $y = ae^{-bt}$ for the population of each country by letting $t = 15$ correspond to 2015. Use the model to predict the population of each country in 2035.

(b) You can see that the populations of the United States and the United Kingdom are growing at different rates. What constant in the equation $y = ae^{bt}$ gives the growth rate? Discuss the relationship between the different growth rates and the magnitude of the constant.

31. Website Growth The number y of hits a new website receives each month can be modeled by $y = 4080e^{kt}$, where t represents the number of months the website has been operating. In the website's third month, there were 10,000 hits. Find the value of k, and use this value to predict the number of hits the website will receive after 24 months.

32. Population The population P (in thousands) of Tallahassee, Florida, from 2000 through 2014 can be modeled by $P = 150.9e^{kt}$, where t represents the year, with $t = 0$ corresponding to 2000. In 2005, the population of Tallahassee was about 163,075. *(Source: U.S. Census Bureau)*

(a) Find the value of k. Is the population increasing or decreasing? Explain.

(b) Use the model to predict the populations of Tallahassee in 2020 and 2025. Are the results reasonable? Explain.

(c) According to the model, during what year will the population reach 200,000?

33. Bacteria Growth The number of bacteria in a culture is increasing according to the law of exponential growth. After 3 hours there are 100 bacteria, and after 5 hours there are 400 bacteria. How many bacteria will there be after 6 hours?

34. Bacteria Growth The number of bacteria in a culture is increasing according to the law of exponential growth. The initial population is 250 bacteria, and the population after 10 hours is double the population after 1 hour. How many bacteria will there be after 6 hours?

35. Depreciation A laptop computer that costs \$575 new has a book value of \$275 after 2 years.

(a) Find the linear model $V = mt + b$.

(b) Find the exponential model $V = ae^{kt}$.

(c) Use a graphing utility to graph the two models in the same viewing window. Which model depreciates faster in the first 2 years?

(d) Find the book values of the computer after 1 year and after 3 years using each model.

(e) Explain the advantages and disadvantages of using each model to a buyer and a seller.

36. Learning Curve The management at a plastics factory has found that the maximum number of units a worker can produce in a day is 30. The learning curve for the number N of units produced per day after a new employee has worked t days is modeled by $N = 30(1 - e^{kt})$. After 20 days on the job, a new employee produces 19 units.

(a) Find the learning curve for this employee. (*Hint:* First, find the value of k.)

(b) How many days does the model predict will pass before this employee is producing 25 units per day?

37. Carbon Dating The ratio of carbon-14 to carbon-12 in a piece of wood discovered in a cave is $R = 1/8^{14}$. Estimate the age of the piece of wood.

38. Carbon Dating The ratio of carbon-14 to carbon-12 in a piece of paper buried in a tomb is $R = 1/13^{11}$. Estimate the age of the piece of paper.

39. IQ Scores The IQ scores for a sample of students at a small college roughly follow the normal distribution

$$y = 0.0266e^{-(x-100)^2/450}, \quad 70 \le x \le 115$$

where x is the IQ score.

(a) Use a graphing utility to graph the function.

(b) From the graph in part (a), estimate the average IQ score of a student.

40. Education The amount of time (in hours per week) a student utilizes a math-tutoring center roughly follows the normal distribution

$$y = 0.7979e^{-(x-5.4)^2/0.5}, \quad 4 \le x \le 7$$

where x is the number of hours.

(a) Use a graphing utility to graph the function.

(b) From the graph in part (a), estimate the average number of hours per week a student uses the tutoring center.

41. Cell Sites A cell site is a site where electronic communications equipment is placed in a cellular network for the use of mobile phones. The numbers y of cell sites from 1985 through 2014 can be modeled by

$$y = \frac{320{,}110}{1 + 374e^{-0.252t}}$$

where t represents the year, with $t = 5$ corresponding to 1985. *(Source: CTIA-The Wireless Association)*

(a) Use the model to find the numbers of cell sites in the years 1998, 2003, and 2006.

(b) Use a graphing utility to graph the function.

(c) Use the graph to determine the year in which the number of cell sites reached 270,000.

(d) Confirm your answer to part (c) algebraically.

42. Population The population P (in thousands) of a city from 2000 through 2016 can be modeled by

$$P = \frac{2632}{1 + 0.083e^{0.050t}}$$

where t represents the year, with $t = 0$ corresponding to 2000.

(a) Use the model to find the populations of the city in the years 2000, 2005, 2010, and 2015.

(b) Use a graphing utility to graph the function.

(c) Use the graph to determine the year in which the population reached 2.2 million.

(d) Confirm your answer to part (c) algebraically.

43. Population Growth A conservation organization released 100 animals of an endangered species into a game preserve. The preserve has a carrying capacity of 1000 animals. The growth of the pack is modeled by the logistic curve

$$p(t) = \frac{1000}{1 + 9e^{-0.1656t}}$$

where t is measured in months (see figure).

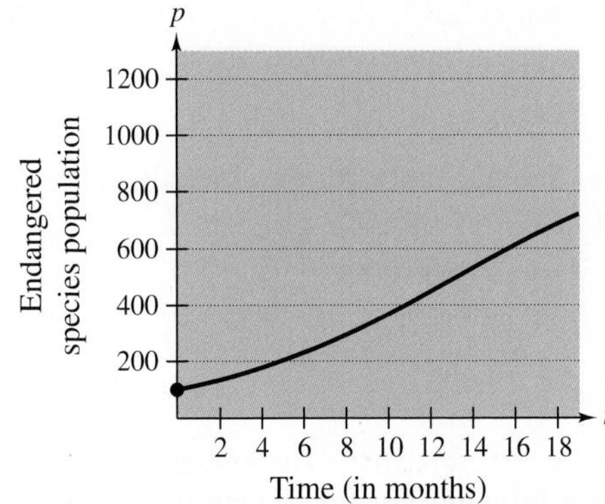

(a) Estimate the population after 5 months.

(b) After how many months is the population 500?

(c) Use a graphing utility to graph the function. Use the graph to determine the horizontal asymptotes, and interpret the meaning of the asymptotes in the context of the problem.

44. Sales After discontinuing all advertising for a tool kit in 2010, the manufacturer noted that sales began to drop according to the model

$$S = \frac{500{,}000}{1 + 0.1e^{kt}}$$

where S represents the number of units sold and t represents the year, with $t = 0$ corresponding to 2010 (see figure). In 2014, 300,000 units were sold.

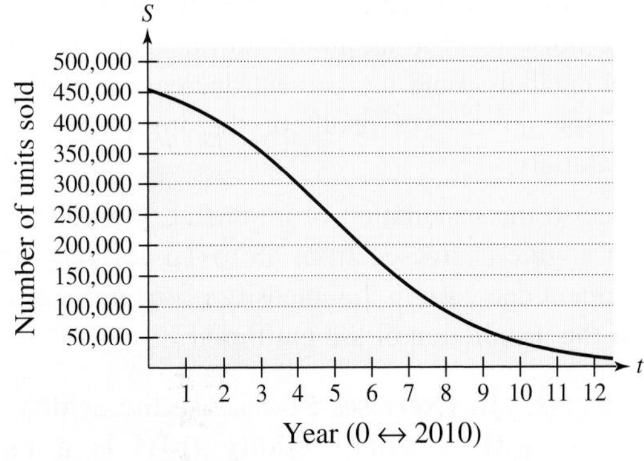

(a) Use the graph to estimate sales in 2020.

(b) Complete the model by solving for k.

(c) Use the model to estimate sales in 2020. Compare your results with that of part (a).

Geology In Exercises 45 and 46, use the Richter scale

$$R = \log \frac{I}{I_0}$$

for measuring the magnitude R of an earthquake.

45. Find the intensity I of an earthquake measuring R on the Richter scale (let $I_0 = 1$).
    (a) Peru in 2015: $R = 7.6$
    (b) Pakistan in 2015: $R = 5.6$
    (c) Indonesia in 2015: $R = 6.6$

46. Find the magnitude R of each earthquake of intensity I (let $I_0 = 1$).
    (a) $I = 199{,}500{,}000$
    (b) $I = 48{,}275{,}000$
    (c) $I = 17{,}000$

Intensity of Sound In Exercises 47–50, use the following information for determining sound intensity. The number of decibels β of a sound with an intensity of I watts per square meter is given by $\beta = 10 \log(I/I_0)$, where I_0 is an intensity of 10^{-12} watt per square meter, corresponding roughly to the faintest sound that can be heard by the human ear. In Exercises 47 and 48, find the number of decibels β of the sound.

47. (a) $I = 10^{-10}$ watt per m^2 (quiet room)
    (b) $I = 10^{-5}$ watt per m^2 (busy street corner)
    (c) $I = 10^{-8}$ watt per m^2 (quiet radio)
    (d) $I = 10^{-3}$ watt per m^2 (loud car horn)

48. (a) $I = 10^{-11}$ watt per m^2 (rustle of leaves)
    (b) $I = 10^{2}$ watt per m^2 (jet at 30 meters)
    (c) $I = 10^{-4}$ watt per m^2 (door slamming)
    (d) $I = 10^{-6}$ watt per m^2 (normal conversation)

49. Due to the installation of noise suppression materials, the noise level in an auditorium decreased from 93 to 80 decibels. Find the percent decrease in the intensity of the noise as a result of the installation of these materials.

50. Due to the installation of a muffler, the noise level of an engine decreased from 88 to 72 decibels. Find the percent decrease in the intensity of the noise as a result of the installation of the muffler.

pH Levels In Exercises 51–56, use the acidity model $\text{pH} = -\log[\text{H}^+]$, where acidity (pH) is a measure of the hydrogen ion concentration $[\text{H}^+]$ (measured in moles of hydrogen per liter) of a solution.

51. Find the pH when $[\text{H}^+] = 2.3 \times 10^{-5}$.

52. Find the pH when $[\text{H}^+] = 1.13 \times 10^{-5}$.

53. Compute $[\text{H}^+]$ for a solution in which $\text{pH} = 5.8$.

54. Compute $[\text{H}^+]$ for a solution in which $\text{pH} = 3.2$.

55. Apple juice has a pH of 2.9 and drinking water has a pH of 8.0. The hydrogen ion concentration of the apple juice is how many times the concentration of drinking water?

56. The pH of a solution decreases by one unit. By what factor does the hydrogen ion concentration increase?

57. **Forensics** At 8:30 A.M., a coroner went to the home of a person who had died during the night. In order to estimate the time of death, the coroner took the person's temperature twice. At 9:00 A.M. the temperature was 85.7°F, and at 11:00 A.M. the temperature was 82.8°F. From these two temperatures, the coroner was able to determine that the time elapsed since death and the body temperature were related by the formula

$$t = -10 \ln \frac{T - 70}{98.6 - 70}$$

where t is the time in hours elapsed since the person died and T is the temperature (in degrees Fahrenheit) of the person's body. (This formula comes from a general cooling principle called *Newton's Law of Cooling*. It uses the assumptions that the person had a normal body temperature of 98.6°F at death and that the room temperature was a constant 70°F.) Use the formula to estimate the time of death of the person.

58. **Home Mortgage** A \$120,000 home mortgage for 30 years at $7\frac{1}{2}\%$ has a monthly payment of \$839.06. Part of the monthly payment covers the interest charge on the unpaid balance, and the remainder of the payment reduces the principal. The amount paid toward the interest is

$$u = M - \left(M - \frac{Pr}{12}\right)\left(1 + \frac{r}{12}\right)^{12t}$$

and the amount paid toward the reduction of the principal is

$$v = \left(M - \frac{Pr}{12}\right)\left(1 + \frac{r}{12}\right)^{12t}.$$

In these formulas, P is the amount of the mortgage, r is the interest rate (in decimal form), M is the monthly payment, and t is the time in years.

(a) Use a graphing utility to graph each function in the same viewing window. (The viewing window should show all 30 years of mortgage payments.)
(b) In the early years of the mortgage, is the greater part of the monthly payment paid toward the interest or the principal? Approximate the time when the monthly payment is evenly divided between interest and principal reduction.
(c) Repeat parts (a) and (b) for a repayment period of 20 years ($M = \$966.71$). What can you conclude?

59. Home Mortgage The total interest u paid on a home mortgage of P dollars at interest rate r (in decimal form) for t years is

$$u = P\left[\frac{rt}{1 - \left(\frac{1}{1 + r/12}\right)^{12t}} - 1\right].$$

Consider a \$120,000 home mortgage at $7\frac{1}{2}\%$.

(a) Use a graphing utility to graph the total interest function.

(b) Approximate the length of the mortgage for which the total interest paid is the same as the size of the mortgage. Is it possible that some people are paying twice as much in interest charges as the size of the mortgage?

60. Car Speed The table shows the time t (in seconds) required for a car to attain a speed of s miles per hour from a standing start.

DATA

Speed, s	Time, t
30	3.4
40	5.0
50	7.0
60	9.3
70	12.0
80	15.8
90	20.0

Spreadsheet at LarsonPrecalculus.com

Two models for these data are given below.

$$t_1 = 40.757 + 0.556s - 15.817 \ln s$$

$$t_2 = 1.2259 + 0.0023s^2$$

(a) Use the *regression* feature of a graphing utility to find a linear model t_3 and an exponential model t_4 for the data.

(b) Use the graphing utility to graph the data and each model in the same viewing window.

(c) Create a table comparing the data with estimates obtained from each model.

(d) Use the results of part (c) to find the sum of the absolute values of the differences between the data and the estimated values found using each model. Based on the four sums, which model do you think best fits the data? Explain.

Exploration

True or False? **In Exercises 61–64, determine whether the statement is true or false. Justify your answer.**

61. The domain of a logistic growth function cannot be the set of real numbers.

62. A logistic growth function will always have an x-intercept.

63. The graph of $f(x) = \dfrac{4}{1 + 6e^{-2x}} + 5$ is the graph of $g(x) = \dfrac{4}{1 + 6e^{-2x}}$ shifted to the right five units.

64. The graph of a Gaussian model will never have an x-intercept.

65. Writing Use your school's library, the Internet, or some other reference source to write a paper describing John Napier's work with logarithms.

66. HOW DO YOU SEE IT? Identify each model as exponential growth, exponential decay, Gaussian, linear, logarithmic, logistic growth, quadratic, or none of the above. Explain your reasoning.

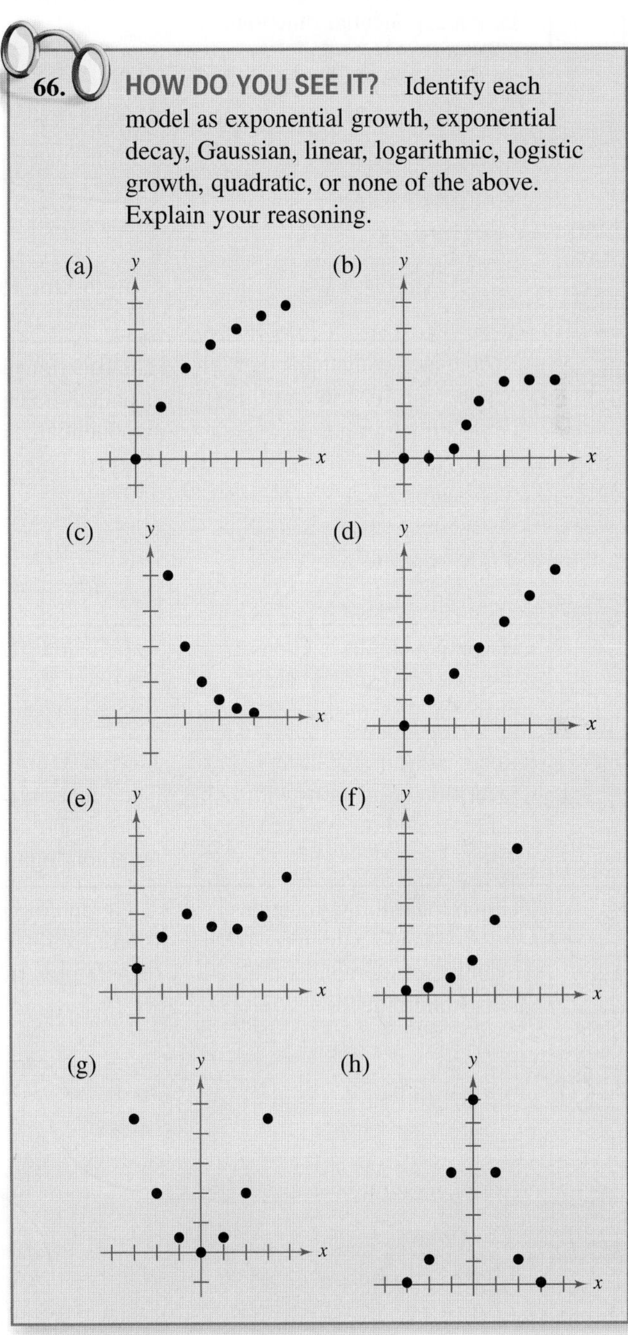

Project: Sales per Share To work an extended application analyzing the sales per share for Kohl's Corporation from 1999 through 2014, visit this text's website at *LarsonPrecalculus.com.* *(Source: Kohl's Corporation)*

Chapter Summary

	What Did You Learn?	Explanation/Examples	Review Exercises
Section 5.1	Recognize and evaluate exponential functions with base *a* *(p. 358)*.	The exponential function f with base a is denoted by $f(x) = a^x$, where $a > 0$, $a \neq 1$, and x is any real number.	1–6
	Graph exponential functions and use a One-to-One Property *(p. 359)*.	**One-to-One Property:** For $a > 0$ and $a \neq 1$, $a^x = a^y$ if and only if $x = y$.	7–20
	Recognize, evaluate, and graph exponential functions with base *e* *(p. 362)*.	The function $f(x) = e^x$ is called the natural exponential function.	21–28
	Use exponential functions to model and solve real-life problems *(p. 363)*.	Exponential functions are used in compound interest formulas (see Example 8) and in radioactive decay models (see Example 9).	29–32
Section 5.2	Recognize and evaluate logarithmic functions with base *a* *(p. 369)*.	For $x > 0$, $a > 0$, and $a \neq 1$, $y = \log_a x$ if and only if $x = a^y$. The function $f(x) = \log_a x$ is the logarithmic function with base a. The logarithmic function with base 10 is called the common logarithmic function. It is denoted by $\log_{10}$ or log.	33–44
	Graph logarithmic functions *(p. 371)*, and recognize, evaluate, and graph natural logarithmic functions *(p. 373)*.	The graph of $g(x) = \log_a x$ is a reflection of the graph of $f(x) = a^x$ in the line $y = x$. The function $g(x) = \ln x$, $x > 0$, is called the natural logarithmic function. Its graph is a reflection of the graph of $f(x) = e^x$ in the line $y = x$.	45–56
	Use logarithmic functions to model and solve real-life problems *(p. 375)*.	A logarithmic function can model human memory. (See Example 11.)	57, 58

	What Did You Learn?	Explanation/Examples	Review Exercises
Section 5.3	Use the change-of-base formula to rewrite and evaluate logarithmic expressions *(p. 379)*.	Let a, b, and x be positive real numbers such that $a \neq 1$ and $b \neq 1$. Then $\log_a x$ can be converted to a different base as follows. **Base b**: $\log_a x = \dfrac{\log_b x}{\log_b a}$ **Base 10**: $\log_a x = \dfrac{\log x}{\log a}$ **Base e**: $\log_a x = \dfrac{\ln x}{\ln a}$	59–62
	Use properties of logarithms to evaluate, rewrite, expand, or condense logarithmic expressions *(pp. 380–381)*.	Let a be a positive number such that $a \neq 1$, let n be a real number, and let u and v be positive real numbers. **1. Product Property:** $\log_a(uv) = \log_a u + \log_a v$ $\ln(uv) = \ln u + \ln v$ **2. Quotient Property:** $\log_a(u/v) = \log_a u - \log_a v$ $\ln(u/v) = \ln u - \ln v$ **3. Power Property:** $\log_a u^n = n \log_a u$, $\ln u^n = n \ln u$	63–78
	Use logarithmic functions to model and solve real-life problems *(p. 382)*.	Logarithmic functions can help you find an equation that relates the periods of several planets and their distances from the sun. (See Example 7.)	79, 80
Section 5.4	Solve simple exponential and logarithmic equations *(p. 386)*.	One-to-One Properties and Inverse Properties of exponential or logarithmic functions are used to solve exponential or logarithmic equations.	81–86
	Solve more complicated exponential equations *(p. 387)* and logarithmic equations *(p. 389)*.	To solve more complicated equations, rewrite the equations to allow the use of the One-to-One Properties or Inverse Properties of exponential or logarithmic functions. (See Examples 2–9.)	87–102
	Use exponential and logarithmic equations to model and solve real-life problems *(p. 391)*.	Exponential and logarithmic equations can help you determine how long it will take to double an investment (see Example 10) and find the year in which an industry had a given amount of sales (see Example 11).	103, 104
Section 5.5	Recognize the five most common types of models involving exponential and logarithmic functions *(p. 396)*.	**1. Exponential growth model:** $y = ae^{bx}$, $b > 0$ **2. Exponential decay model:** $y = ae^{-bx}$, $b > 0$ **3. Gaussian model:** $y = ae^{-(x-b)^2/c}$ **4. Logistic growth model:** $y = \dfrac{a}{1 + be^{-rx}}$ **5. Logarithmic models:** $y = a + b \ln x$, $y = a + b \log x$	105–110
	Use exponential growth and decay functions to model and solve real-life problems *(p. 397)*.	An exponential growth function can help you model a population of fruit flies (see Example 2), and an exponential decay function can help you estimate the age of a fossil (see Example 3).	111, 112
	Use Gaussian functions *(p. 400)*, logistic growth functions *(p. 401)*, and logarithmic functions *(p. 402)* to model and solve real-life problems.	A Gaussian function can help you model SAT mathematics scores for college-bound seniors. (See Example 4.) A logistic growth function can help you model the spread of a flu virus. (See Example 5.) A logarithmic function can help you find the intensity of an earthquake given its magnitude. (See Example 6.)	113–115

Review Exercises

See CalcChat.com for tutorial help and worked-out solutions to odd-numbered exercises.

5.1 Evaluating an Exponential Function In Exercises 1–6, evaluate the function at the given value of x. Round your result to three decimal places.

1. $f(x) = 0.3^x, \quad x = 1.5$
2. $f(x) = 30^x, \quad x = \sqrt{3}$
3. $f(x) = 2^x, \quad x = \frac{2}{3}$
4. $f(x) = \left(\frac{1}{2}\right)^{2x}, \quad x = \pi$
5. $f(x) = 7(0.2^x), \quad x = -\sqrt{11}$
6. $f(x) = -14(5^x), \quad x = -0.8$

Graphing an Exponential Function In Exercises 7–12, use a graphing utility to construct a table of values for the function. Then sketch the graph of the function.

7. $f(x) = 4^{-x} + 4$
8. $f(x) = 2.65^{x-1}$
9. $f(x) = 5^{x-2} + 4$
10. $f(x) = 2^{x-6} - 5$
11. $f(x) = \left(\frac{1}{2}\right)^{-x} + 3$
12. $f(x) = \left(\frac{1}{8}\right)^{x+2} - 5$

Using a One-to-One Property In Exercises 13–16, use a One-to-One Property to solve the equation for x.

13. $\left(\frac{1}{3}\right)^{x-3} = 9$
14. $3^{x+3} = \frac{1}{81}$
15. $e^{3x-5} = e^7$
16. $e^{8-2x} = e^{-3}$

Transforming the Graph of an Exponential Function In Exercises 17–20, describe the transformation of the graph of f that yields the graph of g.

17. $f(x) = 5^x, \quad g(x) = 5^x + 1$
18. $f(x) = 6^x, \quad g(x) = 6^{x+1}$
19. $f(x) = 3^x, \quad g(x) = 1 - 3^x$
20. $f(x) = \left(\frac{1}{2}\right)^x, \quad g(x) = -\left(\frac{1}{2}\right)^{x+2}$

Evaluating the Natural Exponential Function In Exercises 21–24, evaluate $f(x) = e^x$ at the given value of x. Round your result to three decimal places.

21. $x = 3.4$
22. $x = -2.5$
23. $x = \frac{3}{5}$
24. $x = \frac{2}{7}$

Graphing a Natural Exponential Function In Exercises 25–28, use a graphing utility to construct a table of values for the function. Then sketch the graph of the function.

25. $h(x) = e^{-x/2}$
26. $h(x) = 2 - e^{-x/2}$
27. $f(x) = e^{x+2}$
28. $s(t) = 4e^{t-1}$

29. **Waiting Times** The average time between new posts on a message board is 3 minutes. The probability F of waiting less than t minutes until the next post is approximated by the model $F(t) = 1 - e^{-t/3}$. A message has just been posted. Find the probability that the next post will be within (a) 1 minute, (b) 2 minutes, and (c) 5 minutes.

30. **Depreciation** After t years, the value V of a car that originally cost \$23,970 is given by $V(t) = 23{,}970\left(\frac{3}{4}\right)^t$.
    (a) Use a graphing utility to graph the function.
    (b) Find the value of the car 2 years after it was purchased.
    (c) According to the model, when does the car depreciate most rapidly? Is this realistic? Explain.
    (d) According to the model, when will the car have no value?

Compound Interest In Exercises 31 and 32, complete the table by finding the balance A when P dollars is invested at rate r for t years and compounded n times per year.

n	1	2	4	12	365	Continuous
A						

31. $P = \$5000, \quad r = 3\%, \quad t = 10$ years
32. $P = \$4500, \quad r = 2.5\%, \quad t = 30$ years

5.2 Writing a Logarithmic Equation In Exercises 33–36, write the exponential equation in logarithmic form. For example, the logarithmic form of $2^3 = 8$ is $\log_2 8 = 3$.

33. $3^3 = 27$
34. $25^{3/2} = 125$
35. $e^{0.8} = 2.2255 \ldots$
36. $e^0 = 1$

Evaluating a Logarithm In Exercises 37–40, evaluate the logarithm at the given value of x without using a calculator.

37. $f(x) = \log x, \quad x = 1000$
38. $g(x) = \log_9 x, \quad x = 3$
39. $g(x) = \log_2 x, \quad x = \frac{1}{4}$
40. $f(x) = \log_3 x, \quad x = \frac{1}{81}$

Using a One-to-One Property In Exercises 41–44, use a One-to-One Property to solve the equation for x.

41. $\log_4(x + 7) = \log_4 14$
42. $\log_8(3x - 10) = \log_8 5$
43. $\ln(x + 9) = \ln 4$
44. $\log(3x - 2) = \log 7$

Sketching the Graph of a Logarithmic Function In Exercises 45–48, find the domain, x-intercept, and vertical asymptote of the logarithmic function and sketch its graph.

45. $g(x) = \log_7 x$
46. $f(x) = \log \dfrac{x}{3}$
47. $f(x) = 4 - \log(x + 5)$
48. $f(x) = \log(x - 3) + 1$

Evaluating a Logarithmic Function **In Exercises 49–52, use a calculator to evaluate the function at the given value of x. Round your result to three decimal places, if necessary.**

49. $f(x) = \ln x,\ x = 22.6$ **50.** $f(x) = \ln x,\ x = e^{-12}$

51. $f(x) = \frac{1}{2}\ln x,\ x = \sqrt{e}$

52. $f(x) = 5 \ln x,\ x = 0.98$

Graphing a Natural Logarithmic Function **In Exercises 53–56, find the domain, x-intercept, and vertical asymptote of the logarithmic function and sketch its graph.**

53. $f(x) = \ln x + 6$ **54.** $f(x) = \ln x - 5$

55. $h(x) = \ln(x - 6)$ **56.** $f(x) = \ln(x + 4)$

57. Astronomy The formula $M = m - 5\log(d/10)$ gives the distance d (in parsecs) from Earth to a star with apparent magnitude m and absolute magnitude M. The star Rasalhague has an apparent magnitude of 2.08 and an absolute magnitude of 1.3. Find the distance from Earth to Rasalhague.

58. Snow Removal The number of miles s of roads cleared of snow is approximated by the model

$$s = 25 - \frac{13\ln(h/12)}{\ln 3}, \quad 2 \le h \le 15$$

where h is the depth (in inches) of the snow. Use this model to find s when $h = 10$ inches.

5.3 Using the Change-of-Base Formula **In Exercises 59–62, evaluate the logarithm using the change-of-base formula (a) with common logarithms and (b) with natural logarithms. Round your results to three decimal places.**

59. $\log_2 6$ **60.** $\log_{12} 200$

61. $\log_{1/2} 5$ **62.** $\log_4 0.75$

Using Properties of Logarithms **In Exercises 63–66, use the properties of logarithms to write the logarithm in terms of $\log_2 3$ and $\log_2 5$.**

63. $\log_2 \frac{5}{3}$ **64.** $\log_2 45$

65. $\log_2 \frac{9}{5}$ **66.** $\log_2 \frac{20}{9}$

Expanding a Logarithmic Expression **In Exercises 67–72, use the properties of logarithms to expand the expression as a sum, difference, and/or constant multiple of logarithms. (Assume all variables are positive.)**

67. $\log 7x^2$ **68.** $\log 11x^3$

69. $\log_3 \dfrac{9}{\sqrt{x}}$ **70.** $\log_7 \dfrac{\sqrt[3]{x}}{19}$

71. $\ln x^2y^2z$ **72.** $\ln\left(\dfrac{y-1}{3}\right)^2,\ y > 1$

Condensing a Logarithmic Expression **In Exercises 73–78, condense the expression to the logarithm of a single quantity.**

73. $\ln 7 + \ln x$

74. $\log_2 y - \log_2 3$

75. $\log x - \frac{1}{2}\log y$

76. $3\ln x + 2\ln(x + 1)$

77. $\frac{1}{2}\log_3 x - 2\log_3(y + 8)$

78. $5\ln(x - 2) - \ln(x + 2) - 3\ln x$

79. Climb Rate The time t (in minutes) for a small plane to climb to an altitude of h feet is modeled by

$$t = 50\log[18{,}000/(18{,}000 - h)]$$

where 18,000 feet is the plane's absolute ceiling.

(a) Determine the domain of the function in the context of the problem.

(b) Use a graphing utility to graph the function and identify any asymptotes.

(c) As the plane approaches its absolute ceiling, what can be said about the time required to increase its altitude?

(d) Find the time it takes for the plane to climb to an altitude of 4000 feet.

80. Human Memory Model Students in a learning theory study took an exam and then retested monthly for 6 months with an equivalent exam. The data obtained in the study are given by the ordered pairs (t, s), where t is the time (in months) after the initial exam and s is the average score for the class. Use the data to find a logarithmic equation that relates t and s.

(1, 84.2), (2, 78.4), (3, 72.1),
(4, 68.5), (5, 67.1), (6, 65.3)

5.4 Solving a Simple Equation **In Exercises 81–86, solve for x.**

81. $5^x = 125$

82. $6^x = \frac{1}{216}$

83. $e^x = 3$

84. $\log x - \log 5 = 0$

85. $\ln x = 4$

86. $\ln x = -1.6$

Solving an Exponential Equation **In Exercises 87–90, solve the exponential equation algebraically. Approximate the result to three decimal places.**

87. $e^{4x} = e^{x^2+3}$

88. $e^{3x} = 25$

89. $2^x - 3 = 29$

90. $e^{2x} - 6e^x + 8 = 0$

Solving a Logarithmic Equation **In Exercises 91–98, solve the logarithmic equation algebraically. Approximate the result to three decimal places.**

91. $\ln 3x = 8.2$

92. $4 \ln 3x = 15$

93. $\ln x + \ln(x - 3) = 1$

94. $\ln(x + 2) - \ln x = 2$

95. $\log_8(x - 1) = \log_8(x - 2) - \log_8(x + 2)$

96. $\log_6(x + 2) - \log_6 x = \log_6(x + 5)$

97. $\log(1 - x) = -1$

98. $\log(-x - 4) = 2$

Using Technology **In Exercises 99–102, use a graphing utility to graphically solve the equation. Approximate the result to three decimal places. Verify your result algebraically.**

99. $25e^{-0.3x} = 12$

100. $2 = 5 - e^{x+7}$

101. $2 \ln(x + 3) - 3 = 0$

102. $2 \ln x - \ln(3x - 1) = 0$

103. Compound Interest You deposit \$8500 in an account that pays 1.5% interest, compounded continuously. How long will it take for the money to triple?

104. Meteorology The speed of the wind S (in miles per hour) near the center of a tornado and the distance d (in miles) the tornado travels are related by the model $S = 93 \log d + 65$. On March 18, 1925, a large tornado struck portions of Missouri, Illinois, and Indiana with a wind speed at the center of about 283 miles per hour. Approximate the distance traveled by this tornado.

5.5 Matching a Function with Its Graph **In Exercises 105–110, match the function with its graph. [The graphs are labeled (a), (b), (c), (d), (e), and (f).]**

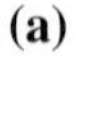

(a)

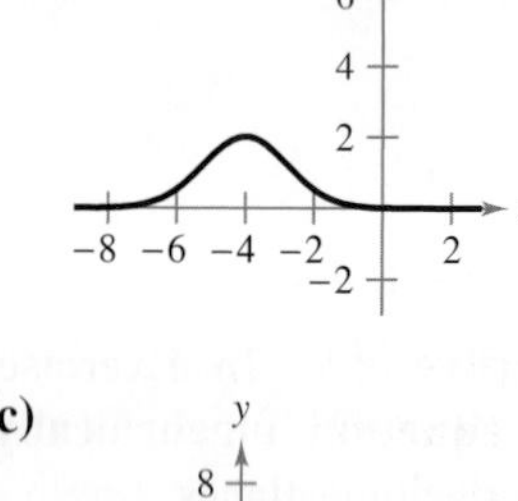

(b)

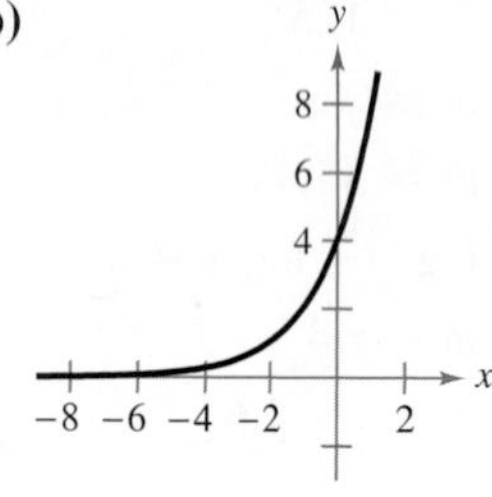

(c)

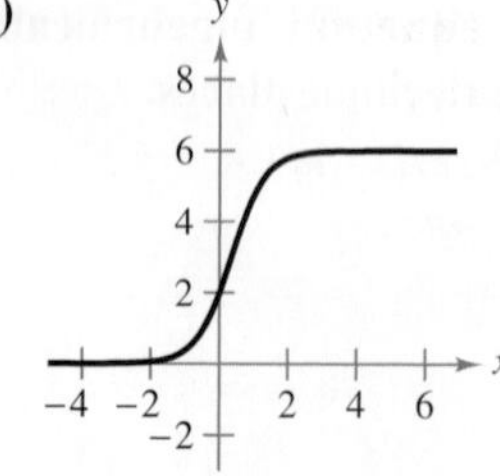

(d)

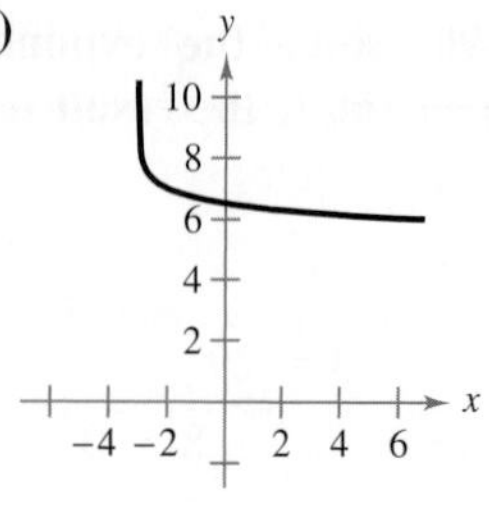

(e)

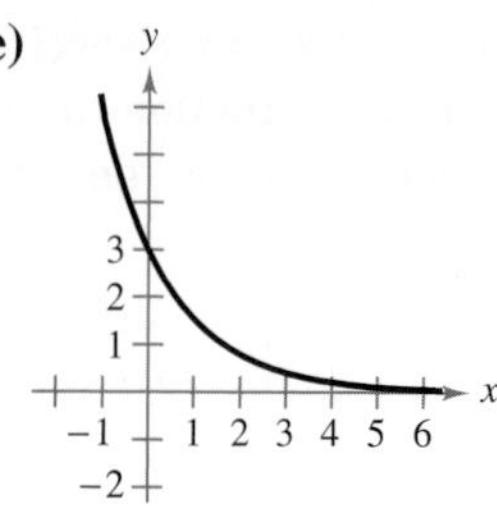

(f)

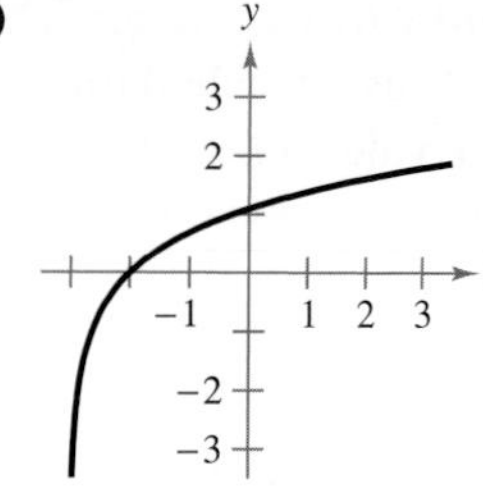

105. $y = 3e^{-2x/3}$

106. $y = 4e^{2x/3}$

107. $y = \ln(x + 3)$

108. $y = 7 - \log(x + 3)$

109. $y = 2e^{-(x+4)^2/3}$

110. $y = \dfrac{6}{1 + 2e^{-2x}}$

111. Finding an Exponential Model Find the exponential model $y = ae^{bx}$ that fits the points $(0, 2)$ and $(4, 3)$.

112. Wildlife Population A species of bat is in danger of becoming extinct. Five years ago, the total population of the species was 2000. Two years ago, the total population of the species was 1400. What was the total population of the species one year ago?

113. Test Scores The test scores for a biology test follow the normal distribution

$$y = 0.0499e^{-(x-71)^2/128}, \quad 40 \le x \le 100$$

where x is the test score. Use a graphing utility to graph the equation and estimate the average test score.

114. Typing Speed In a typing class, the average number N of words per minute typed after t weeks of lessons is

$$N = 157/(1 + 5.4e^{-0.12t}).$$

Find the time necessary to type (a) 50 words per minute and (b) 75 words per minute.

115. Sound Intensity The relationship between the number of decibels β and the intensity of a sound I (in watts per square meter) is

$$\beta = 10 \log(I/10^{-12}).$$

Find the intensity I for each decibel level β.

(a) $\beta = 60$ (b) $\beta = 135$ (c) $\beta = 1$

Exploration

116. Graph of an Exponential Function Consider the graph of $y = e^{kt}$. Describe the characteristics of the graph when k is positive and when k is negative.

True or False? **In Exercises 117 and 118, determine whether the equation is true or false. Justify your answer.**

117. $\log_b b^{2x} = 2x$

118. $\ln(x + y) = \ln x + \ln y$

Chapter Test

See CalcChat.com for tutorial help and worked-out solutions to odd-numbered exercises.

Take this test as you would take a test in class. When you are finished, check your work against the answers given in the back of the book.

In Exercises 1–4, evaluate the expression. Round your result to three decimal places.

1. $0.7^{2.5}$ **2.** $3^{-\pi}$ **3.** $e^{-7/10}$ **4.** $e^{3.1}$

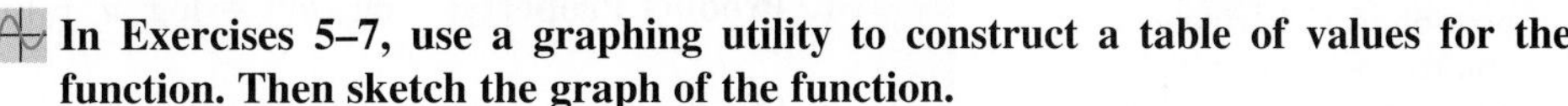

In Exercises 5–7, use a graphing utility to construct a table of values for the function. Then sketch the graph of the function.

5. $f(x) = 10^{-x}$ **6.** $f(x) = -6^{x-2}$ **7.** $f(x) = 1 - e^{2x}$

8. Evaluate (a) $\log_7 7^{-0.89}$ and (b) $4.6 \ln e^2$.

In Exercises 9–11, find the domain, *x*-intercept, and vertical asymptote of the logarithmic function and sketch its graph.

9. $f(x) = 4 + \log x$ **10.** $f(x) = \ln(x - 4)$ **11.** $f(x) = 1 + \ln(x + 6)$

In Exercises 12–14, evaluate the logarithm using the change-of-base formula. Round your result to three decimal places.

12. $\log_5 35$ **13.** $\log_{16} 0.63$ **14.** $\log_{3/4} 24$

In Exercises 15–17, use the properties of logarithms to expand the expression as a sum, difference, and/or constant multiple of logarithms. (Assume all variables are positive.)

15. $\log_2 3a^4$ **16.** $\ln \dfrac{\sqrt{x}}{7}$ **17.** $\log \dfrac{10x^2}{y^3}$

In Exercises 18–20, condense the expression to the logarithm of a single quantity.

18. $\log_3 13 + \log_3 y$ **19.** $4 \ln x - 4 \ln y$

20. $3 \ln x - \ln(x + 3) + 2 \ln y$

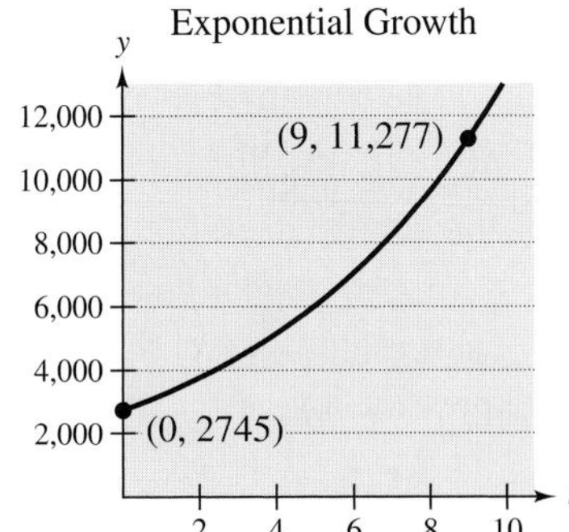

Figure for 27

In Exercises 21–26, solve the equation algebraically. Approximate the result to three decimal places, if necessary.

21. $5^x = \dfrac{1}{25}$ **22.** $3e^{-5x} = 132$

23. $\dfrac{1025}{8 + e^{4x}} = 5$ **24.** $\ln x = \dfrac{1}{2}$

25. $18 + 4 \ln x = 7$ **26.** $\log x + \log(x - 15) = 2$

27. Find the exponential growth model that fits the points shown in the graph.

28. The half-life of radioactive actinium (^{227}Ac) is 21.77 years. What percent of a present amount of radioactive actinium will remain after 19 years?

29. A model that can predict a child's height H (in centimeters) based on the child's age is $H = 70.228 + 5.104x + 9.222 \ln x$, $\frac{1}{4} \le x \le 6$, where x is the child's age in years. *(Source: Snapshots of Applications in Mathematics)*

(a) Construct a table of values for the model. Then sketch the graph of the model.

(b) Use the graph from part (a) to predict the height of a four-year-old child. Then confirm your prediction algebraically.

Proofs in Mathematics

Each of the three properties of logarithms listed below can be proved by using properties of exponential functions.

SLIDE RULES

William Oughtred (1574–1660) is credited with inventing the slide rule. The slide rule is a computational device with a sliding portion and a fixed portion. A slide rule enables you to perform multiplication by using the Product Property of Logarithms. There are other slide rules that allow for the calculation of roots and trigonometric functions. Mathematicians and engineers used slide rules until the hand-held calculator came into widespread use in the 1970s.

Properties of Logarithms *(p. 380)*

Let a be a positive number such that $a \neq 1$, let n be a real number, and let u and v be positive real numbers.

	Logarithm with Base a	Natural Logarithm
1. Product Property:	$\log_a(uv) = \log_a u + \log_a v$	$\ln(uv) = \ln u + \ln v$
2. Quotient Property:	$\log_a \frac{u}{v} = \log_a u - \log_a v$	$\ln \frac{u}{v} = \ln u - \ln v$
3. Power Property:	$\log_a u^n = n \log_a u$	$\ln u^n = n \ln u$

Proof

Let

$$x = \log_a u \quad \text{and} \quad y = \log_a v.$$

The corresponding exponential forms of these two equations are

$$a^x = u \quad \text{and} \quad a^y = v.$$

To prove the Product Property, multiply u and v to obtain

$$\begin{aligned} uv &= a^x a^y \\ &= a^{x+y}. \end{aligned}$$

The corresponding logarithmic form of $uv = a^{x+y}$ is $\log_a(uv) = x + y$. So,

$$\log_a(uv) = \log_a u + \log_a v.$$

To prove the Quotient Property, divide u by v to obtain

$$\begin{aligned} \frac{u}{v} &= \frac{a^x}{a^y} \\ &= a^{x-y}. \end{aligned}$$

The corresponding logarithmic form of $\frac{u}{v} = a^{x-y}$ is $\log_a \frac{u}{v} = x - y$. So,

$$\log_a \frac{u}{v} = \log_a u - \log_a v.$$

To prove the Power Property, substitute a^x for u in the expression $\log_a u^n$.

$$\begin{aligned} \log_a u^n &= \log_a(a^x)^n && \text{Substitute } a^x \text{ for } u. \\ &= \log_a a^{nx} && \text{Property of Exponents} \\ &= nx && \text{Inverse Property} \\ &= n \log_a u && \text{Substitute } \log_a u \text{ for } x. \end{aligned}$$

So, $\log_a u^n = n \log_a u$. ■

P.S. Problem Solving

1. **Graphical Reasoning** Graph the exponential function $y = a^x$ for $a = 0.5, 1.2$, and 2.0. Which of these curves intersects the line $y = x$? Determine all positive numbers a for which the curve $y = a^x$ intersects the line $y = x$.

2. **Graphical Reasoning** Use a graphing utility to graph each of the functions $y_1 = e^x$, $y_2 = x^2$, $y_3 = x^3$, $y_4 = \sqrt{x}$, and $y_5 = |x|$. Which function increases at the greatest rate as x approaches ∞?

3. **Conjecture** Use the result of Exercise 2 to make a conjecture about the rate of growth of $y_1 = e^x$ and $y = x^n$, where n is a natural number and x approaches ∞.

4. **Implication of "Growing Exponentially"** Use the results of Exercises 2 and 3 to describe what is implied when it is stated that a quantity is growing exponentially.

5. **Exponential Function** Given the exponential function

$$f(x) = a^x$$

show that

(a) $f(u + v) = f(u) \cdot f(v)$ and (b) $f(2x) = [f(x)]^2$.

6. **Hyperbolic Functions** Given that

$$f(x) = \frac{e^x + e^{-x}}{2} \quad \text{and} \quad g(x) = \frac{e^x - e^{-x}}{2}$$

show that

$$[f(x)]^2 - [g(x)]^2 = 1.$$

7. **Graphical Reasoning** Use a graphing utility to compare the graph of the function $y = e^x$ with the graph of each function. [$n!$ (read "n factorial") is defined as $n! = 1 \cdot 2 \cdot 3 \cdot \cdot \cdot (n - 1) \cdot n$.]

(a) $y_1 = 1 + \dfrac{x}{1!}$

(b) $y_2 = 1 + \dfrac{x}{1!} + \dfrac{x^2}{2!}$

(c) $y_3 = 1 + \dfrac{x}{1!} + \dfrac{x^2}{2!} + \dfrac{x^3}{3!}$

8. **Identifying a Pattern** Identify the pattern of successive polynomials given in Exercise 7. Extend the pattern one more term and compare the graph of the resulting polynomial function with the graph of $y = e^x$. What do you think this pattern implies?

9. **Finding an Inverse Function** Graph the function

$$f(x) = e^x - e^{-x}.$$

From the graph, the function appears to be one-to-one. Assume that f has an inverse function and find $f^{-1}(x)$.

10. **Finding a Pattern for an Inverse Function** Find a pattern for $f^{-1}(x)$ when

$$f(x) = \frac{a^x + 1}{a^x - 1}$$

where $a > 0$, $a \neq 1$.

11. **Determining the Equation of a Graph** Determine whether the graph represents equation (a), (b), or (c). Explain your reasoning.

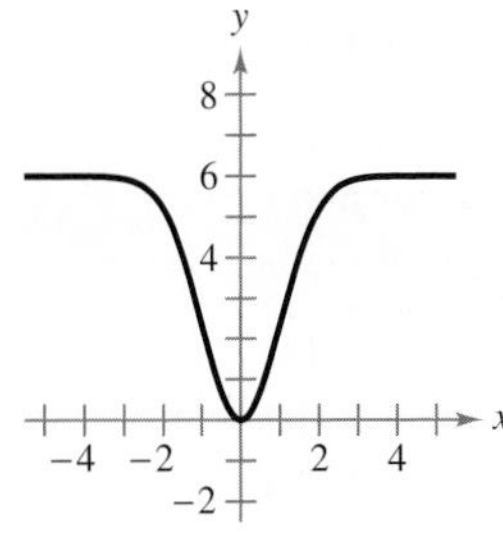

(a) $y = 6e^{-x^2/2}$

(b) $y = \dfrac{6}{1 + e^{-x/2}}$

(c) $y = 6(1 - e^{-x^2/2})$

12. **Simple and Compound Interest** You have two options for investing \$500. The first earns 7% interest compounded annually, and the second earns 7% simple interest. The figure shows the growth of each investment over a 30-year period.

(a) Determine which graph represents each type of investment. Explain your reasoning.

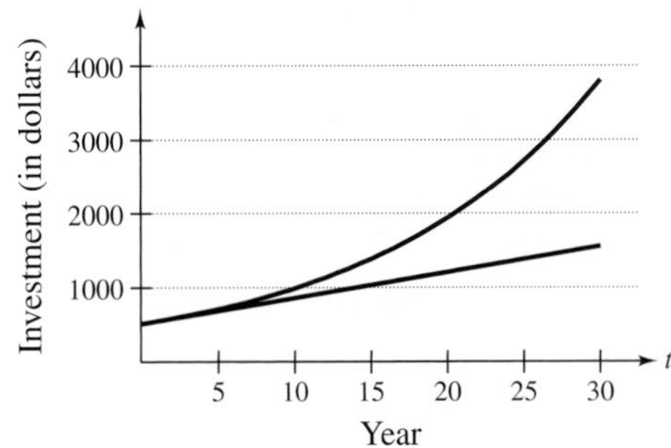

(b) Verify your answer in part (a) by finding the equations that model the investment growth and by graphing the models.

(c) Which option would you choose? Explain.

13. **Radioactive Decay** Two different samples of radioactive isotopes are decaying. The isotopes have initial amounts of c_1 and c_2 and half-lives of k_1 and k_2, respectively. Find an expression for the time t required for the samples to decay to equal amounts.

14. Bacteria Decay A lab culture initially contains 500 bacteria. Two hours later, the number of bacteria decreases to 200. Find the exponential decay model of the form

$$B = B_0a^{kt}$$

that approximates the number of bacteria B in the culture after t hours.

15. Colonial Population The table shows the colonial population estimates of the American colonies for each decade from 1700 through 1780. *(Source: U.S. Census Bureau)*

Year	Population
1700	250,900
1710	331,700
1720	466,200
1730	629,400
1740	905,600
1750	1,170,800
1760	1,593,600
1770	2,148,100
1780	2,780,400

DATA Spreadsheet at LarsonPrecalculus.com

Let y represent the population in the year t, with $t = 0$ corresponding to 1700.

(a) Use the *regression* feature of a graphing utility to find an exponential model for the data.

(b) Use the *regression* feature of the graphing utility to find a quadratic model for the data.

(c) Use the graphing utility to plot the data and the models from parts (a) and (b) in the same viewing window.

(d) Which model is a better fit for the data? Would you use this model to predict the population of the United States in 2020? Explain your reasoning.

16. Ratio of Logarithms Show that

$$\frac{\log_a x}{\log_{a/b} x} = 1 + \log_a \frac{1}{b}.$$

17. Solving a Logarithmic Equation Solve

$$(\ln x)^2 = \ln x^2.$$

18. Graphical Reasoning Use a graphing utility to compare the graph of each function with the graph of $y = \ln x$.

(a) $y_1 = x - 1$

(b) $y_2 = (x - 1) - \frac{1}{2}(x - 1)^2$

(c) $y_3 = (x - 1) - \frac{1}{2}(x - 1)^2 + \frac{1}{3}(x - 1)^3$

19. Identifying a Pattern Identify the pattern of successive polynomials given in Exercise 18. Extend the pattern one more term and compare the graph of the resulting polynomial function with the graph of $y = \ln x$. What do you think the pattern implies?

20. Finding Slope and y-Intercept Take the natural log of each side of each equation below.

$$y = ab^x, \quad y = ax^b$$

(a) What are the slope and y-intercept of the line relating x and $\ln y$ for $y = ab^x$?

(b) What are the slope and y-intercept of the line relating $\ln x$ and $\ln y$ for $y = ax^b$?

Ventilation Rate In Exercises 21 and 22, use the model

$$y = 80.4 - 11 \ln x, \quad 100 \le x \le 1500$$

which approximates the minimum required ventilation rate in terms of the air space per child in a public school classroom. In the model, x is the air space (in cubic feet) per child and y is the ventilation rate (in cubic feet per minute) per child.

21. Use a graphing utility to graph the model and approximate the required ventilation rate when there are 300 cubic feet of air space per child.

22. In a classroom designed for 30 students, the air conditioning system can move 450 cubic feet of air per minute.

(a) Determine the ventilation rate per child in a full classroom.

(b) Estimate the air space required per child.

(c) Determine the minimum number of square feet of floor space required for the room when the ceiling height is 30 feet.

Using Technology In Exercises 23–26, (a) use a graphing utility to create a scatter plot of the data, (b) decide whether the data could best be modeled by a linear model, an exponential model, or a logarithmic model, (c) explain why you chose the model you did in part (b), (d) use the *regression* feature of the graphing utility to find the model you chose in part (b) for the data and graph the model with the scatter plot, and (e) determine how well the model you chose fits the data.

23. (1, 2.0), (1.5, 3.5), (2, 4.0), (4, 5.8), (6, 7.0), (8, 7.8)

24. (1, 4.4), (1.5, 4.7), (2, 5.5), (4, 9.9), (6, 18.1), (8, 33.0)

25. (1, 7.5), (1.5, 7.0), (2, 6.8), (4, 5.0), (6, 3.5), (8, 2.0)

26. (1, 5.0), (1.5, 6.0), (2, 6.4), (4, 7.8), (6, 8.6), (8, 9.0)

6 Topics in Analytic Geometry

Satellite Orbit (Exercise 62, page 490)

Microphone Pickup Pattern (Exercise 69, page 484)

Nuclear Cooling Towers (page 444)

Halley's Comet (page 438)

Suspension Bridge (Exercise 72, page 432)

6.1 Lines

One practical application of the inclination of a line is in measuring heights indirectly. For example, in Exercise 86 on page 424, you will use the inclination of a line to determine the change in elevation from the base to the top of an incline railway.

- **Find the inclination of a line.**
- **Find the angle between two lines.**
- **Find the distance between a point and a line.**

Inclination of a Line

In Section P.4, you learned that the slope of a line is the ratio of the change in y to the change in x. In this section, you will look at the slope of a line in terms of the angle of inclination of the line.

Every nonhorizontal line must intersect the x-axis. The angle formed by such an intersection determines the **inclination** of the line, as specified in the definition below.

> **Definition of Inclination**
>
> The **inclination** of a nonhorizontal line is the positive angle θ (less than π) measured counterclockwise from the x-axis to the line. (See figures below.)

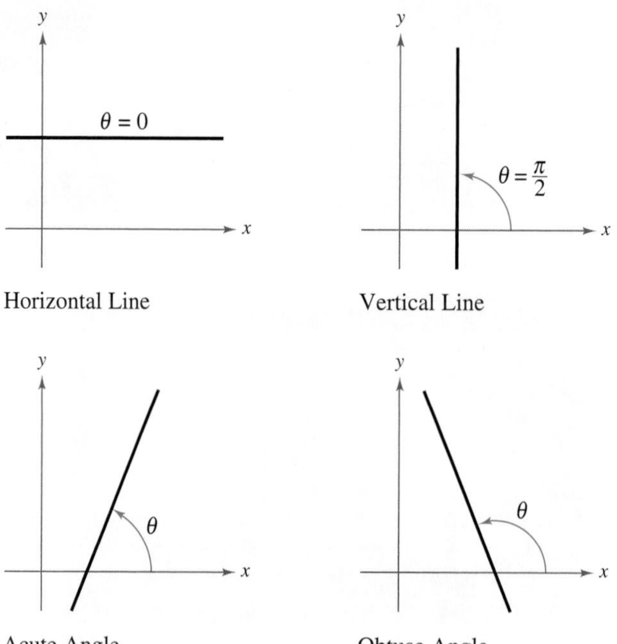

Horizontal Line — Vertical Line — Acute Angle — Obtuse Angle

The inclination of a line is related to its slope in the manner described below.

> **Inclination and Slope**
>
> If a nonvertical line has inclination θ and slope m, then
>
> $m = \tan \theta.$

For a proof of this relation between inclination and slope, see Proofs in Mathematics on page 500.

Note that if $m \geq 0$, then $\theta = \arctan m$ because $0 \leq \theta < \pi/2$. On the other hand, if $m < 0$, then $\theta = \pi + \arctan m$ because $\pi/2 < \theta < \pi$.

(a)

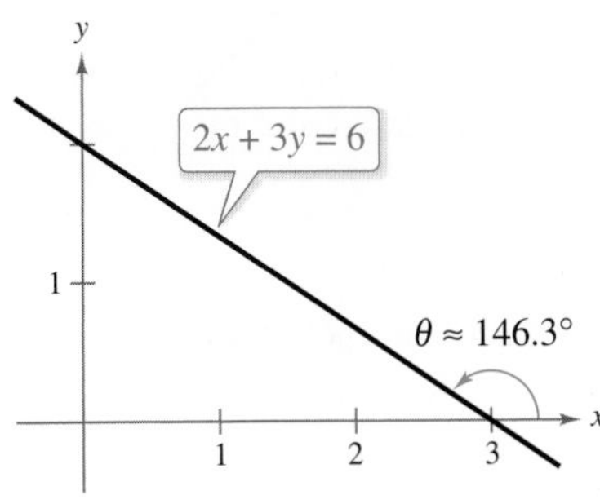

(b)

Figure 6.1

EXAMPLE 1 Finding Inclinations of Lines

Find the inclination of (a) $x - y = 2$ and (b) $2x + 3y = 6$.

Solution

a. The slope of this line is $m = 1$. So, use $\tan\theta = 1$ to determine its inclination. Note that $m \geq 0$. This means that

$$\theta = \arctan 1 = \pi/4 \text{ radian} = 45°$$

as shown in Figure 6.1(a).

b. The slope of this line is $m = -\frac{2}{3}$. So, use $\tan\theta = -\frac{2}{3}$ to determine its inclination. Note that $m < 0$. This means that

$$\theta = \pi + \arctan\left(-\tfrac{2}{3}\right) \approx \pi + (-0.5880) \approx 2.5536 \text{ radians} \approx 146.3°$$

as shown in Figure 6.1(b).

Checkpoint Audio-video solution in English & Spanish at LarsonPrecalculus.com

Find the inclination of (a) $4x - 5y = 7$ and (b) $x + y = -1$.

The Angle Between Two Lines

When two distinct lines intersect and are nonperpendicular, their intersection forms two pairs of opposite angles. One pair is acute and the other pair is obtuse. The smaller of these angles is the **angle between the two lines.** If two lines have inclinations θ_1 and θ_2, where $\theta_1 < \theta_2$ and $\theta_2 - \theta_1 < \pi/2$, then the angle between the two lines is $\theta = \theta_2 - \theta_1$, as shown in Figure 6.2. Use the formula for the tangent of the difference of two angles

$$\tan\theta = \tan(\theta_2 - \theta_1) = \frac{\tan\theta_2 - \tan\theta_1}{1 + \tan\theta_1 \tan\theta_2}$$

to obtain a formula for the tangent of the angle between two lines.

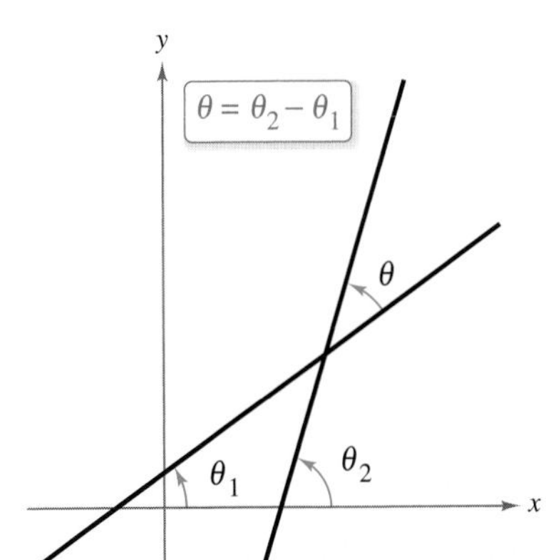

Figure 6.2

Angle Between Two Lines

If two nonperpendicular lines have slopes m_1 and m_2, then the tangent of the angle between the two lines is

$$\tan\theta = \left|\frac{m_2 - m_1}{1 + m_1 m_2}\right|.$$

EXAMPLE 2 Finding the Angle Between Two Lines

Find the angle between $2x - y = 4$ and $3x + 4y = 12$.

Solution The two lines have slopes of $m_1 = 2$ and $m_2 = -\frac{3}{4}$, respectively. So, the tangent of the angle between the two lines is

$$\tan\theta = \left|\frac{m_2 - m_1}{1 + m_1 m_2}\right| = \left|\frac{(-3/4) - 2}{1 + (2)(-3/4)}\right| = \left|\frac{-11/4}{-2/4}\right| = \frac{11}{2}$$

and the angle is $\theta = \arctan \frac{11}{2} \approx 1.3909$ radians $\approx 79.7°$, as shown in Figure 6.3.

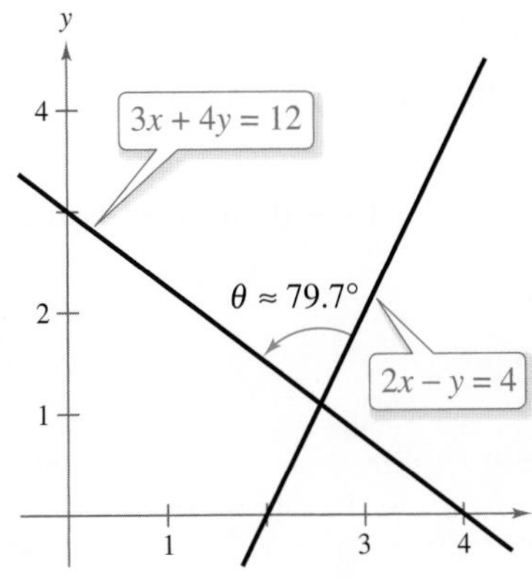

Figure 6.3

Checkpoint Audio-video solution in English & Spanish at LarsonPrecalculus.com

Find the angle between $-4x + 5y = 10$ and $3x + 2y = -5$.

The Distance Between a Point and a Line

Finding the distance between a line and a point not on the line is an application of perpendicular lines. This distance is the length of the perpendicular line segment joining the point and the line, shown as d at the right.

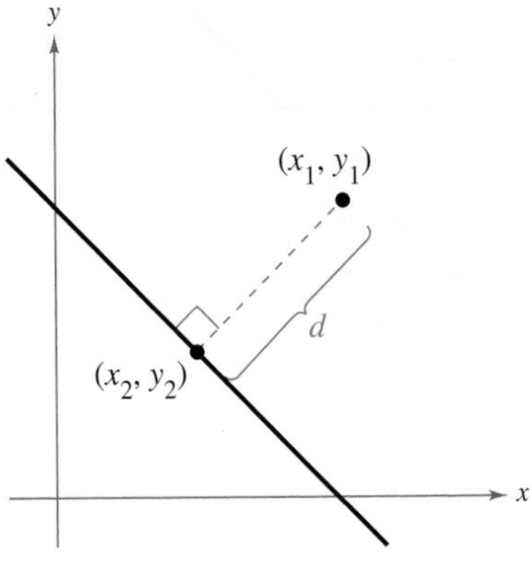

Recall from Section P.3 that d is given by

$$d = \sqrt{(x_2 - x_1)^2 + (y_2 - y_1)^2}.$$

This formula can be written in terms of the coordinates x_1 and y_1 and the coefficients A, B, and C in the general form of the equation of a line, $Ax + By + C = 0$.

Distance Between a Point and a Line

The distance between the point (x_1, y_1) and the line $Ax + By + C = 0$ is

$$d = \frac{|Ax_1 + By_1 + C|}{\sqrt{A^2 + B^2}}.$$

For a proof of this formula for the distance between a point and a line, see Proofs in Mathematics on page 500.

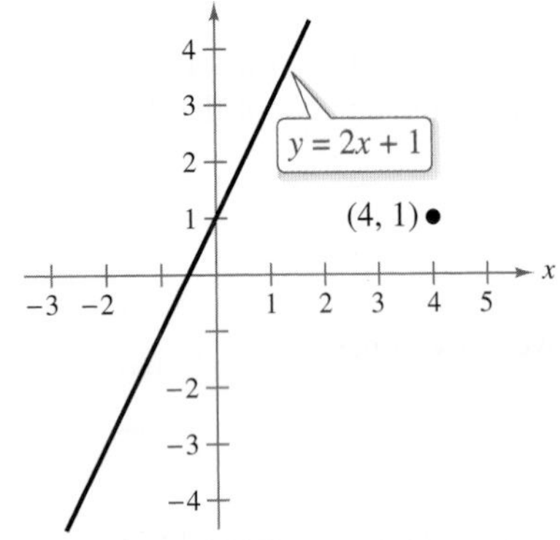

Figure 6.4

EXAMPLE 3 Finding the Distance Between a Point and a Line

Find the distance between the point (4, 1) and the line $y = 2x + 1$. (See Figure 6.4.)

Solution The general form of the equation is $-2x + y - 1 = 0$. So, the distance between the point and the line is

$$d = \frac{|-2(4) + 1(1) + (-1)|}{\sqrt{(-2)^2 + 1^2}} = \frac{8}{\sqrt{5}} \approx 3.58 \text{ units.}$$

✓ **Checkpoint** Audio-video solution in English & Spanish at LarsonPrecalculus.com

Find the distance between the point $(5, -1)$ and the line $y = -3x + 2$.

EXAMPLE 4 Finding the Distance Between a Point and a Line

See LarsonPrecalculus.com for an interactive version of this type of example.

Find the distance between the point $(2, -1)$ and the line $7x + 5y = -13$. (See Figure 6.5.)

Solution The general form of the equation is $7x + 5y + 13 = 0$. So, the distance between the point and the line is

$$d = \frac{|7(2) + 5(-1) + 13|}{\sqrt{7^2 + 5^2}} = \frac{22}{\sqrt{74}} \approx 2.56 \text{ units.}$$

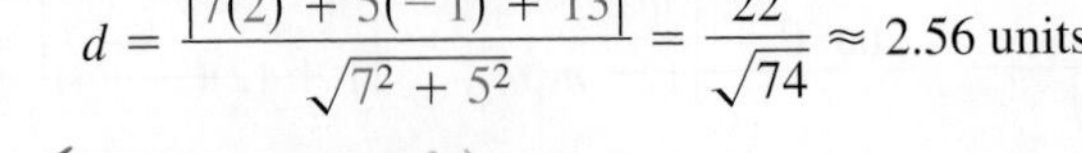

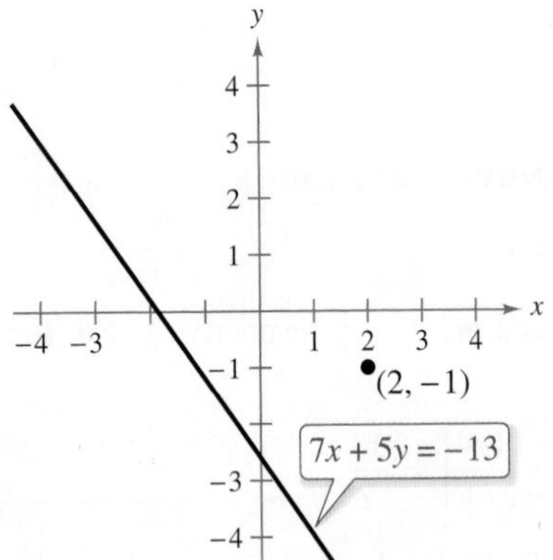

Figure 6.5

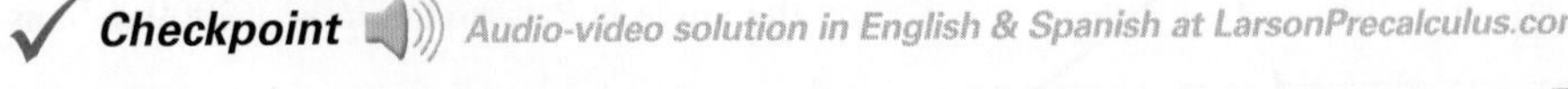

✓ **Checkpoint** Audio-video solution in English & Spanish at LarsonPrecalculus.com

Find the distance between the point (3, 2) and the line $-3x + 5y = -2$.

EXAMPLE 5 An Application of Two Distance Formulas

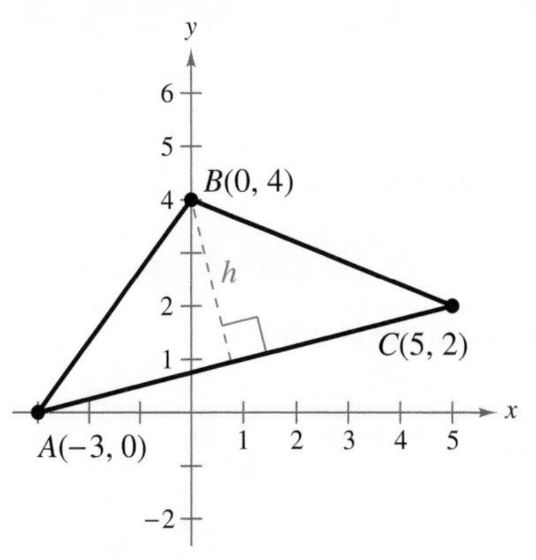

Figure 6.6

Figure 6.6 shows a triangle with vertices $A(-3, 0)$, $B(0, 4)$, and $C(5, 2)$.

a. Find the altitude h from vertex B to side AC.

b. Find the area of the triangle.

Solution

a. First, find the equation of line AC.

$$\textit{Slope: } m = \frac{2-0}{5-(-3)} = \frac{2}{8} = \frac{1}{4}$$

$$\textit{Equation: } \quad y - 0 = \frac{1}{4}(x+3) \qquad \text{Point-slope form}$$

$$4y = x + 3 \qquad \text{Multiply each side by 4.}$$

$$x - 4y + 3 = 0 \qquad \text{General form}$$

Then, use the formula for the distance between line AC and the point $(0, 4)$ to find the altitude.

$$\text{Altitude} = h = \frac{|1(0) + (-4)(4) + 3|}{\sqrt{1^2 + (-4)^2}} = \frac{13}{\sqrt{17}} \text{ units.}$$

b. Use the formula for the distance between two points to find the length of the base AC.

$$b = \sqrt{[5-(-3)]^2 + (2-0)^2} \qquad \text{Distance Formula}$$

$$= \sqrt{8^2 + 2^2} \qquad \text{Simplify.}$$

$$= 2\sqrt{17} \text{ units} \qquad \text{Simplify.}$$

So, the area of the triangle is

$$A = \frac{1}{2}bh \qquad \text{Formula for the area of a triangle}$$

$$= \frac{1}{2}(2\sqrt{17})\left(\frac{13}{\sqrt{17}}\right) \qquad \text{Substitute for } b \text{ and } h.$$

$$= 13 \text{ square units.} \qquad \text{Simplify.}$$

✓ ***Checkpoint***  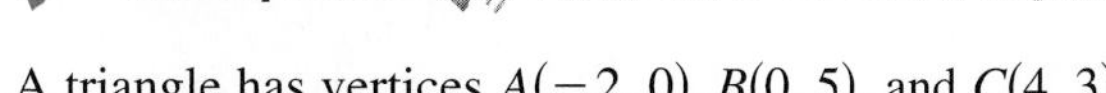

A triangle has vertices $A(-2, 0)$, $B(0, 5)$, and $C(4, 3)$.

a. Find the altitude from vertex B to side AC.

b. Find the area of the triangle.

Summarize (Section 6.1)

1. Explain how to find the inclination of a line *(page 418)*. For an example of finding the inclinations of lines, see Example 1.
2. Explain how to find the angle between two lines *(page 419)*. For an example of finding the angle between two lines, see Example 2.
3. Explain how to find the distance between a point and a line *(page 420)*. For examples of finding the distances between points and lines, see Examples 3–5.

6.1 Exercises

See CalcChat.com for tutorial help and worked-out solutions to odd-numbered exercises.

Vocabulary: Fill in the blanks.

1. The ________ of a nonhorizontal line is the positive angle θ (less than π) measured counterclockwise from the x-axis to the line.
2. If a nonvertical line has inclination θ and slope m, then $m =$ ________.
3. If two nonperpendicular lines have slopes m_1 and m_2, then the tangent of the angle between the two lines is $\tan\theta =$ ________.
4. The distance between the point (x_1, y_1) and the line $Ax + By + C = 0$ is $d =$ ________.

Skills and Applications

Finding the Slope of a Line **In Exercises 5–16, find the slope of the line with inclination θ.**

5. $\theta = \frac{\pi}{6}$ radian
6. $\theta = \frac{\pi}{4}$ radian
7. $\theta = \frac{3\pi}{4}$ radians
8. $\theta = \frac{2\pi}{3}$ radians
9. $\theta = \frac{\pi}{3}$ radians
10. $\theta = \frac{5\pi}{6}$ radians
11. $\theta = 0.39$ radian
12. $\theta = 0.63$ radian
13. $\theta = 1.27$ radians
14. $\theta = 1.35$ radians
15. $\theta = 1.81$ radians
16. $\theta = 2.88$ radians

Finding the Inclination of a Line **In Exercises 17–24, find the inclination θ (in radians and degrees) of the line with slope m.**

17. $m = 1$
18. $m = \sqrt{3}$
19. $m = \frac{2}{3}$
20. $m = \frac{1}{4}$
21. $m = -1$
22. $m = -\sqrt{3}$
23. $m = -\frac{3}{2}$
24. $m = -\frac{5}{9}$

Finding the Inclination of a Line **In Exercises 25–34, find the inclination θ (in radians and degrees) of the line passing through the points.**

25. $(\sqrt{3}, 2), (0, 1)$
26. $(1, 2\sqrt{3}), (0, \sqrt{3})$
27. $(-\sqrt{3}, -1), (0, -2)$
28. $(3, \sqrt{3}), (6, -2\sqrt{3})$
29. $(6, 1), (10, 8)$
30. $(12, 8), (-4, -3)$
31. $(-2, 20), (10, 0)$
32. $(0, 100), (50, 0)$
33. $(\frac{1}{4}, \frac{3}{2}), (\frac{1}{3}, \frac{1}{2})$
34. $(\frac{2}{5}, -\frac{3}{4}), (-\frac{11}{10}, -\frac{1}{4})$

Finding the Inclination of a Line **In Exercises 35–44, find the inclination θ (in radians and degrees) of the line.**

35. $2x + 2y - 5 = 0$
36. $x - \sqrt{3}y + 1 = 0$
37. $3x - 3y + 1 = 0$
38. $\sqrt{3}x - y + 2 = 0$
39. $x + \sqrt{3}y + 2 = 0$
40. $-2\sqrt{3}x - 2y = 0$
41. $6x - 2y + 8 = 0$
42. $2x - 6y - 12 = 0$
43. $4x + 5y - 9 = 0$
44. $5x + 3y = 0$

Finding the Angle Between Two Lines **In Exercises 45–54, find the angle θ (in radians and degrees) between the lines.**

45. $3x + y = 3$
    $x - y = 2$

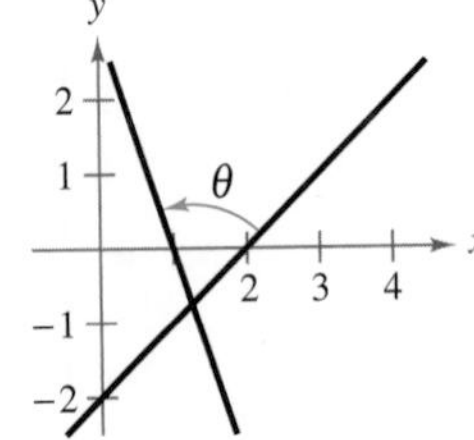

46. $x + 3y = 2$
    $x - 2y = -3$

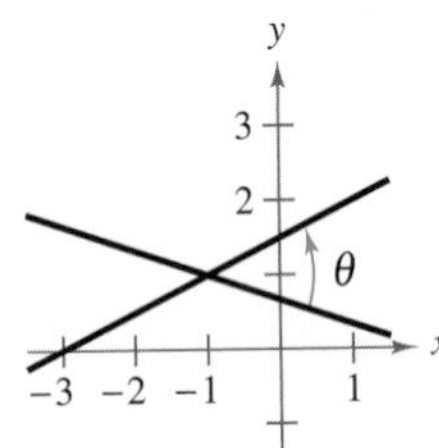

47. $x - y = 0$
    $3x - 2y = -1$

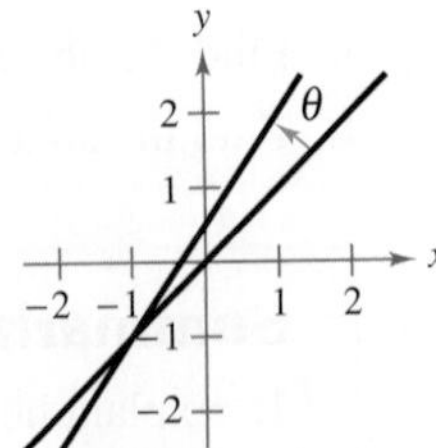

48. $2x - y = 2$
    $4x + 3y = 24$

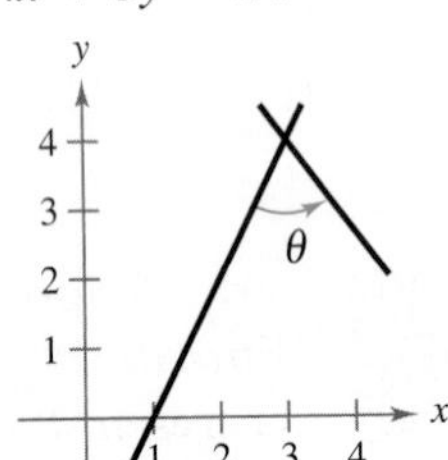

49. $x - 2y = 7$
    $6x + 2y = 5$
50. $5x + 2y = 16$
    $3x - 5y = -1$
51. $x + 2y = 8$
    $x - 2y = 2$
52. $3x - 5y = 3$
    $3x + 5y = 12$
53. $0.05x - 0.03y = 0.21$
    $0.07x + 0.02y = 0.16$
54. $0.02x - 0.05y = -0.19$
    $0.03x + 0.04y = 0.52$

Angle Measurement **In Exercises 55–58, find the slope of each side of the triangle, and use the slopes to find the measures of the interior angles.**

55.

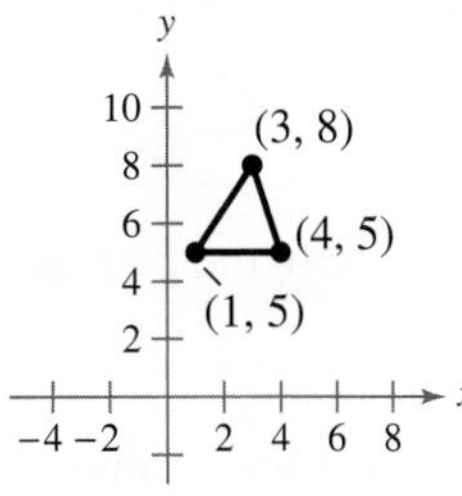

56.

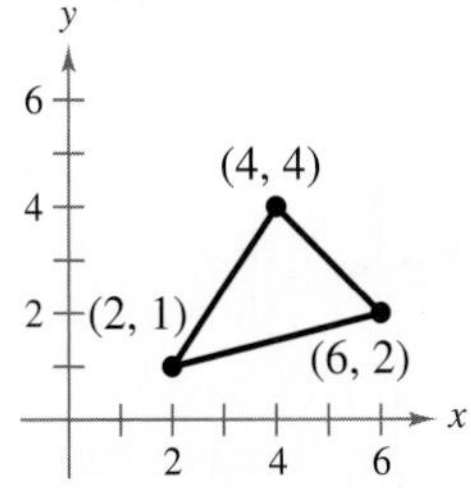

57.

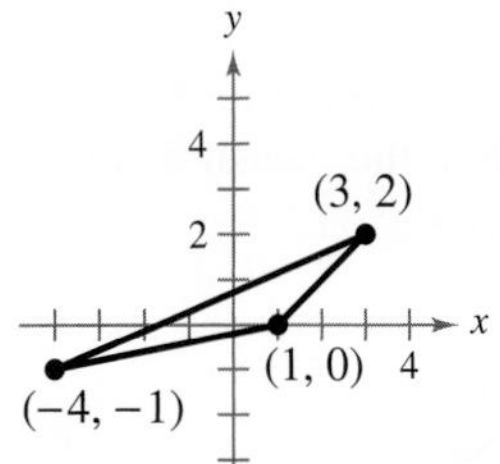

58.

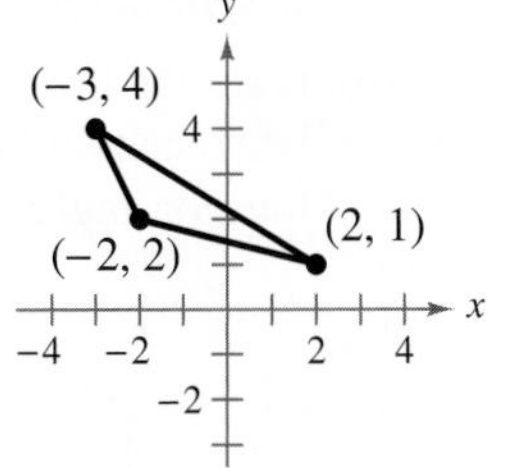

Finding the Distance Between a Point and a Line **In Exercises 59–72, find the distance between the point and the line.**

	Point	Line
59.	$(1, 2)$	$y = x + 2$
60.	$(3, 1)$	$y = x + 3$
61.	$(2, 3)$	$y = 2x - 3$
62.	$(1, 5)$	$y = 4x + 5$
63.	$(-2, 4)$	$y = -x + 6$
64.	$(3, -3)$	$y = -3x - 4$
65.	$(1, -2)$	$y = 3x - 6$
66.	$(-3, 7)$	$y = -4x + 3$
67.	$(2, 3)$	$3x + y = 1$
68.	$(2, 1)$	$-2x + y = 2$
69.	$(6, 2)$	$-3x + 4y = -5$
70.	$(1, -4)$	$2x - 3y = -5$
71.	$(-2, 4)$	$4x + 3y = 5$
72.	$(-3, -5)$	$-3x - 4y = 4$

An Application of Two Distance Formulas **In Exercises 73–78, the points represent the vertices of a triangle. (a) Draw triangle ABC in the coordinate plane, (b) find the altitude from vertex B of the triangle to side AC, and (c) find the area of the triangle.**

73. $A(-1, 0), B(0, 3), C(3, 1)$

74. $A(-4, 0), B(0, 5), C(3, 3)$

75. $A(-3, 0), B(0, -2), C(2, 3)$

76. $A(-2, 0), B(0, -3), C(5, 1)$

77. $A(1, 1), B(2, 4), C(3, 5)$

78. $A(-3, -2), B(-1, -4), C(3, -1)$

Finding the Distance Between Parallel Lines **In Exercises 79 and 80, find the distance between the parallel lines.**

79. $x + y = 1$
$x + y = 5$

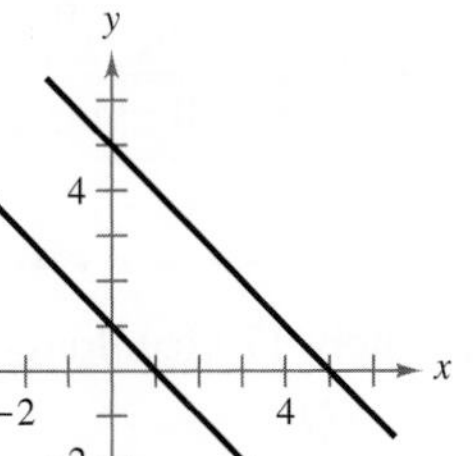

80. $3x - 4y = 1$
$3x - 4y = 10$

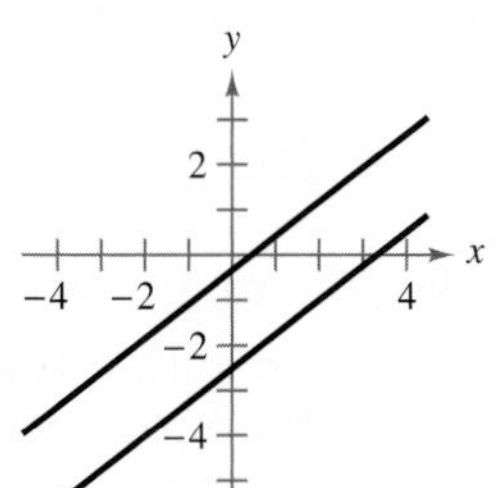

81. Road Grade A straight road rises with an inclination of 0.1 radian from the horizontal (see figure). Find the slope of the road and the change in elevation over a two-mile section of the road.

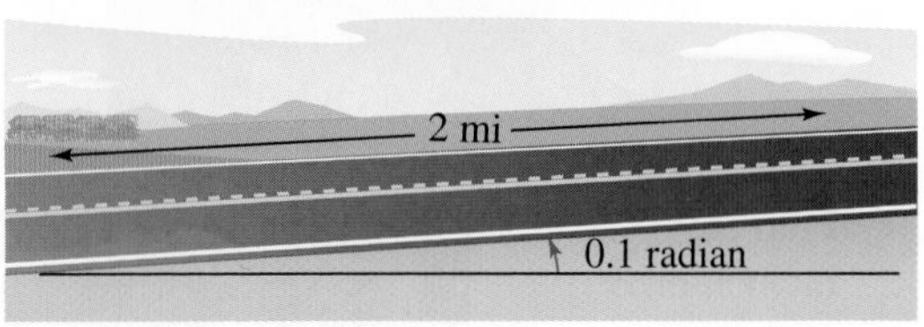

82. Road Grade A straight road rises with an inclination of 0.2 radian from the horizontal. Find the slope of the road and the change in elevation over a one-mile section of the road.

83. Pitch of a Roof A roof has a rise of 3 feet for every horizontal change of 5 feet (see figure). Find the inclination θ of the roof.

84. Conveyor Design A moving conveyor rises 1 meter for every 3 meters of horizontal travel (see figure).

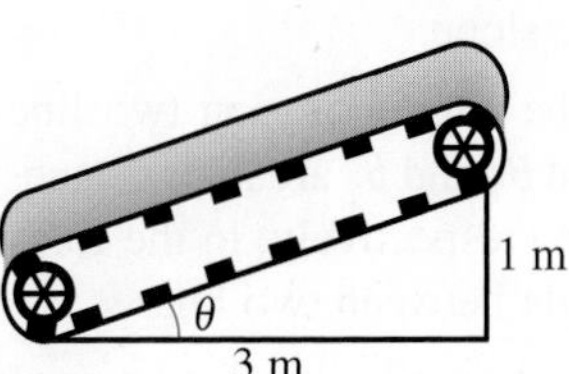

(a) Find the inclination θ of the conveyor.

(b) The conveyor runs between two floors in a factory. The distance between the floors is 5 meters. Find the length of the conveyor.

85. Truss Find the angles α and β shown in the drawing of the roof truss.

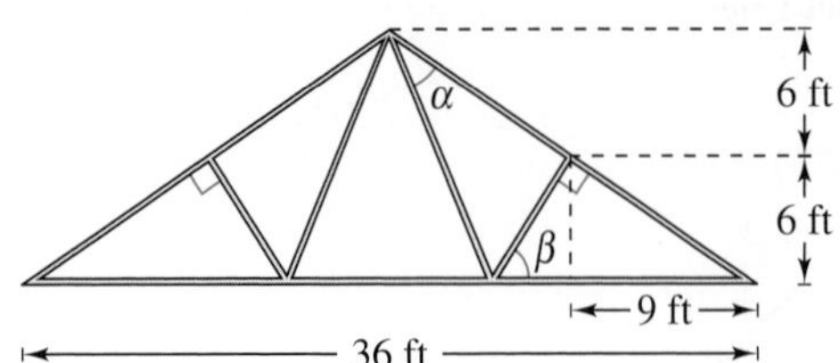

86. Incline Railway

An incline railway is approximately 170 feet long with a 36% uphill grade (see figure).

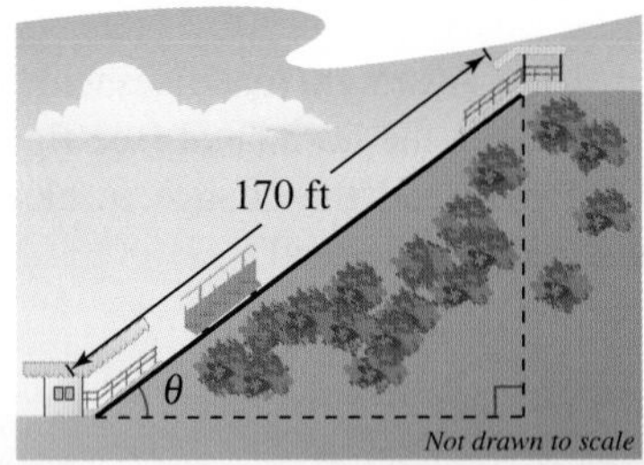

(a) Find the inclination θ of the railway.

(b) Find the change in elevation from the base to the top of the railway.

(c) Using the origin of a rectangular coordinate system as the base of the inclined plane, find an equation of the line that models the railway track.

(d) Sketch a graph of the equation you found in part (c).

Exploration

True or False? **In Exercises 87–90, determine whether the statement is true or false. Justify your answer.**

87. A line that has an inclination of 0 radians has a slope of 0.

88. A line that has an inclination greater than $\frac{\pi}{2}$ radians has a negative slope.

89. To find the angle between two lines whose angles of inclination θ_1 and θ_2 are known, substitute θ_1 and θ_2 for m_1 and m_2, respectively, in the formula for the tangent of the angle between two lines.

90. The inclination of a line is the angle between the line and the x-axis.

91. Writing Explain why the inclination of a line can be an angle that is greater than $\frac{\pi}{2}$, but the angle between two lines cannot be greater than $\frac{\pi}{2}$.

92. HOW DO YOU SEE IT? Use the pentagon shown below.

(a) Describe how to use the formula for the distance between a point and a line to find the area of the pentagon.

(b) Describe how to use the formula for the angle between two lines to find the measures of the interior angles of the pentagon.

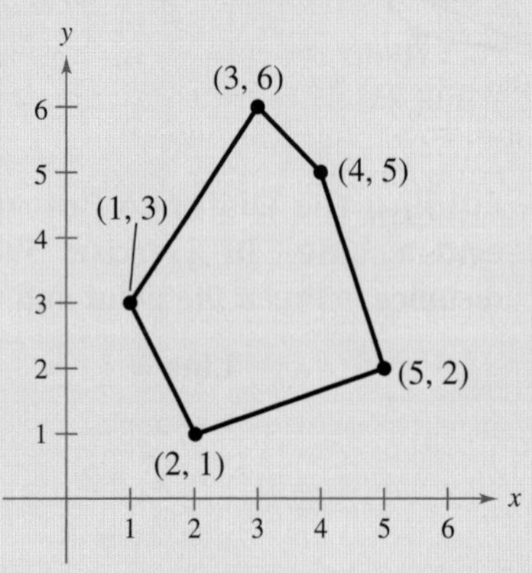

93. Think About It Consider a line with slope m and x-intercept $(0, 4)$.

(a) Write the distance d between the origin and the line as a function of m.

(b) Graph the function in part (a).

(c) Find the slope that yields the maximum distance between the origin and the line.

(d) Find the asymptote of the graph in part (b) and interpret its meaning in the context of the problem.

94. Think About It Consider a line with slope m and y-intercept $(0, 4)$.

(a) Write the distance d between the point $(3, 1)$ and the line as a function of m.

(b) Graph the function in part (a).

(c) Find the slope that yields the maximum distance between the point and the line.

(d) Is it possible for the distance to be 0? If so, what is the slope of the line that yields a distance of 0?

(e) Find the asymptote of the graph in part (b) and interpret its meaning in the context of the problem.

6.2 Introduction to Conics: Parabolas

Parabolas have many real-life applications and are often used to model and solve engineering problems. For example, in Exercise 72 on page 432, you will use a parabola to model the cables of the Golden Gate Bridge.

- **Recognize a conic as the intersection of a plane and a double-napped cone.**
- **Write equations of parabolas in standard form.**
- **Use the reflective property of parabolas to write equations of tangent lines.**

Conics

The earliest basic descriptions of conic sections took place during the classical Greek period, 500 to 336 B.C. This early Greek study was largely concerned with the geometric properties of conics. It was not until the early 17th century that the broad applicability of conics became apparent and played a prominent role in the early development of calculus.

A **conic section** (or simply **conic**) is the intersection of a plane and a double-napped cone. Notice in Figure 6.7 that in the formation of the four basic conics, the intersecting plane does not pass through the vertex of the cone. When the plane does pass through the vertex, the resulting figure is a **degenerate conic,** as shown in Figure 6.8.

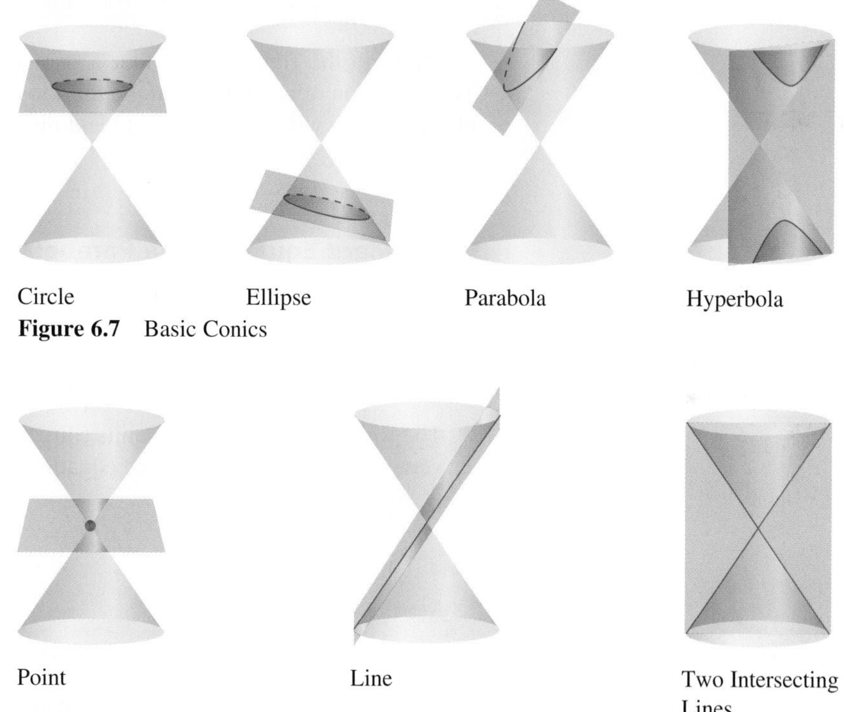

Figure 6.7 Basic Conics

Figure 6.8 Degenerate Conics

There are several ways to approach the study of conics. You could begin by defining conics in terms of the intersections of planes and cones, as the Greeks did, or you could define them algebraically, in terms of the general second-degree equation

$$Ax^2 + Bxy + Cy^2 + Dx + Ey + F = 0.$$

However, you will study a third approach, in which each of the conics is defined as a *locus* (collection) of points satisfying a given geometric property. For example, in Section P.3, you saw how the definition of a circle as *the collection of all points* (x, y) *that are equidistant from a fixed point* (h, k) led to the standard form of the equation of a circle

$$(x - h)^2 + (y - k)^2 = r^2. \quad \text{Equation of a circle}$$

Recall that the center of a circle is at (h, k) and that the radius of the circle is r.

topseller/Shutterstock.com

Parabolas

Recall that the graph of the quadratic function

$$f(x) = ax^2 + bx + c$$

is a parabola that opens upward or downward. The definition of a parabola below is more general in the sense that it is independent of the orientation of the parabola.

Definition of a Parabola

A **parabola** is the set of all points (x, y) in a plane that are equidistant from a fixed line, the **directrix,** and a fixed point, the **focus,** not on the line. (See figure.) The **vertex** is the midpoint between the focus and the directrix. The **axis** of the parabola is the line passing through the focus and the vertex.

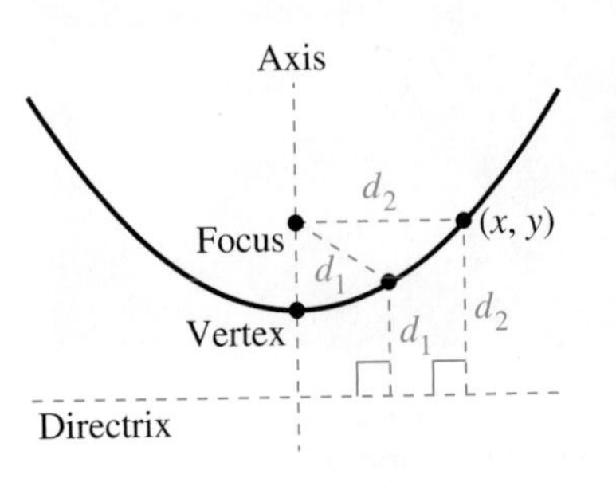

Note in the figure above that a parabola is symmetric with respect to its axis. The definition of a parabola can be used to derive the **standard form of the equation of a parabola** with vertex at (h, k) and directrix parallel to the x-axis or to the y-axis, stated below.

Standard Equation of a Parabola

The **standard form of the equation of a parabola** with vertex at (h, k) is

$(x - h)^2 = 4p(y - k), \quad p \neq 0$ Vertical axis; directrix: $y = k - p$

$(y - k)^2 = 4p(x - h), \quad p \neq 0.$ Horizontal axis; directrix: $x = h - p$

The focus lies on the axis p units (directed distance) from the vertex. If the vertex is at the origin, then the equation takes one of two forms.

$x^2 = 4py$ Vertical axis

$y^2 = 4px$ Horizontal axis

See the figures below.

For a proof of the standard form of the equation of a parabola, see Proofs in Mathematics on page 501.

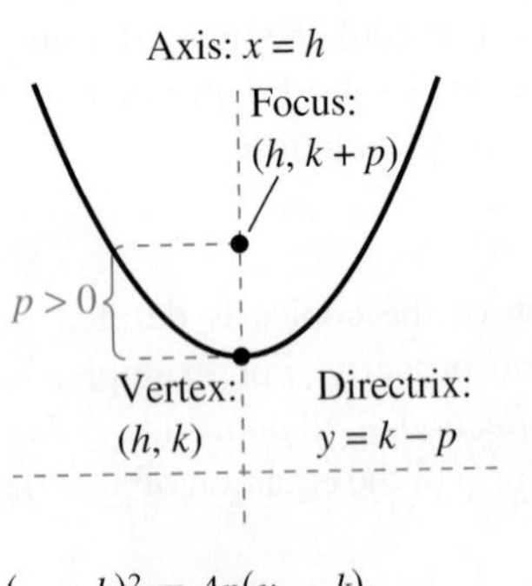

$(x - h)^2 = 4p(y - k)$
Vertical axis: $p > 0$

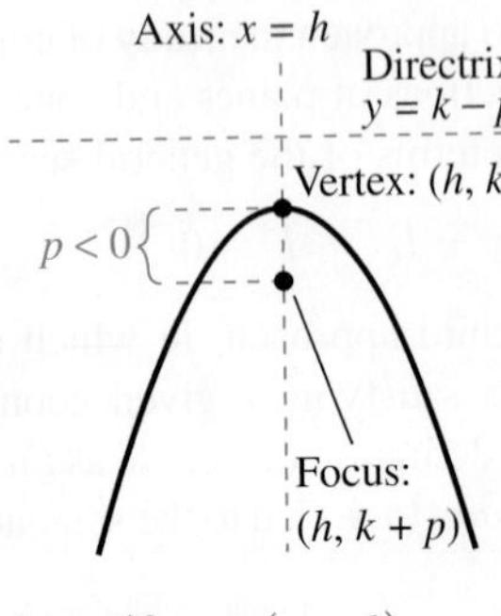

$(x - h)^2 = 4p(y - k)$
Vertical axis: $p < 0$

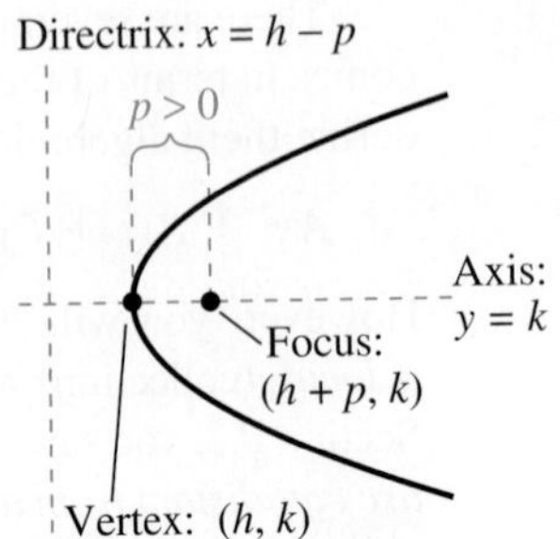

$(y - k)^2 = 4p(x - h)$
Horizontal axis: $p > 0$

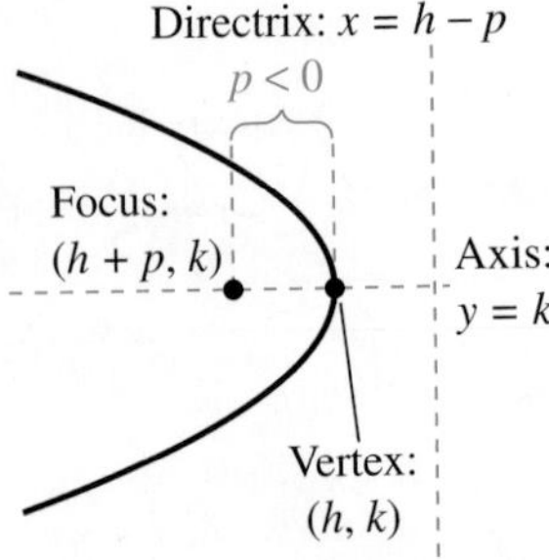

$(y - k)^2 = 4p(x - h)$
Horizontal axis: $p < 0$

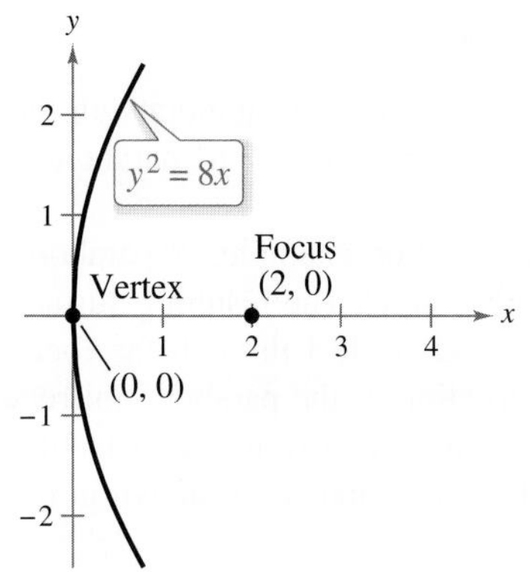

Figure 6.9

EXAMPLE 1 Finding the Standard Equation of a Parabola

Find the standard form of the equation of the parabola with vertex at the origin and focus $(2, 0)$.

Solution The axis of the parabola is horizontal, passing through $(0, 0)$ and $(2, 0)$, as shown in Figure 6.9. The equation is of the form $y^2 = 4px$, where $p = 2$. So, the standard form of the equation is $y^2 = 8x$. You can use a graphing utility to confirm this equation. Let $y_1 = \sqrt{8x}$ to graph the upper portion of the parabola and let $y_2 = -\sqrt{8x}$ to graph the lower portion of the parabola.

✓ **Checkpoint** Audio-video solution in English & Spanish at LarsonPrecalculus.com

Find the standard form of the equation of the parabola with vertex at the origin and focus $\left(0, \frac{3}{8}\right)$.

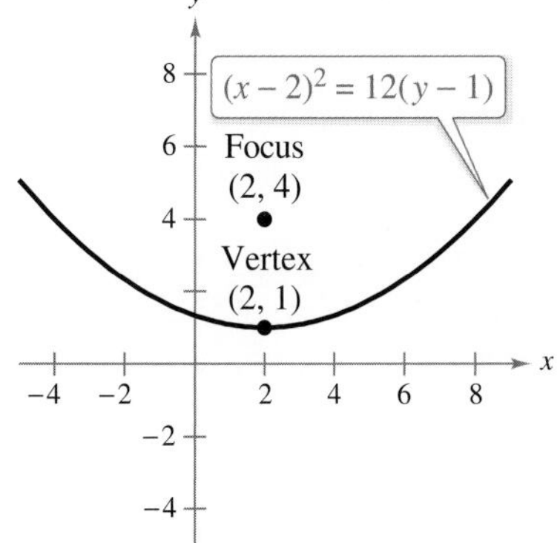

Figure 6.10

EXAMPLE 2 Finding the Standard Equation of a Parabola

See LarsonPrecalculus.com for an interactive version of this type of example.

Find the standard form of the equation of the parabola with vertex $(2, 1)$ and focus $(2, 4)$.

Solution The axis of the parabola is vertical, passing through $(2, 1)$ and $(2, 4)$. The equation is of the form

$$(x - h)^2 = 4p(y - k)$$

where $h = 2$, $k = 1$, and $p = 4 - 1 = 3$. So, the standard form of the equation is

$$(x - 2)^2 = 12(y - 1).$$

Figure 6.10 shows the graph of this parabola.

✓ **Checkpoint** Audio-video solution in English & Spanish at LarsonPrecalculus.com

Find the standard form of the equation of the parabola with vertex $(2, -3)$ and focus $(4, -3)$.

ALGEBRA HELP The technique of completing the square is used to write the equation in Example 3 in standard form. To review completing the square, see Section P.2.

EXAMPLE 3 Finding the Focus of a Parabola

Find the focus of the parabola $y = -\frac{1}{2}x^2 - x + \frac{1}{2}$.

Solution Convert to standard form by completing the square.

$$y = -\tfrac{1}{2}x^2 - x + \tfrac{1}{2} \qquad \text{Write original equation.}$$

$$-2y = x^2 + 2x - 1 \qquad \text{Multiply each side by } -2.$$

$$1 - 2y = x^2 + 2x \qquad \text{Add 1 to each side.}$$

$$1 + 1 - 2y = x^2 + 2x + 1 \qquad \text{Complete the square.}$$

$$2 - 2y = x^2 + 2x + 1 \qquad \text{Combine like terms.}$$

$$-2(y - 1) = (x + 1)^2 \qquad \text{Write in standard form.}$$

Comparing this equation with

$$(x - h)^2 = 4p(y - k)$$

shows that $h = -1$, $k = 1$, and $p = -\frac{1}{2}$. The parabola opens downward, as shown in Figure 6.11, because p is negative. So, the focus of the parabola is $(h, k + p) = \left(-1, \frac{1}{2}\right)$.

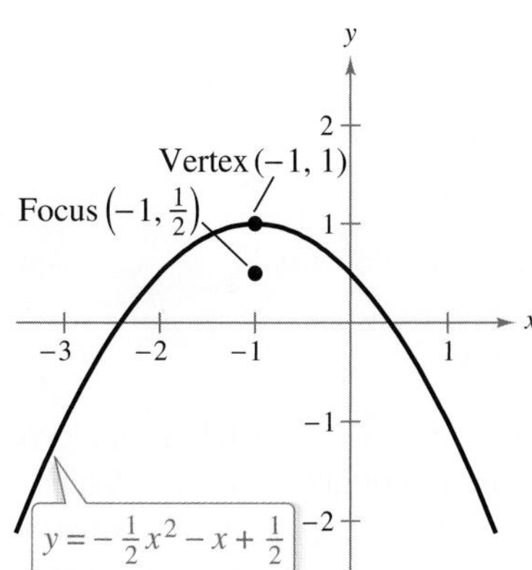

Figure 6.11

✓ **Checkpoint** Audio-video solution in English & Spanish at LarsonPrecalculus.com

Find the focus of the parabola $x = \frac{1}{4}y^2 + \frac{3}{2}y + \frac{13}{4}$.

One important application of parabolas is in astronomy. Radio telescopes use parabolic dishes to collect radio waves from space.

The Reflective Property of Parabolas

A line segment that passes through the focus of a parabola and has endpoints on the parabola is a **focal chord.** The focal chord perpendicular to the axis of the parabola is called the **latus rectum.**

Parabolas occur in a wide variety of applications. For example, a parabolic reflector can be formed by revolving a parabola about its axis. The resulting surface has the property that all incoming rays parallel to the axis reflect through the focus of the parabola. This is the principle behind the construction of the parabolic mirrors used in reflecting telescopes. Conversely, the light rays emanating from the focus of a parabolic reflector used in a flashlight are all parallel to one another, as shown in the figure below.

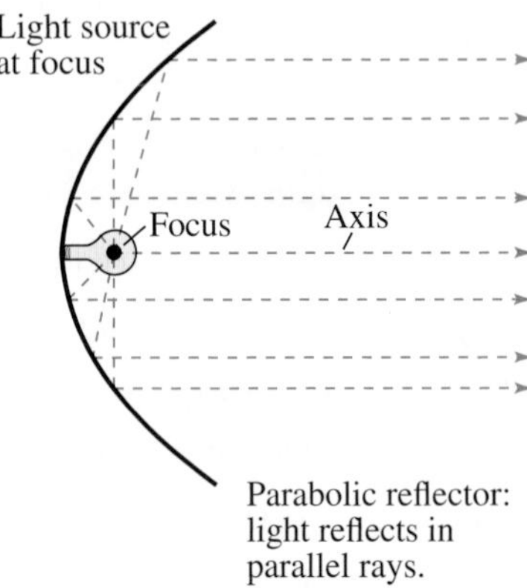

A line is **tangent** to a parabola at a point on the parabola when the line intersects, but does not cross, the parabola at the point. Tangent lines to parabolas have special properties related to the use of parabolas in constructing reflective surfaces.

Reflective Property of a Parabola

The tangent line to a parabola at a point P makes equal angles with the following two lines (see figure below).

1. The line passing through P and the focus
2. The axis of the parabola

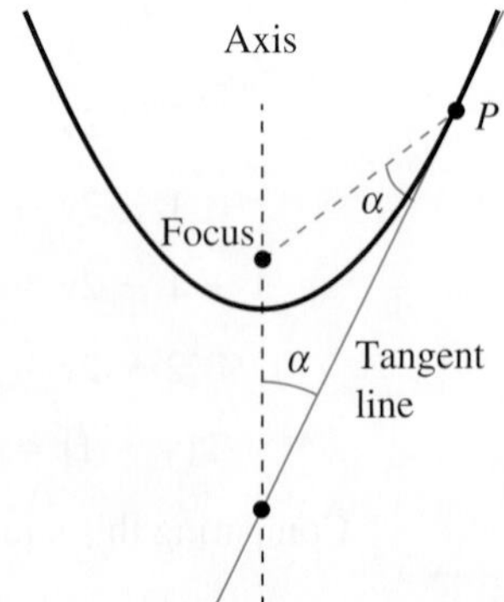

Example 4 shows how to find an equation of a tangent line to a parabola at a given point. Finding slopes and equations of tangent lines are important topics in calculus. If you take a calculus course, you will study techniques for finding slopes and equations of tangent lines to parabolas and other curves.

EXAMPLE 4 Finding the Tangent Line at a Point on a Parabola

Find an equation of the tangent line to the parabola $y = x^2$ at the point $(1, 1)$.

Solution For this parabola, the vertex is at the origin, the axis is vertical, and $p = \frac{1}{4}$, so the focus is $\left(0, \frac{1}{4}\right)$, as shown in the figure below.

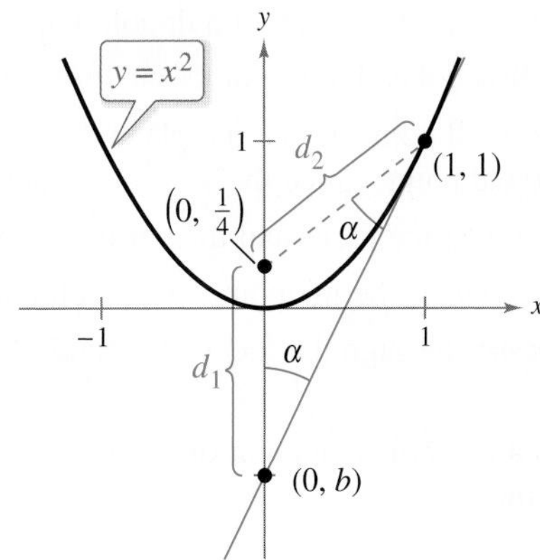

To find the y-intercept $(0, b)$ of the tangent line, equate the lengths of the two sides of the isosceles triangle shown in the figure:

$$d_1 = \frac{1}{4} - b$$

and

$$d_2 = \sqrt{(1 - 0)^2 + \left(1 - \frac{1}{4}\right)^2} = \frac{5}{4}.$$

Note that $d_1 = \frac{1}{4} - b$ rather than $b - \frac{1}{4}$. The order of subtraction for the distance is important because the distance must be positive. Setting $d_1 = d_2$ produces

$$\frac{1}{4} - b = \frac{5}{4}$$

$$b = -1.$$

So, the slope of the tangent line is

$$m = \frac{1 - (-1)}{1 - 0} = 2$$

and the equation of the tangent line in slope-intercept form is

$$y = 2x - 1.$$

TECHNOLOGY Use a graphing utility to confirm the result of Example 4. Graph

$$y_1 = x^2 \quad \text{and} \quad y_2 = 2x - 1$$

in the same viewing window and verify that the line touches the parabola at the point $(1, 1)$.

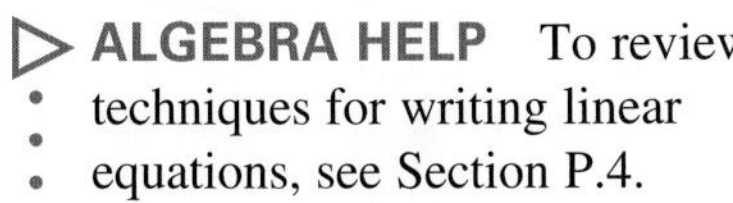

ALGEBRA HELP To review techniques for writing linear equations, see Section P.4.

✓ ***Checkpoint*** *Audio-video solution in English & Spanish at LarsonPrecalculus.com*

Find an equation of the tangent line to the parabola $y = 3x^2$ at the point $(1, 3)$.

Summarize (Section 6.2)

1. List the four basic conic sections and the degenerate conics. Use sketches to show how to form each basic conic section and degenerate conic from the intersection of a plane and a double-napped cone *(page 425)*.
2. State the definition of a parabola and the standard form of the equation of a parabola *(page 426)*. For examples involving writing equations of parabolas in standard form, see Examples 1–3.
3. State the reflective property of a parabola *(page 428)*. For an example of using this property to write an equation of a tangent line, see Example 4.

6.2 Exercises

See **CalcChat.com** for tutorial help and worked-out solutions to odd-numbered exercises.

Vocabulary: Fill in the blanks.

1. A ________ is the intersection of a plane and a double-napped cone.
2. When a plane passes through the vertex of a double-napped cone, the intersection is a ________ ________.
3. A ________ of points is a collection of points satisfying a given geometric property.
4. A ________ is the set of all points (x, y) in a plane that are equidistant from a fixed line, called the ________, and a fixed point, called the ________, not on the line.
5. The line that passes through the focus and the vertex of a parabola is the ________ of the parabola.
6. The ________ of a parabola is the midpoint between the focus and the directrix.
7. A line segment that passes through the focus of a parabola and has endpoints on the parabola is a ________ ________.
8. A line is ________ to a parabola at a point on the parabola when the line intersects, but does not cross, the parabola at the point.

Skills and Applications

Matching **In Exercises 9–12, match the equation with its graph. [The graphs are labeled (a), (b), (c), and (d).]**

(a)

(b)

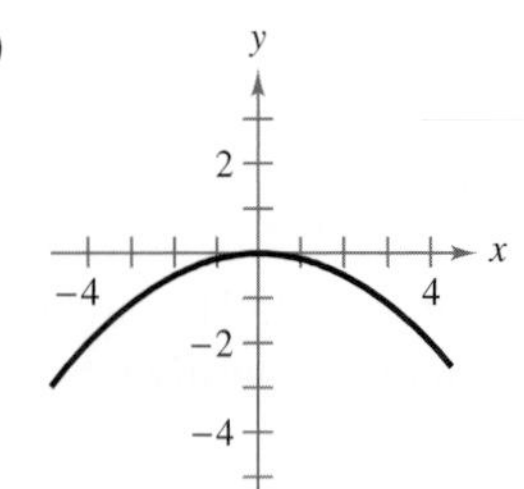

(c)

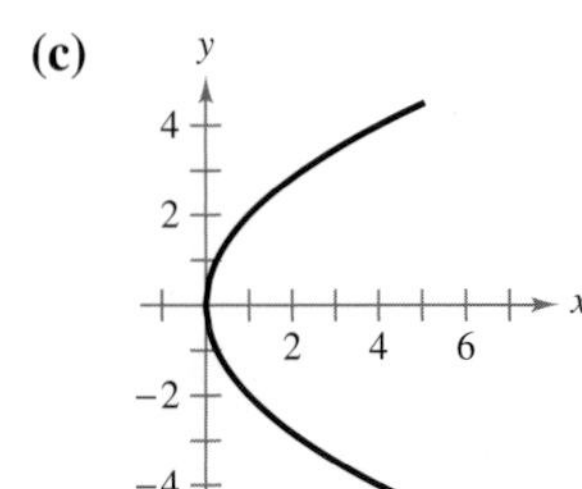

(d)

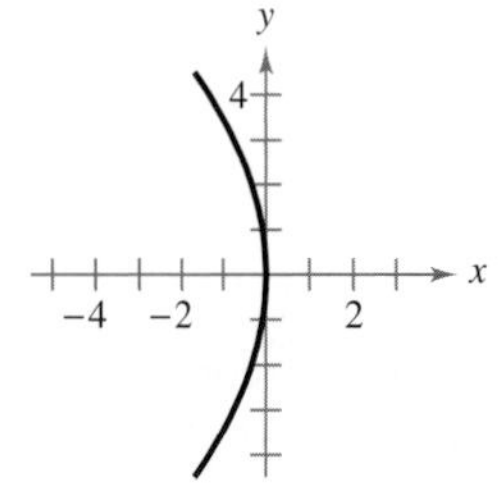

9. $y^2 = 4x$
10. $x^2 = 2y$
11. $x^2 = -8y$
12. $y^2 = -12x$

Finding the Standard Equation of a Parabola **In Exercises 13–26, find the standard form of the equation of the parabola with the given characteristic(s) and vertex at the origin.**

13.

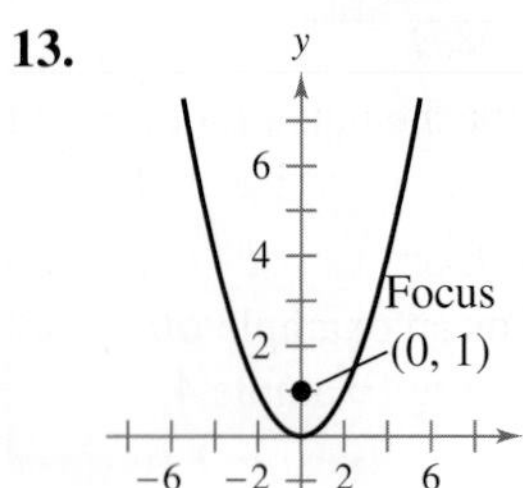

14.

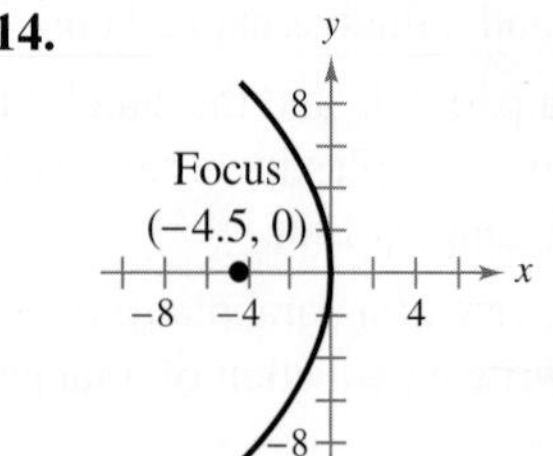

15. Focus: $\left(0, \frac{1}{2}\right)$
16. Focus: $\left(\frac{3}{2}, 0\right)$
17. Focus: $(-2, 0)$
18. Focus: $(0, -1)$
19. Directrix: $y = 2$
20. Directrix: $y = -4$
21. Directrix: $x = -1$
22. Directrix: $x = 3$
23. Vertical axis; passes through the point $(4, 6)$
24. Vertical axis; passes through the point $(-3, -3)$
25. Horizontal axis; passes through the point $(-2, 5)$
26. Horizontal axis; passes through the point $(3, -2)$

Finding the Standard Equation of a Parabola **In Exercises 27–36, find the standard form of the equation of the parabola with the given characteristics.**

27.

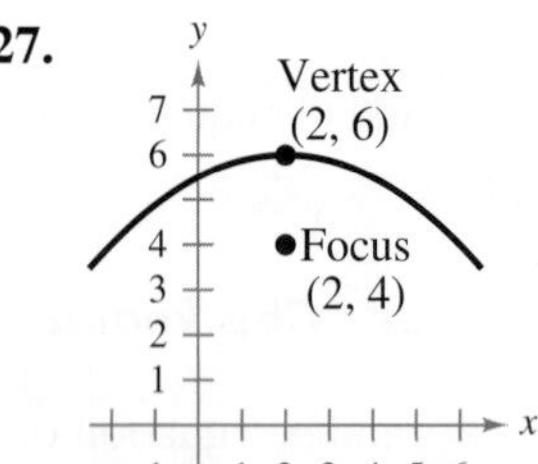

28.

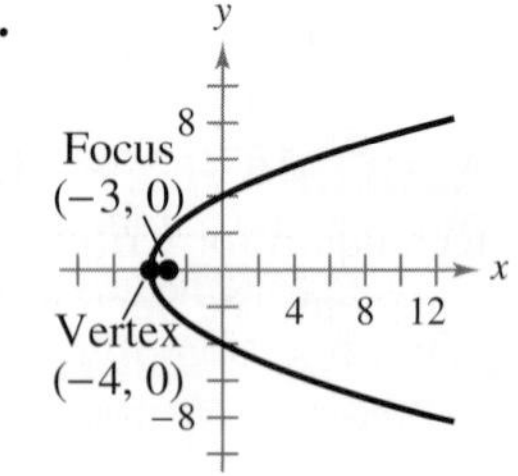

29. Vertex: $(6, 3)$; focus: $(4, 3)$
30. Vertex: $(1, -8)$; focus: $(3, -8)$
31. Vertex: $(0, 2)$; directrix: $y = 4$
32. Vertex: $(1, 2)$; directrix: $y = -1$
33. Focus: $(2, 2)$; directrix: $x = -2$
34. Focus: $(0, 0)$; directrix: $y = 8$
35. Vertex: $(3, -3)$; vertical axis; passes through the point $(0, 0)$
36. Vertex: $(-1, 6)$; horizontal axis; passes through the point $(-9, 2)$

Finding the Vertex, Focus, and Directrix of a Parabola **In Exercises 37–50, find the vertex, focus, and directrix of the parabola. Then sketch the parabola.**

37. $y = \frac{1}{2}x^2$
38. $y = -4x^2$
39. $y^2 = -6x$
40. $y^2 = 3x$
41. $x^2 + 12y = 0$
42. $x + y^2 = 0$
43. $(x - 1)^2 + 8(y + 2) = 0$
44. $(x + 5) + (y - 1)^2 = 0$
45. $(y + 7)^2 = 4\left(x - \frac{3}{2}\right)$
46. $\left(x + \frac{1}{2}\right)^2 = 4(y - 1)$
47. $y = \frac{1}{4}(x^2 - 2x + 5)$
48. $x = \frac{1}{4}(y^2 + 2y + 33)$
49. $y^2 + 6y + 8x + 25 = 0$
50. $x^2 - 4x - 4y = 0$

Finding the Vertex, Focus, and Directrix of a Parabola **In Exercises 51–54, find the vertex, focus, and directrix of the parabola. Use a graphing utility to graph the parabola.**

51. $x^2 + 4x - 6y = -10$
52. $x^2 - 2x + 8y = -9$
53. $y^2 + x + y = 0$
54. $y^2 - 4x - 4 = 0$

Finding the Tangent Line at a Point on a Parabola **In Exercises 55–60, find an equation of the tangent line to the parabola at the given point.**

55.

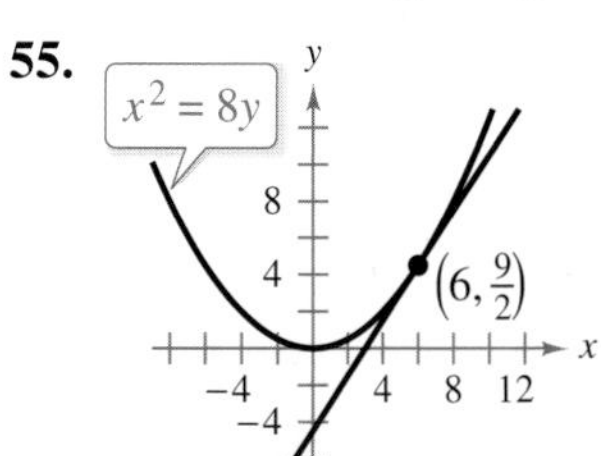

56.

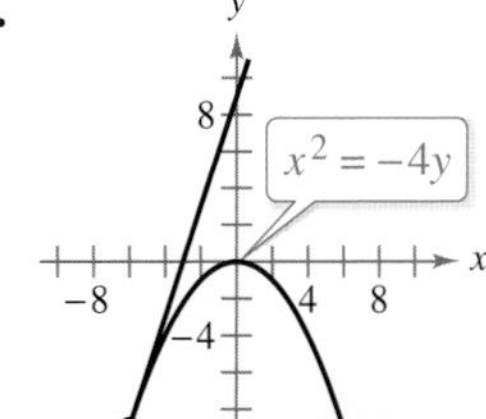

57. $x^2 = 2y, \ (4, 8)$
58. $x^2 = 2y, \ \left(-3, \frac{9}{2}\right)$
59. $y = -2x^2, \ (-1, -2)$
60. $y = -2x^2, \ (2, -8)$

61. Flashlight The light bulb in a flashlight is at the focus of the parabolic reflector, 1.5 centimeters from the vertex of the reflector (see figure). Write an equation for a cross section of the flashlight's reflector with its focus on the positive x-axis and its vertex at the origin.

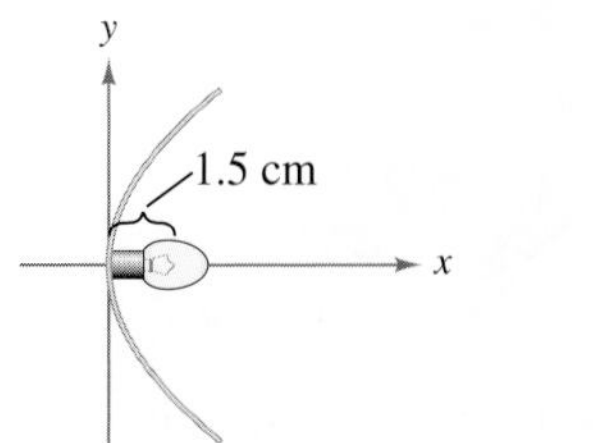

Figure for 61

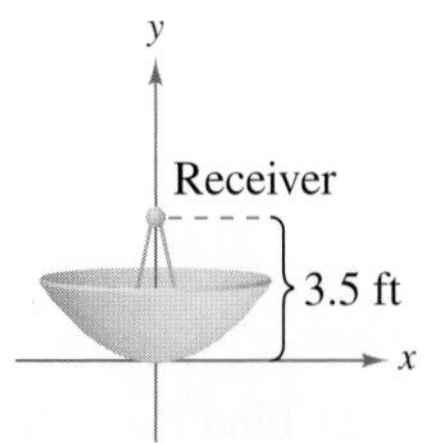

Figure for 62

62. Satellite Dish The receiver of a parabolic satellite dish is at the focus of the parabola (see figure). Write an equation for a cross section of the satellite dish.

63. Highway Design Highway engineers use a parabolic curve to design an entrance ramp from a straight street to an interstate highway (see figure). Write an equation of the parabola.

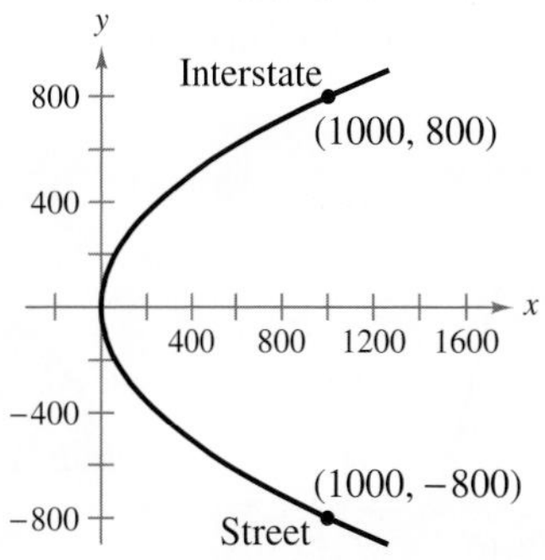

64. Road Design Roads are often designed with parabolic surfaces to allow rain to drain off. A particular road is 32 feet wide and 0.4 foot higher in the center than it is on the sides (see figure).

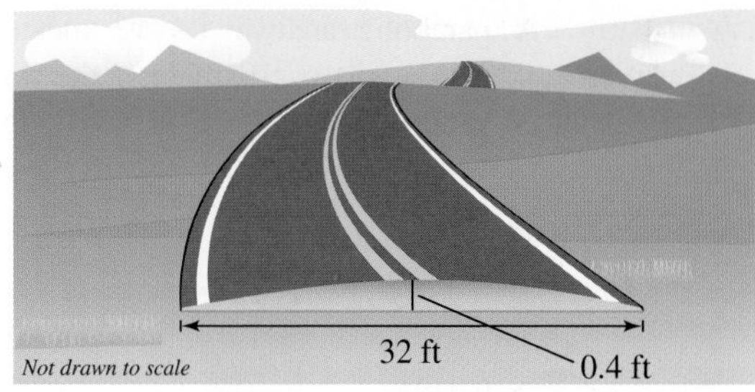

(a) Write an equation of the parabola with its vertex at the origin that models the road surface.

(b) How far from the center of the road is the road surface 0.1 foot lower than the center?

65. Beam Deflection A simply supported beam is 64 feet long and has a load at the center (see figure). The deflection of the beam at its center is 1 inch. The shape of the deflected beam is parabolic.

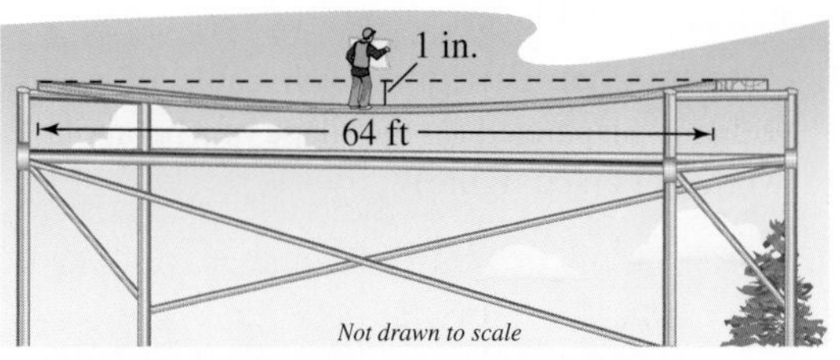

(a) Write an equation of the parabola with its vertex at the origin that models the shape of the beam.

(b) How far from the center of the beam is the deflection $\frac{1}{2}$ inch?

66. Beam Deflection Repeat Exercise 65 when the length of the beam is 36 feet and the deflection of the beam at its center is 2 inches.

67. **Fluid Flow** Water is flowing from a horizontal pipe 48 feet above the ground. The falling stream of water has the shape of a parabola whose vertex $(0, 48)$ is at the end of the pipe (see figure). The stream of water strikes the ocean at the point $(10\sqrt{3}, 0)$. Write an equation for the path of the water.

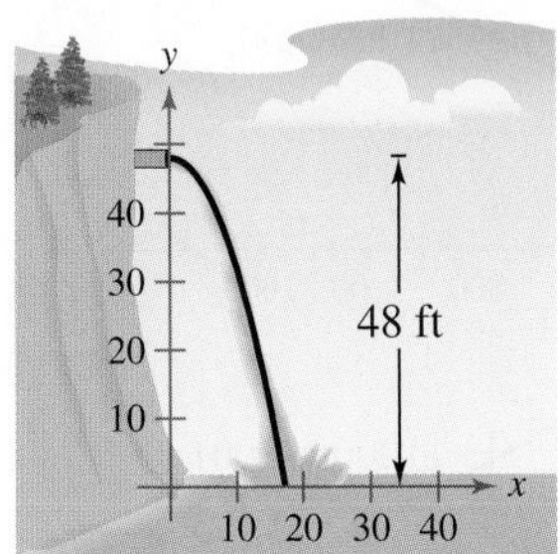

Figure for 67

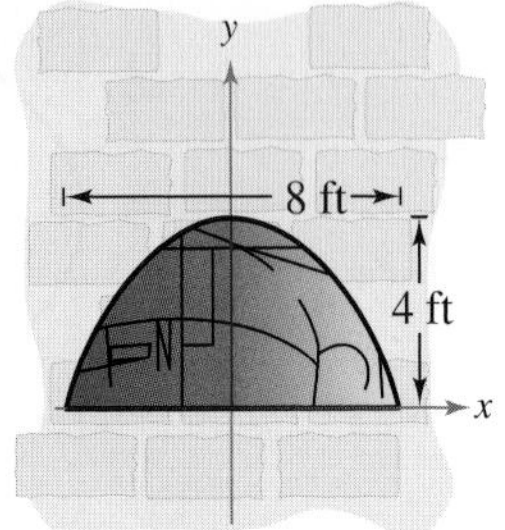

Figure for 68

68. **Window Design** A church window is bounded above by a parabola (see figure). Write an equation of the parabola.

69. **Archway** A parabolic archway is 12 meters high at the vertex. At a height of 10 meters, the width of the archway is 8 meters (see figure). How wide is the archway at ground level?

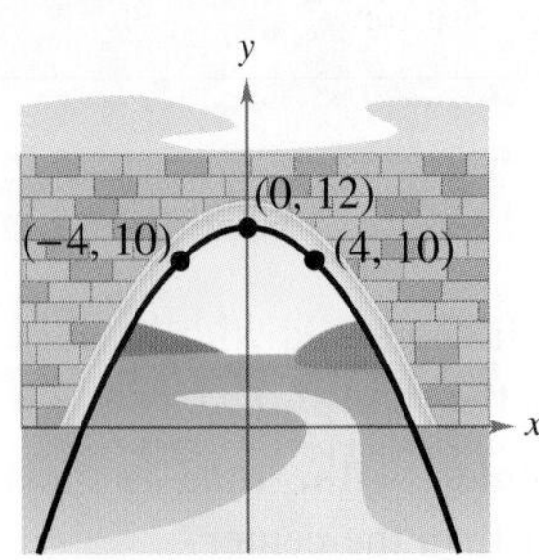

Figure for 69

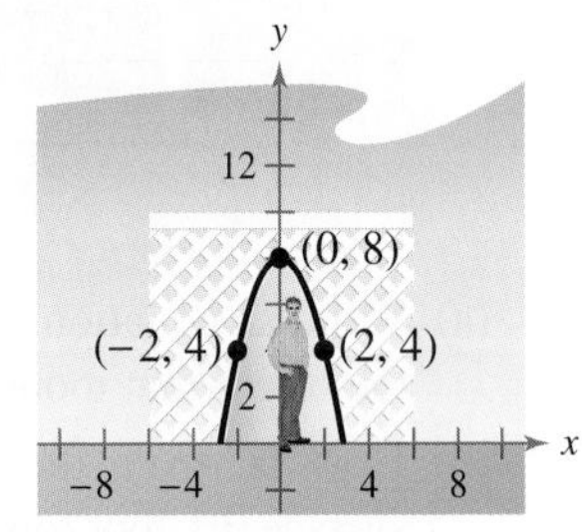

Figure for 70

70. **Lattice Arch** A parabolic lattice arch is 8 feet high at the vertex. At a height of 4 feet, the width of the lattice arch is 4 feet (see figure). How wide is the lattice arch at ground level?

71. **Suspension Bridge** Each cable of a suspension bridge is suspended (in the shape of a parabola) between two towers (see figure).

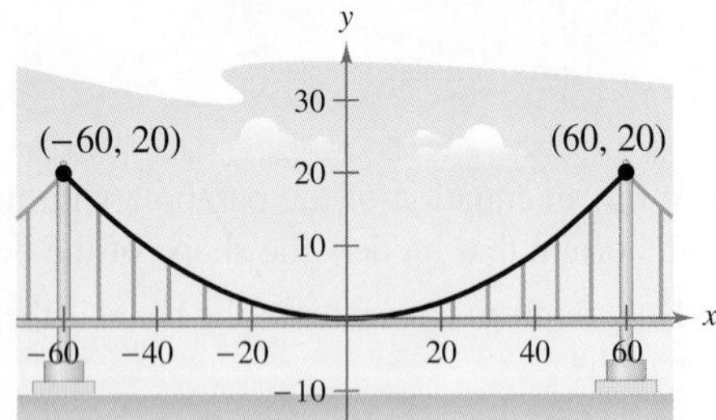

(a) Find the coordinates of the focus.

(b) Write an equation that models the cables.

72. **Suspension Bridge** Each cable of the Golden Gate Bridge is suspended (in the shape of a parabola) between two towers that are 1280 meters apart. The top of each tower is 152 meters above the roadway. The cables touch the roadway at the midpoint between the towers.

(a) Sketch the bridge on a rectangular coordinate system with the cables touching the roadway at the origin. Label the coordinates of the known points.

(b) Write an equation that models the cables.

(c) Complete the table by finding the height y of the cables over the roadway at a distance of x meters from the point where the cables touch the roadway.

Distance, x	Height, y
0	
100	
250	
400	
500	

73. **Satellite Orbit** A satellite in a 100-mile-high circular orbit around Earth has a velocity of approximately 17,500 miles per hour. When this velocity is multiplied by $\sqrt{2}$, the satellite has the minimum velocity necessary to escape Earth's gravity and follow a parabolic path with the center of Earth as the focus (see figure).

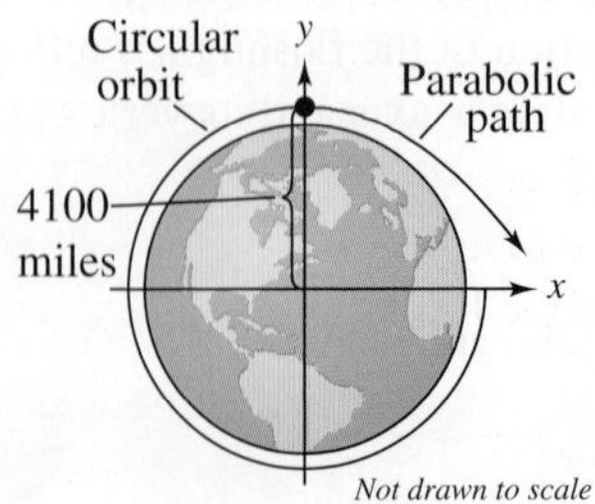

(a) Find the escape velocity of the satellite.

(b) Write an equation for the parabolic path of the satellite. (Assume that the radius of Earth is 4000 miles.)

74. Path of a Softball The path of a softball is modeled by

$$-12.5(y - 7.125) = (x - 6.25)^2$$

where x and y are measured in feet, with $x = 0$ corresponding to the position from which the ball was thrown.

(a) Use a graphing utility to graph the trajectory of the softball.

(b) Use the *trace* feature of the graphing utility to approximate the highest point and the range of the trajectory.

Projectile Motion **In Exercises 75 and 76, consider the path of an object projected horizontally with a velocity of v feet per second at a height of s feet, where the model for the path is**

$$x^2 = -\frac{v^2}{16}(y - s).$$

In this model (in which air resistance is disregarded), y is the height (in feet) of the projectile and x is the horizontal distance (in feet) the projectile travels.

75. A ball is thrown from the top of a 100-foot tower with a velocity of 28 feet per second.

(a) Write an equation for the parabolic path.

(b) How far does the ball travel horizontally before it strikes the ground?

76. A cargo plane is flying at an altitude of 500 feet and a speed of 255 miles per hour. A supply crate is dropped from the plane. How many *feet* will the crate travel horizontally before it hits the ground?

Exploration

True or False? **In Exercises 77–79, determine whether the statement is true or false. Justify your answer.**

77. It is possible for a parabola to intersect its directrix.

78. A tangent line to a parabola always intersects the directrix.

79. When the vertex and focus of a parabola are on a horizontal line, the directrix of the parabola is vertical.

80. Slope of a Tangent Line Let (x_1, y_1) be the coordinates of a point on the parabola $x^2 = 4py$. The equation of the line tangent to the parabola at the point is

$$y - y_1 = \frac{x_1}{2p}(x - x_1).$$

What is the slope of the tangent line?

81. Think About It Explain what each equation represents, and how equations (a) and (b) are equivalent.

(a) $y = a(x - h)^2 + k, \quad a \neq 0$

(b) $(x - h)^2 = 4p(y - k), \quad p \neq 0$

(c) $(y - k)^2 = 4p(x - h), \quad p \neq 0$

82. HOW DO YOU SEE IT?

In parts (a)–(d), describe how a plane could intersect the double-napped cone to form each conic section (see figure).

(a) Circle (b) Ellipse

(c) Parabola (d) Hyperbola

83. Think About It The graph of $x^2 + y^2 = 0$ is a degenerate conic. Sketch the graph of this equation and identify the degenerate conic. Describe the intersection of the plane and the double-napped cone for this conic.

84. Graphical Reasoning Consider the parabola $x^2 = 4py$.

(a) Use a graphing utility to graph the parabola for $p = 1$, $p = 2$, $p = 3$, and $p = 4$. Describe the effect on the graph when p increases.

(b) Find the focus for each parabola in part (a).

(c) For each parabola in part (a), find the length of the latus rectum (see figure). How can the length of the latus rectum be determined directly from the standard form of the equation of the parabola?

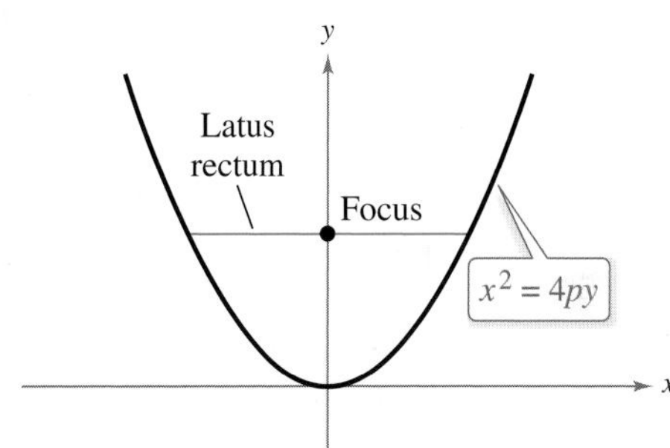

(d) How can you use the result of part (c) as a sketching aid when graphing parabolas?

85. Geometry The area of the shaded region in the figure is $A = \frac{8}{3}p^{1/2}b^{3/2}$.

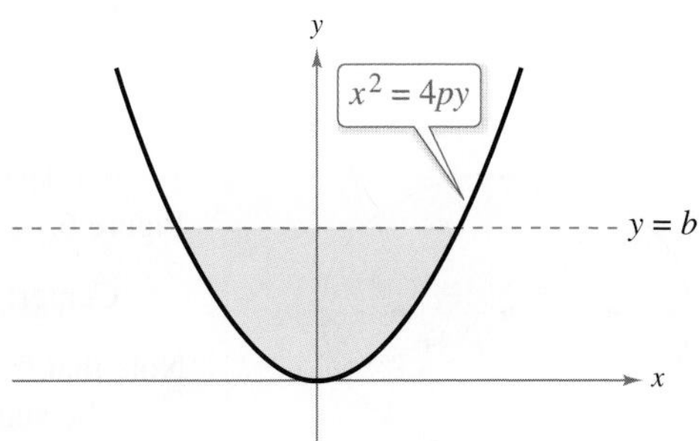

(a) Find the area when $p = 2$ and $b = 4$.

(b) Give a geometric explanation of why the area approaches 0 as p approaches 0.

6.3 Ellipses

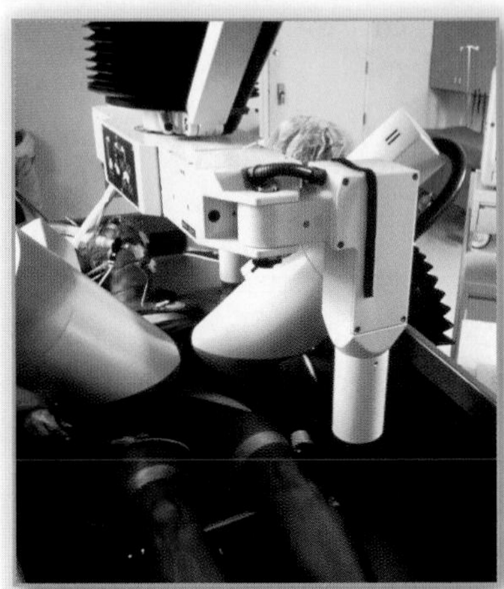

Ellipses have many real-life applications. For example, Exercise 55 on page 441 shows how a lithotripter machine uses the focal properties of an ellipse to break up kidney stones.

- **Write equations of ellipses in standard form and sketch ellipses.**
- **Use properties of ellipses to model and solve real-life problems.**
- **Find eccentricities of ellipses.**

Introduction

Another type of conic is an **ellipse.** It is defined below.

> **Definition of an Ellipse**
>
> An **ellipse** is the set of all points (x, y) in a plane, the sum of whose distances from two distinct fixed points **(foci)** is constant. See Figure 6.12.

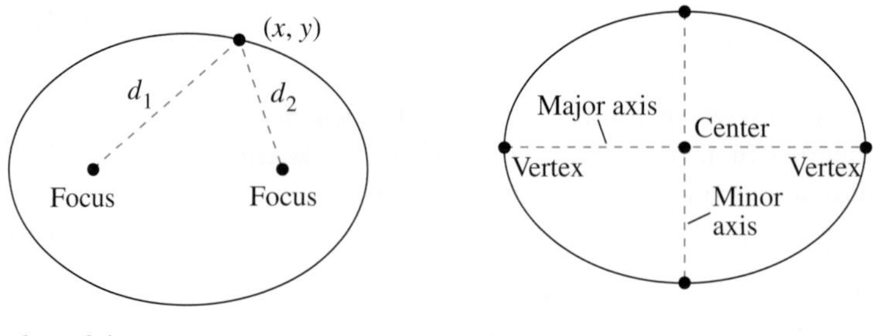

$d_1 + d_2$ is constant.
Figure 6.12

Figure 6.13

The line through the foci intersects the ellipse at two points **(vertices).** The chord joining the vertices is the **major axis,** and its midpoint is the **center** of the ellipse. The chord perpendicular to the major axis at the center is the **minor axis** of the ellipse. (See Figure 6.13.)

To visualize the definition of an ellipse, imagine two thumbtacks placed at the foci, as shown in the figure below. When the ends of a fixed length of string are fastened to the thumbtacks and the string is drawn taut with a pencil, the path traced by the pencil is an ellipse.

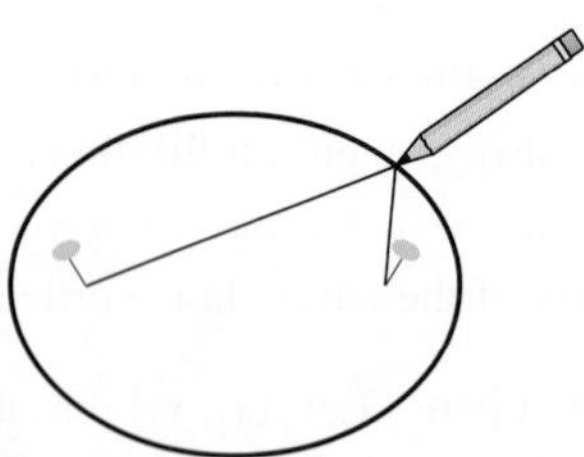

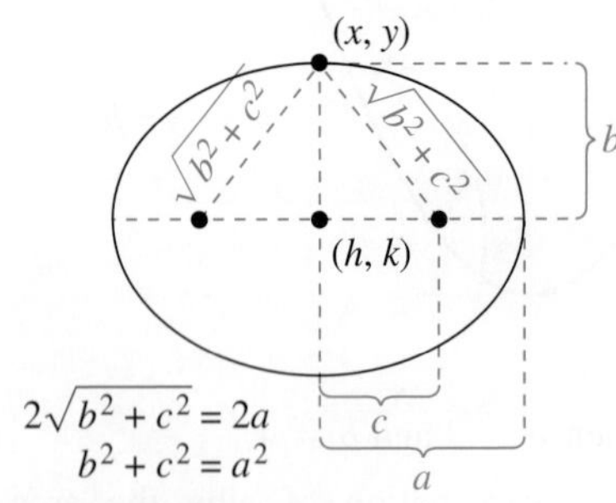

Figure 6.14

To derive the standard form of the equation of an ellipse, consider the ellipse in Figure 6.14 with the points listed below.

Center: (h, k) Vertices: $(h \pm a, k)$ Foci: $(h \pm c, k)$

Note that the center is also the midpoint of the segment joining the foci.

The sum of the distances from any point on the ellipse to the two foci is constant. Using a vertex point, this constant sum is

$$(a + c) + (a - c) = 2a \qquad \text{Length of major axis}$$

which is the length of the major axis.

Southern Illinois University -/Getty Images

Now, if you let (x, y) be *any* point on the ellipse, then the sum of the distances between (x, y) and the two foci must also be $2a$. That is,

$$\sqrt{[x - (h - c)]^2 + (y - k)^2} + \sqrt{[x - (h + c)]^2 + (y - k)^2} = 2a$$

which, after expanding and regrouping, reduces to

$$(a^2 - c^2)(x - h)^2 + a^2(y - k)^2 = a^2(a^2 - c^2).$$

From Figure 6.14,

$$b^2 + c^2 = a^2$$

$$b^2 = a^2 - c^2$$

which implies that the equation of the ellipse is

$$b^2(x - h)^2 + a^2(y - k)^2 = a^2b^2$$

$$\frac{(x - h)^2}{a^2} + \frac{(y - k)^2}{b^2} = 1.$$

You would obtain a similar equation in the derivation by starting with a vertical major axis. A summary of these results is given below.

REMARK Consider the equation of the ellipse

$$\frac{(x - h)^2}{a^2} + \frac{(y - k)^2}{b^2} = 1.$$

If you let $a = b = r$, then the equation can be rewritten as

$$(x - h)^2 + (y - k)^2 = r^2$$

which is the standard form of the equation of a circle with radius r. Geometrically, when $a = b$ for an ellipse, the major and minor axes are of equal length, and so the graph is a circle.

Standard Equation of an Ellipse

The **standard form of the equation of an ellipse** with center (h, k) and major and minor axes of lengths $2a$ and $2b$, respectively, where $0 < b < a$, is

$$\frac{(x - h)^2}{a^2} + \frac{(y - k)^2}{b^2} = 1 \qquad \text{Major axis is horizontal.}$$

$$\frac{(x - h)^2}{b^2} + \frac{(y - k)^2}{a^2} = 1. \qquad \text{Major axis is vertical.}$$

The foci lie on the major axis, c units from the center, with

$$c^2 = a^2 - b^2.$$

If the center is at the origin, then the equation takes one of the forms below.

$$\frac{x^2}{a^2} + \frac{y^2}{b^2} = 1 \qquad \text{Major axis is horizontal.}$$

$$\frac{x^2}{b^2} + \frac{y^2}{a^2} = 1 \qquad \text{Major axis is vertical.}$$

The figures below show generalized horizontal and vertical orientations for ellipses.

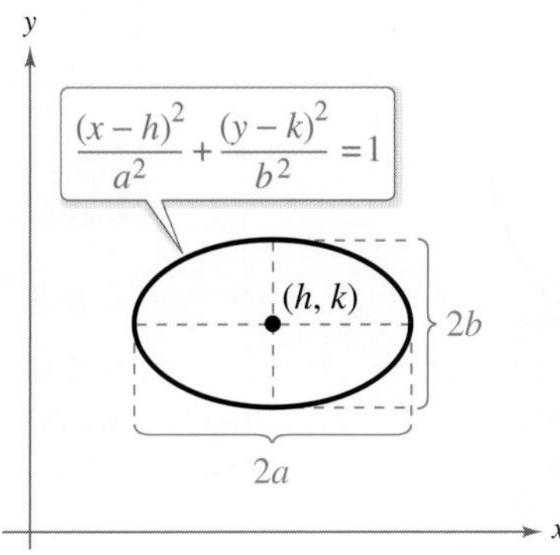

Major axis is horizontal.

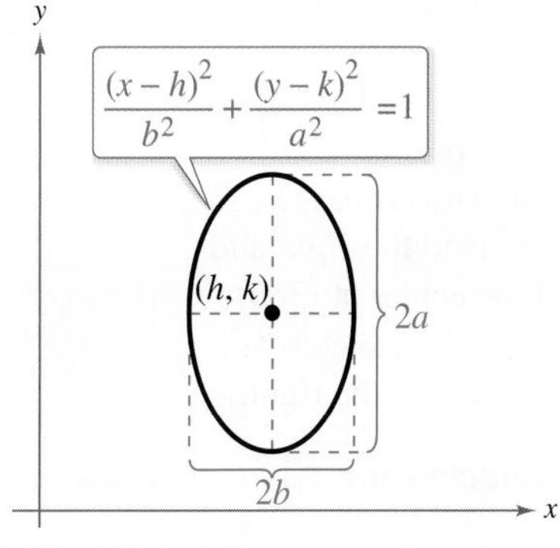

Major axis is vertical.

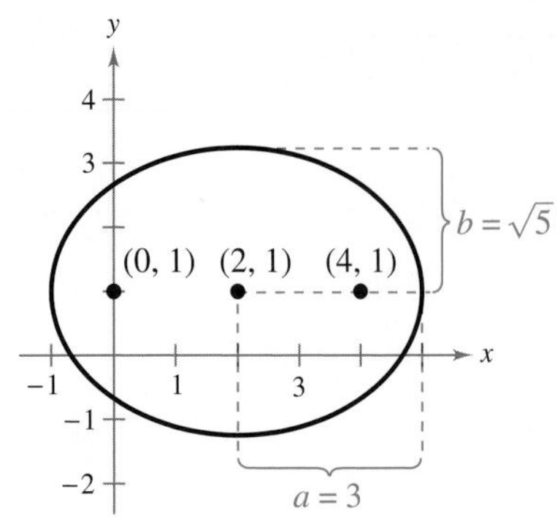

Figure 6.15

EXAMPLE 1 Finding the Standard Equation of an Ellipse

Find the standard form of the equation of the ellipse with foci $(0, 1)$ and $(4, 1)$ and major axis of length 6, as shown in Figure 6.15.

Solution The foci occur at $(0, 1)$ and $(4, 1)$, so the center of the ellipse is $(2, 1)$ and the distance from the center to one of the foci is $c = 2$. Because $2a = 6$, you know that $a = 3$. Now, from $c^2 = a^2 - b^2$, you have

$$b = \sqrt{a^2 - c^2} = \sqrt{3^2 - 2^2} = \sqrt{5}.$$

The major axis is horizontal, so the standard form of the equation is

$$\frac{(x-2)^2}{3^2} + \frac{(y-1)^2}{(\sqrt{5})^2} = 1.$$

This equation simplifies to

$$\frac{(x-2)^2}{9} + \frac{(y-1)^2}{5} = 1.$$

✓ **Checkpoint** Audio-video solution in English & Spanish at LarsonPrecalculus.com

Find the standard form of the equation of the ellipse with foci $(2, 0)$ and $(2, 6)$ and major axis of length 8.

EXAMPLE 2 Sketching an Ellipse

Sketch the ellipse $4x^2 + y^2 = 36$ and identify the center and vertices.

Algebraic Solution

$4x^2 + y^2 = 36$ Write original equation.

$\frac{4x^2}{36} + \frac{y^2}{36} = \frac{36}{36}$ Divide each side by 36.

$\frac{x^2}{3^2} + \frac{y^2}{6^2} = 1$ Write in standard form.

The center of the ellipse is $(0, 0)$. The denominator of the y^2-term is greater than the denominator of the x^2-term, so the major axis is vertical. Moreover, $a^2 = 6^2$, so the endpoints of the major axis (the vertices) lie six units *up* and *down* from the center at $(0, 6)$ and $(0, -6)$. Similarly, the denominator of the x^2-term is $b^2 = 3^2$, so the endpoints of the minor axis (the *co-vertices*) lie three units to the *right* and *left* of the center at $(3, 0)$ and $(-3, 0)$. A sketch of the ellipse is at the right.

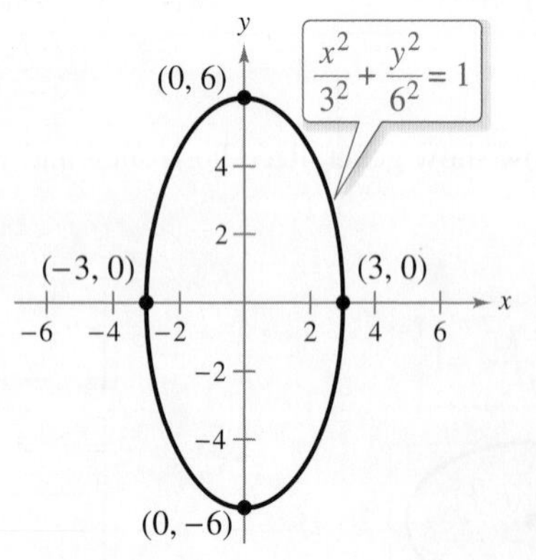

Graphical Solution

Solve the equation of the ellipse for y.

$$4x^2 + y^2 = 36$$
$$y^2 = 36 - 4x^2$$
$$y = \pm\sqrt{36 - 4x^2}$$

Then, use a graphing utility to graph

$$y_1 = \sqrt{36 - 4x^2} \quad \text{and} \quad y_2 = -\sqrt{36 - 4x^2}$$

in the same viewing window, as shown in the figure below. Be sure to use a square setting.

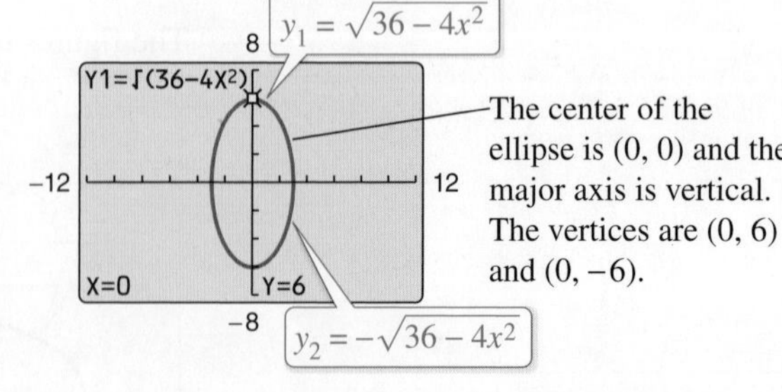

  Audio-video solution in English & Spanish at LarsonPrecalculus.com

Sketch the ellipse $x^2 + 9y^2 = 81$ and identify the center and vertices. ■

EXAMPLE 3 Sketching an Ellipse

Find the center, vertices, and foci of the ellipse $x^2 + 4y^2 + 6x - 8y + 9 = 0$. Then sketch the ellipse.

Solution Begin by writing the original equation in standard form. In the third step, note that you add 9 and 4 to *both* sides of the equation when completing the squares.

$$x^2 + 4y^2 + 6x - 8y + 9 = 0 \quad \text{Write original equation.}$$

$$(x^2 + 6x + \square) + 4(y^2 - 2y + \square) = -9 \quad \text{Group terms and factor 4 out of } y\text{-terms.}$$

$$(x^2 + 6x + 9) + 4(y^2 - 2y + 1) = -9 + 9 + 4(1) \quad \text{Complete the squares.}$$

$$(x + 3)^2 + 4(y - 1)^2 = 4 \quad \text{Write in completed square form.}$$

$$\frac{(x + 3)^2}{2^2} + \frac{(y - 1)^2}{1^2} = 1 \quad \text{Write in standard form.}$$

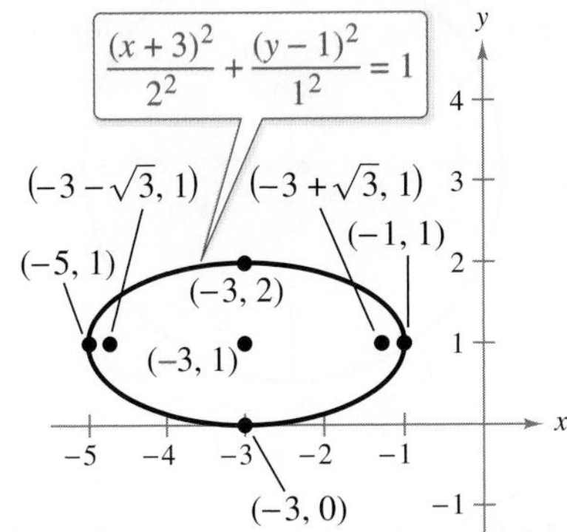

Figure 6.16

From this standard form, it follows that the center is $(h, k) = (-3, 1)$. The denominator of the x-term is $a^2 = 2^2$, so the endpoints of the major axis lie two units to the right and left of the center and the vertices are $(-1, 1)$ and $(-5, 1)$. Similarly, the denominator of the y-term is $b^2 = 1^2$, so the covertices lie one unit up and down from the center at $(-3, 2)$ and $(-3, 0)$. Now, from $c^2 = a^2 - b^2$, you have

$$c = \sqrt{2^2 - 1^2} = \sqrt{3}.$$

So, the foci of the ellipse are $(-3 + \sqrt{3}, 1)$ and $(-3 - \sqrt{3}, 1)$. Figure 6.16 shows the ellipse.

Find the center, vertices, and foci of the ellipse $9x^2 + 4y^2 + 36x - 8y + 4 = 0$. Then sketch the ellipse.

EXAMPLE 4 Sketching an Ellipse

See LarsonPrecalculus.com for an interactive version of this type of example.

Find the center, vertices, and foci of the ellipse $4x^2 + y^2 - 8x + 4y - 8 = 0$. Then sketch the ellipse.

Solution Complete the square to write the original equation in standard form.

$$4x^2 + y^2 - 8x + 4y - 8 = 0 \quad \text{Write original equation.}$$

$$4(x^2 - 2x + \square) + (y^2 + 4y + \square) = 8 \quad \text{Group terms and factor 4 out of } x\text{-terms.}$$

$$4(x^2 - 2x + 1) + (y^2 + 4y + 4) = 8 + 4(1) + 4 \quad \text{Complete the squares.}$$

$$4(x - 1)^2 + (y + 2)^2 = 16 \quad \text{Write in completed square form.}$$

$$\frac{(x - 1)^2}{2^2} + \frac{(y + 2)^2}{4^2} = 1 \quad \text{Write in standard form.}$$

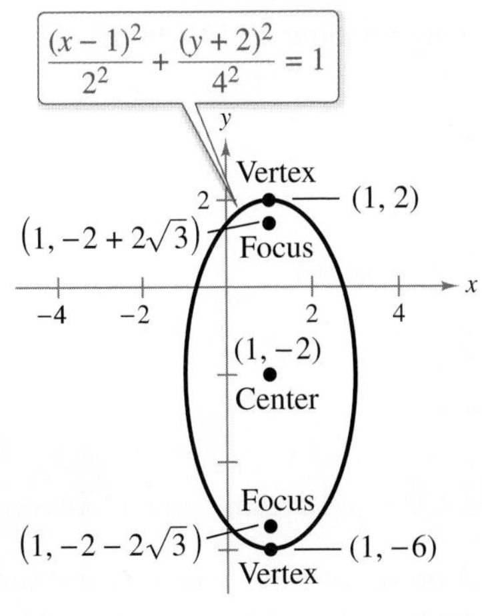

Figure 6.17

The major axis is vertical, $h = 1$, $k = -2$, $a = 4$, $b = 2$, and

$$c = \sqrt{a^2 - b^2} = \sqrt{16 - 4} = \sqrt{12} = 2\sqrt{3}.$$

So, the center is $(h, k) = (1, -2)$, the vertices are $(1, -6)$ and $(1, 2)$, and the foci are $(1, -2 - 2\sqrt{3})$ and $(1, -2 + 2\sqrt{3})$. Figure 6.17 shows the ellipse.

Find the center, vertices, and foci of the ellipse $5x^2 + 9y^2 + 10x - 54y + 41 = 0$. Then sketch the ellipse. ■

Application

Ellipses have many practical and aesthetic uses. For example, machine gears, supporting arches, and acoustic designs often involve elliptical shapes. The orbits of satellites and planets are also ellipses. Example 5 investigates the elliptical orbit of the moon about Earth.

EXAMPLE 5 An Application Involving an Elliptical Orbit

The moon travels about Earth in an elliptical orbit with the center of Earth at one focus, as shown in the figure at the right. The major and minor axes of the orbit have lengths of 768,800 kilometers and 767,641 kilometers, respectively. Find the greatest and least distances (the *apogee* and *perigee,* respectively) from Earth's center to the moon's center. Then use a graphing utility to graph the orbit of the moon.

REMARK Note in Example 5 that Earth is *not* the center of the moon's orbit.

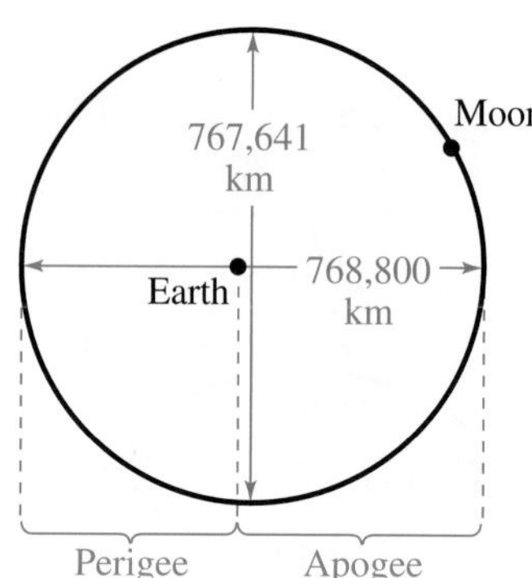

Solution Because $2a = 768{,}800$ and $2b = 767{,}641$, you have $a = 384{,}400$ and $b = 383{,}820.5$, which implies that

$$c = \sqrt{a^2 - b^2} = \sqrt{384{,}400^2 - 383{,}820.5^2} \approx 21{,}099.$$

So, the greatest distance between the center of Earth and the center of the moon is

$$a + c \approx 384{,}400 + 21{,}099 = 405{,}499 \text{ kilometers}$$

and the least distance is

$$a - c \approx 384{,}400 - 21{,}099 = 363{,}301 \text{ kilometers.}$$

To use a graphing utility to graph the orbit of the moon, first let $a = 384{,}400$ and $b = 383{,}820.5$ in the standard form of an equation of an ellipse centered at the origin, and then solve for y.

$$\frac{x^2}{384{,}400^2} + \frac{y^2}{383{,}820.5^2} = 1 \implies y = \pm 383{,}820.5\sqrt{1 - \frac{x^2}{384{,}400^2}}$$

Graph the upper and lower portions in the same viewing window, as shown below.

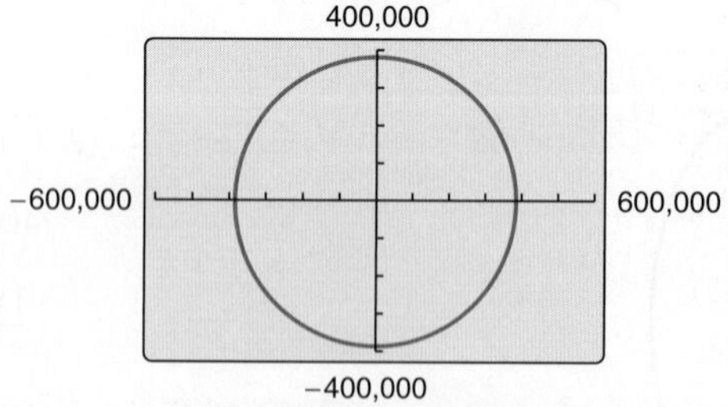

In Exercise 56, you will investigate the elliptical orbit of Halley's comet about the sun. Halley's comet is visible from Earth approximately every 76.1 years. The comet's latest appearance was in 1986.

✓ **Checkpoint** *Audio-video solution in English & Spanish at LarsonPrecalculus.com*

Encke's comet travels about the sun in an elliptical orbit with the center of the sun at one focus. The major and minor axes of the orbit have lengths of approximately 4.420 astronomical units and 2.356 astronomical units, respectively. (An astronomical unit is about 93 million miles.) Find the greatest and least distances (the *aphelion* and *perihelion,* respectively). from the sun's center to the comet's center. Then use a graphing utility to graph the orbit of the comet.

Eccentricity

It was difficult for early astronomers to detect that the orbits of the planets are ellipses because the foci of the planetary orbits are relatively close to their centers, and so the orbits are nearly circular. You can measure the "ovalness" of an ellipse by using the concept of **eccentricity.**

> **Definition of Eccentricity**
>
> The **eccentricity** e of an ellipse is the ratio $e = \dfrac{c}{a}$.

Note that $0 < e < 1$ for *every* ellipse.

To see how this ratio describes the shape of an ellipse, note that because the foci of an ellipse are located along the major axis between the vertices and the center, it follows that $0 < c < a$. For an ellipse that is nearly circular, the foci are close to the center and the ratio c/a is close to 0, as shown in Figure 6.18. On the other hand, for an elongated ellipse, the foci are close to the vertices and the ratio c/a is close to 1, as shown in Figure 6.19.

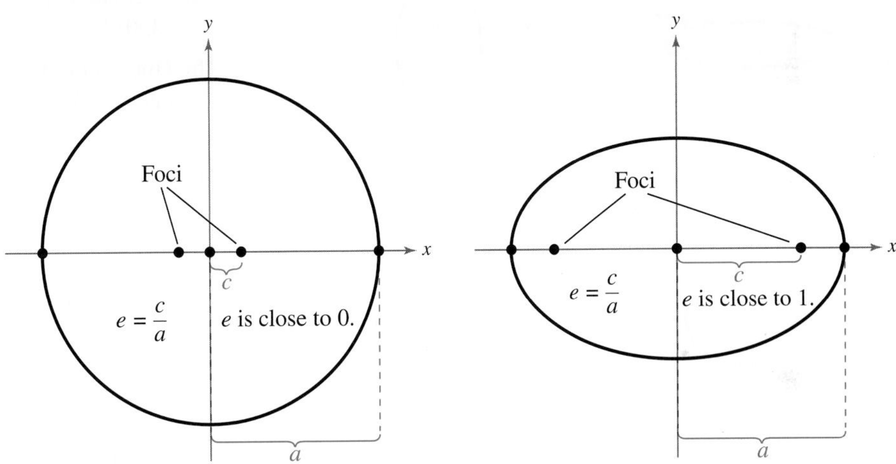

Figure 6.18 **Figure 6.19**

The orbit of the moon has an eccentricity of $e \approx 0.0549$. The eccentricities of the eight planetary orbits are listed below.

Mercury:	$e \approx 0.2056$	Jupiter:	$e \approx 0.0489$
Venus:	$e \approx 0.0067$	Saturn:	$e \approx 0.0565$
Earth:	$e \approx 0.0167$	Uranus:	$e \approx 0.0457$
Mars:	$e \approx 0.0935$	Neptune:	$e \approx 0.0113$

The time it takes Saturn to orbit the sun is about 29.5 Earth years.

Summarize (Section 6.3)

1. State the definition of an ellipse and the standard form of the equation of an ellipse *(page 434)*. For examples involving the equations and graphs of ellipses, see Examples 1–4.
2. Describe a real-life application of an ellipse *(page 438, Example 5)*.
3. State the definition of the eccentricity of an ellipse and explain how eccentricity describes the shape of an ellipse *(page 439)*.

6.3 Exercises

See CalcChat.com for tutorial help and worked-out solutions to odd-numbered exercises.

Vocabulary: Fill in the blanks.

1. An ________ is the set of all points (x, y) in a plane, the sum of whose distances from two distinct fixed points, called ________, is constant.
2. The chord joining the vertices of an ellipse is the ________ ________, and its midpoint is the ________ of the ellipse.
3. The chord perpendicular to the major axis at the center of an ellipse is the ________ ________ of the ellipse.
4. You can measure the "ovalness" of an ellipse by using the concept of ________.

Skills and Applications

Matching **In Exercises 5–8, match the equation with its graph. [The graphs are labeled (a), (b), (c), and (d).]**

(a)

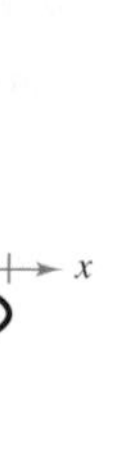

(b)

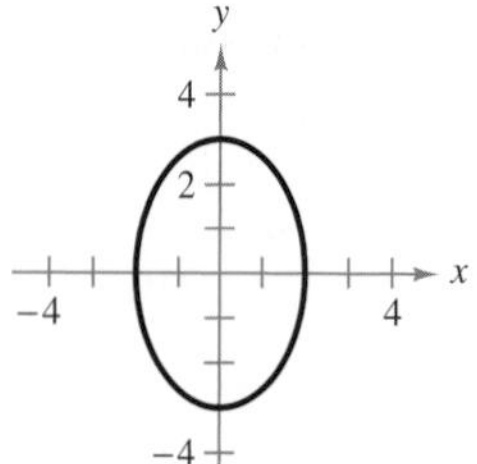

(c)

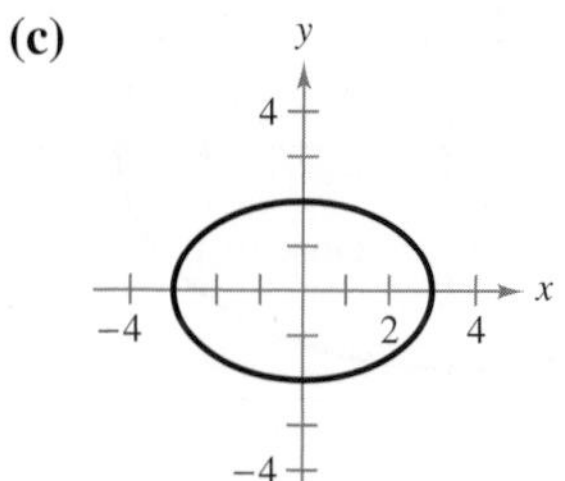

(d)

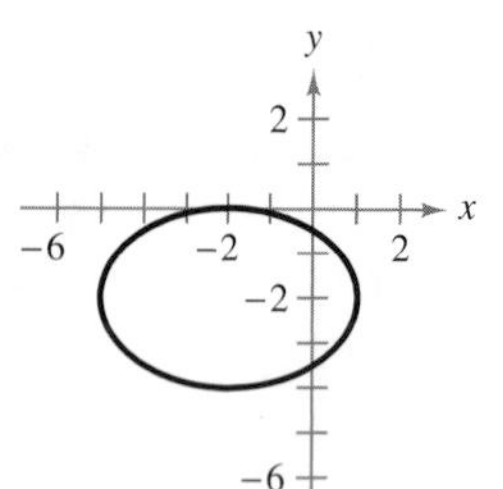

5. $\dfrac{x^2}{4} + \dfrac{y^2}{9} = 1$

6. $\dfrac{x^2}{9} + \dfrac{y^2}{4} = 1$

7. $\dfrac{(x-2)^2}{16} + (y+1)^2 = 1$

8. $\dfrac{(x+2)^2}{9} + \dfrac{(y+2)^2}{4} = 1$

An Ellipse Centered at the Origin **In Exercises 9–18, find the standard form of the equation of the ellipse with the given characteristics and center at the origin.**

9.

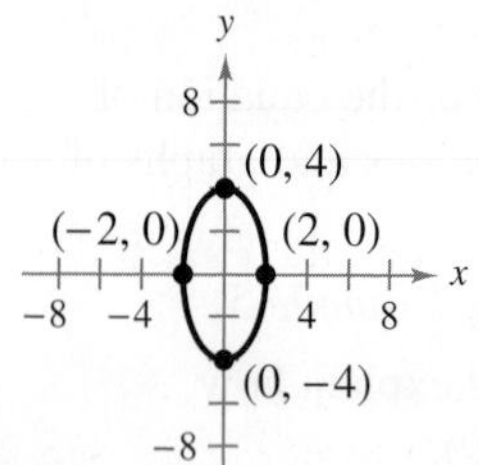

10.

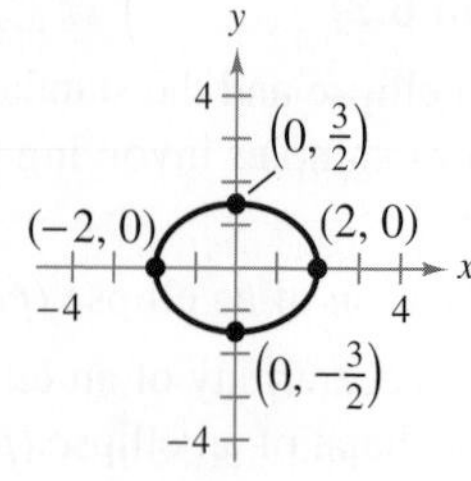

11. Vertices: $(\pm 7, 0)$; foci: $(\pm 2, 0)$
12. Vertices: $(0, \pm 8)$; foci: $(0, \pm 4)$
13. Foci: $(\pm 4, 0)$; major axis of length 10
14. Foci: $(0, \pm 3)$; major axis of length 8
15. Vertical major axis; passes through the points $(0, 6)$ and $(3, 0)$
16. Horizontal major axis; passes through the points $(5, 0)$ and $(0, 2)$
17. Vertices: $(\pm 6, 0)$; passes through the point $(4, 1)$
18. Vertices: $(0, \pm 8)$; passes through the point $(3, 4)$

Finding the Standard Equation of an Ellipse **In Exercises 19–30, find the standard form of the equation of the ellipse with the given characteristics.**

19.

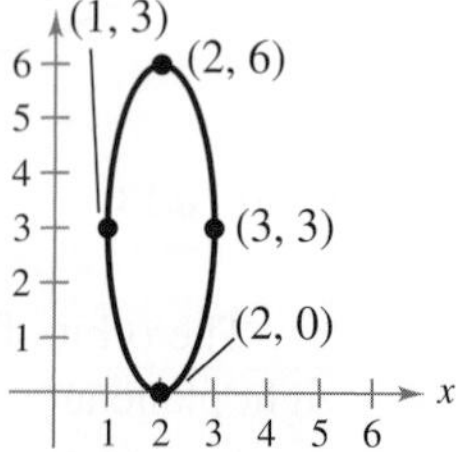

20.

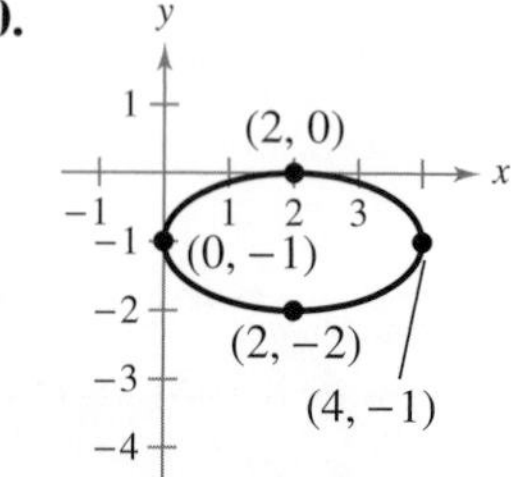

21. Vertices: $(2, 0)$, $(10, 0)$; minor axis of length 4
22. Vertices: $(3, 1)$, $(3, 11)$; minor axis of length 2
23. Foci: $(0, 0)$, $(4, 0)$; major axis of length 6
24. Foci: $(0, 0)$, $(0, 8)$; major axis of length 16
25. Center: $(1, 3)$; vertex: $(-2, 3)$; minor axis of length 4
26. Center: $(2, -1)$; vertex: $\left(2, \frac{1}{2}\right)$; minor axis of length 2
27. Center: $(1, 4)$; $a = 2c$; vertices: $(1, 0)$, $(1, 8)$
28. Center: $(3, 2)$; $a = 3c$; foci: $(1, 2)$, $(5, 2)$
29. Vertices: $(0, 2)$, $(4, 2)$; endpoints of the minor axis: $(2, 3)$, $(2, 1)$
30. Vertices: $(5, 0)$, $(5, 12)$; endpoints of the minor axis: $(1, 6)$, $(9, 6)$

Sketching an Ellipse **In Exercises 31–46, find the center, vertices, foci, and eccentricity of the ellipse. Then sketch the ellipse.**

31. $\dfrac{x^2}{25} + \dfrac{y^2}{16} = 1$

32. $\dfrac{x^2}{16} + \dfrac{y^2}{81} = 1$

33. $9x^2 + y^2 = 36$

34. $x^2 + 16y^2 = 64$

35. $\dfrac{(x-4)^2}{16} + \dfrac{(y+1)^2}{25} = 1$

36. $\dfrac{(x+3)^2}{12} + \dfrac{(y-2)^2}{16} = 1$

37. $\dfrac{(x+5)^2}{9/4} + (y-1)^2 = 1$

38. $(x+2)^2 + \dfrac{(y+4)^2}{1/4} = 1$

39. $9x^2 + 4y^2 + 36x - 24y + 36 = 0$

40. $9x^2 + 4y^2 - 54x + 40y + 37 = 0$

41. $x^2 + 5y^2 - 8x - 30y - 39 = 0$

42. $3x^2 + y^2 + 18x - 2y - 8 = 0$

43. $6x^2 + 2y^2 + 18x - 10y + 2 = 0$

44. $x^2 + 4y^2 - 6x + 20y - 2 = 0$

45. $12x^2 + 20y^2 - 12x + 40y - 37 = 0$

46. $36x^2 + 9y^2 + 48x - 36y + 43 = 0$

Graphing an Ellipse **In Exercises 47–50, use a graphing utility to graph the ellipse. Find the center, foci, and vertices.**

47. $5x^2 + 3y^2 = 15$

48. $3x^2 + 4y^2 = 12$

49. $x^2 + 9y^2 - 10x + 36y + 52 = 0$

50. $4x^2 + 3y^2 - 8x + 18y + 19 = 0$

51. **Using Eccentricity** Find an equation of the ellipse with vertices $(\pm 5, 0)$ and eccentricity $e = \dfrac{4}{5}$.

52. **Using Eccentricity** Find an equation of the ellipse with vertices $(0, \pm 8)$ and eccentricity $e = \dfrac{1}{2}$.

53. **Architecture** Statuary Hall is an elliptical room in the United States Capitol in Washington, D.C. The room is also called the Whispering Gallery because a person standing at one focus of the room can hear even a whisper spoken by a person standing at the other focus. The dimensions of Statuary Hall are 46 feet wide by 97 feet long.
    (a) Find an equation of the shape of the room.
    (b) Determine the distance between the foci.

54. **Architecture** A mason is building a semielliptical fireplace arch that has a height of 2 feet at the center and a width of 6 feet along the base (see figure). The mason draws the semiellipse on the wall by the method shown on page 434. Find the positions of the thumbtacks and the length of the string.

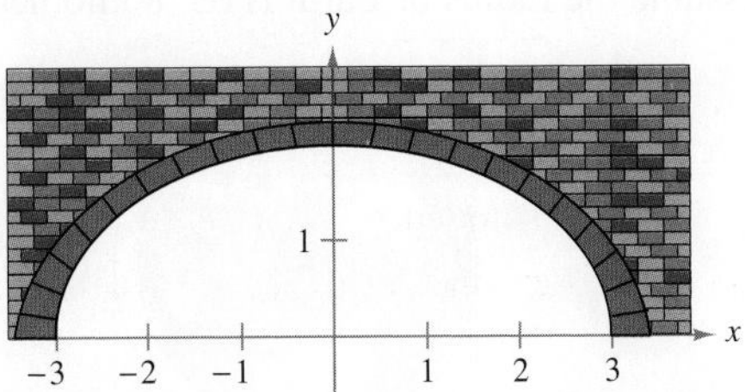

55. **Lithotripter**

A lithotripter machine uses an elliptical reflector to break up kidney stones nonsurgically. A spark plug in the reflector generates energy waves at one focus of an ellipse. The reflector directs these waves toward the kidney stone, positioned at the other focus of the ellipse, with enough energy to break up the stone (see figure). The lengths of the major and minor axes of the ellipse are 280 millimeters and 160 millimeters, respectively. How far is the spark plug from the kidney stone?

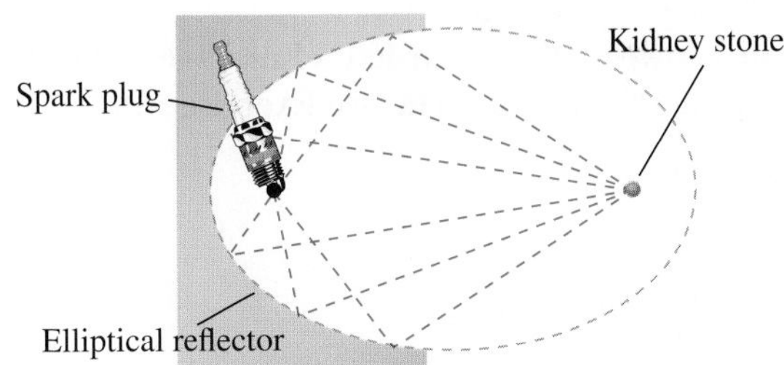

56. **Astronomy** Halley's comet has an elliptical orbit with the center of the sun at one focus. The eccentricity of the orbit is approximately 0.967. The length of the major axis of the orbit is approximately 35.88 astronomical units. (An astronomical unit is about 93 million miles.)
    (a) Find an equation of the orbit. Place the center of the orbit at the origin and place the major axis on the x-axis.
    (b) Use a graphing utility to graph the equation of the orbit.
    (c) Find the greatest and least distances (the aphelion and perihelion, respectively) from the sun's center to the comet's center.

57. Astronomy The first artificial satellite to orbit Earth was Sputnik I (launched by the former Soviet Union in 1957). Its highest point above Earth's surface was 939 kilometers, and its lowest point was 215 kilometers (see figure). The center of Earth was at one focus of the elliptical orbit. Find the eccentricity of the orbit. (Assume the radius of Earth is 6378 kilometers.)

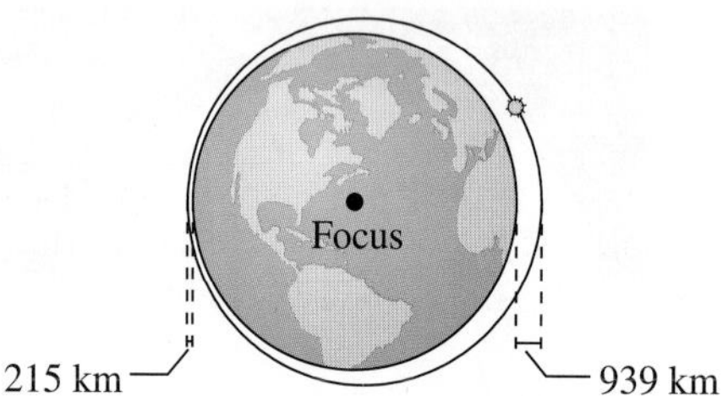

58. Geometry A line segment through a focus of an ellipse with endpoints on the ellipse and perpendicular to the major axis is called a **latus rectum** of the ellipse. An ellipse has two latera recta. Knowing the length of the latera recta is helpful in sketching an ellipse because it yields other points on the curve (see figure). Show that the length of each latus rectum is $2b^2/a$.

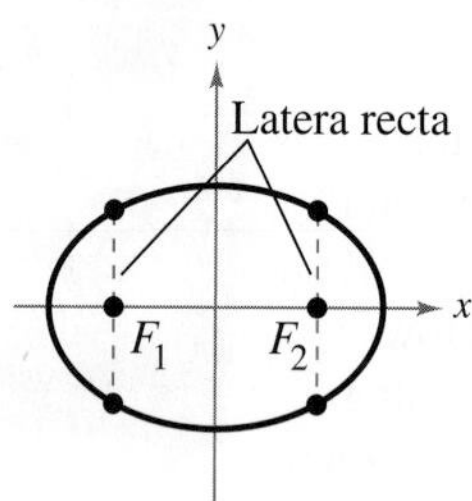

Using Latera Recta **In Exercises 59–62, sketch the ellipse using the latera recta (see Exercise 58).**

59. $\dfrac{x^2}{9} + \dfrac{y^2}{16} = 1$

60. $\dfrac{x^2}{4} + \dfrac{y^2}{1} = 1$

61. $5x^2 + 3y^2 = 15$

62. $9x^2 + 4y^2 = 36$

Exploration

True or False? **In Exercises 63 and 64, determine whether the statement is true or false. Justify your answer.**

63. The graph of $x^2 + 4y^4 - 4 = 0$ is an ellipse.

64. It is easier to distinguish the graph of an ellipse from the graph of a circle when the eccentricity of the ellipse is close to 1.

65. Think About It Find an equation of an ellipse such that for any point on the ellipse, the sum of the distances from the point to the points (2, 2) and (10, 2) is 36.

66. Think About It At the beginning of this section, you learned that an ellipse can be drawn using two thumbtacks, a string of fixed length (greater than the distance between the two thumbtacks), and a pencil. When the ends of the string are fastened to the thumbtacks and the string is drawn taut with the pencil, the path traced by the pencil is an ellipse.

(a) What is the length of the string in terms of a?

(b) Explain why the path is an ellipse.

67. Conjecture Consider the ellipse

$$\frac{x^2}{a^2} + \frac{y^2}{b^2} = 1, \quad a + b = 20.$$

(a) The area of the ellipse is given by $A = \pi ab$. Write the area of the ellipse as a function of a.

(b) Find the equation of an ellipse with an area of 264 square centimeters.

(c) Complete the table using your equation from part (a). Then make a conjecture about the shape of the ellipse with maximum area.

a	8	9	10	11	12	13
A						

(d) Use a graphing utility to graph the area function and use the graph to support your conjecture in part (c).

68. HOW DO YOU SEE IT? Without performing any calculations, order the eccentricities of the ellipses from least to greatest.

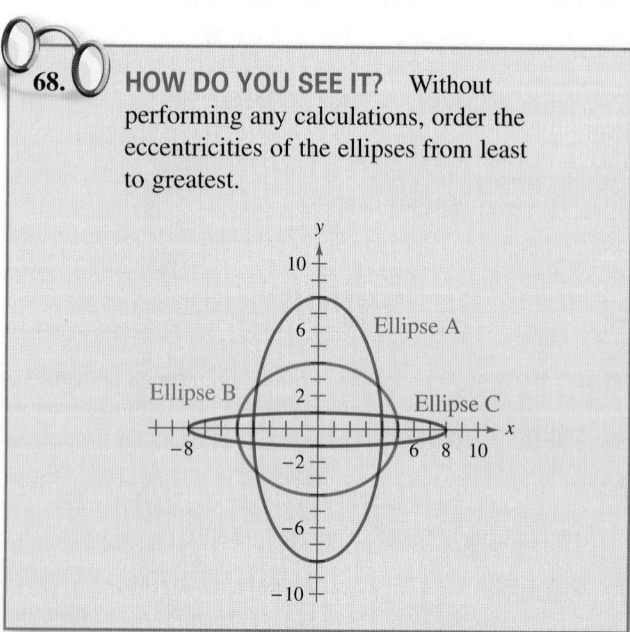

69. Proof Show that $a^2 = b^2 + c^2$ for the ellipse

$$\frac{x^2}{a^2} + \frac{y^2}{b^2} = 1$$

where $a > 0, b > 0$, and the distance from the center of the ellipse (0, 0) to a focus is c.

6.4 Hyperbolas

Hyperbolas have many types of real-life applications. For example, in Exercise 53 on page 451, you will investigate the use of hyperbolas in long distance radio navigation for aircraft and ships.

- **Write equations of hyperbolas in standard form.**
- **Find asymptotes of and sketch hyperbolas.**
- **Use properties of hyperbolas to solve real-life problems.**
- **Classify conics from their general equations.**

Introduction

The definition of a **hyperbola** is similar to that of an ellipse. For an ellipse, the *sum* of the distances between the foci and a point on the ellipse is constant. For a hyperbola, the absolute value of the *difference* of the distances between the foci and a point on the hyperbola is constant.

> **Definition of a Hyperbola**
>
> A **hyperbola** is the set of all points (x, y) in a plane for which the absolute value of the difference of the distances from two distinct fixed points **(foci)** is constant. See Figure 6.20.

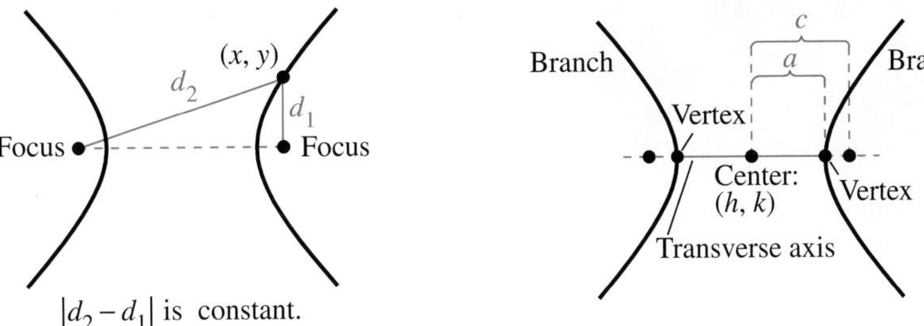

Figure 6.20

Figure 6.21

The graph of a hyperbola has two disconnected parts **(branches).** The line through the foci intersects the hyperbola at two points **(vertices).** The line segment connecting the vertices is the **transverse axis,** and its midpoint is the **center** of the hyperbola.

Consider the hyperbola in Figure 6.21 with the points listed below.

Center: (h, k) Vertices: $(h \pm a, k)$ Foci: $(h \pm c, k)$

Note that the center is also the midpoint of the segment joining the foci.

The absolute value of the difference of the distances from *any* point on the hyperbola to the two foci is constant. Using a vertex point, this constant value is

$$|[2a + (c - a)] - (c - a)| = |2a| = 2a \qquad \text{Length of transverse axis}$$

which is the length of the transverse axis. Now, if you let (x, y) be *any* point on the hyperbola, then

$$|d_2 - d_1| = 2a$$

(see Figure 6.20). You would obtain the same result for a hyperbola with a vertical transverse axis.

The development of the standard form of the equation of a hyperbola is similar to that of an ellipse. Note in the definition on the next page that a, b, and c are related differently for hyperbolas than for ellipses. For a hyperbola, the distance between the foci and the center is greater than the distance between the vertices and the center.

Standard Equation of a Hyperbola

The **standard form of the equation of a hyperbola** with center (h, k) is

$$\frac{(x-h)^2}{a^2} - \frac{(y-k)^2}{b^2} = 1 \qquad \text{Transverse axis is horizontal.}$$

$$\frac{(y-k)^2}{a^2} - \frac{(x-h)^2}{b^2} = 1. \qquad \text{Transverse axis is vertical.}$$

The vertices are a units from the center, and the foci are c units from the center. Moreover, $c^2 = a^2 + b^2$. If the center is at the origin, then the equation takes one of the forms below.

$$\frac{x^2}{a^2} - \frac{y^2}{b^2} = 1 \qquad \text{Transverse axis is horizontal.}$$

$$\frac{y^2}{a^2} - \frac{x^2}{b^2} = 1 \qquad \text{Transverse axis is vertical.}$$

Nuclear cooling towers such as those shown above are in the shapes of hyperboloids. The vertical cross sections of these cooling towers are hyperbolas.

The figures below show generalized horizontal and vertical orientations for hyperbolas.

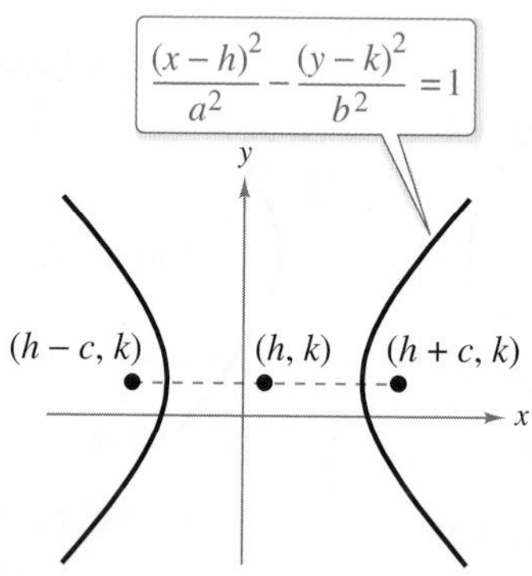

Transverse axis is horizontal.

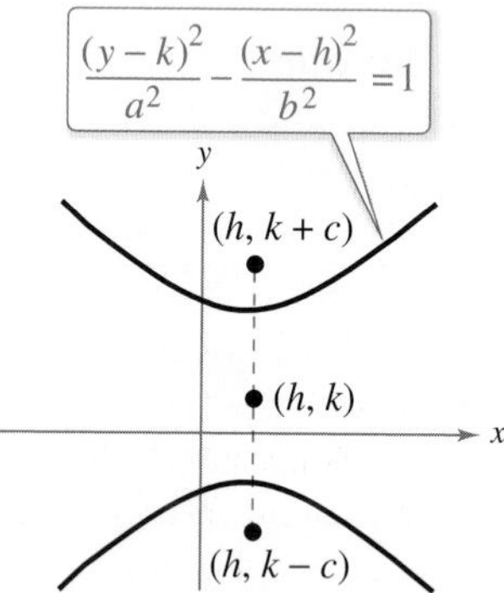

Transverse axis is vertical.

EXAMPLE 1 Finding the Standard Equation of a Hyperbola

Find the standard form of the equation of the hyperbola with vertices $(0, 2)$ and $(4, 2)$ and foci $(-1, 2)$ and $(5, 2)$, as shown in Figure 6.22.

Solution The foci occur at $(-1, 2)$ and $(5, 2)$, so the center of the hyperbola is $(2, 2)$. Furthermore, $c = 5 - 2 = 3$ and $a = 4 - 2 = 2$, and it follows that

$$b = \sqrt{c^2 - a^2} = \sqrt{3^2 - 2^2} = \sqrt{9 - 4} = \sqrt{5}.$$

The hyperbola has a horizontal transverse axis, so the standard form of the equation is

$$\frac{(x-2)^2}{2^2} - \frac{(y-2)^2}{(\sqrt{5})^2} = 1.$$

This equation simplifies to

$$\frac{(x-2)^2}{4} - \frac{(y-2)^2}{5} = 1.$$

Figure 6.22

Checkpoint Audio-video solution in English & Spanish at LarsonPrecalculus.com

Find the standard form of the equation of the hyperbola with vertices $(2, -4)$ and $(2, 2)$ and foci $(2, -5)$ and $(2, 3)$.

Asymptotes of a Hyperbola

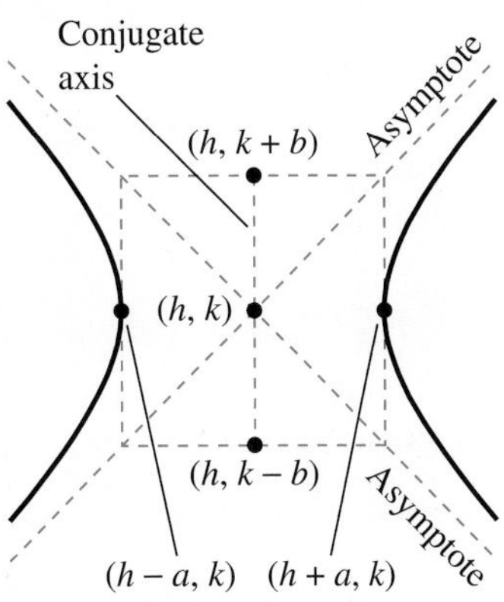

Every hyperbola has two *asymptotes* that intersect at the center of the hyperbola, as shown in Figure 6.23. The asymptotes pass through the vertices of a rectangle of dimensions $2a$ by $2b$, with its center at (h, k). The **conjugate axis** of a hyperbola is the line segment of length $2b$ joining $(h, k + b)$ and $(h, k - b)$ when the transverse axis is horizontal (as in Figure 6.23), and joining $(h + b, k)$ and $(h - b, k)$ when the transverse axis is vertical.

Figure 6.23

Asymptotes of a Hyperbola

The equations of the asymptotes of a hyperbola are

$$y = k \pm \frac{b}{a}(x - h) \qquad \text{Asymptotes for horizontal transverse axis}$$

$$y = k \pm \frac{a}{b}(x - h). \qquad \text{Asymptotes for vertical transverse axis}$$

EXAMPLE 2 Sketching a Hyperbola

Sketch the hyperbola $4x^2 - y^2 = 16$.

Algebraic Solution

Divide each side of the original equation by 16, and write the equation in standard form.

$$\frac{x^2}{2^2} - \frac{y^2}{4^2} = 1 \qquad \text{Write in standard form.}$$

The center of the hyperbola is $(0, 0)$. The x^2-term is postive, so the transverse axis is horizontal. The vertices occur at $(-2, 0)$ and $(2, 0)$, and the endpoints of the conjugate axis occur at $(0, -4)$ and $(0, 4)$. Use the vertices and the endpoints of the conjugate axis to sketch the rectangle shown in Figure 6.24. Sketch the asymptotes $y = 2x$ and $y = -2x$ through the opposite corners of the rectangle. Now, from $c^2 = a^2 + b^2$, you have $c = \sqrt{2^2 + 4^2} = \sqrt{20} = 2\sqrt{5}$. So, the foci of the hyperbola are $(-2\sqrt{5}, 0)$ and $(2\sqrt{5}, 0)$. Figure 6.25 shows a sketch of the hyperbola.

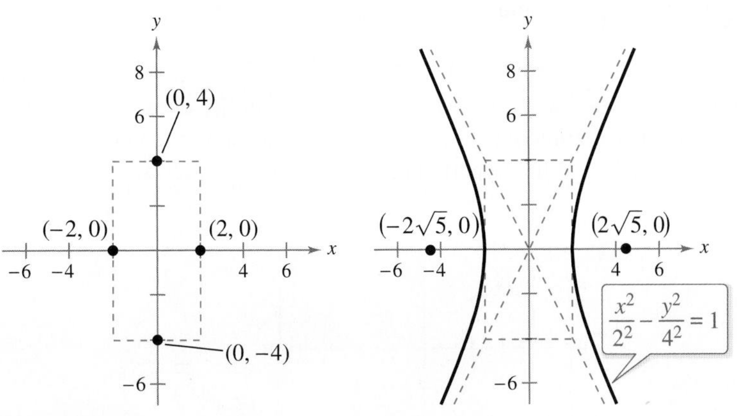

Figure 6.24

Figure 6.25

Graphical Solution

Solve the equation of the hyperbola for y.

$$4x^2 - y^2 = 16$$

$$4x^2 - 16 = y^2$$

$$\pm\sqrt{4x^2 - 16} = y$$

Then use a graphing utility to graph

$$y_1 = \sqrt{4x^2 - 16}$$

and

$$y_2 = -\sqrt{4x^2 - 16}$$

in the same viewing window, as shown in the figure below. Be sure to use a square setting.

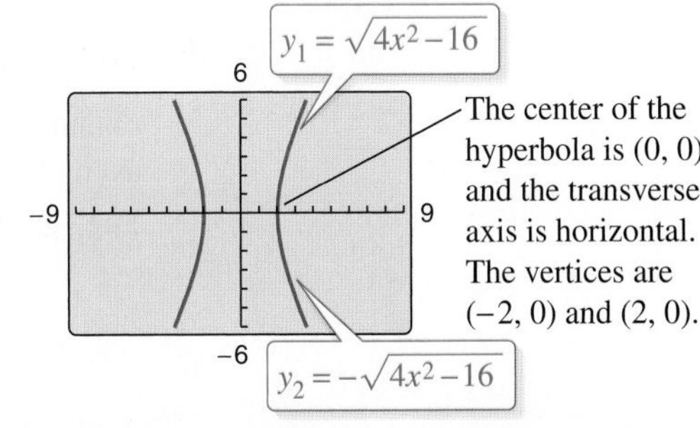

✓ **Checkpoint** Audio-video solution in English & Spanish at LarsonPrecalculus.com

Sketch the hyperbola $4y^2 - 9x^2 = 36$.

EXAMPLE 3 Sketching a Hyperbola

Sketch the hyperbola $4x^2 - 3y^2 + 8x + 16 = 0$.

Solution

$4x^2 - 3y^2 + 8x + 16 = 0$	Write original equation.
$(4x^2 + 8x) - 3y^2 = -16$	Group terms.
$4(x^2 + 2x) - 3y^2 = -16$	Factor 4 out of x-terms.
$4(x^2 + 2x + 1) - 3y^2 = -16 + 4(1)$	Complete the square.
$4(x + 1)^2 - 3y^2 = -12$	Write in completed square form.
$-\dfrac{(x + 1)^2}{3} + \dfrac{y^2}{4} = 1$	Divide each side by -12.
$\dfrac{y^2}{2^2} - \dfrac{(x + 1)^2}{(\sqrt{3})^2} = 1$	Write in standard form.

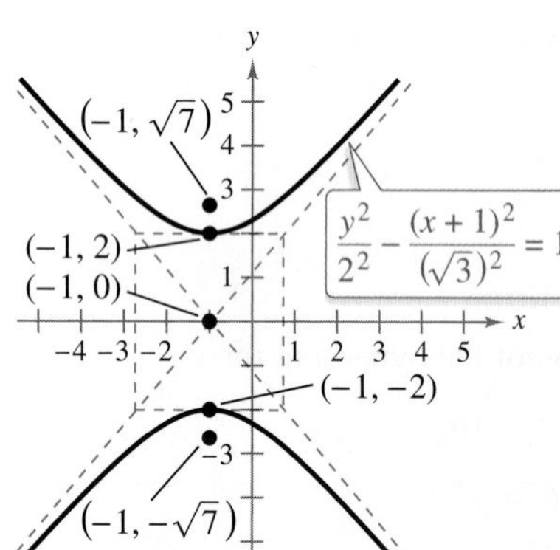

Figure 6.26

The center of the hyperbola is $(-1, 0)$. The y^2-term is positive, so the transverse axis is vertical. The vertices occur at $(-1, 2)$ and $(-1, -2)$, and the endpoints of the conjugate axis occur at $(-1 - \sqrt{3}, 0)$ and $(-1 + \sqrt{3}, 0)$. Draw a rectangle through the vertices and the endpoints of the conjugate axes. Sketch the asymptotes by drawing lines through the opposite corners of the rectangle. Using $a = 2$ and $b = \sqrt{3}$, the equations of the asymptotes are

$$y = \frac{2}{\sqrt{3}}(x + 1) \quad \text{and} \quad y = -\frac{2}{\sqrt{3}}(x + 1).$$

Finally, from $c^2 = a^2 + b^2$, you have $c = \sqrt{2^2 + (\sqrt{3})^2} = \sqrt{7}$. So, the foci of the hyperbola are $(-1, \sqrt{7})$ and $(-1, -\sqrt{7})$. Figure 6.26 shows a sketch of the hyperbola.

✓ Checkpoint *Audio-video solution in English & Spanish at LarsonPrecalculus.com*

Sketch the hyperbola $9x^2 - 4y^2 + 8y - 40 = 0$.

TECHNOLOGY To use a graphing utility to graph a hyperbola, graph the upper and lower portions in the same viewing window. For instance, to graph the hyperbola in Example 3, first solve for y to get

$$y_1 = 2\sqrt{1 + \frac{(x + 1)^2}{3}} \quad \text{and} \quad y_2 = -2\sqrt{1 + \frac{(x + 1)^2}{3}}.$$

Use a viewing window in which $-9 \le x \le 9$ and $-6 \le y \le 6$. You should obtain the graph shown below. Notice that the graphing utility does not draw the asymptotes. However, by graphing the asymptotes in the same viewing window, you can see that the values of the hyperbola approach the asymptotes.

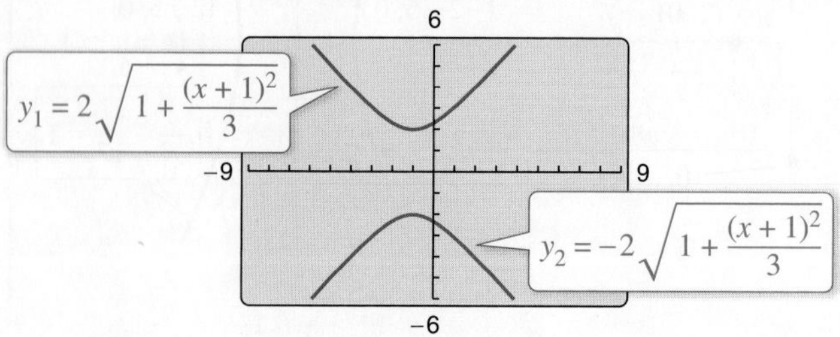

EXAMPLE 4 Using Asymptotes to Find the Standard Equation

See LarsonPrecalculus.com for an interactive version of this type of example.

Find the standard form of the equation of the hyperbola with vertices $(3, -5)$ and $(3, 1)$ and asymptotes

$$y = 2x - 8 \quad \text{and} \quad y = -2x + 4$$

as shown in Figure 6.27.

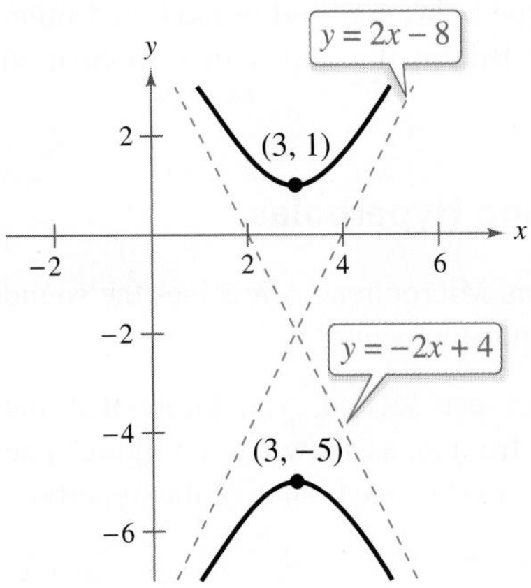

Figure 6.27

Solution The center of the hyperbola is $(3, -2)$. Furthermore, the hyperbola has a vertical transverse axis with $a = 3$. The slopes of the asymptotes are

$$m_1 = 2 = \frac{a}{b} \quad \text{and} \quad m_2 = -2 = -\frac{a}{b}$$

and $a = 3$, so

$$2 = \frac{a}{b} \implies 2 = \frac{3}{b} \implies b = \frac{3}{2}.$$

The standard form of the equation of the hyperbola is

$$\frac{(y+2)^2}{3^2} - \frac{(x-3)^2}{\left(\frac{3}{2}\right)^2} = 1.$$

✓ Checkpoint *Audio-video solution in English & Spanish at LarsonPrecalculus.com*

Find the standard form of the equation of the hyperbola with vertices $(3, 2)$ and $(9, 2)$ and asymptotes

$$y = -2 + \tfrac{2}{3}x \quad \text{and} \quad y = 6 - \tfrac{2}{3}x.$$

As with ellipses, the *eccentricity* of a hyperbola is

$$e = \frac{c}{a}. \qquad \text{Eccentricity}$$

You know that $c > a$ for a hyperbola, so it follows that $e > 1$. When the eccentricity is large, the branches of the hyperbola are nearly flat, as shown in Figure 6.28. When the eccentricity is close to 1, the branches of the hyperbola are more narrow, as shown in Figure 6.29.

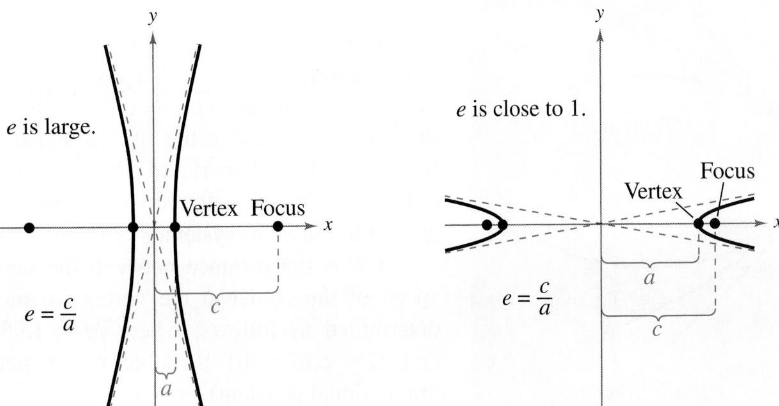

Figure 6.28

Figure 6.29

Applications

The next example shows how the properties of hyperbolas are used in radar and other detection systems. The United States and Great Britain developed this application during World War II.

EXAMPLE 5 An Application Involving Hyperbolas

Two microphones, 1 mile apart, record an explosion. Microphone A receives the sound 2 seconds before microphone B. Where did the explosion occur?

Solution Assuming sound travels at 1100 feet per second, you know that the explosion took place 2200 feet farther from B than from A, as shown in the figure. The locus of all points that are 2200 feet closer to A than to B is one branch of the hyperbola of the form

$$\frac{x^2}{a^2} - \frac{y^2}{b^2} = 1$$

where

$$a = \frac{2200}{2} = 1100.$$

Because

$$c = \frac{5280}{2} = 2640$$

it follows that

$$b^2 = c^2 - a^2$$

$$= 2640^2 - 1100^2$$

$$= 5{,}759{,}600.$$

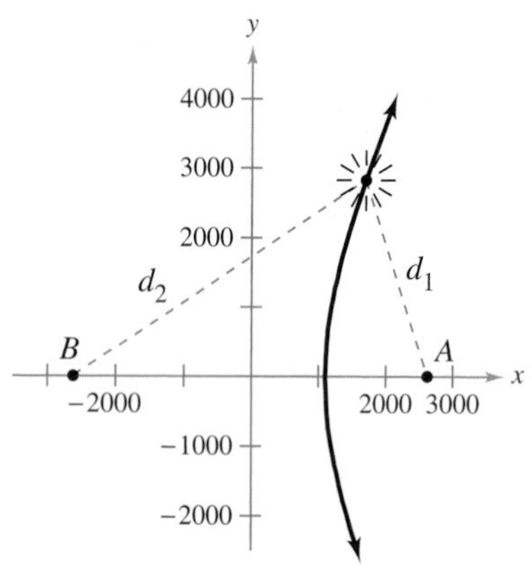

$2c = 1 \text{ mi} = 5280 \text{ ft}$
$|d_2 - d_1| = 2a = 2200 \text{ ft}$

So, the explosion occurred somewhere on the right branch of the hyperbola

$$\frac{x^2}{1{,}210{,}000} - \frac{y^2}{5{,}759{,}600} = 1.$$

✓ **Checkpoint**  Audio-video solution in English & Spanish at LarsonPrecalculus.com

Repeat Example 5 when microphone A receives the sound 4 seconds before microphone B.

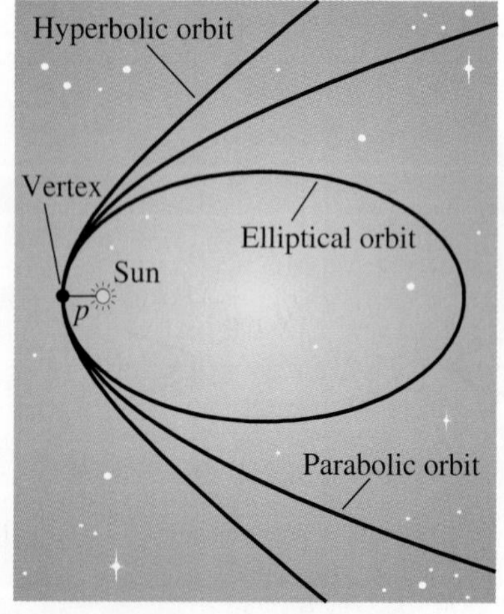

Figure 6.30

Another interesting application of conic sections involves the orbits of comets in our solar system. Comets can have elliptical, parabolic, or hyperbolic orbits. The center of the sun is a focus of each of these orbits, and each orbit has a vertex at the point where the comet is closest to the sun, as shown in Figure 6.30. Undoubtedly, many comets with parabolic or hyperbolic orbits have not been identified. You get to see such comets only *once.* Comets with elliptical orbits, such as Halley's comet, are the only ones that remain in our solar system.

If p is the distance between the vertex and the focus (in meters), and v is the speed of the comet at the vertex (in meters per second), then the type of orbit is determined as follows, where $M = 1.989 \times 10^{30}$ kilograms (the mass of the sun) and $G \approx 6.67 \times 10^{-11}$ cubic meter per kilogram-second squared (the universal gravitational constant).

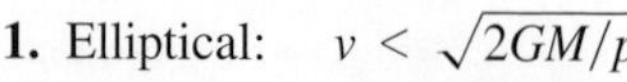

1. Elliptical: $v < \sqrt{2GM/p}$
2. Parabolic: $v = \sqrt{2GM/p}$
3. Hyperbolic: $v > \sqrt{2GM/p}$

General Equations of Conics

Classifying a Conic from Its General Equation

The graph of $Ax^2 + Cy^2 + Dx + Ey + F = 0$ is one of the following.

1. *Circle:* $A = C$ — $A \neq 0$
2. *Parabola:* $AC = 0$ — $A = 0$ or $C = 0$, but not both.
3. *Ellipse:* $AC > 0$ — $A \neq C$ and A and C have like signs.
4. *Hyperbola:* $AC < 0$ — A and C have unlike signs.

The test above is valid when the graph is a conic. The test does not apply to equations such as $x^2 + y^2 = -1$, whose graph is not a conic.

EXAMPLE 6 Classifying Conics from General Equations

a. For the equation $4x^2 - 9x + y - 5 = 0$, you have

$AC = 4(0) = 0.$ Parabola

So, the graph is a parabola.

b. For the equation $4x^2 - y^2 + 8x - 6y + 4 = 0$, you have

$AC = 4(-1) < 0.$ Hyperbola

So, the graph is a hyperbola.

c. For the equation $2x^2 + 4y^2 - 4x + 12y = 0$, you have

$AC = 2(4) > 0.$ Ellipse

So, the graph is an ellipse.

d. For the equation $2x^2 + 2y^2 - 8x + 12y + 2 = 0$, you have

$A = C = 2.$ Circle

So, the graph is a circle.

Caroline Herschel (1750–1848) was the first woman to be credited with discovering a comet. During her long life, this German astronomer discovered a total of eight comets.

 ✓ **Checkpoint** 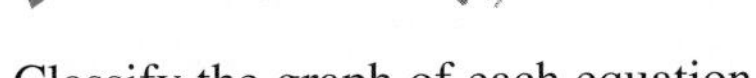 *Audio-video solution in English & Spanish at LarsonPrecalculus.com*

Classify the graph of each equation.

a. $3x^2 + 3y^2 - 6x + 6y + 5 = 0$

b. $2x^2 - 4y^2 + 4x + 8y - 3 = 0$

c. $3x^2 + y^2 + 6x - 2y + 3 = 0$

d. $2x^2 + 4x + y - 2 = 0$

Summarize (Section 6.4)

1. State the definition of a hyperbola and the standard form of the equation of a hyperbola *(page 443)*. For an example of finding the standard form of the equation of a hyperbola, see Example 1.
2. Explain how to find asymptotes of and sketch a hyperbola *(page 445)*. For examples involving asymptotes and graphs of hyperbolas, see Examples 2–4.
3. Describe a real-life application of a hyperbola *(page 448, Example 5)*.
4. Explain how to classify a conic from its general equation *(page 449)*. For an example of classifying conics from their general equations, see Example 6.

6.4 Exercises

See CalcChat.com for tutorial help and worked-out solutions to odd-numbered exercises.

Vocabulary: Fill in the blanks.

1. A ________ is the set of all points (x, y) in a plane for which the absolute value of the difference of the distances from two distinct fixed points, called ________, is constant.
2. The graph of a hyperbola has two disconnected parts called ________.
3. The line segment connecting the vertices of a hyperbola is the ________ ________, and its midpoint is the ________ of the hyperbola.
4. Every hyperbola has two ________ that intersect at the center of the hyperbola.

Skills and Applications

Matching **In Exercises 5–8, match the equation with its graph. [The graphs are labeled (a), (b), (c), and (d).]**

(a)

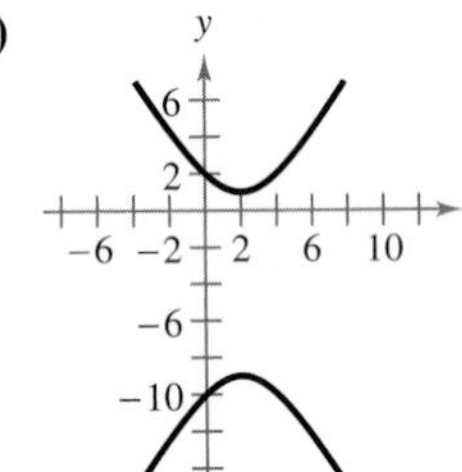

(b)

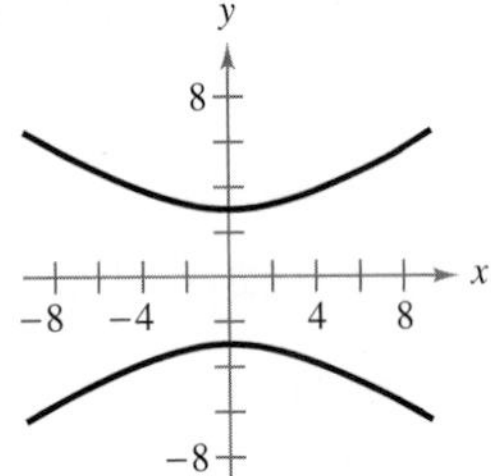

(c)

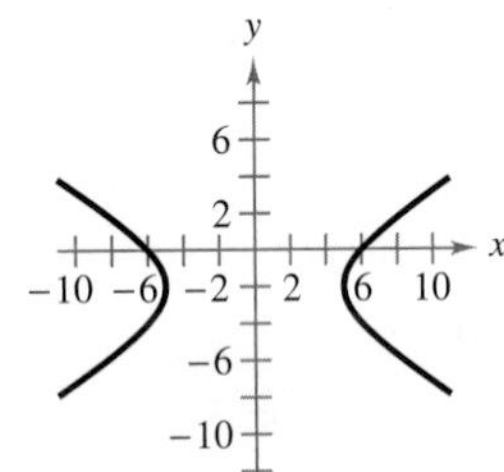

(d)

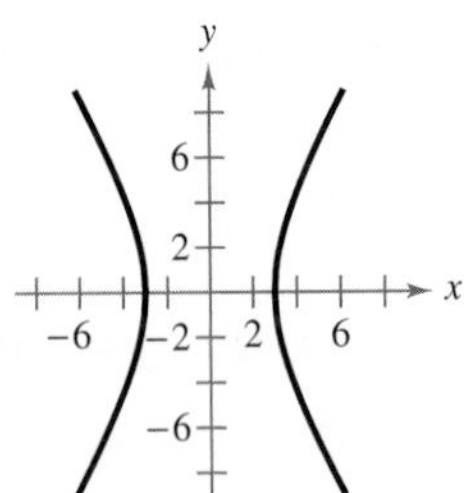

5. $\dfrac{y^2}{9} - \dfrac{x^2}{25} = 1$
6. $\dfrac{x^2}{9} - \dfrac{y^2}{25} = 1$
7. $\dfrac{x^2}{25} - \dfrac{(y + 2)^2}{9} = 1$
8. $\dfrac{(y + 4)^2}{25} - \dfrac{(x - 2)^2}{9} = 1$

Finding the Standard Equation of a Hyperbola **In Exercises 9–18, find the standard form of the equation of the hyperbola with the given characteristics.**

9. Vertices: $(0, \pm 2)$; foci: $(0, \pm 4)$
10. Vertices: $(\pm 4, 0)$; foci: $(\pm 6, 0)$
11. Vertices: $(2, 0), (6, 0)$; foci: $(0, 0), (8, 0)$
12. Vertices: $(2, 3), (2, -3)$; foci: $(2, 6), (2, -6)$
13. Vertices: $(4, 1), (4, 9)$; foci: $(4, 0), (4, 10)$
14. Vertices: $(-1, 1), (3, 1)$; foci: $(-2, 1), (4, 1)$
15. Vertices: $(2, 3), (2, -3)$; passes through the point $(0, 5)$
16. Vertices: $(-2, 1), (2, 1)$; passes through the point $(5, 4)$
17. Vertices: $(0, -3), (4, -3)$; passes through the point $(-4, 5)$
18. Vertices: $(1, -3), (1, -7)$; passes through the point $(5, -11)$

Sketching a Hyperbola **In Exercises 19–32, find the center, vertices, foci, and the equations of the asymptotes of the hyperbola. Then sketch the hyperbola using the asymptotes as an aid.**

19. $x^2 - y^2 = 1$
20. $\dfrac{x^2}{9} - \dfrac{y^2}{16} = 1$
21. $\dfrac{1}{36}y^2 - \dfrac{1}{100}x^2 = 1$
22. $\dfrac{1}{144}x^2 - \dfrac{1}{169}y^2 = 1$
23. $2y^2 - \dfrac{x^2}{2} = 2$
24. $\dfrac{y^2}{3} - 3x^2 = 3$
25. $\dfrac{(x - 1)^2}{4} - \dfrac{(y + 2)^2}{1} = 1$
26. $\dfrac{(x + 3)^2}{144} - \dfrac{(y - 2)^2}{25} = 1$
27. $\dfrac{(y + 6)^2}{1/9} - \dfrac{(x - 2)^2}{1/4} = 1$
28. $\dfrac{(y - 1)^2}{1/4} - \dfrac{(x + 3)^2}{1/16} = 1$
29. $9x^2 - y^2 - 36x - 6y + 18 = 0$
30. $x^2 - 9y^2 + 36y - 72 = 0$
31. $4x^2 - y^2 + 8x + 2y - 1 = 0$
32. $16y^2 - x^2 + 2x + 64y + 64 = 0$

Graphing a Hyperbola **In Exercises 33–38, use a graphing utility to graph the hyperbola and its asymptotes. Find the center, vertices, and foci.**

33. $2x^2 - 3y^2 = 6$
34. $6y^2 - 3x^2 = 18$
35. $25y^2 - 9x^2 = 225$
36. $25x^2 - 4y^2 = 100$
37. $9y^2 - x^2 + 2x + 54y + 62 = 0$
38. $9x^2 - y^2 + 54x + 10y + 55 = 0$

Finding the Standard Equation of a Hyperbola In Exercises 39–48, find the standard form of the equation of the hyperbola with the given characteristics.

39. Vertices: $(\pm 1, 0)$; asymptotes: $y = \pm 5x$
40. Vertices: $(0, \pm 3)$; asymptotes: $y = \pm 3x$
41. Foci: $(0, \pm 8)$; asymptotes: $y = \pm 4x$
42. Foci: $(\pm 10, 0)$; asymptotes: $y = \pm \frac{3}{4}x$
43. Vertices: $(1, 2)$, $(3, 2)$;
asymptotes: $y = x$, $y = 4 - x$
44. Vertices: $(3, 0)$, $(3, 6)$;
asymptotes: $y = 6 - x$, $y = x$
45. Vertices: $(3, 0)$, $(3, 4)$;
asymptotes: $y = \frac{2}{3}x$, $y = 4 - \frac{2}{3}x$
46. Vertices: $(-4, 1)$, $(0, 1)$;
asymptotes: $y = x + 3$, $y = -x - 1$
47. Foci: $(-1, -1)$, $(9, -1)$;
asymptotes: $y = \frac{3}{4}x - 4$, $y = -\frac{3}{4}x + 2$
48. Foci: $(9, \pm 2\sqrt{10})$;
asymptotes: $y = 3x - 27$, $y = -3x + 27$

49. **Art** A cross section of a sculpture can be modeled by a hyperbola (see figure).

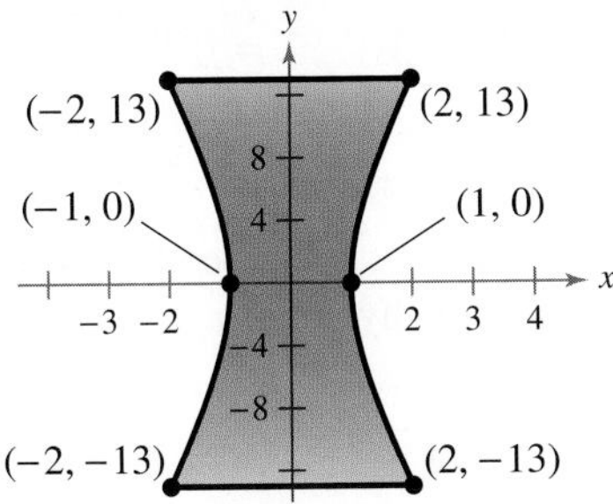

(a) Write an equation that models the curved sides of the sculpture.

(b) Each unit in the coordinate plane represents 1 foot. Find the width of the sculpture at a height of 18 feet.

50. **Clock** The base of a clock has the shape of a hyperbola (see figure).

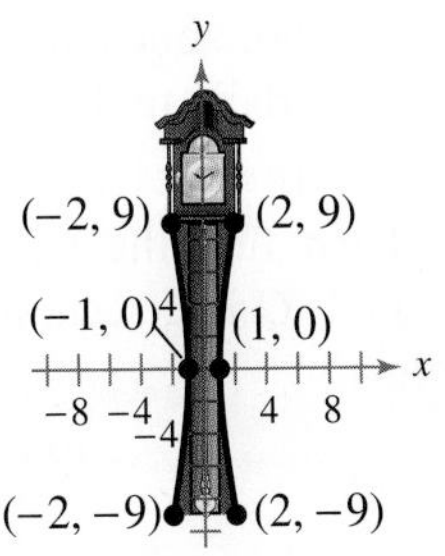

(a) Write an equation of the cross section of the base.

(b) Each unit in the coordinate plane represents $\frac{1}{2}$ foot. Find the width of the base 4 inches from the bottom.

51. **Sound Location** You and a friend live 4 miles apart. You hear a clap of thunder from lightning 18 seconds before your friend hears it. Where did the lightning occur? (Assume sound travels at 1100 feet per second.)

52. **Sound Location** Listening station A and listening station B are located at $(3300, 0)$ and $(-3300, 0)$, respectively. Station A detects an explosion 4 seconds before station B. (Assume the coordinate system is measured in feet and sound travels at 1100 feet per second.)

(a) Where did the explosion occur?

(b) Station C is located at $(3300, 1100)$ and detects the explosion 1 second after station A. Find the coordinates of the explosion.

53. **Navigation**

Long-distance radio navigation for aircraft and ships uses synchronized pulses transmitted by widely separated transmitting stations. These pulses travel at the speed of light (186,000 miles per second). The difference in the times of arrival of these pulses at an aircraft or ship is constant on a hyperbola having the transmitting stations as foci.

Assume that two stations 300 miles apart are positioned on a rectangular coordinate system with coordinates $(-150, 0)$ and $(150, 0)$ and that a ship is traveling on a hyperbolic path with coordinates $(x, 75)$ (see figure).

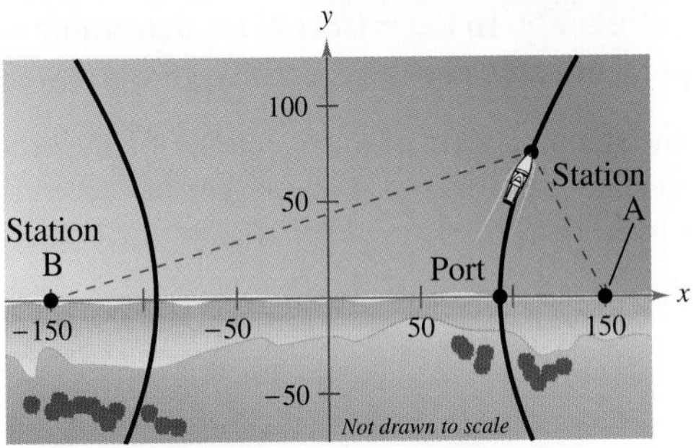

(a) Find the x-coordinate of the position of the ship when the time difference between the pulses from the transmitting stations is 1000 microseconds (0.001 second).

(b) Determine the distance between the port and station A.

(c) Find a linear equation that approximates the ship's path as it travels far away from the shore.

Melissa Madia/Shutterstock.com

54. Hyperbolic Mirror A hyperbolic mirror (used in some telescopes) has the property that a light ray directed at focus A is reflected to focus B (see figure). Find the vertex of the mirror when its mount at the top edge of the mirror has coordinates (24, 24).

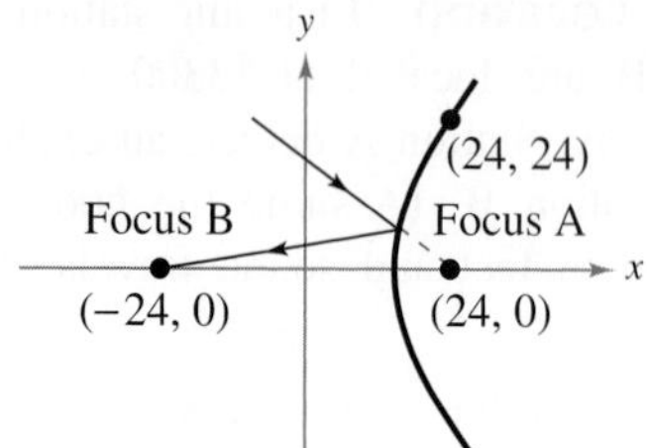

Classifying a Conic from a General Equation **In Exercises 55–66, classify the graph of the equation as a circle, a parabola, an ellipse, or a hyperbola.**

55. $9x^2 + 4y^2 - 18x + 16y - 119 = 0$

56. $x^2 + y^2 - 4x - 6y - 23 = 0$

57. $4x^2 - y^2 - 4x - 3 = 0$

58. $y^2 - 6y - 4x + 21 = 0$

59. $y^2 - 4x^2 + 4x - 2y - 4 = 0$

60. $y^2 + 12x + 4y + 28 = 0$

61. $4x^2 + 25y^2 + 16x + 250y + 541 = 0$

62. $4y^2 - 2x^2 - 4y - 8x - 15 = 0$

63. $25x^2 - 10x - 200y - 119 = 0$

64. $4y^2 + 4x^2 - 24x + 35 = 0$

65. $100x^2 + 100y^2 - 100x + 400y + 409 = 0$

66. $9x^2 + 4y^2 - 90x + 8y + 228 = 0$

Exploration

True or False? **In Exercises 67–69, determine whether the statement is true or false. Justify your answer.**

67. In the standard form of the equation of a hyperbola, the larger the ratio of b to a, the larger the eccentricity of the hyperbola.

68. If the asymptotes of the hyperbola

$$\frac{x^2}{a^2} - \frac{y^2}{b^2} = 1$$

where $a, b > 0$, intersect at right angles, then $a = b$.

69. The graph of

$$x^2 - y^2 + 4x - 4y = 0$$

is a hyperbola.

70. **Think About It** Write an equation whose graph is the bottom half of the hyperbola

$$9x^2 - 54x - 4y^2 + 8y + 41 = 0.$$

71. **Writing** Explain how to use a rectangle to sketch the asymptotes of a hyperbola.

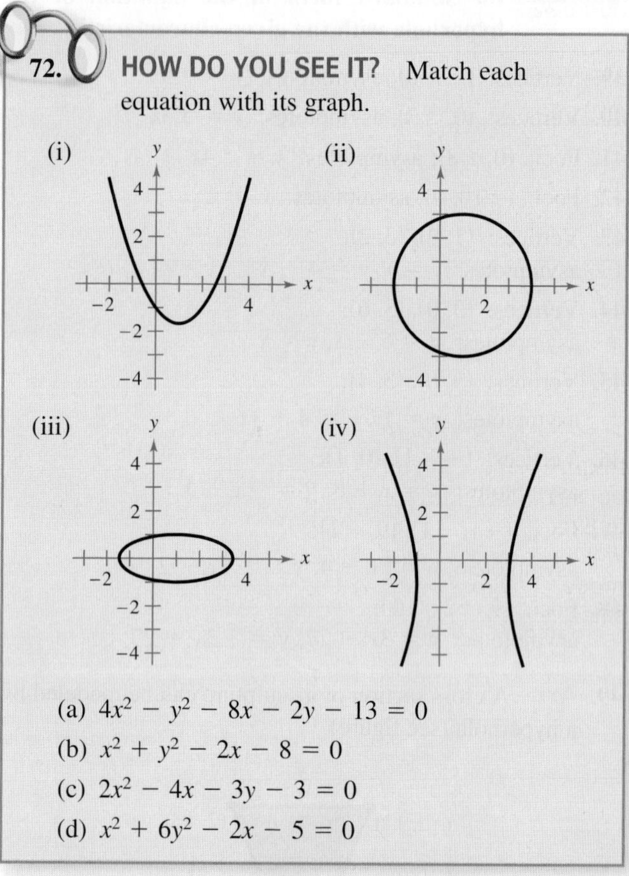

(a) $4x^2 - y^2 - 8x - 2y - 13 = 0$

(b) $x^2 + y^2 - 2x - 8 = 0$

(c) $2x^2 - 4x - 3y - 3 = 0$

(d) $x^2 + 6y^2 - 2x - 5 = 0$

73. **Error Analysis** Describe the error in finding the asymptotes of the hyperbola

$$\frac{(y+5)^2}{9} - \frac{(x-3)^2}{4} = 1.$$

$$y = k \pm \frac{b}{a}(x - h)$$

$$= -5 \pm \tfrac{2}{3}(x - 3)$$

The asymptotes are

$y = \frac{2}{3}x - 7$ and $y = -\frac{2}{3}x - 3$.

74. **Think About It** Consider a hyperbola centered at the origin with a horizontal transverse axis. Use the definition of a hyperbola to derive its standard form.

75. **Points of Intersection** Sketch the circle $x^2 + y^2 = 4$. Then find the values of C so that the parabola $y = x^2 + C$ intersects the circle at the given number of points.

(a) 0 points

(b) 1 point

(c) 2 points

(d) 3 points

(e) 4 points

6.5 Rotation of Conics

Rotated conics can model objects in real life. For example, in Exercise 63 on page 460, you will use a rotated parabola to model the cross section of a satellite dish.

- **Rotate the coordinate axes to eliminate the xy-term in equations of conics.**
- **Use the discriminant to classify conics.**

Rotation

In the preceding section, you classified conics whose equations were written in the general form

$$Ax^2 + Cy^2 + Dx + Ey + F = 0.$$

The graphs of such conics have axes that are parallel to one of the coordinate axes. Conics whose axes are rotated so that they are not parallel to either the x-axis or the y-axis have general equations that contain an xy-term.

$$Ax^2 + Bxy + Cy^2 + Dx + Ey + F = 0 \qquad \text{Equation in } xy\text{-plane}$$

To eliminate this xy-term, use a procedure called **rotation of axes.** The objective is to rotate the x- and y-axes until they are parallel to the axes of the conic. The rotated axes are denoted as the x'-axis and the y'-axis, as shown in Figure 6.31. After the rotation, the equation of the conic in the $x'y'$-plane will have the form

$$A'(x')^2 + C'(y')^2 + D'x' + E'y' + F' = 0. \qquad \text{Equation in } x'y'\text{-plane}$$

This equation has no $x'y'$-term, so you can obtain a standard form by completing the square. The theorem below identifies how much to rotate the axes to eliminate the xy-term and also the equations for determining the new coefficients A', C', D', E', and F'.

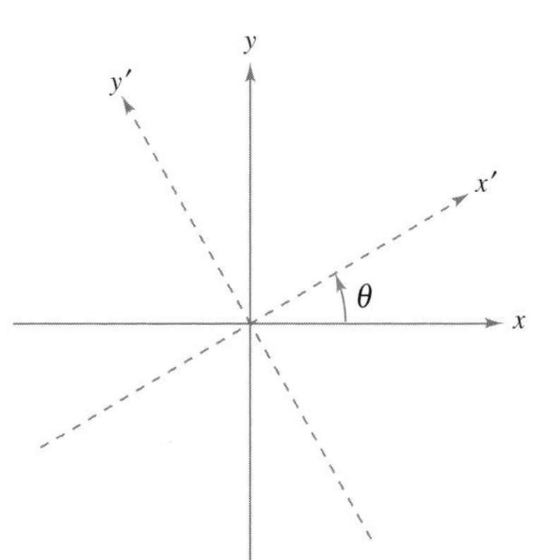

Figure 6.31

Rotation of Axes to Eliminate an *xy*-Term

The general second-degree equation

$$Ax^2 + Bxy + Cy^2 + Dx + Ey + F = 0$$

where $B \neq 0$, can be rewritten as

$$A'(x')^2 + C'(y')^2 + D'x' + E'y' + F' = 0$$

by rotating the coordinate axes through an angle θ, where

$$\cot 2\theta = \frac{A - C}{B}.$$

The coefficients of the new equation are obtained by making the substitutions

$$x = x' \cos \theta - y' \sin \theta$$

and

$$y = x' \sin \theta + y' \cos \theta.$$

Remember that the substitutions

$$x = x' \cos \theta - y' \sin \theta$$

and

$$y = x' \sin \theta + y' \cos \theta$$

should eliminate the $x'y'$-term in the rotated system. Use this as a check of your work. If you obtain an equation of a conic in the $x'y'$-plane that contains an $x'y'$-term, you know that you have made a mistake.

EXAMPLE 1 Rotation of Axes for a Hyperbola

Rotate the axes to eliminate the xy-term in the equation $xy - 1 = 0$. Then write the equation in standard form and sketch its graph.

Solution Because $A = 0$, $B = 1$, and $C = 0$, you have

$$\cot 2\theta = \frac{A - C}{B} = \frac{0 - 0}{1} = 0 \implies 2\theta = \frac{\pi}{2} \implies \theta = \frac{\pi}{4}.$$

Obtain the equation in the $x'y'$-system by substituting

$$\begin{aligned} x &= x' \cos \frac{\pi}{4} - y' \sin \frac{\pi}{4} \\ &= x'\left(\frac{1}{\sqrt{2}}\right) - y'\left(\frac{1}{\sqrt{2}}\right) \\ &= \frac{x' - y'}{\sqrt{2}} \end{aligned}$$

and

$$\begin{aligned} y &= x' \sin \frac{\pi}{4} + y' \cos \frac{\pi}{4} \\ &= x'\left(\frac{1}{\sqrt{2}}\right) + y'\left(\frac{1}{\sqrt{2}}\right) \\ &= \frac{x' + y'}{\sqrt{2}} \end{aligned}$$

into the original equation. So, you have

$$xy - 1 = 0$$

$$\left(\frac{x' - y'}{\sqrt{2}}\right)\left(\frac{x' + y'}{\sqrt{2}}\right) - 1 = 0$$

which simplifies to

$$\frac{(x')^2 - (y')^2}{2} - 1 = 0$$

$$\frac{(x')^2}{(\sqrt{2})^2} - \frac{(y')^2}{(\sqrt{2})^2} = 1. \quad \text{Write in standard form.}$$

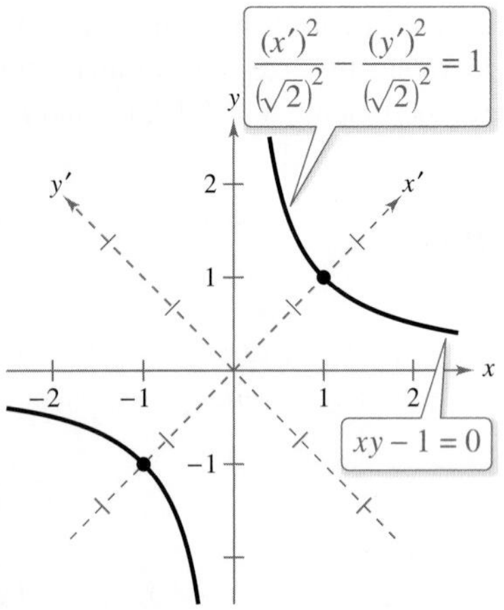

Vertices:
In $x'y'$-system: $(\sqrt{2}, 0), (-\sqrt{2}, 0)$
In xy-system: $(1, 1), (-1, -1)$
Figure 6.32

In the $x'y'$-system, this is the equation of a hyperbola centered at the origin with vertices $(\pm\sqrt{2}, 0)$, as shown in Figure 6.32. Note that to find the coordinates of the vertices in the xy-system, substitute the coordinates $(\pm\sqrt{2}, 0)$ into the equations

$$x = \frac{x' - y'}{\sqrt{2}} \quad \text{and} \quad y = \frac{x' + y'}{\sqrt{2}}.$$

This substitution yields the vertices $(1, 1)$ and $(-1, -1)$ in the xy-system. Note also that the asymptotes of the hyperbola have equations

$$y' = \pm x'$$

which correspond to the original x- and y-axes.

✓ Checkpoint *Audio-video solution in English & Spanish at LarsonPrecalculus.com*

Rotate the axes to eliminate the xy-term in the equation $xy + 6 = 0$. Then write the equation in standard form and sketch its graph.

EXAMPLE 2 Rotation of Axes for an Ellipse

Rotate the axes to eliminate the xy-term in the equation

$$7x^2 - 6\sqrt{3}xy + 13y^2 - 16 = 0.$$

Then write the equation in standard form and sketch its graph.

Solution Because $A = 7$, $B = -6\sqrt{3}$, and $C = 13$, you have

$$\cot 2\theta = \frac{A - C}{B} = \frac{7 - 13}{-6\sqrt{3}} = \frac{1}{\sqrt{3}}$$

which implies that $\theta = \frac{\pi}{6}$. Obtain the equation in the $x'y'$-system by substituting

$$\begin{aligned} x &= x' \cos\frac{\pi}{6} - y' \sin\frac{\pi}{6} \\ &= x'\left(\frac{\sqrt{3}}{2}\right) - y'\left(\frac{1}{2}\right) \\ &= \frac{\sqrt{3}x' - y'}{2} \end{aligned}$$

and

$$\begin{aligned} y &= x' \sin\frac{\pi}{6} + y' \cos\frac{\pi}{6} \\ &= x'\left(\frac{1}{2}\right) + y'\left(\frac{\sqrt{3}}{2}\right) \\ &= \frac{x' + \sqrt{3}y'}{2} \end{aligned}$$

into the original equation. So, you have

$$7x^2 - 6\sqrt{3}xy + 13y^2 - 16 = 0$$

$$7\left(\frac{\sqrt{3}x' - y'}{2}\right)^2 - 6\sqrt{3}\left(\frac{\sqrt{3}x' - y'}{2}\right)\left(\frac{x' + \sqrt{3}y'}{2}\right) + 13\left(\frac{x' + \sqrt{3}y'}{2}\right)^2 - 16 = 0$$

which simplifies to

$$\begin{aligned} 4(x')^2 + 16(y')^2 - 16 &= 0 \\ 4(x')^2 + 16(y')^2 &= 16 \\ \frac{(x')^2}{4} + \frac{(y')^2}{1} &= 1 \\ \frac{(x')^2}{2^2} + \frac{(y')^2}{1^2} &= 1. \end{aligned}$$

Write in standard form.

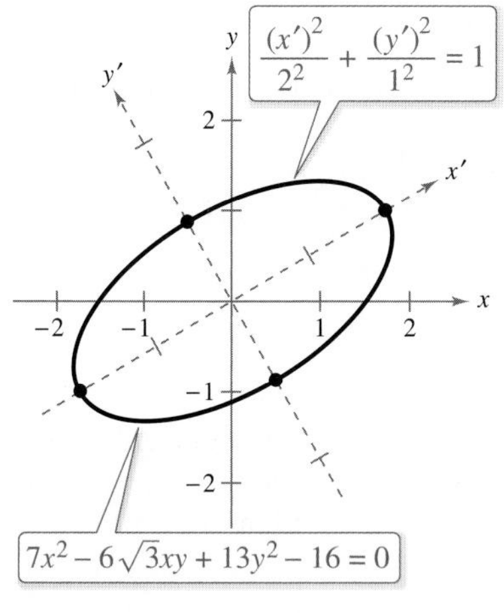

Vertices:
In $x'y'$-system: $(\pm 2, 0)$
In xy-system: $(\sqrt{3}, 1)$, $(-\sqrt{3}, -1)$
Figure 6.33

In the $x'y'$-system, this is the equation of an ellipse centered at the origin with vertices $(\pm 2, 0)$, as shown in Figure 6.33.

Checkpoint Audio-video solution in English & Spanish at LarsonPrecalculus.com

Rotate the axes to eliminate the xy-term in the equation

$$12x^2 + 16\sqrt{3}xy + 28y^2 - 36 = 0.$$

Then write the equation in standard form and sketch its graph.

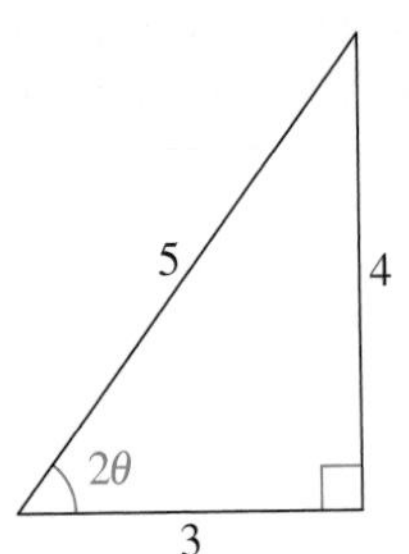

Figure 6.34

EXAMPLE 3 Rotation of Axes for a Parabola

See LarsonPrecalculus.com for an interactive version of this type of example.

Rotate the axes to eliminate the xy-term in the equation

$$x^2 - 4xy + 4y^2 + 5\sqrt{5}y + 1 = 0.$$

Then write the equation in standard form and sketch its graph.

Solution Because $A = 1$, $B = -4$, and $C = 4$, you have

$$\cot 2\theta = \frac{A - C}{B} = \frac{1 - 4}{-4} = \frac{3}{4}.$$

Use this information to draw a right triangle, as shown in Figure 6.34. From the figure, $\cos 2\theta = \frac{3}{5}$. To find the values of $\sin \theta$ and $\cos \theta$, use the half-angle formulas

$$\sin \theta = \sqrt{\frac{1 - \cos 2\theta}{2}} \quad \text{and} \quad \cos \theta = \sqrt{\frac{1 + \cos 2\theta}{2}}.$$

So,

$$\sin \theta = \sqrt{\frac{1 - \frac{3}{5}}{2}} = \sqrt{\frac{1}{5}} = \frac{1}{\sqrt{5}} \quad \text{and} \quad \cos \theta = \sqrt{\frac{1 + \frac{3}{5}}{2}} = \sqrt{\frac{4}{5}} = \frac{2}{\sqrt{5}}.$$

Consequently, use the substitutions

$$x = x'\cos\theta - y'\sin\theta = x'\left(\frac{2}{\sqrt{5}}\right) - y'\left(\frac{1}{\sqrt{5}}\right) = \frac{2x' - y'}{\sqrt{5}}$$

and

$$y = x'\sin\theta + y'\cos\theta = x'\left(\frac{1}{\sqrt{5}}\right) + y'\left(\frac{2}{\sqrt{5}}\right) = \frac{x' + 2y'}{\sqrt{5}}.$$

Substituting these expressions into the original equation, you have

$$x^2 - 4xy + 4y^2 + 5\sqrt{5}y + 1 = 0$$

$$\left(\frac{2x' - y'}{\sqrt{5}}\right)^2 - 4\left(\frac{2x' - y'}{\sqrt{5}}\right)\left(\frac{x' + 2y'}{\sqrt{5}}\right) + 4\left(\frac{x' + 2y'}{\sqrt{5}}\right)^2 + 5\sqrt{5}\left(\frac{x' + 2y'}{\sqrt{5}}\right) + 1 = 0$$

which simplifies to

$$5(y')^2 + 5x' + 10y' + 1 = 0$$

$$5[(y')^2 + 2y'] = -5x' - 1 \qquad \text{Group terms.}$$

$$5[(y')^2 + 2y' + 1] = -5x' - 1 + 5(1) \qquad \text{Complete the square.}$$

$$5(y' + 1)^2 = -5x' + 4 \qquad \text{Write in completed square form.}$$

$$(y' + 1)^2 = (-1)\left(x' - \frac{4}{5}\right). \qquad \text{Write in standard form.}$$

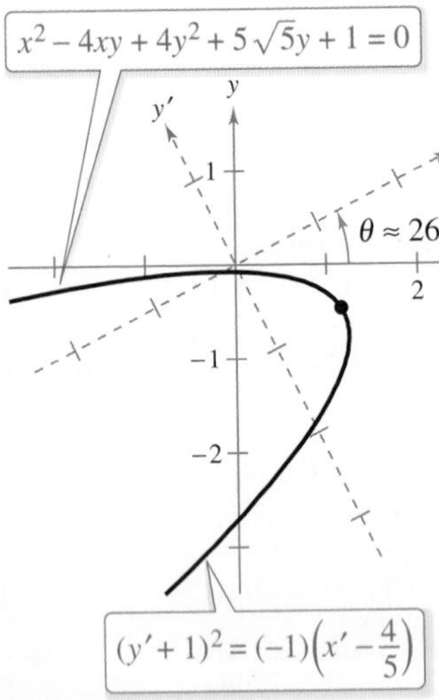

$x^2 - 4xy + 4y^2 + 5\sqrt{5}y + 1 = 0$

$\theta \approx 26.6°$

$(y' + 1)^2 = (-1)\left(x' - \frac{4}{5}\right)$

Vertex:

In $x'y'$-system: $\left(\frac{4}{5}, -1\right)$

In xy-system: $\left(\frac{13}{5\sqrt{5}}, -\frac{6}{5\sqrt{5}}\right)$

Figure 6.35

In the $x'y'$-system, this is the equation of a parabola with vertex $\left(\frac{4}{5}, -1\right)$. Its axis is parallel to the x'-axis in the $x'y'$-system, and $\theta = \sin^{-1}(1/\sqrt{5}) \approx 0.4636$ radian $\approx 26.6°$, as shown in Figure 6.35.

✓ **Checkpoint** *Audio-video solution in English & Spanish at LarsonPrecalculus.com*

Rotate the axes to eliminate the xy-term in the equation

$$4x^2 + 4xy + y^2 - 2\sqrt{5}x + 4\sqrt{5}y - 30 = 0.$$

Then write the equation in standard form and sketch its graph.

Invariants Under Rotation

In the rotation of axes theorem stated at the beginning of this section, the constant term is the same in both equations, that is, $F' = F$. Such quantities are **invariant under rotation.** The next theorem lists this and other rotation invariants.

Rotation Invariants

The rotation of the coordinate axes through an angle θ that transforms the equation $Ax^2 + Bxy + Cy^2 + Dx + Ey + F = 0$ into the form

$$A'(x')^2 + C'(y')^2 + D'x' + E'y' + F' = 0$$

has the rotation invariants listed below.

1. $F = F'$

2. $A + C = A' + C'$

3. $B^2 - 4AC = (B')^2 - 4A'C'$

You can use the results of this theorem to classify the graph of a second-degree equation *with* an xy-term in much the same way you do for a second-degree equation *without* an xy-term. Note that $B' = 0$, so the invariant $B^2 - 4AC$ reduces to

$$B^2 - 4AC = -4A'C'. \qquad \text{Discriminant}$$

This quantity is the **discriminant** of the equation

$$Ax^2 + Bxy + Cy^2 + Dx + Ey + F = 0.$$

Now, from the classification procedure given in Section 6.4, you know that the value of $A'C'$ determines the type of graph for the equation

$$A'(x')^2 + C'(y')^2 + D'x' + E'y' + F' = 0.$$

Consequently, the value of $B^2 - 4AC$ will determine the type of graph for the original equation, as given in the classification below.

REMARK When there is an xy-term in the equation of a conic, you should realize that the conic is rotated. Before rotating the axes, you should use the discriminant to classify the conic. Use this classification to verify the types of graphs in Examples 1–3.

Classification of Conics by the Discriminant

The graph of the equation $Ax^2 + Bxy + Cy^2 + Dx + Ey + F = 0$ is, except in degenerate cases, determined by its discriminant as follows.

1. *Ellipse or circle:* $B^2 - 4AC < 0$

2. *Parabola:* $B^2 - 4AC = 0$

3. *Hyperbola:* $B^2 - 4AC > 0$

For example, in the general equation

$$3x^2 + 7xy + 5y^2 - 6x - 7y + 15 = 0$$

you have $A = 3$, $B = 7$, and $C = 5$. So, the discriminant is

$$\begin{aligned} B^2 - 4AC &= 7^2 - 4(3)(5) \\ &= 49 - 60 \\ &= -11. \end{aligned}$$

The graph of the equation is an ellipse or a circle because $-11 < 0$.

EXAMPLE 4 Rotation and Graphing Utilities

For each equation, classify the graph of the equation, use the Quadratic Formula to solve for y, and then use a graphing utility to graph the equation.

a. $2x^2 - 3xy + 2y^2 - 2x = 0$ **b.** $x^2 - 6xy + 9y^2 - 2y + 1 = 0$

c. $3x^2 + 8xy + 4y^2 - 7 = 0$

Solution

a. Because $B^2 - 4AC = 9 - 16 < 0$, the graph is an ellipse or a circle.

$$2x^2 - 3xy + 2y^2 - 2x = 0 \qquad \text{Write original equation.}$$

$$2y^2 - 3xy + (2x^2 - 2x) = 0 \qquad \text{Quadratic form } ay^2 + by + c = 0$$

$$y = \frac{-(-3x) \pm \sqrt{(-3x)^2 - 4(2)(2x^2 - 2x)}}{2(2)}$$

$$y = \frac{3x \pm \sqrt{x(16 - 7x)}}{4}$$

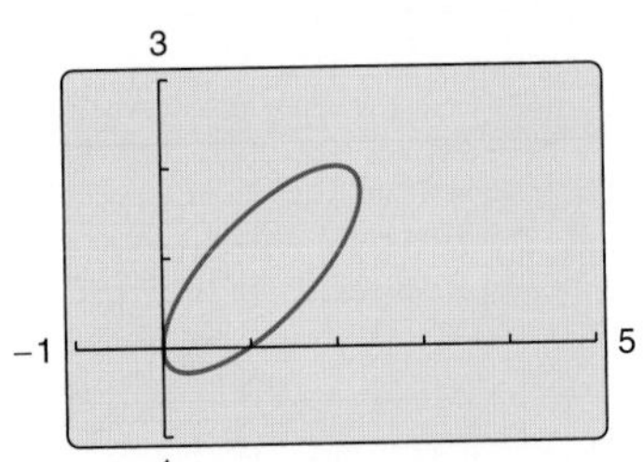

Figure 6.36

Graph both of the equations to obtain the ellipse shown in Figure 6.36.

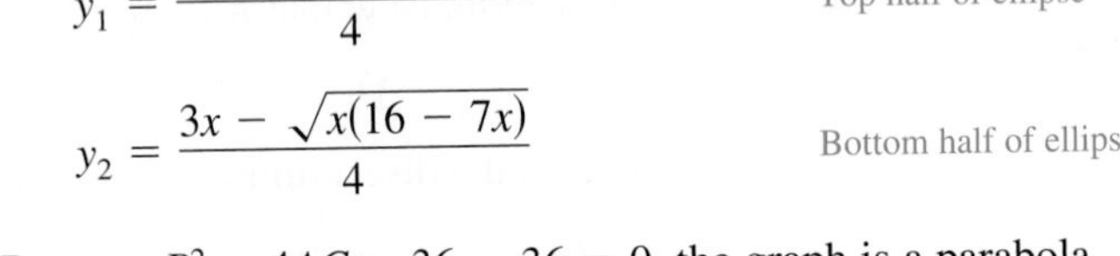

$$y_1 = \frac{3x + \sqrt{x(16 - 7x)}}{4} \qquad \text{Top half of ellipse}$$

$$y_2 = \frac{3x - \sqrt{x(16 - 7x)}}{4} \qquad \text{Bottom half of ellipse}$$

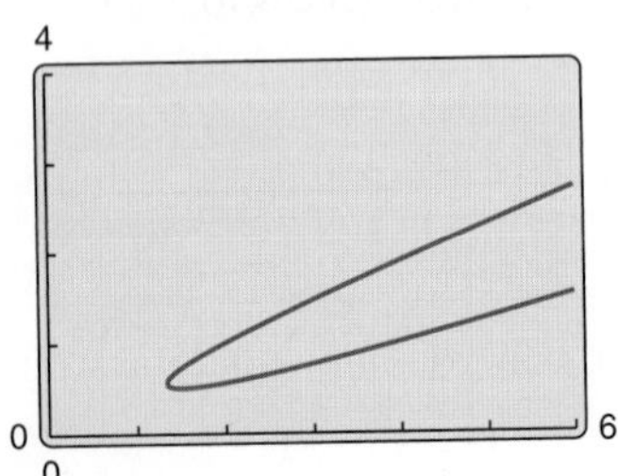

Figure 6.37

b. Because $B^2 - 4AC = 36 - 36 = 0$, the graph is a parabola.

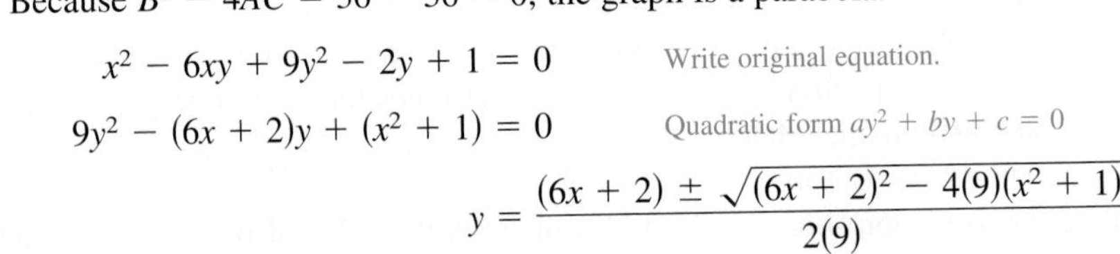

$$x^2 - 6xy + 9y^2 - 2y + 1 = 0 \qquad \text{Write original equation.}$$

$$9y^2 - (6x + 2)y + (x^2 + 1) = 0 \qquad \text{Quadratic form } ay^2 + by + c = 0$$

$$y = \frac{(6x + 2) \pm \sqrt{(6x + 2)^2 - 4(9)(x^2 + 1)}}{2(9)}$$

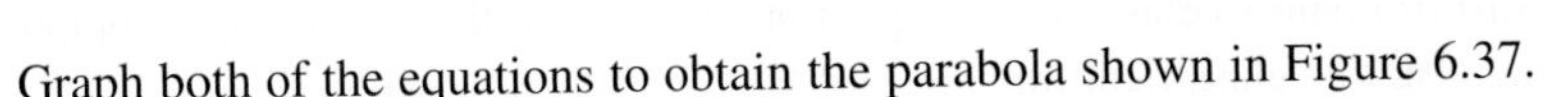

Graph both of the equations to obtain the parabola shown in Figure 6.37.

c. Because $B^2 - 4AC = 64 - 48 > 0$, the graph is a hyperbola.

$$3x^2 + 8xy + 4y^2 - 7 = 0 \qquad \text{Write original equation.}$$

$$4y^2 + 8xy + (3x^2 - 7) = 0 \qquad \text{Quadratic form } ay^2 + by + c = 0$$

$$y = \frac{-8x \pm \sqrt{(8x)^2 - 4(4)(3x^2 - 7)}}{2(4)}$$

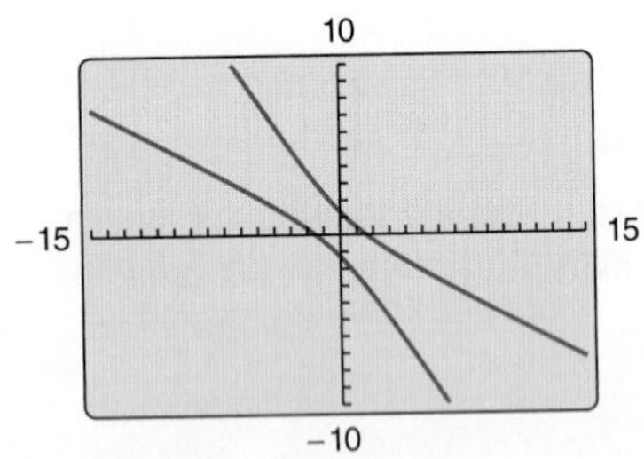

Figure 6.38

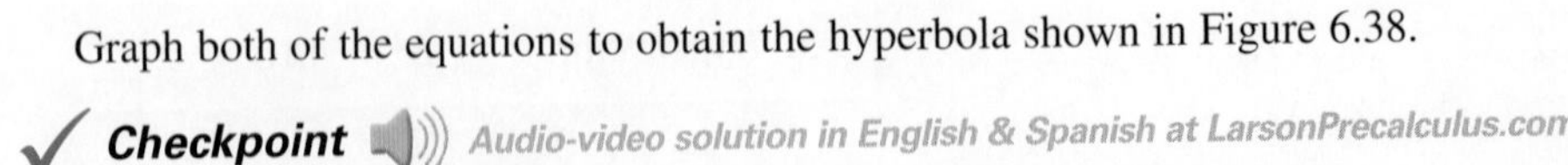

Graph both of the equations to obtain the hyperbola shown in Figure 6.38.

✓ **Checkpoint** *Audio-video solution in English & Spanish at LarsonPrecalculus.com*

Classify the graph of the equation $2x^2 - 8xy + 8y^2 + 3x + 5 = 0$, use the Quadratic Formula to solve for y, and then use a graphing utility to graph the equation.

Summarize (Section 6.5)

1. Explain how to rotate coordinate axes to eliminate the xy-term in the equation of a conic *(page 453)*. For examples of rotating coordinate axes to eliminate the xy-term in equations of conics, see Examples 1–3.
2. Explain how to use the discriminant to classify conics *(page 457)*. For an example of using the discriminant to classify conics, see Example 4.

6.5 Exercises

See CalcChat.com for tutorial help and worked-out solutions to odd-numbered exercises.

Vocabulary: Fill in the blanks.

1. The procedure used to eliminate the xy-term in a general second-degree equation is called ________ of ________.
2. After rotating the coordinate axes through an angle θ, the general second-degree equation in the $x'y'$-plane will have the form ________.
3. Quantities that are equal in both the original equation of a conic and the equation of the rotated conic are ________ ________ ________.
4. The quantity $B^2 - 4AC$ is the ________ of the equation $Ax^2 + Bxy + Cy^2 + Dx + Ey + F = 0$.

Skills and Applications

Finding a Point in a Rotated Coordinate System In Exercises 5–12, the $x'y'$-coordinate system is rotated θ degrees from the xy-coordinate system. The coordinates of a point in the xy-coordinate system are given. Find the coordinates of the point in the rotated coordinate system.

5. $\theta = 90°, (2, 0)$
6. $\theta = 90°, (4, 1)$
7. $\theta = 30°, (1, 3)$
8. $\theta = 30°, (2, 4)$
9. $\theta = 45°, (2, 1)$
10. $\theta = 45°, (4, 4)$
11. $\theta = 60°, (1, 2)$
12. $\theta = 60°, (3, 1)$

Rotation of Axes In Exercises 13–24, rotate the axes to eliminate the xy-term in the equation. Then write the equation in standard form. Sketch the graph of the resulting equation, showing both sets of axes.

13. $xy + 3 = 0$
14. $xy - 4 = 0$
15. $xy + 2x - y + 4 = 0$
16. $xy - 8x - 4y = 0$
17. $5x^2 - 6xy + 5y^2 - 12 = 0$
18. $2x^2 + xy + 2y^2 - 8 = 0$
19. $13x^2 + 6\sqrt{3}xy + 7y^2 - 16 = 0$
20. $7x^2 - 6\sqrt{3}xy + 13y^2 - 64 = 0$
21. $x^2 + 2xy + y^2 + \sqrt{2}x - \sqrt{2}y = 0$
22. $3x^2 - 2\sqrt{3}xy + y^2 + 2x + 2\sqrt{3}y = 0$
23. $9x^2 + 24xy + 16y^2 + 90x - 130y = 0$
24. $9x^2 + 24xy + 16y^2 + 80x - 60y = 0$

Using a Graphing Utility In Exercises 25–30, use a graphing utility to graph the conic. Determine the angle θ through which the axes are rotated. Explain how you used the graphing utility to obtain the graph.

25. $x^2 - 4xy + 2y^2 = 6$
26. $3x^2 + 5xy - 2y^2 = 10$
27. $14x^2 + 16xy + 9y^2 = 44$
28. $24x^2 + 18xy + 12y^2 = 34$
29. $2x^2 + 4xy + 2y^2 + \sqrt{26}x + 3y = -15$
30. $4x^2 - 12xy + 9y^2 + \sqrt{6}x - 29y = 91$

Matching In Exercises 31–36, match the equation with its graph. [The graphs are labeled (a), (b), (c), (d), (e), and (f).]

(a)

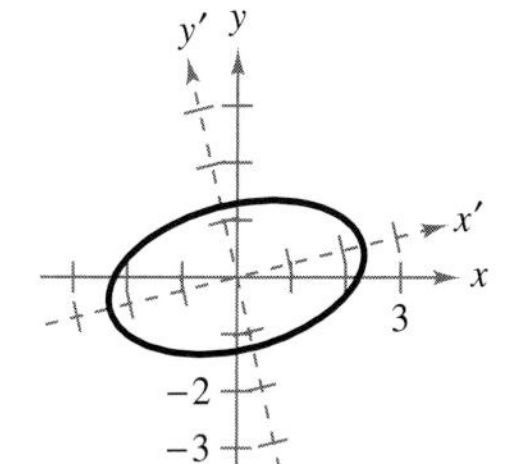

(b)

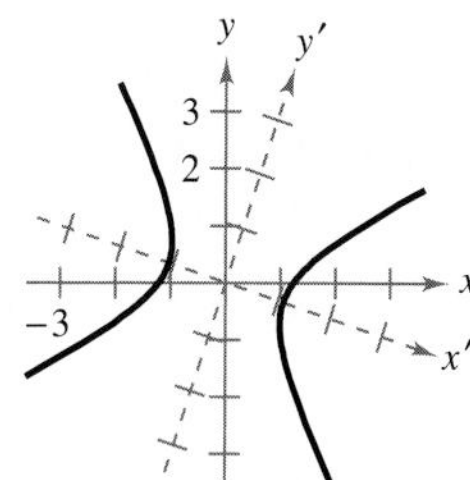

(c)

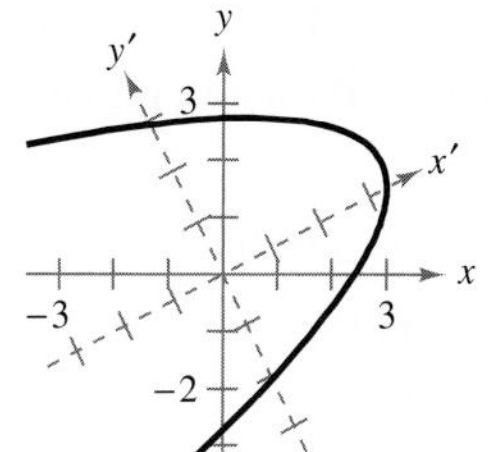

(d)

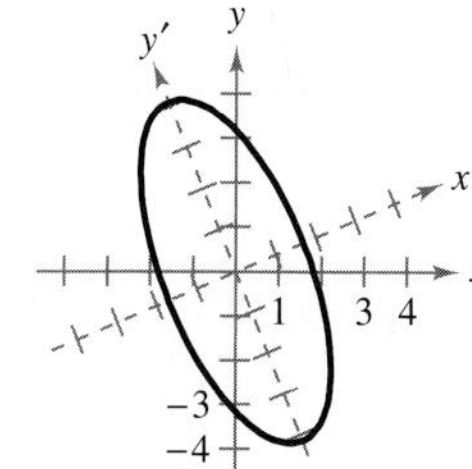

(e)

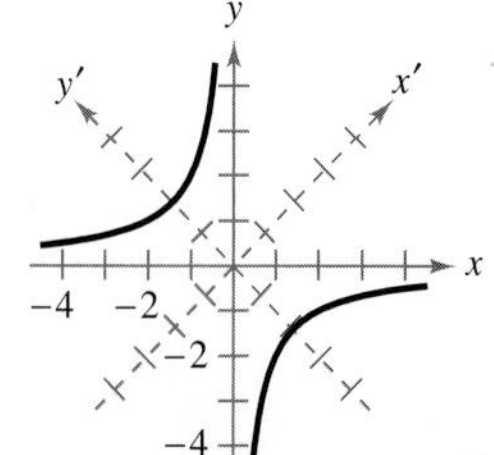

(f)

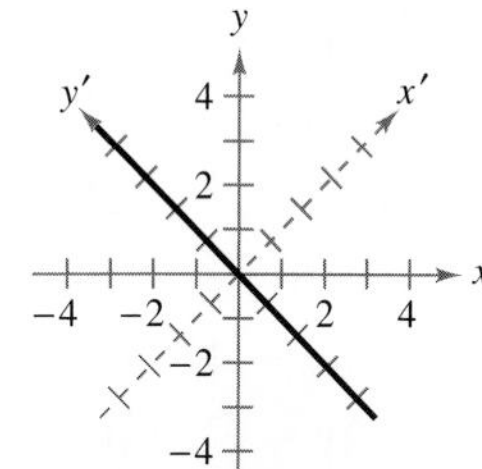

31. $xy + 2 = 0$
32. $x^2 - xy + 3y^2 - 5 = 0$
33. $3x^2 + 2xy + y^2 - 10 = 0$
34. $x^2 - 4xy + 4y^2 + 10x - 30 = 0$
35. $x^2 + 2xy + y^2 = 0$
36. $-2x^2 + 3xy + 2y^2 + 3 = 0$

Rotation and Graphing Utilities In Exercises 37–44, (a) use the discriminant to classify the graph of the equation, (b) use the Quadratic Formula to solve for y, and (c) use a graphing utility to graph the equation.

37. $16x^2 - 8xy + y^2 - 10x + 5y = 0$
38. $x^2 - 4xy - 2y^2 - 6 = 0$
39. $12x^2 - 6xy + 7y^2 - 45 = 0$
40. $2x^2 + 4xy + 5y^2 + 3x - 4y - 20 = 0$
41. $x^2 - 6xy - 5y^2 + 4x - 22 = 0$
42. $36x^2 - 60xy + 25y^2 + 9y = 0$
43. $x^2 + 4xy + 4y^2 - 5x - y - 3 = 0$
44. $x^2 + xy + 4y^2 + x + y - 4 = 0$

Sketching the Graph of a Degenerate Conic In Exercises 45–54, sketch the graph of the degenerate conic.

45. $y^2 - 16x^2 = 0$
46. $y^2 - 25x^2 = 0$
47. $15x^2 - 2xy - y^2 = 0$
48. $32x^2 - 4xy - y^2 = 0$
49. $x^2 - 2xy + y^2 = 0$
50. $x^2 + 4xy + 4y^2 = 0$
51. $x^2 + y^2 + 2x - 4y + 5 = 0$
52. $x^2 + y^2 - 2x + 6y + 10 = 0$
53. $x^2 + 2xy + y^2 - 1 = 0$
54. $4x^2 + 4xy + y^2 - 1 = 0$

Finding Points of Intersection In Exercises 55–62, find any points of intersection of the graphs of the equations algebraically and then verify using a graphing utility.

55. $x^2 - 4y^2 - 20x - 64y - 172 = 0$
$16x^2 + 4y^2 - 320x + 64y + 1600 = 0$
56. $x^2 - y^2 - 12x + 16y - 64 = 0$
$x^2 + y^2 - 12x - 16y + 64 = 0$
57. $x^2 + 4y^2 - 2x - 8y + 1 = 0$
$-x^2 + 2x - 4y - 1 = 0$
58. $-16x^2 - y^2 + 24y - 80 = 0$
$16x^2 + 25y^2 - 400 = 0$
59. $x^2 + y^2 - 4 = 0$
$3x - y^2 = 0$
60. $4x^2 + 9y^2 - 36y = 0$
$x^2 + 9y - 27 = 0$
61. $-x^2 - y^2 - 8x + 20y - 7 = 0$
$x^2 + 9y^2 + 8x + 4y + 7 = 0$
62. $x^2 + 2y^2 - 4x + 6y - 5 = 0$
$x^2 - 4x - y + 4 = 0$

63. **Satellite Dish** The parabolic cross section of a satellite dish is modeled by a portion of the graph of the equation

$$x^2 - 2xy - 27\sqrt{2}x + y^2 + 9\sqrt{2}y + 378 = 0$$

where all measurements are in feet.

(a) Rotate the axes to eliminate the xy-term in the equation. Then write the equation in standard form.

(b) A receiver is located at the focus of the cross section. Find the distance from the vertex of the cross section to the receiver.

64. **HOW DO YOU SEE IT?** Match each graph with the discriminant of its corresponding equation.

(a) -7
(b) 0
(c) 1

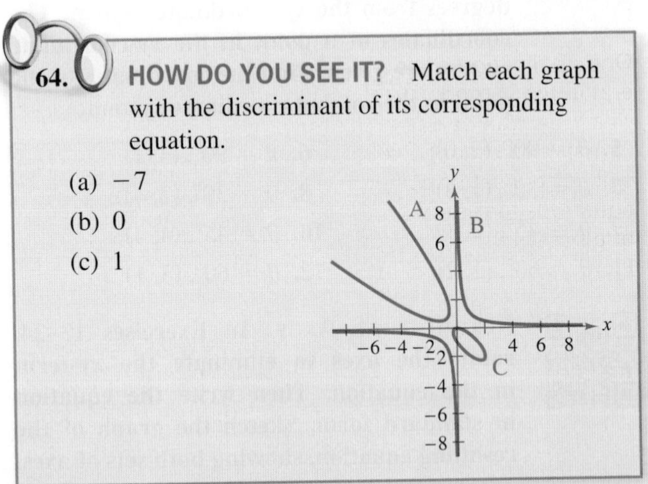

Exploration

True or False? In Exercises 65 and 66, determine whether the statement is true or false. Justify your answer.

65. The graph of the equation

$$x^2 + xy + ky^2 + 6x + 10 = 0$$

where k is any constant less than $\frac{1}{4}$, is a hyperbola.

66. After a rotation of axes is used to eliminate the xy-term from an equation of the form

$$Ax^2 + Bxy + Cy^2 + Dx + Ey + F = 0$$

the coefficients of the x^2- and y^2-terms remain A and C, respectively.

67. **Rotating a Circle** Show that the equation

$$x^2 + y^2 = r^2$$

is invariant under rotation of axes.

68. **Finding Lengths of Axes** Find the lengths of the major and minor axes of the ellipse in Exercise 19.

6.6 Parametric Equations

One application of parametric equations is modeling the path of an object. For example, in Exercise 93 on page 469, you will write a set of parametric equations that models the path of a baseball.

- **Evaluate sets of parametric equations for given values of the parameter.**
- **Sketch curves represented by sets of parametric equations.**
- **Rewrite sets of parametric equations as single rectangular equations by eliminating the parameter.**
- **Find sets of parametric equations for graphs.**

Plane Curves

Up to this point, you have been representing a graph by a single equation involving *two* variables such as x and y. In this section, you will study situations in which it is useful to introduce a *third* variable to represent a curve in the plane.

To see the usefulness of this procedure, consider the path of an object propelled into the air at an angle of 45°. When the initial velocity of the object is 48 feet per second, it can be shown that the object follows the parabolic path

$$y = -\frac{x^2}{72} + x. \qquad \text{Rectangular equation}$$

However, this equation does not tell the whole story. Although it does tell you *where* the object has been, it does not tell you *when* the object was at a given point (x, y) on the path. To determine this time, you can introduce a third variable t, called a **parameter.** It is possible to write both x and y as functions of t to obtain the **parametric equations**

$$x = 24\sqrt{2}t \qquad \text{Parametric equation for } x$$

$$y = -16t^2 + 24\sqrt{2}t. \qquad \text{Parametric equation for } y$$

This set of equations shows that at time $t = 0$, the object is at the point $(0, 0)$. Similarly, at time $t = 1$, the object is at the point $\left(24\sqrt{2}, 24\sqrt{2} - 16\right)$, and so on, as shown in the figure below.

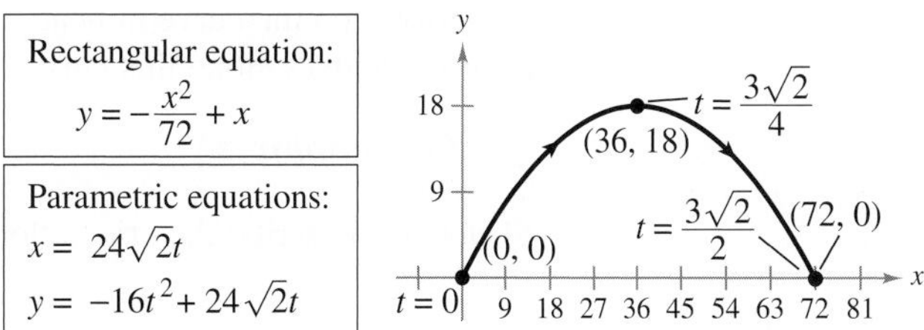

Curvilinear Motion: Two Variables for Position, One Variable for Time

For this motion problem, x and y are continuous functions of t, and the resulting path is a **plane curve.** (Recall that a *continuous function* is one whose graph has no breaks, holes, or gaps.)

Definition of Plane Curve

If f and g are continuous functions of t on an interval I, then the set of ordered pairs $(f(t), g(t))$ is a **plane curve** C. The equations

$$x = f(t) \quad \text{and} \quad y = g(t)$$

are **parametric equations** for C, and t is the **parameter.**

Sketching a Plane Curve

One way to sketch a curve represented by a pair of parametric equations is to plot points in the xy-plane. You determine each set of coordinates (x, y) from a value chosen for the parameter t. Plotting the resulting points in the order of *increasing* values of t traces the curve in a specific direction. This is called the **orientation** of the curve.

EXAMPLE 1 Sketching a Curve

See LarsonPrecalculus.com for an interactive version of this type of example.

Sketch and describe the orientation of the curve given by the parametric equations

$$x = t^2 - 4 \quad \text{and} \quad y = \frac{t}{2}, \quad -2 \le t \le 3.$$

REMARK When using a value of t to find x, be sure to use the same value of t to find the corresponding value of y. Organizing your results in a table, as shown in Example 1, can be helpful.

Solution Using values of t in the specified interval, the parametric equations yield the values of x and y shown in the table. By plotting the points (x, y) in the order of increasing values of t, you obtain the curve shown in the figure.

t	x	y
-2	0	-1
-1	-3	$-\frac{1}{2}$
0	-4	0
1	-3	$\frac{1}{2}$
2	0	1
3	5	$\frac{3}{2}$

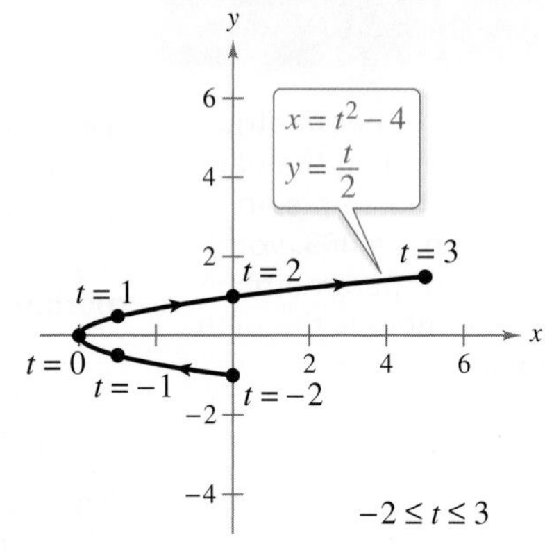

The arrows on the curve indicate its orientation as t increases from -2 to 3. So, when a particle moves along this curve, it starts at $(0, -1)$ and ends at $\left(5, \frac{3}{2}\right)$.

✓ **Checkpoint** Audio-video solution in English & Spanish at LarsonPrecalculus.com

Sketch and describe the orientation of the curve given by the parametric equations

$$x = 2t \quad \text{and} \quad y = 4t^2 + 2, \quad -2 \le t \le 2.$$

Note that the graph in Example 1 does not define y as a function of x. This points out one benefit of parametric equations—they can represent graphs that are not necessarily graphs of functions.

Two different sets of parametric equations can have the same graph. For example, the set of parametric equations

$$x = 4t^2 - 4 \quad \text{and} \quad y = t, \quad -1 \le t \le \frac{3}{2}$$

has the same graph as the set of parametric equations given in Example 1 (see Figure 6.39). However, comparing the values of t in the two graphs shows that the second graph is traced out more *rapidly* (considering t as time) than the first graph. So, in applications, different parametric representations can represent various *speeds* at which objects travel along a given path.

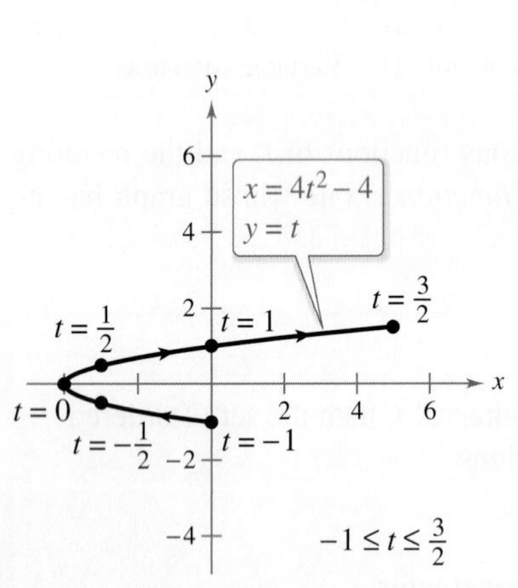

Figure 6.39

Eliminating the Parameter

Sketching a curve represented by a pair of parametric equations can sometimes be simplified by finding a rectangular equation (in x and y) that has the same graph. This process is called **eliminating the parameter,** and is illustrated below using the parametric equations from Example 1.

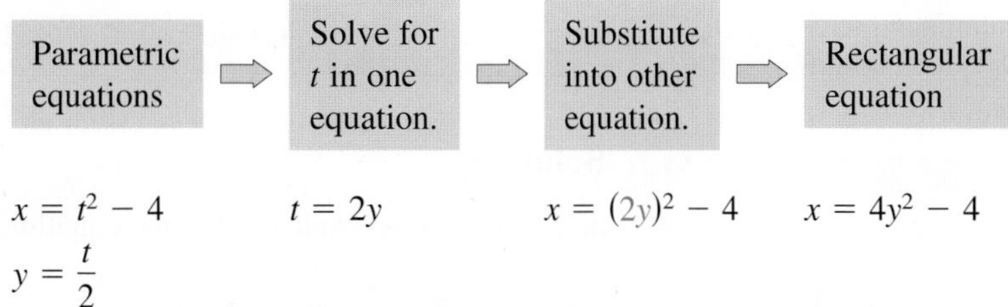

$x = t^2 - 4$ $\quad$ $t = 2y$ $\quad$ $x = (2y)^2 - 4$ $\quad$ $x = 4y^2 - 4$

$y = \frac{t}{2}$

The equation $x = 4y^2 - 4$ represents a parabola with a horizontal axis and vertex at $(-4, 0)$. You graphed a portion of this parabola in Example 1.

When converting equations from parametric to rectangular form, you may need to alter the domain of the rectangular equation so that its graph matches the graph of the parametric equations. Example 2 demonstrates such a situation.

EXAMPLE 2 Eliminating the Parameter

Sketch the curve represented by the equations

$$x = \frac{1}{\sqrt{t+1}} \quad \text{and} \quad y = \frac{t}{t+1}$$

by eliminating the parameter and adjusting the domain of the resulting rectangular equation.

Solution Solve for t in the equation for x.

$$x^2 = \frac{1}{t+1} \implies t + 1 = \frac{1}{x^2} \implies t = \frac{1}{x^2} - 1 = \frac{1 - x^2}{x^2}$$

Then substitute for t in the equation for y to obtain the rectangular equation

$$y = \frac{t}{t+1} = \frac{\frac{1-x^2}{x^2}}{\frac{1-x^2}{x^2} + 1} = \frac{\frac{1-x^2}{x^2}}{\frac{1-x^2}{x^2} + 1} \cdot \frac{x^2}{x^2} = \frac{1-x^2}{1 - x^2 + x^2} = \frac{1-x^2}{1} = 1 - x^2.$$

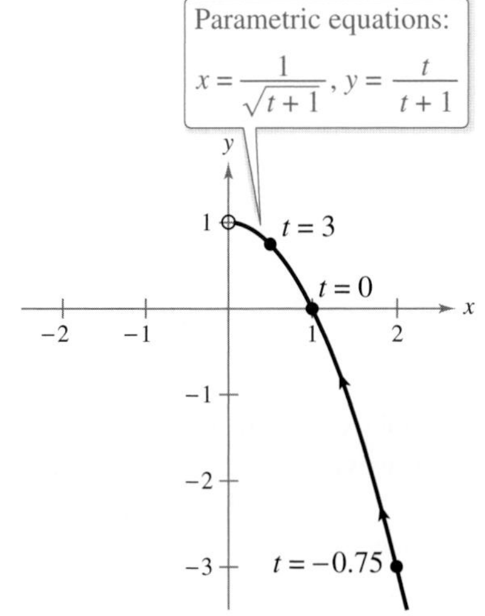

Figure 6.40

This rectangular equation shows that the curve is a parabola that opens downward and has its vertex at $(0, 1)$. Also, this rectangular equation is defined for all values of x. The parametric equation for x, however, is defined only when

$$\sqrt{t+1} > 0 \implies t + 1 > 0 \implies t > -1.$$

This implies that you should restrict the domain of x to positive values. Figure 6.40 shows a sketch of the curve.

✓ **Checkpoint** Audio-video solution in English & Spanish at LarsonPrecalculus.com

Sketch the curve represented by the equations

$$x = \frac{1}{\sqrt{t-1}} \quad \text{and} \quad y = \frac{t+1}{t-1}$$

by eliminating the parameter and adjusting the domain of the resulting rectangular equation.

It is not necessary for the parameter in a set of parametric equations to represent time. The next example uses an *angle* as the parameter.

REMARK To eliminate the parameter in equations involving trigonometric functions, you may need to use fundamental trigonometric identities, as shown in Example 3.

EXAMPLE 3 Eliminating Angle Parameters

Sketch the curve represented by each set of equations by eliminating the parameter.

a. $x = 3\cos\theta$ and $y = 4\sin\theta$, $0 \le \theta < 2\pi$

b. $x = 1 + 3\sec\theta$ and $y = -3 + \tan\theta$, $\pi/2 < \theta < 3\pi/2$

Solution

a. Solve for $\cos\theta$ and $\sin\theta$ in the equations.

$$\cos\theta = \frac{x}{3} \quad \text{and} \quad \sin\theta = \frac{y}{4}$$ Solve for $\cos\theta$ and $\sin\theta$.

Then use the identity $\sin^2\theta + \cos^2\theta = 1$ to form an equation involving only x and y.

$$\cos^2\theta + \sin^2\theta = 1$$ Pythagorean identity

$$\left(\frac{x}{3}\right)^2 + \left(\frac{y}{4}\right)^2 = 1$$ Substitute $\frac{x}{3}$ for $\cos\theta$ and $\frac{y}{4}$ for $\sin\theta$.

$$\frac{x^2}{9} + \frac{y^2}{16} = 1$$ Rectangular equation

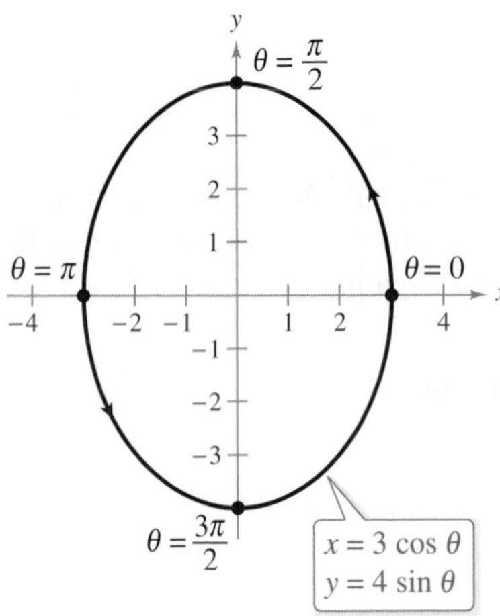

Figure 6.41

The graph of this rectangular equation is an ellipse centered at $(0, 0)$, with vertices $(0, 4)$ and $(0, -4)$, and minor axis of length $2b = 6$, as shown in Figure 6.41. Note that the elliptic curve is traced out *counterclockwise* as θ increases on the interval $[0, 2\pi)$.

b. Solve for $\sec\theta$ and $\tan\theta$ in the equations.

$$\sec\theta = \frac{x-1}{3} \quad \text{and} \quad \tan\theta = y + 3$$ Solve for $\sec\theta$ and $\tan\theta$.

Then use the identity $\sec^2\theta - \tan^2\theta = 1$ to form an equation involving only x and y.

$$\sec^2\theta - \tan^2\theta = 1$$ Pythagorean identity

$$\left(\frac{x-1}{3}\right)^2 - (y+3)^2 = 1$$ Substitute $\frac{x-1}{3}$ for $\sec\theta$ and $y + 3$ for $\tan\theta$.

$$\frac{(x-1)^2}{9} - \frac{(y+3)^2}{1} = 1$$ Rectangular equation

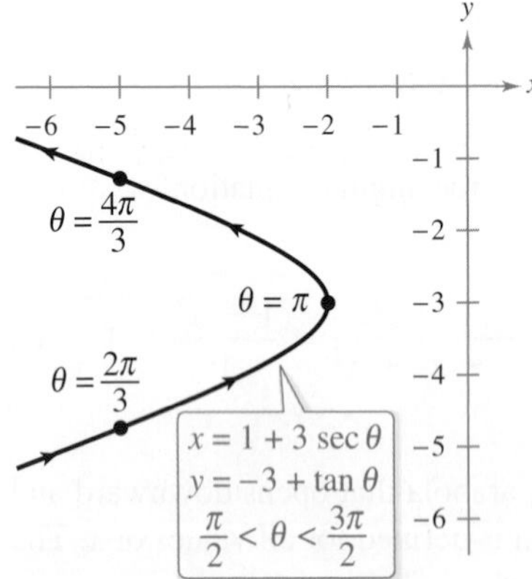

Figure 6.42

The graph of this rectangular equation is a hyperbola centered at $(1, -3)$ with a horizontal transverse axis of length $2a = 6$. However, the restriction on θ corresponds to a restriction on the domain of x to $x \le -2$, which corresponds to the *left* branch of the hyperbola only. Figure 6.42 shows the graph.

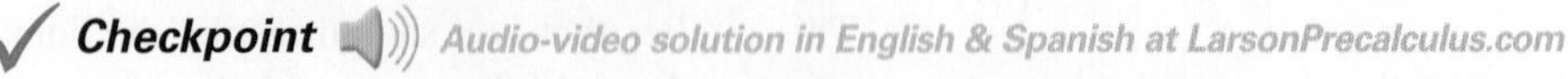

Sketch the curve represented by each set of equations by eliminating the parameter.

a. $x = 5\cos\theta$ and $y = 3\sin\theta$, $0 \le \theta < 2\pi$

b. $x = -1 + \tan\theta$ and $y = 2 + 2\sec\theta$, $\pi/2 < \theta < 3\pi/2$

REMARK It is important to realize that eliminating the parameter is primarily an aid to curve sketching, as demonstrated in Examples 2 and 3.

When parametric equations represent the path of a moving object, the graph of the corresponding rectangular equation is not sufficient to describe the object's motion. You still need the parametric equations to tell you the *position, direction,* and *speed* at a given time.

Finding Parametric Equations for a Graph

You have been studying techniques for sketching the graph represented by a set of parametric equations. Now consider the *reverse* problem—that is, how can you find a set of parametric equations for a given graph or a given physical description? From the discussion after Example 1, you know that such a representation is not unique. That is, the equations

$$x = 4t^2 - 4 \quad \text{and} \quad y = t, \quad -1 \le t \le \frac{3}{2}$$

produced the same graph as the equations

$$x = t^2 - 4 \quad \text{and} \quad y = \frac{t}{2}, \quad -2 \le t \le 3.$$

Example 4 further demonstrates this.

EXAMPLE 4 **Finding Parametric Equations for a Graph**

Find a set of parametric equations to represent the graph of $y = 1 - x^2$, using each parameter.

a. $t = x$

b. $t = 1 - x$

Solution

a. Letting $t = x$, you obtain the parametric equations

$$x = t \quad \text{and} \quad y = 1 - x^2 = 1 - t^2.$$

Figure 6.43 shows the curve represented by the parametric equations.

b. Letting $t = 1 - x$, you obtain the parametric equations

$$x = 1 - t \quad \text{and} \quad y = 1 - x^2 = 1 - (1 - t)^2 = 2t - t^2.$$

Figure 6.44 shows the curve represented by the parametric equations. Note that the graphs in Figures 6.43 and 6.44 have opposite orientations.

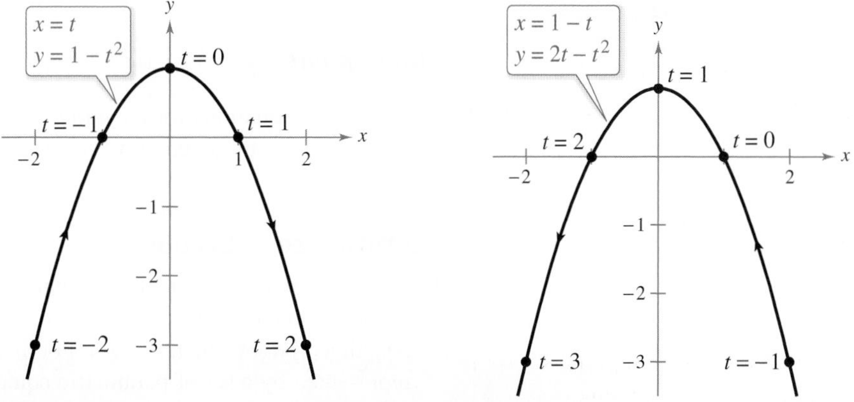

Figure 6.43

Figure 6.44

✓ **Checkpoint** Audio-video solution in English & Spanish at LarsonPrecalculus.com

Find a set of parametric equations to represent the graph of $y = x^2 + 2$, using each parameter.

a. $t = x$ **b.** $t = 2 - x$

A **cycloid** is a curve traced by a point P on a circle as the circle rolls along a straight line in a plane.

EXAMPLE 5 Parametric Equations for a Cycloid

Write parametric equations for a cycloid traced by a point P on a circle of radius a units as the circle rolls along the x-axis given that P is at a minimum when $x = 0$.

Solution Let the parameter θ be the measure of the circle's rotation, and let the point $P(x, y)$ begin at the origin. When $\theta = 0$, P is at the origin; when $\theta = \pi$, P is at a maximum point $(\pi a, 2a)$; and when $\theta = 2\pi$, P is back on the x-axis at $(2\pi a, 0)$. From the figure below, $\angle APC = \pi - \theta$. So, you have

$$\sin\theta = \sin(\pi - \theta) = \sin(\angle APC) = \frac{AC}{a} = \frac{BD}{a}$$

$$\cos\theta = -\cos(\pi - \theta) = -\cos(\angle APC) = -\frac{AP}{a}$$

REMARK In Example 5, $\widehat{PD}$ represents the arc of the circle between points P and D.

which implies that $BD = a\sin\theta$ and $AP = -a\cos\theta$. The circle rolls along the x-axis, so you know that $OD = \widehat{PD} = a\theta$. Furthermore, $BA = DC = a$, so you have

$$x = OD - BD = a\theta - a\sin\theta$$

and

$$y = BA + AP = a - a\cos\theta.$$

The parametric equations are $x = a(\theta - \sin\theta)$ and $y = a(1 - \cos\theta)$.

TECHNOLOGY Use a graphing utility in *parametric* mode to obtain a graph similar to the one in Example 5 by graphing

$$X_{1T} = T - \sin T$$

and

$$Y_{1T} = 1 - \cos T.$$

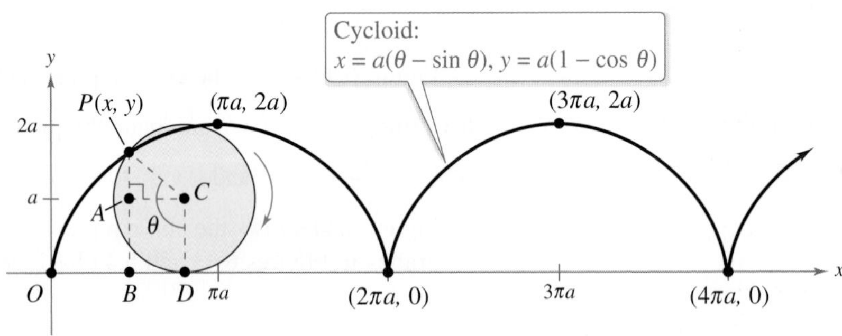

Checkpoint Audio-video solution in English & Spanish at LarsonPrecalculus.com

Write parametric equations for a cycloid traced by a point P on a circle of radius a as the circle rolls along the x-axis given that P is at a maximum when $x = 0$.

Summarize (Section 6.6)

1. Explain how to evaluate a set of parametric equations for given values of the parameter and sketch a curve represented by a set of parametric equations *(pages 461 and 462)*. For an example of sketching a curve represented by a set of parametric equations, see Example 1.
2. Explain how to rewrite a set of parametric equations as a single rectangular equation by eliminating the parameter *(page 463)*. For examples of sketching curves by eliminating the parameter, see Examples 2 and 3.
3. Explain how to find a set of parametric equations for a graph *(page 465)*. For examples of finding sets of parametric equations for graphs, see Examples 4 and 5.

6.6 Exercises

See CalcChat.com for tutorial help and worked-out solutions to odd-numbered exercises.

Vocabulary: Fill in the blanks.

1. If f and g are continuous functions of t on an interval I, then the set of ordered pairs $(f(t), g(t))$ is a ________ ________ C.
2. The ________ of a curve is the direction in which the curve is traced for increasing values of the parameter.
3. The process of converting a set of parametric equations to a corresponding rectangular equation is called ________ the ________.
4. A curve traced by a point on the circumference of a circle as the circle rolls along a straight line in a plane is a ________.

Skills and Applications

5. **Sketching a Curve** Consider the parametric equations $x = \sqrt{t}$ and $y = 3 - t$.
   (a) Create a table of x- and y-values using $t = 0, 1, 2, 3$, and 4.
   (b) Plot the points (x, y) generated in part (a), and sketch a graph of the parametric equations.
   (c) Sketch the graph of $y = 3 - x^2$. How do the graphs differ?
6. **Sketching a Curve** Consider the parametric equations $x = 4\cos^2\theta$ and $y = 2\sin\theta$.
   (a) Create a table of x- and y-values using $\theta = -\pi/2, -\pi/4, 0, \pi/4$, and $\pi/2$.
   (b) Plot the points (x, y) generated in part (a), and sketch a graph of the parametric equations.
   (c) Sketch the graph of $x = -y^2 + 4$. How do the graphs differ?

Sketching a Curve In Exercises 7–12, sketch and describe the orientation of the curve given by the parametric equations.

7. $x = t, \quad y = -5t$
8. $x = 2t - 1, \quad y = t + 4$
9. $x = t^2, \quad y = 3t$
10. $x = \sqrt{t}, \quad y = 2t - 1$
11. $x = 3\cos\theta, \quad y = 2\sin^2\theta, \quad 0 \le \theta \le \pi$
12. $x = \cos\theta, \quad y = 2\sin\theta, \quad 0 \le \theta \le 2\pi$

Sketching a Curve In Exercises 13–38, (a) sketch the curve represented by the parametric equations (indicate the orientation of the curve) and (b) eliminate the parameter and write the resulting rectangular equation whose graph represents the curve. Adjust the domain of the rectangular equation, if necessary.

13. $x = t, \quad y = 4t$
14. $x = t, \quad y = -\frac{1}{2}t$
15. $x = -t + 1, \quad y = -3t$
16. $x = 3 - 2t, \quad y = 2 + 3t$
17. $x = \frac{1}{4}t, \quad y = t^2$
18. $x = t, \quad y = t^3$
19. $x = t^2, \quad y = -2t$
20. $x = -t^2, \quad y = \dfrac{t}{3}$
21. $x = \sqrt{t}, \quad y = 1 - t$
22. $x = \sqrt{t + 2}, \quad y = t - 1$
23. $x = \sqrt{t} - 3, \quad y = t^3$
24. $x = \sqrt{t - 1}, \quad y = \sqrt[3]{t - 1}$
25. $x = t + 1$, $y = \dfrac{t}{t + 1}$
26. $x = t - 1$, $y = \dfrac{t}{t - 1}$
27. $x = 4\cos\theta$, $y = 2\sin\theta$
28. $x = 2\cos\theta$, $y = 3\sin\theta$
29. $x = 1 + \cos\theta$, $y = 1 + 2\sin\theta$
30. $x = 2 + 5\cos\theta$, $y = -6 + 4\sin\theta$
31. $x = 2\sec\theta, \quad y = \tan\theta, \quad \pi/2 \le \theta \le 3\pi/2$
32. $x = 3\cot\theta, \quad y = 4\csc\theta, \quad 0 \le \theta \le \pi$
33. $x = 3\cos\theta$, $y = 3\sin\theta$
34. $x = 6\sin 2\theta$, $y = 6\cos 2\theta$
35. $x = e^t, \quad y = e^{3t}$
36. $x = e^{-t}, \quad y = e^{3t}$
37. $x = t^3, \quad y = 3\ln t$
38. $x = \ln 2t, \quad y = 2t^2$

Graphing a Curve In Exercises 39–48, use a graphing utility to graph the curve represented by the parametric equations.

39. $x = t$, $y = \sqrt{t}$
40. $x = t + 1$, $y = \sqrt{2 - t}$
41. $x = 2t$, $y = |t + 1|$
42. $x = |t + 2|$, $y = 3 - t$
43. $x = 4 + 3\cos\theta$, $y = -2 + \sin\theta$
44. $x = 4 + 3\cos\theta$, $y = -2 + 2\sin\theta$
45. $x = 2\csc\theta$, $y = 4\cot\theta$
46. $x = \sec\theta$, $y = \tan\theta$
47. $x = \frac{1}{2}t$, $t = \ln(t^2 + 1)$
48. $x = 10 - 0.01e^t$, $y = 0.4t^2$

Comparing Plane Curves **In Exercises 49 and 50, determine how the plane curves differ from each other.**

49. (a) $x = t$, $y = 2t + 1$ (b) $x = \cos\theta$, $y = 2\cos\theta + 1$

(c) $x = e^{-t}$, $y = 2e^{-t} + 1$ (d) $x = e^{t}$, $y = 2e^{t} + 1$

50. (a) $x = t$, $y = t^2 - 1$ (b) $x = t^2$, $y = t^4 - 1$

(c) $x = \sin t$, $y = \sin^2 t - 1$ (d) $x = e^{t}$, $y = e^{2t} - 1$

Eliminating the Parameter **In Exercises 51–54, eliminate the parameter and obtain the standard form of the rectangular equation.**

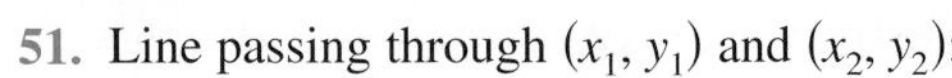

51. Line passing through (x_1, y_1) and (x_2, y_2):
$x = x_1 + t(x_2 - x_1),\quad y = y_1 + t(y_2 - y_1)$

52. Circle: $x = h + r\cos\theta,\quad y = k + r\sin\theta$

53. Ellipse with horizontal major axis:
$x = h + a\cos\theta,\quad y = k + b\sin\theta$

54. Hyperbola with horizontal transverse axis:
$x = h + a\sec\theta,\quad y = k + b\tan\theta$

Finding Parametric Equations for a Graph **In Exercises 55–62, use the results of Exercises 51–54 to find a set of parametric equations to represent the graph of the line or conic.**

55. Line: passes through $(0, 0)$ and $(3, 6)$

56. Line: passes through $(3, 2)$ and $(-6, 3)$

57. Circle: center: $(3, 2)$; radius: 4

58. Circle: center: $(-2, -5)$; radius: 7

59. Ellipse: vertices: $(\pm 5, 0)$; foci: $(\pm 4, 0)$

60. Ellipse: vertices: $(7, 3), (-1, 3)$; foci: $(5, 3), (1, 3)$

61. Hyperbola: vertices: $(1, 0), (9, 0)$; foci: $(0, 0), (10, 0)$

62. Hyperbola: vertices: $(4, 1), (8, 1)$; foci: $(2, 1), (10, 1)$

Finding Parametric Equations for a Graph **In Exercises 63–66, use the results of Exercises 51 and 54 to find a set of parametric equations to represent the section of the graph of the line or conic. (*Hint:* Adjust the domain of the standard form of the rectangular equation to determine the appropriate interval for the parameter.)**

63. Line segment between $(0, 0)$ and $(-5, 2)$

64. Line segment between $(1, -4)$ and $(9, 0)$

65. Left branch of the hyperbola with vertices $(\pm 3, 0)$ and foci $(\pm 5, 0)$

66. Right branch of the hyperbola with vertices $(-4, 3)$ and $(6, 3)$ and foci $(-12, 3)$ and $(14, 3)$

Finding Parametric Equations for a Graph **In Exercises 67–78, find a set of parametric equations to represent the graph of the rectangular equation using (a) $t = x$ and (b) $t = 2 - x$.**

67. $y = 3x - 2$

68. $y = 2 - x$

69. $x = 2y + 1$

70. $x = 3y - 2$

71. $y = x^2 + 1$

72. $y = 6x^2 - 5$

73. $y = 1 - 2x^2$

74. $y = 2 - 5x^2$

75. $y = \dfrac{1}{x}$

76. $y = \dfrac{1}{2x}$

77. $y = e^{x}$

78. $y = e^{2x}$

Graphing a Curve **In Exercises 79–86, use a graphing utility to graph the curve represented by the parametric equations.**

79. Cycloid: $x = 4(\theta - \sin\theta),\quad y = 4(1 - \cos\theta)$

80. Cycloid: $x = \theta + \sin\theta,\quad y = 1 - \cos\theta$

81. Prolate cycloid: $x = 2\theta - 4\sin\theta,\quad y = 2 - 4\cos\theta$

82. Epicycloid: $x = 8\cos\theta - 2\cos 4\theta$
$y = 8\sin\theta - 2\sin 4\theta$

83. Hypocycloid: $x = 3\cos^3\theta,\quad y = 3\sin^3\theta$

84. Curtate cycloid: $x = 8\theta - 4\sin\theta,\quad y = 8 - 4\cos\theta$

85. Witch of Agnesi: $x = 2\cot\theta,\quad y = 2\sin^2\theta$

86. Folium of Descartes: $x = \dfrac{3t}{1 + t^3},\quad y = \dfrac{3t^2}{1 + t^3}$

Matching **In Exercises 87–90, match the parametric equations with the correct graph and describe the domain and range. [The graphs are labeled (a)–(d).]**

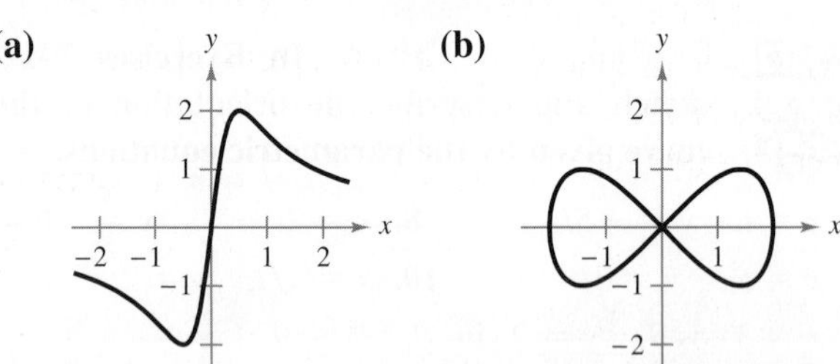

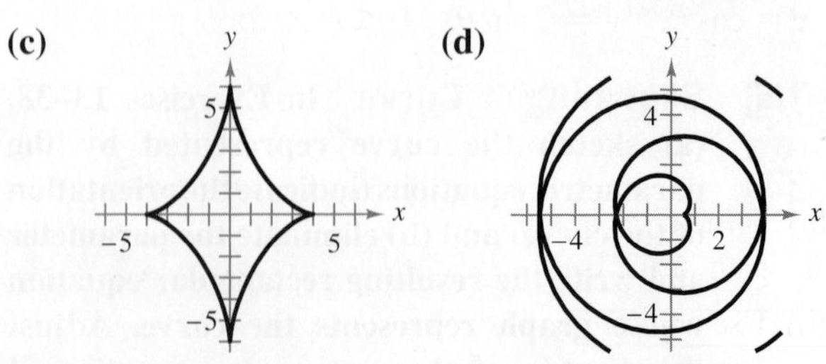

87. Lissajous curve: $x = 2\cos\theta,\quad y = \sin 2\theta$

88. Evolute of ellipse: $x = 4\cos^3\theta,\quad y = 6\sin^3\theta$

89. Involute of circle: $x = \frac{1}{2}(\cos\theta + \theta\sin\theta)$
$y = \frac{1}{2}(\sin\theta - \theta\cos\theta)$

90. Serpentine curve: $x = \frac{1}{2}\cot\theta,\quad y = 4\sin\theta\cos\theta$

Projectile Motion **Consider a projectile launched at a height of h feet above the ground at an angle of θ with the horizontal. The initial velocity is v_0 feet per second, and the path of the projectile is modeled by the parametric equations**

$$x = (v_0 \cos \theta)t$$

and

$$y = h + (v_0 \sin \theta)t - 16t^2.$$

In Exercises 91 and 92, use a graphing utility to graph the paths of a projectile launched from ground level at each value of θ and v_0. For each case, use the graph to approximate the maximum height and the range of the projectile.

91. (a) $\theta = 60°$, $v_0 = 88$ feet per second
(b) $\theta = 60°$, $v_0 = 132$ feet per second
(c) $\theta = 45°$, $v_0 = 88$ feet per second
(d) $\theta = 45°$, $v_0 = 132$ feet per second

92. (a) $\theta = 15°$, $v_0 = 50$ feet per second
(b) $\theta = 15°$, $v_0 = 120$ feet per second
(c) $\theta = 10°$, $v_0 = 50$ feet per second
(d) $\theta = 10°$, $v_0 = 120$ feet per second

93. Path of a Baseball

The center field fence in a baseball stadium is 7 feet high and 408 feet from home plate. A baseball player hits a baseball at a point 3 feet above the ground. The ball leaves the bat at an angle of θ degrees with the horizontal at a speed of 100 miles per hour (see figure).

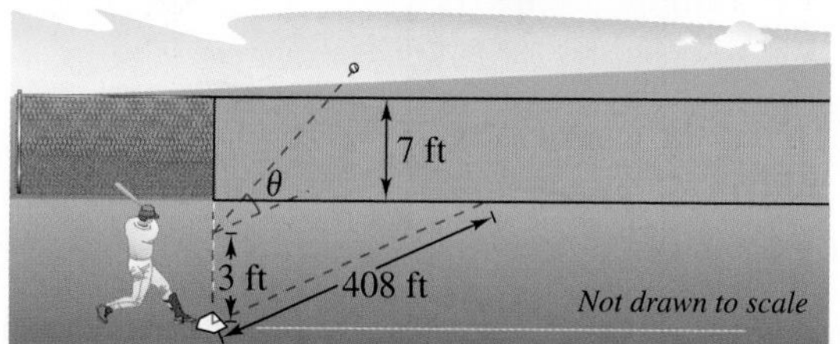

(a) Write a set of parametric equations that model the path of the baseball. (See Exercises 91 and 92.)

(b) Use a graphing utility to graph the path of the baseball when $\theta = 15°$. Is the hit a home run?

(c) Use the graphing utility to graph the path of the baseball when $\theta = 23°$. Is the hit a home run?

(d) Find the minimum angle required for the hit to be a home run.

94. Path of an Arrow An archer releases an arrow from a bow at a point 5 feet above the ground. The arrow leaves the bow at an angle of 15° with the horizontal and at an initial speed of 225 feet per second.

(a) Write a set of parametric equations that model the path of the arrow. (See Exercises 91 and 92.)

(b) Assuming the ground is level, find the distance the arrow travels before it hits the ground. (Ignore air resistance.)

(c) Use a graphing utility to graph the path of the arrow and approximate its maximum height.

(d) Find the total time the arrow is in the air.

95. Path of a Football A quarterback releases a pass at a height of 7 feet above the playing field, and a receiver catches the football at a height of 4 feet, 30 yards directly downfield. The pass is released at an angle of 35° with the horizontal.

(a) Write a set of parametric equations for the path of the football. (See Exercises 91 and 92.)

(b) Find the speed of the football when it is released.

(c) Use a graphing utility to graph the path of the football and approximate its maximum height.

(d) Find the time the receiver has to position himself after the quarterback releases the football.

96. Projectile Motion Eliminate the parameter t in the parametric equations

$$x = (v_0 \cos \theta)t$$

and

$$y = h + (v_0 \sin \theta)t - 16t^2$$

for the motion of a projectile to show that the rectangular equation is

$$y = -\frac{16 \sec^2 \theta}{v_0^2}x^2 + (\tan \theta)x + h.$$

97. Path of a Projectile The path of a projectile is given by the rectangular equation

$$y = 7 + x - 0.02x^2.$$

(a) Find the values of h, v_0, and θ. Then write a set of parametric equations that model the path. (See Exercise 96.)

(b) Use a graphing utility to graph the rectangular equation for the path of the projectile. Confirm your answer in part (a) by sketching the curve represented by the parametric equations.

(c) Use the graphing utility to approximate the maximum height of the projectile and its range.

98. Path of a Projectile Repeat Exercise 97 for a projectile with a path given by the rectangular equation

$$y = 6 + x - 0.08x^2.$$

99. Curtate Cycloid A wheel of radius a units rolls along a straight line without slipping. The curve traced by a point P that is b units from the center ($b < a$) is called a **curtate cycloid** (see figure). Use the angle θ shown in the figure to find a set of parametric equations for the curve.

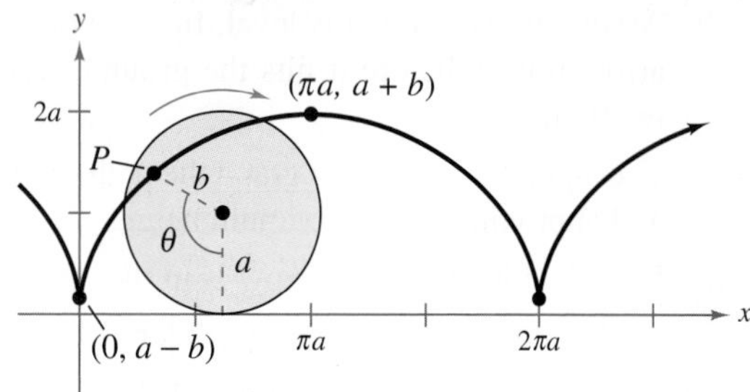

100. Epicycloid A circle of radius one unit rolls around the outside of a circle of radius two units without slipping. The curve traced by a point on the circumference of the smaller circle is called an **epicycloid** (see figure). Use the angle θ shown in the figure to find a set of parametric equations for the curve.

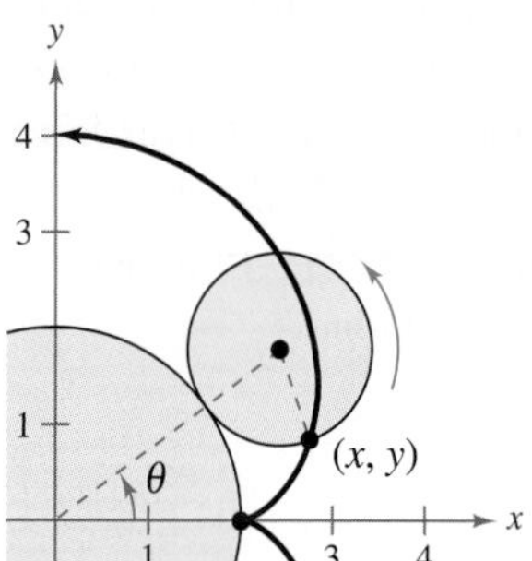

Exploration

True or False? **In Exercises 101–104, determine whether the statement is true or false. Justify your answer.**

101. The two sets of parametric equations

$x = t, \ y = t^2 + 1$ and $x = 3t, \ y = 9t^2 + 1$

correspond to the same rectangular equation.

102. The graphs of the parametric equations

$x = t^2, \ y = t^2$ and $x = t, \ y = t$

both represent the line $y = x$, so they are the same plane curve.

103. If y is a function of t and x is a function of t, then y must be a function of x.

104. The parametric equations

$x = at + h$ and $y = bt + k$

where $a \neq 0$ and $b \neq 0$, represent a circle centered at (h, k) when $a = b$.

105. Writing Write a short paragraph explaining why parametric equations are useful.

106. Writing Explain what is meant by the orientation of a plane curve.

107. Error Analysis Describe the error in finding the rectangular equation for the parametric equations

$x = \sqrt{t - 1}$ and $y = 2t$.

$x = \sqrt{t - 1} \implies t = x^2 + 1$

$y = 2(x^2 + 1) = 2x^2 + 2$

108. HOW DO YOU SEE IT? The graph of the parametric equations $x = t$ and $y = t^2$ is shown below. Determine whether the graph would change for each set of parametric equations. If so, how would it change?

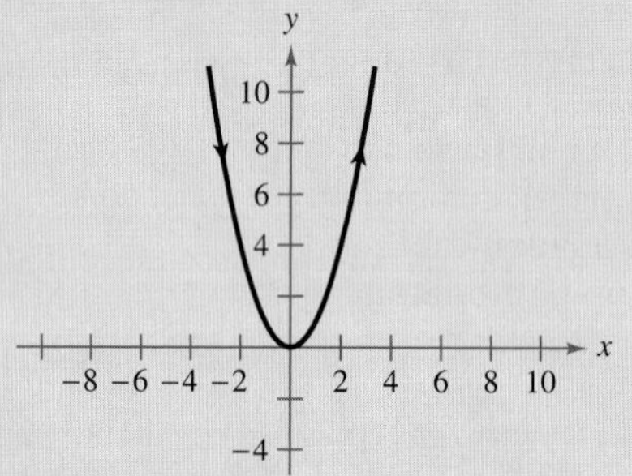

(a) $x = -t, \ y = t^2$

(b) $x = t + 1, \ y = t^2$

(c) $x = t, \ y = t^2 + 1$

109. Think About It The graph of the parametric equations $x = t^3$ and $y = t - 1$ is shown below. Would the graph change for the parametric equations $x = (-t)^3$ and $y = -t - 1$? If so, how would it change?

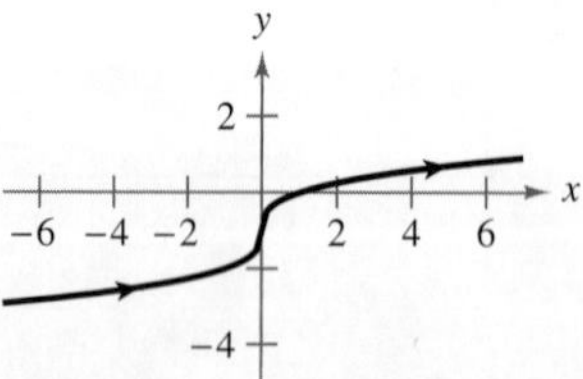

110. Think About It The graph of the parametric equations $x = t^2$ and $y = t + 1$ is shown below. Would the graph change for the parametric equations $x = (t + 1)^2$ and $y = t + 2$? If so, how would it change?

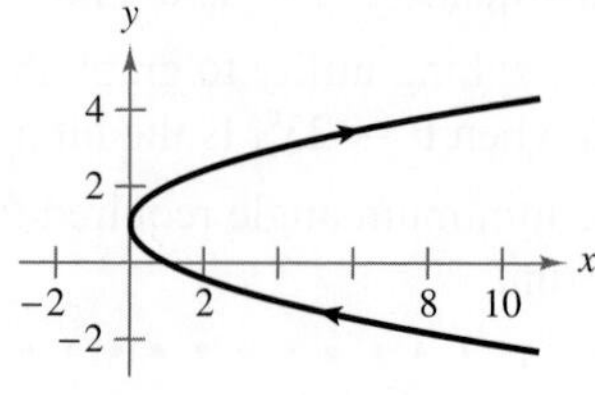

6.7 Polar Coordinates

Polar coordinates are often useful tools in mathematical modeling. For example, in Exercise 109 on page 476, you will use polar coordinates to write an equation that models the position of a passenger car on a Ferris wheel.

- **Plot points in the polar coordinate system.**
- **Convert points from rectangular to polar form and vice versa.**
- **Convert equations from rectangular to polar form and vice versa.**

Introduction

So far, you have been representing graphs of equations as collections of points (x, y) in the rectangular coordinate system, where x and y represent the directed distances from the coordinate axes to the point (x, y). In this section, you will study a different system called the **polar coordinate system.**

To form the polar coordinate system in the plane, fix a point O, called the **pole** (or **origin**), and construct from O an initial ray called the **polar axis,** as shown in the figure at the right. Then each point P in the plane can be assigned **polar coordinates** (r, θ), where r and θ are defined below.

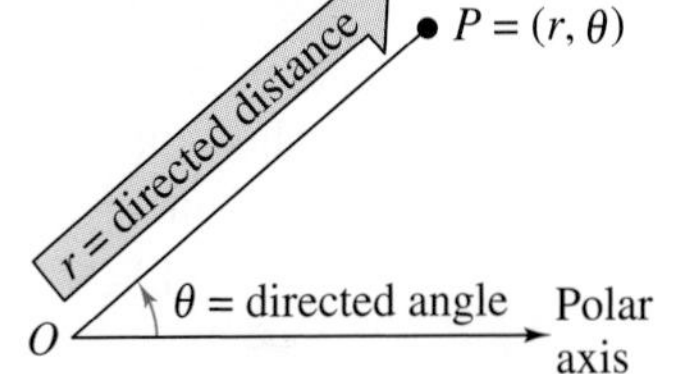

1. r = *directed distance* from O to P
2. θ = *directed angle,* counterclockwise from the polar axis to segment $\overline{OP}$

EXAMPLE 1 Plotting Points in the Polar Coordinate System

Plot each point given in polar coordinates.

a. $(2, \pi/3)$ **b.** $(3, -\pi/6)$ **c.** $(3, 11\pi/6)$

Solution

a. The point $(r, \theta) = (2, \pi/3)$ lies two units from the pole on the terminal side of the angle $\theta = \pi/3$, as shown in Figure 6.45.

b. The point $(r, \theta) = (3, -\pi/6)$ lies three units from the pole on the terminal side of the angle $\theta = -\pi/6$, as shown in Figure 6.46.

c. The point $(r, \theta) = (3, 11\pi/6)$ coincides with the point $(3, -\pi/6)$, as shown in Figure 6.47.

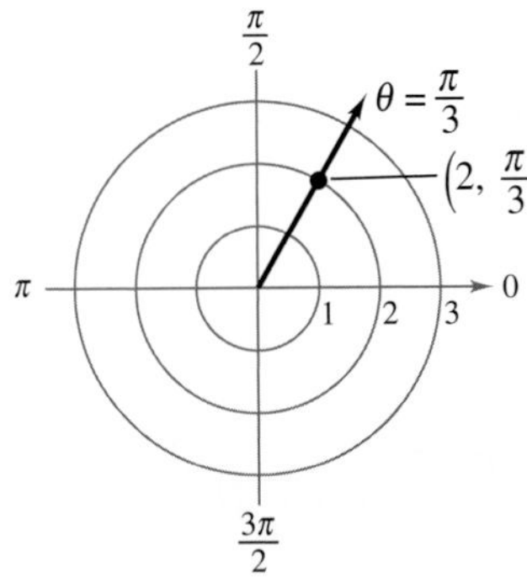

Figure 6.45

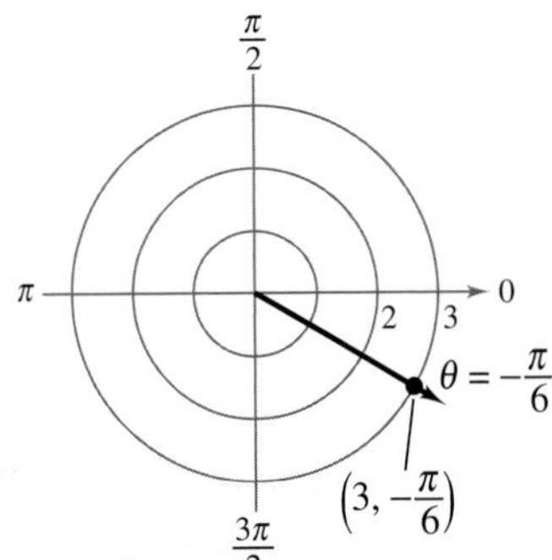

Figure 6.46

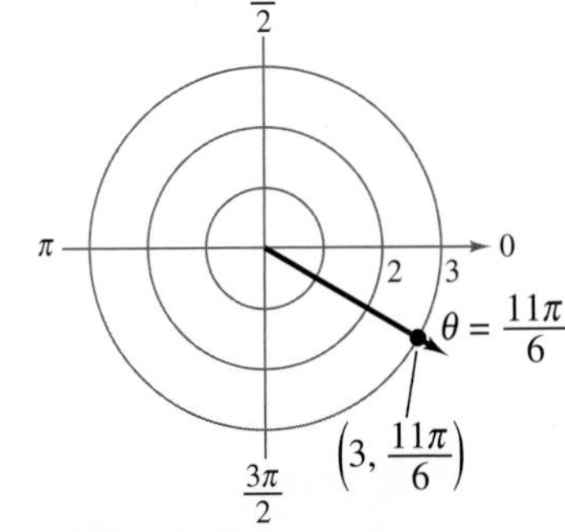

Figure 6.47

✓ **Checkpoint** Audio-video solution in English & Spanish at LarsonPrecalculus.com

Plot each point given in polar coordinates.

a. $(3, \pi/4)$ **b.** $(2, -\pi/3)$ **c.** $(2, 5\pi/3)$

In rectangular coordinates, each point (x, y) has a unique representation. This is not true for polar coordinates. For example, the coordinates

$$(r, \theta) \quad \text{and} \quad (r, \theta + 2\pi)$$

represent the same point, as illustrated in Example 1. Another way to obtain multiple representations of a point is to use negative values for r. Because r is a *directed distance,* the coordinates

$$(r, \theta) \quad \text{and} \quad (-r, \theta + \pi)$$

represent the same point. In general, the point (r, θ) can be represented by

$$(r, \theta) = (r, \theta \pm 2n\pi) \quad \text{or} \quad (r, \theta) = (-r, \theta \pm (2n + 1)\pi)$$

where n is any integer. Moreover, the pole is represented by $(0, \theta)$, where θ is any angle.

EXAMPLE 2 Multiple Representations of Points

Plot the point

$$\left(3, -\frac{3\pi}{4}\right)$$

and find three additional polar representations of this point, using

$$-2\pi < \theta < 2\pi.$$

Solution The point is shown below. Three other representations are

$$\left(3, -\frac{3\pi}{4} + 2\pi\right) = \left(3, \frac{5\pi}{4}\right), \qquad \text{Add } 2\pi \text{ to } \theta.$$

$$\left(-3, -\frac{3\pi}{4} - \pi\right) = \left(-3, -\frac{7\pi}{4}\right), \qquad \text{Replace } r \text{ with } -r \text{ and subtract } \pi \text{ from } \theta.$$

and

$$\left(-3, -\frac{3\pi}{4} + \pi\right) = \left(-3, \frac{\pi}{4}\right). \qquad \text{Replace } r \text{ with } -r \text{ and add } \pi \text{ to } \theta.$$

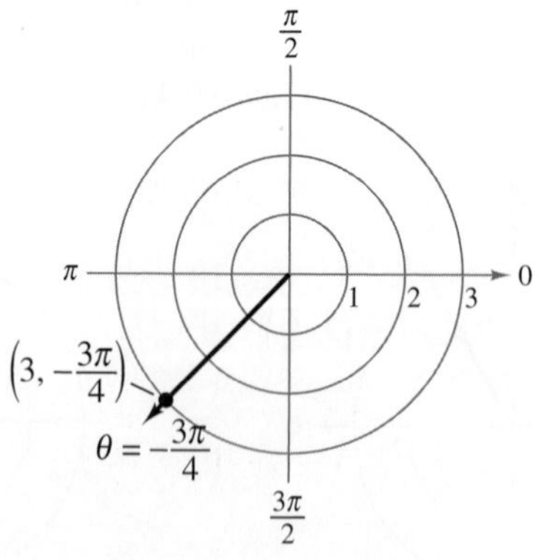

$$\left(3, -\frac{3\pi}{4}\right) = \left(3, \frac{5\pi}{4}\right) = \left(-3, -\frac{7\pi}{4}\right) = \left(-3, \frac{\pi}{4}\right)$$

✓ **Checkpoint** Audio-video solution in English & Spanish at LarsonPrecalculus.com

Plot the point

$$\left(-1, \frac{3\pi}{4}\right)$$

and find three additional polar representations of this point, using $-2\pi < \theta < 2\pi$.

Coordinate Conversion

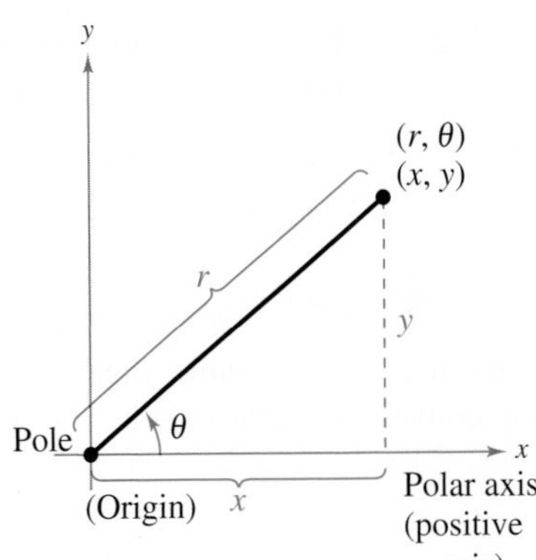

Figure 6.48

To establish the relationship between polar and rectangular coordinates, let the polar axis coincide with the positive x-axis and the pole with the origin, as shown in Figure 6.48. Because (x, y) lies on a circle of radius r, it follows that $r^2 = x^2 + y^2$. Moreover, for $r > 0$, the definitions of the trigonometric functions imply that

$$\tan \theta = \frac{y}{x}, \quad \cos \theta = \frac{x}{r}, \quad \text{and} \quad \sin \theta = \frac{y}{r}.$$

Show that the same relationships hold for $r < 0$.

Coordinate Conversion

The polar coordinates (r, θ) and the rectangular coordinates (x, y) are related as follows.

Polar-to-Rectangular	**Rectangular-to-Polar**
$x = r \cos \theta$	$\tan \theta = \dfrac{y}{x}$
$y = r \sin \theta$	$r^2 = x^2 + y^2$

EXAMPLE 3 Polar-to-Rectangular Conversion

Convert $\left(\sqrt{3}, \dfrac{\pi}{6}\right)$ to rectangular coordinates.

Solution Substitute $r = \sqrt{3}$ and $\theta = \pi/6$ to find the x- and y-coordinates.

$$x = r \cos \theta = \sqrt{3} \cos \frac{\pi}{6} = \sqrt{3}\left(\frac{\sqrt{3}}{2}\right) = \frac{3}{2}$$

$$y = r \sin \theta = \sqrt{3} \sin \frac{\pi}{6} = \sqrt{3}\left(\frac{1}{2}\right) = \frac{\sqrt{3}}{2}$$

The rectangular coordinates are $(x, y) = \left(\dfrac{3}{2}, \dfrac{\sqrt{3}}{2}\right)$. (See Figure 6.49.)

Figure 6.49

✓ **Checkpoint** *Audio-video solution in English & Spanish at LarsonPrecalculus.com*

Convert $(2, \pi)$ to rectangular coordinates.

EXAMPLE 4 Rectangular-to-Polar Conversion

Convert $(-1, 1)$ to polar coordinates.

Solution The point $(x, y) = (-1, 1)$ lies in the second quadrant.

$$\tan \theta = \frac{y}{x} = \frac{1}{-1} = -1 \quad \Longrightarrow \quad \theta = \pi + \arctan(-1) = \frac{3\pi}{4}$$

The angle θ lies in the same quadrant as (x, y), so use the positive value for r.

$$r = \sqrt{x^2 + y^2} = \sqrt{(-1)^2 + (1)^2} = \sqrt{2}$$

So, *one* set of polar coordinates is $(r, \theta) = \left(\sqrt{2}, 3\pi/4\right)$, as shown in Figure 6.50.

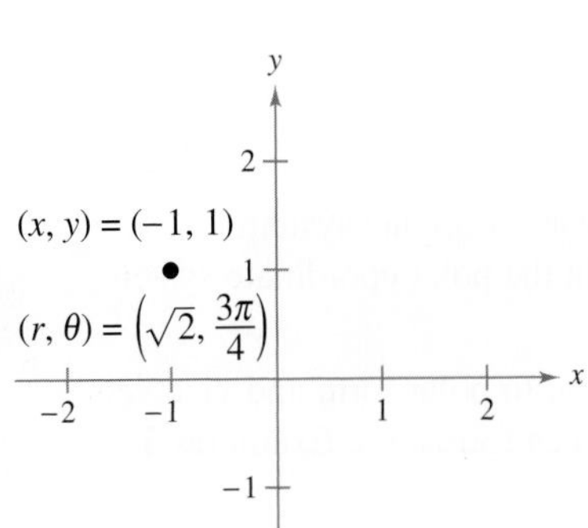

Figure 6.50

✓ **Checkpoint** *Audio-video solution in English & Spanish at LarsonPrecalculus.com*

Convert $(0, 2)$ to polar coordinates.

Equation Conversion

To convert a rectangular equation to polar form, replace x with $r \cos \theta$ and y with $r \sin \theta$. For example, here is how to write the rectangular equation $y = x^2$ in polar form.

$$y = x^2 \qquad \text{Rectangular equation}$$

$$r \sin \theta = (r \cos \theta)^2 \qquad \text{Polar equation}$$

$$r = \sec \theta \tan \theta \qquad \text{Solve for } r.$$

Converting a polar equation to rectangular form requires considerable ingenuity. Example 5 demonstrates several polar-to-rectangular conversions that enable you to sketch the graphs of some polar equations.

EXAMPLE 5 Converting Polar Equations to Rectangular Form

See LarsonPrecalculus.com for an interactive version of this type of example.

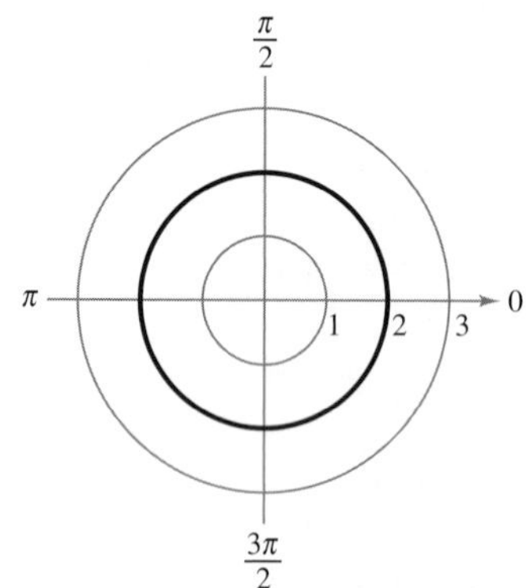

Figure 6.51

a. The graph of the polar equation $r = 2$ consists of all points that are two units from the pole. In other words, this graph is a circle centered at the origin with a radius of 2, as shown in Figure 6.51. Confirm this by converting to rectangular form, using the relationship $r^2 = x^2 + y^2$.

$$\underbrace{r = 2}_{\text{Polar equation}} \implies r^2 = 2^2 \implies \underbrace{x^2 + y^2 = 2^2}_{\text{Rectangular equation}}$$

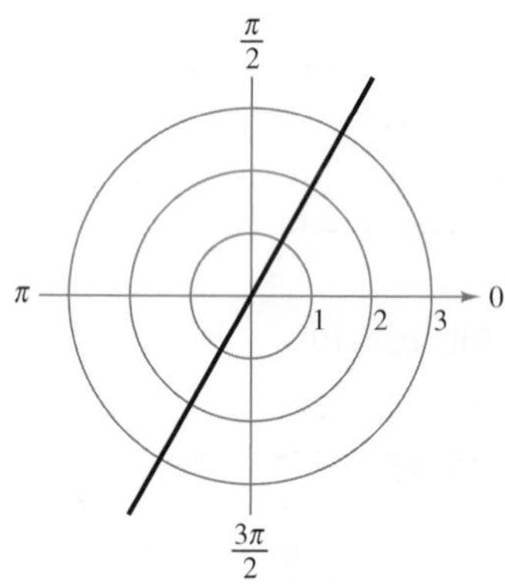

Figure 6.52

b. The graph of the polar equation $\theta = \pi/3$ consists of all points on the line that makes an angle of $\pi/3$ with the polar axis and passes through the pole, as shown in Figure 6.52. To convert to rectangular form, use the relationship $\tan \theta = y/x$.

$$\underbrace{\theta = \pi/3}_{\text{Polar equation}} \implies \tan \theta = \sqrt{3} \implies \underbrace{y = \sqrt{3}x}_{\text{Rectangular equation}}$$

Figure 6.53

c. The graph of the polar equation $r = \sec \theta$ is not evident by inspection, so convert to rectangular form using the relationship $r \cos \theta = x$.

$$\underbrace{r = \sec \theta}_{\text{Polar equation}} \implies r \cos \theta = 1 \implies \underbrace{x = 1}_{\text{Rectangular equation}}$$

The graph is a vertical line, as shown in Figure 6.53.

Describe the graph of each polar equation and find the corresponding rectangular equation.

a. $r = 7$ **b.** $\theta = \pi/4$ **c.** $r = 6 \sin \theta$

Summarize (Section 6.7)

1. Explain how to plot the point (r, θ) in the polar coordinate system *(page 471)*. For examples of plotting points in the polar coordinate system, see Examples 1 and 2.
2. Explain how to convert points from rectangular to polar form and vice versa *(page 473)*. For examples of converting between forms, see Examples 3 and 4.
3. Explain how to convert equations from rectangular to polar form and vice versa *(page 474)*. For an example of converting polar equations to rectangular form, see Example 5.

6.7 Exercises

See CalcChat.com for tutorial help and worked-out solutions to odd-numbered exercises.

Vocabulary: Fill in the blanks.

1. The origin of the polar coordinate system is called the ________.
2. For the point (r, θ), r is the ________ ________ from O to P and θ is the ________ ________, counterclockwise from the polar axis to the line segment $\overline{OP}$.
3. To plot the point (r, θ), use the ________ coordinate system.
4. The polar coordinates (r, θ) and the rectangular coordinates (x, y) are related as follows:

$x =$ ________ $y =$ ________ $\tan\theta =$ ________ $r^2 =$ ________

Skills and Applications

Plotting a Point in the Polar Coordinate System **In Exercises 5–18, plot the point given in polar coordinates and find three additional polar representations of the point, using $-2\pi < \theta < 2\pi$.**

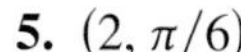

5. $(2, \pi/6)$
6. $(3, 5\pi/4)$
7. $(4, -\pi/3)$
8. $(1, -3\pi/4)$
9. $(2, 3\pi)$
10. $(4, 5\pi/2)$
11. $(-2, 2\pi/3)$
12. $(-3, 11\pi/6)$
13. $(0, 7\pi/6)$
14. $(0, -7\pi/2)$
15. $(\sqrt{2}, 2.36)$
16. $(2\sqrt{2}, 4.71)$
17. $(-3, -1.57)$
18. $(-5, -2.36)$

Polar-to-Rectangular Conversion **In Exercises 19–28, a point is given in polar coordinates. Convert the point to rectangular coordinates.**

19. $(0, \pi)$
20. $(0, -\pi)$
21. $(3, \pi/2)$
22. $(3, 3\pi/2)$
23. $(2, 3\pi/4)$
24. $(1, 5\pi/4)$
25. $(-2, 7\pi/6)$
26. $(-3, 5\pi/6)$
27. $(-3, -\pi/3)$
28. $(-2, -4\pi/3)$

Using a Graphing Utility to Find Rectangular Coordinates **In Exercises 29–38, use a graphing utility to find the rectangular coordinates of the point given in polar coordinates. Round your results to two decimal places.**

29. $(2, 7\pi/8)$
30. $(3/2, 6\pi/5)$
31. $(1, 5\pi/12)$
32. $(4, 7\pi/9)$
33. $(-2.5, 1.1)$
34. $(-2, 5.76)$
35. $(2.5, -2.9)$
36. $(8.75, -6.5)$
37. $(-3.1, 7.92)$
38. $(-2.04, -5.3)$

Rectangular-to-Polar Conversion **In Exercises 39–50, a point is given in rectangular coordinates. Convert the point to polar coordinates. (There are many correct answers.)**

39. $(1, 1)$
40. $(2, 2)$
41. $(-3, -3)$
42. $(-4, -4)$
43. $(3, 0)$
44. $(-6, 0)$
45. $(0, -5)$
46. $(0, 8)$
47. $(-\sqrt{3}, -\sqrt{3})$
48. $(-\sqrt{3}, \sqrt{3})$
49. $(\sqrt{3}, -1)$
50. $(-1, \sqrt{3})$

Using a Graphing Utility to Find Polar Coordinates **In Exercises 51–58, use a graphing utility to find one set of polar coordinates of the point given in rectangular coordinates. Round your results to two decimal places.**

51. $(3, -2)$
52. $(6, 3)$
53. $(-5, 2)$
54. $(7, -2)$
55. $(-\sqrt{3}, -4)$
56. $(5, -\sqrt{2})$
57. $\left(\frac{5}{2}, \frac{4}{3}\right)$
58. $\left(-\frac{7}{9}, -\frac{3}{4}\right)$

Converting a Rectangular Equation to Polar Form **In Exercises 59–78, convert the rectangular equation to polar form. Assume $a > 0$.**

59. $x^2 + y^2 = 9$
60. $x^2 + y^2 = 16$
61. $y = x$
62. $y = -x$
63. $x = 10$
64. $y = -2$
65. $3x - y + 2 = 0$
66. $3x + 5y - 2 = 0$
67. $xy = 16$
68. $2xy = 1$
69. $x = a$
70. $y = a$
71. $x^2 + y^2 = a^2$
72. $x^2 + y^2 = 9a^2$
73. $x^2 + y^2 - 2ax = 0$
74. $x^2 + y^2 - 2ay = 0$
75. $(x^2 + y^2)^2 = x^2 - y^2$
76. $(x^2 + y^2)^2 = 9(x^2 - y^2)$
77. $y^3 = x^2$
78. $y^2 = x^3$

Converting a Polar Equation to Rectangular Form In Exercises 79–100, convert the polar equation to rectangular form.

79. $r = 5$
80. $r = -7$
81. $\theta = 2\pi/3$
82. $\theta = -5\pi/3$
83. $\theta = \pi/2$
84. $\theta = 3\pi/2$
85. $r = 4 \csc \theta$
86. $r = 2 \csc \theta$
87. $r = -3 \sec \theta$
88. $r = -\sec \theta$
89. $r = -2 \cos \theta$
90. $r = 4 \sin \theta$
91. $r^2 = \cos \theta$
92. $r^2 = 2 \sin \theta$
93. $r^2 = \sin 2\theta$
94. $r^2 = \cos 2\theta$
95. $r = 2 \sin 3\theta$
96. $r = 3 \cos 2\theta$
97. $r = \dfrac{2}{1 + \sin \theta}$
98. $r = \dfrac{1}{1 - \cos \theta}$
99. $r = \dfrac{6}{2 - 3 \sin \theta}$
100. $r = \dfrac{5}{\sin \theta - 4 \cos \theta}$

Converting a Polar Equation to Rectangular Form In Exercises 101–108, describe the graph of the polar equation and find the corresponding rectangular equation.

101. $r = 6$
102. $r = 8$
103. $\theta = \pi/6$
104. $\theta = 3\pi/4$
105. $r = 3 \sec \theta$
106. $r = 2 \csc \theta$
107. $r = 2 \sin \theta$
108. $r = -6 \cos \theta$

109. **Ferris Wheel**

The center of a Ferris wheel lies at the pole of the polar coordinate system, where the distances are in feet. Passengers enter a car at $(30, -\pi/2)$. It takes 45 seconds for the wheel to complete one clockwise revolution.

(a) Write a polar equation that models the possible positions of a passenger car.

(b) Passengers enter a car. Find and interpret their coordinates after 15 seconds of rotation.

(c) Convert the point in part (b) to rectangular coordinates. Interpret the coordinates.

110. **Ferris Wheel** Repeat Exercise 109 when the distance from a passenger car to the center is 35 feet and it takes 60 seconds to complete one clockwise revolution.

Exploration

True or False? In Exercises 111 and 112, determine whether the statement is true or false. Justify your answer.

111. If $\theta_1 = \theta_2 + 2\pi n$ for some integer n, then (r, θ_1) and (r, θ_2) represent the same point in the polar coordinate system.

112. If $|r_1| = |r_2|$, then (r_1, θ) and (r_2, θ) represent the same point in the polar coordinate system.

113. **Error Analysis** Describe the error in converting the rectangular coordinates $(1, -\sqrt{3})$ to polar form.

$$\tan \theta = -\sqrt{3}/1 \implies \theta = \frac{2\pi}{3}$$

$$r = \sqrt{1^2 + (-\sqrt{3})^2} = \sqrt{4} = 2$$

$$(r, \theta) = \left(2, \frac{2\pi}{3}\right)$$

114. **HOW DO YOU SEE IT?** Use the polar coordinate system shown below.

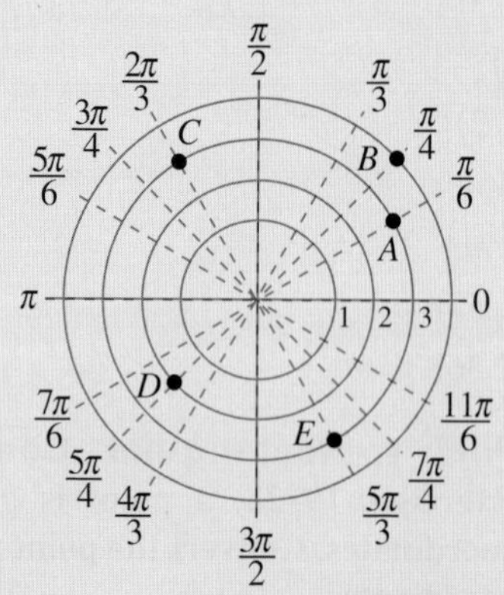

(a) Identify the polar coordinates of points A–E.

(b) Which points lie on the graph of $r = 3$?

(c) Which points lie on the graph of $\theta = \pi/4$?

115. **Think About It**

(a) Convert the polar equation

$$r = 2(h \cos \theta + k \sin \theta)$$

to rectangular form and verify that it represents a circle.

(b) Use the result of part (a) to convert

$$r = \cos \theta + 3 \sin \theta$$

to rectangular form and find the center and radius of the circle it represents.

6.8 Graphs of Polar Equations

Graphs of polar equations are often useful visual tools in mathematical modeling. For example, in Exercise 69 on page 484, you will use the graph of a polar equation to analyze the pickup pattern of a microphone.

- **Graph polar equations by point plotting.**
- **Use symmetry, zeros, and maximum r-values to sketch graphs of polar equations.**
- **Recognize special polar graphs.**

Introduction

In previous chapters, you sketched graphs in the rectangular coordinate system. You began with the basic point-plotting method. Then you used sketching aids such as symmetry, intercepts, asymptotes, periods, and shifts to further investigate the natures of graphs. This section approaches curve sketching in the polar coordinate system similarly, beginning with a demonstration of point plotting.

EXAMPLE 1 Graphing a Polar Equation by Point Plotting

Sketch the graph of the polar equation $r = 4 \sin \theta$.

Solution The sine function is periodic, so to obtain a full range of r-values, consider values of θ in the interval $0 \le \theta \le 2\pi$, as shown in the table below.

θ	0	$\frac{\pi}{6}$	$\frac{\pi}{3}$	$\frac{\pi}{2}$	$\frac{2\pi}{3}$	$\frac{5\pi}{6}$	π	$\frac{7\pi}{6}$	$\frac{3\pi}{2}$	$\frac{11\pi}{6}$	2π
r	0	2	$2\sqrt{3}$	4	$2\sqrt{3}$	2	0	-2	-4	-2	0

By plotting these points, it appears that the graph is a circle of radius 2 whose center is at the point $(x, y) = (0, 2)$, as shown in the figure below.

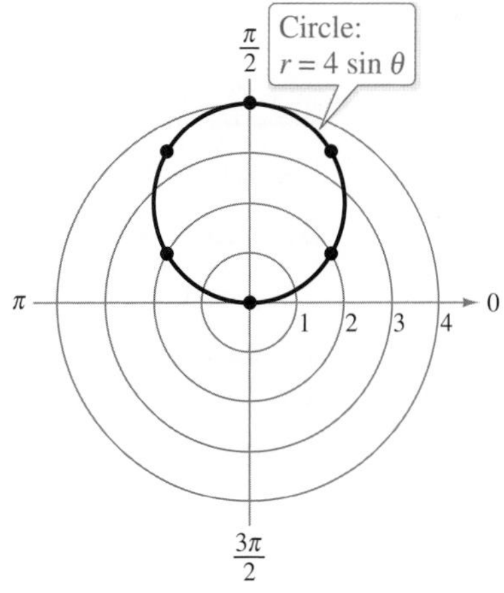

Checkpoint Audio-video solution in English & Spanish at LarsonPrecalculus.com

Sketch the graph of the polar equation $r = 6 \cos \theta$.

One way to confirm the graph in Example 1 is to convert the polar equation to rectangular form and then sketch the graph of the rectangular equation. You can also use a graphing utility set to *polar* mode and graph the polar equation, or use a graphing utility set to *parametric* mode and graph a parametric representation.

nikkytok/Shutterstock.com

Symmetry, Zeros, and Maximum *r*-Values

Note in Example 1 that as θ increases from 0 to 2π, the graph is traced twice. Moreover, note that the graph is *symmetric with respect to the line* $\theta = \pi/2$. Had you known about this symmetry and retracing ahead of time, you could have used fewer points. The figures below show the three important types of symmetry to consider in polar curve sketching.

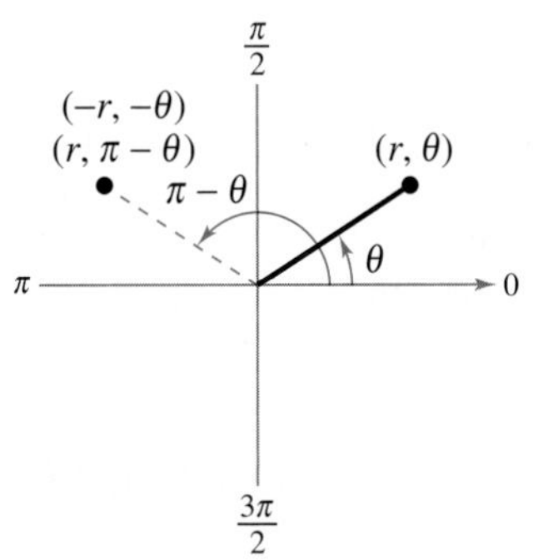

Symmetry with Respect to the Line $\theta = \dfrac{\pi}{2}$

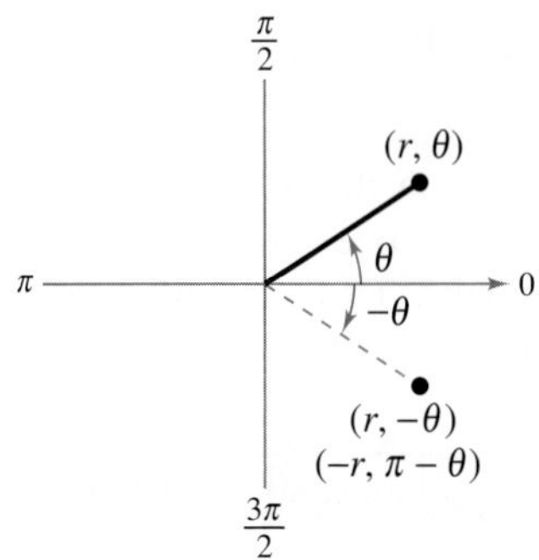

Symmetry with Respect to the Polar Axis

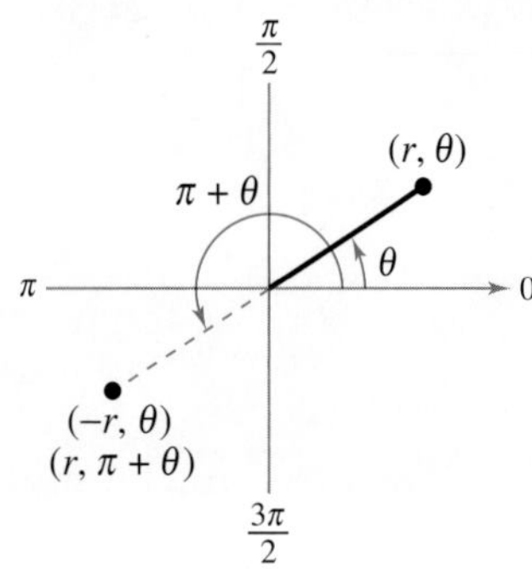

Symmetry with Respect to the Pole

Tests for Symmetry in Polar Coordinates

The graph of a polar equation is symmetric with respect to the following when the given substitution yields an equivalent equation.

1. The line $\theta = \pi/2$: Replace (r, θ) with $(r, \pi - \theta)$ or $(-r, -\theta)$.
2. The polar axis: Replace (r, θ) with $(r, -\theta)$ or $(-r, \pi - \theta)$.
3. The pole: Replace (r, θ) with $(r, \pi + \theta)$ or $(-r, \theta)$.

EXAMPLE 2 Using Symmetry to Sketch a Polar Graph

Use symmetry to sketch the graph of $r = 3 + 2 \cos \theta$.

Solution Replacing (r, θ) with $(r, -\theta)$ produces

$r = 3 + 2 \cos(-\theta) = 3 + 2 \cos \theta$. $\quad \cos(-\theta) = \cos \theta$

So, the curve is symmetric with respect to the polar axis. Plotting the points in the table below and using polar axis symmetry, you obtain the graph shown in Figure 6.54. This graph is called a **limaçon.**

θ	0	$\frac{\pi}{6}$	$\frac{\pi}{3}$	$\frac{\pi}{2}$	$\frac{2\pi}{3}$	$\frac{5\pi}{6}$	π
r	5	$3 + \sqrt{3}$	4	3	2	$3 - \sqrt{3}$	1

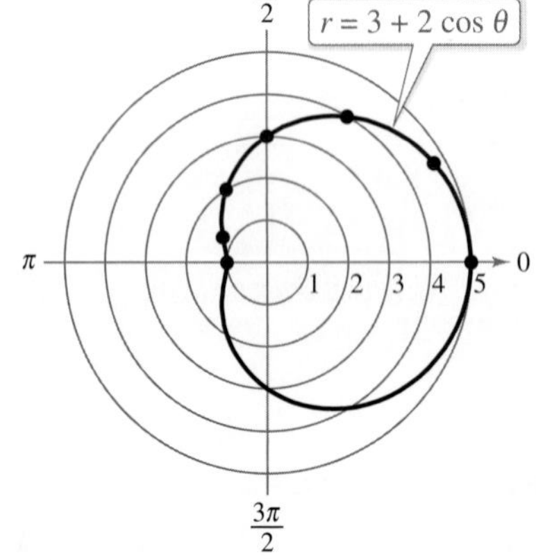

Figure 6.54

✓ ***Checkpoint*** Audio-video solution in English & Spanish at LarsonPrecalculus.com

Use symmetry to sketch the graph of $r = 3 + 2 \sin \theta$.

Example 2 uses the property that the cosine function is *even.* Recall from Section 1.2 that the cosine function is even because $\cos(-\theta) = \cos \theta$, and the sine function is odd because $\sin(-\theta) = -\sin \theta$.

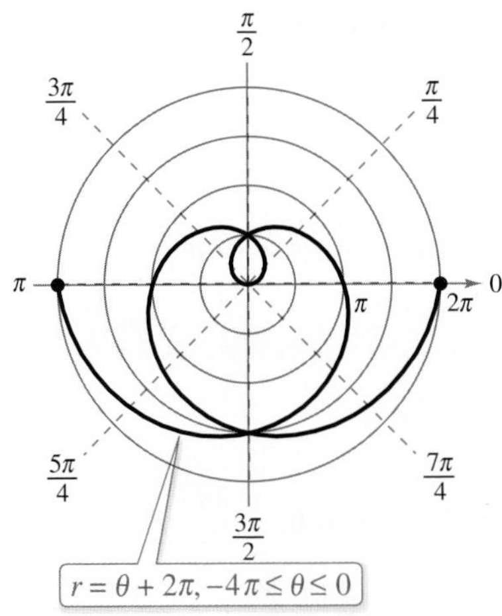

Figure 6.55

The tests for symmetry in polar coordinates listed on the preceding page are sufficient to guarantee symmetry, but a graph may have symmetry even though its equation does not satisfy the tests. For example, Figure 6.55 shows the graph of

$$r = \theta + 2\pi$$

to be symmetric with respect to the line $\theta = \pi/2$, and yet the corresponding test fails to reveal this. That is, neither of the replacements below yields an equivalent equation.

Original Equation	**Replacement**	**New Equation**
$r = \theta + 2\pi$	(r, θ) with $(r, \pi - \theta)$	$r = -\theta + 3\pi$
$r = \theta + 2\pi$	(r, θ) with $(-r, -\theta)$	$-r = -\theta + 2\pi$

The equations $r = 4 \sin \theta$ and $r = 3 + 2 \cos \theta$, discussed in Examples 1 and 2, are of the form

$$r = f(\sin \theta) \quad \text{and} \quad r = g(\cos \theta)$$

respectively. Graphs of equations of these forms have symmetry in polar coordinates as listed below.

Quick Tests for Symmetry in Polar Coordinates

1. The graph of $r = f(\sin \theta)$ is symmetric with respect to the line $\theta = \dfrac{\pi}{2}$.
2. The graph of $r = g(\cos \theta)$ is symmetric with respect to the polar axis.

Two additional aids to sketching graphs of polar equations involve knowing the θ-values for which $|r|$ is maximum and knowing the θ-values for which $r = 0$. For instance, in Example 1, the maximum value of $|r|$ for $r = 4 \sin \theta$ is $|r| = 4$, and this occurs when $\theta = \pi/2$. Moreover, $r = 0$ when $\theta = 0$.

EXAMPLE 3 Sketching a Polar Graph

Sketch the graph of $r = 1 - 2 \cos \theta$.

Solution From the equation $r = 1 - 2 \cos \theta$, you obtain the following features of the graph.

Symmetry: With respect to the polar axis

Maximum value of $|r|$: $r = 3$ when $\theta = \pi$

Zero of r: $r = 0$ when $\theta = \pi/3$

The table shows several θ-values in the interval $[0, \pi]$. Plot the corresponding points and sketch the graph, as shown in Figure 6.56.

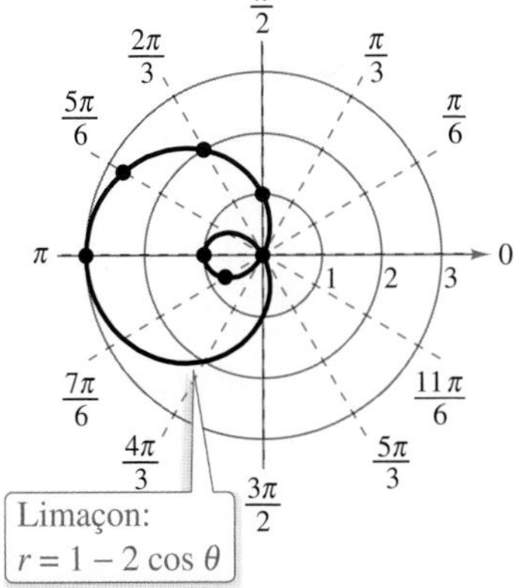

Figure 6.56

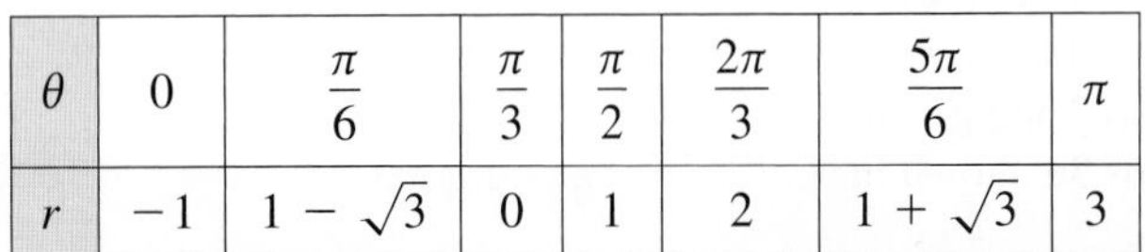

θ	0	$\dfrac{\pi}{6}$	$\dfrac{\pi}{3}$	$\dfrac{\pi}{2}$	$\dfrac{2\pi}{3}$	$\dfrac{5\pi}{6}$	π
r	-1	$1 - \sqrt{3}$	0	1	2	$1 + \sqrt{3}$	3

Note that the negative r-values determine the *inner loop* of the graph in Figure 6.56. This graph, like the graph in Example 2, is a limaçon.

✓ **Checkpoint** Audio-video solution in English & Spanish at LarsonPrecalculus.com

Sketch the graph of $r = 1 + 2 \sin \theta$.

Some curves reach their zeros and maximum r-values at more than one point, as shown in Example 4.

EXAMPLE 4 Sketching a Polar Graph

See LarsonPrecalculus.com for an interactive version of this type of example.

Sketch the graph of $r = 2 \cos 3\theta$.

Solution

Symmetry: With respect to the polar axis

Maximum value of $|r|$: $|r| = 2$ when $3\theta = 0, \pi, 2\pi, 3\pi$ or $\theta = 0, \frac{\pi}{3}, \frac{2\pi}{3}, \pi$

Zeros of r: $r = 0$ when $3\theta = \frac{\pi}{2}, \frac{3\pi}{2}, \frac{5\pi}{2}$ or $\theta = \frac{\pi}{6}, \frac{\pi}{2}, \frac{5\pi}{6}$

θ	0	$\frac{\pi}{12}$	$\frac{\pi}{6}$	$\frac{\pi}{4}$	$\frac{\pi}{3}$	$\frac{5\pi}{12}$	$\frac{\pi}{2}$
r	2	$\sqrt{2}$	0	$-\sqrt{2}$	-2	$-\sqrt{2}$	0

Plot these points and use the specified symmetry, zeros, and maximum values to obtain the graph, as shown in the figures below. This graph is called a **rose curve,** and each loop on the graph is called a *petal*. Note how the entire curve is traced as θ increases from 0 to π.

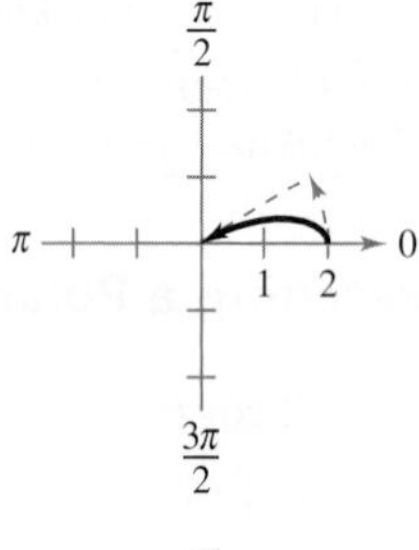

$0 \le \theta \le \frac{\pi}{6}$

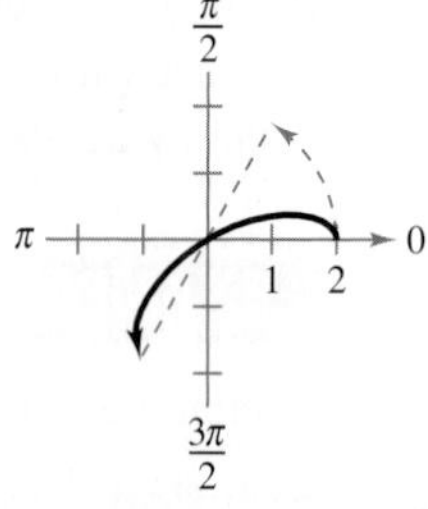

$0 \le \theta \le \frac{\pi}{3}$

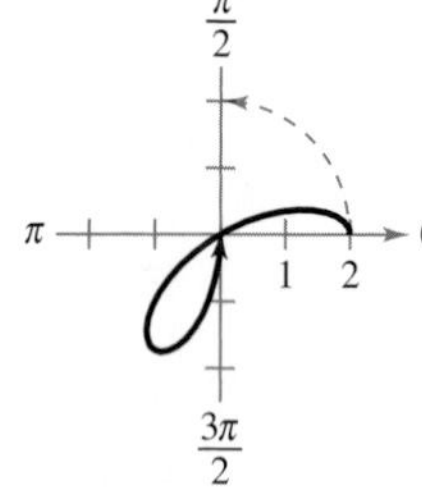

$0 \le \theta \le \frac{\pi}{2}$

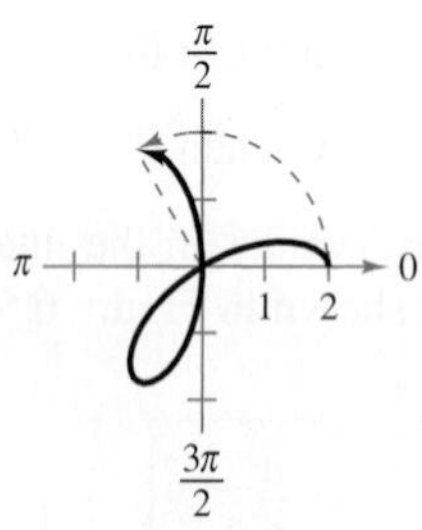

$0 \le \theta \le \frac{2\pi}{3}$

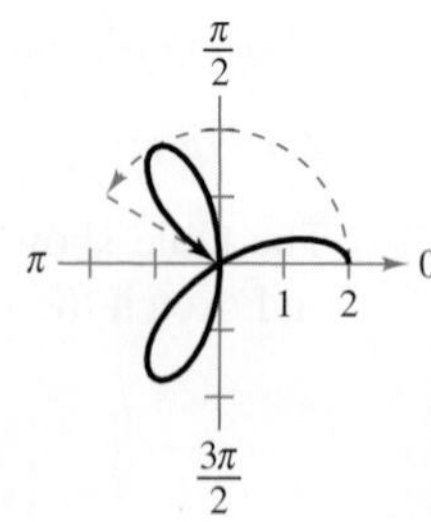

$0 \le \theta \le \frac{5\pi}{6}$

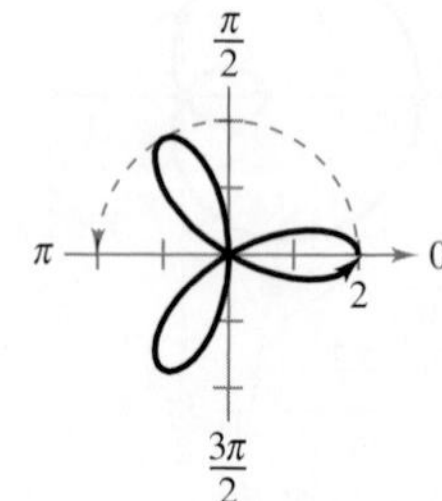

$0 \le \theta \le \pi$

TECHNOLOGY Use a graphing utility in *polar* mode to verify the graph of $r = 2 \cos 3\theta$ shown in Example 4.

✓ ***Checkpoint*** *Audio-video solution in English & Spanish at LarsonPrecalculus.com*

Sketch the graph of $r = 2 \sin 3\theta$.

Special Polar Graphs

Several important types of graphs have equations that are simpler in polar form than in rectangular form. For example, the circle with the polar equation $r = 4 \sin \theta$ in Example 1 has the more complicated rectangular equation $x^2 + (y - 2)^2 = 4$. Several types of graphs that have simpler polar equations are shown below.

Limaçons

$r = a \pm b \cos \theta,\ r = a \pm b \sin \theta \quad (a > 0, b > 0)$

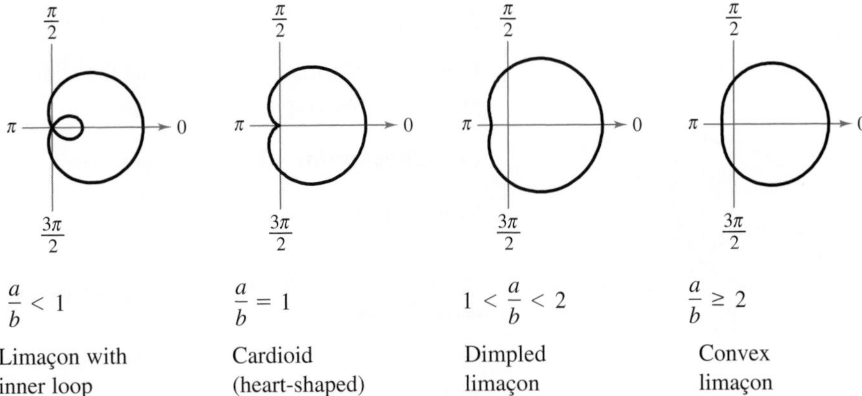

$\frac{a}{b} < 1$	$\frac{a}{b} = 1$	$1 < \frac{a}{b} < 2$	$\frac{a}{b} \geq 2$
Limaçon with inner loop	Cardioid (heart-shaped)	Dimpled limaçon	Convex limaçon

Rose Curves

n petals when n is odd, $2n$ petals when n is even $\quad (n \geq 2)$

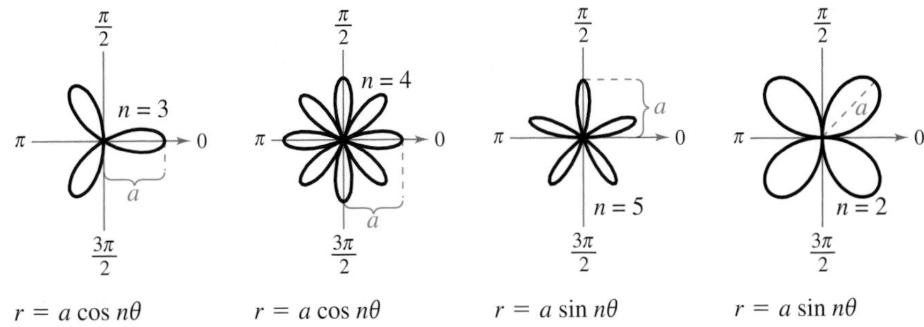

$r = a \cos n\theta$	$r = a \cos n\theta$	$r = a \sin n\theta$	$r = a \sin n\theta$

Circles and Lemniscates

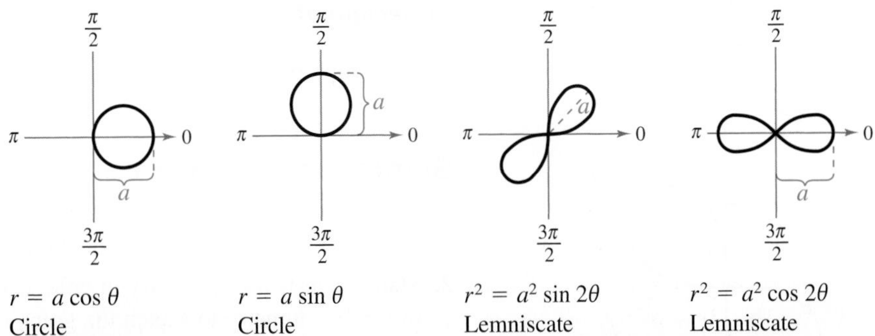

$r = a \cos \theta$	$r = a \sin \theta$	$r^2 = a^2 \sin 2\theta$	$r^2 = a^2 \cos 2\theta$
Circle	Circle	Lemniscate	Lemniscate

The quick tests for symmetry presented on page 479 can be especially useful when graphing many of the curves shown above. For example, limaçons have the form $r = f(\sin \theta)$ or the form $r = g(\cos \theta)$, so you know that a limaçon will be either symmetric with respect to the line $\theta = \pi/2$ or symmetric with respect to the polar axis.

EXAMPLE 5 Sketching a Rose Curve

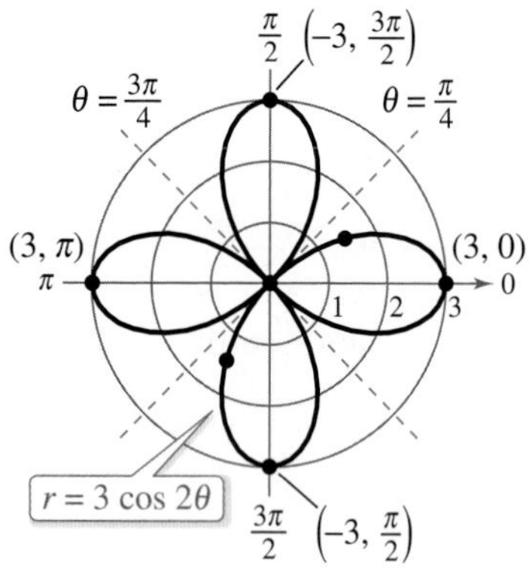

Sketch the graph of $r = 3 \cos 2\theta$.

Solution

Type of curve: Rose curve with $2n = 4$ petals

Symmetry: With respect to the line $\theta = \pi/2$, the polar axis, and the pole

Maximum value of $|r|$: $|r| = 3$ when $\theta = 0, \pi/2, \pi, 3\pi/2$

Zeros of r: $r = 0$ when $\theta = \pi/4, 3\pi/4$

Using this information and plotting the additional points included in the table at the left, you obtain the graph shown in Figure 6.57.

θ	r
0	3
$\frac{\pi}{6}$	$\frac{3}{2}$
$\frac{\pi}{4}$	0
$\frac{\pi}{3}$	$-\frac{3}{2}$

Figure 6.57

✓ **Checkpoint** Audio-video solution in English & Spanish at LarsonPrecalculus.com

Sketch the graph of $r = 3 \cos 3\theta$.

EXAMPLE 6 Sketching a Lemniscate

Sketch the graph of $r^2 = 9 \sin 2\theta$.

Solution

Type of curve: Lemniscate

Symmetry: With respect to the pole

Maximum value of $|r|$: $|r| = 3$ when $\theta = \pi/4$

Zeros of r: $r = 0$ when $\theta = 0, \pi/2$

When $\sin 2\theta < 0$, this equation has no solution points. So, restrict the values of θ to those for which $\sin 2\theta \geq 0$.

$$0 \leq \theta \leq \frac{\pi}{2} \quad \text{or} \quad \pi \leq \theta \leq \frac{3\pi}{2}$$

Using symmetry, you need to consider only the first of these two intervals. By finding a few additional points (included in the table at the left), you obtain the graph shown in Figure 6.58.

θ	$r = \pm 3\sqrt{\sin 2\theta}$
0	0
$\frac{\pi}{12}$	$\pm\frac{3}{\sqrt{2}}$
$\frac{\pi}{4}$	± 3
$\frac{5\pi}{12}$	$\pm\frac{3}{\sqrt{2}}$
$\frac{\pi}{2}$	0

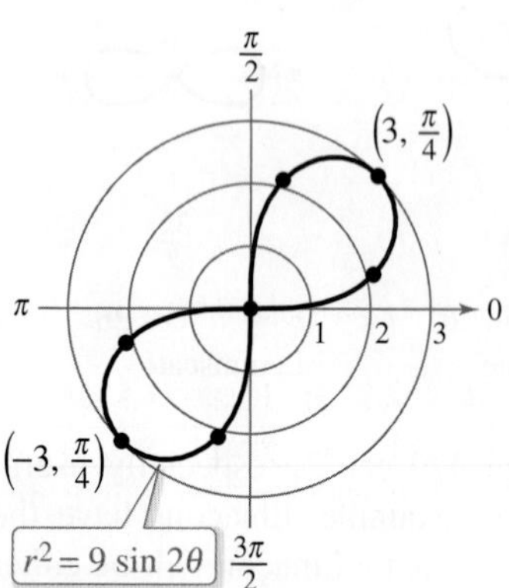

Figure 6.58

✓ **Checkpoint** Audio-video solution in English & Spanish at LarsonPrecalculus.com

Sketch the graph of $r^2 = 4 \cos 2\theta$.

Summarize (Section 6.8)

1. Explain how to graph a polar equation by point plotting *(page 477)*. For an example of graphing a polar equation by point plotting, see Example 1.
2. State the tests for symmetry in polar coordinates *(page 478)*. For an example of using symmetry to sketch the graph of a polar equation, see Example 2.
3. Explain how to use zeros and maximum r-values to sketch the graph of a polar equation *(page 479)*. For examples of using zeros and maximum r-values to sketch graphs of polar equations, see Examples 3 and 4.
4. State and give examples of the special polar graphs discussed in this lesson *(page 481)*. For examples of sketching special polar graphs, see Examples 5 and 6.

6.8 Exercises

See CalcChat.com for tutorial help and worked-out solutions to odd-numbered exercises.

Vocabulary: Fill in the blanks.

1. The graph of $r = f(\sin \theta)$ is symmetric with respect to the line ________.
2. The graph of $r = g(\cos \theta)$ is symmetric with respect to the ________ ________.
3. The equation $r = 2 + \cos \theta$ represents a ________ ________.
4. The equation $r = 2 \cos \theta$ represents a ________.
5. The equation $r^2 = 4 \sin 2\theta$ represents a ________.
6. The equation $r = 1 + \sin \theta$ represents a ________.

Skills and Applications

Identifying Types of Polar Graphs In Exercises 7–12, identify the type of polar graph.

7.
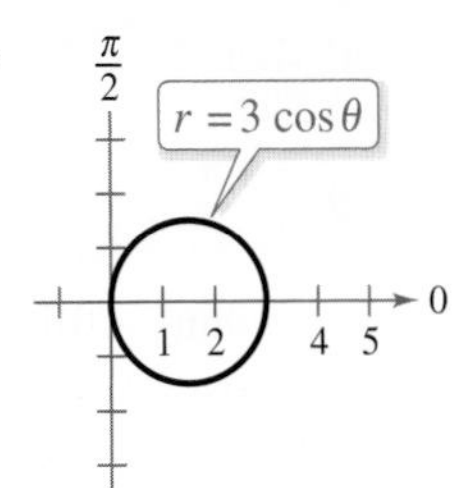

8.
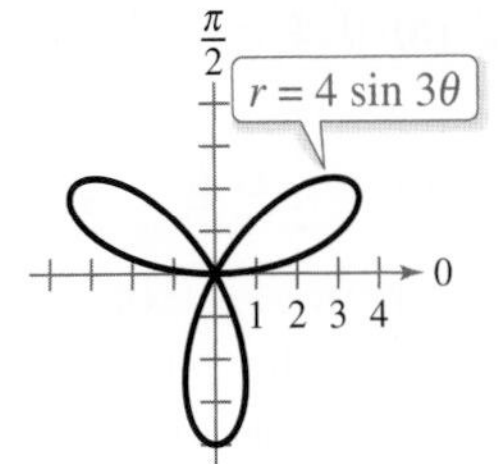

9.
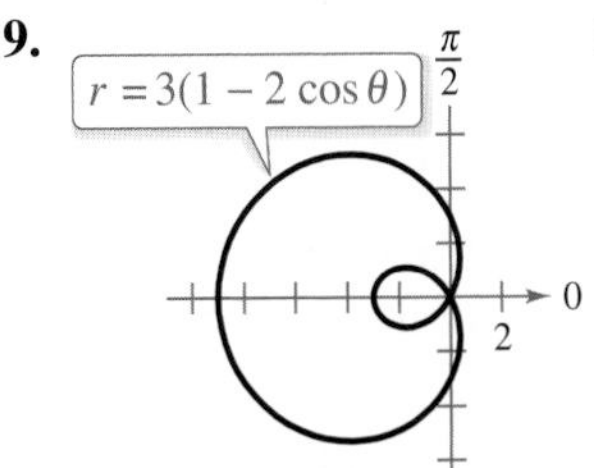

10.
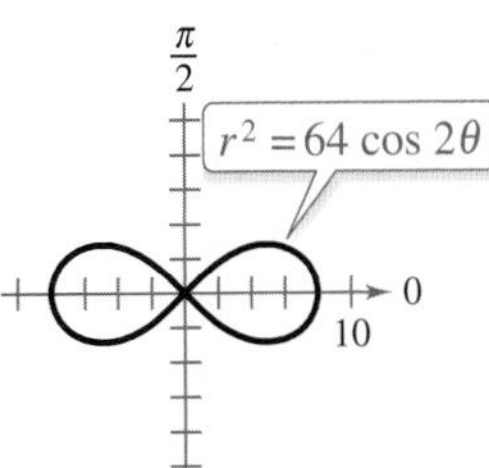

11.
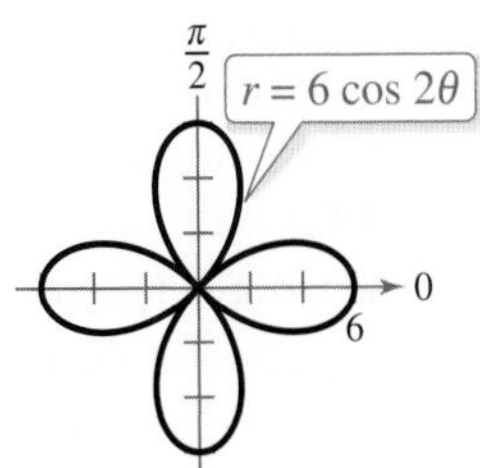

12.
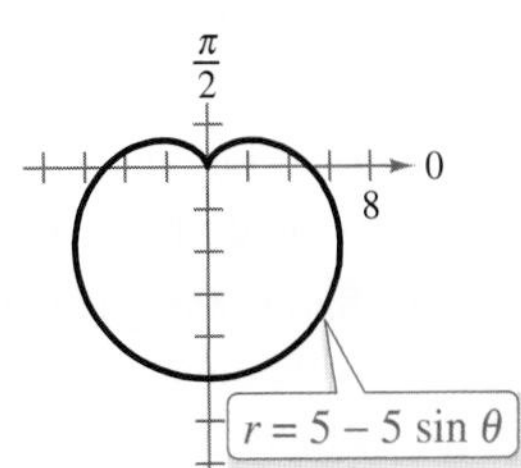

Testing for Symmetry In Exercises 13–18, test for symmetry with respect to the line $\theta = \pi/2$, the polar axis, and the pole.

13. $r = 6 + 3 \cos \theta$
14. $r = 9 \cos 3\theta$
15. $r = \dfrac{2}{1 + \sin \theta}$
16. $r = \dfrac{3}{2 + \cos \theta}$
17. $r^2 = 36 \cos 2\theta$
18. $r^2 = 25 \sin 2\theta$

Finding the Maximum Value of $|r|$ and Zeros of r In Exercises 19–22, find the maximum value of $|r|$ and any zeros of r.

19. $r = 10 - 10 \sin \theta$
20. $r = 6 + 12 \cos \theta$
21. $r = 4 \cos 3\theta$
22. $r = 3 \sin 2\theta$

Sketching the Graph of a Polar Equation In Exercises 23–48, sketch the graph of the polar equation using symmetry, zeros, maximum r-values, and any other additional points.

23. $r = 5$
24. $r = -8$
25. $r = \pi/4$
26. $r = -2\pi/3$
27. $r = 3 \sin \theta$
28. $r = 4 \cos \theta$
29. $r = 3(1 - \cos \theta)$
30. $r = 4(1 - \sin \theta)$
31. $r = 4(1 + \sin \theta)$
32. $r = 6(1 + \cos \theta)$
33. $r = 5 + 2 \cos \theta$
34. $r = 5 - 2 \sin \theta$
35. $r = 1 - 3 \sin \theta$
36. $r = 2 - 5 \cos \theta$
37. $r = 3 - 6 \cos \theta$
38. $r = 4 + 6 \sin \theta$
39. $r = 5 \sin 2\theta$
40. $r = 2 \cos 2\theta$
41. $r = 6 \cos 3\theta$
42. $r = 3 \sin 3\theta$
43. $r = 2 \sec \theta$
44. $r = 5 \csc \theta$
45. $r = \dfrac{3}{\sin \theta - 2 \cos \theta}$
46. $r = \dfrac{6}{2 \sin \theta - 3 \cos \theta}$
47. $r^2 = 9 \cos 2\theta$
48. $r^2 = 16 \sin \theta$

Graphing a Polar Equation In Exercises 49–58, use a graphing utility to graph the polar equation.

49. $r = 9/4$
50. $r = -5/2$
51. $r = 5\pi/8$
52. $r = -\pi/10$
53. $r = 8 \cos \theta$
54. $r = \cos 2\theta$
55. $r = 3(2 - \sin \theta)$
56. $r = 2 \cos(3\theta - 2)$
57. $r = 8 \sin \theta \cos^2 \theta$
58. $r = 2 \csc \theta + 5$

Finding an Interval In Exercises 59–64, use a graphing utility to graph the polar equation. Find an interval for θ for which the graph is traced only once.

59. $r = 3 - 8 \cos \theta$
60. $r = 5 + 4 \cos \theta$
61. $r = 2 \cos(3\theta/2)$
62. $r = 3 \sin(5\theta/2)$
63. $r^2 = 16 \sin 2\theta$
64. $r^2 = 1/\theta$

Asymptote of a Graph of a Polar Equation In Exercises 65–68, use a graphing utility to graph the polar equation and show that the given line is an asymptote of the graph.

	Name of Graph	Polar Equation	Asymptote
65.	Conchoid	$r = 2 - \sec\theta$	$x = -1$
66.	Conchoid	$r = 2 + \csc\theta$	$y = 1$
67.	Hyperbolic spiral	$r = \dfrac{3}{\theta}$	$y = 3$
68.	Strophoid	$r = 2\cos 2\theta \sec\theta$	$x = -2$

69. Microphone

The pickup pattern of a microphone is modeled by the polar equation

$r = 5 + 5\cos\theta$

where $|r|$ measures how sensitive the microphone is to sounds coming from the angle θ.

(a) Sketch the graph of the model and identify the type of polar graph.

(b) At what angle is the microphone most sensitive to sound?

70. Area The total area of the region bounded by the lemniscate $r^2 = a^2\cos 2\theta$ is a^2.

(a) Sketch the graph of $r^2 = 16\cos 2\theta$.

(b) Find the area of one loop of the graph from part (a).

Exploration

True or False? In Exercises 71 and 72, determine whether the statement is true or false. Justify your answer.

71. The graph of $r = 10\sin 5\theta$ is a rose curve with five petals.

72. A rose curve is always symmetric with respect to the line $\theta = \pi/2$.

73. Graphing a Polar Equation Consider the equation $r = 3\sin k\theta$.

(a) Use a graphing utility to graph the equation for $k = 1.5$. Find the interval for θ over which the graph is traced only once.

(b) Use the graphing utility to graph the equation for $k = 2.5$. Find the interval for θ over which the graph is traced only once.

(c) Is it possible to find an interval for θ over which the graph is traced only once for any rational number k? Explain.

74. HOW DO YOU SEE IT? Match each polar equation with its graph.

(i) (ii) (iii)

(a) $r = 5\sin\theta$

(b) $r = 2 + 5\sin\theta$

(c) $r = 5\cos 2\theta$

75. Sketching the Graph of a Polar Equation Sketch the graph of $r = 10\cos\theta$ over each interval. Describe the part of the graph obtained in each case.

(a) $0 \le \theta \le \dfrac{\pi}{2}$ (b) $\dfrac{\pi}{2} \le \theta \le \pi$

(c) $-\dfrac{\pi}{2} \le \theta \le \dfrac{\pi}{2}$ (d) $\dfrac{\pi}{4} \le \theta \le \dfrac{3\pi}{4}$

76. Graphical Reasoning Use a graphing utility to graph the polar equation $r = 6[1 + \cos(\theta - \phi)]$ for each value of ϕ. Use the graphs to describe the effect of the angle ϕ. Write the equation as a function of $\sin\theta$ for part (c).

(a) $\phi = 0$ (b) $\phi = \pi/4$ (c) $\phi = \pi/2$

77. Rotating Polar Graphs The graph of $r = f(\theta)$ is rotated about the pole through an angle ϕ. Show that the equation of the rotated graph is $r = f(\theta - \phi)$.

78. Rotating Polar Graphs Consider the graph of $r = f(\sin\theta)$.

(a) Show that when the graph is rotated counterclockwise $\pi/2$ radians about the pole, the equation of the rotated graph is $r = f(-\cos\theta)$.

(b) Show that when the graph is rotated counterclockwise π radians about the pole, the equation of the rotated graph is $r = f(-\sin\theta)$.

(c) Show that when the graph is rotated counterclockwise $3\pi/2$ radians about the pole, the equation of the rotated graph is $r = f(\cos\theta)$.

Rotating Polar Graphs In Exercises 79 and 80, use the results of Exercises 77 and 78.

79. Write an equation for the limaçon $r = 2 - \sin\theta$ after it rotated through each angle.

(a) $\dfrac{\pi}{4}$ (b) $\dfrac{\pi}{2}$ (c) π (d) $\dfrac{3\pi}{2}$

80. Write an equation for the rose curve $r = 2\sin 2\theta$ after it is rotated through each angle.

(a) $\dfrac{\pi}{6}$ (b) $\dfrac{\pi}{2}$ (c) $\dfrac{2\pi}{3}$ (d) π

6.9 Polar Equations of Conics

Polar equation of conics can model the orbits of planets and satellites. For example, in Exercise 62 on page 490, you will use a polar equation to model the parabolic path of a satellite.

- **Define conics in terms of eccentricity, and write and graph polar equations of conics.**
- **Use equations of conics in polar form to model real-life problems.**

Alternative Definition and Polar Equations of Conics

In Sections 6.3 and 6.4, you learned that the rectangular equations of ellipses and hyperbolas take simpler forms when the origin lies at their *centers*. There are many important applications of conics in which it is more convenient to use a *focus* as the origin. In these cases, it is convenient to use polar coordinates.

To begin, consider an alternative definition of a conic that uses the concept of *eccentricity*.

Alternative Definition of a Conic

The locus of a point in the plane that moves such that its distance from a fixed point (focus) is in a constant ratio to its distance from a fixed line (directrix) is a **conic.** The constant ratio is the *eccentricity* of the conic and is denoted by e. Moreover, the conic is an **ellipse** when $0 < e < 1$, a **parabola** when $e = 1$, and a **hyperbola** when $e > 1$. (See the figures below.)

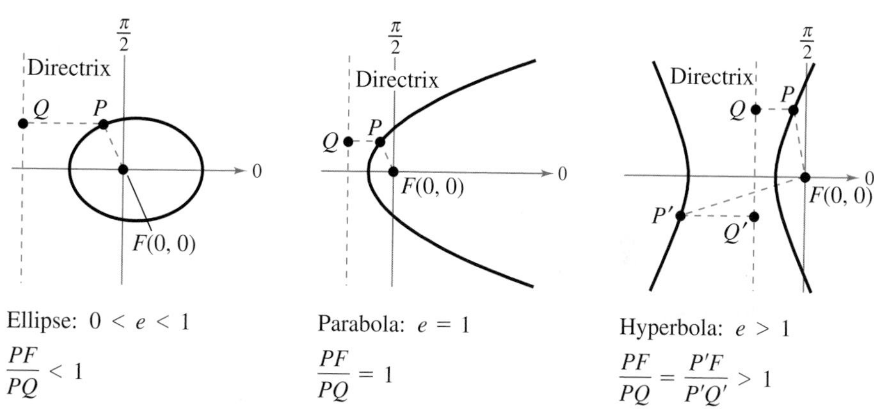

Ellipse: $0 < e < 1$ — $\frac{PF}{PQ} < 1$

Parabola: $e = 1$ — $\frac{PF}{PQ} = 1$

Hyperbola: $e > 1$ — $\frac{PF}{PQ} = \frac{P'F}{P'Q'} > 1$

In the figures, note that for each type of conic, a focus is at the pole. The benefit of locating a focus of a conic at the pole is that the equation of the conic takes on a simpler form.

Polar Equations of Conics

The graph of a polar equation of the form

1. $r = \frac{ep}{1 \pm e\cos\theta}$ or **2.** $r = \frac{ep}{1 \pm e\sin\theta}$

is a conic, where $e > 0$ is the eccentricity and $|p|$ is the distance between the focus (pole) and the directrix.

For a proof of the polar equations of conics, see Proofs in Mathematics on page 502.

An equation of the form

$$r = \frac{ep}{1 \pm e \cos \theta} \qquad \text{Vertical directrix}$$

corresponds to a conic with a vertical directrix and symmetry with respect to the polar axis. An equation of the form

$$r = \frac{ep}{1 \pm e \sin \theta} \qquad \text{Horizontal directrix}$$

corresponds to a conic with a horizontal directrix and symmetry with respect to the line $\theta = \pi/2$. Moreover, the converse is also true—that is, any conic with a focus at the pole and having a horizontal or vertical directrix can be represented by one of these equations.

EXAMPLE 1 Identifying a Conic from Its Equation

See LarsonPrecalculus.com for an interactive version of this type of example.

Identify the type of conic represented by the equation

$$r = \frac{15}{3 - 2 \cos \theta}.$$

Algebraic Solution

To identify the type of conic, rewrite the equation in the form

$$r = \frac{ep}{1 \pm e \cos \theta}.$$

$$r = \frac{15}{3 - 2 \cos \theta} \qquad \text{Write original equation.}$$

$$= \frac{5}{1 - (2/3) \cos \theta} \qquad \text{Divide numerator and denominator by 3.}$$

Because $e = \frac{2}{3} < 1$, the graph is an ellipse.

Graphical Solution

Use a graphing utility in *polar* mode and be sure to use a square setting, as shown in the figure below.

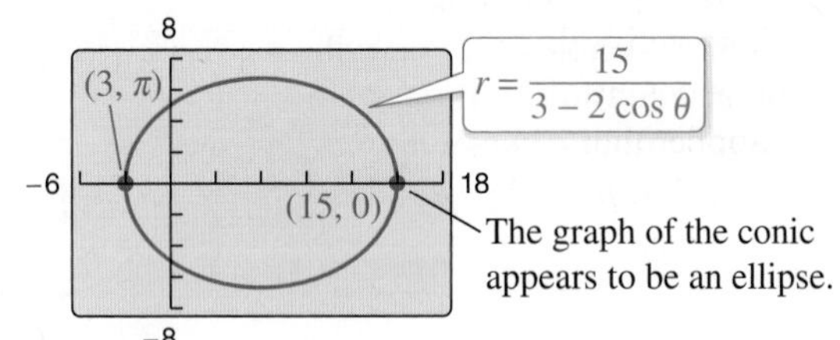

✓ Checkpoint Audio-video solution in English & Spanish at LarsonPrecalculus.com

Identify the type of conic represented by the equation

$$r = \frac{8}{2 - 3 \sin \theta}.$$

For the ellipse in Example 1, the major axis is horizontal and the vertices lie at $(r, \theta) = (15, 0)$ and $(r, \theta) = (3, \pi)$. So, the length of the major axis is $2a = 18$. To find the length of the *minor* axis, use the definition of eccentricity $e = c/a$ and the relation $a^2 = b^2 + c^2$ for ellipses to conclude that

$$b^2 = a^2 - c^2 = a^2 - (ea)^2 = a^2(1 - e^2). \qquad \text{Ellipse}$$

Because $a = 18/2 = 9$ and $e = 2/3$, you have

$$b^2 = 9^2\left[1 - \left(\tfrac{2}{3}\right)^2\right] = 45$$

which implies that $b = \sqrt{45} = 3\sqrt{5}$. So, the length of the minor axis is $2b = 6\sqrt{5}$.

A similar analysis holds for hyperbolas. Using $e = c/a$ and the relation $c^2 = a^2 + b^2$ for hyperbolas yields

$$b^2 = c^2 - a^2 = (ea)^2 - a^2 = a^2(e^2 - 1). \qquad \text{Hyperbola}$$

EXAMPLE 2 Sketching a Conic from Its Polar Equation

Identify the type of conic represented by $r = \dfrac{32}{3 + 5 \sin \theta}$ and sketch its graph.

Solution Dividing the numerator and denominator by 3, you have

$$r = \frac{32/3}{1 + (5/3) \sin \theta}.$$

Because $e = \frac{5}{3} > 1$, the graph is a hyperbola. The transverse axis of the hyperbola lies on the line $\theta = \pi/2$, and the vertices occur at $(r, \theta) = (4, \pi/2)$ and $(r, \theta) = (-16, 3\pi/2)$. The length of the transverse axis is 12, so $a = 6$. To find b, write

$$b^2 = a^2(e^2 - 1) = 6^2\left[\left(\tfrac{5}{3}\right)^2 - 1\right] = 64$$

which implies that $b = 8$. Use a and b to determine that the asymptotes of the hyperbola are $y = 10 \pm \frac{3}{4}x$. Figure 6.59 shows the graph.

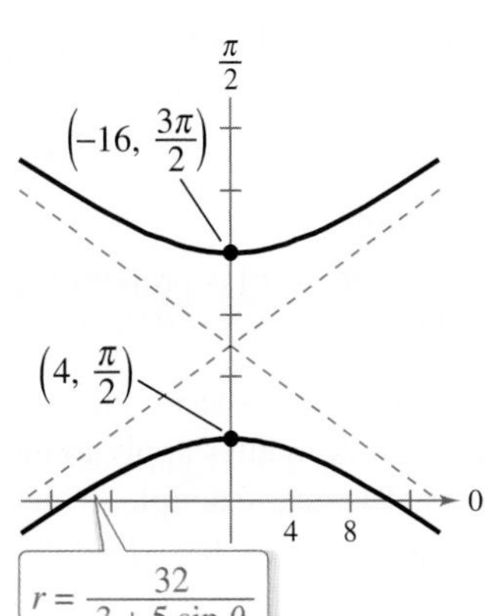

Figure 6.59

Checkpoint Audio-video solution in English & Spanish at LarsonPrecalculus.com

Identify the conic $r = \dfrac{3}{2 - 4 \sin \theta}$ and sketch its graph.

In the next example, you will find a polar equation of a specified conic. To do this, let p be the distance between the pole and the directrix.

TECHNOLOGY Use a graphing utility set in *polar* mode to verify the four orientations listed at the right. Remember that e must be positive, but p can be positive or negative.

1. Horizontal directrix above the pole: $r = \dfrac{ep}{1 + e \sin \theta}$
2. Horizontal directrix below the pole: $r = \dfrac{ep}{1 - e \sin \theta}$
3. Vertical directrix to the right of the pole: $r = \dfrac{ep}{1 + e \cos \theta}$
4. Vertical directrix to the left of the pole: $r = \dfrac{ep}{1 - e \cos \theta}$

EXAMPLE 3 Finding the Polar Equation of a Conic

Find a polar equation of the parabola whose focus is the pole and whose directrix is the line $y = 3$.

Solution The directrix is horizontal and above the pole, so use an equation of the form

$$r = \frac{ep}{1 + e \sin \theta}.$$

Moreover, the eccentricity of a parabola is $e = 1$ and the distance between the pole and the directrix is $p = 3$, so you have the equation

$$r = \frac{3}{1 + \sin \theta}.$$

Figure 6.60 shows the parabola.

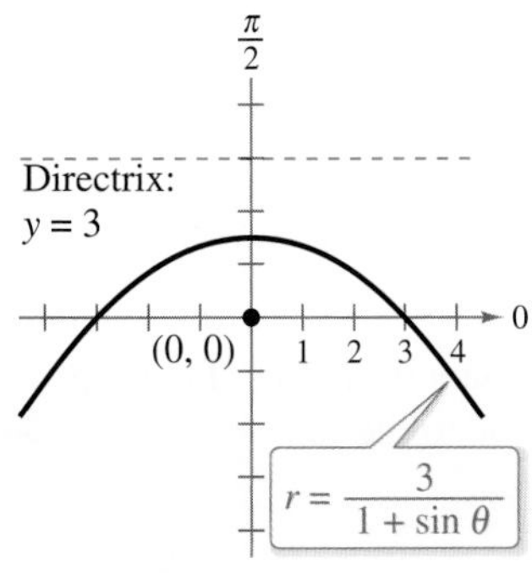

Figure 6.60

Checkpoint Audio-video solution in English & Spanish at LarsonPrecalculus.com

Find a polar equation of the parabola whose focus is the pole and whose directrix is the line $x = -2$.

Application

Kepler's Laws (listed below), named after the German astronomer Johannes Kepler (1571–1630), can be used to describe the orbits of the planets about the sun.

1. Each planet moves in an elliptical orbit with the sun at one focus.
2. A ray from the sun to a planet sweeps out equal areas in equal times.
3. The square of the period (the time it takes for a planet to orbit the sun) is proportional to the cube of the mean distance between the planet and the sun.

Although Kepler stated these laws on the basis of observation, Isaac Newton (1642–1727) later validated them. In fact, Newton showed that these laws apply to the orbits of all heavenly bodies, including comets and satellites. The next example, which involves the comet named after the English mathematician and physicist Edmund Halley (1656–1742), illustrates this.

If you use Earth as a reference with a period of 1 year and a distance of 1 astronomical unit (about 93 million miles), then the proportionality constant in Kepler's third law is 1. For example, Mars has a mean distance to the sun of $d \approx 1.524$ astronomical units. Solve for its period P in $d^3 = P^2$ to find that the period of Mars is $P \approx 1.88$ years.

EXAMPLE 4 Halley's Comet

Halley's comet has an elliptical orbit with an eccentricity of $e \approx 0.967$. The length of the major axis of the orbit is approximately 35.88 astronomical units. Find a polar equation for the orbit. How close does Halley's comet come to the sun?

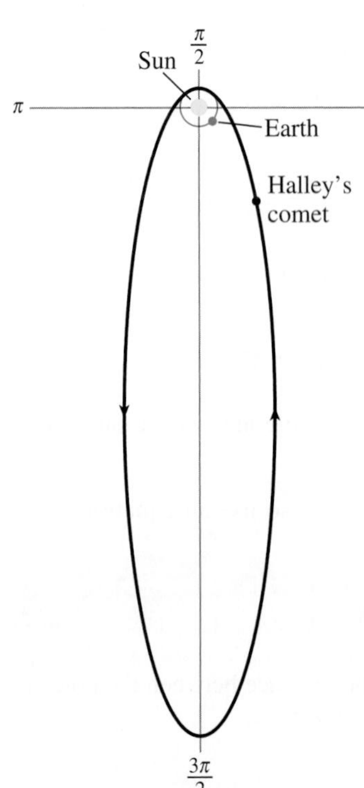

Figure 6.61

Solution Using a vertical major axis, as shown in Figure 6.61, choose an equation of the form $r = ep/(1 + e \sin \theta)$. The vertices of the ellipse occur when $\theta = \pi/2$ and $\theta = 3\pi/2$, and the length of the major axis is the sum of the r-values of the vertices. That is,

$$2a = \frac{0.967p}{1 + 0.967} + \frac{0.967p}{1 - 0.967} \approx 29.79p \approx 35.88.$$

So, $p \approx 1.204$ and $ep \approx (0.967)(1.204) \approx 1.164$. Substituting this value for ep in the equation, you have

$$r = \frac{1.164}{1 + 0.967 \sin \theta}$$

where r is measured in astronomical units. To find the closest point to the sun (a focus), substitute $\theta = \pi/2$ into this equation to obtain

$$r = \frac{1.164}{1 + 0.967 \sin(\pi/2)} \approx 0.59 \text{ astronomical unit} \approx 55{,}000{,}000 \text{ miles}.$$

✓ **Checkpoint** Audio-video solution in English & Spanish at LarsonPrecalculus.com

Encke's comet has an elliptical orbit with an eccentricity of $e \approx 0.847$. The length of the major axis of the orbit is approximately 4.420 astronomical units. Find a polar equation for the orbit. How close does Encke's comet come to the sun?

Summarize (Section 6.9)

1. State the definition of a conic in terms of eccentricity *(page 485)*. For examples of writing and graphing polar equations of conics, see Examples 2 and 3.
2. Describe a real-life application of an equation of a conic in polar form *(page 488, Example 4)*.

6.9 Exercises

See CalcChat.com for tutorial help and worked-out solutions to odd-numbered exercises.

Vocabulary

In Exercises 1–3, fill in the blanks.

1. The locus of a point in the plane that moves such that its distance from a fixed point (focus) is in a constant ratio to its distance from a fixed line (directrix) is a ________.
2. The constant ratio is the ________ of the conic and is denoted by ________.
3. An equation of the form $r = \dfrac{ep}{1 - e\cos\theta}$ has a ________ directrix to the ________ of the pole.
4. Match the conic with its eccentricity.

   (a) $0 < e < 1$ (b) $e = 1$ (c) $e > 1$

   (i) Parabola (ii) Hyperbola (iii) Ellipse

Skills and Applications

Identifying a Conic **In Exercises 5–8, write the polar equation of the conic for each value of e. Identify the type of conic represented by each equation. Verify your answers with a graphing utility.**

(a) $e = 1$ (b) $e = 0.5$ (c) $e = 1.5$

5. $r = \dfrac{2e}{1 + e\cos\theta}$
6. $r = \dfrac{2e}{1 - e\cos\theta}$
7. $r = \dfrac{2e}{1 - e\sin\theta}$
8. $r = \dfrac{2e}{1 + e\sin\theta}$

Matching **In Exercises 9–12, match the polar equation with its graph. [The graphs are labeled (a), (b), (c), and (d).]**

(a)

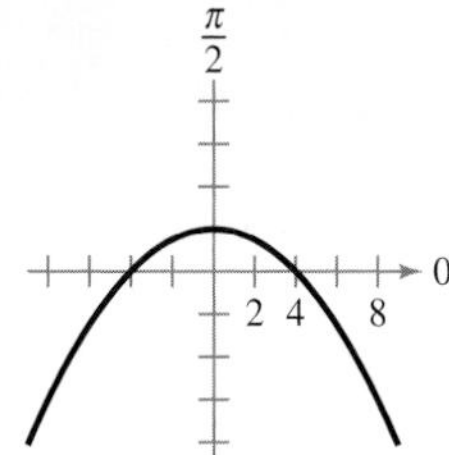

(b)

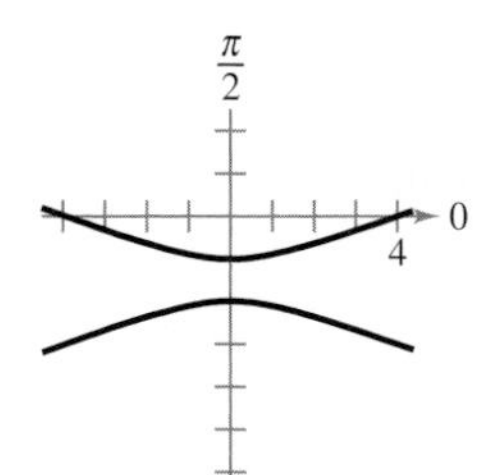

(c)

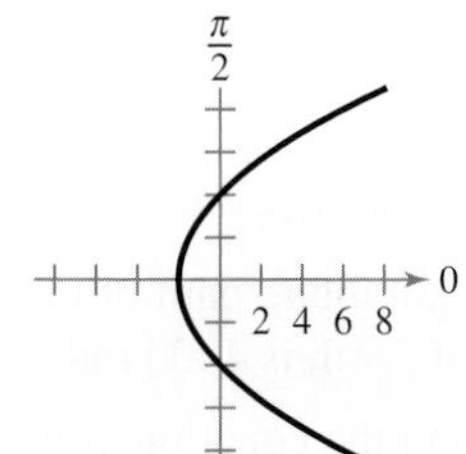

(d)

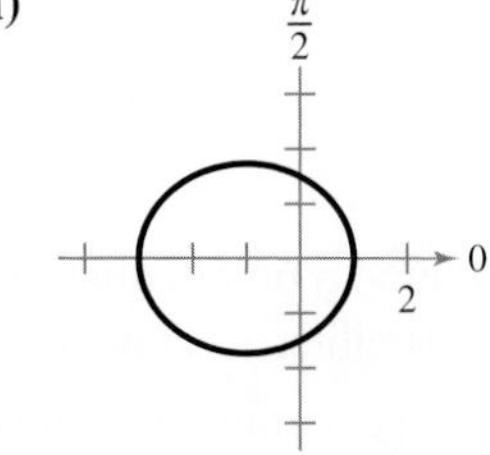

9. $r = \dfrac{4}{1 - \cos\theta}$
10. $r = \dfrac{3}{2 + \cos\theta}$
11. $r = \dfrac{4}{1 + \sin\theta}$
12. $r = \dfrac{4}{1 - 3\sin\theta}$

Sketching a Conic **In Exercises 13–24, identify the conic represented by the equation and sketch its graph.**

13. $r = \dfrac{3}{1 - \cos\theta}$
14. $r = \dfrac{7}{1 + \sin\theta}$
15. $r = \dfrac{5}{1 - \sin\theta}$
16. $r = \dfrac{6}{1 + \cos\theta}$
17. $r = \dfrac{2}{2 - \cos\theta}$
18. $r = \dfrac{4}{4 + \sin\theta}$
19. $r = \dfrac{6}{2 + \sin\theta}$
20. $r = \dfrac{6}{3 - 2\sin\theta}$
21. $r = \dfrac{3}{2 + 4\sin\theta}$
22. $r = \dfrac{5}{-1 + 2\cos\theta}$
23. $r = \dfrac{3}{2 - 6\cos\theta}$
24. $r = \dfrac{3}{2 + 6\sin\theta}$

Graphing a Polar Equation **In Exercises 25–32, use a graphing utility to graph the polar equation. Identify the conic.**

25. $r = \dfrac{-1}{1 - \sin\theta}$
26. $r = \dfrac{-5}{2 + 4\sin\theta}$
27. $r = \dfrac{3}{-4 + 2\cos\theta}$
28. $r = \dfrac{4}{1 - 2\cos\theta}$
29. $r = \dfrac{4}{3 - \cos\theta}$
30. $r = \dfrac{10}{1 + \cos\theta}$
31. $r = \dfrac{14}{14 + 17\sin\theta}$
32. $r = \dfrac{12}{2 - \cos\theta}$

Graphing a Rotated Conic In Exercises 33–36, use a graphing utility to graph the rotated conic.

33. $r = \dfrac{3}{1 - \cos[\theta - (\pi/4)]}$ (See Exercise 13.)

34. $r = \dfrac{4}{4 + \sin[\theta - (\pi/3)]}$ (See Exercise 18.)

35. $r = \dfrac{6}{2 + \sin[\theta + (\pi/6)]}$ (See Exercise 19.)

36. $r = \dfrac{3}{2 + 6\sin[\theta + (2\pi/3)]}$ (See Exercise 24.)

Finding the Polar Equation of a Conic In Exercises 37–52, find a polar equation of the indicated conic with the given characteristics and focus at the pole.

	Conic	Eccentricity	Directrix
37.	Parabola	$e = 1$	$x = -1$
38.	Parabola	$e = 1$	$y = -4$
39.	Ellipse	$e = \frac{1}{2}$	$x = 3$
40.	Ellipse	$e = \frac{3}{4}$	$y = -2$
41.	Hyperbola	$e = 2$	$x = 1$
42.	Hyperbola	$e = \frac{3}{2}$	$y = -2$

	Conic	Vertex or Vertices
43.	Parabola	$(2, 0)$
44.	Parabola	$(10, \pi/2)$
45.	Parabola	$(5, \pi)$
46.	Parabola	$(1, -\pi/2)$
47.	Ellipse	$(2, 0), (10, \pi)$
48.	Ellipse	$(2, \pi/2), (4, 3\pi/2)$
49.	Ellipse	$(20, 0), (4, \pi)$
50.	Hyperbola	$(2, 0), (8, 0)$
51.	Hyperbola	$(1, 3\pi/2), (9, 3\pi/2)$
52.	Hyperbola	$(4, \pi/2), (1, \pi/2)$

53. **Astronomy** The planets travel in elliptical orbits with the sun at one focus. Assume that the focus is at the pole, the major axis lies on the polar axis, and the length of the major axis is $2a$ (see figure). Show that the polar equation of the orbit is $r = a(1 - e^2)/(1 - e\cos\theta)$, where e is the eccentricity.

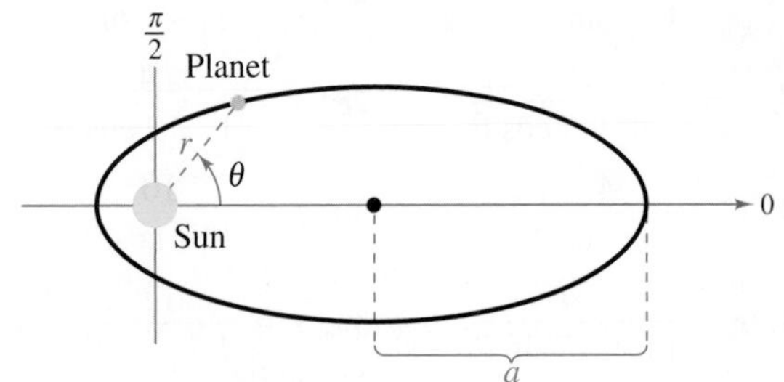

54. **Astronomy** Use the result of Exercise 53 to show that the minimum distance *(perihelion)* from the sun to the planet is

$r = a(1 - e)$

and the maximum distance *(aphelion)* is

$r = a(1 + e)$.

Planetary Motion In Exercises 55–60, use the results of Exercises 53 and 54 to find (a) the polar equation of the planet's orbit and (b) the perihelion and aphelion.

55. Earth $a \approx 9.2957 \times 10^7$ miles, $e \approx 0.0167$
56. Saturn $a \approx 1.4335 \times 10^9$ kilometers, $e \approx 0.0565$
57. Venus $a \approx 1.0821 \times 10^8$ kilometers, $e \approx 0.0067$
58. Mercury $a \approx 3.5984 \times 10^7$ miles, $e \approx 0.2056$
59. Mars $a \approx 1.4162 \times 10^8$ miles, $e \approx 0.0935$
60. Jupiter $a \approx 7.7857 \times 10^8$ kilometers, $e \approx 0.0489$

61. **Error Analysis** Describe the error.

For the polar equation $r = \dfrac{3}{2 + \sin\theta}$, $e = 1$.

So, the equation represents a parabola.

62. **Satellite Orbit**

A satellite in a 100-mile-high circular orbit around Earth has a velocity of approximately 17,500 miles per hour. If this velocity is multiplied by $\sqrt{2}$, then the satellite will have the minimum velocity necessary to escape Earth's gravity and will follow a parabolic path with the center of Earth as the focus (see figure).

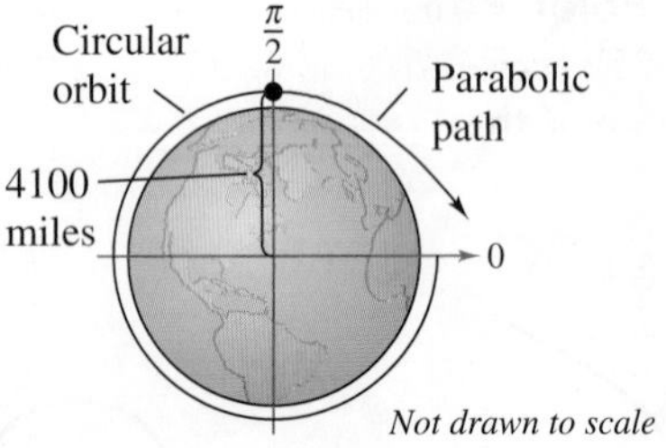

(a) Find a polar equation of the parabolic path of the satellite. Assume the radius of Earth is 4000 miles.

(b) Use a graphing utility to graph the equation you found in part (a).

(c) Find the distance between the surface of the Earth and the satellite when $\theta = 30°$.

(d) Find the distance between the surface of Earth and the satellite when $\theta = 60°$.

Exploration

True or False? **In Exercises 63–66, determine whether the statement is true or false. Justify your answer.**

63. For values of $e > 1$ and $0 \le \theta \le 2\pi$, the graphs of

$$r = \frac{ex}{1 - e\cos\theta} \quad \text{and} \quad r = \frac{e(-x)}{1 + e\cos\theta}$$

are the same.

64. The graph of

$$r = \frac{4}{-3 - 3\sin\theta}$$

has a horizontal directrix above the pole.

65. The conic represented by

$$r^2 = \frac{16}{9 - 4\cos\left(\theta + \frac{\pi}{4}\right)}$$

is an ellipse.

66. The conic represented by

$$r = \frac{6}{3 - 2\cos\theta}$$

is a parabola.

67. Verifying a Polar Equation Show that the polar equation of the ellipse represented by

$$\frac{x^2}{a^2} + \frac{y^2}{b^2} = 1 \quad \text{is} \quad r^2 = \frac{b^2}{1 - e^2\cos^2\theta}.$$

68. Verifying a Polar Equation Show that the polar equation of the hyperbola represented by

$$\frac{x^2}{a^2} - \frac{y^2}{b^2} = 1 \quad \text{is} \quad r^2 = \frac{-b^2}{1 - e^2\cos^2\theta}.$$

Writing a Polar Equation **In Exercises 69–74, use the results of Exercises 67 and 68 to write the polar form of the equation of the conic.**

69. $\dfrac{x^2}{169} + \dfrac{y^2}{144} = 1$

70. $\dfrac{x^2}{25} + \dfrac{y^2}{16} = 1$

71. $\dfrac{x^2}{9} - \dfrac{y^2}{16} = 1$

72. $\dfrac{x^2}{36} - \dfrac{y^2}{4} = 1$

73. Hyperbola
One focus: $(5, 0)$
Vertices: $(4, 0), (4, \pi)$

74. Ellipse
One focus: $(4, 0)$
Vertices: $(5, 0), (5, \pi)$

75. Writing Explain how the graph of each equation differs from the conic represented by $r = \dfrac{5}{1 - \sin\theta}$. (See Exercise 15.)

(a) $r = \dfrac{5}{1 - \cos\theta}$ (b) $r = \dfrac{5}{1 + \sin\theta}$

(c) $r = \dfrac{5}{1 + \cos\theta}$ (d) $r = \dfrac{5}{1 - \sin[\theta - (\pi/4)]}$

76. HOW DO YOU SEE IT? The graph of

$$r = \frac{e}{1 - e\sin\theta}$$

is shown for different values of e. Determine which graph matches each value of e.

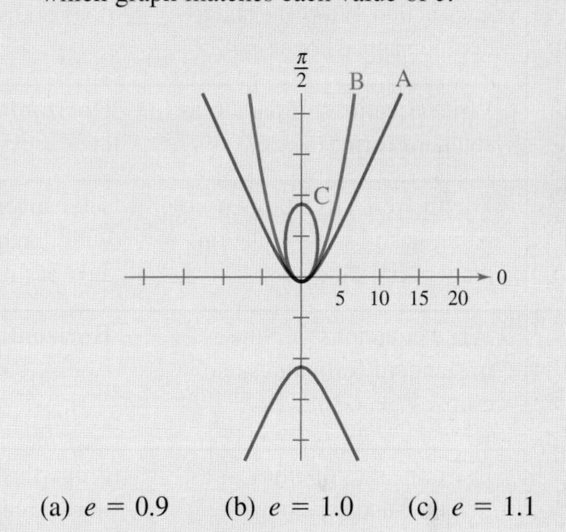

(a) $e = 0.9$ (b) $e = 1.0$ (c) $e = 1.1$

77. Reasoning

(a) Identify the type of conic represented by

$$r = \frac{4}{1 - 0.4\cos\theta}$$

without graphing the equation.

(b) Without graphing the equations, describe how the graph of each equation below differs from the polar equation given in part (a).

$$r_1 = \frac{4}{1 + 0.4\cos\theta} \qquad r_2 = \frac{4}{1 - 0.4\sin\theta}$$

(c) Use a graphing utility to verify your results in part (b).

78. Reasoning The equation

$$r = \frac{ep}{1 \pm e\sin\theta}$$

represents an ellipse with $e < 1$. What happens to the lengths of both the major axis and the minor axis when the value of e remains fixed and the value of p changes? Use an example to explain your reasoning.

Chapter Summary

	What Did You Learn?	Explanation/Examples		Review Exercises
Section 6.1	Find the inclination of a line *(p. 418).*	If a nonvertical line has inclination θ and slope m, then $m = \tan\theta$.		1–4
	Find the angle between two lines *(p. 419).*	If two nonperpendicular lines have slopes m_1 and m_2, then the tangent of the angle between the two lines is $\tan\theta = \lvert(m_2 - m_1)/(1 + m_1 m_2)\rvert$.		5–8
	Find the distance between a point and a line *(p. 420).*	The distance between the point (x_1, y_1) and the line $Ax + By + C = 0$ is $d = \lvert Ax_1 + By_1 + C\rvert/\sqrt{A^2 + B^2}$.		9, 10
Section 6.2	Recognize a conic as the intersection of a plane and a double-napped cone *(p. 425).*	In the formation of the four basic conics, the intersecting plane does not pass through the vertex of the cone. (See Figure 6.7.)		11, 12
	Write equations of parabolas in standard form *(p. 426).*	**Horizontal Axis** $(y - k)^2 = 4p(x - h),\ p \neq 0$	**Vertical Axis** $(x - h)^2 = 4p(y - k),\ p \neq 0$	13–16, 19, 20
	Use the reflective property of parabolas to write equations of tangent lines *(p. 428).*	The tangent line to a parabola at a point P makes equal angles with (1) the line passing through P and the focus and (2) the axis of the parabola.		17, 18
Section 6.3	Write equations of ellipses in standard form and sketch ellipses *(p. 435).*	**Horizontal Major Axis** $\dfrac{(x - h)^2}{a^2} + \dfrac{(y - k)^2}{b^2} = 1$	**Vertical Major Axis** $\dfrac{(x - h)^2}{b^2} + \dfrac{(y - k)^2}{a^2} = 1$	21–24, 27–30
	Use properties of ellipses to model and solve real-life problems *(p. 438).*	Properties of ellipses can be used to find distances from Earth's center to the moon's center in the moon's orbit. (See Example 5.)		25, 26
	Find eccentricities *(p. 439).*	The eccentricity e of an ellipse is the ratio $e = c/a$.		27–30
Section 6.4	Write equations of hyperbolas in standard form *(p. 444),* and find asymptotes of and sketch hyperbolas *(p. 445).*	**Horizontal Transverse Axis** $\dfrac{(x - h)^2}{a^2} - \dfrac{(y - k)^2}{b^2} = 1$	**Vertical Transverse Axis** $\dfrac{(y - k)^2}{a^2} - \dfrac{(x - h)^2}{b^2} = 1$	31–38
	Use properties of hyperbolas to solve real-life problems *(p. 448).*	Properties of hyperbolas can be used in radar and other detection systems. (See Example 5.)		39, 40
	Classify conics from their general equations *(p. 449).*	The graph of $Ax^2 + Cy^2 + Dx + Ey + F = 0$ is, except in degenerate cases, a circle $(A = C)$, a parabola $(AC = 0)$, an ellipse $(A \neq C$ and $AC > 0)$, or a hyperbola $(AC < 0)$.		41–44
Section 6.5	Rotate the coordinate axes to eliminate the xy-term in equations of conics *(p. 453).*	The equation $Ax^2 + Bxy + Cy^2 + Dx + Ey + F = 0$, where $B \neq 0$, can be rewritten as $A'(x')^2 + C'(y')^2 + D'x' + E'y' + F' = 0$ by rotating the coordinate axes through an angle θ, where $\cot 2\theta = (A - C)/B$.		45–48
	Use the discriminant to classify conics *(p. 457).*	The graph of $Ax^2 + Bxy + Cy^2 + Dx + Ey + F = 0$ is, except in degenerate cases, an ellipse or a circle $(B^2 - 4AC < 0)$, a parabola $(B^2 - 4AC = 0)$, or a hyperbola $(B^2 - 4AC > 0)$.		49–52

Section	What Did You Learn?	Explanation/Examples	Review Exercises
Section 6.6	Evaluate sets of parametric equations for given values of the parameter *(p. 461)*.	If f and g are continuous functions of t on an interval I, then the set of ordered pairs $(f(t), g(t))$ is a plane curve C. The equations $x = f(t)$ and $y = g(t)$ are parametric equations for C, and t is the parameter.	53, 54
	Sketch curves represented by sets of parametric equations *(p. 462)*.	One way to sketch a curve represented by a pair of parametric equations is to plot points in the xy-plane. You determine each set of coordinates (x, y) from a value chosen for the parameter t.	55–60
	Rewrite sets of parametric equations as single rectangular equations by eliminating the parameter *(p. 463)*.	To eliminate the parameter in a pair of parametric equations, solve for t in one equation and substitute the expression for t into the other equation. The result is the corresponding rectangular equation.	55–60
	Find sets of parametric equations for graphs *(p. 465)*.	A set of parametric equations that represent a graph is not unique. (See Example 4.)	61–66
Section 6.7	Plot points in the polar coordinate system *(p. 471)*.	$P = (r, \theta)$; r = directed distance; θ = directed angle; O; Polar axis	67–70
	Convert points *(p. 473)* and equations *(p. 474)* from rectangular to polar form and vice versa.	**Polar Coordinates (r, θ) and Rectangular Coordinates (x, y)** Polar-to-Rectangular: $x = r\cos\theta$, $y = r\sin\theta$ Rectangular-to-Polar: $\tan\theta = \frac{y}{x}$, $r^2 = x^2 + y^2$	71–90
Section 6.8	Graph polar equations by point plotting *(p. 477)*.	Graphing a polar equation by point plotting is similar to graphing a rectangular equation. (See Example 1.)	91–100
	Use symmetry, zeros, and maximum r-values to sketch graphs of polar equations *(p. 478)*.	The graph of a polar equation is symmetric with respect to the following when the given substitution yields an equivalent equation. **1.** The line $\theta = \frac{\pi}{2}$: Replace (r, θ) with $(r, \pi - \theta)$ or $(-r, -\theta)$. **2.** The polar axis: Replace (r, θ) with $(r, -\theta)$ or $(-r, \pi - \theta)$. **3.** The pole: Replace (r, θ) with $(r, \pi + \theta)$ or $(-r, \theta)$. Additional aids to graphing polar equations are the θ-values for which $\lvert r\rvert$ is maximum and the θ-values for which $r = 0$.	91–100
	Recognize special polar graphs *(p. 481)*.	Several important types of graphs, such as limaçons, rose curves, circles, and lemniscates, have equations that are simpler in polar form than in rectangular form.	101–104
Section 6.9	Define conics in terms of eccentricity, and write and graph polar equations of conics *(p. 485)*.	**Ellipse:** $0 < e < 1$ **Parabola:** $e = 1$ **Hyperbola:** $e > 1$ The graph of a polar equation of the form **(1)** $r = (ep)/(1 \pm e\cos\theta)$ or **(2)** $r = (ep)/(1 \pm e\sin\theta)$ is a conic, where $e > 0$ is the eccentricity and $\lvert p\rvert$ is the distance between the focus (pole) and the directrix.	105–112
	Use equations of conics in polar form to model real-life problems *(p. 488)*.	The equation of a conic in polar form can be used to model the orbit of Halley's comet. (See Example 4.)	113, 114

Review Exercises

See CalcChat.com for tutorial help and worked-out solutions to odd-numbered exercises.

6.1 Finding the Inclination of a Line In Exercises 1–4, find the inclination θ (in radians and degrees) of the line with the given characteristics.

1. Passes through the points $(-1, 2)$ and $(2, 5)$
2. Passes through the points $(3, 4)$ and $(-2, 7)$
3. Equation: $5x + 2y + 4 = 0$
4. Equation: $2x - 5y - 7 = 0$

Finding the Angle Between Two Lines In Exercises 5–8, find the angle θ (in radians and degrees) between the lines.

5. $4x + y = 2$
   $-5x + y = -1$
6. $-5x + 3y = 3$
   $-2x + 3y = 1$
7. $2x - 7y = 8$
   $\frac{2}{5}x + y = 0$
8. $0.03x + 0.05y = 0.16$
   $0.07x - 0.02y = 0.15$

Finding the Distance Between a Point and a Line In Exercises 9 and 10, find the distance between the point and the line.

	Point	Line
9.	$(4, 3)$	$y = 2x - 1$
10.	$(-2, 1)$	$y = -4x + 2$

6.2 Forming a Conic Section In Exercises 11 and 12, state the type of conic formed by the intersection of the plane and the double-napped cone.

11. 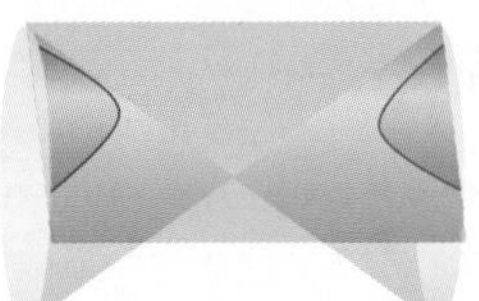

12. 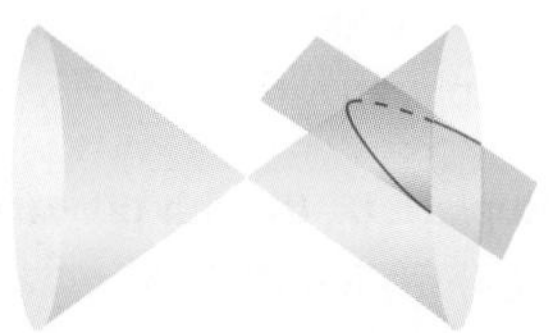

Finding the Standard Equation of a Parabola In Exercises 13–16, find the standard form of the equation of the parabola with the given characteristics. Then sketch the parabola.

13. Vertex: $(0, 0)$
    Focus: $(0, 3)$
14. Vertex: $(4, 0)$
    Focus: $(0, 0)$
15. Vertex: $(0, 2)$
    Directrix: $x = -3$
16. Vertex: $(-3, -3)$
    Directrix: $y = 0$

Finding the Tangent Line at a Point on a Parabola In Exercises 17 and 18, find an equation of the tangent line to the parabola at the given point.

17. $y = 2x^2$, $(-1, 2)$
18. $x^2 = -2y$, $(-4, -8)$

19. **Architecture** A parabolic archway is 10 meters high at the vertex. At a height of 8 meters, the width of the archway is 6 meters (see figure). How wide is the archway at ground level?

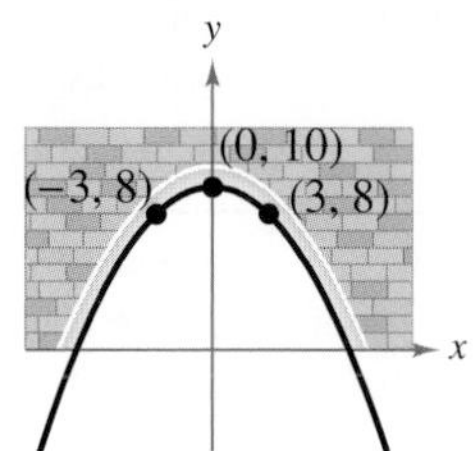

Figure for 19

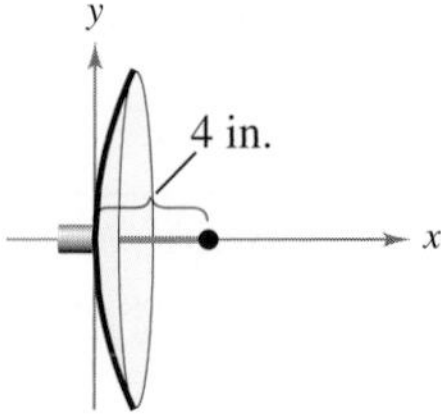

Figure for 20

20. **Parabolic Microphone** The receiver of a parabolic microphone is at the focus of the parabolic reflector, 4 inches from the vertex (see figure). Write an equation for a cross section of the reflector with its focus on the positive x-axis and its vertex at the origin.

6.3 Finding the Standard Equation of an Ellipse In Exercises 21–24, find the standard form of the equation of the ellipse with the given characteristics.

21. Vertices: $(2, 0)$, $(2, 16)$; minor axis of length 6
22. Vertices: $(0, 3)$, $(10, 3)$; minor axis of length 4
23. Vertices: $(0, 1)$, $(4, 1)$; endpoints of the minor axis: $(2, 0)$, $(2, 2)$
24. Vertices: $(-4, -1)$, $(-4, 11)$; endpoints of the minor axis: $(-6, 5)$, $(-2, 5)$

25. **Architecture** A mason is building a semielliptical arch that has a height of 4 feet and a width of 10 feet. Where should the foci be placed in order to sketch the arch?

26. **Wading Pool** You are building a wading pool that is in the shape of an ellipse. An equation for the elliptical shape of the pool is $(x^2/324) + (y^2/196) = 1$, where x and y are measured in feet. Find the longest distance across the pool, the shortest distance, and the distance between the foci.

Sketching an Ellipse In Exercises 27–30, find the center, vertices, foci, and eccentricity of the ellipse. Then sketch the ellipse.

27. $\dfrac{(x + 2)^2}{64} + \dfrac{(y - 5)^2}{36} = 1$
28. $\dfrac{(x - 4)^2}{25} + \dfrac{(y + 3)^2}{49} = 1$
29. $16x^2 + 9y^2 - 32x + 72y + 16 = 0$
30. $4x^2 + 25y^2 + 16x - 150y + 141 = 0$

6.4 Finding the Standard Equation of a Hyperbola **In Exercises 31–34, find the standard form of the equation of the hyperbola with the given characteristics.**

31. Vertices: $(0, \pm 6)$
Foci: $(0, \pm 8)$

32. Vertices: $(5, 2), (-5, 2)$
Foci: $(6, 2), (-6, 2)$

33. Foci: $(\pm 5, 0)$
Asymptotes: $y = \pm \frac{3}{4}x$

34. Foci: $(0, \pm 13)$
Asymptotes: $y = \pm \frac{5}{12}x$

Sketching a Hyperbola **In Exercises 35–38, find the center, vertices, foci, and the equations of the asymptotes of the hyperbola. Then sketch the hyperbola using the asymptotes as an aid.**

35. $\dfrac{(x-4)^2}{49} - \dfrac{(y+2)^2}{25} = 1$

36. $\dfrac{(y-3)^2}{9} - x^2 = 1$

37. $9x^2 - 16y^2 - 18x - 32y - 151 = 0$

38. $-4x^2 + 25y^2 - 8x + 150y + 121 = 0$

39. Sound Location Two microphones, 2 miles apart, record an explosion. Microphone A receives the sound 6 seconds before microphone B. Where did the explosion occur? (Assume sound travels at 1100 feet per second.)

40. Navigation Radio transmitting station A is located 200 miles east of transmitting station B. A ship is in an area to the north and 40 miles west of station A. Synchronized radio pulses transmitted at 186,000 miles per second by the two stations are received 0.0005 second sooner from station A than from station B. How far north is the ship?

Classifying a Conic from a General Equation **In Exercises 41–44, classify the graph of the equation as a circle, a parabola, an ellipse, or a hyperbola.**

41. $5x^2 - 2y^2 + 10x - 4y + 17 = 0$

42. $-4y^2 + 5x + 3y + 7 = 0$

43. $3x^2 + 2y^2 - 12x + 12y + 29 = 0$

44. $4x^2 + 4y^2 - 4x + 8y - 11 = 0$

6.5 Rotation of Axes **In Exercises 45–48, rotate the axes to eliminate the xy-term in the equation. Then write the equation in standard form. Sketch the graph of the resulting equation, showing both sets of axes.**

45. $xy + 5 = 0$

46. $x^2 - 4xy + y^2 + 9 = 0$

47. $5x^2 - 2xy + 5y^2 - 12 = 0$

48. $4x^2 + 8xy + 4y^2 + 7\sqrt{2}x + 9\sqrt{2}y = 0$

Rotation and Graphing Utilities **In Exercises 49–52, (a) use the discriminant to classify the graph of the equation, (b) use the Quadratic Formula to solve for y, and (c) use a graphing utility to graph the equation.**

49. $16x^2 - 24xy + 9y^2 - 30x - 40y = 0$

50. $13x^2 - 8xy + 7y^2 - 45 = 0$

51. $x^2 - 10xy + y^2 + 1 = 0$

52. $x^2 + y^2 + 2xy + 2\sqrt{2}x - 2\sqrt{2}y + 2 = 0$

6.6 Sketching a Curve **In Exercises 53 and 54, (a) create a table of x- and y-values for the parametric equations using $t = -2, -1, 0, 1$, and 2, and (b) plot the points (x, y) generated in part (a) and sketch the graph of the parametric equations.**

53. $x = 3t - 2$ and $y = 7 - 4t$

54. $x = \dfrac{1}{4}t$ and $y = \dfrac{6}{t+3}$

Sketching a Curve **In Exercises 55–60, (a) sketch the curve represented by the parametric equations (indicate the orientation of the curve) and (b) eliminate the parameter and write the resulting rectangular equation whose graph represents the curve. Adjust the domain of the rectangular equation, if necessary. Verify your result with a graphing utility.**

55. $x = 2t$
$y = 4t$

56. $x = 1 + 4t$
$y = 2 - 3t$

57. $x = t^2$
$y = \sqrt{t}$

58. $x = t + 4$
$y = t^2$

59. $x = 3 \cos \theta$
$y = 3 \sin \theta$

60. $x = 3 + 3 \cos \theta$
$y = 2 + 5 \sin \theta$

Finding Parametric Equations for a Graph **In Exercises 61–66, find a set of parametric equations to represent the graph of the rectangular equation using (a) $t = x$, (b) $t = x + 1$, and (c) $t = 3 - x$.**

61. $y = 2x + 3$

62. $y = 4 - 3x$

63. $y = x^2 + 3$

64. $y = 2 - x^2$

65. $y = 1 - 4x^2$

66. $y = 2x^2 + 2$

6.7 Plotting a Point in the Polar Coordinate System **In Exercises 67–70, plot the point given in polar coordinates and find three additional polar representations of the point, using $-2\pi < \theta < 2\pi$.**

67. $\left(4, \dfrac{5\pi}{6}\right)$

68. $\left(-3, -\dfrac{\pi}{4}\right)$

69. $(-7, 4.19)$

70. $(\sqrt{3}, 2.62)$

Polar-to-Rectangular Conversion **In Exercises 71–74, a point is given in polar coordinates. Convert the point to rectangular coordinates.**

71. $\left(0, \frac{\pi}{2}\right)$ **72.** $\left(2, \frac{5\pi}{4}\right)$

73. $\left(-1, \frac{\pi}{3}\right)$ **74.** $\left(3, -\frac{3\pi}{4}\right)$

Rectangular-to-Polar Conversion **In Exercises 75–78, a point is given in rectangular coordinates. Convert the point to polar coordinates. (There are many correct answers.)**

75. $(3, 3)$ **76.** $(3, -4)$

77. $(-\sqrt{5}, \sqrt{5})$ **78.** $(-\sqrt{2}, -\sqrt{2})$

Converting a Rectangular Equation to Polar Form **In Exercises 79–84, convert the rectangular equation to polar form.**

79. $x^2 + y^2 = 81$ **80.** $x^2 + y^2 = 48$

81. $x = 5$ **82.** $y = 4$

83. $xy = 5$ **84.** $xy = -2$

Converting a Polar Equation to Rectangular Form **In Exercises 85–90, convert the polar equation to rectangular form.**

85. $r = 4$ **86.** $r = 12$

87. $r = 3\cos\theta$ **88.** $r = 8\sin\theta$

89. $r^2 = \sin\theta$ **90.** $r^2 = 4\cos 2\theta$

6.8 Sketching the Graph of a Polar Equation **In Exercises 91–100, sketch the graph of the polar equation using symmetry, zeros, maximum r-values, and any other additional points.**

91. $r = 6$ **92.** $r = 11$

93. $r = -2(1 + \cos\theta)$ **94.** $r = 1 - 4\cos\theta$

95. $r = 4\sin 2\theta$ **96.** $r = \cos 5\theta$

97. $r = 2 + 6\sin\theta$ **98.** $r = 5 - 5\cos\theta$

99. $r^2 = 9\sin\theta$ **100.** $r^2 = \cos 2\theta$

Identifying Types of Polar Graphs **In Exercises 101–104, identify the type of polar graph and use a graphing utility to graph the equation.**

101. $r = 3(2 - \cos\theta)$ **102.** $r = 5(1 - 2\cos\theta)$

103. $r = 8\cos 3\theta$ **104.** $r^2 = 2\sin 2\theta$

6.9 Sketching a Conic **In Exercises 105–108, identify the conic represented by the equation and sketch its graph.**

105. $r = \dfrac{1}{1 + 2\sin\theta}$ **106.** $r = \dfrac{6}{1 + \sin\theta}$

107. $r = \dfrac{4}{5 - 3\cos\theta}$ **108.** $r = \dfrac{16}{4 + 5\cos\theta}$

Finding the Polar Equation of a Conic **In Exercises 109–112, find a polar equation of the indicated conic with the given characteristics and focus at the pole.**

	Conic	Vertex or Vertices
109.	Parabola	$(2, \pi)$
110.	Parabola	$(2, \pi/2)$
111.	Ellipse	$(5, 0), (1, \pi)$
112.	Hyperbola	$(1, 0), (7, 0)$

113. Explorer 18 In 1963, the United States launched Explorer 18. Its low and high points above the surface of Earth were 110 miles and 122,800 miles, respectively. The center of Earth was at one focus of the orbit (see figure). Find the polar equation of the orbit and the distance between the surface of Earth and the satellite when $\theta = \pi/3$. Assume Earth has a radius of 4000 miles.

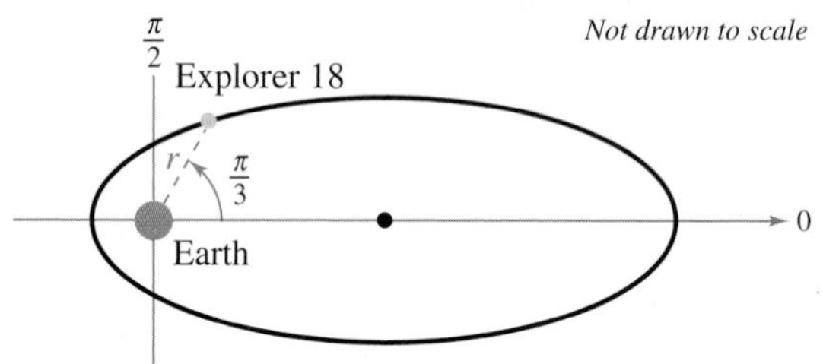

114. Asteroid An asteroid takes a parabolic path with Earth as its focus. It is about 6,000,000 miles from Earth at its closest approach. Write the polar equation of the path of the asteroid with its vertex at $\theta = \pi/2$. Find the distance between the asteroid and Earth when $\theta = -\pi/3$.

Exploration

True or False? **In Exercises 115–117, determine whether the statement is true or false. Justify your answer.**

115. The graph of $\frac{1}{4}x^2 - y^4 = 1$ is a hyperbola.

116. Only one set of parametric equations can represent the line $y = 3 - 2x$.

117. There is a unique polar coordinate representation of each point in the plane.

118. Think About It Consider an ellipse with the major axis horizontal and 10 units in length. The number b in the standard form of the equation of the ellipse must be less than what real number? Explain the change in the shape of the ellipse as b approaches this number.

119. Think About It What is the relationship between the graphs of the rectangular and polar equations?

(a) $x^2 + y^2 = 25, \quad r = 5$

(b) $x - y = 0, \quad \theta = \dfrac{\pi}{4}$

Chapter Test

See CalcChat.com for tutorial help and worked-out solutions to odd-numbered exercises.

Take this test as you would take a test in class. When you are finished, check your work against the answers given in the back of the book.

1. Find the inclination θ (in radians and degrees) of $4x - 7y + 6 = 0$.
2. Find the angle θ (in radians and degrees) between $3x + y = 6$ and $5x - 2y = -4$.
3. Find the distance between the point $(2, 9)$ and the line $y = 3x + 4$.

In Exercises 4–7, identify the conic and write the equation in standard form. Find the center, vertices, foci, and the equations of the asymptotes, if applicable. Then sketch the conic.

4. $y^2 - 2x + 2 = 0$
5. $x^2 - 4y^2 - 4x = 0$
6. $9x^2 + 16y^2 + 54x - 32y - 47 = 0$
7. $2x^2 + 2y^2 - 8x - 4y + 9 = 0$

8. Find the standard form of the equation of the parabola with vertex $(3, -4)$ and focus $(6, -4)$.
9. Find the standard form of the equation of the hyperbola with foci $(0, \pm 2)$ and asymptotes $y = \pm\frac{1}{9}x$.
10. Rotate the axes to eliminate the xy-term in the equation $xy + 1 = 0$. Then write the equation in standard form. Sketch the graph of the resulting equation, showing both sets of axes.
11. Sketch the curve represented by the parametric equations $x = 2 + 3\cos\theta$ and $y = 2\sin\theta$. Eliminate the parameter and write the resulting rectangular equation.
12. Find a set of parametric equations to represent the graph of the rectangular equation $y = 3 - x^2$ using (a) $t = x$ and (b) $t = x + 2$.
13. Convert the polar coordinates $\left(-2, \dfrac{5\pi}{6}\right)$ to rectangular coordinates.
14. Convert the rectangular coordinates $(2, -2)$ to polar coordinates and find three additional polar representations of the point, using $-2\pi < \theta < 2\pi$.
15. Convert the rectangular equation $x^2 + y^2 = 64$ to polar form.

In Exercises 16–19, identify the type of graph represented by the polar equation. Then sketch the graph.

16. $r = \dfrac{4}{1 + \cos\theta}$
17. $r = \dfrac{4}{2 + \sin\theta}$
18. $r = 2 + 3\sin\theta$
19. $r = 2\sin 4\theta$

20. Find a polar equation of the ellipse with focus at the pole, eccentricity $e = \frac{1}{4}$, and directrix $y = 4$.
21. A straight road rises with an inclination of 0.15 radian from the horizontal. Find the slope of the road and the change in elevation over a one-mile section of the road.
22. A baseball is hit at a point 3 feet above the ground toward the left field fence. The fence is 10 feet high and 375 feet from home plate. The path of the baseball can be modeled by the parametric equations $x = (115\cos\theta)t$ and $y = 3 + (115\sin\theta)t - 16t^2$. Does the baseball go over the fence when it is hit at an angle of $\theta = 30°$? Does the baseball go over the fence when $\theta = 35°$?

Cumulative Test for Chapters 4–6

See CalcChat.com for tutorial help and worked-out solutions to odd-numbered exercises.

Take this test as you would take a test in class. When you are finished, check your work against the answers given in the back of the book.

1. Write the complex number $6 - \sqrt{-49}$ in standard form.

In Exercises 2–4, perform the operation and write the result in standard form.

2. $6i - (2 + \sqrt{-81})$ **3.** $(5i - 2)^2$ **4.** $(\sqrt{3} + i)(\sqrt{3} - i)$

5. Write the quotient in standard form: $\dfrac{8i}{10 + 2i}$.

In Exercises 6 and 7, find all the zeros of the function.

6. $f(x) = x^3 + 2x^2 + 4x + 8$

7. $f(x) = x^4 + 4x^3 - 21x^2$

8. Find a polynomial function with real coefficients that has -6, -3, and $4 + \sqrt{5}i$ as its zeros. (There are many correct answers.)

9. Write the complex number $z = -2 + 2i$ in trigonometric form.

10. Find the product of $[4(\cos 30° + i \sin 30°)]$ and $[6(\cos 120° + i \sin 120°)]$. Leave the result in trigonometric form.

11. Plot $3 - 2i$ and its complex conjugate. Write the conjugate as a complex number.

12. Find the distance between $2 + 5i$ and $3 - i$ in the complex plane.

In Exercises 13 and 14, use DeMoivre's Theorem to find the power of the complex number. Write the result in standard form.

13. $\left[2\left(\cos \dfrac{2\pi}{3} + i \sin \dfrac{2\pi}{3}\right)\right]^4$ **14.** $(-\sqrt{3} - i)^6$

15. Find the three cube roots of 1.

16. Find all solutions of the equation $x^4 - 81i = 0$ and represent the solutions graphically.

In Exercises 17 and 18, describe the transformations of the graph of f that yield the graph of g.

17. $f(x) = \left(\frac{2}{5}\right)^x$, $g(x) = -\left(\frac{2}{5}\right)^{-x+3}$

18. $f(x) = 2.2^x$, $g(x) = -2.2^x + 4$

In Exercises 19–22, use a calculator to evaluate the expression. Round your result to three decimal places.

19. $\log 98$ **20.** $\log \frac{6}{7}$

21. $\ln \sqrt{31}$ **22.** $\ln(\sqrt{30} - 4)$

In Exercises 23–25, evaluate the logarithm using the change-of-base formula. Round your result to three decimal places.

23. $\log_5 4.3$ **24.** $\log_3 0.149$ **25.** $\log_{1/2} 17$

26. Use the properties of logarithms to expand $\ln\left(\dfrac{x^2 - 25}{x^4}\right)$, where $x > 5$.

27. Condense $2 \ln x - \frac{1}{2} \ln(x + 5)$ to the logarithm of a single quantity.

In Exercises 28–31, solve the equation algebraically. Approximate the result to three decimal places.

28. $6e^{2x} = 72$

29. $4^{x-5} + 21 = 30$

30. $\log_2 x + \log_2 5 = 6$

31. $\ln 4x - \ln 2 = 8$

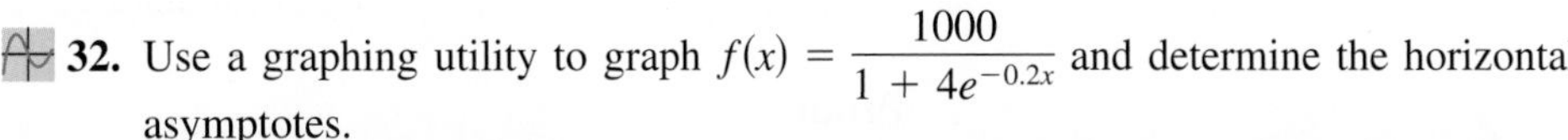

32. Use a graphing utility to graph $f(x) = \dfrac{1000}{1 + 4e^{-0.2x}}$ and determine the horizontal asymptotes.

33. The number N of bacteria in a culture is given by the model $N = 175e^{kt}$, where t is the time in hours. Given that $N = 420$ when $t = 8$, estimate the time required for the population to double in size.

34. The population P (in millions) of Texas from 2001 through 2014 can be approximated by the model $P = 20.913e^{0.0184t}$, where t represents the year, with $t = 1$ corresponding to 2001. According to this model, when will the population reach 32 million? *(Source: U.S. Census Bureau)*

35. Find the angle θ (in radians and degrees) between $2x + y = 3$ and $x - 3y = -6$.

36. Find the distance between the point $(6, -3)$ and the line $y = 2x - 4$.

In Exercises 37–40, identify the conic and write the equation in standard form. Find the center, vertices, foci, and the equations of the asymptotes, if applicable. Then sketch the conic.

37. $9x^2 + 4y^2 - 36x + 8y + 4 = 0$

38. $4x^2 - y^2 - 4 = 0$

39. $x^2 + y^2 + 2x - 6y - 12 = 0$

40. $y^2 + 2x + 2 = 0$

41. Find the standard form of the equation of the ellipse with vertices $(0, 0)$ and $(0, 4)$ and endpoints of the minor axis $(1, 2)$ and $(-1, 2)$.

42. Find the standard form of the equation of the hyperbola with foci $(0, \pm 5)$ and asymptotes $y = \pm \frac{4}{3}x$.

43. Rotate the axes to eliminate the xy-term in the equation

$$x^2 + xy + y^2 + 2x - 3y - 30 = 0.$$

Then write the equation in standard form. Sketch the graph of the resulting equation, showing both sets of axes.

44. Sketch the curve represented by the parametric equations $x = 3 + 4\cos\theta$ and $y = \sin\theta$. Eliminate the parameter and write the resulting rectangular equation.

45. Find a set of parametric equations to represent the graph of the rectangular equation $y = 1 - x$ using (a) $t = x$ and (b) $t = 2 - x$.

46. Plot the point $(-2, -3\pi/4)$ (given in polar coordinates) and find three additional polar representations of the point, using $-2\pi < \theta < 2\pi$.

47. Convert the rectangular equation $x^2 + y^2 - 16y = 0$ to polar form.

48. Convert the polar equation $r = \dfrac{2}{4 - 5\cos\theta}$ to rectangular form.

In Exercises 49 and 50, identify the conic and sketch its graph.

49. $r = \dfrac{4}{2 + \cos\theta}$

50. $r = \dfrac{8}{1 + \sin\theta}$

51. Match each polar equation with its graph at the left.

(a) $r = 2 + 3\sin\theta$ (b) $r = 3\sin\theta$ (c) $r = 3\sin 2\theta$

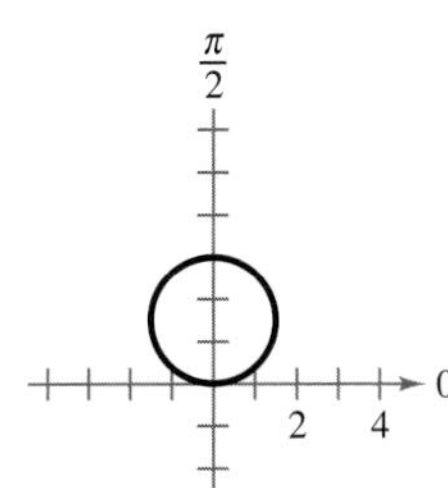

(i)

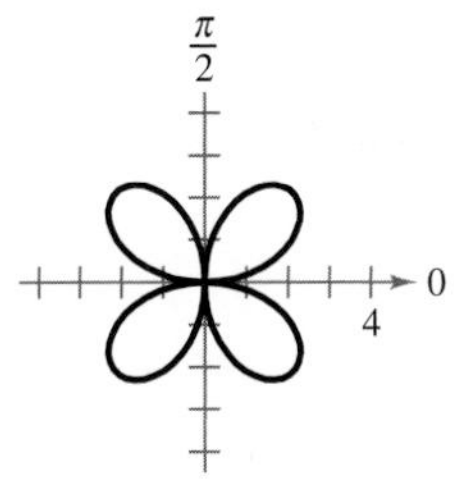

(ii)

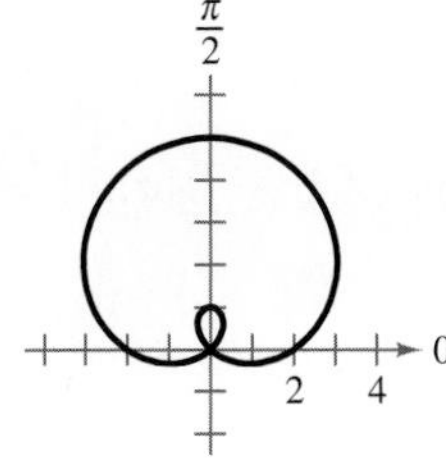

(iii)

Figure for 51

Proofs in Mathematics

Inclination and Slope *(p. 418)*

If a nonvertical line has inclination θ and slope m, then $m = \tan \theta$.

Proof

If $m = 0$, then the line is horizontal and $\theta = 0$. So, the result is true for horizontal lines because $m = 0 = \tan 0$.

If the line has a positive slope, then it will intersect the x-axis. Label this point $(x_1, 0)$, as shown in the figure. If (x_2, y_2) is a second point on the line, then the slope is

$$m = \frac{y_2 - 0}{x_2 - x_1} = \frac{y_2}{x_2 - x_1} = \tan \theta.$$

The case in which the line has a negative slope can be proved in a similar manner.

Distance Between a Point and a Line *(p. 420)*

The distance between the point (x_1, y_1) and the line $Ax + By + C = 0$ is

$$d = \frac{|Ax_1 + By_1 + C|}{\sqrt{A^2 + B^2}}.$$

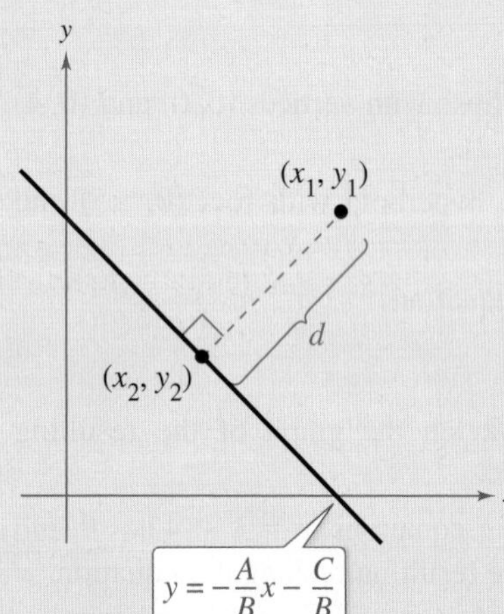

Proof

For simplicity, assume that the given line is neither horizontal nor vertical (see figure). Writing the equation $Ax + By + C = 0$ in slope-intercept form

$$y = -\frac{A}{B}x - \frac{C}{B}$$

shows that the line has a slope of $m = -A/B$. So, the slope of the line passing through (x_1, y_1) and perpendicular to the given line is B/A, and its equation is $y - y_1 = (B/A)(x - x_1)$. These two lines intersect at the point (x_2, y_2), where

$$x_2 = \frac{B^2x_1 - ABy_1 - AC}{A^2 + B^2} \quad \text{and} \quad y_2 = \frac{-ABx_1 + A^2y_1 - BC}{A^2 + B^2}.$$

Finally, the distance between (x_1, y_1) and (x_2, y_2) is

$$\begin{aligned}
d &= \sqrt{(x_2 - x_1)^2 + (y_2 - y_1)^2} \\
&= \sqrt{\left(\frac{B^2x_1 - ABy_1 - AC}{A^2 + B^2} - x_1\right)^2 + \left(\frac{-ABx_1 + A^2y_1 - BC}{A^2 + B^2} - y_1\right)^2} \\
&= \sqrt{\frac{A^2(Ax_1 + By_1 + C)^2 + B^2(Ax_1 + By_1 + C)^2}{(A^2 + B^2)^2}} \\
&= \sqrt{\frac{(Ax_1 + By_1 + C)^2(A^2 + B^2)}{(A^2 + B^2)^2}} \\
&= \sqrt{\frac{(Ax_1 + By_1 + C)^2}{A^2 + B^2}} \\
&= \frac{|Ax_1 + By_1 + C|}{\sqrt{A^2 + B^2}}.
\end{aligned}$$

PARABOLIC PATHS

There are many natural occurrences of parabolas in real life. For example, Italian astronomer and mathematician Galileo Galilei discovered in the 17th century that an object projected upward and obliquely to the pull of gravity travels in a parabolic path. Examples of this include the path of a jumping dolphin and the path of water molecules from a drinking water fountain.

Standard Equation of a Parabola *(p. 426)*

The **standard form of the equation of a parabola** with vertex at (h, k) is

$(x - h)^2 = 4p(y - k), \quad p \neq 0$ Vertical axis; directrix: $y = k - p$

$(y - k)^2 = 4p(x - h), \quad p \neq 0.$ Horizontal axis; directrix: $x = h - p$

The focus lies on the axis p units (directed distance) from the vertex. If the vertex is at the origin, then the equation takes one of two forms.

$x^2 = 4py$ Vertical axis

$y^2 = 4px$ Horizontal axis

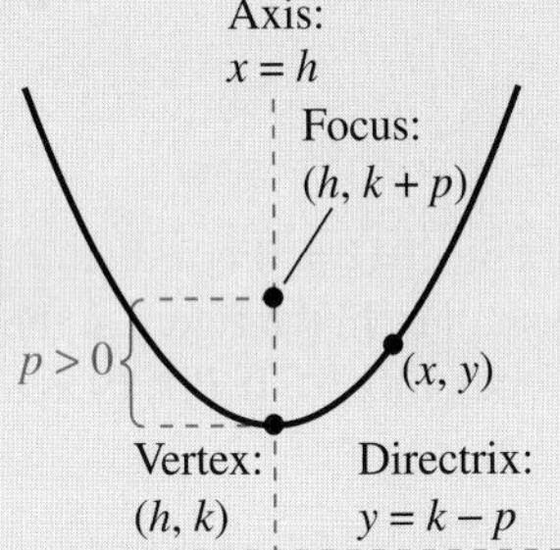

Parabola with vertical axis

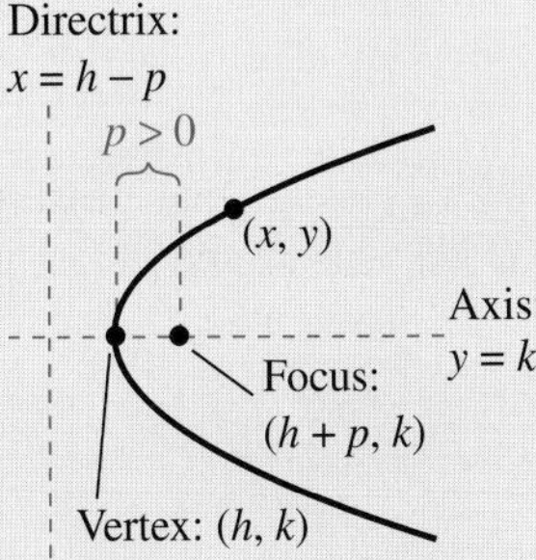

Parabola with horizontal axis

Proof

First, examine the case in which the directrix is parallel to the x-axis and the focus lies above the vertex, as shown in the top figure. If (x, y) is any point on the parabola, then, by definition, it is equidistant from the focus $(h, k + p)$ and the directrix $y = k - p$. Apply the Distance Formula to obtain

$$\sqrt{(x - h)^2 + [y - (k + p)]^2} = y - (k - p)$$

$$(x - h)^2 + [y - (k + p)]^2 = [y - (k - p)]^2$$

$$(x - h)^2 + y^2 - 2y(k + p) + (k + p)^2 = y^2 - 2y(k - p) + (k - p)^2$$

$$(x - h)^2 + y^2 - 2ky - 2py + k^2 + 2pk + p^2 = y^2 - 2ky + 2py + k^2 - 2pk + p^2$$

$$(x - h)^2 - 2py + 2pk = 2py - 2pk$$

$$(x - h)^2 = 4p(y - k).$$

Next, examine the case in which the directrix is parallel to the y-axis and the focus lies to the right of the vertex, as shown in the bottom figure. If (x, y) is any point on the parabola, then, by definition, it is equidistant from the focus $(h + p, k)$ and the directrix $x = h - p$. Apply the Distance Formula to obtain

$$\sqrt{[x - (h + p)]^2 + (y - k)^2} = x - (h - p)$$

$$[x - (h + p)]^2 + (y - k)^2 = [x - (h - p)]^2$$

$$x^2 - 2x(h + p) + (h + p)^2 + (y - k)^2 = x^2 - 2x(h - p) + (h - p)^2$$

$$x^2 - 2hx - 2px + h^2 + 2ph + p^2 + (y - k)^2 = x^2 - 2hx + 2px + h^2 - 2ph + p^2$$

$$-2px + 2ph + (y - k)^2 = 2px - 2ph$$

$$(y - k)^2 = 4p(x - h).$$

Note that if a parabola is centered at the origin, then the two equations above would simplify to $x^2 = 4py$ and $y^2 = 4px$, respectively. The cases in which the focus lies (1) below the vertex and (2) to the left of the vertex can be proved in manners similar to the above. ■

Polar Equations of Conics ***(p. 485)***

The graph of a polar equation of the form

1. $r = \dfrac{ep}{1 \pm e \cos \theta}$

or

2. $r = \dfrac{ep}{1 \pm e \sin \theta}$

is a conic, where $e > 0$ is the eccentricity and $|p|$ is the distance between the focus (pole) and the directrix.

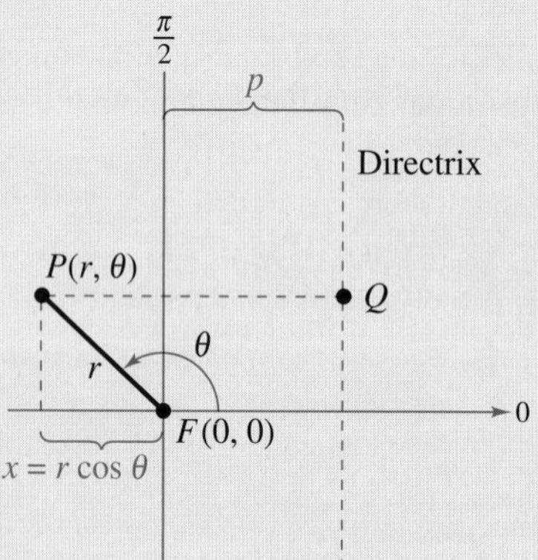

Proof

A proof for

$$r = \frac{ep}{1 + e \cos \theta}$$

with $p > 0$ is shown here. The proofs of the other cases are similar. In the figure at the left, consider a vertical directrix, p units to the right of the focus $F(0, 0)$. If $P(r, \theta)$ is a point on the graph of

$$r = \frac{ep}{1 + e \cos \theta}$$

then the distance between P and the directrix is

$$\begin{aligned} PQ &= |p - x| \\ &= |p - r \cos \theta| \\ &= \left| p - \left(\frac{ep}{1 + e \cos \theta} \right) \cos \theta \right| \\ &= \left| p \left(1 - \frac{e \cos \theta}{1 + e \cos \theta} \right) \right| \\ &= \left| \frac{p}{1 + e \cos \theta} \right| \\ &= \left| \frac{r}{e} \right|. \end{aligned}$$

Moreover, the distance between P and the pole is $PF = |r|$, so the ratio of PF to PQ is

$$\begin{aligned} \frac{PF}{PQ} &= \frac{|r|}{\left| \frac{r}{e} \right|} \\ &= |e| \\ &= e \end{aligned}$$

and, by definition, the graph of the equation must be a conic. ■

P.S. Problem Solving

1. **Mountain Climbing** Several mountain climbers are located in a mountain pass between two peaks. The angles of elevation to the two peaks are 0.84 radian and 1.10 radians. A range finder shows that the distances to the peaks are 3250 feet and 6700 feet, respectively (see figure).

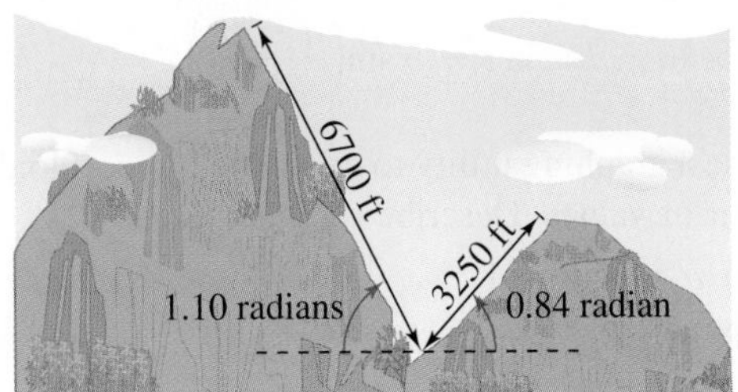

   (a) Find the angle between the two lines.

   (b) Approximate the amount of vertical climb that is necessary to reach the summit of each peak.

2. **Finding the Equation of a Parabola** Find the general equation of a parabola that has the x-axis as the axis of symmetry and the focus at the origin.

3. **Area** Find the area of the square whose vertices lie on the graph of the ellipse, as shown below.

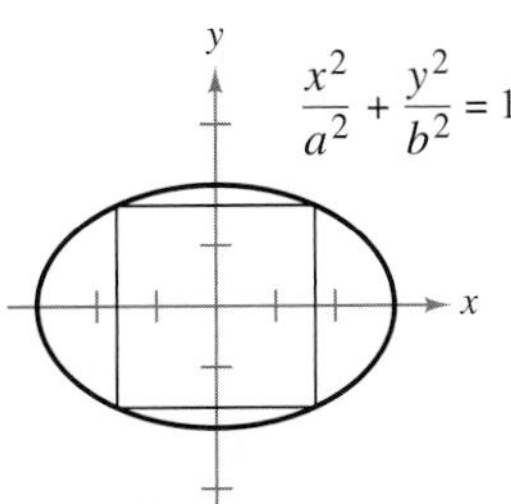

4. **Involute** The *involute* of a circle can be described by the endpoint P of a string that is held taut as it is unwound from a spool (see figure below). The spool does not rotate. Show that the parametric equations

   $x = r(\cos\theta + \theta\sin\theta)$

   and

   $y = r(\sin\theta - \theta\cos\theta)$

   represent the involute of a circle.

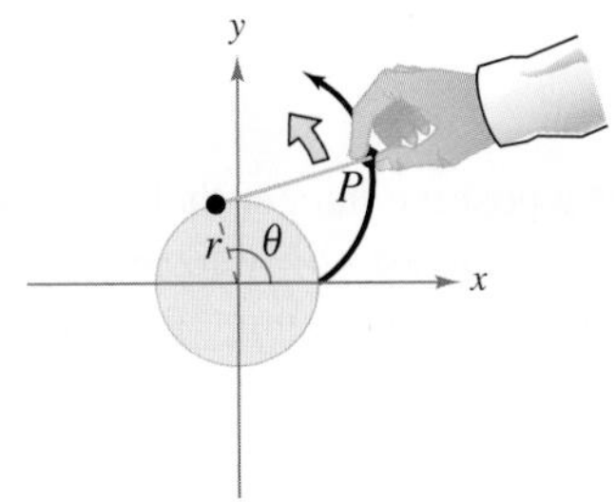

5. **Tour Boat** A tour boat travels between two islands that are 12 miles apart (see figure). There is enough fuel for a 20-mile trip.

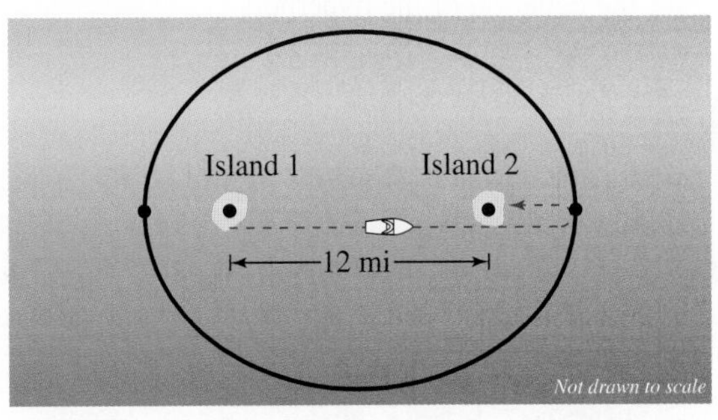

   (a) Explain why the region in which the boat can travel is bounded by an ellipse.

   (b) Let (0, 0) represent the center of the ellipse. Find the coordinates of each island.

   (c) The boat travels from Island 1, past Island 2 to a vertex of the ellipse, and then to Island 2. How many miles does the boat travel? Use your answer to find the coordinates of the vertex.

   (d) Use the results from parts (b) and (c) to write an equation of the ellipse that bounds the region in which the boat can travel.

6. **Finding an Equation of a Hyperbola** Find an equation of the hyperbola such that for any point on the hyperbola, the absolute value of the difference of its distances from the points $(2, 2)$ and $(10, 2)$ is 6.

7. **Proof** Prove that the graph of the equation

   $$Ax^2 + Cy^2 + Dx + Ey + F = 0$$

   is one of the following (except in degenerate cases).

Conic	Condition
(a) Circle	$A = C, A \neq 0$
(b) Parabola	$A = 0$ or $C = 0$, but not both.
(c) Ellipse	$AC > 0, A \neq C$
(d) Hyperbola	$AC < 0$

8. **Projectile Motion** The two sets of parametric equations below model projectile motion.

   $x_1 = (v_0\cos\theta)t, \quad y_1 = (v_0\sin\theta)t$

   $x_2 = (v_0\cos\theta)t, \quad y_2 = h + (v_0\sin\theta)t - 16t^2$

   (a) Under what circumstances is it appropriate to use each model?

   (b) Eliminate the parameter for each set of equations.

   (c) In which case is the path of the moving object not affected by a change in the velocity v? Explain.

9. Proof Prove that

$$c^2 = a^2 + b^2$$

for the equation of the hyperbola

$$\frac{x^2}{a^2} - \frac{y^2}{b^2} = 1$$

where the distance from the center of the hyperbola $(0, 0)$ to a focus is c.

10. Proof Prove that the angle θ used to eliminate the xy-term in $Ax^2 + Bxy + Cy^2 + Dx + Ey + F = 0$ by a rotation of axes is given by

$$\cot 2\theta = \frac{A - C}{B}.$$

11. Orientation of an Ellipse As t increases, the ellipse given by the parametric equations

$$x = \cos t$$

and

$$y = 2 \sin t$$

is traced *counterclockwise*. Find a set of parametric equations that represent the same ellipse traced *clockwise*.

12. Writing Use a graphing utility to graph the polar equation

$$r = \cos 5\theta + n \cos \theta$$

for the integers $n = -5$ to $n = 5$ using $0 \le \theta \le \pi$. As you graph these equations, you should see the graph's shape change from a heart to a bell. Write a short paragraph explaining what values of n produce the heart portion of the curve and what values of n produce the bell portion.

13. Strophoid The curve given by the polar equation

$$r = 2 \cos 2\theta \sec \theta$$

is called a **strophoid.**

(a) Find a rectangular equation of the strophoid.

(b) Find a pair of parametric equations that represent the strophoid.

(c) Use a graphing utility to graph the strophoid.

14. Think About It

(a) Show that the distance between the points (r_1, θ_1) and (r_2, θ_2) is $\sqrt{r_1^2 + r_2^2 - 2r_1r_2\cos(\theta_1 - \theta_2)}$.

(b) Simplify the Distance Formula for $\theta_1 = \theta_2$. Is the simplification what you expected? Explain.

(c) Simplify the Distance Formula for $\theta_1 - \theta_2 = 90°$. Is the simplification what you expected? Explain.

15. Hypocycloid A **hypocycloid** has the parametric equations

$$x = (a - b) \cos t + b \cos\left(\frac{a - b}{b}t\right)$$

and

$$y = (a - b) \sin t - b \sin\left(\frac{a - b}{b}t\right).$$

Use a graphing utility to graph the hypocycloid for each pair of values. Describe each graph.

(a) $a = 2, b = 1$

(b) $a = 3, b = 1$

(c) $a = 4, b = 1$

(d) $a = 10, b = 1$

(e) $a = 3, b = 2$

(f) $a = 4, b = 3$

16. Butterfly Curve The graph of the polar equation

$$r = e^{\cos \theta} - 2 \cos 4\theta + \sin^5\left(\frac{\theta}{12}\right)$$

is called the *butterfly curve,* as shown in the figure.

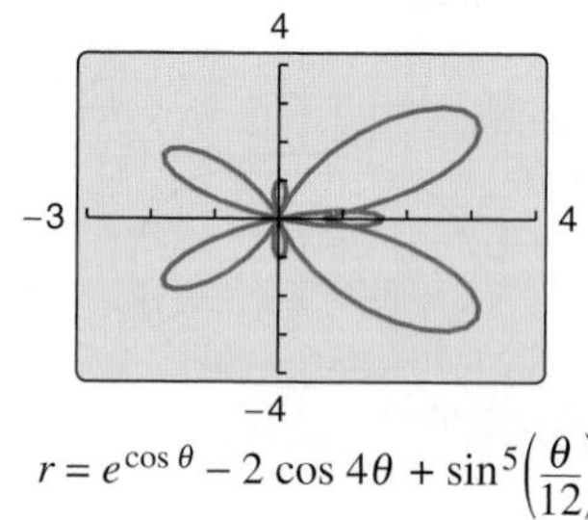

$r = e^{\cos \theta} - 2 \cos 4\theta + \sin^5\left(\frac{\theta}{12}\right)$

(a) The graph shown was produced using $0 \le \theta \le 2\pi$. Does this show the entire graph? Explain.

(b) Approximate the maximum r-value of the graph. Does this value change when you use $0 \le \theta \le 4\pi$ instead of $0 \le \theta \le 2\pi$? Explain.

17. Rose Curves The rose curves described in this chapter are of the form

$$r = a \cos n\theta$$

or

$$r = a \sin n\theta$$

where n is a positive integer that is greater than or equal to 2. Use a graphing utility to graph $r = a \cos n\theta$ and $r = a \sin n\theta$ for some noninteger values of n. Describe the graphs.

Answers to Odd-Numbered Exercises and Tests

Chapter P

Section P.1 *(page 12)*

1. irrational **3.** absolute value **5.** terms

7. (a) 5, 1, 2 (b) 0, 5, 1, 2 (c) $-9, 5, 0, 1, -4, 2, -11$
(d) $-9, -\frac{7}{2}, 5, \frac{2}{3}, 0, 1, -4, 2, -11$ (e) $\sqrt{2}$

9. (a) 1 (b) 1 (c) $-13, 1, -6$
(d) $2.01, 0.\overline{6}, -13, 1, -6$ (e) $0.010110111 \ldots$

11. (a) 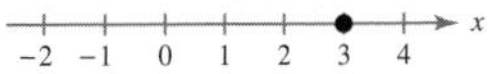

(b) 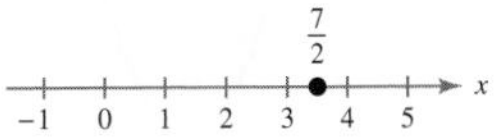

(c) 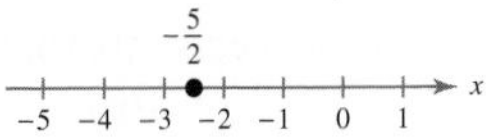

(d) 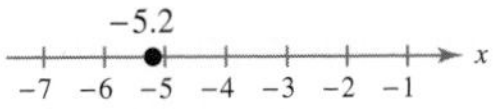

13. 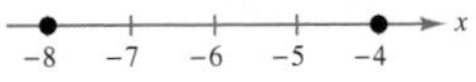

$-4 > -8$

15. 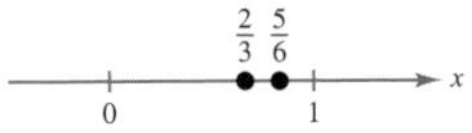

$\frac{5}{6} > \frac{2}{3}$

17. (a) $x \leq 5$ denotes the set of all real numbers less than or equal to 5.
(b) 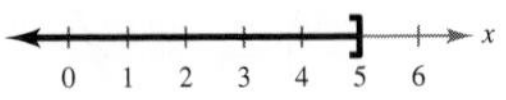

(c) Unbounded

19. (a) $-2 < x < 2$ denotes the set of all real numbers greater than -2 and less than 2.
(b) 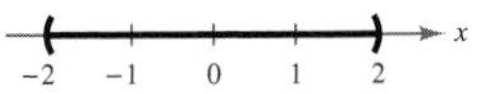

(c) Bounded

21. (a) $[4, \infty)$ denotes the set of all real numbers greater than or equal to 4.
(b) 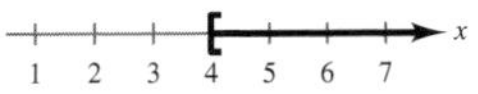

(c) Unbounded

23. (a) $[-5, 2)$ denotes the set of all real numbers greater than or equal to -5 and less than 2.
(b) 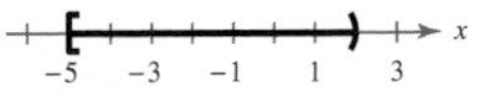

(c) Bounded

	Inequality	**Interval**
25.	$y \geq 0$	$[0, \infty)$
27.	$10 \leq t \leq 22$	$[10, 22]$

29. 10 **31.** 5 **33.** -1 **35.** 25 **37.** -1

39. $|-4| = |4|$ **41.** $-|-6| < |-6|$ **43.** 51 **45.** $\frac{5}{2}$

47. $|x - 5| \leq 3$ **49.** \$2524.0 billion; \$458.5 billion

51. \$2450.0 billion; \$1087.0 billion

53. $7x$ and 4 are the terms; 7 is the coefficient.

55. $6x^3$ and $-5x$ are the terms; 6 and -5 are the coefficients.

57. $3\sqrt{3}x^2$ and 1 are the terms; $3\sqrt{3}$ is the coefficient.

59. (a) -10 (b) -6 **61.** (a) 2 (b) 6

63. (a) Division by 0 is undefined. (b) 0

65. Multiplicative Inverse Property

67. Associative and Commutative Properties of Multiplication

69. $\dfrac{5x}{12}$ **71.** $\dfrac{x}{4}$

73. False. Zero is nonnegative, but not positive.

75. True. The product of two negative numbers is positive.

77. (a)

n	0.0001	0.01	1	100	10,000
$\dfrac{5}{n}$	50,000	500	5	0.05	0.0005

(b) (i) The value of $5/n$ approaches infinity as n approaches 0.
(ii) The value of $5/n$ approaches 0 as n increases without bound.

Section P.2 *(page 24)*

1. equation **3.** extraneous **5.** 4 **7.** -9 **9.** 12

11. No solution **13.** $-\frac{96}{23}$ **15.** 4 **17.** 3

19. No solution. The variable is divided out.

21. No solution. The solution is extraneous.

23. 5 **25.** $0, -\frac{1}{2}$ **27.** -5 **29.** $-\frac{1}{2}, 3$ **31.** $\pm\frac{3}{4}$

33. $-\frac{20}{3}, -4$ **35.** ± 7 **37.** $\pm 3\sqrt{3} \approx 5.20$ **39.** $-3, 11$

41. $\dfrac{1 \pm 3\sqrt{2}}{2} \approx 2.62, -1.62$ **43.** $4, -8$ **45.** $-2 \pm \sqrt{2}$

47. $-1 \pm \dfrac{\sqrt{2}}{2}$ **49.** $\dfrac{-5 \pm \sqrt{89}}{4}$ **51.** $\dfrac{1}{(x-1)^2 + 2^2}$

53. $\dfrac{1}{\sqrt{2^2 - (x-1)^2}}$ **55.** $\dfrac{1}{2}, -1$ **57.** $-\dfrac{5}{3}$ **59.** $\dfrac{7}{4} \pm \dfrac{\sqrt{41}}{4}$

61. $\dfrac{2}{3} \pm \dfrac{\sqrt{7}}{3}$ **63.** $1 \pm \sqrt{3}$ **65.** $2 \pm \dfrac{\sqrt{6}}{2}$

67. $6 \pm \sqrt{11}$ **69.** $0.976, -0.643$ **71.** $2.137, 18.063$

73. $1 \pm \sqrt{2}$ **75.** $-10, 6$ **77.** $\frac{1}{2} \pm \sqrt{3}$

79. $\dfrac{3}{4} \pm \dfrac{\sqrt{97}}{4}$ **81.** $-\dfrac{1}{2} \pm \dfrac{\sqrt{21}}{6}$ **83.** $\dfrac{1}{3} \pm \dfrac{\sqrt{31}}{3}$

85. $0, \pm 3$ **87.** $-2, \pm 2\sqrt{2}$ **89.** $\pm 1, \pm\sqrt{3}$ **91.** 20

93. $-\frac{55}{2}$ **95.** 1 **97.** 4 **99.** 9 **101.** 1

103. $8, -3$ **105.** $-\dfrac{1}{2} - \dfrac{\sqrt{17}}{2}, 3$ **107.** 63.7 in.

109. False. See Example 14 on page 23.

111. True. There is no value that satisfies this equation.

113. Yes. Both equations have the same solution of $x = 11$.

Section P.3 *(page 36)*

1. Cartesian **3.** Midpoint Formula **5.** graph **7.** y-axis

9. A: $(2, 6)$, B: $(-6, -2)$, C: $(4, -4)$, D: $(-3, 2)$

11. 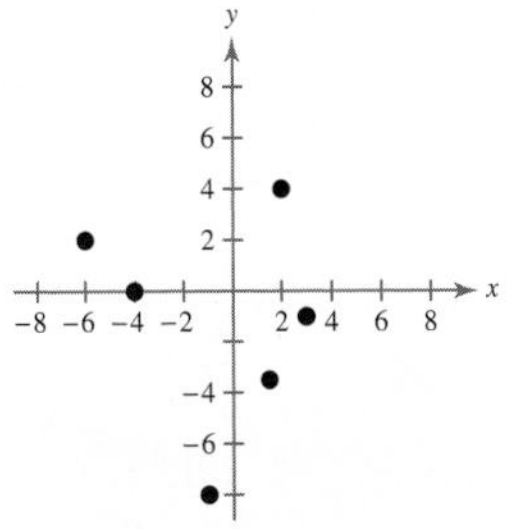

13. $(-3, 4)$ **15.** Quadrant IV **17.** Quadrant II

19. Quadrant II or IV

CHAPTER P

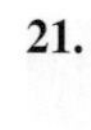

21.

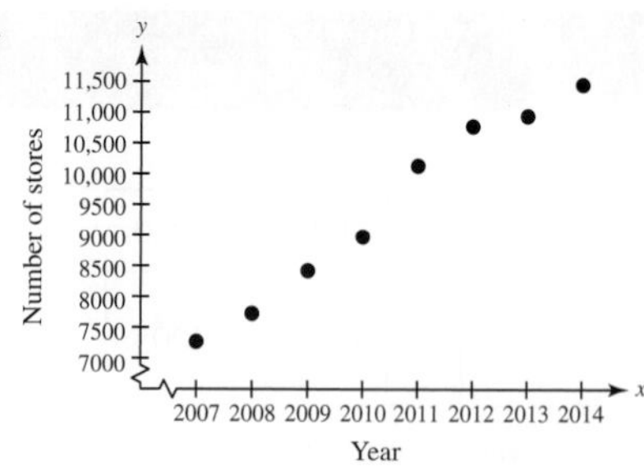

23. 13 **25.** $\sqrt{61}$ **27.** (a) 5, 12, 13 (b) $5^2 + 12^2 = 13^2$

29. $(\sqrt{5})^2 + (\sqrt{45})^2 = (\sqrt{50})^2$

31. Distances between the points: $\sqrt{29}$, $\sqrt{58}$, $\sqrt{29}$

33. (a)

(b) 10 (c) (5, 4)

35. (a)

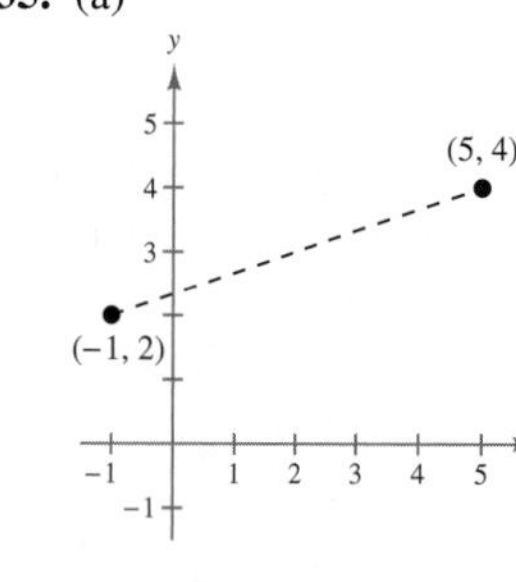

(b) $2\sqrt{10}$ (c) (2, 3)

37. $30\sqrt{41} \approx 192$ km **39.** \$40,560.5 million

41. (a) Yes (b) Yes **43.** (a) Yes (b) No

45. (a) No (b) Yes

47.

x	-1	0	1	2	$\frac{5}{2}$
y	7	5	3	1	0
(x, y)	$(-1, 7)$	$(0, 5)$	$(1, 3)$	$(2, 1)$	$(\frac{5}{2}, 0)$

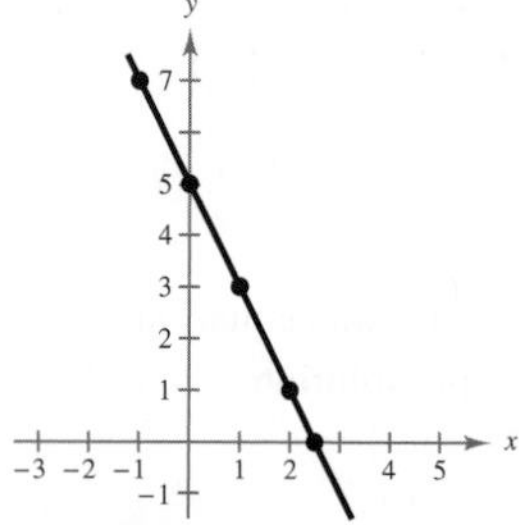

49.

x	-1	0	1	2	3
y	4	0	-2	-2	0
(x, y)	$(-1, 4)$	$(0, 0)$	$(1, -2)$	$(2, -2)$	$(3, 0)$

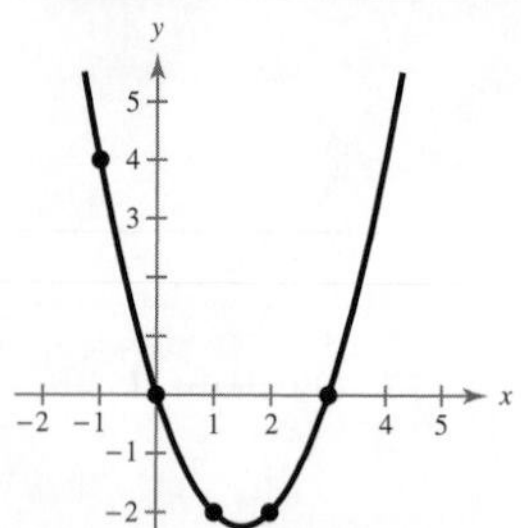

51. x-intercept: $(\frac{6}{5}, 0)$
y-intercept: $(0, -6)$

53. x-intercept: $(-4, 0)$
y-intercept: $(0, 2)$

55. x-intercept: $(\frac{7}{3}, 0)$
y-intercept: $(0, 7)$

57. x-intercepts: $(0, 0)$, $(2, 0)$
y-intercept: $(0, 0)$

59. x-intercept: $(6, 0)$
y-intercepts: $(0, \pm\sqrt{6})$

61. y-axis symmetry **63.** Origin symmetry

65. Origin symmetry **67.** x-axis symmetry

69.

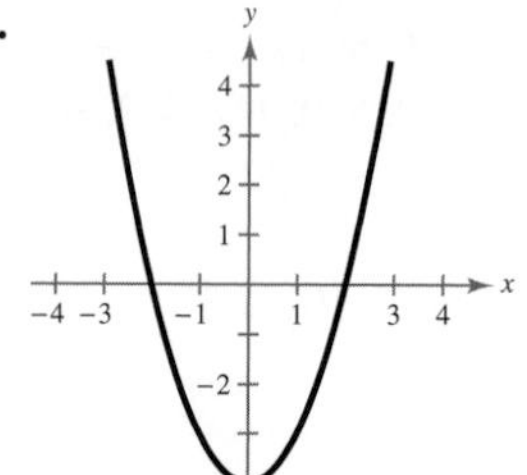

71.

73. x-intercept: $(\frac{1}{3}, 0)$
y-intercept: $(0, 1)$
No symmetry

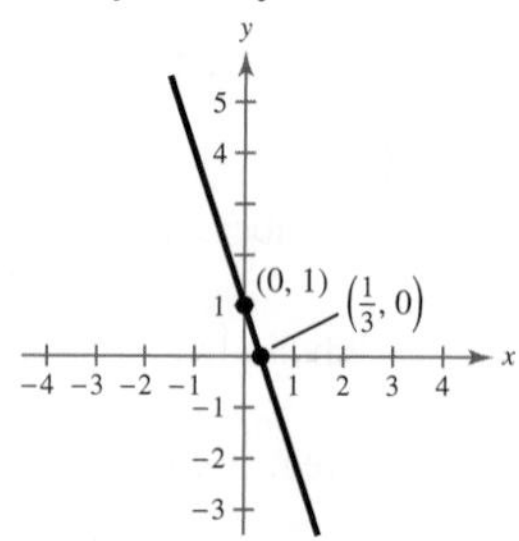

75. x-intercepts: $(0, 0)$, $(2, 0)$
y-intercept: $(0, 0)$
No symmetry

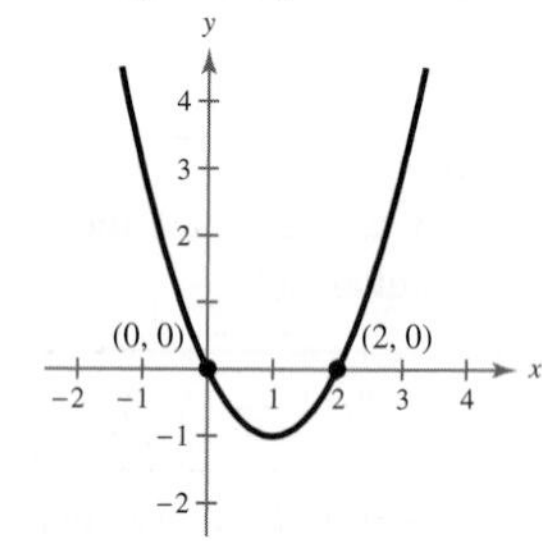

77. x-intercept: $(\sqrt[3]{-3}, 0)$
y-intercept: $(0, 3)$
No symmetry

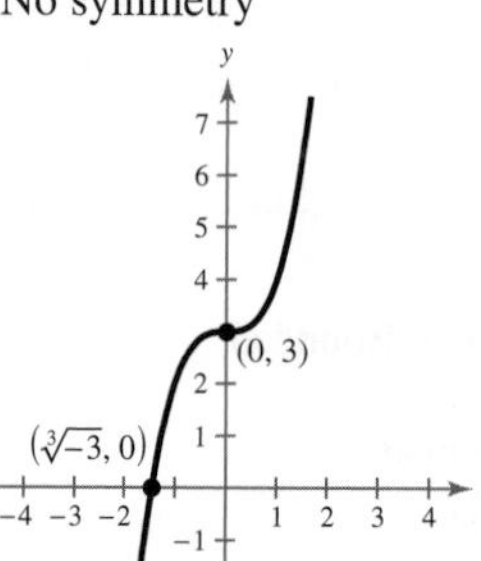

79. x-intercept: $(3, 0)$
y-intercept: None
No symmetry

81. x-intercept: $(6, 0)$
y-intercept: $(0, 6)$
No symmetry

83. $x^2 + y^2 = 49$ **85.** $(x - 3)^2 + (y - 8)^2 = 169$

87. $(x + 3)^2 + (y + 3)^2 = 61$

89. Center: $(0, 0)$; Radius: 5 **91.** Center: $\left(\frac{1}{2}, \frac{1}{2}\right)$; Radius: $\frac{3}{2}$

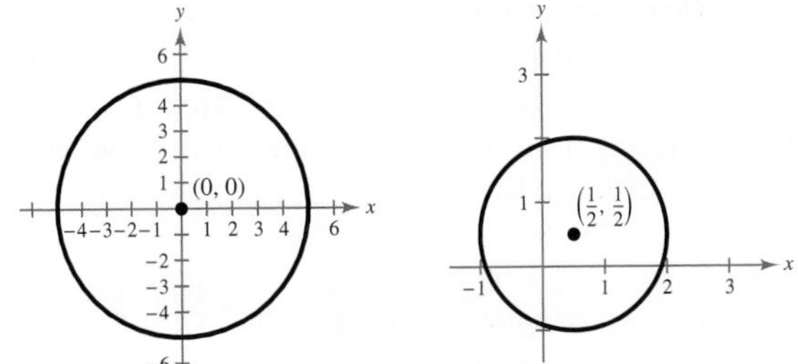

93.

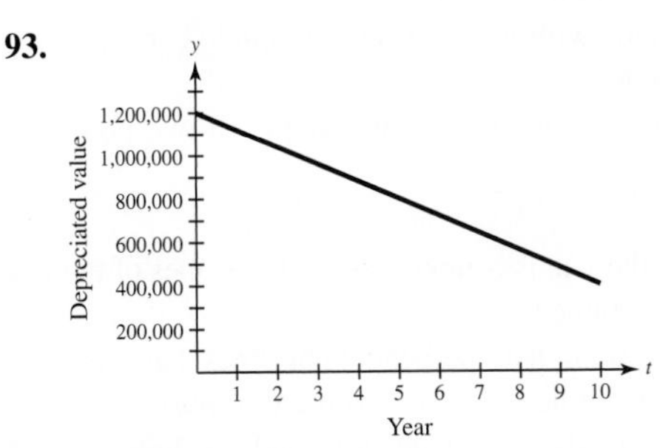

95. (a)

x	5	10	20	30	40
y	414.8	103.7	25.93	11.52	6.48

x	50	60	70	80	90	100
y	4.15	2.88	2.12	1.62	1.28	1.04

(b)

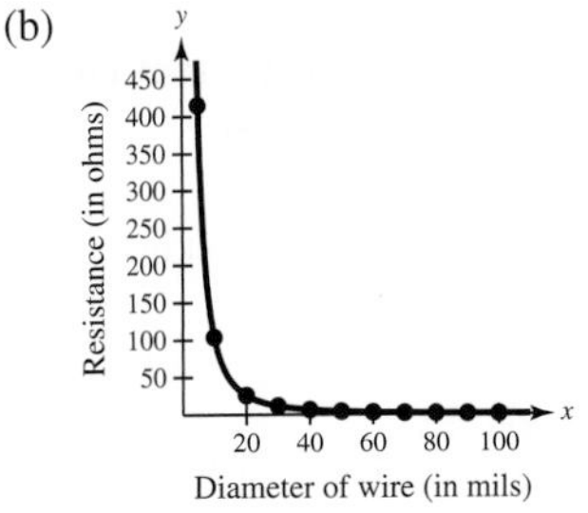

When $x = 85.5$, the resistance is 1.4 ohms.

(c) 1.42 ohms

(d) As the diameter of the copper wire increases, the resistance decreases.

97. False. The Midpoint Formula would be used 15 times.

99. False. $y = x$ is symmetric with respect to the origin.

101. *Sample answer:* When the x-values are much larger or smaller than the y-values.

103. Use the Midpoint Formula to prove that the diagonals of the parallelogram bisect each other.

$$\left(\frac{b+a}{2}, \frac{c+0}{2}\right) = \left(\frac{a+b}{2}, \frac{c}{2}\right)$$

$$\left(\frac{a+b+0}{2}, \frac{c+0}{2}\right) = \left(\frac{a+b}{2}, \frac{c}{2}\right)$$

105. $(2x_m - x_1, 2y_m - y_1)$ (a) $(7, 0)$ (b) $(9, -3)$

Section P.4 *(page 49)*

1. linear **3.** point-slope **5.** perpendicular

7. Linear extrapolation **9.** (a) L_2 (b) L_3 (c) L_1

11.

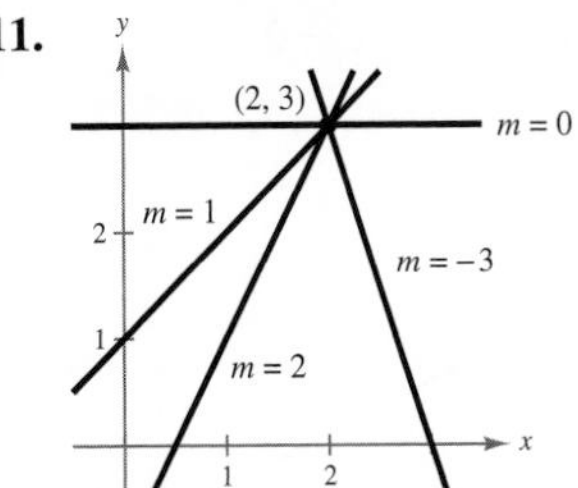

13. $\frac{3}{2}$

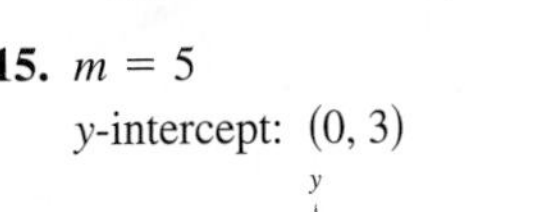

15. $m = 5$
y-intercept: $(0, 3)$

17. $m = -\frac{3}{4}$
y-intercept: $(0, -1)$

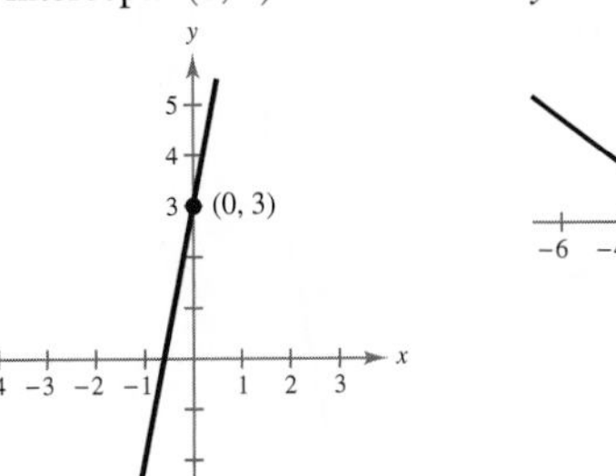

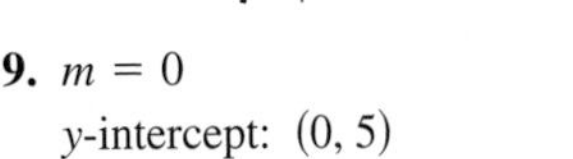

19. $m = 0$
y-intercept: $(0, 5)$

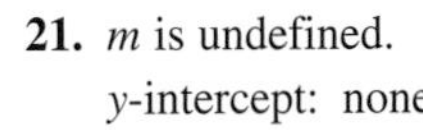

21. m is undefined.
y-intercept: none

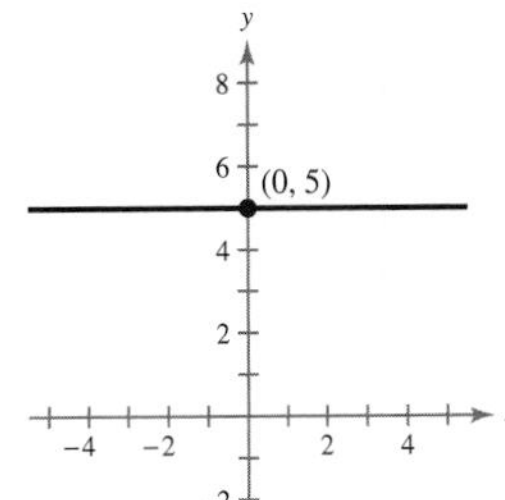

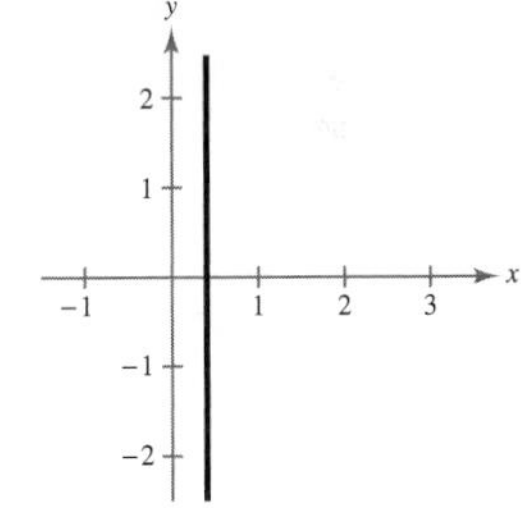

23. $m = \frac{7}{6}$
y-intercept: $(0, -5)$

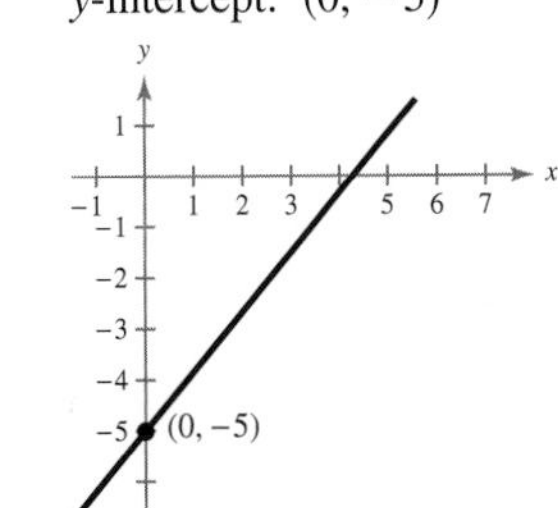

25. $m = -\frac{3}{2}$ **27.** $m = 2$ **29.** $m = 0$

31. m is undefined. **33.** $m = 0.15$

35. $(-1, 7), (0, 7), (4, 7)$ **37.** $(-4, 6), (-3, 8), (-2, 10)$

39. $(-2, 7), \left(0, \frac{19}{3}\right), (1, 6)$ **41.** $(-4, -5), (-4, 0), (-4, 2)$

43. $y = 3x - 2$ **45.** $y = -2x$

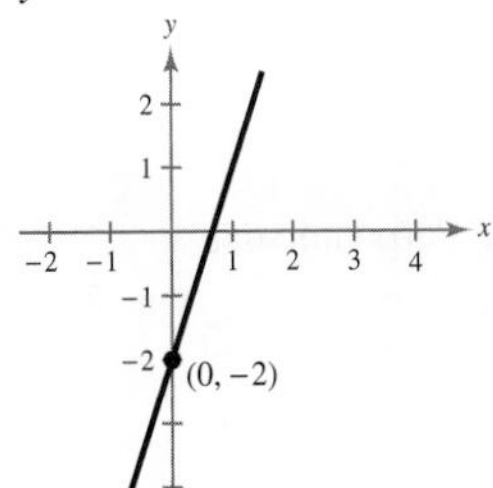

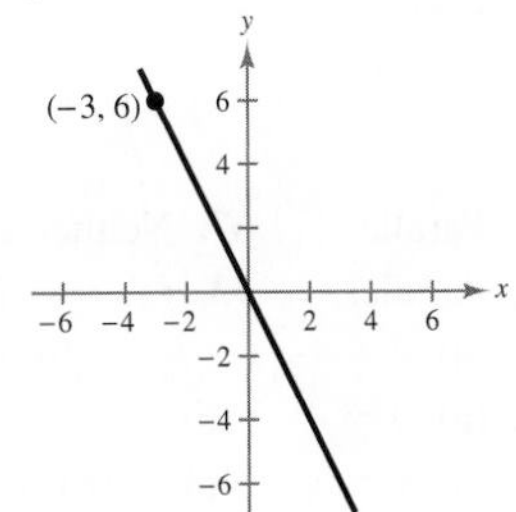

CHAPTER P

47. $y = -\frac{1}{3}x + \frac{4}{3}$ **49.** $y = -\frac{1}{2}x - 2$

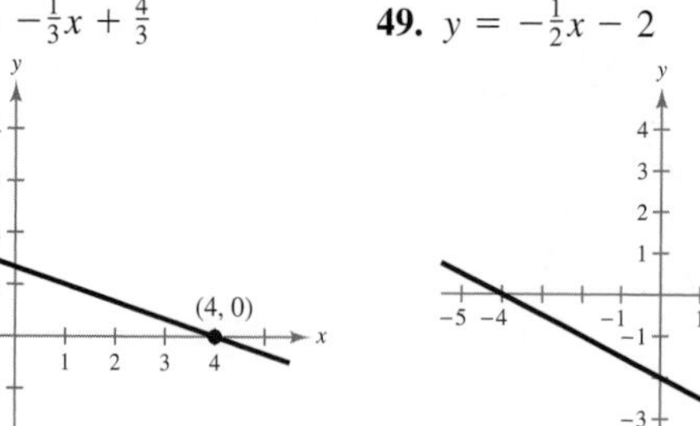

51. $y = \frac{5}{2}$

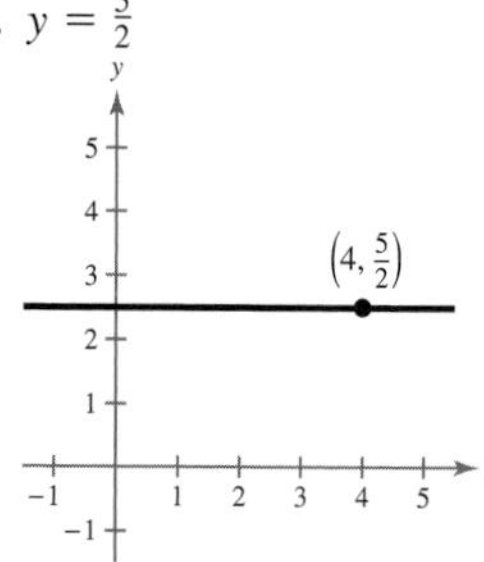

53. $y = 5x + 27.3$

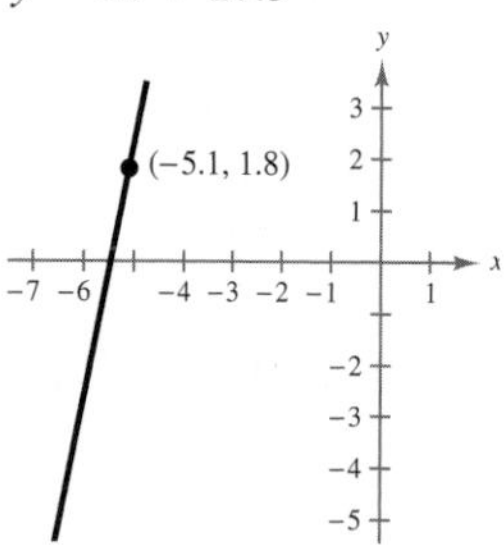

55. $y = -\frac{3}{5}x + 2$

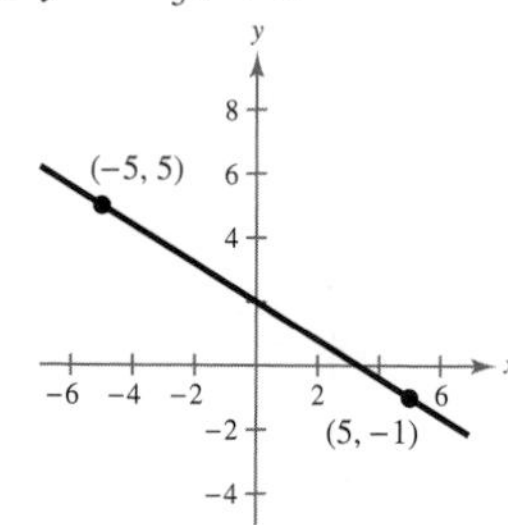

57. $x = -7$

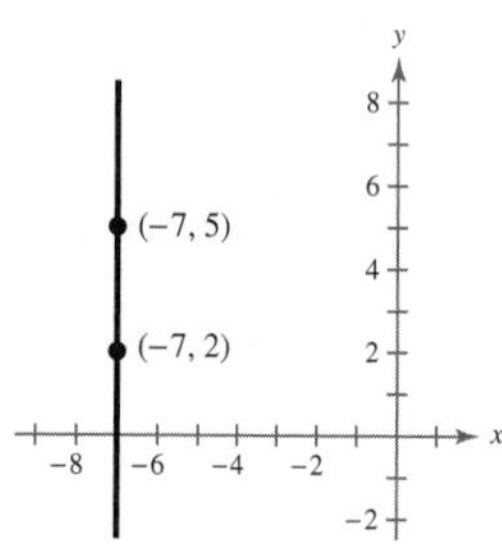

59. $y = -\frac{1}{2}x + \frac{3}{2}$

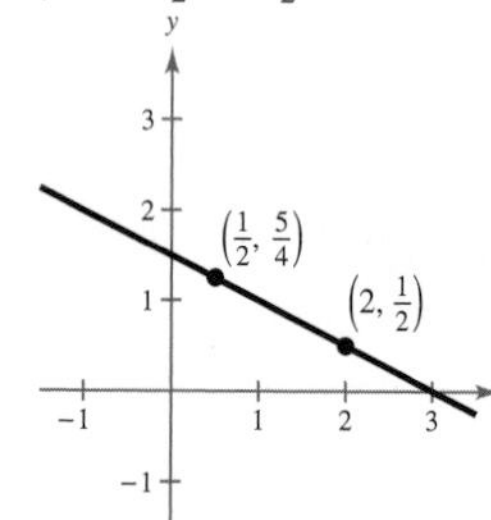

61. $y = 0.4x + 0.2$

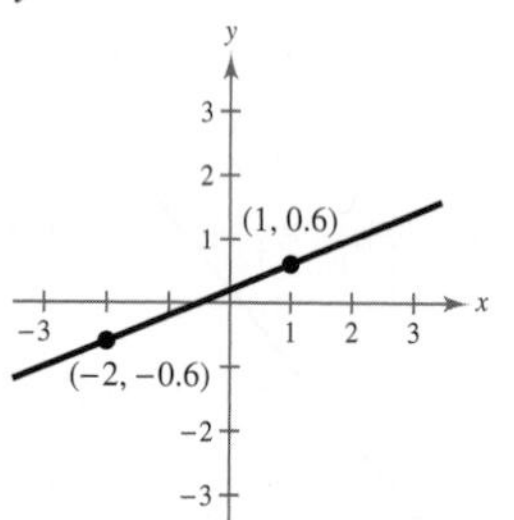

63. $y = -1$

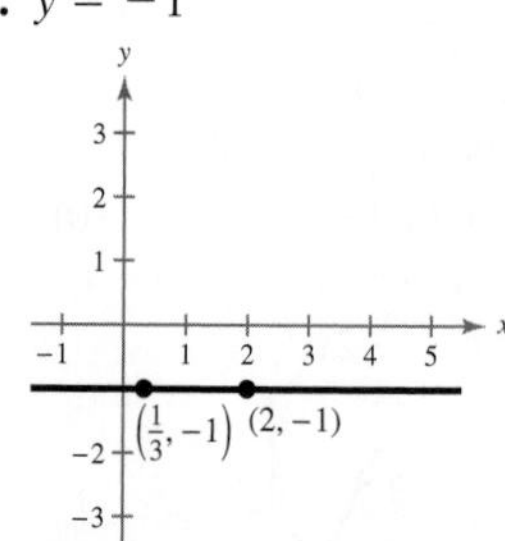

65. Parallel **67.** Neither **69.** Perpendicular

71. Parallel **73.** (a) $y = 2x - 3$ (b) $y = -\frac{1}{2}x + 2$

75. (a) $y = -\frac{3}{4}x + \frac{3}{8}$ (b) $y = \frac{4}{3}x + \frac{127}{72}$

77. (a) $y = 4$ (b) $x = -2$

79. (a) $y = x + 4.3$ (b) $y = -x + 9.3$

81. $5x + 3y - 15 = 0$ **83.** $12x + 3y + 2 = 0$

85. $x + y - 3 = 0$

87. (a) Sales increasing 135 units/yr
(b) No change in sales
(c) Sales decreasing 40 units/yr

89. 12 ft **91.** $V(t) = -150t + 5400, \quad 16 \le t \le 21$

93. C-intercept: fixed initial cost; Slope: cost of producing an additional laptop bag

95. $V(t) = -175t + 875, \quad 0 \le t \le 5$

97. $F = 1.8C + 32$ or $C = \frac{5}{9}F - \frac{160}{9}$

99. (a) $C = 21t + 42{,}000$ (b) $R = 45t$
(c) $P = 24t - 42{,}000$ (d) 1750 h

101. False. The slope with the greatest magnitude corresponds to the steepest line.

103. Find the slopes of the lines containing each two points and use the relationship $m_1 = -\dfrac{1}{m_2}$.

105. The scale on the y-axis is unknown, so the slopes of the lines cannot be determined.

107. No. The slopes of two perpendicular lines have opposite signs (assume that neither line is vertical or horizontal).

109. The line $y = 4x$ rises most quickly, and the line $y = -4x$ falls most quickly. The greater the magnitude of the slope (the absolute value of the slope), the faster the line rises or falls.

111. $3x - 2y - 1 = 0$ **113.** $80x + 12y + 139 = 0$

Section P.5 *(page 62)*

1. domain; range; function **3.** implied domain

5. Function **7.** Not a function

9. (a) Function
(b) Not a function, because the element 1 in A corresponds to two elements, -2 and 1, in B.
(c) Function
(d) Not a function, because not every element in A is matched with an element in B.

11. Not a function **13.** Function

15. Function **17.** Function

19. (a) -2 (b) -14 (c) $3x + 1$

21. (a) 15 (b) $4t^2 - 19t + 27$ (c) $4t^2 - 3t - 10$

23. (a) 1 (b) 2.5 (c) $3 - 2|x|$

25. (a) $-\dfrac{1}{9}$ (b) Undefined (c) $\dfrac{1}{y^2 + 6y}$

27. (a) 1 (b) -1 (c) $\dfrac{|x - 1|}{x - 1}$

29. (a) -1 (b) 2 (c) 6

31.

x	-2	-1	0	1	2
$f(x)$	1	4	5	4	1

33.

x	-2	-1	0	1	2
$f(x)$	5	$\frac{9}{2}$	4	1	0

35. 5 **37.** $\frac{4}{3}$ **39.** ± 9 **41.** $0, \pm 1$ **43.** $-1, 2$

45. $0, \pm 2$ **47.** All real numbers x

49. All real numbers y such that $y \ge -6$

51. All real numbers x except $x = 0, -2$

53. All real numbers s such that $s \ge 1$ except $s = 4$

55. All real numbers x such that $x > 0$

57. (a) The maximum volume is 1024 cubic centimeters.

(b) Yes, V is a function of x.

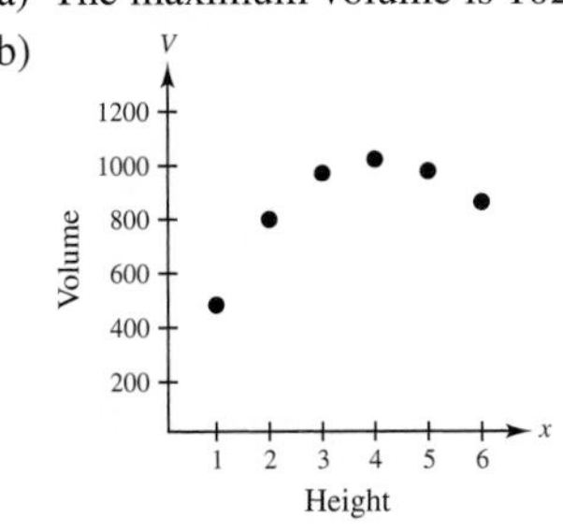

(c) $V = x(24 - 2x)^2, \quad 0 < x < 12$

59. $A = \dfrac{P^2}{16}$ **61.** No, the ball will be at a height of 18.5 feet.

63. $A = \dfrac{x^2}{2(x-2)}, \quad x > 2$

65. 2008: 67.36%
2009: 70.13%
2010: 72.90%
2011: 75.67%
2012: 79.30%
2013: 81.25%
2014: 83.20%

67. (a) $C = 12.30x + 98{,}000$ (b) $R = 17.98x$
(c) $P = 5.68x - 98{,}000$

69. (a)

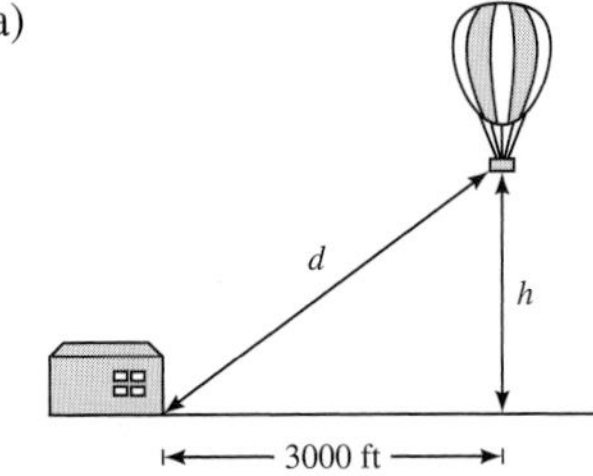

(b) $h = \sqrt{d^2 - 3000^2}, \quad d \geq 3000$

71. (a) $R = \dfrac{240n - n^2}{20}, \quad n \geq 80$

(b)

n	90	100	110	120	130	140	150
$R(n)$	\$675	\$700	\$715	\$720	\$715	\$700	\$675

The revenue is maximum when 120 people take the trip.

73. $2 + h, \quad h \neq 0$ **75.** $3x^2 + 3xh + h^2 + 3, \quad h \neq 0$

77. $-\dfrac{x+3}{9x^2}, \quad x \neq 3$ **79.** $\dfrac{\sqrt{5x} - 5}{x - 5}$

81. $g(x) = cx^2; c = -2$ **83.** $r(x) = \dfrac{c}{x}; c = 32$

85. False. A function is a special type of relation.

87. False. The range is $[-1, \infty)$.

89. The domain of $f(x)$ includes $x = 1$ and the domain of $g(x)$ does not because you cannot divide by 0. So, the functions do not have the same domain.

91. No; x is the independent variable, f is the name of the function.

93. (a) Yes. The amount you pay in sales tax will increase as the price of the item purchased increases.
(b) No. The length of time that you study will not necessarily determine how well you do on an exam.

Section P.6 *(page 74)*

1. Vertical Line Test **3.** decreasing

5. average rate of change; secant

7. Domain: $(-2, 2]$; range: $[-1, 8]$
(a) -1 (b) 0 (c) -1 (d) 8

9. Domain: $(-\infty, \infty)$; range: $(-2, \infty)$
(a) 0 (b) 1 (c) 2 (d) 3

11. Function **13.** Not a function **15.** -6 **17.** $-\frac{5}{2}, 6$

19. -3 **21.** $0, \pm\sqrt{6}$ **23.** $\pm 3, 4$ **25.** $\frac{1}{2}$

27. (a)

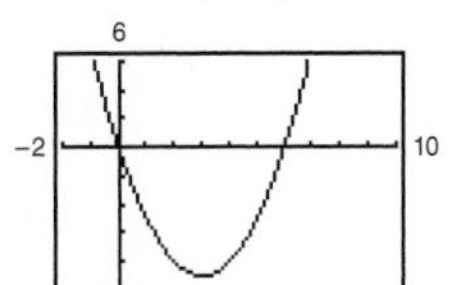

0, 6
(b) 0, 6

29. (a)
-5.5
(b) $-\frac{11}{2}$

31. (a)
0.3333
(b) $\frac{1}{3}$

33. Decreasing on $(-\infty, \infty)$

35. Increasing on $(1, \infty)$; Decreasing on $(-\infty, -1)$

37. Increasing on $(1, \infty)$; Decreasing on $(-\infty, -1)$
Constant on $(-1, 1)$

39. Increasing on $(-\infty, -1)$, $(0, \infty)$; Decreasing on $(-1, 0)$

41.

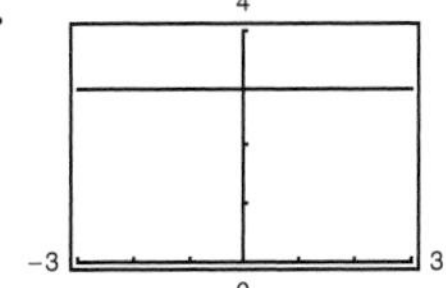

Constant on $(-\infty, \infty)$

43.
Decreasing on $(-\infty, 0)$
Increasing on $(0, \infty)$

45.

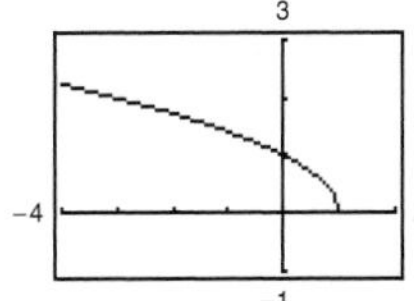

Decreasing on $(-\infty, 1)$

47.
Increasing on $(0, \infty)$

49. Relative minimum: $(-1.5, -2.25)$

51. Relative maximum: $(0, 15)$
Relative minimum: $(4, -17)$

53. Relative minimum: $(0.33, -0.38)$

55.

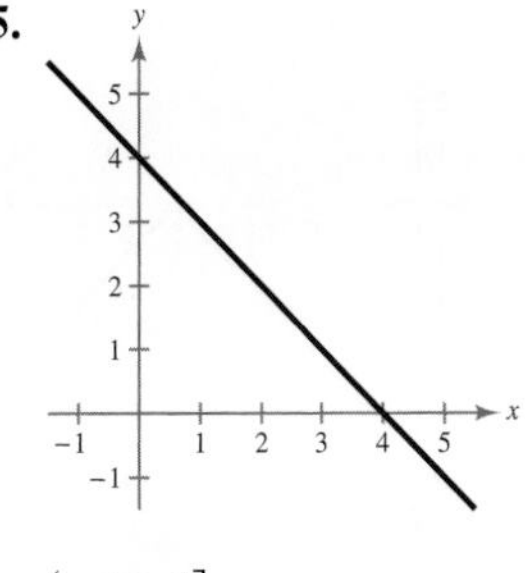

$(-\infty, 4]$

57.
$[-3, 3]$

59.

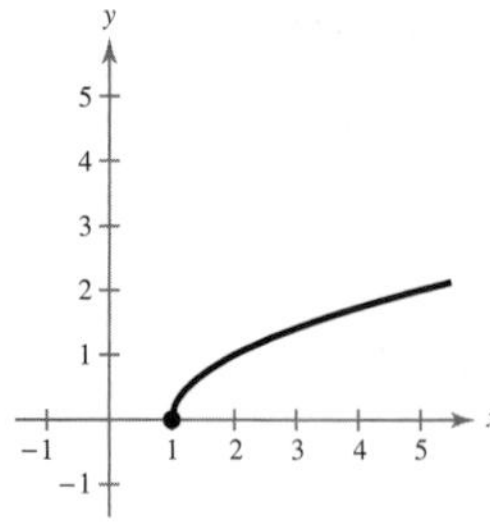

$[1, \infty)$

61. -2 **63.** -1

65. (a)

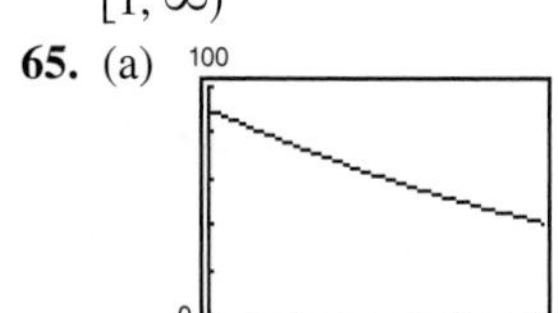

(b) About -6.14; The amount the U.S. federal government spent on research and development for defense decreased by about $6.14 billion each year from 2010 to 2014.

67. (a) $s = -16t^2 + 64t + 6$

(b)

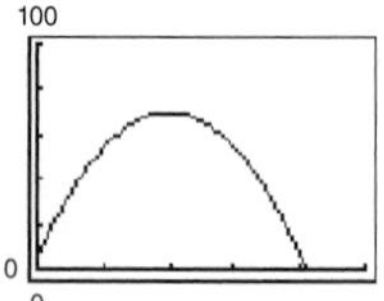

(c) 16 ft/sec

(d) The slope of the secant line is positive. (e) $y = 16t + 6$

(f)

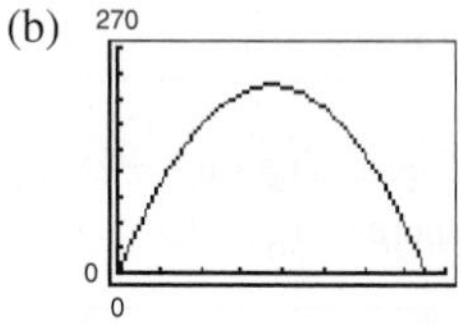

69. (a) $s = -16t^2 + 120t$

(b)

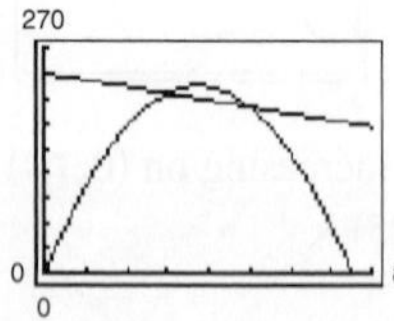

(c) -8 ft/sec

(d) The slope of the secant line is negative.

(e) $y = -8t + 240$

(f) 270

0 8

0

71. Even; y-axis symmetry **73.** Neither; no symmetry

75. Neither; no symmetry

77.

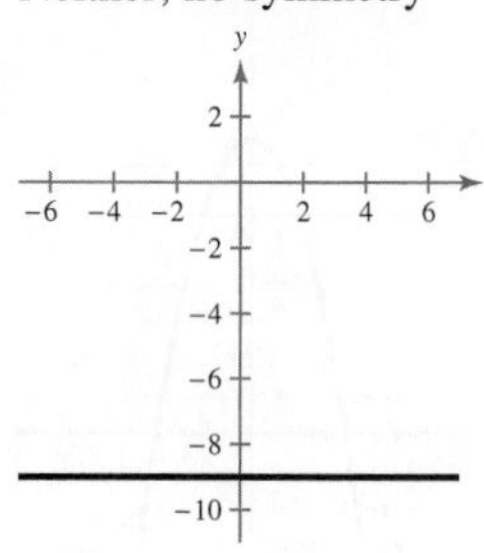

Even

79.

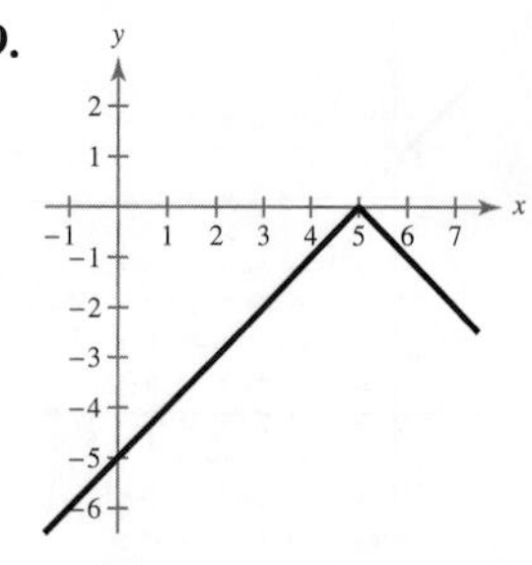

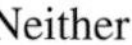

Neither

81.

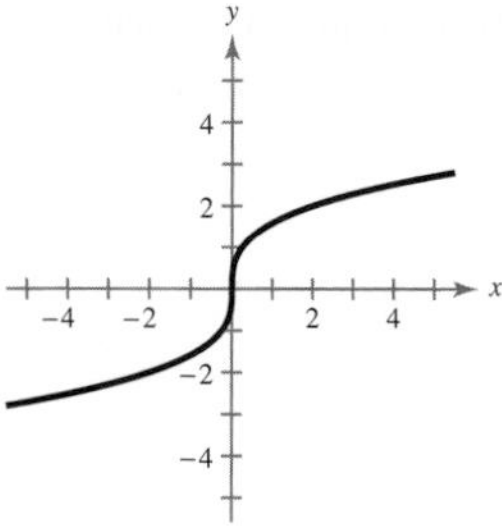

Odd

83. $h = 3 - 4x + x^2$ **85.** $L = 2 - \sqrt[3]{2y}$

87. The negative symbol should be divided out of each term, which yields $f(-x) = -(2x^3 + 5)$. So, the function is neither even nor odd.

89. (a) Ten thousands (b) Ten millions (c) Tens (d) Ones

91. False. The function $f(x) = \sqrt{x^2 + 1}$ has a domain of all real numbers.

93. True. A graph that is symmetric with respect to the y-axis cannot be increasing on its entire domain.

95. (a) $\left(\frac{5}{3}, -7\right)$ (b) $\left(\frac{5}{3}, 7\right)$

97. (a)

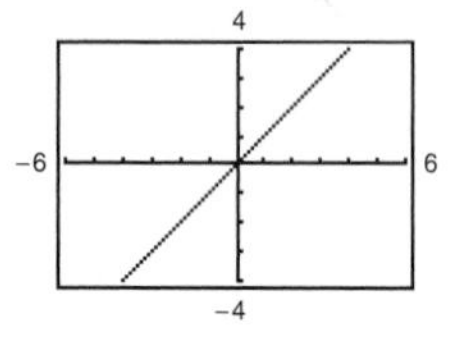

(b)

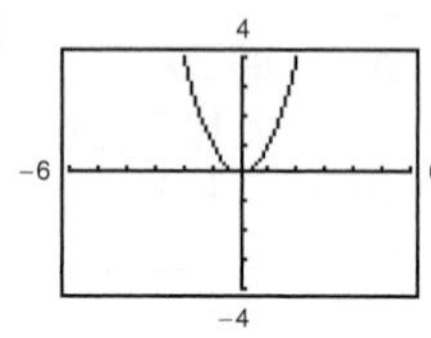

(c)

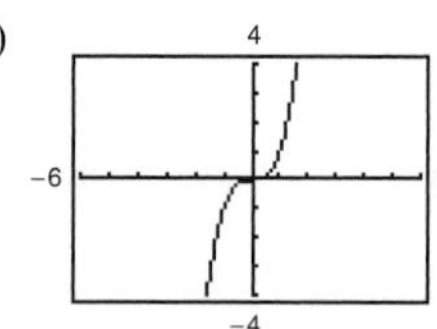

(d)

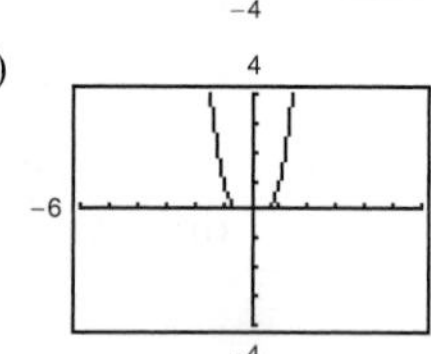

(e)

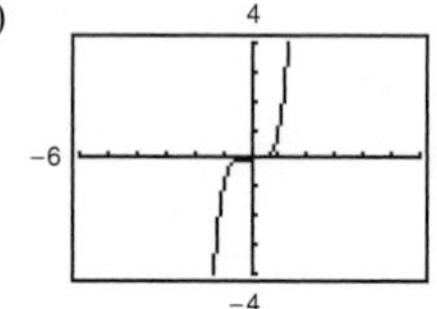

(f)

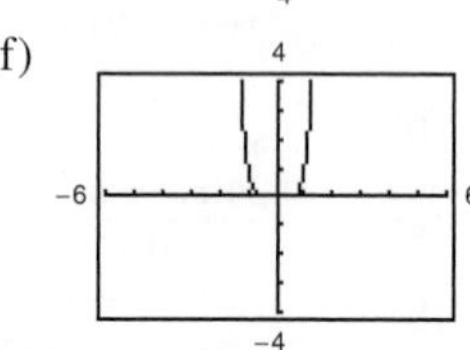

All the graphs pass through the origin. The graphs of the odd powers of x are symmetric with respect to the origin, and the graphs of the even powers are symmetric with respect to the y-axis. As the powers increase, the graphs become flatter in the interval $-1 < x < 1$.

99. (a) Even. The graph is a reflection in the x-axis.
(b) Even. The graph is a reflection in the y-axis.
(c) Even. The graph is a downward shift of f.
(d) Neither. The graph is a right shift of f.

Section P.7 *(page 83)*

1. Greatest integer function **3.** Reciprocal function
5. Square root function **7.** Absolute value function
9. Linear function

11. (a) $f(x) = -2x + 6$
(b)

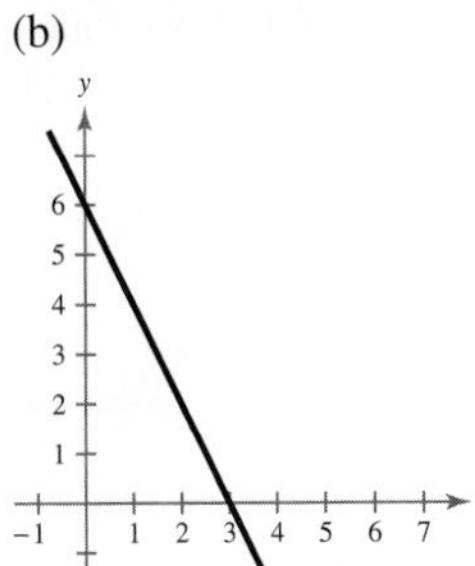

13. (a) $f(x) = \frac{2}{3}x - 2$
(b)

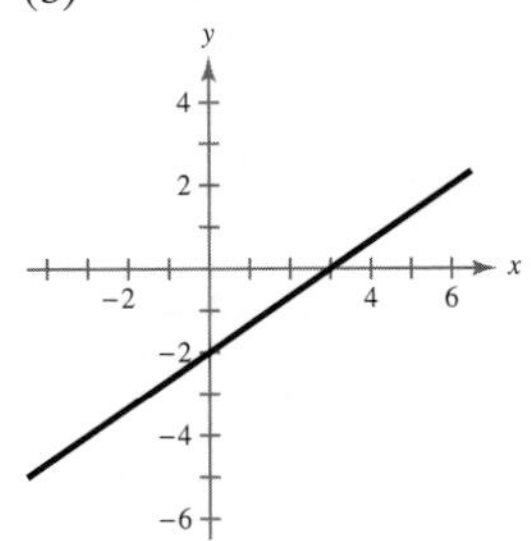

15.

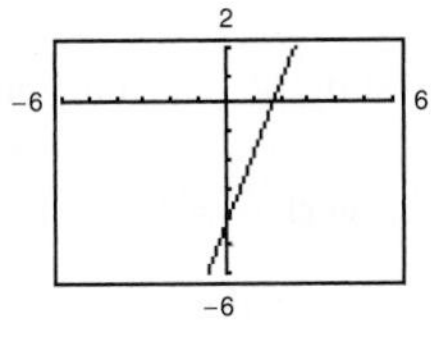

17.

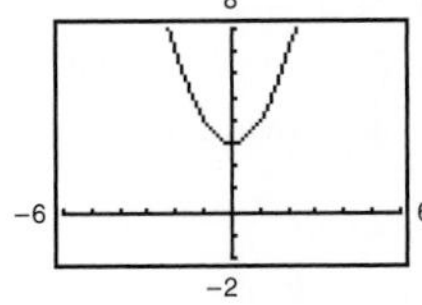

19.

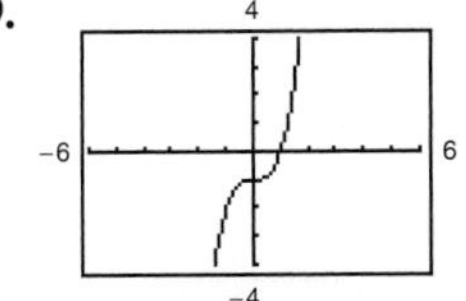

21.

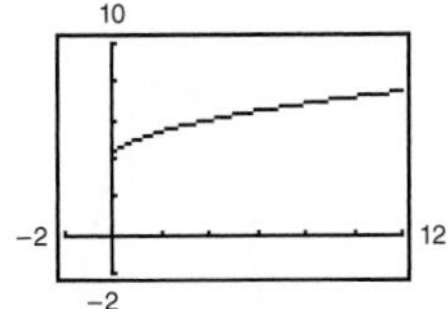

23.

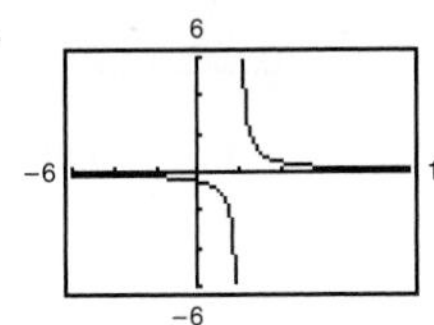

25.

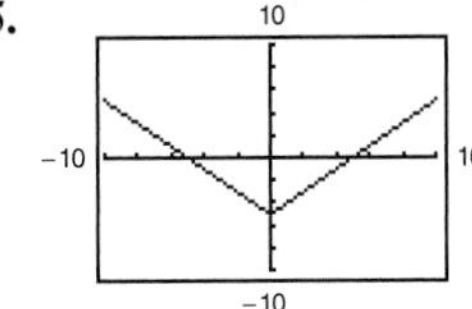

27. (a) 2 (b) 2 (c) -4 (d) 3

29. (a) 1 (b) -4 (c) 3 (d) 2

31.

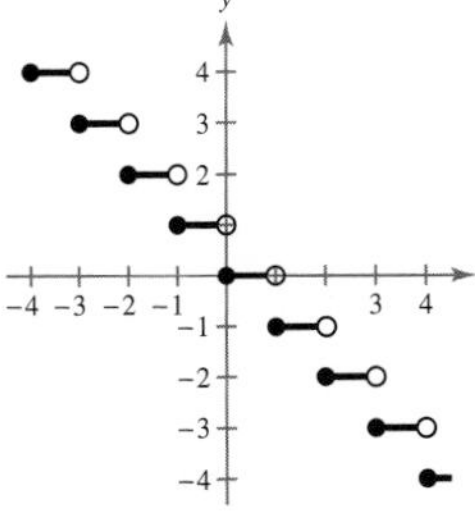

33.

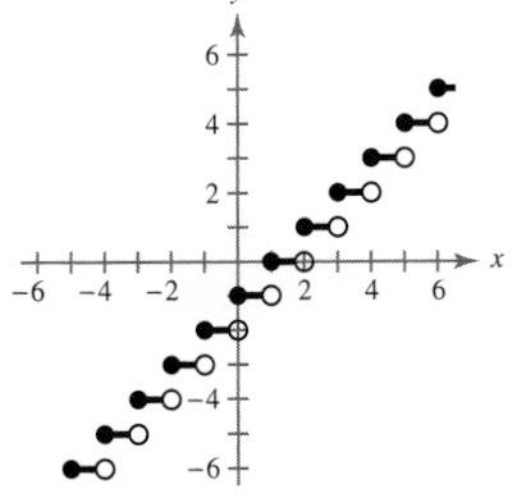

35.

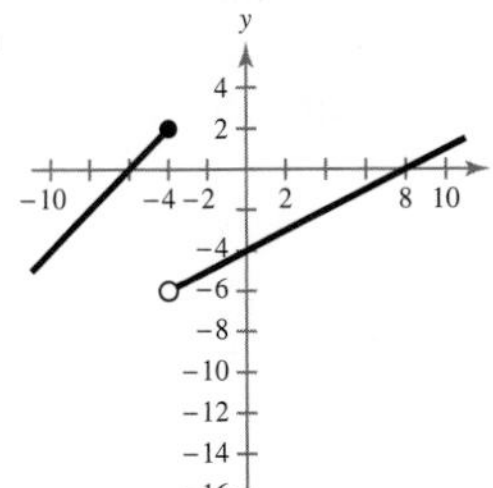

37.

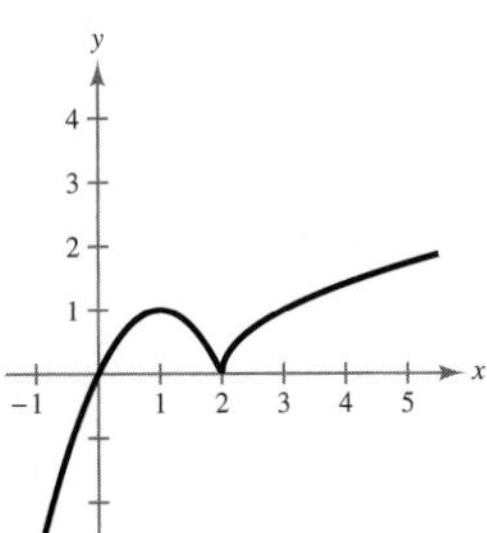

39.

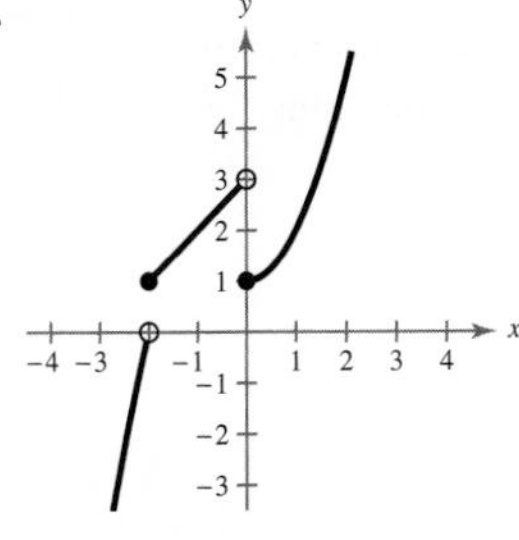

41. (a)

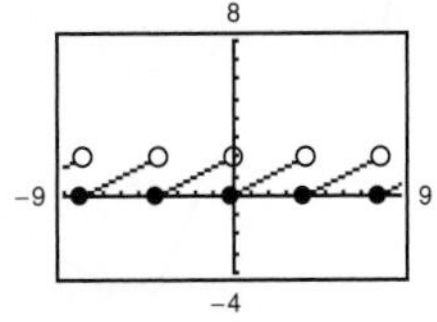

(b) Domain: $(-\infty, \infty)$
Range: $[0, 2)$

43. (a) $W(30) = 420$; $W(40) = 560$; $W(45) = 665$; $W(50) = 770$

(b) $W(h) = \begin{cases} 14h, & 0 < h \le 36 \\ 21(h - 36) + 504, & h > 36 \end{cases}$

(c) $W(h) = \begin{cases} 16h, & 0 < h \le 40 \\ 24(h - 40) + 640, & h > 40 \end{cases}$

45.

Interval	Input Pipe	Drain Pipe 1	Drain Pipe 2
[0, 5]	Open	Closed	Closed
[5, 10]	Open	Open	Closed
[10, 20]	Closed	Closed	Closed
[20, 30]	Closed	Closed	Open
[30, 40]	Open	Open	Open
[40, 45]	Open	Closed	Open
[45, 50]	Open	Open	Open
[50, 60]	Open	Open	Closed

47. $f(t) = \begin{cases} t, & 0 \le t \le 2 \\ 2t - 2, & 2 < t \le 8 \\ \frac{1}{2}t + 10, & 8 < t \le 9 \end{cases}$

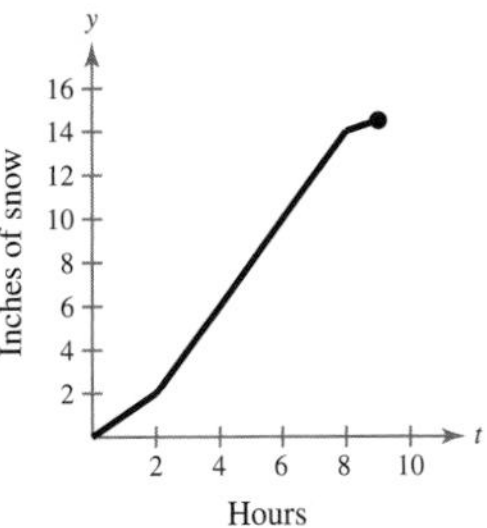

Total accumulation = 14.5 in.

49. False. A piecewise-defined function is a function that is defined by two or more equations over a specified domain. That domain may or may not include x- and y-intercepts.

Section P.8 *(page 90)*

1. rigid **3.** vertical stretch; vertical shrink

5. (a) (b)

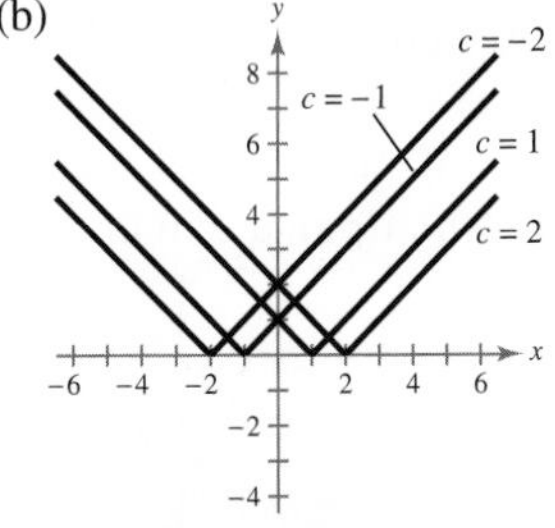

7. (a)

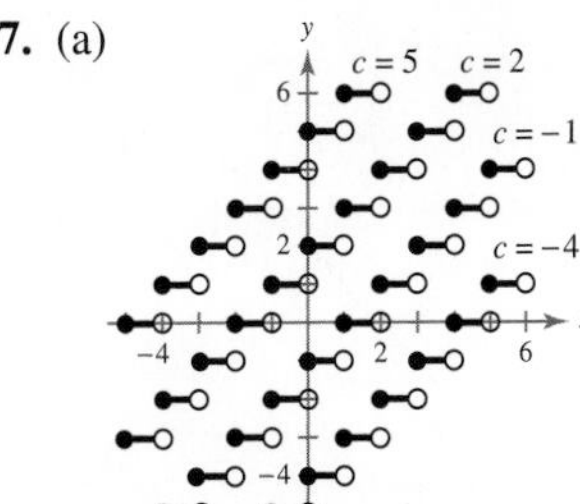

(b)

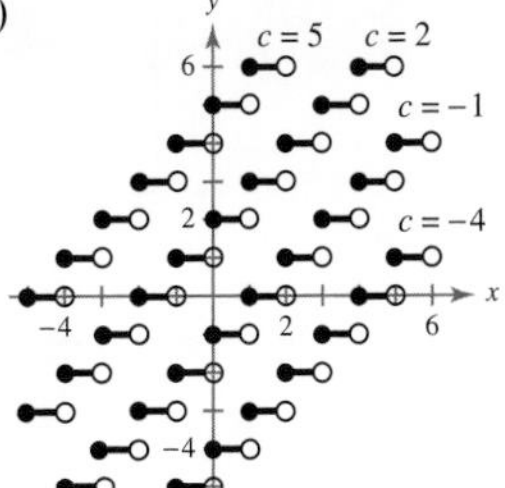

9. (a) (b)

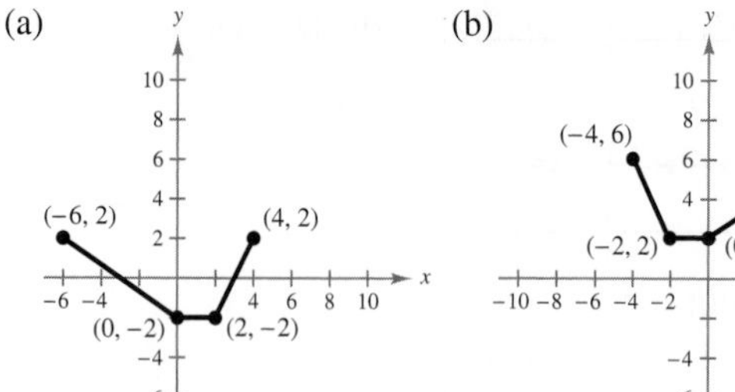

(c) (d)

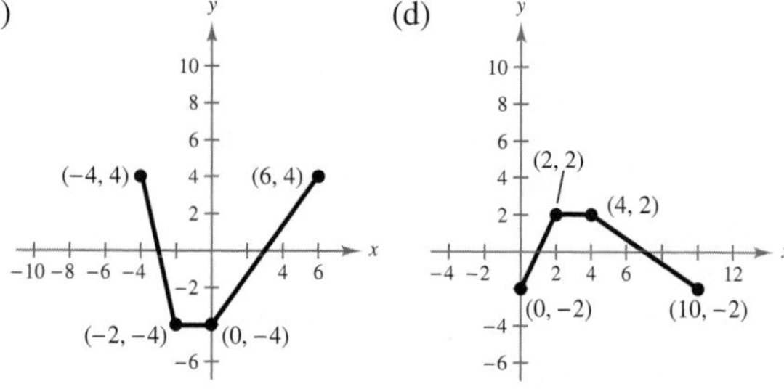

(e) (f)

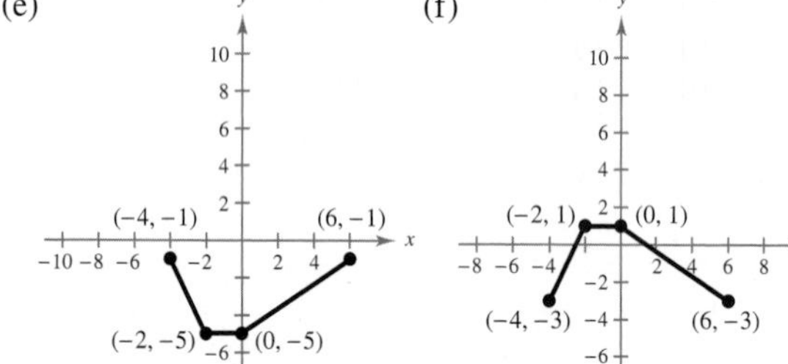

(g)

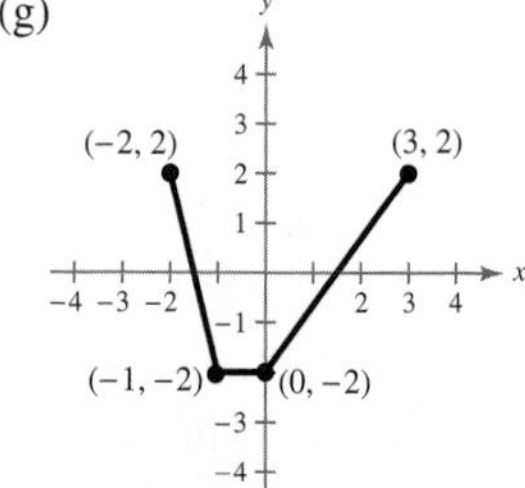

11. (a) $y = x^2 - 1$ (b) $y = -(x + 1)^2 + 1$

13. (a) $y = -|x + 3|$ (b) $y = |x - 2| - 4$

15. Right shift of $y = x^3$; $y = (x - 2)^3$

17. Reflection in the x-axis of $y = x^2$; $y = -x^2$

19. Reflection in the x-axis and upward shift of $y = \sqrt{x}$; $y = 1 - \sqrt{x}$

21. (a) $f(x) = x^2$

(b) Upward shift of six units

(c) (d) $g(x) = f(x) + 6$

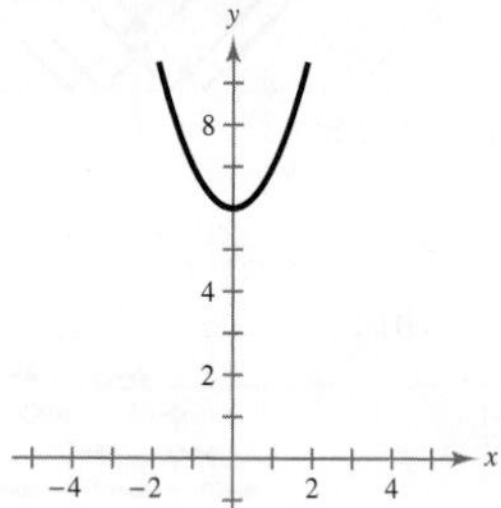

23. (a) $f(x) = x^3$

(b) Reflection in the x-axis and a right shift of two units

(c) (d) $g(x) = -f(x - 2)$

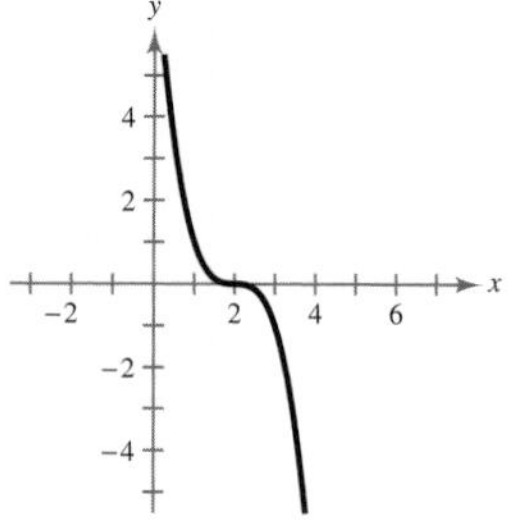

25. (a) $f(x) = x^2$

(b) Reflection in the x-axis, a left shift of one unit, and a downward shift of three units

(c) (d) $g(x) = -3 - f(x + 1)$

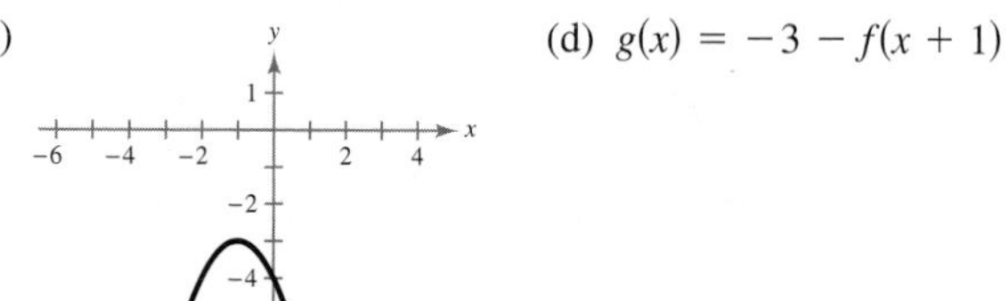

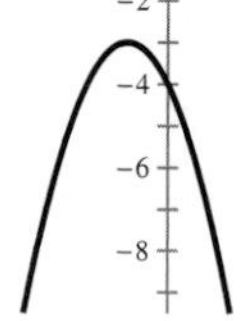

27. (a) $f(x) = |x|$

(b) Right shift of one unit and an upward shift of two units

(c) (d) $g(x) = f(x - 1) + 2$

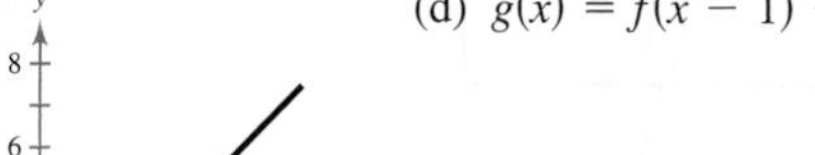

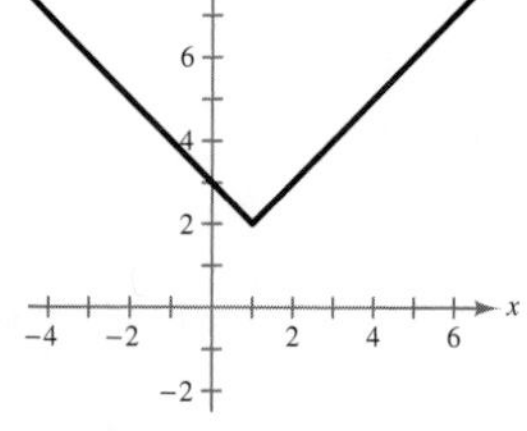

29. (a) $f(x) = \sqrt{x}$ (b) Vertical stretch

(c) (d) $g(x) = 2f(x)$

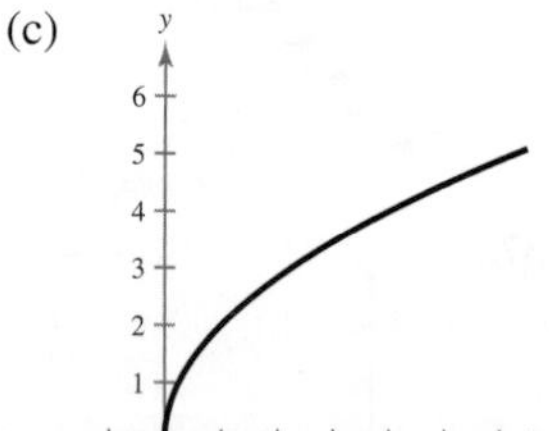

31. (a) $f(x) = [\![x]\!]$

(b) Vertical stretch and a downward shift of one unit

(c) (d) $g(x) = 2f(x) - 1$

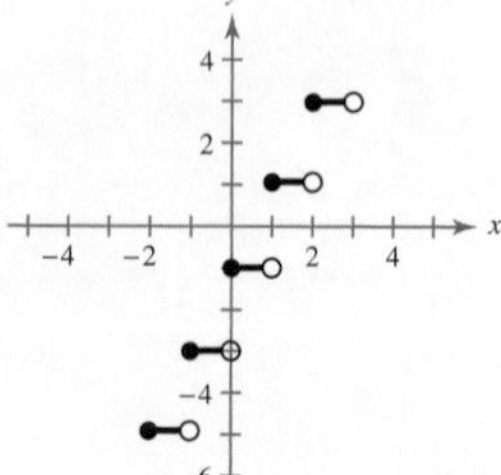

33. (a) $f(x) = |x|$ (b) Horizontal shrink
(c) (d) $g(x) = f(2x)$

35. (a) $f(x) = x^2$
(b) Reflection in the x-axis, a vertical stretch, and an upward shift of one unit
(c) (d) $g(x) = -2f(x) + 1$

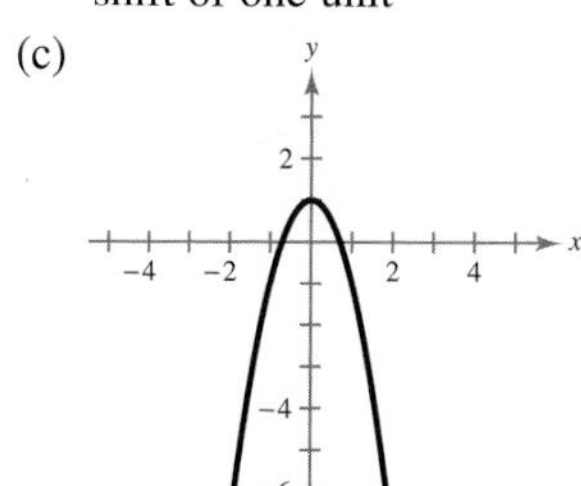

37. (a) $f(x) = |x|$
(b) Vertical stretch, a right shift of one unit, and an upward shift of two units
(c) (d) $g(x) = 3f(x - 1) + 2$

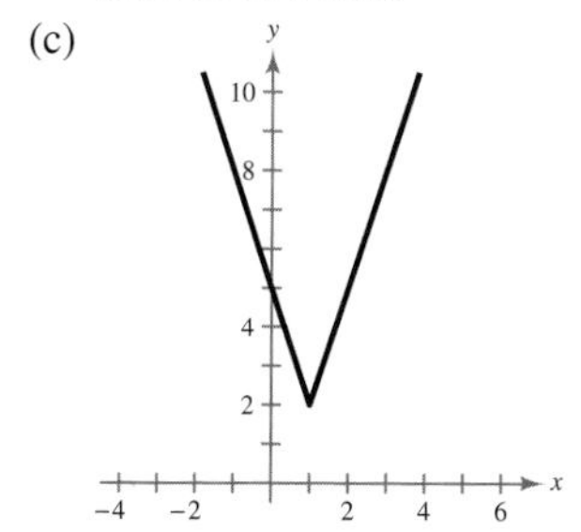

39. $g(x) = (x - 3)^2 - 7$ **41.** $g(x) = (x - 13)^3$
43. $g(x) = -|x| - 12$ **45.** $g(x) = -\sqrt{-x + 6}$
47. (a) $y = -3x^2$ (b) $y = 4x^2 + 3$
49. (a) $y = -\frac{1}{2}|x|$ (b) $y = 3|x| - 3$
51. Vertical stretch of $y = x^3$; $y = 2x^3$
53. Reflection in the x-axis and vertical shrink of $y = x^2$; $y = -\frac{1}{2}x^2$
55. Reflection in the y-axis and vertical shrink of $y = \sqrt{x}$; $y = \frac{1}{2}\sqrt{-x}$
57. $y = -(x - 2)^3 + 2$ **59.** $y = -\sqrt{x} - 3$
61. (a) (b) $H\left(\dfrac{x}{1.6}\right) = 0.00001132x^3$; Horizontal stretch
63. False. The graph of $y = f(-x)$ is a reflection of the graph of $f(x)$ in the y-axis.
65. True. $|-x| = |x|$ **67.** $(-2, 0), (-1, 1), (0, 2)$
69. The equation should be $g(x) = (x - 1)^3$.
71. (a) $g(t) = \frac{3}{4}f(t)$ (b) $g(t) = f(t) + 10{,}000$
(c) $g(t) = f(t - 2)$

Section P.9 *(page 99)*

1. addition; subtraction; multiplication; division
3.

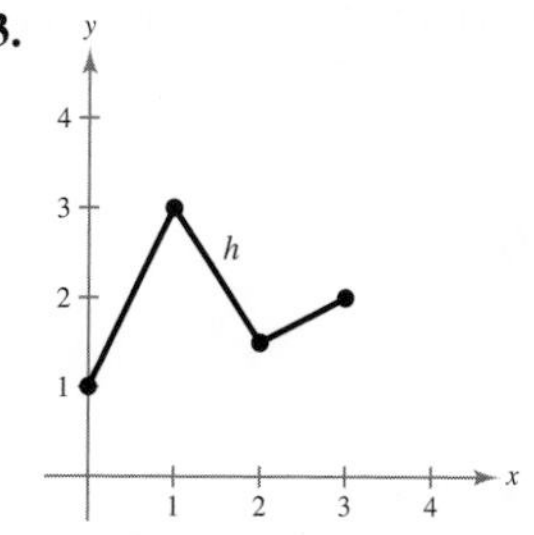

5. (a) $2x$ (b) 4 (c) $x^2 - 4$
(d) $\dfrac{x + 2}{x - 2}$; all real numbers x except $x = 2$
7. (a) $x^2 + 4x - 5$ (b) $x^2 - 4x + 5$ (c) $4x^3 - 5x^2$
(d) $\dfrac{x^2}{4x - 5}$; all real numbers x except $x = \dfrac{5}{4}$
9. (a) $x^2 + 6 + \sqrt{1 - x}$ (b) $x^2 + 6 - \sqrt{1 - x}$
(c) $(x^2 + 6)\sqrt{1 - x}$
(d) $\dfrac{(x^2 + 6)\sqrt{1 - x}}{1 - x}$; all real numbers x such that $x < 1$
11. (a) $\dfrac{x^4 + x^3 + x}{x + 1}$ (b) $\dfrac{-x^4 - x^3 + x}{x + 1}$ (c) $\dfrac{x^4}{x + 1}$
(d) $\dfrac{1}{x^2(x + 1)}$; all real numbers x except $x = 0, -1$
13. 7 **15.** 5 **17.** $-9t^2 + 3t + 5$ **19.** 306
21. $\frac{8}{23}$ **23.** -9
25. **27.**

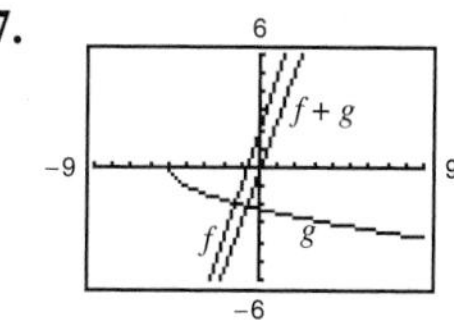

$f(x), g(x)$ $f(x), f(x)$

29. (a) $x + 5$ (b) $x + 5$ (c) $x - 6$
31. (a) $(x - 1)^2$ (b) $x^2 - 1$ (c) $x - 2$
33. (a) x (b) x (c) $x^9 + 3x^6 + 3x^3 + 2$
35. (a) $\sqrt{x^2 + 4}$ (b) $x + 4$
Domains of f and $g \circ f$: all real numbers x such that $x \geq -4$
Domains of g and $f \circ g$: all real numbers x
37. (a) x^2 (b) x^2
Domains of $f, g, f \circ g$, and $g \circ f$: all real numbers x
39. (a) $|x + 6|$ (b) $|x| + 6$
Domains of $f, g, f \circ g$, and $g \circ f$: all real numbers x
41. (a) $\dfrac{1}{x + 3}$ (b) $\dfrac{1}{x} + 3$
Domains of f and $g \circ f$: all real numbers x except $x = 0$
Domain of g: all real numbers x
Domain of $f \circ g$: all real numbers x except $x = -3$
43. (a) (b)

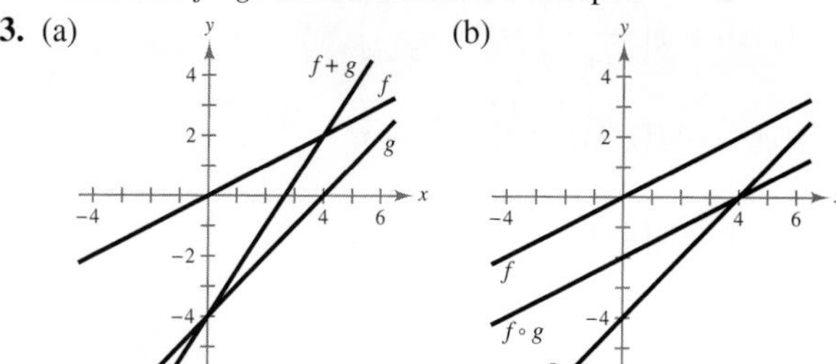

CHAPTER P

45. (a) 3 (b) 0 **47.** (a) 0 (b) 4

49. $f(x) = x^2, g(x) = 2x + 1$ **51.** $f(x) = \sqrt[3]{x}, g(x) = x^2 - 4$

53. $f(x) = \frac{1}{x}, g(x) = x + 2$ **55.** $f(x) = \frac{x+3}{4+x}, g(x) = -x^2$

57. (a) $T = \frac{3}{4}x + \frac{1}{15}x^2$

(b)

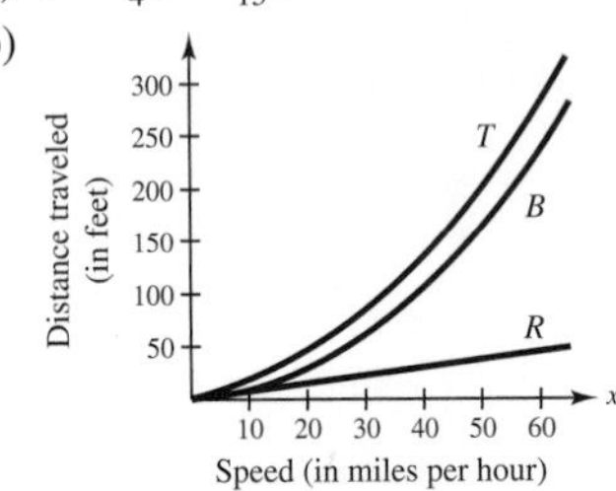

(c) The braking function $B(x)$; As x increases, $B(x)$ increases at a faster rate than $R(x)$.

59. (a) $c(t) = \frac{b(t) - d(t)}{p(t)} \times 100$

(b) $c(16)$ is the percent change in the population due to births and deaths in the year 2016.

61. (a) $r(x) = \frac{x}{2}$ (b) $A(r) = \pi r^2$

(c) $(A \circ r)(x) = \pi\left(\frac{x}{2}\right)^2$;

$(A \circ r)(x)$ represents the area of the circular base of the tank on the square foundation with side length x.

63. $g(f(x))$ represents 3 percent of an amount over \$500,000.

65. False. $(f \circ g)(x) = 6x + 1$ and $(g \circ f)(x) = 6x + 6$

67. (a) $O(M(Y)) = 2\left(6 + \frac{1}{2}Y\right) = 12 + Y$

(b) Middle child is 8 years old; youngest child is 4 years old.

69. Proof

71. (a) *Sample answer:* $f(x) = x + 1, g(x) = x + 3$

(b) *Sample answer:* $f(x) = x^2, g(x) = x^3$

73. (a) Proof

(b) $\frac{1}{2}[f(x) + f(-x)] + \frac{1}{2}[f(x) - f(-x)]$
$= \frac{1}{2}[f(x) + f(-x) + f(x) - f(-x)]$
$= \frac{1}{2}[2f(x)]$
$= f(x)$

(c) $f(x) = (x^2 + 1) + (-2x)$

$k(x) = \frac{-1}{(x+1)(x-1)} + \frac{x}{(x+1)(x-1)}$

Section P.10 *(page 108)*

1. inverse **3.** range; domain **5.** one-to-one

7. $f^{-1}(x) = \frac{1}{6}x$ **9.** $f^{-1}(x) = \frac{x-1}{3}$

11. $f^{-1}(x) = \sqrt{x+4}$ **13.** $f^{-1}(x) = \sqrt[3]{x-1}$

15. $f(g(x)) = f(4x + 9) = \frac{(4x+9) - 9}{4} = x$

$g(f(x)) = g\left(\frac{x-9}{4}\right) = 4\left(\frac{x-9}{4}\right) + 9 = x$

17. $f(g(x)) = f\left(\sqrt[3]{4x}\right) = \frac{\left(\sqrt[3]{4x}\right)^3}{4} = x$

$g(f(x)) = g\left(\frac{x^3}{4}\right) = \sqrt[3]{4\left(\frac{x^3}{4}\right)} = x$

19.

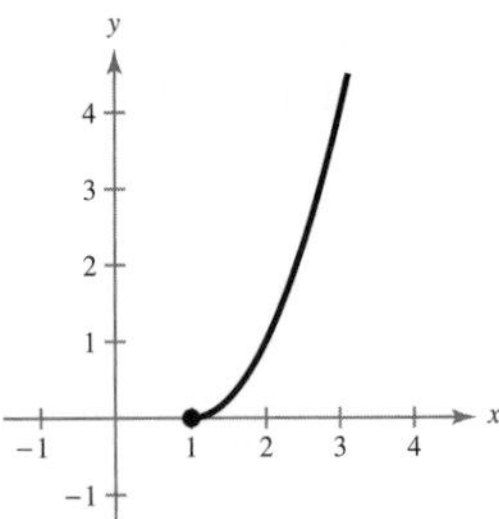

21. (a) $f(g(x)) = f(x + 5) = (x + 5) - 5 = x$
$g(f(x)) = g(x - 5) = (x - 5) + 5 = x$

(b)

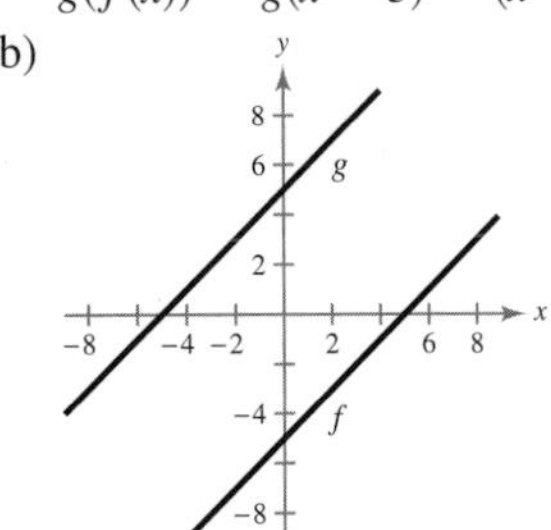

23. (a) $f(g(x)) = f\left(\frac{x-1}{7}\right) = 7\left(\frac{x-1}{7}\right) + 1 = x$

$g(f(x)) = g(7x + 1) = \frac{(7x+1) - 1}{7} = x$

(b)

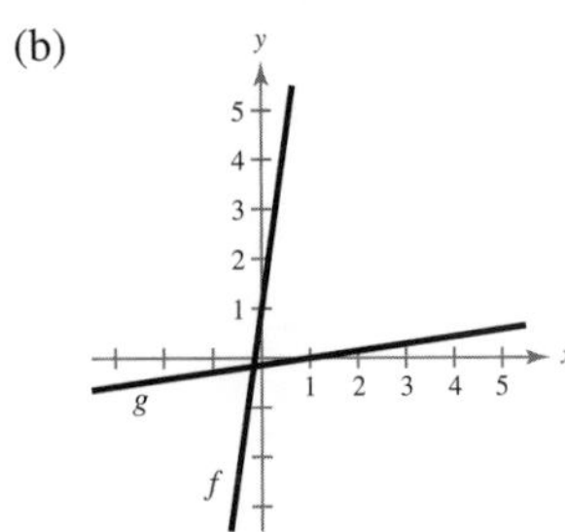

25. (a) $f(g(x)) = f\left(\sqrt[3]{x}\right) = \left(\sqrt[3]{x}\right)^3 = x$
$g(f(x)) = g(x^3) = \sqrt[3]{(x^3)} = x$

(b)

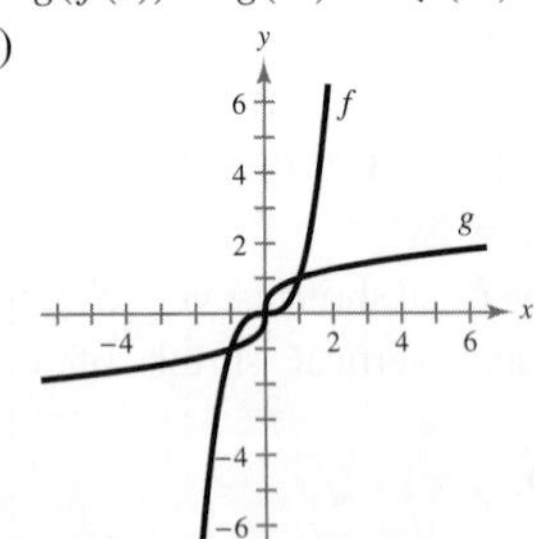

27. (a) $f(g(x)) = f(x^2 - 5) = \sqrt{(x^2 - 5) + 5} = x, x \geq 0$
$g(f(x)) = g\left(\sqrt{x+5}\right) = \left(\sqrt{x+5}\right)^2 - 5 = x$

(b)

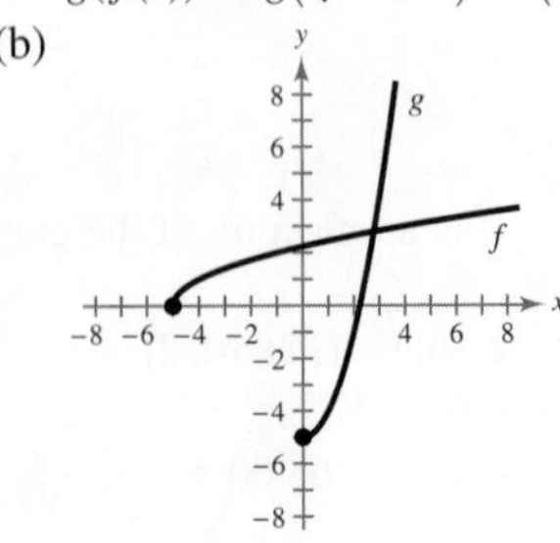

29. (a) $f(g(x)) = f\left(\frac{1}{x}\right) = \frac{1}{(1/x)} = x$

$g(f(x)) = g\left(\frac{1}{x}\right) = \frac{1}{(1/x)} = x$

(b)

31. (a) $f(g(x)) = f\left(-\frac{5x+1}{x-1}\right) = \frac{-\left(\frac{5x+1}{x-1}\right) - 1}{-\left(\frac{5x+1}{x-1}\right) + 5}$

$= \frac{-5x - 1 - x + 1}{-5x - 1 + 5x - 5} = x$

$g(f(x)) = g\left(\frac{x-1}{x+5}\right) = \frac{-5\left(\frac{x-1}{x+5}\right) - 1}{\frac{x-1}{x+5} - 1}$

$= \frac{-5x + 5 - x - 5}{x - 1 - x - 5} = x$

(b)

33. No **35.**

x	3	5	7	9	11	13
$f^{-1}(x)$	-1	0	1	2	3	4

37. Yes **39.** No

41. The function does not have an inverse function.

43. The function does not have an inverse function.

45. (a) $f^{-1}(x) = \sqrt[5]{x+2}$

(b)

(c) The graph of f^{-1} is the reflection of the graph of f in the line $y = x$.

(d) The domains and ranges of f and f^{-1} are all real numbers x.

47. (a) $f^{-1}(x) = \sqrt{4 - x^2}$, $0 \le x \le 2$

(b)

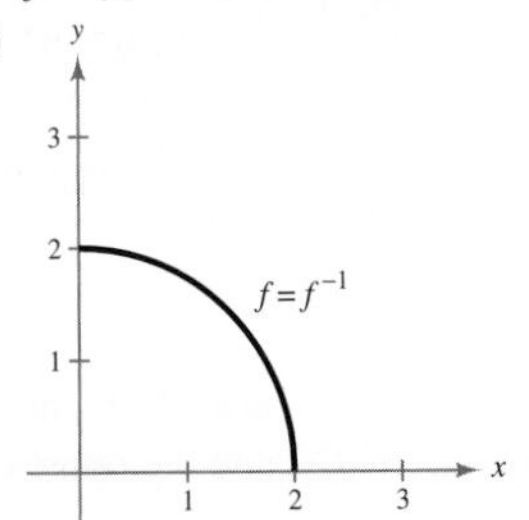

(c) The graph of f^{-1} is the same as the graph of f.

(d) The domains and ranges of f and f^{-1} are all real numbers x such that $0 \le x \le 2$.

49. (a) $f^{-1}(x) = \frac{4}{x}$

(b)

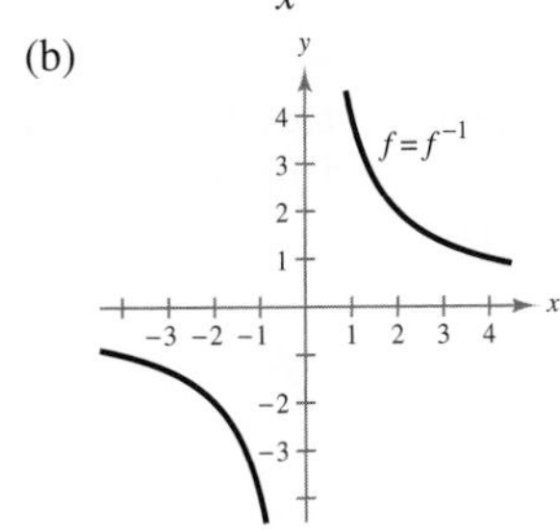

(c) The graph of f^{-1} is the same as the graph of f.

(d) The domains and ranges of f and f^{-1} are all real numbers x except $x = 0$.

51. (a) $f^{-1}(x) = \frac{2x+1}{x-1}$

(b)

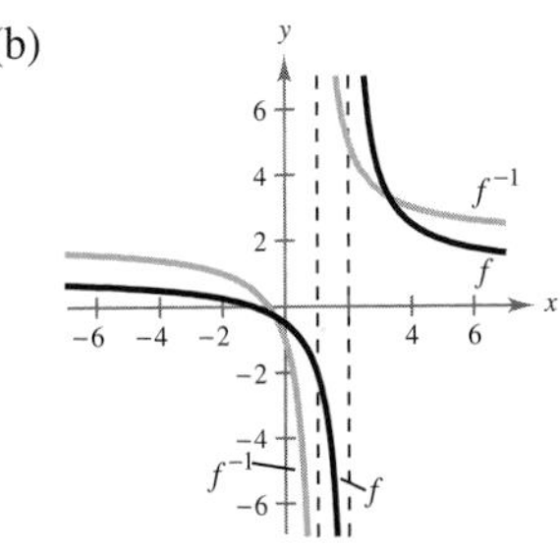

(c) The graph of f^{-1} is the reflection of the graph of f in the line $y = x$.

(d) The domain of f and the range of f^{-1} are all real numbers x except $x = 2$. The domain of f^{-1} and the range of f are all real numbers x except $x = 1$.

53. (a) $f^{-1}(x) = x^3 + 1$

(b)

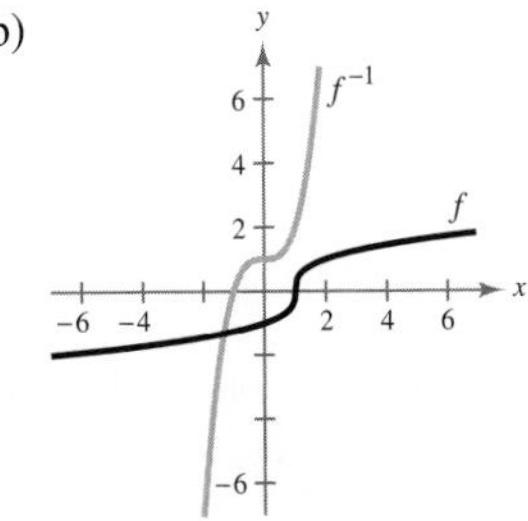

(c) The graph of f^{-1} is the reflection of the graph of f in the line $y = x$.

(d) The domains and ranges of f and f^{-1} are all real numbers x.

55. No inverse function **57.** $g^{-1}(x) = 6x - 1$

59. No inverse function **61.** $f^{-1}(x) = \sqrt{x} - 3$

63. No inverse function **65.** No inverse function

CHAPTER P

67. $f^{-1}(x) = \dfrac{x^2 - 3}{2},\quad x \geq 0$ **69.** $f^{-1}(x) = \dfrac{5x - 4}{6 - 4x}$

71. $f^{-1}(x) = x - 2$
The domain of f and the range of f^{-1} are all real numbers x such that $x \geq -2$. The domain of f^{-1} and the range of f are all real numbers x such that $x \geq 0$.

73. $f^{-1}(x) = \sqrt{x} - 6$
The domain of f and the range of f^{-1} are all real numbers x such that $x \geq -6$. The domain of f^{-1} and the range of f are all real numbers x such that $x \geq 0$.

75. $f^{-1}(x) = \dfrac{\sqrt{-2(x - 5)}}{2}$
The domain of f and the range of f^{-1} are all real numbers x such that $x \geq 0$. The domain of f^{-1} and the range of f are all real numbers x such that $x \leq 5$.

77. $f^{-1}(x) = x + 3$
The domain of f and the range of f^{-1} are all real numbers x such that $x \geq 4$. The domain of f^{-1} and the range of f are all real numbers x such that $x \geq 1$.

79. 32 **81.** 472 **83.** $2\sqrt[3]{x + 3}$

85. $\dfrac{x + 1}{2}$ **87.** $\dfrac{x + 1}{2}$

89. (a) $y = \dfrac{x - 10}{0.75}$
x = hourly wage; y = number of units produced
(b) 19 units

91. False. $f(x) = x^2$ has no inverse function.

93.

x	1	3	4	6
y	1	2	6	7

x	1	2	6	7
$f^{-1}(x)$	1	3	4	6

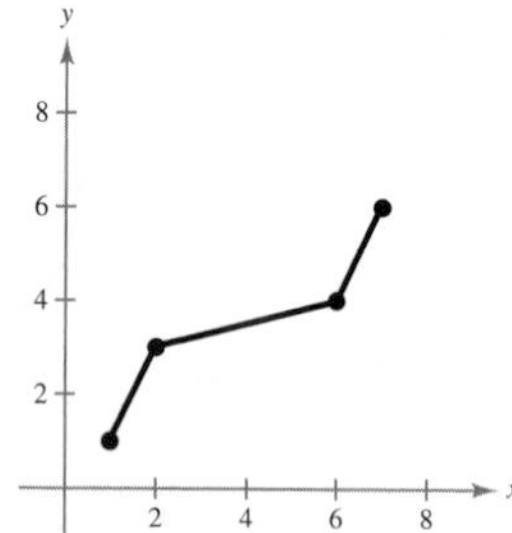

95. Proof **97.** $k = \frac{1}{4}$

99.

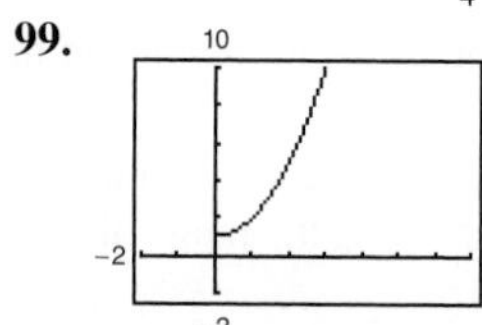

There is an inverse function $f^{-1}(x) = \sqrt{x - 1}$ because the domain of f is equal to the range of f^{-1} and the range of f is equal to the domain of f^{-1}.

101. This situation could be represented by a one-to-one function if the runner does not stop to rest. The inverse function would represent the time in hours for a given number of miles completed.

Review Exercises *(page 114)*

1. (a) 11 (b) 11 (c) 11 (d) 11, $-\frac{8}{9}$, $\frac{5}{2}$, 0.4 (e) $\sqrt{6}$

3. Number line: 0, $\frac{1}{8}$, $\frac{1}{4}$, $\frac{3}{8}$, $\frac{1}{2}$, $\frac{5}{8}$, $\frac{3}{4}$, $\frac{7}{8}$, 1, $\frac{9}{8}$, $\frac{5}{4}$

$\frac{5}{4} > \frac{7}{8}$

5. 122 **7.** $|x - 7| \geq 4$ **9.** (a) -1 (b) -3

11. Additive Identity Property

13. Associative Property of Addition

15. Commutative Property **17.** 0 **19.** 32

21. $\dfrac{x}{10}$ **23.** 6 **25.** -30 **27.** $-\dfrac{7}{2}, 4$ **29.** $-8, -18$

31. $0, \frac{12}{5}$ **33.** 5 **35.** $-124, 126$ **37.** $-5, 15$

39. (a)

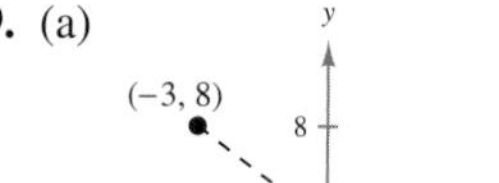

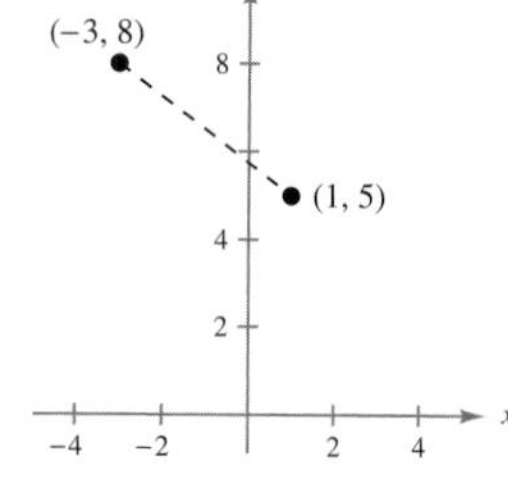

(b) 5 (c) $\left(-1, \frac{13}{2}\right)$

41.

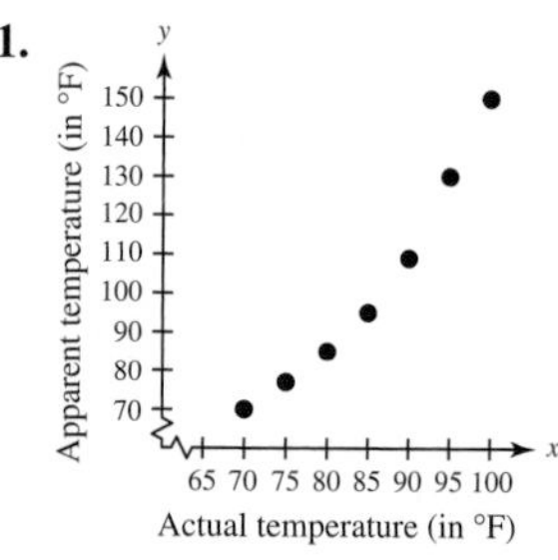

43.

x	1	2	3	4	5
y	-4	-2	0	2	4

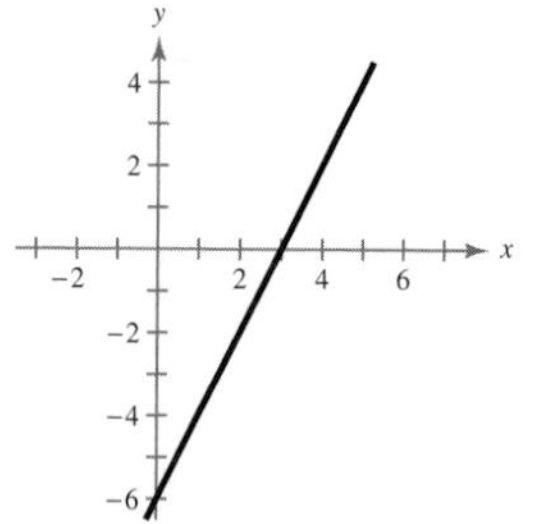

45.

x	-3	-2	-1	0	1
y	3	0	-1	0	3

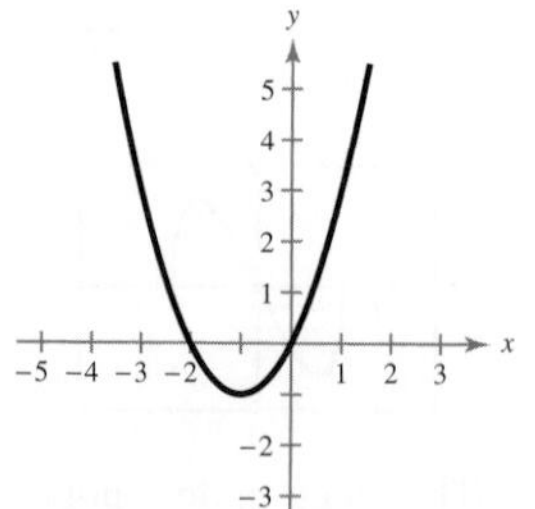

47. x-intercepts: $(1, 0), (5, 0)$
y-intercept: $(0, 5)$

49. No symmetry

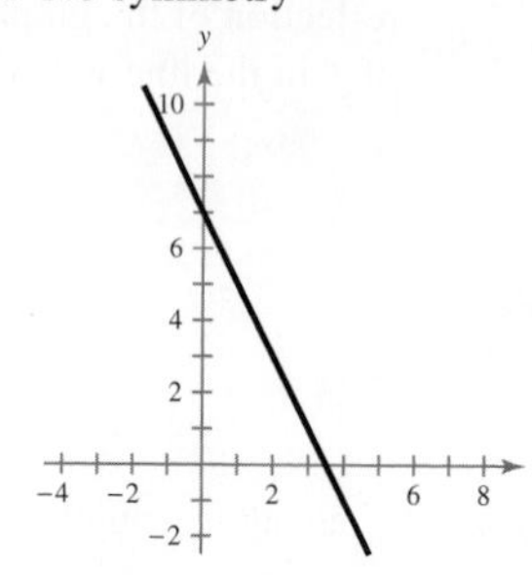

51. y-axis symmetry

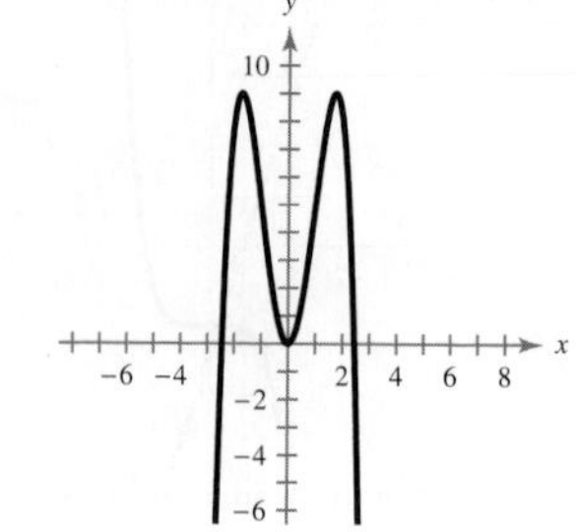

53. $(x - 2)^2 + (y + 3)^2 = 13$

55. Center: $(0, 0)$; Radius: 3

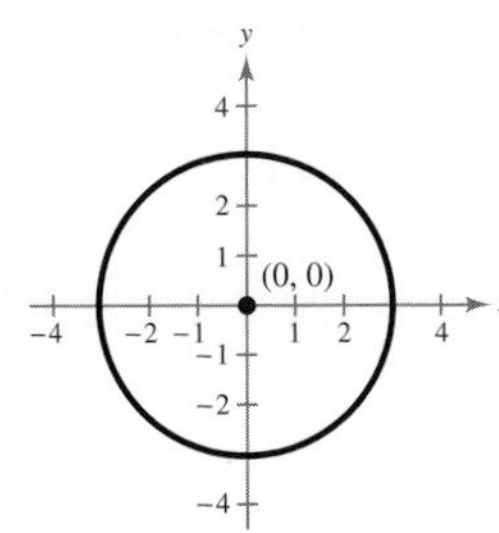

57. $m = -\frac{1}{2}$
y-intercept: $(0, 1)$

59. $m = 0$
y-intercept: $(0, 1)$

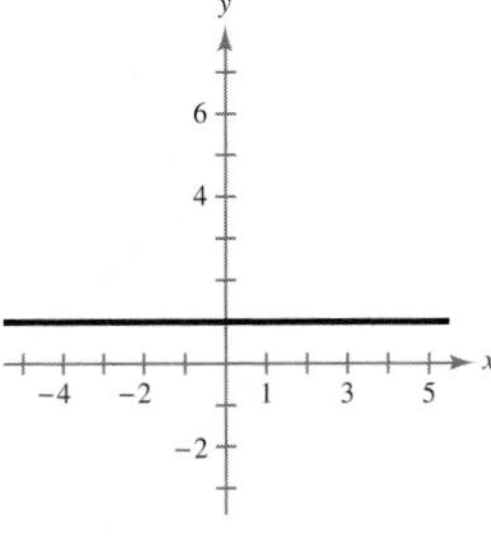

61. $m = -1$

63. $y = \frac{1}{3}x - 7$

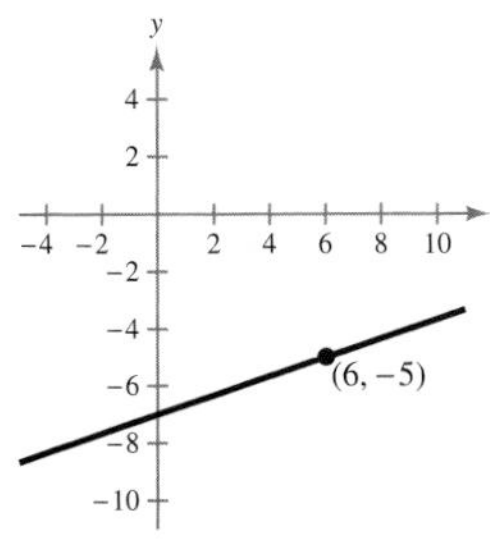

65. $y = \frac{1}{2}x + 7$

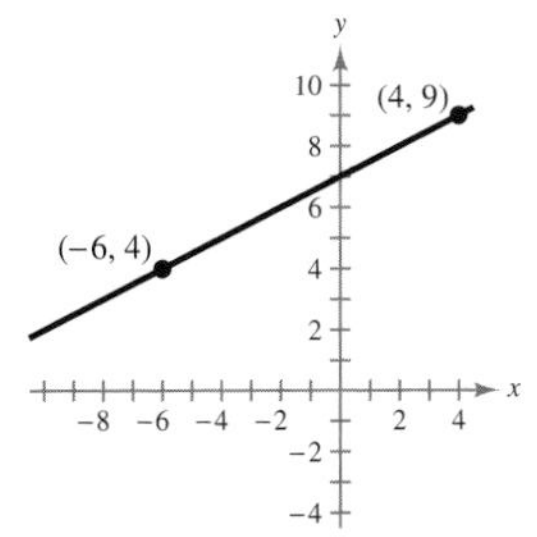

67. (a) $y = \frac{5}{4}x - \frac{23}{4}$ (b) $y = -\frac{4}{5}x + \frac{2}{5}$

69. $S = 0.80L$ **71.** Not a function **73.** Function

75. (a) 16 (b) $(t + 1)^{4/3}$ (c) 81 (d) $x^{4/3}$

77. All real numbers x such that $-5 \leq x \leq 5$

79. (a) 16 ft/sec (b) 1.5 sec **81.** Function **83.** $-\frac{1}{2}$

85.
Increasing on $(0, \infty)$
Decreasing on $(-\infty, -1)$
Constant on $(-1, 0)$

87. Relative maximum: $(1, 2)$ **89.** 4

91. Neither even nor odd; no symmetry

93. Odd; origin symmetry

95. (a) $f(x) = -3x$
(b)

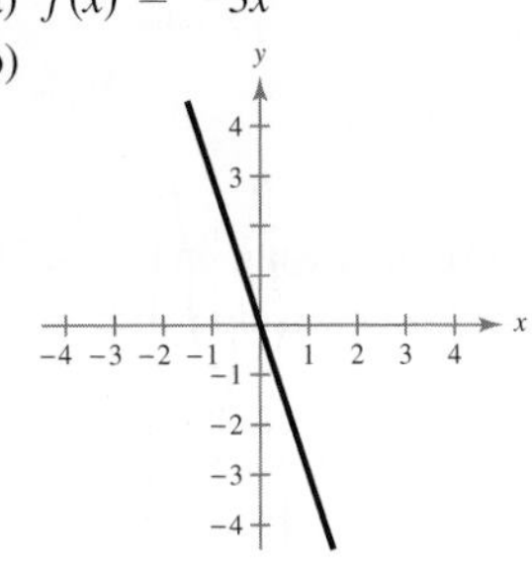

97.

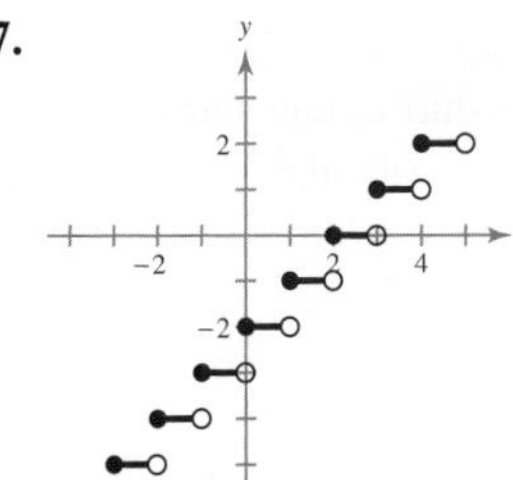

99.

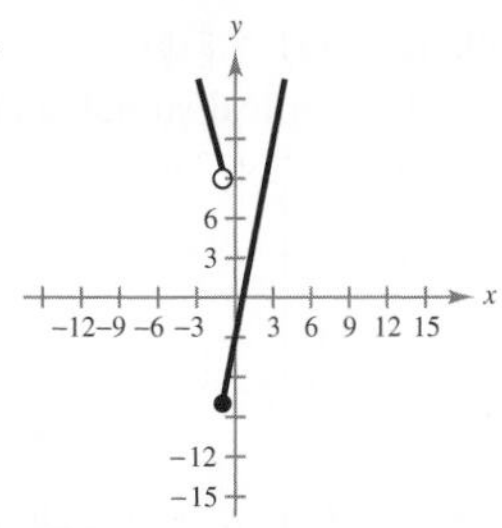

101. (a) $f(x) = x^2$ (b) Downward shift of nine units
(c)

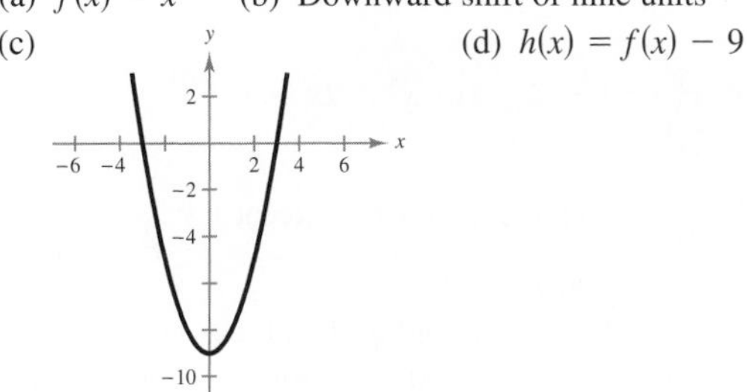

(d) $h(x) = f(x) - 9$

103. (a) $f(x) = \sqrt{x}$
(b) Reflection in the x-axis and an upward shift of four units
(c)

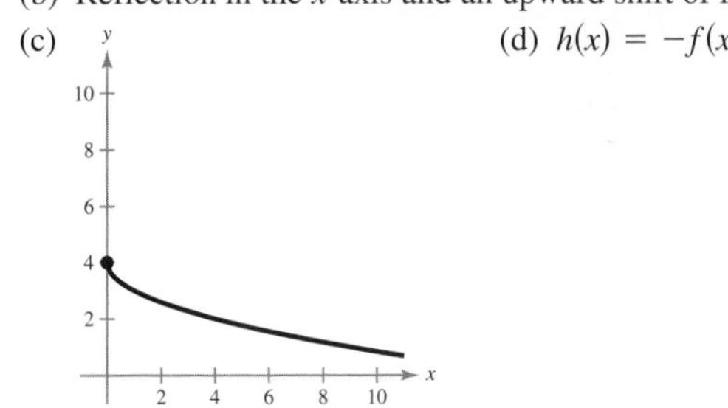

(d) $h(x) = -f(x) + 4$

105. (a) $f(x) = x^2$
(b) Reflection in the x-axis, a left shift of two units, and an upward shift of three units
(c)

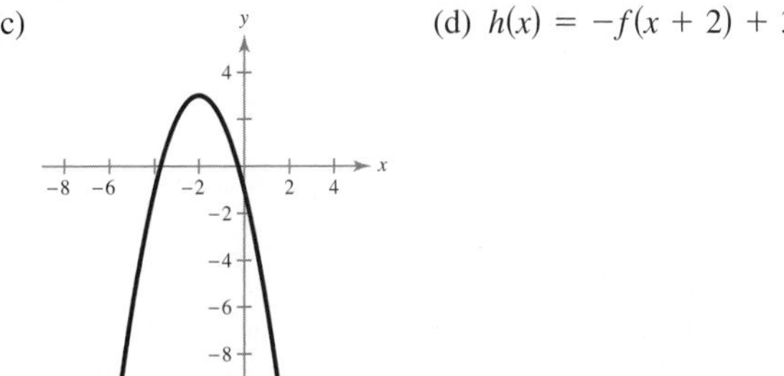

(d) $h(x) = -f(x + 2) + 3$

107. (a) $f(x) = [\![x]\!]$
(b) Reflection in the x-axis and an upward shift of six units
(c)

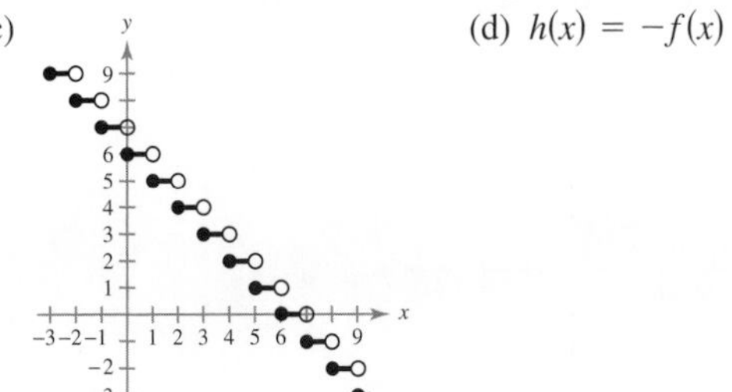

(d) $h(x) = -f(x) + 6$

CHAPTER P

109. (a) $f(x) = [\![x]\!]$
(b) Vertical stretch and a right shift of nine units
(c)

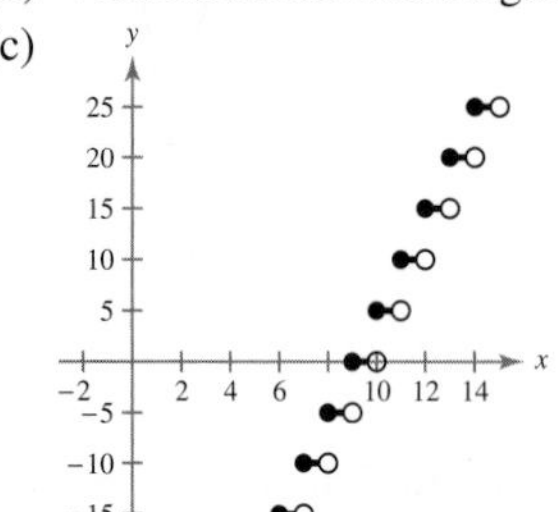

(d) $h(x) = 5f(x - 9)$

111. (a) $x^2 + 2x + 2$ (b) $x^2 - 2x + 4$
(c) $2x^3 - x^2 + 6x - 3$
(d) $\dfrac{x^2 + 3}{2x - 1}$; all real numbers x except $x = \dfrac{1}{2}$

113. (a) $x - \frac{8}{3}$ (b) $x - 8$
Domains of $f, g, f \circ g$, and $g \circ f$: all real numbers x

115. $(f \circ g)(x) = 0.95x - 100$; $(f \circ g)(x)$ represents the 5% discount before the \$100 rebate.

117. $f^{-1}(x) = 5x + 4$

$$f(f^{-1}(x)) = \frac{5x + 4 - 4}{5} = x$$

$$f^{-1}(f(x)) = 5\left(\frac{x - 4}{5}\right) + 4 = x$$

119.

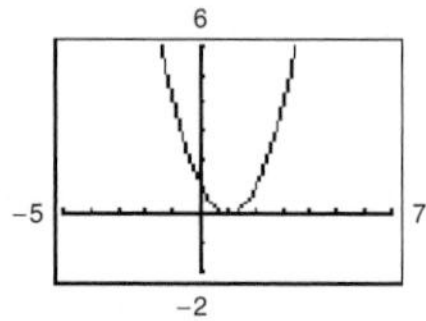

The function does not have an inverse function.

121. (a) $f^{-1}(x) = 2x + 6$
(b)

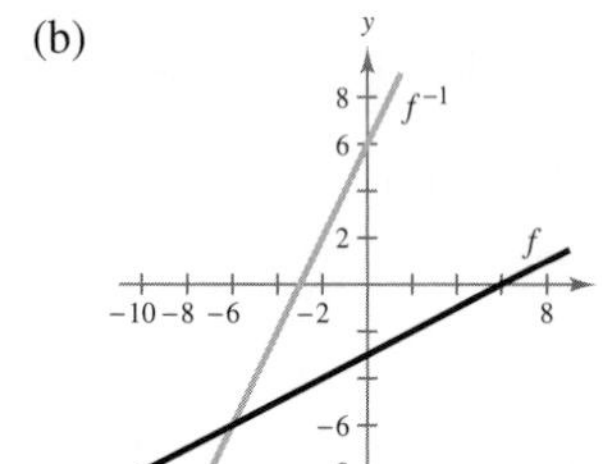

(c) The graphs are reflections of each other in the line $y = x$.
(d) Both f and f^{-1} have domains and ranges that are all real numbers x.

123. $x > 4$; $f^{-1}(x) = \sqrt{\dfrac{x}{2}} + 4, x \neq 0$

125. False. The graph is reflected in the x-axis, shifted 9 units to the left, and then shifted 13 units down.

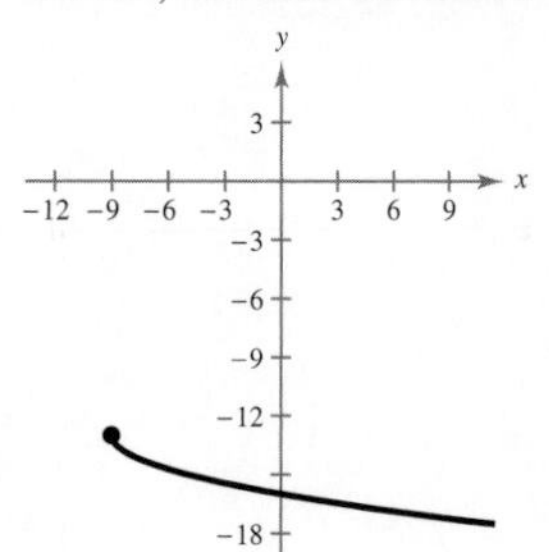

Chapter Test *(page 117)*

1. $-\frac{10}{3} < -\frac{5}{3}$ **2.** 3 **3.** Additive Identity Property
4. $\frac{128}{11}$ **5.** $-3, 5$ **6.** No solution **7.** $-2, \frac{8}{3}$

8.

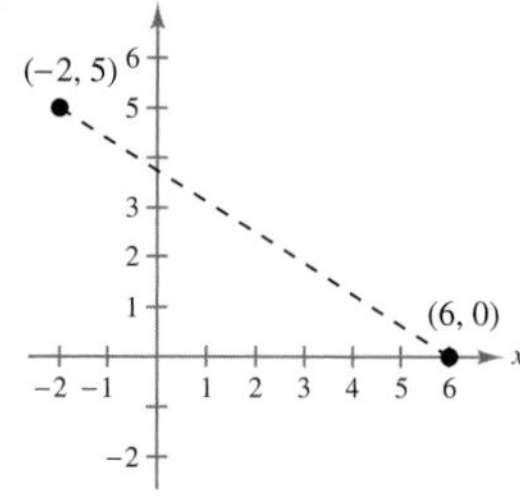

Distance: $\sqrt{89}$;
Midpoint: $\left(2, \frac{5}{2}\right)$

9. No symmetry

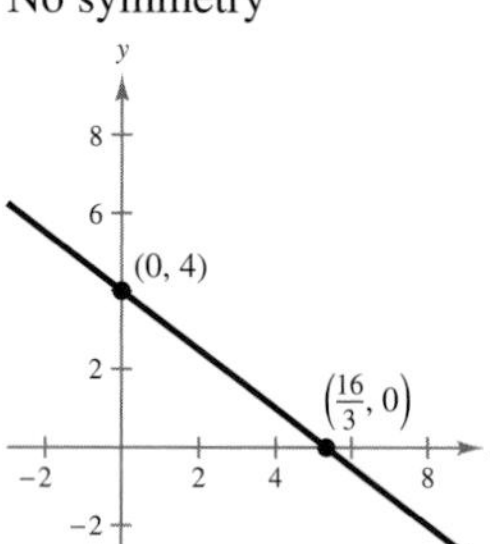

10. y-axis symmetry

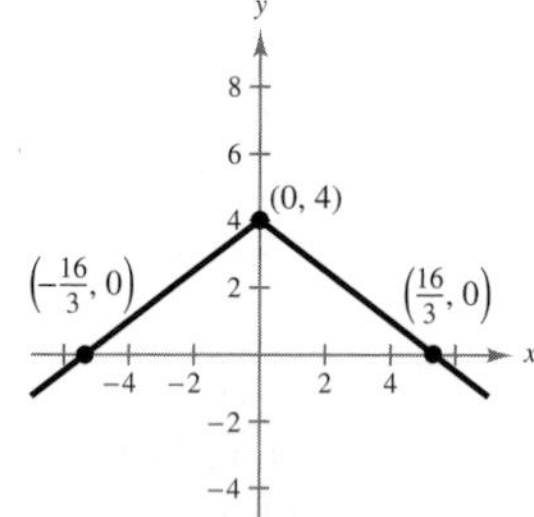

11. Origin symmetry

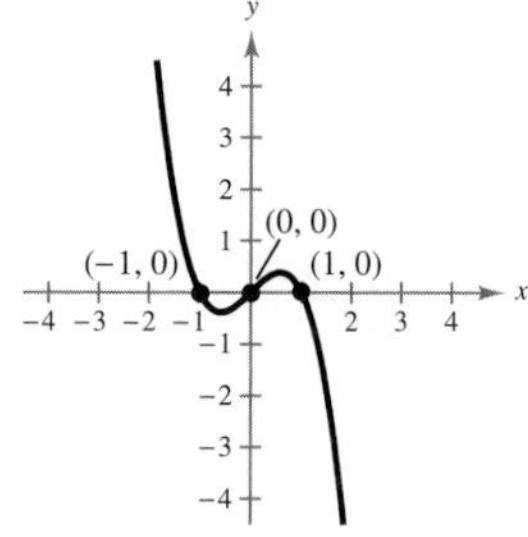

12. Center: $(3, 0)$; Radius: 3

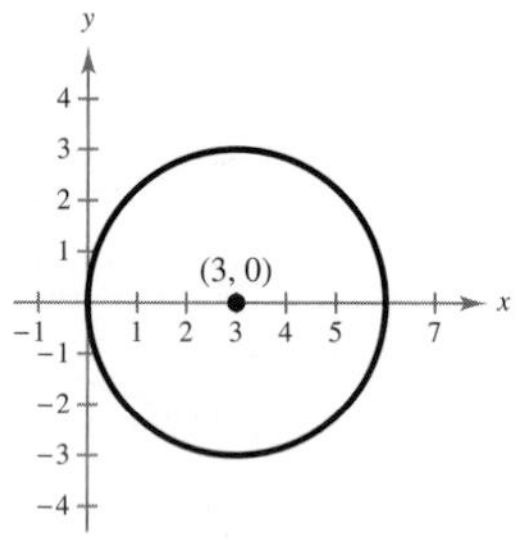

13. $y = -4x - 3$

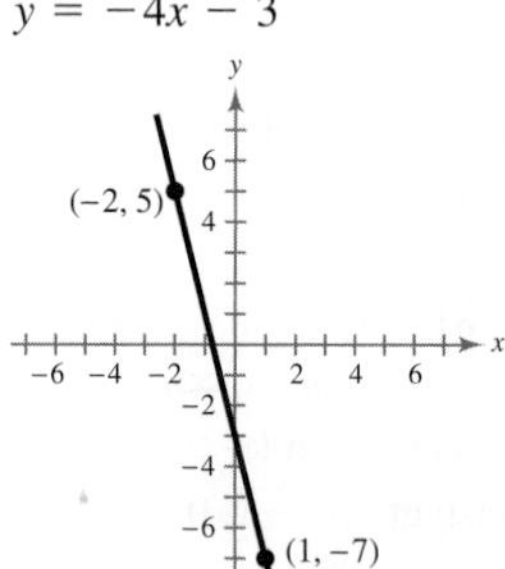

14. $y = \frac{5}{3}x - \frac{1}{3}$

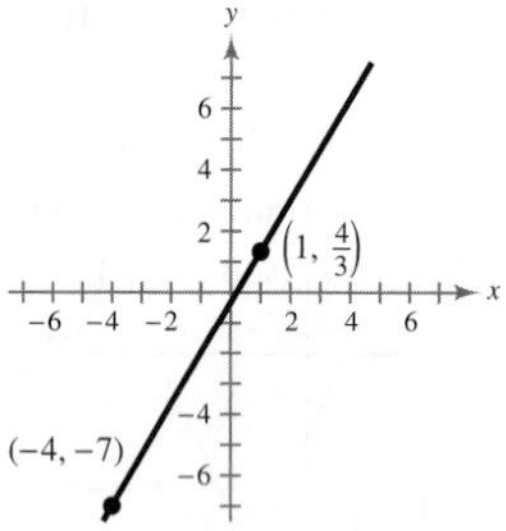

15. (a) $y = -\frac{5}{2}x + 4$ (b) $y = \frac{2}{5}x + 4$

16. (a) -9 (b) 1 (c) $|x - 4| - 15$

17. (a) 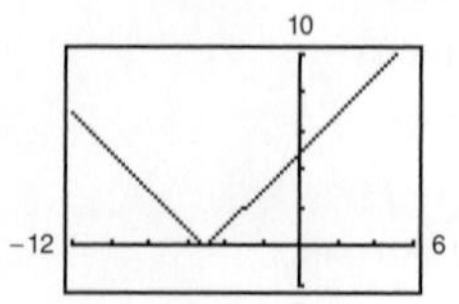
(b) Increasing on $(-5, \infty)$
Decreasing on $(-\infty, -5)$
(c) Neither

18. (a) 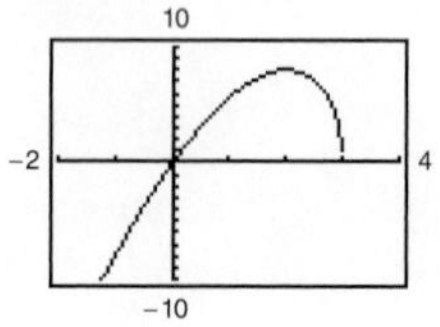
(b) Increasing on $(-\infty, 2)$
Decreasing on $(2, 3)$
(c) Neither

19. (a)

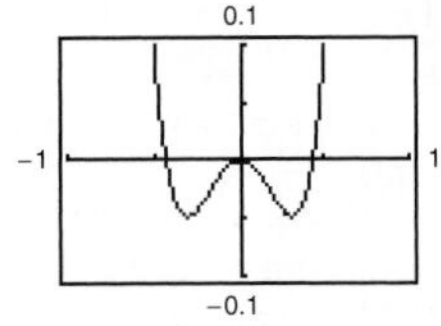

(b) Increasing on $(-0.31, 0)$, $(0.31, \infty)$
Decreasing on $(-\infty, -0.31)$, $(0, 0.31)$

(c) Even

20. (a) $f(x) = [\![x]\!]$

(b) Vertical stretch

(c)

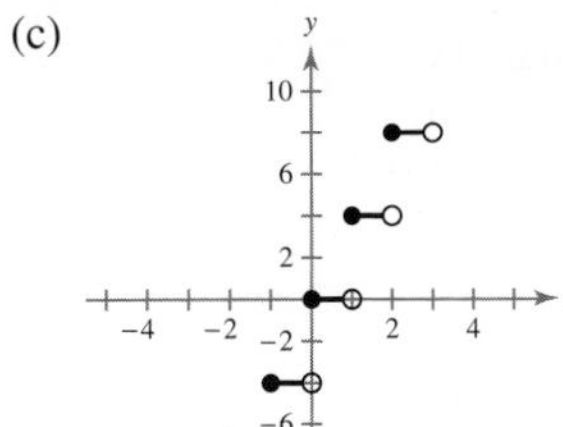

21. (a) $f(x) = \sqrt{x}$

(b) Reflection in the x-axis, left shift of five units, and an upward shift of eight units

(c)

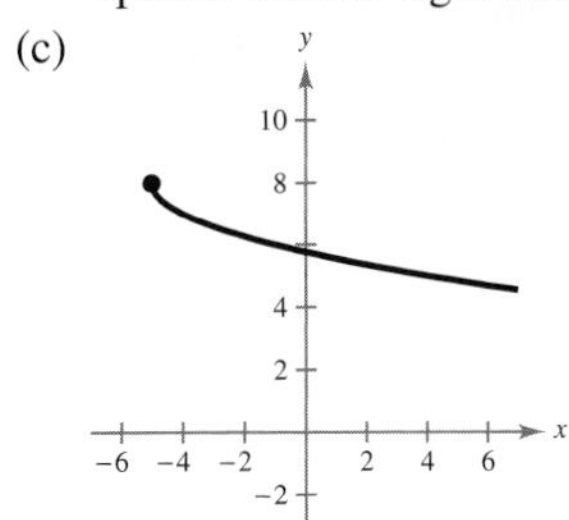

22. (a) $f(x) = x^3$

(b) Vertical stretch, a reflection in the x-axis, a right shift of five units, and an upward shift of three units

(c)

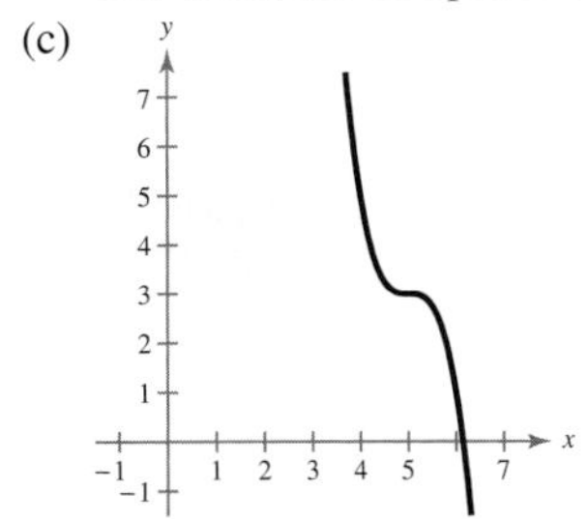

23. (a) $2x^2 - 4x - 2$

(b) $4x^2 + 4x - 12$

(c) $-3x^4 - 12x^3 + 22x^2 + 28x - 35$

(d) $\dfrac{3x^2 - 7}{-x^2 - 4x + 5}$

(e) $3x^4 + 24x^3 + 18x^2 - 120x + 68$

(f) $-9x^4 + 30x^2 - 16$

24. (a) $\dfrac{1 + 2x^{3/2}}{x}$ (b) $\dfrac{1 - 2x^{3/2}}{x}$ (c) $\dfrac{2\sqrt{x}}{x}$

(d) $\dfrac{1}{2x^{3/2}}$ (e) $\dfrac{\sqrt{x}}{2x}$ (f) $\dfrac{2\sqrt{x}}{x}$

25. $f^{-1}(x) = \sqrt[3]{x - 8}$ **26.** No inverse function

27. $f^{-1}(x) = \left(\frac{1}{3}x\right)^{2/3}, \quad x \geq 0$

Problem Solving *(page 119)*

1. (a) $W_1 = 2000 + 0.07S$

(b) $W_2 = 2300 + 0.05S$

(c)

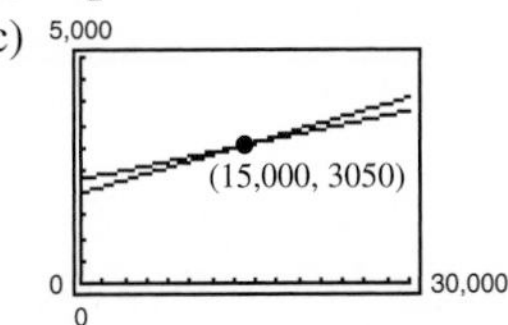

Both jobs pay the same monthly salary when sales equal \$15,000.

(d) No. Job 1 would pay \$3400 and job 2 would pay \$3300.

3. (a) The function will be even.

(b) The function will be odd.

(c) The function will be neither even nor odd.

5. $f(x) = a_{2n}x^{2n} + a_{2n-2}x^{2n-2} + \cdots + a_2x^2 + a_0$

$f(-x) = a_{2n}(-x)^{2n} + a_{2n-2}(-x)^{2n-2} + \cdots + a_2(-x)^2 + a_0$

$= f(x)$

7. (a) $81\frac{2}{3}$ h

(b) $25\frac{5}{7}$ mi/h

(c) $y = \dfrac{-180}{7}x + 3400$

Domain: $0 \leq x \leq \dfrac{1190}{9}$

Range: $0 \leq y \leq 3400$

(d)

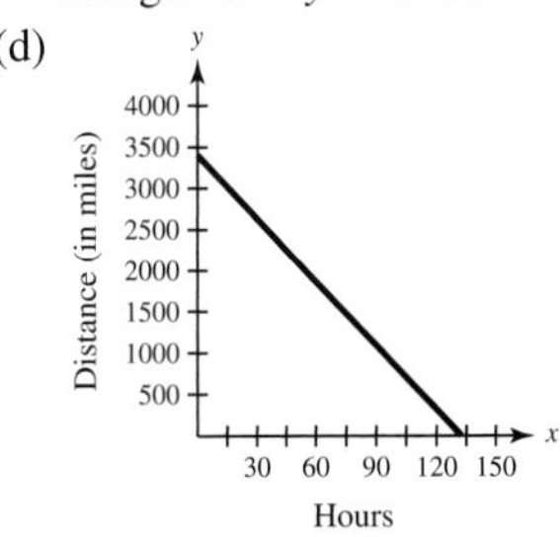

9. (a) $(f \circ g)(x) = 4x + 24$

(b) $(f \circ g)^{-1}(x) = \frac{1}{4}x - 6$

(c) $f^{-1}(x) = \frac{1}{4}x$; $g^{-1}(x) = x - 6$

(d) $(g^{-1} \circ f^{-1})(x) = \frac{1}{4}x - 6$; They are the same.

(e) $(f \circ g)(x) = 8x^3 + 1$; $(f \circ g)^{-1}(x) = \frac{1}{2}\sqrt[3]{x - 1}$;
$f^{-1}(x) = \sqrt[3]{x - 1}$; $g^{-1}(x) = \frac{1}{2}x$;
$(g^{-1} \circ f^{-1})(x) = \frac{1}{2}\sqrt[3]{x - 1}$

(f) Answers will vary.

(g) $(f \circ g)^{-1}(x) = (g^{-1} \circ f^{-1})(x)$

11. (a)

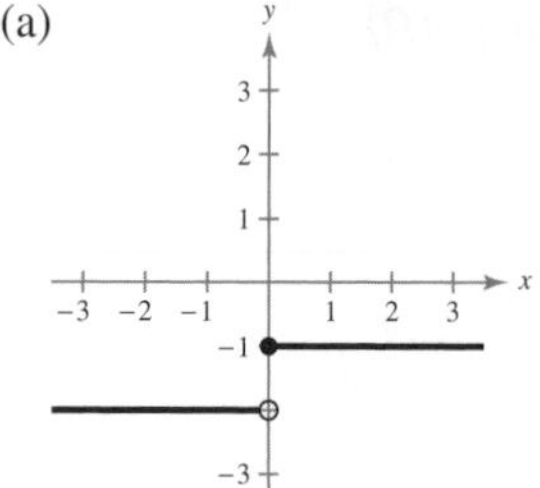

(b)

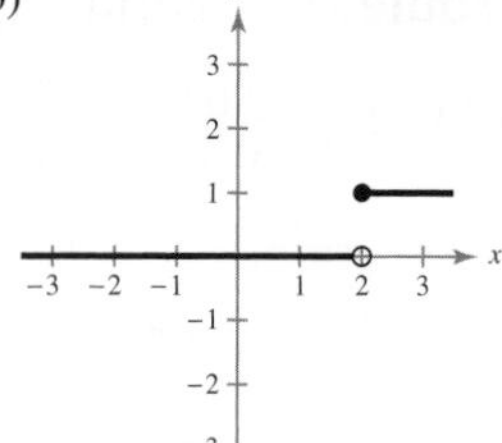

(c)

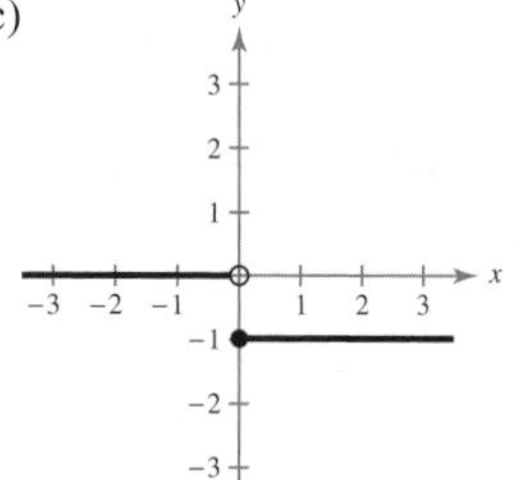

(d)

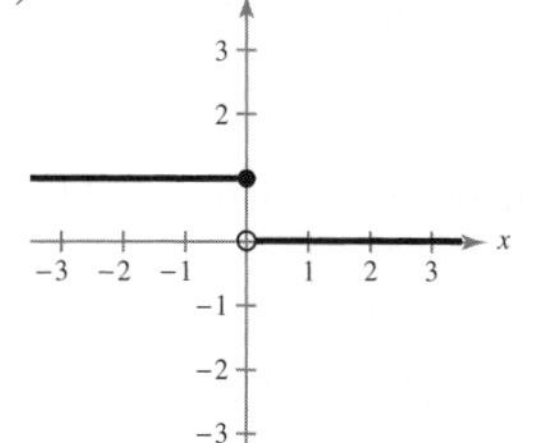

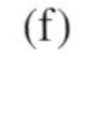

(e)

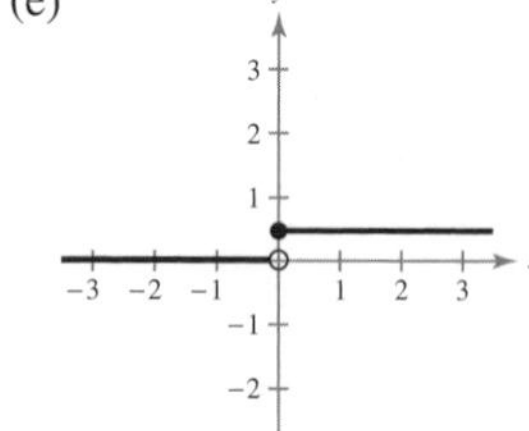

(f)

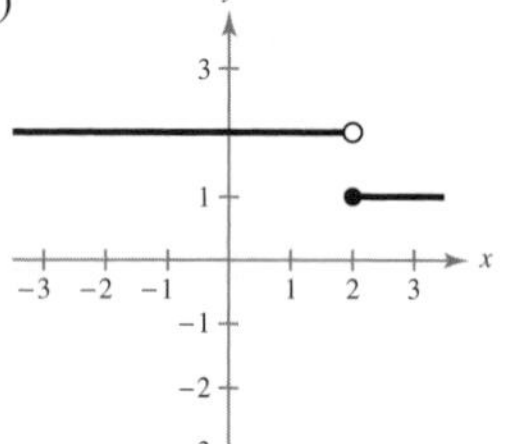

13. Proof

15. (a)

x	-4	-2	0	4
$f(f^{-1}(x))$	-4	-2	0	4

(b)

x	-3	-2	0	1
$(f + f^{-1})(x)$	5	1	-3	-5

(c)

x	-3	-2	0	1
$(f \cdot f^{-1})(x)$	4	0	2	6

(d)

x	-4	-3	0	4
$\lvert f^{-1}(x)\rvert$	2	1	1	3

Chapter 1

Section 1.1 *(page 129)*

1. coterminal **3.** complementary; supplementary

5. angular **7.** 1 rad **9.** -3 rad

11. (a) Quadrant I (b) Quadrant II

13. (a) (b)

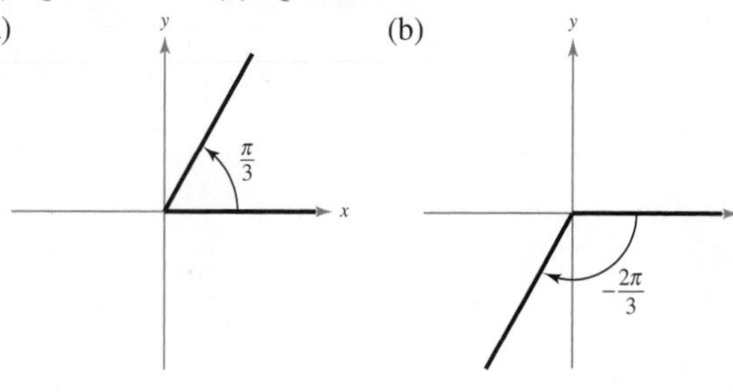

15. *Sample answers:* (a) $\frac{13\pi}{6}, -\frac{11\pi}{6}$ (b) $\frac{7\pi}{6}, -\frac{17\pi}{6}$

17. (a) Complement: $\frac{5\pi}{12}$; Supplement: $\frac{11\pi}{12}$

(b) Complement: none; Supplement: $\frac{\pi}{12}$

19. (a) Complement: $\frac{\pi}{2} - 1 \approx 0.57$;

Supplement: $\pi - 1 \approx 2.14$

(b) Complement: none; Supplement: $\pi - 2 \approx 1.14$

21. $210°$ **23.** $-60°$

25. (a) Quadrant II (b) Quadrant IV

27. (a)

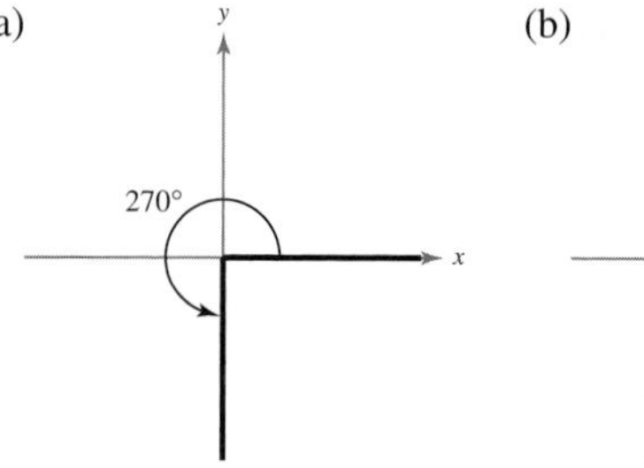

(b)

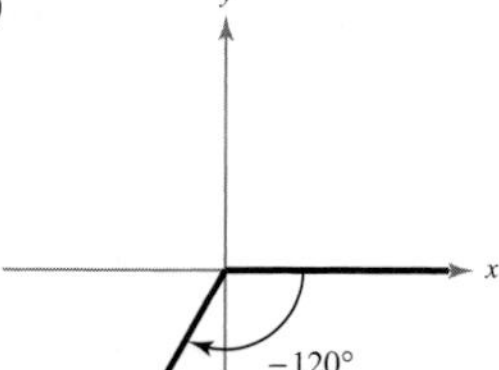

29. (a) $480°, -240°$ (b) $150°, -570°$

31. (a) Complement: $72°$; Supplement: $162°$

(b) Complement: $5°$; Supplement: $95°$

33. (a) Complement: $66°$; Supplement: $156°$

(b) Complement: none; Supplement: $54°$

35. (a) $\frac{2\pi}{3}$ (b) $-\frac{\pi}{9}$ **37.** (a) $270°$ (b) $-210°$

39. 0.785 **41.** -0.009 **43.** $81.818°$ **45.** $-756.000°$

47. (a) $54.75°$ (b) $-128.5°$

49. (a) $240°\ 36'$ (b) $-145°\ 48'$

51. 10π in. ≈ 31.42 in. **53.** $\frac{15}{8}$ rad **55.** 4 rad

57. About 18.85 in.^{2} **59.** $20°$ should be multiplied by $\frac{\pi \text{ rad}}{180 \text{ deg}}$.

61. About 592 mi **63.** About $23.87°$

65. (a) 8π rad/min ≈ 25.13 rad/min

(b) 200π ft/min ≈ 628.3 ft/min

67. (a) About 35.70 mi/h (b) About 739.50 revolutions/min

69.

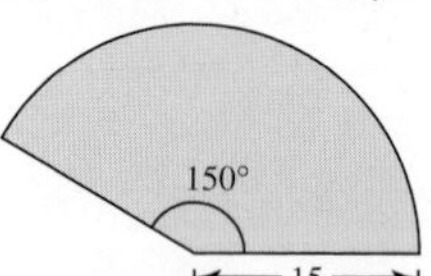

$A = 93.75\pi \text{ m}^2 \approx 294.52 \text{ m}^2$

71. False. $\frac{180°}{\pi}$ is in degree measure.

73. True. Let α and β represent coterminal angles, and let n represent an integer.

$$\alpha = \beta + n(360°)$$
$$\alpha - \beta = n(360°)$$

75. When θ is constant, the length of the arc is proportional to the radius ($s = r\theta$).

77. The speed increases. The linear velocity is proportional to the radius.

79. Proof

Section 1.2 (page 137)

1. unit circle **3.** period

5. $\sin t = \frac{5}{13}$ $\csc t = \frac{13}{5}$
$\cos t = \frac{12}{13}$ $\sec t = \frac{13}{12}$
$\tan t = \frac{5}{12}$ $\cot t = \frac{12}{5}$

7. $\sin t = -\frac{3}{5}$ $\csc t = -\frac{5}{3}$
$\cos t = -\frac{4}{5}$ $\sec t = -\frac{5}{4}$
$\tan t = \frac{3}{4}$ $\cot t = \frac{4}{3}$

9. $(0, 1)$ **11.** $\left(-\frac{\sqrt{3}}{2}, \frac{1}{2}\right)$

13. $\sin \frac{\pi}{4} = \frac{\sqrt{2}}{2}$
$\cos \frac{\pi}{4} = \frac{\sqrt{2}}{2}$
$\tan \frac{\pi}{4} = 1$

15. $\sin\left(-\frac{\pi}{6}\right) = -\frac{1}{2}$
$\cos\left(-\frac{\pi}{6}\right) = \frac{\sqrt{3}}{2}$
$\tan\left(-\frac{\pi}{6}\right) = -\frac{\sqrt{3}}{3}$

17. $\sin\left(-\frac{7\pi}{4}\right) = \frac{\sqrt{2}}{2}$
$\cos\left(-\frac{7\pi}{4}\right) = \frac{\sqrt{2}}{2}$
$\tan\left(-\frac{7\pi}{4}\right) = 1$

19. $\sin \frac{11\pi}{6} = -\frac{1}{2}$
$\cos \frac{11\pi}{6} = \frac{\sqrt{3}}{2}$
$\tan \frac{11\pi}{6} = -\frac{\sqrt{3}}{3}$

21. $\sin\left(-\frac{3\pi}{2}\right) = 1$
$\cos\left(-\frac{3\pi}{2}\right) = 0$
$\tan\left(-\frac{3\pi}{2}\right)$ is undefined.

23. $\sin \frac{2\pi}{3} = \frac{\sqrt{3}}{2}$ $\csc \frac{2\pi}{3} = \frac{2\sqrt{3}}{3}$
$\cos \frac{2\pi}{3} = -\frac{1}{2}$ $\sec \frac{2\pi}{3} = -2$
$\tan \frac{2\pi}{3} = -\sqrt{3}$ $\cot \frac{2\pi}{3} = -\frac{\sqrt{3}}{3}$

25. $\sin \frac{4\pi}{3} = -\frac{\sqrt{3}}{2}$ $\csc \frac{4\pi}{3} = -\frac{2\sqrt{3}}{3}$
$\cos \frac{4\pi}{3} = -\frac{1}{2}$ $\sec \frac{4\pi}{3} = -2$
$\tan \frac{4\pi}{3} = \sqrt{3}$ $\cot \frac{4\pi}{3} = \frac{\sqrt{3}}{3}$

27. $\sin\left(-\frac{5\pi}{3}\right) = \frac{\sqrt{3}}{2}$ $\csc\left(-\frac{5\pi}{3}\right) = \frac{2\sqrt{3}}{3}$
$\cos\left(-\frac{5\pi}{3}\right) = \frac{1}{2}$ $\sec\left(-\frac{5\pi}{3}\right) = 2$
$\tan\left(-\frac{5\pi}{3}\right) = \sqrt{3}$ $\cot\left(-\frac{5\pi}{3}\right) = \frac{\sqrt{3}}{3}$

29. $\sin\left(-\frac{\pi}{2}\right) = -1$ $\csc\left(-\frac{\pi}{2}\right) = -1$
$\cos\left(-\frac{\pi}{2}\right) = 0$ $\sec\left(-\frac{\pi}{2}\right)$ is undefined.
$\tan\left(-\frac{\pi}{2}\right)$ is undefined. $\cot\left(-\frac{\pi}{2}\right) = 0$

31. 0 **33.** $\frac{1}{2}$ **35.** $-\frac{1}{2}$ **37.** (a) $-\frac{1}{2}$ (b) -2

39. (a) $-\frac{1}{5}$ (b) -5 **41.** (a) $\frac{4}{5}$ (b) $-\frac{4}{5}$

43. 0.5646 **45.** 0.4142 **47.** -1.0009

49. (a) 0.50 ft (b) About 0.04 ft (c) About -0.49 ft

51. False. $\sin(-t) = -\sin(t)$ means that the function is odd, not that the sine of a negative angle is a negative number.

53. True. The tangent function has a period of π.

55. (a) y-axis symmetry (b) $\sin t_1 = \sin(\pi - t_1)$
(c) $\cos(\pi - t_1) = -\cos t_1$

57. The calculator was in degree mode instead of radian mode.

59. (a)

Circle of radius 1 centered at $(0, 0)$

(b) The t-values represent the central angle in radians. The x- and y-values represent the location in the coordinate plane.

(c) $-1 \le x \le 1, -1 \le y \le 1$

61. It is an odd function.

Section 1.3 (page 146)

1. (a) v (b) iv (c) vi (d) iii (e) i (f) ii

3. complementary

5. $\sin \theta = \frac{3}{5}$ $\csc \theta = \frac{5}{3}$
$\cos \theta = \frac{4}{5}$ $\sec \theta = \frac{5}{4}$
$\tan \theta = \frac{3}{4}$ $\cot \theta = \frac{4}{3}$

7. $\sin \theta = \frac{9}{41}$ $\csc \theta = \frac{41}{9}$
$\cos \theta = \frac{40}{41}$ $\sec \theta = \frac{41}{40}$
$\tan \theta = \frac{9}{40}$ $\cot \theta = \frac{40}{9}$

9. $\sin \theta = \frac{\sqrt{2}}{2}$ $\csc \theta = \sqrt{2}$
$\cos \theta = \frac{\sqrt{2}}{2}$ $\sec \theta = \sqrt{2}$
$\tan \theta = 1$ $\cot \theta = 1$

11. $\sin \theta = \frac{8}{17}$ $\csc \theta = \frac{17}{8}$
$\cos \theta = \frac{15}{17}$ $\sec \theta = \frac{17}{15}$
$\tan \theta = \frac{8}{15}$ $\cot \theta = \frac{15}{8}$
The triangles are similar, and corresponding sides are proportional.

13. $\sin \theta = \frac{1}{3}$ $\csc \theta = 3$
$\cos \theta = \frac{2\sqrt{2}}{3}$ $\sec \theta = \frac{3\sqrt{2}}{4}$
$\tan \theta = \frac{\sqrt{2}}{4}$ $\cot \theta = 2\sqrt{2}$
The triangles are similar, and corresponding sides are proportional.

15.

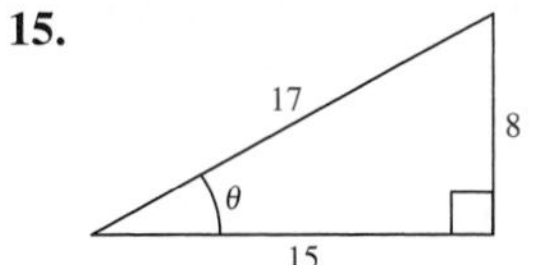

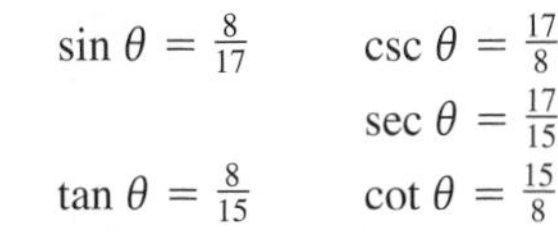

$\sin \theta = \frac{8}{17}$ $\csc \theta = \frac{17}{8}$
$\sec \theta = \frac{17}{15}$
$\tan \theta = \frac{8}{15}$ $\cot \theta = \frac{15}{8}$

17.

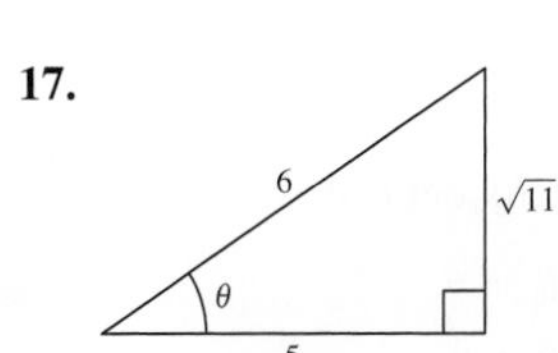

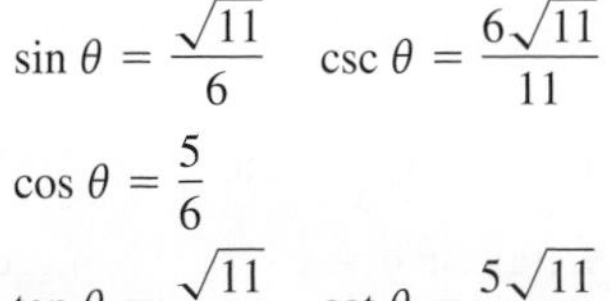

$\sin \theta = \frac{\sqrt{11}}{6}$ $\csc \theta = \frac{6\sqrt{11}}{11}$
$\cos \theta = \frac{5}{6}$
$\tan \theta = \frac{\sqrt{11}}{5}$ $\cot \theta = \frac{5\sqrt{11}}{11}$

19.

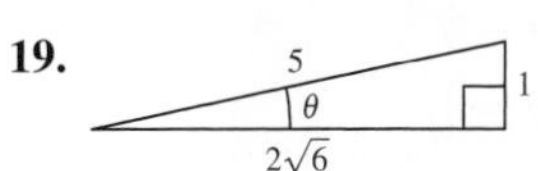

$\csc \theta = 5$
$\cos \theta = \frac{2\sqrt{6}}{5}$ $\sec \theta = \frac{5\sqrt{6}}{12}$
$\tan \theta = \frac{\sqrt{6}}{12}$ $\cot \theta = 2\sqrt{6}$

21. $\sqrt{10}$ θ 1 3

$\sin\theta = \frac{\sqrt{10}}{10}$ $\csc\theta = \sqrt{10}$

$\cos\theta = \frac{3\sqrt{10}}{10}$ $\sec\theta = \frac{\sqrt{10}}{3}$

$\tan\theta = \frac{1}{3}$

23. $\frac{\pi}{6}$; $\frac{\sqrt{3}}{3}$ **25.** $45°$; $\frac{\sqrt{2}}{2}$ **27.** $45°$; $\sqrt{2}$

29. (a) 0.3420 (b) 0.3420 **31.** (a) 0.2455 (b) 4.0737

33. (a) 0.9964 (b) 1.0036 **35.** (a) 3.2205 (b) 0.3105

37. (a) $\frac{1}{2}$ (b) $\frac{\sqrt{3}}{2}$ (c) $\sqrt{3}$ (d) $\frac{\sqrt{3}}{3}$

39. (a) $\frac{2\sqrt{2}}{3}$ (b) $2\sqrt{2}$ (c) 3 (d) 3

41. (a) $\frac{1}{3}$ (b) $\sqrt{10}$ (c) $\frac{1}{3}$ (d) $\frac{\sqrt{10}}{10}$

43–51. Answers will vary. **53.** (a) $30° = \frac{\pi}{6}$ (b) $30° = \frac{\pi}{6}$

55. (a) $60° = \frac{\pi}{3}$ (b) $45° = \frac{\pi}{4}$

57. (a) $60° = \frac{\pi}{3}$ (b) $45° = \frac{\pi}{4}$

59. $x = 9, y = 9\sqrt{3}$ **61.** $x = \frac{32\sqrt{3}}{3}, r = \frac{64\sqrt{3}}{3}$

63. About 443.2 m; about 323.3 m **65.** $30° = \frac{\pi}{6}$

67. (a) About 219.9 ft (b) About 160.9 ft

69. $(x_1, y_1) = (28\sqrt{3}, 28)$
$(x_2, y_2) = (28, 28\sqrt{3})$

71. $\sin 20° \approx 0.34$, $\cos 20° \approx 0.94$, $\tan 20° \approx 0.36$, $\csc 20° \approx 2.92$, $\sec 20° \approx 1.06$, $\cot 20° \approx 2.75$

73. (a) About 519.33 ft (b) About 1174.17 ft (c) About 173.11 ft/min

75. True. $\csc x = \frac{1}{\sin x}$ **77.** False. $\frac{\sqrt{2}}{2} + \frac{\sqrt{2}}{2} \neq 1$

79. False. $1.7321 \neq 0.0349$

81. Yes, $\tan\theta$ is equal to opp/adj. You can find the value of the hypotenuse by the Pythagorean Theorem. Then you can find $\sec\theta$, which is equal to hyp/adj.

83.

θ	0.1	0.2	0.3	0.4	0.5
$\sin\theta$	0.0998	0.1987	0.2955	0.3894	0.4794

(a) θ is greater.

(b) As $\theta \to 0$, $\sin\theta \to 0$ and $\frac{\theta}{\sin\theta} \to 1$.

Section 1.4 *(page 156)*

1. $\frac{y}{r}$ **3.** $\frac{y}{x}$ **5.** $\cos\theta$ **7.** zero; defined

9. (a) $\sin\theta = \frac{3}{5}$ $\csc\theta = \frac{5}{3}$
$\cos\theta = \frac{4}{5}$ $\sec\theta = \frac{5}{4}$
$\tan\theta = \frac{3}{4}$ $\cot\theta = \frac{4}{3}$

(b) $\sin\theta = \frac{15}{17}$ $\csc\theta = \frac{17}{15}$
$\cos\theta = -\frac{8}{17}$ $\sec\theta = -\frac{17}{8}$
$\tan\theta = -\frac{15}{8}$ $\cot\theta = -\frac{8}{15}$

11. (a) $\sin\theta = -\frac{1}{2}$ $\csc\theta = -2$
$\cos\theta = -\frac{\sqrt{3}}{2}$ $\sec\theta = -\frac{2\sqrt{3}}{3}$
$\tan\theta = \frac{\sqrt{3}}{3}$ $\cot\theta = \sqrt{3}$

(b) $\sin\theta = -\frac{\sqrt{17}}{17}$ $\csc\theta = -\sqrt{17}$
$\cos\theta = \frac{4\sqrt{17}}{17}$ $\sec\theta = \frac{\sqrt{17}}{4}$
$\tan\theta = -\frac{1}{4}$ $\cot\theta = -4$

13. $\sin\theta = \frac{12}{13}$ $\csc\theta = \frac{13}{12}$
$\cos\theta = \frac{5}{13}$ $\sec\theta = \frac{13}{5}$
$\tan\theta = \frac{12}{5}$ $\cot\theta = \frac{5}{12}$

15. $\sin\theta = -\frac{2\sqrt{29}}{29}$ $\csc\theta = -\frac{\sqrt{29}}{2}$
$\cos\theta = -\frac{5\sqrt{29}}{29}$ $\sec\theta = -\frac{\sqrt{29}}{5}$
$\tan\theta = \frac{2}{5}$ $\cot\theta = \frac{5}{2}$

17. $\sin\theta = \frac{4}{5}$ $\csc\theta = \frac{5}{4}$
$\cos\theta = -\frac{3}{5}$ $\sec\theta = -\frac{5}{3}$
$\tan\theta = -\frac{4}{3}$ $\cot\theta = -\frac{3}{4}$

19. Quadrant I **21.** Quadrant II

23. $\sin\theta = \frac{15}{17}$ $\csc\theta = \frac{17}{15}$
$\cos\theta = \frac{8}{17}$ $\sec\theta = \frac{17}{8}$
$\cot\theta = \frac{8}{15}$

25. $\csc\theta = \frac{5}{3}$
$\cos\theta = -\frac{4}{5}$ $\sec\theta = -\frac{5}{4}$
$\tan\theta = -\frac{3}{4}$ $\cot\theta = -\frac{4}{3}$

27. $\sin\theta = -\frac{\sqrt{10}}{10}$ $\csc\theta = -\sqrt{10}$
$\cos\theta = \frac{3\sqrt{10}}{10}$ $\sec\theta = \frac{\sqrt{10}}{3}$
$\tan\theta = -\frac{1}{3}$

29. $\sin\theta = 1$
$\sec\theta$ is undefined.
$\tan\theta$ is undefined. $\cot\theta = 0$

31. $\sin\theta = 0$ $\csc\theta$ is undefined.
$\cos\theta = -1$ $\sec\theta = -1$
$\tan\theta = 0$ $\cot\theta$ is undefined.

33. $\sin\theta = \frac{\sqrt{2}}{2}$ $\csc\theta = \sqrt{2}$
$\cos\theta = -\frac{\sqrt{2}}{2}$ $\sec\theta = -\sqrt{2}$
$\tan\theta = -1$ $\cot\theta = -1$

35. $\sin\theta = \frac{2\sqrt{5}}{5}$ $\csc\theta = \frac{\sqrt{5}}{2}$
$\cos\theta = \frac{\sqrt{5}}{5}$ $\sec\theta = \sqrt{5}$
$\tan\theta = 2$ $\cot\theta = \frac{1}{2}$

37. 0 **39.** Undefined **41.** 1 **43.** Undefined **45.** 0

47. $\theta' = 20°$

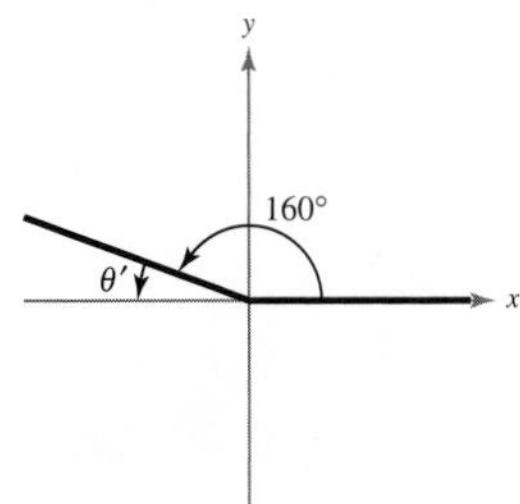

49. $\theta' = 55°$

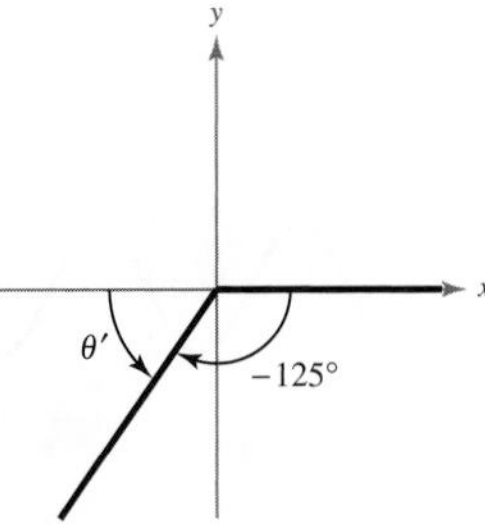

51. $\theta' = \dfrac{\pi}{3}$

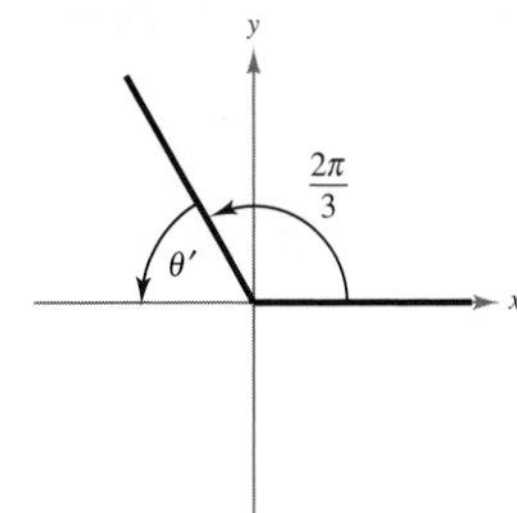

53. $\theta' = 2\pi - 4.8$

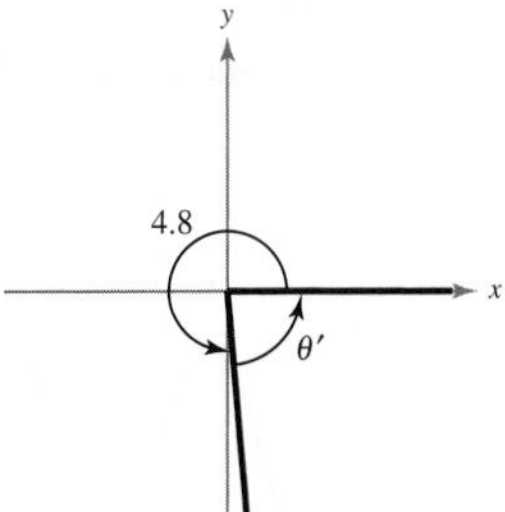

55. $\sin 225° = -\dfrac{\sqrt{2}}{2}$
$\cos 225° = -\dfrac{\sqrt{2}}{2}$
$\tan 225° = 1$

57. $\sin 750° = \dfrac{1}{2}$
$\cos 750° = \dfrac{\sqrt{3}}{2}$
$\tan 750° = \dfrac{\sqrt{3}}{3}$

59. $\sin(-120°) = -\dfrac{\sqrt{3}}{2}$
$\cos(-120°) = -\dfrac{1}{2}$
$\tan(-120°) = \sqrt{3}$

61. $\sin \dfrac{2\pi}{3} = \dfrac{\sqrt{3}}{2}$
$\cos \dfrac{2\pi}{3} = -\dfrac{1}{2}$
$\tan \dfrac{2\pi}{3} = -\sqrt{3}$

63. $\sin\left(-\dfrac{\pi}{6}\right) = -\dfrac{1}{2}$
$\cos\left(-\dfrac{\pi}{6}\right) = \dfrac{\sqrt{3}}{2}$
$\tan\left(-\dfrac{\pi}{6}\right) = -\dfrac{\sqrt{3}}{3}$

65. $\sin \dfrac{11\pi}{4} = \dfrac{\sqrt{2}}{2}$
$\cos \dfrac{11\pi}{4} = -\dfrac{\sqrt{2}}{2}$
$\tan \dfrac{11\pi}{4} = -1$

67. $\sin\left(-\dfrac{17\pi}{6}\right) = -\dfrac{1}{2}$
$\cos\left(-\dfrac{17\pi}{6}\right) = -\dfrac{\sqrt{3}}{2}$
$\tan\left(-\dfrac{17\pi}{6}\right) = \dfrac{\sqrt{3}}{3}$

69. $\dfrac{4}{5}$ **71.** $-\dfrac{\sqrt{13}}{12}$ **73.** $\dfrac{8\sqrt{39}}{39}$ **75.** 0.1736

77. −0.3420 **79.** −28.6363 **81.** 1.4142 **83.** 0.3640

85. −2.6131 **87.** −0.6052 **89.** 1.8382

91. (a) $30° = \dfrac{\pi}{6}, 150° = \dfrac{5\pi}{6}$ (b) $210° = \dfrac{7\pi}{6}, 330° = \dfrac{11\pi}{6}$

93. (a) $60° = \dfrac{\pi}{3}, 300° = \dfrac{5\pi}{3}$ (b) $60° = \dfrac{\pi}{3}, 300° = \dfrac{5\pi}{3}$

95. (a) $45° = \dfrac{\pi}{4}, 225° = \dfrac{5\pi}{4}$ (b) $150° = \dfrac{5\pi}{6}, 330° = \dfrac{11\pi}{6}$

97. (a) 12 mi (b) 6 mi (c) About 6.9 mi

99. (a) $B = 24.593 \sin(0.495t - 2.262) + 57.387$
$F = 39.071 \sin(0.448t - 1.366) + 32.204$

(b) February: $B \approx 33.9°, F \approx 14.5°$
April: $B \approx 50.5°, F \approx 48.3°$
May: $B \approx 62.6°, F \approx 62.2°$
July: $B \approx 80.3°, F \approx 70.5°$
September: $B \approx 77.4°, F \approx 50.1°$
October: $B \approx 68.2°, F \approx 33.3°$
December: $B \approx 44.8°, F \approx 2.4°$

(c)

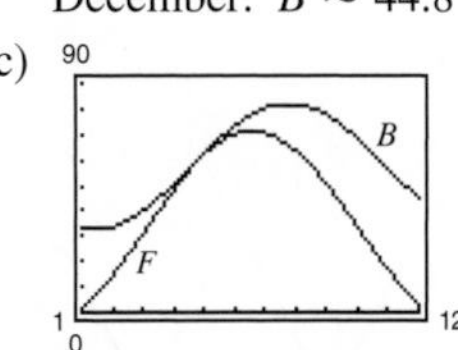

Answers will vary.

101. (a) About 270.63 ft (b) About 307.75 ft
(c) About 270.63 ft

103. False. In each of the four quadrants, the signs of the secant function and the cosine function are the same because these functions are reciprocals of each other.

105. Answers will vary.

Section 1.5 *(page 166)*

1. cycle **3.** phase shift **5.** Period: $\dfrac{2\pi}{5}$; Amplitude: 2

7. Period: 4; Amplitude: $\dfrac{3}{4}$ **9.** Period: $\dfrac{8\pi}{5}$; Amplitude: $\dfrac{1}{2}$

11. Period: 24; Amplitude: $\frac{5}{3}$

13. The period of g is one-fifth the period of f.

15. g is a reflection of f in the x-axis.

17. g is a shift of f π units to the right.

19. g is a shift of f three units up.

21. The graph of g has twice the amplitude of the graph of f.

23. The graph of g is a horizontal shift of the graph of f π units to the right.

25.

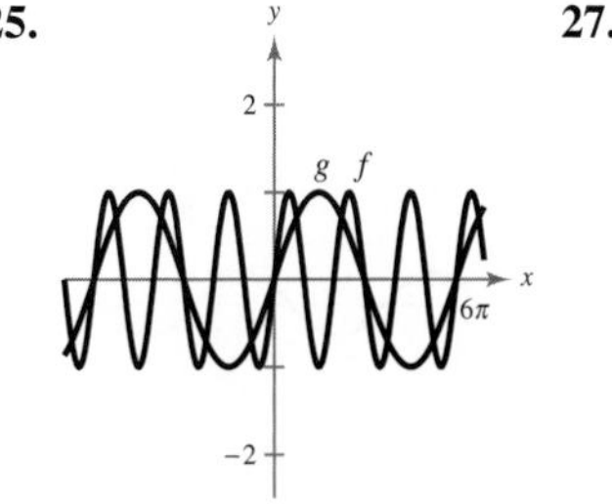

27.

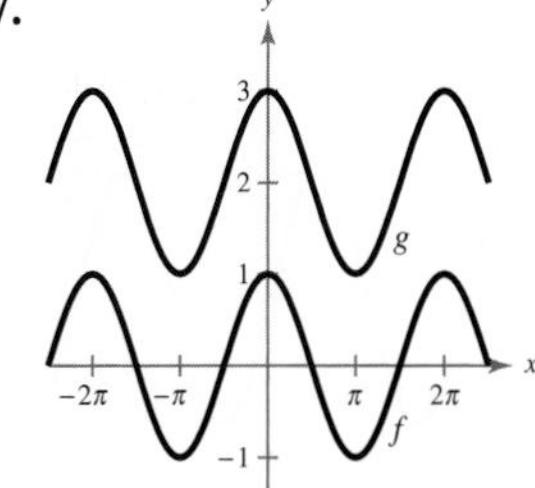

29.

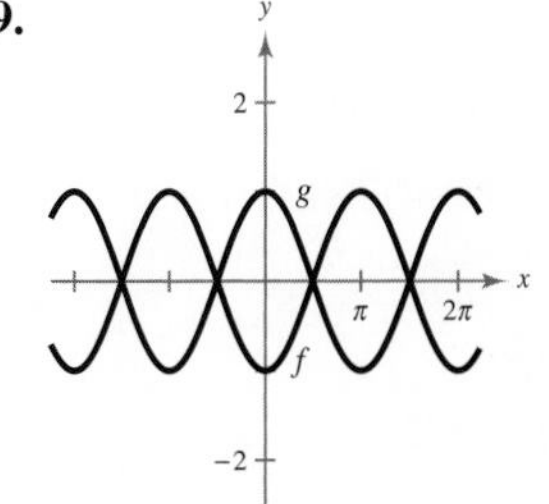

31.

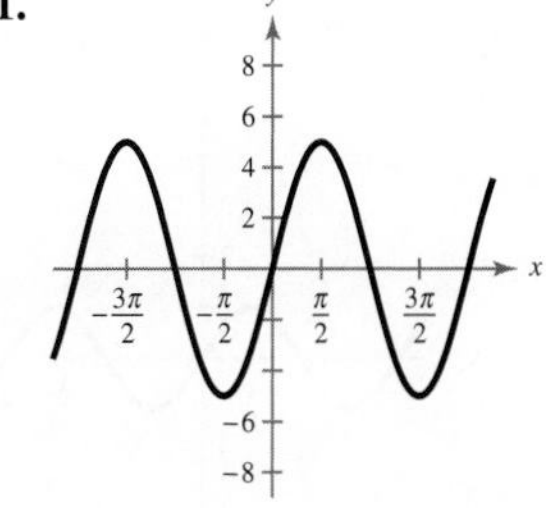

33.

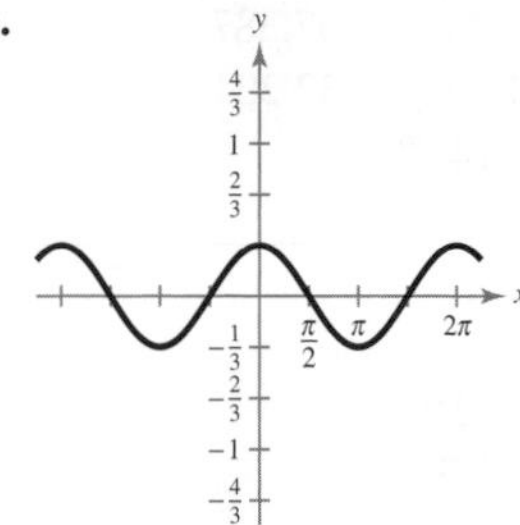

35.

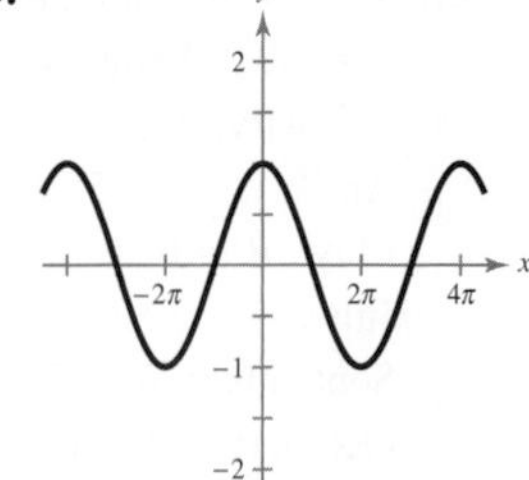

37.

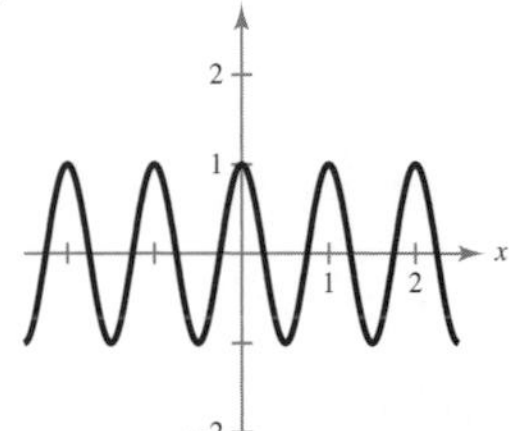

39.

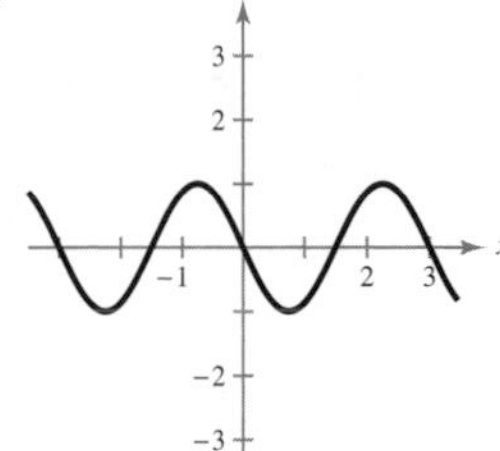

41.

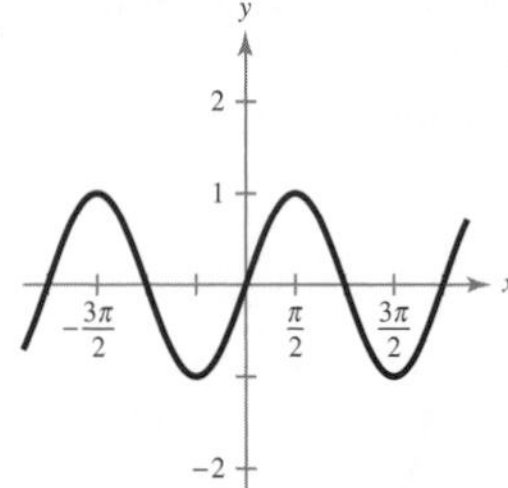

43.

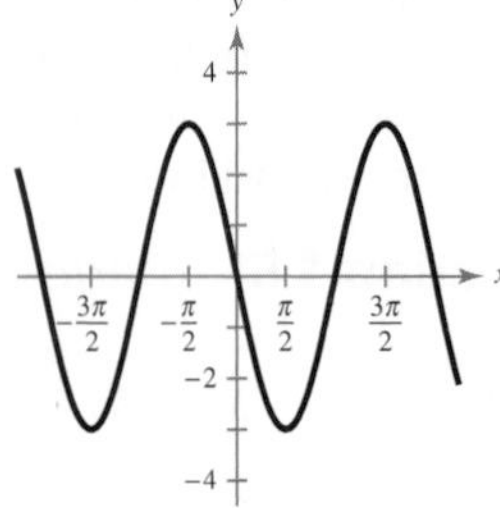

45.

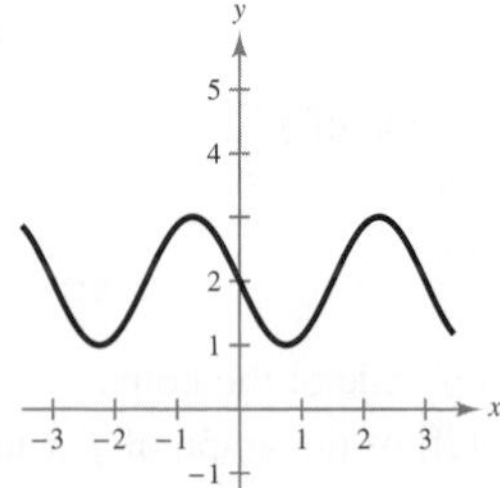

47.

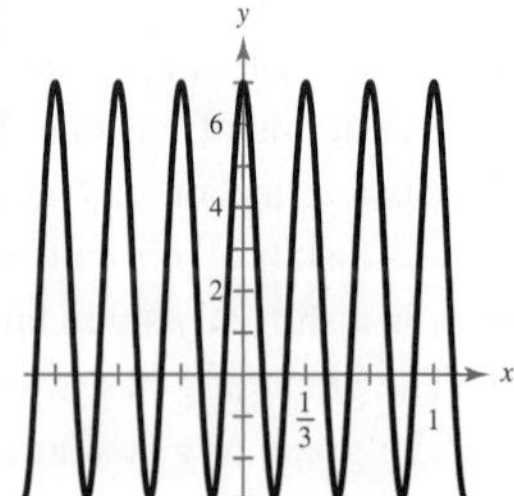

49.

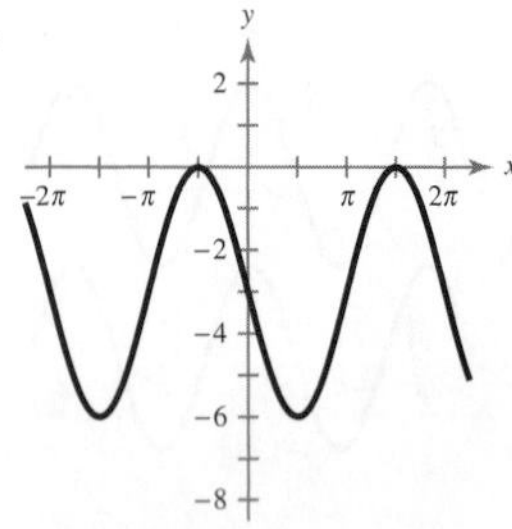

51.

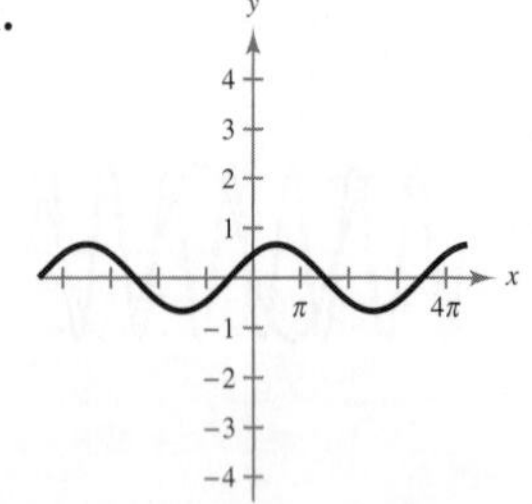

53. (a) Horizontal shrink and a phase shift $\pi/4$ unit right
(b)

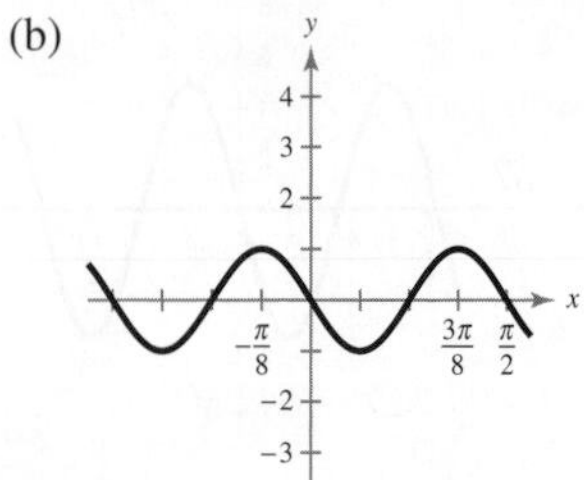

(c) $g(x) = f(4x - \pi)$

55. (a) Shift two units up and a phase shift $\pi/2$ units right
(b)

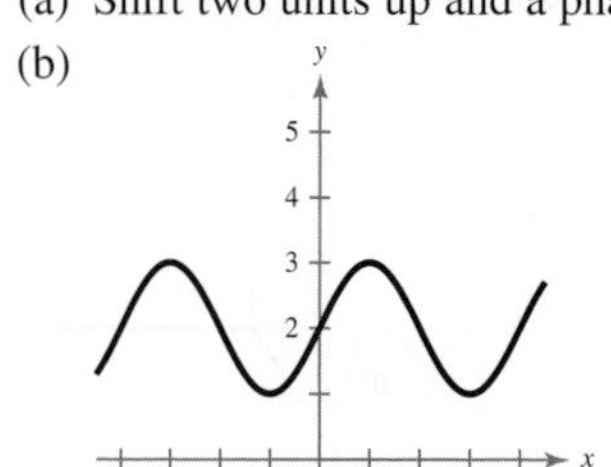

(c) $g(x) = f\left(x - \dfrac{\pi}{2}\right) + 2$

57. (a) Horizontal shrink, a vertical stretch, a shift three units down, and a phase shift $\pi/4$ unit right
(b)

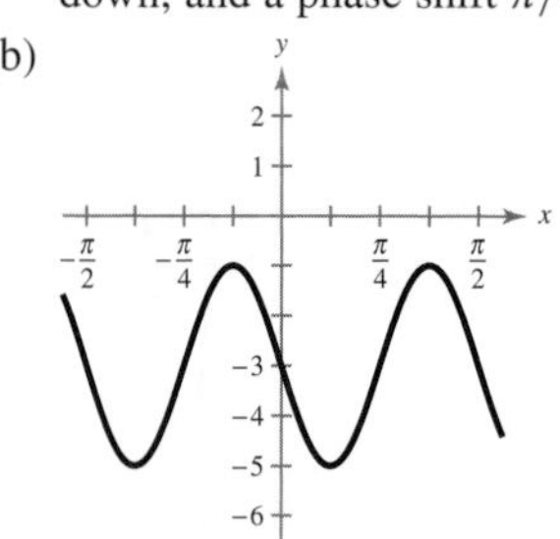

(c) $g(x) = 2f(4x - \pi) - 3$

59.

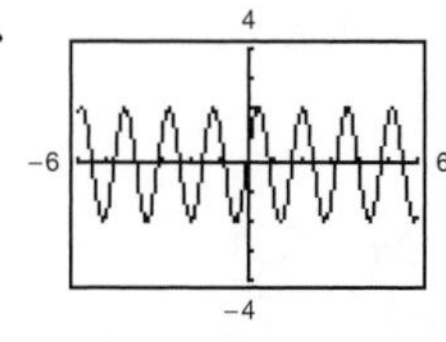

61.

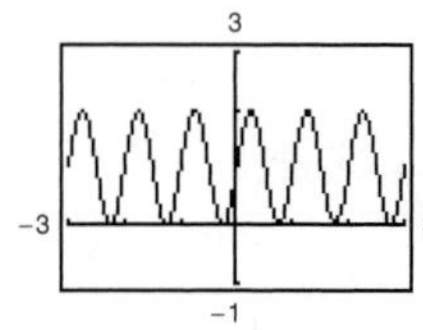

63.

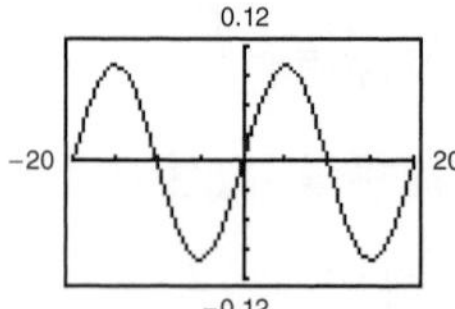

65. $a = 2, d = 1$ **67.** $a = -4, d = 4$

69. $a = -3, b = 2, c = 0$ **71.** $a = 2, b = 1, c = -\dfrac{\pi}{4}$

73.

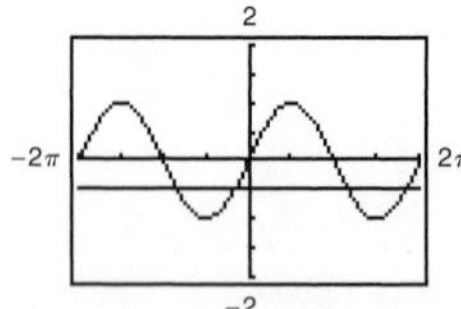

$x = -\dfrac{\pi}{6}, -\dfrac{5\pi}{6}, \dfrac{7\pi}{6}, \dfrac{11\pi}{6}$

75. $y = 1 + 2\sin(2x - \pi)$ **77.** $y = \cos(2x + 2\pi) - \frac{3}{2}$

79. (a) 4 sec
(b) 15 cycles/min
(c)

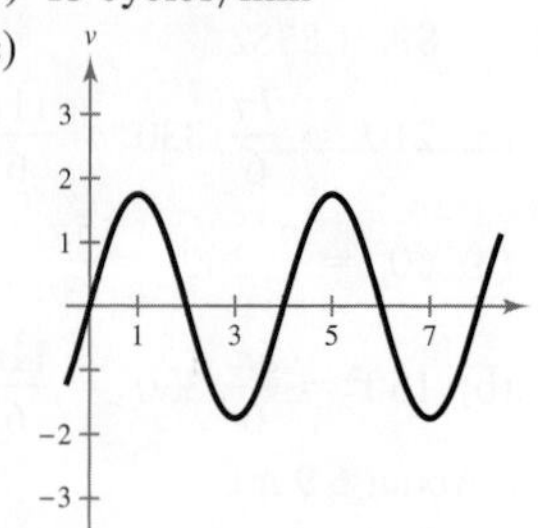

81. (a) $\frac{6}{5}$ sec (b) 50 heartbeats/min

83. (a) and (c)

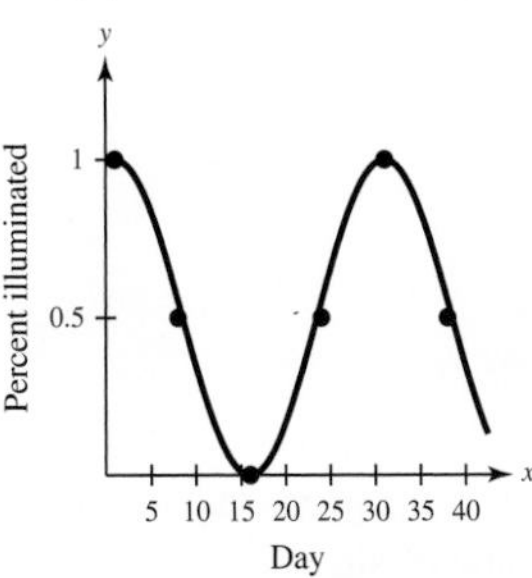

The model fits the data well.

(b) $y = 0.5 \cos\left(\frac{\pi x}{15} - \frac{\pi}{15}\right) + 0.5$

(d) 30 days (e) 25%

85. (a) 20 sec; It takes 20 seconds to complete one revolution on the Ferris wheel.

(b) 50 ft; The diameter of the Ferris wheel is 100 feet.

(c)

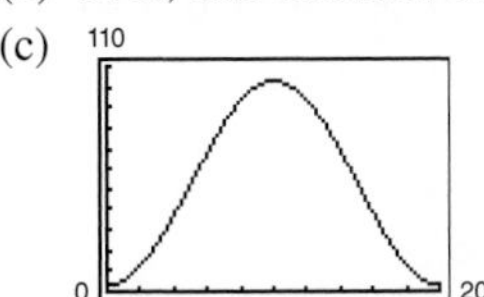

87. False. The graph of g is shifted one period to the left.

89. True. $-\sin\left(x + \frac{\pi}{2}\right) = -\cos x$

91.

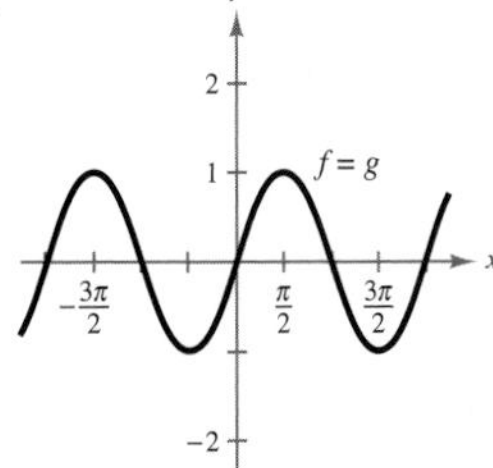

Conjecture: $\sin x = \cos\left(x - \frac{\pi}{2}\right)$

93.

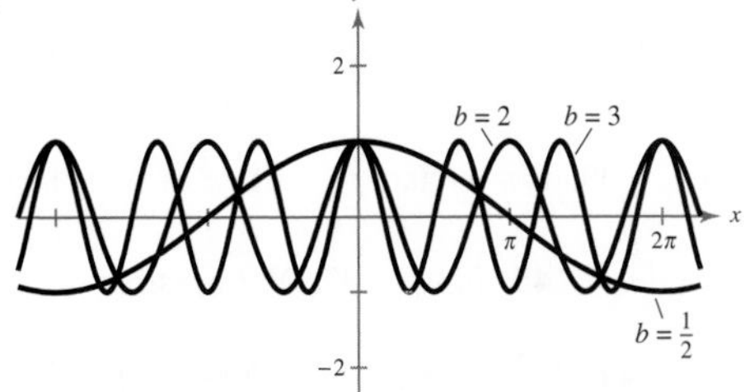

The value of b affects the period of the graph.

$b = \frac{1}{2} \rightarrow \frac{1}{2}$ cycle

$b = 2 \rightarrow 2$ cycles

$b = 3 \rightarrow 3$ cycles

95. (a) 0.4794, 0.4794 (b) 0.8417, 0.8415 (c) 0.5, 0.5

(d) 0.8776, 0.8776 (e) 0.5417, 0.5403

(f) 0.7074, 0.7071

The error increases as x moves farther away from 0.

Section 1.6 *(page 177)*

1. odd; origin **3.** reciprocal **5.** π

7. $(-\infty, -1] \cup [1, \infty)$ **9.** e, π **10.** c, 2π **11.** a, 1

12. d, 2π **13.** f, 4 **14.** b, 4

15.

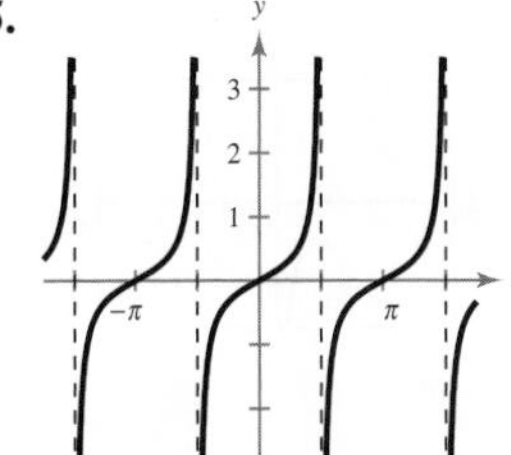

17.

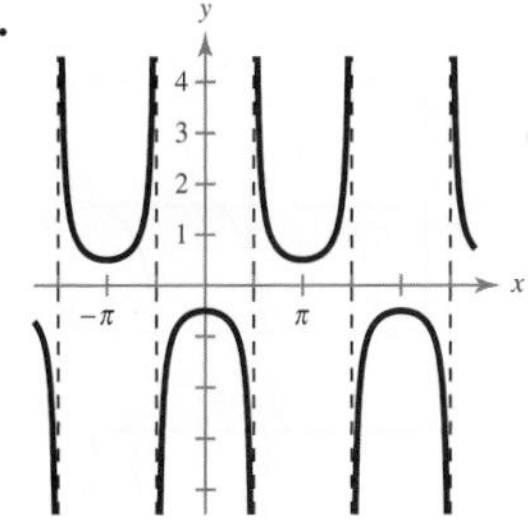

19.

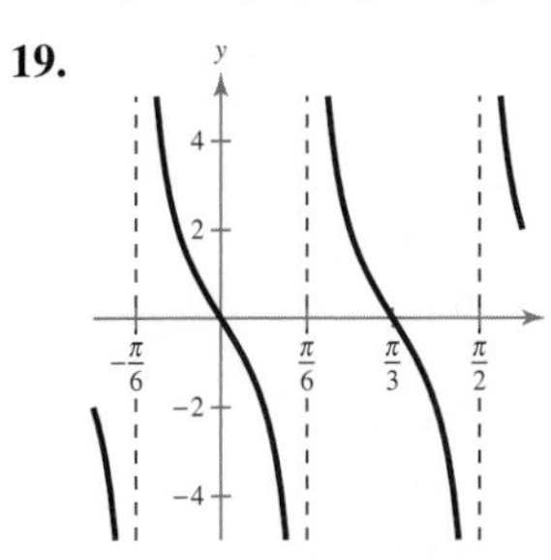

21.

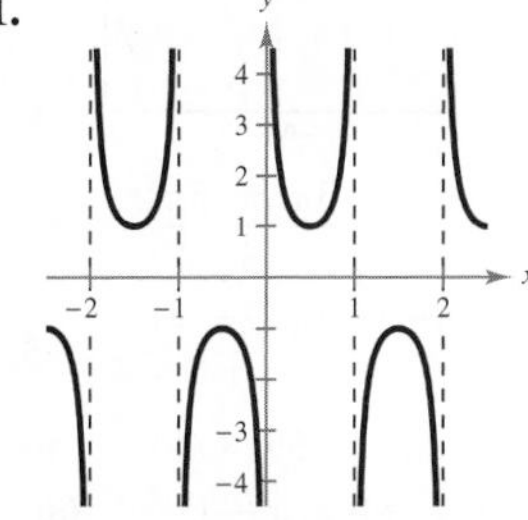

23.

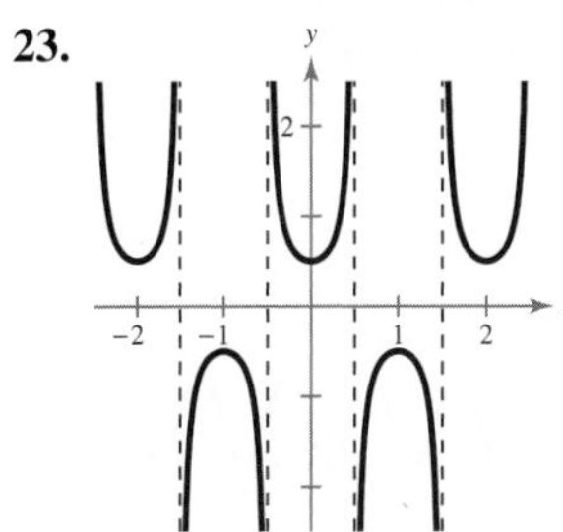

25.

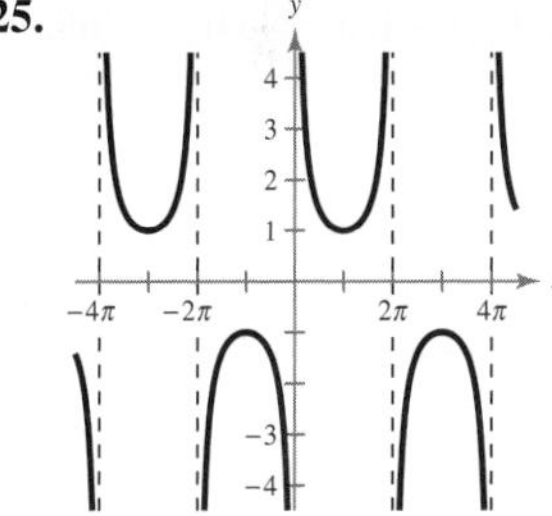

27.

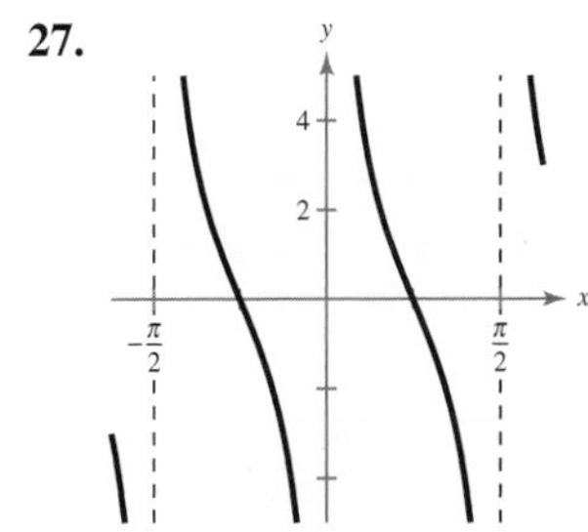

29.

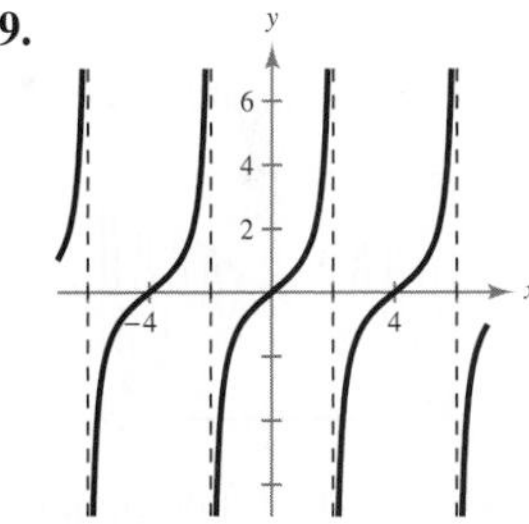

31.

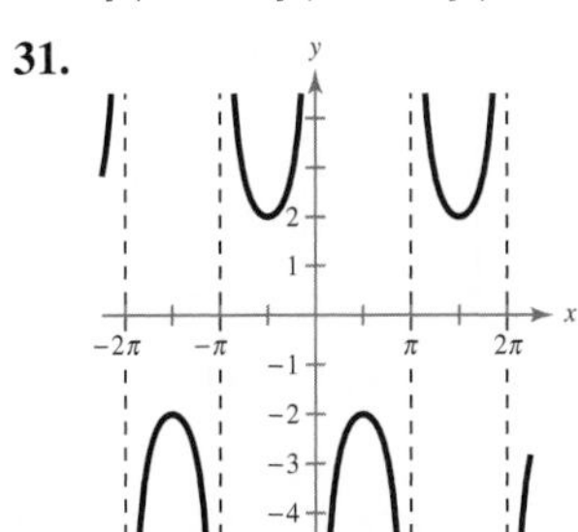

33.

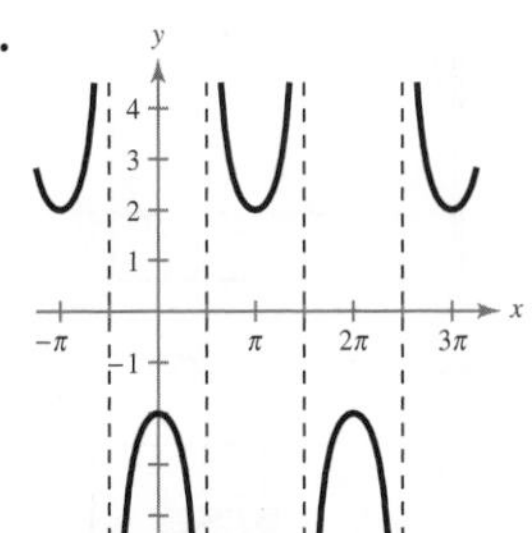

35.

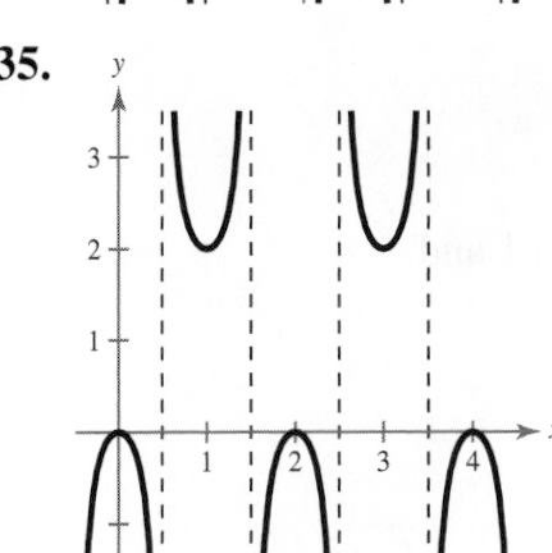

37.

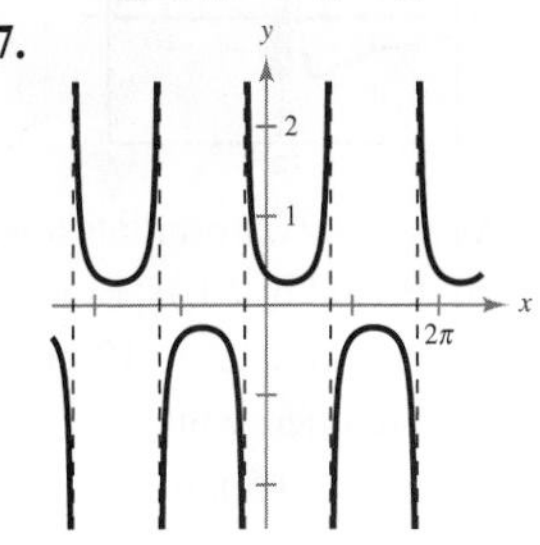

39. 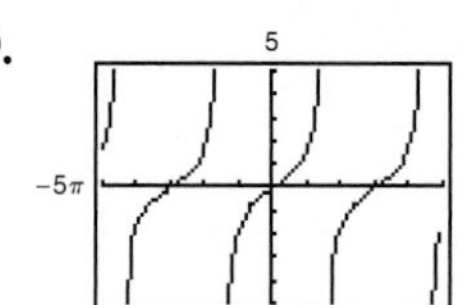

41. 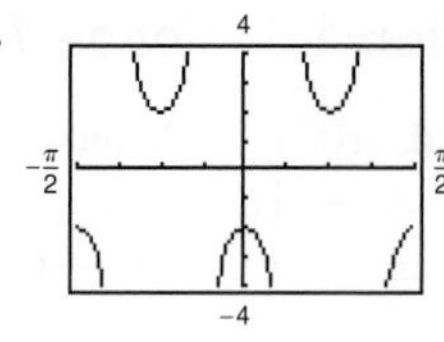

43. 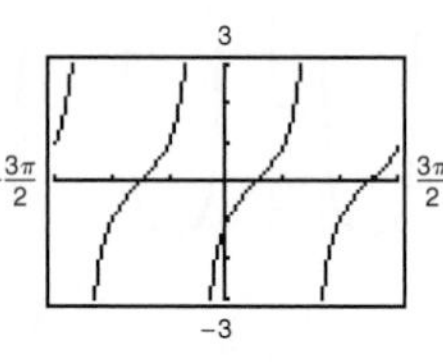

45. 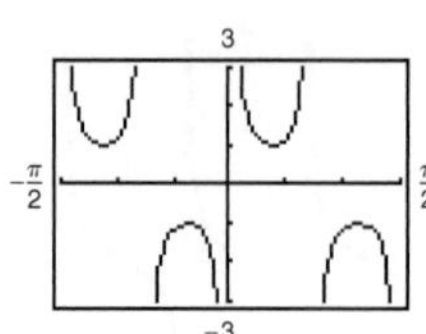

47. 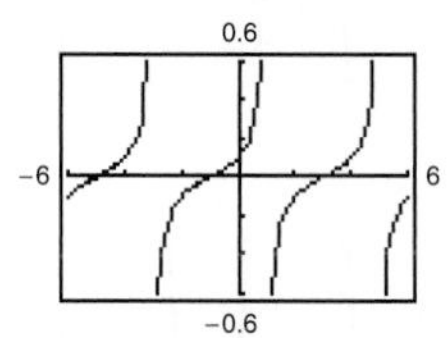

49. $-\frac{7\pi}{4}, -\frac{3\pi}{4}, \frac{\pi}{4}, \frac{5\pi}{4}$ **51.** $-\frac{7\pi}{6}, -\frac{\pi}{6}, \frac{5\pi}{6}, \frac{11\pi}{6}$

53. $-\frac{4\pi}{3}, -\frac{2\pi}{3}, \frac{2\pi}{3}, \frac{4\pi}{3}$ **55.** $-\frac{7\pi}{4}, -\frac{5\pi}{4}, \frac{\pi}{4}, \frac{3\pi}{4}$

57. Even **59.** Odd **61.** Odd **63.** Even

65. d, $f \to 0$ as $x \to 0$. **66.** a, $f \to 0$ as $x \to 0$.

67. b, $g \to 0$ as $x \to 0$. **68.** c, $g \to 0$ as $x \to 0$.

69. 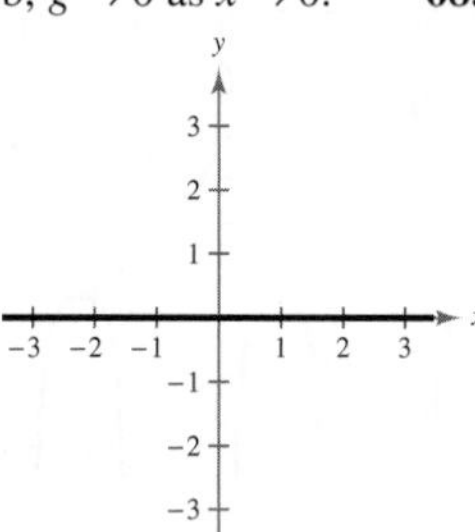

The functions are equal.

71. 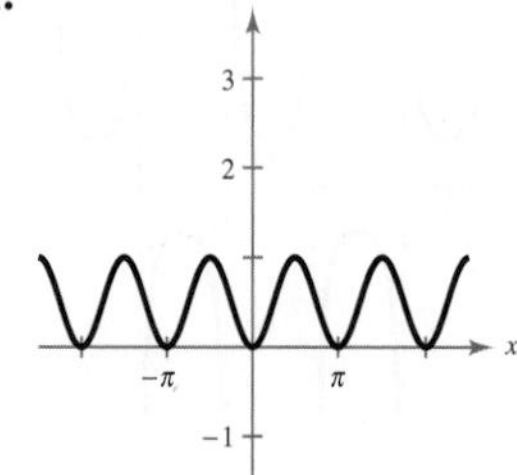

The functions are equal.

73. 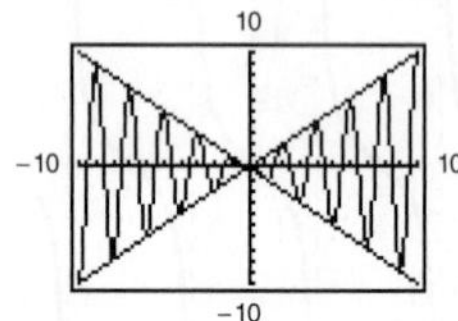

As $x \to \infty$, $g(x)$ oscillates.

75. 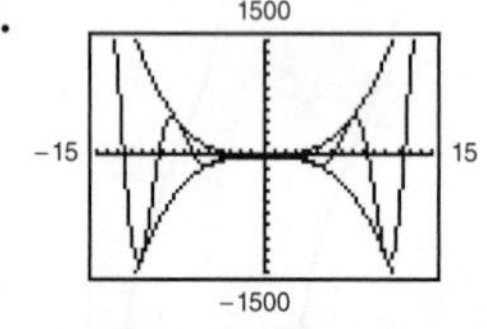

As $x \to \infty$, $f(x)$ oscillates.

77. 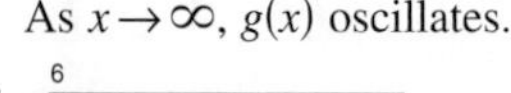

As $x \to 0$, $y \to \infty$.

79. 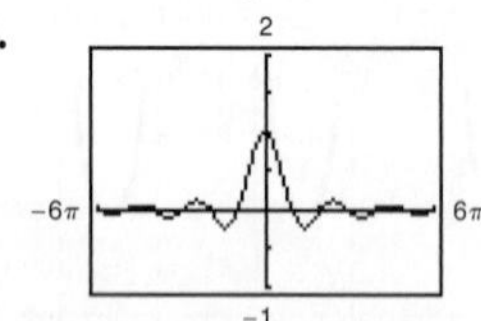

As $x \to 0$, $g(x) \to 1$.

81. 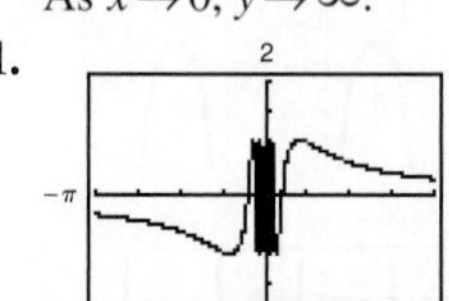

As $x \to 0$, $f(x)$ oscillates between 1 and -1.

83. (a) Period of $H(t)$: 12 mo
Period of $L(t)$: 12 mo
(b) Summer; winter
(c) About 0.5 mo

85. $d = 27 \sec x$

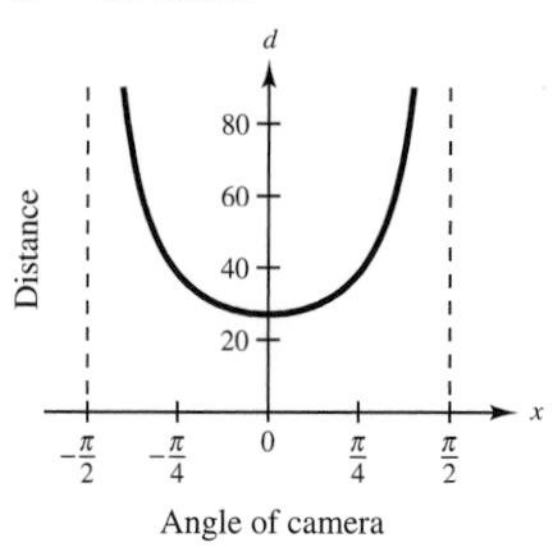

87. True. For a given value of x, the y-coordinate of $\csc x$ is the reciprocal of the y-coordinate of $\sin x$.

89. (a) 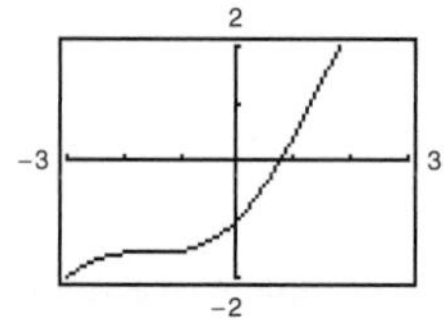

0.7391

(b) 1, 0.5403, 0.8576, 0.6543, 0.7935, 0.7014, 0.7640, 0.7221, 0.7504, 0.7314, . . . ; 0.7391

91. (a) $f(x) \to \infty$ (b) $f(x) \to -\infty$
(c) $f(x) \to \infty$ (d) $f(x) \to -\infty$

93. (a) $f(x) \to -\infty$ (b) $f(x) \to \infty$
(c) $f(x) \to -\infty$ (d) $f(x) \to \infty$

Section 1.7 *(page 186)*

1. $y = \sin^{-1} x;\ -1 \le x \le 1$

3. $y = \tan^{-1} x;\ -\infty < x < \infty;\ -\frac{\pi}{2} < y < \frac{\pi}{2}$ **5.** $\frac{\pi}{6}$

7. $\frac{\pi}{3}$ **9.** $\frac{\pi}{6}$ **11.** Not possible **13.** $-\frac{\pi}{3}$

15. $\frac{2\pi}{3}$ **17.** $-\frac{\pi}{3}$

19. 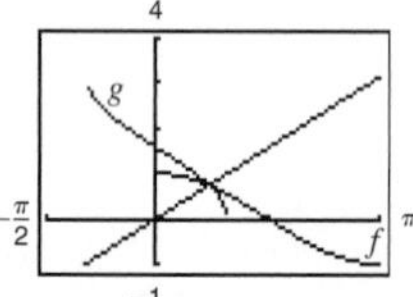

21. 1.19 **23.** -0.85 **25.** -1.25 **27.** Not possible

29. 1.99 **31.** 0.74 **33.** 1.07 **35.** -1.50

37. $-\frac{\pi}{3}, -\frac{\sqrt{3}}{3}, 1$ **39.** $\theta = \arctan \frac{x}{4}$ **41.** $\theta = \arcsin \frac{x+2}{5}$

43. $\theta = \arccos \frac{x+3}{2x}$ **45.** 0.3 **47.** Not possible

49. $\frac{\pi}{4}$ **51.** $\frac{3}{5}$ **53.** $\frac{\sqrt{5}}{5}$ **55.** $\frac{13}{12}$ **57.** $-\frac{5}{3}$

59. $-\frac{\sqrt{5}}{2}$ **61.** 2 **63.** $\sqrt{1-4x^2}$ **65.** $\frac{1}{x}$

67. $\sqrt{1-x^2}$ **69.** $\frac{\sqrt{9-x^2}}{x}$ **71.** $\frac{\sqrt{x^2+a^2}}{x}$

73. 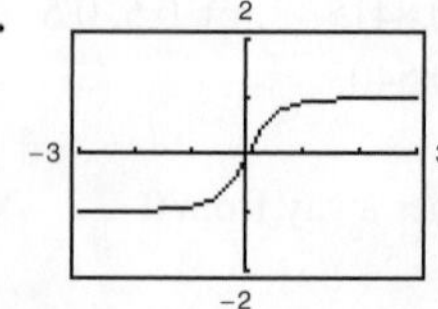

Asymptotes: $y = \pm 1$

75. $\dfrac{9}{\sqrt{x^2 + 81}}$ **77.** $\dfrac{|x - 1|}{\sqrt{x^2 - 2x + 10}}$

79.

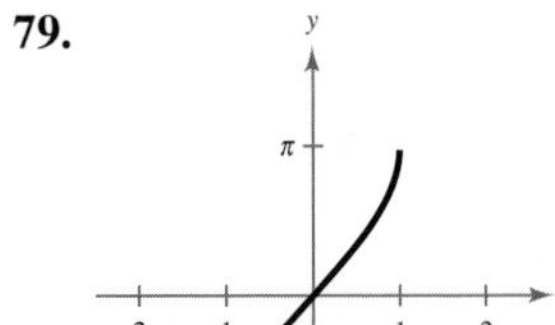

Vertical stretch

81.

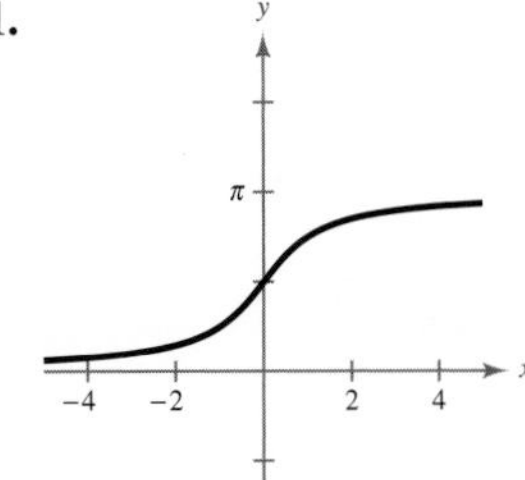

Shift $\dfrac{\pi}{2}$ units up

83.

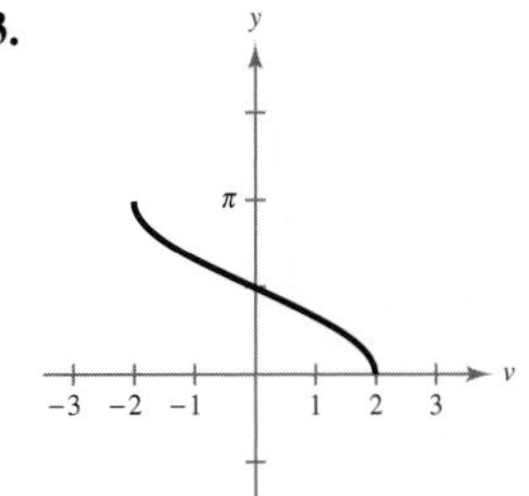

Horizontal stretch

85.

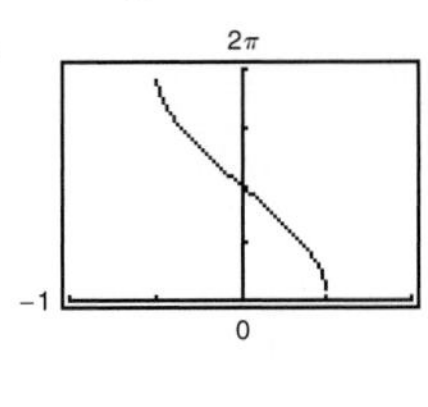

87.

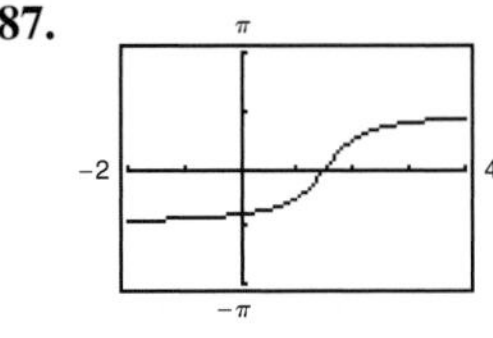

89.

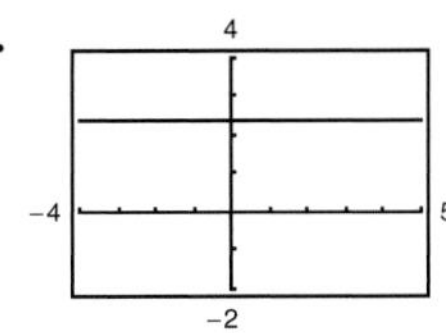

91. $3\sqrt{2}\sin\left(2t + \dfrac{\pi}{4}\right)$

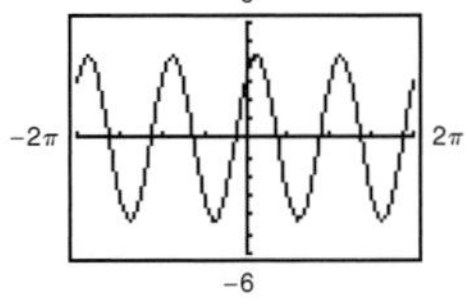

The graph implies that the identity is true.

93. $\dfrac{\pi}{2}$ **95.** $\dfrac{\pi}{2}$ **97.** π

99. (a) $\theta = \arcsin \dfrac{5}{s}$ (b) About 0.13, about 0.25

101. (a) About 32.9° (b) About 6.5 m

103. (a)

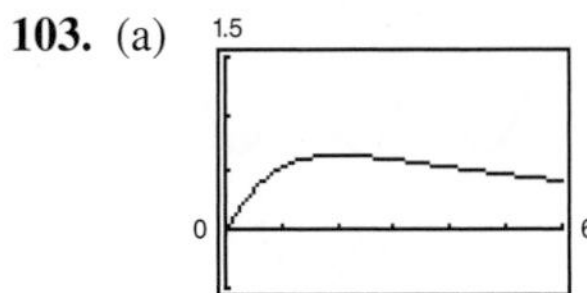

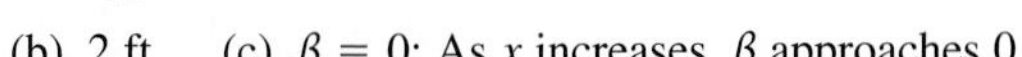

(b) 2 ft (c) $\beta = 0$; As x increases, β approaches 0.

105. (a) $\theta = \arctan \dfrac{x}{20}$ (b) About 14.0°, about 31.0°

107. True. $-\dfrac{\pi}{4}$ is in the range of the arctangent function.

109. False. $\sin^{-1} x$ is the inverse of $\sin x$, not the reciprocal.

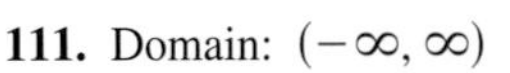

111. Domain: $(-\infty, \infty)$
Range: $(0, \pi)$

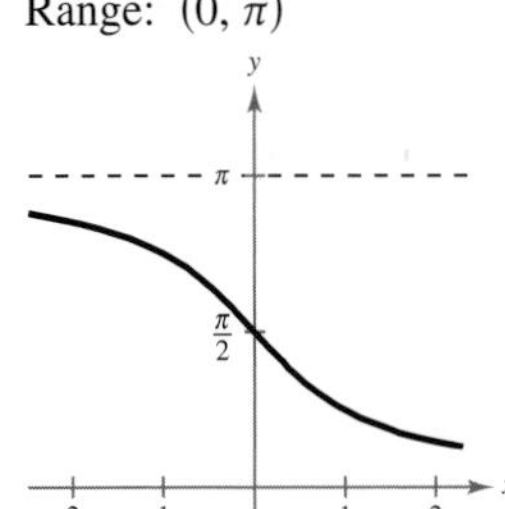

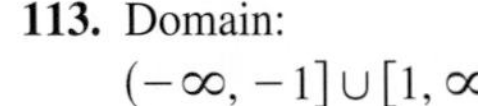

113. Domain:
$(-\infty, -1] \cup [1, \infty)$
Range:
$[-\pi/2, 0) \cup (0, \pi/2]$

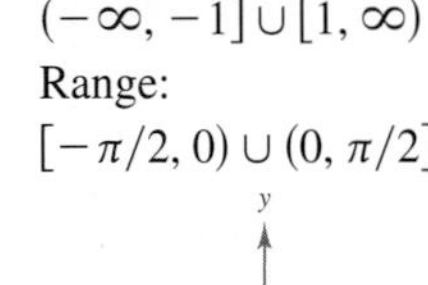

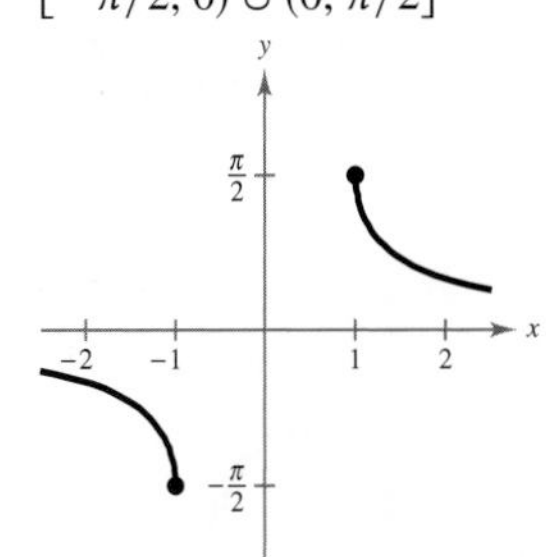

115. $\dfrac{\pi}{4}$ **117.** $\dfrac{3\pi}{4}$ **119.** $-\dfrac{\pi}{2}$ **121.** 1.17

123. -0.12 **125.** 0.19

127. (a) $\dfrac{\pi}{4}$ (b) $\dfrac{\pi}{2}$ (c) About 1.25 (d) About 2.03

129. (a) $f \circ f^{-1}$ $f^{-1} \circ f$

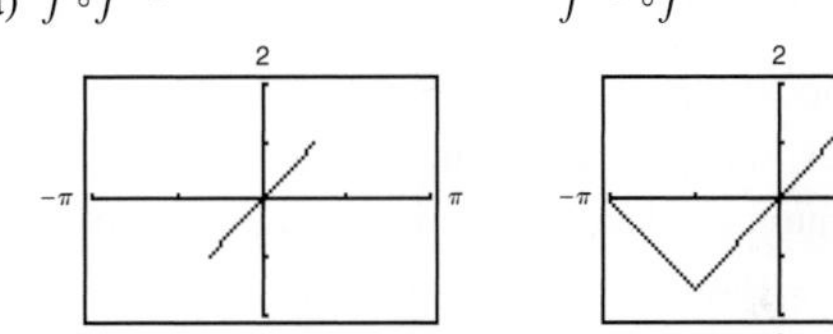

(b) The domains and ranges of the functions are restricted. The graphs of $f \circ f^{-1}$ and $f^{-1} \circ f$ differ because of the domains and ranges of f and f^{-1}.

Section 1.8 *(page 196)*

1. bearing **3.** period

5. $a \approx 10.39$
$b = 6$
$B = 30°$

7. $b \approx 14.21$
$c \approx 14.88$
$A = 17.2°$

9. $c = 5$
$A \approx 36.87°$
$B \approx 53.13°$

11. $a \approx 52.88$
$A \approx 73.46°$
$B \approx 16.54°$

13. 3.00 **15.** 2.50 **17.** About 214.45 ft

19. About 19.7 ft **21.** About 20.5 ft **23.** About 11.8 km

25. About 56.3° **27.** About 75.97°

29. About 3.23 mi or about 17,054 ft

31. (a) $l = 250$ ft, $A \approx 36.87°$, $B \approx 53.13°$
(b) About 4.87 sec

33. About 508 mi north, about 650 mi east

35. (a) About 104.95 nmi south, about 58.18 nmi west
(b) S 36.7° W; about 130.9 nmi

37. N 56.31° W **39.** (a) N 58° E (b) About 68.82 m

41. About 35.3° **43.** About 29.4 in. **45.** $d = 4 \sin \pi t$

47. $d = 3 \cos \dfrac{4\pi t}{3}$ **49.** $\omega = 524\pi$

51. (a) 9 (b) $\frac{3}{5}$ (c) 9 (d) $\frac{5}{12}$

53. (a) $\frac{1}{4}$ (b) 3 (c) 0 (d) $\frac{1}{6}$

55. (a)

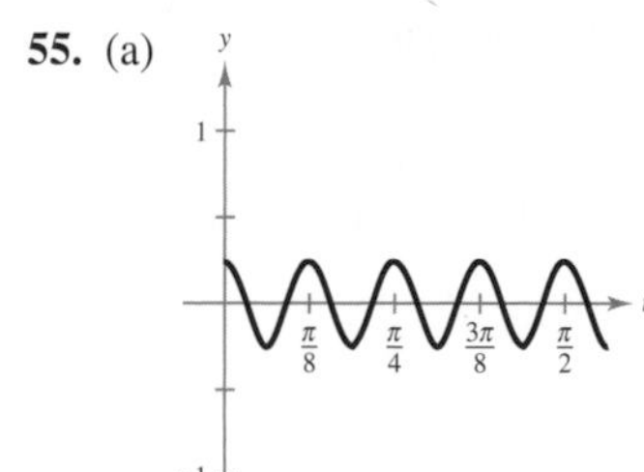

(b) $\frac{\pi}{8}$ (c) $\frac{\pi}{32}$

57. (a)

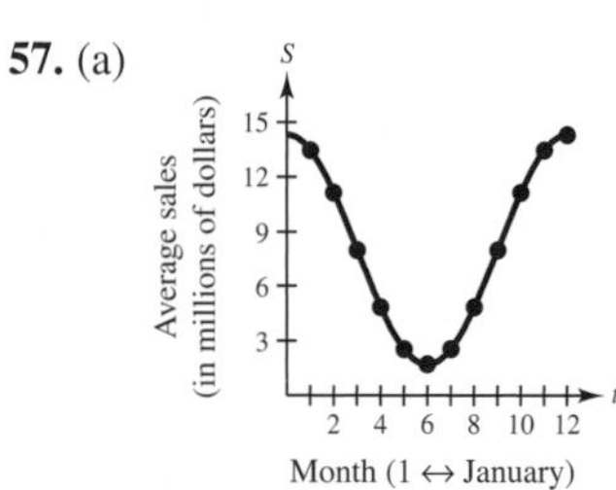

(b) $S = 8 + 6.3 \cos \frac{\pi}{6}t$ or $S = 8 + 6.3 \sin\left(\frac{\pi}{6}t + \frac{\pi}{2}\right)$

The model is a good fit.

(c) 12. Yes, sales of outerwear are seasonal.

(d) Maximum displacement from average sales of $8 million

59. False. The scenario does not create a right triangle because the tower is not vertical.

Review Exercises *(page 202)*

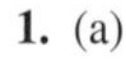

1. (a)

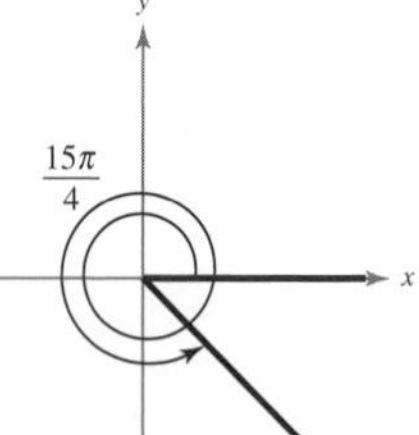

(b) Quadrant IV

(c) $\frac{23\pi}{4}, -\frac{\pi}{4}$

3. (a)

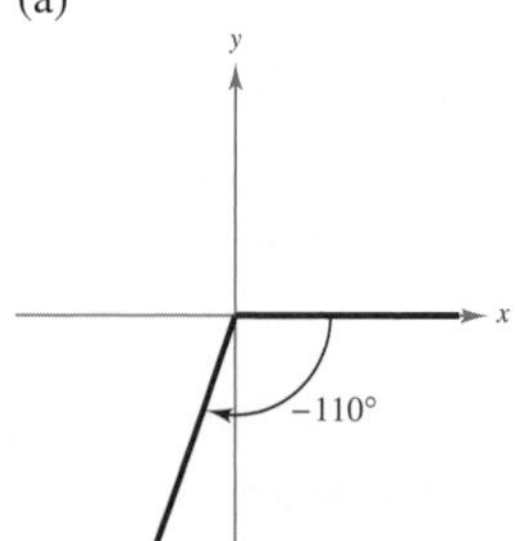

(b) Quadrant III

(c) $250°, -470°$

5. 7.854 **7.** -0.279 **9.** $54°$ **11.** $-200.535°$

13. $198° 24'$ **15.** About 48.17 in. **17.** About 523.60 in.^{2}

19. $\left(-\frac{1}{2}, \frac{\sqrt{3}}{2}\right)$ **21.** $\left(-\frac{\sqrt{3}}{2}, -\frac{1}{2}\right)$

23. $\sin \frac{3\pi}{4} = \frac{\sqrt{2}}{2}$ $\csc \frac{3\pi}{4} = \sqrt{2}$

$\cos \frac{3\pi}{4} = -\frac{\sqrt{2}}{2}$ $\sec \frac{3\pi}{4} = -\sqrt{2}$

$\tan \frac{3\pi}{4} = -1$ $\cot \frac{3\pi}{4} = -1$

25. $\frac{\sqrt{2}}{2}$ **27.** $-\frac{\sqrt{3}}{2}$ **29.** 3.2361 **31.** -75.3130

33. $\sin \theta = \frac{4\sqrt{41}}{41}$ $\csc \theta = \frac{\sqrt{41}}{4}$

$\cos \theta = \frac{5\sqrt{41}}{41}$ $\sec \theta = \frac{\sqrt{41}}{5}$

$\tan \theta = \frac{4}{5}$ $\cot \theta = \frac{5}{4}$

35. 0.6494 **37.** 3.6722

39. (a) 3 (b) $\frac{2\sqrt{2}}{3}$ (c) $\frac{3\sqrt{2}}{4}$ (d) $\frac{\sqrt{2}}{4}$

41. About 73.3 m

43. $\sin \theta = \frac{4}{5}$ $\csc \theta = \frac{5}{4}$

$\cos \theta = \frac{3}{5}$ $\sec \theta = \frac{5}{3}$

$\tan \theta = \frac{4}{3}$ $\cot \theta = \frac{3}{4}$

45. $\sin \theta = \frac{4}{5}$ $\csc \theta = \frac{5}{4}$

$\cos \theta = \frac{3}{5}$ $\sec \theta = \frac{5}{3}$

$\tan \theta = \frac{4}{3}$ $\cot \theta = \frac{3}{4}$

47. $\sin \theta = -\frac{\sqrt{11}}{6}$ $\csc \theta = -\frac{6\sqrt{11}}{11}$

$\cos \theta = \frac{5}{6}$

$\tan \theta = -\frac{\sqrt{11}}{5}$ $\cot \theta = -\frac{5\sqrt{11}}{11}$

49. $\sin \theta = \frac{\sqrt{21}}{5}$ $\csc \theta = \frac{5\sqrt{21}}{21}$

$\sec \theta = -\frac{5}{2}$

$\tan \theta = -\frac{\sqrt{21}}{2}$ $\cot \theta = -\frac{2\sqrt{21}}{21}$

51. $\theta' = 84°$

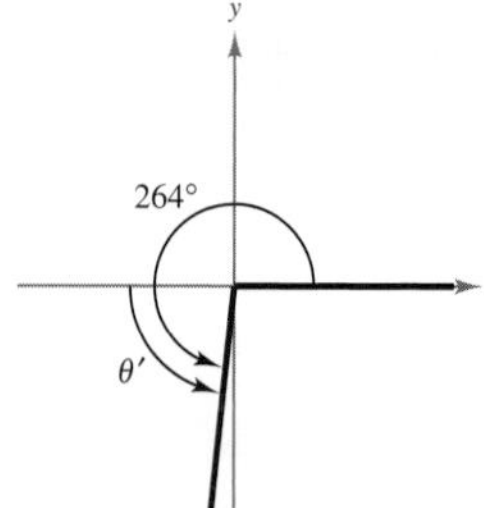

53. $\theta' = \frac{\pi}{5}$

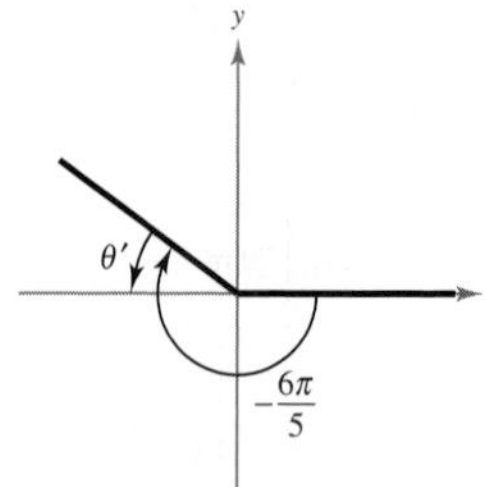

55. $\sin(-150°) = -\frac{1}{2}$; $\cos(-150°) = -\frac{\sqrt{3}}{2}$; $\tan(-150°) = \frac{\sqrt{3}}{3}$

57. $\sin \frac{\pi}{3} = \frac{\sqrt{3}}{2}$; $\cos \frac{\pi}{3} = \frac{1}{2}$; $\tan \frac{\pi}{3} = \sqrt{3}$

59. 0.9613 **61.** -0.4452

63.

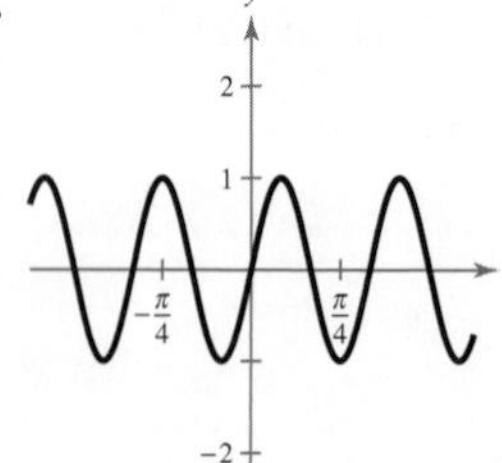

65.

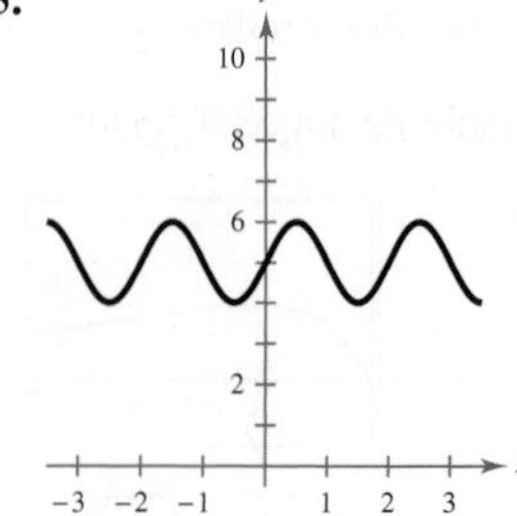

67.

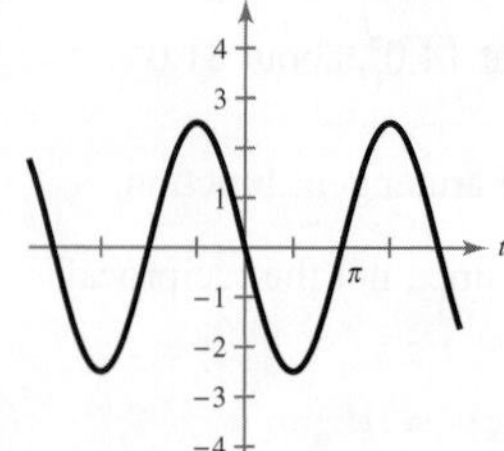

69. (a) $y = 2 \sin 528\pi x$ (b) 264 cycles/sec

71.

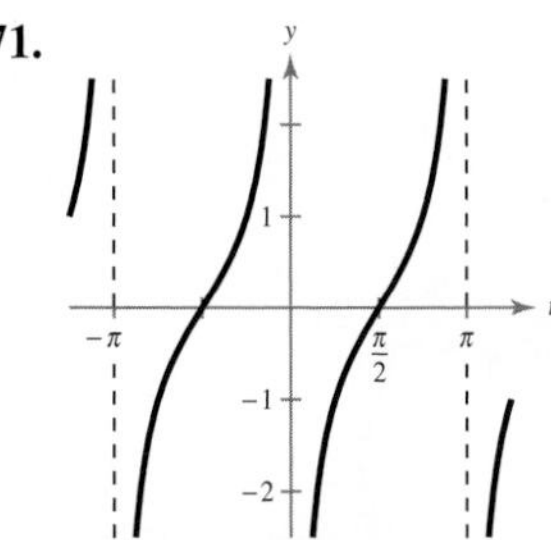

73.

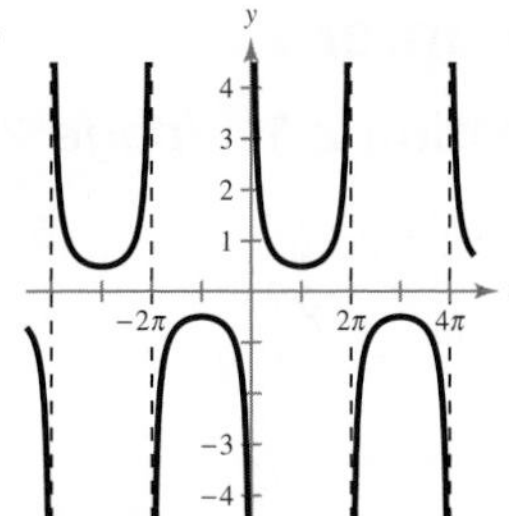

75.

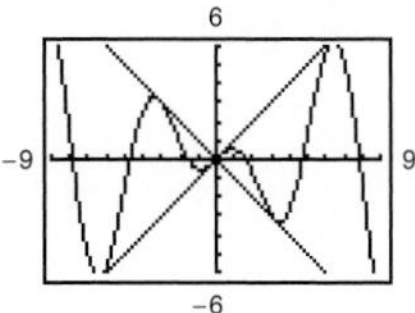

As $x \to +\infty$, $f(x)$ oscillates.

77. $-\frac{\pi}{2}$ **79.** $\frac{\pi}{6}$ **81.** -0.92 **83.** 0.06

85.

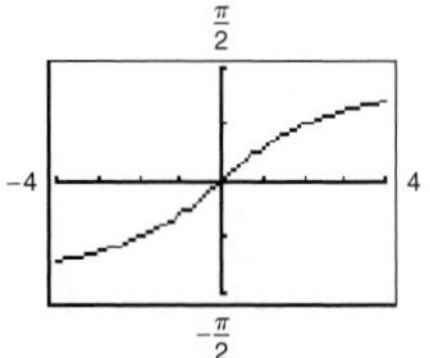

87. $\frac{4}{5}$ **89.** $\frac{13}{5}$ **91.** $\frac{\sqrt{4-x^2}}{x}$

93.

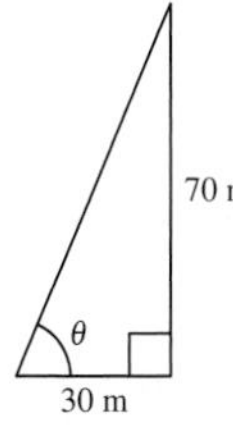

$\theta \approx 66.8°$

95. About 1221 mi, about 85.6°

97. False. For each θ there corresponds exactly one value of y.

99. The function is undefined because $\sec\theta = 1/\cos\theta$.

101. The ranges of the other four trigonometric functions are $(-\infty, \infty)$ or $(-\infty, -1] \cup [1, \infty)$.

103. Answers will vary.

Chapter Test *(page 205)*

1. (a)

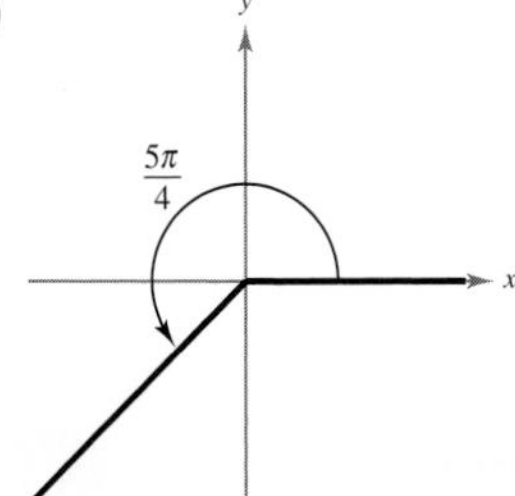

(b) $\frac{13\pi}{4}, -\frac{3\pi}{4}$

(c) 225°

2. 3500 rad/min **3.** About 709.04 ft^2

4. $\sin\theta = \frac{3\sqrt{13}}{13}$ $\csc\theta = \frac{\sqrt{13}}{3}$

$\cos\theta = \frac{2\sqrt{13}}{13}$ $\sec\theta = \frac{\sqrt{13}}{2}$

$\cot\theta = \frac{2}{3}$

5. $\sin\theta = \frac{3\sqrt{10}}{10}$ $\csc\theta = \frac{\sqrt{10}}{3}$

$\cos\theta = -\frac{\sqrt{10}}{10}$ $\sec\theta = -\sqrt{10}$

$\tan\theta = -3$ $\cot\theta = -\frac{1}{3}$

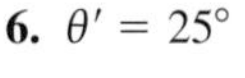

6. $\theta' = 25°$

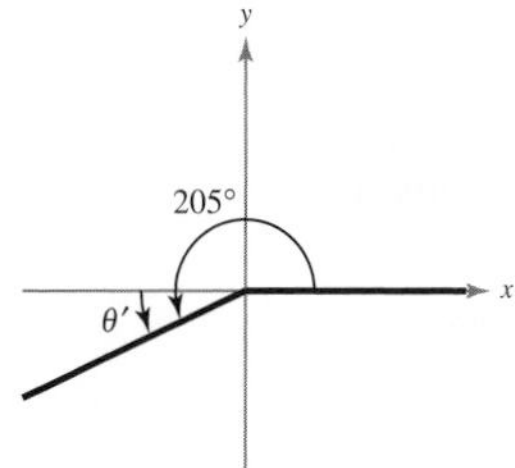

7. Quadrant III **8.** 150°, 210°

9. $\sin\theta = -\frac{4}{5}$
$\tan\theta = -\frac{4}{3}$
$\csc\theta = -\frac{5}{4}$
$\sec\theta = \frac{5}{3}$
$\cot\theta = -\frac{3}{4}$

10. $\sin\theta = \frac{21}{29}$
$\cos\theta = -\frac{20}{29}$
$\tan\theta = -\frac{21}{20}$
$\csc\theta = \frac{29}{21}$
$\cot\theta = -\frac{20}{21}$

11.

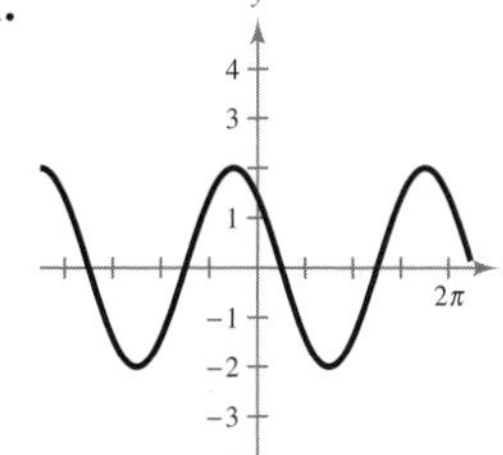

12.

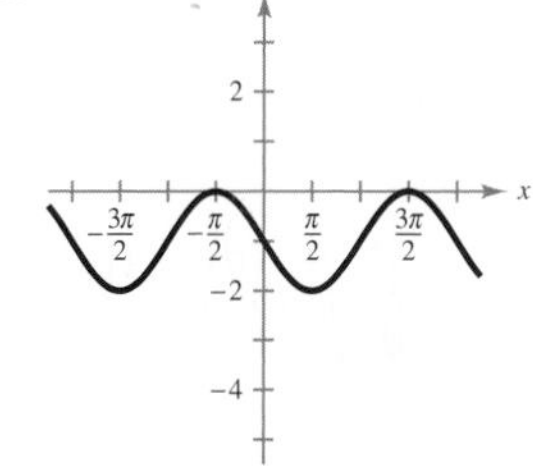

13.

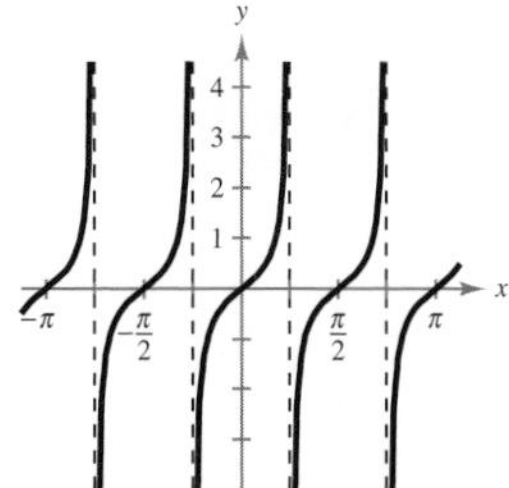

14.

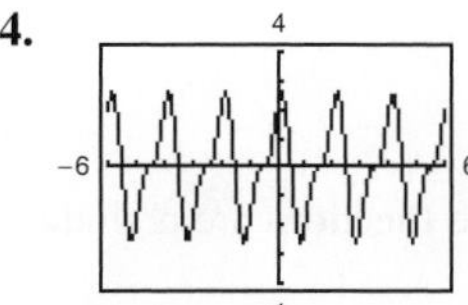

Period: 2

15.

160
0
32
−80

Not periodic;
As $x \to \infty$, y oscillates.

16. $a = -2, b = \frac{1}{2}, c = -\frac{\pi}{4}$ **17.** $\frac{\sqrt{55}}{3}$

CHAPTER 1

18.

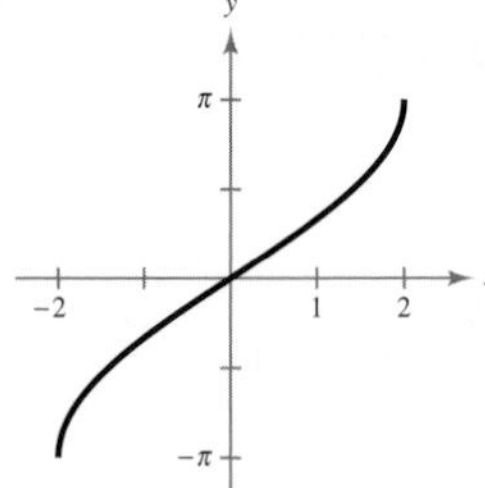

19. About 309.3° **20.** $d = -6 \cos \pi t$

Problem Solving *(page 207)*

1. (a) $\frac{11\pi}{2}$ rad or 990° (b) About 816.42 ft

3. $h = 51 - 50 \sin\left(8\pi t + \frac{\pi}{2}\right)$

5. (a) About 4767 ft (b) About 3705 ft

(c) $w \approx 2183$ ft, $\tan 63° = \frac{w + 3705}{3000}$

7. (a)

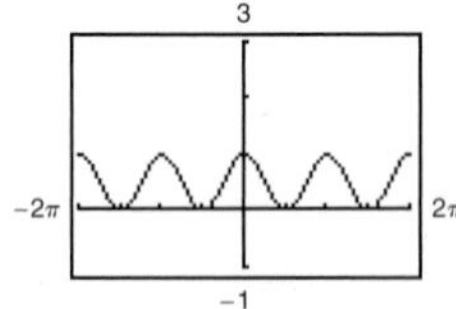

Even

(b)

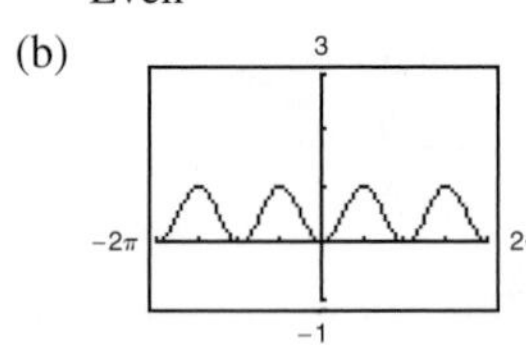

Even

9. (a)

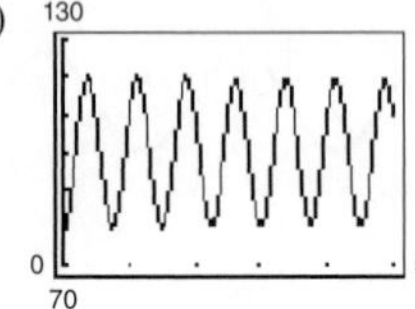

(b) Period $= \frac{3}{4}$ sec; Answers will vary.

(c) 20 mm; Answers will vary. (d) 80 beats/min

(e) Period $= \frac{15}{16}$ sec; $\frac{32\pi}{15}$

11. (a)

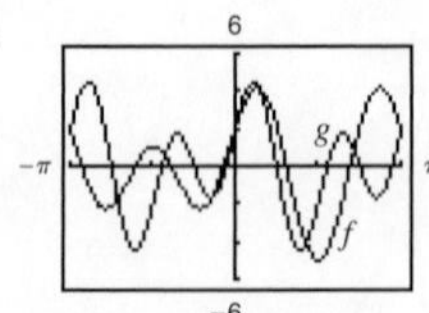

(b) Period of f: 2π

Period of g: π

(c) Yes, because the sine and cosine functions are periodic.

13. (a) About 40.5°

(b) $x \approx 1.71$ ft; $y \approx 3.46$ ft

(c) About 1.75 ft

(d) As you move closer to the rock, d must get smaller and smaller. The angles θ_1 and θ_2 will decrease along with the distance y, so d will decrease.

Chapter 2

Section 2.1 *(page 215)*

1. $\tan u$ **3.** $\cot u$ **5.** 1

7. $\sin x = \frac{\sqrt{21}}{5}$ $\csc x = \frac{5\sqrt{21}}{21}$

$\cos x = -\frac{2}{5}$ $\sec x = -\frac{5}{2}$

$\tan x = -\frac{\sqrt{21}}{2}$ $\cot x = -\frac{2\sqrt{21}}{21}$

9. $\sin \theta = -\frac{3}{4}$ $\csc \theta = -\frac{4}{3}$

$\cos \theta = \frac{\sqrt{7}}{4}$ $\sec \theta = \frac{4\sqrt{7}}{7}$

$\tan \theta = -\frac{3\sqrt{7}}{7}$ $\cot \theta = -\frac{\sqrt{7}}{3}$

11. $\sin x = \frac{2\sqrt{13}}{13}$ $\csc x = \frac{\sqrt{13}}{2}$

$\cos x = \frac{3\sqrt{13}}{13}$ $\sec x = \frac{\sqrt{13}}{3}$

$\tan x = \frac{2}{3}$ $\cot x = \frac{3}{2}$

13. c **14.** b **15.** f **16.** a **17.** e **18.** d

19. $\cos \theta$ **21.** $\sin^2 x$ **23.** $\sec x + 1$ **25.** $\sin^4 x$

27. $\csc^2 x(\cot x + 1)$ **29.** $(3 \sin x + 1)(\sin x - 2)$

31. $(\csc x - 1)(\csc x + 2)$ **33.** $\sec \theta$ **35.** $\cos^2 \phi$

37. $\sec \beta$ **39.** $\sin^2 x$ **41.** $1 + 2 \sin x \cos x$

43. $2 \csc^2 x$ **45.** $-2 \tan x$ **47.** $-\cot x$ **49.** $1 + \cos y$

51. $\cos x$ **53.** $\sin x$ **55.** $3 \sin \theta$ **57.** $2 \tan \theta$

59. $2 \cos \theta = \sqrt{2}$; $\sin \theta = \pm\frac{\sqrt{2}}{2}$; $\cos \theta = \frac{\sqrt{2}}{2}$

61. $6 \cos \theta = 3$; $\sin \theta = \pm\frac{\sqrt{3}}{2}$; $\cos \theta = \frac{1}{2}$ **63.** $0 \le \theta \le \pi$

65. $0 \le \theta < \frac{\pi}{2}, \frac{3\pi}{2} < \theta < 2\pi$ **67.** $\mu = \tan \theta$

69. True. $\csc x = \frac{1}{\sin x}$, $\sec x = \frac{1}{\cos x}$, $\tan x = \frac{\sin x}{\cos x}$, $\cot x = \frac{\cos x}{\sin x}$

71. $\infty, 0$ **73.** $\cos(-\theta) = \cos \theta$

75. $\cos \theta = \pm\sqrt{1 - \sin^2 \theta}$

$\tan \theta = \pm\frac{\sin \theta}{\sqrt{1 - \sin^2 \theta}}$

$\cot \theta = \pm\frac{\sqrt{1 - \sin^2 \theta}}{\sin \theta}$

$\sec \theta = \pm\frac{1}{\sqrt{1 - \sin^2 \theta}}$

$\csc \theta = \frac{1}{\sin \theta}$

77. $\frac{\sin \theta}{\cos \theta}$

Section 2.2 *(page 222)*

1. identity **3.** $\tan u$ **5.** $\sin u$ **7.** $-\csc u$

9–41. Answers will vary. **43.** $\cot(-x) = -\cot x$

45. (a)

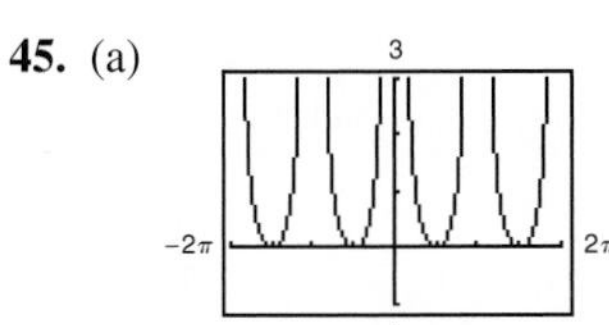

(b)

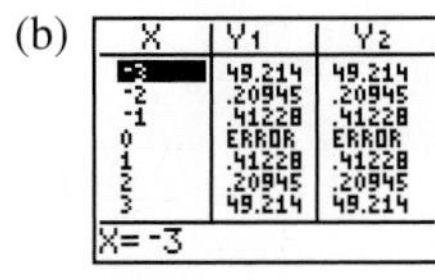

X	Y1	Y2
-3	49.214	49.214
-2	.20945	.20945
-1	.41228	.41228
0	ERROR	ERROR
1	.41228	.41228
2	.20945	.20945
3	49.214	49.214

X=-3

Identity

(c) Answers will vary.

47. (a)

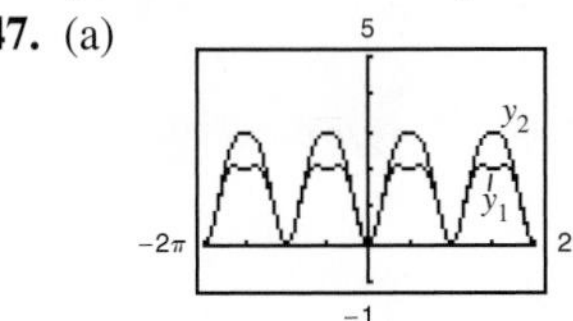

(b)

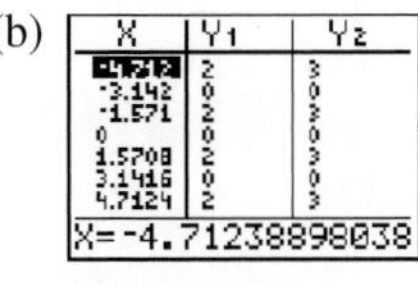

X	Y1	Y2
-4.712	2	3
-3.142	0	0
-1.571	2	3
0	0	0
1.5708	2	3
3.1416	0	0
4.7124	2	3

X=-4.71238898038

Not an identity

(c) Answers will vary.

49. (a) (b)

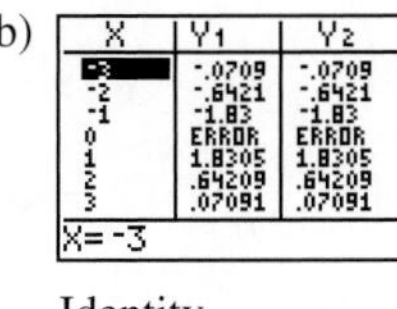

X	Y1	Y2
-3	-.0709	-.0709
-2	-.6421	-.6421
-1	-1.83	-1.83
0	ERROR	ERROR
1	1.8305	1.8305
2	.64209	.64209
3	.07091	.07091

X=-3

Identity

(c) Answers will vary.

51–53. Answers will vary. **55.** 1

57–61. Answers will vary.

63. False. $\tan x^2 = \tan(x \cdot x) \neq \tan^2 x = (\tan x)(\tan x)$

65. False. An identity is an equation that is true for all real values of θ.

67. The equation is not an identity because $\sin\theta = \pm\sqrt{1 - \cos^2\theta}$.

Sample answer: $\frac{7\pi}{4}$

69. The equation is not an identity because $1 - \cos^2\theta = \sin^2\theta$.

Sample answer: $-\frac{\pi}{2}$

Section 2.3 *(page 232)*

1. isolate **3.** quadratic **5–9.** Answers will vary.

11. $\frac{\pi}{3} + 2n\pi, \frac{2\pi}{3} + 2n\pi$ **13.** $\frac{2\pi}{3} + 2n\pi, \frac{4\pi}{3} + 2n\pi$

15. $\frac{\pi}{6} + n\pi, \frac{5\pi}{6} + n\pi$ **17.** $\frac{\pi}{3} + n\pi, \frac{2\pi}{3} + n\pi$

19. $n\pi, \frac{3\pi}{2} + 2n\pi$ **21.** $\frac{n\pi}{2}$ **23.** $n\pi, \frac{\pi}{6} + n\pi, \frac{5\pi}{6} + n\pi$

25. $\pi + 2n\pi, \frac{\pi}{3} + 2n\pi, \frac{5\pi}{3} + 2n\pi$

27. $\frac{\pi}{3} + 2n\pi, \pi + 2n\pi, \frac{5\pi}{3} + 2n\pi$ **29.** $\frac{\pi}{4}, \frac{5\pi}{4}$

31. $\frac{\pi}{2}, \frac{3\pi}{2}, \frac{2\pi}{3}, \frac{4\pi}{3}$ **33.** $\frac{\pi}{3}, \frac{2\pi}{3}, \frac{4\pi}{3}, \frac{5\pi}{3}$ **35.** No solution

37. $\frac{\pi}{2}$ **39.** $\frac{\pi}{6} + n\pi, \frac{5\pi}{6} + n\pi$ **41.** $\frac{\pi}{12} + \frac{n\pi}{3}$

43. $\frac{\pi}{2} + 4n\pi, \frac{7\pi}{2} + 4n\pi$ **45.** $\frac{\pi}{3} + 2n\pi$ **47.** $3 + 4n$

49. 3.553, 5.872 **51.** 1.249, 4.391 **53.** 0.739

55. 0.955, 2.186, 4.097, 5.328 **57.** 1.221, 1.921, 4.362, 5.062

59. $\arctan(-4) + n\pi, \arctan 3 + n\pi$

61. $\frac{\pi}{4} + n\pi, \arctan 5 + n\pi$ **63.** $\frac{\pi}{3} + 2n\pi, \frac{5\pi}{3} + 2n\pi$

65. $\arctan\frac{1}{3} + n\pi, \arctan\left(-\frac{1}{3}\right) + n\pi$

67. $\arccos\frac{1}{4} + 2n\pi, -\arccos\frac{1}{4} + 2n\pi$

69. $\frac{\pi}{2} + 2n\pi, \arcsin\left(-\frac{1}{4}\right) + 2n\pi, \arcsin\frac{1}{4} + 2n\pi$

71. 0.3398, 0.8481, 2.2935, 2.8018

73. 1.9357, 2.7767, 5.0773, 5.9183

75. −1.154, 0.534 **77.** 1.110

79. (a)

(b) $\frac{\pi}{3} \approx 1.0472$
$\frac{5\pi}{3} \approx 5.2360$
0
$\pi \approx 3.1416$

Maximum: (1.0472, 1.25)
Maximum: (5.2360, 1.25)
Minimum: (0, 1)
Minimum: (3.1416, −1)

81. (a)

(b) $\frac{\pi}{4} \approx 0.7854$
$\frac{5\pi}{4} \approx 3.9270$

Maximum: (0.7854, 1.4142)
Minimum: (3.9270, −1.4142)

83. (a)

(b) $\frac{\pi}{4} \approx 0.7854$
$\frac{5\pi}{4} \approx 3.9270$
$\frac{3\pi}{4} \approx 2.3562$
$\frac{7\pi}{4} \approx 5.4978$

Maximum: (0.7854. 0.5)
Maximum: (3.9270, 0.5)
Minimum: (2.3562, −0.5)
Minimum: (5.4978, −0.5)

85. 1

87. (a) All real numbers x except $x = 0$
(b) y-axis symmetry; Horizontal asymptote: $y = 0$
(c) y approaches 1.
(d) Four solutions: $\pm\pi, \pm 2\pi$

89. 0.04 sec, 0.43 sec, 0.83 sec

91. January, November, December

93. (a) and (c)

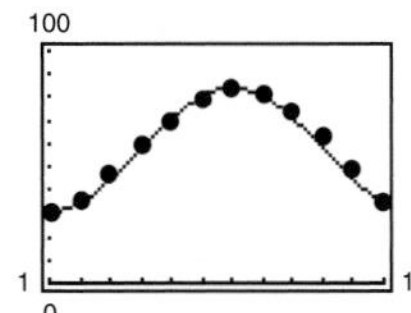

The model fits the data well.

(b) $C = 26.55\cos\left(\frac{\pi t}{6} - \frac{7\pi}{6}\right) + 57.55$

(d) 57.55°F

(e) Above 72°F: June through September
Below 72°F: October through May

95. (a)

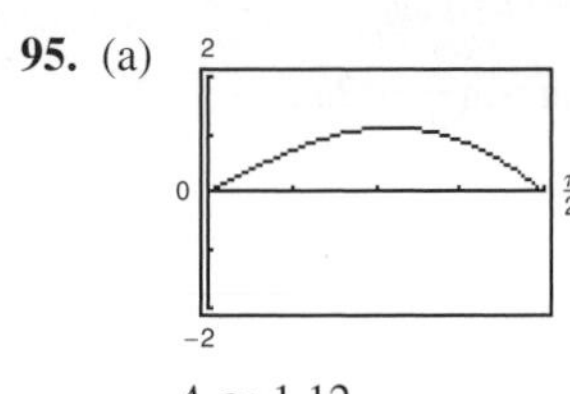

$A \approx 1.12$

(b) $0.6 < x < 1.1$

97. 1

99. True. The first equation has a smaller period than the second equation, so it will have more solutions in the interval $[0, 2\pi)$.

101. The equation would become $\cos^2 x = 2$; this is not the correct method to use when solving equations.

103. (a)

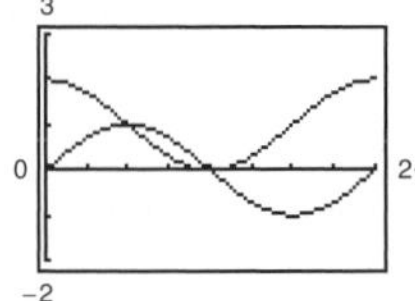

Graphs intersect when $x = \frac{\pi}{2}$ and $x = \pi$.

(b)

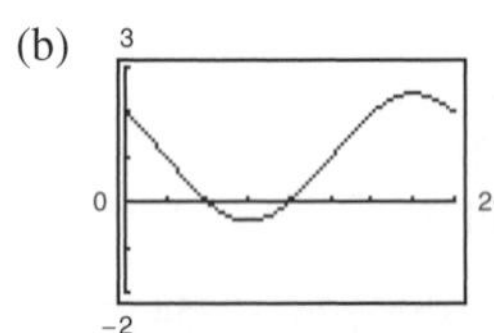

x-intercepts: $\left(\frac{\pi}{2}, 0\right), (\pi, 0)$

(c) Yes; Answers will vary.

Section 2.4 *(page 240)*

1. $\sin u \cos v - \cos u \sin v$ **3.** $\dfrac{\tan u + \tan v}{1 - \tan u \tan v}$

5. $\cos u \cos v + \sin u \sin v$

7. (a) $\dfrac{\sqrt{2} - \sqrt{6}}{4}$ (b) $\dfrac{\sqrt{2} + 1}{2}$

9. (a) $\dfrac{\sqrt{6} + \sqrt{2}}{4}$ (b) $\dfrac{\sqrt{2} - \sqrt{3}}{2}$

11. $\sin \frac{11\pi}{12} = \frac{\sqrt{2}}{4}(\sqrt{3} - 1)$

$\cos \frac{11\pi}{12} = -\frac{\sqrt{2}}{4}(\sqrt{3} + 1)$

$\tan \frac{11\pi}{12} = -2 + \sqrt{3}$

13. $\sin \frac{17\pi}{12} = -\frac{\sqrt{2}}{4}(\sqrt{3} + 1)$

$\cos \frac{17\pi}{12} = \frac{\sqrt{2}}{4}(1 - \sqrt{3})$

$\tan \frac{17\pi}{12} = 2 + \sqrt{3}$

15. $\sin 105° = \frac{\sqrt{2}}{4}(\sqrt{3} + 1)$

$\cos 105° = \frac{\sqrt{2}}{4}(1 - \sqrt{3})$

$\tan 105° = -2 - \sqrt{3}$

17. $\sin(-195°) = \frac{\sqrt{2}}{4}(\sqrt{3} - 1)$

$\cos(-195°) = -\frac{\sqrt{2}}{4}(\sqrt{3} + 1)$

$\tan(-195°) = -2 + \sqrt{3}$

19. $\sin \frac{13\pi}{12} = \frac{\sqrt{2}}{4}(1 - \sqrt{3})$

$\cos \frac{13\pi}{12} = -\frac{\sqrt{2}}{4}(1 + \sqrt{3})$

$\tan \frac{13\pi}{12} = 2 - \sqrt{3}$

21. $\sin\left(-\frac{5\pi}{12}\right) = -\frac{\sqrt{2}}{4}(1 + \sqrt{3})$

$\cos\left(-\frac{5\pi}{12}\right) = \frac{\sqrt{2}}{4}(\sqrt{3} - 1)$

$\tan\left(-\frac{5\pi}{12}\right) = -2 - \sqrt{3}$

23. $\sin 285° = -\frac{\sqrt{2}}{4}(\sqrt{3} + 1)$

$\cos 285° = \frac{\sqrt{2}}{4}(\sqrt{3} - 1)$

$\tan 285° = -(2 + \sqrt{3})$

25. $\sin(-165°) = -\frac{\sqrt{2}}{4}(\sqrt{3} - 1)$

$\cos(-165°) = -\frac{\sqrt{2}}{4}(1 + \sqrt{3})$

$\tan(-165°) = 2 - \sqrt{3}$

27. $\sin 1.8$ **29.** $\sin 75°$ **31.** $\tan \frac{7\pi}{15}$ **33.** $\cos(3x - 2y)$

35. $\frac{\sqrt{3}}{2}$ **37.** $-\frac{1}{2}$ **39.** 0 **41.** $-\frac{13}{85}$ **43.** $-\frac{13}{84}$

45. $\frac{85}{36}$ **47.** $\frac{3}{5}$ **49.** $-\frac{44}{117}$ **51.** $-\frac{125}{44}$ **53.** 1 **55.** 0

57–63. Answers will vary. **65.** $-\sin \theta$ **67.** $-\sec \theta$

69. $\frac{\pi}{6}, \frac{5\pi}{6}$ **71.** $\frac{5\pi}{4}, \frac{7\pi}{4}$ **73.** $0, \frac{\pi}{3}, \pi, \frac{5\pi}{3}$

75. $\frac{\pi}{4}, \frac{7\pi}{4}$ **77.** $\frac{\pi}{2}, \pi, \frac{3\pi}{2}$

79. (a) $y = \frac{5}{12}\sin(2t + 0.6435)$ (b) $\frac{5}{12}$ ft (c) $\frac{1}{\pi}$ cycle/sec

81. True. $\sin(u \pm v) = \sin u \cos v \pm \cos u \sin v$

83. False. $\sin \alpha \cos \beta = -\cos \alpha \sin \beta$

85. The denominator should be $1 + \tan x \tan \frac{\pi}{4} = 1 + \tan x$.

87–89. Answers will vary.

91. (a) $\sqrt{2}\sin\left(\theta + \frac{\pi}{4}\right)$ (b) $\sqrt{2}\cos\left(\theta - \frac{\pi}{4}\right)$

93. (a) $13\sin(3\theta + 0.3948)$ (b) $13\cos(3\theta - 1.1760)$

95. $\sqrt{2}\sin \theta + \sqrt{2}\cos \theta$ **97.** 15°

99.

No, $y_1 \neq y_2$ because their graphs are different.

101. Proof

Section 2.5 *(page 250)*

1. $2 \sin u \cos u$ **3.** $\frac{1}{2}[\sin(u + v) + \sin(u - v)]$

5. $\pm\sqrt{\dfrac{1 - \cos u}{2}}$ **7.** $n\pi, \dfrac{\pi}{3} + 2n\pi, \dfrac{5\pi}{3} + 2n\pi$ **9.** $\dfrac{2n\pi}{3}$

11. $\dfrac{n\pi}{2}$ **13.** $\dfrac{\pi}{6} + n\pi, \dfrac{\pi}{2} + n\pi, \dfrac{5\pi}{6} + n\pi$ **15.** $3 \sin 2x$

17. $3 \cos 2x$ **19.** $4 \cos 2x$

21. $\sin 2u = -\frac{24}{25}, \cos 2u = \frac{7}{25}, \tan 2u = -\frac{24}{7}$

23. $\sin 2u = \frac{15}{17}, \cos 2u = \frac{8}{17}, \tan 2u = \frac{15}{8}$

25. $8 \cos^4 x - 8 \cos^2 x + 1$ **27.** $\frac{1}{8}(3 + 4 \cos 2x + \cos 4x)$

29. $\dfrac{1}{8}(3 - 4 \cos 4x + \cos 8x)$ **31.** $\dfrac{(3 - 4 \cos 4x + \cos 8x)}{(3 + 4 \cos 4x + \cos 8x)}$

33. $\frac{1}{8}(1 - \cos 8x)$

35. $\sin 75^\circ = \frac{1}{2}\sqrt{2 + \sqrt{3}}$

$\cos 75^\circ = \frac{1}{2}\sqrt{2 - \sqrt{3}}$

$\tan 75^\circ = 2 + \sqrt{3}$

37. $\sin 112^\circ 30' = \frac{1}{2}\sqrt{2 + \sqrt{2}}$

$\cos 112^\circ 30' = -\frac{1}{2}\sqrt{2 - \sqrt{2}}$

$\tan 112^\circ 30' = -1 - \sqrt{2}$

39. $\sin \dfrac{\pi}{8} = \dfrac{1}{2}\sqrt{2 - \sqrt{2}}$

$\cos \dfrac{\pi}{8} = \dfrac{1}{2}\sqrt{2 + \sqrt{2}}$

$\tan \dfrac{\pi}{8} = \sqrt{2} - 1$

41. (a) Quadrant I

(b) $\sin \dfrac{u}{2} = \dfrac{3}{5}, \cos \dfrac{u}{2} = \dfrac{4}{5}, \tan \dfrac{u}{2} = \dfrac{3}{4}$

43. (a) Quadrant II

(b) $\sin \dfrac{u}{2} = \dfrac{\sqrt{26}}{26}, \cos \dfrac{u}{2} = -\dfrac{5\sqrt{26}}{26}, \tan \dfrac{u}{2} = -\dfrac{1}{5}$

45. π **47.** $\dfrac{\pi}{3}, \pi, \dfrac{5\pi}{3}$ **49.** $\dfrac{1}{2}(\cos 2\theta - \cos 8\theta)$

51. $\frac{1}{2}(\cos(-2\theta) + \cos 6\theta)$ **53.** $2 \cos 4\theta \sin \theta$

55. $2 \cos 4x \cos 2x$ **57.** $\dfrac{\sqrt{6}}{2}$ **59.** $-\sqrt{2}$

61. $0, \dfrac{\pi}{4}, \dfrac{\pi}{2}, \dfrac{3\pi}{4}, \pi, \dfrac{5\pi}{4}, \dfrac{3\pi}{2}, \dfrac{7\pi}{4}$ **63.** $\dfrac{\pi}{6}, \dfrac{5\pi}{6}$

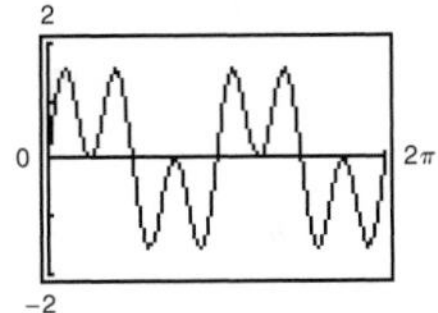

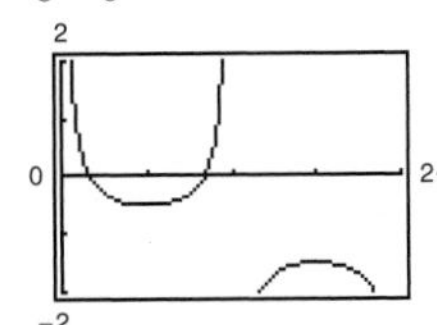

65–69. Answers will vary.

71. (a) $\cos \theta = \dfrac{M^2 - 2}{M^2}$ (b) $\dfrac{\pi}{3}$ (c) 0.4482

(d) 1520 mi/h; 3420 mi/h

73. $x = 2r(1 - \cos \theta)$

75. True. $\sin(-2x) = 2 \sin(-x) \cos(-x) = -2 \sin x \cos x$.

77. Answers will vary.

Review Exercises *(page 254)*

1. $\cot x$ **3.** $\cos x$

5. $\sin \theta = -\dfrac{\sqrt{21}}{5}$ $\csc \theta = -\dfrac{5\sqrt{21}}{21}$

$\cos \theta = -\dfrac{2}{5}$ $\sec \theta = -\dfrac{5}{2}$

$\tan \theta = \dfrac{\sqrt{21}}{2}$ $\cot \theta = \dfrac{2\sqrt{21}}{21}$

7. $\sin^2 x$ **9.** 1 **11.** $\tan u \sec u$ **13.** $\cot^2 x$

15. $-2 \tan^2 \theta$ **17.** $5 \cos \theta$ **19–25.** Answers will vary.

27. $\dfrac{\pi}{3} + 2n\pi, \dfrac{2\pi}{3} + 2n\pi$ **29.** $\dfrac{\pi}{6} + n\pi$

31. $\dfrac{\pi}{3} + n\pi, \dfrac{2\pi}{3} + n\pi$ **33.** $0, \dfrac{\pi}{2}, \pi, \dfrac{3\pi}{2}$ **35.** $0, \dfrac{\pi}{2}, \pi$

37. $\dfrac{\pi}{8}, \dfrac{3\pi}{8}, \dfrac{9\pi}{8}, \dfrac{11\pi}{8}$ **39.** $\dfrac{\pi}{2}$

41. $0, \dfrac{\pi}{8}, \dfrac{3\pi}{8}, \dfrac{5\pi}{8}, \dfrac{7\pi}{8}, \dfrac{9\pi}{8}, \dfrac{11\pi}{8}, \dfrac{13\pi}{8}, \dfrac{15\pi}{8}$

43. $n\pi$, $\arctan 2 + n\pi$

45. $\arctan(-3) + n\pi$, $\arctan 2 + n\pi$

47. $\sin 75^\circ = \dfrac{\sqrt{2}}{4}(1 + \sqrt{3})$ **49.** $\sin \dfrac{25\pi}{12} = \dfrac{\sqrt{2}}{4}(\sqrt{3} - 1)$

$\cos 75^\circ = \dfrac{\sqrt{2}}{4}(\sqrt{3} - 1)$ $\cos \dfrac{25\pi}{12} = \dfrac{\sqrt{2}}{4}(\sqrt{3} + 1)$

$\tan 75^\circ = 2 + \sqrt{3}$ $\tan \dfrac{25\pi}{12} = 2 - \sqrt{3}$

51. $\sin 15^\circ$ **53.** $-\frac{24}{25}$ **55.** -1

57–59. Answers will vary. **61.** $\dfrac{\pi}{4}, \dfrac{7\pi}{4}$

63. $\sin 2u = \frac{24}{25}$

$\cos 2u = -\frac{7}{25}$

$\tan 2u = -\frac{24}{7}$

65. Answers will vary. **67.** $\dfrac{1 - \cos 6x}{1 + \cos 6x}$

69. $\sin(-75^\circ) = -\frac{1}{2}\sqrt{2 + \sqrt{3}}$

$\cos(-75^\circ) = \frac{1}{2}\sqrt{2 - \sqrt{3}}$

$\tan(-75^\circ) = -2 - \sqrt{3}$

71. (a) Quadrant II

(b) $\sin \dfrac{u}{2} = \dfrac{2\sqrt{5}}{5}, \cos \dfrac{u}{2} = -\dfrac{\sqrt{5}}{5}, \tan \dfrac{u}{2} = -2$

73. (a) Quadrant I

(b) $\sin \dfrac{u}{2} = \dfrac{3\sqrt{14}}{14}, \cos \dfrac{u}{2} = \dfrac{\sqrt{70}}{14}, \tan \dfrac{u}{2} = \dfrac{3\sqrt{5}}{5}$

75. $\frac{1}{2}[\sin 10\theta - \sin(-2\theta)]$

77. $2 \cos \dfrac{11\theta}{2} \cos \dfrac{\theta}{2}$ **79.** $\theta = 15^\circ$ or $\dfrac{\pi}{12}$

81. False. If $(\pi/2) < \theta < \pi$, then $\theta/2$ lies in Quadrant I.

83. True. $4 \sin(-x) \cos(-x) = 4(-\sin x) \cos x$

$= -4 \sin x \cos x$

$= -2(2 \sin x \cos x)$

$= -2 \sin 2x$

85. Yes. *Sample answer:* $\sin x = \frac{1}{2}$ has an infinite number of solutions.

CHAPTER 2

Chapter Test (page 256)

1. $\sin\theta = \dfrac{2}{5}$ $\quad\csc\theta = \dfrac{5}{2}$

$\cos\theta = -\dfrac{\sqrt{21}}{5}$ $\quad\sec\theta = -\dfrac{5\sqrt{21}}{21}$

$\tan\theta = -\dfrac{2\sqrt{21}}{21}$ $\quad\cot\theta = -\dfrac{\sqrt{21}}{2}$

2. 1 **3.** 1 **4.** $\csc\theta\sec\theta$ **5–10.** Answers will vary.

11. $2(\sin 5\theta + \sin\theta)$ **12.** $-2\sin\theta$

13. $0, \dfrac{3\pi}{4}, \pi, \dfrac{7\pi}{4}$ **14.** $\dfrac{\pi}{6}, \dfrac{\pi}{2}, \dfrac{5\pi}{6}, \dfrac{3\pi}{2}$ **15.** $\dfrac{\pi}{6}, \dfrac{5\pi}{6}, \dfrac{7\pi}{6}, \dfrac{11\pi}{6}$

16. $\dfrac{\pi}{6}, \dfrac{5\pi}{6}, \dfrac{3\pi}{2}$ **17.** 0, 2.596 **18.** $\dfrac{\sqrt{2}-\sqrt{6}}{4}$

19. $\sin 2u = -\frac{20}{29}$, $\cos 2u = -\frac{21}{29}$, $\tan 2u = \frac{20}{21}$

20. Day 30 to day 310

21. 0.26 min, 0.58 min, 0.89 min, 1.20 min, 1.52 min, 1.83 min

Problem Solving (page 259)

1. $\sin\theta = \pm\sqrt{1-\cos^2\theta}$

$\tan\theta = \pm\dfrac{\sqrt{1-\cos^2\theta}}{\cos\theta}$

$\csc\theta = \pm\dfrac{1}{\sqrt{1-\cos^2\theta}}$

$\sec\theta = \dfrac{1}{\cos\theta}$

$\cot\theta = \pm\dfrac{\cos\theta}{\sqrt{1-\cos^2\theta}}$

3. Answers will vary. **5.** $u + v = w$; Proof

7. (a) $A = 100\sin\dfrac{\theta}{2}\cos\dfrac{\theta}{2}$

(b) $A = 50\sin\theta$; $\theta = \dfrac{\pi}{2}$

9. (a) $F = \dfrac{0.6W\cos\theta}{\sin 12°}$

(b)

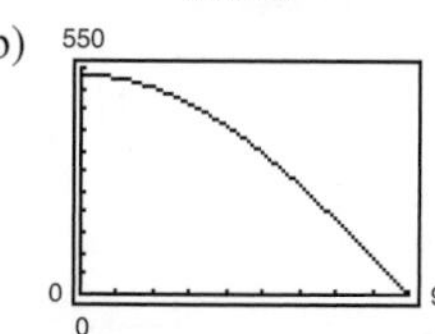

(c) Maximum: $\theta = 0°$
Minimum: $\theta = 90°$

11. (a) High tides: 6:12 A.M., 6:36 P.M.
Low tides: 12:00 A.M., 12:24 P.M.

(b) The water depth never falls below 7 feet.

(c) 70 0 0 24

13. (a) $n = \dfrac{1}{2}\left(\cot\dfrac{\theta}{2} + \sqrt{3}\right)$

(b) $\theta \approx 76.5°$

15. (a) $\dfrac{\pi}{6} \le x \le \dfrac{5\pi}{6}$

(b) $\dfrac{2\pi}{3} \le x \le \dfrac{4\pi}{3}$

(c) $\dfrac{\pi}{2} < x < \pi, \dfrac{3\pi}{2} < x < 2\pi$

(d) $0 \le x \le \dfrac{\pi}{4}, \dfrac{5\pi}{4} \le x < 2\pi$

Chapter 3

Section 3.1 (page 268)

1. oblique **3.** angles; side

5. $A = 30°, a \approx 14.14, c \approx 27.32$

7. $C = 105°, a \approx 5.94, b \approx 6.65$

9. $B = 60.9°, b \approx 19.32, c \approx 6.36$

11. $B = 42°\,4', a \approx 22.05, b \approx 14.88$

13. $C = 80°, a \approx 5.82, b \approx 9.20$

15. $C = 83°, a \approx 0.62, b \approx 0.51$

17. $B \approx 21.55°, C \approx 122.45°, c \approx 11.49$

19. $B \approx 9.43°, C \approx 25.57°, c \approx 10.53$

21. $A \approx 10°\,11', C \approx 154°\,19', c \approx 11.03$

23. $B \approx 48.74°, C \approx 21.26°, c \approx 48.23$ **25.** No solution

27. Two solutions:
$B \approx 72.21°, C \approx 49.79°, c \approx 10.27$
$B \approx 107.79°, C \approx 14.21°, c \approx 3.30$

29. No solution **31.** $B = 45°, C = 90°, c \approx 1.41$

33. (a) $b \le 5, b = \dfrac{5}{\sin 36°}$ (b) $5 < b < \dfrac{5}{\sin 36°}$

(c) $b > \dfrac{5}{\sin 36°}$

35. (a) $b < 80$ (b) Not possible (c) $b \ge 80$

37. 22.1 **39.** 94.4 **41.** 218.0 **43.** 22.3

45. (a) $\dfrac{h}{\sin 30°} = \dfrac{40}{\sin 56°}$ (b) About 24.1 m

47. From Pine Knob: about 42.4 km
From Colt Station: about 15.5 km

49. About 16.1°

51. (a)

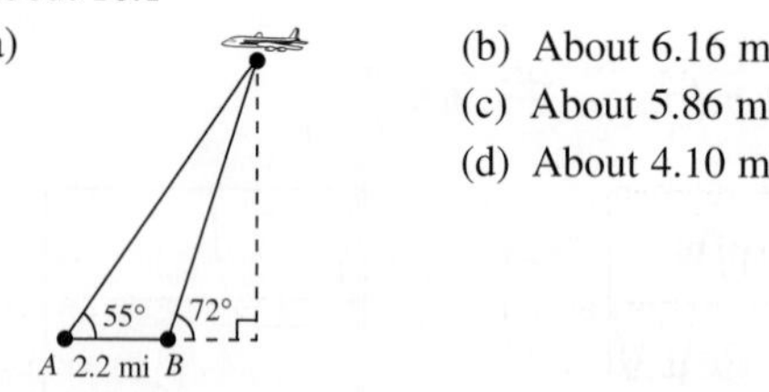

(b) About 6.16 mi
(c) About 5.86 mi
(d) About 4.10 mi

53. $d = \dfrac{2\sin\theta}{\sin(\phi - \theta)}$

55. True. If an angle of a triangle is obtuse, then the other two angles must be acute.

57. False. When just three angles are known, the triangle cannot be solved.

59. The formula is $A = \frac{1}{2}ab\sin C$, so solve the triangle to find the value of a.

61. Yes. $A = 40°, b \approx 11.9, c \approx 15.6$; Yes; *Sample answer:* Use the definitions of cosine and tangent.

Section 3.2 *(page 275)*

1. $b^2 = a^2 + c^2 - 2ac \cos B$ **3.** standard
5. $A \approx 38.62°, B \approx 48.51°, C \approx 92.87°$
7. $A \approx 26.38°, B \approx 36.34°, C \approx 117.28°$
9. $B \approx 23.79°, C \approx 126.21°, a \approx 18.59$
11. $B \approx 29.44°, C \approx 100.56°, a \approx 23.38$
13. $A \approx 30.11°, B \approx 43.16°, C \approx 106.73°$
15. $A \approx 132.77°, B \approx 31.91°, C \approx 15.32°$
17. $B \approx 27.46°, C \approx 32.54°, a \approx 11.27$
19. $A \approx 141° 45', C \approx 27° 40', b \approx 11.87$
21. $A = 27° 10', C = 27° 10', b \approx 65.84$
23. $A \approx 33.80°, B \approx 103.20°, c \approx 0.54$
25. 12.07; 5.69; 135° **27.** 13.86; 68.2°; 111.8°
29. 16.96; 77.2°; 102.8°
31. Yes; $A \approx 102.44°, C \approx 37.56°, b \approx 5.26$
33. No; No solution **35.** No; $C = 103°, a \approx 0.82, b \approx 0.71$
37. 23.53 **39.** 5.35 **41.** 0.24 **43.** 1514.14
45. About 373.3 m **47.** About 103.9 ft
49. About 131.1 ft, about 118.6 ft
51.

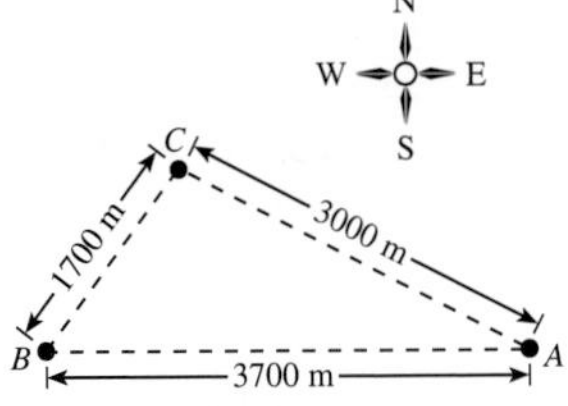

N 37.1° E, S 63.1° E
53. 41.2°, 52.9°
55. (a) $d = \sqrt{9s^2 - 108s + 1296}$ (b) About 15.87 mi/h
57. About 46,837.5 ft^2 **59.** $83,336.37
61. False. For s to be the average of the lengths of the three sides of the triangle, s would be equal to $(a + b + c)/3$.
63. $c^2 = a^2 + b^2$; The Pythagorean Theorem is a special case of the Law of Cosines.
65. The Law of Cosines can be used to solve the single-solution case of SSA. There is no method that can solve the no-solution case of SSA.
67. Proof

Section 3.3 *(page 287)*

1. directed line segment **3.** vector
5. standard position **7.** multiplication; addition
9. Equivalent; **u** and **v** have the same magnitude and direction.
11. Not equivalent; **u** and **v** do not have the same direction.
13. Equivalent; **u** and **v** have the same magnitude and direction.
15. $\mathbf{v} = \langle 1, 3 \rangle; \|\mathbf{v}\| = \sqrt{10}$ **17.** $\mathbf{v} = \langle 0, 5 \rangle; \|\mathbf{v}\| = 5$
19. $\mathbf{v} = \langle -8, 6 \rangle; \|\mathbf{v}\| = 10$ **21.** $\mathbf{v} = \langle -9, -12 \rangle; \|\mathbf{v}\| = 15$
23. $\mathbf{v} = \langle 16, -26 \rangle; \|\mathbf{v}\| = 2\sqrt{233}$
25.

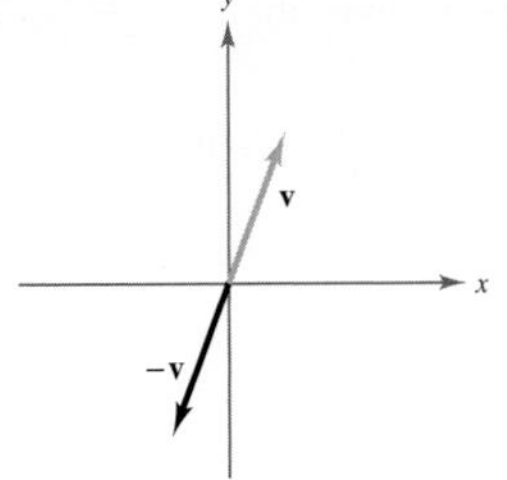

27.

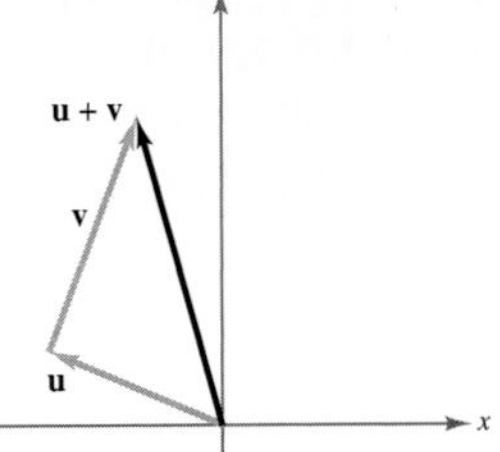

29.

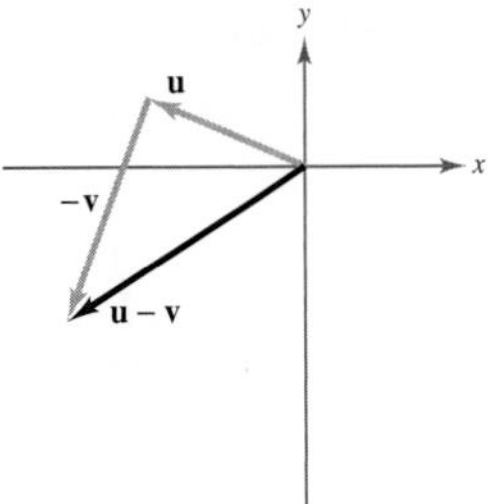

31. (a) $\langle 3, 4 \rangle$

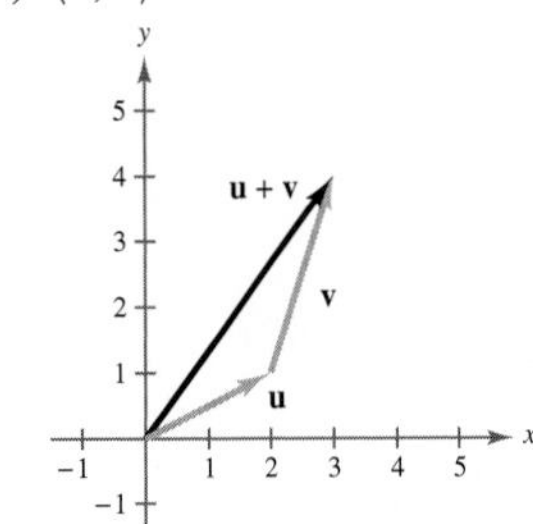

(b) $\langle 1, -2 \rangle$

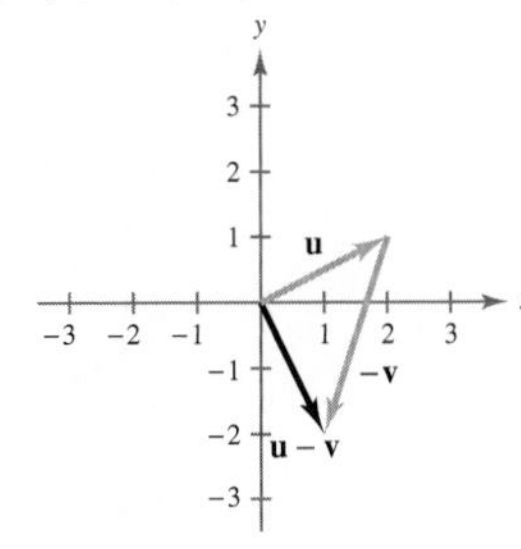

(c) $\langle 1, -7 \rangle$

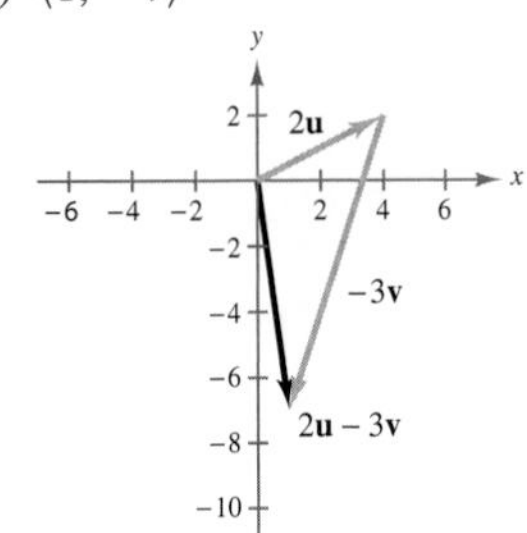

33. (a) $\langle -5, 3 \rangle$ (b) $\langle -5, 3 \rangle$

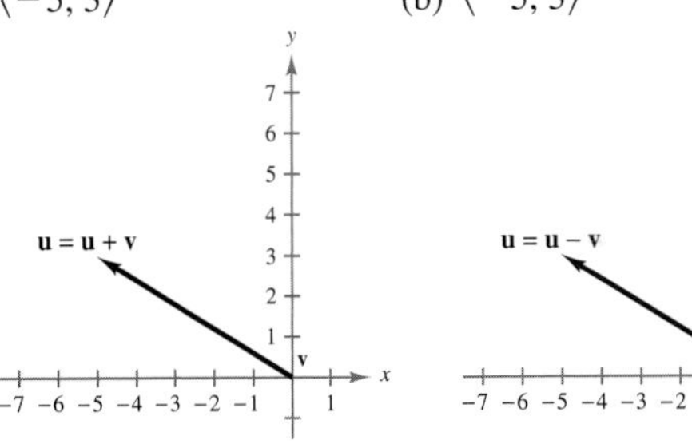

(c) $\langle -10, 6 \rangle$

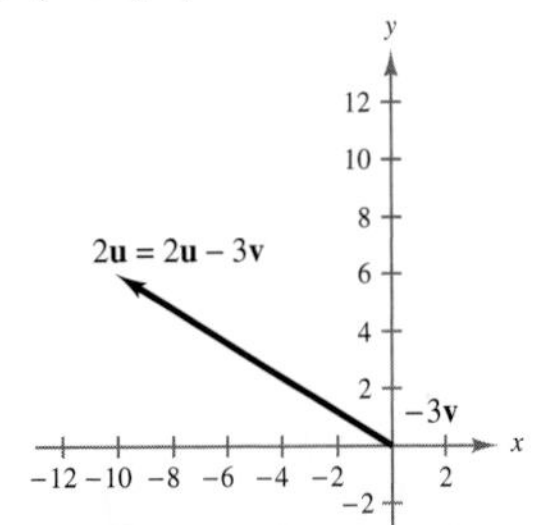

CHAPTER 3

35. (a) $\langle 1, -9\rangle$

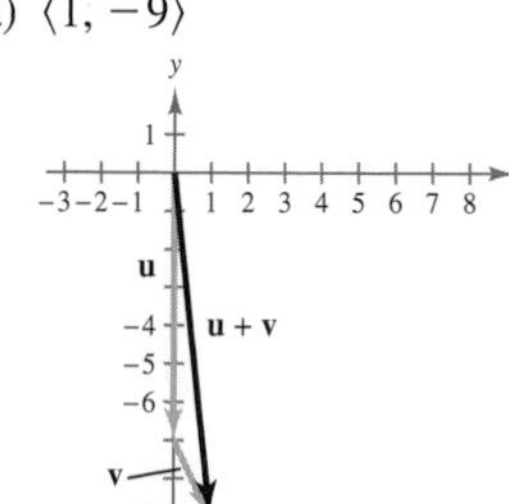

(b) $\langle -1, -5\rangle$

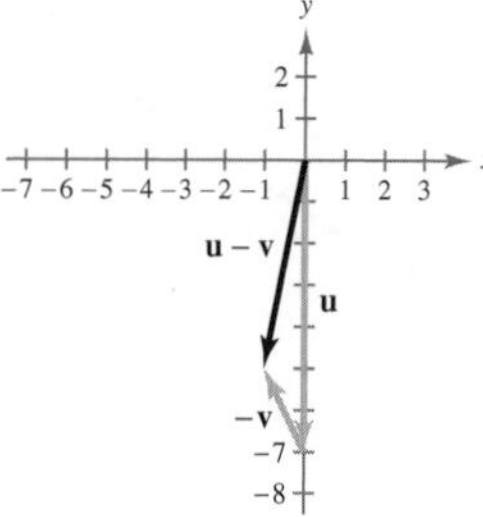

(c) $\langle -3, -8\rangle$

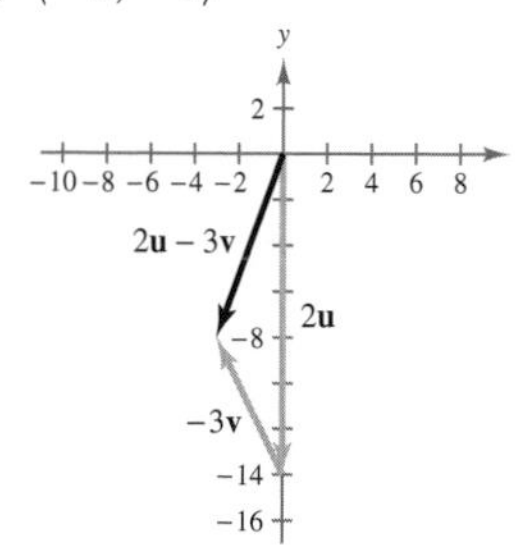

37. 10 **39.** $9\sqrt{5}$ **41.** $\langle 1, 0\rangle$ **43.** $\left\langle -\frac{\sqrt{2}}{2}, \frac{\sqrt{2}}{2}\right\rangle$

45. $\left\langle \frac{\sqrt{37}}{37}, -\frac{6\sqrt{37}}{37}\right\rangle$ **47.** $\mathbf{w} = \langle -6, 8\rangle$

49. $\mathbf{w} = \left\langle \frac{18\sqrt{29}}{29}, \frac{45\sqrt{29}}{29}\right\rangle$ **51.** $5\mathbf{i} - 3\mathbf{j}$ **53.** $-6\mathbf{i} + 3\mathbf{j}$

55. $\mathbf{v} = \langle 3, -\frac{3}{2}\rangle$ **57.** $\mathbf{v} = \langle 4, 3\rangle$

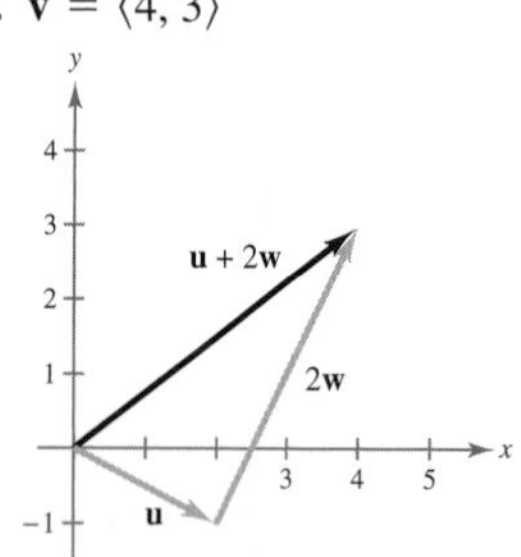

59. $\mathbf{v} = \langle 0, -5\rangle$

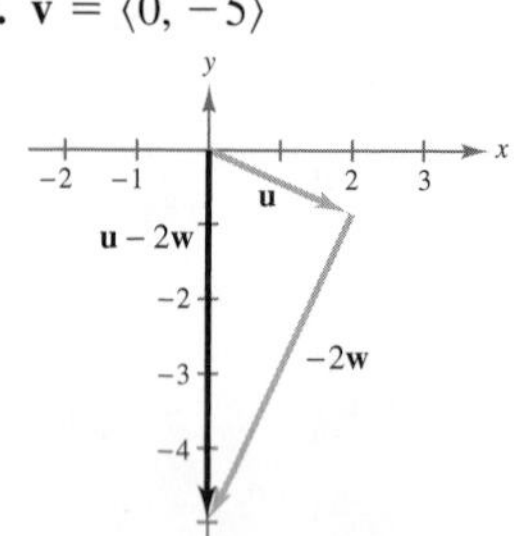

61. $\|\mathbf{v}\| = 6\sqrt{2}; \theta = 315°$ **63.** $\|\mathbf{v}\| = 3; \theta = 60°$

65. $\mathbf{v} = \langle 3, 0\rangle$ **67.** $\mathbf{v} = \left\langle -\frac{7\sqrt{3}}{4}, \frac{7}{4}\right\rangle$

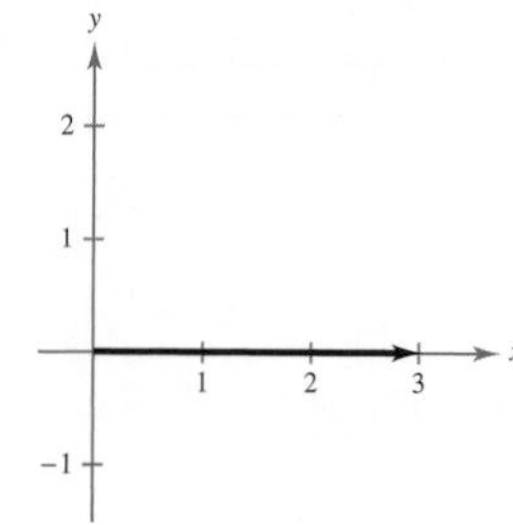

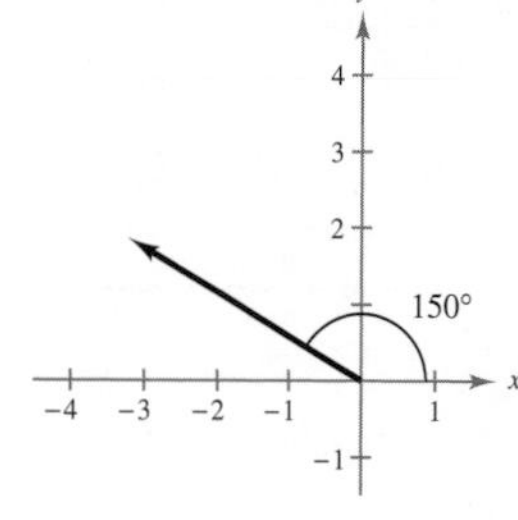

69. $\mathbf{v} = \langle \frac{9}{5}, \frac{12}{5}\rangle$

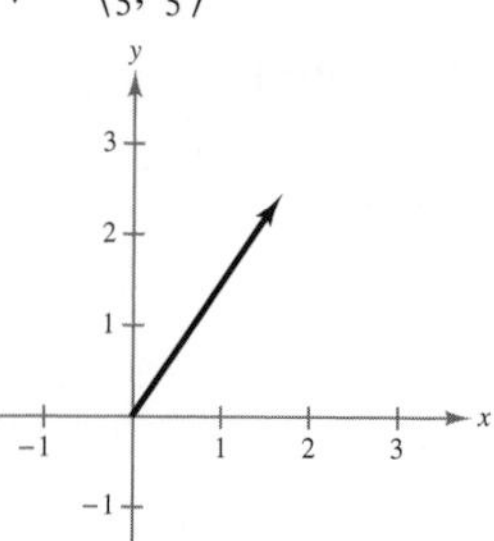

71. $\langle 2, 4 + 2\sqrt{3}\rangle$ **73.** 90° **75.** About 62.7°

77. Vertical $\approx$ 125.4 ft/sec, horizontal $\approx$ 1193.4 ft/sec

79. About 12.8°; about 398.32 N

81. About 71.3°; about 228.5 lb

83. $T_L \approx$ 15,484 lb **85.** $\sqrt{2}$ lb; 1 lb **87.** About 3154.4 lb
$T_R \approx$ 19,786 lb

89. About 20.8 lb **91.** About 19.5°

93. N 21.4° E; about 138.7 km/h

95. True. See Example 1. **97.** True. $a = b = 0$

99. $u_1 = 6 - (-3) = 9$ and $u_2 = -1 - 4 = -5$, so $\mathbf{u} = \langle 9, -5\rangle$.

101. Proof **103.** $\langle 1, 3\rangle$ or $\langle -1, -3\rangle$

105. (a) $5\sqrt{5 + 4\cos\theta}$

(b)

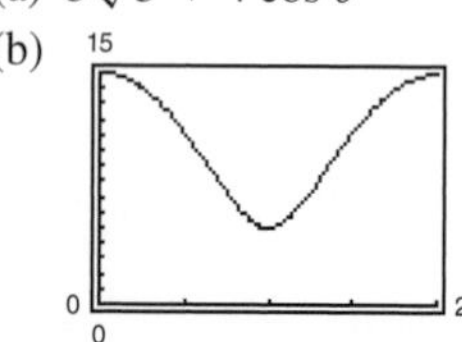

(c) Range: $[5, 15]$
Maximum is 15 when $\theta = 0$.
Minimum is 5 when $\theta = \pi$.

(d) The magnitudes of $\mathbf{F}_1$ and $\mathbf{F}_2$ are not the same.

107. (a) and (b) Answers will vary.

Section 3.4 *(page 297)*

1. dot product **3.** $\frac{\mathbf{u} \cdot \mathbf{v}}{\|\mathbf{u}\|\|\mathbf{v}\|}$ **5.** $\left(\frac{\mathbf{u} \cdot \mathbf{v}}{\|\mathbf{v}\|^2}\right)\mathbf{v}$

7. -19 **9.** 0 **11.** 6 **13.** 18; scalar

15. $\langle 24, -12\rangle$; vector **17.** $\mathbf{0}$; vector

19. $\sqrt{10} - 1$; scalar **21.** -12; scalar **23.** 17

25. $5\sqrt{41}$ **27.** 6 **29.** $\frac{\pi}{2}$ **31.** About 2.50 **33.** 0

35. About 0.93 **37.** $\frac{5\pi}{12}$ **39.** About 91.33° **41.** 90°

43. 26.57°, 63.43°, 90° **45.** 41.63°, 53.13°, 85.24°

47. -20 **49.** $12{,}500\sqrt{3}$ **51.** Not orthogonal

53. Orthogonal **55.** Not orthogonal

57. $\frac{1}{37}\langle 84, 14\rangle, \frac{1}{37}\langle -10, 60\rangle$ **59.** $\langle 0, 0\rangle, \langle 4, 2\rangle$ **61.** $\langle 3, 2\rangle$

63. $\langle 0, 0\rangle$ **65.** $\langle -5, 3\rangle, \langle 5, -3\rangle$

67. $\frac{2}{3}\mathbf{i} + \frac{1}{2}\mathbf{j}, -\frac{2}{3}\mathbf{i} - \frac{1}{2}\mathbf{j}$ **69.** 32

71. (a) \$35,727.50
This value gives the total amount paid to the employees.
(b) Multiply $\mathbf{v}$ by 1.02.

73. (a) Force $= 30{,}000 \sin d$

(b)

d	0°	1°	2°	3°	4°	5°
Force	0	523.6	1047.0	1570.1	2092.7	2614.7

d	6°	7°	8°	9°	10°
Force	3135.9	3656.1	4175.2	4693.0	5209.4

(c) About 29,885.8 lb

75. 735 N-m **77.** About 779.4 ft-lb

79. About 10,282,651.78 N-m **81.** About 1174.62 ft-lb

83. False. Work is represented by a scalar.

85. The dot product is the scalar 0. **87.** 4

89. 1; $\mathbf{u} \cdot \mathbf{u} = \|\mathbf{u}\|^2$

91. (a) **u** and **v** are parallel. (b) **u** and **v** are orthogonal.

93. Proof

Review Exercises *(page 302)*

1. $C = 72°, b \approx 12.21, c \approx 12.36$

3. $A = 26°, a \approx 24.89, c \approx 56.23$

5. $C = 66°, a \approx 2.53, b \approx 9.11$

7. $B = 108°, a \approx 11.76, c \approx 21.49$

9. $A \approx 20.41°, C \approx 9.59°, a \approx 20.92$

11. $B \approx 39.48°, C \approx 65.52°, c \approx 48.24$

13. 19.1 **15.** 47.2 **17.** 3204.5 **19.** About 31.01 ft

21. $A \approx 27.81°, B \approx 54.75°, C \approx 97.44°$

23. $A \approx 16.99°, B \approx 26.00°, C \approx 137.01°$

25. $A \approx 29.92°, B \approx 86.18°, C \approx 63.90°$

27. $A = 36°, C = 36°, b \approx 17.80$

29. $A \approx 45.76°, B \approx 91.24°, c \approx 21.42$

31. No; $A \approx 77.52°, B \approx 38.48°, a \approx 14.12$

33. Yes; $A \approx 28.62°, B \approx 33.56°, C \approx 117.82°$

35. No; $A = 161°, b \approx 102.18, c \approx 59.89$

37. About 8.57 ft, about 25.27 ft **39.** About 615.1 m

41. 7.64 **43.** 8.36

45. Equivalent; **u** and **v** have the same magnitude and direction.

47. $\langle 7, -5 \rangle, \sqrt{74}$ **49.** $\langle 7, -7 \rangle; 7\sqrt{2}$

51. (a) $\langle -4, 3 \rangle$

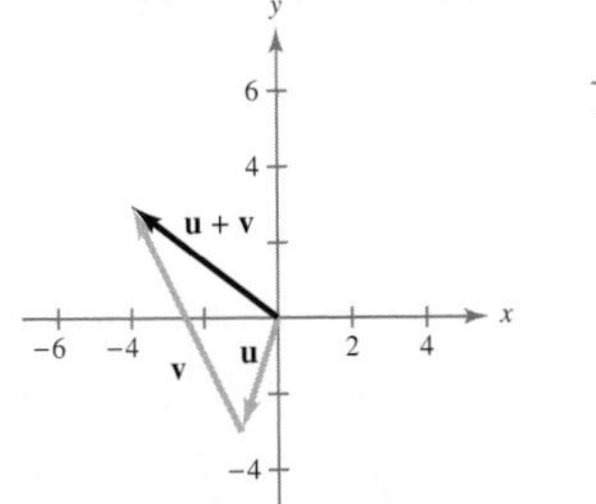

(b) $\langle 2, -9 \rangle$

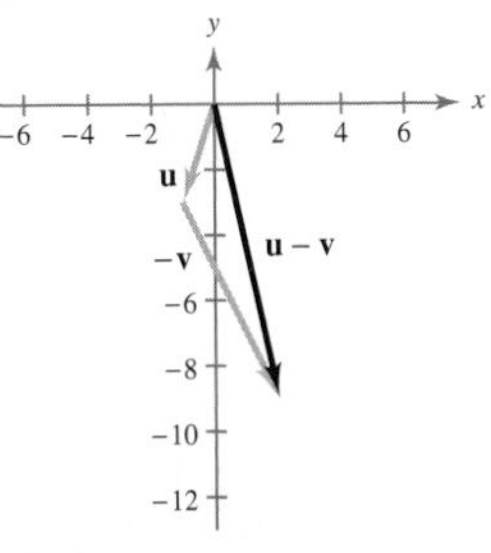

(c) $\langle -4, -12 \rangle$

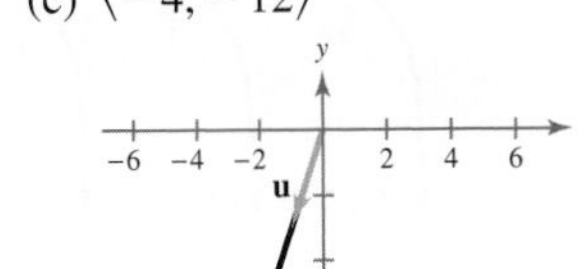

(d) $\langle -14, 3 \rangle$

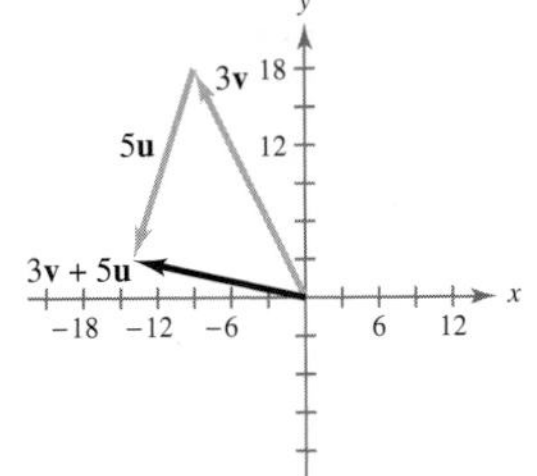

53. (a) $\langle -1, 6 \rangle$

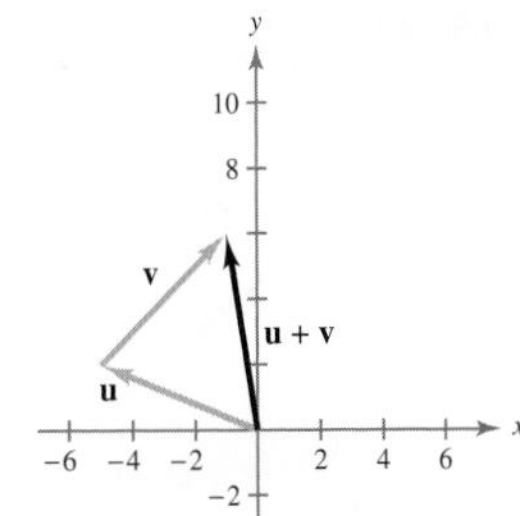

(b) $\langle -9, -2 \rangle$

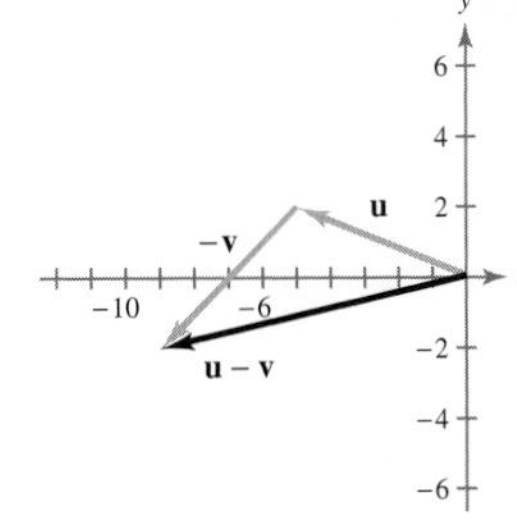

(c) $\langle -20, 8 \rangle$

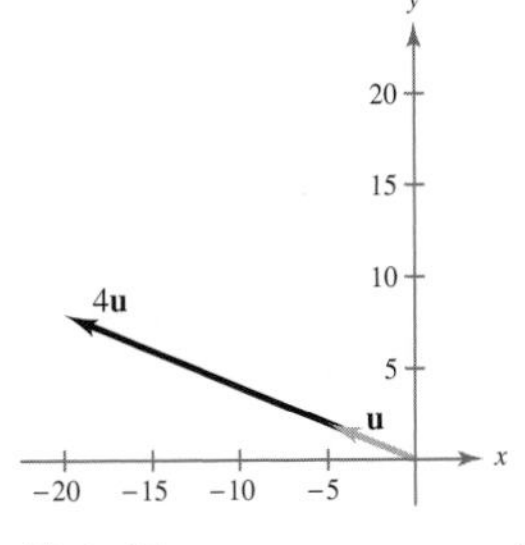

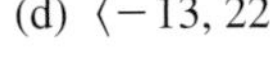

(d) $\langle -13, 22 \rangle$

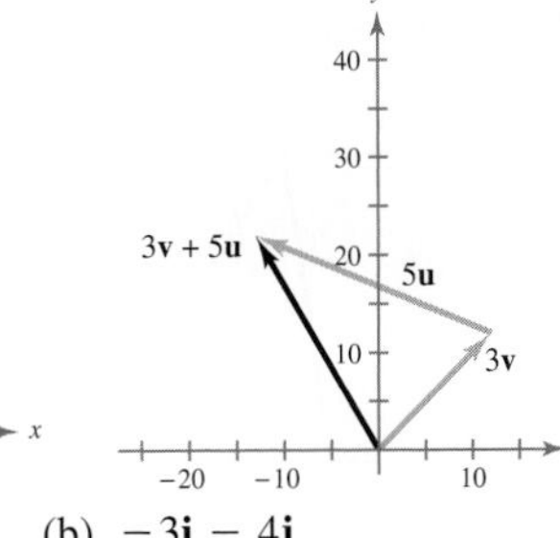

55. (a) $7\mathbf{i} + 2\mathbf{j}$

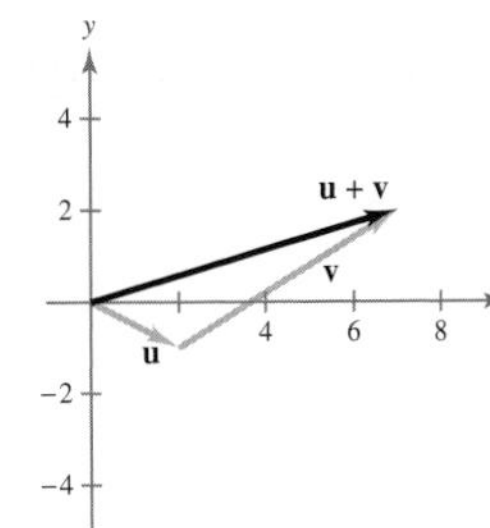

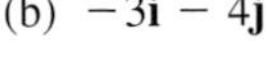

(b) $-3\mathbf{i} - 4\mathbf{j}$

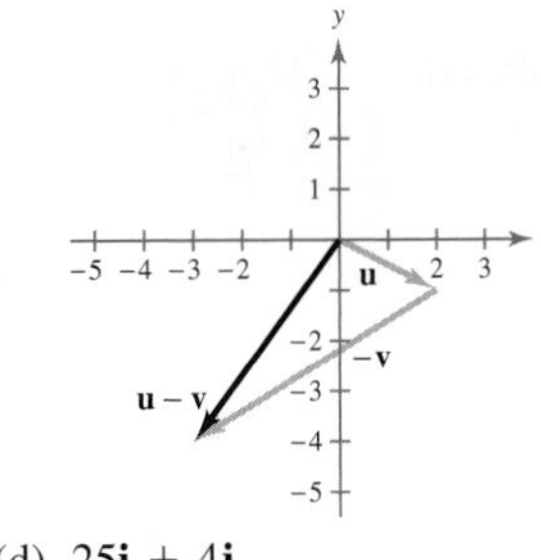

(c) $8\mathbf{i} - 4\mathbf{j}$

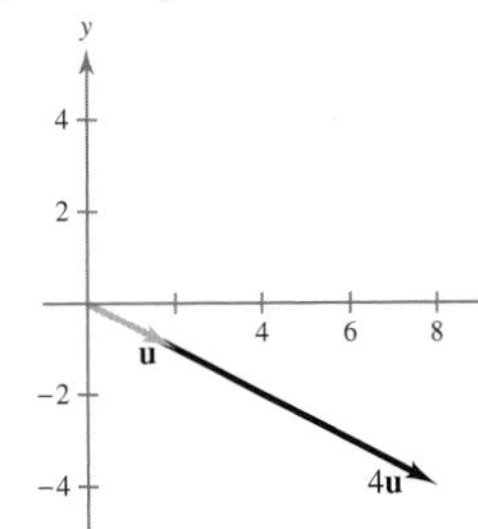

(d) $25\mathbf{i} + 4\mathbf{j}$

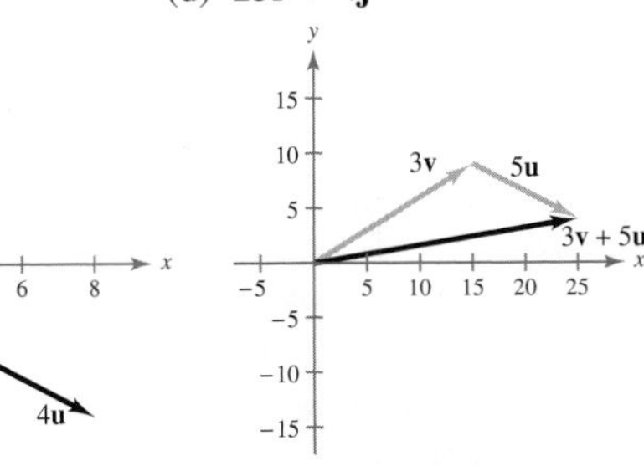

57. (a) $3\mathbf{i} + 6\mathbf{j}$

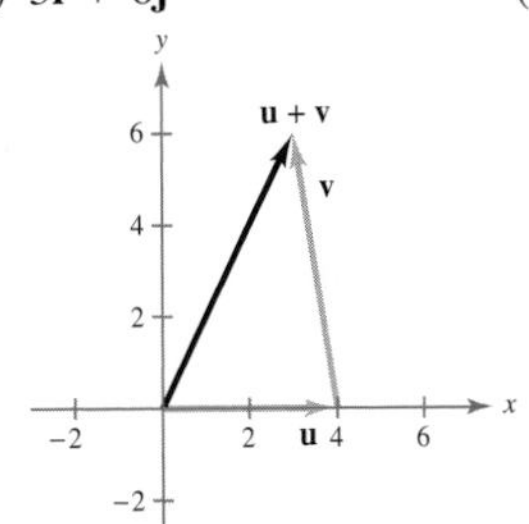

(b) $5\mathbf{i} - 6\mathbf{j}$

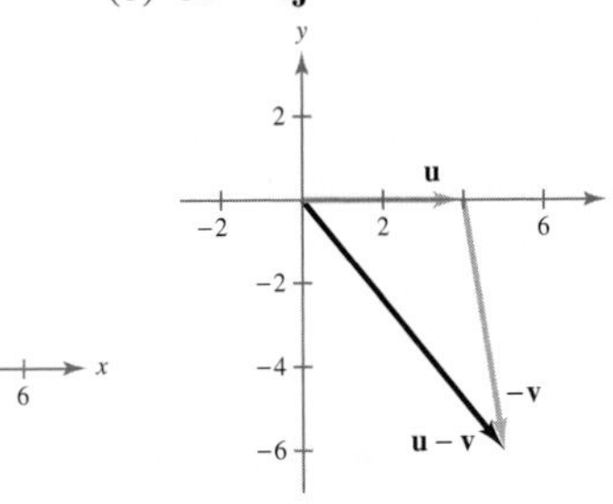

(c) $16\mathbf{i}$

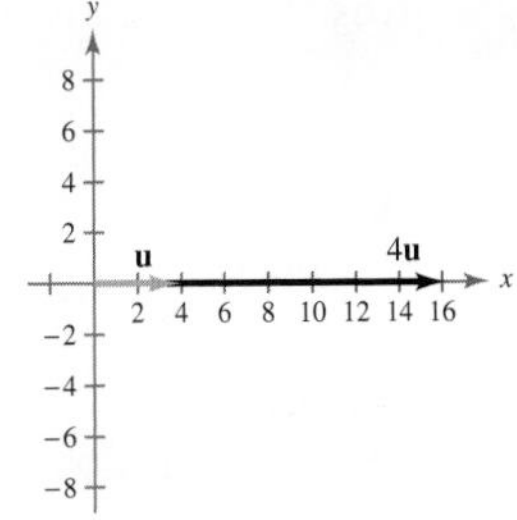

(d) $17\mathbf{i} + 18\mathbf{j}$

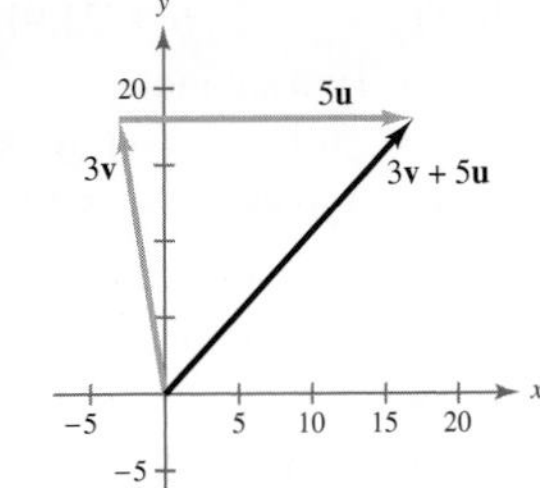

59. $-\mathbf{i} + 5\mathbf{j}$ **61.** $6\mathbf{i} + 4\mathbf{j}$

63. $\langle 30, 9\rangle$ **65.** $\langle 22, -7\rangle$

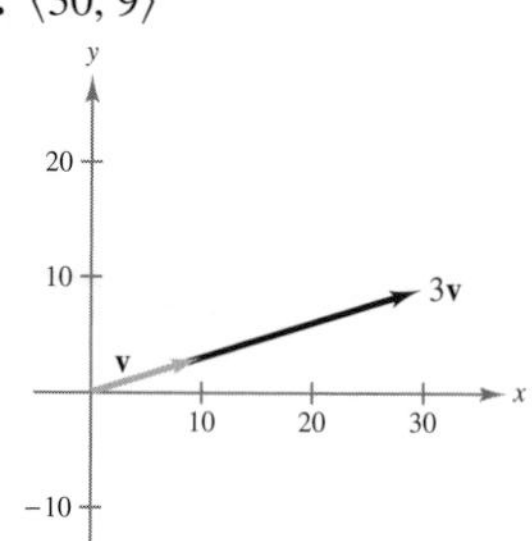

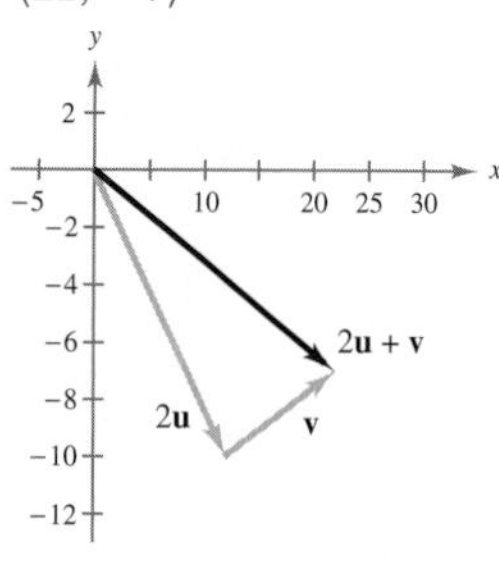

67. $\langle -10, -37\rangle$

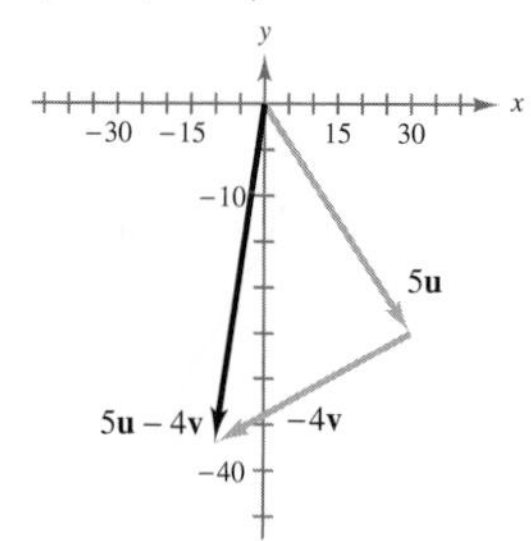

69. $\|\mathbf{v}\| = \sqrt{41}$; $\theta = 38.7°$ **71.** $\|\mathbf{v}\| = 3\sqrt{2}$; $\theta = 225°$

73. $\mathbf{v} = \langle -4, 4\sqrt{3}\rangle$

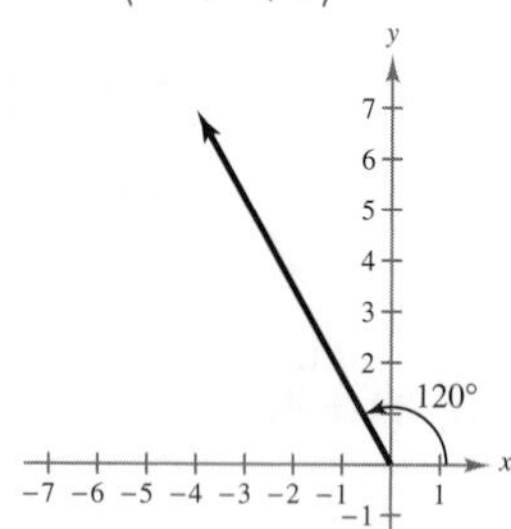

75. 180 lb each **77.** 45 **79.** -2 **81.** 40; scalar

83. $4 - 2\sqrt{5}$; scalar **85.** $\langle 72, -36\rangle$; scalar

87. 38; scalar **89.** 165° **91.** About 160.5°

93. Orthogonal **95.** Not orthogonal

97. $-\frac{13}{17}\langle 4, 1\rangle, \frac{16}{17}\langle -1, 4\rangle$ **99.** $\frac{5}{2}\langle -1, 1\rangle, \frac{9}{2}\langle 1, 1\rangle$

101. 48 **103.** 72,000 ft-lb

105. True. Using the Law of Cosines, there is exactly one solution.

107. True. sin 90° is defined in the Law of Sines.

109. A and C

111. For $k > 0$, the direction is the same and the magnitude is k times as great.

For $k < 0$, the result is a vector in the opposite direction and the magnitude is $|k|$ times as great.

113. Direction and magnitude

Chapter Test *(page 305)*

1. No; $C = 88°, b \approx 27.81, c \approx 29.98$

2. No; $A = 42°, b \approx 21.91, c \approx 10.95$

3. No; Two solutions:

$B \approx 29.12°, C \approx 126.88°, c \approx 22.03$

$B \approx 150.88°, C \approx 5.12°, c \approx 2.46$

4. Yes; $A \approx 19.12°, B \approx 23.49°, C \approx 137.39°$

5. No; No solution

6. Yes; $A \approx 21.90°, B \approx 37.10°, c \approx 78.15$

7. 2052.5 m^2 **8.** 606.3 mi; 29.1° **9.** $\langle 14, -23\rangle$

10. $\left\langle \frac{18\sqrt{34}}{17}, -\frac{30\sqrt{34}}{17}\right\rangle$

11. $\langle -4, 12\rangle$ **12.** $\langle 8, 2\rangle$

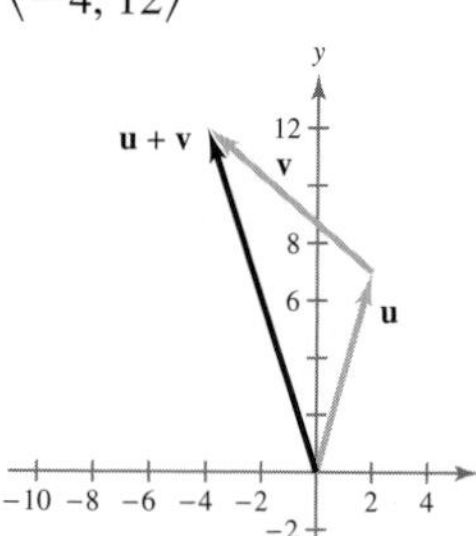

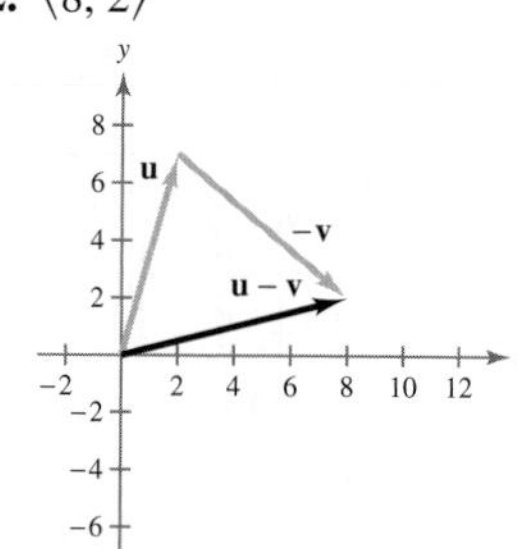

13. $\langle 28, 20\rangle$ **14.** $\langle -4, 38\rangle$

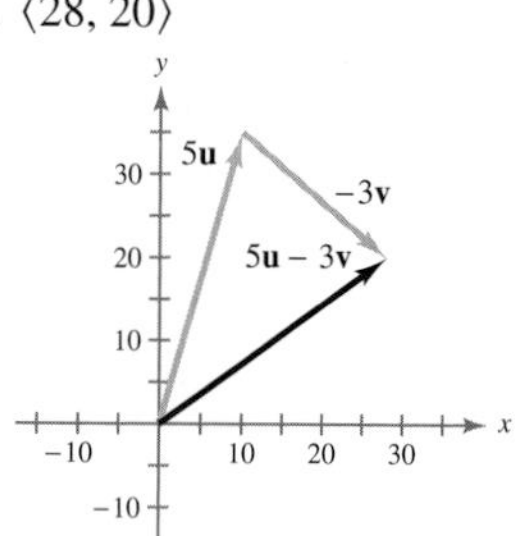

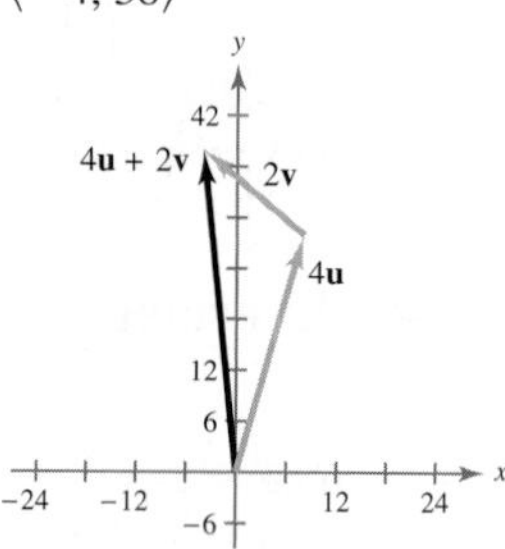

15. $\left\langle \frac{24}{25}, -\frac{7}{25}\right\rangle$ **16.** About 14.9°; about 205.15 lb **17.** -1

18. 135° **19.** Orthogonal **20.** Orthogonal

21. $\frac{37}{26}\langle 5, 1\rangle; \frac{29}{26}\langle -1, 5\rangle$ **22.** About 104 lb

Cumulative Test for Chapters 1–3 *(page 306)*

1. (a)

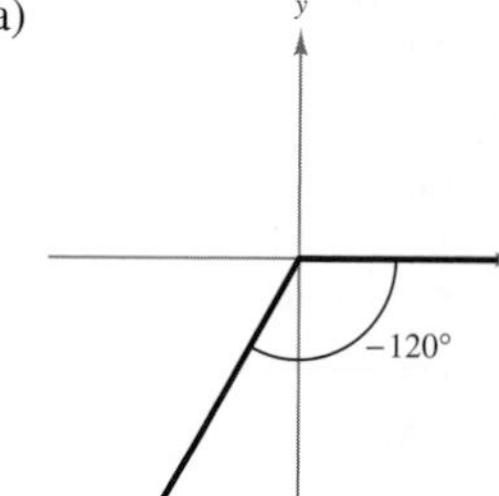

(b) 240°

(c) $-\frac{2\pi}{3}$

(d) 60°

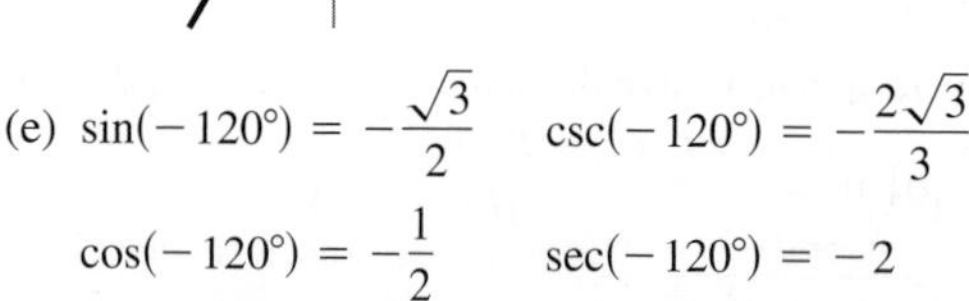

(e) $\sin(-120°) = -\frac{\sqrt{3}}{2}$ $\csc(-120°) = -\frac{2\sqrt{3}}{3}$

$\cos(-120°) = -\frac{1}{2}$ $\sec(-120°) = -2$

$\tan(-120°) = \sqrt{3}$ $\cot(-120°) = \frac{\sqrt{3}}{3}$

2. $-83.079°$ **3.** $\frac{20}{29}$

4.

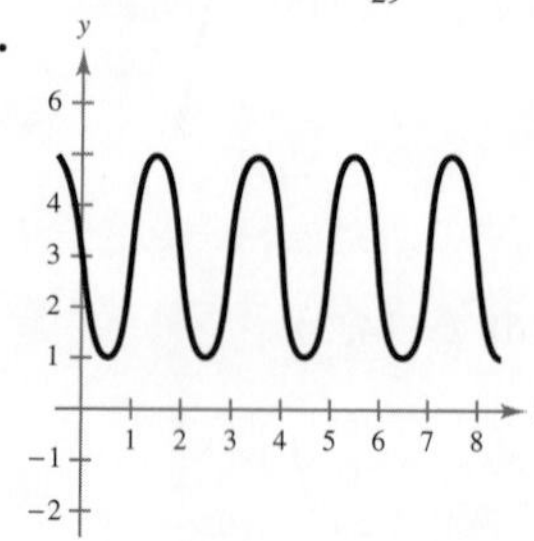

5.

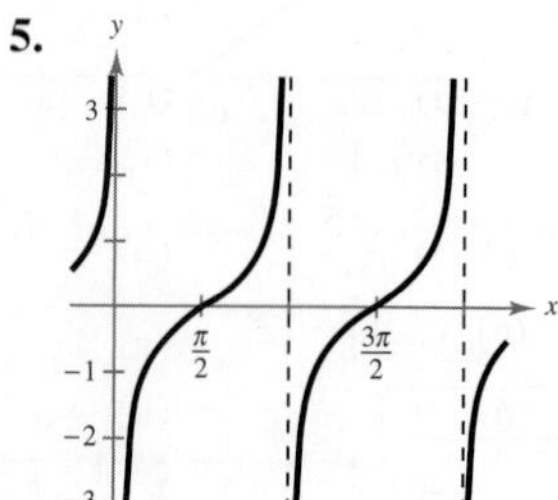

6. 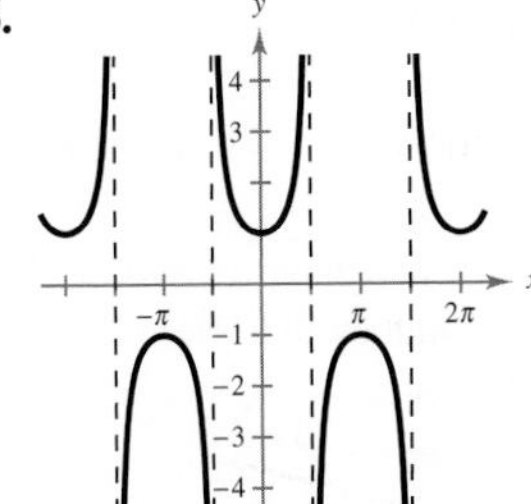

7. $a = -3, b = \pi, c = 0$

8. 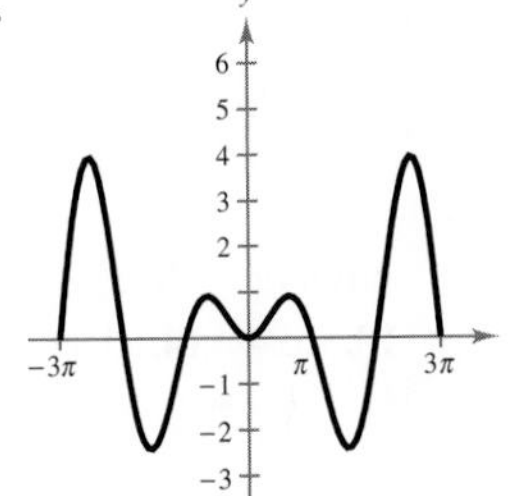

9. 4.9
10. $\frac{3}{4}$
11. $\sqrt{1 - 4x^2}$
12. 1
13. $2 \tan \theta$
14–16. Answers will vary.
17. $\frac{\pi}{3}, \frac{\pi}{2}, \frac{3\pi}{2}, \frac{5\pi}{3}$

18. $\frac{\pi}{6}, \frac{5\pi}{6}, \frac{7\pi}{6}, \frac{11\pi}{6}$ 19. $\frac{3\pi}{2}$ 20. $\frac{16}{63}$ 21. $\frac{4}{3}$

22. $\frac{\sqrt{5}}{5}$ 23. $\frac{5}{2}\left(\sin \frac{5\pi}{2} - \sin \pi\right)$ 24. $-2 \sin 8x \sin x$

25. No; $B \approx 26.39°, C \approx 123.61°, c \approx 14.99$
26. Yes; $B \approx 52.48°, C \approx 97.52°, a \approx 5.04$
27. No; $B = 60°, a \approx 5.77, c \approx 11.55$
28. Yes; $A \approx 26.28°, B \approx 49.74°, C \approx 103.98°$
29. No; $C = 109°, a \approx 14.96, b \approx 9.27$
30. Yes; $A \approx 6.88°, B \approx 93.12°, c \approx 9.86$
31. 41.48 in.^{2} 32. 599.09 m^2 33. $7\mathbf{i} + 8\mathbf{j}$
34. $\left\langle \frac{\sqrt{2}}{2}, \frac{\sqrt{2}}{2} \right\rangle$ 35. -5 36. $-\frac{1}{13}\langle 1, 5\rangle; \frac{21}{13}\langle 5, -1\rangle$
37. About 395.8 rad/min; about 8312.7 in./min
38. 42π yd$^2 \approx 131.95$ yd^2 39. 5 ft 40. About 22.6°
41. $d = 4 \cos \frac{\pi}{4}t$ 42. About 32.6°; about 543.9 km/h
43. 425 ft-lb

Problem Solving (page 311)

1. About 2.01 ft
3. (a) 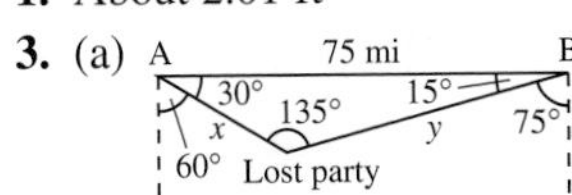

(b) Station A: about 27.45 mi; Station B: about 53.03 mi
(c) About 11.03 mi; S 21.7° E

5. (a) (i) $\sqrt{2}$ (ii) $\sqrt{5}$ (iii) 1
(iv) 1 (v) 1 (vi) 1
(b) (i) 1 (ii) $3\sqrt{2}$ (iii) $\sqrt{13}$
(iv) 1 (v) 1 (vi) 1
(c) (i) $\frac{\sqrt{5}}{2}$ (ii) $\sqrt{13}$ (iii) $\frac{\sqrt{85}}{2}$
(iv) 1 (v) 1 (vi) 1
(d) (i) $2\sqrt{5}$ (ii) $5\sqrt{2}$ (iii) $5\sqrt{2}$
(iv) 1 (v) 1 (vi) 1
7. (a) 0° (b) 180°
(c) No. The magnitude is greatest when the angle is 0°.
9. Proof 11. Answers will vary.

13. (a) $\mathbf{u} = \langle 0, -120\rangle, \mathbf{v} = \langle 40, 0\rangle$
(b) 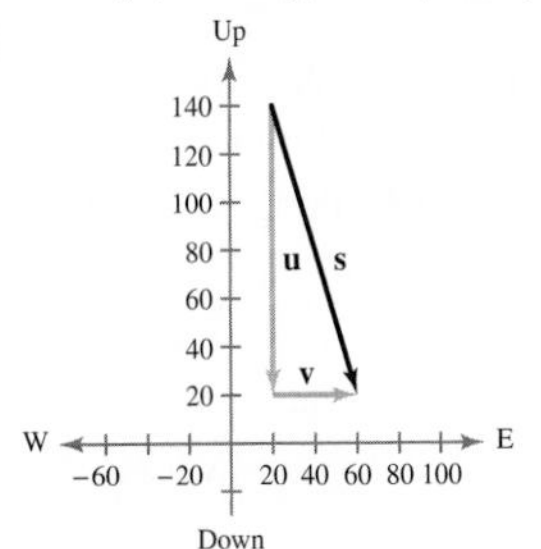
(c) About 126.5 miles per hour; The magnitude gives the actual rate of the skydiver's fall.
(d) About 71.57°
(e) 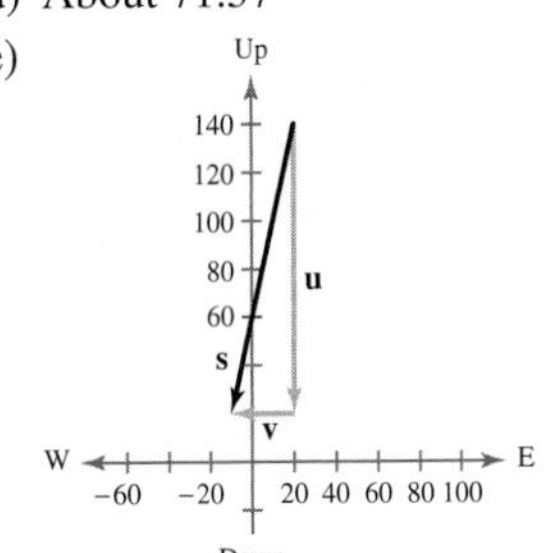
About 123.7 mi/h

Chapter 4

Section 4.1 (page 319)

1. real 3. pure imaginary 5. principal square
7. $a = 9, b = 8$ 9. $a = 8, b = 4$ 11. $2 + 5i$
13. $1 - 2\sqrt{3}i$ 15. $2\sqrt{10}i$ 17. 23 19. $-1 - 6i$
21. $0.2i$ 23. $7 + 4i$ 25. 1 27. $3 - 3\sqrt{2}i$
29. $-14 + 20i$ 31. $5 + i$ 33. $108 + 12i$ 35. 11
37. $-13 + 84i$ 39. $9 - 2i, 85$ 41. $-1 + \sqrt{5}i, 6$
43. $-2\sqrt{5}i, 20$ 45. $\sqrt{6}, 6$ 47. $\frac{8}{41} + \frac{10}{41}i$ 49. $\frac{12}{13} + \frac{5}{13}i$
51. $-4 - 9i$ 53. $-\frac{120}{1681} - \frac{27}{1681}i$ 55. $-\frac{1}{2} - \frac{5}{2}i$
57. $\frac{62}{949} + \frac{297}{949}i$ 59. $-2\sqrt{3}$ 61. -15 63. $7\sqrt{2}i$
65. $(21 + 5\sqrt{2}) + (7\sqrt{5} - 3\sqrt{10})i$ 67. $1 \pm i$
69. $-2 \pm \frac{1}{2}i$ 71. $-2 \pm \frac{\sqrt{5}}{2}i$ 73. $2 \pm \sqrt{2}i$
75. $\frac{5}{7} \pm \frac{5\sqrt{13}}{7}i$ 77. $-1 + 6i$ 79. $-14i$
81. $-432\sqrt{2}i$ 83. i 85. 81
87. (a) $z_1 = 9 + 16i, z_2 = 20 - 10i$ (b) $z = \frac{11{,}240}{877} + \frac{4630}{877}i$
89. False. *Sample answer:* $(1 + i) + (3 + i) = 4 + 2i$
91. True.
$$x^4 - x^2 + 14 = 56$$
$$(-i\sqrt{6})^4 - (-i\sqrt{6})^2 + 14 \stackrel{?}{=} 56$$
$$36 + 6 + 14 \stackrel{?}{=} 56$$
$$56 = 56$$
93. $i, -1, -i, 1, i, -1, -i, 1$; The pattern repeats the first four results. Divide the exponent by 4.
When the remainder is 1, the result is i.
When the remainder is 2, the result is -1.
When the remainder is 3, the result is $-i$.
When the remainder is 0, the result is 1.
95. $\sqrt{-6}\sqrt{-6} = \sqrt{6}i\sqrt{6}i = 6i^2 = -6$ 97. Proof

Section 4.2 *(page 326)*

1. Fundamental Theorem; Algebra **3.** complex conjugates
5. 3 **7.** 4 **9.** Two imaginary solutions
11. One repeated real solution **13.** Two real solutions
15. $\pm\sqrt{5}$ **17.** $-1 \pm \sqrt{3}$ **19.** 4 **21.** $-1 \pm 2i$
23. $\frac{1}{2} \pm i$ **25.** $\pm\sqrt{7}, \pm i$ **27.** $\pm\sqrt{6}, \pm i$
29. (a)

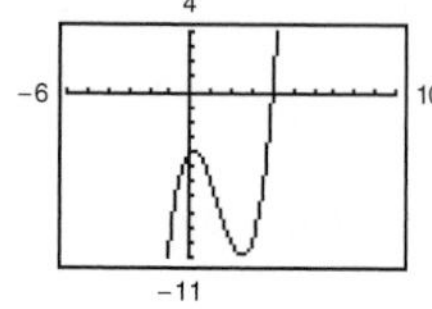

(b) $4, \pm i$
(c) They are the same.

31. (a)

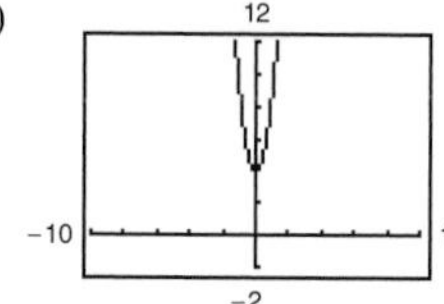

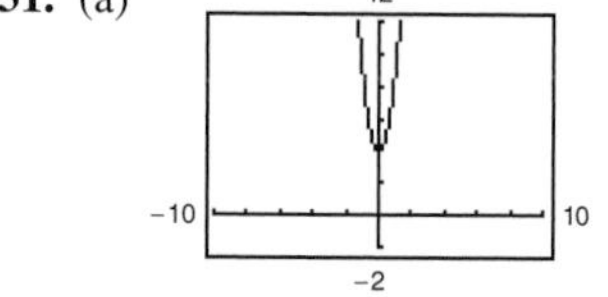

(b) $\pm\sqrt{2}i$
(c) They are the same.

33. $(x + 6i)(x - 6i); \pm 6i$
35. $(x + 3)(x - 3)(x + 3i)(x - 3i); \pm 3, \pm 3i$
37. $(x - 1 - 4i)(x - 1 + 4i); 1 \pm 4i$
39. $(x - 3 + \sqrt{19})(x - 3 - \sqrt{19}); 3 \pm \sqrt{19}$
41. $(x + 3)(x + \sqrt{3})(x - \sqrt{3}); -3, \pm\sqrt{3}$
43. $(2x - 1)(x + 3\sqrt{2}i)(x - 3\sqrt{2}i); \frac{1}{2}, \pm 3\sqrt{2}i$
45. $x(x - 6)(x + 4i)(x - 4i); 0, 6, \pm 4i$
47. $(x + i)(x - i)(x + 3i)(x - 3i); \pm i, \pm 3i$
49. $1, \pm 2i$ **51.** $\pm 2i, 1, -\frac{1}{2}$ **53.** $-4, 3 \pm i$
55. $2, 3 \pm 2i$ **57.** $\pm 2i, 1 \pm \sqrt{3}i$
59. $f(x) = x^3 - x^2 + 25x - 25$
61. $f(x) = x^4 - 6x^3 + 14x^2 - 16x + 8$
63. $f(x) = 3x^4 - 17x^3 + 25x^2 + 23x - 22$
65. $f(x) = -x^3 + x^2 - 4x + 4$
67. $f(x) = 2x^4 + 2x^3 - 2x^2 + 2x - 4$
69. $f(x) = x^3 + x^2 - 2x + 12$
71. $f(x) = x^3 - 6x^2 + 4x + 40$
73. $f(x) = x^3 - 6x^2 + 17x - 18$
75. $f(x) = \frac{1}{2}x^4 + \frac{1}{2}x^3 - 2x^2 + x - 6$
77. (a)

t	0	0.5	1	1.5	2	2.5	3
h	0	20	32	36	32	20	0

No

(b) When $h = 64$, the resulting equation yields imaginary roots. So, the ball will not reach a height of 64 feet.
(c)

y = 64
$y = -16t^2 + 48t$

The graphs do not intersect, so the ball does not reach 64 feet.
(d) The results show that it is not possible for the ball to reach a height of 64 feet.

79. (a) $P = -0.0001x^2 + 60x - 150{,}000$
(b) \$8,600,000 (c) \$115
(d) Not possible; When $P = 10{,}000{,}000$, the solutions are imaginary.
81. False. The most imaginary zeros it can have is two, and the Linear Factorization Theorem guarantees that there are three linear factors, so one zero must be real.
83. Answers will vary.
85. (a) $x^2 + b$ (b) $x^2 - 2ax + a^2 + b^2$
87. r_1, r_2, r_3 **89.** $5 + r_1, 5 + r_2, 5 + r_3$
91. The zeros cannot be determined.

Section 4.3 *(page 334)*

1. real **3.** absolute value **5.** reflections **7.** c **8.** f
9. h **10.** a **11.** b **12.** g **13.** e **14.** d
15.

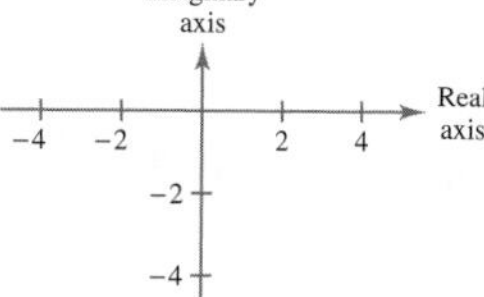

7

17.

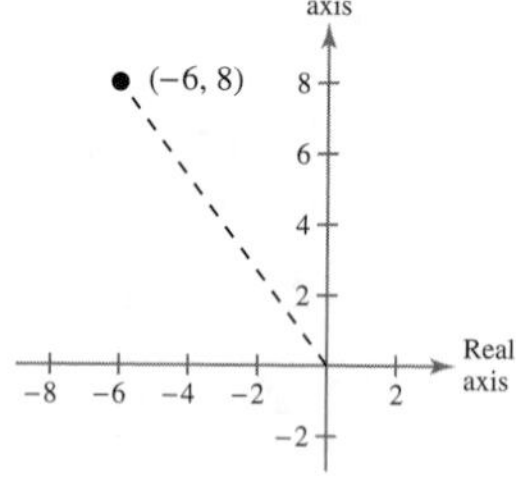

10

19.

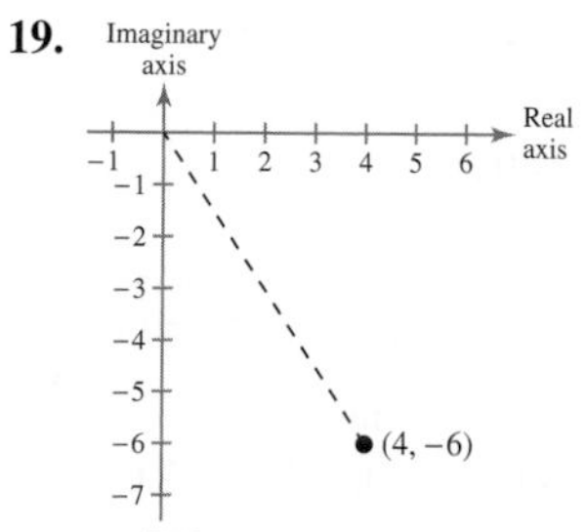

$2\sqrt{13}$

21. $5 + 6i$ **23.** $10 + 4i$ **25.** $6 + 5i$ **27.** $-5 + 7i$
29. $-2 - 2i$ **31.** $10 - 3i$ **33.** $-6i$ **35.** $-3 + 3i$
37.

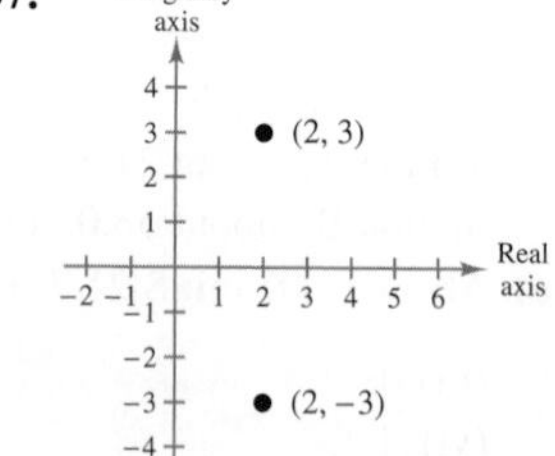

$2 - 3i$

39.

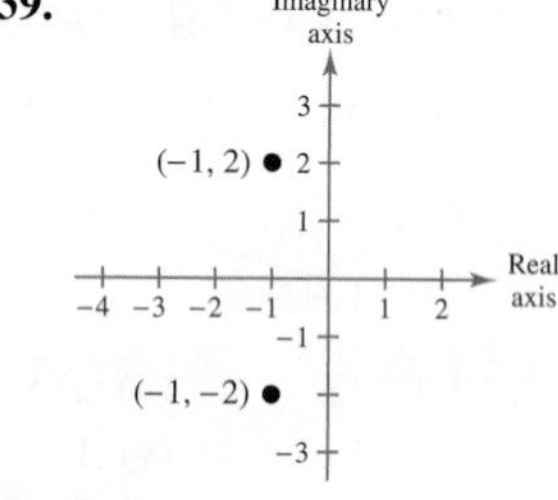

$-1 + 2i$

41. $2\sqrt{2} \approx 2.83$ **43.** $\sqrt{109} \approx 10.44$
45. $(4, 3)$ **47.** $\left(\frac{9}{2}, -\frac{3}{2}\right)$
49. (a) Ship A: $3 + 4i$, Ship B: $-5 + 2i$
(b) *Sample answer:* Find the modulus of the difference of the complex numbers.
51. False. The modulus is always real.
53. False. $|1 + i| + |1 - i| = 2\sqrt{2}$ and $|(1 + i) + (1 - i)| = 2$
55. A circle; The modulus represents the distance from the origin.
57. Isosceles triangle; The moduli are equal.

Section 4.4 *(page 340)*

1. trigonometric form; modulus; argument

3. $3\left(\cos\frac{\pi}{2} + i\sin\frac{\pi}{2}\right)$ **5.** $3\sqrt{2}\left(\cos\frac{5\pi}{4} + i\sin\frac{5\pi}{4}\right)$

7.

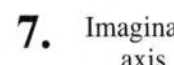

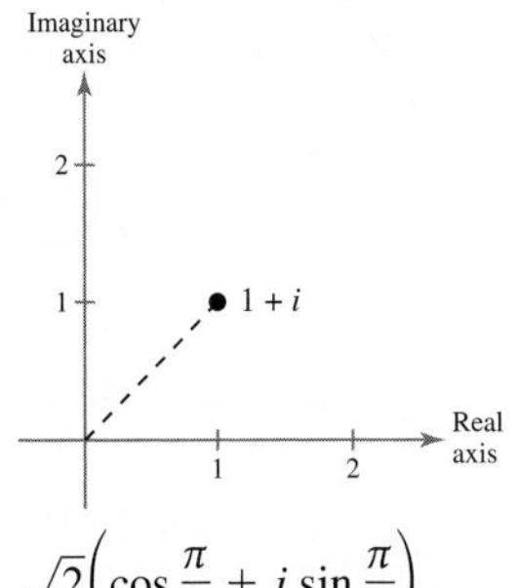

$\sqrt{2}\left(\cos\frac{\pi}{4} + i\sin\frac{\pi}{4}\right)$

9.

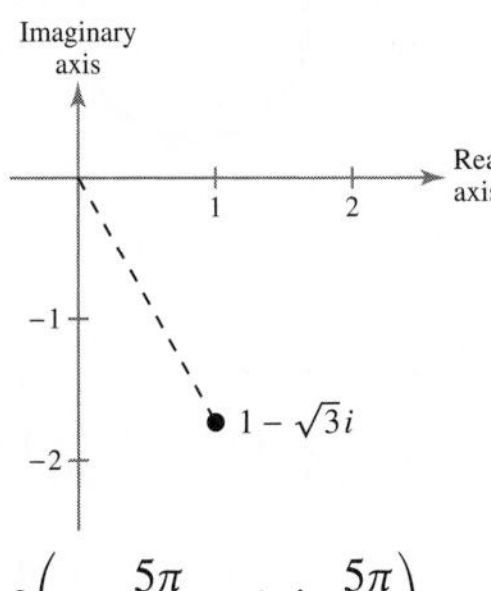

$2\left(\cos\frac{5\pi}{3} + i\sin\frac{5\pi}{3}\right)$

11.

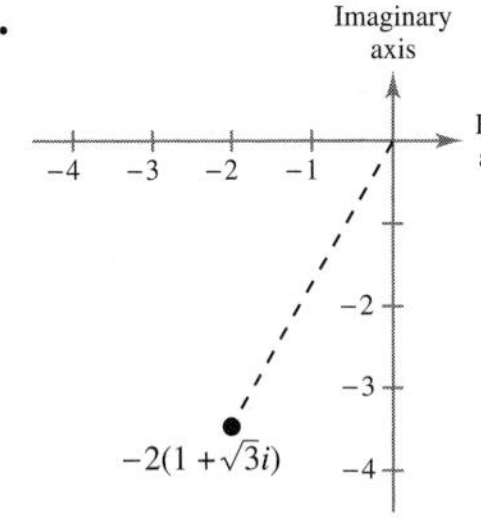

$4\left(\cos\frac{4\pi}{3} + i\sin\frac{4\pi}{3}\right)$

13.

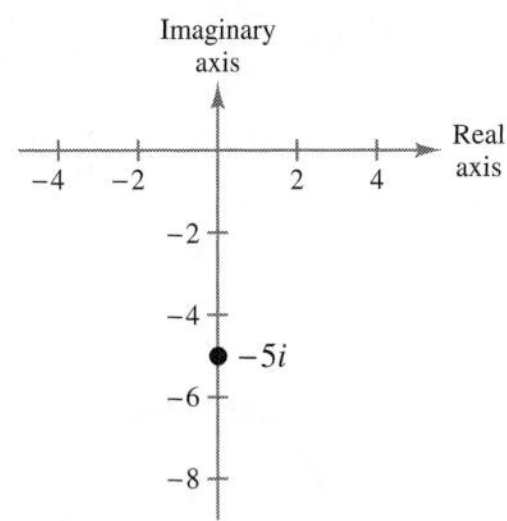

$5\left(\cos\frac{3\pi}{2} + i\sin\frac{3\pi}{2}\right)$

15.

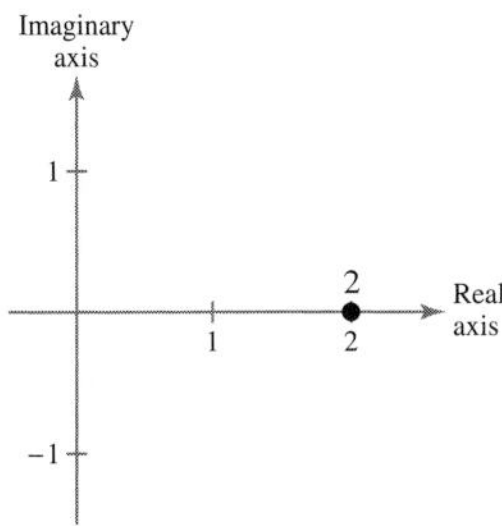

$2(\cos 0 + i\sin 0)$

17.

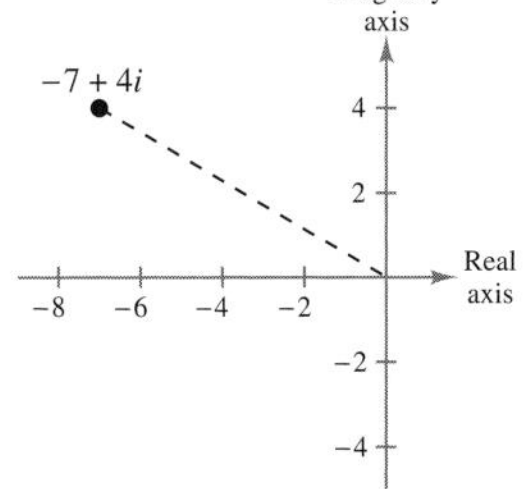

$\sqrt{65}(\cos 2.62 + i\sin 2.62)$

19.

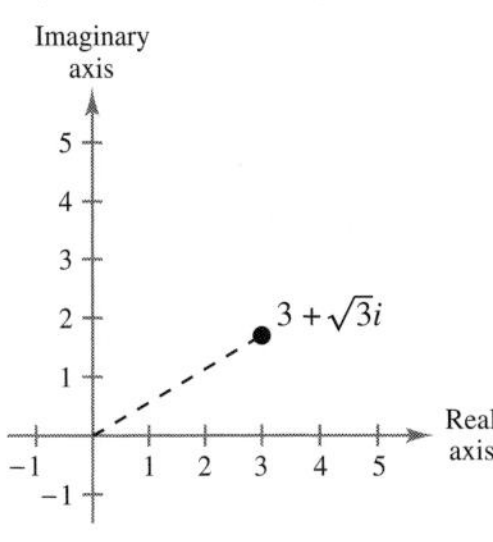

$2\sqrt{3}\left(\cos\frac{\pi}{6} + i\sin\frac{\pi}{6}\right)$

21.

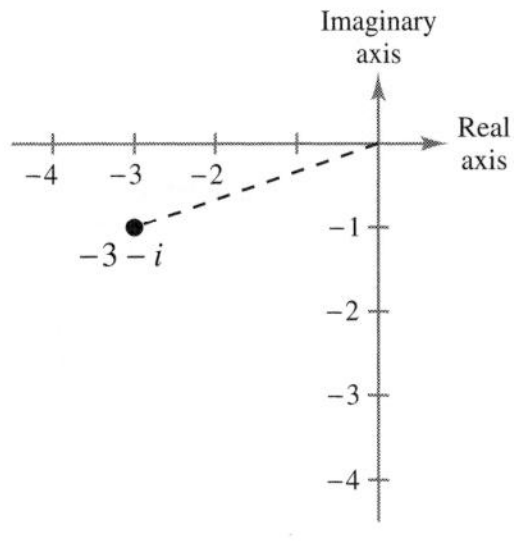

$\sqrt{10}(\cos 3.46 + i\sin 3.46)$

23.

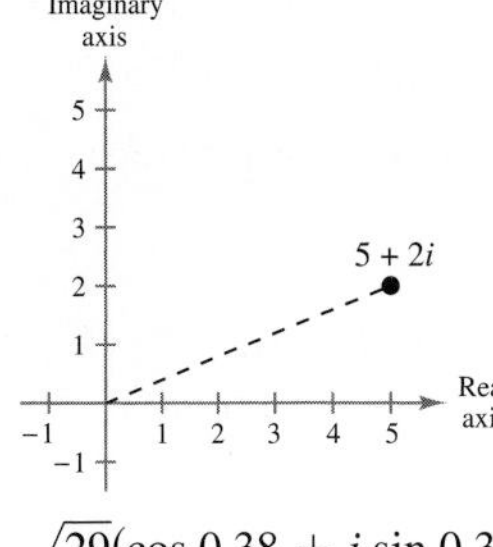

$\sqrt{29}(\cos 0.38 + i\sin 0.38)$

25.

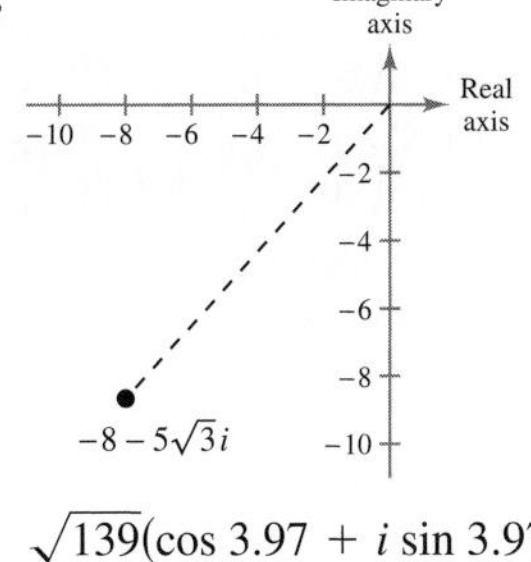

$\sqrt{139}(\cos 3.97 + i\sin 3.97)$

27. $1 + \sqrt{3}i$

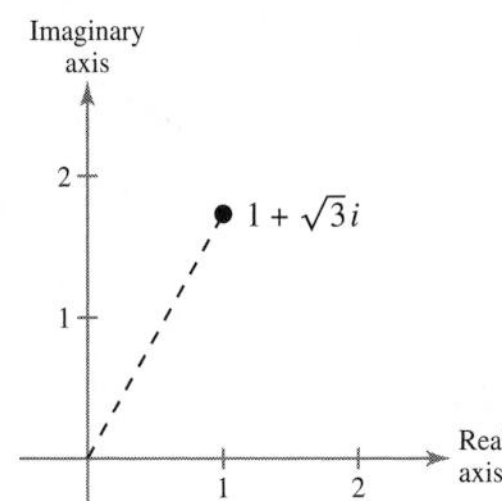

29. $-\frac{9\sqrt{2}}{8} + \frac{9\sqrt{2}}{8}i$

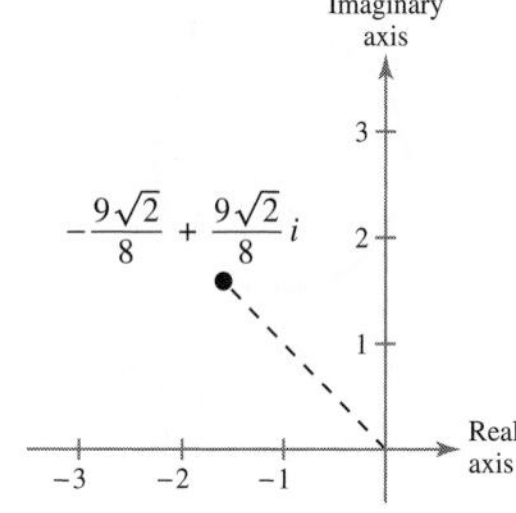

31. $6 - 2\sqrt{3}i$

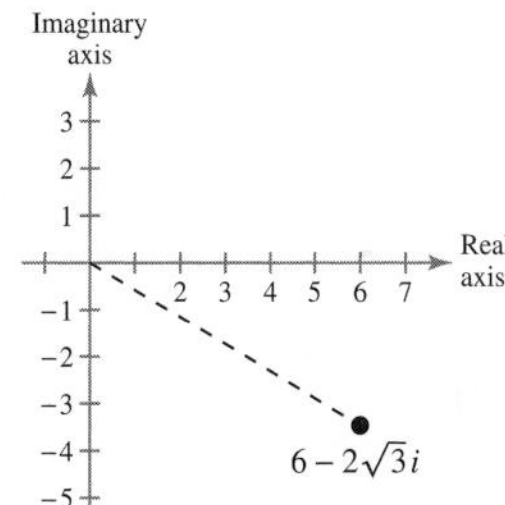

33. 7

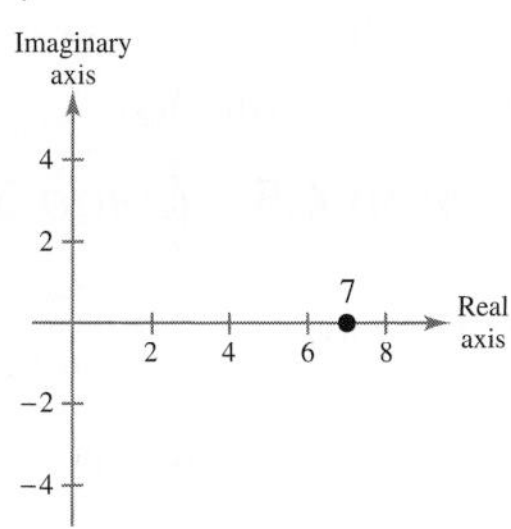

35. $-4.7347 - 1.6072i$

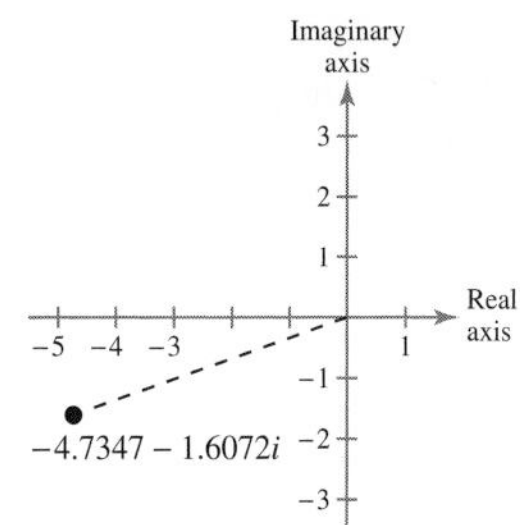

37. $4.6985 + 1.7101i$ **39.** $-1.8126 + 0.8452i$

41. $12\left(\cos\frac{\pi}{3} + i\sin\frac{\pi}{3}\right)$ **43.** $\frac{10}{9}(\cos 150° + i\sin 150°)$

45. $\cos 50° + i\sin 50°$ **47.** $\frac{1}{3}(\cos 30° + i\sin 30°)$

49. $\cos\frac{2\pi}{3} + i\sin\frac{2\pi}{3}$ **51.** $6(\cos 330° + i\sin 330°)$

53. (a) $\left[2\sqrt{2}\left(\cos\frac{\pi}{4} + i\sin\frac{\pi}{4}\right)\right]\left[\sqrt{2}\left(\cos\frac{7\pi}{4} + i\sin\frac{7\pi}{4}\right)\right]$
(b) $4(\cos 0 + i\sin 0) = 4$ (c) 4

55. (a) $\left[2\left(\cos\frac{3\pi}{2} + i\sin\frac{3\pi}{2}\right)\right]\left[\sqrt{2}\left(\cos\frac{\pi}{4} + i\sin\frac{\pi}{4}\right)\right]$
(b) $2\sqrt{2}\left(\cos\frac{7\pi}{4} + i\sin\frac{7\pi}{4}\right) = 2 - 2i$
(c) $-2i - 2i^2 = -2i + 2 = 2 - 2i$

57. (a) $[5(\cos 0.93 + i\sin 0.93)] \div \left[2\left(\cos\frac{5\pi}{3} + i\sin\frac{5\pi}{3}\right)\right]$
(b) $\frac{5}{2}(\cos 1.97 + i\sin 1.97) \approx -0.982 + 2.299i$
(c) About $-0.982 + 2.299i$

59. (a) $[5(\cos 0 + i\sin 0)] \div \left[\sqrt{13}(\cos 0.98 + i\sin 0.98)\right]$
(b) $\frac{5}{\sqrt{13}}(\cos 5.30 + i\sin 5.30) \approx 0.769 - 1.154i$
(c) $\frac{10}{13} - \frac{15}{13}i \approx 0.769 - 1.154i$

61. -1 **63.** $-10\sqrt{2} + 10\sqrt{2}i$

65.

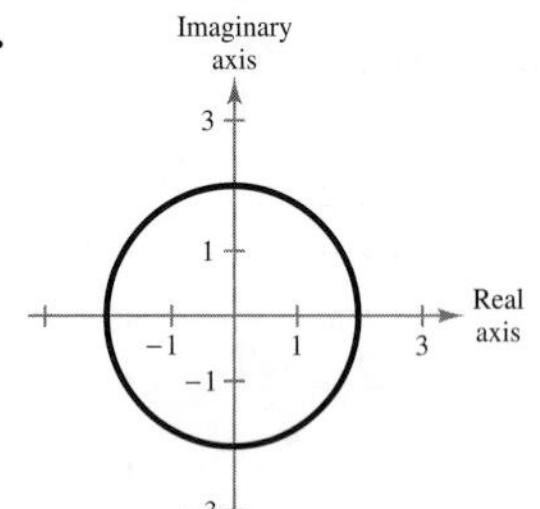

67.

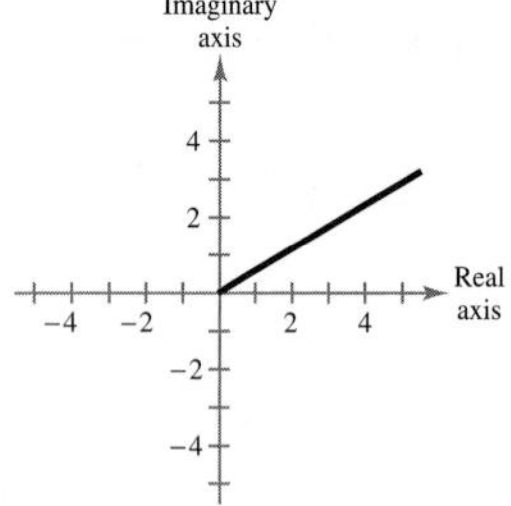

69. (a) $E = 24(\cos 30° + i \sin 30°)$ volts
(b) $E = 12\sqrt{3} + 12i$ volts (c) $|E| = 24$ volts

71. True. $z_1 z_2 = r_1 r_2[\cos(\theta_1 + \theta_2) + i \sin(\theta_1 + \theta_2)] = 0$ if and only if $r_1 = 0$ and/or $r_2 = 0$.

73. Answers will vary.

Section 4.5 *(page 346)*

1. DeMoivre's **3.** $\dfrac{2\pi}{n}$ **5.** $\dfrac{125}{2} + \dfrac{125\sqrt{3}}{2}i$ **7.** -1

9. $608.0204 + 144.6936i$ **11.** $\dfrac{81}{2} + \dfrac{81\sqrt{3}}{2}i$

13. $32.3524 - 120.7407i$ **15.** $32i$ **17.** 27

19. $-4 - 4i$ **21.** $8i$ **23.** $1024 - 1024\sqrt{3}i$

25. $-597 - 122i$ **27.** $-43\sqrt{5} + 4i$

29.

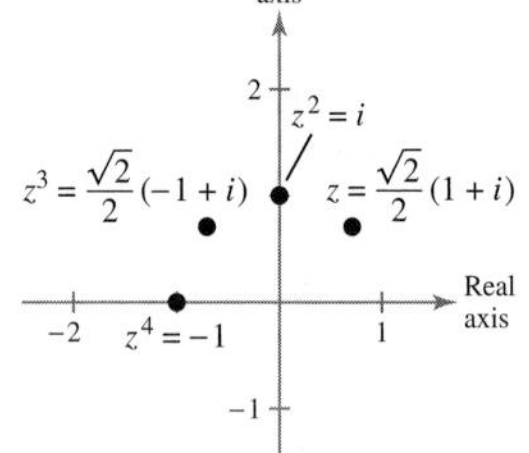

The absolute value of each is 1, and the consecutive powers of z are each 45° apart.

31. $1 + i, -1 - i$ **33.** $-\dfrac{\sqrt{6}}{2} + \dfrac{\sqrt{6}}{2}i, \dfrac{\sqrt{6}}{2} - \dfrac{\sqrt{6}}{2}i$

35. $-1.5538 + 0.6436i, 1.5538 - 0.6436i$

37. $\dfrac{\sqrt{6}}{2} + \dfrac{\sqrt{2}}{2}i, -\dfrac{\sqrt{6}}{2} - \dfrac{\sqrt{2}}{2}i$

39. (a) $\sqrt{5}(\cos 60° + i \sin 60°)$
$\sqrt{5}(\cos 240° + i \sin 240°)$
(b) $\dfrac{\sqrt{5}}{2} + \dfrac{\sqrt{15}}{2}i, -\dfrac{\sqrt{5}}{2} - \dfrac{\sqrt{15}}{2}i$
(c)

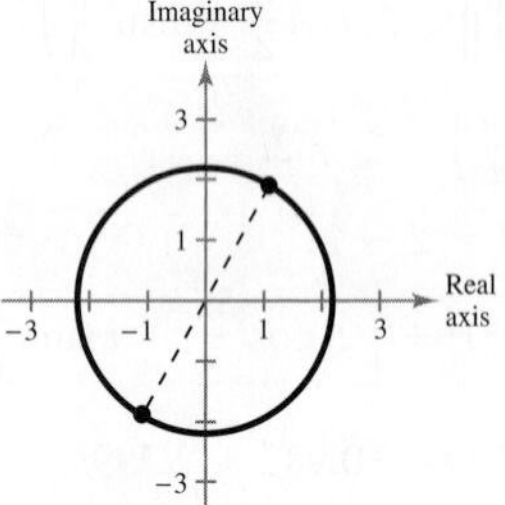

41. (a) $2\left(\cos \dfrac{2\pi}{9} + i \sin \dfrac{2\pi}{9}\right)$
$2\left(\cos \dfrac{8\pi}{9} + i \sin \dfrac{8\pi}{9}\right)$
$2\left(\cos \dfrac{14\pi}{9} + i \sin \dfrac{14\pi}{9}\right)$
(b) $1.5321 + 1.2856i$, $-1.8794 + 0.6840i$, $0.3473 - 1.9696i$

(c)

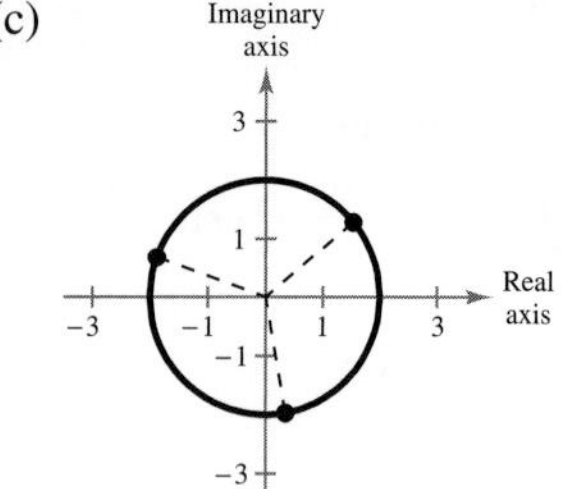

43. (a) $3\left(\cos \dfrac{\pi}{30} + i \sin \dfrac{\pi}{30}\right)$
$3\left(\cos \dfrac{13\pi}{30} + i \sin \dfrac{13\pi}{30}\right)$
$3\left(\cos \dfrac{5\pi}{6} + i \sin \dfrac{5\pi}{6}\right)$
$3\left(\cos \dfrac{37\pi}{30} + i \sin \dfrac{37\pi}{30}\right)$
$3\left(\cos \dfrac{49\pi}{30} + i \sin \dfrac{49\pi}{30}\right)$
(b) $2.9836 + 0.3136i$, $0.6237 + 2.9344i$, $-2.5981 + 1.5i$, $-2.2294 - 2.0074i$, $1.2202 - 2.7406i$
(c)

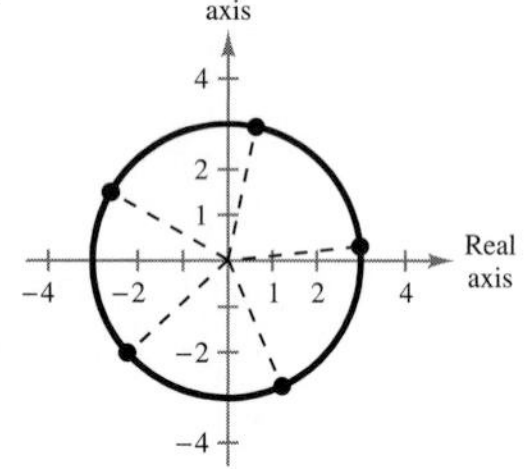

45. (a) $3\left(\cos \dfrac{\pi}{8} + i \sin \dfrac{\pi}{8}\right)$
$3\left(\cos \dfrac{5\pi}{8} + i \sin \dfrac{5\pi}{8}\right)$
$3\left(\cos \dfrac{9\pi}{8} + i \sin \dfrac{9\pi}{8}\right)$
$3\left(\cos \dfrac{13\pi}{8} + i \sin \dfrac{13\pi}{8}\right)$
(b) $2.7716 + 1.1481i$, $-1.1481 + 2.7716i$, $-2.7716 - 1.1481i$, $1.1481 - 2.7716i$
(c)

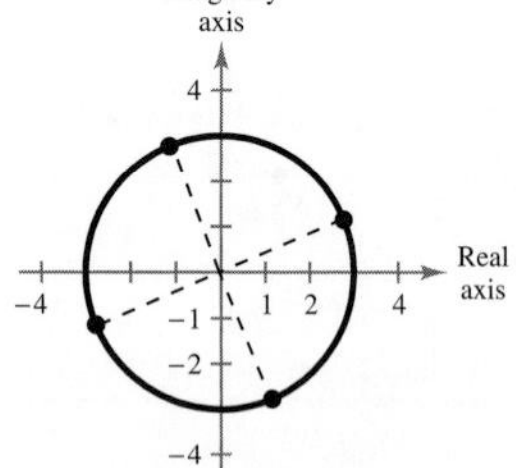

47. (a) $5\left(\cos \dfrac{4\pi}{9} + i \sin \dfrac{4\pi}{9}\right)$
$5\left(\cos \dfrac{10\pi}{9} + i \sin \dfrac{10\pi}{9}\right)$
$5\left(\cos \dfrac{16\pi}{9} + i \sin \dfrac{16\pi}{9}\right)$
(b) $0.8682 + 4.9240i$, $-4.6985 - 1.7101i$, $3.8302 - 3.2139i$

(c)

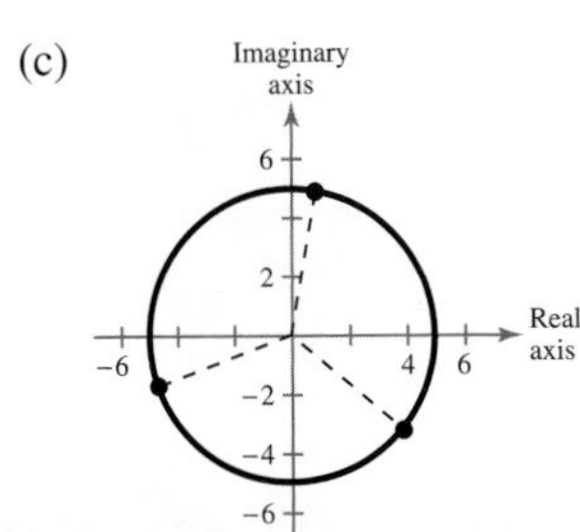

49. (a) $2(\cos 0 + i \sin 0)$

$2\left(\cos \frac{\pi}{2} + i \sin \frac{\pi}{2}\right)$

$2(\cos \pi + i \sin \pi)$

$2\left(\cos \frac{3\pi}{2} + i \sin \frac{3\pi}{2}\right)$

(b) $2, 2i, -2, -2i$

(c)

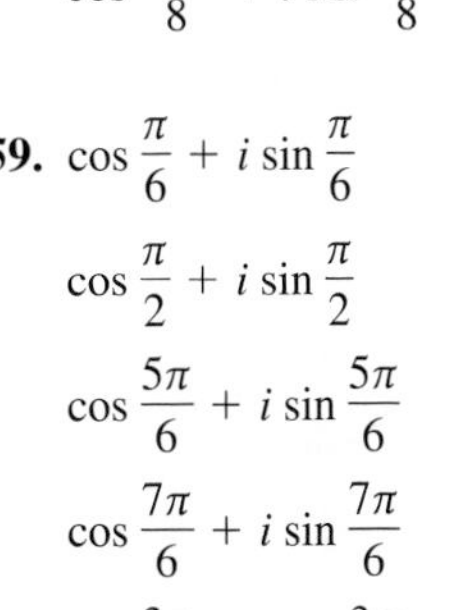

51. (a) $\cos 0 + i \sin 0$

$\cos \frac{2\pi}{5} + i \sin \frac{2\pi}{5}$

$\cos \frac{4\pi}{5} + i \sin \frac{4\pi}{5}$

$\cos \frac{6\pi}{5} + i \sin \frac{6\pi}{5}$

$\cos \frac{8\pi}{5} + i \sin \frac{8\pi}{5}$

(b) $1, 0.3090 + 0.9511i,$ $-0.8090 + 0.5878i,$ $-0.8090 - 0.5878i,$ $0.3090 - 0.9511i$

(c)

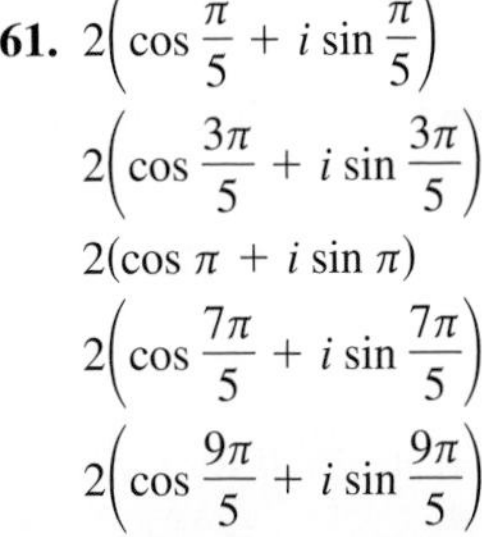

53. (a) $5\left(\cos \frac{\pi}{3} + i \sin \frac{\pi}{3}\right)$

$5(\cos \pi + i \sin \pi)$

$5\left(\cos \frac{5\pi}{3} + i \sin \frac{5\pi}{3}\right)$

(b) $\frac{5}{2} + \frac{5\sqrt{3}}{2}i, -5,$ $\frac{5}{2} - \frac{5\sqrt{3}}{2}i$

(c)

55. (a) $\sqrt{2}\left(\cos \frac{7\pi}{20} + i \sin \frac{7\pi}{20}\right)$

$\sqrt{2}\left(\cos \frac{3\pi}{4} + i \sin \frac{3\pi}{4}\right)$

$\sqrt{2}\left(\cos \frac{23\pi}{20} + i \sin \frac{23\pi}{20}\right)$

$\sqrt{2}\left(\cos \frac{31\pi}{20} + i \sin \frac{31\pi}{20}\right)$

$\sqrt{2}\left(\cos \frac{39\pi}{20} + i \sin \frac{39\pi}{20}\right)$

(b) $0.6420 + 1.2601i,$ $-1 + i,$ $-1.2601 - 0.6420i,$ $0.2212 - 1.3968i,$ $1.3968 - 0.2212i$

(c)

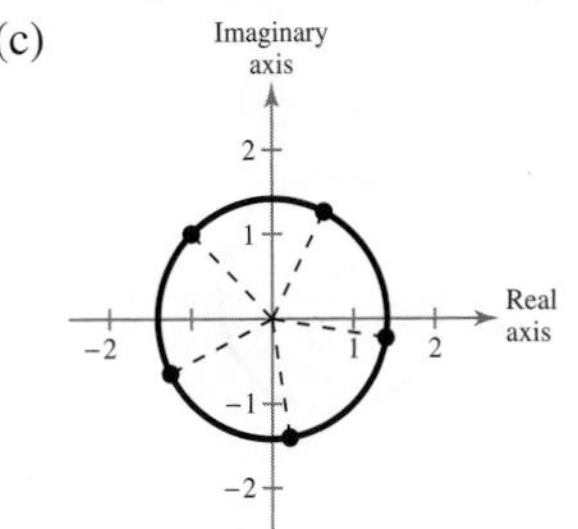

57. $\cos \frac{3\pi}{8} + i \sin \frac{3\pi}{8}$

$\cos \frac{7\pi}{8} + i \sin \frac{7\pi}{8}$

$\cos \frac{11\pi}{8} + i \sin \frac{11\pi}{8}$

$\cos \frac{15\pi}{8} + i \sin \frac{15\pi}{8}$

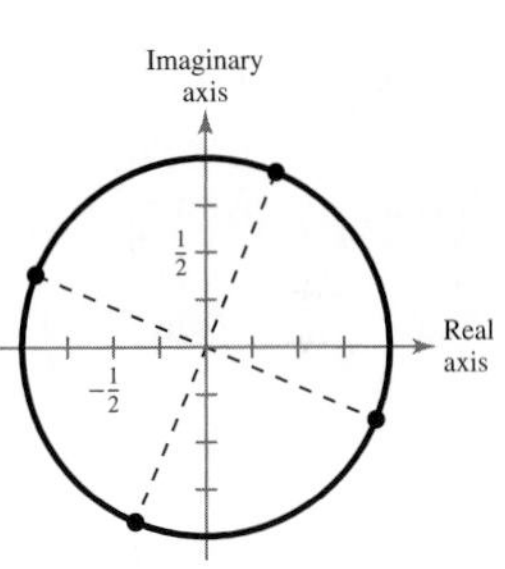

59. $\cos \frac{\pi}{6} + i \sin \frac{\pi}{6}$

$\cos \frac{\pi}{2} + i \sin \frac{\pi}{2}$

$\cos \frac{5\pi}{6} + i \sin \frac{5\pi}{6}$

$\cos \frac{7\pi}{6} + i \sin \frac{7\pi}{6}$

$\cos \frac{3\pi}{2} + i \sin \frac{3\pi}{2}$

$\cos \frac{11\pi}{6} + i \sin \frac{11\pi}{6}$

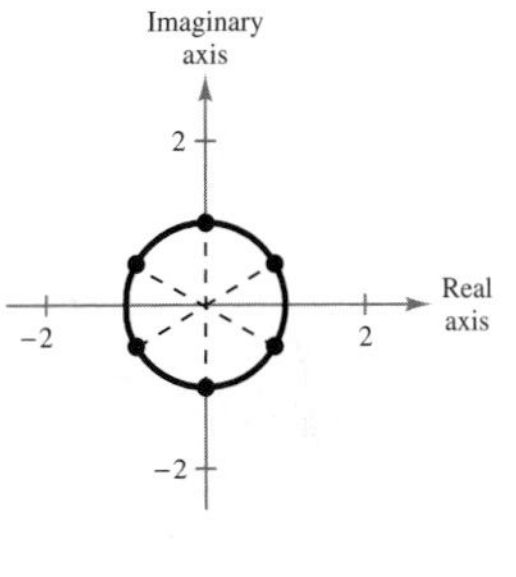

61. $2\left(\cos \frac{\pi}{5} + i \sin \frac{\pi}{5}\right)$

$2\left(\cos \frac{3\pi}{5} + i \sin \frac{3\pi}{5}\right)$

$2(\cos \pi + i \sin \pi)$

$2\left(\cos \frac{7\pi}{5} + i \sin \frac{7\pi}{5}\right)$

$2\left(\cos \frac{9\pi}{5} + i \sin \frac{9\pi}{5}\right)$

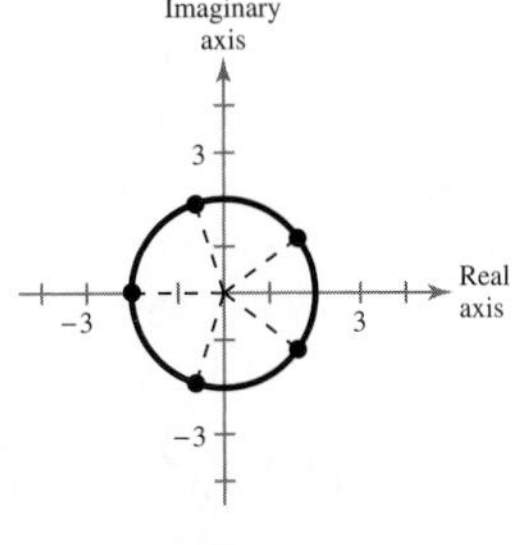

63. $3(\cos 0 + i \sin 0)$

$3\left(\cos \frac{2\pi}{3} + i \sin \frac{2\pi}{3}\right)$

$3\left(\cos \frac{4\pi}{3} + i \sin \frac{4\pi}{3}\right)$

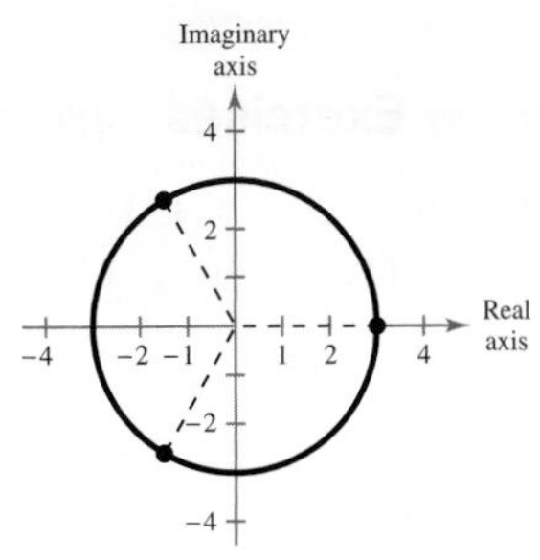

65. $2\left(\cos\frac{3\pi}{8} + i\sin\frac{3\pi}{8}\right)$
$2\left(\cos\frac{7\pi}{8} + i\sin\frac{7\pi}{8}\right)$
$2\left(\cos\frac{11\pi}{8} + i\sin\frac{11\pi}{8}\right)$
$2\left(\cos\frac{15\pi}{8} + i\sin\frac{15\pi}{8}\right)$

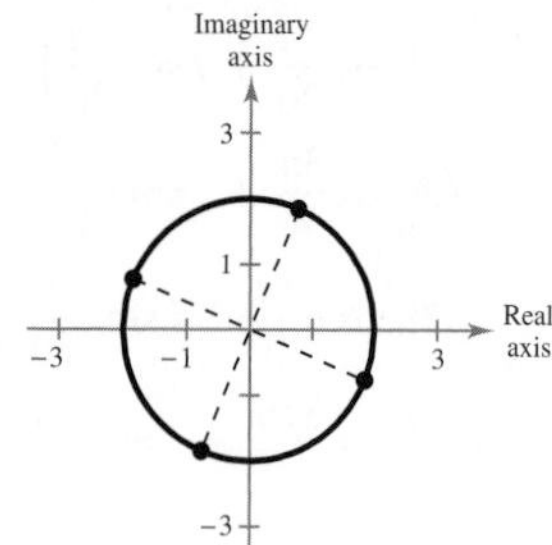

67. $2\left(\cos\frac{\pi}{8} + i\sin\frac{\pi}{8}\right)$
$2\left(\cos\frac{5\pi}{8} + i\sin\frac{5\pi}{8}\right)$
$2\left(\cos\frac{9\pi}{8} + i\sin\frac{9\pi}{8}\right)$
$2\left(\cos\frac{13\pi}{8} + i\sin\frac{13\pi}{8}\right)$

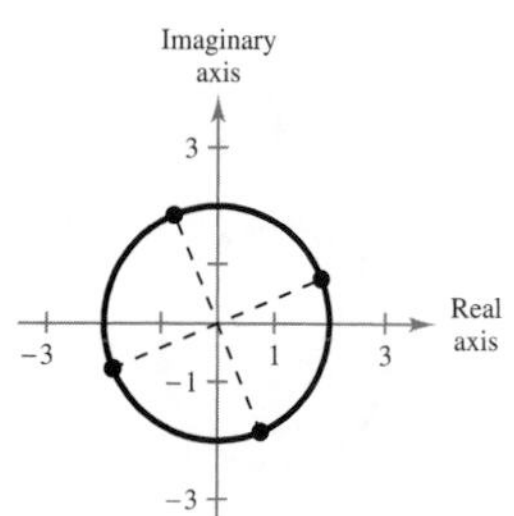

69. $\sqrt[6]{2}\left(\cos\frac{7\pi}{12} + i\sin\frac{7\pi}{12}\right)$
$\sqrt[6]{2}\left(\cos\frac{5\pi}{4} + i\sin\frac{5\pi}{4}\right)$
$\sqrt[6]{2}\left(\cos\frac{23\pi}{12} + i\sin\frac{23\pi}{12}\right)$

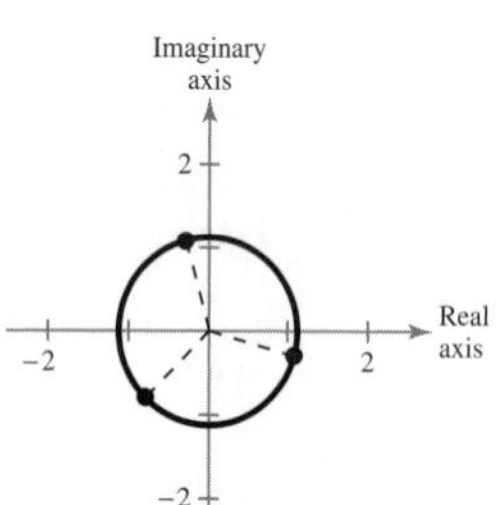

71. $\sqrt[12]{2}\left(\cos\frac{5\pi}{24} + i\sin\frac{5\pi}{24}\right)$
$\sqrt[12]{2}\left(\cos\frac{13\pi}{24} + i\sin\frac{13\pi}{24}\right)$
$\sqrt[12]{2}\left(\cos\frac{7\pi}{8} + i\sin\frac{7\pi}{8}\right)$
$\sqrt[12]{2}\left(\cos\frac{29\pi}{24} + i\sin\frac{29\pi}{24}\right)$
$\sqrt[12]{2}\left(\cos\frac{37\pi}{24} + i\sin\frac{37\pi}{24}\right)$
$\sqrt[12]{2}\left(\cos\frac{15\pi}{8} + i\sin\frac{15\pi}{8}\right)$

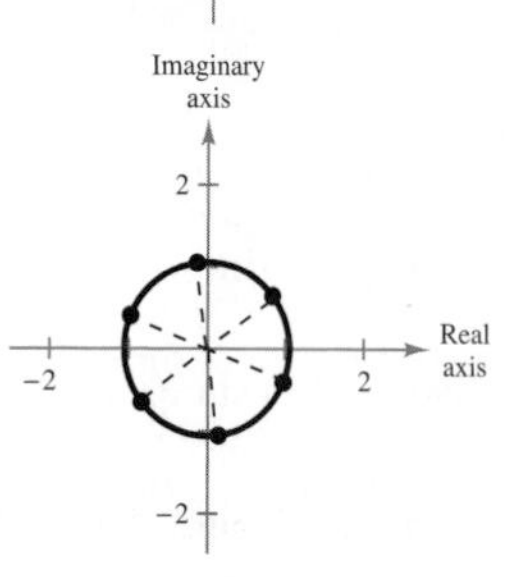

73. (a) Prisoner set (b) Escape set
(c) Escape set (d) Prisoner set

75. False. They are equally spaced around the circle centered at the origin with radius $\sqrt[n]{r}$.

77. True. $(i)^3 = -i$ and $(-i)^3 = i$

79–81. Answers will vary.

83. (a) $i, -2i$ (b) $\left(-1 \pm \sqrt{2}\right)i$
(c) $\frac{\sqrt{2}}{2} - \left(1 + \frac{\sqrt{6}}{2}\right)i, -\frac{\sqrt{2}}{2} + \left(-1 + \frac{\sqrt{6}}{2}\right)i$

85. Answers will vary.

Review Exercises *(page 350)*

1. $6 + 2i$ **3.** $-1 + 3i$ **5.** $-3 - 3i$ **7.** $15 + 6i$
9. 50 **11.** $\frac{4}{5} + \frac{8}{5}i$ **13.** $\frac{17}{26} + \frac{7}{26}i$ **15.** $\frac{21}{13} - \frac{1}{13}i$
17. $1 \pm 3i$ **19.** $-\frac{1}{2} \pm \frac{\sqrt{6}}{2}i$ **21.** $-10 + i$
23. i **25.** 6 **27.** Two real solutions
29. Two imaginary solutions **31.** 0, 2 **33.** $\frac{3}{2} \pm \frac{\sqrt{11}i}{2}$
35. $-\frac{3}{4} \pm \frac{\sqrt{39}i}{4}$ **37.** $\pm i, \pm\sqrt{7}i$ **39.** 16.8°C
41. $\frac{1}{2}\left(2x + 1 - \sqrt{5}i\right)\left(2x + 1 + \sqrt{5}i\right); -\frac{1}{2} \pm \frac{\sqrt{5}}{2}i$
43. $(2x - 3)(x + 5i)(x - 5i); \frac{3}{2}, \pm 5i$
45. $\left(2x + \sqrt{5}\right)\left(2x - \sqrt{5}\right)\left(x + \sqrt{2}i\right)\left(x - \sqrt{2}i\right); \pm\frac{\sqrt{5}}{2}, \pm\sqrt{2}i$
47. $-7, 2$ **49.** $2, \pm 4i$ **51.** $-3, \frac{1}{2}, 2 \pm i$
53. $f(x) = 3x^4 - 14x^3 + 17x^2 - 42x + 24$
55. $f(x) = x^4 + 27x^2 + 50$
57. $f(x) = 2x^3 - 14x^2 + 24x - 20$

59.

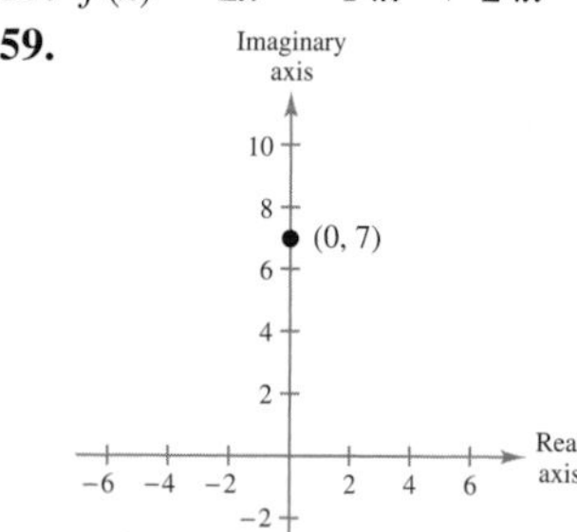

7

61.

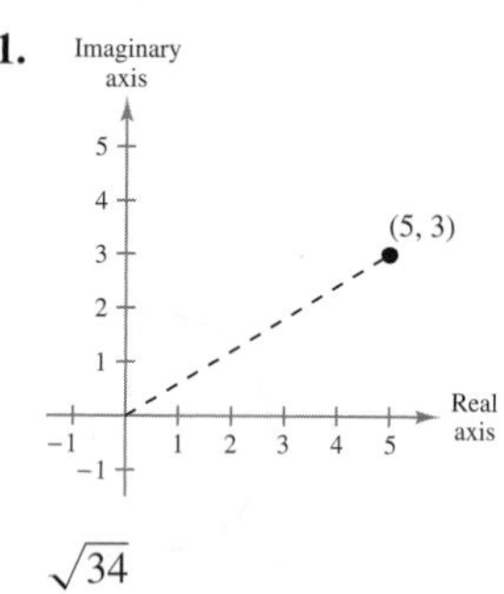

$\sqrt{34}$

63. $3 + i$ **65.** $-2 + i$

67.

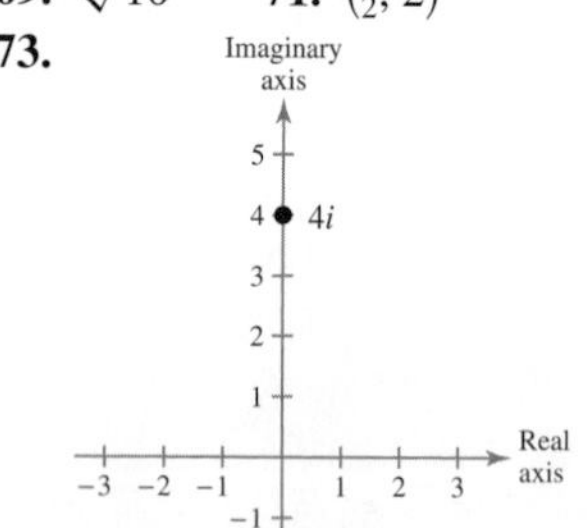

$3 - i$

69. $\sqrt{10}$ **71.** $\left(\frac{5}{2}, 2\right)$

73.

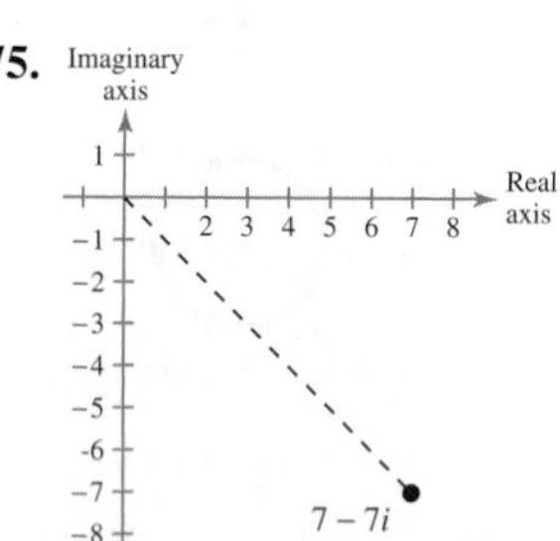

$4\left(\cos\frac{\pi}{2} + i\sin\frac{\pi}{2}\right)$

75.

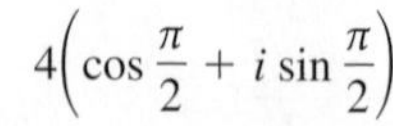

$7\sqrt{2}\left(\cos\frac{7\pi}{4} + i\sin\frac{7\pi}{4}\right)$

77.

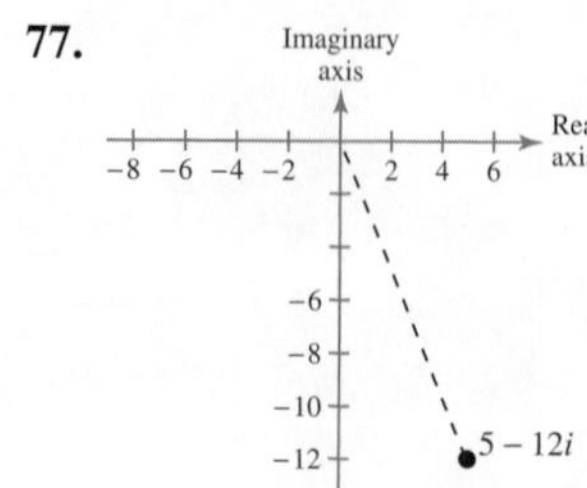

$13(\cos 5.107 + i\sin 5.107)$

79. $\sqrt{3} + i$

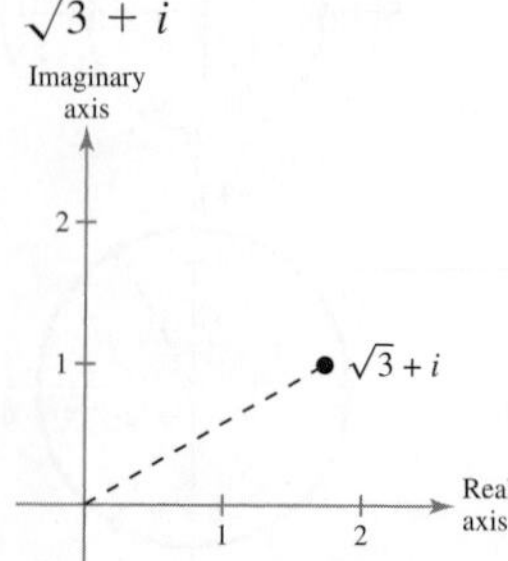

81. $1 - i$

83. $-\sqrt{3} + i$

85. $4\left(\cos\frac{7\pi}{12} + i\sin\frac{7\pi}{12}\right)$ **87.** $\frac{2}{3}(\cos 45° + i\sin 45°)$

89. $-8i$ **91.** $\frac{625}{2} + \frac{625\sqrt{3}}{2}i$ **93.** $2035 - 828i$

95. $-8 - 8i$

97. (a) $3\left(\cos\frac{\pi}{4} + i\sin\frac{\pi}{4}\right)$

$3\left(\cos\frac{7\pi}{12} + i\sin\frac{7\pi}{12}\right)$

$3\left(\cos\frac{11\pi}{12} + i\sin\frac{11\pi}{12}\right)$

$3\left(\cos\frac{5\pi}{4} + i\sin\frac{5\pi}{4}\right)$

$3\left(\cos\frac{19\pi}{12} + i\sin\frac{19\pi}{12}\right)$

$3\left(\cos\frac{23\pi}{12} + i\sin\frac{23\pi}{12}\right)$

(b) $\frac{3\sqrt{2}}{2} + \frac{3\sqrt{2}}{2}i,$
$-0.7765 + 2.8978i,$
$-2.8978 + 0.7765i,$
$-\frac{3\sqrt{2}}{2} - \frac{3\sqrt{2}}{2}i,$
$0.7765 - 2.8978i,$
$2.8978 - 0.7765i$

(c)

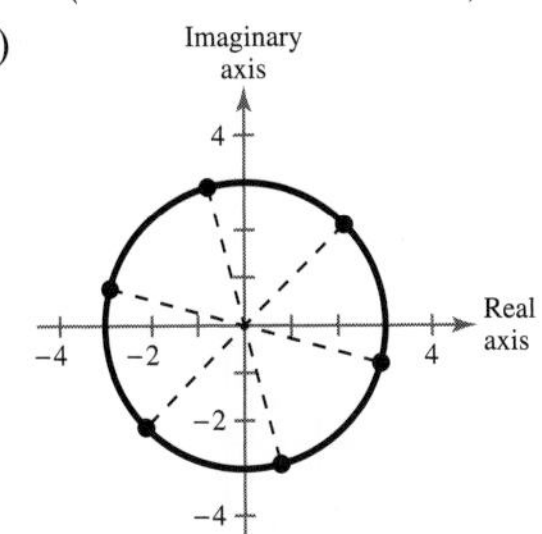

99. (a) $2(\cos 0 + i\sin 0)$

$2\left(\cos\frac{2\pi}{3} + i\sin\frac{2\pi}{3}\right)$

$2\left(\cos\frac{4\pi}{3} + i\sin\frac{4\pi}{3}\right)$

(b) $2, -1 + \sqrt{3}i,$
$-1 - \sqrt{3}i$

(c)

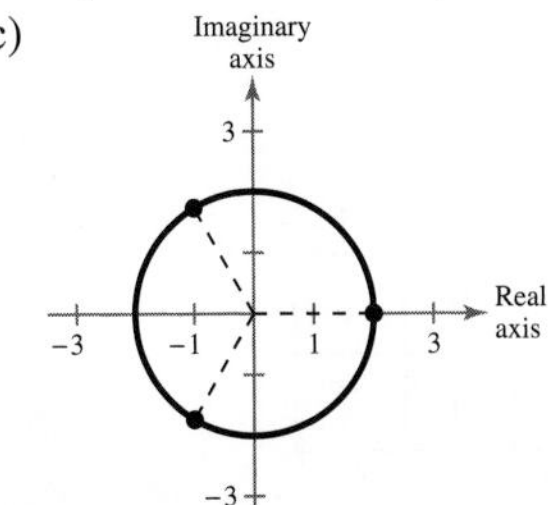

101. $3\left(\cos\frac{\pi}{4} + i\sin\frac{\pi}{4}\right) = \frac{3\sqrt{2}}{2} + \frac{3\sqrt{2}}{2}i$

$3\left(\cos\frac{3\pi}{4} + i\sin\frac{3\pi}{4}\right) = -\frac{3\sqrt{2}}{2} + \frac{3\sqrt{2}}{2}i$

$3\left(\cos\frac{5\pi}{4} + i\sin\frac{5\pi}{4}\right) = -\frac{3\sqrt{2}}{2} - \frac{3\sqrt{2}}{2}i$

$3\left(\cos\frac{7\pi}{4} + i\sin\frac{7\pi}{4}\right) = \frac{3\sqrt{2}}{2} - \frac{3\sqrt{2}}{2}i$

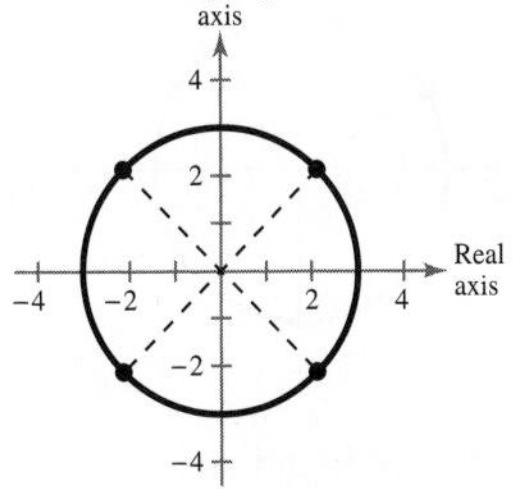

103. $2\left(\cos\frac{\pi}{2} + i\sin\frac{\pi}{2}\right) = 2i$

$2\left(\cos\frac{7\pi}{6} + i\sin\frac{7\pi}{6}\right) = -\sqrt{3} - i$

$2\left(\cos\frac{11\pi}{6} + i\sin\frac{11\pi}{6}\right) = \sqrt{3} - i$

105. False. $\sqrt{-18}\sqrt{-2} = 3\sqrt{2}i\sqrt{2}i = 6i^2 = -6$ and $\sqrt{(-18)(-2)} = \sqrt{36} = 6.$

107. False. A fourth-degree polynomial with real coefficients has four zeros, and complex zeros occur in conjugate pairs.

109. (a) $4(\cos 60° + i\sin 60°)$ (b) -64
$4(\cos 180° + i\sin 180°)$
$4(\cos 300° + i\sin 300°)$

111. $z_1z_2 = -4, \frac{z_1}{z_2} = -\cos 2\theta - i\sin 2\theta$

Chapter Test *(page 353)*

1. $-5 + 10i$ **2.** -14 **3.** $-65 + 72i$ **4.** 43

5. $\frac{8}{5} - \frac{16}{5}i$ **6.** $\frac{1}{2} \pm \frac{\sqrt{5}}{2}i$ **7.** 5

8. $(x - 6)(x + \sqrt{5}i)(x - \sqrt{5}i); 6, \pm\sqrt{5}i$

9. $(x + \sqrt{6})(x - \sqrt{6})(x + 2i)(x - 2i); \pm\sqrt{6}, \pm 2i$

10. $\pm 2, \pm\sqrt{2}i$ **11.** $\frac{3}{2}, 2 \pm i$

12. $f(x) = x^4 - 2x^3 + 9x^2 - 18x$

13. $f(x) = x^4 - 6x^3 + 16x^2 - 18x + 7$

14. No. If $a + bi, b \neq 0$, is a zero, its complex conjugate $a - bi$ is also a zero.

15. 5 **16.** $4\sqrt{2}\left(\cos\frac{7\pi}{4} + i\sin\frac{7\pi}{4}\right)$ **17.** $-3 + 3\sqrt{3}i$

18. $-\frac{6561}{2} - \frac{6561\sqrt{3}}{2}i$ **19.** $5832i$

CHAPTER 4

20. $4(\cos 0 + i \sin 0)$

$4\left(\cos \frac{\pi}{2} + i \sin \frac{\pi}{2}\right)$

$4(\cos \pi + i \sin \pi)$

$4\left(\cos \frac{3\pi}{2} + i \sin \frac{3\pi}{2}\right)$

21. $3\left(\cos \frac{\pi}{6} + i \sin \frac{\pi}{6}\right)$

$3\left(\cos \frac{5\pi}{6} + i \sin \frac{5\pi}{6}\right)$

$3\left(\cos \frac{3\pi}{2} + i \sin \frac{3\pi}{2}\right)$

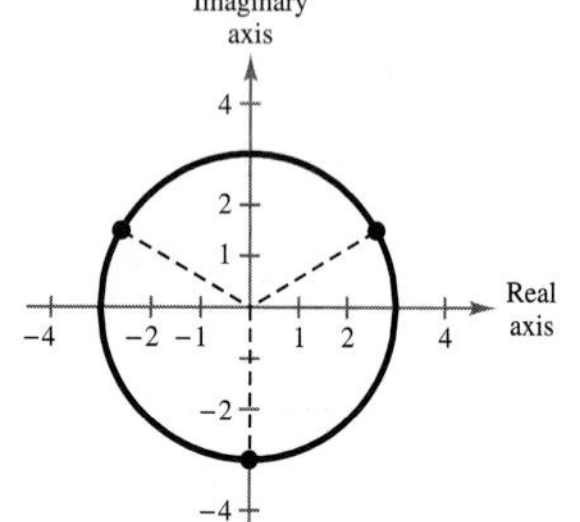

22. No. When $h = 125$, the solutions are imaginary.

Problem Solving *(page 355)*

1. (a) $z^3 = 8$ for all three complex numbers.
(b) $z^3 = 27$ for all three complex numbers.
(c) The cube roots of a positive real number a are:
$\sqrt[3]{a}$, $\frac{-\sqrt[3]{a} + \sqrt[3]{a}\sqrt{3}i}{2}$, and $\frac{-\sqrt[3]{a} - \sqrt[3]{a}\sqrt{3}i}{2}$.

3. (a) *Sample answer:* $f(x) = -2x^3 + 3x^2 + 11x - 6$
(b)

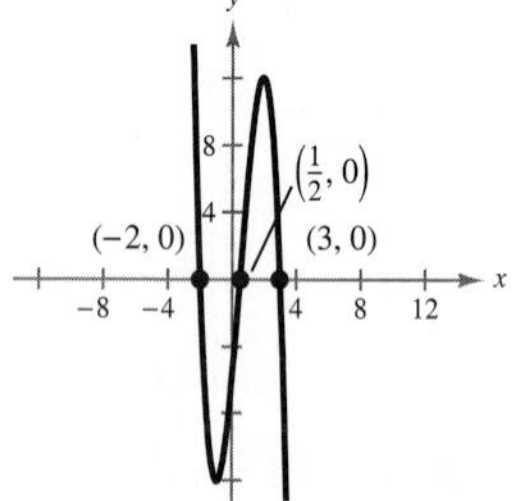

(c) Infinitely many

5. Proof **7.** (a) $0 < k < 4$ (b) $k < 0$ (c) $k > 4$

9. c; f has $(0, 0)$ as an intercept, the degree of g is 2, and k has $(-1, 0)$ as an intercept.

11. (a) $-2, 1, 4$ (b) The graph touches the x-axis at $x = 1$.
(c) 4; There are at least four real zeros (1 is repeated); no; Imaginary zeros occur in conjugate pairs.
(d) Positive. The graph rises to the left and to the right.
(e) *Sample answer:* $f(x) = x^4 - 4x^3 - 3x^2 + 14x - 8$
(f)

13. Answers will vary. **15.** Answers will vary. Circle

Chapter 5

Section 5.1 *(page 366)*

1. algebraic **3.** One-to-One **5.** $A = P\left(1 + \frac{r}{n}\right)^{nt}$

7. 0.863 **9.** 1.552 **11.** 1767.767

13. d **14.** c **15.** a **16.** b

17.

x	-2	-1	0	1	2
$f(x)$	0.020	0.143	1	7	49

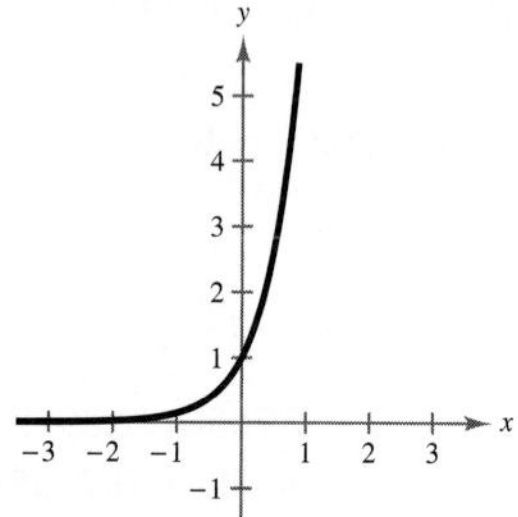

19.

x	-2	-1	0	1	2
$f(x)$	0.063	0.25	1	4	16

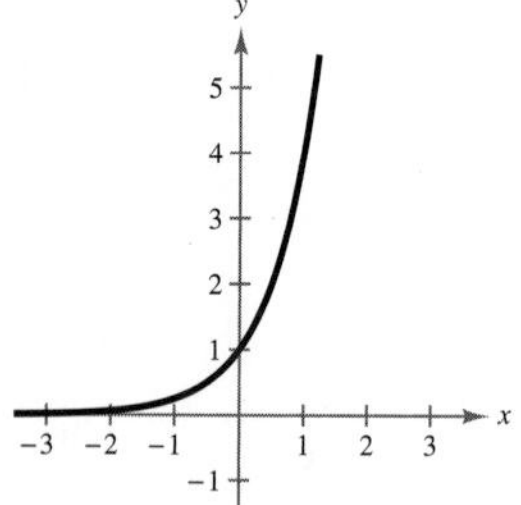

21.

x	-2	-1	0	1	2
$f(x)$	0.016	0.063	0.25	1	4

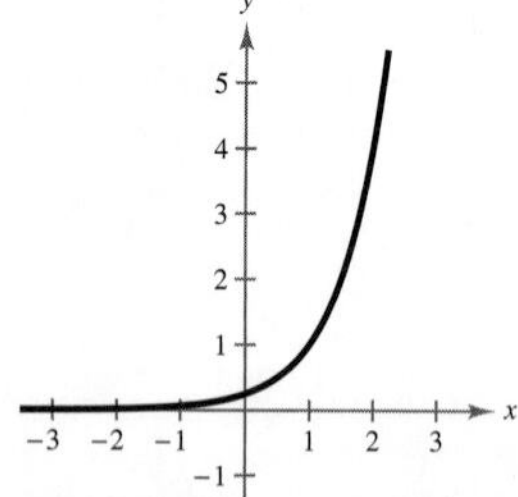

23.

x	-3	-2	-1	0	1
$f(x)$	3.25	3.5	4	5	7

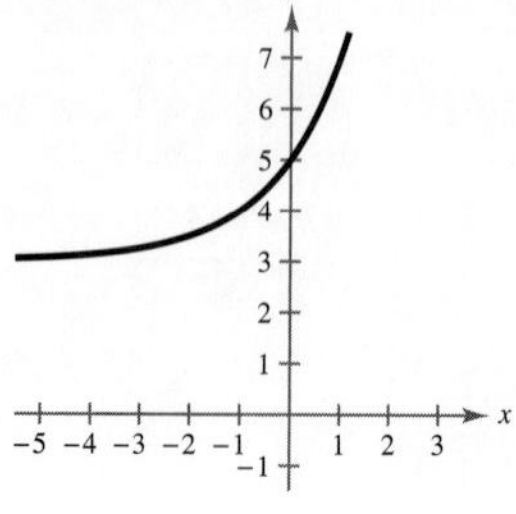

25. 2 **27.** -5 **29.** Shift the graph of f one unit up.

31. Reflect the graph of f in the y-axis and shift three units to the right.

33. 6.686 **35.** 7166.647

37.

x	-8	-7	-6	-5	-4
$f(x)$	0.055	0.149	0.406	1.104	3

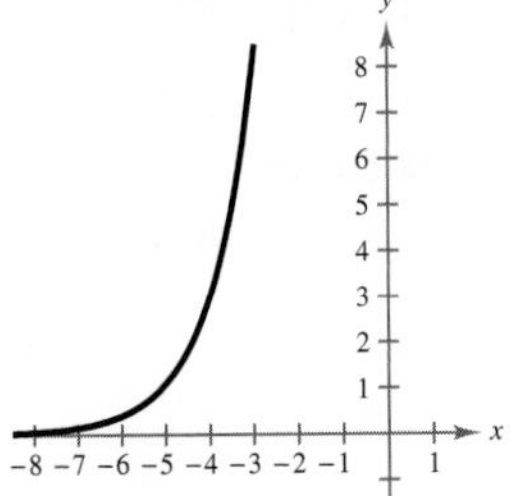

39.

x	-2	-1	0	1	2
$f(x)$	4.037	4.100	4.271	4.736	6

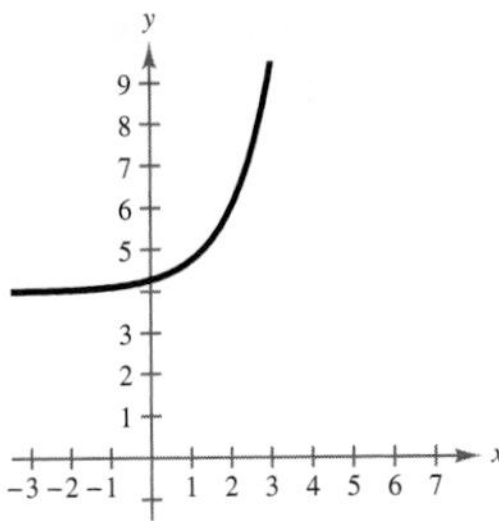

41.

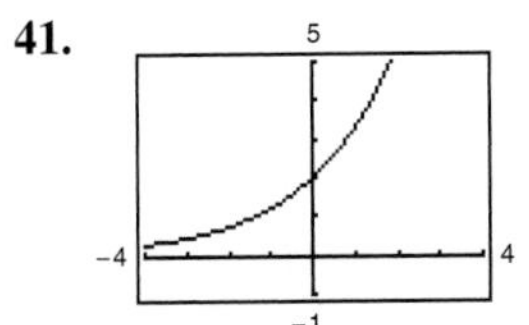

43.

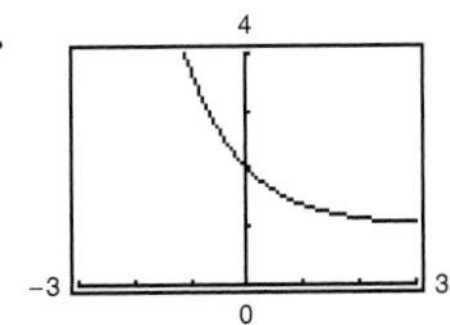

45. $\frac{1}{3}$ **47.** 3, -1

49.

n	1	2	4	12
A	\$1828.49	\$1830.29	\$1831.19	\$1831.80

n	365	Continuous
A	\$1832.09	\$1832.10

51.

n	1	2	4	12
A	\$5477.81	\$5520.10	\$5541.79	\$5556.46

n	365	Continuous
A	\$5563.61	\$5563.85

53.

t	10	20	30
A	\$17,901.90	\$26,706.49	\$39,841.40

t	40	50
A	\$59,436.39	\$88,668.67

55.

t	10	20	30
A	\$22,986.49	\$44,031.56	\$84,344.25

t	40	50
A	\$161,564.86	\$309,484.08

57. \$104,710.29 **59.** \$44.23

61. (a)

(b)

t	25	26	27	28
P (in millions)	350.281	352.107	353.943	355.788

t	29	30	31	32
P (in millions)	357.643	359.508	361.382	363.266

t	33	34	35	36
P (in millions)	365.160	367.064	368.977	370.901

t	37	38	39	40
P (in millions)	372.835	374.779	376.732	378.697

t	41	42	43	44
P (in millions)	380.671	382.656	384.651	386.656

t	45	46	47	48
P (in millions)	388.672	390.698	392.735	394.783

t	49	50	51	52
P (in millions)	396.841	398.910	400.989	403.080

t	53	54	55
P (in millions)	405.182	407.294	409.417

(c) 2064

63. (a) 16 g (b) 1.85 g

(c)

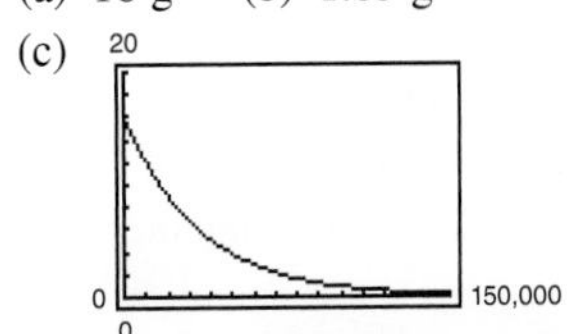

65. (a) $V(t) = 49{,}810\left(\frac{7}{8}\right)^t$ (b) \$29,197.71

67. True. As $x \to -\infty$, $f(x) \to -2$ but never reaches -2.

69. $f(x) = h(x)$ **71.** $f(x) = g(x) = h(x)$

73.

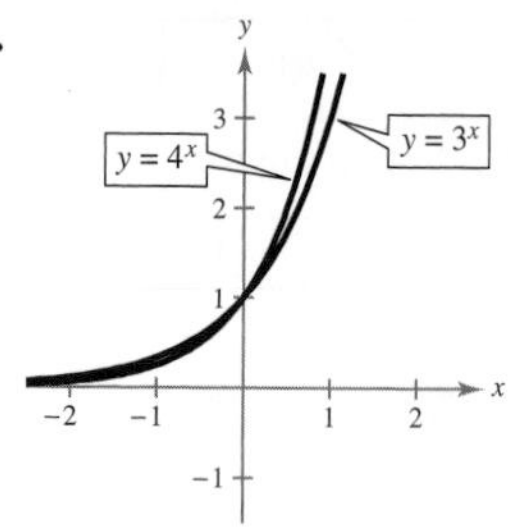

(a) $x < 0$ (b) $x > 0$

75.

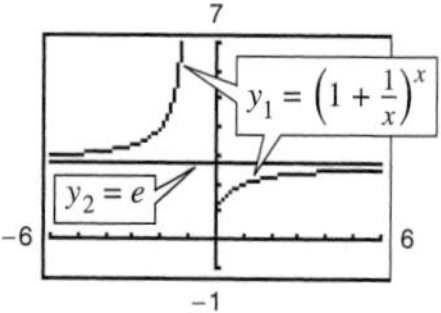

As the x-value increases, y_1 approaches the value of e.

77. (a)

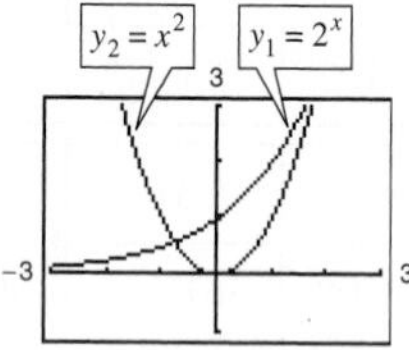

(b)

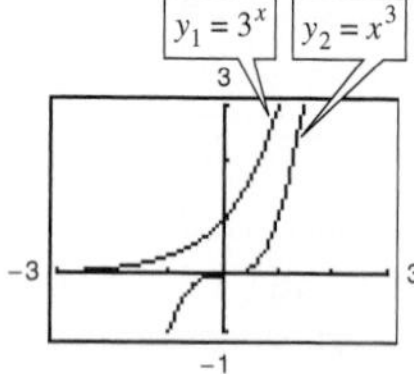

In both viewing windows, the constant raised to a variable power increases more rapidly than the variable raised to a constant power.

79. c, d

Section 5.2 *(page 376)*

1. logarithmic **3.** natural; e **5.** $x = y$ **7.** $4^2 = 16$
9. $12^1 = 12$ **11.** $\log_5 125 = 3$ **13.** $\log_4 \frac{1}{64} = -3$
15. 6 **17.** 0 **19.** -2 **21.** -0.058 **23.** 1.097
25. 1 **27.** 0 **29.** 5 **31.** ± 2

33.

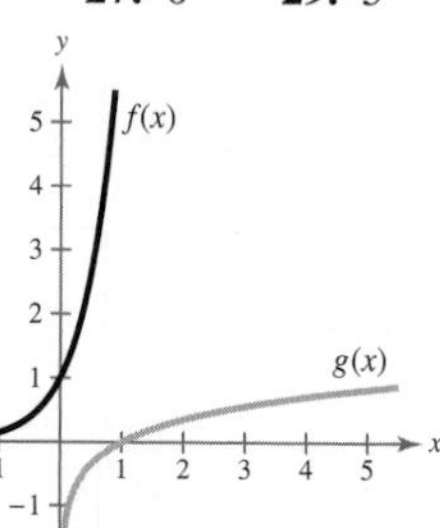

35.

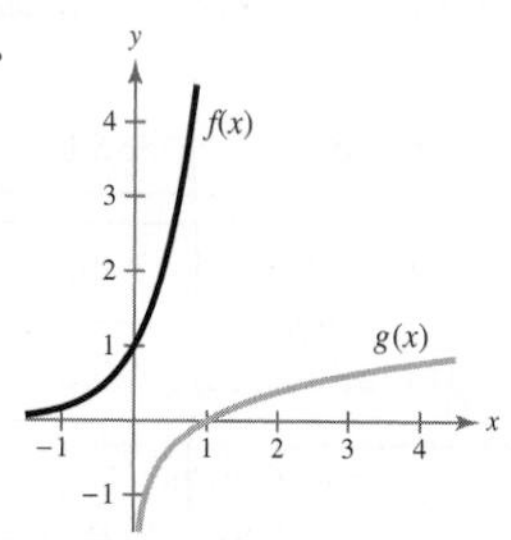

37. a; Upward shift of two units
38. d; Right shift of one unit
39. b; Reflection in the y-axis and a right shift of one unit
40. c; Reflection in the x-axis

41.

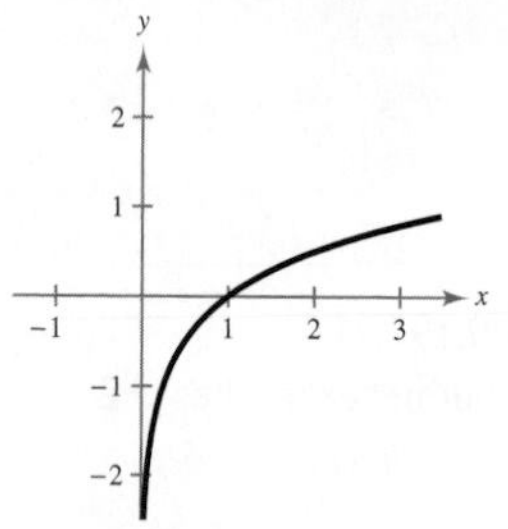

Domain: $(0, \infty)$
x-intercept: $(1, 0)$
Vertical asymptote: $x = 0$

43.

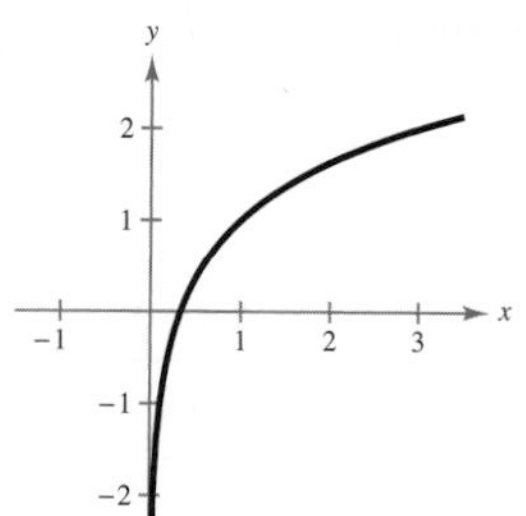

Domain: $(0, \infty)$
x-intercept: $\left(\frac{1}{3}, 0\right)$
Vertical asymptote: $x = 0$

45.

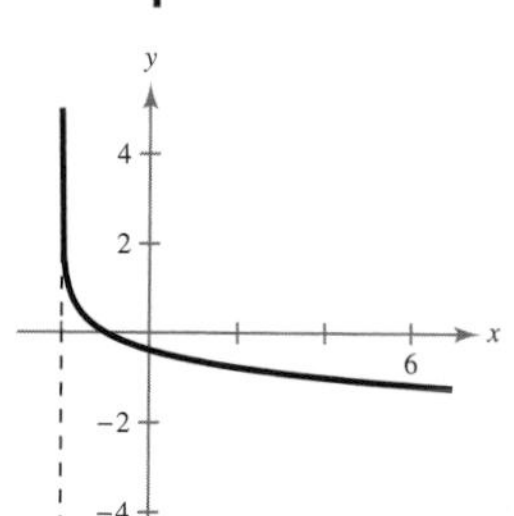

Domain: $(-2, \infty)$
x-intercept: $(-1, 0)$
Vertical asymptote: $x = -2$

47.

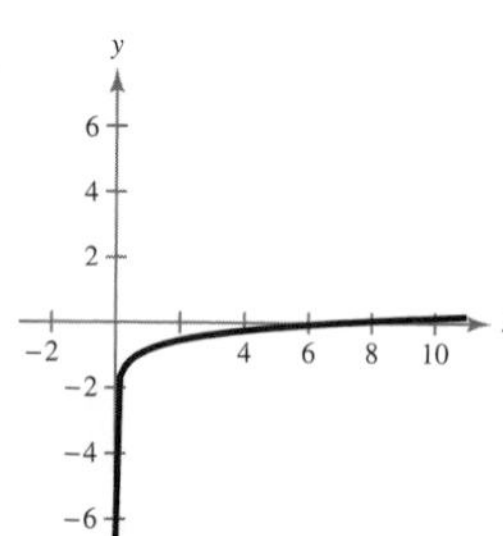

Domain: $(0, \infty)$
x-intercept: $(7, 0)$
Vertical asymptote: $x = 0$

49. $e^{-0693\ldots} = \frac{1}{2}$ **51.** $e^{5.521\ldots} = 250$
53. $\ln 7.3890\ldots = 2$ **55.** $\ln \frac{1}{2} = -4x$ **57.** 2.913
59. 6.438 **61.** 4 **63.** 0 **65.** 1

67.

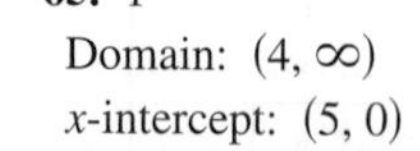

Domain: $(4, \infty)$
x-intercept: $(5, 0)$
Vertical asymptote: $x = 4$

69.

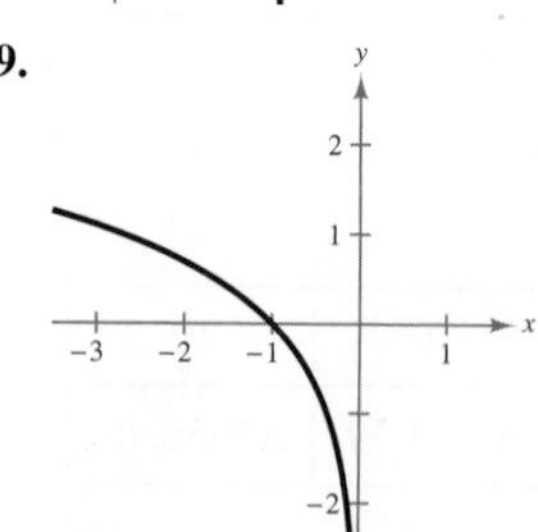

Domain: $(-\infty, 0)$
x-intercept: $(-1, 0)$
Vertical asymptote: $x = 0$

71.

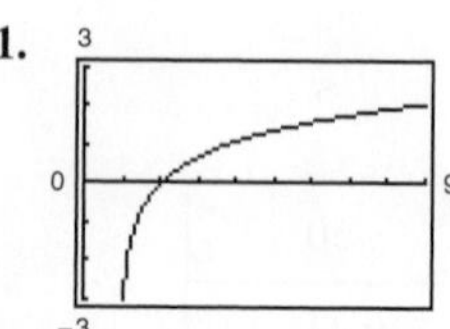

73.

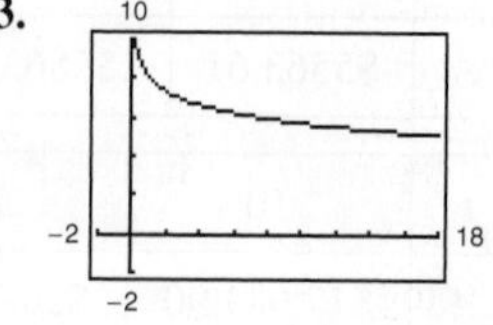

75. 8 **77.** $-2, 3$
79. (a) 30 yr; 10 yr
(b) \$323,179; \$199,109; \$173,179; \$49,109
(c) $x = 750$; The monthly payment must be greater than \$750.

81. (a)

r	0.005	0.010	0.015	0.020	0.025	0.030
t	138.6	69.3	46.2	34.7	27.7	23.1

As the rate of increase r increases, the time t in years for the population to double decreases.

(b) 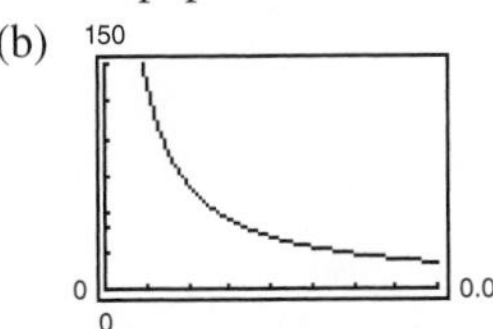

83. (a) (b) 80 (c) 68.1 (d) 62.3

85. False. Reflecting $g(x)$ in the line $y = x$ will determine the graph of $f(x)$.

87. (a) $g(x)$; The natural log function grows at a slower rate than the square root function.

(b) 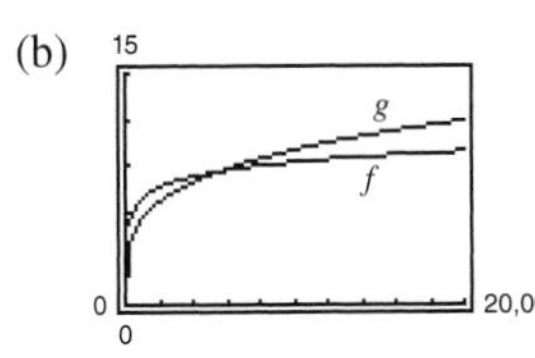

$g(x)$; The natural log function grows at a slower rate than the fourth root function.

89. $y = \log_2 x$, so y is a logarithmic function of x.

91. (a)

x	1	5	10	10^2
$f(x)$	0	0.322	0.230	0.046

x	10^4	10^6
$f(x)$	0.00092	0.0000138

(b) 0

(c) 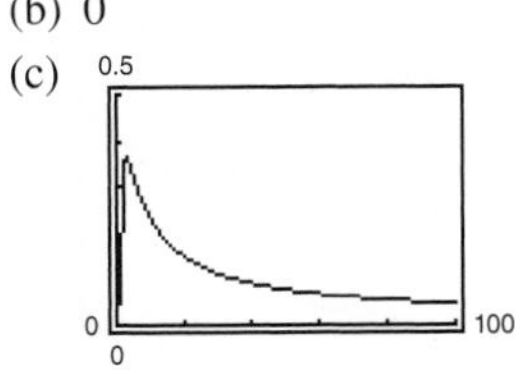

Section 5.3 *(page 383)*

1. change-of-base **3.** $\dfrac{1}{\log_b a}$ **5.** (a) $\dfrac{\log 16}{\log 5}$ (b) $\dfrac{\ln 16}{\ln 5}$

7. (a) $\dfrac{\log \frac{3}{10}}{\log x}$ (b) $\dfrac{\ln \frac{3}{10}}{\ln x}$ **9.** 2.579 **11.** -0.606

13. $\log_3 5 + \log_3 7$ **15.** $\log_3 7 - 2 \log_3 5$

17. $1 + \log_3 7 - \log_3 5$ **19.** 2 **21.** $-\frac{1}{3}$

23. -2 is not in the domain of $\log_2 x$. **25.** $\frac{3}{4}$

27. 7 **29.** 2 **31.** $\frac{3}{2}$ **33.** 1.1833 **35.** -1.6542

37. 1.9563 **39.** -2.7124 **41.** $\ln 7 + \ln x$

43. $4 \log_8 x$ **45.** $1 - \log_5 x$ **47.** $\frac{1}{2} \ln z$

49. $\ln x + \ln y + 2 \ln z$ **51.** $\ln z + 2 \ln(z - 1)$

53. $\frac{1}{2} \log_2 (a + 2) + \frac{1}{2} \log_2 (a - 2) - \log_2 7$

55. $2 \log_5 x - 2 \log_5 y - 3 \log_5 z$

57. $\frac{1}{3} \ln y + \frac{1}{3} \ln z - \frac{2}{3} \ln x$ **59.** $\frac{3}{4} \ln x + \frac{1}{4} \ln(x^2 + 3)$

61. $\ln 3x$ **63.** $\log_7 (z - 2)^{2/3}$ **65.** $\log_3 \dfrac{5}{x^3}$

67. $\log x(x + 1)^2$ **69.** $\log \dfrac{xz^3}{y^2}$ **71.** $\ln \dfrac{x}{(x + 1)(x - 1)}$

73. $\ln \sqrt{\dfrac{x(x + 3)^2}{x^2 - 1}}$ **75.** $\log_8 \dfrac{\sqrt[3]{y(y + 4)^2}}{y - 1}$

77. $\log_2 \frac{32}{4} = \log_2 32 - \log_2 4$; Property 2

79. $\beta = 10(\log I + 12)$; 60 dB **81.** 70 dB

83. $\ln y = \frac{1}{4} \ln x$ **85.** $\ln y = -\frac{1}{4} \ln x + \ln \frac{5}{2}$

87. $\ln y = -0.14 \ln x + 5.7$

89. (a) and (b) (c) 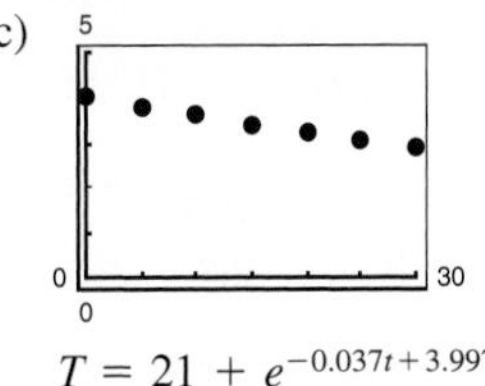

$T = 21 + e^{-0.037t + 3.997}$

The results are similar.

(d)

$$T = 21 + \frac{1}{0.001t + 0.016}$$

91. False; $\ln 1 = 0$ **93.** False; $\ln(x - 2) \neq \ln x - \ln 2$

95. False; $u = v^2$

97. $f(x) = \dfrac{\log x}{\log 2} = \dfrac{\ln x}{\ln 2}$ **99.** $f(x) = \dfrac{\log x}{\log \frac{1}{4}} = \dfrac{\ln x}{\ln \frac{1}{4}}$

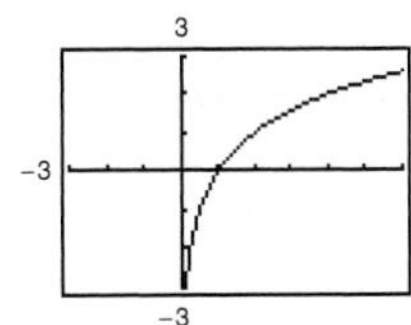

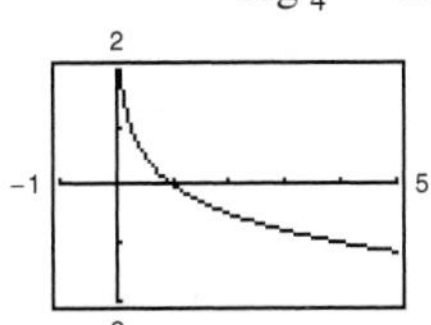

101. The Power Property cannot be used because $\ln e$ is raised to the second power, not just e.

103. No; $\dfrac{x}{x - 3} > 0$ when $x < 0$.

105.

$\ln 1 = 0$ | $\ln 9 \approx 2.1972$
$\ln 2 \approx 0.6931$ | $\ln 10 \approx 2.3025$
$\ln 3 \approx 1.0986$ | $\ln 12 \approx 2.4848$
$\ln 4 \approx 1.3862$ | $\ln 15 \approx 2.7080$
$\ln 5 \approx 1.6094$ | $\ln 16 \approx 2.7724$
$\ln 6 \approx 1.7917$ | $\ln 18 \approx 2.8903$
$\ln 8 \approx 2.0793$ | $\ln 20 \approx 2.9956$

Section 5.4 (page 393)

1. (a) $x = y$ (b) $x = y$ (c) x (d) x
3. (a) Yes (b) No (c) Yes
5. (a) Yes (b) No (c) No **7.** 2
9. 2 **11.** $\ln 2 \approx 0.693$ **13.** $e^{-1} \approx 0.368$ **15.** 64
17. (3, 8) **19.** 2, −1 **21.** $\dfrac{\ln 5}{\ln 3} \approx 1.465$
23. $\ln 39 \approx 3.664$ **25.** $\dfrac{\ln 80}{2 \ln 3} \approx 1.994$
27. $2 - \dfrac{\ln 400}{\ln 3} \approx -3.454$ **29.** $\dfrac{1}{3}\log\dfrac{3}{2} \approx 0.059$
31. $\dfrac{\ln 12}{3} \approx 0.828$ **33.** 0 **35.** $\dfrac{\ln \frac{8}{3}}{3 \ln 2} + \dfrac{1}{3} \approx 0.805$
37. $-\dfrac{\ln 2}{\ln 3 - \ln 2} \approx -1.710$ **39.** $0, \dfrac{\ln 4}{\ln 5} \approx 0.861$
41. $\ln 5 \approx 1.609$ **43.** $\ln \frac{4}{5} \approx -0.223$
45. $\dfrac{\ln 4}{365 \ln\left(1 + \dfrac{0.065}{365}\right)} \approx 21.330$ **47.** $e^{-3} \approx 0.050$
49. $\dfrac{e^{2.1}}{6} \approx 1.361$ **51.** $e^{-2} \approx 0.135$ **53.** $2(3^{11/6}) \approx 14.988$
55. No solution **57.** No solution **59.** No solution
61. 2 **63.** 3.328 **65.** −0.478 **67.** 20.086
69. 1.482 **71.** (a) 27.73 yr (b) 43.94 yr **73.** −1, 0
75. 1 **77.** $e^{-1} \approx 0.368$ **79.** $e^{-1/2} \approx 0.607$
81. (a) $y = 100$ and $y = 0$; The range falls between 0% and 100%.
(b) Males: 69.51 in. Females: 64.49 in.
83. 5 years **85.** 2011 **87.** About 3.039 min
89. $\log_b uv = \log_b u + \log_b v$
True by Property 1 in Section 5.3.
91. $\log_b(u - v) = \log_b u - \log_b v$
False.
$1.95 \approx \log(100 - 10) \neq \log 100 - \log 10 = 1$
93. Yes. See Exercise 57.
95. For $rt < \ln 2$ years, double the amount you invest. For $rt > \ln 2$ years, double your interest rate or double the number of years, because either of these will double the exponent in the exponential function.
97. (a) 7% (b) 7.25% (c) 7.19% (d) 7.45%
The investment plan with the greatest effective yield and the highest balance after 5 years is plan (d).

Section 5.5 (page 403)

1. $y = ae^{bx}$; $y = ae^{-bx}$ **3.** normally distributed
5. (a) $P = \dfrac{A}{e^{rt}}$ (b) $t = \dfrac{\ln\left(\dfrac{A}{P}\right)}{r}$ **7.** 19.8 yr; $1419.07
9. 8.9438%; $1834.37 **11.** $6376.28; 15.4 yr
13. $303,580.52
15. (a) 7.27 yr (b) 6.96 yr (c) 6.93 yr (d) 6.93 yr
17. (a)

r	2%	4%	6%	8%	10%	12%
t	54.93	27.47	18.31	13.73	10.99	9.16

(b)

r	2%	4%	6%	8%	10%	12%
t	55.48	28.01	18.85	14.27	11.53	9.69

19.

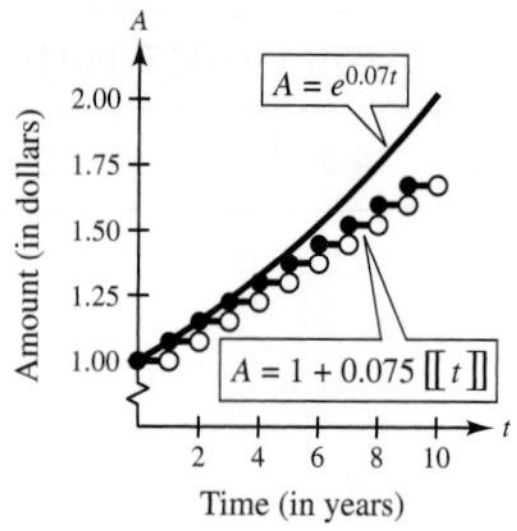

Continuous compounding
21. 6.48 g **23.** 2.26 g **25.** $y = e^{0.7675x}$
27. $y = 5e^{-0.4024x}$
29. (a)

Year	Population
1980	104,752
1990	143,251
2000	195,899
2010	267,896

(b) 2019
(c) *Sample answer:* No; As t increases, the population increases rapidly.
31. $k = 0.2988$; About 5,309,734 hits **33.** About 800 bacteria
35. (a) $V = -150t + 575$ (b) $V = 575e^{-0.3688t}$
(c)

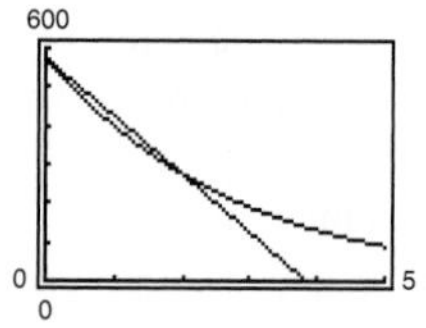

The exponential model depreciates faster.
(d) Linear model: $425; $125
Exponential model: $397.65; $190.18
(e) Answers will vary.
37. About 12,180 yr old
39. (a)

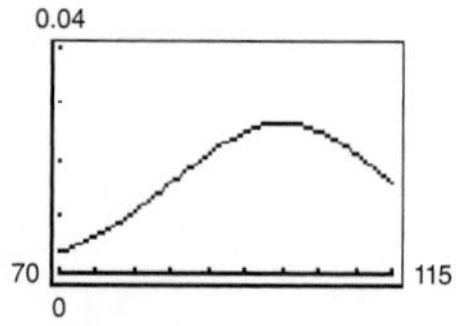

(b) 100
41. (a) 1998: 63,922 sites
2003: 149,805 sites
2006: 208,705 sites
(b)

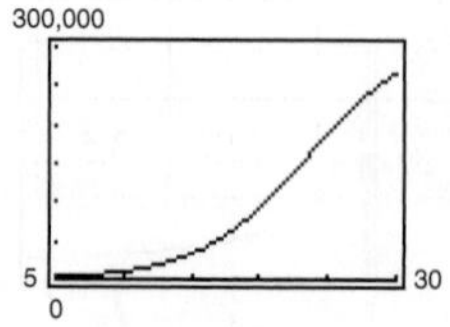

(c) and (d) 2010
43. (a) 203 animals (b) 13 mo
(c)

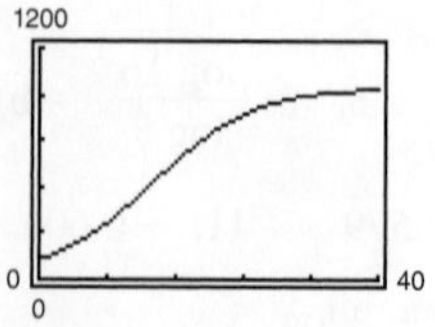

Horizontal asymptotes: $p = 0, p = 1000$. The population size will approach 1000 as time increases.
45. (a) $10^{7.6} \approx 39{,}810{,}717$
(b) $10^{5.6} \approx 398{,}107$ (c) $10^{6.6} \approx 3{,}981{,}072$
47. (a) 20 dB (b) 70 dB (c) 40 dB (d) 90 dB
49. 95% **51.** 4.64 **53.** 1.58×10^{-6} moles/L
55. $10^{5.1}$ **57.** 3:00 A.M.

59. (a)
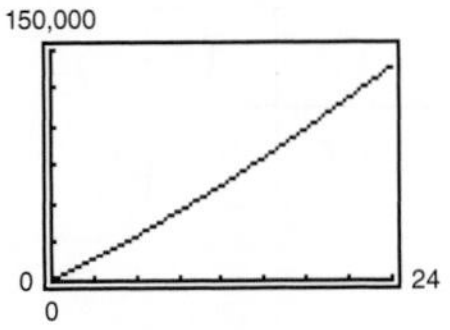

(b) $t \approx 21$ yr; Yes

61. False. The domain can be the set of real numbers for a logistic growth function.

63. False. The graph of $f(x)$ is the graph of $g(x)$ shifted five units up.

65. Answers will vary.

Review Exercises *(page 410)*

1. 0.164 **3.** 1.587 **5.** 1456.529

7.

x	-1	0	1	2	3
$f(x)$	8	5	4.25	4.063	4.016

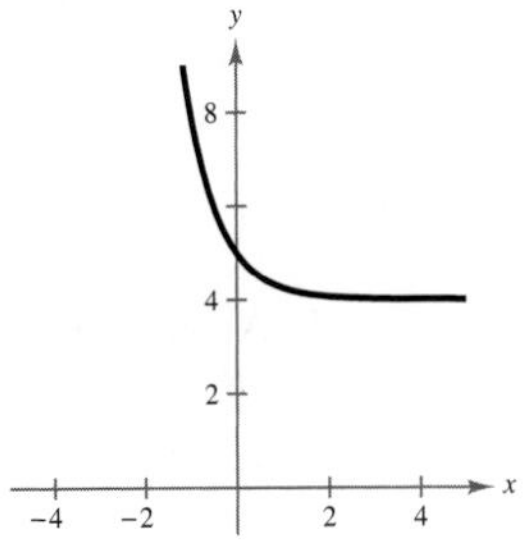

9.

x	-1	0	1	2	3
$f(x)$	4.008	4.04	4.2	5	9

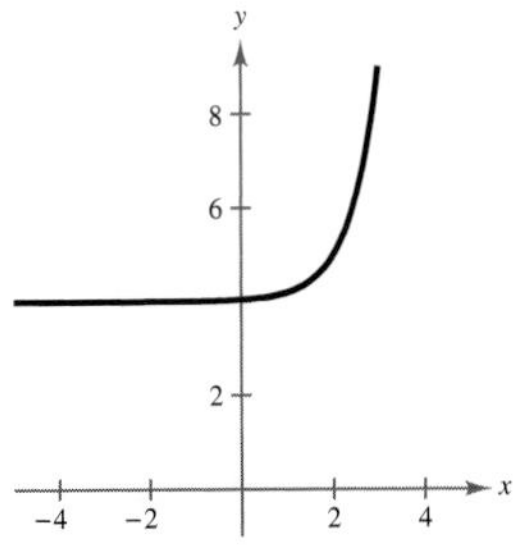

11.

x	-2	-1	0	1	2
$f(x)$	3.25	3.5	4	5	7

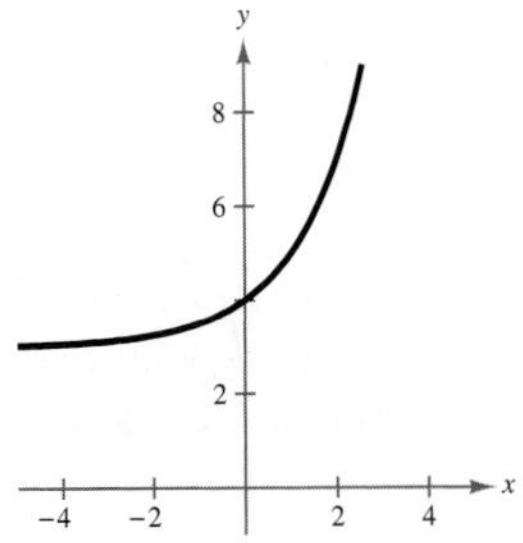

13. 1 **15.** 4 **17.** Shift the graph of f one unit up.

19. Reflect f in the x-axis and shift one unit up.

21. 29.964 **23.** 1.822

25.

x	-2	-1	0	1	2
$h(x)$	2.72	1.65	1	0.61	0.37

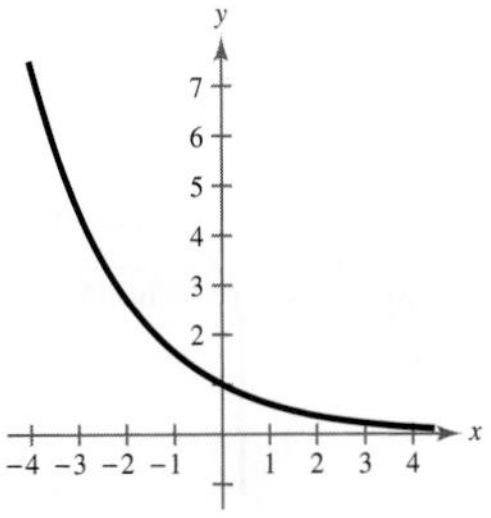

27.

x	-3	-2	-1	0	1
$f(x)$	0.37	1	2.72	7.39	20.09

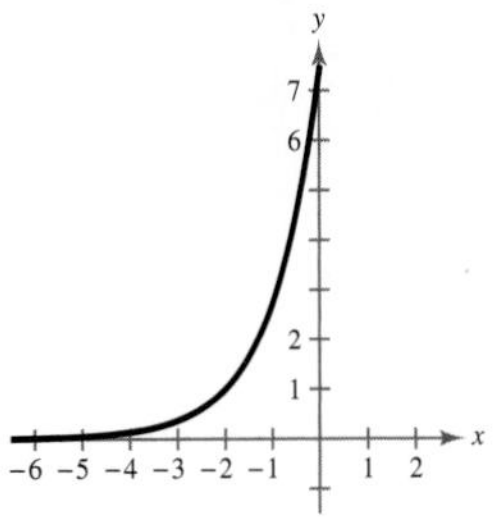

29. (a) 0.283 (b) 0.487 (c) 0.811

31.

n	1	2	4	12
A	\$6719.58	\$6734.28	\$6741.74	\$6746.77

n	365	Continuous
A	\$6749.21	\$6749.29

33. $\log_3 27 = 3$ **35.** $\ln 2.2255\ldots = 0.8$ **37.** 3

39. -2 **41.** 7 **43.** -5

45. Domain: $(0, \infty)$
x-intercept: $(1, 0)$
Vertical asymptote: $x = 0$

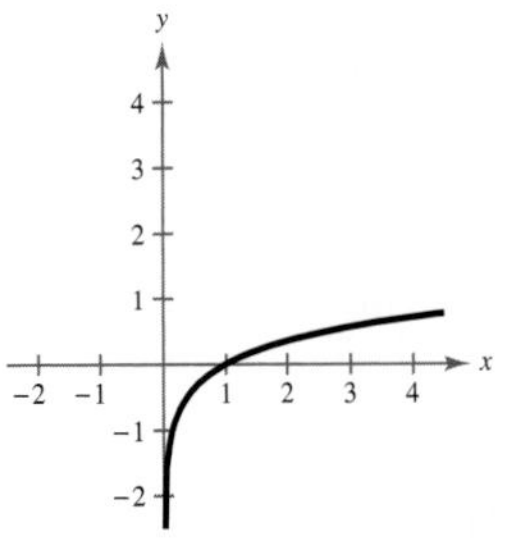

47. Domain: $(-5, \infty)$
x-intercept: $(9995, 0)$
Vertical asymptote: $x = -5$

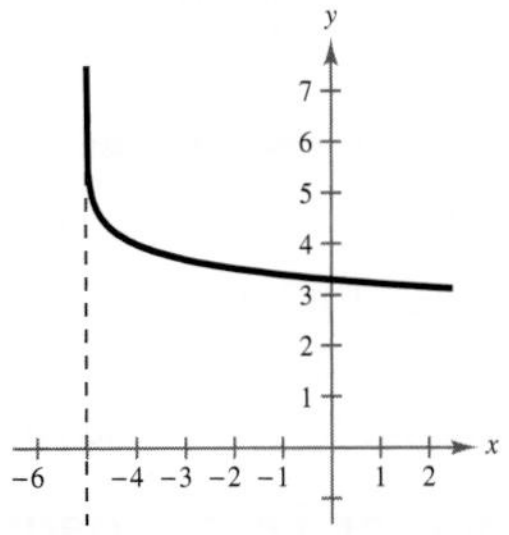

49. 3.118 **51.** 0.25

53. Domain: $(0, \infty)$
x-intercept: $(e^{-6}, 0)$
Vertical asymptote: $x = 0$

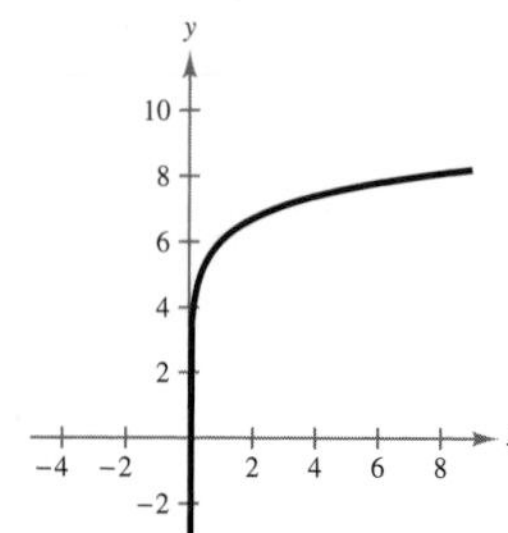

55. Domain: $(6, \infty)$
x-intercept: $(7, 0)$
Vertical asymptote: $x = 6$

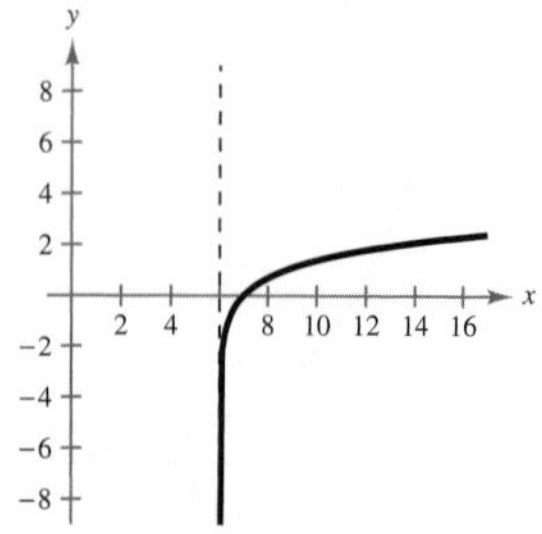

57. About 14.32 parsecs **59.** (a) and (b) 2.585
61. (a) and (b) -2.322 **63.** $\log_2 5 - \log_2 3$
65. $2 \log_2 3 - \log_2 5$ **67.** $\log 7 + 2 \log x$
69. $2 - \frac{1}{2} \log_3 x$ **71.** $2 \ln x + 2 \ln y + \ln z$
73. $\ln 7x$ **75.** $\log \dfrac{x}{\sqrt{y}}$ **77.** $\log_3 \dfrac{\sqrt{x}}{(y + 8)^2}$
79. (a) $0 \le h < 18{,}000$
(b)

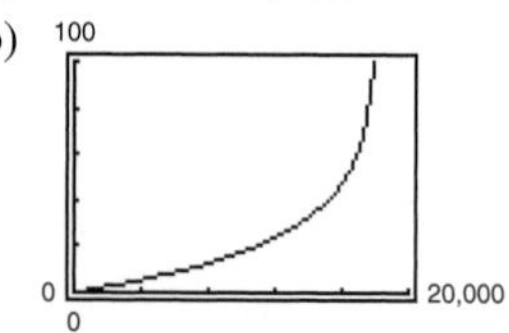

Vertical asymptote: $h = 18{,}000$
(c) The plane is climbing at a slower rate, so the time required increases.
(d) 5.46 min
81. 3 **83.** $\ln 3 \approx 1.099$ **85.** $e^4 \approx 54.598$
87. 1, 3 **89.** $\dfrac{\ln 32}{\ln 2} = 5$ **91.** $\frac{1}{3}e^{8.2} \approx 1213.650$
93. $\dfrac{3}{2} + \dfrac{\sqrt{9 + 4e}}{2} \approx 3.729$ **95.** No solution **97.** 0.900
99. 2.447 **101.** 1.482 **103.** 73.2 yr **105.** e
106. b **107.** f **108.** d **109.** a **110.** c
111. $y = 2e^{0.1014x}$
113.

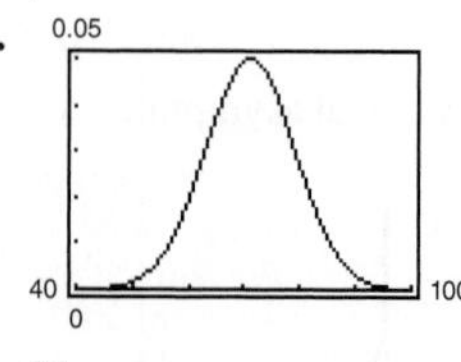

71
115. (a) 10^{-6} W/m^2 (b) $10\sqrt{10}$ W/m^2
(c) 1.259×10^{-12} W/m^2
117. True by the inverse properties.

Chapter Test *(page 413)*

1. 0.410 **2.** 0.032 **3.** 0.497 **4.** 22.198

5.

x	-1	$-\frac{1}{2}$	0	$\frac{1}{2}$	1
$f(x)$	10	3.162	1	0.316	0.1

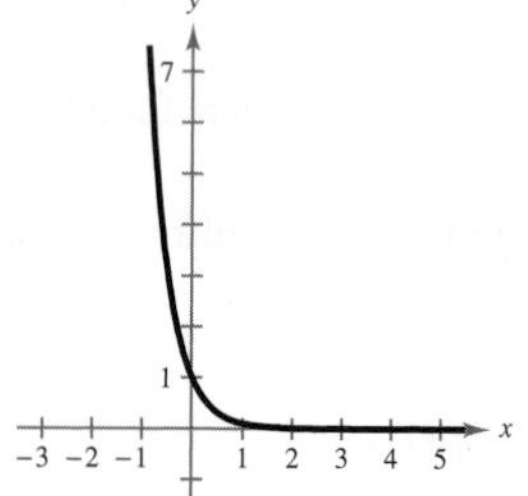

6.

x	-1	0	1	2	3
$f(x)$	-0.005	-0.028	-0.167	-1	-6

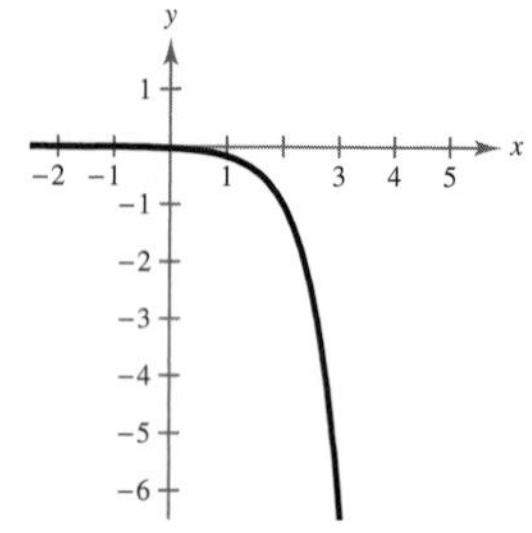

7.

x	-1	$-\frac{1}{2}$	0	$\frac{1}{2}$	1
$f(x)$	0.865	0.632	0	-1.718	-6.389

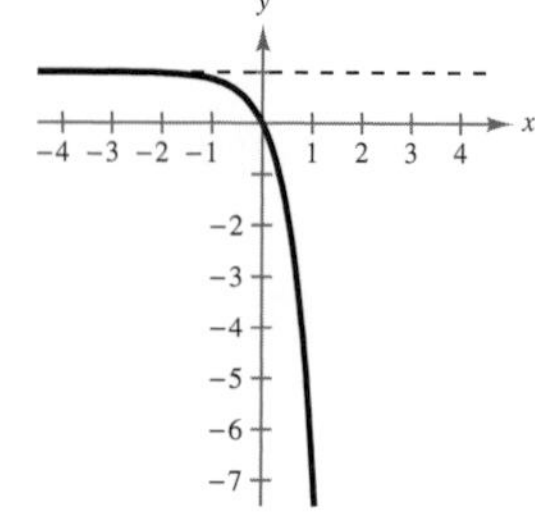

8. (a) -0.89 (b) 9.2
9. Domain: $(0, \infty)$
x-intercept: $(10^{-4}, 0)$
Vertical asymptote: $x = 0$

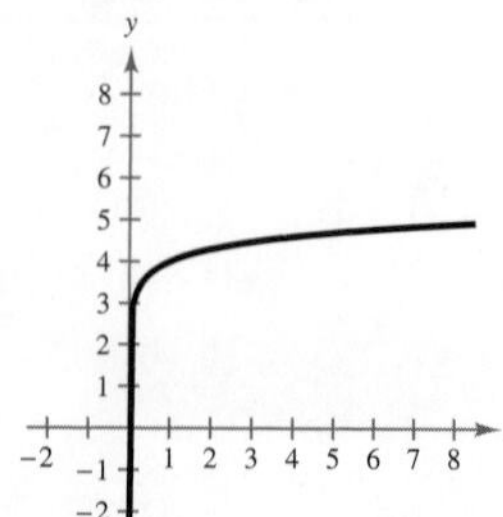

10. Domain: $(4, \infty)$
x-intercept: $(5, 0)$
Vertical asymptote: $x = 4$

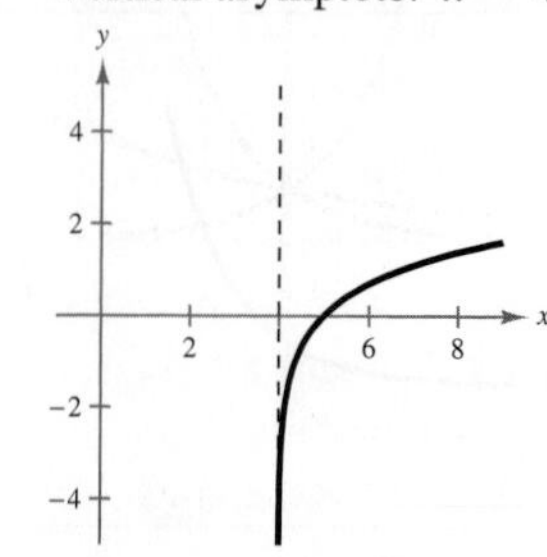

11. Domain: $(-6, \infty)$
x-intercept: $(e^{-1} - 6, 0)$
Vertical asymptote: $x = -6$

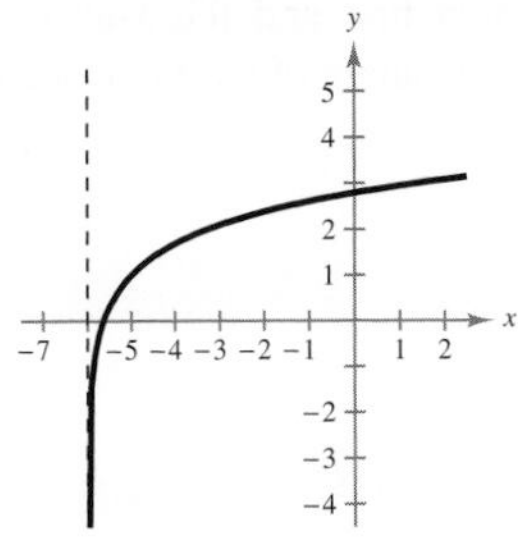

12. 2.209 **13.** -0.167 **14.** -11.047
15. $\log_2 3 + 4 \log_2 a$ **16.** $\frac{1}{2} \ln x - \ln 7$
17. $1 + 2 \log x - 3 \log y$ **18.** $\log_3 13y$ **19.** $\ln \dfrac{x^4}{y^4}$
20. $\ln \dfrac{x^3y^2}{x + 3}$ **21.** -2 **22.** $\dfrac{\ln 44}{-5} \approx -0.757$
23. $\dfrac{\ln 197}{4} \approx 1.321$ **24.** $e^{1/2} \approx 1.649$
25. $e^{-11/4} \approx 0.0639$ **26.** 20
27. $y = 2745e^{0.1570t}$ **28.** 55%
29. (a)

x	$\frac{1}{4}$	1	2	4	5	6
H	58.720	75.332	86.828	103.43	110.59	117.38

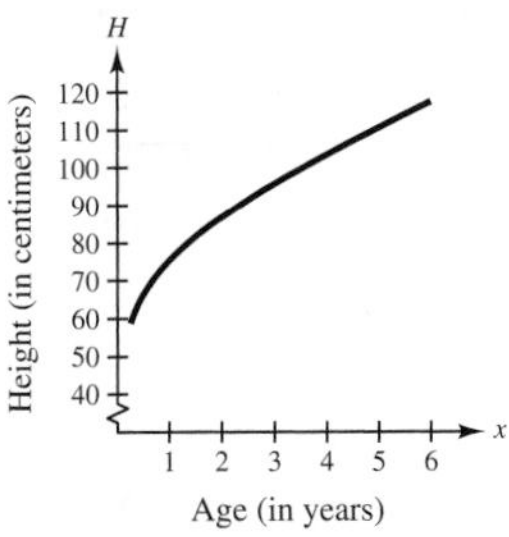

(b) 103 cm; 103.43 cm

Problem Solving *(page 415)*

1.

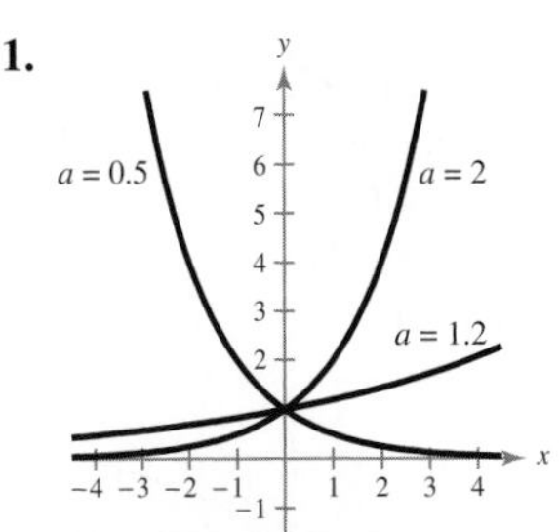

$y = 0.5^x$ and $y = 1.2^x$
$0 < a \le e^{1/e}$

3. As $x \to \infty$, the graph of e^x increases at a greater rate than the graph of x^n.

5. Answers will vary.

7. (a)

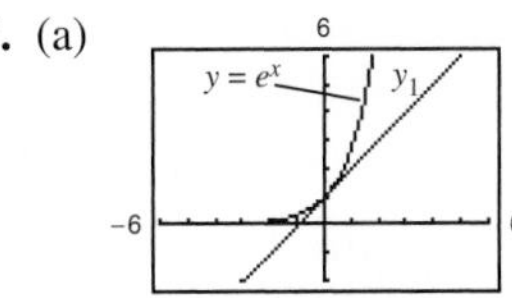

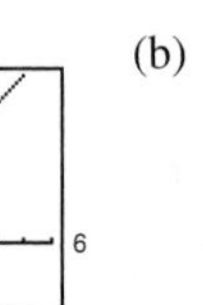

(b)

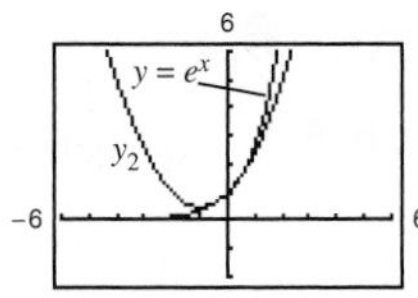

(c)

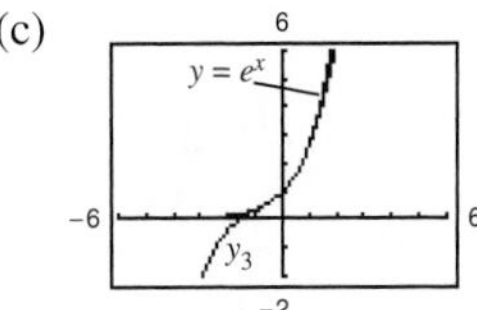

9.

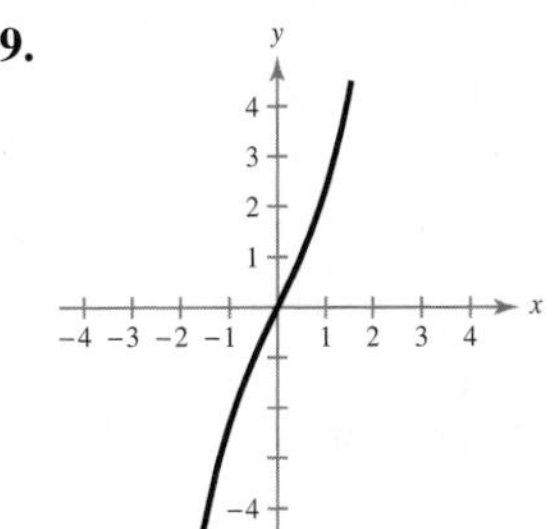

$f^{-1}(x) = \ln\left(\dfrac{x + \sqrt{x^2 + 4}}{2}\right)$

11. c **13.** $t = \dfrac{\ln c_1 - \ln c_2}{\left(\dfrac{1}{k_2} - \dfrac{1}{k_1}\right)\ln \dfrac{1}{2}}$

15. (a) $y_1 = 252{,}606(1.0310)^t$
(b) $y_2 = 400.88t^2 - 1464.6t + 291{,}782$
(c) 2,900,000 ... 0 ... 85 ... 200,000
(d) The exponential model is a better fit. No, because the model is rapidly approaching infinity.

17. $1, e^2$

19. $y_4 = (x - 1) - \frac{1}{2}(x - 1)^2 + \frac{1}{3}(x - 1)^3 - \frac{1}{4}(x - 1)^4$

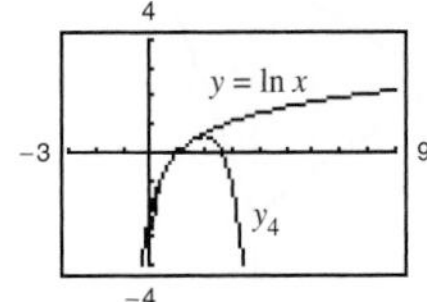

The pattern implies that
$\ln x = (x - 1) - \frac{1}{2}(x - 1)^2 + \frac{1}{3}(x - 1)^3 - \cdots.$

21. 30 ... 100 ... 1500 ... 0

17.7 ft³/min

23. (a)

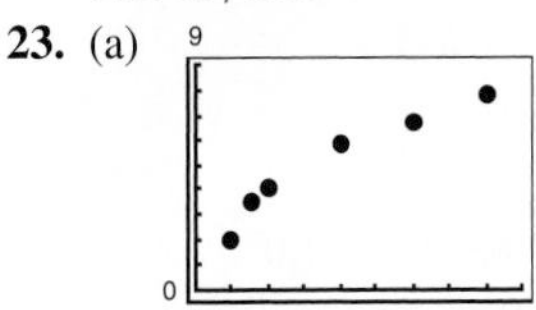

(b)–(e) Answers will vary.

25. (a)

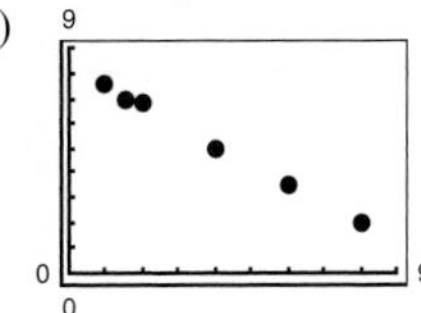

(b)–(e) Answers will vary.

Chapter 6

Section 6.1 *(page 422)*

1. inclination **3.** $\left|\dfrac{m_2 - m_1}{1 + m_1 m_2}\right|$ **5.** $\dfrac{\sqrt{3}}{3}$ **7.** -1

9. $\sqrt{3}$ **11.** 0.4111 **13.** 3.2236 **15.** -4.1005

17. $\dfrac{\pi}{4}$ rad, 45° **19.** 0.5880 rad, 33.7° **21.** $\dfrac{3\pi}{4}$ rad, 135°

23. 2.1588 rad, 123.7° **25.** $\dfrac{\pi}{6}$ rad, 30° **27.** $\dfrac{5\pi}{6}$ rad, 150°

29. 1.0517 rad, 60.3° **31.** 2.1112 rad, 121.0°

33. 1.6539 rad, 94.8° **35.** $\dfrac{3\pi}{4}$ rad, 135° **37.** $\dfrac{\pi}{4}$ rad, 45°

39. $\dfrac{5\pi}{6}$ rad, 150° **41.** 1.2490 rad, 71.6°

43. 2.4669 rad, 141.3° **45.** 1.1071 rad, 63.4°

47. 0.1974 rad, 11.3° **49.** 1.4289 rad, 81.9°

51. 0.9273 rad, 53.1° **53.** 0.8187 rad, 46.9°

55. $(1, 5) \leftrightarrow (4, 5)$: slope $= 0$
$(4, 5) \leftrightarrow (3, 8)$: slope $= -3$
$(3, 8) \leftrightarrow (1, 5)$: slope $= \frac{3}{2}$
$(1, 5)$: 56.3°; $(4, 5)$: 71.6°; $(3, 8)$: 52.1°

57. $(-4, -1) \leftrightarrow (3, 2)$: slope $= \frac{3}{7}$
$(3, 2) \leftrightarrow (1, 0)$: slope $= 1$
$(1, 0) \leftrightarrow (-4, -1)$: slope $= \frac{1}{5}$
$(-4, -1)$: 11.9°; $(3, 2)$: 21.8°; $(1, 0)$: 146.3°

59. $\dfrac{\sqrt{2}}{2} \approx 0.7071$ **61.** $\dfrac{2\sqrt{5}}{5} \approx 0.8944$

63. $2\sqrt{2} \approx 2.8284$ **65.** $\dfrac{\sqrt{10}}{10} \approx 0.3162$

67. $\dfrac{4\sqrt{10}}{5} \approx 2.5298$ **69.** 1 **71.** $\dfrac{1}{5}$

73. (a)

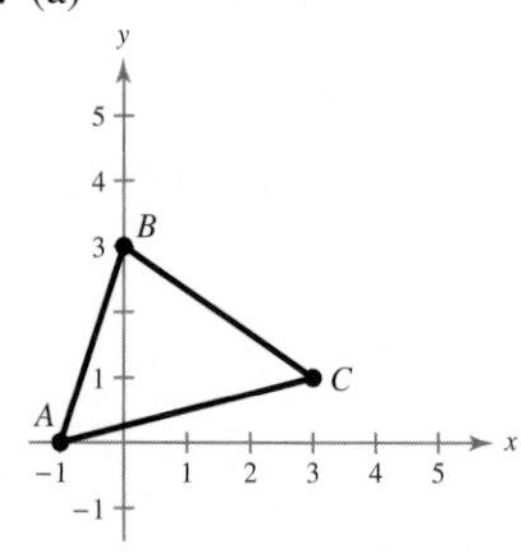

(b) $\dfrac{11\sqrt{17}}{17}$ (c) $\dfrac{11}{2}$

75. (a)

(b) $\dfrac{19\sqrt{34}}{34}$ (c) $\dfrac{19}{2}$

77. (a)

(b) $\dfrac{\sqrt{5}}{5}$ (c) 1

79. $2\sqrt{2}$ **81.** 0.1003, 1054 ft **83.** $\theta \approx 31.0°$

85. $\alpha \approx 33.69°$; $\beta \approx 56.31°$ **87.** True. $\tan 0 = 0$

89. False. Substitute $\tan \theta_1$ and $\tan \theta_2$ for m_1 and m_2 in the formula for the angle between two lines.

91. The inclination of a line measures the angle of intersection (measured counterclockwise) of a line and the x-axis. The angle between two lines is the acute angle of their intersection, which must be less than $\pi/2$.

93. (a) $d = \dfrac{4}{\sqrt{m^2 + 1}}$

(b)

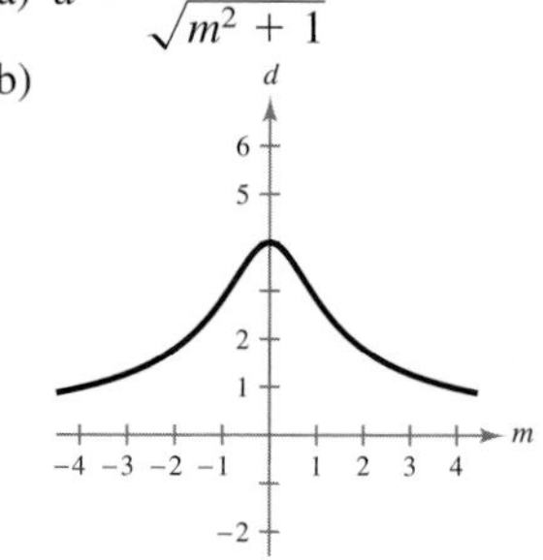

(c) $m = 0$

(d) The graph has a horizontal asymptote of $d = 0$. As the slope becomes larger, the distance between the origin and the line, $y = mx + 4$, becomes smaller and approaches 0.

Section 6.2 *(page 430)*

1. conic **3.** locus **5.** axis **7.** focal chord **9.** c

10. a **11.** b **12.** d **13.** $x^2 = 4y$ **15.** $x^2 = 2y$

17. $y^2 = -8x$ **19.** $x^2 = -8y$ **21.** $y^2 = 4x$

23. $x^2 = \frac{8}{3}y$ **25.** $y^2 = -\frac{25}{2}x$ **27.** $(x - 2)^2 = -8(y - 6)$

29. $(y - 3)^2 = -8(x - 6)$ **31.** $x^2 = -8(y - 2)$

33. $(y - 2)^2 = 8x$ **35.** $(x - 3)^2 = 3(y + 3)$

37. Vertex: $(0, 0)$
Focus: $\left(0, \frac{1}{2}\right)$
Directrix: $y = -\frac{1}{2}$

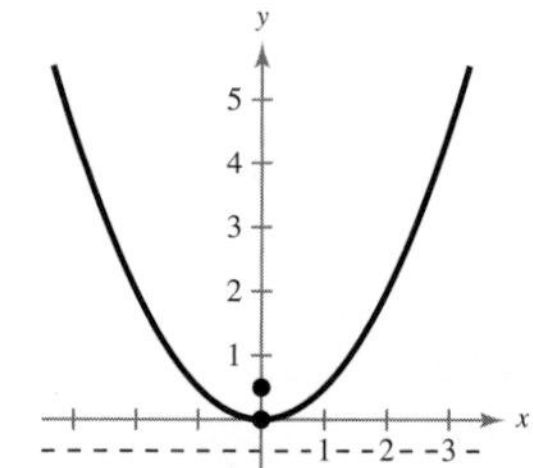

39. Vertex: $(0, 0)$
Focus: $\left(-\frac{3}{2}, 0\right)$
Directrix: $x = \frac{3}{2}$

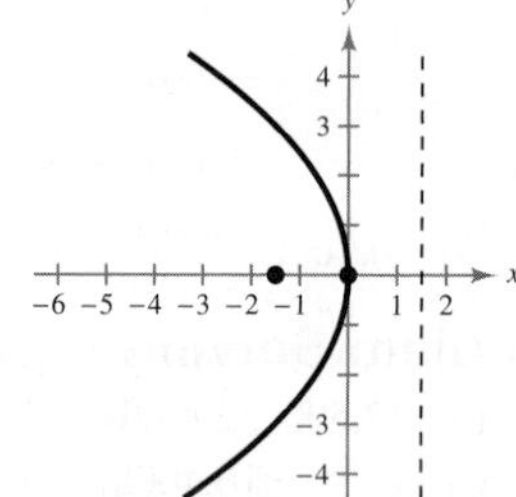

41. Vertex: $(0, 0)$
Focus: $(0, -3)$
Directrix: $y = 3$

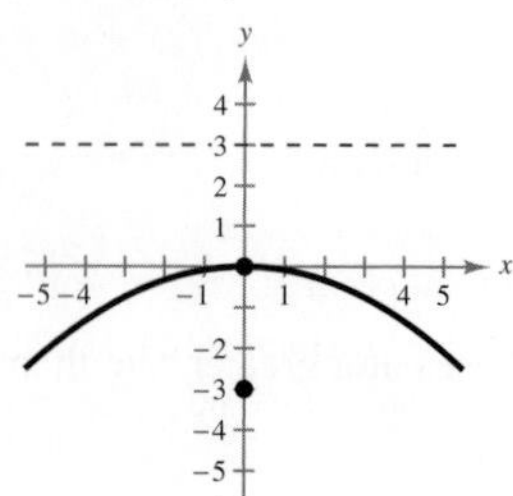

43. Vertex: $(1, -2)$
Focus: $(1, -4)$
Directrix: $y = 0$

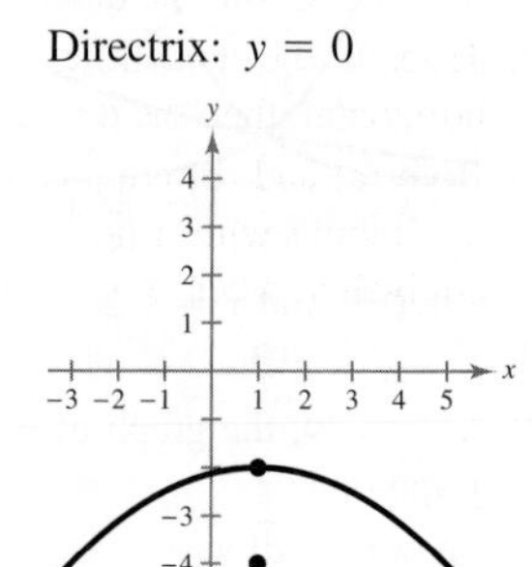

45. Vertex: $\left(\frac{3}{2}, -7\right)$
Focus: $\left(\frac{5}{2}, -7\right)$
Directrix: $x = \frac{1}{2}$

47. Vertex: $(1, 1)$
Focus: $(1, 2)$
Directrix: $y = 0$

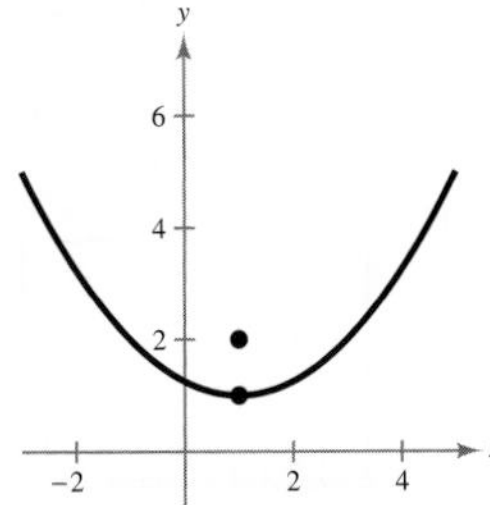

49. Vertex: $(-2, -3)$
Focus: $(-4, -3)$
Directrix: $x = 0$

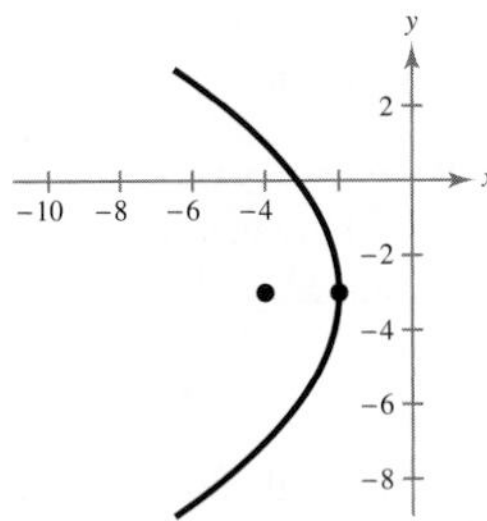

51. Vertex: $(-2, 1)$
Focus: $\left(-2, \frac{5}{2}\right)$
Directrix: $y = -\frac{1}{2}$

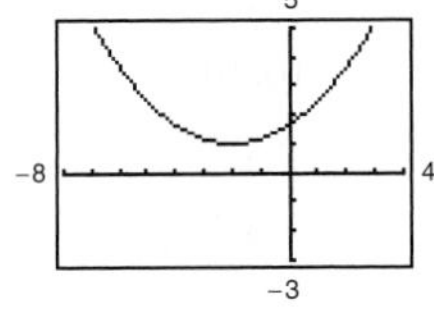

53. Vertex: $\left(\frac{1}{4}, -\frac{1}{2}\right)$
Focus: $\left(0, -\frac{1}{2}\right)$
Directrix: $x = \frac{1}{2}$

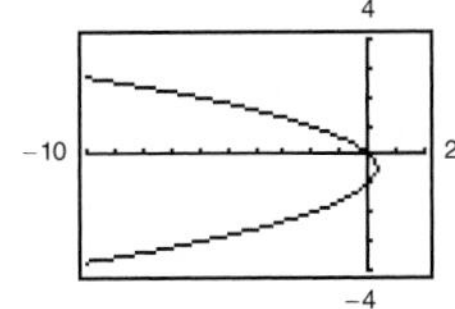

55. $y = \frac{3}{2}x - \frac{9}{2}$ **57.** $y = 4x - 8$ **59.** $y = 4x + 2$
61. $y^2 = 6x$ **63.** $y^2 = 640x$
65. (a) $x^2 = 12{,}288y$ (in feet) (b) About 22.6 ft
67. $x^2 = -\frac{25}{4}(y - 48)$ **69.** About 19.6 m
71. (a) $(0, 45)$ (b) $y = \frac{1}{180}x^2$
73. (a) $17{,}500\sqrt{2}$ mi/h $\approx 24{,}750$ mi/h
(b) $x^2 = -16{,}400(y - 4100)$
75. (a) $x^2 = -49(y - 100)$ (b) 70 ft
77. False. If the graph crossed the directrix, then there would exist points closer to the directrix than the focus.
79. True. If the axis (line connecting the vertex and focus) is horizontal, then the directrix must be vertical.
81. Both (a) and (b) are parabolas with vertical axes, while (c) is a parabola with a horizontal axis. Equations (a) and (b) are equivalent when $p = 1/(4a)$.
83.

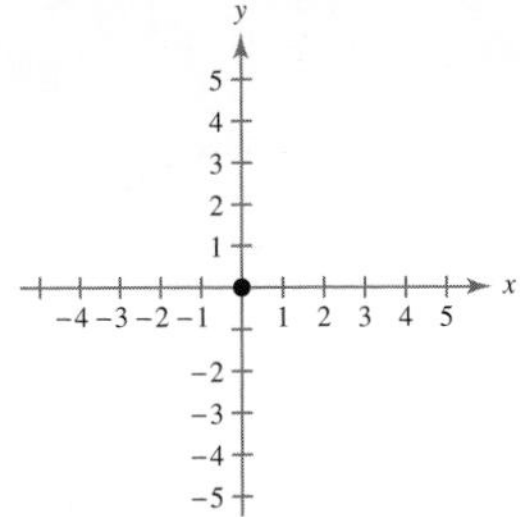

Single point $(0, 0)$; A single point is formed when a plane intersects only the vertex of the cone.

85. (a) $\frac{64\sqrt{2}}{3} \approx 30.17$
(b) As p approaches zero, the parabola becomes narrower and narrower, thus the area becomes smaller and smaller.

Section 6.3 *(page 440)*

1. ellipse; foci **3.** minor axis
5. b **6.** c **7.** a **8.** d
9. $\frac{x^2}{4} + \frac{y^2}{16} = 1$ **11.** $\frac{x^2}{49} + \frac{y^2}{45} = 1$ **13.** $\frac{x^2}{25} + \frac{y^2}{9} = 1$
15. $\frac{x^2}{9} + \frac{y^2}{36} = 1$ **17.** $\frac{x^2}{36} + \frac{5y^2}{9} = 1$
19. $\frac{(x-2)^2}{1} + \frac{(y-3)^2}{9} = 1$ **21.** $\frac{(x-6)^2}{16} + \frac{y^2}{4} = 1$
23. $\frac{(x-2)^2}{9} + \frac{y^2}{5} = 1$ **25.** $\frac{(x-1)^2}{9} + \frac{(y-3)^2}{4} = 1$
27. $\frac{(x-1)^2}{12} + \frac{(y-4)^2}{16} = 1$ **29.** $\frac{(x-2)^2}{4} + \frac{(y-2)^2}{1} = 1$

31. Center: $(0, 0)$
Vertices: $(\pm 5, 0)$
Foci: $(\pm 3, 0)$
Eccentricity: $\frac{3}{5}$

33. Center: $(0, 0)$
Vertices: $(0, \pm 6)$
Foci: $\left(0, \pm 4\sqrt{2}\right)$
Eccentricity: $\frac{2\sqrt{2}}{3}$

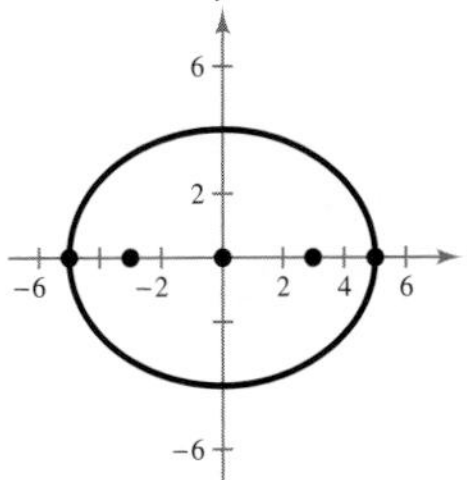

35. Center: $(4, -1)$
Vertices: $(4, -6), (4, 4)$
Foci: $(4, 2), (4, -4)$
Eccentricity: $\frac{3}{5}$

37. Center: $(-5, 1)$
Vertices: $\left(-\frac{7}{2}, 1\right), \left(-\frac{13}{2}, 1\right)$
Foci: $\left(-5 \pm \frac{\sqrt{5}}{2}, 1\right)$
Eccentricity: $\frac{\sqrt{5}}{3}$

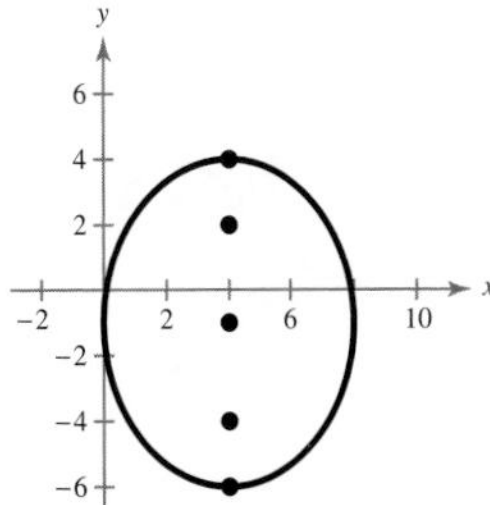

39. Center: $(-2, 3)$
Vertices: $(-2, 6), (-2, 0)$
Foci: $\left(-2, 3 \pm \sqrt{5}\right)$
Eccentricity: $\frac{\sqrt{5}}{3}$

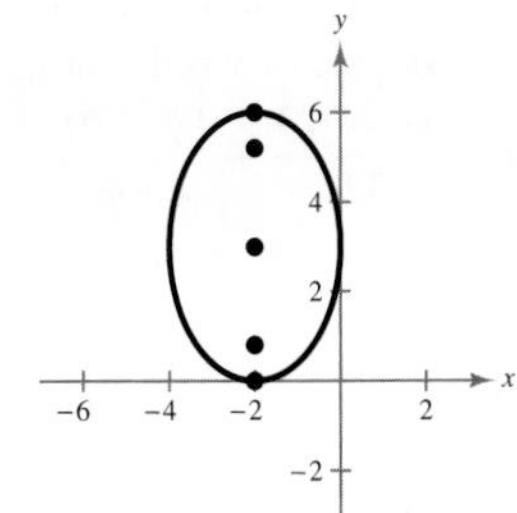

41. Center: $(4, 3)$

Vertices: $(14, 3), (-6, 3)$

Foci: $\left(4 \pm 4\sqrt{5}, 3\right)$

Eccentricity: $\dfrac{2\sqrt{5}}{5}$

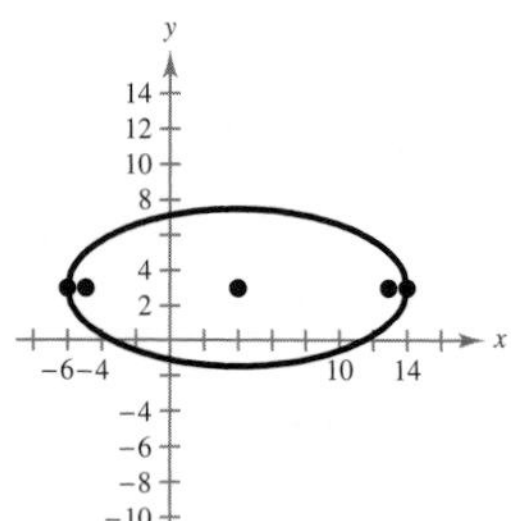

43. Center: $\left(-\dfrac{3}{2}, \dfrac{5}{2}\right)$

Vertices: $\left(-\dfrac{3}{2}, \dfrac{5}{2} \pm 2\sqrt{3}\right)$

Foci: $\left(-\dfrac{3}{2}, \dfrac{5}{2} \pm 2\sqrt{2}\right)$

Eccentricity: $\dfrac{\sqrt{6}}{3}$

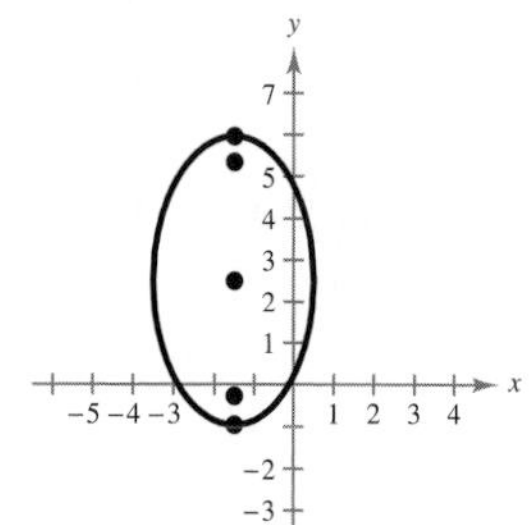

45. Center: $\left(\dfrac{1}{2}, -1\right)$

Vertices: $\left(\dfrac{1}{2} \pm \sqrt{5}, -1\right)$

Foci: $\left(\dfrac{1}{2} \pm \sqrt{2}, -1\right)$

Eccentricity: $\dfrac{\sqrt{10}}{5}$

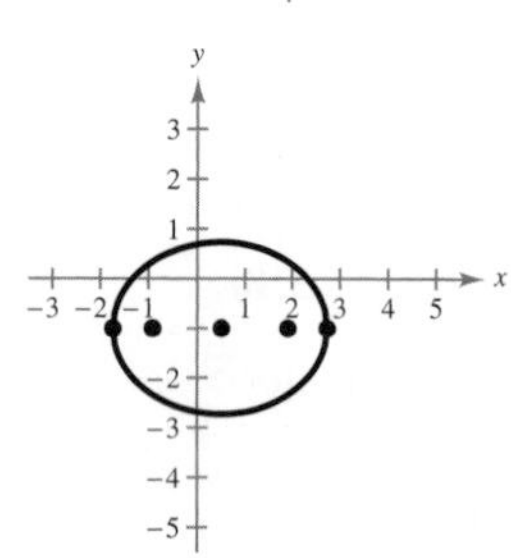

47.

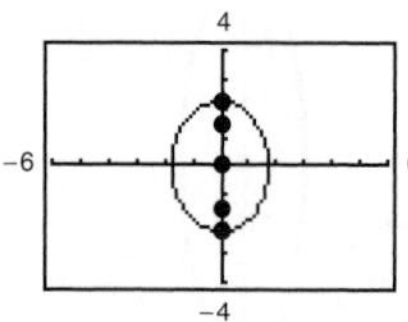

Center: $(0, 0)$

Vertices: $\left(0, \pm\sqrt{5}\right)$

Foci: $\left(0, \pm\sqrt{2}\right)$

49.

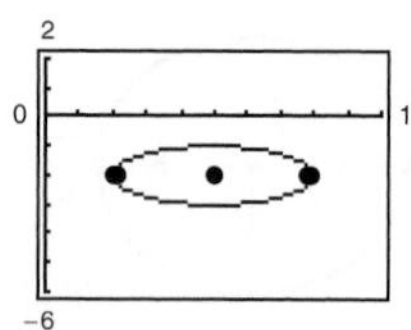

Center: $(5, -2)$

Vertices: $(2, -2), (8, -2)$

Foci: $\left(5 \pm 2\sqrt{2}, -2\right)$

51. $\dfrac{x^2}{25} + \dfrac{y^2}{9} = 1$

53. (a) $\dfrac{x^2}{2352.25} + \dfrac{y^2}{529} = 1$ (b) About 85.4 ft

55. About 229.8 mm **57.** $e \approx 0.0520$

59.

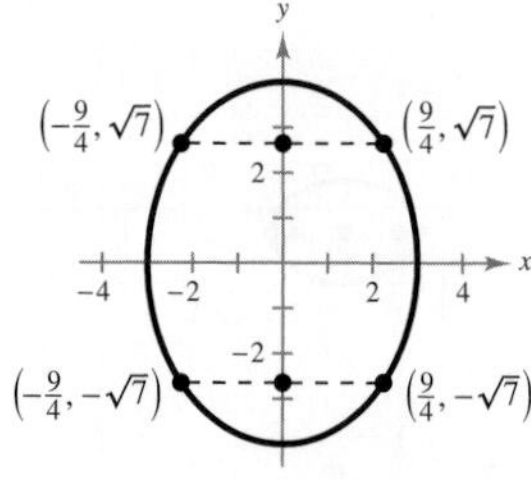

61.

63. False. The graph of $(x^2/4) + y^4 = 1$ is not an ellipse. The degree of y is 4, not 2.

65. $\dfrac{(x-6)^2}{324} + \dfrac{(y-2)^2}{308} = 1$

67. (a) $A = \pi a(20 - a)$ (b) $\dfrac{x^2}{196} + \dfrac{y^2}{36} = 1$

(c)

a	8	9	10	11	12	13
A	301.6	311.0	314.2	311.0	301.6	285.9

$a = 10$, circle

(d)

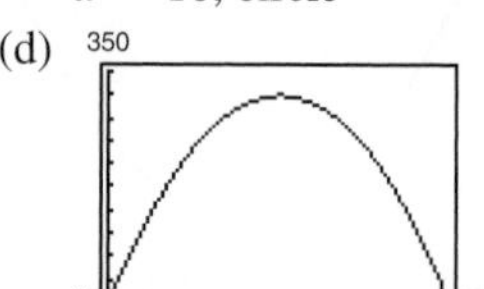

The maximum occurs at $a = 10$.

69. Proof

Section 6.4 *(page 450)*

1. hyperbola; foci **3.** transverse axis; center

5. b **6.** d **7.** c **8.** a **9.** $\dfrac{y^2}{4} - \dfrac{x^2}{12} = 1$

11. $\dfrac{(x-4)^2}{4} - \dfrac{y^2}{12} = 1$ **13.** $\dfrac{(y-5)^2}{16} - \dfrac{(x-4)^2}{9} = 1$

15. $\dfrac{y^2}{9} - \dfrac{4(x-2)^2}{9} = 1$ **17.** $\dfrac{(x-2)^2}{4} - \dfrac{(y+3)^2}{8} = 1$

19. Center: $(0, 0)$
Vertices: $(\pm 1, 0)$
Foci: $\left(\pm\sqrt{2}, 0\right)$
Asymptotes: $y = \pm x$

21. Center: $(0, 0)$
Vertices: $(0, \pm 6)$
Foci: $\left(0 \pm 2\sqrt{34}\right)$
Asymptotes: $y = \pm\frac{3}{5}x$

23. Center: $(0, 0)$
Vertices: $(0, \pm 1)$
Foci: $\left(0, \pm\sqrt{5}\right)$
Asymptotes: $y = \pm\frac{1}{2}x$

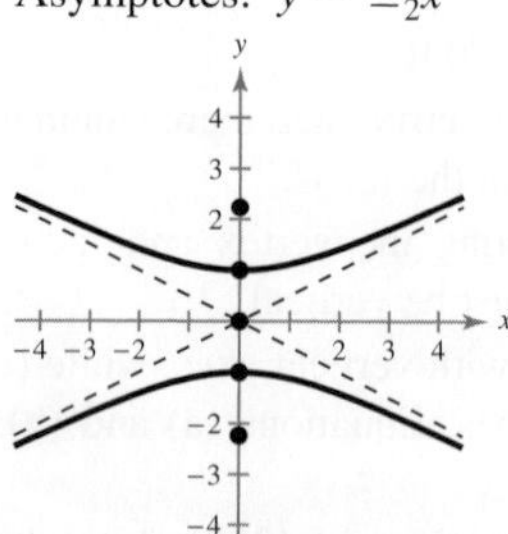

25. Center: $(1, -2)$
Vertices: $(3, -2), (-1, -2)$
Foci: $\left(1 \pm \sqrt{5}, -2\right)$
Asymptotes:
$y = -2 \pm \frac{1}{2}(x - 1)$

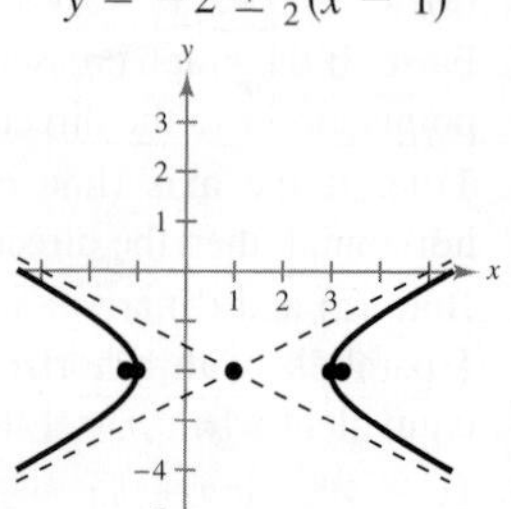

27. Center: $(2, -6)$

Vertices: $\left(2, -\frac{17}{3}\right), \left(2, -\frac{19}{3}\right)$

Foci: $\left(2, -6 \pm \frac{\sqrt{13}}{6}\right)$

Asymptotes: $y = -6 \pm \frac{2}{3}(x - 2)$

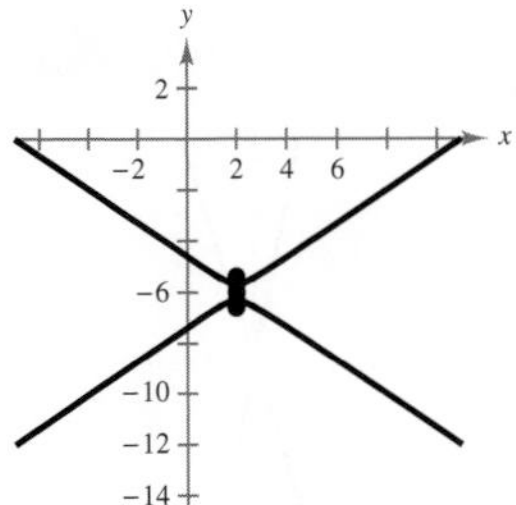

29. Center: $(2, -3)$
Vertices: $(3, -3), (1, -3)$
Foci: $\left(2 \pm \sqrt{10}, -3\right)$
Asymptotes:
$y = -3 \pm 3(x - 2)$

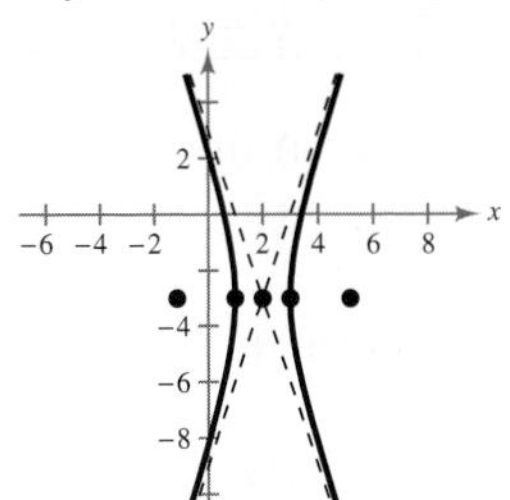

31. Center: $(-1, 1)$
Vertices: $(-2, 1), (0, 1)$
Foci: $\left(-1 \pm \sqrt{5}, 1\right)$
Asymptotes:
$y = 1 \pm 2(x + 1)$

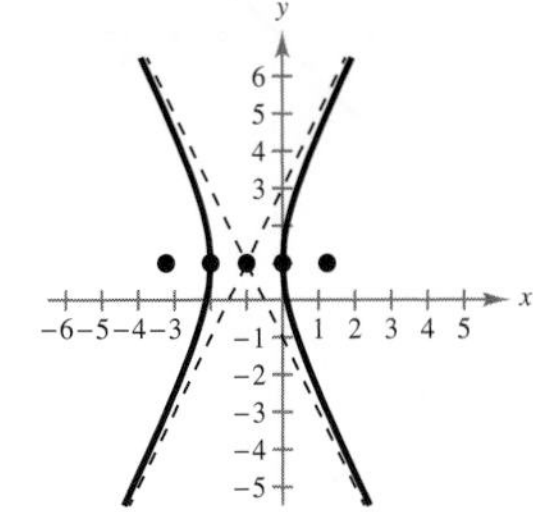

33.

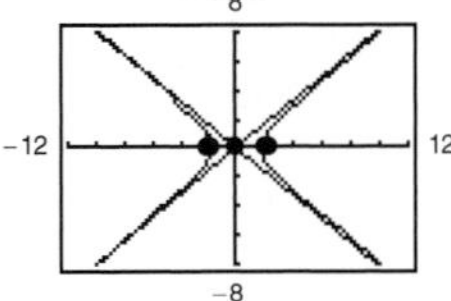

Center: $(0, 0)$
Vertices: $\left(\pm\sqrt{3}, 0\right)$
Foci: $\left(\pm\sqrt{5}, 0\right)$

35.

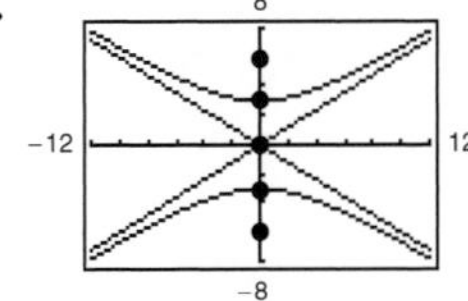

Center: $(0, 0)$
Vertices: $(0, \pm 3)$
Foci: $\left(0, \pm\sqrt{34}\right)$

37.

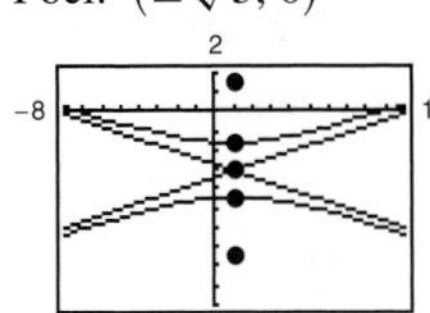

Center: $(1, -3)$
Vertices: $\left(1, -3 \pm \sqrt{2}\right)$
Foci: $\left(1, -3 \pm 2\sqrt{5}\right)$

39. $\frac{x^2}{1} - \frac{y^2}{25} = 1$ **41.** $\frac{17y^2}{1024} - \frac{17x^2}{64} = 1$

43. $\frac{(x - 2)^2}{1} - \frac{(y - 2)^2}{1} = 1$ **45.** $\frac{(y - 2)^2}{4} - \frac{(x - 3)^2}{9} = 1$

47. $\frac{(x - 4)^2}{16} - \frac{(y + 1)^2}{9} = 1$

49. (a) $\frac{x^2}{1} - \frac{y^2}{169/3} = 1$ (b) About 2.403 ft

51. $\frac{x^2}{98{,}010{,}000} - \frac{y^2}{13{,}503{,}600} = 1$

53. (a) $x \approx 110.3$ mi (b) 57.0 mi (c) $y = \frac{27\sqrt{19}}{93}x$

55. Ellipse **57.** Hyperbola **59.** Hyperbola

61. Ellipse **63.** Parabola **65.** Circle

67. True. For a hyperbola, $c^2 = a^2 + b^2$. The larger the ratio of b to a, the larger the eccentricity of the hyperbola, $e = c/a$.

69. False. The graph is two intersecting lines.

71. Draw a rectangle through the vertices and the endpoints of the conjugate axis. Sketch the asymptotes by drawing lines through the opposite corners of the rectangle.

73. The equations of the asymptotes should be $y = k \pm \frac{a}{b}(x - h)$.

75.

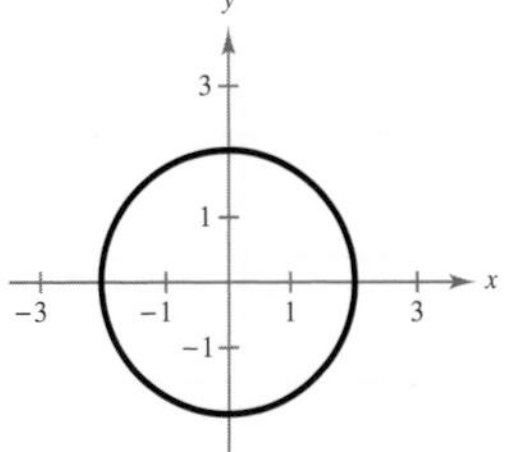

(a) $C > 2$ and $C < -\frac{17}{4}$
(b) $C = 2$
(c) $-2 < C < 2, C = -\frac{17}{4}$
(d) $C = -2$
(e) $-\frac{17}{4} < C < -2$

Section 6.5 *(page 459)*

1. rotation; axes **3.** invariant under rotation **5.** $(0, -2)$

7. $\left(\frac{3 + \sqrt{3}}{2}, \frac{3\sqrt{3} - 1}{2}\right)$ **9.** $\left(\frac{3\sqrt{2}}{2}, -\frac{\sqrt{2}}{2}\right)$

11. $\left(\frac{2\sqrt{3} + 1}{2}, \frac{2 - \sqrt{3}}{2}\right)$ **13.** $\frac{(y')^2}{6} - \frac{(x')^2}{6} = 1$

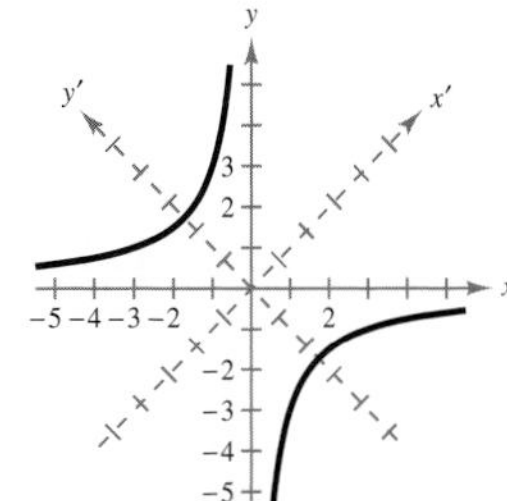

15. $\frac{\left(y' + \frac{3\sqrt{2}}{2}\right)^2}{12} - \frac{\left(x' + \frac{\sqrt{2}}{2}\right)^2}{12} = 1$

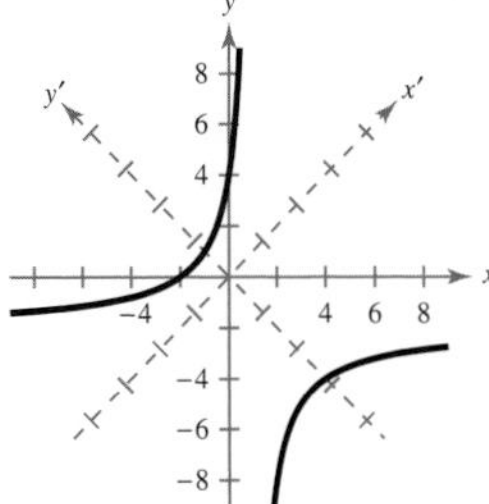

17. 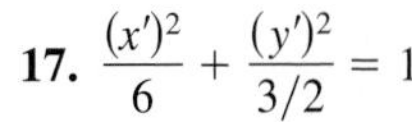 $\frac{(x')^2}{6} + \frac{(y')^2}{3/2} = 1$

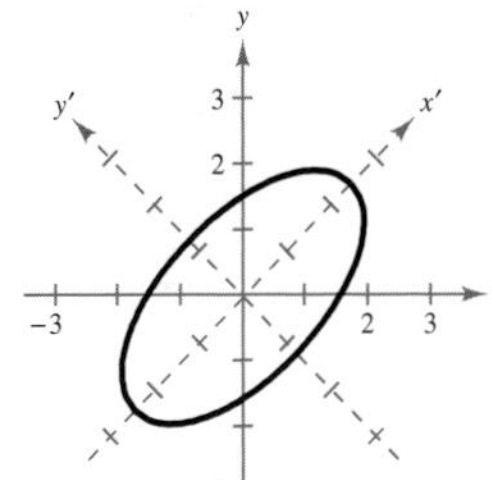

19. 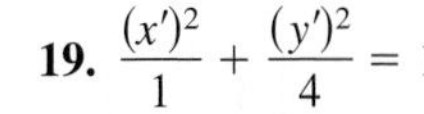 $\frac{(x')^2}{1} + \frac{(y')^2}{4} = 1$

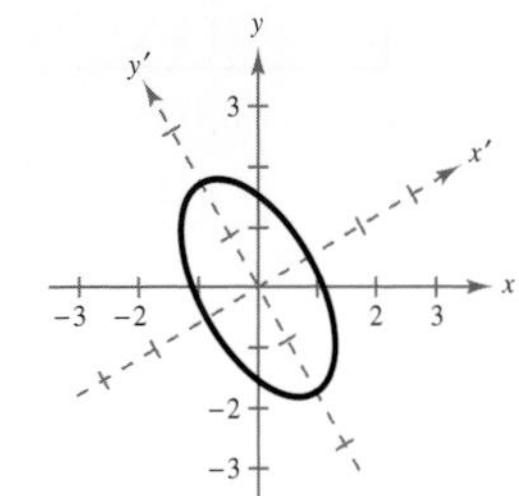

CHAPTER 6

21. $(x')^2 = y'$

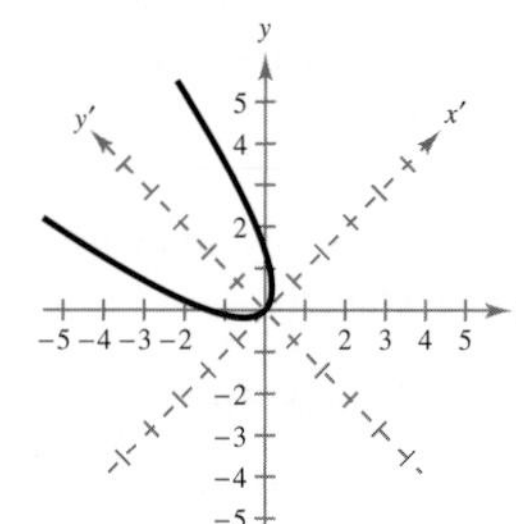

23. $(x' - 1)^2 = 6\left(y' + \frac{1}{6}\right)$

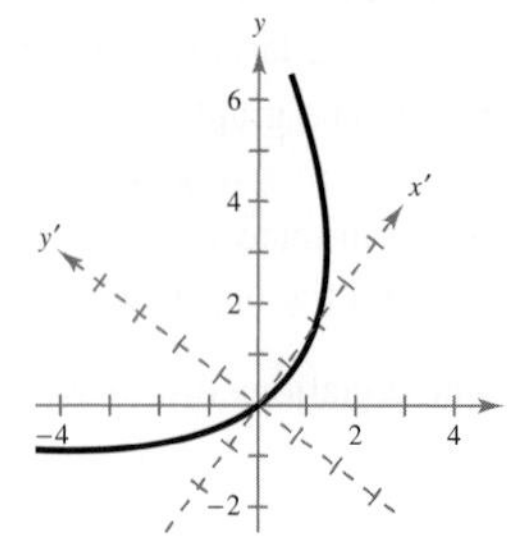

25.

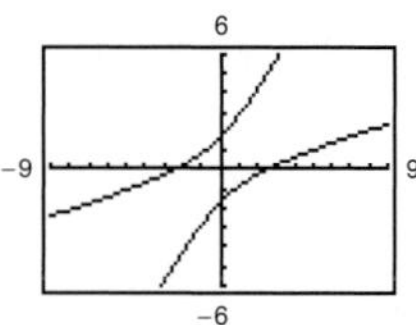

$\theta \approx 37.98°$

27.

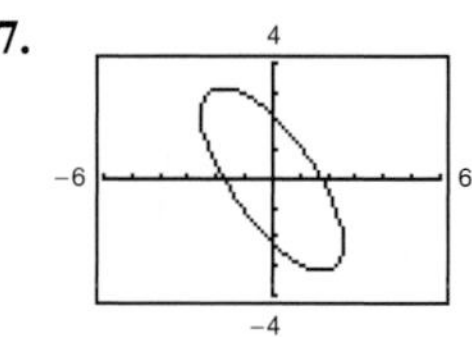

$\theta \approx 36.32°$

29.

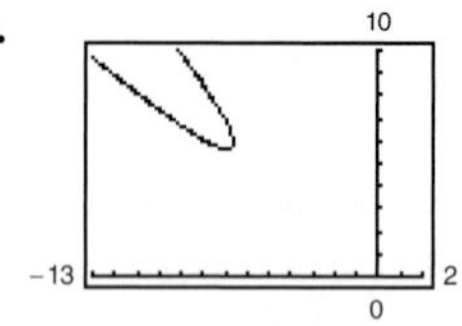

$\theta = 45°$

31. e **32.** a **33.** d **34.** c **35.** f **36.** b

37. (a) Parabola

(b) $y = \dfrac{(8x - 5) \pm \sqrt{(8x - 5)^2 - 4(16x^2 - 10x)}}{2}$

(c)

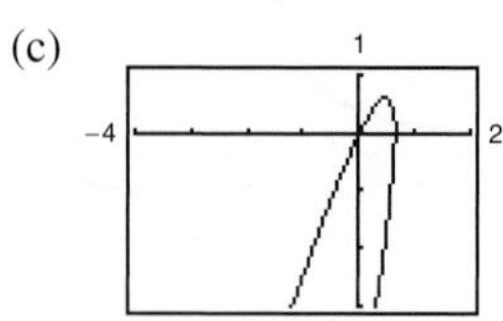

39. (a) Ellipse

(b) $y = \dfrac{6x \pm \sqrt{36x^2 - 28(12x^2 - 45)}}{14}$

(c)

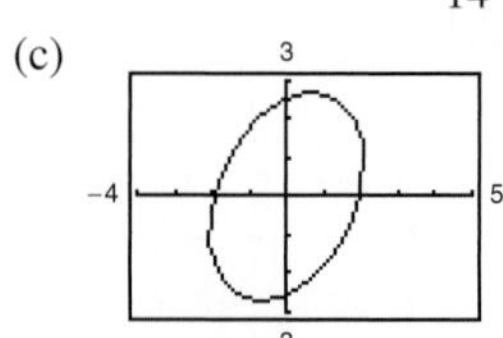

41. (a) Hyperbola

(b) $y = \dfrac{6x \pm \sqrt{36x^2 + 20(x^2 + 4x - 22)}}{-10}$

(c)

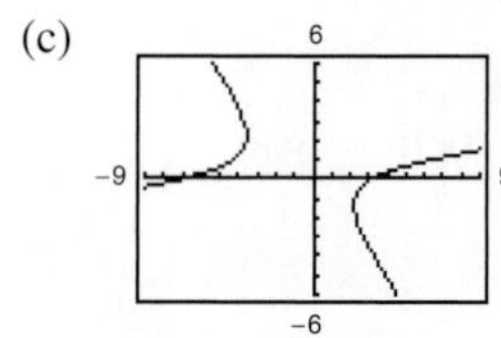

43. (a) Parabola

(b) $y = \dfrac{-(4x - 1) \pm \sqrt{(4x - 1)^2 - 16(x^2 - 5x - 3)}}{8}$

(c)

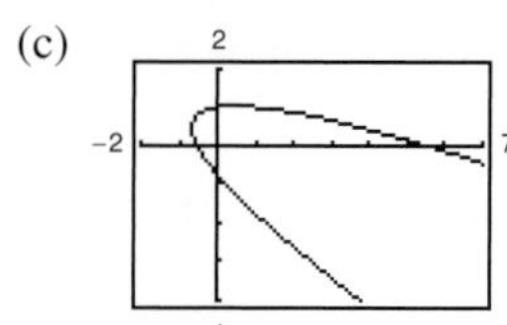

45.

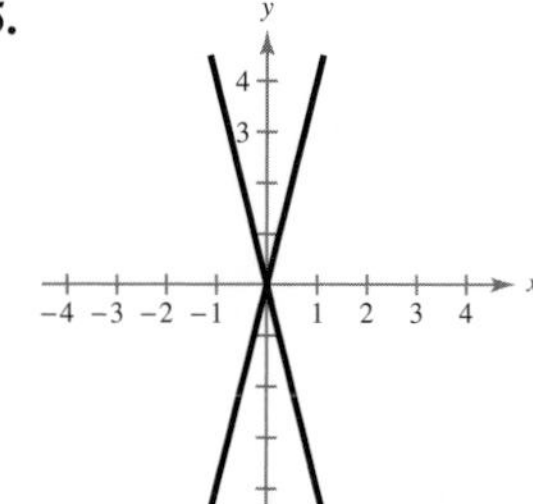

47.

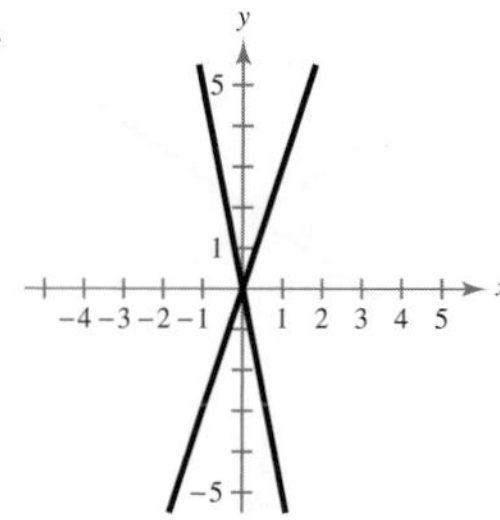

49.

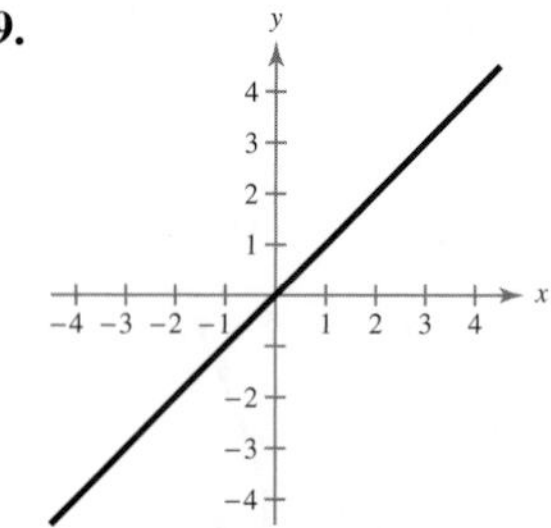

51.

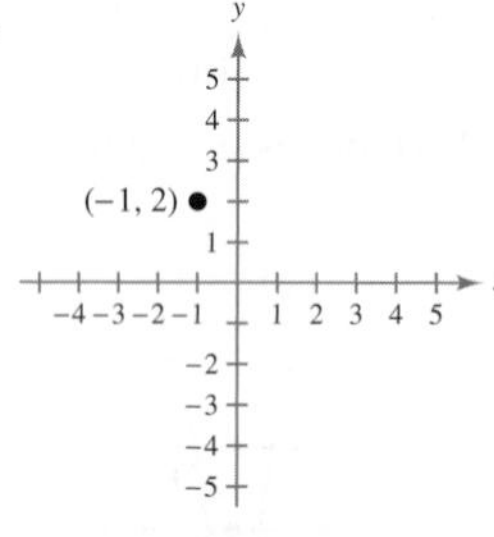

53.

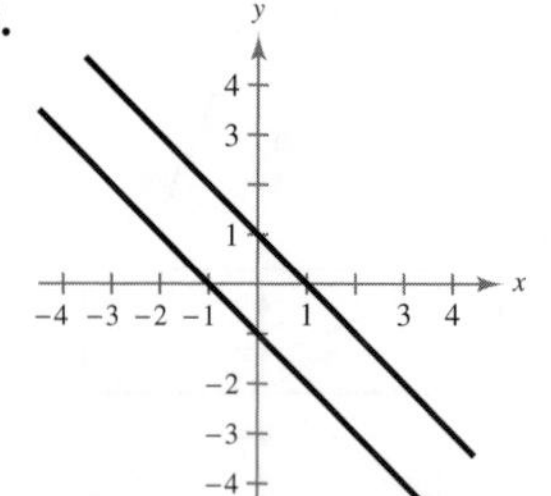

55. $(14, -8), (6, -8)$

57. $(1, 0)$

59. $\left(1, \sqrt{3}\right), \left(1, -\sqrt{3}\right)$

61. $(-7, 0), (-1, 0)$

63. (a) $(y' + 9)^2 = 9(x' - 12)$

(b) 2.25 ft

65. True. The graph of the equation can be classified by finding the discriminant. For a graph to be a hyperbola, the discriminant must be greater than zero. If $k \geq \frac{1}{4}$, then the discriminant would be less than or equal to zero.

67. Answers will vary.

Section 6.6 *(page 467)*

1. plane curve **3.** eliminating; parameter

5. (a)

t	0	1	2	3	4
x	0	1	$\sqrt{2}$	$\sqrt{3}$	2
y	3	2	1	0	-1

(b)

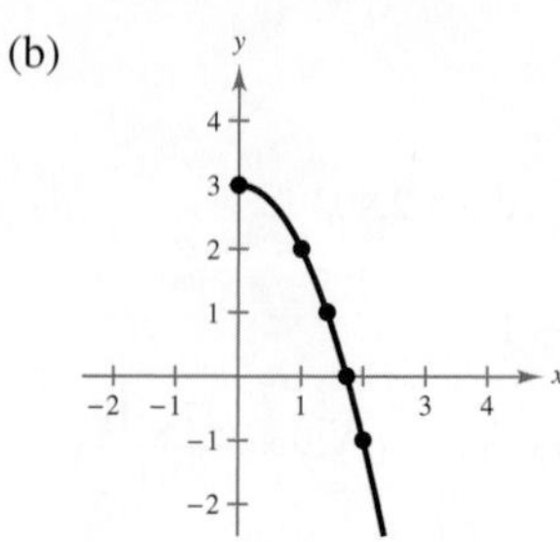

(c)

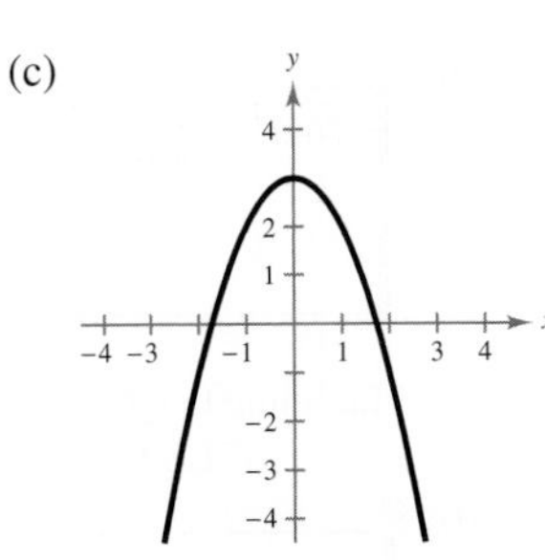

The graph of $y = 3 - x^2$ shows the entire parabola rather than just the right half.

7.

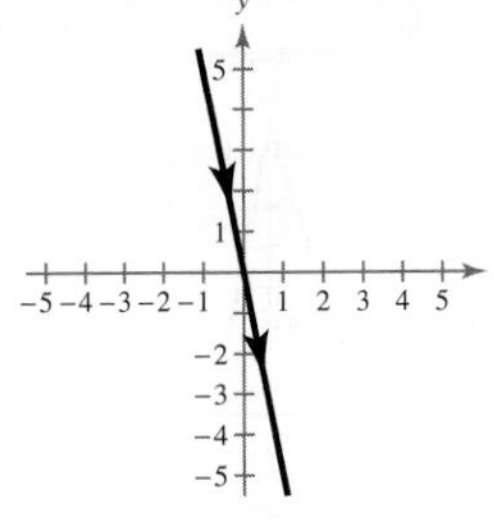

The curve is traced from left to right.

9.

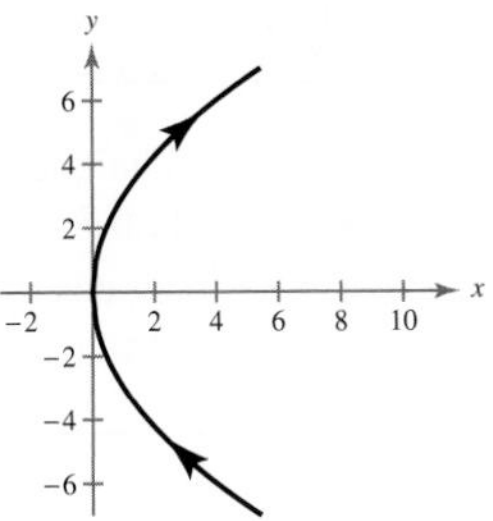

The curve is traced clockwise.

11.

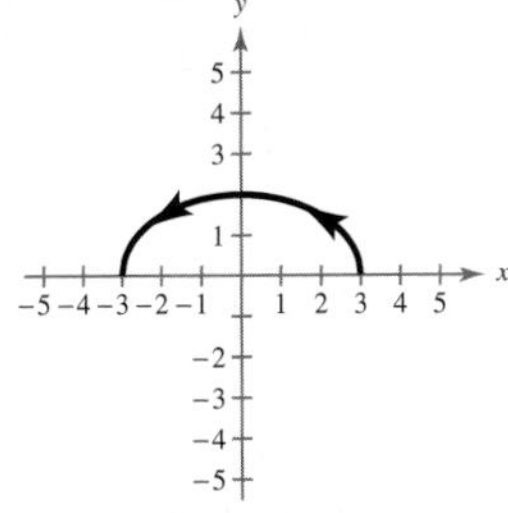

The curve is traced from right to left.

13. (a)

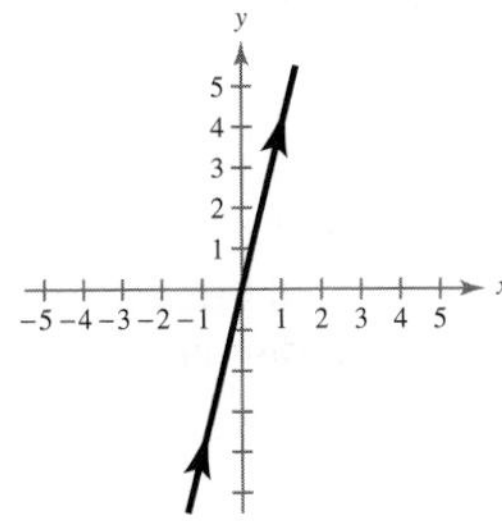

(b) $y = 4x$

15. (a)

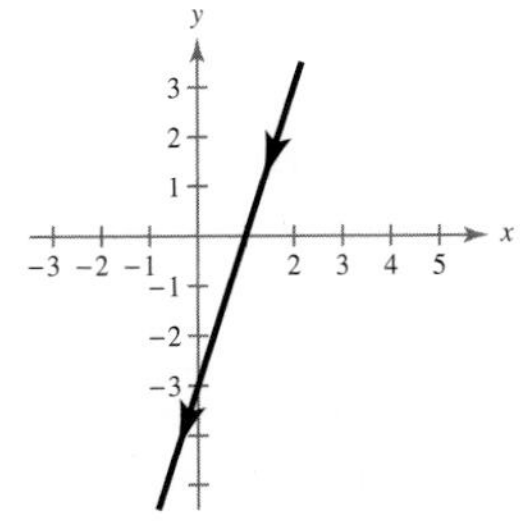

(b) $y = 3x - 3$

17. (a)

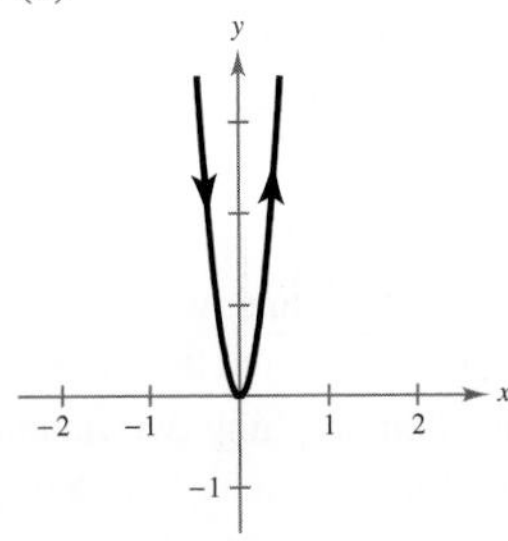

(b) $y = 16x^2$

19. (a)

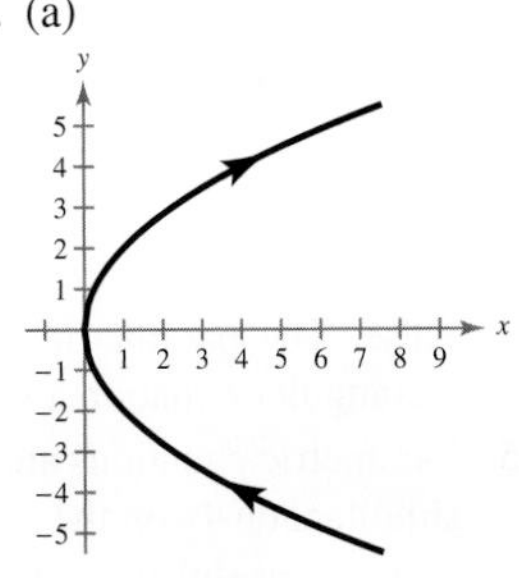

(b) $y = \pm 2\sqrt{x}$

21. (a)

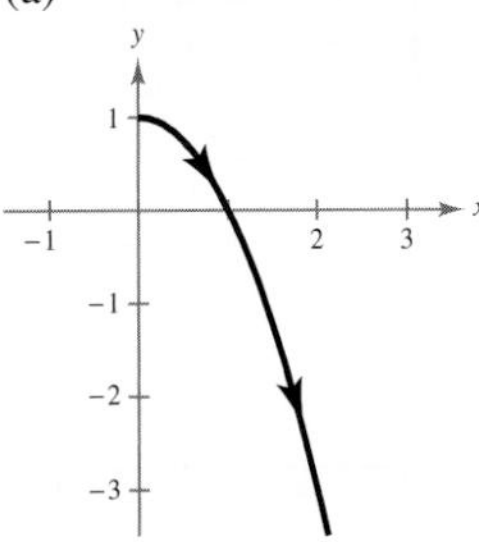

(b) $y = 1 - x^2,\ x \geq 0$

23. (a)

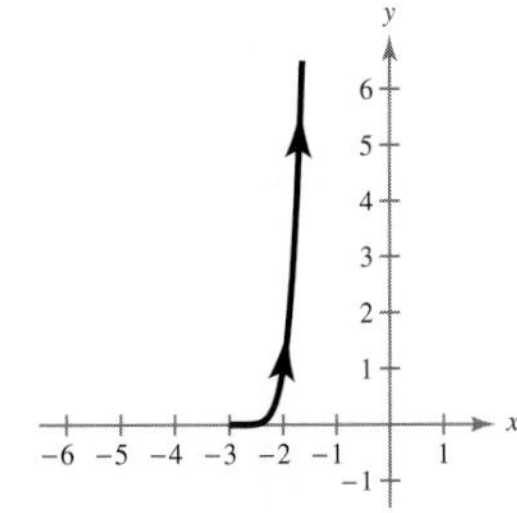

(b) $y = (x + 3)^6,\ x \geq -3$

25. (a)

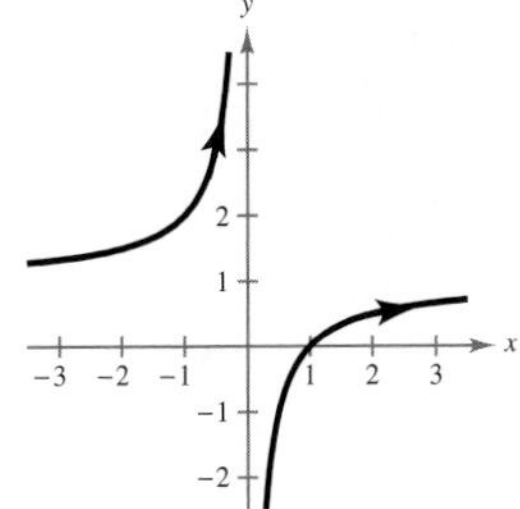

(b) $y = \dfrac{(x - 1)}{x}$

27. (a)

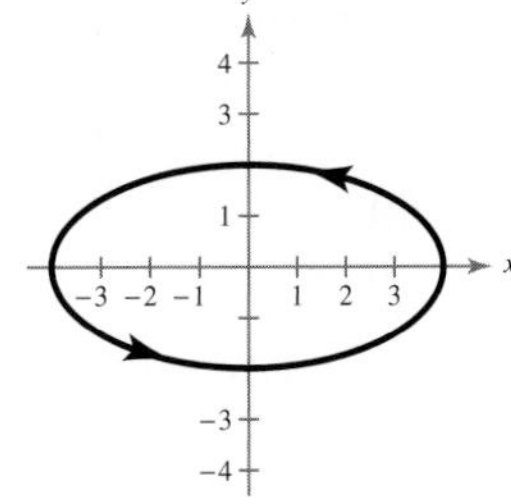

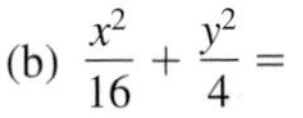

(b) $\dfrac{x^2}{16} + \dfrac{y^2}{4} = 1$

29. (a)

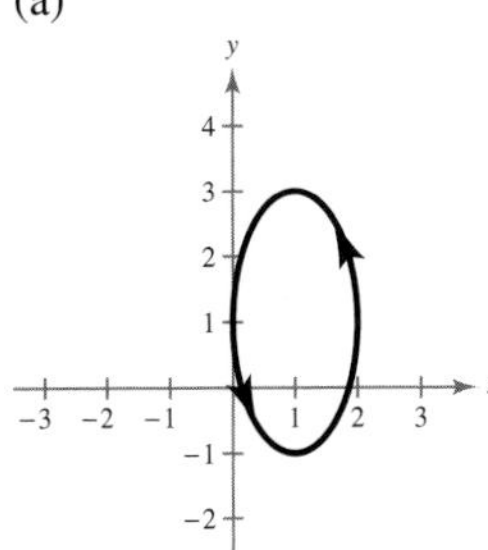

(b) $\dfrac{(x - 1)^2}{1} + \dfrac{(y - 1)^2}{4} = 1$

31. (a)

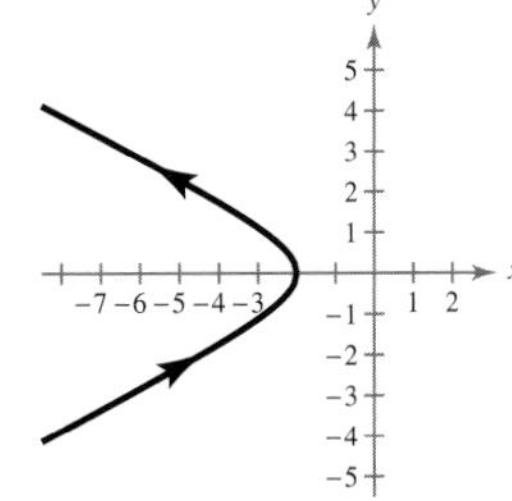

(b) $\dfrac{x^2}{4} - y^2 = 1,\ x \leq -2$

33. (a)

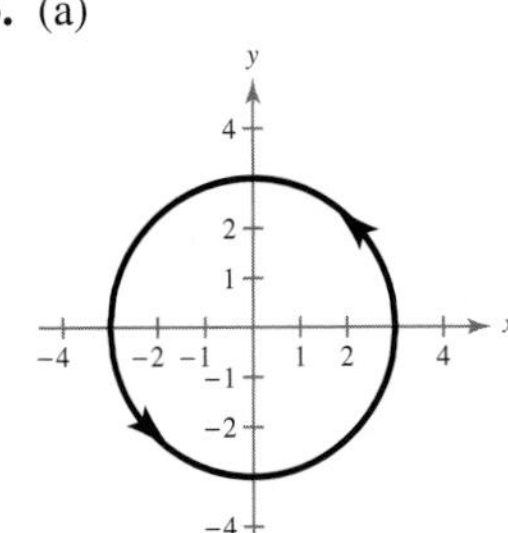

(b) $x^2 + y^2 = 9$

35. (a)

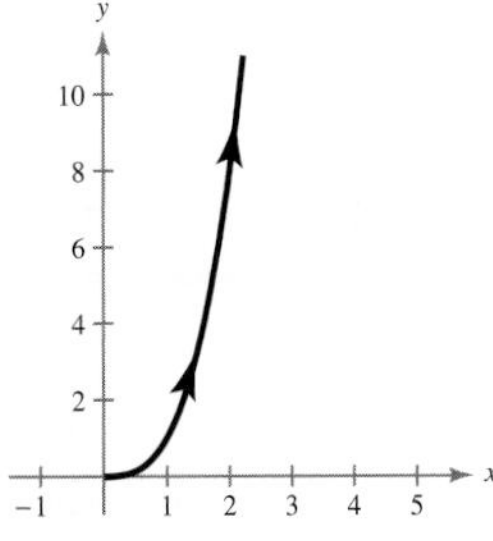

(b) $y = x^3,\ x > 0$

37. (a)

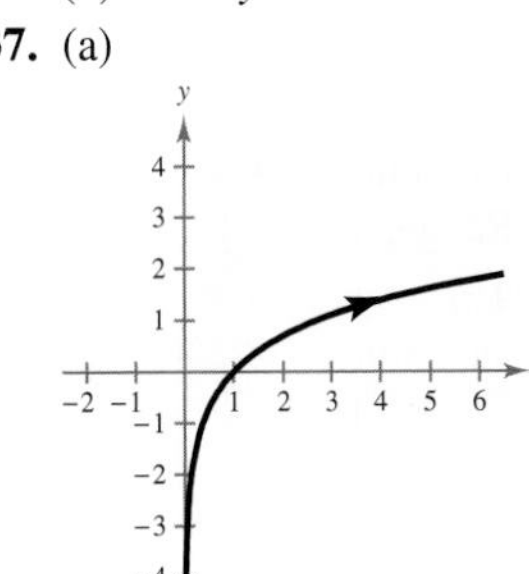

(b) $y = \ln x$

39.

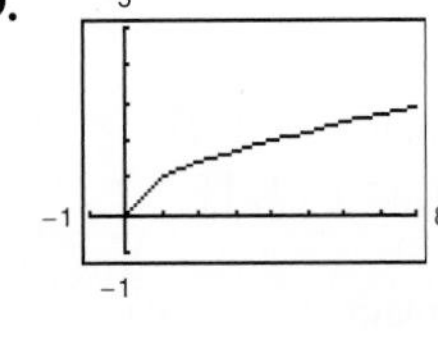

41.

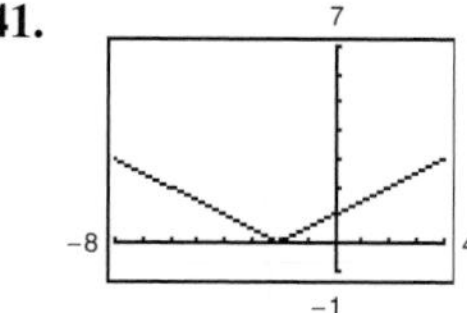

43.

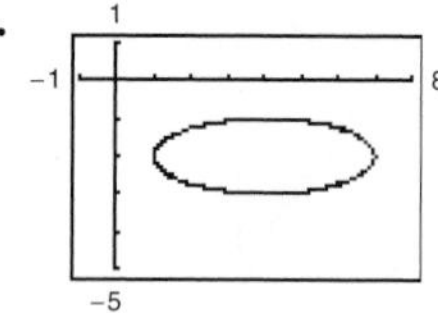

45.

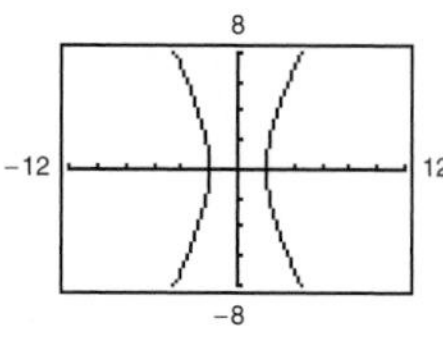

47.

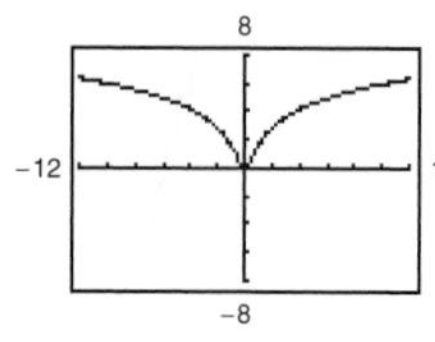

49. Each curve represents a portion of the line $y = 2x + 1$.

Domain	Orientation
(a) $(-\infty, \infty)$	Left to right
(b) $[-1, 1]$	Depends on θ
(c) $(0, \infty)$	Right to left
(d) $(0, \infty)$	Left to right

51. $y - y_1 = m(x - x_1)$

53. $\dfrac{(x-h)^2}{a^2} + \dfrac{(y-k)^2}{b^2} = 1$

55. $x = 3t$
$y = 6t$

57. $x = 3 + 4\cos\theta$
$y = 2 + 4\sin\theta$

59. $x = 5\cos\theta$
$y = 3\sin\theta$

61. $x = 5 + 4\sec\theta$
$y = 3\tan\theta$

63. $x = -5t$
$y = 2t$
$0 \le t \le 1$

65. $x = 3\sec\theta$
$y = 4\tan\theta$
$\dfrac{\pi}{2} < \theta < \dfrac{3\pi}{2}$

67. (a) $x = t,\ y = 3t - 2$ (b) $x = -t + 2,\ y = -3t + 4$

69. (a) $x = t,\ y = \frac{1}{2}(t - 1)$ (b) $x = -t + 2,\ y = -\frac{1}{2}(t - 1)$

71. (a) $x = t,\ y = t^2 + 1$ (b) $x = -t + 2,\ y = t^2 - 4t + 5$

73. (a) $x = t,\ y = 1 - 2t^2$
(b) $x = 2 - t,\ y = -2t^2 + 8t - 7$

75. (a) $x = t,\ y = \dfrac{1}{t}$ (b) $x = -t + 2,\ y = -\dfrac{1}{t - 2}$

77. (a) $x = t,\ y = e^t$ (b) $x = -t + 2,\ y = e^{-t+2}$

79.

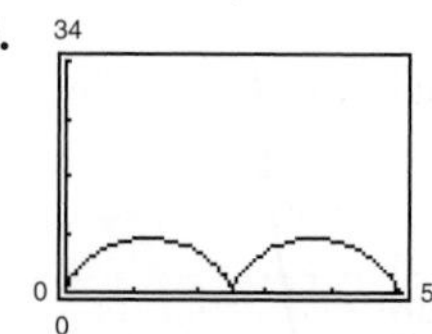

81.

83.

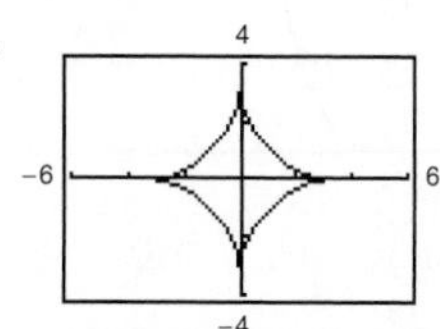

85.

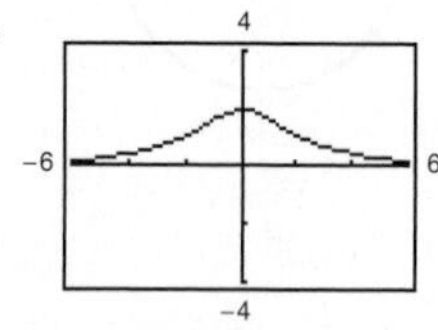

87. b
Domain: $[-2, 2]$
Range: $[-1, 1]$

88. c
Domain: $[-4, 4]$
Range: $[-6, 6]$

89. d
Domain: $(-\infty, \infty)$
Range: $(-\infty, \infty)$

90. a
Domain: $(-\infty, \infty)$
Range: $[-2, 2]$

91. (a)

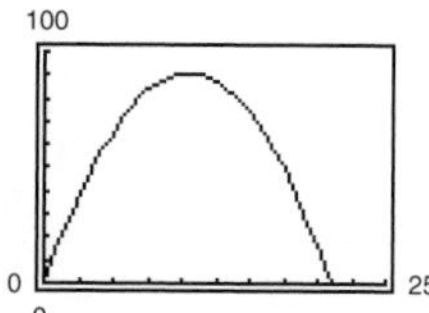

Maximum height: 90.7 ft
Range: 209.6 ft

(b)

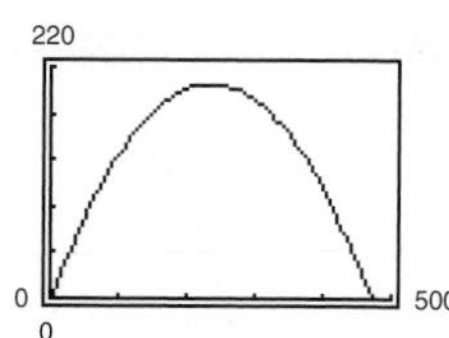

Maximum height: 204.2 ft
Range: 471.6 ft

(c)

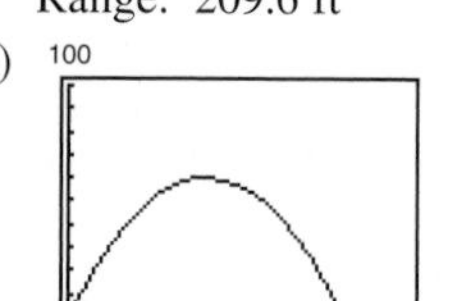

Maximum height: 60.5 ft
Range: 242.0 ft

(d)

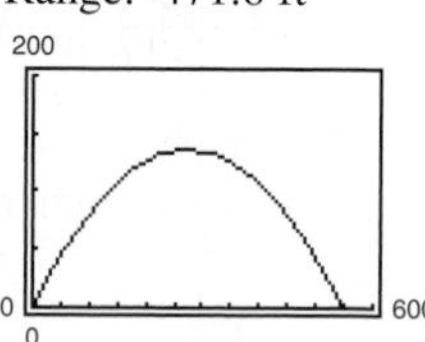

Maximum height: 136.1 ft
Range: 544.5 ft

93. (a) $x = (146.67\cos\theta)t$
$y = 3 + (146.67\sin\theta)t - 16t^2$

(b)

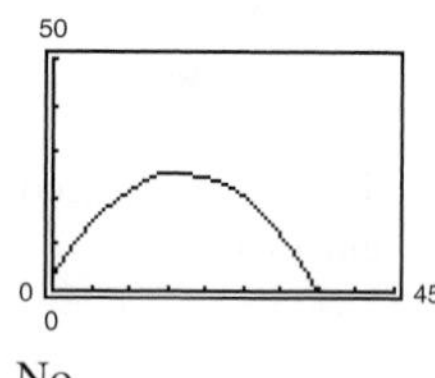

No

(c)

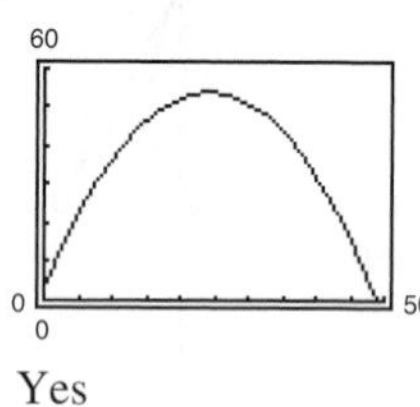

Yes

(d) $19.3°$

95. (a) $x = (\cos 35°)v_0 t$
$y = 7 + (\sin 35°)v_0 t - 16t^2$
(b) About 54.09 ft/sec
(c) 22.04 ft (d) About 2.03 sec

97. (a) $h = 7,\ v_0 = 40,\ \theta = 45°$
$x = (40\cos 45°)t$
$y = 7 + (40\sin 45°)t - 16t^2$
(b)
(c) Maximum height: 19.5 ft
Range: 56.2 ft

99. $x = a\theta - b\sin\theta$
$y = a - b\cos\theta$

101. True
$x = t$
$y = t^2 + 1 \Rightarrow y = x^2 + 1$
$x = 3t$
$y = 9t^2 + 1 \Rightarrow y = x^2 + 1$

103. False. The parametric equations $x = t^2$ and $y = t$ give the rectangular equation $x = y^2$, so y is not a function of x.

105. Parametric equations are useful when graphing two functions simultaneously on the same coordinate system. For example, they are useful when tracking the path of an object so that the position and the time associated with that position can be determined.

107. The parametric equation for x is defined only when $t \ge 1$, so the domain of the rectangular equation is $x \ge 0$.

109. Yes. The orientation would change.

Section 6.7 *(page 475)*

1. pole **3.** polar

5. $\left(2, -\frac{11\pi}{6}\right), \left(-2, -\frac{5\pi}{6}\right), \left(-2, \frac{7\pi}{6}\right)$

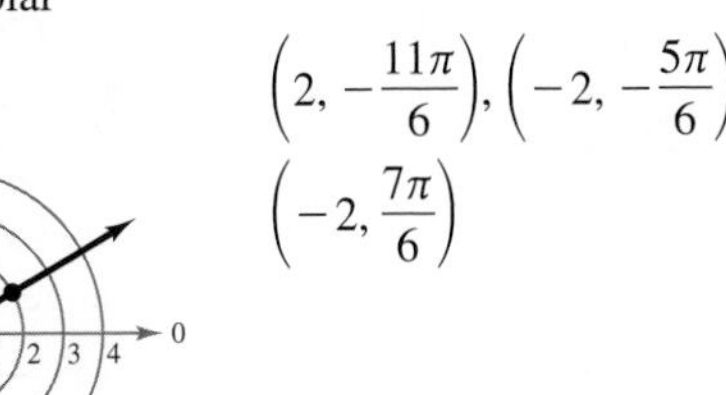

7. $\left(4, \frac{5\pi}{3}\right), \left(-4, \frac{2\pi}{3}\right), \left(-4, -\frac{4\pi}{3}\right)$

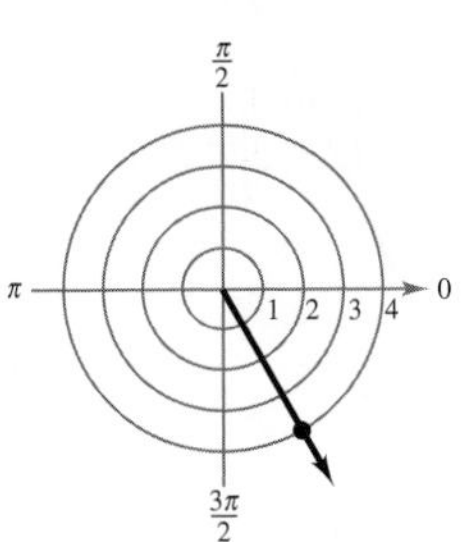

9. $(2, \pi), (-2, 0), (2, -\pi)$

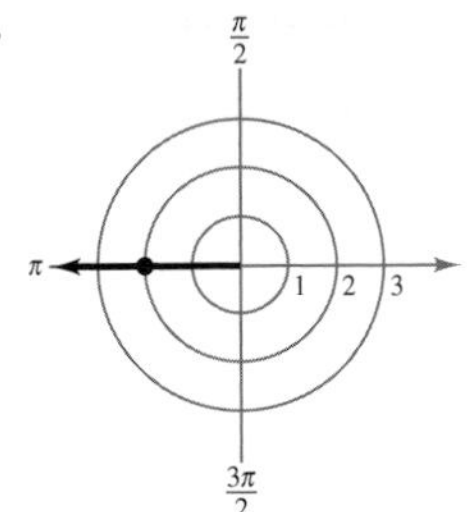

11. $\left(-2, -\frac{4\pi}{3}\right), \left(2, -\frac{\pi}{3}\right), \left(2, \frac{5\pi}{3}\right)$

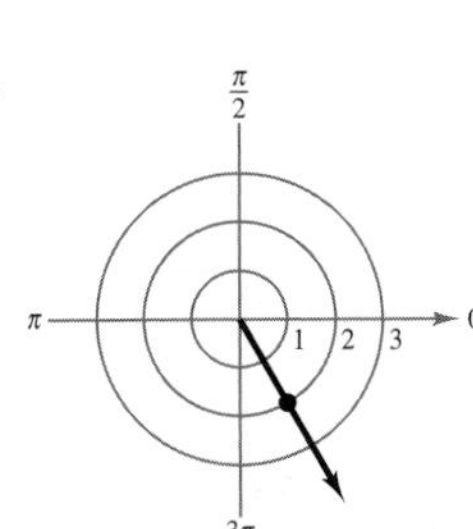

13. *Sample answer:* $\left(0, -\frac{11\pi}{6}\right), \left(0, -\frac{5\pi}{6}\right), \left(0, \frac{\pi}{6}\right)$

15. $(\sqrt{2}, -3.92), (-\sqrt{2}, -0.78), (-\sqrt{2}, 5.50)$

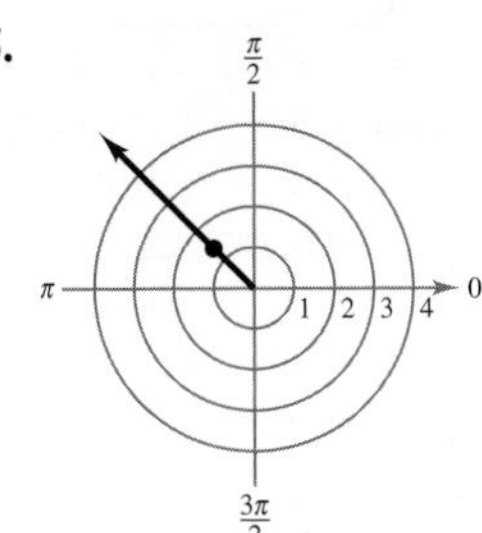

17. $(-3, 4.71), (3, 1.57), (3, -4.71)$

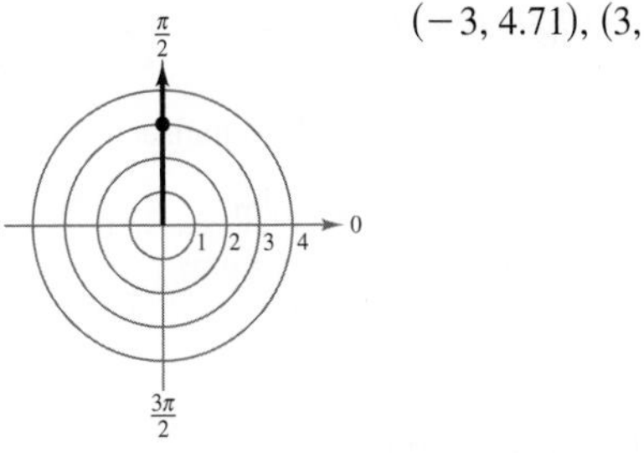

19. $(0, 0)$ **21.** $(0, 3)$ **23.** $(-\sqrt{2}, \sqrt{2})$ **25.** $(\sqrt{3}, 1)$

27. $\left(-\frac{3}{2}, \frac{3\sqrt{3}}{2}\right)$ **29.** $(-1.85, 0.77)$ **31.** $(0.26, 0.97)$

33. $(-1.13, -2.23)$ **35.** $(-2.43, -0.60)$

37. $(0.20, -3.09)$ **39.** $\left(\sqrt{2}, \frac{\pi}{4}\right)$ **41.** $\left(3\sqrt{2}, \frac{5\pi}{4}\right)$

43. $(3, 0)$ **45.** $\left(5, \frac{3\pi}{2}\right)$ **47.** $\left(\sqrt{6}, \frac{5\pi}{4}\right)$ **49.** $\left(2, \frac{11\pi}{6}\right)$

51. $(3.61, 5.70)$ **53.** $(5.39, 2.76)$ **55.** $(4.36, -1.98)$

57. $(2.83, 0.49)$ **59.** $r = 3$ **61.** $\theta = \frac{\pi}{4}$

63. $r = 10 \sec \theta$ **65.** $r = \frac{-2}{3 \cos \theta - \sin \theta}$

67. $r^2 = 16 \sec \theta \csc \theta = 32 \csc 2\theta$ **69.** $r = a \sec \theta$

71. $r = a$ **73.** $r = 2a \cos \theta$ **75.** $r^2 = \cos 2\theta$

77. $r = \cot^2 \theta \csc \theta$ **79.** $x^2 + y^2 = 25$

81. $\sqrt{3}x + y = 0$ **83.** $x = 0$ **85.** $y = 4$

87. $x = -3$ **89.** $x^2 + y^2 + 2x = 0$

91. $x^2 + y^2 - x^{2/3} = 0$ **93.** $(x^2 + y^2)^2 = 2xy$

95. $(x^2 + y^2)^2 = 6x^2y - 2y^3$ **97.** $x^2 + 4y - 4 = 0$

99. $4x^2 - 5y^2 - 36y - 36 = 0$

101. The graph consists of all points six units from the pole; $x^2 + y^2 = 36$

103. The graph consists of all points on the line that makes an angle of $\pi/6$ with the polar axis and passes through the pole; $-\sqrt{3}x + 3y = 0$

105. The graph is a vertical line; $x - 3 = 0$

107. The graph is a circle with center $(0, 1)$ and radius 1; $x^2 + (y - 1)^2 = 1$

109. (a) $r = 30$

(b) $\left(30, \frac{5\pi}{6}\right)$; 30 represents the distance of the passenger car from the center, and $\frac{5\pi}{6} = 150°$ represents the angle to which the car has rotated.

(c) $(-25.98, 15)$; The car is about 25.98 feet to the left of the center and 15 feet above the center.

111. True. Because r is a directed distance, the point (r, θ) can be represented as $(r, \theta \pm 2\pi n)$.

113. $(1, -\sqrt{3})$ is in Quadrant IV, so $\theta = -\frac{\pi}{3}$.

115. (a) $(x - h)^2 + (y - k)^2 = h^2 + k^2$; $r = \sqrt{h^2 + k^2}$

(b) $\left(x - \frac{1}{2}\right)^2 + \left(y - \frac{3}{2}\right)^2 = \frac{5}{2}$

Center: $\left(\frac{1}{2}, \frac{3}{2}\right)$

Radius: $\frac{\sqrt{10}}{2}$

CHAPTER 6

Section 6.8 *(page 483)*

1. $\theta = \dfrac{\pi}{2}$ **3.** convex limaçon **5.** lemniscate

7. Circle **9.** Limaçon with inner loop

11. Rose curve with 4 petals **13.** Polar axis

15. $\theta = \dfrac{\pi}{2}$ **17.** $\theta = \dfrac{\pi}{2}$, polar axis, pole

19. Maximum: $|r| = 20$ when $\theta = \dfrac{3\pi}{2}$

Zero: $r = 0$ when $\theta = \dfrac{\pi}{2}$

21. Maximum: $|r| = 4$ when $\theta = 0, \dfrac{\pi}{3}, \dfrac{2\pi}{3}$

Zeros: $r = 0$ when $\theta = \dfrac{\pi}{6}, \dfrac{\pi}{2}, \dfrac{5\pi}{6}$

23.

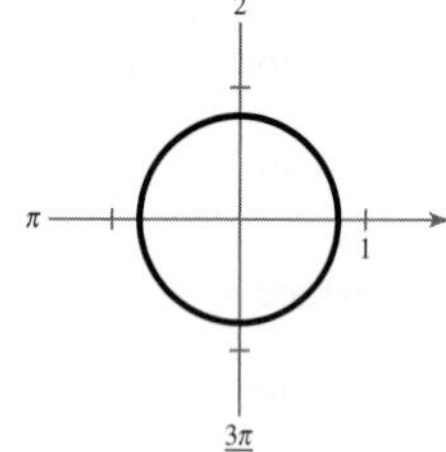

25.

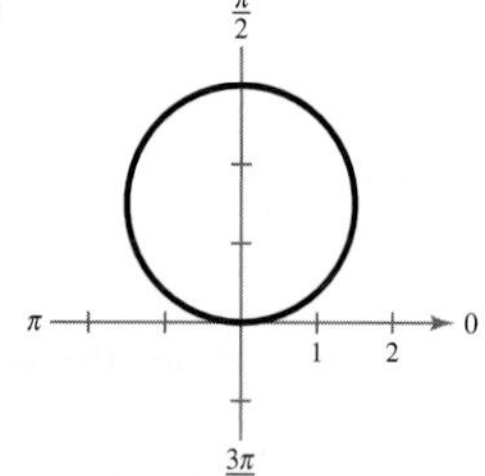

27.

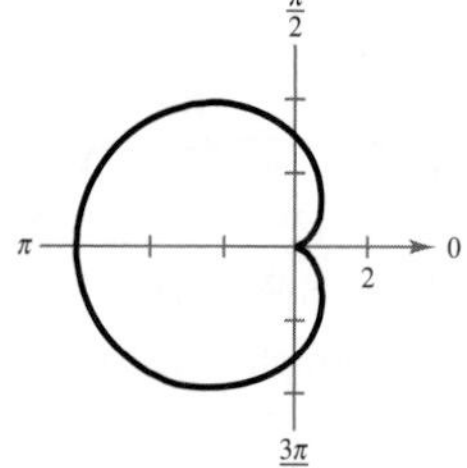

29.

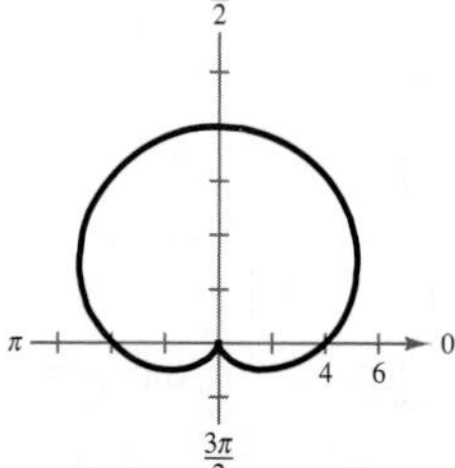

31.

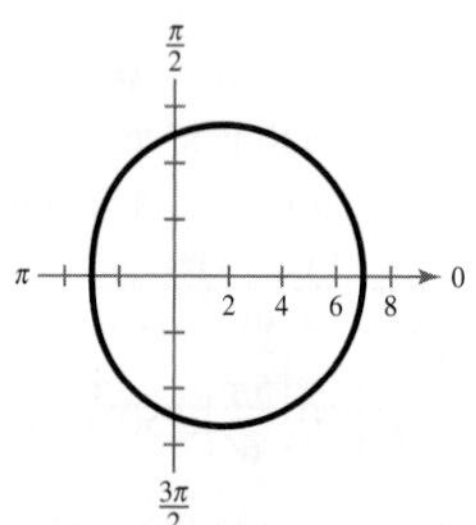

33.

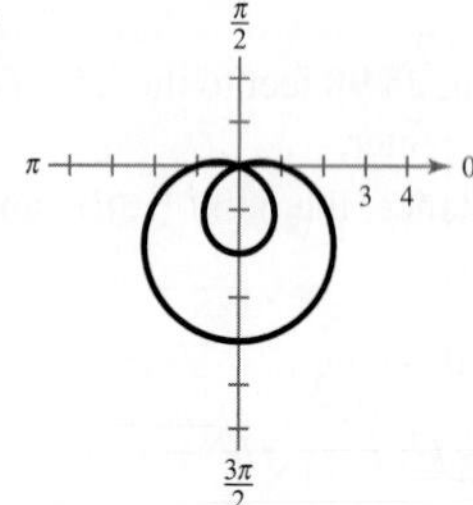

35.

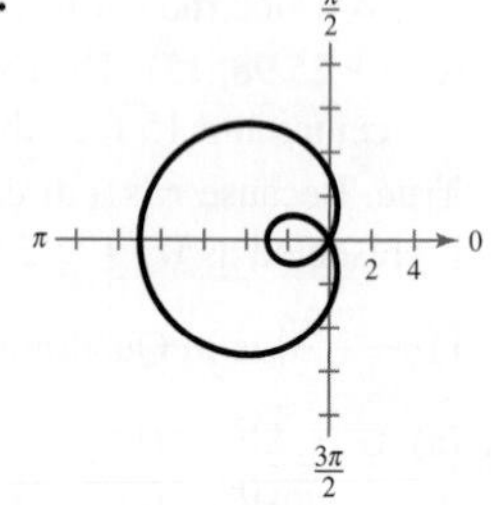

37.

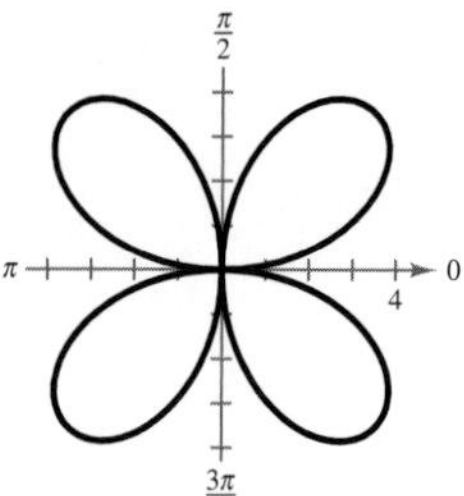

39.

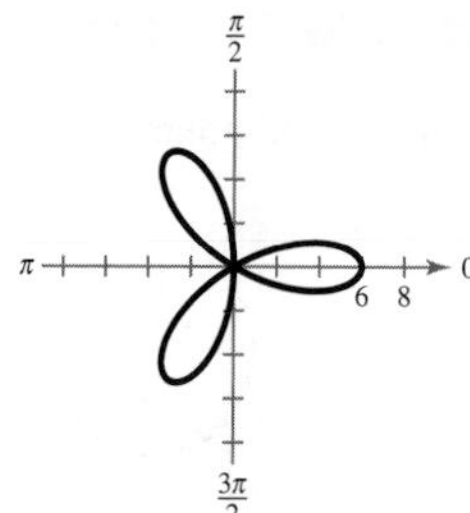

41.

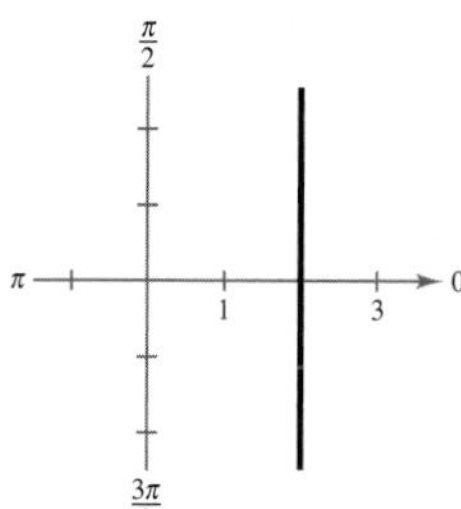

43.

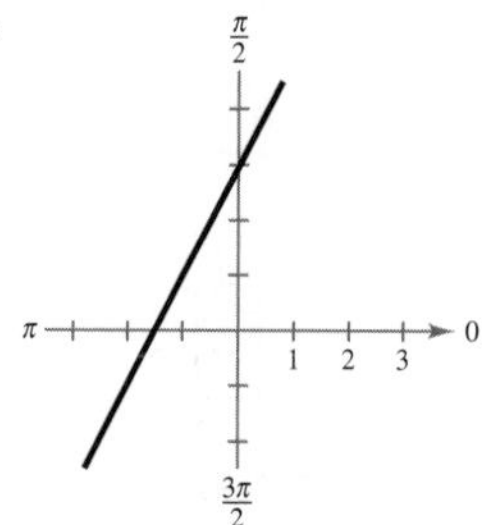

45.

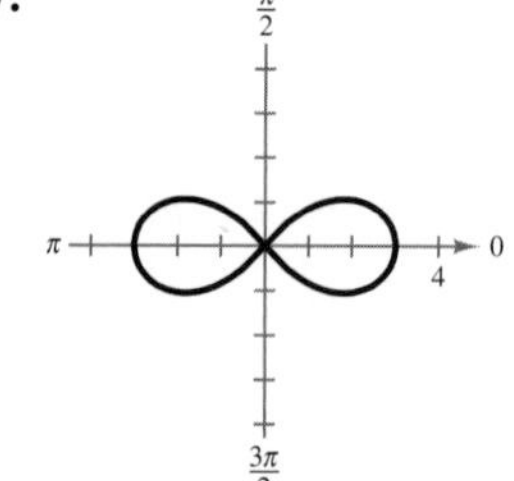

47.

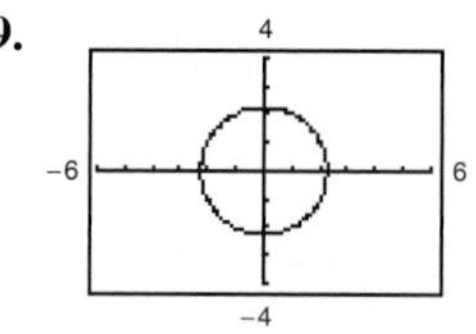

49.

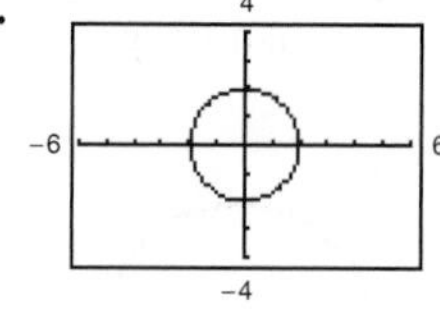

51.

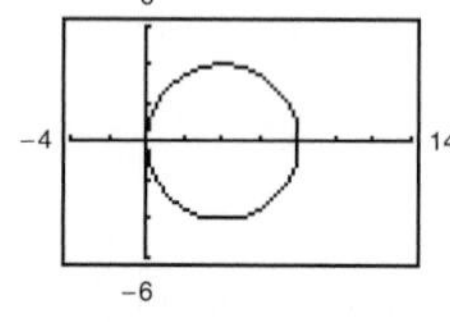

53.

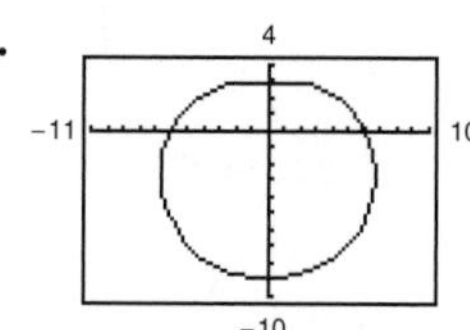

55.

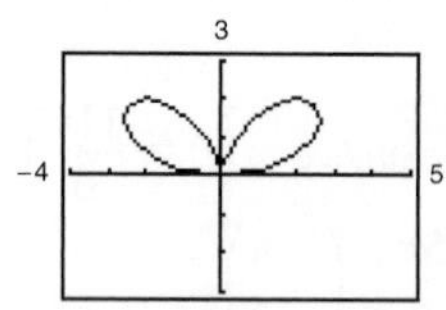

57.

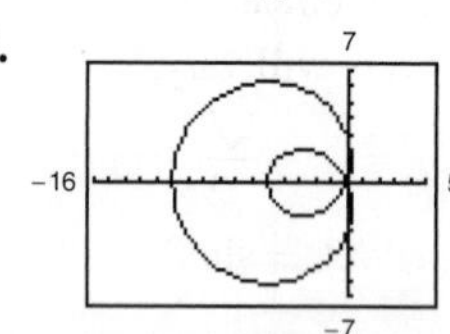

59.

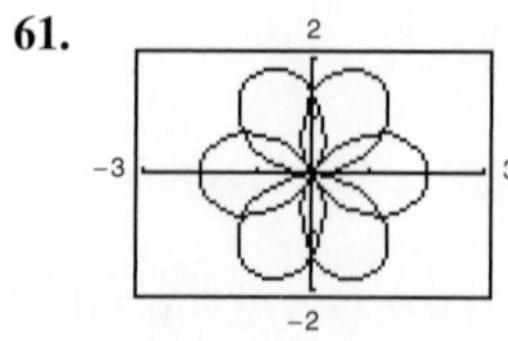

$0 \le \theta < 2\pi$

61.

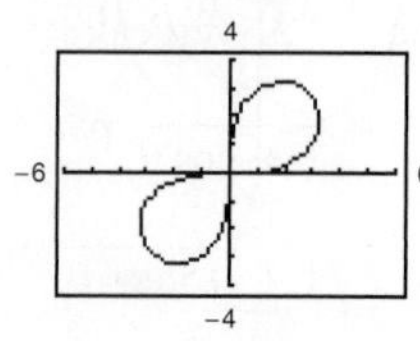

$0 \le \theta < 4\pi$

63.

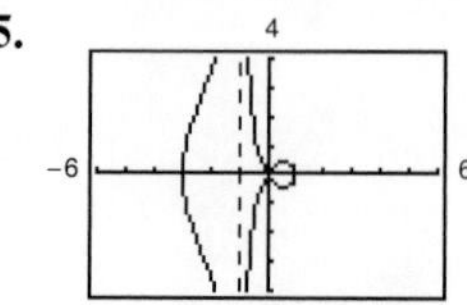

$0 \le \theta < \dfrac{\pi}{2}$

65.

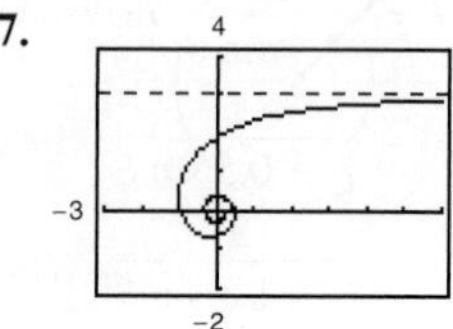

67.

69. (a)

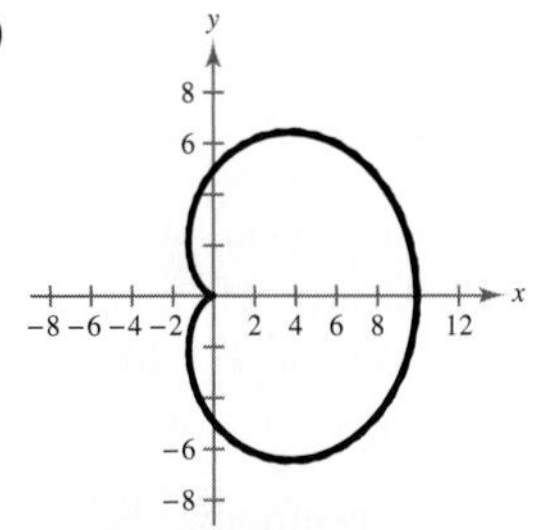

Cardioid

(b) 0 radians

71. True. The equation is of the form $r = a \sin n\theta$, where n is odd.

73. (a)

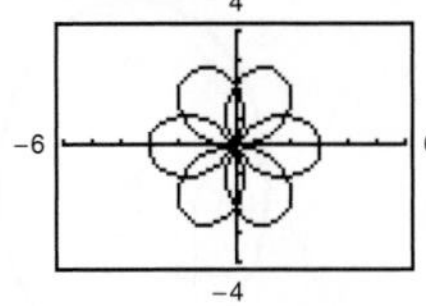

$0 \le \theta < 4\pi$

(b)

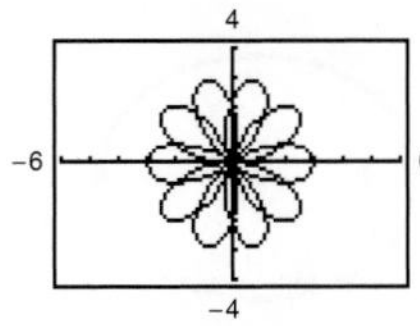

$0 \le \theta < 4\pi$

(c) Yes. Explanations will vary.

75. (a)

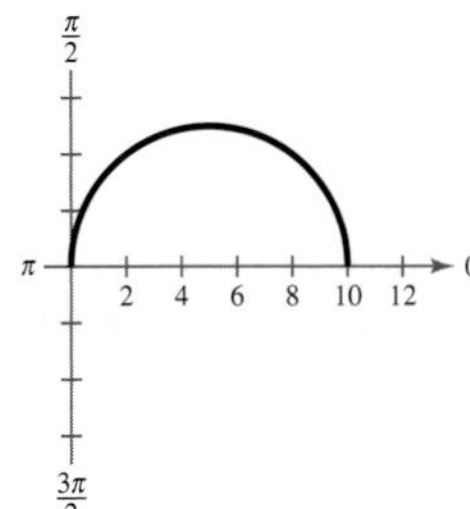

Upper half of circle

(b)

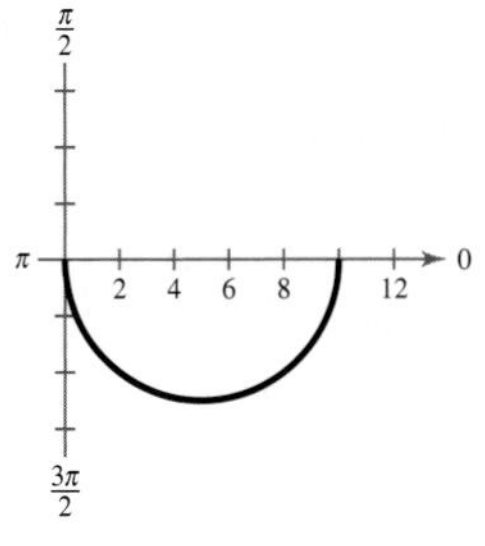

Lower half of circle

(c)

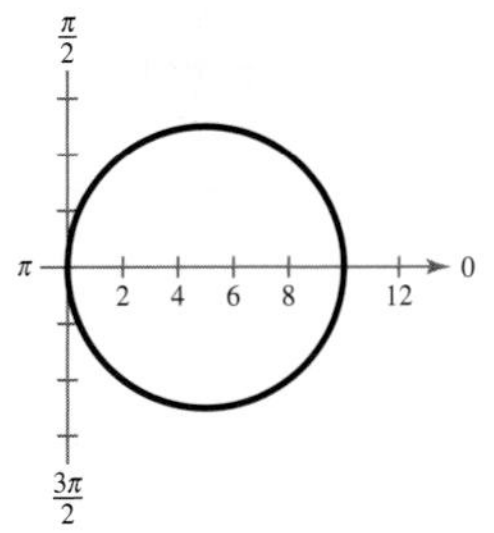

Full circle

(d)

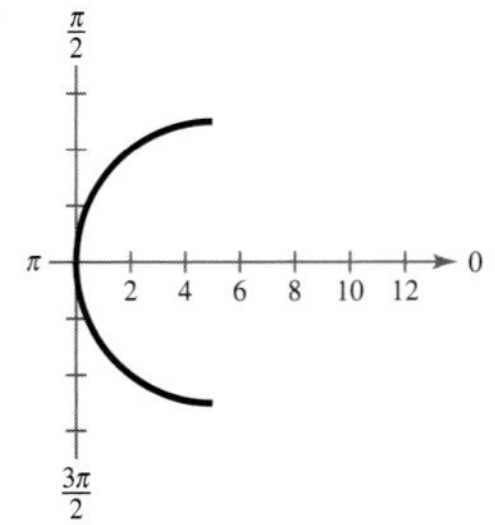

Left half of circle

77. Answers will vary.

79. (a) $r = 2 - \dfrac{\sqrt{2}}{2}(\sin\theta - \cos\theta)$ (b) $r = 2 + \cos\theta$

(c) $r = 2 + \sin\theta$ (d) $r = 2 - \cos\theta$

Section 6.9 *(page 489)*

1. conic **3.** vertical; left

5. (a) $r = \dfrac{2}{1 + \cos\theta}$, parabola

(b) $r = \dfrac{1}{1 + 0.5\cos\theta}$, ellipse

(c) $r = \dfrac{3}{1 + 1.5\cos\theta}$, hyperbola

7. (a) $r = \dfrac{2}{1 - \sin\theta}$, parabola

(b) $r = \dfrac{1}{1 - 0.5\sin\theta}$, ellipse

(c) $r = \dfrac{3}{1 - 1.5\sin\theta}$, hyperbola

9. c **10.** d **11.** a **12.** b

13. Parabola

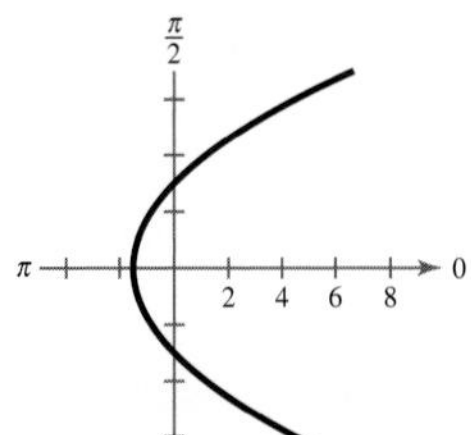

15. Parabola

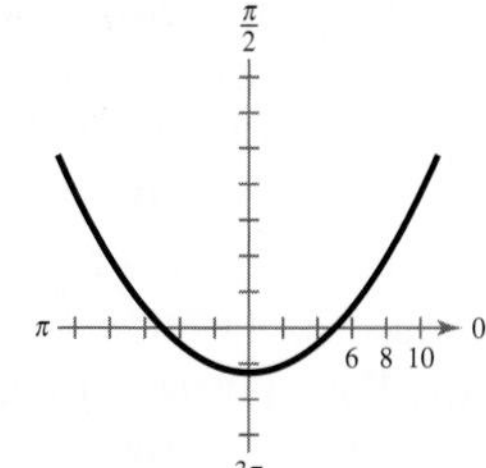

17. Ellipse

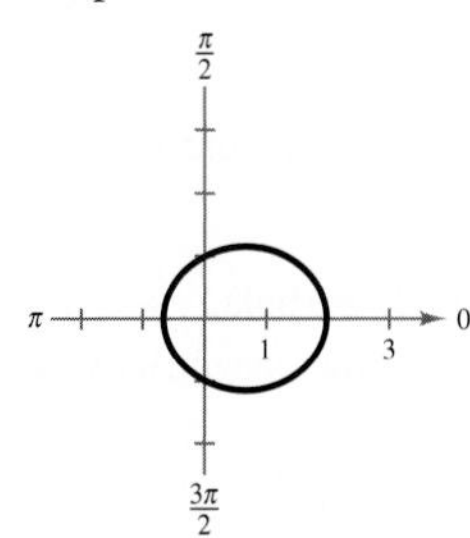

19. Ellipse

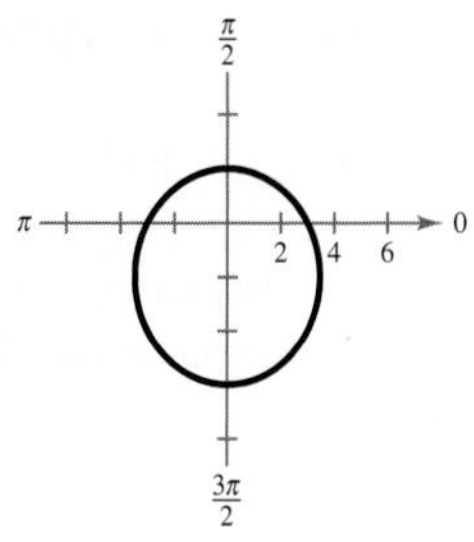

21. Hyperbola

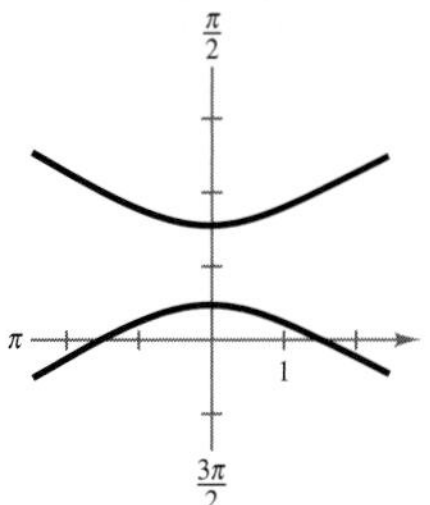

23. Hyperbola

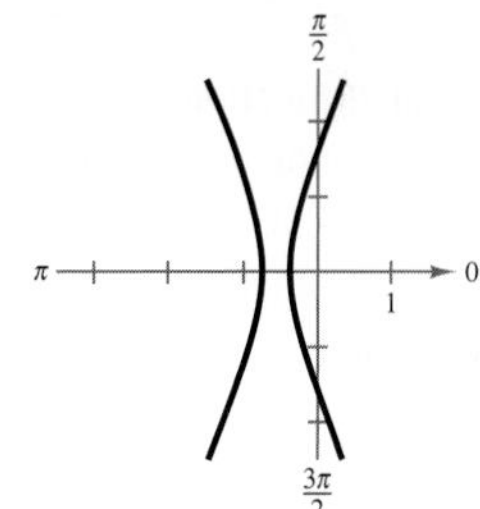

25.

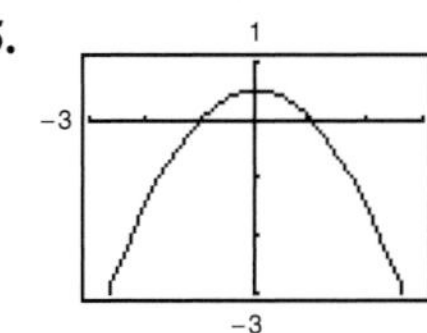

Parabola

27.

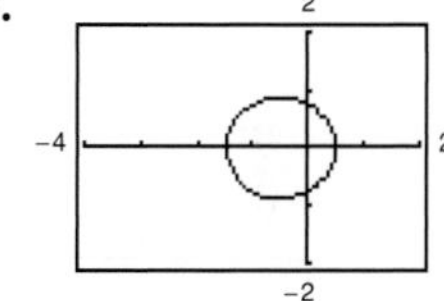

Ellipse

29.

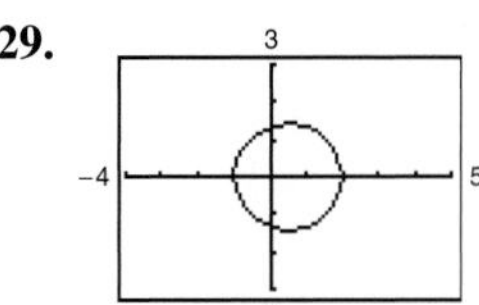

Ellipse

31.

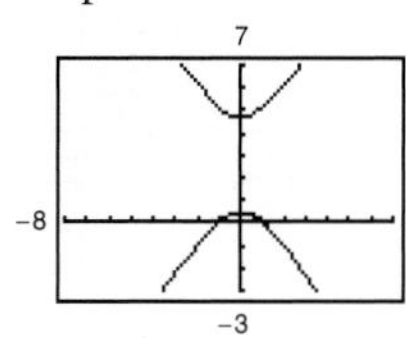

Hyperbola

33.

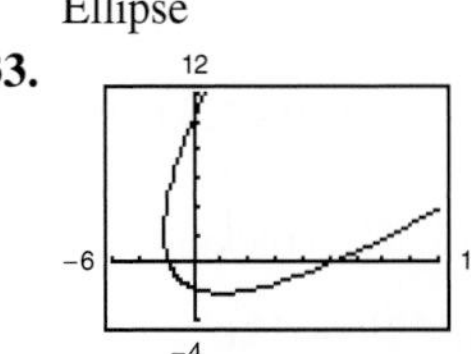

35.

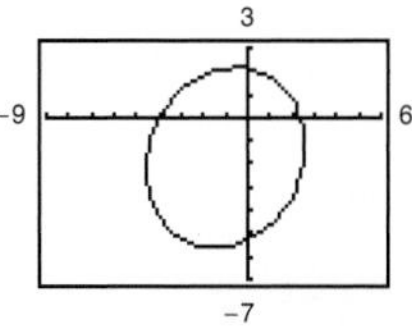

37. $r = \dfrac{1}{1 - \cos\theta}$ **39.** $r = \dfrac{3}{2 + \cos\theta}$

41. $r = \dfrac{2}{1 + 2\cos\theta}$ **43.** $r = \dfrac{4}{1 + \cos\theta}$

45. $r = \dfrac{10}{1 - \cos\theta}$ **47.** $r = \dfrac{10}{3 + 2\cos\theta}$

49. $r = \dfrac{20}{3 - 2\cos\theta}$ **51.** $r = \dfrac{9}{4 - 5\sin\theta}$

53. Answers will vary.

55. $r = \dfrac{9.2931 \times 10^7}{1 - 0.0167 \cos \theta}$

Perihelion: 9.1405×10^7 mi

Aphelion: 9.4509×10^7 mi

57. $r = \dfrac{1.0821 \times 10^8}{1 - 0.0067 \cos \theta}$

Perihelion: 1.0748×10^8 km

Aphelion: 1.0894×10^8 km

59. $r = \dfrac{1.4038 \times 10^8}{1 - 0.0935 \cos \theta}$

Perihelion: 1.2838×10^8 mi

Aphelion: 1.5486×10^8 mi

61. $\dfrac{3}{2 + \sin \theta} = \dfrac{3/2}{1 + (1/2) \sin \theta}$, so $e = \dfrac{1}{2}$ and the conic is an ellipse.

63. True. The graphs represent the same hyperbola.

65. True. The conic is an ellipse because the eccentricity is less than 1.

67. Answers will vary. **69.** $r^2 = \dfrac{24{,}336}{169 - 25 \cos^2 \theta}$

71. $r^2 = \dfrac{144}{25 \cos^2 \theta - 9}$ **73.** $r^2 = \dfrac{144}{25 \cos^2 \theta - 16}$

75. The original equation graphs as a parabola that opens upward.
(a) The parabola opens to the right.
(b) The parabola opens downward.
(c) The parabola opens to the left.
(d) The parabola has been rotated.

77. (a) Ellipse
(b) The given polar equation, r, has a vertical directrix to the left of the pole. The equation r_1 has a vertical directrix to the right of the pole, and the equation r_2 has a horizontal directrix below the pole.
(c)

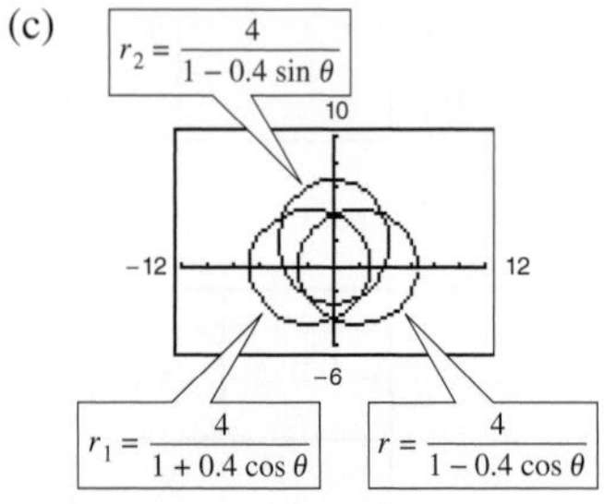

Review Exercises (page 494)

1. $\dfrac{\pi}{4}$ rad, 45° **3.** 1.9513 rad, 111.8° **5.** 0.4424 rad, 25.3°

7. 0.6588 rad, 37.7° **9.** $\dfrac{4\sqrt{5}}{5}$ **11.** Hyperbola

13. $x^2 = 12y$ **15.** $(y - 2)^2 = 12x$

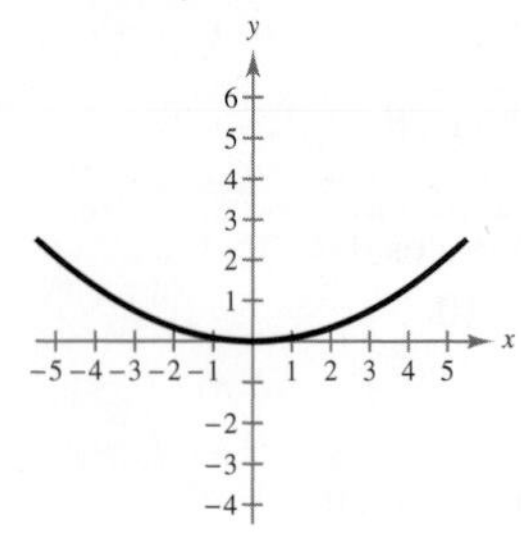

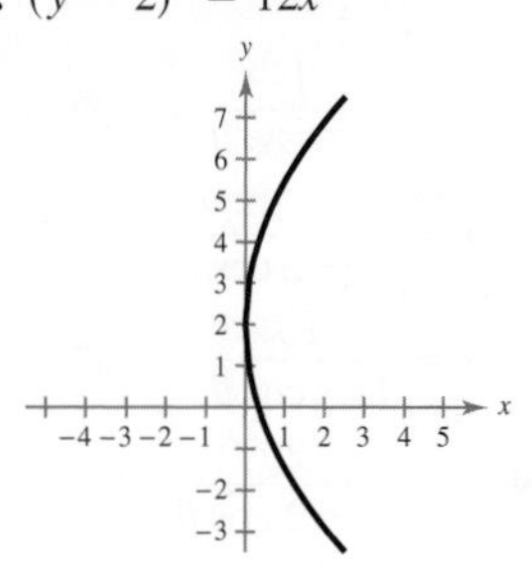

17. $y = -4x - 2$ **19.** $6\sqrt{5}$ m

21. $\dfrac{(x-2)^2}{9} + \dfrac{(y-8)^2}{64} = 1$ **23.** $\dfrac{(x-2)^2}{4} + \dfrac{(y-1)^2}{1} = 1$

25. 3 feet from the center of the arch at ground level.

27. Center: $(-2, 5)$
Vertices: $(-10, 5), (6, 5)$
Foci: $\left(-2 \pm 2\sqrt{7}, 5\right)$
Eccentricity: $e = \dfrac{\sqrt{7}}{4}$

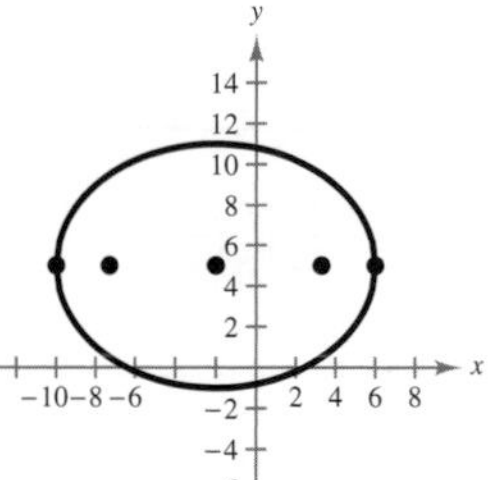

29. Center: $(1, -4)$
Vertices: $(1, 0), (1, -8)$
Foci: $\left(1, -4 \pm \sqrt{7}\right)$
Eccentricity: $\dfrac{\sqrt{7}}{4}$

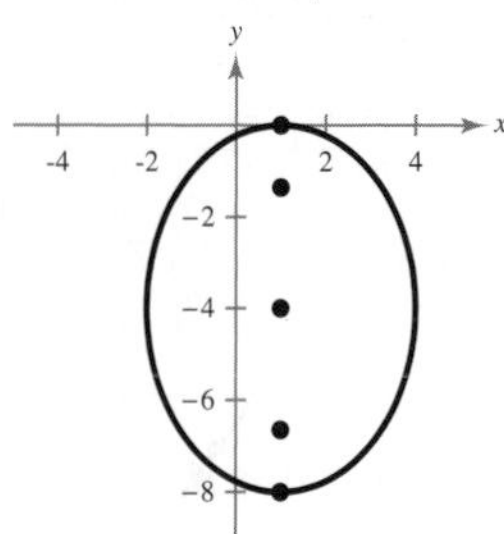

31. $\dfrac{y^2}{36} - \dfrac{x^2}{28} = 1$ **33.** $\dfrac{x^2}{16} - \dfrac{y^2}{9} = 1$

35. Center: $(4, -2)$
Vertices:
$(-3, -2), (11, -2)$
Foci: $\left(4 \pm \sqrt{74}, -2\right)$
Asymptotes:
$y = -2 \pm \frac{5}{7}(x - 4)$

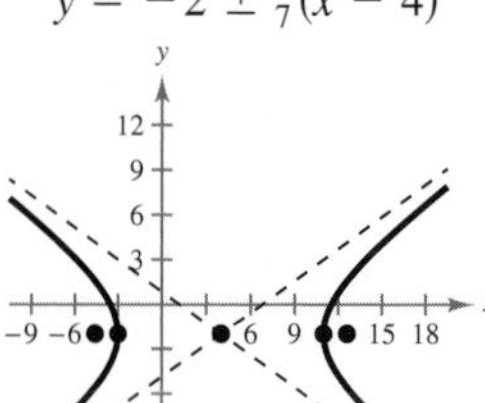

37. Center: $(1, -1)$
Vertices:
$(5, -1), (-3, -1)$
Foci: $(6, -1), (-4, -1)$
Asymptotes:
$y = -1 \pm \frac{3}{4}(x - 1)$

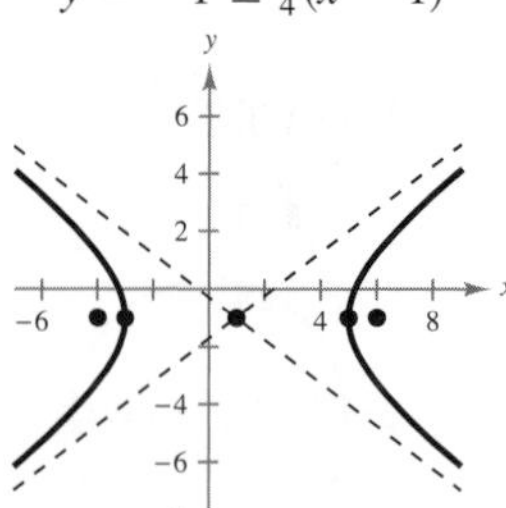

39. $\dfrac{x^2}{10{,}890{,}000} - \dfrac{y^2}{16{,}988{,}400} = 1$

41. Hyperbola **43.** Ellipse

45. $\dfrac{(y')^2}{10} - \dfrac{(x')^2}{10} = 1$ **47.** $\dfrac{(x')^2}{3} + \dfrac{(y')^2}{2} = 1$

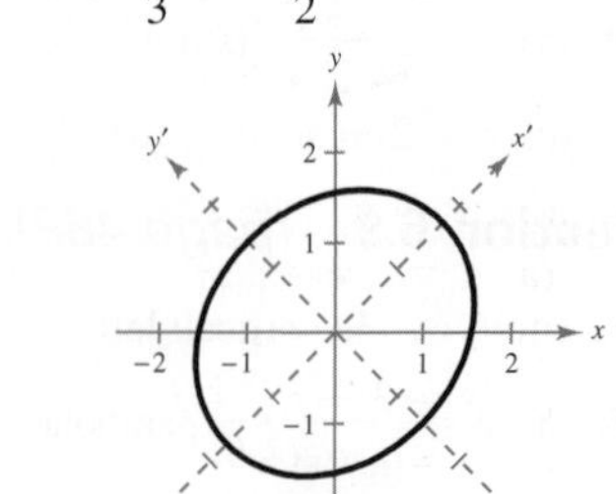

49. (a) Parabola
(b) $y = \dfrac{24x + 40 \pm \sqrt{(24x + 40)^2 - 36(16x^2 - 30x)}}{18}$

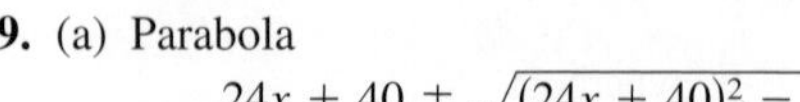

(c)

51. (a) Hyperbola (b) $y = \dfrac{10x \pm \sqrt{100x^2 - 4(x^2 + 1)}}{2}$

(c)

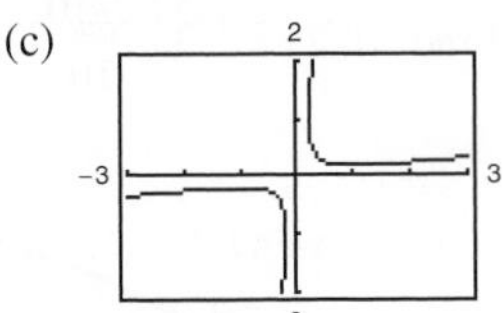

53. (a)

t	-2	-1	0	1	2
x	-8	-5	-2	1	4
y	15	11	7	3	-1

(b)

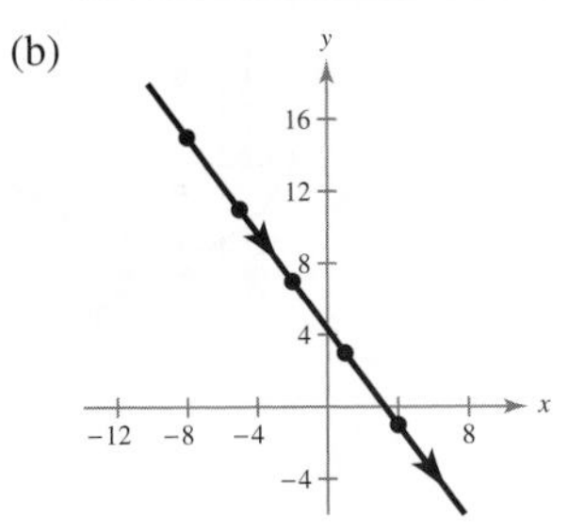

55. (a)

(b) $y = 2x$

57. (a)

(b) $y = \sqrt[4]{x}$

59. (a)

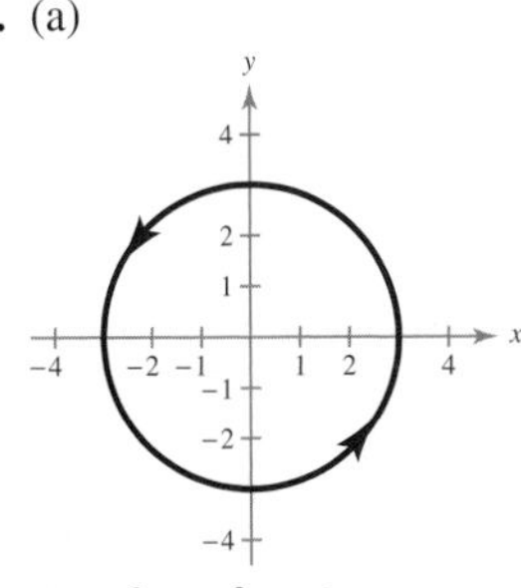

(b) $x^2 + y^2 = 9$

61. (a) $x = t,\ y = 2t + 3$
(b) $x = t - 1,\ y = 2t + 1$
(c) $x = 3 - t,\ y = 9 - 2t$

63. (a) $x = t,\ y = t^2 + 3$
(b) $x = t - 1,\ y = t^2 - 2t + 4$
(c) $x = 3 - t,\ y = t^2 - 6t + 12$

65. (a) $x = t,\ y = 1 - 4t^2$
(b) $x = t - 1,\ y = -4t^2 + 8t - 3$
(c) $x = 3 - t,\ y = -4t^2 + 24t - 35$

67.

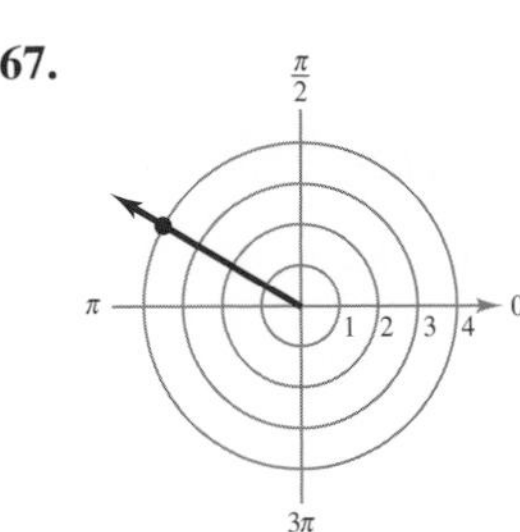

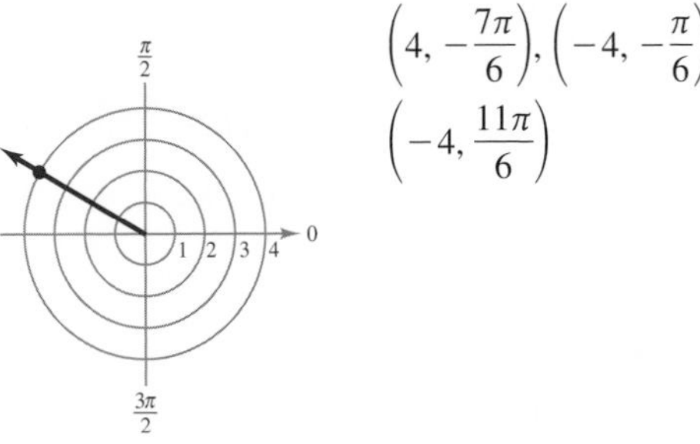

$\left(4, -\frac{7\pi}{6}\right), \left(-4, -\frac{\pi}{6}\right), \left(-4, \frac{11\pi}{6}\right)$

69.

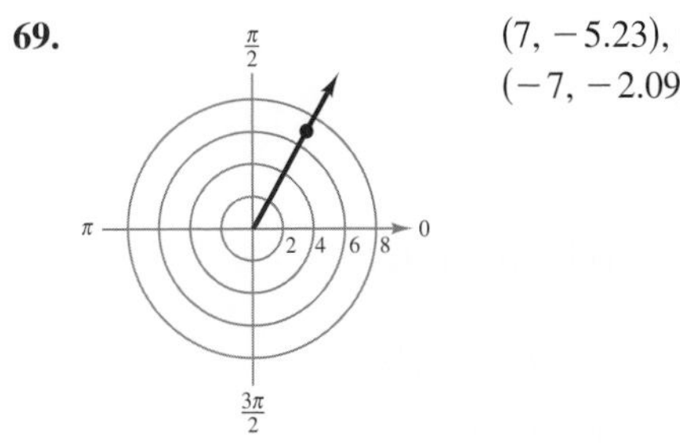

$(7, -5.23), (7, 1.05), (-7, -2.09)$

71. $(0, 0)$ **73.** $\left(-\frac{1}{2}, -\frac{\sqrt{3}}{2}\right)$ **75.** $\left(3\sqrt{2}, \frac{\pi}{4}\right)$

77. $\left(\sqrt{10}, \frac{3\pi}{4}\right)$ **79.** $r = 9$ **81.** $r = 5 \sec \theta$

83. $r^2 = 10 \csc 2\theta$ **85.** $x^2 + y^2 = 16$ **87.** $x^2 + y^2 = 3x$

89. $x^2 + y^2 = y^{2/3}$

91. Symmetry: $\theta = \frac{\pi}{2}$, polar axis, pole
Maximum value of $|r|$:
$|r| = 6$ for all values of θ
No zeros of r

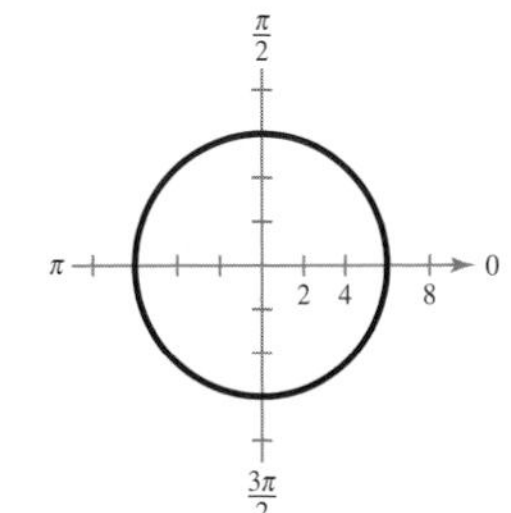

93. Symmetry: polar axis
Maximum value of $|r|$:
$|r| = 4$ when $\theta = 0$
Zeros of r: $r = 0$ when $\theta = \pi$

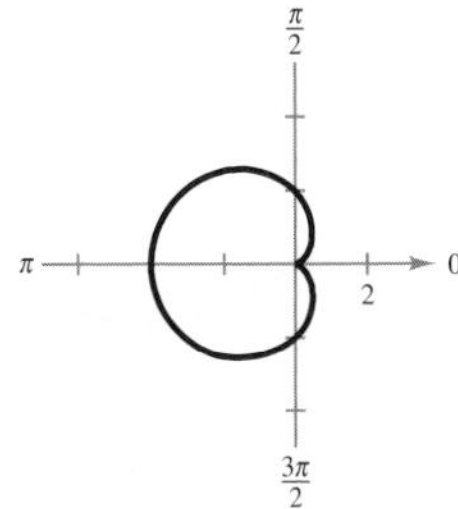

95. Symmetry: $\theta = \frac{\pi}{2}$, polar axis, pole
Maximum value of $|r|$:
$|r| = 4$ when $\theta = \frac{\pi}{4}, \frac{3\pi}{4}, \frac{5\pi}{4}, \frac{7\pi}{4}$
Zeros of r:

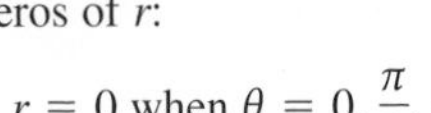

$r = 0$ when $\theta = 0, \frac{\pi}{2}, \pi, \frac{3\pi}{2}$

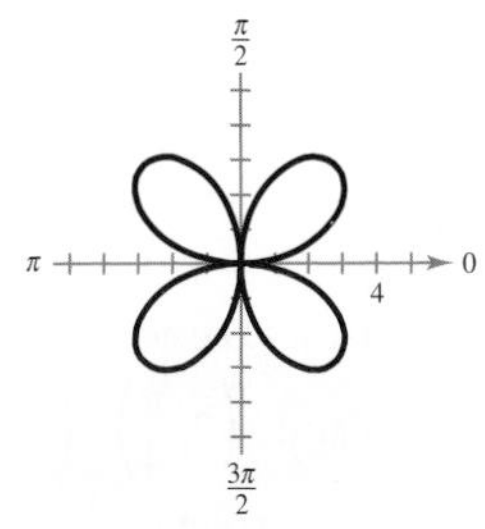

CHAPTER 6

97. Symmetry: $\theta = \dfrac{\pi}{2}$

Maximum value of $|r|$: $|r| = 8$ when $\theta = \dfrac{\pi}{2}$

Zeros of r: $r = 0$ when $\theta = 3.4814, 5.9433$

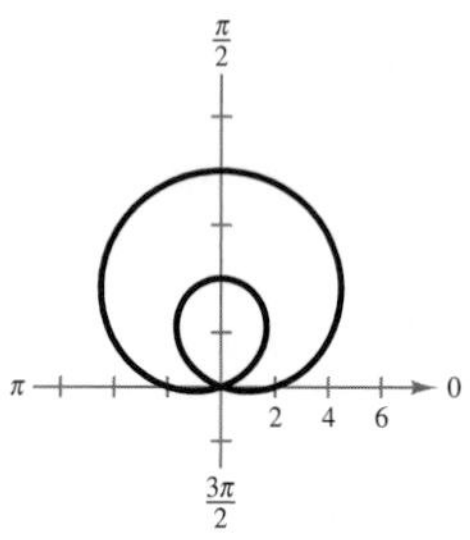

99. Symmetry: $\theta = \dfrac{\pi}{2}$, polar axis, pole

Maximum value of $|r|$: $|r| = 3$ when $\theta = \dfrac{\pi}{2}$

Zeros of r: $r = 0$ when $\theta = 0, \pi$

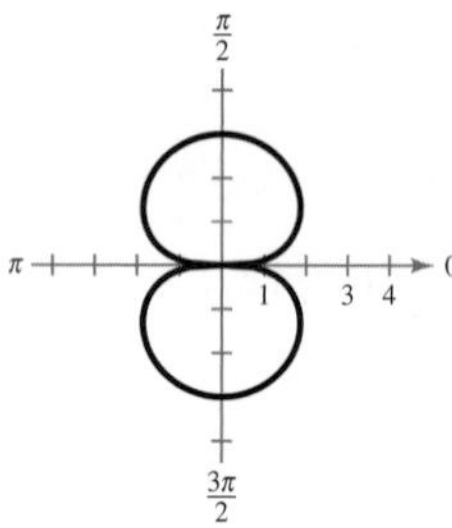

101. Limaçon

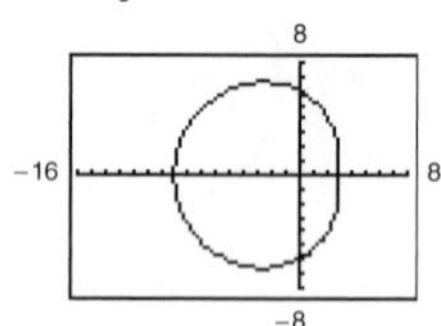

103. Rose curve

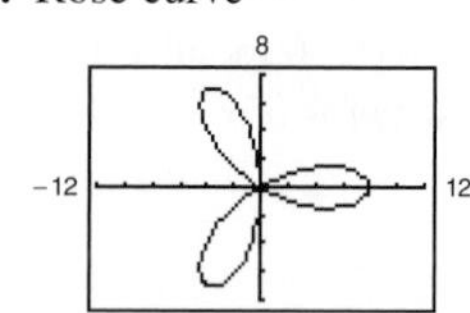

105. Hyperbola

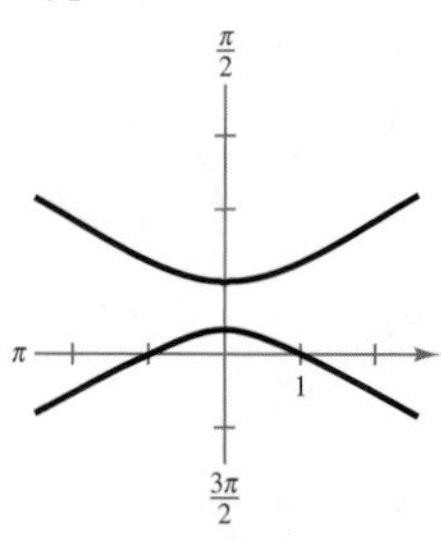

107. Ellipse

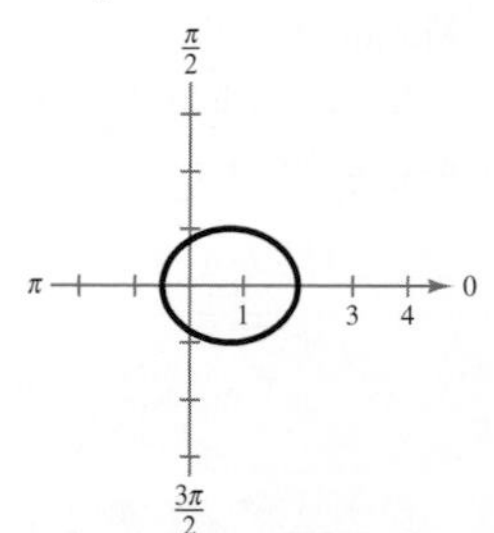

109. $r = \dfrac{4}{1 - \cos\theta}$ **111.** $r = \dfrac{5}{3 - 2\cos\theta}$

113. $r = \dfrac{7961.93}{1 - 0.937\cos\theta}$; 10,980.11 mi

115. False. The equation of a hyperbola is a second-degree equation.

117. False. $\left(2, \dfrac{\pi}{4}\right)$, $\left(-2, \dfrac{5\pi}{4}\right)$, and $\left(2, \dfrac{9\pi}{4}\right)$ all represent the same point.

119. (a) The graphs are the same.
(b) The graphs are the same.

Chapter Test *(page 497)*

1. 0.5191 rad, 29.7° **2.** 0.7023 rad, 40.2° **3.** $\dfrac{\sqrt{10}}{10}$

4. Parabola: $y^2 = 2(x - 1)$
Vertex: $(1, 0)$
Focus: $\left(\frac{3}{2}, 0\right)$

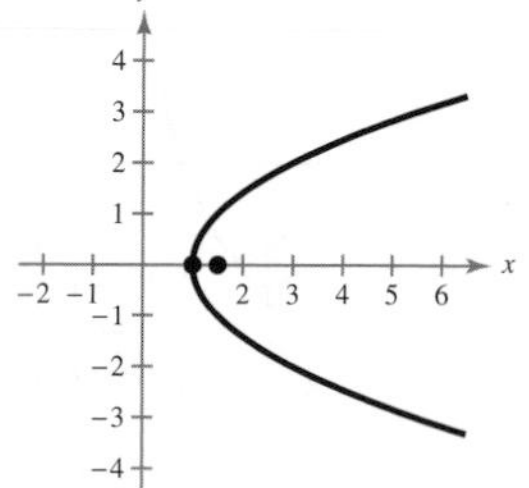

5. Hyperbola: $\dfrac{(x - 2)^2}{4} - y^2 = 1$
Center: $(2, 0)$
Vertices: $(0, 0), (4, 0)$
Foci: $\left(2 \pm \sqrt{5}, 0\right)$
Asymptotes: $y = \pm\dfrac{1}{2}(x - 2)$

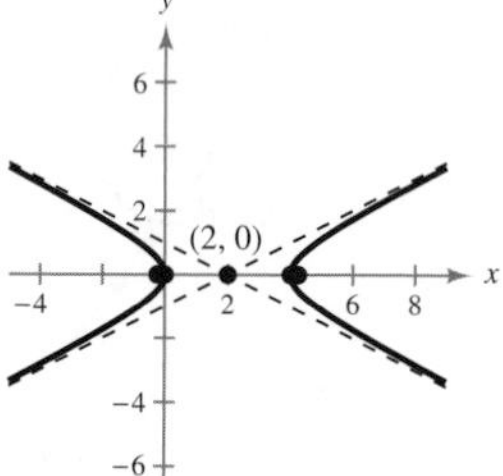

6. Ellipse:
$\dfrac{(x + 3)^2}{16} + \dfrac{(y - 1)^2}{9} = 1$
Center: $(-3, 1)$
Vertices: $(1, 1), (-7, 1)$
Foci: $\left(-3 \pm \sqrt{7}, 1\right)$

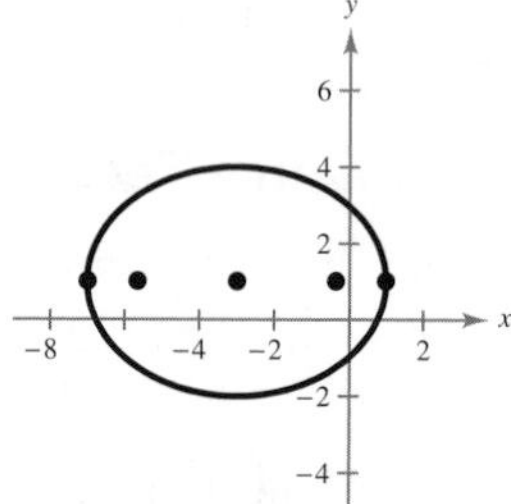

7. Circle:
$(x - 2)^2 + (y - 1)^2 = \frac{1}{2}$
Center: $(2, 1)$

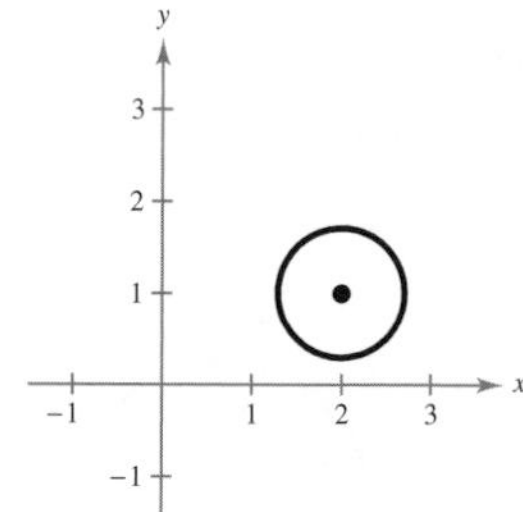

8. $(y + 4)^2 = 12(x - 3)$ **9.** $\dfrac{y^2}{2/41} - \dfrac{x^2}{162/41} = 1$

10. (a) $\dfrac{(y')^2}{2} - \dfrac{(x')^2}{2} = 1$ **11.**
(b)

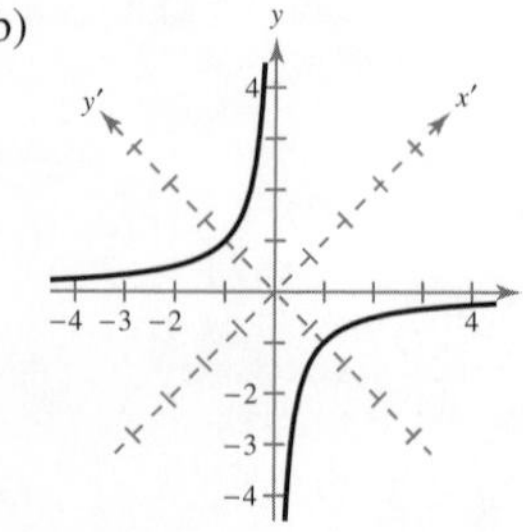

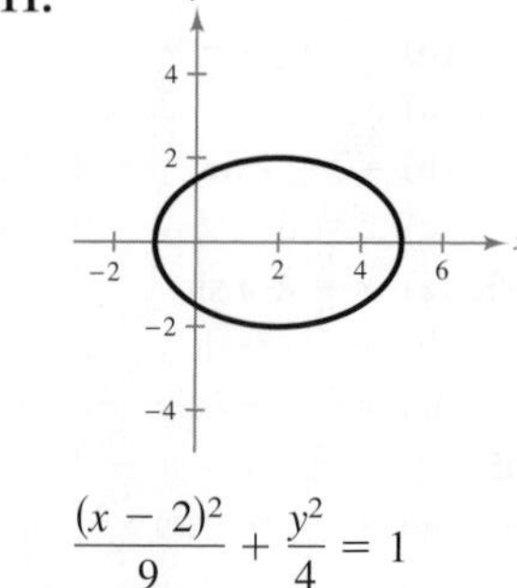

$\dfrac{(x - 2)^2}{9} + \dfrac{y^2}{4} = 1$

12. (a) $x = t$, $y = 3 - t^2$ (b) $x = t - 2$, $y = -t^2 + 4t - 1$

13. $\left(\sqrt{3}, -1\right)$

14. $\left(2\sqrt{2}, \dfrac{7\pi}{4}\right)$, $\left(-2\sqrt{2}, \dfrac{3\pi}{4}\right)$, $\left(2\sqrt{2}, -\dfrac{\pi}{4}\right)$, $\left(-2\sqrt{2}, -\dfrac{5\pi}{4}\right)$

15. $r = 8$

16. Parabola

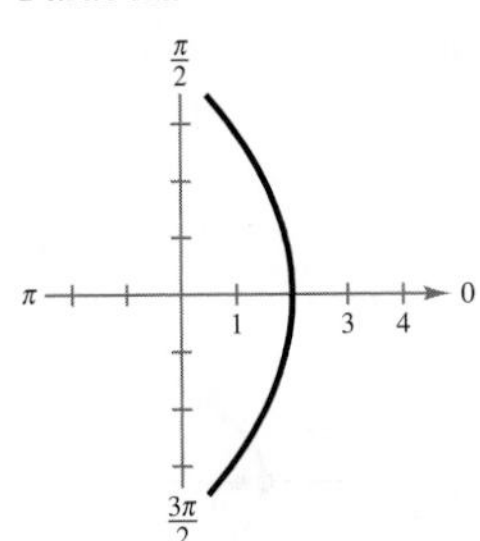

17. Ellipse

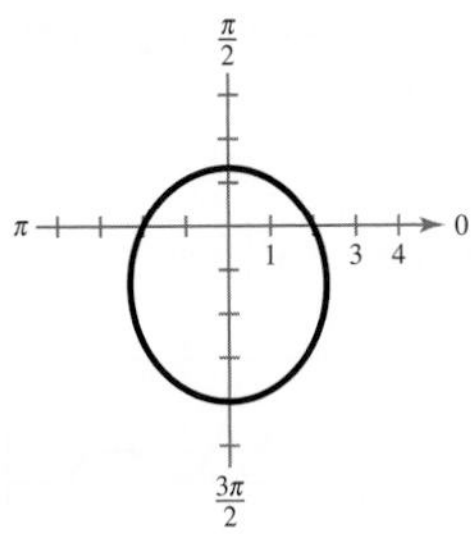

18. Limaçon with inner loop

19. Rose curve

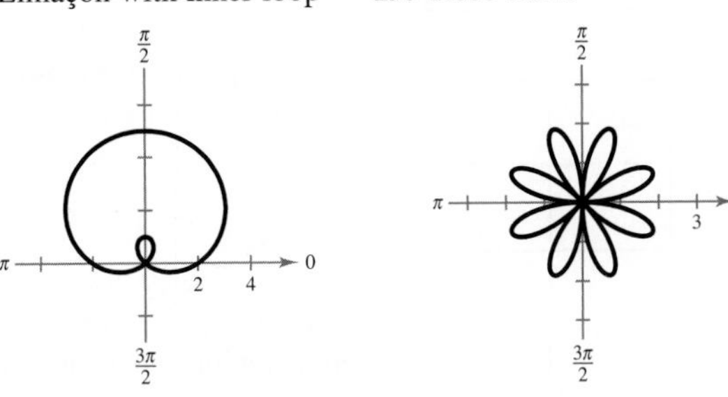

20. Answers will vary. For example: $r = \dfrac{1}{1 + 0.25 \sin \theta}$

21. Slope: 0.1511; Change in elevation: 789 ft

22. No; Yes

Cumulative Test for Chapters 4–6 *(page 498)*

1. $6 - 7i$ **2.** $-2 - 3i$ **3.** $-21 - 20i$ **4.** 4

5. $\frac{2}{13} + \frac{10}{13}i$ **6.** $-2, \pm 2i$ **7.** $-7, 0, 3$

8. $f(x) = x^4 + x^3 - 33x^2 + 45x + 378$

9. $2\sqrt{2}\left(\cos \dfrac{3\pi}{4} + i \sin \dfrac{3\pi}{4}\right)$ **10.** $24(\cos 150° + i \sin 150°)$

11.

Imaginary axis, Real axis; points (3, 2), (3, −2)

$3 + 2i$

12. $\sqrt{37}$ **13.** $-8 + 8\sqrt{3}i$ **14.** -64

15. $\cos 0 + i \sin 0$

$\cos \dfrac{2\pi}{3} + i \sin \dfrac{2\pi}{3}$

$\cos \dfrac{4\pi}{3} + i \sin \dfrac{4\pi}{3}$

16. $3\left(\cos \dfrac{\pi}{8} + i \sin \dfrac{\pi}{8}\right)$

$3\left(\cos \dfrac{5\pi}{8} + i \sin \dfrac{5\pi}{8}\right)$

$3\left(\cos \dfrac{9\pi}{8} + i \sin \dfrac{9\pi}{8}\right)$

$3\left(\cos \dfrac{13\pi}{8} + i \sin \dfrac{13\pi}{8}\right)$

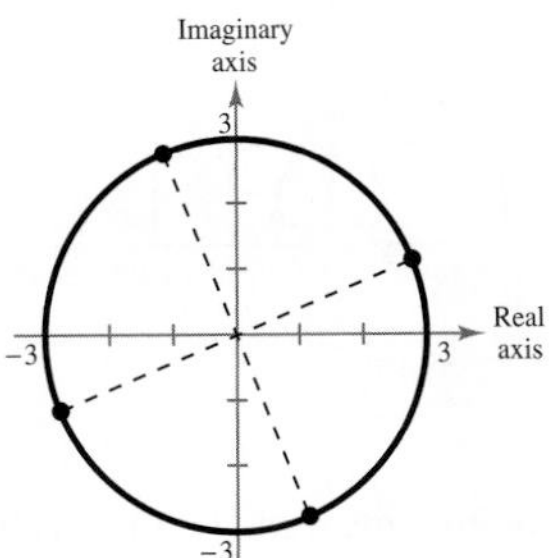

17. Reflect f in the x-axis and y-axis, and shift three units to the right.

18. Reflect f in the x-axis, and shift four units up.

19. 1.991 **20.** -0.067 **21.** 1.717 **22.** 0.390

23. 0.906 **24.** -1.733 **25.** -4.087

26. $\ln(x + 5) + \ln(x - 5) - 4 \ln x$ **27.** $\ln \dfrac{x^2}{\sqrt{x + 5}}, \; x > 0$

28. $\dfrac{\ln 12}{2} \approx 1.242$ **29.** $\dfrac{\ln 9}{\ln 4} + 5 \approx 6.585$ **30.** $\dfrac{64}{5} = 12.8$

31. $\frac{1}{2}e^8 \approx 1490.479$

32.

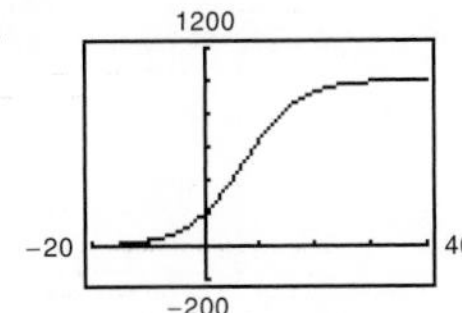

Horizontal asymptotes: $y = 0$, $y = 1000$

33. 6.3 h **34.** 2023 **35.** About 1.43 rad; about 81.87°

36. $\dfrac{11\sqrt{5}}{5}$

37. Ellipse;

$\dfrac{(x - 2)^2}{4} + \dfrac{(y + 1)^2}{9} = 1$

Center: $(2, -1)$

Vertices: $(2, 2), (2, -4)$

Foci: $\left(2, -1 \pm \sqrt{5}\right)$

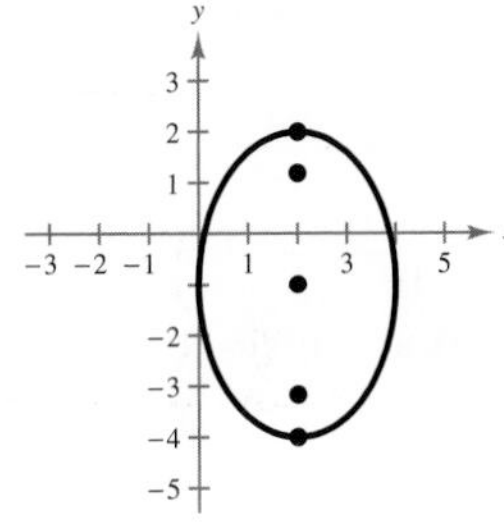

38. Hyperbola; $x^2 - \dfrac{y^2}{4} = 1$

Center: $(0, 0)$

Vertices: $(1, 0), (-1, 0)$

Foci: $\left(\sqrt{5}, 0\right), \left(-\sqrt{5}, 0\right)$

Asymptotes: $y = \pm 2x$

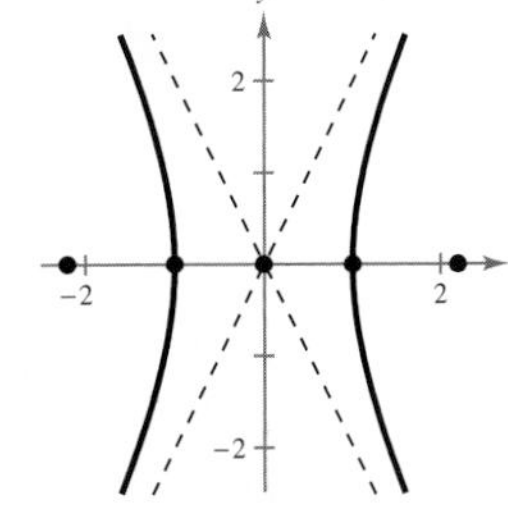

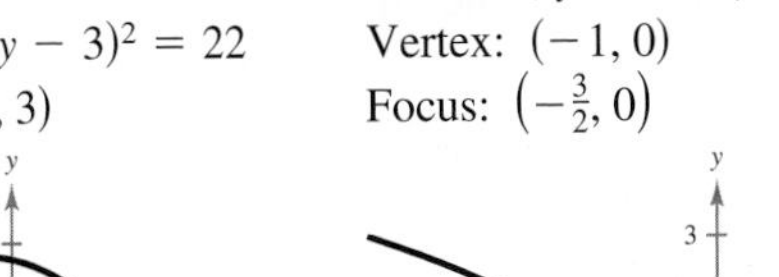

39. Circle;

$(x + 1)^2 + (y - 3)^2 = 22$

Center: $(-1, 3)$

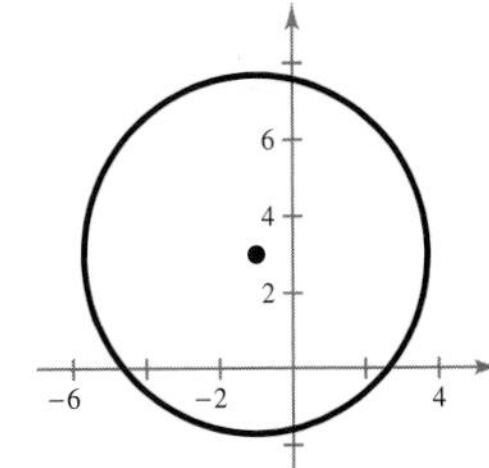

40. Parabola; $y^2 = -2(x + 1)$

Vertex: $(-1, 0)$

Focus: $\left(-\frac{3}{2}, 0\right)$

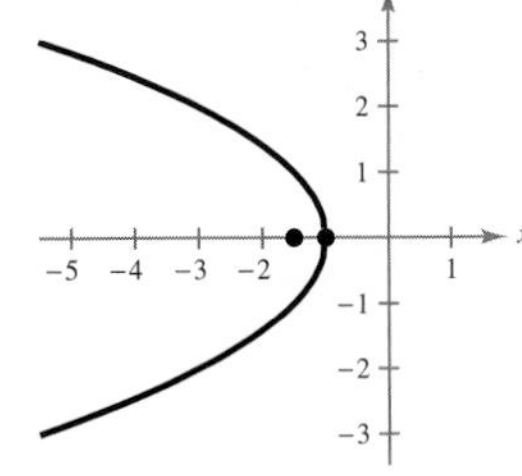

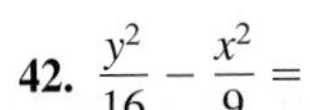

41. $x^2 + \dfrac{(y - 2)^2}{4} = 1$ **42.** $\dfrac{y^2}{16} - \dfrac{x^2}{9} = 1$

43. $\dfrac{\left(x' - \frac{\sqrt{2}}{6}\right)^2}{218/9} + \dfrac{\left(y' - \frac{5\sqrt{2}}{2}\right)^2}{218/3} = 1$

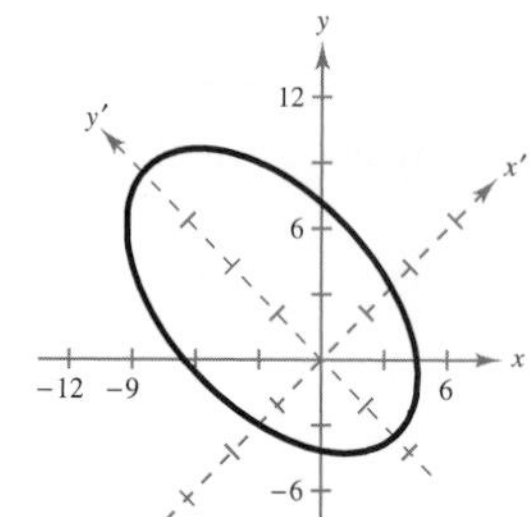

44.

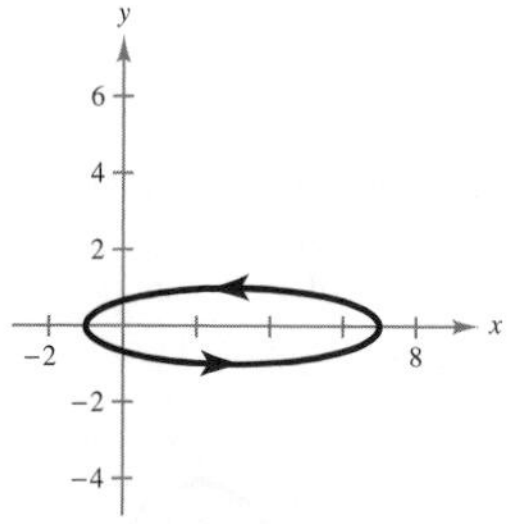

$\dfrac{(x-3)^2}{16} + y^2 = 1$

45. (a) $x = t,\ y = 1 - t$
(b) $x = 2 - t,\ y = t - 1$

46.

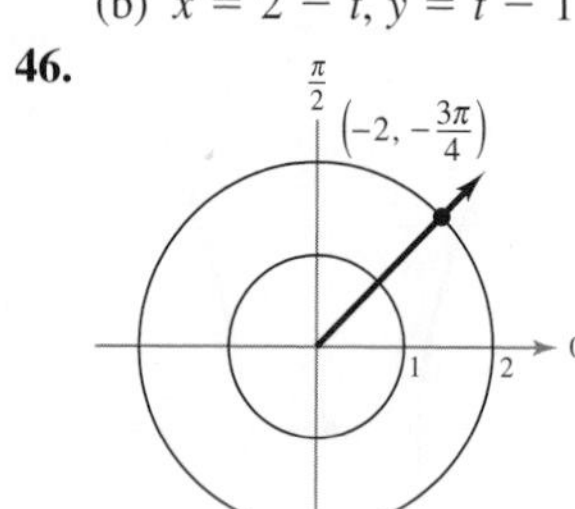

$\left(-2, \frac{5\pi}{4}\right), \left(2, -\frac{7\pi}{4}\right), \left(2, \frac{\pi}{4}\right)$

47. $r = 16 \sin \theta$ **48.** $9x^2 - 16y^2 + 20x + 4 = 0$

49. Ellipse **50.** Parabola

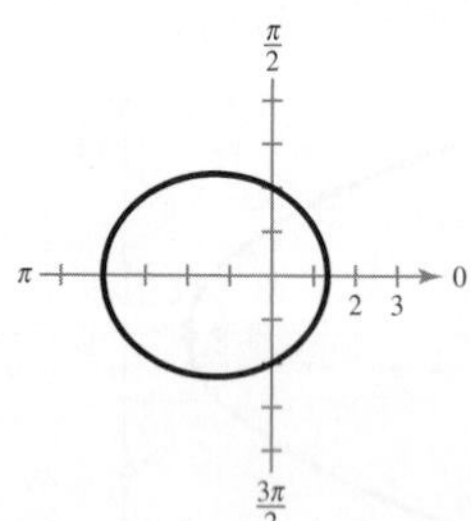

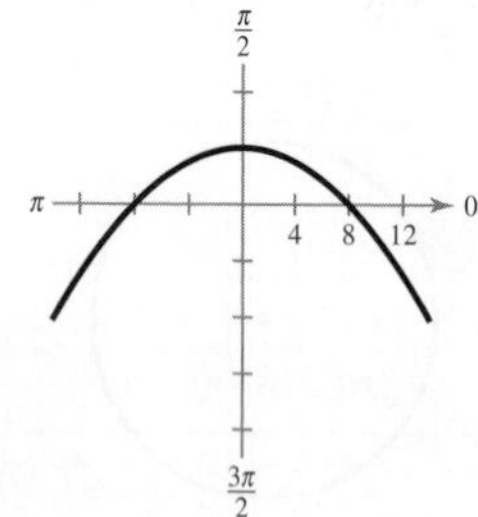

51. (a) iii (b) i (c) ii

Problem Solving *(page 503)*

1. (a) 1.2016 rad (b) 2420 ft, 5971 ft **3.** $A = \dfrac{4a^2b^2}{a^2 + b^2}$

5. (a) Because $d_1 + d_2 \leq 20$, by definition, the outer bound that the boat can travel is an ellipse. The islands are the foci.
(b) Island 1: $(-6, 0)$; Island 2: $(6, 0)$
(c) 20 mi; Vertex: $(10, 0)$
(d) $\dfrac{x^2}{100} - \dfrac{y^2}{64} = 1$

7–9. Proofs

11. Answers will vary. *Sample answer:*
$x = \cos(-t)$
$y = 2 \sin(-t)$

13. (a) $y^2 = x^2\left(\dfrac{2-x}{2+x}\right)$ (b) $x = \dfrac{2 - 2t^2}{1 + t^2},\ y = \dfrac{t(2 - 2t^2)}{1 + t^2}$
(c)

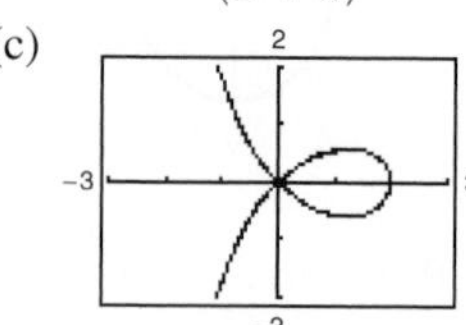

15. (a) 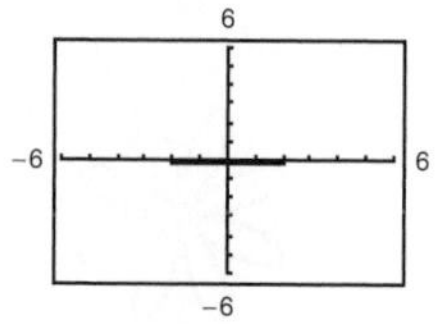 The graph is a line between -2 and 2 on the x-axis.

(b) The graph is a three-sided figure with counterclockwise orientation.

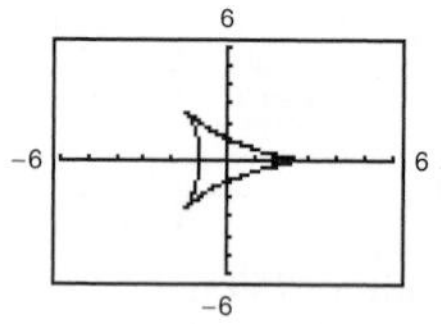

(c) The graph is a four-sided figure with counterclockwise orientation.

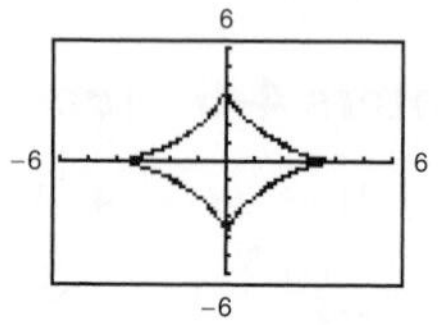

(d) The graph is a 10-sided figure with counterclockwise orientation.

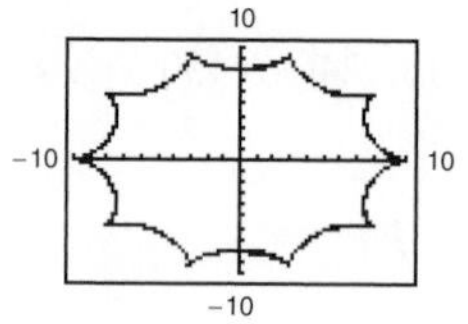

(e) The graph is a three-sided figure with clockwise orientation.

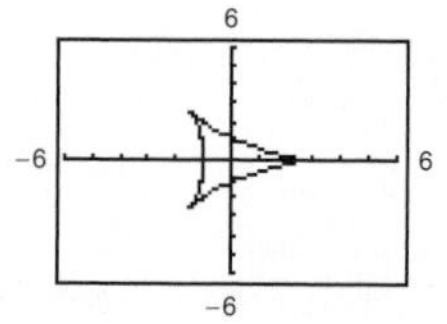

(f)

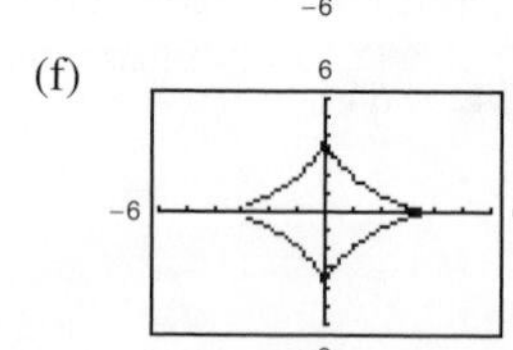

The graph is a four-sided figure with clockwise orientation.

17.

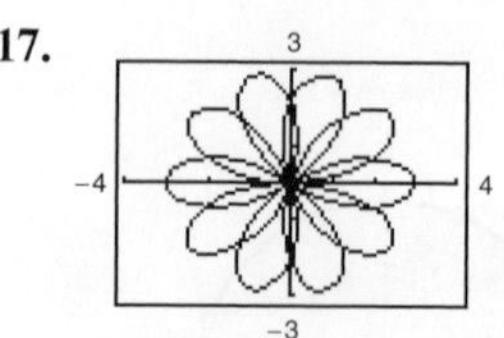

$r = 3 \sin\left(\dfrac{5\theta}{2}\right)$

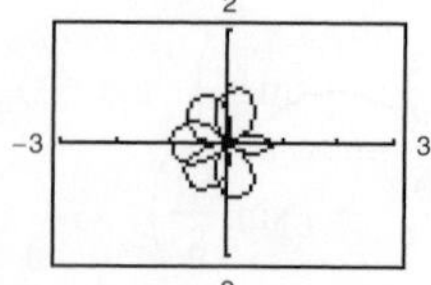

$r = -\cos\left(\sqrt{2}\theta\right),$
$-2\pi \leq \theta \leq 2\pi$

Sample answer: If n is a rational number, then the curve has a finite number of petals. If n is an irrational number, then the curve has an infinite number of petals.

Technology

Chapter P *(page 27)*

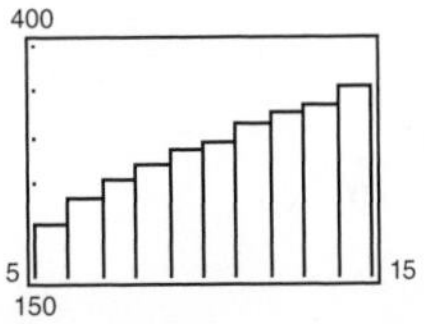

(page 45)

The lines appear perpendicular on the square setting.

(page 58)

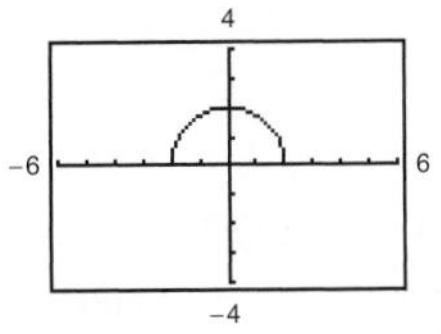

Domain: $[-2, 2]$

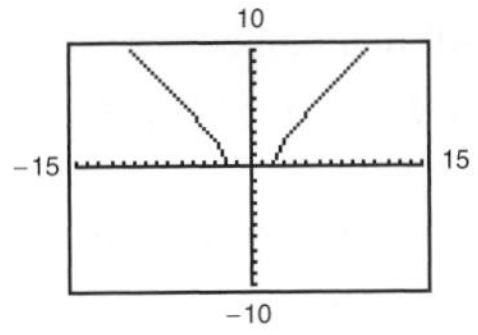

Domain: $(-\infty, -2] \cup [2, \infty)$
Yes, for -2 and 2.

(page 81)

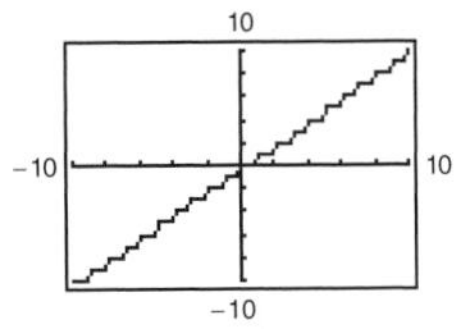

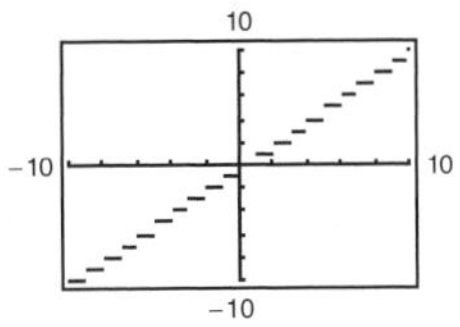

The graph in *dot* mode illustrates that the range is the set of all integers.

Chapter 1 *(page 160)*

No graph is visible. $-\pi \le x \le \pi$ and $-0.5 \le y \le 0.5$ displays a good view of the graph.

Chapter 5 *(page 397)*

$S = 0.00036(2.130)^t$

The exponential regression model has the same coefficient as the model in Example 1. However, the model given in Example 1 contains the natural exponential function.

Chapter 6 *(page 466)*

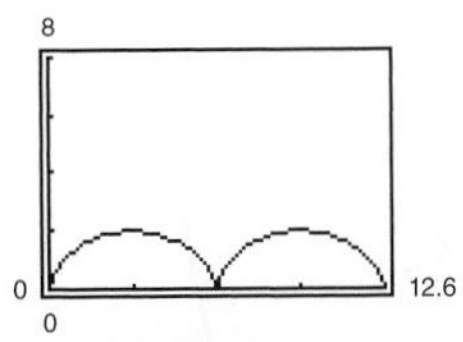

Checkpoints

Chapter P

Section P.1

1. (a) Natural numbers: $\left\{\frac{6}{3}, 8\right\}$ (b) Whole numbers: $\left\{\frac{6}{3}, 8\right\}$
(c) Integers: $\left\{\frac{6}{3}, -1, 8, -22\right\}$
(d) Rational numbers: $\left\{-\frac{1}{4}, \frac{6}{3}, -7.5, -1, 8, -22\right\}$
(e) Irrational numbers: $\left\{-\pi, \frac{1}{2}\sqrt{2}\right\}$

2.

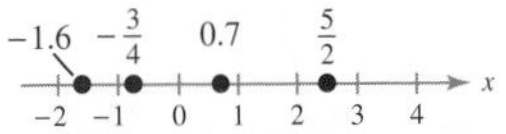

3. (a) $1 > -5$ (b) $\frac{3}{2} < 7$ (c) $-\frac{2}{3} > -\frac{3}{4}$

4. (a) The inequality $x > -3$ denotes all real numbers greater than -3.
(b) The inequality $0 < x \le 4$ denotes all real numbers between 0 and 4, not including 0, but including 4.

5. The interval consists of all real numbers greater than or equal to -2 and less than 5.

6. $-2 \le x < 4$ **7.** (a) 1 (b) $-\frac{3}{4}$ (c) $\frac{2}{3}$ (d) -0.7

8. (a) 1 (b) -1

9. (a) $|-3| < |4|$ (b) $-|-4| = -|4|$
(c) $|-3| > -|-3|$

10. (a) 58 (b) 12 (c) 12

11. Terms: $-2x, 4$; Coefficients: $-2, 4$ **12.** -5

13. (a) Commutative Property of Addition
(b) Associative Property of Multiplication
(c) Distributive Property

14. (a) $\dfrac{x}{10}$ (b) $\dfrac{x}{2}$

Section P.2

1. (a) -4 (b) 8 **2.** $-\frac{18}{5}$ **3.** No solution

4. $-1, \frac{5}{2}$ **5.** (a) $\pm 2\sqrt{3}$ (b) $1 \pm \sqrt{10}$ **6.** $2 \pm \sqrt{5}$

7. $\dfrac{5}{3} \pm \dfrac{\sqrt{31}}{3}$ **8.** $-\dfrac{1}{3} \pm \dfrac{\sqrt{31}}{3}$ **9.** $\dfrac{4}{3}$ **10.** $0, \pm\dfrac{2\sqrt{3}}{3}$

11. (a) $5, \pm\sqrt{2}$ (b) $0, -\frac{3}{2}, 6$ **12.** -9 **13.** $-59, 69$

14. $-2, 6$

Section P.3

1.

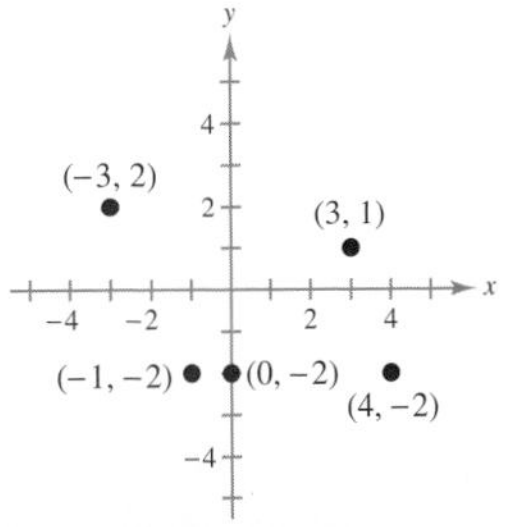

2.

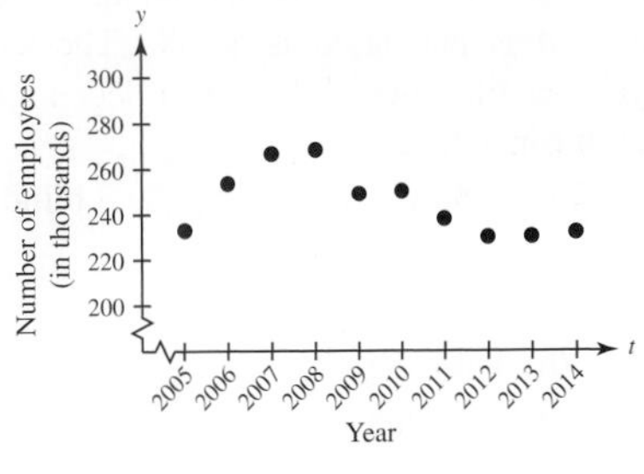

CHECKPOINTS

3. $\sqrt{37} \approx 6.08$

4. $d_1 = \sqrt{45}, d_2 = \sqrt{20}, d_3 = \sqrt{65}$
$(\sqrt{45})^2 + (\sqrt{20})^2 = (\sqrt{65})^2$

5. $(1, -1)$ **6.** $\sqrt{709} \approx 27$ yd **7.** \$4.8 billion

8. (a)

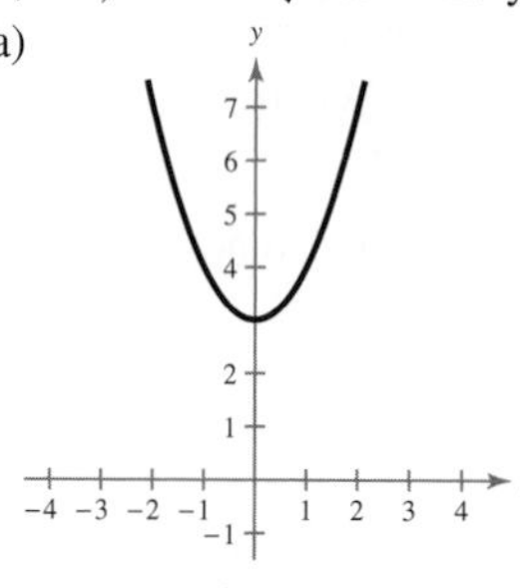

(b)

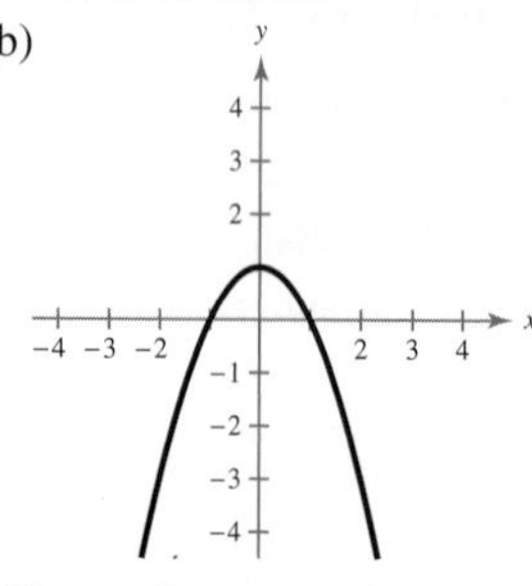

9 x-intercepts: $(0, 0), (-5, 0)$, y-intercept: $(0, 0)$ **10.** x-axis symmetry

11.

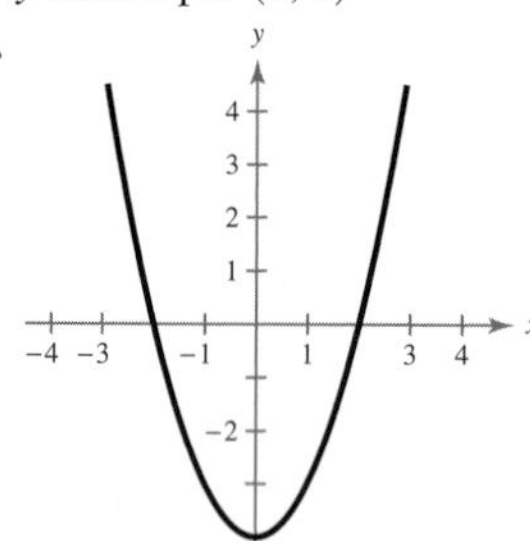

12.

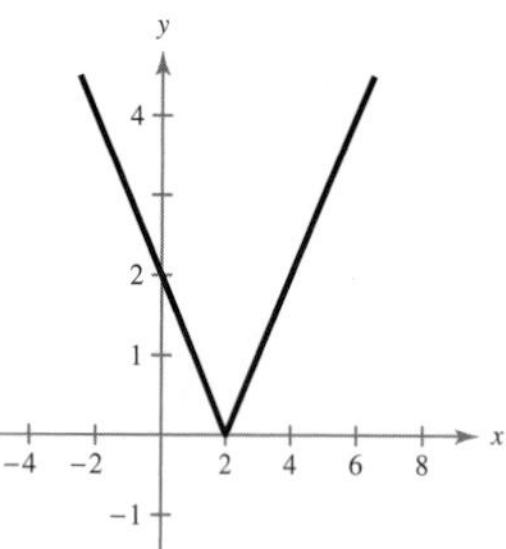

13. $(x + 3)^2 + (y + 5)^2 = 25$

Section P.4

1. (a)

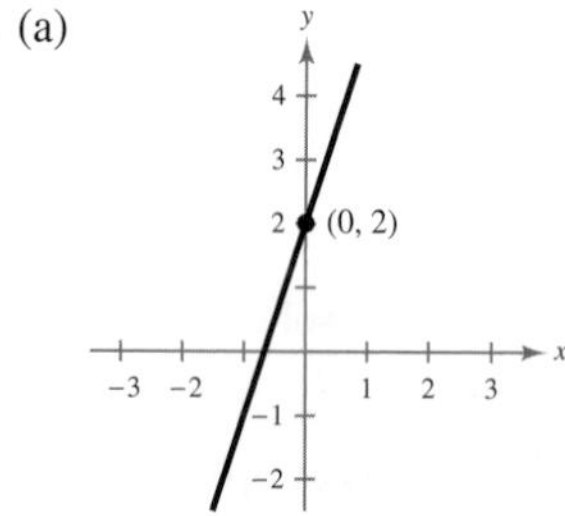

(b)

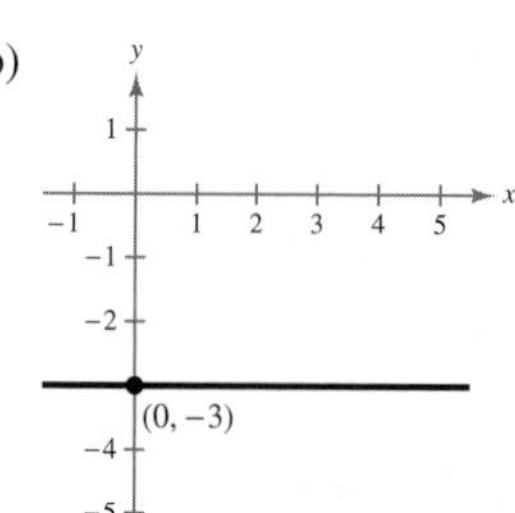

(c)

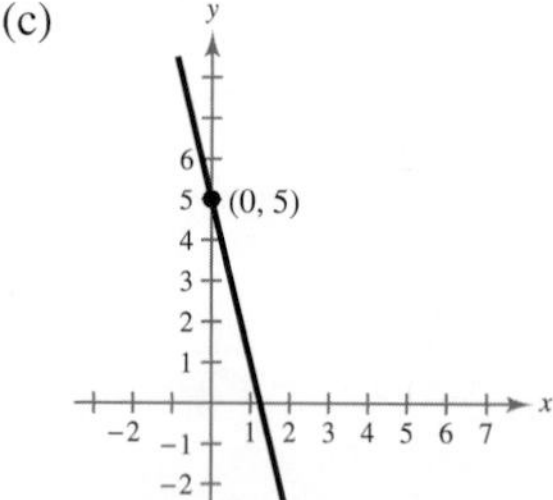

2. (a) 2 (b) $-\frac{3}{2}$ (c) Undefined (d) 0

3. (a) $y = 2x - 13$ (b) $y = -\frac{2}{3}x + \frac{5}{3}$ (c) $y = 1$

4. (a) $y = \frac{5}{3}x + \frac{23}{3}$ (b) $y = -\frac{3}{5}x - \frac{7}{5}$ **5.** Yes

6. The y-intercept, $(0, 1500)$, tells you that the initial value of a copier at the time it is purchased is \$1500. The slope, $m = -300$, tells you that the value of the copier decreases by \$300 each year after it is purchased.

7. $y = -4125x + 24{,}750$ **8.** $y = 0.7t + 4.4$; \$9.3 billion

Section P.5

1. (a) Not a function (b) Function

2. (a) Not a function (b) Function

3. (a) -2 (b) -38 (c) $-3x^2 + 6x + 7$

4. $f(-2) = 5, f(2) = 1, f(3) = 2$ **5.** ± 4 **6.** $-4, 3$

7. (a) $\{-2, -1, 0, 1, 2\}$
(b) All real numbers x except $x = 3$
(c) All real numbers r such that $r > 0$
(d) All real numbers x such that $x \geq 16$

8. (a) $S(r) = 10\pi r^2$ (b) $S(h) = \frac{5}{8}\pi h^2$ **9.** No

10. 2009: 772 2013: 1277
2010: 841 2014: 1437
2011: 910 2015: 1597
2012: 1117

11. $2x + h + 2, h \neq 0$

Section P.6

1. (a) All real numbers x except $x = -3$
(b) $f(0) = 3; f(3) = -6$ (c) $(-\infty, 3]$

2. Function

3. (a) $x = -8, x = \frac{3}{2}$ (b) $t = 25$ (c) $x = \pm\sqrt{2}$

4.

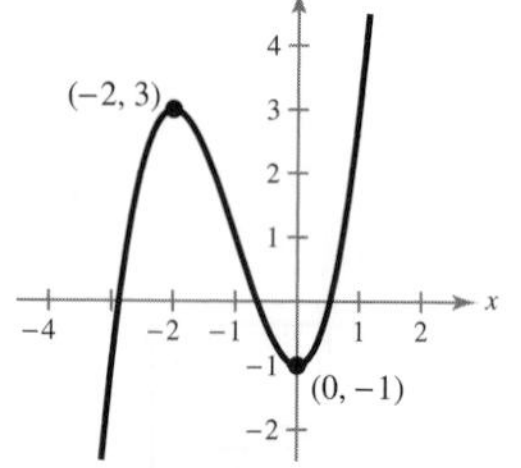

Increasing on $(-\infty, -2)$ and $(0, \infty)$
Decreasing on $(-2, 0)$

5. $(-0.88, 6.06)$ **6.** (a) -3 (b) 0

7. (a) 20 ft/sec (b) $\frac{140}{3}$ ft/sec

8. (a) Neither; No symmetry (b) Even; y-axis symmetry
(c) Odd; Origin symmetry

Section P.7

1. $f(x) = -\frac{5}{2}x + 1$ **2.** $f(-\frac{3}{2}) = 0, f(1) = 3, f(-\frac{5}{2}) = -1$

3.

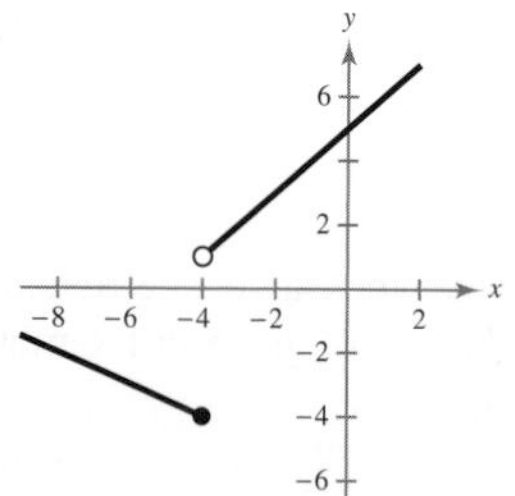

Section P.8

1. (a)

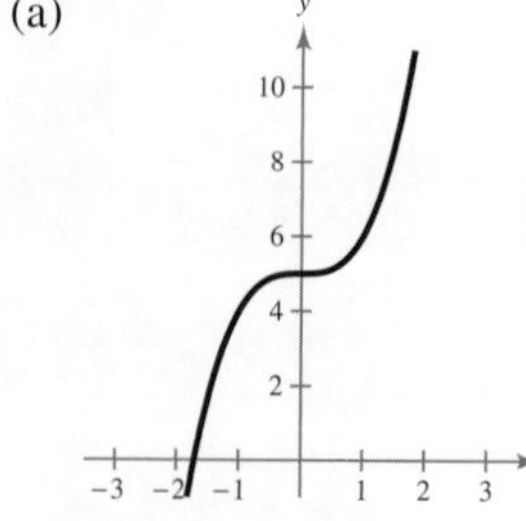

(b)

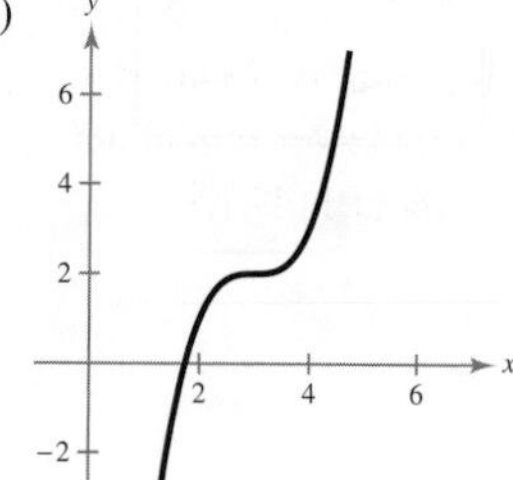

2. $j(x) = -(x + 3)^4$

3. (a) The graph of g is a reflection in the x-axis of the graph of f.
   (b) The graph of h is a reflection in the y-axis of the graph of f.
4. (a) The graph of g is a vertical stretch of the graph of f.
   (b) The graph of h is a vertical shrink of the graph of f.
5. (a) The graph of g is a horizontal shrink of the graph of f.
   (b) The graph of h is a horizontal stretch of the graph of f.

Section P.9

1. $x^2 - x + 1; 3$ 2. $x^2 + x - 1; 11$ 3. $x^2 - x^3; -18$

4. $\left(\frac{f}{g}\right)(x) = \frac{\sqrt{x-3}}{\sqrt{16-x^2}}$; Domain: $[3, 4)$

   $\left(\frac{g}{f}\right)(x) = \frac{\sqrt{16-x^2}}{\sqrt{x-3}}$; Domain: $(3, 4]$

5. (a) $8x^2 + 7$ (b) $16x^2 + 80x + 101$ (c) 9
6. All real numbers x 7. $f(x) = \frac{1}{5}\sqrt[3]{x}$, $g(x) = 8 - x$
8. (a) $(N \circ T)(t) = 32t^2 + 36t + 204$ (b) About 4.5 h

Section P.10

1. $f^{-1}(x) = 5x, f(f^{-1}(x)) = \frac{1}{5}(5x) = x, f^{-1}(f(x)) = 5\left(\frac{1}{5}x\right) = x$
2. $g(x) = 7x + 4$
3.

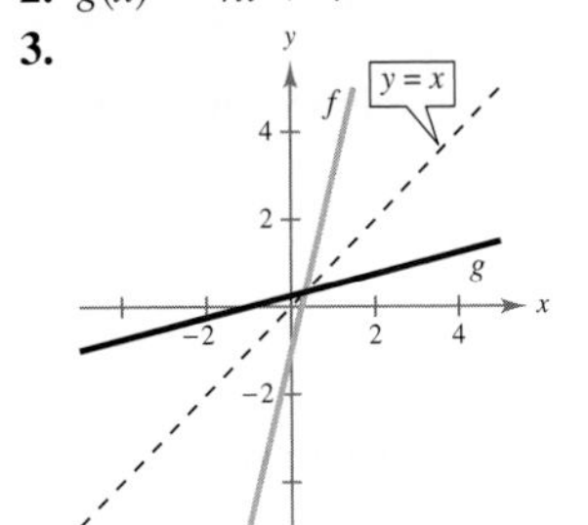

4.

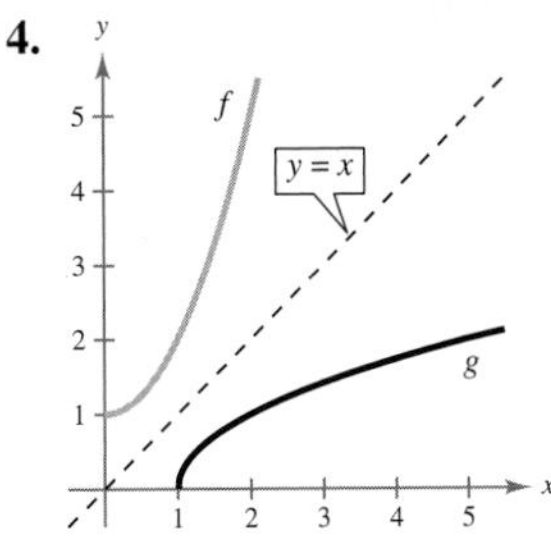

5. (a) Yes (b) No 6. $f^{-1}(x) = \frac{5-2x}{x+3}$
7. $f^{-1}(x) = x^3 - 10$

Chapter 1

Section 1.1

1. (a) $\frac{\pi}{4}, -\frac{7\pi}{4}$ (b) $\frac{5\pi}{3}, -\frac{7\pi}{3}$
2. (a) Complement: $\frac{\pi}{3}$; Supplement: $\frac{5\pi}{6}$

   (b) Complement: none; Supplement: $\frac{\pi}{6}$
3. (a) $\frac{\pi}{3}$ (b) $\frac{16\pi}{9}$ 4. (a) 30° (b) 300°
5. 24π in. $\approx$ 75.40 in. 6. About 0.84 cm/sec
7. (a) 4800π rad/min (b) About 60.319 in./min
8. About 1117 ft^2

Section 1.2

1. (a) $\sin\frac{\pi}{2} = 1$ $\csc\frac{\pi}{2} = 1$

   $\cos\frac{\pi}{2} = 0$ $\sec\frac{\pi}{2}$ is undefined.

   $\tan\frac{\pi}{2}$ is undefined. $\cot\frac{\pi}{2} = 0$

   (b) $\sin 0 = 0$ $\csc 0$ is undefined.

   $\cos 0 = 1$ $\sec 0 = 1$

   $\tan 0 = 0$ $\cot 0$ is undefined.

   (c) $\sin\left(-\frac{5\pi}{6}\right) = -\frac{1}{2}$ $\csc\left(-\frac{5\pi}{6}\right) = -2$

   $\cos\left(-\frac{5\pi}{6}\right) = -\frac{\sqrt{3}}{2}$ $\sec\left(-\frac{5\pi}{6}\right) = -\frac{2\sqrt{3}}{3}$

   $\tan\left(-\frac{5\pi}{6}\right) = \frac{\sqrt{3}}{3}$ $\cot\left(-\frac{5\pi}{6}\right) = \sqrt{3}$

   (d) $\sin\left(-\frac{3\pi}{4}\right) = -\frac{\sqrt{2}}{2}$ $\csc\left(-\frac{3\pi}{4}\right) = -\sqrt{2}$

   $\cos\left(-\frac{3\pi}{4}\right) = -\frac{\sqrt{2}}{2}$ $\sec\left(-\frac{3\pi}{4}\right) = -\sqrt{2}$

   $\tan\left(-\frac{3\pi}{4}\right) = 1$ $\cot\left(-\frac{3\pi}{4}\right) = 1$
2. (a) 0 (b) $-\frac{\sqrt{3}}{2}$ (c) 0.3
3. (a) 0.78183148 (b) 1.0997502

Section 1.3

1. $\sin\theta = \frac{1}{2}$ $\csc\theta = 2$

   $\cos\theta = \frac{\sqrt{3}}{2}$ $\sec\theta = \frac{2\sqrt{3}}{3}$

   $\tan\theta = \frac{\sqrt{3}}{3}$ $\cot\theta = \sqrt{3}$
2. $\cot 45° = 1$, $\sec 45° = \sqrt{2}$, $\csc 45° = \sqrt{2}$
3. $\tan 60° = \sqrt{3}$, $\tan 30° = \frac{\sqrt{3}}{3}$ 4. 1.7650691
5. (a) 0.28 (b) 0.2917 6. (a) $\frac{1}{2}$ (b) $\sqrt{5}$
7. Answers will vary. 8. About 40 ft
9. 60° 10. About 17.6 ft; about 17.2 ft

Section 1.4

1. $\sin\theta = \frac{3\sqrt{13}}{13}$, $\cos\theta = -\frac{2\sqrt{13}}{13}$, $\tan\theta = -\frac{3}{2}$
2. $\cos\theta = -\frac{3}{5}$, $\tan\theta = -\frac{4}{3}$
3. $-1; 0$ 4. (a) 33° (b) $\frac{4\pi}{9}$ (c) $\frac{\pi}{5}$
5. (a) $-\frac{\sqrt{2}}{2}$ (b) $-\frac{1}{2}$ (c) $-\frac{\sqrt{3}}{3}$ 6. (a) $-\frac{3}{5}$ (b) $\frac{4}{3}$
7. (a) −1.8040478 (b) −1.0428352 (c) 0.8090170

Section 1.5

1.
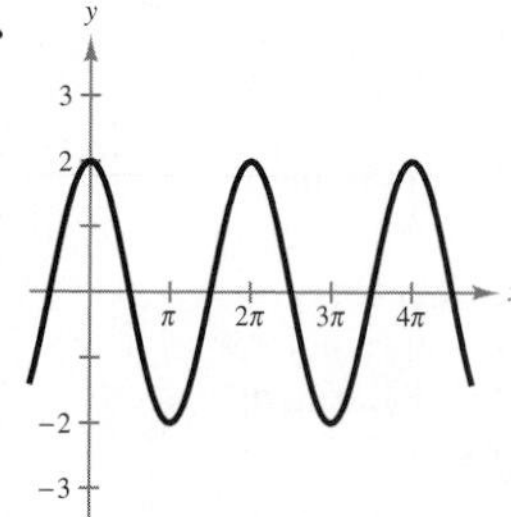

2.
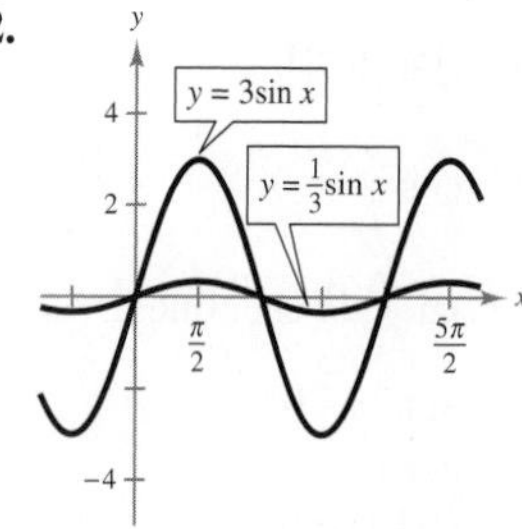

3.
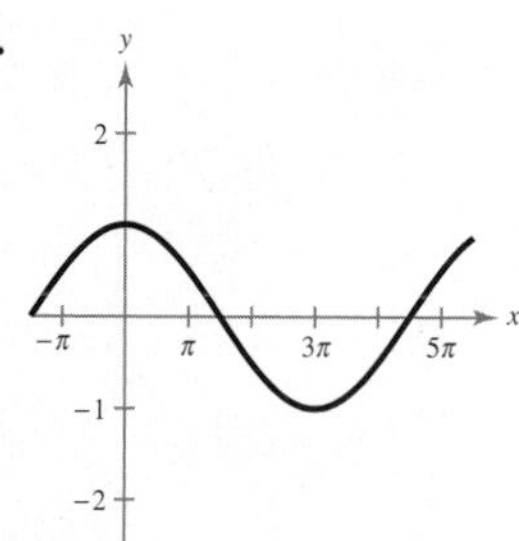

4. Period: 2π
Amplitude: 2
$\left[\frac{\pi}{2}, \frac{5\pi}{2}\right]$ corresponds to one cycle.
Key points: $\left(\frac{\pi}{2}, 2\right), (\pi, 0), \left(\frac{3\pi}{2}, -2\right), (2\pi, 0), \left(\frac{5\pi}{2}, 2\right)$

5.
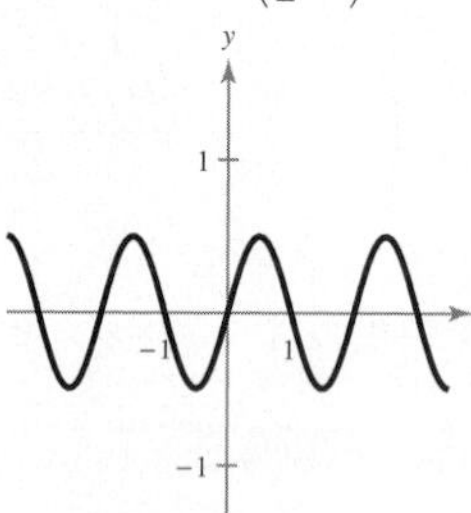

6.
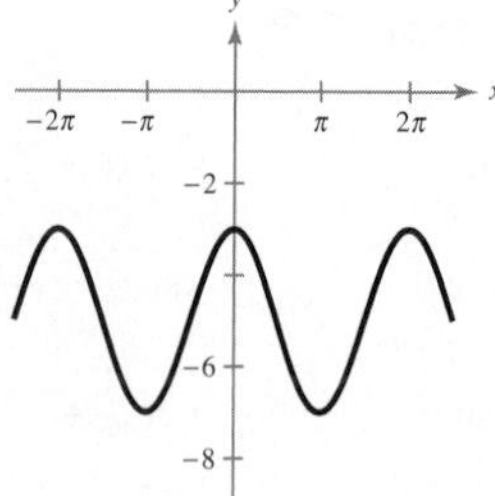

7. $y = 5.6 \sin(0.524t - 0.524) + 5.7$

Section 1.6

1.
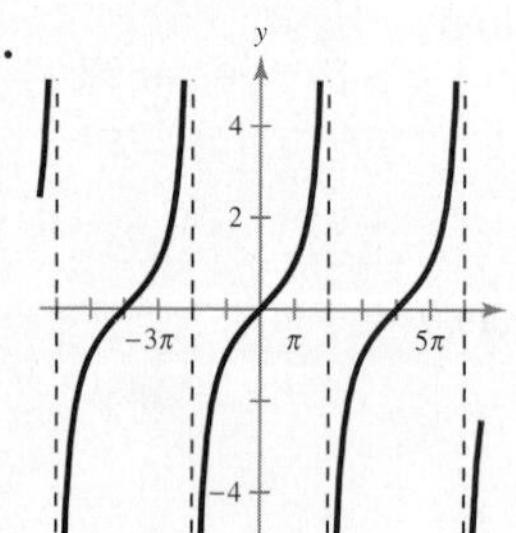

2.
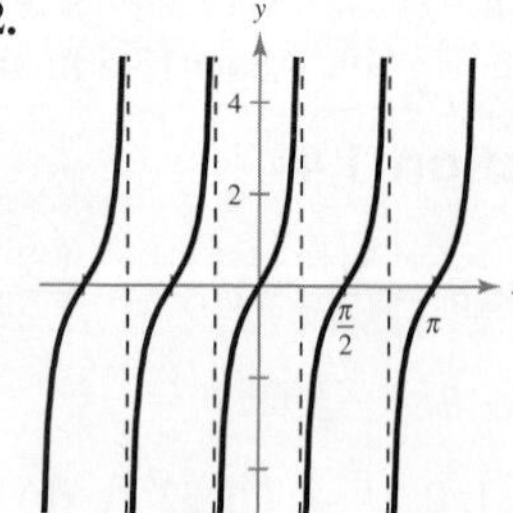

3.
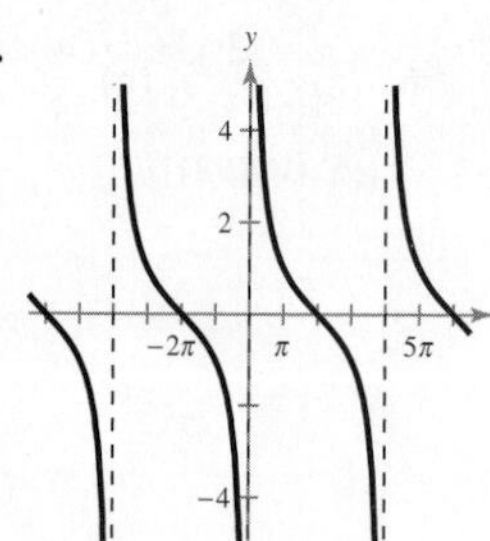

4.
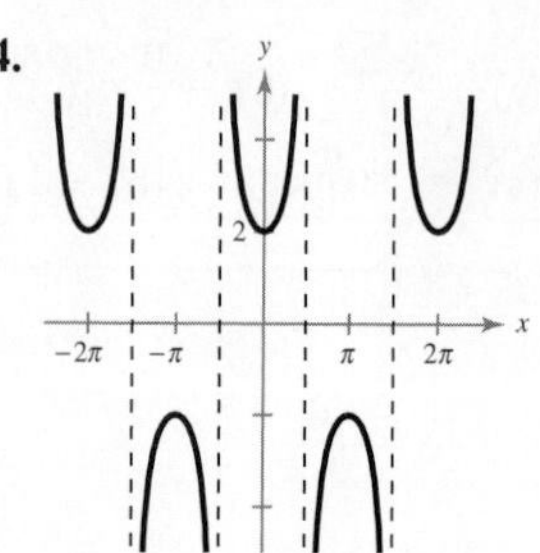

5.
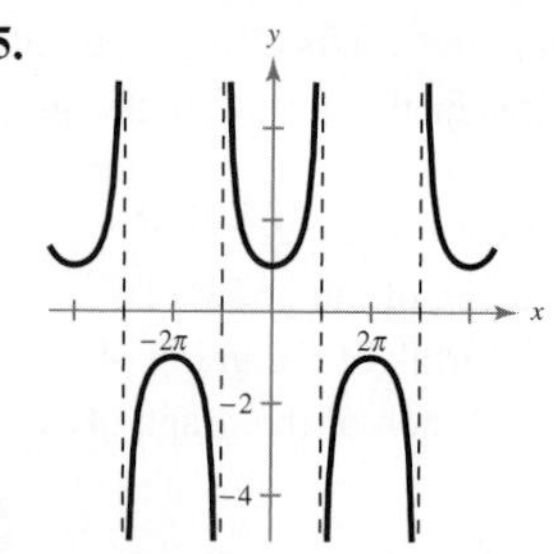

6.
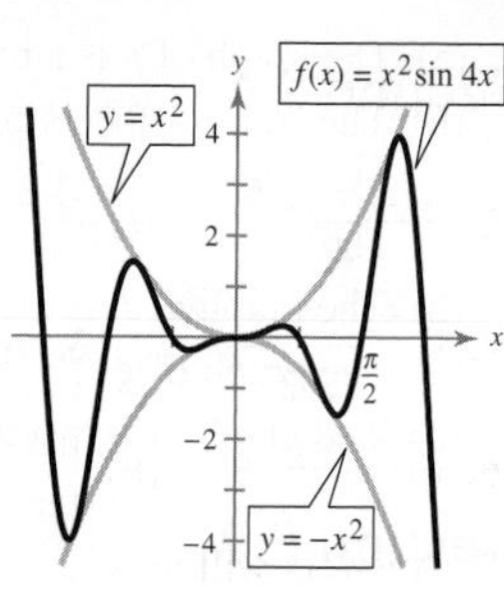

Section 1.7

1. (a) $\frac{\pi}{2}$
(b) Not possible
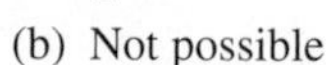

2.
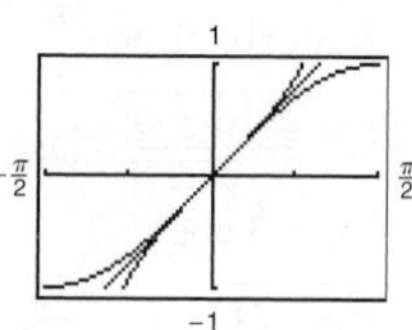

3. π
4. (a) 1.3670516 (b) Not possible (c) 1.9273001
5. (a) -14 (b) $-\frac{\pi}{4}$ (c) 0.54 6. $\frac{4}{5}$ 7. $\sqrt{x^2 + 1}$

Section 1.8

1. $a \approx 5.46, c \approx 15.96, B = 70°$ 2. About 15.8 ft
3. About 15.1 ft 4. 3.58°
5. Bearing: N 53° W, Distance: about 3.2 nmi
6. $d = 6 \sin \frac{2\pi}{3} t$
7. (a) 4 (b) 3 cycles per unit of time (c) 4 (d) $\frac{1}{12}$

Chapter 2

Section 2.1

1. $\sin x = -\frac{\sqrt{10}}{10}, \cos x = -\frac{3\sqrt{10}}{10}, \tan x = \frac{1}{3}, \csc x = -\sqrt{10}$
$\sec x = -\frac{\sqrt{10}}{3}, \cot x = 3$
2. $-\sin x$
3. (a) $(1 + \cos\theta)(1 - \cos\theta)$ (b) $(2\csc\theta - 3)(\csc\theta - 2)$
4. $(\tan x + 1)(\tan x + 2)$ 5. $\sin x$ 6. $2\sec^2\theta$
7. $1 + \sin\theta$ 8. $3\cos\theta$

Section 2.2

1–7. Answers will vary.

Section 2.3

1. $\frac{\pi}{4} + 2n\pi, \frac{3\pi}{4} + 2n\pi$
2. $\frac{\pi}{3} + 2n\pi, \frac{2\pi}{3} + 2n\pi, \frac{4\pi}{3} + 2n\pi, \frac{5\pi}{3} + 2n\pi$ 3. $n\pi$
4. $\frac{\pi}{6}, \frac{\pi}{2}, \frac{5\pi}{6}$ 5. $\frac{\pi}{4} + n\pi, \frac{3\pi}{4} + n\pi$ 6. $0, \frac{3\pi}{2}$
7. $\frac{\pi}{6} + n\pi, \frac{\pi}{3} + n\pi$ 8. $\frac{\pi}{2} + 2n\pi$
9. $\arctan\frac{3}{4} + n\pi, \arctan(-2) + n\pi$ 10. 0.4271, 2.7145
11. $\theta \approx 54.7356°$

Section 2.4

1. $\dfrac{\sqrt{2}+\sqrt{6}}{4}$ 2. $\dfrac{\sqrt{2}+\sqrt{6}}{4}$ 3. $-\dfrac{63}{65}$
4. $\dfrac{x+\sqrt{1-x^2}}{\sqrt{2}}$ 5. Answers will vary.
6. (a) $-\cos\theta$ (b) $\dfrac{\tan\theta - 1}{\tan\theta + 1}$ 7. $\dfrac{\pi}{3}, \dfrac{5\pi}{3}$
8. Answers will vary.

Section 2.5

1. $\dfrac{\pi}{3} + 2n\pi, \pi + 2n\pi, \dfrac{5\pi}{3} + 2n\pi$
2. $\sin 2\theta = \frac{24}{25}, \cos 2\theta = \frac{7}{25}, \tan 2\theta = \frac{24}{7}$
3. $4\cos^3 x - 3\cos x$ 4. $\dfrac{\cos 4x - 4\cos 2x + 3}{\cos 4x + 4\cos 2x + 3}$
5. $\dfrac{-\sqrt{2-\sqrt{3}}}{2}$ 6. $\dfrac{\pi}{3}, \pi, \dfrac{5\pi}{3}$ 7. $\dfrac{1}{2}\sin 8x + \dfrac{1}{2}\sin 2x$
8. $\dfrac{\sqrt{2}}{2}$ 9. $\dfrac{\pi}{6} + \dfrac{2n\pi}{3}, \dfrac{\pi}{2} + \dfrac{2n\pi}{3}, n\pi$ 10. $45°$

Chapter 3

Section 3.1

1. $C = 105°, b \approx 45.25$ cm, $c \approx 61.82$ cm 2. 13.40 m
3. $B \approx 12.39°, C \approx 136.61°, c \approx 16.01$ in.
4. $\sin B \approx 3.0311 > 1$
5. Two solutions:
$B \approx 70.4°, C \approx 51.6°, c \approx 4.16$ ft
$B \approx 109.6°, C \approx 12.4°, c \approx 1.14$ ft
6. About 213 yd^2 7. About 1856.59 m

Section 3.2

1. $A \approx 26.38°, B \approx 36.34°, C \approx 117.28°$
2. $B \approx 59.66°, C \approx 40.34°, a \approx 18.26$ m 3. About 202 ft
4. N 15.37° E 5. About 19.90 in.^2

Section 3.3

1. $\|\overrightarrow{PQ}\| = \|\overrightarrow{RS}\| = \sqrt{10}$, $\text{slope}_{\overrightarrow{PQ}} = \text{slope}_{\overrightarrow{RS}} = \frac{1}{3}$
$\overrightarrow{PQ}$ and $\overrightarrow{RS}$ have the same magnitude and direction, so they are equivalent.
2. $\mathbf{v} = \langle -5, 6\rangle, \|\mathbf{v}\| = \sqrt{61}$
3. (a) $\langle 4, 6\rangle$ (b) $\langle -2, 2\rangle$ (c) $\langle -7, 2\rangle$
4. (a) $3\sqrt{17}$ (b) $2\sqrt{13}$ (c) $5\sqrt{13}$
5. $\left\langle \dfrac{6}{\sqrt{37}}, -\dfrac{1}{\sqrt{37}} \right\rangle$ 6. $\left\langle \dfrac{6}{\sqrt{5}}, \dfrac{-12}{\sqrt{5}} \right\rangle$ 7. $-6\mathbf{i} - 3\mathbf{j}$
8. $11\mathbf{i} - 14\mathbf{j}$ 9. (a) 135° (b) About 209.74°
10. $\langle -96.59, -25.88\rangle$ 11. About 2405 lb
12. $\|\mathbf{v}\| \approx 451.8$ mi/h, $\theta \approx 305.1°$

Section 3.4

1. (a) -6 (b) 37 (c) 0
2. (a) $\langle -36, 108\rangle$ (b) 43 (c) $2\sqrt{10}$ 3. 45°
4. Yes 5. $\frac{1}{17}\langle 64, 16\rangle; \frac{1}{17}\langle -13, 52\rangle$ 6. About 38.8 lb
7. About 1212 ft-lb

Chapter 4

Section 4.1

1. (a) $12 - i$ (b) $-2 + 7i$ (c) i (d) 0
2. (a) $-15 + 10i$ (b) $18 - 6i$ (c) 41 (d) $12 + 16i$
3. (a) 45 (b) 29 4. $\frac{3}{5} + \frac{4}{5}i$ 5. $-2\sqrt{7}$
6. $-\dfrac{7}{8} \pm \dfrac{\sqrt{23}}{8}i$

Section 4.2

1. 3
2. (a) Two real solutions
(b) One repeated real solution
(c) Two imaginary solutions
3. $2 \pm i$ 4. $\pm\sqrt{2}, \pm 3i$ 5. $\frac{2}{3}, \pm 4i$
6. $f(x) = x^4 + 45x^2 - 196$
7. $f(x) = 2x^3 - 10x^2 + 18x - 10$

Section 4.3

1.
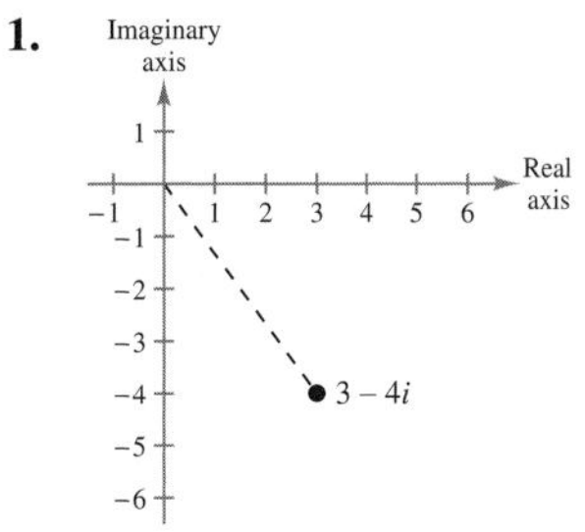

5
2. $4 + 3i$ 3. $1 - 5i$
4.
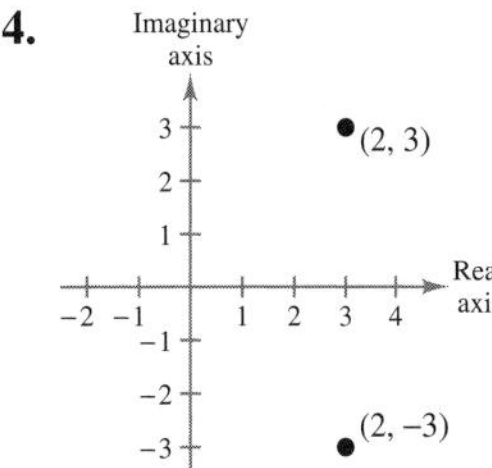

$2 + 3i$
5. $\sqrt{82} \approx 9.06$ units 6. $\left(\frac{7}{2}, -2\right)$

Section 4.4

1. $6\sqrt{2}\left(\cos\dfrac{7\pi}{4} + i\sin\dfrac{7\pi}{4}\right)$ 2. $5(\cos 53.1° + i\sin 53.1°)$
3. $-\sqrt{3} + i$ 4. $-4 + 4\sqrt{3}i$ 5. $12i$
6. $\dfrac{\sqrt{3}}{2} + \dfrac{1}{2}i$ 7. -8

Section 4.5

1. -4 2. $1, i, -1, -i$
3. $\sqrt[3]{3} + \sqrt[3]{3}i, -1.9701 + 0.5279i, 0.5279 - 1.9701i$

Chapter 5

Section 5.1

1. 0.0528248

2.

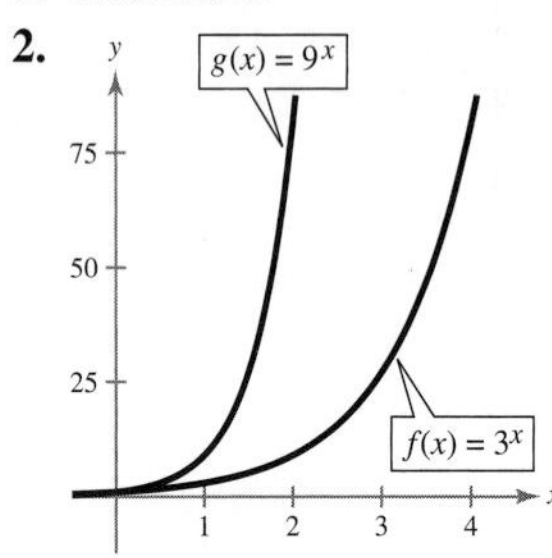

3.

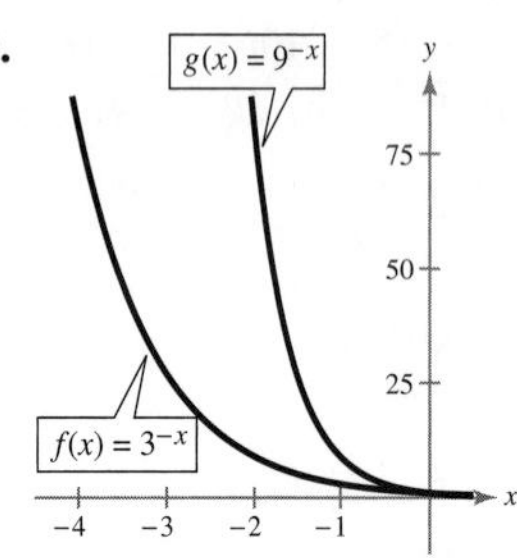

4. (a) 2 (b) 3

5. (a) Shift the graph of f two units to the right.
(b) Shift the graph of f three units up.
(c) Reflect the graph of f in the y-axis and shift three units down.

6. (a) 1.3498588 (b) 0.3011942 (c) 492.7490411

7.

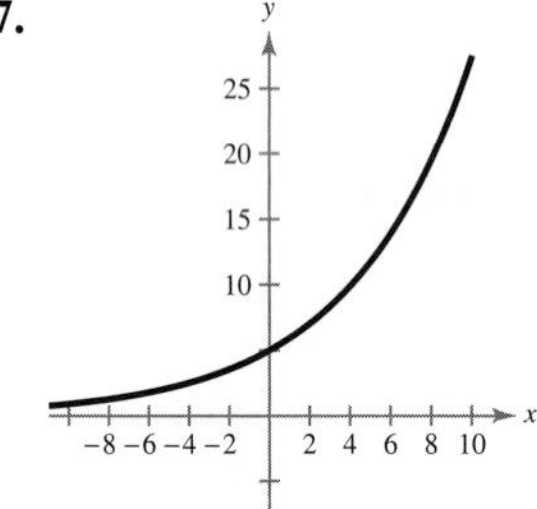

8. (a) \$7927.75 (b) \$7935.08 (c) \$7938.78

9. About 9.970 lb; about 0.275 lb

Section 5.2

1. (a) 0 (b) -3 (c) 3

2. (a) 2.4393327 (b) Error or complex number
(c) -0.3010300

3. (a) 1 (b) 3 (c) 0 **4.** ± 3

5.

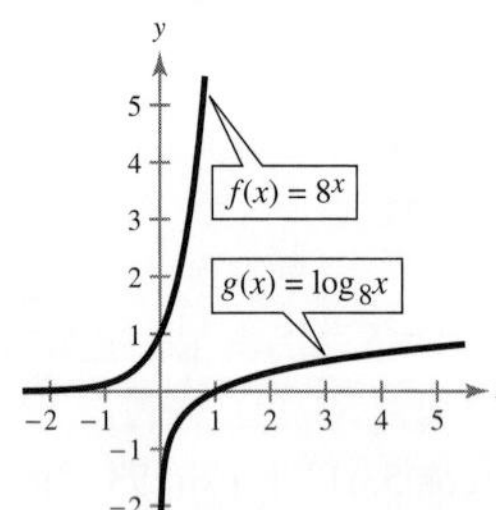

6.

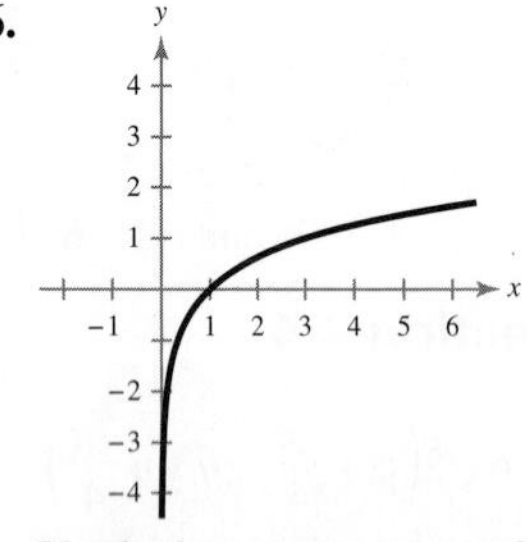

Vertical asymptote: $x = 0$

7. (a)

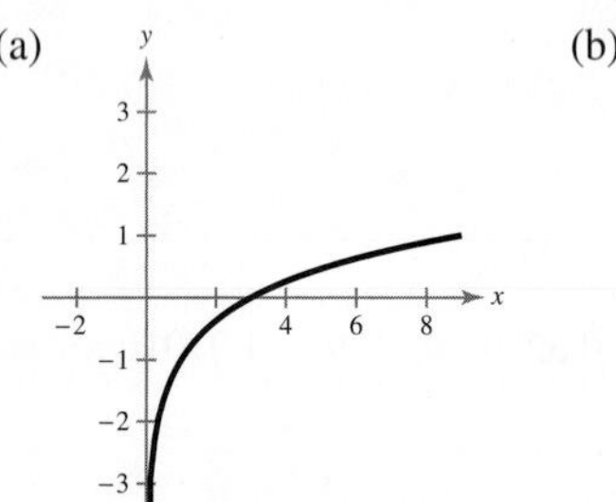

(b)

8. (a) -4.6051702 (b) 1.3862944 (c) 1.3169579
(d) Error or complex number

9. (a) $\frac{1}{3}$ (b) 0 (c) $\frac{3}{4}$ (d) 7 **10.** $(-3, \infty)$

11. (a) 70.84 (b) 61.18 (c) 59.61

Section 5.3

1. 3.5850 **2.** 3.5850

3. (a) $\log 3 + 2 \log 5$ (b) $2 \log 3 - 3 \log 5$ **4.** 4

5. $\log_3 4 + 2 \log_3 x - \dfrac{1}{2} \log_3 y$ **6.** $\log \dfrac{(x + 3)^2}{(x - 2)^4}$

7. $\ln y = \frac{2}{3} \ln x$

Section 5.4

1. (a) 9 (b) 216 (c) $\ln 5$ (d) $-\frac{1}{2}$

2. (a) $4, -2$ (b) 1.723 **3.** 3.401 **4.** -4.778

5. 1.099, 1.386 **6.** (a) $e^{2/3}$ (b) 7 (c) 12

7. 0.513 **8.** $\frac{32}{3}$ **9.** 10

10. About 13.2 years; It takes longer for your money to double at a lower interest rate.

11. 2010

Section 5.5

1. 2018 **2.** 400 bacteria **3.** About 38,000 yr

4.

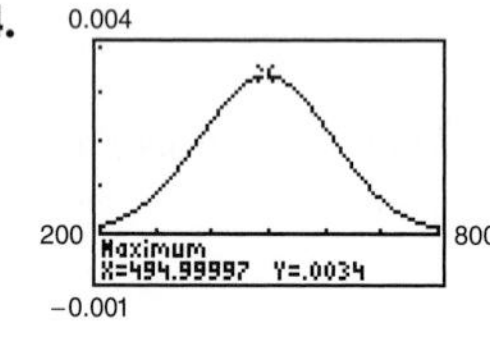

495

5. About 7 days **6.** (a) 1,000,000 (b) About 80,000,000

Chapter 6

Section 6.1

1. (a) $0.6747 \approx 38.7°$ (b) $\dfrac{3\pi}{4} \approx 135°$ **2.** $1.4841 \approx 85.0°$

3. $\dfrac{12}{\sqrt{10}} \approx 3.79$ units **4.** $\dfrac{3}{\sqrt{34}} \approx 0.51$ unit

5. (a) $\dfrac{8}{\sqrt{5}} \approx 3.58$ units (b) 12 square units

Section 6.2

1. $x^2 = \frac{3}{2}y$ **2.** $(y + 3)^2 = 8(x - 2)$ **3.** $(2, -3)$

4. $y = 6x - 3$

Section 6.3

1. $\dfrac{(x - 2)^2}{7} + \dfrac{(y - 3)^2}{16} = 1$

2.

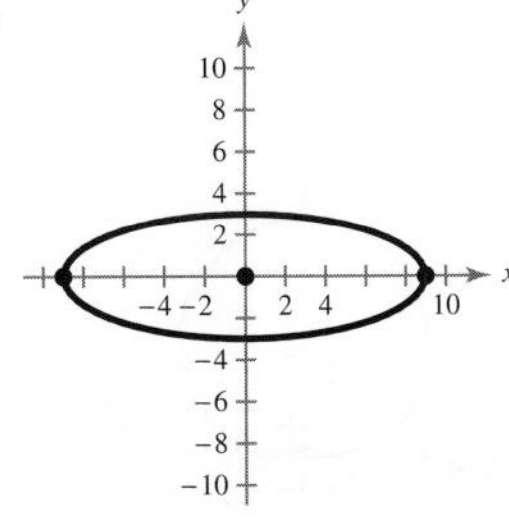

Center: $(0, 0)$
Vertices: $(-9, 0), (9, 0)$

3. Center: $(-2, 1)$
Vertices: $(-2, -2), (-2, 4)$
Foci: $\left(-2, 1 \pm \sqrt{5}\right)$

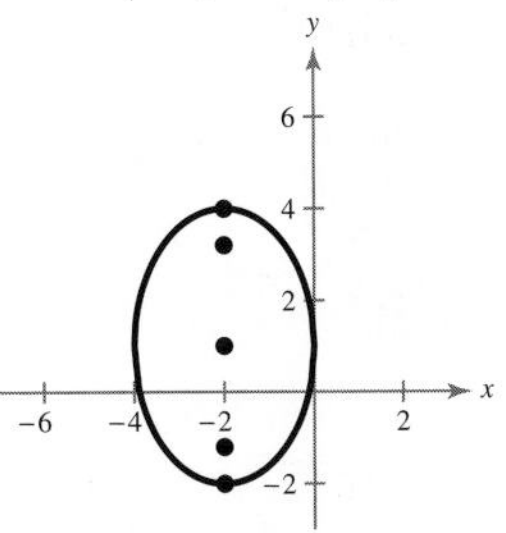

4. Center: $(-1, 3)$
Vertices: $(-4, 3), (2, 3)$
Foci: $(-3, 3), (1, 3)$

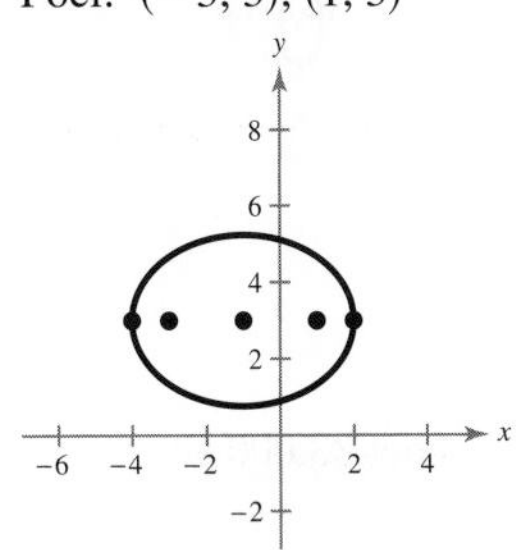

5. Aphelion:
4.080 astronomical units
Perihelion:
0.340 astronomical unit

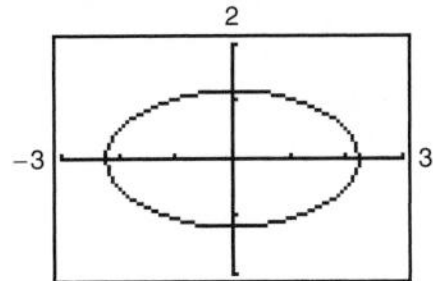

Section 6.4

1. $\dfrac{(y+1)^2}{9} - \dfrac{(x-2)^2}{7} = 1$

2.

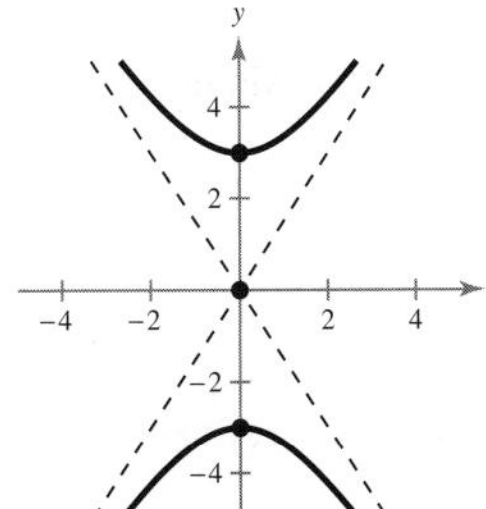

3.

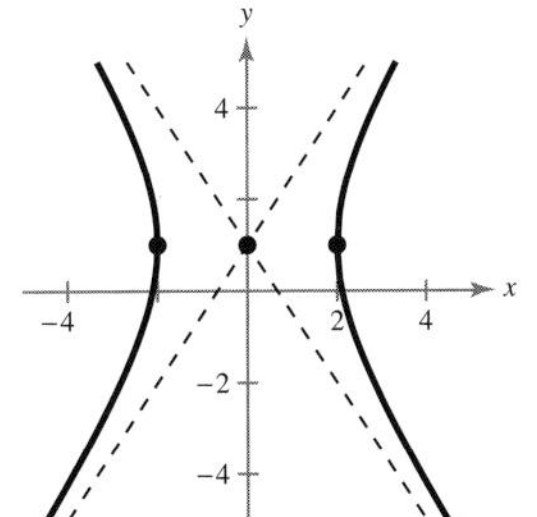

4. $\dfrac{(x-6)^2}{9} - \dfrac{(y-2)^2}{4} = 1$

5. The explosion occurred somewhere on the right branch of the hyperbola

$$\frac{x^2}{4{,}840{,}000} - \frac{y^2}{2{,}129{,}600} = 1.$$

6. (a) Circle (b) Hyperbola (c) Ellipse (d) Parabola

Section 6.5

1. $\dfrac{(y')^2}{12} - \dfrac{(x')^2}{12} = 1$

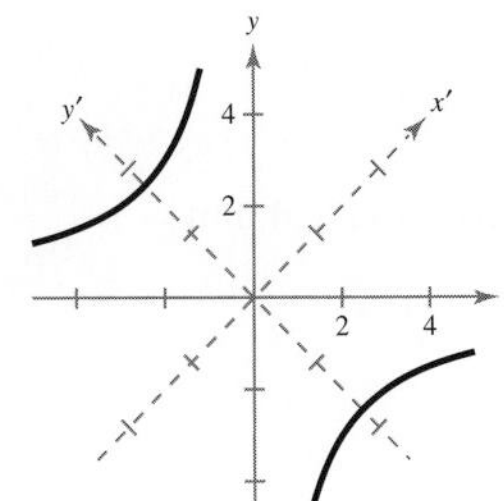

2. $\dfrac{(x')^2}{1} - \dfrac{(y')^2}{9} = 1$

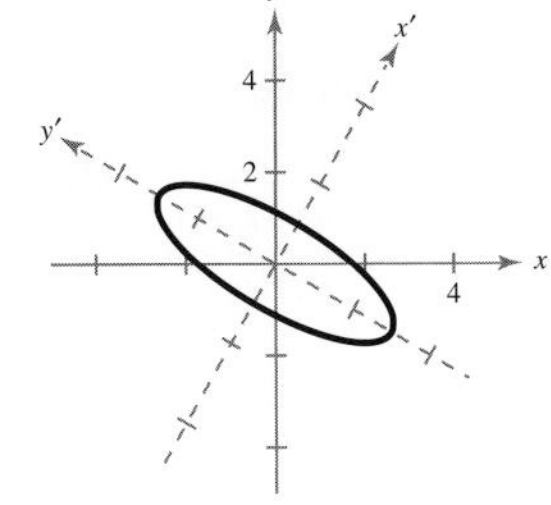

3. $(x')^2 = -2(y' - 3)$

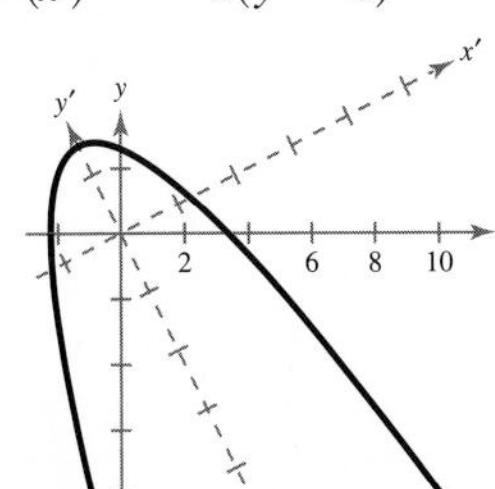

4. Parabola;
$y = \dfrac{2x \pm \sqrt{-6x - 10}}{4}$

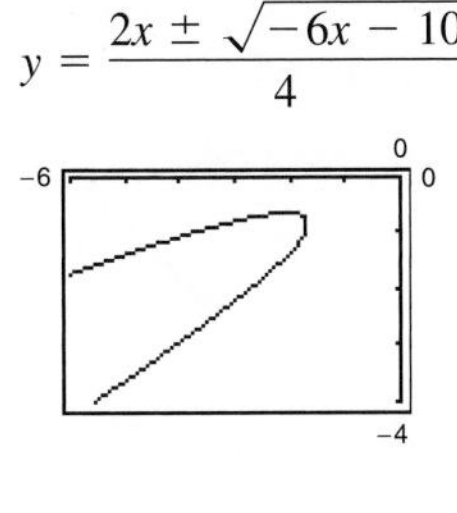

Section 6.6

1.

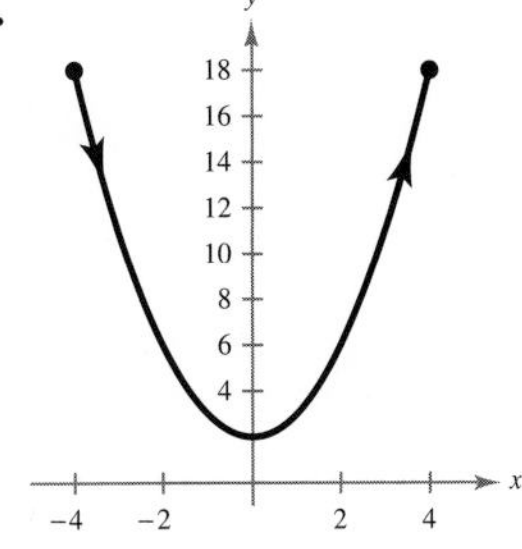

The curve starts at $(-4, 18)$ and ends at $(4, 18)$.

2.

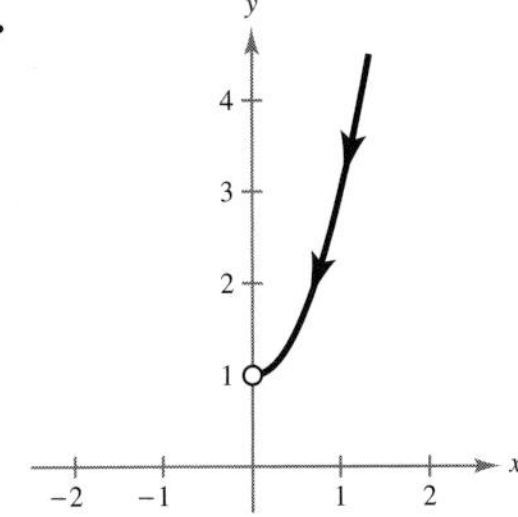

3. (a)

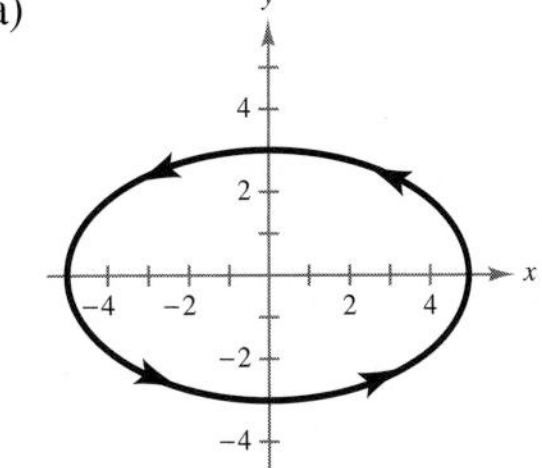

(b)

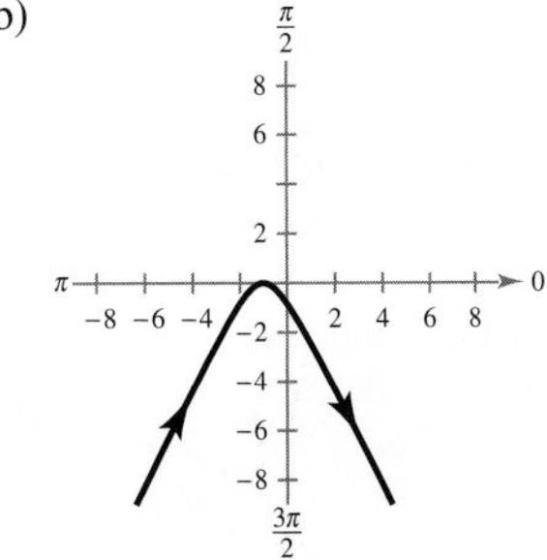

4. (a) $x = t,\ y = t^2 + 2$ (b) $x = 2 - t,\ y = t^2 - 4t + 6$

5. $x = a(\theta + \sin\theta),\ y = a(1 + \cos\theta)$

Section 6.7

1. (a)

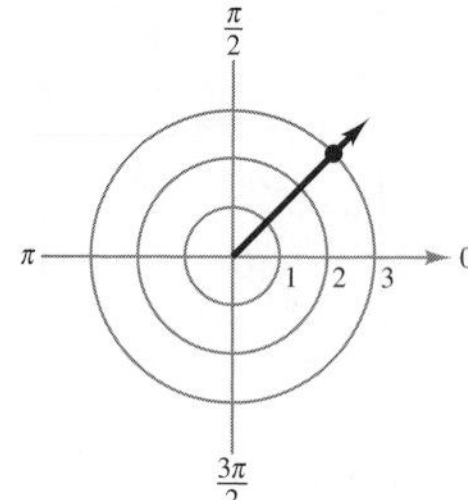

(b)

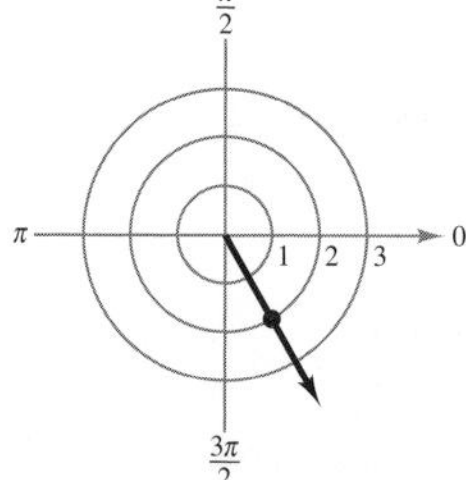

(c)

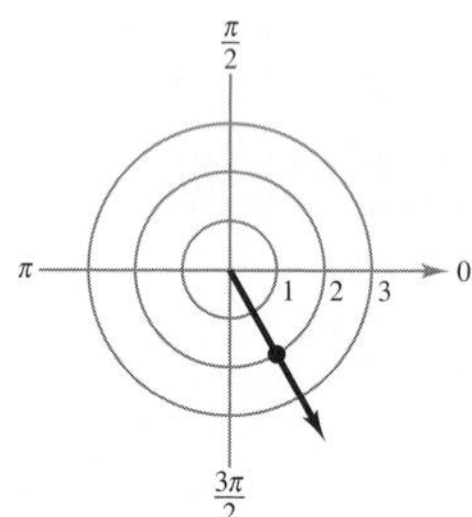

2.

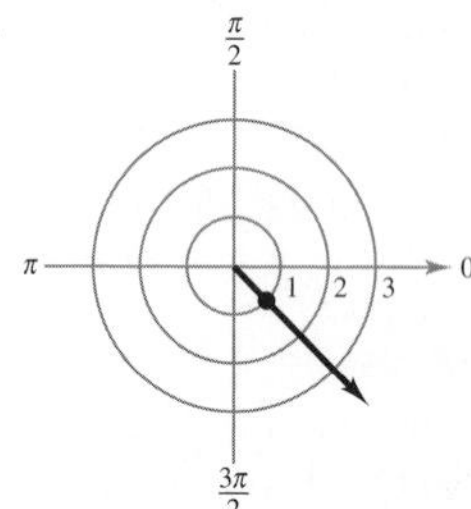

$\left(-1, -\frac{5\pi}{4}\right), \left(1, \frac{7\pi}{4}\right), \left(1, -\frac{\pi}{4}\right)$

3. $(-2, 0)$ 4. $\left(2, \frac{\pi}{2}\right)$

5. (a) The graph consists of all points seven units from the pole; $x^2 + y^2 = 49$
   (b) The graph consists of all points on the line that makes an angle of $\pi/4$ with the polar axis and passes through the pole; $y = x$
   (c) The graph is a circle with center $(0, 3)$ and radius 3; $x^2 + (y - 3)^2 = 9$

Section 6.8

1.

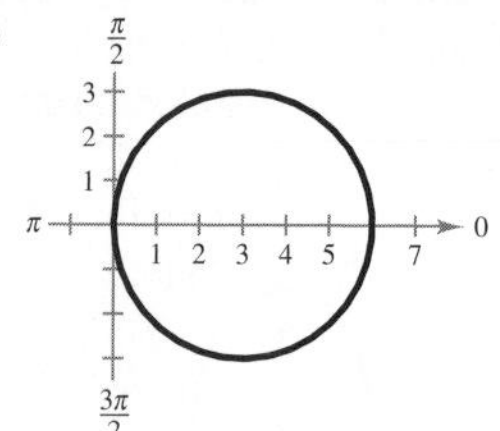

2.

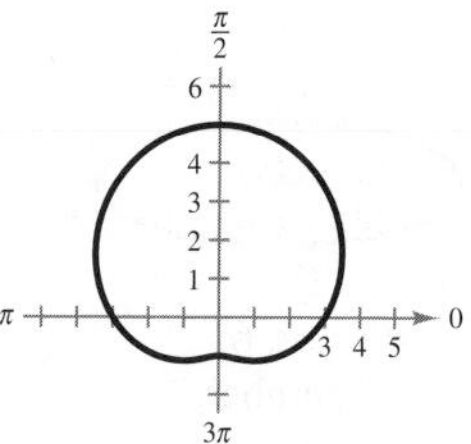

3.

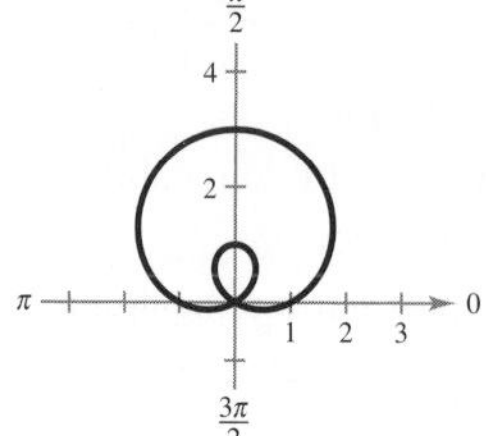

4.

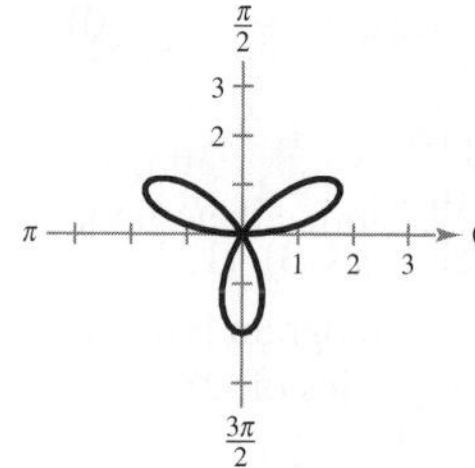

5.

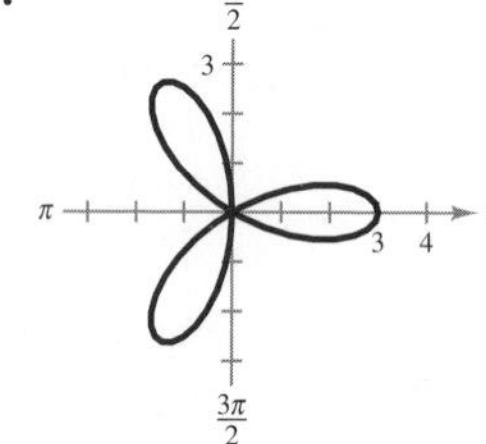

6.

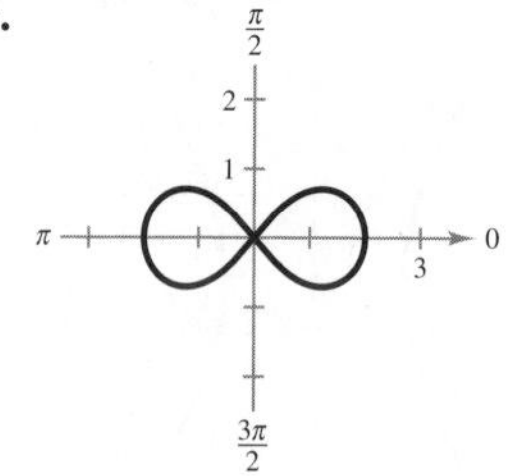

Section 6.9

1. Hyperbola
2. Hyperbola

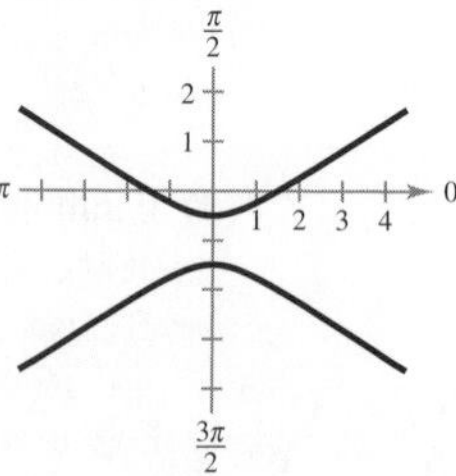

3. $r = \dfrac{2}{1 - \cos\theta}$
4. $r = \dfrac{0.625}{1 + 0.847 \sin\theta}$; about 0.340 astronomical unit

Index

D

E

F

Q

R

S

T